Environment, Colonization, and the Baltic Crusader States

ENVIRONMENTAL HISTORIES OF THE NORTH ATLANTIC WORLD

VOLUME 2

Previously published volumes in this series are listed at the back of the book.

ENVIRONMENT, COLONIZATION, AND THE BALTIC CRUSADER STATES

TERRA SACRA I

Edited by

Aleksander Pluskowski

BREPOLS

British Library Cataloguing in Publication Data

A catalogue record for this book is available from the British Library.

D/2019/0095/49
ISBN: 978-2-503-55132-6

Printed in the EU on acid-free paper

Contents

Prussia

Livonia and Prussia: A Comparison

Appendices

LIST OF ILLUSTRATIONS

Aleksander Pluskowski, Kaspars Kļaviņš, Eva Eihmane, Agris Dzenis, and Juhan Kreem

Aleksander Pluskowski

Prussia

Aleksander Pluskowski, Marc Jarzebowski, Małgorzata Karczewska, and Maciej Karczewski

Alexander Brown

Daniel Makowiecki, Mirosława Zabilska-Kunek, Krish Seetah, Marc Jarzebowski, and Aleksander Pluskowski

Aleksander Pluskowski, Monika Badura, and Marc Jarzebowski

Rowena Banerjea

Marc Jarzebowski, Aleksander Pluskowski, Tomasz Nowakiewicz, and Maciej Karczewski

Aleksander Pluskowski

Livonia and Prussia: A Comparison

Aleksander Pluskowski

Appendices

Alexander Brown

Stuart Black

Preface

This book is the end result of a four-year project funded by the European Research Council on the environmental impact of the crusades in the eastern Baltic and the societies created by the process of military conquest, colonization, and religious conversion. Adopting a multi-scalar, multi-proxy methodology, synthesizing a range of archaeological, environmental and historical sources, this study demonstrates how indigenous societies had already begun to significantly transform the environment in the pre-crusade period. Following the establishment of a militarised Christian theocratic regime in the regions that became known as Livonia and Prussia, environmental exploitation was reorganized alongside territorial reconfiguration. Fortified convents were at the centre of this reorganization and formed a network that connected the conquered territories within a hierarchy centred on the headquarters of the Teutonic Order at Marienburg, today Malbork in Poland. The Teutonic Order, which dominated the theocratic regime, promoted itself as the defender of Latin Christendom in north-eastern Europe. Environmental exploitation therefore served to secure and maintain the political and military authority of the Order, and is at the heart of understanding the success and longevity of the Baltic 'crusader states'. However, despite the theocracy's monolithic image and centralized administration, its impact on the conquered territories was variable. This can be synchronized with the regionally uneven and protracted process of colonization which followed the military conquests, as well as reflecting varied relations with the indigenous population. This study provides an important contribution to our understanding of the development of Europe in the formative period of the Middle Ages, the crusading movement, and how environmental change can be used as a critical proxy of colonization.

Aleksander Pluskowski is Associate Professor of Medieval Archaeology at the Department of Archaeology at the University of Reading, UK. His research focuses on the relationship between nature and culture in medieval Europe, particularly within frontier societies associated with crusading, colonization, cultural encounters, and religious transformations. Alongside numerous articles, his publications include *Wolves and the Wilderness in the Middle Ages* (2006), *The Ritual Killing and Burial of Animals: European Perspectives* (2011), and *The Archaeology of the Prussian Crusade: Holy War and Colonisation* (2013).

Acknowledgements

A considerable number of people in various countries have contributed to the success of this project. This has included the core team, its partners, collaborators and steering panel, as well as undergraduate and postgraduate students in multiple institutions who have been involved at various levels, particularly in excavations, and during the various stages of post-excavation analysis.

Project Core Team

The core team consisted of Aleksander Pluskowski (principal investigator), Alexander Brown (palynology), Monika Badura (plant macro-fossils), Krish Seetah (zooarchaeology), Daniel Makowiecki (zooarchaeology), Rowena Banerjea (geoarchaeology), Lisa-Marie Shillito (geoarchaeology), Kevin Hayward (petrography), Marc Jarzebowski (written sources), Kaspars Kļaviņš (written sources), and Juhan Kreem (written sources). All geophysics work for the project was carried out by David Thornley.

Project Partners

This project would not have been possible without the tireless support and collegiality of our partners across the eastern Baltic and in the UK. They are listed by country and grouped according to institution, with co-directors of excavations indicated by *.

Estonia

Heiki Valk*, Eve Rannamäe and Martin Malve (Institute of History and Archaeology, University of Tartu), Lembi Lõugas (Archaeological Research Collection, Tallinn University) and Siim Veski (Department of Earth Sciences, Tallinn University of Technology).

Latvia

Gundars Kalniņš* (Cēsis Castle Museum), Zigrīda Apala* (Institute of Latvian History, University of Latvia), OskarsUšpelis* (independent), Eva Eihmane (independent), Laimdota Kalniņa and Normunds Stivriņš (Faculty of Geography and Earth Sciences, University of Latvia), Māris Zunde (Dendrochronological laboratory, Institute of Latvian History, University of Latvia), Agris Dzenis (independent), Uldis Kalējs (Architectural Investigation Group Ltd, Riga), Arnis Mugurēvičs (Faculty of Veterinary Medicine, Latvia University of Agriculture).

Lithuania

Linas Daugnora and Vladas Žulkus (Institute of Baltic Sea Region History and Archaeology, University of Klaipėda), Roman Shiroukhov (Department of Archaeology, Vilnius University).

Poland

Zbigniew Sawicki* and Waldek Jaszczyński (Castle Museum in Malbork); Marcin Wiewióra, Adam Chęć, and Dariusz Poliński (Institute of Archaeology, University of Nicolaus Copernicus in Toruń); Maciej Karczewski* and Małgorzata Karczewska* (Institute of History, University of Białystok); Maria Kasprzycka, Mirosław Marcinkowski*, and Joanna Fonferek (Archaeology and History Museum in Elbląg); Katarzyna Pińska (independent); and Mirosława Zabilska-Kunek (Institute of Archaeology, University of Rzeszów).

United Kingdom

Gundula Müldner, Stuart Black, and Charlotte Scull (Department of Archaeology, University of Reading); Gary King and Chris Gerrard (Department of Archaeology, University of Durham); Ivy Yeh and Piers Mitchell (Department of Archaeology, University of Cambridge); and Alaina Schmisseur (Department of Archaeology, University of York).

Steering Panel

The project has benefited from guidance provided by Grenville Astill (Department of Archaeology, University of Reading), József Laszlovszky (Central European University, Budapest), Mark Maltby (Department of Archaeology, Anthropology and Forensic Sciences, Bournemouth University), and Anders Andrén (Institute of Archaeology and Classical History, Stockholm University).

Additional Thanks

Numerous people have been involved at various stages of the project, particularly during excavations and periods of data collection in Poland, Estonia, and Latvia. It is impossible to thank everyone individually, but a special thanks goes to Magnus Elander, who photographed our fieldwork, sites, and associated landscapes, prompting us to think about presenting our results in a different way. We would also like to thank Roberta Gilchrist, Nick Branch, and Dominik Fleitmann (Department of Archaeology, University of Reading) for their continuing advice and support. We would like to extend our thanks to the staff of the British Library; the libraries of the Castle Museum in Malbork, the Museum of Warmia, and Masuria in Olsztyn; the University of Nicolaus Copernicus in Toruń; and the Warburg Institute in London for their help with accessing maps and relevant source material. We would like to thank Kristin Ismail-Meyer, Philippe Rentzel, Christine Pümpin, and David Brönnimann at the IPNA in Basel for access to their reference collections of thin sections from plant materials, waterlogged and stable contexts. We would like to particularly thank Seweryn Szczepański (Wojciech Kętrzyński Research Centre in Olsztyn) for his comments on the entirety of the text, for his continuing support and collegiality, as well as Yannick Devos and Cristiano Nicosia for additionally reviewing the geoarchaeological chapters. We would also like to thank all the participants and hosts of the Terra Sacra conference which took place in Riga and Cēsis on 17–21 September 2014. Finally, we would like to thank Rosie Bonté, Shannon Cunningham, Lauren Davis, Martine Maguire-Weltecke and the staff of Brepols for all their work in the preparation of the project's two volumes.

Funding

The members of the project team and partners are extremely grateful to the European Research Council for enabling the realization of the research programme 'The Ecology of Crusading: The Environmental Impact of Conquest, Colonisation and Religious Conversion in the Medieval Baltic' from 2010–14. The research leading to the results in this book has received funding from the European Union's Seventh Framework Programme (FP7/2007–13) under grant agreement number 263735. Data from parallel projects has been included where appropriate, and with permission. Eve Rannamäe's contribution was supported by the European Union through the European Regional Development Fund (Centre of Excellence CECT) and the institutional research funding IUT20-7 of the Estonian Ministry of Education and Research, and part of Monika Badura's contribution was carried out within the parallel project 2012/05/B/ST10/00243, financed by the National Science Centre in Poland.

Glossary

Place Names and Other Terminology

Modern place names are used when referring to a monument or archaeological site, and chosen variants of medieval names are used for historical references. Each time a major place name is used for the first time in a chapter, its equivalent is provided in brackets. It is anticipated that readers will be comfortable between this interchange, whilst at the same time distinguishing archaeological from historical data. There are many variants of regional names applied to the many territories of the eastern Baltic, which are discussed in further detail by Bojtár.[1] The glossary below provides alphabetically organized lists of major place names referred to in the text, with English spellings that are largely based on German or Latin renderings. This is followed by a list of key terms associated with the Teutonic Order. These have all been anglicized wherever possible for the sake of consistency, although some German terms — such as *Ordensland* — have been retained and used where appropriate, including the first use of a term in each chapter.

[1] E. Bojtár (1999), *Foreword to the Past: A Cultural History of the Baltic People* (Budapest: Central European University Press).

Principal Territories in Livonia

German	Latvian /Estonian	Version Used
Harrien	Harjumaa	Harria
Jerwen	Järvamaa	Jervia
Kurland	Kurzeme	Curonia/Courland
Letgallen	Latgale	Latgalia
Ösel	Saaremaa	Osilia/Ösel
Sakala	Sakala	Saccala
Selonien	Sēlija	Selonia
Semgallen	Zemgale	Semigallia
Ugaunien	Ugandi	Ugaunia
Wiek	Läänemaa	Maritima
Wierland	Virumaa	Vironia

Principal Territories in and adjacent to Prussia

German	Polish /Lithuanian	Version Used
Dobriner Land	Ziemia Dobrzyńska	Dobrin Land
Ermland	Warmia	Ermland/Warmia
Galindien	Galindia	Galindia
Kuja	Kujawy	Kuyavia
Kulmerland	Ziemia Chełmińska	Kulmerland
Löbauer Land	Ziemia Lubawska	Lubavia
Masowien	Mazowsze	Masovia
Masuren	Mazury	Masuria
Memmelland	Klaipėdos apskritis	Memmelland
Nadrauen	Nadrowia	Nadruvia
Natangen	Natangia	Natangia
Ostpreussen	Prusy Wschodnie	East Prussia
Pogesanien	Pogezania	Pogesania
Pomesanien	Pomezania	Pomesania
Pommerellen	Pomorze Wschodnie	Pomerelia
Pommern	Pomorze	Pomerania
Preussen	Prusy	Prussia
Rus	Ruś	Rus'
Samland	Sambia	Sambia
Sassen	Sasinia	Sasna
Schalauen	Skalowia	Scalovia
Sudauen	Jaćwież /Sudawia	Sudovia

Major Place Names in Livonia Mentioned in the Text

German	Estonian	Latvian
Adsel	Koivaliina	Gaujiena
Alt-Dahlen		Vecdole
Arrasch		Āraiši
Ascheraden		Aizkraukle
Babbit		Babīte
Dorpat	Tartu	
Dünaburg		Daugavpils
Dünamünde		Daugavgrīva
Dundagen		Dundaga
Ermes	Härgmäe	Ērģeme
Fellin	Viljandi	
Frauenburg		Saldus
Goldingen		Kuldīga
Heiligenberg		Svētkalns
Helmet	Helme	
Holme		Mārtiņsala
Jürgensburg		Zaube
Kalverholm		Klīversala
Kandau		Kandava
Karkus	Karksi	
Kirchholm		Salaspils
Kokenhusen		Koknese
Kreutzburg		Krustpils
Lennewarden		Lielvārde
Marienburg		Alūksne
Marienhausen		Viļaka
Mesothen		Mežotne
Mitau		Jelgava
Ösel	Saaremaa	
Odenpäh	Otepää	
Pernau	Pärnu	
Reval	Tallinn	
Riga		Rīga
Rodenpois		Ropaži
Ronneburg		Rauna
Rositten		Rēzekne
Sabell		Sabile
Segewold		Sigulda
Seßwegen		Cesvaine
Tarwast	Tarvastu	
Treiden		Turaida
Trikaten		Trikāta
Üxküll		Ikšķile
Walk	Valga	Valka
Wenden	Võnnu	Cēsis
Windau		Ventspils
Wolmar		Valmiera

Major Place Names in Prussia Mentioned in the Text

German	Polish	Russian/Lithuanian
Allenstein	Olsztyn	
Angerburg	Węgorzewo	
Balga	Bałga	Wiesiołoje (Russia)
Barten	Barciany	
Bäslack	Bezławki	
Birgelau	Bierzgłowo	
Bischöflich Papau	Papowo Biskupie	
Brandenburg	Pokarmin	Ushakovo (Russia)
Briesen	Wąbrzeźno	
Christburg	Dzierzgoń	
Danzig	Gdańsk	
Dirschau	Tczew	
Eckersberg	Okartowo	
Elbing	Elbląg	
Engelsburg	Pokrzywno	
Eylau	Pruska Iława /Iławka	Bagrationovsk (Russia)
Fischhausen	Rybaki	Primorsk (Russia)
Frauenburg	Frombork	
Graudenz	Grudziądz	
Heilsberg	Lidzbark Warmiński	
Heiliger Berg	Święta Góra	
Königsberg	Królewiec	Kaliningrad (Russia)
Kulm	Chełmno	
Lochstädt		Pavlovo (Russia)
Lötzen	Giżycko	
Lyck	Ełk	
Marienburg	Malbork	
Marienwerder	Kwidzyn	
Memel	Kłajpeda	Klaipėda (Lithuania)
Mewe	Gniew	
Mohrungen	Morąg	
Nessau	Mała Nieszawka	
Osterode	Ostróda	
Ragnit	Ragneta	Neman (Russia)
Rastenburg	Kętrzyn	
Rehden	Radzyń Chełmiński	
Rhein	Ryn	
Schönsee	Kowalewo Pomorskie	
Schwetz	Świecie	
Seehesten	Szestno	
Strasburg	Brodnica	
Thorn	Toruń	
Tilsit	Tylża	Sovetsk (Russia)
Weißenberg	Biała Góra	

German	English	Other (Where Applicable)
Bauleiter	Master Mason	Also (Ger.) Muwermeister (Lat.) Magister lapidum /Magister laterum
Bernsteinmeister	Amber Master	Also (Lat.) Magister sucinum
Großkomtur	Grand Commander	Also (Lat.) Magnus commendator
Haus	House	Also (Lat.) Domus
Hochmeister	Grand Master	Also (Lat.) Magister generalis
Komturei	Commandery	Also (Ger.) Kommende; (Pol.) Komturstwo
Komtur	Commander	Also (Lat.) Commendator
Landmeister	Provincial Master	Also (Lat.) Magister provincialis
Marschall	Marshal	Also (Lat.) Summus marescalus
Pfleger	Procurator	Also (Lat.) Procuratoris; (Pol.) Prokurator
Reise	Campaign	In relation to fourteenth-century Lithuania
Ritter	Knight	Also (Lat.) Miles
Spittler	Hospitaller	Also (Lat.) Hospitalarius
Trappier	Quartermaster	Also (Lat.) Trapearius
Tressler	Treasurer	Also (Lat.) Thesaurarius
Vogt	Advocate	Also (Lat.) Advocatus
Vorburg	Outer bailey or ward	Also (Lat.) Subcastrum
Vorwerk	Specialised farm	Also (Lat.) Allodium; (Pol.) Folwark
Waldmeister	Forest Master	Also (Lat.) Magister silvarum

Terminology of the Teutonic Order

The Hospital of St Mary of the Germans in Jerusalem has been abbreviated to Teutonic Order, interchangeable with Teutonic Knights and both further abbreviated to 'the Order'. Teutonic is a Latinized rendering of 'German', and whilst the term 'German Order' is also encountered in scholarly literature, the Latinized form is more typically used within international scholarship. The locally formed military order operating in Livonia in 1202–36 was the *Fratres militiae Christi Livoniae*, more commonly known as the Livonian Brothers of the Sword, or the Sword Brothers. In 1237, the remnants of the Sword Brothers were formally merged with the Teutonic Order. The combined territories were previously referred to in German historiography as *der Deutschordensstaat* or simply *der Ordensstaat* — the Order's state — variously referred to in the literature as the Teutonic Order's monastic, theocratic, or crusader state. However, the term *Ordensland* or 'Order's land', also featured in earlier scholarship, is increasingly being used to distinguish this polity from neighbouring medieval 'states' which had different characteristics. This is the term used throughout this book to refer to the combined territories of the Teutonic Order in the eastern Baltic, although the term 'crusader states' is used to draw a comparison with similar polities, particularly in the Latin East. The term *Ordensland* acknowledges that alongside these territories were domains run by bishops who had variable associations with the Order (alongside the exceptional secular Duchy of Estonia), and whilst there were key differences between the two groups the general regime in the medieval eastern Baltic can be characterized as a theocracy. The principal terms associated with the Order's organizational structure are listed in the table above.

Other

Royal Polish lands subdivided into *starostwa* (sing. *starostwo*) are translated as elderships. *Paganus* and the German term *Heidin*, which features in the narratives of the Baltic Crusades, have been translated as both 'pagan' and 'heathen'. However, 'pagan' also remains widely used within modern scholarship of the Conversion Period, where it functions as a shorthand for the societies encountered by Christian missionaries and crusaders. It remains a problematic term insofar as it assumes a contrast between two pristine religious systems: 'Christian' and 'pagan', ultimately derived from contemporary Christian essentialist and derogatory designations of non-Christian peoples.[2]

2 W. Kilbride (2000), 'Why I feel Cheated by the Term "Christianisation"', *Archaeological Review from Cambridge* 17/2, 1–17.

Abbreviations

AMH	*Das Ausgabenbuch des Marienburger Hauskomturs*
CTP	Peter von Dusburg, *Chronicon terra Prussiae* (*Chronicle of Prussia*)
CS	Arnold von Lübeck, *Chronica Slavorum* (*Chronicle of the Slavs*)
GAD	*Das große Ämterbuch des Deutschen Ordens* (*Great Book of Offices of the Teutonic Order*)
Est.	Estonian
Ger.	German
HCL	*Heinrici Cronicon Lyvoniae* (*Henry's Chronicle of Livonia*) (following subdivisions in Brundage 2003)
HHC	Hermann Helewegh, *Das rothe Buch inter Archiepiscopalia* (*Red Book of the Archbishops*)
HWC	Hermann von Wartberge, *Chronicon Livoniale* (*Chronicle of Livonia*)
KP	Nicolaus von Jeroschin, *Di Kronike von Pruzinlant* (*Chronicle of Prussia*)
Lat.	Latin
Latv.	Latvian
LGU	*Livländische Güterurkunden (aus den Jahren 1207 bis 1545)*
Lith.	Lithuanian
LR	*Die Libri Redituum der Stadt Riga* (*The Income Book of the City of Riga*)
LRC	*Livländische Reimchronik* (*Livonian Rhymed Chronicle*)
LUB	*Liv-, Est- und Kurländisches Urkundenbuch nebst regesten* (*Book of Certificates and Register for Livonia, Estonia and Courland*)
LVVA	Latvijas Valsts vēstures Arhīvs (Latvian State Historical Archives, Riga)
LW	*Martin Luthers Werke: kritische Gesamtausgabe* (*Martin Luther's Works*)
MA	*Das Marienburger Ämterbuch* (*The Marienburg Book of Offices*)
MLG	Middle Low German
MTB	*Das Marienburger Tresslerbuch* (*The Marienburg Treasurer's Book*)
OBA	*Ordensbriefarchiv* in the twentieth Hauptabteilung (Historcial State Archive Königsberg) at the Geheimes Staatsarchiv Preußischer Kulturbesitz at Berlin-Dahlen
OF	*Ordensfolianten* in the twentieth Hauptabteilung at the Geheimes Staatsarchiv Preußischer Kulturbesitz, Berlin-Dahlen (as above)
Plu.	Plural
Pol.	Polish
PrUB	*Preußisches Urkundenbuch* (*Prussian Records Book*)
Prus.	Old Prussian
Rus.	Russian
Sing.	Singular

Trees felled by beavers in Masuria, Poland.
Photo by M. Elander.

1. Introduction: *Terra Sacra* in the Eastern Baltic

Aleksander Pluskowski

In the last few years of the twelfth century, the missionary activities of a German bishop in lands that became referred to in Latin as 'Livonia' rapidly developed into a series of crusades which saw the creation of a local military order, the Sword Brothers, inspired by the precedent set in the Holy Land. From its very onset in the eastern Baltic, crusading, targeting indigenous societies branded by Christian commentators as pagan, was facilitated by the acquisition of territory. As the initial conquest of Livonia was nearing completion, the Teutonic Order and its allies unleashed a relentless holy war against tribal groups in the south-eastern Baltic, collectively known as Prussians. In 1237, following a catastrophic defeat in the autumn of the previous year, the remnants of the Sword Brothers were officially merged with the Teutonic Order. The following decades saw the subjugation and reorganization of tribal territories across the eastern Baltic into new Christian polities run by a theocratic elite: the Order, individual bishops, and their cathedral chapters, although north Estonia came under the control of the Danish crown. This was punctuated by episodes of indigenous military resistance, which in the case of Estonia continued into the first half of the fourteenth century (for historical overviews, see Boockmann 1981; Christiansen 1997; Urban 2003; Biskup et al. 2009). The theocracy organized the construction of castles, encouraged colonists, developed towns and introduced Christianity. Only the Lithuanians were able to resist the incursions of crusading armies and remained independent, defiant, and pagan: opponents in an incessant war sustained by the Order throughout the fourteenth century.

The impact of this process on the environment would have been profound as infrastructure and provisioning networks were reorganized. The process of colonization was accompanied by a general increase in cultivation alongside deforestation, although land management strategies varied between commanderies. In most regions, the theocratic authorities encountered landscapes that had already been modified to varying degrees in previous centuries. At the same time, the reorganization of territory and the ongoing war with Lithuania saw the creation of new frontiers and heartlands, which included a distinctive borderland that became known as the 'Great Wilderness'. Levels of biodiversity remained high in this sparsely populated region into the sixteenth century, whilst they noticeably decreased elsewhere due to pre-crusade exploitation. Local livestock husbandry regimes typically continued, although in some areas, such as the Kulmerland, there is evidence for an increase in the size of stock most likely related to improvements in fodder and the introduction of new animals. The process of urbanization created new ecosystems alongside expansive hinterlands that became connected to interregional mercantile networks dominated by the Hanse. The Teutonic Order's own network of houses in Livonia and Prussia also enabled resources to be moved within, and in some cases between, its domains. Luxury commodities such as amber and fur, as well as fishing and access to local resources, came under intensive regulation. The exploitation of resources was ultimately directed to sustaining the theocracy, particularly through the generation of surpluses that could be exported. This was most intensive in Prussia, where a substantial trade in grain and timber developed after the crusades.

Since many aspects of the natural world were sacred to the Baltic tribes, this impact would also be synonymous with the cultural changes that created a new world — a Latin European world — at this newly created frontier between Catholic and Eastern Orthodox 'Christendoms' (Lind 2004). Conquered in the name of the Virgin Mary and with papal authorization, the Baltic crusader states were not replicas of the colonists' homelands but a unique, composite culture. Today, the memories of this culture are preserved in the most spectacular historical monuments in north-eastern Europe: castles, churches, and preserved town centres, as well as the more poignant monuments of the conquered indigenous societies which have survived as striking earthworks. They are typically experienced and understood in isolation, disconnected from their broader spatial context which defined the daily lives of their inhabitants in the past.

The results synthesized within this book are the product of a four-year project entitled 'The Ecology of Crusading: The Environmental Impact of Conquest, Colonisation and Religious Conversion in the Medieval Baltic', funded by the European Research Council (2010–14). The project was motivated by the largely neglected study of environmental change as a means of furthering our understanding of the pivotal cultural transformations in the eastern Baltic, following the crusades in the thirteenth century. Historians and archaeologists, particularly from the 1990s, had engaged with aspects of environmental exploitation in the medieval eastern Baltic, but focused on very specific topics, with data fragmented between disciplines and with a particularistic geographical scope; the Teutonic Order's territories in Prussia and Livonia have been rarely examined as a whole (e.g., Dygo 2003; Ekdahl 1991; Gancewski 2000; Makowiecki 2007; Mugurēvičs 2002; Samsonowicz 1990; Stančikaitė et al. 2009). At the same time, developments in environmental archaeology and history over the last decade have emphasized the fundamental importance of an ecological perspective for understanding the expansion of European society at this formative period in its history (Ervynck 2004; Chudziak 2008; Lagerås 2007; Pluskowski 2010; Hoffman 2014). This refers to the complex relationships between humans and other species — plants and animals — in their shared physical surroundings. The term *Terra Sacra* has been adopted to emphasize the clash of sacred associations embedded in the landscapes of the eastern Baltic. The indigenous network of sacred groves, trees, rivers, lakes, hills, and cemeteries, juxtaposed with the sacralization of space associated with crusading, where the purpose of a holy war could not simply or even primarily be territorial conquest, but the recovery of a sacral domain, of a *terra sancta* — in this case dedicated to the Virgin Mary and ruled on her behalf by a militarised theocracy (Tamm 2013, 445). This transition from an indigenous landscape of localized sacrality to a monolithic Europeanised network of cathedrals, churches, and shrines was in fact very gradual, and the two co-existed alongside each other throughout the Middle Ages. More than a passive backdrop, this provides an essential context for changes in the exploitation of the physical environment.

From Crusading Frontier to Theocratic Polity

Geopolitical boundaries in the eastern Baltic region have changed extensively between the relatively coarsely defined Late Iron Age territories (Fig. 1.1) and present day Estonia, Latvia, Lithuania, the Russian Kaliningrad Oblast, north-eastern Poland, and western Belarus — the latter encompassing part of the territory of Sudovia (or Yotvingia) which was not settled by the Teutonic Order until the late fifteenth century. Indigenous social organization prior to the crusades is widely referred to as 'tribal' (for a recent discussion of this and related organizational terminology, see Bojtár 2009; Nowakiewicz 2010; Valk 2014a). There is evidence for changes in the nature of this organization from the eleventh century, resulting in increasing social stratification linked to the emergence of a distinct, regional aristocratic class (Žulkus, 2002; Šnē, 2005, 2006). Strongholds functioned as central economic entities with their own arable, pasture, and access to various natural resources (see papers in Valk 2014b). The external written sources describe populations in hillforts under the leadership of powerful families led by *nobiles* or *seniors* as they were defined in external Latin sources (Blomkvist 2009, 444), with *konagis* and *rīkīs* as the Prussian equivalent (Długokęcki 2009b, 42; Vykypěl 2011, 249). Political centralization is most evident in Lithuania with the formation of the Grand Duchy (Rowell 1994; Žulkus 2002) but is also evident in neighbouring regions, particularly in response to the incursions of crusading armies. In the southern Estonian province of Saccala, Lembitu emerges as a short-lived leader whilst in Prussia, the Natangian Herkus (or Henry) Monte is one of the leading figures of the Great Prussian Uprising (1260–74). Similar coordination was evident in the failed Estonian uprising in 1343. However, the general lack of a unified supra-regional political structure, as well as the widely documented in-fighting between tribal groups, prevented a cohesive and consistent military response to the crusades (Eihmane 2009). In both Livonia and Prussia, the series of military conquests, treaties, and annexations resulted in the dissolution of the political efficacy of tribal polities, although the individual groups themselves remained linked to their earlier lands, except in

Map 1.1: A map of the modern geopolitical situation in the eastern Baltic region. Map by author.

parts of eastern Prussia which saw significant depopulation and even resettlement (Urban 2000, 332–33).

Whilst the crusades in the Holy Land continue to attract the most attention from historians, the other two major theatres of Christian holy war in the Middle Ages — the eastern Baltic and Iberian Peninsula — have seen a steady increase in regional and international scholarship. By default, this project is firmly situated within the 'Pluralist School' of crusader studies (Riley-Smith 1999), recognizing the military campaigns against the Baltic tribes as penitential wars sanctioned by the papacy, with spiritual rewards (and financial deficits) that were eventually placed on a level with the Holy Land (Fonnesberg-Schmid 2007). Traditional historiography in the Baltic countries viewed the crusades as wars of economic and political aggression, tied to the idea of an ethnically driven struggle promoted by nationalists in the nineteenth and early twentieth centuries (Kreem 2001b; Ekdahl 2004; see below). The continuing debate over the motivations of soldiers, knights, merchants, and colonists to travel to the eastern Baltic and participate in the crusading ventures of bishops, the Teutonic Order, and its secular allies has illuminated a diverse spectrum of agendas. It is no longer useful to separate the crusading movement in this region from the promotion of interregional commerce, exemplified from the thirteenth century by the increasingly sophisticated merchant confederation which came to be known as the Hanse (see papers in Murray 2001; 2009; 2012). The crusading period saw the amalgamation of the interests of *ecclesia* and *mercatura* (Blomkvist 2005, 685) and the emergence of a trading network defined by a common language (Middle Low German), economic paradigm, and broadly shared material culture (Gaimster 2005). At the same time, thirteenth-century *excitatoria* presented the eastern Baltic as a sort of paradise, a land of abundant resources, which required taming and exploiting (Tamm 2009, 21–23). When compared to the diminishing Christian foothold on the Holy Land, all of this generated a climate which made the eastern Baltic an attractive location for colonists who spoke the language of the Christian authorities and merchants.

The impact of the crusades can be characterized as a dramatic shift in regional power relations, resulting in the most extensive and lasting changes to the definition and control of territory in north-eastern Europe. This consisted of a shift from the mosaic of tribal polities to a cohesive polity dominated by a single institution — the Teutonic Order — with authority over extensive areas in the hands of bishops and their cathedral chapters (Maps 1.2–1.3). Bishops were initially able to exercise relatively more sovereignty in Livonia, and the Archbishops of Riga oversaw the dioceses of both Livonia and Prussia (with the exception of Reval (Est. Tallin); Kala 2007, 57), but their power was successfully challenged by the Order in 1268, and the office was incorporated into the Order for a short period with a papal ratification in 1397. In the 1420s this papal order was first suspended and then revoked, but henceforth the Archbishop of Riga never aspired to exercise authority over the Order. In Prussia, the political and military power of the Order was strongest, and of the four cathedral chapters, only Ermland (Pol. Warmia) remained independent from the Order. The Bishops of Ermland, from the mid-fourteenth century designated as 'prince-bishops' by the Holy Roman Emperor, increasingly cherished notions of sovereignty, as expressed in the minting of coinage (Paszkiewicz 2009, 48–49), as well as in the architectural vocabulary of their fortified houses, particularly at Frauenburg (Pol. Frombork) (Dygo 1989, 69). Ermland was itself subordinate to the Archdiocese of Riga from 1245 to 1488, and this relationship remained *de jure* until 1566, when the Archdiocese was abolished (Hubatsch 1973, 144). Despite its aspirations, the Ermland episcopate co-operated with the Order; in 1387 Bishop Heinrich imposed ecclesiastical punishments on anyone who impeded pilgrims (i.e. crusaders) who came to the Order's aid (Ehlers 2001, 37), and in 1410 the forces of the episcopate formed a component of the Order's army.

Given how crusading in the Baltic was intended to safeguard and promote missionary activity, it is perhaps surprising that few monasteries were established on the Order's territories between the thirteenth and fifteenth centuries. By the early decades of the fifteenth century in the Prussian *Ordensland*, there were six Dominican and seven Franciscan houses, five Cistercian nunneries, three Augustinian eremite foundations, and individual convents of Bridgettines, Norbertine nuns, and Carthusians, with some towns containing Beguine communities (Radzimiński 2009, 174–75). In Livonia, there were comparatively fewer monastic houses, with three Dominican and seven Franciscan friaries (of which six were founded in the later fifteenth century, one of which included a community of Poor Clares), three Cistercian monasteries, four Cistercian nunneries, at least two Beguine houses, and individual Augustinian eremite, Antonine, and Bridgettine foundations (Radzimiński 2013, 266–67). Virtually all monasteries were in towns, and whilst the mendicants played a role in the Christianization of the multicultural urban populations, it was left to the theocratic authorities to convert indigenous rural communities (see below and Chapter 2). The Order itself was not engaged in missionary activity, and in contrast to the episcopates it did not actively promote this on its territories. Secular power also remained limited. In Prussia, private residences of knights were permitted, even encouraged, on the Order's lands, but their numbers remained small and the group mustered no effective political agency until the fifteenth century (Jóźwiak 2009a). Indigenous Prussian vassals expected to provide military service, the so-called greater free (*große Freie*) and lesser free (*kleine Freie*, defined as *equites* in Ermland) formed a specific type of knightly class (Dobrosielska and Radzicki 2013). The situation in Livonia was different, where local power was sometimes devolved to loyal *seniors*, which included the so-called Curonian 'kings' (*Könige*). These individuals exercised a higher degree of political freedom but were ultimately subordinated to the Teutonic Order's hierarchy.

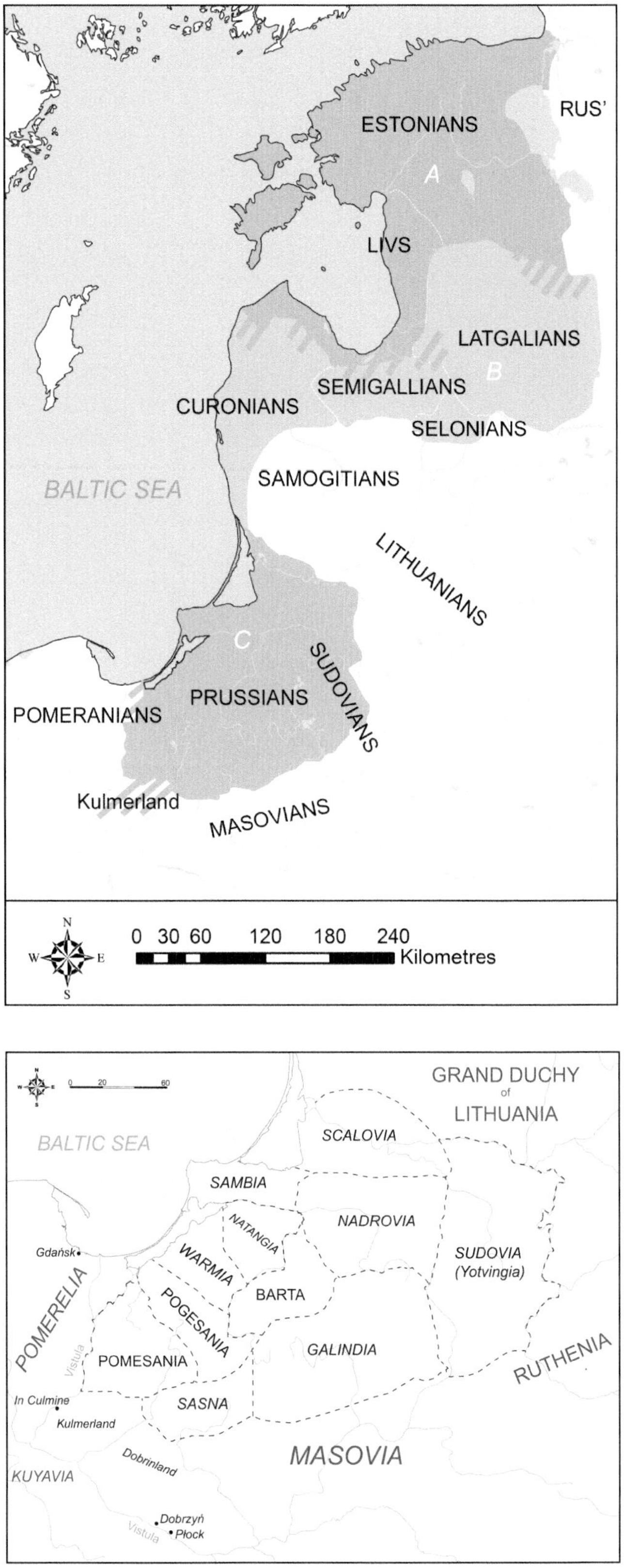

Map 1.2: (a) Principal indigenous territories in the eastern Baltic as documented by the Teutonic Order in the crusading period. Linguistic groupings: A = Finno-Ugrians, B = Transitional Balts, C = Prussians; (b) Prussian territories on the basis of external written sources. Map by author.

Crusader State?

The Teutonic Order's lands in Prussia and Livonia, which came under the single jurisdiction of the Grand Master (*Hochmeister*) in Marienburg (Pol. Malbork)

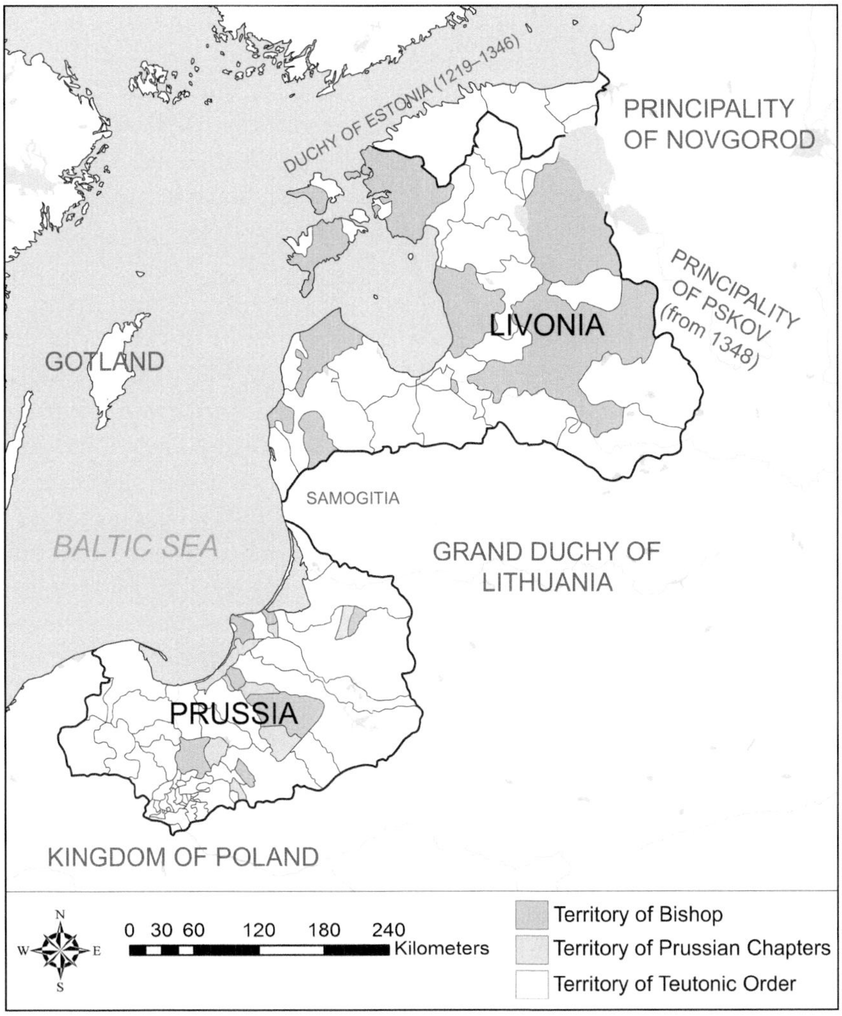

Map 1.3: The Baltic *Ordensland*, showing territorial boundaries in Prussia from *c.* 1400 (after Biskup et al. 2009, with the Kulmerland after Poliński 2003, Map 2), and Livonia *c.* 1450 (after Benninghoven 1963), including the outline of the former Danish Duchy of Estonia, which was sold to the Teutonic Order in 1346. Map by author.

from 1309, has been variously referred to in the scholarly literature as a monastic or theocratic state, as the *Ordensstaat* or Order's state, although given the ongoing discussion on the definition of a 'state', the more preferred term is *Ordensland*, already in scholarly use from the latter decades of the nineteenth century (Treitschke 1862; see also Wenskus 1970). Typically applied to Prussia, the term is used here for all the Order's territories in the eastern Baltic. The name *Terra Mariana* (Land of Mary) was used for Livonia (Mänd 2009, 195), but the Virgin was considered to be the sovereign of both regions and featured widely in the religious and political symbolism adopted by the Teutonic Order (Dygo 1989; Jóźwiak and Trupinda 2007), as well as within its liturgy (Gancarczyk 2010, 273). Some scholars, such as William Urban (2000), consider the use of the term 'crusading' as only applicable to the period of active holy war supported by the papacy throughout parts of the thirteenth century. The development of the *Ordensland* in the fourteenth century has also been compared by some to a secular polity, divorced from its earlier crusading ideology. Certainly, the relocation of the Order's headquarters from Venice to Marienburg in 1309, initiated by Siegfried von Feuchtwangen, was a conscious choice to abandon the eastern Mediterranean front (Morton 2009, 143). Moreover, the Order moved beyond its self-defining roles — holy war, hospitality, and medical care (Sterns 1969, 203) — to include land and resource management on a much bigger scale than in the Holy Land. The narratives produced by and for the Order from the 1290s through to 1390s, particularly the anonymous *Livländische Reimchronik* (*Livonian Rhymed Chronicle*; hereafter *LRC*), *Di Kronike von Pruzinlant* (*The Chronicle of Prussia*; hereafter *KP*) by Nicolaus von Jeroschin and Wigand of Marburg's *Chronica nova Prutenica* (*New Prussian Chronicle*; see Zonenberg 1994), have been interpreted as promoting the knightly identity of the Order in the eastern Baltic, with little or no theological justification (Fischer 2007, 266; Zonenberg 1994, 76), although Jeroschin does refer to the war with Lithuania as *gotis werc* — 'the work of God' (Ehlers 2001, 33). Marienburg, in the last decades of the fourteenth century, has also been compared to a secular court (see Jóźwiak and Trupinda 2007), as has the status of the Livonian Master from the fifteenth century, culminating in the reign of Plettenberg (Kreem 2001a; 2002; 2013).

However, the Order successfully promoted its wars against Lithuania in terms that would appeal to aspiring crusaders, coinciding with a period of internal reform and spiritual revivalism in the first half of the fourteenth century (Fischer 1991, 190), in part reflected by the importance allocated to religious spaces within its fortified convents (Jóźwiak and Trupinda 2012, 177–84), and the commissioning of iconography that framed the struggle with paganism in apocalyptic terms (Kutzner

Figure 1.1: A drawing of a wall painting at the procurator's castle in Lochstädt (Rus. Pawlowo), showing the archangel Michael as a crusader, slaying the apocalyptic dragon (after Domasłowski 2010, 112; *Schloß Lochstedt und seine Malereien*, pl. 3, Berlin, 1910).

1995; 2010) (Fig. 1.1). Links with the Holy Land were emphasized until at least the early fifteenth century, particularly in the Order's architectural iconography and the promotion of the cult of the cross (Dygo 2009a, 359–60), itself an integral part of crusader identity (Schenk 2012), whilst perceptions of pagan opponents in the Baltic were aligned with stereotypes of Muslim aggressors (Murray 2010). The Order's *Reisen* or campaigns across the borderlands into Samogitia and Lithuania throughout the fourteenth century dovetailed with the chivalric culture of European knights. But they were punctuated by religious ceremonial beyond the usual conduct of secular warfare, with participants defined as 'pilgrims' who took vows and could benefit from indulgences offered by the Order (Paravicini 1989; Ehlers 2001; Pluskowski 2013, 161–62, 280). Above all, the Teutonic Order maintained the cult of the Virgin Mary as a central part of its corporate identity into the fifteenth century (Fig. 1.2). In contrast to the emblematic use of saints by secular states, such as the leonine St Mark by the Republic of Venice, the cult of the Virgin pervaded the devotional lifestyle of the Teutonic Order, as well as other social groups within the *Ordensland* (Dygo 1989). Historians see the official conversion of Lithuania to Catholicism as provoking an internal crisis within the Order, precisely due to the loss of its original *raison d'être* (Józwiak 2009b, 332–33). However, the problem was not only external: increasing secular tendencies and a decline of interest in the religious life amongst the brethren were already evident from the reign of Grand Master Winrich von Kniprode (1352–82), with falling standards also noted in some of the Prussian convents in the first half of the fifteenth century (Tandecki 2009b, 414–15). Whilst the Order's campaigns into Samogitia from 1401–09 were framed as crusades against apostates and pagans, it is clear that not all those who participated considered these as penitential wars (Chollet 2015). Moreover, the Order was also evidently investing in expansion away from the 'crusading frontiers' of the east. In the last two decades of the fourteenth century it sought to align its territories with the Holy Roman Empire, culminating in the purchase of the Neumark in 1402, which remained in the Order's possession until 1455 (Tandecki 2009a, 106).

Despite this internal crisis and the increasingly secular character of the Order, in the early fifteenth century, the Knights were still perceived as a functioning military order and defended their crusading role at the Council of Constance (1415–18), where the Polish-Lithuanian delegation argued for the Order to redirect its crusading efforts against Tartars and Turks (Mickūnaitė 2007, 47). Over the course of the fifteenth century the Prussian *Ordensland* was truncated and its remnants secularized in 1525. During this time the Livonian branch of the Order began to operate more independently from its dominant Prussian branch (Kreem 2001a, 13, 34) and held onto its lands until secularization in 1561, but its adherence to a crusading ideology up to that point remains unclear. Not all military activity organized by the Order was framed in crusading terms, but the Grand Master had used traditional crusading rhetoric when seeking papal and Imperial support during the war against Novgorod in 1443–48 (Dirks 1988, 44). Wolter von Plettenberg's constant petitioning for crusade indulgences for wars against the Russian principalities only achieved success near their conclusion, as the papacy sought Eastern Orthodox assistance against the Turks (Housely 1992, 373), but suggested the ideology of holy war was still considered as a viable option for conducting

Figure 1.2: The Virgin and Child at the eastern end of the church at the high castle in Marienburg (Pol. Malbork): (a) a rendition from the lost painting 'the siege of Marienburg', *c.* 1480; (b) the sculpture restored in 2016 (after Kilarski 2007).

warfare in the eastern Baltic in the early sixteenth century (Riley-Smith 2005, 254; Kreem 2013). Following Jensen's (2001) use of 'crusader state' to characterize twelfth-century Denmark, Reisnert's (2009, 61) description of Estonia as 'a crusader state under Danish rule', and most recently Barber's (2012) use of the term for the Latin East, it is adopted here to emphasize the fact that a military order dominated the governance of the medieval eastern Baltic and promoted a crusading ideology that could be readily situated within the chivalric culture of late medieval Latin Christendom.

Multiculturalism and Apartheid

If the Order continued to present itself as the defender of *ecclesia* committed to an ongoing holy war against paganism, to what extent did this affect relations with the conquered populations? The juxtaposition of colonists and colonized in any model of Europeanization is inevitable, but the realities of the ethnic dialectic in the *Ordensland* remain debated. Historians have challenged many earlier assumptions and have distanced themselves from nationalist interpretations of cultural struggles. The *Drang nach Osten* (Push to the East) is no longer accepted as a valid understanding of the settlement of German-speaking colonists in north-eastern Europe over the course of the Middle Ages — there was no ethnically driven agenda linked to insatiable 'land hunger' in the eastern Baltic. Towns were multicultural centres that attracted rural migrants as well as foreign merchants, whether the Russians in Riga or the English in Danzig (Pol. Gdańsk) who invariably clustered together in districts or 'quarters'. Colonists conveniently labelled 'German' could also include Scandinavians and Slavs; Lübeck, the core of the Hanse and source of many colonists heading to the eastern Baltic, was itself populated by Germans and Slavs (Fehring 1990). A variable percentage of the populations of Prussian and Livonian towns included indigenous people; documentary sources indicate native Prussians made up 3% of the population of the old town of Elbing (Pol. Elbląg) in north-west Prussia, and they were also present in the new town (Czaja 2009b, 447), whilst Dorpat (Est. Tartu) in eastern Livonia appears to have been dominated by local migrants from the mid-fourteenth century on the basis of archaeological interpretations (Kalling 1995). Castle communities, effectively miniature urban enclaves, were also multi-ethnic; the Teutonic Order's brethren were supported by a household drawn from colonists and the indigenous population. But untangling the socio-topography of castles at any given time is highly problematic, and these sites are typically considered as a single unit: a cohesive, hierarchical community. Archaeologically, the presence of indigenous populations in the urban colonies of the Baltic crusader states is much more clearly defined in Livonia than in Prussia, where the lack of excavated late medieval indigenous cemeteries and settlements (with the exception of sites such as Równina Dolna and Alt Wehlau) makes it difficult to assess the resilience of native Prussian culture. Invariably a proportion of the indigenous population fulfilled the role of middlemen between the colonies and the countryside, opening the way for cross-ethnic friendships and liaisons (Blomkvist 2009, 448).

Nonetheless, there is evidence for segregation. In Prussia, the *Iura Pruthenorum* (Prussian Law) governed indigenous settlements and restricted the mobility of the native population from the mid-fourteenth century (Matuszewski 1963; Długokęcki 2009c), although its application varied regionally (e.g., Szczepański 2016). In the case of Livonia, archaeologists have consistently argued for a segregated society, with colonists restricted to towns and castles and largely absent from the countryside (Mugurēvičs 1990; Pluskowski and Valk 2016). From the mid-fourteenth century, it was possible to speak of a legal separation of Germans (*Deutsche*) from non-Germans (*Undeutsche*). This designation should not be considered in nationalist terms but rather as social categories, with authority more clearly linked to a developing local German identity that did not exist during the crusading period (Kala 2009, 189). This was also a time when German was replacing Latin as the official language of record-keeping (Kala 2006, 593). Blomkvist (2005, 232) labels this segregation as a 'truly colonial apartheid system', but the extent to which this affected cross-ethnic relations is, again, debatable. The sharpest contrasts were invariably found between towns and villages, and in the fifteenth-century struggles between the townspeople of Riga and the Order, both German and non-German were united. Although indigenous Prussian culture was more heavily Germanized, and by the fifteenth century all inhabitants could be described as 'Prussians' irrespective of what language they spoke (Czacharowski 1983, 72), ethnic distinctions remained, no doubt reinforced by the *Iura Pruthenorum* and the maintenance of distinct and socially charged funerary practices, as well as the persistence of the native Prussian language in some regions into the sixteenth century.

Religious Transformation

A major impact on the identities of the indigenous populations, and a key raison d'etre for the crusades, was Christianization. Religion, after all, distinguished crusaders from their targets and the narratives of the crusades emphasised polarised differences. To Henry of Livonia, for example, technological supremacy acquired a religious dimension: Christians used it to conquer and convert pagans (Blomkvist 2005, 674–76), whilst the native woods were a distinctly pagan landscape (Nielsen 2011). Throughout and into the fourteenth century, the Teutonic Order continued to view the struggle against paganism in apocalyptic terms; an ideology framing the relentless *Reisen* into Lithuanian territories (Kutzner 2010, 241). Archaeological data points to some Christian influences before the crusading period, although the use of cross pendants and the presence of early churches on Saaremaa (Mägi 2002), and perhaps also in Palanga (Žulkus 2002, 202), does not amount to the proliferation of Christianity before the imposition of a diocesan system. The religious conversion of the eastern Baltic has been studied mostly by historians (for a summary, see Pluskowski 2013, 246–93), and it is clear that the process took many centuries to implement. Dioceses were established relatively quickly in Livonia and Prussia, and parishes gradually developed alongside the reorganization of settlement and colonization. Officially, the entire region came under the administration of the Latin Church following the establishment of the Samogitian Bishopric in 1417, but a range of sources indicate that non-Christian cult practices continued, particularly in Lithuania and Livonia. This typically took place in the countryside, reinforcing the urban-rural contrasts outlined above, and towns have been perceived as 'Christian islands in a sea of pagan-Christian-syncretism' (Blomkvist 2009, 454; Valk 2003).

Burial rites and the deposition of grave goods continued in both town and country, and whilst these are not generically reliable indicators of religious allegiance, as they feature widely across Christian Europe into the modern period, for Christians in the eastern Baltic they were a defining feature of the pagan 'other' (Tamm

2009, 30–34). The use of cemeteries as sacred sites — important loci of pre-Christian cult praxis most explicitly outlined in the Treaty of Christburg (1249) — does appear to continue in rural Lithuania and Livonia; in 1428 the Provincial Council of Riga condemned burials in unconsecrated cemeteries in the wilderness (Valk 2004). Indigenous religious practices could be highly localized, as evident in the documented continuation of cremation rites in Curonia described by the French knight Guillebert de Lannoy in 1414 (Laime 2009), and the cremation urn associated with a fifteenth-century coin hoard discovered in the forest of Eylau (Pol. Iława Pruska, Rus. Bagrationovsk) (Chmielewski 1963). This notion of a 'syncretic society' provides an interesting window into the tolerance of indigenous customs or, perhaps more likely, on the limits of theocratic authority alongside local expressions of *Realpolitik*. The Order was consistently accused of failing to Christianize its pagan vassals, with language remaining an insurmountable barrier for successfully conveying even basic Christian concepts (Kosman 1976, 25). Evidence for continued or discontinued perceptions of elements of the landscape as sacred can therefore function as a barometer of tolerance and control, of the impact of the Christian elite beyond the castle walls. However, it can also reveal longer term trends in religious transformation which were already evident before the crusades.

Europeanization and Protocols of Territorial Organization

The transformation of eastern Baltic society as a result of the crusades has in recent decades been framed as part of the process of 'Europeanization' (see Chapter 2). The model of Europeanization sees its development within a Continental (i.e., Western European) core, providing a single concept for integration through cultural subordination using peaceful and violent mechanisms. This was facilitated directly by colonization, defined here as the movement of new ethnic groups into politically subordinated territories resulting in the establishment of new settlements. In the case of the eastern Baltic region, specific cultural elements — namely language, religion, technology, social and economic paradigms — were imported through conquest and trade, and imposed on indigenous populations who reacted in various ways (Blomkvist 1998; 2009). The resulting societies, under the political aegis of the Teutonic Order, emerged from the spectrum of encounters between colonists and colonized, although in general this resulted in the marginalization and subjugation of the indigenous populations. The model of Europeanization has been criticized and even dismissed (see Kļaviņš 2019), but given its pervasive influence it is important to engage with (see Chapter 2).

Medieval Livonia and Prussia have often, and usually in casual rather than analytical terms, been described as 'colonial' societies. Whilst the early territorial gains in the eastern Baltic had initially been ratified by the Emperor, the papacy came to play a more prominent role. Papal authority in Prussia was confirmed by Gregory IX in 1234, and in 1243, Pope Innocent IV had specified that Prussia (including the Kulmerland) was to be governed by the Order on behalf of the Holy See (Dygo 2009b, 64). In Livonia this was never as clearly articulated, where the Archbishops of Riga considered the Order as their vassals (inheriting this status from the Sword Brothers), an ineffective claim that was contested and eventually abandoned by the end of the fourteenth century. Nonetheless, subordination to papal or even Imperial authority was relevant in name only, qualified by the recognition of papal authority with regular lobbying of the curia for political support in the fourteenth and fifteenth centuries (Urban 1978; Eihmane 2012). Moreover, relationships with the papacy were not always positive; in 1354, when the Order refused to give up properties it had seized in Riga's Archbishopric, the pope placed the Knights' territories under interdict for around three decades (Kala 2007, 58). Relationships with the Holy Roman Empire were equally diffuse. There were of course cultural synergies. The Order exercised its power in a comparable way to ecclesiastical lords in Germany, using visible symbols of authority (fortified convents and smaller residences) and territorial control (articulated by intensive regulation and military force) (Eldevik 2003).

The specialized economies developed by the Order within its eastern Baltic territories were also found in neighbouring regions. In Brandenburg, lordships operated *Vorwerke* (arable domain farms), *Meiereien* (beef or dairy-cattle farms), and *Schäfereien* (sheep-farms), which served a variety of markets and were associated with different patterns of landuse, in turn related to the variable topography of the province (Scott 2002, 84). The process of colonization was certainly a familiar tool of settlement in Central Europe, and Lübeck's variable influence can be glimpsed in both Livonia and Prussia, but the Order's Livonian towns were founded using Riga's template, and in Prussia the ubiquitous Kulm Law was a creation of the Order (see below). The Empire was also the principal area for recruiting brethren into the Order, as well

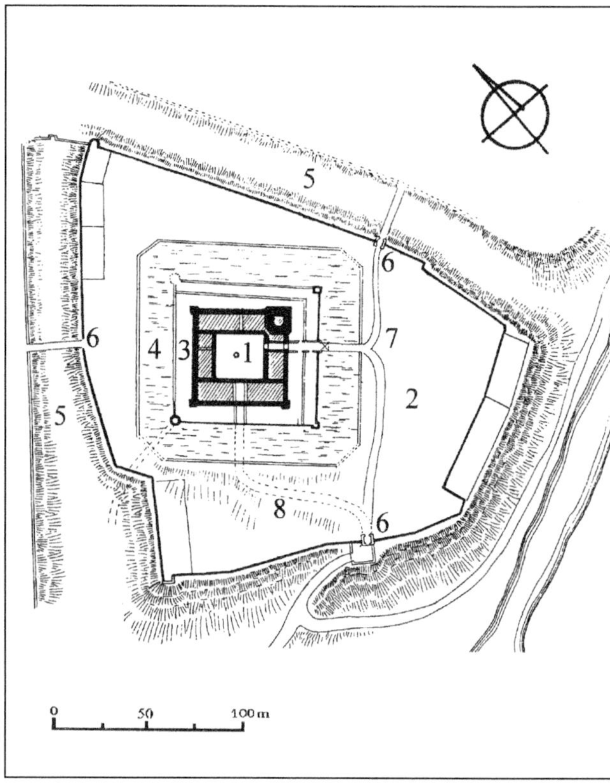

Figure 1.3: (a) Gniew (Ger. Mewe), the typical form of the conventual castle constructed by the Teutonic Order in the eastern Baltic (after Klat and Mykowski 2002, 23), with (b) an adaptation of Conrad Steinbrecht's plan.

as clerics, and the source of many wealthier merchants and peasant colonists. Whilst the Order enjoyed the nominal support of the emperor and became entangled in Imperial politics in the fifteenth century, there was never any implication of vassalage (Biskup 2009, 291). These connections were therefore ones of shared cultural practices, patronage, and commerce, rather than an Imperial core with the Baltic theocracies as politically subservient satellites. The Order's lands were certainly not replicas of Imperial provinces. Even if Gaimster's (2005, 2007) characterization of the consumption of imported ceramics by German mercantile communities in the eastern Baltic as a form of cultural dependency is accepted, in the territories of the Order this appears to have reflected commercial connectivity. There is a visible gradient in the consumption of imported ceramics from the major Hanseatic ports, particularly Elbing, where excavated imports appear in abundant and diverse numbers, through to those located some distance inland, but still connected with the network, such as Thorn and Graudenz, through to less well-connected castles of the Order, such as Strasburg (Pol. Brodnica) where imported wares are found only very sporadically (Grążawski 2009). Indeed, numbers of imported ceramics in the Order's castles are comparatively limited compared to urban assemblages, although data is lacking from the convents of the major Hanseatic centres.

Therefore, with the possible exception of Danish Estonia until 1346, the term 'colonial' is neither appropriate nor useful for understanding the character of the eastern Baltic crusader polities, and the Order was the de facto ruler in the territory it occupied, with its own sovereign government, monetary system, and bureaucracy (see the discussion on crusading and colonialism in Kedar 1992; see also Czaja and Nowak 2013, 17; and Herrmann 2007, 33, for a summary of the rejection of medieval Prussian art and architecture as *Kolonialkunst*). It is more useful to define the variable process of colonization as facilitating gradual 'Germanization'; Mortensen and Bisgaard (2013, 26) even described the medieval Baltic Sea region as 'a cultural province of northern Germany', and in south-east Prussia the introduction of Masovian culture would eventually result in the formation of a distinct Masurian identity (Ekdahl 2004; Długokęcki 2009a, 212; see also Blanke 1999). However, these cultural synergies should not be con-

Figure 1.4: Symbols of theocratic authority: Schillings of (a) Winrich von Kniprode (1351–82); (b) Michael Küchmeister von Sternberg (1414–22), displaying the emblem of the Teutonic Order and the eagle of the Grand Master (Warszawskie Centrum Numizmatyczne); (c) an artig (equivalent of half a schilling) of Livonian Master Wennemar von Brüggenei (1389–1401), minted in Reval (Est. Tallin) (Cēsis Castle Museum).

flated with political subservience. Indigenous leaders who wanted to keep their position had to adopt *select* elements of the incoming German culture and language, evident in both Livonia and Prussia already during the period of active crusading. In this respect, indigenous power structures were accommodated to varying degrees within the new political regime, which was defined by a wholly new system of territorial administration.

The Administration of Conquered Territory

The Teutonic Order's own administrative structure developed rapidly over the course of the thirteenth century, reflecting the dynamics of its activities in the Holy Land, Prussia, Livonia, and elsewhere in Europe, primarily within the Empire. The Order's estates across these regions provided the income and resources required for maintaining its garrisons and infrastructure, and the loss of territories in the Holy Land (particularly Montfort (Starkenberg) in 1271) saw a gradual realignment of investment to the eastern Baltic. Here, the administrative structure developed to cope with extensive territorial gains (for the most recent summary of the administration of territory in both Prussia and Livonia, see Czaja and Radzimiński 2013). In the period of active crusading, the construction of castles and planned towns required immediate and sustained provisioning; food, fuel, and building materials as basics, followed by a suite of secondary products. This was organized from the very onset through the imposition of commanderies (*Komturei*) on the annexed landscape. Commanderies defined a territory managed by a commander (*Komtur*) heading a fortified convent (rendered in Latin as *domus* and German as *hus*/*haus*; see Jóźwiak and Trupinda 2012, 96–98), where military resources and political authority were concentrated. Convents consisted of a nominally regular number of brethren as established in the Order's rule — twelve, representing the apostles — although in reality this figure varied dramatically over time, and the brethren were supported by an extended household of sergeants, priests and chaplains, half-brothers, half-sisters, servants, and administrators, as well as visiting guests (Tandecki 2009b, 406–11). The shape and size of commanderies varied dramatically as a result of their specific historical development, initially linked to the process of military conquest and the subdivision of territory. The delineation of commanderies began first in Livonia, with Wenden (1237, Latv. Cēsis), Leal (1242, Est. Lihula), and Fellin (1248, Est. Viljandi), followed by Goldingen (Latv. Kuldīga) and Ascheraden (Latv. Aizkraukle) in 1252. Riga's commander is first mentioned in 1253, although the convent clearly existed earlier (Tandecki 2013, 174). In Prussia, commanderies were also not clearly defined until the mid-thirteenth century, when the first commanders of Elbing (1246), Christburg (1250, Pol. Dzierzgoń), Balga (1251), Zantyr (1251), and Sambia (1254) were named (Dygo 2009b, 74).

The system of territorial reorganization was a composite drawn from the Order's lands in the Levant and various parts of the Empire, perhaps even informed by the Order's relatively obscure experience in the Transylvanian Burzenland (1211–25) which ended shortly before the Prussian Crusade began. From 1237, the crusading enterprise in the eastern Baltic was dominated by the Teutonic Order. Both Livonia and Prussia were governed by land masters (*Landmeisters*), with the Prussian office dissolved following the relocation of the Grand Master to Marienburg in 1309. The Livonian branch of the Order continued to be presided over by the Land Master (henceforth referred to as the Livonian Master), who was based in Riga and from the late fifteenth century in Wenden, and his deputy, the Marshal (*Landmarschall*) based in Segewold (Latv. Sigulda). A further level of standardization developed by the late thirteenth century with the design of 'conventual' castles (with quadrangular or rectangular plans, referred to by German scholars as *Konventsburgen*), which then typified the majority of the Order's convents in Prussia (Herrmann 2007, 81) and Livonia (Caune and Ose 2004; Arszyński 2013). These fortified monasteries expressed both the identity and lifestyle of the Order as a religious corporation, and as sovereign lords (Fig. 1.3). The coinage issued by the Order became standardized, reflecting the authority of the Grand Master (see Paszkiewicz 2009) (Fig. 1.4), and its aristocratic character was consolidated in 1309, when recruitment began to focus solely on the nobility. The centralization of the administration of the *Ordensland* in the fourteenth century, particularly associated with Grand Master Winrych von Kniprode, resulted in convents maintaining inventories, copies of which were sent to the chancellery in Marienburg and collated (and updated) from 1400 as *Das große Ämterbuch des Deutschen Ordens* — the 'Great Book of Offices' (*GAD*). The main organizational difference between Prussia and Livonia was one of greater centralized control by Marienburg in the former, and more independence exercised by individual commanders in the latter, albeit under the leadership of the *Landmeister*. Moreover, the fifteenth century saw this trend accentuated with the effective privatization of conventual property, a shift from corporate to individual ownership (Kreem 2002). The production of separate Prussian and Livonian literature within the Order — emphasizing regional histories — suggests a level of detachment within the superstructure that nominally came under the singular authority of the grand masters.

Over the course of the fourteenth century, commanderies began to include smaller districts (*Ämter*) called advocateships (*Vogteien*), procuratorships (*Pflegerämter*), and at the lowest level bailiwicks (*Kammerämter*). The first advocates were appointed in Livonia and are documented in Prussia from the 1270s, with the first procurators associated with Lochstädt (today Rus. Pawlowo) and Golub before 1309 (Jóźwiak and Trupinda 2007, 25). Although a few new convents were established in the fourteenth century, the preferred policy for administering newly acquired territories (e.g., Gotland, the Neumark) was to appoint advocates who were directly subservient to the Grand Master. Livonian advocates were typically subordinated to either individual commanders or the Master, and the unusual office of a *Landvogt* is documented briefly in Karkus (Est. Karksi) in the mid-fifteenth century (Tandecki 2013, 172, 175). More specific adaptations to the challenges of managing particular environmental resources resulted in two specialized districts: *Waldämter*, which managed heavily wooded landscapes, and *Fischämter*, which focused on fishing. The abundance of amber in the western Sambian coastline saw the creation of the office of *Bernsteinmeister* (amber master), based in the castle at Lochstädt. Specialized farms (*Vorwerke*) were introduced at the frontline of exploiting the landscape; by *c.* 1400 the Order maintained 212 such farms in Prussia alone (Czaja and Nowak 2013, 20).

These administrative districts were directly run by officials of the Order: advocates (sing. *Vogt*), procurators (or provisors; sing. *Pfleger*), bailiffs (sing. *Kämmerer*), and forest masters (sing. *Waldmeister*) from their castles or residences (Jóźwiak 2001, 223; for the latest summary of Prussia, see Biskup et al. 2009; for Livonia's administration, see Tandecki 2013). These lesser officials served the role of judges, tax collectors, and military organizers, supervising the Order's tenants and provisioning networks and relieving some of the administrative burden of the commanders. Heterogeneity across the *Ordensland* can be explained in terms of the agency of individual grand masters, commanders, procurators, and advocates, some of whom were more ambitious than others and invested more (or less) in the infrastructure they were directly responsible for, through choice or for strategic reasons. The commandery system, from its initial conception to its continued use into the fifteenth century, was a means of organizing settlement and the resources required to sustain the Order's convents. It was a model adopted by the military orders in general, but tailored specifically to the experiences and challenges of governing Livonia and Prussia.

The Organization of Colonization and Vassalage

All new settlements established by Christian authorities in the eastern Baltic were organized according to formulaic town laws, the application of which varied from region to region and over time. In Danish Estonia, the Lübeck Law was used, whilst in the rest of Livonia town privileges were organized using the Riga Law, itself derived from the Lübeck Law (Czaja 2013, 220). Up to the mid-sixteenth century, a total of nineteen towns was founded in Livonia, of which ten were located on the Order's territory. With the exception of Riga, all Livonian towns originated alongside castles. In Prussia, the process of colonization was much more intensive, particularly from the end of the thirteenth century until the 1360s. The most frequently used municipal charter was the Order's own Kulm Law (Lat. *Ius Culmense*, Ger. *Kulmer Handfeste*), derived from the Magdeburg Law, formulated in 1233 within a few years of the start of the Prussian Crusade and reissued in 1251. Of the ninety-six towns established in Prussia (of which seventy-two were ruled by the Order), eighty-nine were organized under the Kulm Law, whilst the Lübeck Law, which gave towns more autonomy, was granted to only seven (Czaja 2009a). The Lübeck Law had been adopted in Pomerelia before its annexation by the Order, whilst the Magdeburg Law became the standard for towns in Poland, Lithuania, and western Ukraine (Ebel 2004). The relationship between the Teutonic Order and its towns was definitively one of overlord and subject, although some of the largest urban centres aspired to a high degree of political autonomy. Relations between the Order and Riga, in particular, became problematic shortly after the former had acquired the lands of the Sword Brothers (see Chapter 8). The town represented the centre of power for the Order in Livonia, although its importance diminished over time; by the end of the thirteenth century some sixty brothers are documented in the Riga convent, compared to thirteen (including the commander) in 1442 (Tandecki 2013, 177). Reval, incorporated into the Order's domain only in the mid-fourteenth century, came to be regarded as a problematic 'military partner' (Kreem 2002, 21, 100), whilst the patricians of the main Hanseatic towns in Prussia worked closely with the Order's quartermasters and agents (Czaja and Nowak 2013, 21). The situation changed only in the early decades of the fifteenth century and the infamous Prussian Confederation in 1440 saw a protracted rebellion against the Order in over a dozen of its key urban centres, and ultimately led to the shrinkage of the Prussian *Ordensland* in 1466.

In contrast to Livonia, within the Order's Prussian territories, alongside towns, hundreds of colonizing villages were also organized under the Kulm Law and Magdeburg Law, with certain indigenous Prussian settlements reorganized under the more restrictive Prussian Law (Lat. *Iura Pruthenorum*, see below). In Pomerelia, the Polish Law of settlement was also defined, and a variant used in Pomesania, Sasna, and part of Pogesania was codified in *c*. 1340 as the Pomesanian Law (Patze 1964; Długokęcki 2009c). These laws prescribed a settlement's layout, its rights of access to fields, woods, rivers and lakes, as well as its juridical and inheritance rights (see Chapter 15). The allocation of fiefs and real estate provided attractive incentives for settlers, and consolidated guarantees of loyalty from indigenous vassals. In return, they were bound to *corvées*, usually in the form of military support and economic services, and they also had to pay taxes and tithes (Czacharowski 1986; Długokęcki 2009c). In both Prussia and Livonia, the Order assimilated native forms of territorial organization such as the Prussian *lauks* (rendered in Latin as *campus* and German as *Feld*), Estonian *vackus*, Latvian (previously Latgalian) *pagasts*, and Curonian-Semigallian *novads*: regionally variable settlement units whose precise character continues to be the subject of debate (Nowakiewicz 2010). At the same time, land divisions widely used in German-speaking regions were imported into the eastern Baltic, specifically the *Hufe* or *mansus* (a unit of land, around 16.5 hectares) which was used when designating land for colonists. The *Hake*, *uncus*, and *aratrum* all designated units of land that could be ploughed (variously defined as 8–12 hectares), with specifications reflecting actual differences between the introduced German, asymmetrical heavy ploughs, and the smaller Slavic/Baltic ard ploughs (Bartlett 1993; for varying definitions of small and large *Haken* used in commanderies in Pomesania, see Szczepański 2015, 105–111).

The regular use of formulaic municipal laws within the annexed territories emphasized the fundamental connection between settlements and their surroundings. The specificity of the charters and privileges handed out by the Order indicates a careful and calculated appraisal of annexed territory, a landscape which they had become intimately familiar with over the many decades of crusading. Effective provisioning underpinned the success of colonization, and required significant logistical organization from the very onset of crusading. In some areas, the theocracy simply procured existing provisioning systems, seizing fields, pasture, livestock, and agricultural surpluses. In other areas colonists had to develop a working relationship with the indigenous population, and where depopulation had occurred necessitated starting from scratch. At the same time, the paradigm of colonization was cohesive; when it came to planned settlements, those established by the Order were the most uniform in terms of land allocation of all medieval German-speaking colonies (Aubin 1966, 463).

Taxation became integrally connected with the use of the landscape; a surplus had to be produced to convert into coin or provide payment in kind. The commodities provided by the Order's vassals extended to all valued natural resources, including animal products (particularly fur), timber, honey, and even amber in some settlements. These derived from the lands attached to individual settlements, which in the case of towns were referred to collectively as its patrimony or *Stadtmark*. However, the staple commodity provided by the Order's vassals was grain. Taxes stipulated by the Kulm Law consisted of a tithe of a measure of wheat and rye, and an annual tax of one-half mark per *Hufe*. Tithes could also be paid in grain and, as a 'religious tax', would have underlined the economic authority of the new theocracy. Indeed, after the alleged mass-baptisms at Holm (Latv. Salaspils) and Üxküll (Latv. Ikšķile) in 1198 following the arrival of a German crusading army, as documented in Henry's *Chronicle of Livonia* (*HCL* II, 7), the converts were obliged to pay a measure of grain per *aratrum* for the expenses of each priest. The collection of taxes and tithes required the building of extensive storage facilities and provided the Order with its most important commodity for export. Trade in Prussia was organized by the *Großschäffer* and his agents in Marienburg and Königsberg (Rus. Kaliningrad), whilst in Livonia it appears to have been supervised by the Master and organized by lesser officials (Kreem 2002, 104). At the 'receiving end' of its West European network, the Order had agents in Bruges and London, closely associated with the Hanse's *kontores* (Samsonowicz 1994).

Historians have suggested the principal economic differences between Prussia and Livonia reflected the development of intensive grain production in the former region. But even the comparatively modest economy of Livonia was largely driven by the production of grain (Kreem 2002, 101–03). In contrast to Prussia, there is a general consensus that in Livonia the agrarian practices of the indigenous population did not change until the fifteenth and sixteenth centuries, coinciding with more intensive grain production, the introduction of serfdom, the restructuring of the manorial economy, and the reduction in the number of local vassals exercising power (Šnē 2008, 94–95). Therefore, an economic model has already been developed by historians for medieval Prussia and Livonia, albeit one which treats these regions as peripheral singularities serving a European core (for a detailed discussion see Chapters 9 and 16). Nonetheless, it is clear that protocols of territorial organization were also protocols of environmental exploitation. Given this inextricable connection and following Karczewski's (2006, 2011) lead in regarding the environmental aspects of eastern Prussian as culturally defining, which can also be understood from a human ecological perspective as 'niche construction' (see Chapter 2), long-term trends in environmental exploitation can be used as valuable and multi-scalar barometers of cultural change.

Cultural Landscapes in the Eastern Baltic: Towards an Environmental Archaeology of the Ordensland

The conquest of the medieval eastern Baltic clearly resulted in significant changes to the organization of the landscape alongside the mobilization of residing and migrant communities; indeed, as outlined above, this was the very foundation of the *Ordensland* and is integral to understanding its development and maintenance at virtually every level (see Chapter 9). But what is meant by 'landscape'? From an archaeological perspective, the term has been discussed for over a century (Johnson 2006). Most recently, the Historical Landscape Characterization protocol (see Chapter 2) distinguishes 'landscape' from 'environment', defining it largely as a product of human perception (Fairclough 2006, 204). This, in turn, places an emphasis on the value attributed to any given space. Given the focus on environmental exploitation in this study, the term 'economic landscape' may seem most appropriate (Scott 2002, 72), and fiscal histories of the Teutonic Order have been written (e.g., Sarnowsky 1993). One of the major impacts on indigenous societies was subordination to a system of taxes and tithes, and the introduction of the regulation of property and land. But this excludes indigenous, non-economic values which endured into the post-medieval period. The persistence of indigenous technologies has a compelling cultural explanation; the use of specific architecture, clothing, tools, and other artefacts was inextricably entangled in customs and habits, in turn linked to social standing within the community — these were most pronounced in the countryside, where contact with colonists was minimized compared with the towns (Blomkvist 2009, 458). In a society where varying degrees of apartheid were a daily reality, emblems were powerful symbols of identity and rank, not least of all the cross of the ruling theocracy which served as a dual religious and political symbol. From the colonists' perspective, the construction of moated systems and canals,

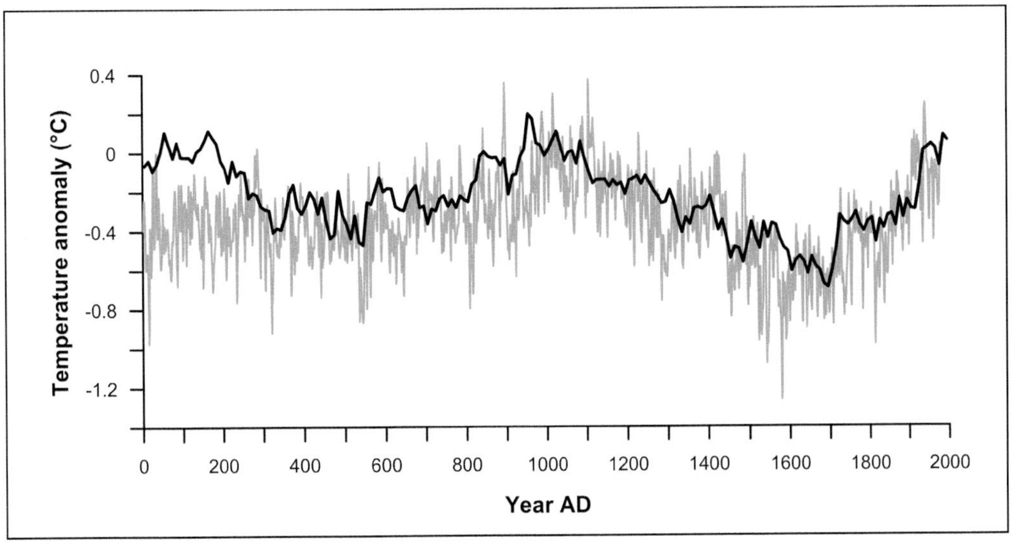

Figure 1.5:
Climate records showing the Medieval Climatic Anomaly (MCA) followed by the Little Ice Age (LIA). After Ljungqvist 2010 (black) and Moberg et al. 2005 (grey). Figure courtesy of D. Fleitmann.

amongst other features, also points to a more complex utilization of the landscape by the Order. Significant investment in religious infrastructure and material culture, as well as the promotion of a crusading ideology, moves us away even further from a purely economic model, despite the Order's much lauded fiscal achievements by the end of the fourteenth century.

A more nuanced understanding subdivides the landscape into physical (the matrix of ecological interactions between people, animals, plants, soils, and water), conceptual (the framework for social, religious, and economic organization), and chronological elements, with the latter two as the main drivers (Palang et al. 2005, 290). The importance of climate is increasingly recognised in studies of cultural change. The crusader states in the eastern Baltic formed at a time traditionally linked to the end of the Medieval Climatic Anomaly (or Medieval Warm Period) and their development is spanned by two climatic minima: the Wolf Minimum (1280–1350) and Sporer Minimum (1460–1550) (Xoplaki et al. 2011) (Fig. 1.5). The use of the term 'setting' does not imply a passive backdrop or stage but rather an interactive space which shapes, and is simultaneously shaped by, human activity. Together, they are referred to as the 'cultural landscape' (Birks et al. 2004) or, 'biocultural landscape' (Pungetti 2013). This definition of human ecology recognizes the complex relationship between the two and forms an important focus of inquiry (see Chapter 2): in a colonizing context, to what extent do settlers adapt to/modify local ecosystems, and how is this related to the imposition and development of cultural norms and worldviews?

Whilst previous studies have typically included an economic role for the Order's castles as a descriptive addendum with sporadic examples of agrarian commodities listed in inventories, here we argue for a fundamental relationship between castle construction, maintenance, administration, and the reorganization of annexed landscapes. At the same time, the theocratic character of the castle communities cannot be understated in how political, military, and religious authority was blended and articulated, although the effectiveness of this authority — particularly in the context of Christianization — can be questioned through its impact on the conceptual landscapes of the eastern Baltic. Recognizing the fundamental connections between colonization, cultural transformation, and environmental exploitation, this study investigates the relationship between the developments of the crusader states in the eastern Baltic, focusing on the territories of the Teutonic Order, and regional environmental change, through four principal questions. These form the basis of the multi-proxy, integrated, and comparative study of select territories in Livonia and Prussia:

1. How were local environments in the eastern Baltic region transformed from the thirteenth to fifteenth centuries with the establishment of castles and their territories?
2. How were wild and domestic plants and animals used by crusading institutions in the eastern Baltic, and to what extent did this reflect ideology and adaptation to local ecology?
3. To what extent did the crusading movement in the eastern Baltic frontier represent a cohesive force of ecological transformation? Is there evidence for a standardized approach to exploiting the conquered landscape?

4. How did the process of Christianization, facilitated by the Baltic Crusades, change the way indigenous groups conceptualized the natural world?

Some time was spent trying to ascertain whether a wholly synthetic study, subdivided chronologically or geographically, would be the most useful way of presenting the findings of such an expansive project. Björn Berglund's (1991) monumental *Ystad Project* provided an initial model, not only for the type of integration of different sets of archaeological and historical data that may be attempted for the eastern Baltic, but also for how this could be presented. However, a more accessible template was provided by the outputs of more recent multiproxy research, such as James Barrett's (2012) *Being an Islander: Production and Identity at Quoygrew, Orkney, AD 900–1600*. In these cases, an approach to presenting data which emulated an archaeological site report was adopted, with a series of individual specialist reports brought together in a cascading series of integrative summaries.

This volume, the first of two complimentary books from the Ecology of Crusading project, is subdivided into Livonia and Prussia, with each geographically defined section containing specialist chapters concluding with an integrated summary, and then a final chapter where both regions are compared. The concluding chapter revisits the research questions set out above. This format enables readers interested in particular aspects to navigate their way around the data quickly and easily. The second volume brings together a series of case-studies from across the eastern Baltic region, following more closely the format adopted by the Culture, Clash or Compromise (CCC) project (Blomkvist 2009; see Chapter 2). Together, they represent the current state of scholarship on the relationship between cultural and environmental changes that occurred during the Middle Ages in north-eastern Europe. They provide an interpretative framework for understanding the relationship between cultural and ecological change associated with colonization, one that can be developed further in the Baltic Sea Region or applied and adapted to the study of other frontier societies; a blueprint for future inter-regional research programmes which strive for holistic understandings of human colonization, cultural confrontation, and transformation.

Works Cited

Please note that the names of Russian authors have been transliterated from Cyrillic.

Primary Sources

GAD Ziesemer, W., ed. 1968. *Das große Ämterbuch des Deutschen Ordens: Mit Unterstützung des Vereins für die Herstellung und Ausschmückung der Marienburg*, Wiesbaden: Sändig (repr. 2009)

HCL Brundage, J. A., trans. 2003. *The Chronicle of Henry of Livonia/Henricus Lettus*, New York: Columbia University Press

KP Fischer, M. 2010. *The Chronicle of Prussia by Nicolaus von Jeroschin: A History of the Teutonic Knights in Prussia, 1190–1331*, Farnham: Ashgate

Matuszewski, J., ed. 1963. *Iura Pruthenorum*, Toruń: Państwowe Wydawnictwo Naukowe

LRC Smith, J. C., and Urban, W., trans. 2006. *The Livonian Rhymed Chronicle*, London: Routledge Curzon

Secondary Sources

Arszyński, M. 2013. 'Zamki i umocnienia zakonu krzyżackiego i biskupów w Inflantach', in *Zakon krzyżacki w Prusach i Inflantach: podziały administracyjne i kościelne w XIII–XVI wieku*, eds R. Czaja and A. Radzimiński, Toruń: Wydawnictwo Naukowe Uniwersytetu Mikołaja Kopernika, 183–213

Aubin, H. 1966. 'The Lands East of the Elbe and German Colonization Eastwards', in *The Cambridge Economic History of Europe from the Decline of the Roman Empire*, I: *Agrarian Life of the Middle Ages*, ed. M. M. Postan, Cambridge: Cambridge University Press, 449–86

Barber, M. 2012. *The Crusader States*, New Haven: Yale University Press

Barrett, J. 2012. *Being an Islander: Production and Identity at Quoygrew, Orkney, AD 900–1600*, Cambridge: McDonald Institute for Archaeological Research

Bartlett, R. 1993. *Making of Europe. Conquest, Colonization and Cultural Change 950–1350* (London: Penguin)

Benninghoven, F. 1963. 'Probleme der Zahl und Standortverteilung der livländischen Streitkräfte im ausgehenden Mittelalter', *Zeitschrift für Ostforschung* 12, 601–22

Berglund, B. E. 1991. *The Cultural Landscape during 6000 Years in Southern Sweden — The Ystad Project*, Copenhagen: Munksgaard

Birks, H. H., et al., eds. 2004. *The Cultural Landscape: Past, Present and Future*, Cambridge: Cambridge University Press (3rd edn)

Biskup, M. 2009. 'Polityka zewnętrzna zakonu krzyżackiego', in *Państwo zakonu krzyżackiego w Prusach: Władza i społeczeństwo*, ed. M. Biskup et al., Warsaw: Państwowe Wydawnictwo Naukowe, 233–331

Biskup, M. et al., ed. 2009. *Państwo zakonu krzyżackiego w Prusach: Władza i społeczeństwo*, Warsaw: Wydawnictwo Naukowe PWN

Blanke, R. 1999. '"Polish-Speaking Germans?" Language and National Identity among the Masurians', *Nationalities Papers: The Journal of Nationalism and Ethnicity* 27/3, 429–53

Blomkvist, N. 1998. 'The Medieval Europeanisation Process of the Baltic Rim Region 1100–1400 AD: Problems for an International Study', in *Culture, Clash or Compromise: The Europeanisation of the Baltic Sea Area 1100–1400 AD*, ed. N. Blomkvist, Visby: Gotland Centre for Baltic Studies, 9–36

——, 2005. *The Discovery of the Baltic: The Reception of a Catholic World-System in the European North (AD 1075–1225)*, Leiden: Brill

——, 2009. '12th–14th Century European Expansion and its Reception in the Baltic North', in *The Reception of Medieval Europe in the Baltic Sea Region*, ed. J. Staecker, Visby: Gotland University Press, 431–84

Bojtár, E. 1999. *Foreword to the Past: A Cultural History of the Baltic People*, Budapest: Central European University Press

Boockmann, H. 1981. *Der Deutsche Orden: Zwölf Kapitel aus seiner Geschichte*, Munich: Beck

Caune A., and Ose, I. 2004. *Latvijas 12. gadsimtā beigu–17. gadsimtā vācu piļu leksikons*, Rīga: Latvijas vēstures institūta apgāds

Chollet, L. 2015. 'Les "Voyages en Prusse" vus de France: la perception de la croisade contre la Lituanie dans quelques sources francophones (fin XIVe–début XVe siècle)', in *Studia z Dziejów Średniowiecza XIX*, eds B. Możejko, M. Smoliński, and S. Szybkowski, Warsaw: Wydawnictwo DIG, 51–80

Chmielewski, S. 1963. 'Czy pruski grób ciałopalny z XV wieku? (Próba interpretacji opisu znaleziska z 1703 r.)', *Rocznik Olsztyński* 5, 295–319

Christiansen, E. 1997. *The Northern Crusades*, London: Penguin

Chudziak, W., ed. 2008. *Człowiek i środowisko przyrodnicze we wczesnym średniowieczu w świetle badań interdyscyplinarnych*, Toruń: Wydawnictwo Naukowe Uniwersytetu Mikołaja Kopernika

Czacharowski, A. 1983. 'Toruń średniowieczny (do roku 1454)', in *Toruń dawny i dzisiejszy: zarys dziejów*, ed. M. Biskup, Warsaw: Państwowe Wydawnictwo Naukowe, 31–131

——, 1986. 'Prawo chełmińskie w życiu stanowym państwa krzyżackiego', in *Niemcy-Polska w średniowieczu: Materiały z konferencji naukowej zorganizowanej przez Instytut Historii UAM w dniach 14–16 XI 1983*, ed. J. Strzelczyk, Poznań: UAM, 275–81

Czaja, R. 2009a. 'Rozwój osadnictwa miejskiego', in *Państwo zakonu krzyżackiego w Prusach: Władza i społeczeństwo*, ed. M. Biskup et al., Warsaw: Państwowe Wydawnictwo Naukowe, 370–83
——, 2009b. 'Struktura społeczna', in *Państwo zakonu krzyżackiego w Prusach: Władza i społeczeństwo*, ed. M. Biskup et al., Warsaw: Państwowe Wydawnictwo Naukowe, 444–50
——, 2013. 'Miasta inflanckie XIII–XVI w.', in *Zakon krzyżacki w Prusach i Inflantach: podziały administracyjne i kościelne w XIII–XVI wieku*, eds R. Czaja and A. Radzimiński, Toruń: Wydawnictwo Naukowe Uniwersytetu Mikołaja Kopernika, 215–40
Czaja, R., and Nowak, Z. H. 2013. 'Państwo zakonu krzyżackiego w Prusach — próba charakterystyki', in *Zakon krzyżacki w Prusach i Inflantach: podziały administracyjne i kościelne w XIII–XVI wieku*, eds R. Czaja and A. Radzimiński, Toruń: Wydawnictwo Naukowe Uniwersytetu Mikołaja Kopernika, 1–27
Czaja, R., and Radzimiński, A., eds. 2013. *Zakon krzyżacki w Prusach i Inflantach: podziały administracyjne i kościelne w XIII–XVI wieku*, Toruń: Wydawnictwo Naukowe Uniwersytetu Mikołaja Kopernika
Dirks, B. 1988. 'Der Krieg des Deutschen Ordens gegen Novgorod 1443–1448', in *Deutschland-Livland-Russland: Ihre Beziehungen vom 15. Bis zum 17. Jahrhundert*, ed. N. Angermann, Lüneburg: Nordostdeutsches Kulturwerk, 29–52
Długokęcki, W. 2009a. 'Kolonizacja ziemi chełmińskiej, Prus i Pomorza Gdańskiego do 1410 r.', in *Państwo zakonu krzyżackiego w Prusach: Władza i społeczeństwo*, ed. M. Biskup et al., Warsaw: Państwowe Wydawnictwo Naukowe, 200–17
——, 2009b. 'Prusy w starożytności i we wczesnym średniowieczu', in *Państwo zakonu krzyżackiego w Prusach: Władza i społeczeństwo*, eds M. Biskup et al., Warsaw: Państwowe Wydawnictwo Naukowe, 25–50
——, 2009c. 'Społeczeństwo wiejskie', in *Państwo zakonu krzyżackiego w Prusach: Władza i społeczeństwo*, ed. M. Biskup et al., Warsaw: Państwowe Wydawnictwo Naukowe, 460–94
Dobrosielska, A., and Radzicki, B. 2013. 'Między Cholinun a Rzymem: Zmiana jako autopojeza systemu społecznego Prusów', *Pruthenia* 8, 7–57
Domasłowski, J. 2010. 'Krzyżacy, biskupi, mieszczanie...malowidła ścienne w państwie zakonnym i krąg fundatowrów do połowy XV wieku', in *Fundacje artystyczne na terenie Państwa Krzyżackiego w Prusach*, ed. B. Pospieszna, Pelplin: Wydawnictwo Bernardinum, 106–41
Dygo, M. 1989. 'The Political Role of the Cult of the Virgin Mary in Teutonic Prussia in the Fourteenth and Fifteenth Centuries', *Journal of Medieval History* 15, 63–80
——, 2003. 'Die Wirtschaftstätigkeit des Deutschen Ordens in Preußen im 14.–15. Jahrhundert', in *Die Ritterorden in der europäischen Wirtschaft des Mittelalters*, eds R. Czaja and J. Sarnowsky, Toruń: Wydawnictwo Uniwersytetu Mikołaja Kopernika, 147–60
——, 2009a. 'Ideologia panowania zakonu niemieckiego w Prusach', in *Państwo zakonu krzyżackiego w Prusach: Władza i społeczeństwo*, ed. M. Biskup et al., Warsaw: Państwowe Wydawnictwo Naukowe, 357–69
——, 2009b. 'Początki i budowa władztwa zakonu krzyżackiego (1226–1309)', in *Państwo zakonu krzyżackiego w Prusach: Władza i społeczeństwo*, ed. M. Biskup et al., Warsaw: Państwowe Wydawnictwo Naukowe, 53–78
Ebel, F. 2004. *Unseren fruntlichen Grus zuvor: Deutsches Recht des Mittelalters im mittel- und osteuropäischen Raum*. Cologne: Böhlau
Ehlers, A. 2001. 'The Crusade of the Teutonic Knights against Lithuania Reconsidered', in *Crusade and Conversion on the Baltic Frontier, 1150–1500*, ed. A. V. Murray, Farnham: Ashgate, 21–44
Eihmane, E. 2009. 'The Baltic Crusades: A Clash of Two Identities', in *The Clash of Cultures on the Medieval Baltic Frontier*, ed. A. V. Murray, Farnham: Ashgate, 37–51
——, 2012. 'Livonia and the Holy See in 13th–Early 15th Centuries: Centre and Periphery of Christendom', in *The Image of the Baltic: A Festschrift for Nils Blomkvist*, eds M. F. Scholz, R. Bohn, and C. Johansson, Visby: Gotland University Press, 23–40
Ekdahl, S. 1991. 'Das Pferd und seine Rolle im Kriegswesen des Deutschen Ordens', in *Das Kriegswesen der Ritterorden im Mittelalter*, ed. Z. H. Nowak, Toruń: Wydawnictwo Uniwersytetu Mikołaja Kopernika, 25–47
——, 2004. 'Crusades and Colonisation in the Baltic: A Historiographic Analysis', *Rocznik Instytutu Polsko-Skandynawskiego* 19, 1–42
Eldevik, J. 2003. 'Ecclesiastical Lordship and the Politics of Submitting Tithes in Medieval Germany: The Thuringian Tithe Dispute in Social Context', *Viator* 34, 40–56
Ervynck, A. 2004. 'The Past is a Distant Planet: Reasons to Integrate Environmental data into Archaeological Interpretations', *Interpreting the Past* 1, 97–118
Fairclough, G. 2006. 'Large Scale, Long Duration, and Broad Perceptions: Scale Issues in Historic Landscape Characterisation', in *Confronting Scale in Archaeology*, eds G. Lock and B. L. Molyneaux, New York: Springer, 203–215
Fehring, G. P. 1990. 'Origins and Development of Slavic and German Lübeck', in *From the Baltic to the Black Sea: Studies in Medieval Archaeology*, eds D. Austin and L. Alcock, London: Routledge, 251–66
Fischer, M. 1991. *Di Himmels Rôte: The Idea of Christian Chivalry in the Chronicles of the Teutonic Order*, Göppingen: Kümmerle
——, 2007. 'Des tuvils kint? The German Order's Perception of its Enemies', *Archiv fur das Studium der neurern Sprachnund Literaturen* 244/159, 260–75
Fonnesberg-Schmid, I. 2007. *The Popes and the Baltic Crusades: 1147–1254*, Leiden: Brill
Gaimster, D. 2005. 'A Parallel History: The Archaeology of Hanseatic Urban Culture in the Baltic *c.* 1200–1600', *World Archaeology* 37/3, 408–23
Gaimster, D. 2007. 'The Baltic Ceramic Market 1200–1600: Measuring Hanseatic Cultural Transfer and Resistance', in *Cultural Exchange in Early Modern Europe*, IV: *Forging European Identities, 1400–1700*, ed. H. Roodenburg, Cambridge: Cambridge University Press, 30–58

Gancarczyk, P. 2010. 'Kultura muzyczna zakonu krzyżackiego w Prusach', in *Fundacje artystyczne na terenie Państwa Krzyżackiego w Prusach*, ed. B. Pospieszna, Pelplin: Wydawnictwo Bernardinum, 269–82

Gancewski, J. 2000. 'Przemiany gospodarcze w południowo-wschodniej części państwa krzyżackiego: Źródła i stan badań', *Komunikaty Mazursko-Warmińskie* 2, 177–86

Grążawski, K. 2009. *Ziemia lubawska na pograniczu słowiańsko-pruskim w VIII–XIII w.: studium nad rozwojem osadnictwa*, Olsztyn: Wydawnictwo Uniwersytetu Warmińsko-Mazurskiego

Herrmann, C. 2007. *Mittelalterliche Architektur im Preußenland: Untersuchungen zur Frage der Kunstlandschaft und -geographie*, Petersberg: Michael Imhof Verlag

Hoffman, R. 2014. *An Environmental History of Medieval Europe*, Cambridge: Cambridge University Press

Housely, N. 1992. *The Later Crusades, 1274–1580: From Lyons to Alcazar*, Oxford: Oxford University Press

Hubatsch, W. 1973. 'Das westliche Preußen und das Ermland zur Zeit des Copernicus', in *Nicolaus Copernicus zum 500. Geburtstag*, eds F. Kaulbach, U. W. Bargenda, and J. Blühdorn, Köln: Böhlau, 138–54

Jensen, C. S. 2001. 'Urban Life and the Crusades in North Germany and the Baltic Lands in the Early Thirteenth Century', in *Crusade and Conversion on the Baltic Frontier, 1150–1500*, ed. A. V. Murray, Aldershot: Ashgate, 75–94

Johnson, M. 2006. *Ideas of Landscape*, Oxford: Blackwell

Jóźwiak, S. 2001. *Centralne i terytorialne organy władzy zakonu krzyżackiego w Prusach w latach 1228–1410: rozwój, przekształcenia, kompetencje*, Toruń: Wydawnictwo Uniwersytetu Mikołaja Kopernika

——, 2009a. Dobra rycerskie', in *Państwo zakonu krzyżackiego w Prusach: Władza i społeczeństwo*, ed. M. Biskup et al., Warsaw: Państwowe Wydawnictwo Naukowe, 217–30

——, 2009b. 'Kryzys władzy terytorialnej', in *Państwo zakonu krzyżackiego w Prusach: Władza i społeczeństwo*, ed. M. Biskup et al., Warsaw: Państwowe Wydawnictwo Naukowe, 332–56

Jóźwiak, S., and Trupinda, J. 2007. *Organizacja życia na zamku krzyżackim w Malborku w czasach wielkich mistrzów (1309–1457)*, Malbork: Muzeum Zamkowe w Malborku

——, 2012. *Krzyżackie zamki komturskie w Prusach*, Toruń: Wydawnictwo Naukowe Uniwersytetu Mikołaja Kopernika

Kala, T. 2006. 'Languages in a Medieval North European City: An Example from Medieval Tallinn', in *Frontiers in the Middle Ages*, eds O. Merisalo and P. Pahta, Louvain-la-Neuve: Fédération Internationale des Institutes d'Études Médiévales, 585–603

——, 2007. 'Northern Estonian Church Life in the Thirteenth and Fourteenth Century: The Religious Environment in which the Pirita Monastery Emerged', *Studies on Art and Architecture (Kunstiteaduslikke Uurimusi)* 4, 57–59

——, 2009. 'Rural Society and Religious Innovation: Acceptance and Rejection of Catholicism among the Native Inhabitants of Medieval Livonia', in *The Clash of Cultures on the Medieval Baltic Frontier*, ed. A. V. Murray, Farnham: Ashgate, 168–90

Kalling, K. 1995. 'Paleoantropoloogilisi andmeid Tartu Jaani kiriku kalmistu 13.14. sajandi matuste kohta: Tartu arheoloogiast ja vanemast ehitusloost', *Tartu Ülikooli Arheoloogia Kabineti toimetised* 8, 47–58

Karczewski, M. 2006. 'Environment, Settlement and Economy of the West Baltic Tribes in the Roman Period', *Archaeologia Lituania*, 7, 54–65

——, M. 2011. *Archeologia środowiska zachodniobałtyjskiego kręgu kulturowego na pojezierzach*, Poznań-Białystok: Bogucki Wydawnictwo Naukowe

Kedar, B. Z. 1992. 'The Crusading Kingdom of Jerusalem: The First European Colonial Society? A Symposium', in *The Horns of Hattin*, ed. B. Z. Kedar, London: Variorum, 341–66

Kilarski, M. 2007. *Odbudowa i konserwacja zespołu zamkowego w Malborku w latach 1945–2000*, Malbork: Muzeum Zamkowe w Malborku

Klat, M., and Mykowski, J. 2002. *Krzyżackie tajemnice: zamki, skarby, odkrycia*, Gdańsk: Oficyna Pomorska

Kļaviņš, K. 2019. 'Reorganizing the Livonian Landscape: Some Issues and Research Perspectives', in *Ecologies of Crusading, Colonization, and Religious Conversion in the Medieval Baltic: Terra Sacra II*, ed. A. Pluskowski, Environmental History of the North Atlantic World, 3, Turnhout, Brepols (forthcoming)

Kosman, M. 1976. 'Zanik pogaństwa w Prusach', *Komunikaty Mazursko-Warmińskie* 1/131, 3–29

Kreem, J. 2001a. 'The Teutonic Order as a Secular Ruler in Livonia: The Privileges and Oath of Reval', in *Crusade and Conversion on the Baltic Frontier, 1150–1500*, ed. A. V. Murray, Aldershot: Ashgate, 215–32.

——, 2001b. 'The Teutonic Order in Livonia: Diverging Historiographic Traditions', in *The Crusades and the Military Orders: Expanding the Frontiers of Medieval Latin Christianity*, eds Z. Hungyadi and J. Laszlovszky, Budapest: Central European University Press, 467–79

——, 2002. *The Town and its Lord: Reval and the Teutonic Order (in the Fifteenth Century)*, Tallinn: Tallinna Linnaarhiiv

——, 2013. 'Crusading Traditions and Chivalric Ideals: The Mentality of the Teutonic Order in Livonia at the Beginning of the Sixteenth Century', *Crusades* 12/1, 233–50

Kutzner, M. 1995. 'Propaganda władzy w sztuce Zakonu Niemieckiego w Prusach', in *Sztuka w kręgu zakonu krzyżackiego w Prusach i Inflantach*, ed. M. Woźniak, Toruń: Uniwersytet Mikołaja Kopernika, 32–85

——, 2010. 'Propaganda władzy Zakonu Niemieckiego w sztuce na terenie Państwa Zakonnego w Prusach', in *Fundacje artystyczne na terenie Państwa Krzyżackiego w Prusach*, ed. B. Pospieszna, Pelplin: Wydawnictwo Bernardinum, 238–59

Lagerås, P. 2007. *The Ecology of Expansion and Abandonment: Medieval and Post-Medieval Land-Use and Settlement Dynamics in a Landscape Perspective*. Lund: National Heritage Board

Laime, S. 2009. 'The Sacred Groves of the Curonian Ķoniņi: Past and Present', *Folklore* 42, 67–80

Lind, J. H. 2004. 'The Concept of "Europeanisation" on the Baltic Rim as Seen from the East', in *The European Frontier: Clashes and Compromises in the Middle Ages*, ed. J. Staecker, Lund: Almqvist and Wiksell, 41–44

Ljungqvist, F. C. 2010. 'A New Reconstruction of Temperature Variability in the Extra-Tropical Northern Hemisphere during the Last Two Millennia', *Geogr. Ann.*, 92 A (3), 339–51

Mägi, M. 2002. *At the Crossroads of Space and Time: Graves, Changing Society and Ideology on Saaremaa (Ösel), 9th–13th centuries AD*, Tallinn: Institute of History

Makowiecki, D. 2007. 'Gospodarka zwierzętami w strefie chełmińskiego pogranicza słowiańsko pruskiego i krzyżackiego w średniowieczu', in *Pogranicze polsko-pruskie i krzyżackie*, ed. K. Grążawski, Włocławek: Lega, 23–37

Mänd, A. 2009. 'Saints' Cults in Medieval Livonia', in *The Clash of Cultures on the Medieval Baltic Frontier*, ed. A. V. Murray, Farnham: Ashgate, 191–223

Mickūnaitė, G. 2007. *Making a Great Ruler: Grand Duke Vytautas of Lithuania*, Budapest: Central European University Press

Moberg, A., Sonechkin, D. M., Holmgren, K., Datsenko, N. M., and Karlén, W. 2005. 'Highly Variable Northern Hemisphere Temperatures Reconstructed from Low- and High-Resolution Proxy Data', *Nature* 433, 613–17

Mortensen, L. B., and Bisgaard, L. 2013. 'Introduction: Late Medieval Urban Civilization and its North European Variant', in *Guilds, Towns and Cultural Transmission in the North, 1300–1500*, eds L. Bisgaard, L. B. Mortensen, and T. Pettitt, Odense: University Press of Southern Denmark, 7–30

Morton, N. E. 2009. *The Teutonic Knights in the Holy Land: 1190–1291*, Woodbridge: Boydell

Mugurēvičs, Ē. 1990. 'Interactions between indigenous and Western Culture in Livonia in the 13th to 16th Centuries', in *From the Baltic to the Black Sea*, eds D. Austin and L. Alcock, London: Routledge, 168–78

——, 2002. 'Forest Animals and Hunting in Medieval Livonia', in *Core-Centre-Periphery: Medieval Europe 2*, eds G. Helmig, B. Scholkmann, and M. Untermann, Basel: Folio Verlag Dr G. Wesselkamp, Archäologische Bodenforschung Basel-Stadt, 177–81

Murray, A. V., ed. 2001. *Crusade and Conversion on the Medieval Baltic Frontier*, Aldershot: Ashgate

——, ed. 2009. *The Clash of Cultures on the Medieval Baltic Frontier*, Farnham: Ashgate

——, 2010. 'The Saracens of the Baltic: Pagan and Christian Lithuanians in the Perception of English and French Crusaders to the Late Medieval Prussia', *Journal of Baltic Studies* 41/4, 413–29

——, ed. 2012. *The Northern-Eastern Frontiers of Medieval Europe: The Expansion of Latin Christendom in the Baltic Lands*, Farnham: Ashgate

Nielsen, T. K. 2011. 'Henry of Livonia on Woods and Wilderness', in *Crusading and Chronicle Writing on the Medieval Baltic Frontier*, eds M. Tamm, L. Kaljundi, and C. S. Jensen, Farnham: Ashgate, 157–78

Nowakiewicz, T. 2010. 'Some Remarks on Settlement Systems of early Medieval Prussians: The Case of Northern Galindia', in *Worlds Apart? Contacts across the Baltic Sea in the Iron Age*, eds U. L. Hansen and A. Bitner-Wróblewska, Copenhagen: Det kgl. Nordiske Oldskriftselskab, 487–504

Palang, H., Printsmann, A., Konsa, M., and Lang, V. 2005. 'Ideology and Tradition in Landscape Change: A Case of the Helme Parish, Estonia', in *Landscape, Law and Justice*, eds T. Peil and M. Jones, Oslo: Novus, 288–97

Paravicini, W. 1989. *Die Preussenreisen des europäischen Adels* (vol. I), Sigmaringen: Thorbecke

Paszkiewicz, B. 2009. *Brakteaty — pieniądz średniowiecznych Prus*, Wrocław: Wydawnictwo Uniwersytetu Wrocławskiego

Patze, H. 1964. 'Die deutsche bäuerliche Gemeinde im Ordensstaat Preussen', *Die Anfänge der Landgemeinde und ihr Wesen* 2, 149–200

Pluskowski, A. G. 2010. 'The Zooarchaeology of Medieval "Christendom": Ideology, the Treatment of Animals and the Making of Medieval Europe', *World Archaeology* 42/2, 201–14

——, 2013. *The Archaeology of the Prussian Crusade: Holy War and Colonisation*, London: Routledge

Pluskowski, A. G., and Valk, H. 2016. 'Conquest and Europeanisation: The Archaeology of the Crusades in Livonia, Prussia and Lithuania', in *The Crusader World*, ed. A. Boas, London: Routledge, 568–592.

Pungetti, G. 2013. 'Biocultural Diversity for Sustainable Ecological, Cultural and Sacred Landscapes: The Biocultural Landscape Approach', in *Landscape Ecology for Sustainable Environment and Culture*, eds. B. Fu and K. Jones, Dordrecht: Springer, 55–76

Radzimiński, A. 2009. 'Geneza oraz ukształtowanie się organizacji kościelnej (1206–1409)', in *Państwo zakonu krzyżackiego w Prusach: Władza i społeczeństwo*, ed. M. Biskup et al., Warsaw: Państwowe Wydawnictwo Naukowe, 143–76

——, 2013. 'Podziały kościelne Inflant', in *Zakon krzyżacki w Prusach i Inflantach: podziały administracyjne i kościelne w XIII–XVI wieku*, eds R. Czaja and A. Radzimiński, Toruń: Wydawnictwo Naukowe Uniwersytetu Mikołaja Kopernika, 241–75

Reisnert, A. 2009. 'The Danish Crusade against Estonia 1219–1346', in *Expansion-Integration? Danish-Baltic Contacts 1147–1410 AD*, eds B. F. Jensen and D. Wille-Jørgensen, Vordingborg: Danmarks Borgcenter, 53–65

Riley-Smith, J. 1999. 'The Crusading Movement and Historians', in *The Oxford History of the Crusades*, ed. J. Riley-Smith, Oxford: Oxford University Press, 1–14

——, 2005. *The Crusades: A History*, London: Continuum

Rowell, S. C. 1994. *Lithuania Ascending: A Pagan Empire within East-Central Europe, 1295–1345*, Cambridge: Cambridge University Press
Samsonowicz, H. 1990. 'Gospodarcza ekspansja Zakonu Krzyżackiego nad Bałtykiem w XIV i na początku XV w.', in *Ekspansja niemieckich zakonów rycerskich w strefie Bałtyku od XIII do połowy XVI wieku*, ed. M. Biskup, Toruń: Uniwersytet Mikołaja Kopernika, 143–58
——, 1994. 'Europejski handel Krzyżaków w XIV–XV wieku', *Komunikaty Mazursko-Warmińskie* 2–3, 137–45
Sarnowsky, J. 1993. *Die Wirtschaftsführung des Deutschen Ordens in Preußen (1382–1454)*, Cologne: Böhlau
Schenk, J. 2012. 'The Cult of the Cross in the Order of the Temple', in *As Ordens Militares: Freires, Guerreiros, Cavaleiros. Actas do VI Encontro sobre Ordens Militares* (vol. I), ed. I. C. Ferreira Fernandes, Palmela: GEsOS / Município de Palmela, Palmela, 207–19
Scott, T. 2002. *Society and Economy in Germany, 1300–1600*, Basingstoke: Palgrave
Šnē, A. 2005. 'Archaeological Evidence of Social Relations and Power in the Late Prehistoric Societies of the Livs', *Suomalaisen tiedeakatemian toimituksia. Sarja Humaniora* 336, 169–85
——, 2006. 'The Economy and Social Power in the late Prehistoric Chiefdoms in Eastern Lativa', *Archaeologia Baltica* 6, 68–78
——, 2008. 'The Medieval Peasantry: On the Social and Religious Position of the Rural Natives in Southern Livonia (13th–15th Centuries)', *Ajalooline Ajakiri* 1/2 (123/24), 89–100
Stančikaitė, M., Šinkūnas, P., Risberg, J., Šeirienė, V., Blažauskas, N., Jarockis, R., Karlsson, S., and Miller, U. 2009. 'Human Activity and the Environment during the Late Iron Age and Middle Ages at the Impiltis Archaeological Site, NW Lithuania', *Quaternary International* 20, 74–90
Sterns, I. 1969. 'The Statutes of the Teutonic Knights: A Study of Religious Chivalry' (unpublished doctoral dissertation, University of Pennsylvania)
Szczepański, S. 2016. *Pomezania pruska: Dzieje osadnictwa w XIII–XV wieku*, Olsztyn: Rozprawy i Materiały Ośrodka Badań Naukowych im. Wojciecha Kętrzyńskiego w Olsztynie
Tamm, M. 2009. 'A New World into Old Words: The Eastern Baltic Region and the Cultural Geography of Medieval Europe', in *The Clash of Cultures on the Medieval Baltic Frontier*, ed. A. V. Murray, Farnham: Ashgate, 11–36.
——, 2013. 'How to Justify a Crusade? The Conquest of Livonia and New Crusade Rhetoric in the Early Thirteenth Century', *Journal of Medieval History* 39/4, 431–55
Tandecki, J. 2009a. 'Rozwój terytorialny państwa zakonnego w Prusach', in *Państwo zakonu krzyżackiego w Prusach: Władza i społeczeństwo*, ed. M. Biskup et al., Warsaw: Państwowe Wydawnictwo Naukowe, 105–09
——, 2009b. 'Zakon krzyżacki', in *Państwo zakonu krzyżackiego w Prusach: Władza i społeczeństwo*, ed. M. Biskup et al., Warsaw: Państwowe Wydawnictwo Naukowe, 405–19
——, 2013. 'Struktury i podziały administracyjne w zakonie krzyżackim w Inflantach', in *Zakon krzyżacki w Prusach i Inflantach: podziały administracyjne i kościelne w XIII–XVI wieku*, eds R. Czaja and A. Radzimiński, Toruń: Wydawnictwo Naukowe Uniwersytetu Mikołaja Kopernika, 167–81
Treitschke, H. von. 1862. 'Das deutsche Ordensland Preußen', *Preußische Jahrbücher* 10, 95–151
Urban, W. 1978. 'The Diplomacy of the Teutonic Knights at the Curia', *Journal of Baltic Studies* 9/2, 116–28
——, 2000. *The Prussian Crusade*, rev. edn, Chicago: Lithuanian Research and Studies Center
——, 2003. *The Teutonic Knights: A Military History*, London: Greenhill Book.
Valk, H. 2003. 'Christianisation in Estonia: a process of dual-faith and syncretism', in *The Cross Goes North: Processes of Conversion in Northern Europe, AD 300–1300*, ed. M. Carver, Woodbridge: Boydell, 571–79
——, 2004. 'Artefacts in Estonian Urban Churchyards: Reflections on Different Traditions and Ideologies', in *Der Ostseeraum und Kontinentaleuropa 1100–1600*, eds D. Kattinger, J. E. Olesen, and H. Wernicke, Schwerin: Helms, 103–12
——, 2014a. 'Strongholds East of the Baltic Sea in the 11th–13th Centuries: The Topic, Concepts, Research, and Terms', in *Strongholds and Power Centres East of the Baltic Sea in the 11th–13th Centuries*, ed. H. Valk, Tartu: Institute of History and Archaeology at the University of Tartu, Department of Archaeology at the Institute of History of Tallinn University, 81–114
——, ed. 2014b. *Strongholds and Power Centres East of the Baltic Sea in the 11th–13th Centuries*, Tartu: Institute of History and Archaeology at the University of Tartu, Department of Archaeology at the Institute of History of Tallinn University
Vykypěl, B. 2011. *Studie k šlechtickým titulům v germánských, slovanských a baltských jazycích*, Praha: Lidové noviny
Wenskus, R. 1970. 'Das Ordensland Preußen als Territorialstaat des 14. Jahrhunderts', in *Der deutsche Territorialstaat im 14. Jahrhundert I*. ed. H. Patze, Frankfurt am Main: Thorbecke, 347–82
Xoplaki, E., Fleitmann, D., Diaz, H., von Gunten, L., and Kiefer, T., eds 2011. 'Medieval Climate Anomaly', *Pages* 19/1, 1–40
Zonenberg, S. 1994. *Kronika Wiganda z Marburga*, Bydgoszcz: Wydawnictwo Uczelniane WSP w Bydgoszczy
Žulkus, V. 2002. 'The Balts: Economy and Society', in *The Neighbours of Poland in the 11th Century*, ed. P. Urbańczyk, Warsaw: Institute of Archaeology and Ethnology, Polish Academy of Sciences, 169–210

2. Methodology and Interpretative Framework

Aleksander Pluskowski, Alexander Brown, Rowena Banerjea, and Kevin Hayward

Introduction: A Multidisciplinary Approach with Interdisciplinary Objectives

Multidisciplinary approaches in medieval archaeology are well established (Gerrard 2003) and are the norm for archaeological site reports that incorporate a range of specialist reports (including written and cartographic sources where appropriate), with a synthetic conclusion. Multidisciplinary applications of environmental archaeology to medieval social contexts are also steadily increasing. This study has adopted an interdisciplinary approach, which, in contrast to multidisciplinarity, is defined by its coherent outcome (van den Besselaar and Heimeriks 2001, 2); the use of an integrated, multiscalar approach differentiates this from previous studies. All specialist analyses were focused on the same set of research objectives outlined in Chapter 1, and datasets gathered since pilot studies in 2006 (Maltby et al. 2009; Pluskowski et al. 2010; Brown and Pluskowski 2011), and more systematically from 2010–14, have been selected with these specific questions in mind. Integration has been facilitated by the involvement of specialists from the start of the project in the articulation of research questions, and by applying the same questions to a clearly defined set of case studies.

Theoretical Framework

There are three main theoretical frameworks that need to be engaged with to situate the key questions posed in this study within contemporary scholarship, and to challenge traditionally held assumptions. These are Europeanization, frontiers, and anthropogenic ecology.

Europeanization

The concept of Europeanization has dominated scholarly debates on the northern crusades in the last few decades, and has been undoubtedly redefined by the Culture-Clash-Compromise (CCC) project (see Blomkvist 2009 for a summary and bibliography of work published by the project which was carried out in 1998–2001). This extended research network compared regional developments across the Baltic Sea between the eleventh and thirteenth centuries, using a modified world-systems model, reframed as the 'Catholic world-system', as its theoretical baseline (Blomkvist 2004; 2005). The term 'Europeanization' was in fact largely applied at the scale of Northern Europe, and in the case of the eastern Baltic it would be more accurate to characterise the prevailing processes as 'Latinization' (Lind 2004) and 'Germanization' (Ekdahl 2004, 7), alongside variable influences from neighbouring Scandinavian, Polish, Lithuanian, and Rus' regions, which in some cases included actual colonization, particularly the various Polish (Masovian, Pomeranian) components in Prussia. One side of the Europeanization model presented the view from the core, or from the representatives of the Catholic system within the conquered territories. Livonia, in particular, was increasingly considered as the most peripheral region of Latin Christendom, with the projected threat of pagan (Lithuanian) and schismatic (Rus') neighbours becoming an established theme in conflicts between Livonian political factions (Selart 2015, 310–11).

However, Blomkvist placed on emphasis on the *reception* of European ideas, moving away from the traditional model of peripheries as passive recipients of ideas from the core. The process was now regarded as a complex dialectic between incoming agents and indigenous reagents, rather than a top-down model of exploiting conquered territory. This bridged the gap between how the eastern Baltic was imagined, and how its societies actually functioned (Tamm 2009, 35). The basic premise of Europeanization recognizes the importation of a 'cultural package' — a series of cultural elements introduced by different social agents — which was then applied and modified to the local context, provoking and affected by a spectrum of reactions from the indigenous population. The encounter was nonetheless a violent one, a clear 'clash of cultures' (e.g., Murray 2009) and following Blomkvist's model, a confrontation of 'world systems', of action and reaction (see Kļaviņš 2019 for a rejection of this). From the socio-economic perspective favoured by this model, this resulted in specific changes to the organization of trade, particularly the rationalization of clusters of independent trading places into central and highly regulated urban hubs, as seen in the case of Riga within the context of the Lower Daugava (Blomkvist 2005, 562). This transformation was sustained due to the connectivity of the Baltic Sea region, bringing 'the rim' closer to the European core; the process envisaged as a 'thickening web' of networks (Mortensen and Bisgaard 2013, 25).

The crusades in the thirteenth century can, on one level, be termed 'a tale of many cities'. At this time the networks converged in key urban hubs, particularly Lübeck, Visby, and Riga; both the Teutonic Order and Sword Brothers had a fixed presence in Lübeck from the end of the 1220s (Jensen 2001, 90), embodying the alliance between religious and mercantile groups, a symbiosis which contributed to, and continued with, the phenomenon of the Hanse (Blomkvist 2005, 526; Tandecki 2009b). Gaimster (1999, 2002, 2005, 2007, 2011, 2014) has consistently argued for the creation of a distinct Hanseatic (i.e., German mercantile collective) identity, based on the association between imported German ceramics across towns in Northern Europe and the presence of German merchants. This alludes to a level of dependence on the core envisaged by the model of Europeanization (Blomkvist 2005, 704). This was above all a commercial dependence, and it can be argued that the cheaply exploitable exports which defined the economies of Prussia, and to a lesser extent Livonia, were driven by core markets (Jahnke 2015), in the same way that *some* communities of German merchants were dependent on products from the core for the maintenance of elements of their own identity. The Order itself promoted self-sustainable convents but depended on recruits from German aristocratic families to sustain its higher ranks (Militzer 1994), facilitated by a shared language, kinship, and patronage links, as well as geographic proximity. The training of German aristocrats in complex territorial administration, a familiar rite of passage from the twelfth century, would have also been highly valued by the Order (Lyon 2008).

The idea of Hanseatic cultural solidarity and exclusivity in towns such as Turku has been challenged, with post-colonial interpretations of the distribution of ceramic types emphasizing the importance of more complex interactions between German merchants and local inhabitants (Immonen 2007; Gaimster 2014, 76). Similar doubts about the correlation between imported ceramics and the presence of colonists have been raised in the case of Viljandi (Ger. Fellin) (Haak and Russow 2013, 70–73), whilst a comparison of imported pottery in west Estonian towns has indicated variability between centres depending on their economic, political, and geographical contexts (Russow 2006). Excavations in Elbląg (Ger. Elbing), however, where the majority of the urban population was German, along with a small proportion of indigenous Prussians, and from the fourteenth century migrants from Polish towns (Tandecki 2009a, 441), have indicated that imports and other luxuries were clearly associated with the patrician class, which in 1417 made up around 11% of the town's population (Czaja 2009, 445). In the early years of the fourteenth century, an extension was added to the town hall, where large quantities of ceramics were uncovered, of which around 90% were imported and included fragments of stoneware jugs from the Rhineland (Nawrolska 2008). In general, finds of imported stoneware and ceramic stove tiles in the Baltic *Ordensland* are associated with urban and castle sites in stark contrast to rural settlements which are almost exclusively dominated by locally produced ceramics, but the socio-topography of all of these categories of site remains to be explored in more detail (Pluskowski 2013, 237–41). What is clear is that both towns and castles, especially the larger convents, provided local people or reagents, as well as migrants, with opportunities. They became multicultural communities exhibiting varying degrees of difference and homogeneity. Where urban and rural cemeteries have been systematically compared, as in the case of Estonia, it appears the majority of urban-dwelling indigenous people adopted the burial rites of the Christian colonists. They were hindered by their relative purchasing power, which would have translated to differences in diet, the ownership of property and luxury/exotic commodities (Naum 2014). If earlier forms of ceramic technology are absent, it is perhaps because new locally produced wares were quickly and cheaply adopted. Though upwardly mobile, the indigenous urban population remained — for a time — a hybridized community, occupying a distinct space between the worlds of the urban Germans and the rural Estonians (Valk 2004, 111).

At the same time, colonists were primed in the climate of the thirteenth-century *excitatoria* to treat the indigenous populations they encountered as 'other', defined most of all by their religious beliefs, burial rites, and cemetery locations (Tamm 2009, 30–31), even though these were clearly concessions to indigenous neophytes, such as the use of village cemeteries which was permitted until the seventeenth century (Laul and Valk 2007, 129–30). The theoretical framework of otherness has been used to understand the tensions visible in multicultural communities in other parts of Latin Christendom (e.g., Iberia; Glick 1992). In the towns and castles of Livonia and Prussia the sense of the other was perhaps not so much pronounced in the sphere of religion, but by differences in language and associated mannerisms, personal presentation (archaeologically most visible through the retention of indigenous forms of jewellery), and the 'secret "language" of the domestic culture' (Blomkvist 2005, 709), reinforced by economic status and ultimately by legal definitions of social status (see Chapter 1). In the Liv quarter in Riga, within the bounds of the town walls, the absence of 'Germanized' material culture, including architecture, may be interpreted as reflecting low socio-economic status or cultural resilience — i.e., indigenous adaptations to an urban environment — or a combination of both. Irrespective, it would have contributed to a sense of visual separation, underlined from the latter half of the fourteenth century by the legal distinction between German and non-German. Similarly, in medieval Reval (Est. Tallinn), native Estonians and Germans interacted with each other whilst levels of segregation were maintained, especially between wealthy patricians and the lower un-German class, in part reflected in the exclusive use of material culture by specific groups (Naum 2014).

To conclude, the relationship between the colonists and the colonized in the revised model of Europeanization, framed in the language of political and economic relations, regularly dovetails with the term 'colonial' (e.g., Bartlett 1993), in contrast to the prevalent historiographic rejection of crusading as colonialism (see Kedar 1992). Whilst rejecting the Baltic Crusades as a colonial process (see Chapters 1, 17), the contribution of this study is to consider whether trends in environmental exploitation can be understood as the result of the implementation of a new world (Catholic, Latin, German, and Hanseatic) system, separating it from former systems. However, despite the acknowledgement of agents and reagents, there is a danger of envisaging the eastern Baltic as a passive periphery, subject to a form of 'ecological imperialism' driven by the core (Crosby 1986; Lahtinen 2006, 187; Mortensen and Bisgaard 2013, 16). This can be addressed by contextualizing trends in environmental exploitation in the eastern Baltic within the theoretical framework of frontiers.

Frontiers

Frontiers in medieval Europe have been largely studied by historians over the last few decades, providing opportunities for examining the complexity of cultural encounters, ethnic identities, multiculturalism, and militarism in regions positioned between culturally distinct polities (Bartlett and Mackay 1989; Berend 1999; Abulafia and Berend 2002; Merisalo and Pahta 2006). This has included a number of studies dedicated to the frontiers of Prussia, before and after the crusades, drawing on archaeological, written, and cartographic sources (Kowalczyk 1999; Grążawski 2003, 2007, 2009; Urban 2001; Marciniak-Kajzer 2006; Karwowska and Andrzejewski 2006; Skrobot 2007; Sawicki et al. 2015). Post-colonial theory has been used to understand the social processes characterizing these ambiguous 'third spaces'; fragmented landscapes characterized by cultural heterogeneity and hybridity, rather than uniformity (Naum 2010, 2012). In particular, the concept of 'creolization' was developed to explain the interactions between colonists and colonized populations in societies with hybrid characteristics (Deetz 1996; Ewan 2000). Creolization has been typically used in the study of colonial societies, although it has also informed some analyses of pre-colonial multicultural encounters (e.g., the provinces of the Roman Empire; Eckardt 2010). In the case of the societies created by the crusades in the eastern Baltic, post-colonial theoretical influences are most evident in the model of Europeanization, but the frontier is better understood as multiple and dynamic, rather than an essentialist dichotomy between core and periphery. Moreover, the model of Europeanization betrays its colonial undertones by treating the whole of the medieval eastern Baltic as a static frontier, when in fact Livonia and Prussia became cores in their own right after the crusades, with their own frontiers. The military conquests of the thirteenth century incrementally extended the area controlled by Latin Christian authorities in north-eastern Europe only from the perspective of Christendom as an entity with loosely shared values, for with the exception of Danish Estonia until the mid-fourteenth century, the result was politically separate dominions. The expansion, organization, and administration of territory was controlled solely by the theocracy. This was itself a protracted and regionally variable process, but one that was instigated from local rather than distant power centres. Moreover, it required a level of collaboration with the local population, whether native or migrant, in order to avoid the creation of 'counter landscapes' (Palang et al. 2005, 289).

The consolidation of the eastern frontier in Prussia took much longer than the formation of Livonia's borders (with the exception of the procurement of Danish Estonia) and can be defined less by the presence of a militarized political authority (as represented by the Order's fortified outposts), and more by stability in settlement reorganization and growth associated with colonization. In the Kulmerland and north Pomesania this began in earnest in the 1280s, at the end of the thirteenth/start of the fourteenth century there was intensive colonization of the commanderies of Elbing, Christburg (Pol. Dzierzgoń), and the area around Lake Geserich (Pol. Jeziorak) and the Eylenz (Pol. Iławka River) Valley. The early decades of the fourteenth century saw increasing colonization in Ermland (Pol. Warmia), the commanderies of Marienburg (Pol. Malbork), Balga, and the northern stretch of Brandenburg (Rus. Ushakovo). Colonization and settlement reorganization also began on the Order's Pomerelian territories at this time. From the mid-fourteenth century the process of colonization intensified in the southern parts of the commanderies of Brandenburg and the eastern part of Königsberg's (Rus. Kaliningrad) commandery, with colonization in Sambia generally limited by a higher density of surviving indigenous settlements and the proximity of the frontier with Lithuania. Colonization in western Sasna began in the 1320s, the establishment of the convent at Osterode (Pol. Ostróda) and the fixing of the border with Masovia facilitated colonization from the 1340s, whilst the gradual colonization of the former tribal territories of eastern Warmia, Barta, and western Galindia began from the 1330s. Settlement gave way to a line of fortified outposts which marked the beginning of the militarized zone of the 'Great Wilderness'. Following significant damage caused by Lithuanian incursions in 1347–48 in this region, the process of colonization began again and intensified in the later decades of the fourteenth century under the direction of the commanders of Balga and Brandenburg. Further east, marked by the outlying castle of Lyck (Pol. Ełk), the process of colonization would begin in the early fifteenth century and would not intensify until after the cessation of hostilities with Poland-Lithuania (Długokęcki 2009). As the frontier shifted, new heartlands were created, and this has direct implications for understanding the management of the landscape. The borders with Lithuania — the southern frontier of Livonia and the eastern frontier of Prussia — remained completely permeable throughout the fourteenth and fifteenth centuries. They were thinly populated or depopulated militarized zones which were the crossing points for numerous raids from both sides (Paravicini 1989).

The differences between Prussia and Livonia are important to note. In Prussia, a number of landscapes which can be considered as frontiers in the pre-crusade period, or during the period of active crusading, were subsequently redefined as heartlands (e.g., the Kulmerland, Marienburg, Elbing). Others, which had been tribal heartlands were redefined as frontiers (Sudovia). The core of annexed Pomerelia remained unchanged, except in relation to the central convent at Marienburg. In Livonia, the theocratic authorities took over and consolidated a number of indigenous heartlands of varying sizes (the Lower Daugava, Wenden (Latv. Cēsis), Fellin (Est. Viljandi), Ösel (Est. Saaremaa)) and prompted the transformation of former cores into frontiers in only a few areas, most clearly in the southern stretches of Curonia (Memmelland) and Semigallia. Contemporary Christian understandings of frontiers also framed them in relation to central places. New cores were represented by the bases of military, political, and religious authorities, but also by a religious landscape with cores defined by cathedrals, churches, and the presence of relics in the chapels of the military orders, with paganism represented at the periphery, mapped onto the physical margins of the landscape (Jensen 2009).

Given this level of regulation of settlement, the actions of the various colonizing groups, whether represented by the Order's convents or peasant villages, can be treated as proxies for the governing theocracy when it comes to linking changes in the landscape with the establishment of convents and their networks of subjects. This is particularly useful where the data for settlement is limited or coarse in resolution. But at the same time, the interregional comparative approach makes it possible to determine the extent to which the transition from a frontier to a heartland, and vice versa, resulted in a comparable ecological profile across the *Ordensland*.

Key variables would have been differences in the landscapes encountered by the colonists, the nature and extent of settlement, relationships with the surviving indigenous population, as well as the aspirations and abilities of individual commanders and other officials charged with territorial management. As with the building of a castle (Johnson 2002, 181), which involved a negotiation between the Teutonic Order's requirements and the practicalities of construction envisaged by architects and masons (Arszyński 2010, 42), the impact on the landscape would be the result of multiple agencies coming together. If trends in environmental exploitation in Prussia can be synchronized with the staggered chronology of colonization outlined above, as well as the reorganization of surviving Prussian communities, then in Livonia similar changes would reflect the activities of the indigenous population, and the degree to which they were mobilized by the theocratic regime. In this respect the definition of a frontier, informed by postcolonial and complexity theory, provides a useful framework for contextualizing the variable character of diachronic trends in environmental exploitation. Despite its projected corporate image, these variables would have compromised aspirations towards homogeneity in lordship across the territories of the Teutonic Order in the eastern Baltic.

Anthropogenic Ecology and Resilience

Postcolonial frontier theory is concerned with the variability of cultural encounters, resulting from confronting unexpected, misunderstood, or incompatible social praxis, ideologies, and norms (Naum 2010, 106). On one level the eastern Baltic frontier can be defined as a place where 'the exclusive presence of Christianity was challenged' (Abulafia 2002, 34), indeed where the whole paradigm of Europeanization as expressed in the actions of crusading institutions was challenged, unsuccessfully at least on a political level. An important aspect of this, if not the most important in the eastern Baltic, was the encounter with the environment, where the ideology of colonization was inextricably tied to protocols of resource exploitation as a means of sustaining the process of conquest and settlement. Since aspects of the environment were not only sacred to the indigenous population, but also fundamentally linked to Late Iron Age social organization, the ecological dimensions of these cultural encounters are synonymous with the transformations that can be broadly framed within the revised model of Europeanization. Andris Šnē (2006, 69) defined Late Iron Age Baltic social relations primarily in terms of economic power, supported by an ideological superstructure of symbolic capital and violence. At the heart of these power relations, before and after the crusading period, was the ownership of land and access to its resources (see Chapter 8; Palang et al. 2005). If a consistent template of environmental exploitation was implemented by the Teutonic Order, then the impact as understood through the ecological datasets considered within this study should be equally consistent. If not, then it is instructive to consider colonization as adaptation. The social mechanisms of adaptation following colonization have been explored from archaeological perspectives, although largely related to prehistoric contexts and pioneer phases of occupation (Rockman and Steele 2003). However, there is a methodological gulf between modern ecology and historical ecology. On the one hand, models of biological adaptation to new ecosystems are primarily concerned with fitness linked to genetic variation within an evolutionary framework (Hoffman and Parsons 1997), far too coarse for contextualizing the nuances of the time period in question, although useful for understanding the depletion of certain species such as the aurochs (see Chapters 6 and 12). On the other hand, an ecosystemic approach to past societies cannot match the specificity required by modern ecology, nor can the latter account for the nuances of historical contexts (Horden and Purcell 2000, 48–49). The term 'ecology' is therefore used in this study to refer to the multiple interactive factors that enable anthropogenic impact on the environment to be measured comparatively over time and space. Most recently Campbell (2016, 22) has conceptualized human-environment interactions in the later Middle Ages as six interrelated components — climate, society, ecology, biology, microbes, and humans — arguing for a holistic, multi-scalar approach to human history.

Within modern ecology, a useful framework that can be adopted (and adapted) for understanding anthropogenic impact is niche construction (NCT), which considers how species modify their environments to influence their own evolution and that of other species (Odling-Smee et al. 2003; Kendal et al. 2011). Again, this has been defined within an evolutionary framework, with cultural inheritance envisaged as a selective evolutionary force, and has been adopted within environmental archaeology in the study of the formation of human habitats (O'Connor and Evans 2005; Wilkinson and Stevens 2013: 241–68). Niche construction results from the interaction of multiple drivers, both indirect (demographic, economic, socio-political, religious, and technological) and direct (physical and biological factors, including climate, soil nutrients, land conversion, biological invasions, and disease) (Nelson et al. 2006). This can be readily aligned with the definition of the cultural landscape, which in the context of multiple phases of occupation and associated modification can be envisaged as a palimpsest (see Chapter 1). Indeed, the long-term expansion of Europe from the tenth century had been framed very broadly in ecological terms already by Alfred Crosby (1986) in the 1980s, and more precisely in relation to the Middle Ages by Richard Hoffman (2014). Increasingly more studies are adopting a holistic approach to the ecology of medieval archaeological sites, articulating the nuanced relationship between changing environmental and cultural contexts (Schreg 2014, 88). Zooarchaeologists have gone so far as to 'biologically profile' medieval sites (Ervynck 2004; Pluskowski 2010), although without detailed socio-topographic comparisons within and across sites (Ervynck 1992), these profiles remain relatively generic. Indeed, the fragmentary nature of palaeoenvironmental data falls short of the expectations of modern ecologists.

The twelfth and thirteenth centuries, the period associated with the high medieval expansion of Europe (Bartlett 1993), can be broadly understood as a time of niche construction driven by urbanization, colonization, and the reclamation of marginal land, linked to a favourable climate — the end stage of the Medieval Climatic Anomaly (MCA). In the eastern Baltic, the establishment of new fortifications, towns, and rural colonies, as well as the reorganization of the landscape, resulted in new ecological niches being created. The concept of ecological inheritance in niche construction, when applied to the frontier, is therefore more complicated and subject to the same vicissitudes which define a 'frontier society'. The laws of settlement adopted by the Teutonic Order provided templates for new niches (see Chapter 1), represented by the systematic management of its subjects, the landscape and provisioning networks. At the same time, the surviving indigenous population was in the process of creating its own ecological niches; adaptations within the framework imposed by the ruling theocracy. The extent to which the colonists' templates were or could be effectively implemented, whether they interrupted or accentuated existing trends in niche construction, is a key question this study will address. Both incoming and native responses can be understood within the context of 'resilience', a term that defines degrees of change in cultural identity and related modes of environmental exploitation (Adger 2000).

Synchronicity of Theoretical Frameworks

The theoretical frameworks of Europeanization, frontiers, niche construction, and resilience can be synchronized to provide a valuable methodological tool for contextualizing the environmental changes associated with the Baltic Crusades. Europeanization recognizes the complexity of interactions between colonists and colonized at the inter-regional scale, within the context of indigenous Baltic, Finno-Ugric, Latin, German, and Hanseatic cultures. Frontiers are situated within this broader model and have defined spatial and temporal parameters, within which they are 'in flux' (Naum 2010, 108). Frontiers share some common characteristics, and when their status changes into heartlands, these characteristics also change. It is important to recognize the dynamism of frontiers and cores within the eastern Baltic. Related to this dynamism is the concept of niche construction, which defines the process of transforming the landscape as a result of colonization following military conquest. Niche construction cannot simply be understood through the lens of economics but is mediated by a host of variables, including the various ideologies and agendas of the colonists, the responses of the indigenous population, as well as the capacity to adapt to unfamiliar or challenging environments — which can be understood as a measure of interrelated social and ecological resilience. Taken as a whole, this moves us away from a straightforward top-down model of exploitation wholly dictated by the ruling theocracy. These theoretical frameworks are concerned with change over time, and it is ultimately the longue durée perspective of environmental change that enables the impact of the crusades to be contextualised.

Scales of Analysis: Regions, Territories, Sites

The CCC project used geography, specifically the landscape, as its integrative framework for comparing societies across the Baltic Sea region. It established three levels of analysis: primary or local, focusing on dwelling sites; secondary, focusing on centralized and distributive systems between areas/groups; and tertiary, focusing on regions and polities (Blomkvist 1998, 29–30). The lowest spatial level for aggregating multidisciplinary data was relatively coarse, defined as the 'region' (Blomkvist 2009, 438). A similar multi-scalar continuum was used by Horden and Purcell (2000) to move from the micro-ecological scale to level of the region. In this study, two spatial levels for aggregation are defined as 'the site', largely consisting of castles built by the Teutonic Order, and their associated districts or territories.

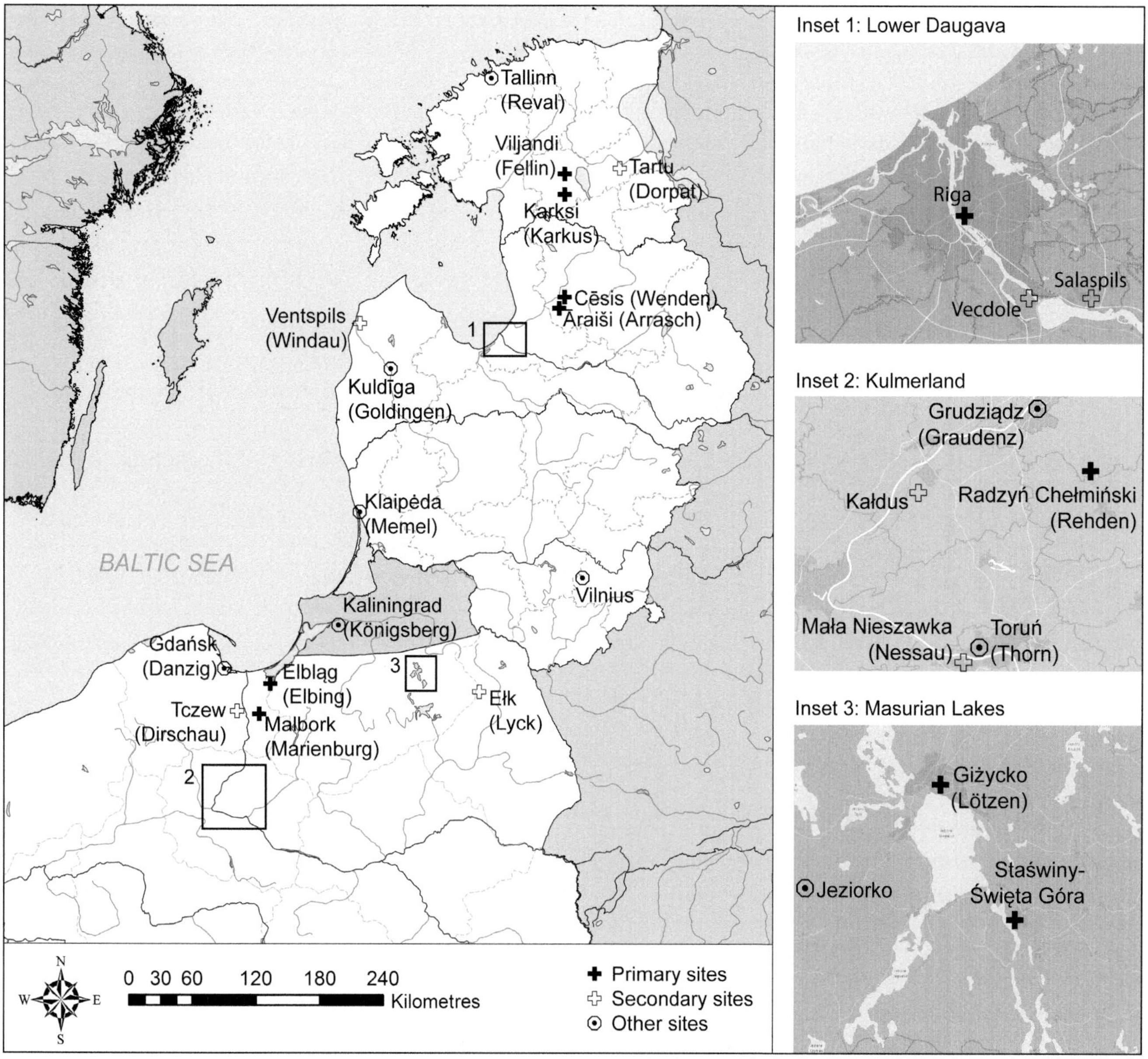

Map 2.1: Map showing the primary and secondary sites included within the study. Map by A. Pluskowski.

In the interests of avoiding particularistic interpretations of one or two castles, extrapolated to the entire medieval theocratic polity, it was decided to include a range of sites from the Teutonic Order's territories in both Livonia and Prussia. These are primarily situated within the boundaries of modern Estonia, Latvia, and north-east Poland, with one site located within western Lithuania (Klaipėda, Ger. Memel) and the full extent of three commanderies falling today within the Russian Kaliningrad Oblast: Brandenburg (Rus. Ushakovo), Königsberg (Rus. Kaliningrad), and Balga. Given the variability in the experiences of crusading and the relationships with the indigenous populations in Prussia and Livonia, as well as the hierarchical structure of the Teutonic Order and the cohesive image it projected, a comparative approach was essential for understanding whether an institutionalized template had been introduced or developed by the Order across its eastern Baltic territories. The choice of sites was dictated by a number of factors, in particular accessibility to ongoing or planned excavations, archives of faunal assemblages, and the potential for recovering palaeoenvironmental data from sites and their associated territories.

Sites included within the project were subdivided into two categories: *primary sites* were the focus of detailed, multi-proxy analyses where data was obtained through excavation, coring and the study of related written sources. *Secondary sites* were associated with at least one or more types of data, typically assemblages of animal bones, plant macro-fossils and parasite remains, as well as written sources. This material was largely derived from previous or unrelated excavations, where the data was made available to the project for analysis. The exception is the Liv quarter in Riga, where the comprehensive sampling strategy employed by the excavation team produced a diverse suite of environmental data from well-dated contexts (see below). The distribution of primary and secondary sites is shown in Map 2.1. This hierarchy of sites enabled archival data to be integrated with material collected systematically using consistent sampling protocols, whilst excavation techniques represented a composite of national traditions.

Sites Associated with the Teutonic Order

The emphasis of this study is on colonizing sites, and the majority of case studies consist of castles of varying hierarchical importance within the *Ordensland*, and in a few instances related settlements. Three basic types of castles were included within the study: convents (centres of commanderies), advocates' residences (centres of advocateships), and procurators' residences (centres of procuratorships). Secondary sites also included episcopal castles for comparative purposes; those garrisoned by vassals (e.g. Vecdole in Latvia, Ger. Alt-Dahlen) and the residences of bishops (e.g. Lidzbark Warmiński in Poland, Ger. Heilsberg). Material derived from castle contexts was sub-divided further into one or more of the chronological phases of 'social time' outlined in Chapter 3.

Castle building defined and facilitated the crusades in the eastern Baltic; the most explicit physical manifestation of the new cultural imports which symbolised and implemented the political reorganization following military conquest. The introduction of stone castles in the early years of the Livonian Crusade was noted as something completely novel, whilst in the early decades of crusading in Prussia *castra* (the term used by Peter von Dusburg) were constructed from timber and earth, emulating the design of contemporary indigenous and neighbouring fortifications. The internal organization of these early *castra* remains unclear as they have left only fragmented archaeological traces, with the exception of the thirteenth-century timber castle uncovered in Elbląg (Fonferek 2019) and, to a lesser extent, parts of the Prussian stronghold at Stary Dzierzgoń (Ger. Alt Christburg) reused by the Teutonic Order (see Pluskowski 2013, Gazda 2018). The narratives of the crusades suggest they had clearly defined military, economic, social, and religious functions, and excavations have shed important light on their early role in organizing resource exploitation. The period of active crusading saw warfare conducted on two levels: physical and spiritual. Castles played a pivotal role in both; as bases for garrisons and as beacons of perceived sanctity, named after the Virgin, Christ, saints and archangels, with relics deposited in their chapels (see chapters 8 and 15). In this respect the spiritual war, articulated through the belief system that supported crusading vows, had clear material components which became embedded within the landscape.

The second chronological phase is associated with the period of castle rebuilding and expansion, when many of the smaller residences were built as the central places within the various *Ämter* created to manage the conquered lands. In Prussia, brick and field stones began to be regularly used in the Teutonic Order's buildings as early as the mid-thirteenth century (in the case of Thorn), and adopted as the standard building materials for the conventual castle from the 1280s (Arszyński 1995, 2013a; Herrmann 2007). In Livonia, both glacial erratics and quarried stone were more typically used, alongside brick and timber (Caune and Ose 2004). Within the range of castle complexes, consisting of a core group of buildings (often referred to as 'high castle'; Ger. Haupthaus, Pol. zamek wysoki) and one or several outer baileys, wards or *Vorburgen* fortified with curtain walls, the policies of grand masters, the ambitions of individual commanders and the levels of funding they could access, are visible. Yet there was a high degree of standardization in their design. Only six castles built by the Teutonic Order (and one episcopal castle) in Prussia were irregular in form, reflecting adaptation to a specific topographic setting which could include pre-existing earthworks (Herrmann 2007, 81). In Livonia, a number of castles and their associated towns were built on interfluvial spurs, resulting in trapezoidal plans (Leighly 1939, 271), although the use of the conventual plan was widespread, perhaps even originating in this region before its adoption in Prussia (Arszyński 2013b, 193). All Livonian and Prussian convents included refectories (where meetings of the chapter usually took place; Jóźwiak and Trupinda 2012, 48), dormitories, infirmaries and separate *camerae* for commanders and other officials, as well as kitchens and stores. Other service areas, such as stables and smithies, were typically located in the outer baileys (or outer wards and sing. Ger. *Vorburg*), although these could also include refectories, chapels, and various *camerae*. Smaller castles included chapels and a combination of residential and service spaces. They were often dominated by a main residential wing attached to a ward enclosed by a wall. The internal arrangement of these spaces varied, but chapels were consistently located in the east-facing parts of castle wings reflecting the alignment of altars (Arszyński 2013a, 66). The terms used for these spaces in contemporary documents varied, and it is often difficult to determine the internal spatial functions of the Order's castles given their history of dismantling and reconstruction, and particularly how these changed over time (Jóźwiak and Trupinda 2012). The specific details of case study sites are discussed in the regional introductions (Chapters 4 and 10).

A small group of castles served additional political functions within the *Ordensland*, where they doubled up as residences of major officials within the Order. The residence of the Grand Master was situated in the Marienburg convent from 1309 to 1457, and relocated to Königsberg from 1457 to 1525. The residence of the Livonian Master was situated in the fortified convent in Riga from 1237, although after it was destroyed the later castle was rebuilt at the north-western edge of the town in 1330. The Master's residence was relocated to Wenden in 1481 until the dissolution of the Livonian branch of the Order in 1561. The presence of major officials did not necessarily result in larger households than found in other convents, with the exception of Marienburg where the size of the complex, its maintenance, and provisioning requirements were significantly larger than any other castle in the *Ordensland* (Jóźwiak and Trupinda 2007). A similar hierarchy operated for the castles of bishops and cathedral chapters within episcopal lands in Livonia and Prussia.

Primary case study sites were selected according to a range of factors: accessibility, preservation, earlier work, associated historical documentation, the suitability of deposits within their territories, and their relative size and importance within the *Ordensland*. They were evaluated through a systematic process which began with a geophysical survey and a programme of on-site coring, assessing soil conditions, as well as the depth and quality of occupation deposits. This informed targeted excavations aimed at recovering the broadest range of environmental data. The territories of these sites were defined in three ways: the boundaries of the pre-crusade territory as reconstructed from archaeological and later written sources, the boundaries of the Teutonic Order's relevant *Ämter* (such as a commandery), and as a unit with ecological integrity, such as a drainage basin or discrete area of woodland.

Sites Associated with Indigenous Settlement

Until recently, the archaeology of pre-crusade settlements (Late Iron Age/early medieval) within modern Poland, Latvia, and Estonia has attracted relatively more scholarly attention than the later medieval archaeology of the crusader states (excluding architectural studies). In this respect and due the constraints of time and funding, the project included a smaller number of indigenous settlements as primary and secondary sites. All the chosen sites have produced significant quantities of data, with two designated as primary sites: the Staświny-Święta Góra complex in north-east Poland defined as *lauks* Staświny, and the Liv quarter in Riga. The most important secondary sites consisted of the Late Iron Age tribal centre at Viljandi in south-central Estonia and the early medieval Kałdus complex in the Kulmerland, north Poland, occupied until the arrival of the Teutonic Order.

Each primary site was the focus of two spatial scales of analysis which linked the settlement in question with its surrounding landscape. These scales are defined as 'on-site' and 'off-site' and characterize the multidisciplinary methodology adopted within this study.

On and Off-Site Analysis

On-site analysis targeted the area within the physical boundaries of a site. In the case of castles this extended to the surrounding moat systems; in the case of open settlements (as at Święta Góra and Biała Góra), these boundaries were not always clearly defined, and an understanding of the spatial context of these sites was dependent on mapping the relative density of occupation deposits, as well as evaluating their topography. The ecological impact of castle construction in the eastern Baltic was primarily associated with the reorganization of natural resource exploitation in variously defined territories. Off-site analysis therefore focused on reconstructed tribal territories or macro-regions associated with archaeological sites for the pre-crusade period, and subsequently on the commandery system introduced by the Teutonic Order. Virtually all convents were associated with planned towns and commanderies were associated with indigenous settlements (primarily in the case of Livonia and eastern Prussia) and newly established colonies (primarily in Prussia). Segregating the role of the castle from its associated settlements has not been straightforward and has been considered on a case-by-case basis. Material recovered directly from occupation deposits within castles could be readily distinguished from landscape changes evident within the territories of vassal settlements. In some cases, it was possible to compare the activities of different types of settlements within the same commandery.

Methods and Categories of Data

Historical Sources

The range of written sources is outlined for Livonia by Kreem (2002) and summarized most recently for Prussia and Livonia by Czaja and Radzmiński (2013, part 5). These can be subdivided into literary and documentary sources (see the list of primary sources). The first category of texts provides some information about the use — and particularly Christian perception — of the landscape during the crusading period in the eastern Baltic (Nielsen 2011). The Latin chronicles of Peter von Dusburg (*Chronicon terrae Prussiae* (*Chronicle of Prussia*), completed *c.* 1326; see *CTP*; Trupinda 1999, 7–16) and Henry (*Henrici Chronicon Livoniae*, *Chronicle of Livonia*, written *c.* 1224–27), were intended for clerical audiences, the papal curia and the bishop of Livonia respectively. Subsequently, the most important narratives are the *Livländische Reimchronik* (*Livonian Rhymed Chronicle*; hereafter *LRC*) written in the last decade of the thirteenth century and covering the period 1180–1290; Nicolaus von Jeroschin's *Die Kronike von Pruzinlant* (*Chronicle of Prussia; hereafter KP*), written in German in the 1330s; Hermann von Wartberg's *Chronicon Livoniale* (*Livonian Chronicle*, hereafter *HWC*), covering the period from 1196–1378; Jan Długosz's Latin translation of Wigand von Marburg's *Chronica nova Prutenica* (*New Prussian Chronicle*), covering events from 1311 until 1394 (Hirsch 1863; Zonenberg 1994), and Długosz's own *Annales seu cronicae incliti Regni Poloniae* (*Annals of the Kingdom of Poland*), written between 1455 and 1480 (Przezdziecki et al. 1863–78; Smith 1997). The first attempt at detailed ethnographic writing can be attributed to Simon Grunau, a Dominican from Elbing, whose descriptions of indigenous customs and lifestyles in his *Preussische Chronik* (*Prussian Chronicle*), written in the early decades of the sixteenth century, have until recently been dismissed as largely fictitious (Perlbach et al. 1876–96; Zonenberg 2009).

In terms of documentary sources, there is a disparity between what is extant for medieval Prussia and Livonia. From the last decades of the fourteenth century, inventory books are available for a number of Prussian convents, compiled as *Das große Ämterbuch des Deutschen Ordens* (*GAD*), along with a detailed inventory from Marienburg — the treasurer's book or *Das Marienburger Tresslerbuch* (*MTB*). The extensive sources for the Prussian *Ordensland* were synthesized in vast corpuses by German scholars in the nineteenth and twentieth centuries: firstly and most famously by Voigt (1827–39); and then the compilation of documents from the mid-fourteenth century in the series *Preussisches Urkundendenbuch* (1882–1975); and the *Scriptores rerum Prussicarum* (Hirsch et al. 1861–74; for a summary, see Jähnig and Michels 2005, 23–48). The archives of the Livonian branch of the Order were largely destroyed, with documentary evidence pertaining to the Order's holdings surviving in the civic archives of Tallinn and Riga from the fifteenth century (Kreem 2002), and along with additional sources relating to castles and their territories compiled as *Liv-, Est- und Kurländisches Urkundenbuch nebst regesten* (*LUB*). Of particular value are the descriptions of land boundaries, which provide detailed insights into the organization of landscapes in medieval Curonia, although they are lacking for many other regions (see Chapter 8). The items listed in castle inventories largely derived from the related *Ämter*, and therefore also contribute to an understanding of territory use during discrete periods of time. Written sources have been incorporated throughout where appropriate, but form the substantial basis of Chapters 8 and 15.

Figure 2.1: One of the most striking examples of in situ animal iconography from a convent is found on the 'Golden Gate' at Malbork Castle (Ger. Marienburg), *c.* 1280. It shows a pig feeding on fallen acorns. According to thirteenth-century Christian iconology, this is most likely a representation of sin, of unclean iniquity (cf. Barber 1993, 84–86). Photo by M. Elander.

Figure 2.2: Religious sculptures from the Teutonic Order's Prussian territories: (a) wooden image of the enthroned Virgin and Child, imported into Prussia (*c*. 1250), perhaps used in the castle chapel of Rogasen (Pol. Rogóźno) then in the church at Grutta (Pol. Gruta); (b) St George, linden, *c*. 1380–90, church in Altmünsterberg (Pol. Stara Kościelnica) (Pospieszna 2010 (I), 98, 103, 107); (c) Crucifix, linden and pine, church in Trunz (Pol. Milejewo), *c*. 1389.

Medieval Artistic Sources

Elite and local patronage within the crusader states of the eastern Baltic promoted a flourishing industry of sculpture, painting and applied art (for the latest survey on Prussia see Pospieszna 2010). The majority of this art, as elsewhere in Europe, was religious and often metaphorical in character (Fig. 2.1) (Jóźwiak and Trupinda 2007, 109–34). The Order also sponsored art emblematizing its identity and heritage, but relatively little of this has survived. The most decorated spaces within the Teutonic Order's castles were chapels, and within the *Ordensland* cathedrals and parish churches, although very little original artwork has been preserved. What has survived provides little direct evidence of the role of the local environment, although artistic media, most often locally sourced wood and brick, and to a lesser extent amber, contribute to an understanding of the utilization of local natural resources (Fig. 2.2). Amber, in particular, was widely recognized as a luxury raw material many centuries before the crusades, with the largest deposits of tertiary amber in the world located in the north-western part of the Sambian Peninsula (Lukashina and Kharin 1999, 27). The Order's subsequent monopoly over its procurement from the coastal zone of its territories necessitated the creation of the office of *Bersteinmeister* (amber master). The largest quantities were stored in his residence at Lochstädt, where 51½ barrels of amber were recorded in 1392 (Mugurēvičs 2003, 94), and raw amber was exported across Europe to established manufacturing centres such as Lübeck, where religious artefacts, particularly rosaries, were produced (Paner 1999, 391). However, there is also evidence for the production of luxury amber artefacts within the Order's territories. With the dissolution of the *Ordensland*, significant international trade in amber would only commence again in the sixteenth century (Budrys 2001, 267). In general, and in contrast to other regions of Europe, the art produced within the Teutonic Order's territories provides very little information on the exploitation of plants, animals, and the organization of the landscape.

Regional Maps and Site Plans

The earliest maps of the eastern Baltic, dating to the later fifteenth century, provide virtually no detailed information on the eastern Baltic region other than indicating its peripheral location in relation to the main spiritual centres of Latin Christendom — Jerusalem and Rome. Sixteenth-century maps, most famously produced by Caspar Henneberger,

Map 2.2: Fragment of Olaus Magnus's *Carta Marina*, showing the eastern Baltic regions. Livonia and Samogitia are pictured as having a frontier defined by woodland which stretches south into eastern Prussia. The map was produced in Venice in 1539, over a decade after the secularization of the Prussian branch of the Teutonic Order. Digital version of map released into public domain by Uppsala University Library.

Map 2.3: Nicolas Sanson's map of Livonia (1663) showing the relative distribution of woodland at a very coarse resolution. DIGAR, The Digital Archive of the National Library of Estonia.

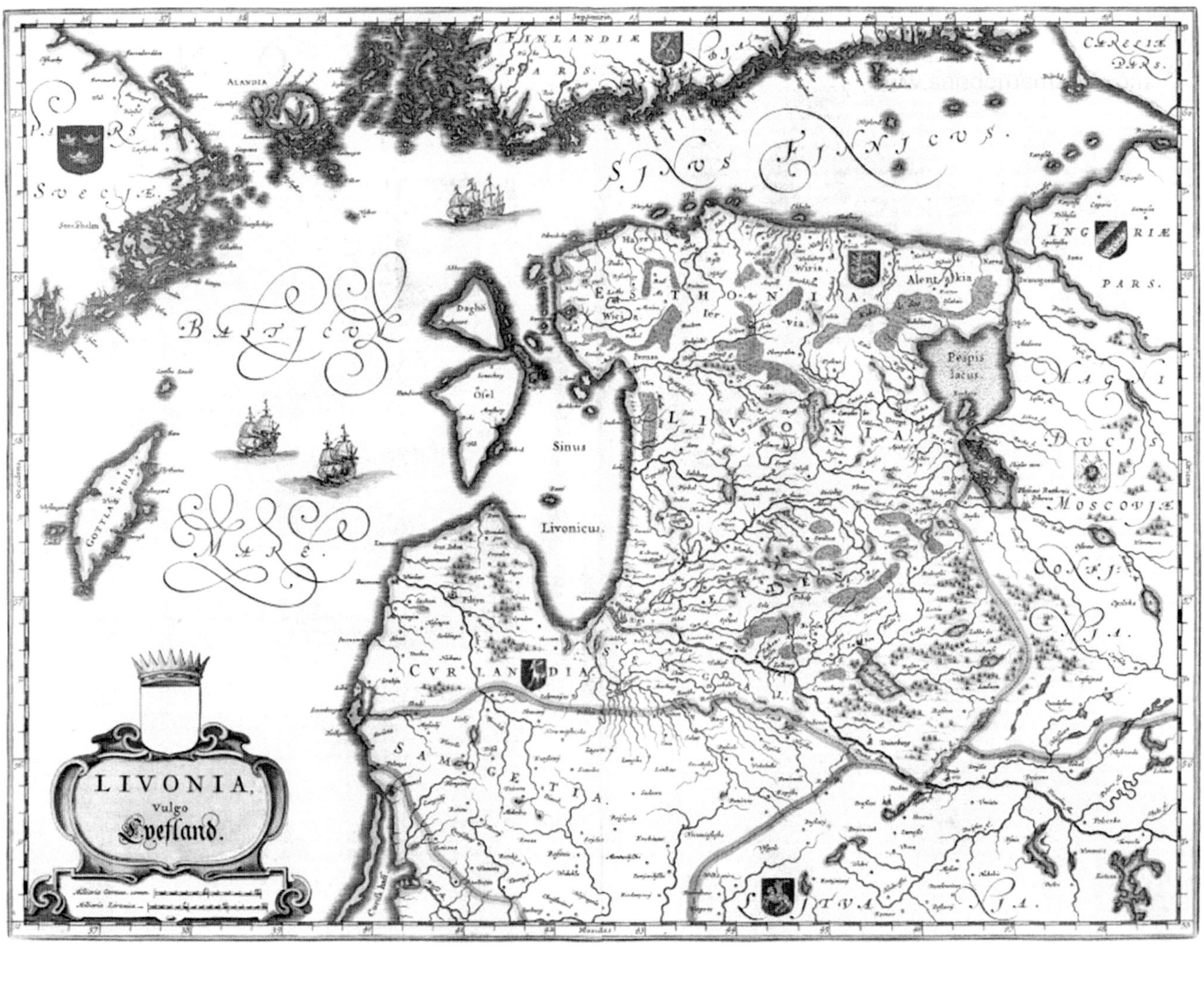

Abraham Ortelius, and Gerardus Mercator, sketched the distribution of major towns and emblematized landuse, particularly hydrology and woodland, which in some cases may have a coarse link with physical ecology and in others are clearly products of the map-maker's imagination (Map 2.2). The Teutonic Order and episcopal powers in the eastern Baltic had not produced maps themselves but did keep a detailed (albeit unevenly preserved) record of the geographical boundaries of properties within their lands. This has enabled the mapping of reorganized administrative units, particularly in the Prussian-Polish borderlands during the fifteenth century. In other cases, the better documented sixteenth/seventeenth-century boundaries of elderships (*starostwa*) and castellanies have been projected back to the late medieval period, on the grounds of demonstrable continuity in administration between the *Ordensland* and the post-medieval secular duchies. These are not without their problems, and the reconstructions of the Order's commanderies and pre-crusade tribal territories in this study are discussed in more detail under the relevant sections (see Chapters 4 and 10). From the later seventeenth century, regional maps were produced more systematically up to the present day. Nicolas Sanson's map of Livonia (1663) shows a clear difference in the distribution of woodland, with large expanses present in the south-east (Map 2.3). Particularly valuable are maps produced between the end of the eighteenth century and 1939/1944, as they provide detailed geographic information before the widespread destruction, depopulation, migration, and urban renewal which characterized the middle decades of the twentieth century. They show, for example, how woodland was largely cleared within the coastal zone between Brandenburg and Balga by the end of the eighteenth century, which today has regenerated (Map. 2.4). Geology and soil maps obtained from research institutions and national mapping agencies vary in resolution across the eastern Baltic countries, and were utilized at both regional and local scales. All relevant maps were digitized and synchronized with contemporary data from Google Earth. The specific location of cores taken from lakes and peat deposits was recorded using a hand-held GPS receiver, whilst acknowledging the spatial relationship between the point at which the core was taken and the size and shape of the associated basin.

Site plans were obtained from a range of sources. The earliest detailed plans of castles were produced in the later seventeenth century, and subsequent plans were made between the eighteenth and twentieth centuries. Antiquarians and archaeologists have been producing plans of varying quality of medieval sites in the eastern Baltic since the late nineteenth century through to the present day. In the process of geophysical surveying, a grid was laid out and linked to the most updated plan available for each site. Each excavation resulted in the production of new site plans using modern surveying techniques, and the augmentation of existing plans.

Geophysical Survey

All geophysical work within the project was conducted by David Thornley (Reading). The aim of these surveys was to identify 'magnetically quiet' areas that would be suitable for siting an archaeological excavation with a view to finding undisturbed anthropogenic deposits. Two types of equipment were used: a Bartington Grad 601–02 dual fluxgate gradiometer and a Ground Penetrating Radar (GPR) (Fig. 2.3). The choice of equipment was dictated by the type and condition of soils (including the level of the water table), as well as the compositional range of buried material at any given site. A gradiometer survey was carried out across the main complex of Karksi Castle (Estonia); gradiometer and GPR surveys of the unexcavated part of the moat and entire outer bailey at Cēsis Castle (Latvia); a gradiometer survey of the entire outer bailey at Āraiši Castle (Ger. Arrasch, Latvia); a gradiometer survey across the entire hilltop area, embankment, and part of the field immediately to the north at Święta Góra (Poland); a GPR survey in a number of areas in the courtyard of the outer bailey of the castle in Elbląg (Poland); gradiometer and GPR surveys at Malbork (Poland) in the courtyard of the high castle, the western range of the outer bailey, in the moat between the outer bailey, and middle castle; and finally a gradiometer survey was carried out across the entire site and the western escarpment down to the former river bed of the Nogat at Biała Góra (Poland).

The data was processed to produce magnetic contrast maps from the gradiometer surveys, and radargrams from the GPR, which highlighted the position of buried archaeological features, including building material (brick and stone), pits, ditches, and metals. This data informed the positioning of small-scale trenches, and the precise location of these trenches was synchronized with the grid established during the course of the geophysical surveys. Areas of rubble, typically composed of brick, stone, and other debris, were identified and, where appropriate, as in the case of the moat at Cēsis Castle, avoided (Fig. 2.4). In all cases the excavations confirmed the findings of the geophysical surveys, which in turn provided useful information on the characterization of future data obtained with using gradiometer and GPR surveys.

Figure 2.3: Gradiometer and Ground Penetrating Radar surveys at the project's sites: (a) Cēsis; (b) Elbląg; (c) Malbork; (d) Karksi. Photos by A. Pluskowski.

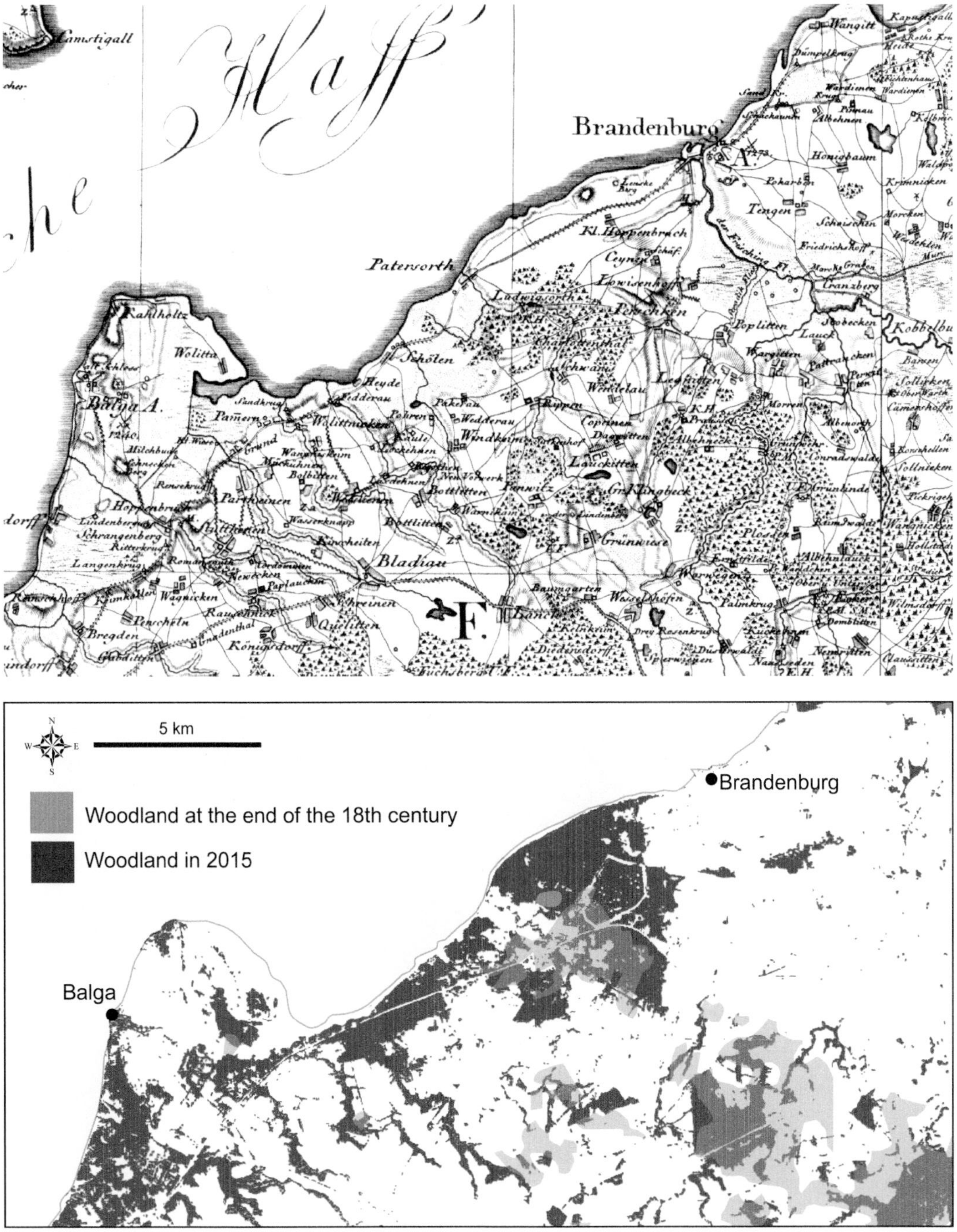

Map 2.4: (a) Part of Friedrich Schrötter's map of East Prussia, 1810 (scale 1:150,000), showing the distribution of woodland between Brandenburg and Balga (Staatsbibliothek zu Berlin - Preußischer Kulturbesitz; Kart. N 1020; image reproduced with permission), compared to (b) the current distribution of woodland (Google, map data: 2015 DigitalGlobe).

Archaeology and Archaeological Sources

Following a geophysical survey, a select number of trenches were opened within specific areas of the case study sites. The location of trenches aimed to uncover intact occupation deposits and maximize the recovery of ecofacts (plant remains, animal bones, and related products, as well as geochemical traces of human activities in the soil) and artefacts. Summaries are provided in the introductions to the sites in the later chapters of this book, whilst a number of reports have already been published on the primary sites of Karksi (Valk et al. 2009; Valk et al. 2012; Valk et al. 2013) and Biała Góra (Pluskowski et al. 2014; Sawicki et al. 2015). Off-site archaeological sources consist of material culture obtained through field walking, irregular, small-scale rescue excavations, and documented landscape features, including earthworks and known archaeological sites. Whilst contributing to an understanding of the chronology and extent of settlement in the pre- and post-cru-

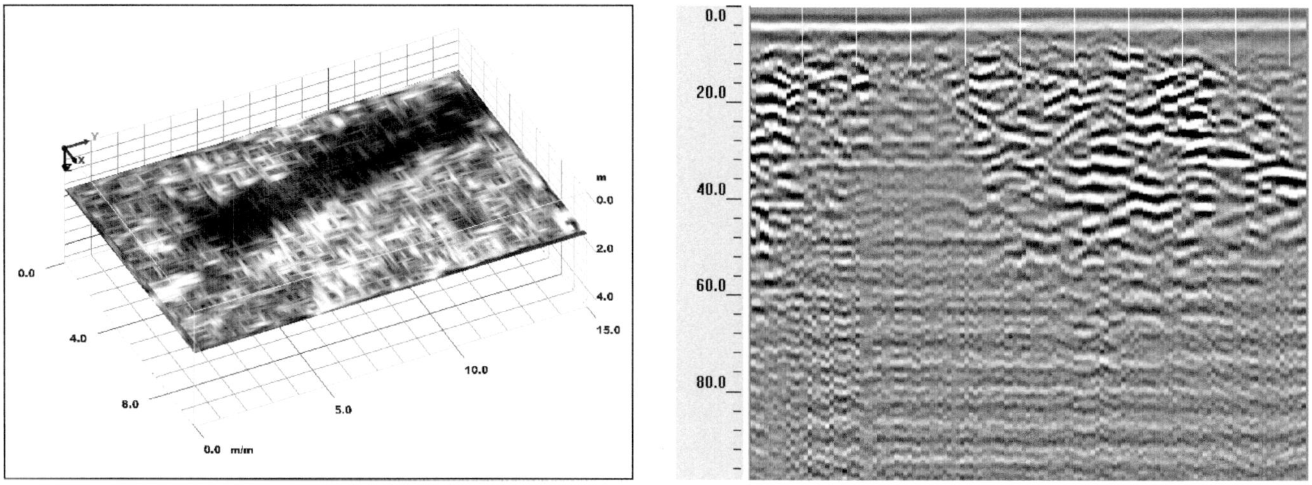

Figure 2.4: Radargram of the moat at Cēsis Castle showing (a) depth slice at 1.72 m; (b) transect at 8.5 m. Areas of rubble and disturbance are marked by strong signal responses and were confirmed by excavation. Figures by D. Thornley.

sade period, they were not usually recorded at the same spatial resolution as the primary, case study sites.

Animal Bones

Animal bones represent one of the most abundant ecofacts recovered from archaeological contexts, particularly from sites with intensive occupation. In the majority of cases they are the end result of human depositional activities and subsequently affected by various taphonomic processes, and do not reflect the actual number of animals present at the site or within its vicinity. However, comparative analysis of assemblages can indicate relative significance across multiple sites and over time. In all but one case study site (Āraiši Castle), a viable sample of animal bone was recovered during excavation. Archaeological faunal assemblages are also routinely archived from earlier excavations, and as a result material that had been recovered from the mid-twentieth century to the present day could be incorporated into the study. This means that despite the paucity of animal bone found during targeted excavations at Āraiši in 2012, a significant dataset could be sampled from the archive of earlier excavations in the 1970s, stored in the National History Museum in Riga and made accessible to the project.

In all primary site excavations sieving was part of the sampling protocol, resulting in more abundant datasets (see above), although as a result a certain percentage of animal bone fragments could not be identified beyond the level of 'class' (e.g., mammal, bird, fish). Animal bones examined from previous excavations had been typically hand collected, which would result in a bias against smaller fragments. This in turn had clear implications for the accuracy of estimated taxonomic abundance (in particular for micro-fauna, including fish), element distribution, and mortality profiles, as well as potentially skewing metric analyses. The routine use of sieving of select contexts in future excavations will build on the assessment of taxonomic abundance as well as element distribution presented within this study. Taphonomic factors varied from site to site and are discussed in more detail in the relevant sections.

Two basic techniques for abundance estimations were employed: Number of Identified Specimens (NISP); and Minimum Number of Individuals (MNI) (Reitz and Wing 2000, 191; Klein and Cruz-Uribe 1984, 24–37) in order to make quantitative estimates of the relative species abundance. These particular methods were employed as they have been used widely (Watson 1979) to analyse animal numbers and therefore facilitate comparisons with other sites (O'Connor 2000, 54–68; Maltby 1979; Sykes 2007). However, while these methods have considerable merit, not least of all because of their widespread usage amongst analysts, a number of caveats must be noted when using both NISP and MNI counts (Lyman 2008). The main concern when using NISP counts is that of biased sampling of material in favour of larger bones (which are easier to recover), of species with a relatively greater number of bones (for example, pigs when compared to cattle) (O'Connor 2000, 56; Davis 1987, 29), and that associated bone groups from a single individual (ABGs) cannot be accounted for when using fragment counts (Fig. 2.5). MNI counts, however, can overrepresent rare taxa (Watson 1979), misrepre-

Figure 2.5: Associated bone groups: (a) a complete skeleton of a cat (*F. silvestris*) uncovered under a plank in the Liv quarter of Riga; (b) a complete skeleton of a horse (*E. caballus*) in situ under a collapsed building in the outer bailey of Cēsis Castle. Photos by A. Pluskowski.

sent individuals through aggregating bone fragments from multiple contexts, and variable means of calculating abundance (through different zoning systems or assigning sides to bone fragments) have been criticized for preventing inter-site comparisons. This latter criticism implies an underlying distrust of analysts' ability to quantify individual animals, whilst simultaneously accepting the accuracy of species and element identification, and would render a large proportion of the existing literature redundant.

For the purposes of this study, both NISP and MNI are used where appropriate, at different scales of site-specific and comparative analysis. In general terms, the NISP count was calculated as a sum of fragments for each species and split by archaeological context. The MNI count for each assemblage was not calculated as a direct proportion of the NISP, but from the Minimum Number of Elements (MNE) by summing the zones of each element to find the most abundant, then taking the number for the most abundant element to represent the MNI for the species overall (Hambleton 1999). The zoning system used in this study for assemblages recorded in the National History Museum in Riga, Cēsis Castle and Malbork Castle, follows the Bournemouth system developed by Mark Maltby and Sheila Hamilton-Dyer. Assemblages from Prussia and related Polish sites were recorded following Makowiecki's (2010) protocol and assemblages from Estonian sites following Rannamäe's (2010, 17–19) system. These recording systems, which are structured around Microsoft Access forms and queries, alongside Microsoft Excel tabulation, are all comparable. Cut marks on bone, where included for detailed butchery analysis, were recorded using Seetah's (2007) protocol. Measurements of animal bone mainly follow von den Driesch (1976) for mammals and birds, and Morales and Rosenlund (1979) for fish. General data for fish was taken from Fishbase (Froese and Pauly 2006). Animal bone was identified according to the reference collections at the University of Nicolaus Copernicus in Toruń, the University of Reading, the Natural History Museum in London, the University of Cambridge, the University of Tartu, Tallinn University, as well as the Anatomy Museum in the Veterinary School of the Estonian University of Life Sciences.

There is a final consideration that needs to be highlighted when discussing the value of quantifying animal bones from an archaeological site. In merging the data from a number of different archaeological contexts dated to the same chronological range, results were being linked to a particular cultural horizon. What is then emphasized is the representation of different species and anatomical elements in relation to each other. When compared, they should be viewed as tendencies, rather than as comparable absolute values. Here the use of MNI becomes problematic, as it is usually impossible to demonstrate whether a fragmented carcass from a single individual has been deposited in multiple features across the site, or whether only a portion has survived in a single feature. This is a taphonomic issue which can be mitigated to a certain degree by including data merged from multiple features of the same occupation phase alongside the recognition of individual depositional events, linked to specific archaeological contexts.

Alongside tabulating and comparing quantities of bones, a range of statistical tools was used to understand the significance of trends in the recorded faunal assemblages, particularly multivariate analysis for comparing species within and across assemblages, chi-squared tests to estimate relative statistical significance (O'Connor 2003, 79), and biodiversity indices (Simpson 1949). Taphonomic variables including fragmentation were noted for the purposes of comparability, but were not the focus of analyses. Animal bone fragments were also sub-sampled for isotopic analyses (see below), uranium-series dating (both conducted at Reading), and AMS dating (by SUERC). There was no scope or possibility to include a programme of aDNA analysis, with the exception of samples of horse bone from the Prussian site of Poganowo which were analysed by Mim Bower (Cambridge), and samples of aurochs, bison, and elk bone from relevant sites in north Poland were provided by Daniel Makowiecki for inclusion within a phylogenetic project based in Białowieża, directed by Rafał Kowalczyk. The results of these analyses are pending.

Isotopic Analyses of Bone

The analysis of stable isotopes has been successfully applied to archaeological bone with the aim of reconstructing dietary patterns and mobility (Lajtha and Michener 1994, xi). Whilst many studies have focused on reconstructing past human diets, isotopic data extracted from the bones of mammals and fish is shedding important light on exploitation and management strategies (Ervynck et al. 2008; Millard et al. 2011; Barrett et al. 2011; Orton et al. 2011; Hamilton and Thomas 2012; Müldner et al. 2014). Studies conducted within the framework of the project primarily focused on stable isotopes of Carbon (C), Nitrogen (N), and Hydrogen (H) from mammal bones derived from sites in both Prussia and Livonia, and were carried out in Reading by Gundula Müldner, Stuart Black, Charlotte Scull, Eilis Weldon, Louise Wakefield, and Alexander Coogan (Scull et al. 2019). Additional isotopic studies of human bone derived from cemeteries in and around Tartu were carried out by Martin Malve, with a parallel project focusing on C and N stable isotopes from cod bones from eleven sites across the Baltic region (Orton et al. 2019).

The principal aim of these studies was to reconstruct dietary patterns in livestock before and after the thirteenth century, with the aim of identifying changes in the provision of fodder and related agrarian regimes at different sites (Scull, Weldon, Wakefield). In the case of Tartu, the study focused on human remains from urban and rural cemeteries and was conducted within the framework of doctoral research (Malve pers comm). The aim was to determine whether there were noticeable dietary differences between urban and rural populations, and whether this could also be linked to the variable presence of colonists. It was also possible to pilot the use of H stable isotopes for provenancing horses from a select number of sites, the results of which will be published separately (Coogan pers comm). Samples of cod bone were provided for a parallel project investigating the impact of the crusades and the emergence of the Hanse on commercial fishing, with all analysis carried out in Cambridge (Orton pers comm).

Palynology (Pollen and Microscopic Charcoal)

The analysis of pollen and spores from peat and lake sediments represents a fundamental aspect of the off-site component of this research, providing key data on changes in vegetation environment and land-use from the Late Iron Age to post-medieval times. The broad chronological focus is critical to placing the evidence for changing land-use during the crusades within the more extensive documented history of conquest and colonization occurring across the Baltic region from the eighth century. This approach provides a fundamental level of comparison between palynological data across a range of temporal and spatial scales.

Coring focused on suitable pollen-bearing deposits located within the historical boundaries of several administrative districts of the Teutonic Order. The choice of administrative districts was chosen primarily on the basis of the suitability of on-site studies, rather than solely on the palaeoenvironmental potential of the landscape. The aim of the coring was to target suitable deposits representative of the likely range of landscape settings and patterns of land-use, both in proximity to castles, urban, and rural sites, and from areas more remote from settlements, including cases where remnant woodland is argued to have been maintained as a managed resource through the medieval period.

Smaller peat and lake basins have been targeted as they derive the majority of their pollen from the local environment and therefore have a finer spatial resolution than large lakes that derive their pollen from a much larger source area, particularly those lakes with large hydrological catchments. Estimates of the relative source area of pollen (RSAP) from different sized basins have been suggested on the basis on computer modelling. The RSAP will vary dependent on factors such as the surrounding vegetation structure (Hellman et al. 2009a, 2009b), the productivity and dispersal of pollen of specific plant taxa (Brostrom et al. 2008), and the influence of pollen from outside the RSAP, which may be as much as 50% of the pollen loading (Soepboer et al. 2008). The RSAP for small basins (*c.* 50 m radius) within open/semi-open cultural landscapes have been estimated at between 1000–2500 m (Hellman et al. 2009b), and provide a useful guide for considering the spatial resolution of the pollen sequences presented here.

In total, forty-seven cores have been sampled from peat mires and lakes across the Baltic, in addition to a small number of cores sampled from on-site archaeological context (Fig. 2.6). Not all cores produced material suitable for analysis; pollen was either poorly preserved or absent entirely, and in other cases sequences had been truncated with sediments of medieval or later date removed altogether. Full analysis was undertaken on twenty-one cores from the following commanderies; Marienburg (four), Rehden (four) (Pol. Radzyń Chełmiński), Elbing (one), Lötzen (three) (Pol. Giżycko), Wenden (five), Fellin (two), and Karkus (two). Sediments sequences were sampled using a Russian auger, a standard piece of coring equipment used by palynologists to collect uncompressed sequences of soft sediment within an enclosed barrel, thereby minimizing the possibility for contamination.

Pollen samples *c.* 1 cm^3 in volume were taken from each cores at set intervals dependent on the individually required sample and chronological resolution, typically ranging from 0.5 to 4 cm. One tablet containing exotic spores (*Lycopodium*) was added to each sample to enable calculation of pollen concentrations. Samples were prepared following standard laboratory techniques (Moore et al. 1991) and mounted in glycerol jelly stained with safranin. A minimum of five hundred pollen of terrestrial species were counted for each level. Pollen percentages are calculated based on terrestrial

Figure 2.6: Coring peat deposits (a) south of Malbork; (b) near Ełk; (c) the Castle Lake in Radzyń Chełmiński; (d) Lake Drużno. Photos 2.6a, 2.6b, 2.6d by A. Pluskowski and A. Brown; Photo 2.6c by M. Elander.

plants. Fern spores, aquatics, and *Sphagnum* are calculated as a percentage of terrestrial pollen plus the sum of the component taxa within the respective category. Identification of indeterminable grains was recorded according to Cushing (1967). Identification of cereal pollen followed the criteria of Andersen (1978), with Poaceae pollen separated into four groups (wild grasses, *Hordeum* group, *Avena-Triticum* group, and *Secale cereale*) based largely on mean annulus and grain diameter, but also surface sculpturing, grain morphology, and the protrudance of the annulus. Key distinguishing features are not always easy to identify where variable preservation of pollen grains occur, in which case caution should be exercised in any identification based on grain and annulus diameter alone, particularly where there is a degree of overlap between the different groups. *Secale cereale* by comparison can be confidentially identified to species level because of its characteristic prolate grain shape. Identification of indeterminable grains was recorded according to Cushing (1967).

Estimating the abundance and spatial extent of particular vegetation environments (e.g., woodland, arable, pasture) around a sampling point represents a major interpretative challenge facing all palynologists. The abundance of individual species in pollen diagrams does not reflect their actual contribution to past physical environments and is largely, though not exclusively, a consequence of differential pollen productivity and dispersal, and the size and therefore pollen source area of mire and lake basins — with pollen source area increasing with basin size. Although interpretation of pollen data is based on an in-depth understanding of the complex taphonomic factors involved in the formation of the palynological record (see Moore et al. 1991), inferences of the likely pollen source area and vegetation composition have traditionally been based on a qualitative or semi-quantitative approach. However, quantitative approaches to reconstructing past vegetation are increasingly applied to palynological data, most often using one of two principal computer-based models (Broström et al. 2005; Gaillard et al. 2008; Hellman et al. 2009a, 2009b). The Multiple Scenario Approach (MSA) uses pollen signals to generate multiple ecologically distinct reconstructions of past vegetation in a GIS output (Bunting and Middleton 2005, 2009a, 2005b), whilst the Landscape Reconstruction Algorithm (LRA), applied here, uses pollen records to produce quantified distance-based estimates of past regional and local vegetation cover within the 'relative pollen source area' of a depositional basin (lakes and bogs) (Sugita 2007a, 2007b). The LRA has been widely tested and verified in numerous palynologically based lake studies across Europe (e.g., Sugita 2007a, 2007b; Hellman et al. 2008; Nielson and Odgaard 2010; Mazier et al. 2012; Fedh et al. 2013; Marquer et al. 2014; Poska et al. 2014).

Quantitative vegetation reconstructions using the LRA were created by Anneli Poska for pollen sequences from the Castle Lake (Radzyń), Rywałd and Rozlewisko, and peat sequences from Benowo and Parpary (Chapter 12). The LRA divides quantitative reconstruction into two stages; Regional Estimates of Vegetation Abundance from Large Sites (REVELS) uses large lakes (≥ 100 ha) to model regional vegetation composition within 10^4–10^5 km^2 (Sugita 2007a), whilst Local Vegetation Estimates (LOVE) requires small lakes (<50 ha), subtracting the background regional component from REVEALS to model local vegetation composition within ≤10^4 ha (Sugita 2007b). Pollen productivity estimates (PPEs) and fall speeds are required as input parameters. The method for obtaining PPEs is summarized by Broström et al. (2008) and has been calculated for various regions of Europe, including southern Estonia (Poska et al. 2011). PPEs do not currently exist for Poland, so PPEs for southern Estonia were applied (Poska et al. 2011, 2014, Table 4), comprising twenty-one mostly wind pollinated taxa that account for more than 90% of the pollen sums in the majority of samples. REVEALS models were created for the Castle Lake (Radzyń) and Lake Rozlewisko and combined to reconstruct regional vegetation abundance as both sites are located less than 100 km apart. LOVE models were produced for hundred-year time-slices from AD 0–1800, dependent on the availability of data, with the proportions of taxa summed within eight groups: coniferous trees, boreal broad-leaved trees, temperate trees, dwarf shrubs, wet meadows, meadows, ruderal, and crops. The data are combined with pollen percentage diagrams to provide more detailed quantitative-based interpretations of the proportions of open and closed land, and particularly the proportions of arable and pastoral land. Quantitative reconstructions were not undertaken on pollen sequences from Latvia because of the absence of existing chronologically robust pollen studies from large lakes from which REVEALS models could be constructed.

Microscopic charcoal was quantified for each pollen sample using the point count method of Clark (1982). The point count method expresses the total charcoal count as an estimate of the area of charcoal on any given pollen slide for a standard 1 cm^3 volume of sediment. Randomly spaced parallel transects were analysed to ensure that a representative portion of the slide was exam-

Figure 2.7: Targeted excavations in the castles at (a) Malbork; (b) Karksi. Photos by A. Pluskowski and A. Brown.

ined. This is important as pollen and microscopic charcoal are not distributed evenly across a microscope slide. Charcoal is identifiable as highly angular, usually opaque black fragments of varying size and shape, sometimes preserving cellular structure. Humified plant matter may appear very dark brown black under the microscope, but unlike charcoal, usually exhibits translucent and rounded edges. As regards the source area for charcoal particles, studies of size-distribution of charcoal from fires show that large particles are distributed close to the burn area, reflecting local fires, whilst microscopic charcoal may be transported longer distances, reflecting regional fires. Charcoal fragments <50 μm and>125 μm are generally considered to correlate strongly with regional and local fires, respectively (Tinner et al. 1998; Blackford 2000; Duffin et al. 2008). Source area will vary depending on factors such as the size of the depositional basin, degree of vegetation openness, and meteorological conditions at the time of the fire (Patterson et al. 1987). Microscopic charcoal may be under-represented in closed or semi-open wooded environments (Blackford 2000). Charcoal quantified in this study is predominantly <50 μm; the results are not plotted on the pollen diagrams as the quantities of microscopic charcoal were typically negligible.

Plant Macrofossils

Plant macrofossil is a term used to cover the remains of plants visible with the naked eye, including seeds, fruits, nuts, mosses, and charcoal. In archaeological contexts, assemblages of plant macrofossils typically include plants derived from the local vegetation environment and those both intentionally or unintentionally transported to a site. The charred and uncharred remains of seeds and fruits are more often the most ubiquitous category of plant macrofossil recorded from archaeological sites, representing a key source of data on past plant exploitation, including: the use of both cultivated and wild plants for nutritional, medicinal, or industrial purposes; evidence for trade, including both the export of agricultural produce and the import of exotic/luxury items; the role of food plants as indicators of social status; and as a tool for reconstructing the ecological conditions of a site and its surrounding environment. The varied site-types examined across the Baltic present a unique opportunity to consider aspects of plant use across both a large geographical area but also between very different functional and socio-economic settings.

Samples for plant macrofossil analysis were taken from a range of occupation contexts, including deposits from occupation surfaces, the fills of pits, moats and latrines, middens, animal dung, and in a single case from a grain silo (Fig. 2.7). These deposits can be broadly classified as either specific or non-specific contexts, reflecting the difference between plant assemblages accumulating from specific processes, such as latrines or granaries, and those accumulating through a range of more-or-less unspecified processes (Greig 1982), most obviously in the case of occupation deposits on floor surfaces and in pits. Taphonomic processes therefore require especially careful thought when interpreting an assemblage.

Bulk samples of a minimum 2 kg volume were taken from a representative range of context revealed through excavation. The size of bulk samples is significantly smaller than for most archaeological excavations, where individual samples of up to forty litres are typically processed. However, the smaller size of bulks employed here was considered sufficient to identify the broad range of botanical remains occurring at a site, and ideally suited for targeted analysis of ecofacts from several sites distributed across the Baltic, each often with multiple trenches containing a large range of features. Samples were processed either in the field or in the laboratory. Those samples processed in the field were wet-sieved through a 0.25, 0.5, and 2 mm mesh to separate plant remains from sediment (Fig. 2.8). Samples processed in the laboratory were soaked for twenty-four hours in solution of potassium hydroxide and washed through 2.0, 0.5, and 0.2 mm mesh sieves. From each sample, 300 cm^3 from each of the fractions was selected all seeds and fruits were collected. Because of a significant number of cereal bran and mosses remains, abundance was recorded on a three-point scale, where '+' means 'remains present' and '+++' 'remains abundant'. The determination of pericarp- and testa structure of cereal bran was followed by Dickson (1987), Holden (1990), and Körber-Grohne (1990). Taxa names have been used after Mirek et al. (2002) and Podbielkowski and Sudnik-Wójcikowska (2003). It has only been possible to include a summary of the plant macrofossils from Prussia in this volume (for a more detailed study see Badura and Możejko 2019), and the richer assemblages from Livonia will be published separately. All plant macro-fossil analysis was directed by Monika Badura (Gdańsk), with a discrete study on the assemblage from Święta Góra by Tom Rose-Jones (Reading).

Invertebrate Analysis

Invertebrate remains, largely consisting of wing cases, were preserved at some sites with waterlogged archaeo-

Figure 2.8: Wet-sieving samples in the field through a mesh at (a) Cēsis; (b) Święta Góra. Photos by G. Kalniņš and A. Pluskowski.

logical contexts and samples were processed by Gary King (Durham). They provide a sensitive indicator of local environmental conditions, particularly local habitat composition (King 2019). Samples were prepared for analysis of the remains of Arthropoda (beetles, flies, and mites) using the paraffin flotation method (Kenward et al. 1980; Schelvis 1990). For each sample, 1000 cm^3 of sediment was processed and washed through a 250 µm mesh sieve. The identification of the invertebrate remains was carried out through comparison with a reference collection and follows the criteria outlined in standard keys (e.g., Balogh and Mahunka 1983; Garcia 2009; Joy 1976; Smith 1989).

Parasites

Samples from waterlogged contexts in the Liv quarter in Riga were analysed for parasites by Ivy Hui-Yuan Yeh and Piers Mitchell (Cambridge; see also Yeh et al. 2014; Yeh and Mitchell 2019). The study of eggs and cysts of parasites from Latrines provide valuable information hygiene and indicators of local dietary habitats and food preparation techniques. In order to analyse the latrine soil for intestinal parasites, 1 gram of the sediment was disaggregated by adding it to 10 ml of distilled water for 1 hour (Anastasiou and Mitchell 2013). This suspension was passed through a column of sequential micro-sieves with mesh size of 300µm, 160µm, and 20µm. Most parasite eggs have dimensions of 25–150µm so will be trapped on the 20µm sieve (Bouchet et al. 1999). After washing the sediment from the 20µm sieve, the suspension was centrifuged to concentrate the sediment, excess water removed, and glycerol added prior to mounting on slides. In order to identify the parasite remains the slides were examined using digital light microscopy. They were identified based upon their shape, size, colour, and special characteristics when compared with standard parasitology sources (Garcia 2009; Gunn and Pitt 2012). Disaggregated latrine soil that passed through the 20µm sieve was collected and analysed for evidence of the organisms that cause dysentery. Cysts of *Giardia duodenalis*, *Entamoeba histolytica*, and *Cryptosporidium parvum* measure around 5–20µm in size (Garcia 2009). Six samples of this fluid was tested with Enzyme Linked Immunosorbent Assay (ELISA) kits specifically designed by TECHLAB© to detect these species of dysentery organisms in faecal specimens (http://www.techlab.com). The test uses monoclonal antibody-peroxidase conjugate specific for proteins uniquely secreted by these organisms.

Phytoliths

Phytoliths are plant microfossils that are formed of biogenic silica; as the plant grows, it takes-up silica from the soil and deposits it within its cell structure. When plant remains decay, phytoliths are left behind in the soil, and generally preserve well in archaeological deposits unless soils and sediments are highly alkaline, particularly

Figure 2.9: Taking micromorphology blocks from archaeological profiles from (a) Cēsis; (b) Elbląg. Photos by G. Kalniņš and A. Pluskowski.

around pH9 (Piperno 2006). Phytoliths can be particularly useful tools in archaeology to examine crop processing and animal husbandry activities (e.g. Harvey and Fuller 2005), as grasses in particular, when certain phytoliths such as conjoined cells from the husk are present, can be identified to species (Rosen 1992; Twiss 1992), and C3 and C4 grasses can be differentiated between (Barboni et al. 1999; Twiss 1992). It is important not to over-interpret phytolith assemblages, and to apply the technique in conjunction with thin-section micromorphology to enable these microfossils to be studied in their depositional context (Shillito 2013). Phytoliths have been extracted from loose sediment samples from Elbląg (Prussia) and Karksi (Livonia), and are studied along-side thin-section micromorphology, macrobotanical, and pollen evidence to form more robust interpretations of occupation deposits and specific activities.

Loose sediment samples were processed using the following methodology: (1) the sample was screened through a 0.5 mm mesh to remove coarse sized particles; (2) approximately 1 gm of dried raw sediment was weighed out; (3) calcium carbonates were dissolved using a dilution of 10% hydrochloric acid and then washed in distilled water three times; (4) clay was removed using a settling procedure and sodium hexametaphosphate (Calgon) as a dispersant. Distilled water was added and the samples left for seventy-five minutes before pouring off the suspense. This was repeated at hourly intervals until the samples were clear. Samples were then transferred into crucibles and left to dry at a temperature of less than 50°C; (5) after drying, samples were placed in a muffle furnace for two hours at 500°C to remove organic matter; (6) phytoliths were then separated from the remaining material using a heavy liquid calibrated to a specific gravity of 2.3. Phytoliths were transferred to centrifuge tubes and washed three times in distilled water. They were then placed in small Pyrex beakers and left to dry; (7) approximately two milligrams of phytoliths per sample were mounted onto microscope slides, using the mounting agent Entellan.

Microscope slides were analysed using a Leica DMLP microscope with an attachable stage using x 100 and x 400 magnifications. A minimum of three hundred single cells were identified and counted. The remainder of the slide was then scanned at x 100 magnification to identify any further multicelled phytolith forms. Phytoliths were further classified as deriving either from woody (dicotyledon) or non-woody.

Micromorphology

Micromorphology enables the in situ study of the sedimentary matrix of the depositional environment which is particularly useful for analysing the formation processes of archaeological deposits and the depositional context of micro-refuse for ascertaining use of settlement space. The technique has been widely applied to understand and interpret the use of settlement space in relation to activities inside and outside of buildings, and middening practices (e.g., Matthews 1995; Matthews et al. 1997; Milek and French 2007; Shillito et al. 2014; Simpson et al. 2006), the development of cultural landscapes (French 2003), and the reworking of archaeological stratigraphy by post-depositional processes (e.g. Canti 2003; Devos et al. 2009; French 2003, 123, 156; Gé et al. 1993; Macphail 1994; Macphail et al. 2003; Weiner 2010).

Samples for micromorphological analysis were collected from profiles from a range of context types and features from sites in Prussia and Livonia in order to examine the development of occupation at the microscopic scale (Fig. 2.9). Samples were collected from 'dark earth'-type deposits or 'cultural layers' at Biała Góra (Pluskowski et al. 2014), and within the stronghold on Riekstu Hill in Cēsis, to understand the composition of the cultural material, and the timescales of the formation of these deposits. Samples from organic-rich, or 'wood lenses', adjacent to the burnt timber fortifications within the stronghold on Riekstu Hill were collected to understand their formation processes and how they relate to the destruction of the timber fortifications. Samples from Święta Góra were collected from pit features to understand the periodicity and intensity of the occupation on this multi-phase site. Samples from the outer bailey of the castle at Elbląg, and from the area of the high castle at Karksi, were collected to understand the very earliest occupation at these key sites, in terms of the activities of the early colonists.

Micromorphology samples were prepared in the Microanalysis Unit at the University of Reading. The procedure followed is the standard protocol for thin section preparation (Murphy 1986). Samples were oven-dried at 40°C and then impregnated with epoxy resin while under vacuum. Slides, 11.5 cm × 7.5 cm, were prepared to the standard geological thickness of 30 μm. Micromorphological investigation was carried out using a Leica DMLP polarizing microscope at magnifications of x 40–x 630 under Plane Polarised Light (PPL), Crossed Polarised Light (XPL), and Oblique Incident Light (OIL). Thin-section description was conducted using the identification and quantification criteria set out by Bullock et al. (1985) and Stoops (2003), with reference to Courty et al. (1989), to Mackenzie and Adams (1994), and Mackenzie and Guilford (1980) for rock and mineral identification, and Fitzpatrick (1993) for further identification of features such as clay coatings. Photomicrographs were taken using a Leica camera attached to the Leica DMLP microscope.

Inorganic Geochemistry

Two methods of inorganic geochemical analysis have been used on sediment samples that have been collected from archaeological sites at in Prussia (Biała Góra, Elbląg, Malbork, and Święta Góra) and Livonia (Cēsis, Karksi, and Riga): X-Ray Fluorescence (XRF), including the use of the portable XRF (pXRF); and Inductively Coupled-Optical Emission Spectroscopy (ICP-OES). ICP-OES was used to analyse sediments from lake and bog sequences immediately adjacent to the castles at Radzyń and Āraiši. In order to investigate the link between sedimentation history and human impact, concentrations (ppm) were determined for eighteen elements using ICP-OES.

Multi-elemental enrichment has been widely applied in archaeology to examine the use of settlement space and to identify specific activities such as metalworking and animal husbandry (Banerjea 2008; Cook et al. 2010; Fernandez et al. 2002; Hutson and Terry 2006; Wilson et al. 2005). In lake and bog sediments it is used to examine the effects of anthropogenic pollution (Belzile et al. 2004) and those associated with precipitation, sedimentation, and erosion processes in lake and bog sequences, such as calcium and titanium (Lomas-Clarke and Barber 2007; O'Connell et al. 2013).

Loose sediment samples for XRF analysis were air dried at 40°C to remove moisture content, and then crushed to pieces of <2 mm in size using a pestle and mortar. The crushed sample was then ground in an agate mill and pressed into pellets using boric acid and a 12-tonne press. The pellets were analysed using an X-Ray spectrometer for a range of major and trace elements. The p-XRF samples were poured into circular plastic containers, closed on one face by a polypropylene film. The powder is gently pushed against the polypropylene film using a mortar, in order to remove air trapped and to permit a direct contact between the samples and the film. The samples are then placed into the portable XRF stand, which is closed by a lead lid. The samples are analysed for a range of major and trace elements using a portable XRF Niton XL3T GOLDD plugged into the stand. The reported values are in parts per million.

For ICP-OES analysis, samples of sediment (0.5 kg dried weight) were taken at intervals of 2–4 cm throughout lake and bog sequences from Radzyń and Āraiši, from the same depths as pollen samples. The sediment was pretreated with nitric acid to digest organic matter, with standards and blanks analysed alongside the samples. Geochemical concentration are expressed as parts per million (ppm) per 0.5 kg of dried sediment.

For both archaeological settlements and lake and bog sequences, an element enrichment index was calculated (*sensu lato*; Belzile et al. 2004), dividing element concentration for each sample by the average element concentration for surface geology samples from the surrounding area resulting in a mean value. Indexes of enrichment three or more times larger than the mean background value were considered potentially highly enriched (Belzile et al. 2004). A three-tiered system to measure enrichment has been developed to enable the impact of geochemical changes and fluctuations to be easily identifiable; concentrations 2.95 times the background mean value were considered high; concentrations between 2.50–2.95 were considered medium level; and concentrations between 1.5–2.5 were considered low level. For lake and bog sequences, the anthropogenic fraction of trace element enrichment (the remaining fraction represents the natural trace element supply) was calculated by using a Ti (titanium) ratio as a soil mineral tracer, whereby the anthropogenic fraction can be calculated by subtracting the background fraction from the total enrichment (Boyle 2001).

Earth Science Methods and Building Materials

The decision to adopt visual earth science and geochemical methods to describe castle building materials from the eastern Baltic, stems from the successful application of these techniques in recent years in sourcing ancient stone, brick, and mortar materials in a number of study areas. High resolution thin-section comparative microscopic analysis, for example, has been successful in determining the character and source of worked stone from Roman Britain (Allen and Fulford 2004; Hayward 2009; Allen 2014) and the north-west provinces (Stribrny 1987). Studies of ICP-AES geochemical preparation and statistical analysis have focused on the concentration of clay in ceramic building materials such medieval floor tiles from Bordesley (Stopford et al. 1991) and medieval and post-medieval roof and floor tiles (Hughes 2009). Both techniques were also successful in establishing the source of Victorian stone and brick building materials in Glasgow (Hayward 2015), where multivariate statistical techniques (Baxter 1994, 2003) have been used to group and discriminate the geochemical data. In light of this, the study of the medieval Baltic and Northern European stone trade and the sourcing of the medieval brick used in the Teutonic Order's castles are areas of study that would benefit greatly from a more considered scientific approach. A pilot study, aimed at documenting and understanding the beginnings of masonry construction in the Order's Prussian and Livonian castles was carried out within the framework of this project, and the results are presented in Volume II (Hayward 2019).

With each castle or construction phase, five representative and well-dated brick, tile, and stone samples were taken to ensure a good representative statistical and geochemical sample size. Overall, eighty-six brick, tile, mortar, and stone samples were taken from four sites in the Vistula Delta region of Prussia (Malbork, Elbląg, Dzierzgoń, and Biała Góra) and two from Livonia (Cēsis, Viljandi). The dimensions and weight of each whole construction brick and tile were recorded with any features which may help ascertain the manufacturing process drawn and photographed. The application of a 1 kg masons hammer and sharp chisel to each building material sub-sample ensured that a small fresh fabric surface was exposed. Sub-samples were examined under a hand lens (Gowland x 10) and long-arm stereo-microscope and their macroscopic texture, colour (Munsell Color Group 1980), and inclusions were recorded. Treatment of dilute Hydrochloric acid determined whether or not the rock, mortar, or brick had a calcareous composition. A 20 mm long × 15 mm thick sample was large enough to undergo thin-section preparation and analysis, with the remaining sub-sample ground up as powder for geochemical analysis. Although destructive, this method ensured that the maximum amount of information could be obtained from the smallest possible sample size thus fulfilling the museums and archives sampling policies.

Forty-four sub-samples underwent thin-section preparation using the rock preparation facilities at the University of Reading (the procedural implications and sample preparation are referred to in Hayward 2009.) Each sample was additionally embedded in coloured araldite resin (CY1301); the process of embedding with the addition of a hardener (HY951) strengthened the mortar samples, whilst the addition of a green colouring agent

(BW1034) highlighted the pore spaces in the sample enabling the overall porosity to be determined. Staining is a necessary process (Adams and Mackenzie 1998) during the production of the more lime rich thin sections. It picks out the variability in colour between ferroan and non-ferroan calcite as well as dolomite, with the addition of Alizarin Red C and Potassium Hexocynoferrate. This high resolution petrographic approach enabled each 30 micron-thick slice of brick, stone, or mortar to be viewed under the polarizing microscope at magnifications greater than × 400 (Leica DMLP). Thin sections can help distinguish between different types of calcite grains, minerals, cements, and microfossils.

The chemical analyses used inductively coupled plasma-atomic emission spectroscopy (ICP-AES) (Thompson and Walsh 1989). The eighty-six samples were powdered and prepared by HF digestion and analysed by ICP-AES and ICP-MS at the Department of Earth Sciences at Royal Holloway (University of London). ICP-AES was used to analyse the ten major elements as oxides and nine trace elements in parts per million (ppm). Finally, the concentration (ppm) of Rare Earth Elements (REE) in each sample were analysed using the sensitive ICP-MS (Inductively Coupled Plasma Mass Spectroscopy) machine.

With the brick and roofing tile, the intention here was to determine what the bulk geochemistry was for the clay raw material in both Prussian and Livonian sites. This would provide a better understanding of possible sources for the manufacture of bricks and roofing tile used in the construction of the Order's houses in both regions and whether clay was supplied locally or brought in from further afield. It was expected that the application of this technique would also indicate whether the clays derived from Quaternary till and glacio-lacustrine deposits; Holocene alluvium or in the case of Livonia from geologically older (Palaeozoic) clays from the underlying bedrock. For the binding mortar and artificial stone, some idea of the constituent elements, particularly the proportion of calcium (calcite in limestone and shell and gypsum), silica (quartz), and magnesium (dolomite). The chemical composition of the different worked stone types used in the Order's Livonian castles could also be ascertained.

Detailed statistical interpretation of the geochemical data including Discriminant Analysis and Principal Component Analysis (Baxter 1994; 2003) was undertaken by Stuart Black (Reading). Principal Component Analysis (PCA) examines the entire dataset to see whether there were any similarities in the composition of the samples undertaken so that Prussian and Livonian bricks, mortar types and rock types may form particular groupings. Discriminant Analysis (DA) examines whether small differences in chemistry between different types of ceramic sources are statistically significant. This was found to be particularly important in fingerprinting the geochemistry of the bricks used in the different Prussian castles.

Wood, Leather, and Other Organics

A number of castles and towns were built on waterlogged soils, with their foundations often stabilised with timber piling. In other cases subterranean springs and raised water levels have ensured the preservation of organic materials within anaerobic conditions. Leather, textile, wooden artefacts and structural timbers were found in waterlogged contexts in the moat and outer bailey at Cēsis Castle, in the Liv quarter in Riga, and within the courtyards of the castles at Karksi and Elbląg. Structural timbers could often be dated by dendrochronology and provided direct evidence of the use of specific tree species for building materials (see Chapter 3). However, an extensive archive of organic material exists from earlier excavations across the eastern Baltic region, much of which has been published and could be incorporated into the study.

Furthermore, a significant body of original documents and manuscripts produced by the Teutonic Order has survived. These consist of pages of skin derived primarily from calves (vellum), sheep, and goats (parchment; a term also used for calf skin). No detailed analysis of this manuscript fabric was possible within the confines of the project, but the provisioning of vellum to meet the needs of individual convents, as well as the main chancellery in Marienburg, represented a new component of livestock management after the creation of the *Ordensland* (this is discussed in more detail in Chapters 6 and 13; see also Pluskowski et al. 2019). Future aDNA studies of the Order's manuscripts, integrated with faunal assemblages recovered from its castles, would shed important light on the provisioning and trade of skin in medieval Prussia, and to a lesser extent Livonia, where comparatively few original documents have survived (cf. Stinson 2011). Another fundamental yet poorly represented natural resource was the fuel required for lighting the Order's castles: oils, tallow (animal fat), and beeswax, the latter also used for sealing documents. The importance attached to 'bee-trees' is clearly expressed in the documentary record (see Chapter 9), even though wax

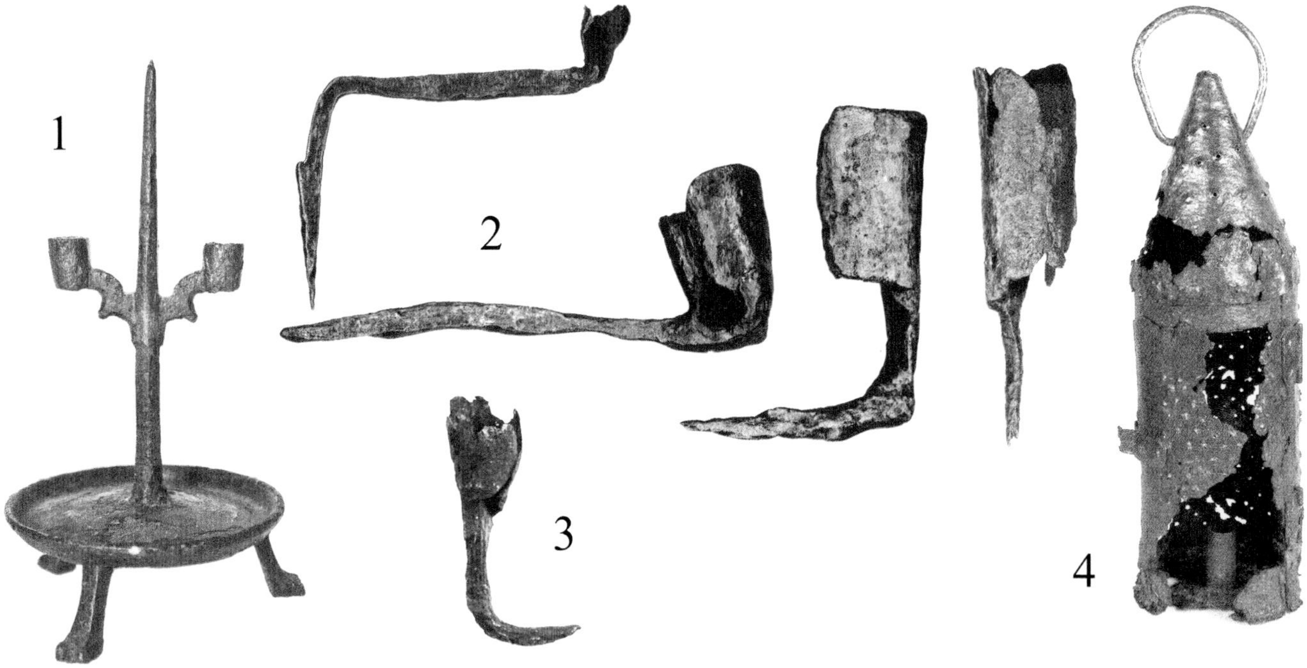

Figure 2.10: Artefacts associated with lighting from the medieval Baltic *Ordensland*: (1) a candlestick, Elbląg (Ger. Elbing), fifteenth century (Fonferek et al. 2012, 114); (2) candlesticks from Turaida Castle (Ger. Treiden) (Graudonis 2003, cat. 20); (3) candlestick from Karksi Castle (Ger. Karkus) (Valk et al. 2012, 78); (4) lantern from Cēsis (Ger. Wenden) (Apala 2007, 52). Not to scale.

is rarely detectable in archaeological contexts. Candles and lamps were used for both practical illumination and liturgical purposes within castle chapels. Indeed, fire played a practical and sacred role within both indigenous and crusader societies in the eastern Baltic. The organic materials themselves rarely survive, but fragments of candle spikes, holders, and lamps have been found in excavations (Fig. 2.10). As with the case of parchment, the provisioning implications (alongside other fuels such as wood, charcoal, and dung) are considered in this study.

Scales of Resolution and Integration

All environmental data collected and synthesized by the project shares a geographic (Cartesian) location and a chronological context (i.e., an associated time period). The resolution of both is variable, but one of the most challenging aspects of the study has been determining the relationship between changes observed in the landscape and within specific sites. The study has adopted a model based on integral associations between individual sites and their territories, as understood from reconstructions of defined territories. However, the scales of resolution vary. With archaeological data recovered from specific contexts, it is possible to make a strong case for a relationship between a castle or town community and the observed exploitation of plants, animals, and soils. With pollen data the modelling of catchment aims to sharpen the resolution of landscape change, otherwise associated with a general intensification of settlement. Where there is enough data to reconstruct both the settlement pattern and its ranking in terms of power relations and vassalage obligations, it is possible to separate the impact of a castle community from that of a town or rural settlements. Whilst castles and their associated towns utilized the landscape in separate ways, the town's privileges in the majority of foundations were granted by the Teutonic Order, and the residing brethren represented the main political authority, elements of which were devolved to town councils which became increasingly powerful. This study aims for a nuanced understanding of the agency of colonizing and indigenous groups but recognizes that such a level of resolution is not always possible. The sovereign authority of the Teutonic Order was challenged significantly in Riga within decades of its foundation, and from the fifteenth century in parts of Prussia, at which point trends in environmental exploitation require recontextualization. This issue is discussed in more detail in relation to the relevant case studies.

The integration of data takes places on multiple levels and is a common feature of archaeological site reports, as well as more wide-ranging studies. Individual datasets were integrated through the process of specialist analysis, and these synthesized datasets were then integrated at varying temporal and spatial scales. This study presents both types of integration with the aim of reaching specialist and generalist audiences. Integration was achieved through the consistent use of two factors that were common to every set of data: a specific geographic location and a temporal context. This enabled the synthesis of datasets at different scales of spatial and chronological resolution. One of the most important objectives of the study was to develop a robust chronological framework which is discussed in more detail in Chapter 3. This framework was subdivided into cultural phases which, in turn, were situated within the synthetic theoretical framework presented at the start of this chapter. This multi-scalar approach ultimately enabled broader, interregional trends to be observed and compared with the nuances of interactions between the colonizing and indigenous populations of the medieval eastern Baltic. The result is an interregional model which can be tested and challenged with future excavations of individual sites alongside palaeoenvironmental studies. Above all it emphasizes connections between multi-scalar levels of analysis, specifically sites and their associated territories.

Works Cited

Primary Sources

Barber, R. W. 1993. *Bestiary: Being an English Version of the Bodleian Library, Oxford M.S. Bodley 764.* Woodbridge: Boydell

CTP Scholz, K. and Wojtecki, D., trans. 1984. *Chronik des Preussenlandes / Peter von Dusburg,* Darmstadt: Wissenschaftliche Buchgesellschaft

GAD Ziesemer, W., ed. 1968. *Das große Ämterbuch des Deutschen Ordens: Mit Unterstützung des Vereins für die Herstellung und Ausschmückung der Marienburg,* Wiesbaden: Sändig (repr. 2009)

HCL Brundage, J. A., trans. 2003. *The Chronicle of Henry of Livonia / Henricus Lettus,* New York: Columbia University Press

Hirsch, T., ed. 1863. *Die Chronik Wigands von Marburg,* Leipzig: S. Hirzel (repr. 1965)

Hirsch, T., Töppen, M., and Strehlke, E., eds. 1861–74. *Scriptores rerum Prussicarum: die Geschichtsquellen der Preussischen Vorzeit bis zum Untergange der Ordensherrschaft.* 5 vols. Leipzig: S. Hirzel (repr. 1968)

HWC Mugurēvičs, Ē., ed. 2005. *Vartberges Hermaņa hronika,* Rīga: Latvijas vēstures institūta apgāds

KP Fischer, M. 2010. *The Chronicle of Prussia by Nicolaus von Jeroschin: A History of the Teutonic Knights in Prussia, 1190–1331,* Farnham: Ashgate

LRC Smith, J. C. and Urban, W., trans. 2006. *The Livonian Rhymed Chronicle,* London: Routledge Curzon

LUB Bunge, F. G., von, ed. 1853–57. *Liv-, Est- und Kurländisches Urkundenbuch nebst regesten* (vols I–III), 13 vols to date, Reval: Kluge and Ströhm

MTB Joachim, E., ed. 1896. *Das Marienburger Tresslerbuch der Jahre 1399–1409,* Königsberg: Thomas and Oppermann

Perlbach, M., Philippi, R., and Wagner, P., eds. 1876–96. *Simon Grunau's Preussische Chronik,* 3 vols, Leipzig: Duncker and Humblot

Przezdziecki, A., Polkowski, I., and Pauli, Z., eds. 1863–78. *Joannis Dlugossii: Opera omnia*, 14 vols, Kraków: Fr. Kluczycki
Smith, P., trans. 1997. *The Annals of Jan Długosz*, Chichester: IM
Voigt, J., ed. 1827–39. *Geschichte Prussens von den ältesten Zeiten bis zum Untergange der Herrschaft des Deutschen Ordens*, Königsberg (repr. 1968)

Secondary Works

Abulafia, D. 2002. 'Introduction: Seven Types of Ambiguity, *c.* 1100–*c.* 1500', in *Medieval Frontiers: Concepts and Practices*, eds D. Abulafia and N. Berend, Aldershot: Ashgate, 1–34
Abulafia, D., and Berend, N., eds. 2002. *Medieval Frontiers: Concepts and Practices*, Aldershot: Ashgate
Adams, A. E., and Mackenzie, W. E. 1998. *A Colour Atlas of Carbonate Sediments and Rocks under the Microscope*, London: Manson
Adger, W. N. 2000. 'Social and Ecological Resilience: Are They Related?', *Prog. in Hu. Geog.* 24/3, 347–64
Allen, J. R. L. 2014. *Whetstones from Roman Silchester (Calleva Atrebatum), North Hampshire, Character, Manufacture, Provenance and Use; Putting an Edge on It*, British Archaeological Reports, 597, Oxford: Archaeopress
Allen, J. R. L., and Fulford, M. G. 2004. 'Early Roman Mosaic Materials in southern Britain with particular reference to Silchester (Calleva Atrebatum): A Regional Geological Perspective', *Britannia* 35, 9–38
Anastasiou, E., and Mitchell, P. D. 2013. 'Simplifying the Process for Extracting Parasitic Worm Eggs from Cesspool and Latrine Sediments: A Trial Comparing the Efficacy of Widely Used Techniques for Disaggregation', *International Journal of Paleopathology* 3, 218–23
Andersen, S. T. 1978. 'Identification of Wild Grasses and Cereal Pollen', *Danmarks Geologiske Undersøgelse*, Årbog, 69–92.
Apala, Z. 2007. 'Cesu pils arheoloģiskās izpētes (1974–2006) rezultāti', in *Cēsīm 800: Quo vadis, Cēsis? Vēsture un mūsdienu nosacījumi pilsētas attīstībai*, ed. V. Rozentāle et al., Cēsis: Cēsu pašvaldības aģentūra, Vidzemes Vēstures un tūrisma centrs, 45–76
Arszyński, M. 1995. *Budownictwo warowne zakonu krzyżackiego w Prusach (1230–1454)*, Toruń: Wydawnictwo Uniwersytetu Mikołaja Kopernika
——, 2010. 'Architektura warowna Zakonu Krzyżackiego w Prusach', in *Fundacje artystyczne na terenie państwa krzyżackiego w Prusach*, ed. B. Pospieszna, Pelplin: Wydawnictwo Bernardinum, 7–45
——, 2013a. 'Zamki i umocnienia zakonu krzyżackiego i hierarchii kościelnej w Prusach', in *Zakon krzyżacki w Prusach i Inflantach: podziały administracyjne i kościelne w XIII–XVI wieku*, eds R. Czaja and A. Radzimiński, Toruń: Wydawnictwo Naukowe Uniwersytetu Mikołaja Kopernika, 55–80
——, 2013b. 'Zamki i umocnienia zakonu krzyżackiego i biskupów w Inflantach', in *Zakon krzyżacki w Prusach i Inflantach: podziały administracyjne i kościelne w XIII–XVI wieku*, eds R. Czaja and A. Radzimiński, Toruń: Wydawnictwo Naukowe Uniwersytetu Mikołaja Kopernika, 183–213
Badura, M., and Możejko, B. 2019. 'The Plant Element in the Diet of the Inhabitants of Gdańsk, Elbląg and Malbork during the Rule of the Teutonic Order: The Historical and Archaeobotanical Perspective', in *Ecologies of Crusading, Colonization, and Religious Conversion in the Medieval Baltic: Terra Sacra II*, ed. A. Pluskowski, Environmental History of the North Atlantic World, 3, Turnhout, Brepols (forthcoming)
Balogh, J., and Mahunka, S. 1983. *The Soil Mites of the World*, I: *Primitive Oribatids of the Palearctic Region*, Budapest: Akadtmiai Kiado
Banerjea, R. Y. 2008. 'Experimental Geochemistry: A Multi-Elemental Characterisation of Known Activity Areas', *Antiquity* 82, 318 Project Gallery
Barber, M. 2012. *The Crusader States*. New Haven: Yale University Press
Barboni, D., Bonnefille, R. A. A., and Meunier, J. D. 1999. 'Phytoliths as Paleoenvironmental Indicators, West Side Awash Valley, Ethiopia', *Palaeogeography, Palaeoclimatology, Palaeoecology* 152, 87–100
Barrett, J., Orton, D., Johnstone, C., Harland, J., Van Neer, W., Ervynck, A., Roberts, C., Locker, A., Amundsen, C., Enghoff, I. B., Hamilton-Dyer, S., Heinrich, D., Hufthammer, A., K. Jones A. K. G., Jonsson, L., Makowiecki, D., Pope. P., O'Connell, T. C., de Roo, T., and Richards, M. 2011. 'Interpreting the Expansion of Sea Fishing in medieval Europe Using Stable Isotope Analysis of Archaeological Cod Bones', *Journal of Archaeological Science* 38/7, 1516–24
Bartlett, R. 1993. *Making of Europe: Conquest, Colonization and Cultural Change, 950–1350*, London: Penguin
Bartlett, R., and Mackay, A., eds. 1989. *Medieval Frontier Societies*, Oxford: Clarendon Press
Baxter, M. 1994. *Exploratory Multivariate Statistics in Archaeology*, Edinburgh: University of Edinburgh Press
——, 2003. *Statistics in Archaeology*, London: Arnold
Belzile, N., Chen, Y., Gunn, J. M., and Dixit, S. S. 2004. 'Sediment Trace Metal Profiles in Lakes of Killarney Park, Canada: From Regional to Continental Influence', *Environmental Pollution* 130, 239–48
Berend, N. 1999. 'Medievalists and the Notion of the Frontier', *The Medieval History Journal* 2, 55–72
Blackford, J. J. 2000. 'Charcoal Fragments in Surface Samples Following a Fire and the Implications for Interpretation of Subfossil Charcoal Data', *Palaeogeography, Palaeoclimatology, Palaeoecology* 164, 33–42

Blomkvist, N. 1998. 'The Medieval Europeanisation Process of the Baltic Rim Region 1100–1400 AD: Problems for an International Study', in *Culture, Clash or Compromise: The Europeanisation of the Baltic Sea Area 1100–1400 AD*, ed. N. Blomkvist, Visby: Gotland Centre for Baltic Studies, 9–36

——, 2004. 'The Medieval Catholic World-System and the Making of Europe', in *The European Frontier: Clashes and Compromises in the Middle Ages*, ed. J. Staecker, Lund: Almqvist and Wiksell, 15–34

——, 2005. *The Discovery of the Baltic: The Reception of a Catholic World-System in the European North (AD 1075–1225)*, Leiden: Brill

——, 2009. '12th–14th-Century European Expansion and its Reception in the Baltic North', in *The Reception of Medieval Europe in the Baltic Sea Region*, ed. J. Staecker, Visby: Gotland University Press, 431–84

Bouchet, F., Lefévre, C., West, D., and Corbertt, D. 1999. 'First Paleo-Parasitological Analysis of a Midden in the Aleutian Islands (Alaska): Results and Limits', *J Parasitol* 85, 369–372

Boyle, J. F. 2001. 'Inorganic Geochemical Methods in Palaeolimnology', in *Tracking Environmental Change Using Lake Sediments*, II: *Physical and Geochemical Methods*, eds W. M. Last and J. P. Smol, Dordrecht: Kluwer, 83–141

Broström, A., Nielsen, A., Gaillard, M-J., Hjelle, K., Florence, M., Binney, H., Bunting, J., Fyfe, R., Meltsov, V., Poska, A., Räsänen, S., Soepboer, W., von Stedingk, H., Suurari, H, and Sugita, S. 2008. 'Pollen Productivity Estimates of Key European Plant Taxa for Quantitative Reconstruction of Past Vegetation: A Review', *Vegetation History and Archaeobotany* 17, 461–78

Broström, A., Sugita, S., Gaillard, M-J., and Pilesjö, P. 2005. 'Estimating the Spatial Scale of Pollen Dispersal in the Cultural Landscape of Southern Sweden', *The Holcoene* 15, 252–62

Brown, A. D., and Pluskowski, A. G. 2011. 'Detecting the Environmental Impact of the Baltic Crusades on a Late-Medieval (13th–15th Century) Frontier Landscape: Palynological Analysis from Malbork Castle and Hinterland, Northern Poland', *Journal of Archaeological Science* 38, 1957–66

Budrys, R. 2001. 'Review of the Literature on Amber', in *Baltic Amber*, ed. A. Butrimas, Vilnius: Vilnius Academy of Fine Arts Press, 261–72

Bullock, P., Fedoroff, N., Jongerius, A., Stoops, G., and Tursina, T. 1985. *Handbook for Thin Section Description*, Wolverhampton: Waine Reasearch

Bunting, J., and Middleton, R. 2005. 'Modelling Pollen Dispersal and Deposition Using HUMPOL Software, Including Simulating Windroses and Irregular Lakes', *Review of Palaeobotany and Palynology*, 134, 185–96

——, 2009a. 'Equifinality and Uncertainty in the Interpretation of Pollen Data: The Multiple Scenario Approach to Reconstruction of Past Vegetation Mosaic', *The Holocene* 19, 799–803

——, 2009b. 'The Multiple Scenario Approach — A Pragmatic Method for Past Vegetation Mosaic Reconstruction', *The Holocene* 19, 797–801

Campbell, B. M. S. 2016. *The Great Transition: Climate, Disease and Society in the Late-Medieval World*, Cambridge: Cambridge University Press

Canti, M. G. 2003. 'Earthworm Activity and Archaeological Stratigraphy: A Review of Products and Processes', *Journal of Archaeological Science* 30, 135–48

Caune A., and Ose I. 2004. *Latvijas 12. gadsimtā beigu - 17. gadsimtā vācu piļu leksikons*, Rīga: Latvijas vēstures institūta apgāds

Clarke, R. 1982. 'Point Count Estimation of Charcoal in Pollen Preparations and Thin Sections of Sediments', *Pollen et Spores* 24, 523–35

Cook, S. R., Banerjea, R. Y., Marshall, L. J., Fulford, M., Clarke, A., and van Zweiten, C. 2010. 'Concentrations of Copper, Zinc and Lead as Indicators of Hearth Usage at the Roman Town of Calleva Atrebatum (Silchester, Hampshire, UK)', *Journal of Archaeological Science* 37, 871–79

Courty, M. A., Goldberg, P., and Macphail, R. 1989. *Soils and Micromorphology in Archaeology*, Cambridge: Cambridge University Press

Crosby, B. 1986. *Ecological Imperialism: The Biological Expansion of Europe, 900–1900*, Cambridge: Cambridge University Press (rev. edn 2015)

Cushing, E. J. 1967. 'Evidence for Differential Pollen Preservation in Late Quaternary Sediments in Minnesota', *Review of Palaeobotany and Palynology* 4, 87–101

Czaja, R. 2009. 'Struktura społeczna', in *Państwo zakonu krzyżackiego w Prusach: Władza i społeczeństwo*, ed. M. Biskup et al., Warsaw: Państwowe Wydawnictwo Naukowe, 444–50

Czaja, R., and Radzimiński, A., eds 2013. *Zakon krzyżacki w Prusach i Inflantach: podziały administracyjne i kościelne w XIII–XVI wieku*, Toruń: Wydawnictwo Naukowe Uniwersytetu Mikołaja Kopernika

Davis, S. 1987. *The Archaeology of Animal Bones*, London: Routledge

Deetz, J. 1996. *In Small Things Forgotten*, London: Anchor

Devos, Y., Vrydaghs, L., Degraeve, A., and Fechner, K. 2009. 'An Archaeopedological and Phytolitarian Study of the "Dark Earth" on the Site of Rue de Dinant (Brussels, Belgium)', *Catena* 78, 270–84

Dickson, C. 1987. 'The Identification of Cereals from Ancient Bran Fragments', *Circaea* 4, 95–102

Długokęcki, W. 2009. 'Kolonizacja ziemi chełmińskiej, Prus i Pomorza Gdańskiego do 1410 r.', in *Państwo zakonu krzyżackiego w Prusach: Władza i społeczeństwo*, ed. M. Biskup et al., Warsaw: Państwowe Wydawnictwo Naukowe, 200–17

Driesch, A. von den, 1976. *A Guide to the Measurement of Animal Bones from Archaeological Sites*, Cambridge, MA: Peabody Museum of Archaeology and Ethnology, Harvard University

Duffin, K. I., Gilson, L., and Willis, K. J. 2008. 'Testing the Sensitivity of Charcoal as an Indicator of Fire Events in Savannah Environments: Quantitative Predictions of Fire Proximity, Area and Intensity', *The Holocene* 18, 279–91

Eckardt, H., ed. 2010. *Roman Diasporas: Archaeological Approaches to Mobility and Diversity in the Roman Empire* (= *Journal of Roman Archaeology*, supp. ser., 78)

Ekdahl, S. 2004. 'Crusades and Colonisation in the Baltic: A Historiographic Analysis', *Rocznik Instytutu Polsko-skandynawskiego* 19, 1–42

Ervynck, A. 1992. 'Medieval Castles as Top-Predators of the Feudal System: An Archaeozoological Approach', *Château Gaillard* 15, 152–59

——, 2004. '*Orant, pugnant, laborant*. The Diet of the Three Orders in the Feudal Society of Medieval North-Western Europe', in *Behaviour behind Bones: The Zooarchaeology of Ritual, Religion, Status and Identity*, eds S. J. O' Day, W. Van Neer, and A. Ervynck, Oxford: Oxbow, 215–23

Ervynck, A., Dobney, K., Lentacker, A., Müldner, G., and Richards, M. 2008. 'An Investigation into the Transition from Forest Dwelling Pigs to Farm Animals in Medieval Flanders, Belgium', in *Pigs and Humans: 10,000 Years of Interaction*, eds U. Albarella et al., Oxford: Oxford University Press, 171–93

Ewan, C. R. 2000. 'From Colonist to Creole: Archaeological Patterns of Spanish Colonization in the New World', *Historical Archaeology* 34/3, 36–45

Fedh, D., Broström, A., Rundgren, M., Lagerås, P., Florence, M., and Zillen, L. 2013. 'The Impact of Land-Use Change on Floristic Diversity at Regional Scale in Southern Sweden 600 BC–AD 2008', *Biogeosciences* 10, 3159–3173

Fernandez, F. G., Terry, R. E., Inomata, T., and Eberl, M. 2002. 'An Ethnoarchaeological Study of Chemical Residues in the Floors and Soils of Q'eqchi' Maya Houses at Las Pozas, Guatemala', *Geoarchaeology* 17, 487–519

FitzPatrick, E. A. 1993. *Soil Microscopy and Micromorphology*, Chichester: Wiley

Fonferek, J. 2019. 'The Environmental Setting of the Earliest Teutonic Order Stronghold in Elbląg', in *Ecologies of Crusading, Colonization, and Religious Conversion in the Medieval Baltic: Terra Sacra II*, ed. A. Pluskowski, Environmental History of the North Atlantic World, 3, Turnhout, Brepols (forthcoming)

Fonferek, J., Marinkowski, M., and Sieńkowska, U. 2012. *Elbląg — życie codzienne w porcie hanzeatyckim*, Elbląg: Muzeum Archeologiczno-Historyczne w Elblągu

French, C. 2003. *Geoarchaeology in Action*, New York: Routledge

Froese, R., and Pauly D., eds. 2006. *FishBase* <www.fishbase.org> (version 05/2006)

Gaillard, M-J., Sugita, S., Bunting, J., Middleton, R., Broström, A., Caseldine, C., Giesecke, T., Hellman, S., Hicks, S., Hjelle, K., Langdon, C., Nielsen, A-B., Poska, A., Stedingk, H., and Veski, S. 2008. 'The Use of Modelling and Simulation Approach in Reconstructing Past Landscapes from Fossil Pollen Data: A Review and Results from the POLLANDCAL Network', *Vegetation History and Archaeobotany* 17, 419–43

Gaimster, D. 1999. 'German Stoneware and Stove-Tiles: Type-Fossils of Hanseatic Urban Culture in the Baltic *c.* 1200–1600', in *The Medieval Town in the Baltic: Hanseatic History and Archaeology*, eds R. Vissak and A. Mäesalu, Tartu: Tartu City Museum, 53–64

——, 2002. 'Tile-Stove Production in the Baltic *c.* 1400–1600: An Index of Hanseatic Cultural and Technological Exchange', in *Centre, Region, Periphery: Medieval Europe Basel*, eds G. Helmig, G. Scholkmann, and M. Untermann, Basel: Folio Verlag Dr G. Wesselkamp, Archäologische Bodenforschung Basel-Stadt, 110–17

——, 2005. 'A Parallel History: The Archaeology of Hanseatic Urban Culture in the Baltic *c.* 1200–1600', *World Archaeology* 37/3, 408–23

——, 2007. 'The Baltic Ceramic Market 1200–1600: Measuring Hanseatic Cultural Transfer and Resistance', in *Cultural Exchange in Early Modern Europe*, IV: *Forging European Identities, 1400–1700*, ed. H. Roodenburg, Cambridge: Cambridge University Press, 30–58

——, 2011. 'Archaeology of a Trade Network: The Hanseatic League, 1200–1500 AD', in *The Archaeology of Medieval Europe*, II: *Twelfth to Sixteenth Centuries*, ed. M. Carver, Aarhus: Aarhus University Press, 340–50

——, 2014. 'The Hanseatic Cultural Signature: Exploring Globalization on the Micro-Scale in Late Medieval Northern Europe', *European Journal of Archaeology* 17/1, 60–81

Garcia, L. S. 2009. *Diagnostic Medical Parasitology*, Washington, DC: ASM Press

Gazda, D. 2018. *Wielokulturowy obiekt warowny na Górze Zamkowej oraz gród cyplowy w Starym Dzierzgoniu. Studia i materiały*, Warsaw: Wydawnictwo Trzecia Strona

Gé, T., Courty, M. A., Matthews, W., and Wattez, J. 1993. 'Sedimentary Formation Processes of Occupation Deposits', in *Formation Processes in Archaeological Context*, eds P. Goldberg, D. T. Nash, and M. D. Petraglia, Madison: Prehistory Press

Gerrard, C. 2003. *Medieval Archaeology: Understanding Traditions and Contemporary Approaches*, London: Routledge

Glick, T. F. 1992. '*Convivencia*: An Introductory Note', in *Convicencia: Jews, Muslims and Christians in Medieval Spain*, eds V. B. Mann, T. F. Glick, and J. D. Dodds, New York: Braziller/Jewish Museum, 1–9

Graudonis, J. 2003. *Turaidas pils*, II: *Atradumi*, Turaida: Turaidas muzejrezervāts

Grążawski, K., ed. 2003. *Pogranicze polsko-pruskie i krzyżackie: Materiały z konferencji naukowej Górzno, 1–2 czerwca 2002*, Włocławek: Włocławskie Wydawnictwo Naukowe

——, ed. 2007. *Pogranicze polsko-pruskie i krzyżackie 2*, Włocławek-Brodnica: Włocławskie Towarzystwo Naukowe

——, 2009. *Ziemia lubawska na pograniczu słowiańsko-pruskim w VIII–XIII w.: studium nad rozwojem osadnictwa*, Olsztyn: Wydawnictwo Uniwersytetu Warmińsko-Mazurskiego

Greig, J. 1982. 'The Interpretation of Pollen Spectra from Urban Archaeological Deposits', in *Environment Archaeology in the Urban Context*, eds A. R. Hall and H. K. Kenward, CBA Research Report, 43, York: Council for British Archaeology

Gunn, A., and Pitt, S. J. 2012. *Parasitology: An Integrated Approach*, Chichester: Wiley Blackwell

Haak, A., and Russow, E. 2013. 'On the Development of the Town of Viljandi in the Light of Earliest Archaeological Find Complexes', *Estonian Journal of Archaeology* 17/1, 57–86

Hambleton, E. 1999. *Animal Husbandry Regimes in Iron Age Britain*, British Archaeological Reports, 282, Oxford: Archaeopress

Hamilton, J., and Thomas, R. 2012. 'Pannage, Pulses and Pigs: Isotopic and Zooarchaeological Evidence for Changing Pig Management Practices in 14th-Century England', *Journal of Medieval Archaeology* 56, 234–59

Harvey E. L., and Fuller D. Q. 2005. 'Investigating Crop Processing using Phytolith Analysis: The Example of Rice and Millets', *Journal of Archaeological Science* 32, 739–52

Hayward, K. M. J. 2009. *Roman Quarrying and Stone Supply on the Periphery — Southern England: A Geological Study of First Century Funerary Monuments and Monumental architecture*, British Archaeological Reports, 500, Oxford: Archaeopress

——, 2015. 'Identifying Geological Materials from Nineteenth-Century Residential and Industrial Properties in South Glasgow: The M74 Excavations', *Scottish Journal of Geology* 51, 81–94

——, 2019. 'Resources for Castle Building in Prussia and Livonia', in *Ecologies of Crusading, Colonization, and Religious Conversion in the Medieval Baltic: Terra Sacra II*, ed. A. Pluskowski, Environmental History of the North Atlantic World, 3, Turnhout, Brepols (forthcoming)

Hellman, S., Gaillard, M-J., Broström, A., and Sugita, S. 2008. 'Effects of the Sampling Design and Selection of Parameter Values on Pollen-Based Quantitative Reconstructions of Regional Vegetation: A Case Study in Southern Sweden Using the REVEALS Model', *Vegetation History and Archaeobotany* 17, 445–59

Hellman, S., Bunting, M.-J., and Gaillard, M-J. 2009a. 'Relevant Source Area of Pollen in a Patchy Cultural Landscape and Signals of Anthropogenic Landscape Disturbance in the Pollen Record: A Simulation Approach', *Review of Palaeobotany and Palynology* 153, 245–58

Hellman, S., Gaillard, M-J., Bunting, J., and Mazier, F., 2009b. 'Estimating the Relevant Source Area of Pollen in the Past Cultural Landscape of Southern Sweden — A Forward Modelling Approach', *Review of Palaeobotany and Palynology* 151, 259–71

Herrmann, C. 2007. *Mittelalterliche Architektur in Preussenland: Untersuchungen zur Frage der Kunstlandschaft und -geographie*, Petersberg: Michael Imhof Verlag

Holden T. 1990. 'Transverse Cell Patterns of Wheat and Rye Bran and their Variation over the Surface of a Single Grain', *Circaea*, 6/2, 97–104.

Hoffman, A. A., and Parsons P. A. 1997. *Extreme Environmental Change and Evolution*, Cambridge: Cambridge University Press

Hoffman, R. 2014. *An Environmental History of Medieval Europe*, Cambridge: Cambridge University Press

Horden, P., and Purcell, N. 2000. *The Corrupting Sea: A Study of Mediterranean History*, Malden: Blackwell

Hughes, M. J. 2009. 'ICP-AES Analysis of Ceramic Building Material', in *Hill Hall; A Singular House Devised by a Tudor Intellectual*, eds P. Drury and R. Simpson, London: Society of Antiquaries, 366–79

Hutson, S. R., and Terry, R. E. 2006. 'Recovering Social and Cultural Dynamics from Plaster Floors: Chemical Analyses at Ancient Chunchucmil, Yucatan, Mexico', *Journal of Archaeological Science* 33, 391–404

Immonen, V. 2007. 'Defining a Culture: The Meaning of Hanseatic in Medieval Turku', *Antiquity* 31/313, 720–32

Jähnig, B., and Michels, G. 2005. *Deutscher Orden und Preußenland: ausgewählte Aufsätze anlässlich des 65. Geburtstages/Udo Arnold*, Marburg: Elwert

Jahnke, C. 2015. 'The Baltic Trade', in *A Companion to the Hanseatic League*, ed. D. J. Harreld, Leiden: Brill, 194–240

Jensen, C. S. 2001. 'Urban Life and the Crusades in North Germany and the Baltic Lands in the Early Thirteenth Century', in *Crusade and Conversion on the Baltic Frontier, 1150–1500*, ed. A. V. Murray, Aldershot: Ashgate, 75–94

Johnson, M. 2002. *Behind the Castle Gate: From Medieval to Renaissance*, Oxford: Blackwell

Joy, N. H. 1976. *A Practical Handbook of British Beetles*, 2 vols, Farningdon: Classey

Jóźwiak, S., and Trupinda, J. 2007. *Organizacja życia na zamku krzyżackim w Malborku w czasach wielkich mistrzów (1309–1457)*, Malbork: Muzeum Zamkowe w Malborku

——, 2012. *Krzyżackie zamki komturskie w Prusach*, Toruń: Wydawnictwo Naukowe Uniwersytetu Mikołaja Kopernika

Karwowska, H., and Andrzejewski, A., eds. 2006. *Stan badań archeologicznych na pograniczu polsko-białoruskim od wczesnego średniowiecza po czasy nowożytne*, Białystok: Muzeum Podlaskie

Kedar, B. Z. 1992. 'The Crusading Kingdom of Jerusalem — The First European Colonial Society? A Symposium', in *The Horns of Hattin*, ed. B. Z. Kedar, London: Variorum, 341–66

Kendal, J., Jamshid, J., Tehrani, J., and Odling-Smee, J. 2011. 'Human Niche Construction in Interdisciplinary Focus', *Philosophical Transactions of the Royal Society B* 366/1566, 785–92

Kenward, H. K., Hall, A. R., and Jones, A. K. G. 1980. 'A Tested Set of Techniques for the Extraction of Plant and Animal Macrofossils from Waterlogged Archaeological Deposits', *Science and Archaeology* 22, 3–15

King, G. A. 2019. 'A Foul Perspective: Analysis of Invertebrate Remains in a Mid-Fourteenth Century Latrine from Riga, Latvia', in *Ecologies of Crusading, Colonization, and Religious Conversion in the Medieval Baltic: Terra Sacra II*, ed. A. Pluskowski, Environmental History of the North Atlantic World, 3, Turnhout, Brepols (forthcoming)

Kļaviņš, K. 2019. 'Reorganizing the Livonian Landscape: Some Issues and Research Perspectives', in *Ecologies of Crusading, Colonization, and Religious Conversion in the Medieval Baltic: Terra Sacra II*, ed. A. Pluskowski, Environmental History of the North Atlantic World, 3, Turnhout, Brepols (forthcoming)

Klein R., and Cruz-Uribe, K. 1984. *The Analysis of Animal Bones from Archaeological Sites*, Chicago: University of Chicago Press

Körber-Grohne, U. 1990. 'Gramineen und Grünlandvegetation vom Neolithikum bis zum Mittelalter in Mitteleuropa', *Bibliotheca Botanica* 139, 1–104

Kowalczyk, E. 1999. 'Szlaki drogowe w puszczy na pograniczu mazowiecko-pruskim w Średniowieczu', *Kwartalnik Historyczny* 1, 3–18

Kreem, J. 2002. *The Town and its Lord: Reval and the Teutonic Order (in the Fifteenth Century)*, Tallinn: Tallinna Linnaarhiiv

Lahtinen, A. 2006. 'Frontier and Borderland', in *Frontiers in the Middle Ages*, eds O. Merisalo and P. Pahta, Louvain-la-Neuve: Fédération Internationale des Institutes d'Études Médiévales, 184–96

Lajtha, K., and Michener, R. 1994. *Stable Isotopes in Ecology and Environmental Science*, Oxford: Blackwell

Laul, S., and Valk, H. 2007. *Siksälä: A Community at the Frontiers. Iron Age and Medieval*, Tartu: University of Tartu

Leighly, J. 1939. 'The Towns of Medieval Livonia', *University of California Publications in Geography* 6/7, 235–314

Lind, J. H. 2004. 'The Concept of "Europeanisation" on the Baltic Rim as Seen from the East', in *The European Frontier: Clashes and Compromises in the Middle Ages*, ed. J. Staecker, Lund: Almqvist and Wiksell, 41–44

Lomas-Clarke, A. H., and Barber, K. E. 2007. 'Human Impact Signals from Peat Bogs — A Combined Palynological and Geochemical Approach', *Vegetation History and Archaeobotany* 16, 419–29

Lukashina, N. P., and Kharin, G. S. 1999. 'Criteria for the Recognition of the Prussian Suite (Formation) and Prussian Horizon', in *Investigations into Amber*, eds B. Kosmowska-Ceranowicz and H. Paner, Gdańsk: Archaeology Museum in Gdańsk, Museum of the Earth, Polish Academy of Sciences, 27–32

Lyman, R. L. 2008. *Quantitative Palaeozoology*, Cambridge: Cambridge University Press

Lyon, J. R. 2008. 'Fathers and Sons: Preparing Noble Youths to be Lords in Twelfth-Century Germany', *Journal of Medieval History* 34/3, 291–310

Mackenzie, W. S., and Adams, A. E. 1994. *A Colour Atlas of Rocks and Minerals in Thin Section*, New York: Wiley

Mackenzie, W. S., and Guilford, C. 1980. *Atlas of Rock Forming Minerals*, Harlow: Longman

Macphail, R. I. 1994. 'The Reworking of Urban Stratigraphy by Human and Natural Processes', in *Urban-Rural Connexions: Perspectives from Environmental Archaeology*, ed. H. K. Kenward, Oxford: Oxbow, 13–43

Macphail, R. I., Galinié, H., and Verhaeghe, F. 2003. 'A Future for Dark Earth?', *Antiquity*, 77, 349–58

Makowiecki, D. 2010. *Wczesnośredniowieczna gospodarka zwierzętami i socjotopografia in Culmine na Pomorzu Nadwiślańskim: Studium archeozoologiczne*, Toruń: Wydawnictwo Naukowe Uniwersytetu Mikołaja Kopernika

Maltby, M. 1979. *Faunal Studies on Urban Sites: The Animal Bones from Exeter, 1971–1975*, Sheffield: Department of Prehistory and Archaeology, University of Sheffield

Maltby, M., Pluskowski, A. G., and Seetah, K. 2009. 'Animal Bones from an Industrial Quarter at Malbork, Poland: Towards an Ecology of a Castle built in Prussia by the Teutonic Order', *Crusades*, 8, 191–212

Marciniak-Kajzer, A. 2006. Wczesnośredniowieczne osadnictwo na pograniczu prusko-mazowieckim w świetle badań grodziska w Leszczu w woj. warmińsko-mazurskim, *Archeologia Polski* 51/1–2, 121–44

Marquer, L., et al. 2014. 'Holocene Changes in Vegetation Composition in Northern Europe: Why Quantitative Pollen-Based Vegetation Reconstructions Matter', *Quaternary Science Reviews* 90, 199–216

Matthews, W. 1995. 'Micromorphological Characteristics of Occupation Deposits and Microstratigraphic Sequences at Abu Salabikh, Southern Iraq', in *Archaeological Sediments and Soils: Analysis, Interpretation and Management*, eds A. J. Barnham and R. I. Macphail, London: Institute of Archaeology, University College, 41–76

Matthews, W., French, C. A. I., Lawrence, T., Cutler, D. F., and Jones, M. K. 1997. 'Microstratigraphic Traces of Site Formation Processes and Human Activities', *World Archaeology* 29/2, 281–308

Mazier, F., Gaillard, M-J., Sugita, S., Trondman, A-K., and Broström, A. 2012. 'Testing the Effect of Site Selection and Parameter on REVEALS-Model Estimates of Plant Abundance using the Czech Quaternary Palynological Database', *Review of Palaeobotany and Palynology* 187, 38–49

Merisalo, O., and Pahta, P., eds. 2006. *Frontiers in the Middle Ages*. Louvain-la-Neuve: Fédération Internationale des Institutes d'Études Médiévales

Milek, K. B., and French, C. A. I. 2007. 'Soils and Sediments in the Settlement and Harbour at Kaupang', in *Kaupang in Skiringssal*, ed. D. Skre, Aarhus: Aarhus University Press, 321–61

Militzer, K. 1994. 'The Recruitment of Brethren for the Teutonic Order in Livonia, 1237–1562', in *The Military Orders: Fighting for the Faith and Caring for the Sick*, ed. M. Barber, Aldershort: Ashgate, 270–77

Millard, A. R., Jiminez-Cano, N., Lebrasseur, O., and Sakai, Y. 2011. 'Isotopic Investigation of Animal Husbandry in the Welsh and English Periods at Dryslwyn Castle, Carmarthenshire, Wales', *International Journal of Osteoarchaeology* 23/6, 640–50

Mirek, Z., Piękoś-Mirkowa, H., and Zając M. 2002. *Flowering Plants and Pteridophytes of Poland: A Checklist*, Kraków, Szafer Institute of Botany Polish Academy of Science

Moore P. D., Webb, J. A., and Collinson, M. E. 1991. *Pollen Analysis*, Oxford: Blackwell (2nd edn)

Morales, A., and Rosenlund, K. 1979. *Fish Bone Measurements*, Copenhagen: Steenstrupia

Mortensen, L. B., and Bisgaard, L. 2013. 'Introduction: Late Medieval Urban Civilization and its North European Variant', in *Guilds, Towns and Cultural Transmission in the North, 1300–1500*, eds L. Bisgaard, L. B. Mortensen, and T. Pettitt, Odense: University Press of Southern Denmark, 7–30

Mugurēvičs, Ē. 2003. 'Viking Age and Medieval Finds of East Baltic Amber in Latvia and the Neighbouring Countries (9th–16th Century)', in *Amber in Archaeology*, eds C. W. Beck, I. B. Loze, and J. M. Todd, Rīga: Institute of the History of Latvia Publishers, 90–95

Müldner, G., Britton, K., and Ervynck, A. 2014. 'Inferring Animal Husbandry Strategies in Coastal Zones through Stable Isotope Analysis: New Evidence from the Flemish Coastal Plain (Belgium, 1st–15th century AD)', *Journal of Archaeological Science* 41, 322–32

Murray, A. V., ed. 2009. *The Clash of Cultures on the Medieval Baltic Frontier*, Farnham: Ashgate

Naum, M. 2010. 'Re-emerging Frontiers: Postcolonial Theory and Historical Archaeology of the Borderlands', *Journal of Archaeological Method and Theory* 17, 101–31

——, 2012. 'Difficult Middles, Hybridity and Ambivalence of a Medieval Frontier: The Cultural Landscape of Lolland and Falster (Denmark)', *Journal of Medieval History* 38/1, 56–75

——, 2014. 'Multi-Ethnicity and Material Exchanges in Late Medieval Tallinn', *European Journal of Archaeology* 17/4, 656–77

Nawrolska, G. 2008. 'A Way of Life: Luxury in a Medieval Town', in *Lübecker Kolloquium zur Stadtarchäologie im Hanseraum*, VI: *Luxus und Lifestyle*, eds C. Kimminus-Schneider, D. Mührenberg, M. Schneider, and C. Sheehan, Lübeck: Schmidt-Römhild, 509–27

Nelson, G. C., Bennett, E., Berhe, A. A., Cassman, K., DeFries, R., Dietz, T., Dobermann, A., Dobson, A., Janetos, A., Levy, M., Marco, D., Nakicenovic, N., O'Neill, B., Norgaard, R., Petschel-Held, G., Ojima, D., Pingali, P., Watson, R., and Zurek, M. 2006. 'Anthropogenic Drivers of Ecosystem Change: An Overview', *Ecology and Society* 11/2, 29

Nielsen, A. B., and Odgaard, B. V. 2010. 'Quantitative Landscape Dynamics in Denmark through the Last Three Millennia Based on the Landscape Reconstruction Algorithm Approach', *Vegetation History and Archaeobotany* 19, 375–87

Nielsen, T. K. 2011. 'Henry of Livonia on Woods and Wilderness', in *Crusading and Chronicle Writing on the Medieval Baltic Frontier*, eds M. Tamm, L. Kaljundi, and C. S. Jensen, Farnham: Ashgate, 157–78

O'Connell, M., Ghilardi, B., and Morrison, L. 2013. 'A 7000-Year Record of Environmental Change, Including Early Farming Impact, Based on Lake-Sediment Geochemistry and Pollen Data from County Sligo, Western Ireland', *Quaternary Research* 81, 35–49

O'Connor. T, P. 2000. *The Archaeology of Animal Bones*, College Station, TX: A & M University Press

——, 2003. *The Analysis of Urban Animal Bone Assemblages: A Handbook for Archaeologists*, York: Council for British Archaeology

O'Connor, T. P., and Evans, J. G. 2005. *Environmental Archaeology*, Stroud: Sutton

Odling-Smee, F. J., Laland, K. N., and Feldman, M. W. 2003. *Niche Construction: The Neglected Process in Evolution*, Princeton: Princeton University Press

Orton, D. C., Makowiecki, D., de Roo, T., Johnstone, C., Harland, J., Jonsson, L., Heinrich, D., Enghoff, I. B., Lõugas, L., Van Neer, W., Ervynck, A., Hufthammer, A. K., Amundsen, C., Jones, A. K. G., Locker, A., Hamilton-Dyer, S., Pope, P., MacKenzie, B. R., Richards, M., O'Connell, T. C., and Barrett, J. H. 2011. 'Stable Isotope Evidence for Late Medieval (14th–15th C) Origins of the Eastern Baltic Cod (*Gadus morhua*) fishery', *PLoS ONE* 6, e27568

Orton, D., Rannamäe, E., Lõugas, L., Makowiecki, D., Hamilton-Dyer, S., Pluskowski, A. G., O'Connell, T., and Barrett, J. 2019. 'The Teutonic Order's Role in the Development of a Medieval Eastern Baltic Cod Fishery', in *Ecologies of Crusading, Colonization, and Religious Conversion in the Medieval Baltic: Terra Sacra II*, ed. A. Pluskowski, Environmental History of the North Atlantic World, 3, Turnhout, Brepols (forthcoming)

Palang, H., Printsmann, A., Konsa, M., and Lang, V. 2005. 'Ideology and Tradition in Landscape Change: A Case of the Helme Parish, Estonia', in *Landscape, Law and Justice*, eds T. Peil and M. Jones, Oslo: Novus, 288–97

Paner, H. 1999. 'Archaeological Evidence for Trade in Gdańsk from the 12th to 17th Century', in *Lübecker Kolloquium zur Stadtarchäologie im Hanseraum*, II: *Der Handel*, eds R. Duncke, M. Gläser, and U. Oltmanns, Lübeck: Schmidt-Römhild, 387–402

Paravicini, W. 1989. *Die Preussenreisen des europäischen Adels* (vol. I), Sigmaringen: Thorbecke

Patterson, W. A., Edwards, K. J., and Maguire, D. J. 1987. 'Microscopic Charcoal as a Fossil Indicator of Fire', *Quaternary Science Reviews* 6, 3–23

Piperno, D. 2006. *Phytoliths: A Comprehensive Guide for Archaeologists and Palaeoecologists*, Lanham, MD: Alta Mira Press

Pluskowski, A. G. 2010. 'The Zooarchaeology of Medieval "Christendom": Ideology, the Treatment of Animals and the Making of Medieval Europe', *World Archaeology* 42/2, 201–14

——, 2013. *The Archaeology of the Prussian Crusade: Holy War and Colonisation*, London: Routledge

Pluskowski, A. G., Sawicki, Z., Shillito, L-M., Badura, M., Makowiecki, D., Zabilska-Kunek, M., Seetah, K., and Brown, A. 2014. 'Biała Góra: The Forgotten Colony in the Medieval Pomeranian-Prussian Borderlands', Antiquity 88/341, 863–82

Pluskowski, A. G., Brown, A. D., Banerjea, R., Makowiecki, D., Seetah, K., Rannamäe, E., Badura, M., Jarzebowski, M., Kreem, J., and Kļaviņš, K., 2019. 'From the Convent to the Commandery: The Pivotal Role of the Environment in Defining the Medieval Baltic *Ordensland*', *Quellen und Studien zur Geschichte des Deutschen Ordens* (forthcoming)

Pluskowski, A. G., Maltby, M., and Seetah, K. 2010. 'Potential Osteoarchaeological Evidence for Riding and the Military Use of Horses at the Castle in Malbork, Poland', *International Journal of Osteoarchaeology* 20/3, 335–43

Pluskowski, A. G., Sawicki, Z., Shillito, L.-M., Badura, M., Makowiecki, D., Zabilska-Kunek, M., Seetah, K., and Brown, A. 2014. 'Biała Góra: The Forgotten Colony in the Medieval Pomeranian-Prussian Borderlands', *Antiquity* 88/341, 863–82

Podbielkowski, Z., and Sudnik-Wójcikowska, B. 2003. *Słownik roślin użytkowych. Państwowe Wydawnictwo Rolnicze i Leśne*, Warsaw: Państwowe Wydawnictwo Rolnicze i Leśne

Poska, A., Meltsov, V., Sugita, S., and Vassiljev, J. 2011. 'Relative Pollen Productivity Estimates of Major Anemophilous Taxa and Relevant Source Area of Pollen in Cultural Landscapes of the Hemi-Boreal Forest Zone (Estonia)', *Review of Palaeobotany and Palynology* 167, 30–39

Poska, A., Saarse, L., Koppel, K., Nielsen, A. B., Avel, E., Vassiljev, J., and Väli, V. 2014. 'The Verijärv Area, South Estonia over the Last Millennium: A High Resolution Quantitative Land-Cover Reconstruction Based on Pollen and Historical Data', *Review of Palaeobotany and Palynology* 207, 5–17

Pospieszna, B., ed. 2010. *Fundacje artystyczne na terenie Państwa Krzyżackiego w Prusach*, 2 vols, Malbork: Muzeum Zamkowe w Malborku

Rannamäe, E. 2010. *A Zooarchaeological Study of Animal Consumption in Medieval Viljandi* (unpublished MA thesis, University of Tartu)

Reitz, E., and Wing, E. 2000. *Zooarchaeology*, Cambridge: Cambridge University Press

Rockman, M., and Steele, J., eds 2003. *The Colonization of Unfamiliar Landscapes: The Archaeology of Adaptation*, London: Routledge

Rosen, A. M. 1992. 'Preliminary Identification of Silica Skeletons from Near Eastern Archaeological Sites: An Anatomical Approach', in *Phytolith Systematics: Emerging Issues*, eds G. Rapp Jr and S. C. Mulholland SC, London: Plenum, 129–47

Russow, E. 2006. *Importkeraamika Lääne-Eesti linnades 13.–17. sajandil*, Tallinn: Russow

Sawicki, Z., Pluskowski, A. G., Brown, A., Badura, M., Makowiecki, D., Shillito, L–M., Zabilska-Kunek, M., and Seetah, K. 2015. 'Survival at the Frontier of Holy War: Political Expansion, Crusading, Environmental Exploitation and the Medieval Colonising Settlement at Biała Góra, North Poland', *European Journal of Archaeology*, 18/2, 282–311

Schelvis, J. 1990. 'The Reconstruction of Local Environments on the Basis of Remains of Oribatid Mites (Acari Oribatida)', *Journal of Archaeological Science* 17, 559–71

Schreg, R. 2014. 'Ecological Approaches to Medieval Rural Archaeology', *European Journal of Archaeology* 17/1, 83–119

Scull, C., Müldner, G., and Makowiecki, D. 2019. 'Carbon and Nitrogen Stable Isotope Evidence for Animal Husbandry and Environmental Change in the Medieval Kulmerland', in *Ecologies of Crusading, Colonization, and Religious Conversion in the Medieval Baltic: Terra Sacra II*, ed. A. Pluskowski, Environmental History of the North Atlantic World, 3, Turnhout, Brepols (forthcoming)

Seetah, K. 2007. '*Butchery as an Analytical Tool: A Comparative Study of the Romano-British and Medieval Periods*' (unpublished doctoral thesis, University of Cambridge)

Selart, A. 2015. *Livonia, Rus' and the Baltic Crusades in the Thirteenth Century*, Leiden: Brill

Shillito, L-M. 2013. 'Grains of Truth or Transparent Blindfolds? A Review of Current Debates in Archaeological Phytolith Analysis', *Vegetation History and Archaeobotany* 22, 71–82

Shillito, L-M., Balazs Major, B., Almond, M., Anderson, E., and Pluskowski, A. 2014. 'Micromorphological and Geochemical Investigation of Formation Processes in the Refectory at the Castle of Margat (Qal'at al-Marqab), Syria', *Journal of Archaeological Science* 50, 451–59

Simpson E. H. 1949. 'Measurement of Diversity', *Nature* 163, 688

Simpson I. A., Guttmann E. B., Cluett, J., and Shepherd, A. 2006. 'Characterizing Anthropic Sediments in North European Neolithic Settlements: An Assessment from Skara Brae, Orkney', *Geoarchaeology* 21/3, 221–35

Skrobot, W. 2007. 'Osadnictwo wczesnośredniowieczne w strefach pogranicznych Pojezierza Iławskiego', in *Pruskie baby kamienne: fenomen kulturowy czy europejska codzienność*, eds J. M. Łapo and G. Białuński, Olsztyn: Ośrodek Badań Naukowych im. Wojciecha Kętrzyńskiego, 116–22

Smith, K. G. V. 1989. *An Introduction to the Immature Stages of British Flies*, London: Royal Entomological Society

Šnē, A. 2006. 'The Economy and Social Power in the Late Prehistoric Chiefdoms in Eastern Lativa', *Archaeologia Baltica* 6, 68–78

Soepboer, W., Vervoort, J. M., Sugita, S., and Lotter, A. F. 2008. 'Evaluating Swiss Pollen Productivity Estimates Using a Simulation Approach', *Vegetation History and Archaeobotany*, 17, 497–506

Stinson, T. 2011. 'Counting Sheep: Potential Applications of DNA Analysis to the Study of Medieval Parchment Production', in *Kodikologie und Paläographie im digitalen Zeitalter 2 - Codicology and Palaeography in the Digital Age 2.*, eds F. Fischer, C. Fritze, and G. Vogele, Norderstedt: Schriften des Instituts für Dokumentologie und Editorik, 191–207

Stoops, G. 2003. *Guidelines for Analysis and Description of Soil Thin Sections*, Madison: Soil Science Society of America

Stopford, J., Hughes M. J., and Leese, M. N. 1991. 'A Scientific Study of Medieval Tiles from Bordesley Abbey, near Redditch (Hereford and Worcester)', *Oxford Journal of Archaeology* 10, 349–60

Stribrny, C. 1987. *CSIR Deutschl. II, 8 Germania superior: Die Herkunft der römischen Werksteine aus Mainz und Umgebung. Vergleichende petrographische und geochemische Untersuchungen an skulptierten Kalksteinen*, Bonn: Habelt

Sugita, S. 2007a. 'Theory of Quantitative Reconstruction of Vegetation I: Pollen from Large Sites REVEALS Regional Vegetation Composition', *The Holocene* 17, 229–41

——, 2007b. 'Theory of Quantitative Reconstruction of Vegetation II: All You Need is LOVE', *The Holocene* 17, 243–57

Sykes, N. 2007. *The Norman Conquest: A Zooarchaeological Perspective*, Oxford: Archaeopress

Tamm, M. 2009. 'A New World into Old Words: The Eastern Baltic Region and the Cultural Geography of Medieval Europe', in *The Clash of Cultures on the Medieval Baltic Frontier*, ed. A. V. Murray, Farnham: Ashgate, 11–36

Tandecki, J. 2009a. 'Liczebność i pochodzenie terytorialne', in *Państwo zakonu krzyżackiego w Prusach. Władza i społeczeństwo*, ed. M. Biskup et al., Warsaw: Państwowe Wydawnictwo Naukowe, 437–44

——, 2009b. 'Rozwój terytorialny państwa zakonnego w Prusach', in *Państwo zakonu krzyżackiego w Prusach. Władza i społeczeństwo*, ed. M. Biskup et al., Warsaw: Państwowe Wydawnictwo Naukowe, 105–09

Thompson, M., and Walsh, J. N. 1989. *A Handbook of Inductively Coupled Plasma Spectrometry*, Glasgow: Blackie

Tinner, W., Conedera, M., Ammann, B., Gaggeler H. W., Gedye, S., Jones, R., and Sagesser, B., 1998. 'Pollen and Charcoal in Lake Sediments Compared with Historically Documented Forest Fires in Southern Switzerland since AD 1920', *The Holocene* 8, 31–42

Trupinda, J. 1999. *Ideologia krucjatowa w kronice Piotra z Dusburga*, Gdańsk: Officina Ferberiana

Twiss, P. C. 1992. 'World Distribution of C3 and C4 Grass Phytoliths', in *Phytolith Systematics: Emerging Issues*, eds G. Rapp Jr and S. C. Mulholland, New York: Plenum Press, 113–28

Urban, W. 2001. 'The Frontier Thesis and the Baltic Crusade', in *Crusade and Conversion on the Baltic Frontier 1150–1500*, ed. A. V. Murray, Aldershot: Ashgate, 45–71

Valk, H. 2004. 'Artefacts in Estonian Urban Churchyards: Reflections on Different Traditions and Ideologies', in *Der Ostseeraum und Kontinentaleuropa 1100–1600*, eds D. Kattinger, J. E. Olesen, and H. Wernicke, Schwerin: T. Helms, 103–12

Valk, H., Pluskowski, A., Brown, A., Rannamäe, E., Malve, M., and Varul, L. 2012. 'The Castle of the Teutonic Order in Karksi: Preliminary Excavation Results', *Arheoloogilised välitööd Eestis/Archaeological Field Work in Estonia*, 2011, 47–56

Valk, H., Pluskowski, A. G., Thornley, D., Brown, A., and Summerfield, C. 2009. 'Fluxgate Gradiometry Survey in the Ruins of Karksi Castle and Palaeoenvironmental Analysis in its Hinterlands', *Arheoloogilised välitööd Eestis/Archaeological Field Work in Estonia*, 2008, 213–19

Valk, H., Rannamäe, E., Brown, A., Pluskowski, A., Badura, M., and Lõugas, L. 2013. 'Thirteenth-Century Cultural Deposits at the Castle of the Teutonic Order in Karksi', *Arheoloogilised välitööd Eestis/Archaeological Field Work in Estonia*, 2012, 73–92

Van den Besselaar, P., and Heimeriks, G. 2001. 'Disciplinary, Multidisciplinary, Interdisciplinary — Concepts and Indicators', in *Proceedings of the 8th International Conference on Scientometrics and Informetrics — ISSI2001*, eds M. Davis and C. S. Wilson, Sydney: UNSW, 1–9.

Watson. J. 1979. 'The Estimation of the Relative Frequency of Mammalian Species: Khirokitia 1972', *Journal of Archaeological Science* 6, 127–37

Weiner, S. 2010. *Microarchaeology: Beyond the Visible Archaeological Record*, Cambridge: Cambridge University Press

Wilkinson, K., and Stevens, C. 2013. *Environmental Archaeology*, Stroud: History Press

Wilson, C. A., Davidson, D. A., Cresser, M. S. 2005. 'An Evaluation of Multi-Element Analysis of Historic Soil Contamination to Differentiate Space Use and Former Function in and around Abandoned Farms', *The Holocene* 15, 1094–99

Yeh, H. Y., and Mitchell, P. 2019. 'Parasites and Baltic Crusading in Medieval Riga: Insights into Disease, Diet and Hygiene', in *Ecologies of Crusading, Colonization, and Religious Conversion in the Medieval Baltic: Terra Sacra II*, ed. A. Pluskowski, Environmental History of the North Atlantic World, 3, Turnhout, Brepols (forthcoming)

Yeh, H. Y., Pluskowski, A., Kalējs, U., and Mitchell, P. D. 2014. 'Intestinal Parasites in a Mid-14th Century Latrine from Riga, Latvia: Fish Tapeworm and the Consumption of Uncooked Fish in the medieval eastern Baltic Region', *Journal of Archaeological Science* 49, 83–89

Zonenberg, S. 1994. *Kronika Wiganda z Marburga*, Bydgoszcz: Wydawnictwo Uczelniane WSP w Bydgoszczy

——, 2009. *Kronika Szymona Grunaua*, Bydgoszcz: Uniwersytet Kazimierza Wielkiego

3. Chronology

Aleksander Pluskowski, Alexander Brown, and Stuart Black

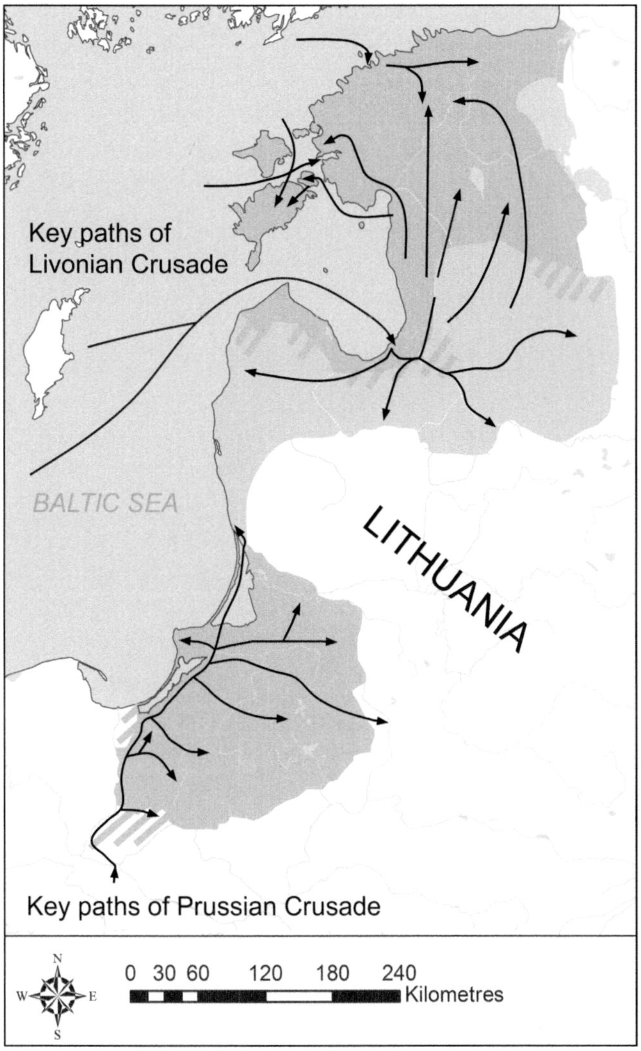

Map 3.1: Map showing the general directions of the Livonian and Prussian crusades, synthesizing a series of military campaigns. Map by A. Pluskowski.

The Baltic Crusades and Multi-Scalar Chronology

The chronology of cultural developments in the medieval eastern Baltic is multi-scalar and can be understood using all three categories of historical time highlighted by Fernand Braudel (2009): the 'long-term' chronology of landscape archaeology; 'social time', the history of groups of people; and 'individual time', the history of events. Medieval archaeologists are well placed to engage with all three (Gerrard 2003, 229), and through interdisciplinarity develop a dialectic of multiple temporalities (Lee 2012). The periods of 'individual time' associated with crusading in the eastern Baltic are relatively short and intensive. The decades leading up to the official issuing of crusading bulls are marked by unsuccessful military incursions into Prussia and the eastern Baltic tribal territories. The events of the Prussian Crusade span a period of fifty-three years (1230–83); the events of the Livonian Crusade cover a longer period, from the issuing of the first crusading bull in 1198 until the official subjugation of the Semigallians in 1290. Within this period, the military conquest of regions was incremental, starting from Riga and the Lower Daugava, branching quickly northwards towards Estonia which was subjugated by *c.* 1227, before campaigns were launched south towards the more volatile Lithuanian frontier (Map 3.1). These time spans include periods of warfare alongside years of political stability and demographic growth.

The war between the Teutonic Order and the Grand Duchy of Lithuania, including Samogitia, started, according to Peter von Dusburg, immediately after the conclusion of the Prussian Crusade in 1283 and continued beyond the official acceptance of Catholicism by the Lithuanian Grand Duke in 1386. As many as 299 *Reisen* were organized by the Order against Lithuania between 1305 and 1403, with regular campaigns from the 1360s intensifying in the following two decades (Paravicini 1989). Samogitia was occupied by the Order in 1398–1409, and subsequently ceded to Lithuania in 1422. Throughout the fifteenth century, there were a series of wars between the Order and Poland-Lithuania (1409–11, 1422, 1431–35, 1454–66), interspersed with lesser conflicts and treaties that eventually resulted in the reduction of the territorial extent of the Prussian *Ordensland* (The 'Peace of Thorn' in 1411, and more decisively in 1466), and a final war in 1519–21 with conflict continuing until the submission of the last Grand Master, Albert von Hohenzollern, to the Polish crown and the secularization of the Prussian branch of the Order in 1525. In Livonia, however, the only major military event after the subjugation of Semigallia was the Estonian uprising in 1343–45. The Livonian branch of the Order was also involved in small-scale conflict with Lithuania and participated in a civil war which culminated in the disastrous battle of Wilkomierz (Lith. Pabaiskas) in 1435. Subsequently, Wolter von Plettenberg led the Order into a war with the emergent Muscovite state in 1501–03. Tensions between Livonia and Russia, as well as the political machinations of Poland-Lithuania, erupted into the Livonian War (1558–83) during which this branch of the Order was secularized by the last Master, Gotthard Kettler, in 1561.

After the end of the period of active crusading in Prussia and Livonia, military activities were regionally and temporally concentrated, although throughout the fourteenth century the borderlands between the Teutonic Order's territories and the Grand Duchy of Lithuania can be broadly defined as a sparsely inhabited, militarized zone. Nonetheless, a good indicator of stability was the regionally variable process of colonization which in Prussia began in earnest in the last decades of the thirteenth century, and which was paralleled in Livonia by a phase of expansion of the Order's network of castles (see Chapter 1). This was also the time when the Order's convents were rebuilt or expanded. Instability on the frontier is exemplified by the procurator's castle at Lyck (Pol. Ełk), the easternmost holding of the Order in Prussia. This was constructed at the start of the fifteenth century on an island in one of the largest lakes in the region (Ger. Lyck-See, Pol. Jezioro Ełckie), in close proximity to Masovian and Lithuanian territory (see Chapter 10). Within the island, the core of the early-fifteenth-century complex consisted of a timber tower with its own moat. Although a settlement was established on the edge of the lake adjacent to the castle, colonization of the surrounding region, the depopulated tribal territory of Sudovia, remained limited until the late fifteenth and sixteenth century. The impact of the fifteenth-century wars in Prussia is matched by the devastation of the Livonian War, resulting in episodes of depopulation and the destruction of some of the Order's houses.

In archaeological terms, the chronology of the crusading period and the *Ordensland* in the eastern Baltic is relatively short. Linking specific archaeological and palaeoenvironmental material to such brief time spans is problematic, and particularly when compared to the temporal resolution of documentary sources, archaeological chronologies may appear coarse. In fact, archaeological data can be dated to varying levels of precision, with style-based dating of artefacts usually providing the lowest resolution chronologies, exemplified by the problems of dating locally produced Baltic ceramics (Karczewska and Karczewski 2004; see also the section on Święta Góra in Chapter 10). The CCC project tackled this issue of multi-scalar chronologies by relying on historical sources for understanding rapid changes and archaeological data for a background chronology of long-term, cultural transformations (Blomkvist 2009). This chapter outlines a phased chronological framework for the medieval eastern Baltic and then discusses the range of dating techniques used to situate the archaeological, palaenvironmental, and historical data from case study sites and their territories within this framework.

A Chronological Framework for the Medieval Eastern Baltic

Archaeological chronologies for the Middle Ages in the eastern Baltic, as with all other divisions of time, are artificial constructs which function as aids for coherent explanations of cultural change. They traditionally begin with the start of the crusades and the appearance of written sources, defined in Estonia as the end stage of the 'Final Iron Age' (between the end of the Viking Age and the crusades), in Latvia and Lithuania as the last phase of the 'Late Iron Age' and in Poland (including Prussia) as the end of the 'Early Medieval' period (Buko 2005; Valk 2014). These are marked by the appearance of new forms of material culture (greyware ceramics, imported stoneware vessels, crossbow bolts, coinage), architecture (timber-framed houses, ecclesiastical and municipal brick and stone buildings), and settlement types (castles and towns) associated with the process of colonization accompanying crusading, as well as the abandonment of the majority of indigenous centres of power (Mugurēvičs 1990; Pluskowski 2013; Pluskowski and Valk 2016; see also Chapters 1 and 2). Trends in settlement development and territorial reorganization, which frame the chronology for the study, can then be subdivided into four broad phases. These correspond to periods of 'social time', defining categories of long-term activities of the various groups involved in the formation of the medieval Baltic *Ordensland*. This timeline is necessarily broad but enables changes in the status of individual regions, characterized using the model of frontiers, to be compared with environmental transformations (Chapter 2). The regional status of each territorial case study is outlined in Chapters 9 (for Livonia) and 16 (for Prussia). Once these long-term trends have been determined, it is possible to incorporate periods of 'individual time' where appropriate, represented by singular archaeological deposits and historical sources.

Phase 1: The Final Phase of Indigenous Society

The first phase is the least precisely defined, as it relates to general trends in social organization across the eastern Baltic which are discussed in more detail for each case study (Chapters 4, and 10). From the eleventh to thirteenth century, the indigenous territories of Prussia and Livonia were ruled by elite groups, engaged in territorial expansion and consolidation. In Estonia this manifested in the formation of regional overlordships, in Latvia in more diffuse clusters of power represented by individuals or families based in key strongholds. In the twelfth century, south-eastern Livonia was draw more closely into the political sphere of neighbouring Rus' principalities, and north-western Estonia into the agendas of the Danish Crown and Church (see Chapter 8). In Prussia, there was a similar reorganization of power from the eleventh century (see Chapter 15). Key strongholds dominated their macro-regions, which have largely been understood in relation to the tribal territories documented in external written sources. The borderlands of Pomerania–Pomesania and Masovia–Lubavia–Sasna, as well as the Kulmerland, were characterised by earlier colonization driven by the expansion of the Polish Piast state in this region from the end of the tenth century, subsequently undermined in the later twelfth century by Prussian raiding. Polish military campaigns, aimed at subduing the Prussians and framed in the rhetoric of holy war, generally failed. In the east, Sudovia was hit by Rus' invasions, and the process of depopulating the 'Great Wilderness' may have begun before the crusades. The final phase of indigenous society is also marked by missionary activity, which in the case of Prussia began with St Adalbert (Pol. Wojciech, d. 997) and St Bruno of Querfurt (d. 1009), and was taken up again by the Cistercians from the 1140s. In Livonia, missionary activities began under the Cistercians in the 1170s, intensified in the 1190s, and, as in the case of the Prussian Cistercian missions, became the springboard for the crusades. The final phase of indigenous society can therefore be characterized as a period of increasing centralization of political authority, although in Prussia and Livonia this did not result in 'state formation', as was the case in Lithuania. This period also saw increasing involvement in the political and cultural spheres of neighbouring Christian regions, both Latin and Eastern Orthodox.

Phase 2: The Period of Active Crusading

The second phase is associated with the period of active crusading in the thirteenth century, from *c.* 1230 to 1283 in Prussia, ending in the conquest of Sudovia, and from *c.* 1200 (with the active involvement of Bishop Albert von Buxthoeven) to 1290 in Livonia, ending in the subjugation of Semigallia. Within this period there are numerous military campaigns, often categorized as discrete crusades against specific regions or tribal groups, as well as marked episodes of organised indigenous resistance (see Chapter 1). From an archaeological perspective, this phase is marked by an initial wave of colonization associated with the crusading process, represented by the construction of *castra*. In Livonia, stone castles were built from the very onset of crusading, but in Prussia there was clearly a rapid assimilation of Prussian, Slavic, and German timber and earth fort-building techniques in the early stages of the crusade. In both regions there is evidence for the reuse of abandoned or destroyed strongholds (Pluskowski 2013, 98). These *castra* were associated with *suburbia* and *preurbia*, effectively fortified towns, some of which may have functioned as outer baileys in their initial phases (Kowalczyk-Heyman 2006). These composite colonies, providing the commercial infrastructure required to sustain crusading efforts and taking advantage of the booming war economy, represented bridgeheads for subsequent crusades. Their earliest mercantile complexes consisted of residences attached to trading posts and company houses of foreign traders, functioning as working centres for trade and providing security and legal protection. The oldest of these outposts were timber-framed structures built on stone foundations as in Elbing (Pol. Elbląg), or in brick as in Thorn (Pol. Toruń); later they would develop into more elaborate complexes (Pilecka 2005). Only a relatively small number of towns were founded during this time by the Teutonic Order and individual bishops, collaborating with magnates and merchants involved in crusading enterprises (Czaja 2009, 178). Riga was founded by Bishop Albert in collaboration with the local Liv communities and German merchants; Polish princes and the Burgrave of Magdeburg, together with the Order, founded Thorn, Kulm (Pol. Chełmno), and Marienwerder (Pol. Kwidzyn), whilst the crusading contingent of Henry III, the burgrave of Meissen, founded Elbing. In Prussia the process of urbanization was largely focused on the Kulmerland and Lower Vistula; in Livonia this was mapped onto former central places: from the Lower Daugava, north-east to Wenden

(Lat. Cēsis), then to Fellin and Reval (Est. Tallin). The crusading period is synonymous with the emergence of the Hanse, which by the mid-thirteenth century had established a well-connected trading system in the Baltic (Jahnke 2015, 198).

In Prussia and the fringes of western Lithuania (mostly southern Curonian territory), all major indigenous political centres were destroyed or abandoned during this period. In Livonia, many were destroyed or abandoned, but a number continued to function throughout the thirteenth century and some even into the fourteenth century, reflecting the incorporation of indigenous power structures into the new system of vassalage (Šnē 2009a; Pluskowski and Valk 2016). In some cases, it is possible to link the destruction or abandonment of specific sites to the historical events of the crusades, of which the best example is the stronghold at Viljandi (Ger. Fellin; see Chapter 4). The physical traces and progress of the Sword Brothers' siege in 1223 have been identified archaeologically, whilst the subsequent refortification of the stronghold and its eventual transformation into a convent by the Livonian branch of the Order are poorly known (Lang and Valk 2011). Opinion remains divided on the date of the associated town's foundation, although it was not a substantial settlement until the end of the thirteenth century (Valk 2005; Haak and Russow 2013). The crusading narratives describe significant losses amongst the indigenous population, but in Livonia and eastern Prussia a large proportion survived, although the class of indigenous Prussian *nobiles* was heavily depleted and resettled to other regions, with some areas becoming depopulated. By the end of the thirteenth century, indigenous territories had been reorganized into Christian polities, and the key urban centres established during the period of active crusading were thriving. The regions had been subdivided into dioceses (four in Prussia, five in Livonia, including the Danish-oriented Bishopric of Reval), the episcopates had established diocesenal centres, and a number of churches had been built and rebuilt. Within this broader phase of cultural transformation, the specific impact of the crusades has to be considered on a regional and even local basis, and this is discussed in more detail under the individual case studies (Chapters 4 and 10).

Phase 3: The Period of Colonization and Consolidation

The third phase can be defined as a sustained process of colonization in Prussia which begins in the 1280s and continues into the later fifteenth century (Czaja 2009), and in Livonia as the consolidation of power through the expansion of existing castles, the construction of new ones, and the growth of existing urban centres in Livonia. Both regions also saw the proliferation of Christian parochial infrastructure, with the aim of serving the spiritual needs of the colonists and facilitating the conversion of the indigenous population. In Prussia, this was most intensive in the Kulmerland and the valley and delta of the Lower Vistula, as well as in the territory of the bishops of Ermland, and consisted of largely peasant communities. Further east, colonization began in earnest from the mid-fourteenth century and on the eastern frontier of Prussia, from the Masovian border in the south to the Memelland (Lith. Klaipėda region) in the north, it remained limited until the later fifteenth and sixteenth centuries. The relocation of the Order's headquarters from Venice to Marienburg (Pol. Malbork) in 1309 led to significant investment in the expansion of the convent over much of the fourteenth century. The Order's annexation of Pomerelia also has clear archaeological dimensions, resulting in the construction of multiple castles as well as the reorganization of Danzig (Pol. Gdańsk) following the siege of 1308, defining a wholly new phase of urban development. By the start of the fifteenth century ninety-four towns had been founded within the Prussian territories of the Order and episcopate, with the majority on the territory of the former established between the 1320s and 1350s, although the process of urbanization would continue into the first half of the fifteenth century with the foundations of the towns of Stuhm (Pol. Sztum), Johannisburg (Pol. Pisz), and Lyck.

In Livonia, whilst there were of course regional differences in the diachronic development of territorial administration (Šnē 2008, 90–92; 2009b, 133), this phase can be defined by investment in castle expansion and urbanization, which was limited to the nineteen towns established by the mid-sixteenth century (ten within the Order's territory), each of them associated with a castle. A small number of indigenous strongholds also continued to be occupied into the fourteenth century. There was no colonization of the countryside by immigrant peasants and knights, most probably due to the prodigious costs of travelling across the Baltic (compared to Prussia, which was more easily accessible) as well as the presence of the surviving indigenous population (Biskup 2013, 161). The indigenous rural settlement pattern appears to have remained more or less intact, except in those southern borderlands where the population density did not reach pre-crusade levels for several centuries. Whilst indigenous people also migrated to towns, this led to some level of cultural segregation between urban and rural communities (one sharpened from the mid-fourteenth century by the policy of legal segregation), although provisioning networks linking the two were clearly functional and necessitated a level of collaboration. The mid-fourteenth century saw the incorporation of Danish Estonia into the Order's territories, which prompted a phase of refortification and the reorganization of land administration. In summary, this third phase can be defined as the period of the established *Ordensland*, which sees the consolidation of the Christian theocratic regime, the creation of a centralized political, commercial, and bureaucratic system in Prussia, paralleled by a connected but more loosely organized system in Livonia, and the generation of significant wealth resulting from the intensive exploitation of natural resources.

Phase 4: Political Decline and Secularization

The fourth phase encompasses a period of fiscal and political decline for the Teutonic Order, traditionally dated to 1410–1525 in Prussia where it was punctuated by numerous conflicts (see above), whilst Livonia saw increasing political autonomy culminating in the reign of Plettenberg (1494–1535), alongside the introduction of the manorial system, which promoted a system of serfdom. This developed more quickly on the lands of the Archbishop of Riga than within the Order's commanderies, but by the fifteenth century the manors of the descendants of crusaders and German immigrants become more numerous than those of native vassals (Šnē 2008, 94). Theocratic rule came to an end in the sixteenth century, with the secularization of the Prussian branch in 1525 and the Livonian branch in 1561. In some sources, such as in the detailed inventories of Prussian castles, the relative impact of 1410 on their function and resources is striking (see Chapter 15). This is most evident from a fiscal perspective, where the last two decades of the fourteenth century saw a dramatic expansion in agriculture and charges levied on the Order's subjects, followed by a sudden decline in the fifteenth century, which was not reversed despite some financial stabilization *c.* 1450 (Sarnowsky 1993, 435–36). In the first half of the fifteenth century, knightly families in Prussia took a more active role in exploiting their estates for trade, and from 1466 a distinctive 'knightly class' emerged in eastern Prussia, whose descendants would eventually come to be known as *Junkers* (Jóźwiak 2009, 458). From an archaeological perspective, this gradual decline is less clearly defined, and in Livonia the first few decades of the sixteenth century are regarded as the zenith of the Livonian Master's rule, independent of Prussia. It is a period associated with lavish investment in the castle at Wenden, followed shortly after by the devastation of the Livonian War. The stabilization of the border with Samogitia/Lithuania stimulated a process of colonization in the borderlands and led to the intensification of settlement from the sixteenth century, as well as a boom in the life of coastal towns, particularly Gdańsk (Pluskowski 2013, 353–55). The final stage is therefore more complex as it encompasses a historically volatile period, episodes of fiscal and demographic stagnation, but also investment in new fortifications in response to the proliferation of artillery and the foundation of new settlements, towns, and religious institutions. It concludes with another reconfiguration of power relations followed by new forms of investment and territorial reorganization. The impact of the wars with Poland, Lithuania, and Muscovite Russia on settlement was devastating, but the extent to which this is visible in long-term environmental trends is not straightforward and must be considered almost on a site-by-site basis (see Chapter 17). The acceptance of Lutheranism by the two branches of the Order effectively ended the system of ecclesiastical organization within the newly created duchies which had defined the medieval *Ordensland*. Lutheran bishops continued to administer these dioceses into the later sixteenth century, and even with the administrative reorganization under Polish rule and the assertion of the Counter Reformation, the resulting administration represented a break from the medieval episcopates.

The time-span considered by this study extends for several centuries on either side of these four phases, enabling them to be contextualized within the longue durée of cultural and ecological trends in the eastern Baltic. These phases of 'social time' are presented as a framework, against which the detailed ecological histories of individual commanderies may be compared.

Creating a Chronology for Case Study Sites and their Territories

The chronological framework for archaeological and palaeoenvironmental data, to complement the information obtained from precisely dated historical sources (see Chapter 2), was developed using multi-proxy techniques, which included a combination of absolute and relative dating. This approach was partly necessitated by the uneven range of materials that could be precisely dated at individual sites and by the need to synchronize environmental transformations with the phases of 'social time' outlined above.

Absolute Dating Methods

Radiocarbon Dating

A total of 204 AMS radiocarbon dates were obtained within the framework of the project, of which the majority (174) were used for dating peat and lake cores, and thirty-five for dating animal bone and seeds from archaeological contexts (Appendix A). Large quantities of radiocarbon dates are required to develop robust chronological frameworks for interpreting the palynological sequences, and particularly to situate this within the broader cultural context and allow for comparison with the archaeological and historical data. The majority of radiocarbon dates (121: 58%) are from pollen sequences and sites in Poland, with a smaller number from Latvia (60: 29%) and Estonia (28: 13%). Although this reflects the greater number of pollen sequences studied from Poland, the geographical distribution of dates is more even when considered in the context of Prussia and Livonia (121: 88).

The underlying principles and methods of radiocarbon dating have been well covered in numerous publications (e.g., Libby 1955; Lowe and Walker 1997; Björck and Wohlfarth 2001) and are considered here more specifically in relation to potential sources of error in dating. The majority of AMS dates were derived from bulk samples of peat and gyttja. Lake sediments produced very few plant macrofossils and in many cases the dating of these failed due to insufficient carbon. Plant macrofossils were equally difficult to retrieve from peat bogs. Limited material was available because of the smaller auger used for coring peat (5 cm dia. compared to 10 cm dia. Russian used for lakes), further constrained by the requirement for large numbers of often closely spaced dates and the need to preserve sediment for additional analytical techniques.

Material for dating was carefully selected to minimise potential sources of error and a number of dates (twenty) were subsequently excluded (Appendix A). Areas of root penetration and bioturbation were identified to avoid sediments potentially contaminated by organic material of younger date. Dating of wood and charcoal has been minimized to avoid problems arising through samples derived from old wood, potentially already centuries old at the time of deposition. Older dates were produced on plant macrofossil samples from securely dated late medieval contexts from Tczew (Ger. Dirschau) and Malbork. The granary at Tczew, constructed at the beginning of the fifteenth century (Jonakowski, 2009), includes a defined layer of charred cereals (see Chapter 13). Two dates on charred *Secale cereale* grains produced reliable fifteenth century dates (Appendix A, Table A.1), but a third date on a single charred *Triticum aestivum* seed produced a date of 1664±33 BP (GU-29831, cal. AD 257–529) and clearly reflects intrusive material, perhaps transported with the grain prior to storage or introduced through animal agencies. Likewise, the construction of the outer moat at Malbork was established as *c.* 1320–40 (Jóźwiak and Trupinda 2007), but a sample of seeds from the base of an organic layer within the moat produced a date of 899±28 BP (GU-31437, cal. AD 1040–1212), predating moat construction by as much as three centuries. In this situation it is likely the sample reflects the erosion and redeposition of material within the moat from a secondary context of earlier pre-crusading date.

Mixing of sediments is also apparent from Lake Nineris (Latvia) where the top four of six samples display a clear inversion, with dates of 972±33 BP (988–89 cm), 905±35 BP (1000–1001 cm), 805±33 BP (1045–1046 cm), and 745±35 (1086–1087 cm), followed by a date of 1546±33 BP (1106–1107 cm). The first two dates come from the top metre of the sequence (950–1050 cm), the third and fourth dates from the second metre (1030–1130 cm), and the fifth and six dates from the third metre (1110–1210 cm). The sediments were extremely organic and highly unconsolidated, and it seems likely mixing occurred during both the initial sampling and subsequent transport of the samples to the laboratory. Further errors in dating were apparent from Lakes Kosilase, Viljandi, Trikāta, and Drużno. Older dates are frequently produced on gyttja and aquatic material, either where lakes are enriched in ancient dissolved carbonates (from carbonate-rich rocks too old to contain 14C), resulting in a hardwater reservoir effect, or because of the slower carbon exchange between the atmosphere and the lake, creating a freshwater reservoir effect. The impact of a freshwater/hardwater reservoir effect is illustrated most clearly at Lake Trikāta where plant macrofossils retrieved from a parallel core typically produced dates 1000–1500 younger than the bulk gyttja samples (Appendix A, Table A.4). The single date from Lake Viljandi (900–01 cm, cal. AD 780–980) also appears several centuries too old given the single grains of buckwheat recorded at 936, 920, and 896 cm, a crop

Figure 3.1: Timber structures uncovered during excavations of the Liv quarter in the south-east corner of the Old Town of Riga in 2012. Photo by U. Kalējs.

not recorded in the eastern Baltic before the thirteenth or fourteenth century (Sillasoo and Hiie 2007), and more frequently present in post-medieval deposits.

An alternative method for dating lake sediments involves the direct dating of pollen concentrates separated from aquatic plant matter, and has been shown to produce viable results, including from hardwater lakes where contamination by old carbon is frequently an issue (e.g., Regnell 1992; Mensing and Southon 1999; Vandergoes and Prior 2003). Several samples of pollen concentrates were sent for dating from Lakes Kosilase, Trikāta, Staświny, and Drużno, using a heavy liquid flotation method modified from Nakagawa et al. (1998). This produced viable results from Lake Staświny but was unsuccessful from Lakes Trikāta and Kosilase where the high organic content of the sediments made it difficult to effectively separate pollen from aquatic plant matter. Despite the low organic content of sediments from Lake Drużno, the single date from 200 cm proved as much as two thousand years too old, producing a Bronze Age date from a level with a characteristic hornbeam peak widely dated in other pollen sequences to the Migration period (Chapter 11).

Radiocarbon dates were calibrated using the IntCal13 calibration dataset (Reimer et al. 2013), with age-depth models constructed for pollen sequences using a Bayesian statistical approach within the Bacon computer program (Blaauw and Christen 2011). The age-depth models produced comparable results that compare favourably with existing palynological studies from across the Baltic, demonstrating the viability of the radiocarbon results.

Uranium-Series Dating (U-series)

During the course of the project a new iteration of U-series dating was developed at the University of Reading, using slices of cortical bone and shell with a

Figure 3.2: Timbers uncovered during excavations in (a–b) the outer bailey of the castle in Elbląg in 2013; (c) Ełk in 2011. Photos by A. Pluskowski.

date range obtained from modelling the series of decaying uranium isotopes. U is taken up by animals during life from groundwater and ingested via food, both of which have a relatively low content of dissolved U (Ivanovich and Harmon, 1992). Dissolved thorium levels and food concentrations, however, are very low, and uptake from solution and food are negligible. The dating method of bone is based on the gradual growth of ^{230}Th from the decay of its parents ^{238}U. Since bone, when newly deposited, contains only small amounts of ^{238}U (and daughter products), virtually all the U found in archaeological bones results from deposition post-death and subsequent growth of the daughter radionuclides (see Hedges and van Iclinken, 1994). A total of sixty-nine animal bone samples were dated using this method, deriving from the castle and hillfort at Cēsis (Latvia), Tartu (Estonia), and Poganowo (Poland), with twenty-four shell samples deriving from cores obtained from Lake Drużno (Poland). The dates with the smallest ranges and margins of error were obtained from cattle long bones, although the majority of dated samples derived from horses. This technique was initially piloted on the horse remains from the 'stable' in the outer bailey of Cēsis Castle and compared with AMS and relative dates taken from the same archaeological layer (see below). Subsequently, it was applied to animal bone from sites across the eastern Baltic. This method proved to be the most accurate method of absolute dating used within the project. The full details of this method are outlined in Appendix B.

Dendrochronology

The dating of timber structures is widely used in the eastern Baltic given how often these survived in waterlogged soils. Timber was the most important building material used by the indigenous societies of the eastern Baltic before the crusades, and was also used by the incoming colonists for building castles, vernacular housing, bridges, palisades, streets, and also for providing stable foundations in flooded soils. Although a number of towns in the *Ordensland* banned the use of timber as a building material, as it was a fire hazard, it continued to be used in the construction of fortifications and houses into the sixteenth century. The most extensive dendrochronological sequence in Prussia comes from the Old Town of Elbląg (Ważny 2002, 2004, 2005); in Livonia, excavations in Riga have provided the largest number of datable timbers (Zunde 2019). In the case of the former, this enabled a potter's workshop, containing the largest discrete assemblage of pottery recovered from a colony established during the crusading period, to be precisely dated from the turn of the 1280s until 1288 (Marcinkowski 2003). In the context of the project, datable timbers were uncovered during excavations in the moat and outer bailey of Cēsis Castle, in the Liv quarter in Riga (Fig. 3.1), within parts of the outer bailey of the castle at Elbląg, and in various parts of the castle island at Ełk (Fig. 3.2). Dates were provided by Māris Zunde for Livonian sites and by Tomasz Ważny for Prussian sites.

Coins

The first coins in the eastern Baltic were struck in Riga in the early thirteenth century, although the majority circulating at this time in Livonia and in the early decades of crusading in Prussia were based on the Cologne penny, the standard preferred by German merchants, particularly from Lübeck. The Teutonic Order minted its own coins (initially bracteates) from the mid-thirteenth century as an integral part of its command economy and visual expression of authority imposed on the conquered territories. As a result, coins are frequently encountered on medieval colonizing sites (castles, towns, villages) in the eastern Baltic, although not typically in large quantities (Fig. 3.3). A fully functioning monetary economy took several decades to develop, slowly replacing the indigenous use of raw precious metals as a form of currency. In Livonia, coins are also primarily found in association with castles, towns and churches or chapels, although they also feature as grave goods in rural cemeteries. However, the autonomy of the various political authorities in this region is represented by the fact that each issued coins, in contrast to Prussia. Ingots, valued by weight, continued to be used alongside coins in the late thirteenth century, whilst the documented use of coins significantly increased in the *Ordensland* from the fourteenth century as payment in raw silver and gold was gradually abandoned (Paszkiewicz, 2009, 55–56). Coins found on medieval sites from the mid-thirteenth century are used for both dating and defining the presence of, or contact with, the Order or Livonian bishops. A chronology has been constructed for the Order's coins in Prussia (Paszkiewicz 2009) and Livonia (Urban 1993); within the context of the project, coins were found at all sites associated with the Order and contributed to dating specific contexts.

Documentary Sources

The most precise sources for the historical chronology of the medieval eastern Baltic are documentary records, which can be associated with a specific day, month, and year. The majority of these survive from the later decades of the fourteenth century, with many deriving from the very last decades of the Teutonic Order's rule in the sixteenth century. Often there are significant gaps in the timelines of documents for individual sites. Where a series of documents exists for an individual castle with a discrete time span, it is possible to determine short-term provisioning, commercial, and other economic trends, which in turn can be used to assign a *relative* function to that castle and its associated estates within the broader

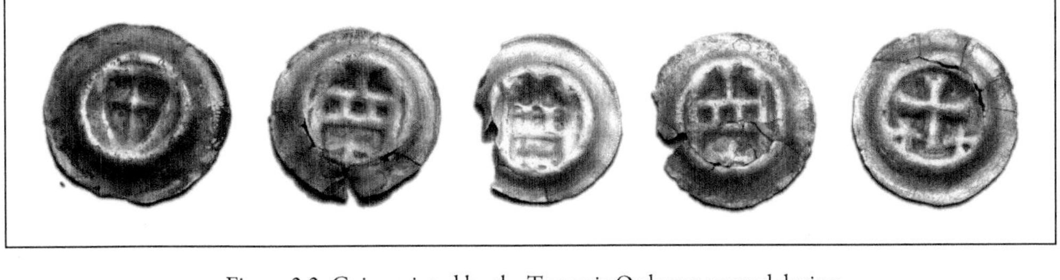

Figure 3.3: Coins minted by the Teutonic Order uncovered during the series of excavations in Biała Góra. Photo by L. Okoński.

Figure 3.4: An example of the 'Wendish bond' from the castle at Malbork.

Figure 3.5: An example of the 'Gothic bond' from the castle at Malbork. Photos by M. Elander.

framework of the *Ordensland*. These trends can, in turn, be situated within the longer-term chronologies constructed for palaeoenvironmental and archaeological data. This type of synchronization has not been possible for every site but has been attempted for a select number of case studies with the highest number of dating proxies. The majority of these documents exist for Prussia, with comparatively little surviving from the Order's Livonian houses.

Relative Dating of Material Culture

Many sites in the eastern Baltic have been dated solely on the basis of the style of excavated material culture. Some artefacts have relatively short time spans associated with them, whilst others remain unchanged in form for several centuries. Moreover, ceramics, which are the most frequently used relative dating proxy, do not yet have a robust chronology associated with them, particularly in the century before and after the crusades. The best known ceramic assemblage with a discrete chronology was derived from a potter's yard excavated at Elbląg and dated by the dendrochronology of the associated timbers (Marcinkowski 2003). Material culture recovered from the project's sites provided broad chronological ranges and was an invaluable tool for monitoring and modifying the direction and pace of excavations. However, the chronological framework for each site was always linked to several dates from at least one type of absolute dating proxy.

Bricks

In both Prussia and Livonia brick was used as a building material from the mid-thirteenth century. In the former region it came to typify the architecture of castles, churches, town halls, houses, and walls, largely due to the absence of suitable building stone. Dating these structures has always involved a reliance on their mention in written sources. From this, the style of laying bricks, where they have survived intact in situ, has been typically used to assign a relative chronology. The thirteenth-century so-called 'Wendish bond', where bricks are laid in sequences of two stretchers and a single header, is defined by large bricks and associated with constructions sponsored by the Cistercians, Dominicans, and Franciscans. This appears in Masovia in the first half of the thirteenth century (Brykowska 2002, 33) and is subsequently used in the territories conquered by the Teutonic Order (Fig. 3.4). The second castle built during the Prussian Crusade — at Thorn — may even have been constructed from brick at the outset (Chudziakowa 1983, 17–19). The Wendish bond was slowly replaced by the 'Dutch or Flemish bond', more popularly called 'Gothic', in the fourteenth and fifteenth centuries (earlier in southern Poland), which consisted of alternating single headers and stretchers, with subsequent courses offset to position headers between stretchers (Fig. 3.5).

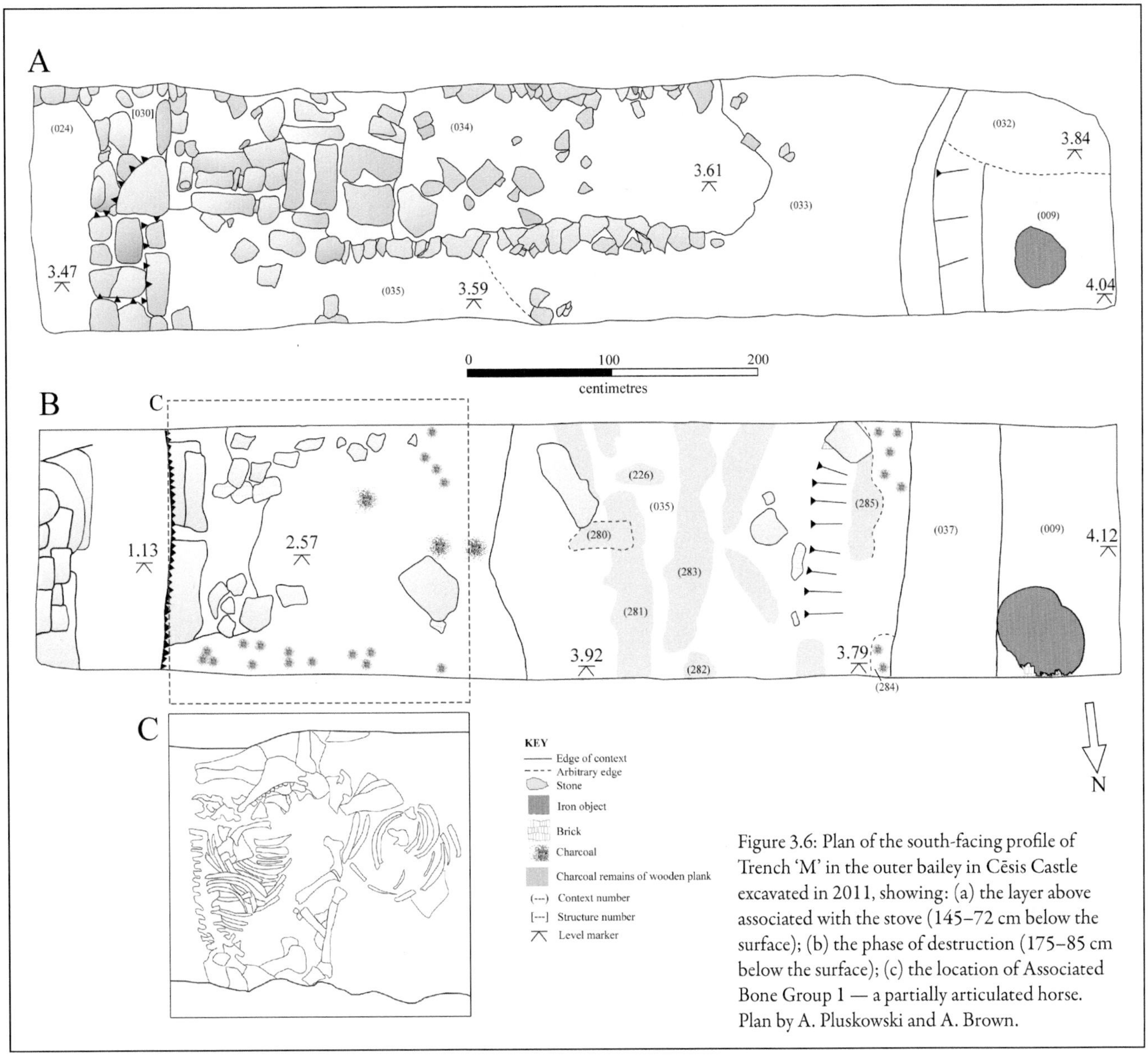

Figure 3.6: Plan of the south-facing profile of Trench 'M' in the outer bailey in Cēsis Castle excavated in 2011, showing: (a) the layer above associated with the stove (145–72 cm below the surface); (b) the phase of destruction (175–85 cm below the surface); (c) the location of Associated Bone Group 1 — a partially articulated horse. Plan by A. Pluskowski and A. Brown.

Bricks made in the thirteenth century tend to be larger than in the fourteenth and fifteenth century. This has been demonstrated where a substantial diachronic assemblage of material is available, as in Gdańsk (Ger. Danzig), where from the fourteenth to fifteenth century the thickness of bricks changed from 10.5 cm to 6–7 cm, their width from 15–16 cm to 12 cm, and their length from 32–33 cm to 26–27 cm (Paner 2001, 499). There is even evidence for the regulation of brick sizes by the Order. In 1423, the Grand Master specified a standard for brick size, translated as 30 × 15 × 7.5 cm. Similar stylistic dating has been applied to ceramic roof tiles, both Prussian and Livonian examples (see Hayward 2019).

There are of course problems with this type of dating. Comparative studies of thirteenth- to sixteenth-century brick structures in Greater Poland demonstrated that establishing a regional chronology was problematic, and that the use of bricks for dating should be focused on individual structures and their immediate vicinity (Żemigała 2008). In some cases, bricks appear to have been reused and recycled, and there is documentary evidence for the dismantling of castles to provide building material for new structures. For example, the Order commissioned the construction of part of the interior walls of Tilsit Castle (Rus. Sovetsk) from elements recovered from the dismantled stronghold at Neuhaus (Clasen-Sandt 1931, 214). In some structures both bonds are found together, as at the episcopal castle at Wąbrzeźno (Ger. Briesen), although along with the irregularity of the brick composition, this lack of standardization has

Figure 3.7: Decorative equestrian artefacts recovered from the collapsed stable (a–b in situ and c–d after conservation). Photo by G. Kalniņš.

been interpreted as reflecting low-quality craftsmanship (Wiewióra 2012). Absolute dating of bricks from the eastern Baltic has only been attempted on material from Malbork and Toruń using thermoluminescence (TL) (Chruścińska et al. 2008). The results confirmed the relative dates established by architectural historians on the basis of style and historical sources. Although promising, this time-consuming and expensive method has yet to be applied to a broader range of structures. Re-hydroxilation was attempted as an alternative method of dating fired ceramics (Wilson et al. 2005), but the initial results indicated several problems with this method and it was abandoned. A study of building fabric, with a particular focus on bricks, was undertaken during the course of this project, and the results are presented in the second volume (Hayward 2019).

Conclusion: Synchronizing Multi-Proxy Dates

The varying precision of the different dating methods presents challenges for developing a cohesive chronological framework across all the case-study sites included within the project, particularly when attempting to synchronize archaeological sites, historical documents, and pollen cores. At the same time, multi-proxy dating allowed increased confidence in the robustness of the results. For example, the dating of the horse remains from the outer bailey in Cēsis Castle in 2011 prompted a change in the initial interpretation linking this with the documented Russian siege of 1577–78 (Fig. 3.6). The recovery of partially articulated horse skeletons during excavations in 1927 and 2005 within the castle's Western Range (alongside human remains), famously detonated during the siege, provided grounds for anticipating a comparable destruction horizon in other parts of the castle. However, AMS dating of the horse bones from the outer bailey resulted in a calibrated date of 407±39 (*c.* 1429–1524 (71.4%), *c.* 1558–1631 24%), a range supported by twelve U-series dates on the same material with a mean of 447±35 (*c.* 1468–1538). Associated fragments of the collapsed building, particularly burnt timbers and ceramic roof tiles, could not be dated. However, a set of decorated horse equipment found in situ could be stylistically dated to the later fifteenth/early sixteenth century (Fig. 3.7).

Whilst the exact context of the destruction of the uncovered stable may only be clarified with further excavation, chronologically this event is most plausibly associated with the final phase of the Livonian branch of the Order, linked to the reign of Plettenberg (1494–1535). Interestingly, the upper levels of the southern tower (the 'Tall Hermann') of the main castle collapsed into the moat at this time. This has been linked to a written reference to a lightning strike in 1538, which damaged a tower which contained an arsenal, including gunpowder. The rebuilding of this tower could be dated by the dendrochronology of the associated timbers within its windows to *c.* 1540. There is no written reference to a destroyed stable or its subsequent rebuilding, except this may be alluded to within a single document in the National Archives of Sweden, which mentions renovation works in a range, block, or wing of the castle that had collapsed in 1545. This record could also refer to the rebuilding of the southern tower. On the basis of the multi-proxy dating, cut marks on three horse bones indicative of meat removal and marrow extraction, as well as mid-shaft charring, could not be readily linked with the documented consumption of 124 horses in the town following a famine in February 1578. This example demonstrates how difficult it is to synchronize archaeological, architectural, and historical data, and how alternative chronologies based on multi-proxy dating can challenge traditional assumptions driven by cultural-historical intellectual paradigms. What is clear is that the remains of the building containing the horse carcasses were levelled and reconstructed immediately after it collapsed, at which point the function of the new building became residential. This too was subsequently destroyed, although whether this phase can be linked to the impact of the Livonian War remains to be verified. At the same time, linking the assemblage with specific events was of limited relevance within the context of the project, as the horse remains (and associated ecofacts) contributed to a broader understanding of the equine population within the Order's sites in Livonia, and can be situated within phase 4 of the chronological framework for the medieval eastern Baltic outlined above (Pluskowski et al. 2018).

In summary, separate archaeological chronologies were constructed for each site and for each pollen core. Documentary sources were situated within separate chronologies, which could be tabulated in the case of the Prussian castles included within this study (see Chapter 15). These were then situated within the broader phases of 'social time' outlined above. The emphasis throughout was on the longue durée perspective, which enabled the impact of the crusades and the creation of the *Ordensland* to be contextualized, that is, the shift from the Late Iron Age to the Middle Ages. Whilst some categories of data with short date ranges (such as written sources) indicated a significant impact within the duration of the theocratic polity, others with a longer date range (such as pollen) demonstrated that significant ecological transformations in the eastern Baltic had already begun several centuries before the arrival of crusading armies. One of the most important challenges encountered during the course of this project has been to integrate these contrasting diachronic scales.

Works Cited

Biskup, M. 2013. 'Władztwa terytorialne w średniowiecznych Inflantach i ich granice', in *Zakon krzyżacki w Prusach i Inflantach: podziały administracyjne i kościelne w XIII–XVI wieku*, eds R. Czaja and A. Radzimiński, Toruń: Wydawnictwo Naukowe Uniwersytetu Mikołaja Kopernika, 153–65

Björck, S., and Wohlfarth, B. W. 2001. '14C Chronostratigraphic Techniques in Palaeolimnology', in *Tracking Environmental Change Using Lake Sediments* (vol. I), eds W. M. Last and J. P. Smol, Amsterdam: Kluuwer, 205–46

Blaauw, M., and Christen, A. 2011. 'Flexible Palaeoclimate Age-Depth Models Using an Auto-Regressive Gamma Process', *Bayesian Analysis* 6, 457–74

Blomkvist, N. 2009. '12th–14th-Century European Expansion and its Reception in the Baltic North', in *The Reception of Medieval Europe in the Baltic Sea Region*, ed. J. Staecker, Visby: Gotland University Press, 431–84

Braudel, F. 2009. 'History and the Social Sciences: The Longue Durée', trans. by Immanuel Wallerstein, *Review* 32/2, 171–203

Brykowska, M. 2002. 'Studia nad wschodnim zasięgiem architektury ceglanej', in *Cegła w architekturze środkowo-wschodniej Europy: historia, metody, badań, konserwacja*, eds M. Arszyński and M. Mierzwiński, Malbork: Muzeum Zamkowe w Malborku, 30–42

Buko, A. 2005. *Archeologia Polski wczesnośredniowiecznej*, Warsaw: Trio (published in English as *The Archaeology of Early Medieval Poland* by Brill in 2008)

Chruścińska, A., Jesionowski, B., Oczkowski, H., and Przegiętka, K. 2008. 'Using the TL Single-Aliquot Regenerative-Dose Protocol for the Verification of the Chronology of the Teutonic Order Castle in Malbork', *Geochronometria* 30/1, 61–67

Chudziakowa, J. 1983. 'Toruń najdawniejszy i wczesnośredniowieczny', in *Toruń dawny i dzisiejszy: Zarys dziejów*, ed. M. Biskup, Warsaw: Państwowe Wydawnictow Naukowe, 11–30

Clasen-Sandt, K. 1931. 'Zur Baugeschichte der Memelburgen Ragnit, Splitter und Tilsit', *Prussia* 29, 196–222

Czaja, R. 2009. 'Urbanizacja kraju', in *Państwo zakonu krzyżackiego w Prusach: Władza i społeczeństwo*, eds M. Biskup et al., Warsaw: Państwowe Wydawnictwo Naukowe, 177–230

Gerrard, C. 2003. *Medieval Archaeology: Understanding Traditions and Contemporary Approaches*, London: Routledge

Haak, A., and Russow, E. 2013. 'On the Development of the Town of Viljandi in the Light of Earliest Archaeological Find Complexes', *Estonian Journal of Archaeology* 17/1, 57–86

Hayward, K. M. J. 2019. 'Resources for Castle Building in Prussia and Livonia', in *Ecologies of Crusading, Colonization, and Religious Conversion in the Medieval Baltic: Terra Sacra II*, ed. by A. Pluskowski, Environmental History of the North Atlantic World, 3, Turnhout, Brepols (forthcoming)

Hedges, R., and van Klinken, G. J., eds 1994. Special issue on bone diagenesis, *Journal of Archaeological Science* 22/2, 147–340

Ivanovich, M., and Harmon, R. S., eds 1992. *Uranium Series Disequilibrium Applications to Environmental Problems*, Oxford: Oxford University Press

Jahnke, C. 2015. 'The Baltic Trade', in *A Companion to the Hanseatic League*, ed. D. J. Harreld, Leiden: Brill, 194–240

Jonakowski, M. 2009. *Sprawozdanie wstępne z przeprowadzonych archeologicznych badań wykopaliskowych na obszarze działki nr 418/14 przy ul. Zamkowej w Tczewie*, unpublished report

Jóźwiak, S. 2009. 'Rycerstwo-szlachta', in *Państwo zakonu krzyżackiego w Prusach: Władza i społeczeństwo*, eds M. Biskup et al., Warsaw: Państwowe Wydawnictwo Naukowe, 450–59

Jóźwiak, S., and Trupinda, J. 2007. *Organizacja życia na zamku krzyżackim w Malborku w czasach wielkich mistrzów (1309–1457)*, Malbork: Muzeum Zamkowe w Malborku

Karczewska, M., and Karczewski, M., eds. 2004. *Ceramika zachodniobałtyjska: nowe źródlła i interpretacje: materiały z konferencji, Białystok 23–24 września 2002 roku*. Białystok: Uniwersytet w Białymstoku, Instytut Historii

Kowalczyk-Heyman, E. 2006. 'Na styku historii z archeologią, czyli o przkłamaniach terminologicznych w tłumaczeniach źródeł średniowiecznych (na przykladzie Kroniki Ziemi Pruskiej Piotra z Dusburga)', *Kwartalnik Historii Kultury Materialnej* 2, 221–26

Lang, V., and Valk, H. 2011. 'An Archaeological Reading of the Chronicle of Henry of Livonia: Events, Traces, Contexts and Interpretations', in *Crusading and Chronicle Writing on the Medieval Baltic Frontier*, eds M. Tamm, L. Kaljundi, and C. S. Jensen, Aldershot: Ashgate, 291–316

Lee, R. E. 2012. 'Introduction: Fernand Braudel, the *Longue Durée* and World-Systems Analysis', in *The Longue Durée and World-Systems Analysis*, ed. R. E. Lee, Albany: SUNY Press, 1–8

Libby, W. 1955. *Radiocarbon Dating*, Chicago: University of Chicago Press (second edition)

Lowe, J. J., and Walker, M. J. C. 1997. *Reconstructing Quaternary Environments*, London: Prentice and Hall (2nd edn)

Marcinkowski, M. 2003. 'Średniowieczny warsztat garncarski ze Starego Miasta w Elblągu', *Pomorania Antiqua* 19, 194–250

Mensing, S. A., and Southon, J. R. 1999. 'A Simple Method to Separate Pollen for AMS Radiocarbon Dating and its Application to Lacustrine and Marine Sediments', *Radiocarbon* 41, 1–8

Mugurēvičs, Ē. 1990. 'Interactions between Indigenous and Western Culture in Livonia in the 13th to 16th Centuries', in *From the Baltic to the Black Sea*, eds D. Austin and L. Alcock, London: Routledge, 168–78

Nakagawa, T., Brugiapaglia, E., Digerfeldt, G., Reille, M., de Beaulieu, J.-L., and Yasuda, Y. 1998. 'Dense-Media Separation as a More Efficient Pollen Extraction Method for Use with Organic Sediment/Deposit Samples: Comparison with the Conventional Method', *Boreas* 27, 15–24

Paner, H. 2001. '10th- to 17th-Century Domestic Architecture in Gdańsk', in *Lübecker Kolloquium zur Stadtarchäologie im Hanseraum*, III: *Der Hausbau*, eds B. Dahmen, M. Gläser, U. Oltmanns, and S. Schindel, Lübeck: Schmidt-Römhild, 491–509

Paravicini, W. 1989. *Die Preussenreisen des europäischen Adels* (vol. I), Sigmaringen: Thorbecke

Paszkiewicz, B. 2009. *Brakteaty — pieniądz średniowiecznych Prus*, Wrocław: Wydawnictwo Uniwersytetu Wrocławskiego

Pilecka, E. 2005. *Średniowieczne Dwory Artusa w Prusach: świadectwo kształtowania się nowej świadomości mieszczańskiej*, Toruń: Wydawnictwo Uniwersytetu Mikołaja Kopernika

Pluskowski, A. G. 2013. *The Archaeology of the Prussian Crusade: Holy War and Colonisation*, London: Routledge

Pluskowski, A. G., and Valk, H. 2016. 'Conquest and Europeanisation: The Archaeology of the Crusades in Livonia, Prussia and Lithuania', in *The Crusader World*, ed. A. Boas, London: Routledge, 568–92

Pluskowski, A., Seetah, K., Maltby, M., Banerjea, R., Black, S., and Kalniņš, G. 2018. 'Late-Medieval Horse Remains at Cēsis Castle, Latvia, and the Teutonic Order's Equestrian Resources in Livonia', *Medieval Archaeology* 62/2, 351–79

Regnéll, J. 1992. 'Preparing Pollen Concentrates for AMS Dating: A Methodological Study from Hardwater Lakes in Southern Sweden', *Boreas* 21, 373–77

Reimer, P. J., Bard, E., Bayliss, A., Beck, J. W., Blackwell, P. G., Bronk Ramsey, C., Buck, C. E., Cheng, H., Edwards, R. L., Friedrich, M., Grootes, P. M., Guilderson, T. P., Haflidason, H., Hajdas, I., Hatté, C., Heaton, T. J., Hoffmann, D. L., Hogg, A. G., Hughen, K. A., Kaiser, K. F., Kromer, B., Manning, S. W., Niu, M., Reimer, R. W., Richards, D. A., Scott, E. M., Southon, J. R., Staff, R. A., Turney, C. S. M., van der Plicht, J. 2013. 'IntCal13 and Marine13 Radiocarbon Age Calibration Curves 0–50,000 Years cal BP', *Radiocarbon* 55/4, 1869–87

Sarnowsky, J. 1993. *Die Wirtschaftsführung des Deutschen Ordens in Preußen (1382–1454)*, Cologne: Böhlau

Sillasoo, Ü., and Hiie, S. 2007. 'An Archaeobotanical Approach to Investigating Food of the Hanseatic Period in Estonia', in *Medieval Food Traditions in Northern Europe*, ed. S. Karg, Copenhagen: National Museum of Denmark, 73–96

Šnē, A. 2008. 'The Medieval Peasantry: On the Social and Religious Position of the Rural Natives in Southern Livonia (13th–15th centuries)', *Ajalooline Ajakiri* 1/2 (123/24), 89–100

——, 2009a. 'The Emergence of Livonia: The Transformation of Social and Political Structures in the Territory of Latvia during the Twelfth and Thirteenth Centuries', in *The Clash of Cultures on the Medieval Baltic Frontier*, ed. A. V. Murray, Farnham: Ashgate, 53–71

——, 2009b. 'The Early Town in Late Prehistoric Latvia', in *The Reception of Medieval Europe in the Baltic Sea Region*, ed. J. Staecker, Visby: Gotland University Press, 127–36

Urban, W. 1993. 'Medieval Livonian Numismatics', *Journal of Baltic Studies* 24/1, 37–52

Valk, H. 2005. 'The Genesis of Viljandi (Fellin): Archaeological Data', in *Riga und der Ostseeraum von der Gründung 1201 bis in die Frühe Neuzeit*, eds I. Misāns and H. Wernicke, Tagungen zur Ostmitteleuropa-forschung, 22, Marburg: Herder Institut, 95–107

——, 2014. 'Strongholds East of the Baltic Sea in the 11th–13th Centuries: The Topic, Concepts, Research, and Terms', in *Strongholds and Power Centres East of the Baltic Sea in the 11th–13th Centuries*, ed. H. Valk, Tartu: Institute of History and Archaeology at the University of Tartu, Department of Archaeology at the Institute of History of Tallinn University, 81–114

Vandergoes, M. J., and Prior, C. A., 2003. 'AMS Dating of Pollen Concentrates — A Methodological Study of Late Quaternary Sediments from South Westland, New Zealand', *Radiocarbon* 45, 479–91

Ważny, T. 2002. 'Baltic Timber in Western Europe: An Exciting Dendrochronological Question', *Dendrochronologia* 20, 313–20

——, 2004. 'Zastosowanie dendrochronologii w badaniach starego miasta w Elblągu', in *Archaeologia et Historia Urbana*, eds R. Czaja, G. Nawrolska, M. Rębkowski, and J. Tandecki, Elbląg: Muzeum w Elblągu, 377–82

——, 2005. 'The Origin, Assortments and Transport of Baltic Timber', in *Constructing Wooden Images*, eds C. Van de Velde, H. Beeckman, J. Van Acker, and F. Verhaeghe, Brussels: Brussels University Press, 115–26

Wiewióra, M. 2012. 'Nowe źródła do dziejów zamku biskupów w Wąbrzeźnie', *Rocznik Grudziądzki* 20, 11–29

Wilson, C. A., Davidson, D. A., and Cresser, M. S. 2005. 'An Evaluation of Multi-Element Analysis of Historic Soil Contamination to Differentiate Space Use and Former Function in and around Abandoned Farms', *The Holocene* 15, 1094–99

Żemigała, M. 2008. *Cegła w budownictwie wielkopolskim w średniowieczu*, Łódź: Instytut Archeologii i Etnologii PAN

LIVONIA

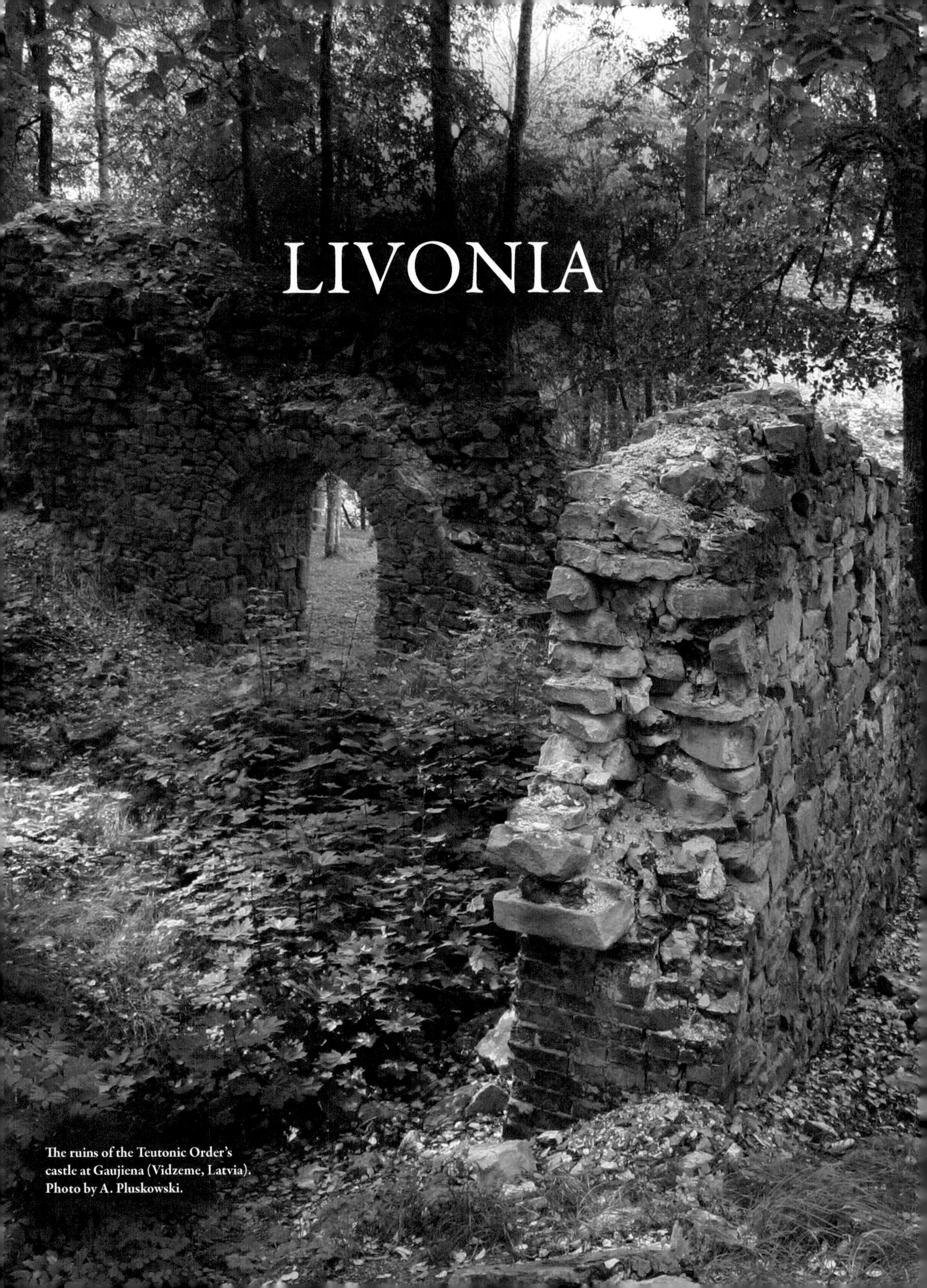

The ruins of the Teutonic Order's castle at Gaujiena (Vidzeme, Latvia). Photo by A. Pluskowski.

4. Sites in Livonia: The Historical and Archaeological Background

Aleksander Pluskowski, Heiki Valk, Juhan Kreem, and Gundars Kalniņš

The term 'Livonia' is used to refer to the lands inhabited by Baltic and Finno-Ugric speaking societies which roughly corresponds to modern Latvia and Estonia, and which were conquered by crusading armies. The Latin name, rendered in German as 'Livland', derives from the Livs settled in the Lower Daugava region, the first area targeted by missionary activity in the last decades of the twelfth century. After officially accepting Christianity, the Livs were the first to ally with the crusaders. The dedication of Livonia to the Virgin Mary was instigated by Bishop Albert, with the aim of making the region more attractive to crusaders and placing it on a par with the Holy Land (Mänd 2009, 195). At the end of the twelfth century, the name Livonia simply referred to the area in which Bishop Meinhard had been active (Tamm 2011), but with the subsequent conquests the name spread in usage over the course of the thirteenth century and was exclusively used for the conglomerate of territories in the region by the fifteenth century (Bojtár 1999, 129). Livonia extended as far north as the borders of the Danish Duchy of Estonia, which was only passed to the Livonian branch of the Teutonic Order after 1346. The use of the name Livonia was then gradually applied to the former Danish territories in northern Estonia. The southern and eastern borders of Livonia remained more or less fixed throughout the Middle Ages, although their geographic definition was blurred in the south by a sparsely populated 'Great Wilderness', which in its northern stretch extended from the Curonian coast through to south eastern Latgale.

As a geographic entity linked to the identity of the ruling branch of the Teutonic Order, whose territory at its greatest extent of 67,000 km^2 was larger than that of its ecclesiastical co-rulers, Livonia lasted until 1561, at which point those lands belonging to the Order which had not been conquered by Muscovite Russia were redefined as duchies, with their leaders becoming vassals of the Polish Crown and Grand Duchy of Lithuania, subsequently the Polish-Lithuanian Commonwealth. The earliest detailed descriptions of the internal subdivisions of Livonia date from this period (the 1580s and 1590s), and have been used to reconstruct the otherwise poorly known medieval administrative districts. This does not mean the divisions of these *Ämter* were not carefully defined — the surviving evidence indicates that boundaries were precisely delineated — but very little information has survived except for the lands around Riga and within a few select commanderies (see Chapter 9). From the thirteenth to early sixteenth century, the internal politics of Livonia were dominated by an ongoing struggle of power between the Order, the Archbishops of Riga and the town of Riga, and the gradual emergence of the *Landtag* added a further level of complexity to the articulation of territorial authority. The region subsequently underwent several periods of reorganization from the sixteenth to eighteenth century. From the 1630s, the name Livonia designated the Swedish province in northern Latvia and southern Estonia, but unofficially the name was used for the whole area of (former) medieval Livonia into the eighteenth century.

The demographic composition of Livonia did not change as much as that of Prussia after the crusades. At the turn of the fourteenth century most of the population was indigenous, with the permanent presence of colonists limited to towns and castles. The ethnic composition of first generation colonists was predominantly German (i.e., originating from regions defined in modern scholarship as the Holy Roman Empire), but this could include Slavs from north Germany, Danes, Swedes (Blomkvist 2005, 232), and small Russian communities also developed in some Livonian towns (Selart 2009). The population of present-day Estonia has been estimated at 150,000–180,000 people in the early thirteenth century, on the eve of the crusades, and up to 250,000–300,000 in the mid-sixteenth century (Palli 1996, 23, 40). Corresponding estimates for Latvia's population are *c.* 200,000 and *c.* 400,000 respectively (Šterns 1997, 2002).

The purpose of this chapter is to present the historical and archaeological background to primary and secondary sites (see Chapter 2) included within this study. Each site is presented in turn with an outline of its known historical development, a summary of its associated territory, which includes what is presently known about the distribution of sacred sites, followed by a summary of the archaeology and how this relates to the present study, concluding with a characterization of the site within the context of a frontier (see Chapter 2).

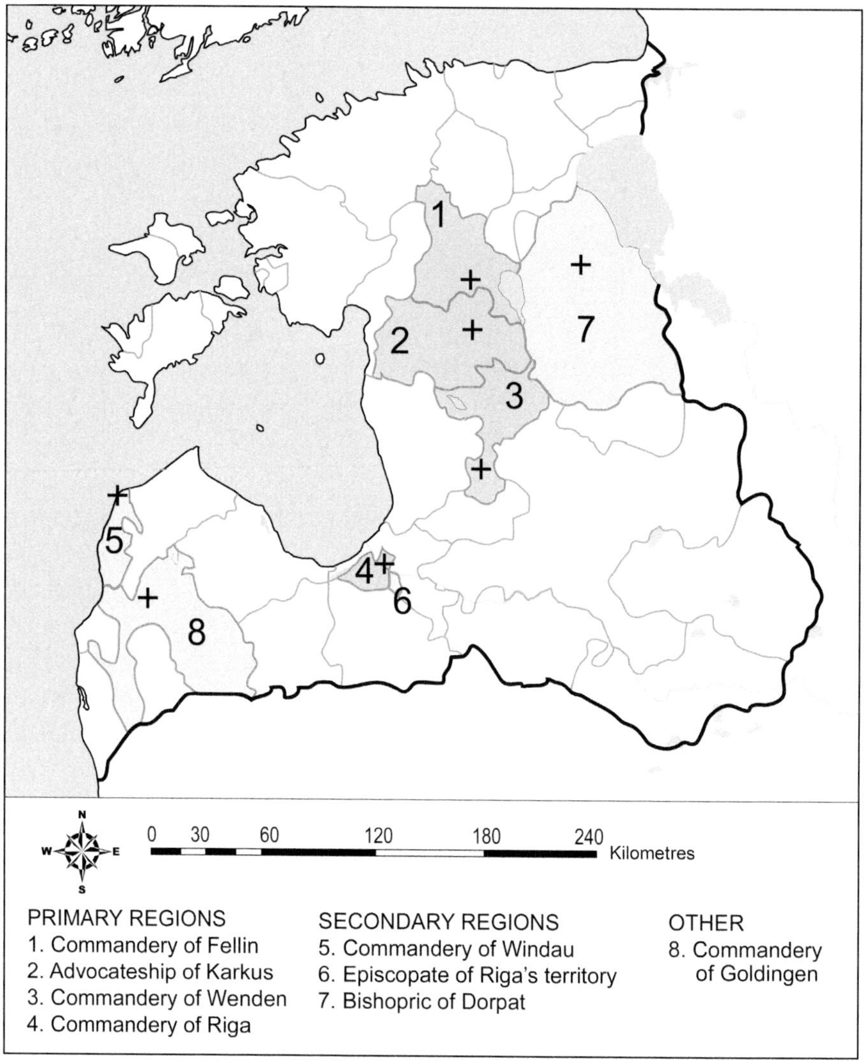

Map 4.1: Primary and secondary case study regions in Livonia, with crosses indicating principal centres of power. Map by A. Pluskowski.

Primary Sites and Regions

The project's primary sites in Livonia are situated within both modern Estonia and Latvia. Their details are outlined below with sites and their territories listed from north to south (Map 4.1).

1. The Commandery of Fellin (Viljandi)

Fellin, the centre of a commandery, was arguably the most important convent of the Order in Livonia, formerly within the Estonian province of Saccala. The castle is situated in the present-day town of Viljandi within the county of Viljandimaa in southern Estonia. The stronghold of Viliende, mentioned repeatedly in Henry's *Chronicle*, emerged as one of the centres of Saccala already before the crusades, and this stronghold was conquered by the Sword Brothers for the first time in 1211 (*HCL* xv, 1). Lembitu, the elder of a politically less-important Saccalian centre, led the military resistance against the crusaders including armies of unified Estonian groups organized according to provinces, and directed raids into neighbouring Rus' territories. He was killed along with Caupo of Turaida during a major clash with a crusading army on 21 September 1217, at the so-called 'Battle of St Matthew's Day' (in German scholarship the 'Battle near Fellin'). A few years later, in 1223, the Sword Brothers besieged Viliende (*HCL* xxvii, 2), the progress of which has left substantial archaeological traces (Lang and Valk 2011, 296–301). The settlement spread across the hills around the southern edges of the stronghold was destroyed by fire, and the ruins of the houses were levelled to provide platforms for siege engines (Valk 2015). After seizing and destroying the stronghold, the Sword Brothers began to strongly refortify it (*HCL* xxviii, 9). What developed into the later Teutonic Order's castle was located on the edge of a

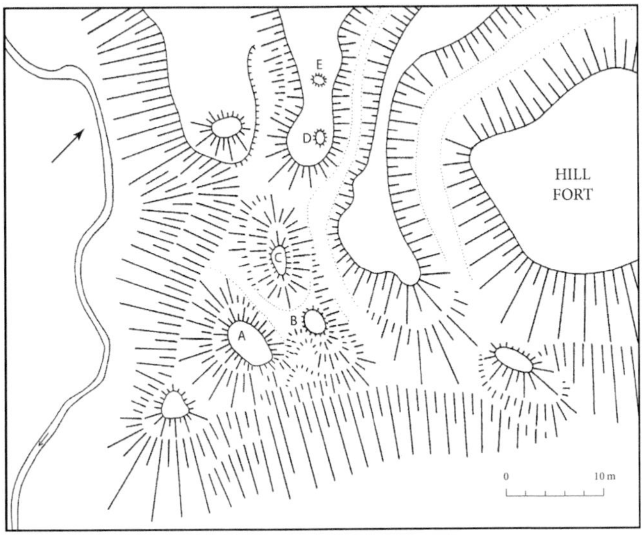

Figure 4.1: Plan of the hills and stronghold (later the Order's castle) at Viljandi. Modified after Valk 2001a.

Figure 4.2: (a) The valley of the second outer bailey of the castle at Viljandi; (b) the view towards the remains of the main convent building. Photos by A. Pluskowski.

plateau bordering the valley of Lake Viljandi, with two adjacent outer baileys located on natural outcrops separated by deep glacial valleys which served as moats, and only accessible with bridges (Figs 4.1–4.2).

In 1225 Henry of Livonia describes the castle as well built and suitable enough for the lengthy stay of William of Modena, the papal legate. By the time the Sword Brothers were merged with the Teutonic Order in 1237, Fellin had developed into an important commandery, prompting the construction of a new castle in *c.* 1250. This castle would become one of the strongest and largest in medieval Livonia (Tuulse 1942, 55–57, 139–49; Haak 2015). The significant investment in the earliest phase of the castle is evidenced by numerous early Gothic column capitals found within the ruins, which were reused in the later convent building (Alttoa 2015). The convent building, measuring 47 × 45 m, the second largest in Livonia after Riga, was constructed in the late thirteenth and/or early fourteenth century (Fig. 4.3). This is partly datable by the use of the 'Prussian bond' in some brick courses. The ambitious construction programme in the high castle area resulted in the removal of all earlier occupation layers, including those of the Late Iron Age stronghold. Moreover, not much of the castle's architecture has survived (Alttoa 2015), but a detailed description from 1599 notes the presence of a high watchtower and three floors (Jakubowski and Kordzikowski 1915, 162–72); the ground floor walls were constructed from stone and their upper storeys were built from brick. According to the visitation report of 1451, Fellin housed the largest convent in Livonia, consisting of twenty-nine knight brothers, five priest brothers, and six sergeant brothers (Biskup and Janosz-Biskupowa 2004, 35–36), where the total household could have numbered around two hundred (Kreem 2015). The convent is reported to be a loca-

Figure 4.3: The remains of the main convent building in Viljandi. Photo by A. Pluskowski.

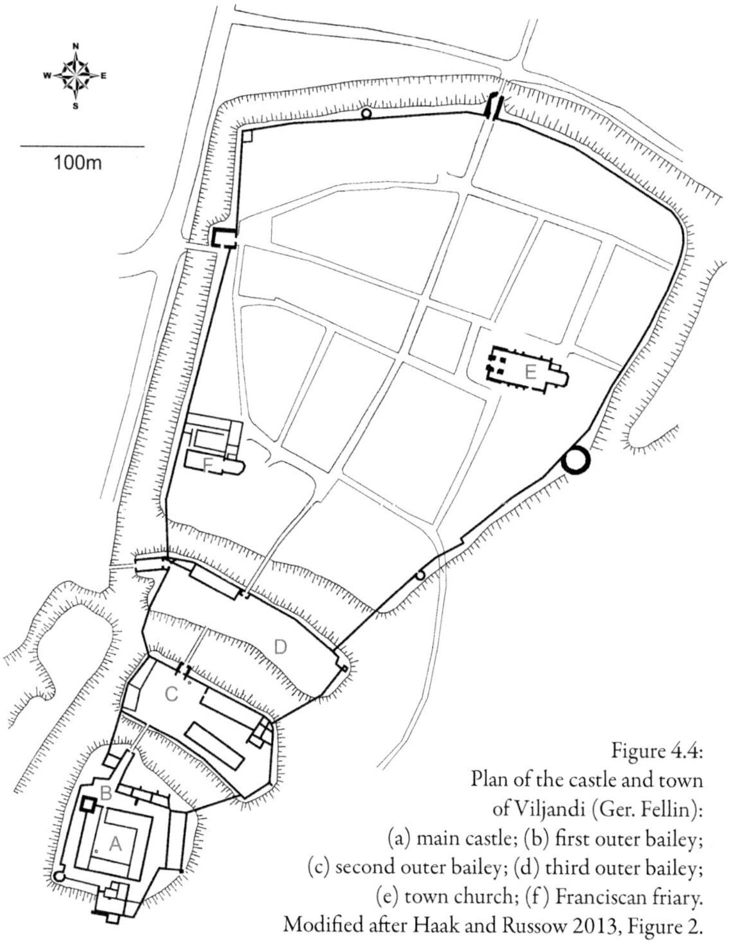

Figure 4.4:
Plan of the castle and town of Viljandi (Ger. Fellin):
(a) main castle; (b) first outer bailey;
(c) second outer bailey; (d) third outer bailey;
(e) town church; (f) Franciscan friary.
Modified after Haak and Russow 2013, Figure 2.

tion of the Order's treasury, as well as the depository of some of the Order's most important privileges. Its commanders (*Komturs*) ranked as some of the highest officials of the Order, and were elevated to the position of Livonian aster eight times, of which seven were directly after serving at the convent. At least two of the Livonian masters, Johann Waldhaus von Heerse (1470–71) and Johann von der Recke (1549–51), made Fellin their main residence for a short time.

The origins of the adjacent town are not fully clear yet. The medieval town area was used as a field in the Late Iron Age, as shown by ard traces. While earlier research located the first urban nucleus at St. John's church (Valk 2005), later analysis of imported stoneware has shown the presence of the earliest settlement at the later market place and in the northern part of the town, whereby town formation intensified only after 1250 (Haak and Russow 2013). With the confirmation of the town's privileges in 1283, the settlement developed its own commercial and civic functions, although the townspeople remained subjects of the Livonian branch of the Order. The physical segregation of the town from the castle complex, as well as the architectural dominance of the convent, reinforced the Order's authority. During the fourteenth century, the town expanded to the north-east. Its strategic location on the main route linking Reval and Riga, as well as by river to Pernau, provided the medieval town of Fellin with vital access to the trading networks of the Baltic, and it is mentioned as a Hanseatic town in 1346. In 1481 the town was attacked and burnt down by Muscovites, but the castle remained in the hands of the Order. The territory came under attack by Tsar Ivan IV Vasilyevich's armies during the Livonian War (1558–83), and Prince Andrei Kurbsky took control of the castle and town in 1561 for over two decades. The castle was then ruined during the Swedish-Polish wars (1600–25/29), and most of its ruins, still standing in the late eighteenth century, were destroyed in the first quarter of the nineteenth century.

Reconstructing the extent of Fellin's commandery is not straightforward. Its bounds are only documented from the early sixteenth century, at a time when its southernmost border was adjacent to the recently established Livonian Master's District. In the north it bordered with the district run by the advocate (*Vogt*) of Jerwen (Est. Järva), and until 1346 the Danish Duchy of Estonia, which was then reorganised into the commandery of Reval, with the rest of the territory split into advocateships. In all likelihood this northern boundary remained more or less fixed throughout the Middle Ages in a landscape dominated by till, fen, and alluvial deposits (Map 4.2). The commandery included the castles of Tarwast (Tarvastu) and Lais (Laiuse), and from the end of the fifteenth century also Oberpahlen (Põltsamaa), controlling key routes linking northern Estonia with western and central Livonia. The town of Fellin was the only urban centre within the commandery, within a region dominated by indigenous rural settlements. Important manors are documented at Fellin, Vastemõisa, and Türi-Alliku, and the convent also owned townhouses in Pernau and Reval. In the mid-sixteenth century the commandery generated an annual contribution of 450 marks to the central coffers of the Order's Livonian branch, which was also the largest single contribution from the whole of Livonia. This sum, however, must have formed only a minor part of the total income of the district. The settlements within the commandery — which generated the agricultural surplus from which the convent derived its wealth — are relatively poorly known archaeologically, in contrast to their associated cemeteries which demonstrate the continuity of non-Christian furnished burial rites which gradually declined in the eighteenth century (Valk 2001b).

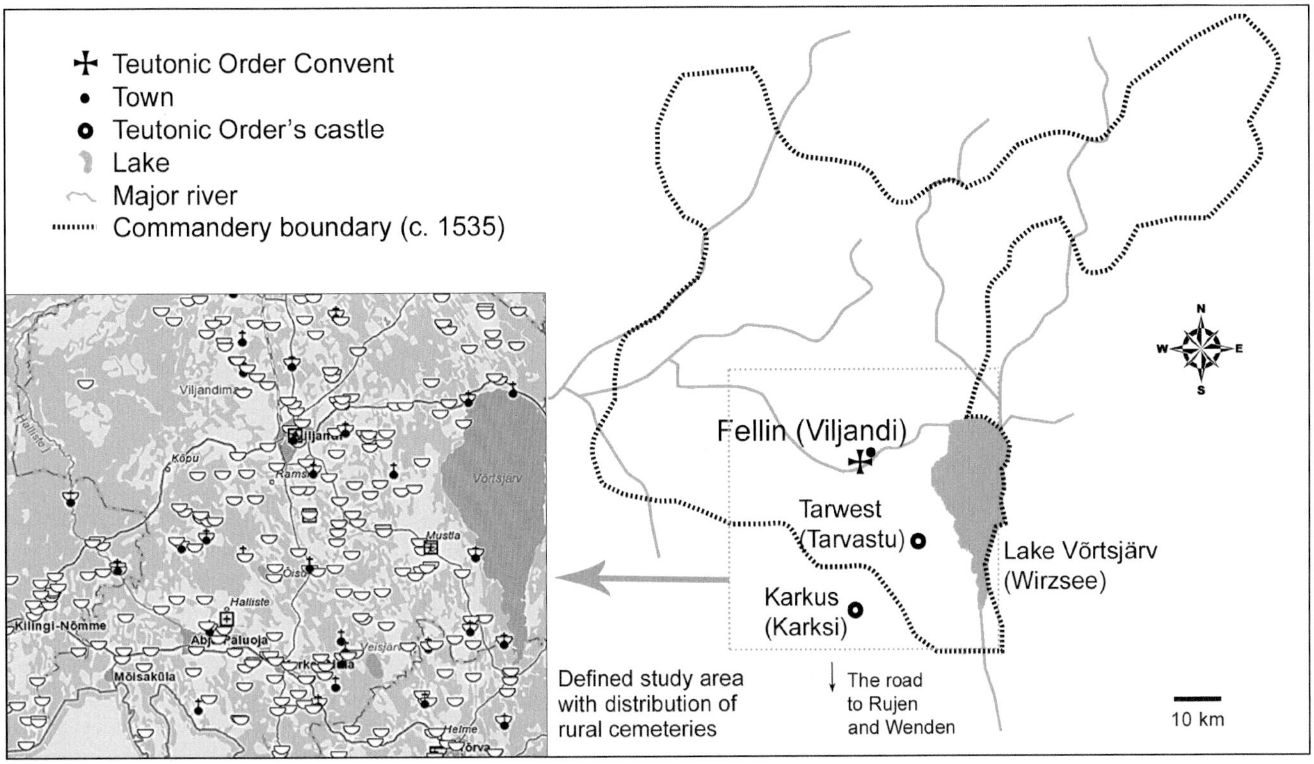

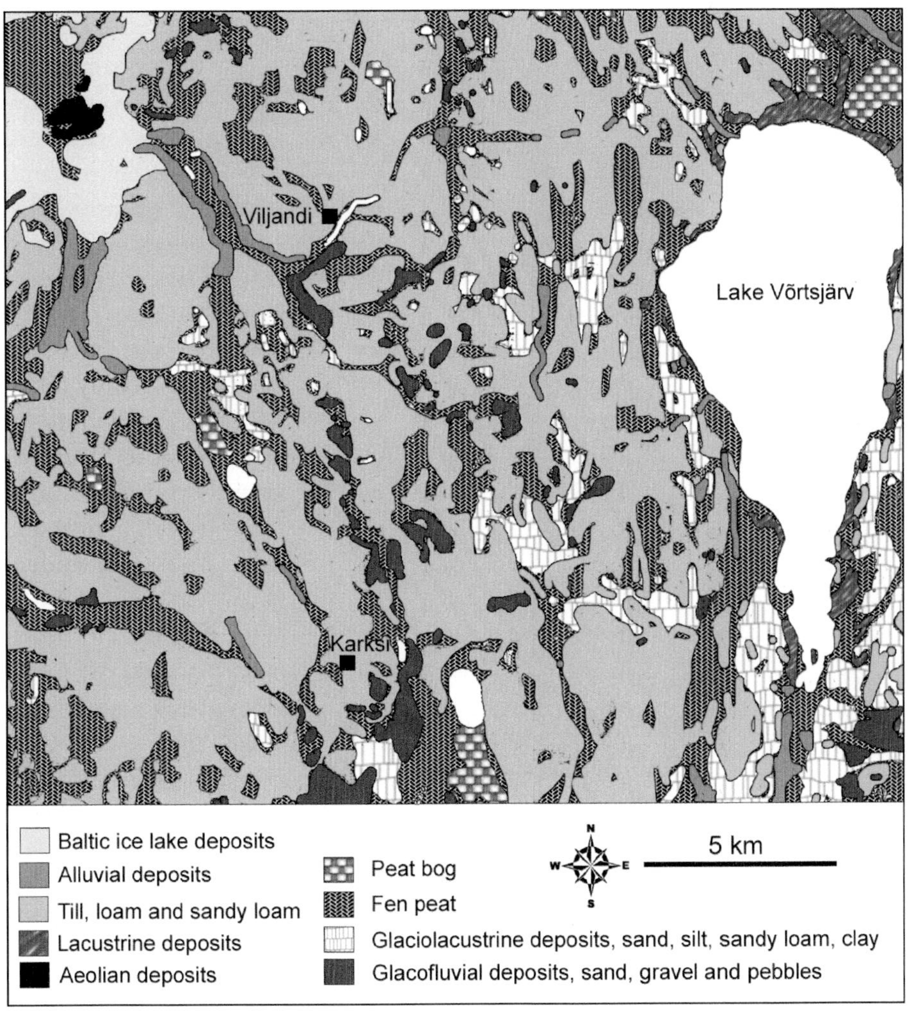

Map 4.2:
(a) The commandery of Fellin, with an inset of rural cemeteries that provides the best representation of the associated extent of settlement (Valk 2019);
(b) a map of the base geology between Viljandi and Karksi (1: 400,000, after Tallinn Geological Institute).

Figure 4.5: Coring Lake Viljandi with Alexander Brown, Siim Veski, and Normunds Stivriņš. Photo by A. Pluskowski.

Sacred Sites

Although a Christian infrastructure was established under the Order's rule, it was less developed than in the neighbouring Bishopric of Dorpat (Est. Tartu), and a high density of indigenous sacred sites has been documented. In the densely populated regions of northern Viljandimaa, the majority of sacred natural sites are associated with villages which were in use in the late sixteenth and late seventeenth century, and which have their own cemetery. Sacred natural sites were also found in association with villages close to parish centres where churchyards were used for burial. Given the association between the oldest settlements and sacred natural sites, this implies a level of continuity in use from at least the period of the Order's rule or, more likely, from the pre-crusade period. Where new settlements were created after the crusading period, it appears that new sacred natural sites were established. Within the vicinity of Fellin, the relatively smaller number of sacred natural sites to the west of the town and convent may reflect settlement discontinuity as in the case of the landscape around Wenden. Evidently the establishment of the Order's power centre here did not significantly disrupt the settlement pattern and its associated network of sacred natural sites. In fact, a tradition identifying a sacred site on the shore of the lake below the castle, coupled with archaeological evidence of a potential late pre-Christian offering site in the modern park area, suggest that surviving local communities continued to attach sacred significance to natural features within the vicinity of the convent, despite the destruction of the indigenous regional *centre* (for details, see Valk 2019).

Archaeology

The ruins and cellars of the high castle were excavated in 1878–79, without any documentation. In 1939 and 1997–2007, small-scale archaeological excavations took place in the high castle area. Substantial occupation layers were only discovered in one excavated area with a date range of 1260s–*c.* 1350 (Haak 2004; 2005). The data included within this project derives from this most recent period of excavation and largely consists of the faunal remains recovered from the various locations in the castle and town. A programme of coring was conducted within the surrounding landscape, including Lake Viljandi (Fig. 4.5). Written sources were relatively more abundant for the convent and underlined its importance within the Order's network in Livonia. The commandery of Fellin is an example of a Late Iron Age heartland that was maintained as a core region by the Order.

Regional Status

The replacement of the overlordship of Viliende with the convent of Fellin did not change the status of the associated territory. This functioned as an important heartland in both the Final Iron Age and in the medieval period.

2. *The Advocateship of Karkus (Karksi)*

Figure 4.6: The ruins of the castle at Karksi overlooking the valley. Photo by A. Pluskowski.

The place name 'Karkus' is first mentioned in written sources in 1248. The earliest reference to the castle, located some 30 km south of Viljandi, is found only much later in the chronicle of Hermann von Wartberge, who reports its destruction by the Lithuanians in 1298; the Lithuanians are said to have also plundered the castle in 1329. The earliest archaeological deposits at the castle date from *c.* 1260 to 1293, with intensive building activities beginning in the 1280s, and there was no evidence the Order's fortifications were constructed on the site of an earlier stronghold (Valk et al. 2013). The ruins of the castle are situated on the bank of the ancient Karksi-Halliste Valley (Fig. 4.6), alongside the historical road linking the convents and towns of Fellin, Wenden, and Riga. The castle buildings also took advantage of the natural topography by using the cutting valleys for the moat system (Tuulse 1942, 244–46; Alttoa 2009).

From the beginning of the fourteenth century, Karkus is associated with an advocate (*Vogt*) in the written sources, although there must have already been an official in charge of the castle and its territory also earlier. Although the advocate of Karkus did not rank amongst the highest officials of the Order in Livonia, the castle's location on the principal strategic route linking northern and southern Livonia ensured it remained an independent administrative unit. The castle's district also included the castle in Helmet (Helme), and in the south its border must have extended to the vicinity of Rujena (Rūjiena, Latvia). A visitation report from 1442 states that the castle owned 1407¼ *Haken* of arable land, out of which around a third were unoccupied at the time. The main building of the castle remained modest in size throughout the Middle Ages, despite the impressive extent of the adjacent outer bailey. Perhaps the comparatively small size of the high castle reflected the size of the household, which was never particularly high. In 1442, there were four knight brothers and a priest brother (Biskup and Janosz-Biskupowa 2002, 185–86); in 1451 there were six knights, a priest brother, and a sergeant brother (Biskup and Janosz-Biskupowa 2004, 37). The same visitation reports from the mid-fifteenth century note a considerable amount of food stored in the castle, quite disproportional to the number of documented inhabitants. The castle must therefore be regarded as a concentration point of supplies, a site used principally for storage and redistribution by the Order. The advocate of Karkus also owned houses in New Pernau (Uus-Pärnu), probably to take advantage of the trading connections of this Baltic port. Such houses could serve as administrative centres, residences and storage depots.

In the 1530s, the castle was incorporated into the Livonian Master's district. In 1534, the advocate Melchior von Galen was deposed and retired to New Pernau. The district was formally incorporated into the 'bailiwick' (*Kammergebiet*) of the Livonian Master, and from that time only a minor official resided at Karkus. In contemporary records the district was still treated separately, and its military contribution, together with the local nobility, is estimated to have been a hundred horses in 1556. Karkus passed into the hands of the Polish crown in 1560, along with the rest of the Order's territory in Livonia. The castle subsequently changed hands several times and was ruined during the Livonian War and the Swedish-Polish wars, losing its military significance during the seventeenth century. The remaining ruins were demolished in the eighteenth century, and only the entrance and some fragments of wall, primar-

Figure 4.7: (a) The ruins of the main castle at Karksi, looking north-west towards the high castle; (b) the remains of the bridge linking the main castle with the outer bailey; (c) the outer bailey, looking north-east. Photo by A. Pluskowski.

ily from the outer bailey, have survived today (Fig. 4.7). The most accurate plan of the castle, which informs our understanding of its medieval layout, derives from a seventeenth-century Swedish map (Fig. 4.8).

Sacred Sites

A limited parochial network was established in the territory of Karkus Castle during the medieval period, where sacred natural sites most probably continued to function in association with settlements. The twin-parish of Karksi-Halliste as documented in the seventeenth century may have existed in some form in the preceding period, but with the exception of the castle chapel and a chapel in front of the castle there is no clear evidence for a medieval church at Karkus. As in the case of the territory around Fellin, Christian and non-Christian sites were in parallel use, and there are even examples of sacred natural sites (specifically healing springs) associated with the Order's castles at Karkus and Helmet. Their origin is obscure, but given the general trends in the use of sacred natural sites in this region of Estonia it is plausible they functioned during the period of the Order's rule (Valk 2019).

Archaeology

In 1939, excavations led by Armin Tuulse within the castle discovered the well-preserved remains of a wooden building and timber constructions, around 1.5 m below the cobblestone pavement. However, no specific information has survived from these investigations. Small-scale rescue excavations were carried out in the outer bailey in 1994 and 2007, recovering material that was dated to the last quarter of the thirteenth/early fourteenth centuries (Valk and Malve 2008). In 2008, the castle was the focus of a gradiometer survey and a programme of prospective coring, which identified areas of undisturbed cultural layers (Valk et al. 2009). This was followed by two seasons of targeted excavation, in 2011 (Valk et al. 2012) and 2012 (Valk et. al 2013) (Fig. 4.8). The excavations recovered a significant quantity of environmental data, alongside a range of artefacts from the earliest phase of the castle through to the first half of the sixteenth century (Fig. 4.9). Little is known about any associated settlement within the vicinity of the castle. Field walking along ploughed fields to the north-east recovered fragments of ceramic building material and pottery, but these were difficult to characterize.

Data included within the project focused on the botanical and faunal remains, as well as the sediments, timbers, and organic artefacts recovered from the castle. In particular, the midden layers of a late-thirteenth-century waste pit, waterlogged by subterranean spring water, provided excellent material for palaeoenvironmental and zooarchaeological analysis, alongside numerous organic artefact finds. Elements of the castle's fabric — stone and mortar — were included within a study of building materials (Hayward 2019). A programme of coring was conducted within the surrounding landscape, with the best pollen sequence recovered from a peat deposit in Äriküla (Valk et. al. 2009). Written sources relating to the site were relatively limited but included useful quantifiable information. The landscape around Karksi had been at the fringes of the overlordship of Viljandi, but the construction of the castle here with access to a major route through Livonia resulted in a more strategically important status for the district.

Regional Status

The area around Karksi, with no evident stronghold in the vicinity, can be considered as peripheral to the overlordship of Viliende in the Final Iron Age. When the Order established their castle here, its strategic location in proximity to a major communication route running through Livonia changed the region's status. No longer on the periphery, it became an administrative centre in its own right (albeit lesser in status compared to Fellin), and this is reflected in associated changes in the surrounding landscape.

3. *The Commandery of Wenden (Cēsis)*

The commandery of Wenden would become the most important centre of power in the eastern Baltic during the final phase of the Order's rule in Livonia, and today the castle complex at Cēsis, within the district of Vidzeme, is the most significant historical monument in Latvia alongside the Old Town of Riga. In 1212, the Sword Brothers established a base at Wenden (on Riekstu Hill/Nussberg) which was described in Henry's *Chronicle* as occupied by a group originally deriving from the Curonian coast called the 'Vends' (Latin *Wendi*), which some scholars have related to the easternmost distribution of the Polabian Slavs, largely concentrated in the Elbe Valley of modern eastern Germany (Apals 2007; Viitso 2009, 273). This interpretation has been recently rejected, and the Vends appear to have been Curonian Livs, part of a north-eastwards migration that began in the tenth century (Apala and Apals 2014, 116–17). Henry describes the encounter at Wenden in positive terms, with the Sword Brothers living alongside the indigenous population (*HCL* XII, 6; XIII, 2). Indeed, the conversion of the Vends in the winter of 1206–07 had presented the crusaders with dependable allies and auxiliaries with a fortified stronghold in the northern extent of the Sword Brothers' territory at the time (Murray 2013, 36). Excavations on the hillfort began in 1884 but commenced in earnest in 1980 and 2010, uncovering the foundations of a 1.55–1.7 m thick stone wall on three sides, with the eastern side associated with the stratified remains of a burnt timber structure. At least four distinct

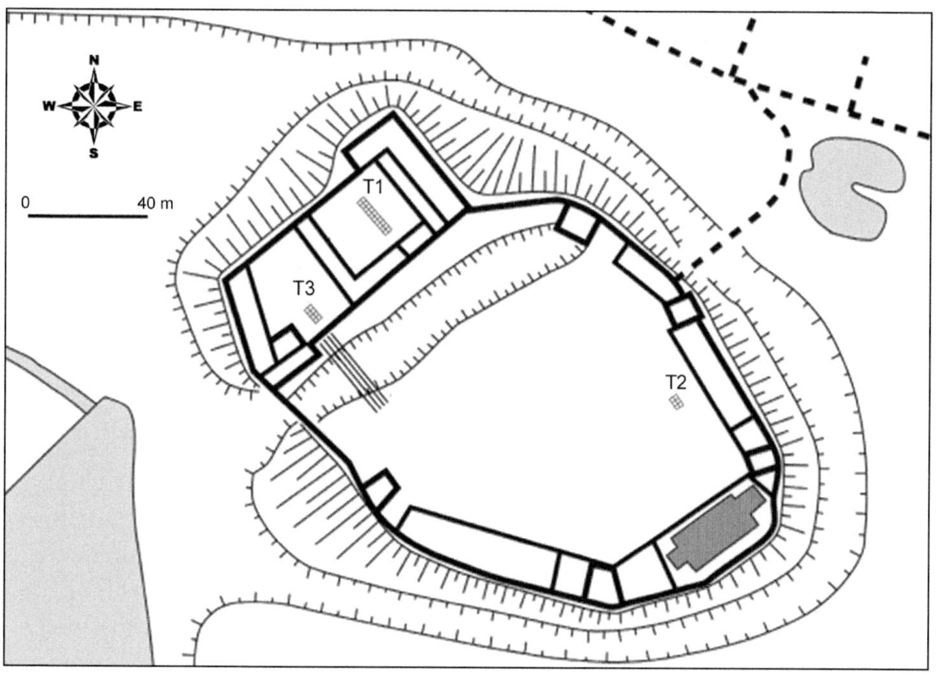

Figure 4.8: Plan of Karksi Castle showing the location of excavations (after Valk et al. 2012, Figure 1).

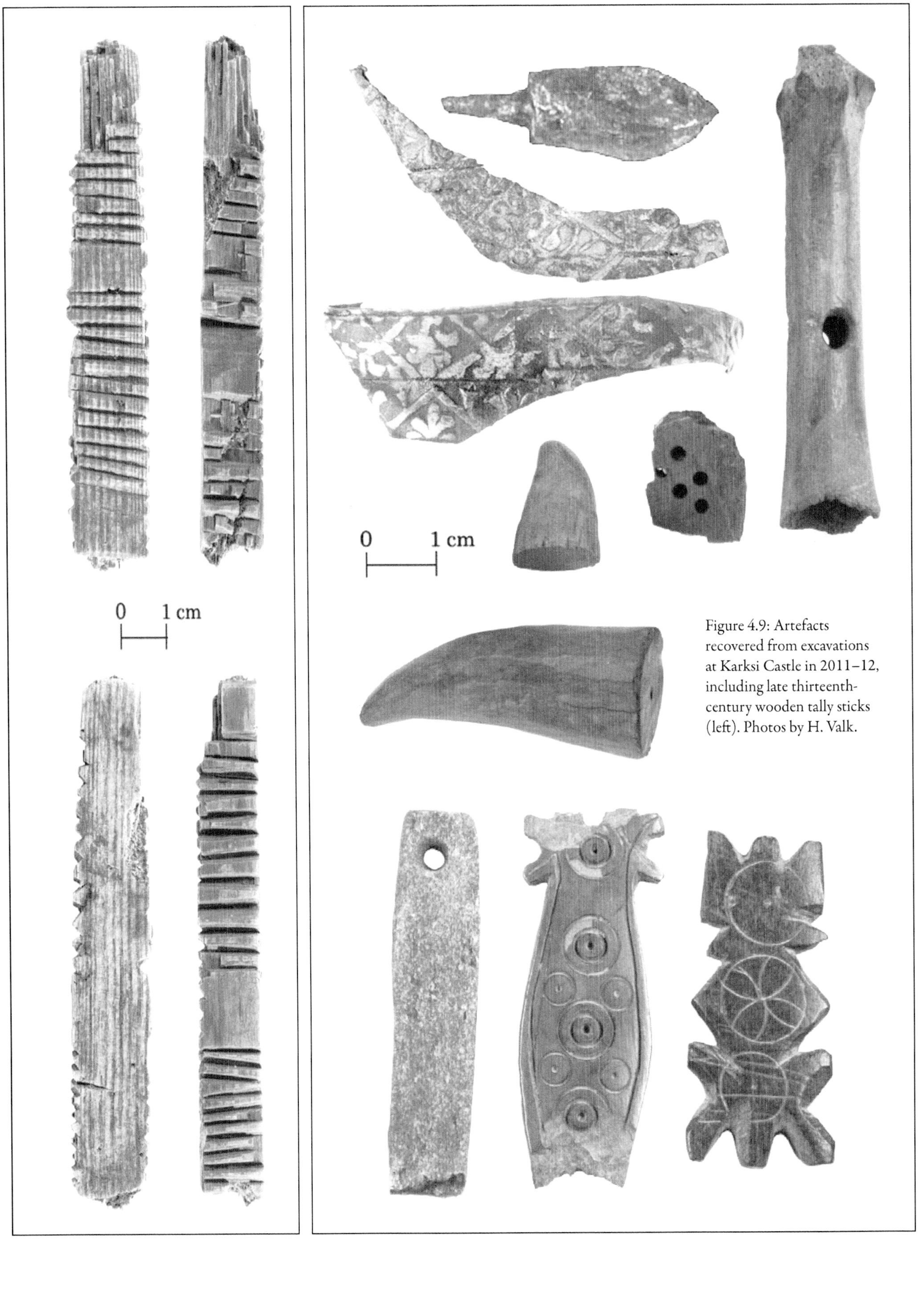

Figure 4.9: Artefacts recovered from excavations at Karksi Castle in 2011–12, including late thirteenth-century wooden tally sticks (left). Photos by H. Valk.

Figure 4.10: The stone walls uncovered during excavations in 2009 on Riekstu Hill in Cēsis. Photo by A. Pluskowski.

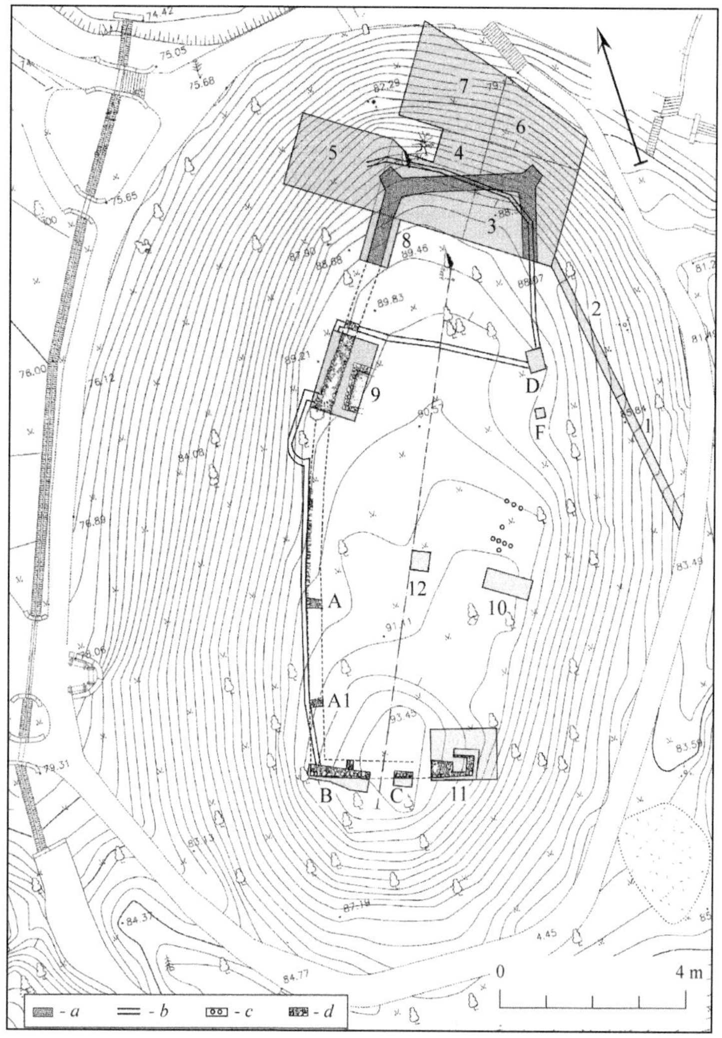

Figure 4.11: Plan of Riekstu Hill showing areas excavated in 1980 (Trenches 1–8) and 2009–10 (Trenches 9–12), as well as the outline of the wall (a, d) (after Apala and Apals 2014, 122).

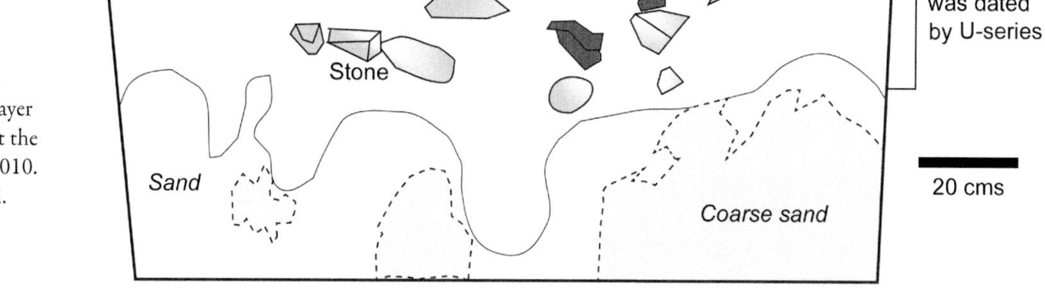

Figure 4.12: The southern profile of the occupation layer uncovered in Trench 12, at the centre of Riekstu Hill in 2010. Drawing by A. Pluskowski.

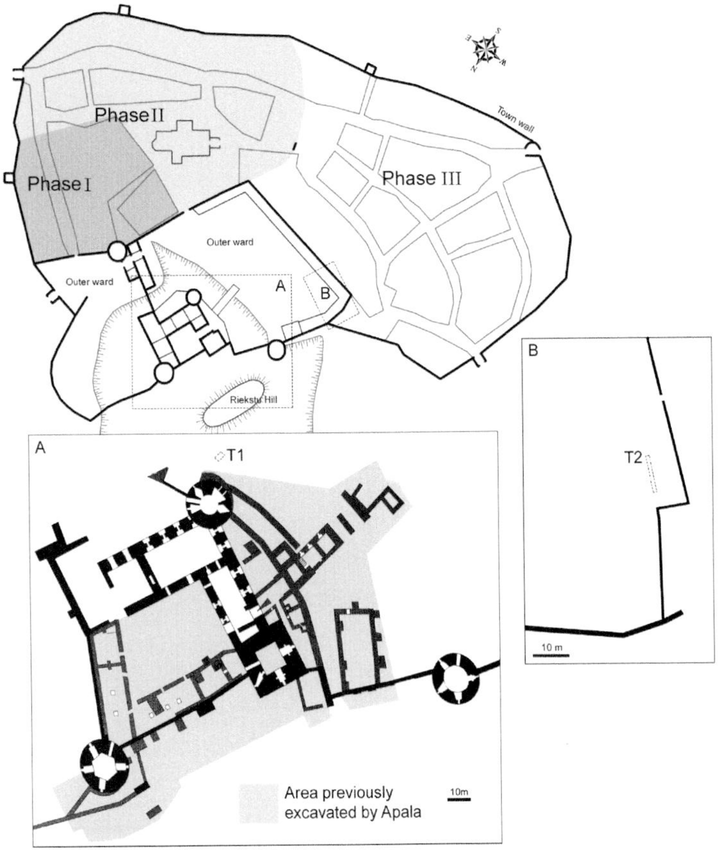

Figure 4.13: (a) Plan of the castle and town of Cēsis (Ger. Wenden), showing the location of the two principal excavations in the castle grounds (T1: moat; T2: outer bailey) (after Apala 2007 and Kalniņš 2014, with town based on Ulrich's plan of 1693 and showing the three hypothetical stages of development), plan created by Zigrīda Apala and modified by G. Kalniņš and A. Pluskowski; (b) the locations of the trenches on an aerial photo; (c) the southern range of the conventual castle, modified further in the sixteenth century. Photos by G. Kalniņš and A. Pluskowski.

occupation phases were identified, the first dating from the first to sixth century AD, the eleventh to twelfth century, the thirteenth century, and the uppermost associated with the nearby castle (Apala 2010a; Apala and Apals 2014). In the centre of the stronghold (Trench 12), a layer was uncovered during excavations in 2009 which microstratigraphic analysis indicated a short and intensive period of occupation. This layer contained pig bones which were U-series dated with mean dates of *c.* 1257 (±38) and 1279 (±14) (Figs 4.10–4.12).

The construction of the adjacent castle, which would become the Order's convent of Wenden, overlapped with the occupation of the hillfort. The castle complex has been the focus of excavations for many decades, but its early phases remain poorly known (Fig. 4.13). The convent began to accrue territory quickly. In June 1212, there was a conflict between the Sword Brothers of Wenden and the Latgalians (or Letts) of Autine (at that time episcopal territory) for fields and bee-trees (trees containing hives; see also Chapter 8). After peace was restored it was agreed that the Latgalians should regain access to their bee-trees, but the Sword Brothers would retain the fields and adequately compensate the Latgalians for the losses that they had incurred (*HCL* xv, 6). The following year Bishop Albert reconfigured the distribution of annexed Lettish territory which had been previously fixed in the autumn of 1211, and exchanged some districts with the Sword Brothers: the latter were granted the whole of Autine, as well as the Bishop's tithe in Ascheraden (Aizkraukle) and two villages (Švābe 1939, 56–57). After this, the details of the territorial expansion of the commandery remain obscure. The town of Wenden is supposed to have received its charter in 1224 (Šterns 1997, 136), and its burghers were obliged to provide military service to the convent (Zeids 1978, 49). The late-seventeenth-century plans of Wenden show that the town, covering an area of approximately 10 hectares, was enclosed by a kilometre-long wall with three gates and four towers, although the development of its morphology remains unclear (Map 4.16).

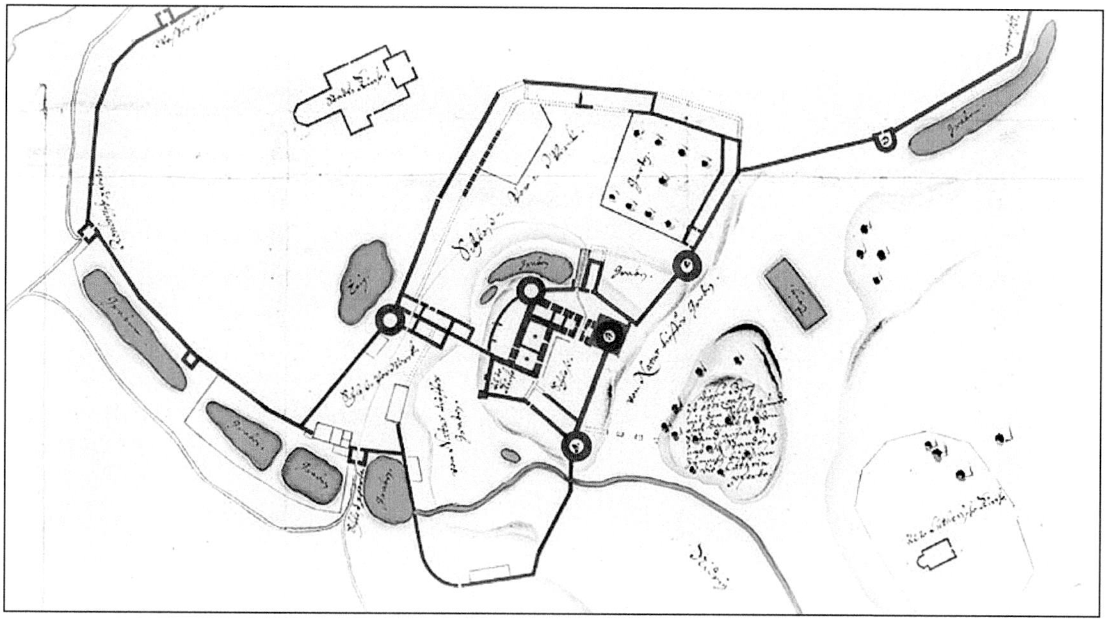

Figure 4.14: Johan Palmstruck's plan of Wenden, *c.* 1685, showing bodies of water within the vicinity of the castle. Swedish Military Archives, Stockholm.

In the latter half of the fourteenth century the convent was significantly expanded, and it remained an important centre for controlling the route linking Riga with the central territories of the Order in Livonia. The Swordbrothers already had a commander in Wenden, and an advocate is mentioned *c.* 1304, and an advocate is mentioned in association with the convent eighteen times between 1255 and 1461. Frank Spede, the advocate of Wenden in 1392–97, then rose to the position of commander of Fellin, and when he died in 1404 he was buried in St John's church in Wenden (Fenske and Militzer 1993, no. 824). During the fourteenth century, the advocates of Wenden wielded significant power and were involved in numerous diplomatic arbitrations. They also owned a house in Riga (mentioned in 1366; *LUB* II, 1033). The most important change in the functioning of the convent of Wenden came in 1413, when it became increasingly used as the residence of the itinerant Livonian Master and the centre for the governance of the Livonian Master's district (*Kammergebiet*). Riga remained the most important centre for the Order in Livonia until problematic relations with the town and Archbishop prompted a permanent relocation to Wenden (Neitmann 1993). At this point, the residential quarters of the castle were upgraded. Written sources relating to the castle's inventories and holdings are rare and typically late. An audit conducted by Polish officials in 1590 provides the most detailed snapshot, but it is important to note that the state of the castle following the damages of the Livonian War cannot be readily projected back to the period of the Order's rule. At this time within the vicinity of the castle there were two ponds, a flour mill, a cattle farm, and three fields, one of which lay fallow as — the audit states — the peasants had been killed by the Muscovites.

The stronghold of Riekstu Hill is the only indigenous central place within this region noted in Henry's *Chronicle*, in proximity to the border of the territories attributed to the Livs in the west, and the Latgalians in the east, broadly marked by Gauja Valley. This Late Iron Age borderland, itself the product of both Liv and Lettish expansion in the eleventh and twelfth century (Zemītis 2014, 447), may explain the limited number of central places encountered by the crusaders in the early thirteenth century. Indeed, the Vends themselves appear to have been assimilated into the neighbouring Lettish population by the end of the thirteenth century, although the tribal name was incorporated into the new regime; the red and white banner carried by the local militia was described in the *Livonian Rhymed Chronicle* as 'according to the Vendish custom', and the name was preserved in the toponym (Murray 2013, 36). The commandery of Wenden, which also adopted the Gauja as a boundary, a clear physical separation with episcopal lands to the west (Fig. 4.15, Map 4.3b), included within it a series of smaller districts governed by advocates, based in the castles of Wolmar (Valmiera), Trikaten (Trikāta), Burtneck (Burtnieki), and Ermes (Ērģeme). There is no surviving evidence documenting the type of official present at Arrasch (Āraiši), which was nonetheless subservient to Wenden and belonged to the *Kammergebiet* of the Livonian Master. It is possible to sketch the extent of these territories, a snapshot of their fullest extent perhaps, based on the earliest map of Polish-Lithuanian castellanies (Map 4.3a). Only

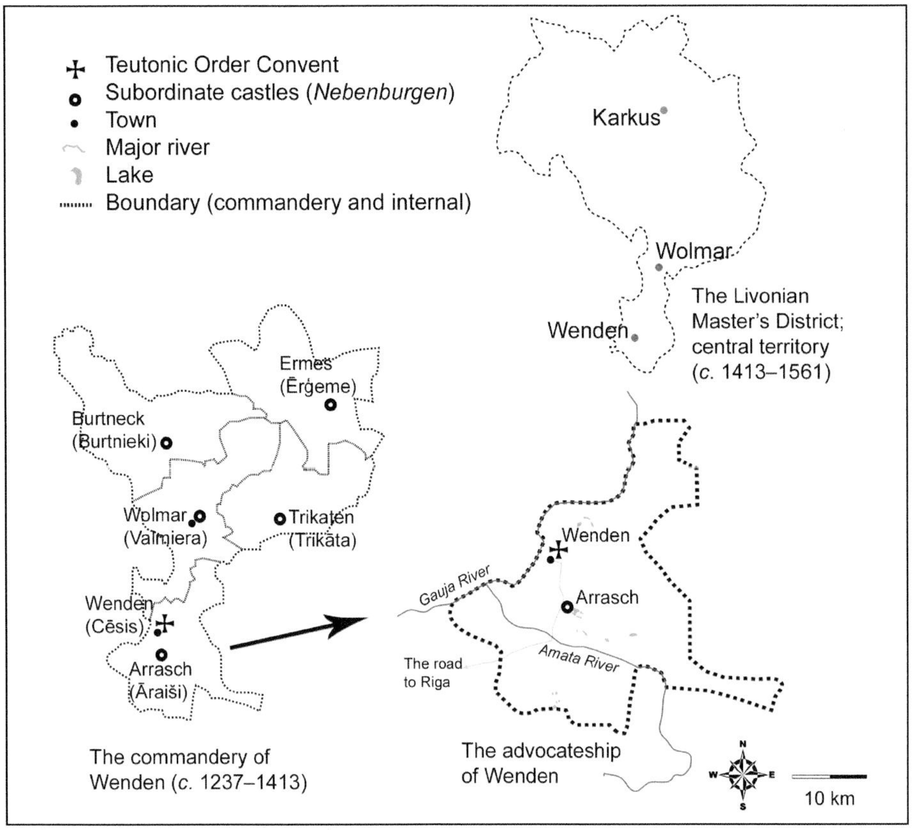

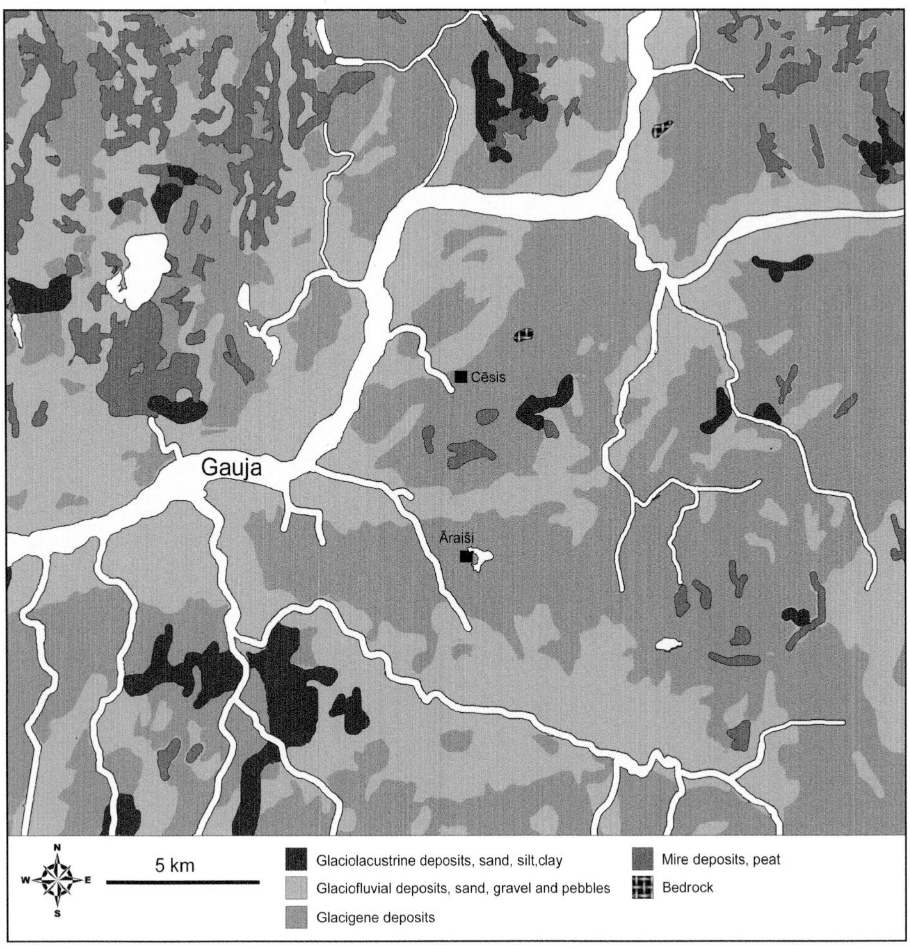

Map 4.3: (a) The commandery of Wenden, its associated territories, and the Livonian Master's district; (b) a map of the superficial geology (1: 200,000, after Latvian Geological Institute).

Wenden and Wolmar were associated with planned towns, and the commandery was dominated by indigenous rural settlements. An audit of 1582 states that ten civil parishes were attached to the castle, with a combination of populated and vacant farmsteads (Map 4.4). For example, in Gaidan (Gaidāna) parish, there were thirty farmsteads of which only ten were occupied, no doubt as a result of the ravages of the Livonian War. In some parishes, due to the depopulation, impoverishment, and reduction in harvests, the dues provided to the castle had drastically fallen, a situation to which the presiding Polish royal official (*Starosta*) responded sympathetically. Uncovered medieval buildings had been constructed from pine with timber-framed superstructures, the architectural technology imported by colonists from Western Europe. The oldest have been discovered in the town square, Rožu laukums, dated by dendrochronology to the 1280s (Apala 2010c; Zunde 2019). However, as with many Livonian towns, Wenden's population was mixed, and included indigenous peoples, German, and Rus' colonists (Caune 2007a).

Sacred Sites

In comparison to other parts of Livonia, relatively little is known about the survival of indigenous sacred sites within the Wenden commandery; however, the presence of boulders cut with crosses may reflect the process of marking out a Christian landscape, most visible in the documented boundaries of Curonia (see Chapter 8). The so-called Cēsu Svētavota ala (Holy Spring Cave) on the outskirts of the town of Cēsis was first documented by folklorists in the second half of the eighteenth century, and its origins remain obscure and speculative (Laime 2009, 332–36); in the southernmost part of the territory Īļaku Svētavots (Īļaki Holy Spring) on the right bank of the Amata River only survives as the place name with no associated folklore. Within the vicinity of Valmiera, place names suggest the presence of potential cult sites, such as Stoķu svētavots (Stoķi Holy Spring), Antonu Velnakmens (Antoni Devil's Stone), Ogkalns Elkukalns (Ogkalns Idols' Hill), and Elku saliņa (Idols' island). In the case of the latter, which included a sacred oak and a spring attributed with healing properties, excavations in 1979 uncovered three hearths (Berga 1980, 29). The largest complex within the vicinity is found around Zilaiskalns Hill, which includes sacred trees, smaller sacred hills, a spring, and a large glacial erratic described as a 'sacrificial stone'. The complex has been associated with the site of Midsummer gatherings in popular tradition. Although some are associated with the Middle Ages in folklore, it has not been

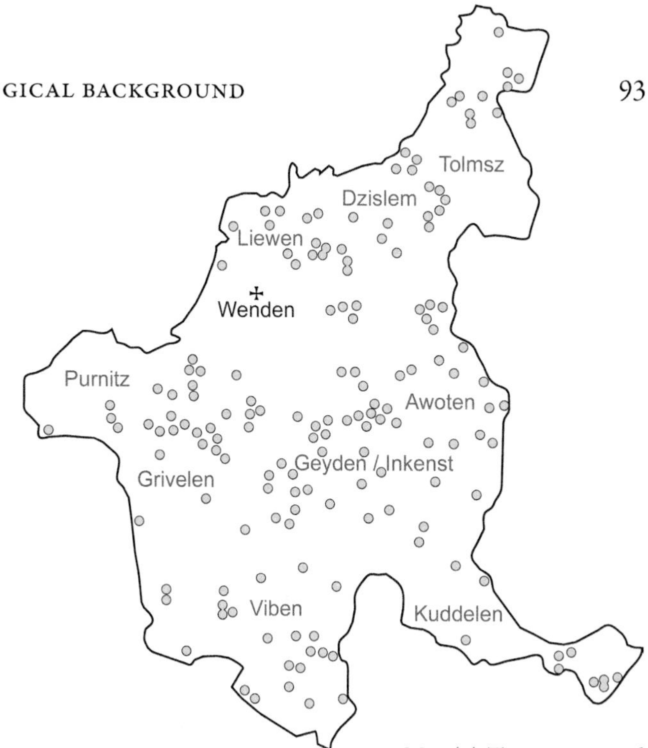

Map 4.4: The reconstructed distribution of known farms in named *pagasts* (districts) recorded in the castellany of Kieś (Ger. Wenden) in the sixteenth century, which would have supplied the convent in the final period of the Teutonic Order's rule. It has not been possible to identify the location of all recorded farms. Map by G. Kalniņš.

Figure 4.15: Gauja Valley with the castle at Turaida (Ger. Treiden) in the foreground — a naturally striking border in the Late Iron Age and Middle Ages (The Turaida Museum Reserve; author <www.aerovision.lv>).

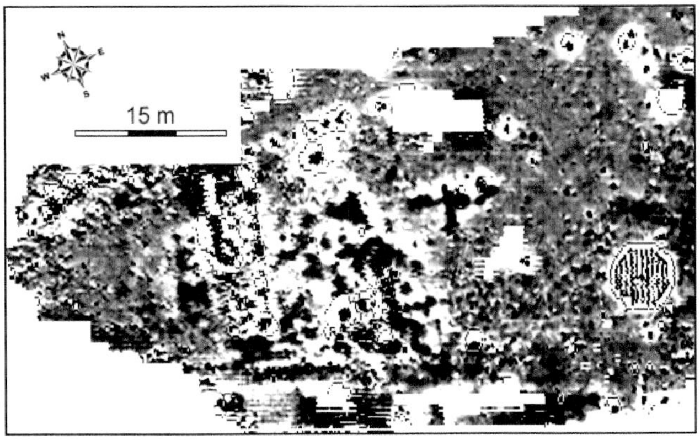

Figure 4.16: Magnetometery map of the outer bailey at Cēsis. Figure by D. Thornley.

possible to date the origins of these sites with any certainty. It is of course entirely plausible their sacred attributions are rooted in pre-Christian indigenous culture, but it is striking how few there are within the territory of Wenden compared to other parts of Vidzeme.

Archaeology

The castle complex has been the subject of over three decades of excavations (1974–2008), directed by Zigrīda Apala (2007; 2010b). Much of the area within the castle was excavated during this time, resulting in the recovery of a substantial quantity of material culture (including, for example, one of the most important assemblages of late medieval armour in Northern Europe) and certain categories of ecofacts, such as animal bones. In particular, the detonated remains of the western range (deliberately destroyed in 1577), along with some of the casualties, were uncovered. Following discussions with Zigrīda Apala and Gundars Kalniņš, it was possible to target parts of the surrounding castle precinct and moat with the aim of recovering a broader suite of palaeoenvironmental data. In 2010, surveys of the southern outer bailey and moat were conducted with a gradiometer and ground penetrating radar. These identified areas with undisturbed cultural deposits that were the focus of targeted excavations in 2011 and 2012 (Fig. 4.20). A range of ecofacts and artefacts were recovered during these excavations, the earliest phases of which could be dated by the dendrochronology of associated timber structures and linked to the expansion of the castle in the later fourteenth century. The soil in the moat was waterlogged at a depth of around 2 m, and in the raised southern part of the outer bailey at a depth of around 2.8 m. The historic centre of Cēsis has also been the subject of various rescue excavations led by Apala, and some material (in particular animal bone) was made available from these for further analysis.

Regional Status

The landscape associated with the commandery of Wenden is complicated to define before it was transformed into a heartland by the Order. It was clearly a core for the Vends but also a borderland which included the Gauja Livs and the western Latgalians. The Gauja remained a border between the territories of the Order and the episcopacy, and whilst it was partially militarized (with, for example, the castles of Segewold and Treiden facing each other across the river), this political tension should not be overemphasized in comparison to external frontiers, particularly with Samogitia and Lithuania.

4. The District of Arrasch (Āraiši)

The southernmost part of the commandery of Wenden was, at one point, sub-divided into a district which came under the jurisdiction of a separate official. A castle was constructed near the edge of the territory, as understood from the boundary of the later castellany, at a site first documented as *Arries* in 1410 (subsequently Arrasch; Latvian Āraiši (Apals 1996, 27)). The castle was built on a hill at the end of a peninsula extending into Lake Āraišu (Arrasch See) (Fig. 4.17). To the north, on a small island, a fortified lake settlement was occupied in the ninth and tenth centuries, a site made famous by the excavations of Jānis Apals and its subsequent partial reconstruction and transformation into a 'museum park', a milestone in European experimental archaeology (Apals 2012) (Fig. 4.18). As a result, the lake settlement is far better known than the castle and its outer bailey, which was also the focus of excavations led by Apals (1995, 1996) in 1972–74 and 1981–92. Nonetheless, the material culture recovered from the outer bailey indicated the indigenous population formed a significant part of the castle community, as evident on other Livonian sites. A substantial faunal assemblage had been collected and was subsequently made available to the project and phased on the basis of six AMS dates. The oldest date range encompassed the latter decades of the thirteenth century (*c.* 1260–1394), but otherwise the castle's structure has been dated to the fourteenth century (Apals 1995, 1996).

There is a gap of at least two centuries before the abandonment of the lake settlement and the construction of the castle, and it is likely the site was chosen

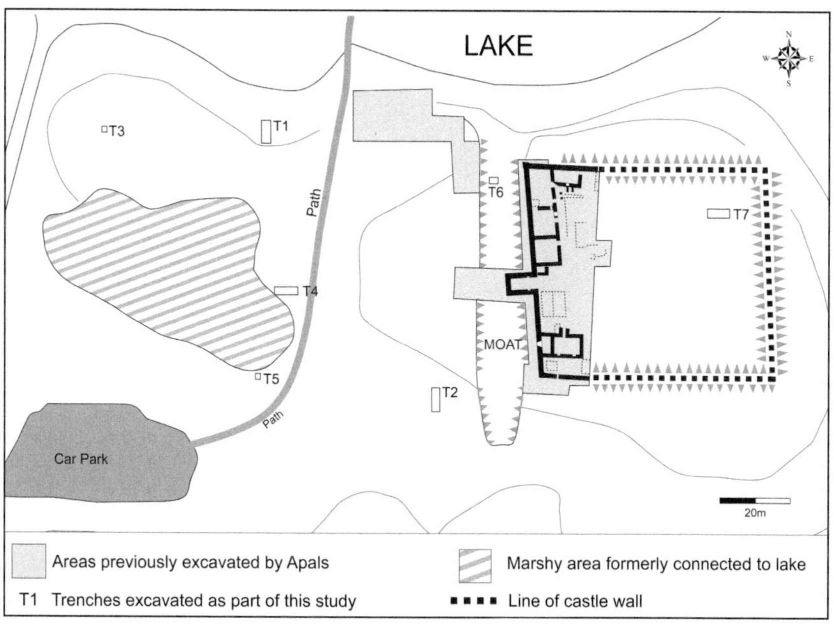

Figure 4.17: Location (a) and plan (b) of the Teutonic Order's castle at Āraiši (Ger. Arrasch) showing locations of previous excavations (after Apals 1995, 33–34) and those carried out as part of this study (Tomsons 2014).

because of its proximity to the route linking Wenden with Riga, and its proximity to the southern border with the commandery of Segewold (Sigulda), the territory of the Livonian Land Marshal from 1432. There is very little information in historical sources about the castle, although during the medieval period it is interpreted as a site for mustering troops and storing arms and armour (Apals 1996, 35). It was maintained as the residence of an official until the Livonian War, when there is a documented siege which ended with the surrender of the garrison. From 1561 to 1629, the castle functioned as a Polish manor (*folwark*), and the site was abandoned by the mid-seventeenth century. The Polish audit of 1582 referring to Aries or stara Kies (Old Wenden, although this name does not feature before the sixteenth century) describes it as almost empty and without lower-ranking subordinates, nor any cultivated lands within its vicinity. However, the arable fields attached to the *folwark* that lay fallow were described as large, and it was anticipated that when the surrounding villages would be repopulated they would be able to cultivate and make good use of them. The medieval settlement pattern in the surrounding landscape, as with much of Wenden's commandery, remains poorly known but would have consisted of small-

Figure 4.18: The reconstructed fortified lake settlement at Āraiši. Photo by A. Brown.

scale, indigenous communities, many of which were destroyed during the Livonian War. This depopulation in the later sixteenth century may indicate a discontinuation in indigenous cult praxis that has resulted in the virtual absence of sacred toponyms within the vicinities of Cēsis and Āraiši. The only surviving place name associated with a sacred site can be found a few kilometres to the south of Āraiši at Drabeši, where a small river at one point included the place-name element Vēlė, which can be interpreted as 'souls' river' (Vaitkevičius 2004, 40, 92). Again, it is extremely difficult to date the antiquity of this name.

Archaeology

In 2011, a gradiometer survey was conducted across the dry areas of the outer bailey, but the results did not point to any clearly defined structures. In 2012, six trenches were opened in the outer bailey and one within the eastern precinct of the castle. Very few artefacts and ecofacts were recovered from this excavation; the most important information pertained to the function of the site, confirming the absence of substantial structures in the central parts of the outer bailey, and demonstrating that the castle consisted of a single wing rather than a set of enclosed ranges. The 1590 audit specifies residential spaces at what was then defined as a *folwark*, with a granary, two cattle sheds, a horse stable, a bath-house, and two threshing barns (Ose 2011). Artefacts recovered from the outer bailey supported Apals' earlier findings that pointed to the presence of an indigenous population within the castle's extended household (Tomsons 2014). More information was obtained from the analysis of a lake core undertaken by Normunds Stivriņš et al. (2015), and from cores taken from nearby peat deposits. The castle's fabric was also evaluated further in 2014 (Hayward 2019).

Regional Status

The territory managed by the castle at Arrasch was located in the southernmost part of Wenden's commandery, and the landscape appears to have operated as a local heartland in the Late Viking Age, at which point it can be characterized as a frontier in the context of Lettish westward expansion. With the establishment of a castle here in proximity to a strategic route through Livonia, the status of Arrasch is comparable to that of Helmet (Est. Helme) or Tarwast (Est. Tarvastu), insofar as it was peripheral centre managing a territory on behalf of a convent.

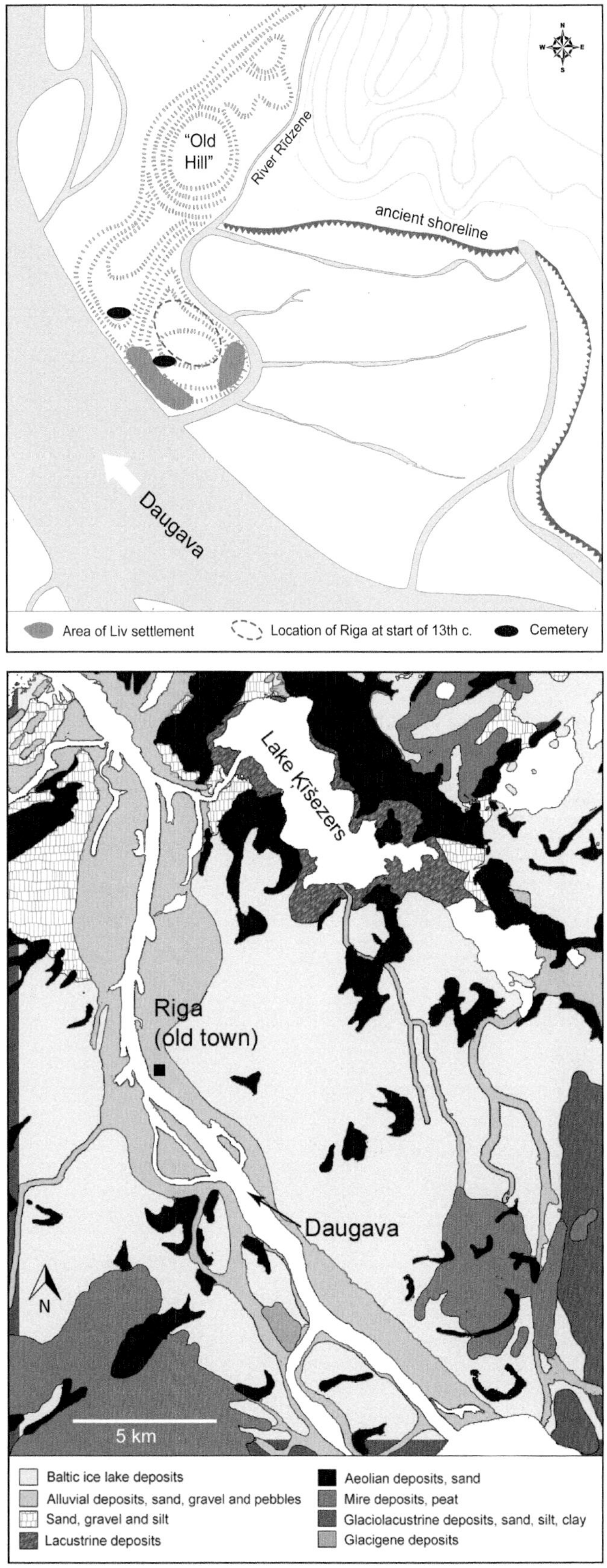

Map 4.5: (a) The vicinity of Riga in the twelfth century (after Caune 2007b, 61); (b) a map of the superficial geology (1: 200,000) (after Latvian Geological Institute).

5. *Riga*

The establishment of Riga, the modern capital of Latvia, is synonymous with the Livonian Crusade. It is the third fortified colony, built under the direction of Bishop Albert in *c.* 1201, after Üxküll and Holm which were constructed under Bishop Meinhard, but quickly became the most important town in Livonia, the base of the Sword Brothers, and the springboard for crusades throughout the thirteenth century, as well as select *Reisen* into northern Samogitia and Lithuania in the fourteenth century. The Late Iron Age of the central Riga district is relatively well known, consisting of two settlements and cemeteries situated on the peninsula bordered by the Daugava in the west and the River Rīdzene on its eastern and southern flanks, as well as stretches of wetlands (Map 4.5). The Lower Daugava, like the Gauja, formed a major territorial boundary between the territories of the Finno-Ugric Livs and the Baltic Semigallians, marked by distinctive sites and material culture. Contact between German merchants and the inhabitants of the Lower Daugava preceded the crusades in the 1180s, and this soon accelerated into a series of violent confrontations resulting in the destruction or procurement of the indigenous strongholds which had previously dominated this strategic riverine zone. The town of Riga was constructed over the course of the thirteenth century (Fig. 4.19), enclosed by a series of walls resulting in the incorporation of the earlier Liv settlements within Riga's bounds by 1225 (Caune 2007b, 156–67). The creation of Riga changed the configuration of the Lower Daugava, creating a new commercial centre at the expense of the existing distribution of trading interests, and many of the settlements stagnated or came under the control of Riga's various authorities (Blomkvist 2005, 531); indeed, this was part of the process of creating a self-governing town which required a newly defined territory with substantial provisioning systems. The Order gained Imperial ratification of its rule over the town in 1274, which resulted in open conflict that was not resolved, at least in the short-term, until 1330 (Czaja 2013, 221, 235).

Ownership of the town's lands (*Stadtmark*, *Marke*, or *patrimonium*) was subdivided between the town, the Archbishops and the Cathedral Chapter, the Sword Brothers, and later the Livonian branch of the Order, and these diachronic vicissitudes are relatively well documented (see Chapter 9). The reorganization of lands in Livonia resulted in the loss of the town's distant territories, and by the end of the thirteenth century its patrimony encompassed *c.* 4900 hectares in the area of the

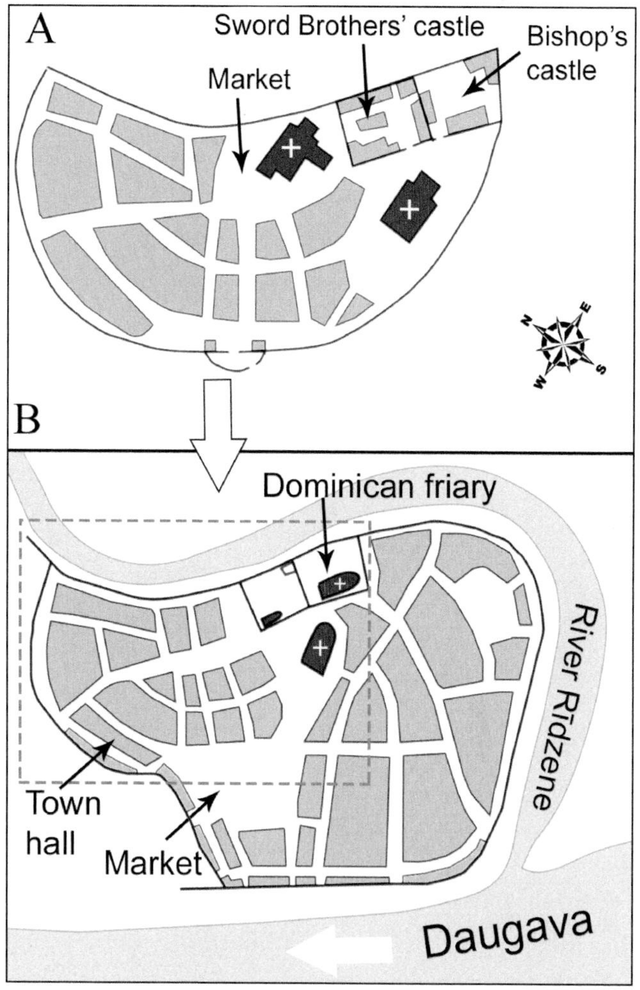

Figure 4.19: The development of Riga during the thirteenth century; (a) *c.* 1201; (b) mid-thirteenth century (after Caune 2007b, 163).

Lower Daugava. The town itself expanded to occupy an area of *c.* 28 hectares and by the fifteenth century had around eight thousand inhabitants, the largest number in Livonia. The mosaic of ownership within the patrimony connected the various social factions of the city with its suburban sprawl, as well as outlying rural settlements, pastures, fields, and woods (Šterns 1998a). At the same time, the town's territory which extended across the Daugava was constrained by neighbouring commanderies and episcopal lands. No sacred sites have been positively identified within the former patrimony of Riga, with the exception of a potential cult site under the cathedral, dated to the twelfth century (Caune 1992). However, given the concentration of theocratic authority in the city, the early emphasis on missionary activity as well as a dominant colonizing population, this is not surprising.

Ethnically, Riga was diverse from its very foundation, with a nominally Christianized Liv population identifi-

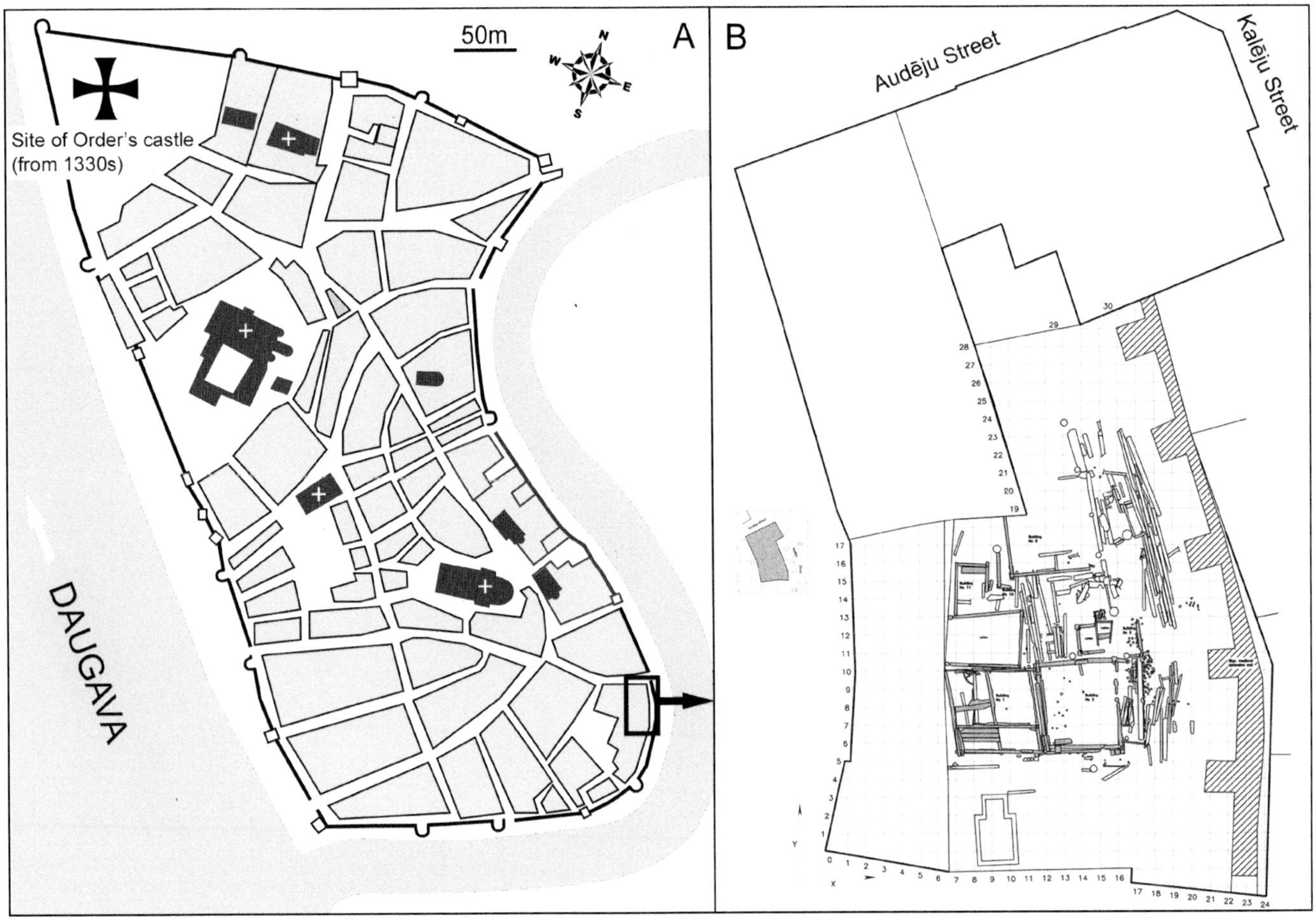

Figure 4.20: Plan of Riga from the end of the thirteenth century showing: (a) the location of the castle and Liv quarter sites included within this study (after Caune 2007b); (b) plan of the excavation in the Liv quarter in the Old Town of Riga. Plan by U. Kalējs, modified by A. Pluskowski.

able by a range of distinctive material culture, including architecture (Smiltniece and Vijups 1998), and other ethnicities noted in documentary sources. *Das Rigische Schuldbuch* (The Debtor's Book) from the end of the thirteenth century indicates the presence of German merchants primarily from west and north Germany, but also from England, Hungary, and other parts of Livonia. The presence of Livs, Latgalians, Estonians, Lithuanians, and Russians is also noted; the latter, who maintained a mercantile presence in Riga but also participated in other civic activities, had their own district (*Russche dorp*) and church dedicated to St Nicholas (Šterns 1998b; Selart 2009, 38–41). Overall, the German character of Riga was due to the dominant role of its resident political, military, mercantile, and religious institutions. Nonetheless, despite the significant presence of the ruling theocracy, the town quickly developed a sense of autonomy, resulting in a protracted, antagonistic relationship with the Order, which culminated in attacks on the convent on a number of occasions. The town continued to struggle against the rule of the Order, pressed into recognizing the authority of both theocratic factions in the Treaty of Kirchholm in 1452 and resulting in the destruction of the castle in 1484. The ultimate solidarity of the town's population against the Order's rule is evident in a proclamation issued by the town council on 8 May 1484. This stated that after the surrender of the Order's castle, everyone — rich and poor, old and young, German and non-German — had the right to take part in razing the castle to the ground, a process completed by mid-August (*SRL* II: 780, 790, 792). The town was defeated militarily by Wolter von Plettenberg in 1491, the Livonian Marshal at the time.

Archaeology

The historic centre of Riga was designated as a UNESCO World Heritage Site in 1997, and it has been the focus of excavations for many decades. Within Riga, two key sites provided material which was incorporated into the project (Fig. 4.20). First, a faunal assemblage recovered from excavations in 2008 the courtyard of the Order's castle derived from early urban (mid-late thirteenth century/start of fourteenth century) and castle phases of occupation into the sixteenth century, confirming the convent had been destroyed in the fifteenth century (Ciglis and Tomsons 2010). Second, a diverse range of ecofacts and sediments from excavations in the Liv quarter in the south-eastern corner of the town (between thirteen Audeju iela and thirty-three Kalēju iela) which took place in 2011–12, were made available for the project. Earlier researchers had defined this area as the city's oldest quarter, where the indigenous Liv settlement was thought to have existed from the twelfth century. The excavations in 2011–12, uncovering an area of 543 m^2, rank amongst the most important in the history of archaeological research in Riga, but the lowest stratigraphic levels were not reached and the date range of the timber buildings began from the early thirteenth century; one of the oldest buildings in Riga was discovered here and dated by dendrochronology to *c.* 1209, constructed within a decade of the town's foundation. In total, the remains of seventeen wooden buildings were uncovered, along with timber streets, several timber latrines, and additional structures such as piling. More than 1400 artefacts were recovered (including jewellery, tools, and domestic artefacts), several hundred objects and fragments of leather, pottery sherds, and over four thousand animal bone fragments. The buildings had been constructed using well-documented indigenous building techniques consisting of logs with cross-joints, in contrast to the timber-framed architecture introduced by the colonists. Virtually all the artefacts could be connected with the Late Iron Age artisanal traditions of the indigenous Liv population, again in contrast to the imported material culture found in other areas of the town and the quarter's character is comparable to the one uncovered between 1959 and 1964 in Albert Square, identified as the other main area of Liv settlement (Caune 2007b, 77–78). The excavation confirmed that this part of Riga would have been densely populated from the thirteenth to fourteenth century. By the mid-thirteenth century, timber-framed structures dominated the townscape but after a fire in 1293, stone and brick were increasingly used for domestic buildings with the introduction of construction regulations (Caune 2007b, 316). However, the consistent integrity of the material culture in the Liv quarter suggests it remained distinct within an otherwise 'Germanized' townscape (Svarāne 1998). Although the focus of analytical studies was on material from within the town, palaeoenvironmental studies within the vicinity of Riga have been conducted by Laimdota Kalniņa (2019), and pollen from the locality was also preserved within one of the latrine fills within the Liv quarter. There is also a significant corpus of documentary sources for Riga which has been included within this study.

Regional Status

The district of Riga can be defined as the edge of a cluster of settlements that controlled the Lower Daugava in the twelfth century. The cores of this extended territory were represented by centres such as Daugmale and their associated territories. This stretch of the river functioned as a broad frontier between Livs, Semigallians and Latgalians, and following the crusades it developed into a clear heartland, with Riga as the most important centre in Livonia.

Secondary Sites

A number of secondary sites were included within the study, which consisted of castle and urban sites associated with the Order, with the Archbishops and the Cathedral Chapter of Riga, and with the bishops of Dorpat.

The Town of Windau (Ventspils)

Faunal remains from excavations in the coastal town of Ventspils (Windau) represented the only archaeological data included within the project from a medieval site

Figure 4.21: Plan showing the urban area in Ventspils (Ger. Windau) occupied from the fourteenth to sixteenth centuries, with the excavations in Maiznieku iela 7/9–11 highlighted. From Vijups 2012, 85, Figure 2.

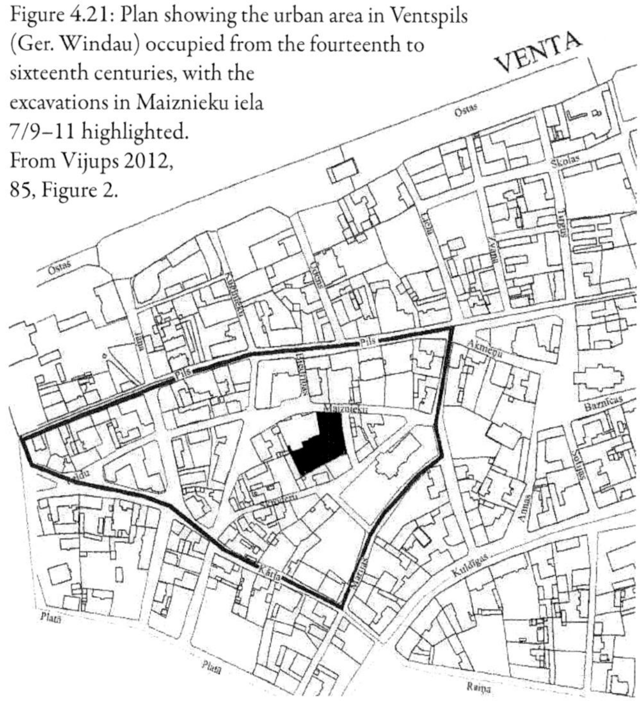

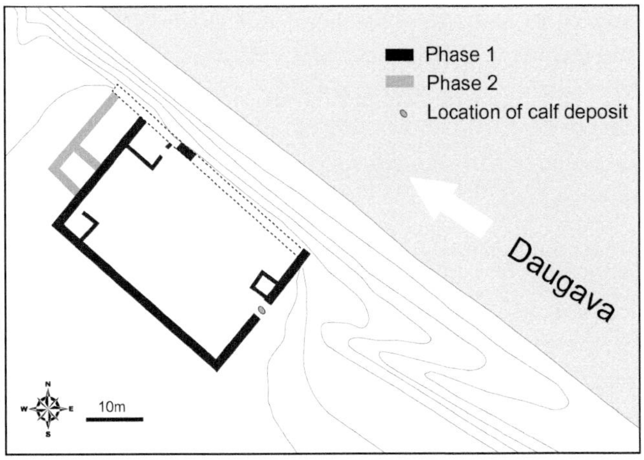

Figure 4.22: Plan of Vecdole Castle (Ger. Alt-Dahlen). After Atgāzis 1999.

in Curonia (Latv. Kurzeme). The town had been established alongside the Order's convent, and whilst the latter had been the focus of excavations, there were problems with accessing material from well-dated occupation deposits. The material derived from occupation layers uncovered during excavations in Maiznieku iela 7/9–11 in the historic centre in 2010, broadly dated from the fourteenth to sixteenth centuries (Vijups 2012) (Fig. 4.21). Further details concerning the convent are presented in Chapter 8.

Castles in the Lower Daugava: Kirchholm (Salaspils) and Alt-Dahlen (Vecdole)

One of the most important medieval sites excavated in Latvia in the second half of the twentieth century was located on Mārtiņsala, an island in the Daugava with a land surface of around 20 hectares. The site consisted of a Liv village with a defensive bank (Sala), a castle attributed to the activity of the first bishop of Livonia, Meinhard (Latv. Salaspils, known from the mid-thirteenth century as Ger. Kirchholm), which subsequently became the property of the Cathedral Chapter of Riga, an associated church and cemetery. The first excavations in 1897 and 1899 recovered over a thousand artefacts, but rescue excavations ahead of the construction of Riga's Hydro-Power Station took place between 1966 and 1974, providing a comprehensive understanding of the site. The results of the excavations were published in a seminal monograph, and the diverse artefactual assemblage has been used as a model for investigating the relationship between the indigenous population and the colonizing elite in medieval Livonia (Mugurēvičs 2008). The village consisted of sunken buildings dating to the twelfth and thirteenth centuries, and standing buildings dating to the fourteenth and fifteenth centuries associated with German timber-framed architecture. The castle, built from stone at the end of the twelfth century, had a second discernible phase from the mid-thirteenth century through to the second half of the fourteenth century. Although no material from the excavation was analysed further by the project, this site and the model of cultural interaction developed by Mugurēvičs represents a fundamental interpretative framework against which other sites could be compared.

The castle of Vecdole (Ger. Alt-Dahlen) was also excavated during the construction of Riga's Hydro-Power Station, although on a much smaller scale than Salaspils. Excavations in 1966–68 focused solely on this enclosure castle located on the island of Dole within the Daugava, just across from Salaspils, and discerned two building phases within a short period of occupation spanning much of the thirteenth century (Atgāzis 1999) (Fig. 4.22). The castle on Dole, first mentioned as *Alt-Dahlen* in 1226 and last mentioned with its destruction in 1298, belonged to the vassals of the bishops of Riga and was destroyed in the war with the Order. Subsequently, a lime kiln functioned at the site in the fourteenth century and the castle does appear to have been rebuilt. The animal bone assemblage from the excavations was made available to the project, and associated with the single phase of occupation in the thirteenth century.

The Bishopric of Dorpat (Tartu)

South-eastern Estonia, corresponding to the tribal territory of Ugaunia (or Ugandi) and associated smaller northern provinces, had been long-involved with political, military, and religious vicissitudes involving the neighbouring Rus' principalities. With the coming of the crusaders in the first decade of the thirteenth century, the Ugaunians ended up fighting a protracted war on two fronts. In 1224, the neighbouring territory of Saccala was officially annexed by the Sword Brothers, and Ugaunia came under the control of Bishop Hermann von Buxhoeveden, who established his principal residence at Dorpat. The bishopric lasted until the town was conquered by the army of Ivan IV in 1558. A series of sites within Tartu, and the village cemeteries in its immediate territory, have been the focus of ongoing research by archaeologists at the University of Tartu. The most important assemblages were represented by human remains derived from a series of cemetery sites in the town and in a series of nearby villages. The results of these will be published in the future and will contribute an important human osteoarchaeological dataset on dietary differences between urban and rural populations in medieval Livonia. The network of sacred sites documented within the Bishopric of Dorpat also provided a useful comparison for the Order's territories around Fellin and Karkus (Valk 2019).

Conclusion

The primary Livonian sites included within the study represent a range of frontier types, with sustained heartlands (Viljandi, Cēsis, and Riga) and transitional peripheries (Karksi, Āraiši) defining the use of these sites and their landscapes from the Late Iron Age to the Middle Ages. The exact diachronic boundaries of their territories are difficult to define, even for the medieval period, when the Order recorded its estates meticulously: the evidence simply has not survived. The precise marking of boundaries appeared after the crusades but is only well known for select fiefs in Livonia and for specific periods of time, particularly in Curonia (see Chapter 9). Future work would particularly benefit from investigating sites and their associated territories in the southernmost borderland of Livonia, as well as the Curonian commandery of Goldingen.

Works Cited

Primary Sources

Biskup, M., and Janosz-Biskupowa, I., eds. 2002. *Visitationen im Deutschen Orden im Mittelalter*, I: *1236–1449*, Quellen und Studien zur Geschichte des Deutschen Ordens, 50, Marbug: Elwert

——, eds. 2004. *Visitationen im Deutschen Orden im Mittelalter*, II: *1450–1519*, Quellen und Studien zur Geschichte des Deutschen Ordens, 50, Marbug: Elwert

HCL Brundage, J. A., trans. 2003. *The Chronicle of Henry of Livonia/Henricus Lettus*, New York: Columbia University Press

Jakubowsky, J., and Kordzikowski, J. 1915. *Polska XVI wieku pod wzglednem geograficno-statystycznym*, XIII: *Inflanty, Cz. 1*, Warsaw: Skład Główny Gebethnera i Wollfa

LUB Bunge, F. G. von, ed. 1853–57. *Liv-, Est- und Kurländisches Urkundenbuch nebst regesten*, 13 vols to date, Reval: Kluge and Ströhm

Fenske, L. and Militzer, K. 1993. Ritterbrüder im livländischen Zweig des Deutschen Ordens, Köln: Böhlau

Ose, I., ed. 2011. 'Cēsu pils 1582. gada revīzija', *Latvijas viduslaiku pilis* 7, 429–32

SRL Fuchs, M., ed. 1853. *Das rothe Buch inter Archiepiscopalia // Scriptores Rerum Livonicarum*, 2 vols, Rīga: Frantzen

Švābe, A. 1939. *Senās Latvijas vēstures avoti*, Rīga: Latvijas Vēstures institūta apgādiens

Secondary Works

Alttoa, K. 2009. 'Karksi keskajal: Mõningaid probleeme', *Viljandi Muuseumi Aastaraamat* 2008, 266–74

——, 2015. 'Viljandi ordulinnuse arhitektuur', in *Viljandi ordulinnus ja Lossimäed läbi aegade*/The Teutonic Order's Castle and Castle Hills in Viljandi through Time, Viljandi: Viljandi Muuseum, 87–110

Apala, Z. 2007. 'Cesu pils arheoloģiskās izpētes (1974–2006) rezultāti', in *Cēsīm 800: Quo vadis, Cēsis? Vēsture un mūsdienu nosacījumi pilsētas attīstībai*, ed. V. Rozentāle et al., Cēsis: Cēsu pašvaldības aģentūra, Vidzemes Vēstures un tūrisma centrs, 45–76

——, 2010a. 'Arheoloģiskie pētījumi Cēsu Rožu laukumā', *Arheologu Pētījumi Latvijā* 2008/09, 68–72

——, 2010b. 'Jaunākie arheoloģiskie pētījumi Cēsu Riekstu kalnā', *Arheologu Pētījumi Latvijā* 2008/09, 35–40

——, 2010c. 'Pabeigta Cēsu viduslaiku pils pagalma arheoloģiskā izpēte', *Arheologu Pētījumi Latvijā* 2008/09, 65–68

Apals, J. 1995. 'Āraišu mūra pils arheoloģiskā izpēte', *Cēsu novada vēsture* 1, 59–76

——, 1996. '*Vēstures avotu ziņas* par *Āraišu mūra pili*', *Arheoloģija un etnogrāfija* 18, 27–37

——, 2007. 'Cēsis un vendi', in *Cēsīm 800: Quo vadis, Cēsis? Vēsture un mūsdienu nosacījumi pilsētas attīstībai*, ed. V. Rozentāle et al., Cēsis: Cēsu pašvaldības aģentūra, Vidzemes Vēstures un tūrisma centrs, 25–44

——, 2012. *Āraišu ezerpils: rakstu izlase un draugu atmiņas*, Rīga: LU Latvijas vēstures institūts

Apala, Z., and Apals, J. 2014. 'The Vendic Hill Fort on Riekstu Kalns in Cēsis', in *Strongholds and Power Centres East of the Baltic Sea in the 11th–13th Centuries*, ed. H. Valk, Tartu: Institute of History and Archaeology at the University of Tartu, Department of Archaeology at the Institute of History of Tallinn University, 115–38

Atgāzis, M. 1999. 'Vecdoles pils nocietinājumi un apbūve', *Latvijas Viduslaiku Pilis* 1, 312–48

Berga, T. 1980. *Zinātniskās atskaites sesijas materiāli par arheologu un etnogrāfu 1979. gada pētījumu rezultātiem*, Rīga: Zinātne

Blomkvist, N. 2005. *The Discovery of the Baltic: The Reception of a Catholic World-System in the European North (AD 1075–1225)*, Leiden: Brill

Bojtár, E. 1999. *Foreword to the Past: A Cultural History of the Baltic People*, Budapest: Central European University Press

Caune, A. 1992. 'Arheoloģiskās liecības par senāko apdzīvotību Rīgas Doma baznīcas apkārtnē', *Latvijas Vēstures Institūta Žurnāls* 3, 21–38

——, 2007a. 'Cēsis 14.–18. gadsimtā: plānojums, apbūve un iedzīvotāji', in *Cēsīm 800: Quo vadis, Cēsis? Vēsture un mūsdienu nosacījumi pilsētas attīstībai*, ed. V. Rozentāle et al., Cēsis: Cēsu pašvaldības aģentūra, Vidzemes Vēstures un tūrisma centrs, 155–66

——, 2007b. *Pētījumi Rīgas archeoloģijā*, Rīga: LVIA

Ciglis, J., and Tomsons, A. 2010. 'Pētījumi Rīgas pils Galvenās pils palgalmā', *Arheologu Pētījumi Latvijā* 2008/09, 96–99

Czaja, R. 2013. 'Miasta inflanckie XIII–XVI w', in *Zakon krzyżacki w Prusach i Inflantach: podziały administracyjne i kościelne w XIII–XVI wieku*, eds R. Czaja and A. Radzimiński, Toruń: Wydawnictwo Naukowe Uniwersytetu Mikołaja Kopernika, 215–40

Haak. A. 2004. 'Archaeological Investigations at Viljandi Castle of the Teutonic Order and in Medieval Viljandi', *Arheoloogilised välitööd Eestis/Archaeological Fieldwork in Estonia* 2003, 107–21

——, 2005. 'Archaeological Investigations at the Late Iron Age Settlement Site of Huntaugu and the Castle of the Teutonic Order in Viljandi', *Arheoloogilised välitööd Eestis/Archaeological Fieldwork in Estonia* 2004, 89–101

——, 2015. 'Arheoloogia andmed Viljandi ordulinnuse kesk- ja uusaja kohta', in *Viljandi ordulinnus ja Lossimäed läbi aegade*/The Teutonic Order's Castle and Castle Hills in Viljandi through Time, Viljandi: Viljandi Muuseum, 27–64

Haak, A., and Russow, E. 2013. 'On the Development of the Town of Viljandi in the Light of Earliest Archaeological Find Complexes', *Estonian Journal of Archaeology* 17/1, 57–86

Hayward, K. M. J. 2019. 'Resources for Castle Building in Prussia and Livonia', in *Ecologies of Crusading, Colonization, and Religious Conversion in the Medieval Baltic: Terra Sacra II*, ed. A. Pluskowski, Environmental History of the North Atlantic World, 3, Turnhout, Brepols (forthcoming)

Kalniņa, L., Ceriņa, A., Kiziks, K., Stankeviča, K., Pujāte, A., and Dručka, A. 2019. 'Evidence of Human Impact and Vegetation Change during the Late Iron Age and the Medieval Livonian Period at Some Sites along the Lower Course of the River Daugava', in *Ecologies of Crusading, Colonization, and Religious Conversion in the Medieval Baltic: Terra Sacra II*, ed. A. Pluskowski, Environmental History of the North Atlantic World, 3, Turnhout, Brepols (forthcoming)

Kalniņš, G. 2014. *Cēsu Pils (Cēsis Castle)*, Cēsis: Cēsis Culture and Tourism Centre

Kreem, J. 2015. 'Saksa ordu ja Viljandi linnus', in *Viljandi ordulinnus ja Lossimäed läbi aegade*/The Teutonic Order's Castle and Castle Hills in Viljandi through Time, Viljandi: Viljandi Muuseum, 65–86

Laime, S. 2009. *Svētā pazeme: Latvijas alu folklore*, Rīga: Zinātne

Lang, V., and Valk, H. 2011. 'An Archaeological Reading of the Chronicle of Henry of Livonia: Events, Traces, Contexts and Interpretations', in *Crusading and Chronicle Writing on the Medieval Baltic Frontier*, eds M. Tamm, L. Kaljundi, and C. S. Jensen, Aldershot: Ashgate, 291–316

Mänd, A. 2009. 'Saints' Cults in Medieval Livonia', in *The Clash of Cultures on the Medieval Baltic Frontier*, ed. A. V. Murray, Farnham: Ashgate, 191–223

Mugurēvičs, Ē., 2008. *Viduslaiku ciems un pils Salaspils novadā*, Rīga: Latvijas vēstures institūta apgāds

Murray, A. 2013. 'The Sword Brothers at War: Observations on the Military Activity of the Knighthood of Christ in the Conquest of Livonia and Estonia (1203–1227)', *Ordines Militares*, 18, 27–38

Neitmann, K. 1993. 'Riga und Wenden als Residenzen des livländischen Landmeisters im 15. Jahrhundert', in *Stadt und Orden: Das Verhältnis des Deutschen Ordens zu den Städten in Livland, Preussen und im Deutschen Reich*, ed. U. Arnold, Marburg: Elwert, 59–93

Palli, H. 1996. *Eesti rahvastiku ajalugu aastani 1712*, Tallinn: Teaduste Akadeemia Kirjastus

Selart, A. 2009. 'Russians in Livonian Towns in the Thirteenth and Fourteenth Centuries', in *Segregation-Integration-Assimilation: Religious and Ethnic Groups in the Medieval Towns of Central and Eastern Europe*, eds D. Keene, B. Nagy, and K. Szende, Farnham: Ashgate, 33–50

Smiltniece, E., and Vijups, A. 1998. 'Rotas Alberta laukuma arheoloģiskajā materiālā', *Senā Rīga* 1, 105–24

Šterns, I. 1997. *Latvijas vēsture, 1290–1500*, Rīga: Daugava

——, 1998a. 'Viduslaiku Rīga ārpus Rīgas', *Senā Rīga* 1, 342–77

——, 1998b. 'Viduslaiku Rīgas Krievu ciema plāns', *Senā Rīga* 1, 22–54

Stivriņš, N., Brown, A., Reitalu, T., Veski, S., Heinsalu, A., Banerjea, R. Y., and Elmi, K. 2015. 'Landscape Change in Central Latvia since the Iron Age: Multiproxy Analysis of the Vegetation Impact of Conflict, Colonization and Economic Expansion during the Last 2,000 years', *Vegetation History and Archaeobotany* 24, 377–91

Svarāne, D. 1998. 'Rīgā, Mārstaļu ielā 3, atrastā 13. gadsimtā guļbūve', *Senā Rīga* 1, 263–72

Tamm, M. 2011. 'Inventing Livonia. The Name and Fame of a New Christian Colony on the Medieval Baltic Frontier', *Zeitschrift für Ostforschung* 60, 186–209

Tomsons, A. 2014. *Pārskats par arheoloģiskajiem izpētes darbiem Āraišu mūra pilī un priekšpilī (valsts aizsardzības nr. 401). 2012. gada 9.-18. Jūlijā*, Rīga: National History Museum (unpublished report)

Tuulse, A. 1942. *Die Burgen in Estland und Lettland*, Dorpat: Aasta

Vaitkevičius, V. 2004. *Studies into the Balts' Sacred Places*, BAR International Series, 1228, Oxford: Archaeopress

Valk, H. 2001a. *Rural Cemeteries of Southern Estonia: 1225–1800 AD*, Tartu: Archaeology Centre, University of Tartu

Valk, H. 2001b. 'Besieging Constructions from 1223 in Viljandi', *Arheoloogilised välitööd Eestis/Archaeological Field Works in Estonia*, 2000, 65–79

——, 2005. 'The Genesis of Viljandi (Fellin): Archaeological Data', in *Riga und der Ostseeraum von der Gründung 1201 bis in die Frühe Neuzeit*, eds I. Misāns and H. Wernicke, Tagungen zur Ostmitteleuropa-Forschung, 22, Marburg: Herder Institut, 95–107

——, 2015. 'Viljandi muinasaeg', in *Viljandi ordulinnus ja Lossimäed läbi aja/The Teutonic Order's Castle and Castle Hills in Viljandi through Time*, Viljandi: Viljandi Muuseum, 5–23

——, 2019. 'Sacred Landscapes of Medieval Livonia: Lands of the Teutonic Order in Southern Estonia', in *Ecologies of Crusading, Colonization, and Religious Conversion in the Medieval Baltic: Terra Sacra II*, ed. A. Pluskowski, Environmental History of the North Atlantic World, 3, Turnhout, Brepols (forthcoming)

Valk, H., and Malve, M. 2008. 'Archaeological Supervision Work in the Outer Bailey of Karksi Castle', *Arheoloogilised välitööd Eestis/Archaeological Field Work in Estonia*, 2007, 121–26

Valk, H., Pluskowski, A., Brown, A., Rannamäe, E., Malve, M., and Varul, L. 2012. 'The Castle of the Teutonic Order in Karksi: Preliminary Excavation Results', *Arheoloogilised välitööd Eestis/Archaeological Field Work in Estonia*, 2011, 47–56

Valk, H., Pluskowski, A., Thornley, D., Brown, A., and Summerfield, C. 2009. 'Fluxgate Gradiometry Survey in the Ruins of Karksi Castle and Palaeoenvironmental Analysis in its Hinterlands', *Arheoloogilised välitööd Eestis/Archaeological Field Work in Estonia*, 2008, 213–19

Valk, H., Rannamäe, E., Brown, A., Pluskowski, A., Badura, M., and Lõugas, L. 2013. 'Thirteenth Century Cultural Deposits at the Castle of the Teutonic Order in Karksi', *Arheoloogilised välitööd Eestis/Archaeological Field Work in Estonia*, 2012, 73–92

Viitso, T-R. 2009. 'Livonian and Leivu: Shared Innovations and Problems', *Linguistica Uralica* 45/4, 269–82

Vijups, A. 2012. 'Ventspils pilsētas senākās daļas aizputināšana ar smiltim 17. gs.: arheoloģisko pētijumu rezultāti', *Ventspils Muzeja Raksti* 7, 79–93

Zeids, T., ed. 1978. *Feodālā Rīga*, Rīga: Zinātne

Zemītis, G. 2014. 'The Hill Fort of Satesele — A Power Centre in the Territory of the Livs in the 11th–13th Centuries', in *Strongholds and Power Centres East of the Baltic Sea in the 11th–13th Centuries*, ed. H. Valk, Tartu: Institute of History and Archaeology at the University of Tartu, Department of Archaeology at the Institute of History of Tallinn University, 445–61

Zunde, M. 2019. 'Timber and its Use from the Late Iron Age to the End of the Medieval Period in Latvia', in *Ecologies of Crusading, Colonization, and Religious Conversion in the Medieval Baltic: Terra Sacra II*, ed. A. Pluskowski, Environmental History of the North Atlantic World, 3, Turnhout, Brepols (forthcoming)

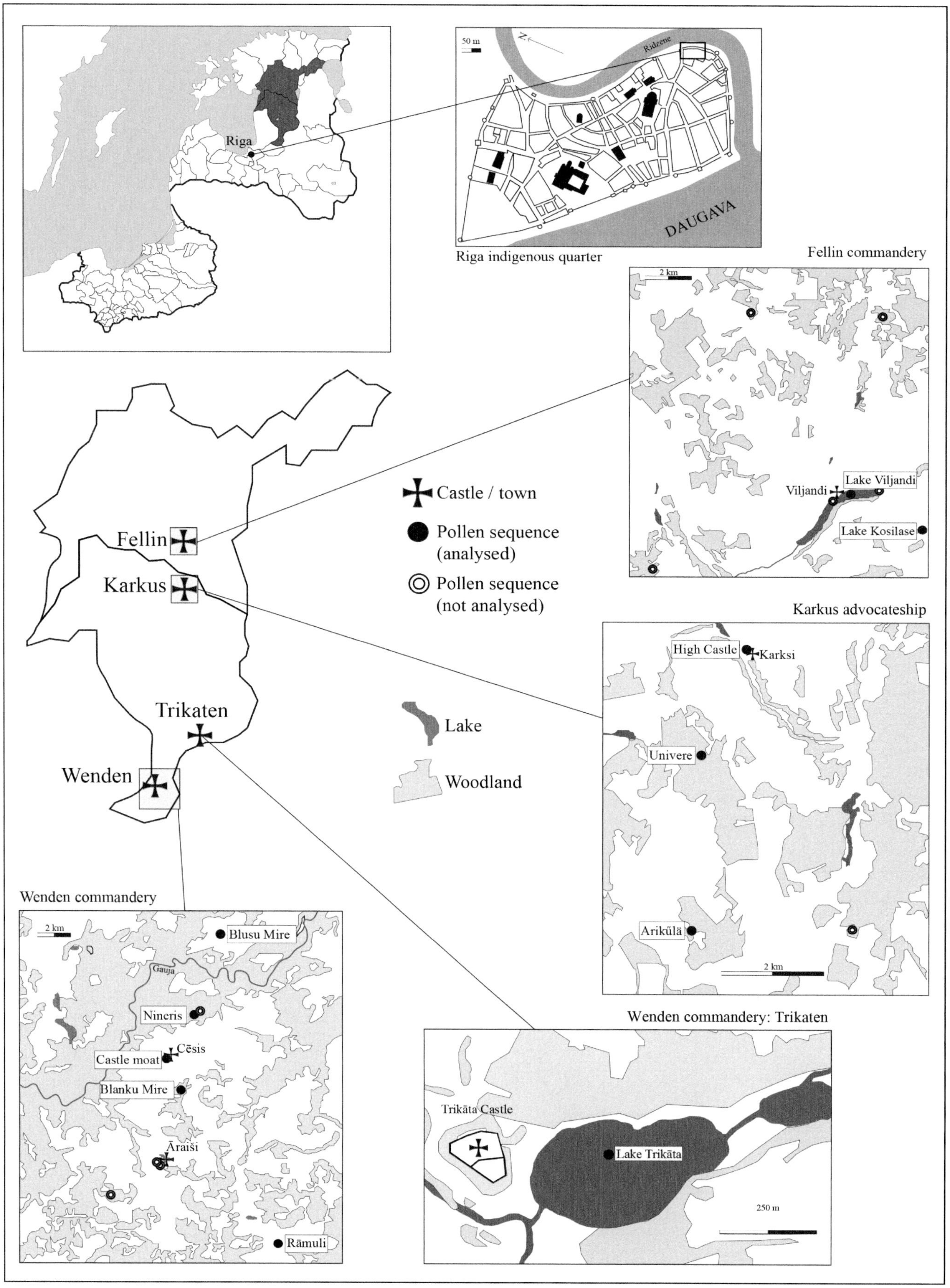

Map 5.1: Location map of pollen studies from Livonia within Riga, Wenden, and Fellin commanderies and Karkus advocateship. Figure by author.

5. Vegetation Change in Livonia: The Palynological Data

Alexander Brown
With contributions by Normunds Stivriņš and Jenny Austin

Introduction

The commanderies of Wenden (Latv. Cēsis), Fellin (Est. Viljandi), the advocateship of Karkus (Est. Karksi), and the Liv quarter in Riga form the focus for palynological studies from Livonia, comprising a significant tract of land extending from present-day central Latvia through to southern Estonia. Palynological data is presented from six peat bogs, three lakes, and cultural deposits from archaeological excavations at Cēsis, Karksi, and Riga (Map 5.1). The studies are ordered in the text geographically from south to north mirroring the broad chronological progression of the crusade, starting with the foundation of Riga, the conquest of the Livs and Latgalians, followed by the conquest of the southern Estonians. The principal aim is to construct detailed vegetation histories for these sites and micro-regions, providing data on how local environments were transformed during the medieval period, set within the broader timeframe of environmental and culture change since the Iron Age. These data are considered in the context of existing palynological studies from across the eastern Baltic, some presented in Volume II, providing an important lens for considering the impact of the crusades on the physical landscape and patterns of land-use across the Eastern Baltic.

Present-Day Biogeography, Relief, and Climate

Both Latvia and Estonia belong biogeographically to the hemiboreal vegetation zone, characterized by mixed boreal coniferous and deciduous (boreo-nemoral) forests dominated by pine (*Pinus sylvestris*), spruce (*Picea abies*), birch (*Betula*), and alder (*Alnus glutinosa*), along with smaller quantities of broad-leaved trees including ash (*Fraxinus*), elm (*Ulmus*), lime (*Tilia*), oak (*Quercus*), and hazel (*Corylus avellana type*). Today, forests form approximately 55% of the land area of Latvia and Estonia (Forest Europe, UNECE, and FAO 2011), with a lesser component covered by agricultural land (*c.* 32%), with peatlands forming a significant component of the land area, totalling 10% in Latvia (Kalniņa et al. 2014) and 6% in Estonia; lakes account for a further 4.6% of the Estonian land-mass (Reintam et al. 2005). The relief of the eastern Baltic has been shaped primarily during the last glaciation and deglaciation, characterized by a largely low-relief averaging no greater than 100 m a.s.l, with the highest points at *c.* 312 m a.s.l in the Vidzeme uplands, north-east Latvia (Zelčs and Markots 2004) and *c.* 318 m a.s.l in the Haanja Heights, south-east Estonia (Niinemets 2010). The climate is transitional with mean January and July temperatures varying between -2.5°C and -6.5°C and +17°C in Estonia (Seepä and Poska 2004) and -3°C--8°C and +16°C –+18°C in Latvia (Heikkilä and Seepä 2010).

Palynological Studies

Riga

Riga Liv quarter (56°56′49″ N, 24°06′45″ E): located in the south-eastern corner of the historic medieval centre of Riga, excavations revealed the exceptionally well-preserved waterlogged remains of timber buildings, latrines, and animal dung, dating to the thirteenth and fourteenth centuries. Samples for palynological analysis were taken from the mid-fourteenth century latrine and from animal manure in contexts sealed by and adjacent to several of the buildings. The site is located adjacent to the medieval city walls; the city itself was built on a marshy peninsula bounded by the Rīdzene River, which ran adjacent the town wall to the west, and the River Daugava to the east.

Wenden Commandery

Blusu Mire (57°17′28″ N, 25°17′20.1″ E): small raised bog (*c.* 45 hectares) located on the present south-eastern edge of Cēsis, *c.* 2.2 km from Cēsis Castle. The bog is chiefly covered in pine, birch, and spruce woodland, especially along the margins, and more sparsely within the centre of the bog, where the surface is dominated by dwarf shrubs and bryophytes dominated by *Sphagnum* species. Coring across the mire indicates a relatively shallow depth of peat to a maximum depth of 1.5 m, but typically no more than 0.75 m. The core was taken from the greatest depth of deposits within the central area of the current mire, comprising moss (*Sphagnum*) peats (0.10–1.06 m) overlying structureless organic matter (1.06–1.50 m). The peat overlies silts, suggesting a former now infilled lake. Pollen was focused on the top 0.72 m of the core covering the last 1500 years; the peat at 1.4 m dated to 4878±33 BP (GU-29816, 3712–3541 cal. BC).

Nineris Mire (57°20′10.4″ N, 25°17′16.5″ E): located adjacent to Lake Nineris and extending a further 800 m to the west, the mire comprises a sequence of ombrotrophic peats several metres deep. The bog surface is currently covered with young pine trees and a surface layer of dwarf shrubs, surrounded on the dry ground by mature pine plantation woodland. The top 2.5 m was sampled, comprising a *Sphagnum*-dominated moss peat, with palynological analysis undertaken to 2.18 m on sediments dating from the fourth century.

Blanku Mire (57°23′12.4″ N, 25°18′05.1″ E): large raised mire (*c.* 300 hectares) 8 km north of Cēsis Castle. The bog is largely open, dominated by dwarf shrubs and bryophytes with a sparse cover of young pine trees; the margins of the bog are surrounded by pine-dominated woodland. Coring across the mire indicates a shallow depth of moss peat to a maximum depth of 2.25 m overlying lacustrine silts. The peat was sampled to a depth of 2 m, *c.* 150 m from the western edge of the mire. Analysis focused on the top 1.1 m covering the last 1500 years, comprising a well humified *Sphagnum* moss peat. The peat at 1.65 m dates to 2575±29 BP (GU-29811, 810–569 cal. BC).

Cēsis Castle moat (57°18′46.7″ N, 25°16′14.3″ E): coring and excavations revealed a deep sequence of waterlogged sediment extending to a depth of *c.* 4 m within the moat fronting the southern range of the castle. Pollen samples were taken from 0.8–2.34 m, avoiding disturbed sediments at the top and overlying a timber drain at *c.* 2.5 m, dendrochronologically dated to AD 1374/75. The sediments underneath the timber drain are both archaeologically and geochemically sterile and were not sampled; these sediments, homogenous in character, most likely accumulated rapidly after construction, eroding into the moat before the banks stabilized with vegetation.

Rāmuli (57°12′49.3″ N, 25°22′43.2″ E): small (4 hectares) basin infill 14 km east-south-east of Cēsis, located within the edge of pine-dominated woodland adjacent to agricultural fields 1 km from the settlement of the same name. The deposits comprise a sequence of structureless herbaceous peats to a depth of 2.35 m. Only the top 1.75 m was sampled, with analysis undertaken on the top 0.5 m, covering the last thousand years.

Lake Trikāta (57°32′28.2″ N, 25°42′49.9″ E): the largest of two lakes (13 hectares), fed by a small stream with an outflow connected with the River Abuls (a tributary of the Gauja) on the western side of the lake. The castle and medieval settlement of Trikāta (Ger. Trikaten) occupy the plateau along the eastern and northern shore. The top 3.4 m of sediment was sampled, consisting of a sequence of organic silts of varying hue from dark olive brown (4.2–5.15 m), olive brown (5.15–5.36 m), very dark greyish brown (5.36–5.42 m), light olive brown (5.42–5.71 m), greyish brown (5.71–5.80 m), dark yellowish-brown (5.80–6.0 m), and very dark brown (6.0–7.4 m); the lake has a water depth of 4 m. The landscape surrounding Trikāta is predominantly agricultural fields, and meadows developed on sandy and podzolic soils, with stands of alder, birch, spruce, and pine surrounding the edges of the lake. Large tract of woodland, largely plantation, 2.5 km to the north-west extending to the Estonian border along the course of the River Gauja, with more fragmented woodland to the south and south-east.

Fellin Commandery

Lake Viljandi (58°21′26.5″ N, 25°36′15.1″ E): large (135 hectares) sinuous lake running north-east to south-west for 4.5 km along the south-eastern edge of Viljandi and fed by two small streams. The lake has a maximum depth of 11 m and a mean depth of 5 m with a catchment of 66.8 km^2 (Vassiljev, Harrison, and Haxeltine 1995). Coring revealed a relatively shallow depth of highly minerogenic sediments (*c.* 1.5 m) overlying reworking Devonian sandstone. The sequence was taken from the northern section of the lake, 200 m from the shoreline adjacent to the castle. The steep eastern shores of the lake are covered with woodland. The landscape around Viljandi is largely agricultural but with more substantial tracts of continuous woodland 20 km to the west in south-western Estonia where there are extensive raised bog formations.

Lake Kosilase (58°20'57.2" N, 25°39'48.6" E): small (3 hectares) isolated basin located 3.5 km east of Lake Viljandi. The lake lacks either inflowing or outflowing streams, limiting the inflow of mineral matter and resulting in a highly organic sequence of Holocene sediment to a depth of 10 m; only the top 3.4 m was sampled. The lake is surrounded by a thin strip of pine woodland within an otherwise largely agricultural landscape, but with more extensive woodland to the north and south.

Karkus Advocateship

Karksi high castle (58°06'20" N, 25°33'50" E): excavations within the high castle revealed waterlogged sediments at the base of a sequence of late medieval cultural deposit. Pollen analysis was focused on two main deposits; a highly organic layer (13) interpreted as dumped fodder, and the overlying layer (12), interpreted as a midden, and including significant quantities of animal bone, fish remains, leather, and other cultural material. Radiocarbon dating of these deposits indicates that they accumulated during the second half of the thirteenth century (cal. AD 1266–90).

Univere, Karksi (58°05'26.5" N, 25°33'24.9" E): small fen deposit 1 km south-west of Karksi Castle, consisting of sequence of organo-mineral deposits (0–0.22 m) overlying herbaceous peats (0.22–1.07 m) and lacustrine silts (1.07 m–) sampled to a depth of 1.25 m. The location is fringed by a large area of woodland to the immediate south-west and a mosaic landscape of open fields and patchy woodland to the north-east.

Äriküla (58°03'26" N, 25°32'27" E): large raised bog *c.* 5 km south-south-east from Karksi Castle, comprising 2.8 m of *Sphagnum-Calluna* peat overlying sandy-clay loam. The bog is currently wooded with planted pine trees, an understorey of birch and a ground flora of dwarf shrubs and bryophytes. Pollen analysis was undertaken on the top 1 m of sediment.

Sequences Assessed but not Analysed

Eight additional cores were assessed but not analysed, in most cases due to poor pollen preservation, slow accumulation rates, and poor chronological resolution of sediments, or through subsequent removal of medieval and later deposits.

Pollen was variably preserved in two sequences from Āraiši Castle, one from a small bog within the outer bailey and a second external to the outer eastern wall. Pollen from both sequences is heavily dominated by local sedge-fen communities that are likely to inhibit the ingress of pollen from the surrounding landscape, masking the wider cultural signal. Pollen was also largely absent in twelfth- to fourteenth-century deposits from the outer bailey sequence, although geochemical analysis has been undertaken. However, pollen was well preserved in a sequence retrieved and analysed by Normunds Stivriņš from Lake Āraišu on the western side of the castle (Stivriņš et al. 2015). A shallow 0.75 m peat sequence from Bille (2.5 km SSE from Āraiši) was found to be heavily truncated with a date at 36 cm of 4167±36 BP (2883–2631 cal. BC).

Lake Nineris, located adjacent to Nineris Mire, was also sampled. The top 3.5 m of sediment was sampled, comprising highly homogenous lake muds, with a maximum water depth of 9.5 m. However, it became apparent subsequent to analysis and detailed radiocarbon dating that the upper 2 m of the Lake Nineris sequence had been disturbed, either during sampling or transport owing to the unconsolidated nature of the sediments. The sequence includes three overlapping cores (9.5–10.5, 10.3–11.3, and 11.1–12.1 m). Six samples were sent for radiocarbon dating, three from the first core, two from the second and one from the third core. The top four radiocarbon dates are inverted (972±33 BP–9.88 m; 905±35 BP–10 m; 805±33 BP–10.45 m; 745±35 BP–10.86 m). In addition, the pollen signal from *c.* 1.10 m appears highly homogenized through mixing, with constant values for all major plant taxa.

In addition, radiocarbon dating of peat deposits from three cores several kilometres to the north, north-west, and west of Viljandi demonstrated that the medieval deposits had been truncated to varying degrees. Pollen was poorly preserved or absent entirely in two further peat sequences from the western and northern margins of the Lake Viljandi.

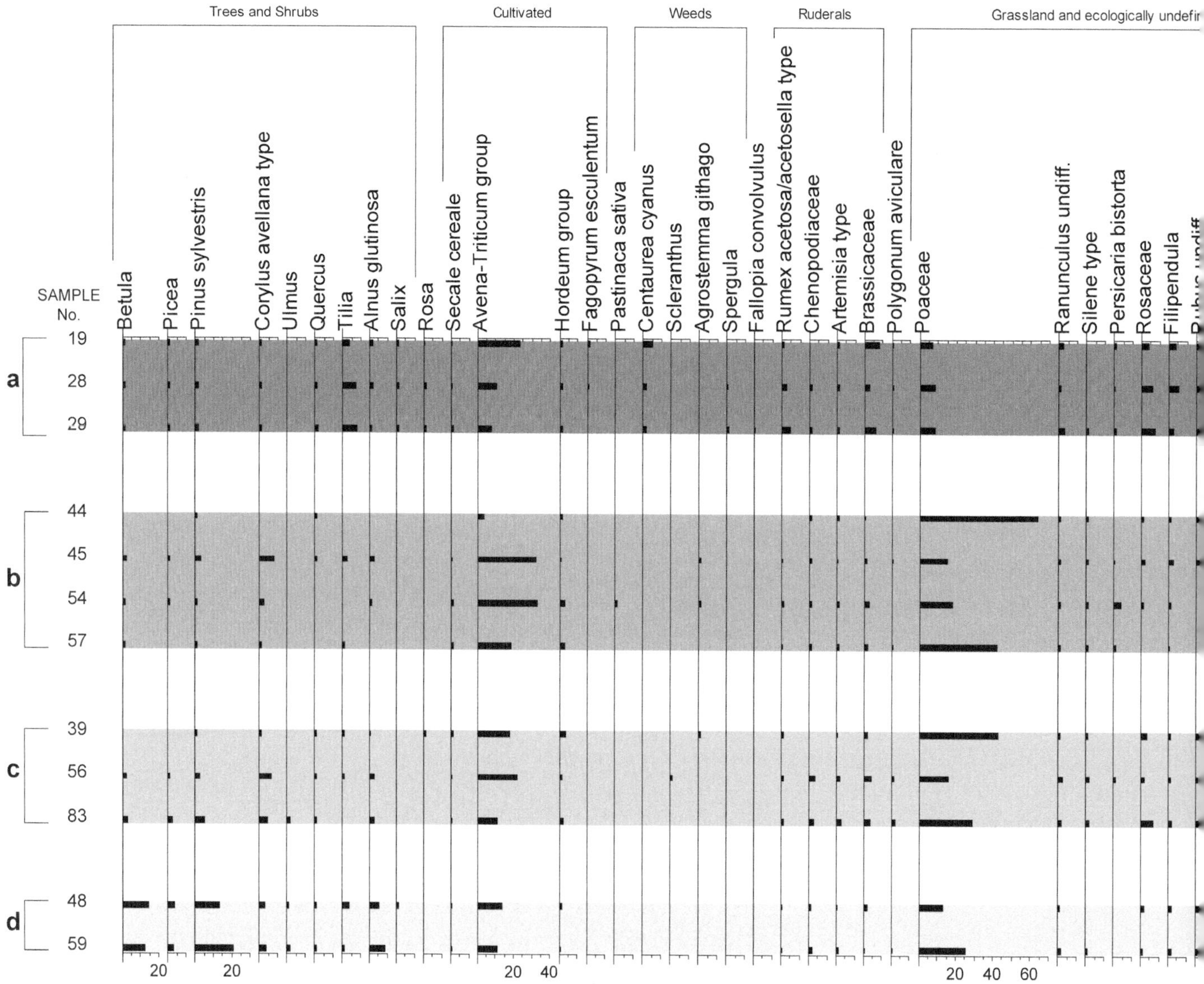

Results

Riga Liv quarter

Latrine (Samples 19, 28, and 29) (Fig. 5.1a)

Three samples from the latrine contained exceptionally well-preserved pollen with a high degree of similarity in pollen assemblage, reflecting a range of habitats within the vicinity of medieval Riga strongly influenced by human land-use. A proportion of the pollen is likely to have entered the latrine within waste food and through faecal matter, but insects are also a major transporter of pollen and would have been attracted to human waste within the latrine. Wood from the latrine was dendrochronologically dated to *c.* 1356; although the deposits may date later than this if the latrine was subsequently emptied. The low levels of arboreal pollen and high diversity of non-arboreal pollen types imply a predominantly open mosaic landscape, although the high values for lime (max 7.2%) do suggest nearby stands of this taxon as its pollen is poorly dispersed and therefore highly localized in extent. Considerable quantities of cereal-type pollen were recorded, principally oat-wheat (6.8–22.8%), with lesser amounts of rye (0.2–0.8%), barley (0.6–0.8%), and other cultivated plants (buckwheat, hemp). Field weeds are generally absent apart from relatively high proportions of cornflower, particularly within the uppermost sample (4.6%), with similarly low proportions for ruderal plants, apart from cabbages (2.4–7.8%); the latter may include pollen of cultivated varieties (e.g., mustards, turnips, cabbages). There are several pollen taxa indicative of meadows and pastures, particularly those characteristic of wet soils (*Ranunculus*, buttercups; *Trifolium*, clovers; *Plantago lanceolata*, ribwort plantain; *Mentha*, type-mint), and including large quantities of *Rubus* undiff., that could reflect collected edible fruits

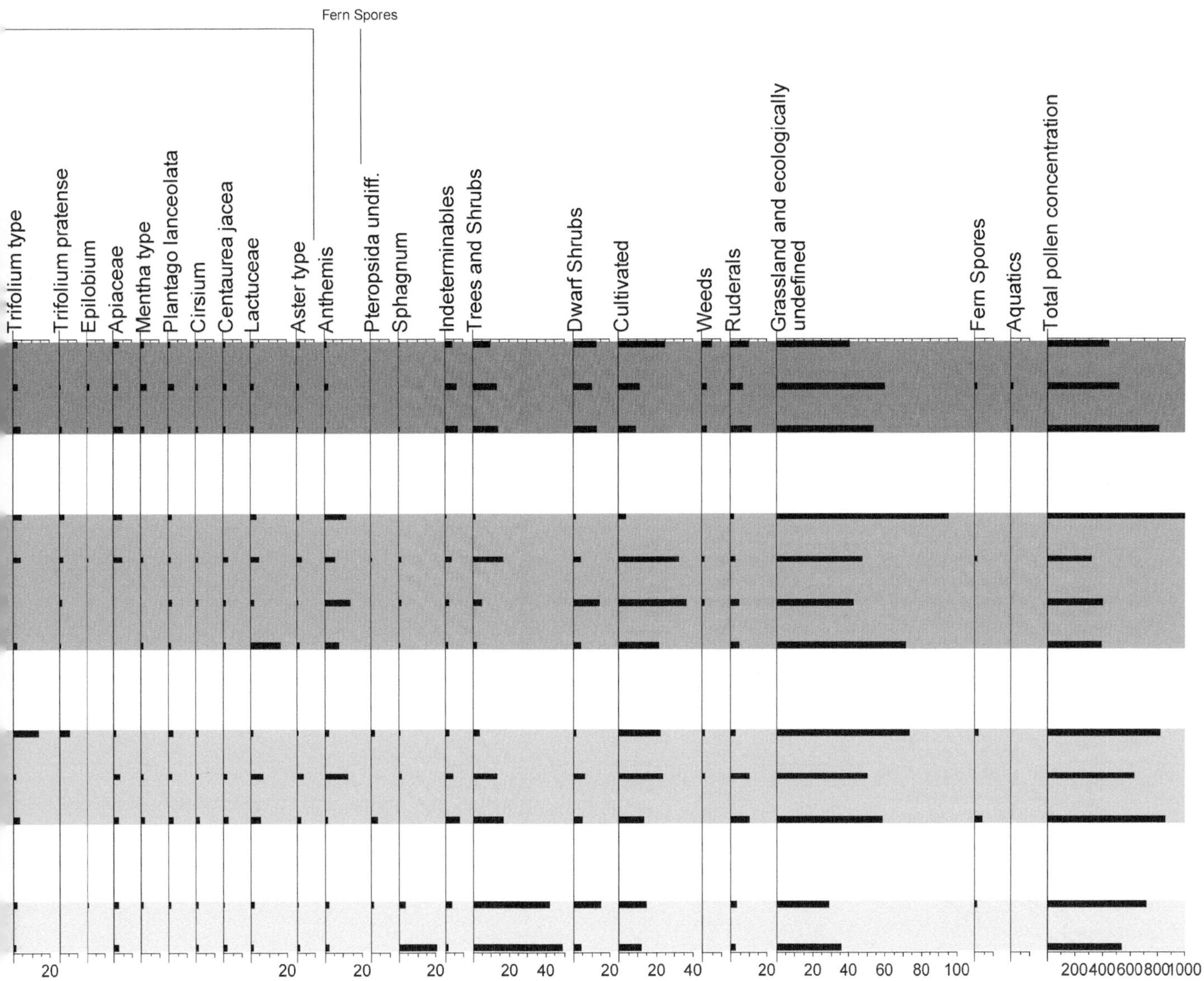

Figure 5.1: Riga Liv quarter, selected taxa percentage pollen diagram: (a) latrine; (b) animal manure; (c) trampled grass, cereal waste/fodder; (d) moss infilling building timbers. Black bars = percentage of taxon. All figures by the author.

(including blackberries, raspberries, dewberries), and represented by large quantities of dewberry seeds in the macrofossil record (Brown et al. 2017). Conversely, areas of dry grassland or dry hummocky vegetation in bogs is suggested by heather/ling (9.5–12.4%), although values for grasses are comparatively low (max. 7.8%).

Animal Manure (Samples 44, 45, 54, and 57) (Fig. 5.1b)

A second class of samples (44, 45, 54, and 57) relate to deposits of animal manure sealed both underneath and in-between houses. Samples 44 and 45 are sealed beneath building 11, and sample 54 beneath building 14, but differ radically in pollen signal. Sample 44 has large quantities of grasses (63%) and smaller quantities of cereal pollen (3.4%), whilst samples 45 and 54 have large quantities of cereal pollen (34% and 36.1%) but significantly lower values for grasses (14.8% and 17.4%); both samples contain a similar range of herbaceous pollen types indicative of meadows and pasture. Sample 57, located between buildings 13, 14, 15, and 17, includes large values for both cereals and grass pollen and range of herbs.

Trampled Grass, Cereal Waste, and/or Fodder (Samples 39, 56, and 83) (Fig. 5.1c)

The pollen signals show a significant degree of similarity with the animal manure samples, but the low levels of phosphorus in samples 39, 56, and 83 strongly suggests they are not manure. These samples may therefore represent accumulations of unconsumed or trampled grass

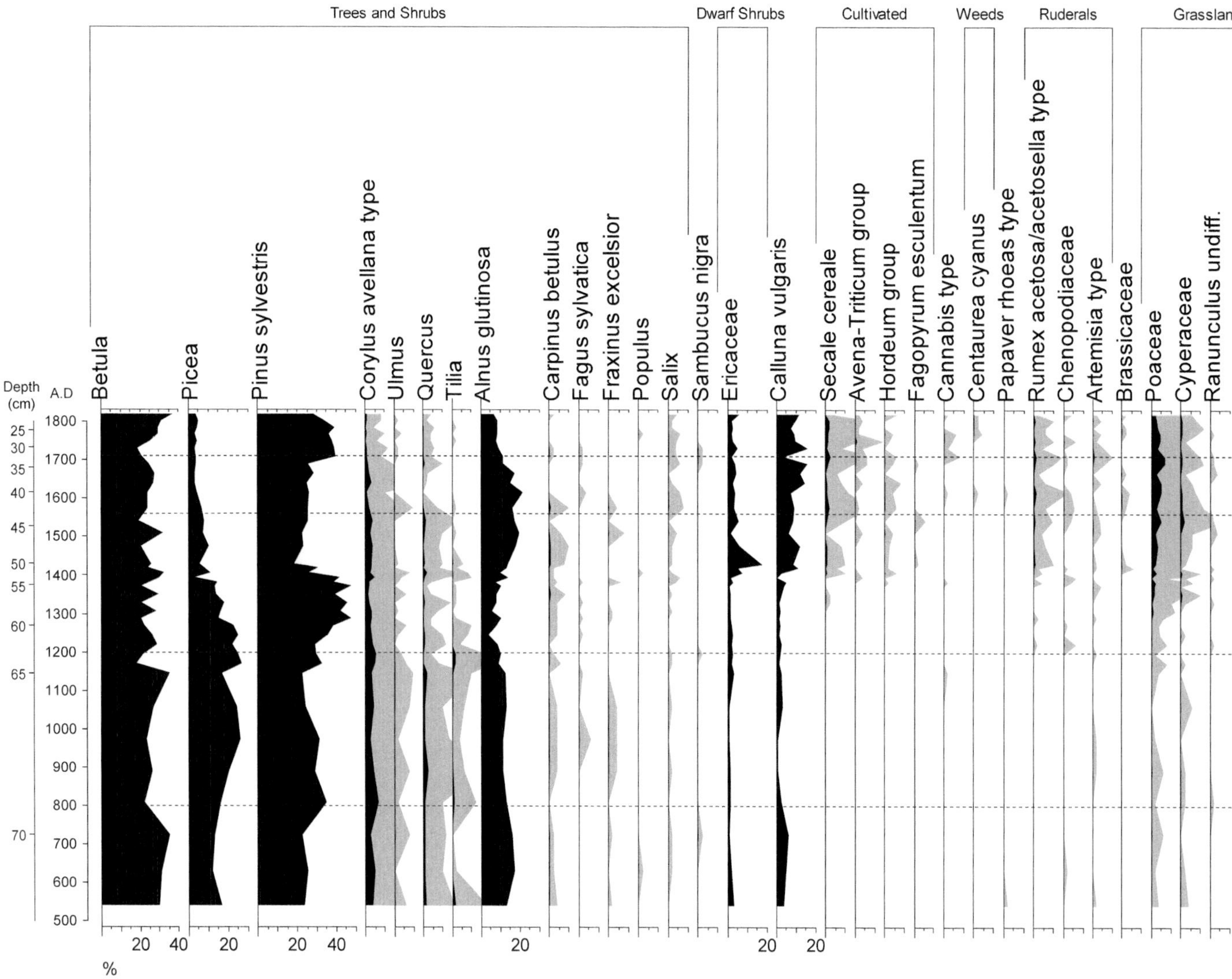

and fodder. All four samples contain high values for pollen of grasses and cultivated plants, and likewise containing a range of herbaceous species indicative of meadows, pastures, and grassland. Sample 39 includes high values for clovers, whilst samples 56 include higher values for pollen of the lettuce family and chamomiles (*Anthemis*).

Moss Infilling Building Timbers (Samples 48 and 59) (Fig. 5.1d)

Two samples were processed from moss infilling the gaps between the planks of buildings (sample 59, building 15), and from with trampled deposits (sample 48). The pollen signals of both samples are distinct from all other samples in containing higher quantities of arboreal pollen, mostly pine and birch. Sample 59 includes the highest values for *Sphagnum* moss spores characteristic of raised bog, though sample 48 includes very little *Sphagnum* but more heather — perhaps suggesting an origin from the drier areas of raised bog. Both samples include high quantities of cereal and grass pollen that is likely to have been incorporated at a later stage.

Wenden Commandery

Blusu Mire (Fig. 5.2)

The vegetation during the Middle and Late Iron Age is dominated by woodland largely comprising birch, spruce, pine, and alder, with lesser hazel, oak, elm, lime, and hornbeam (*Carpinus betulus*). Alder along with birch and pine, the latter two both pioneer species, can tolerate a range of soil conditions including moist peaty soils and are likely to have formed a component of the mire vegetation with a ground flora of dwarf shrubs (*Calluna vulgaris* and Ericaceous plants) and bryophytes (largely *Sphagnum*). There is little sign in the pollen record for human activity at this time; occasional pollen grains of ruderal taxa including mugworts (*Artemisia*) and goosefoots (Chenopodiaceae) are

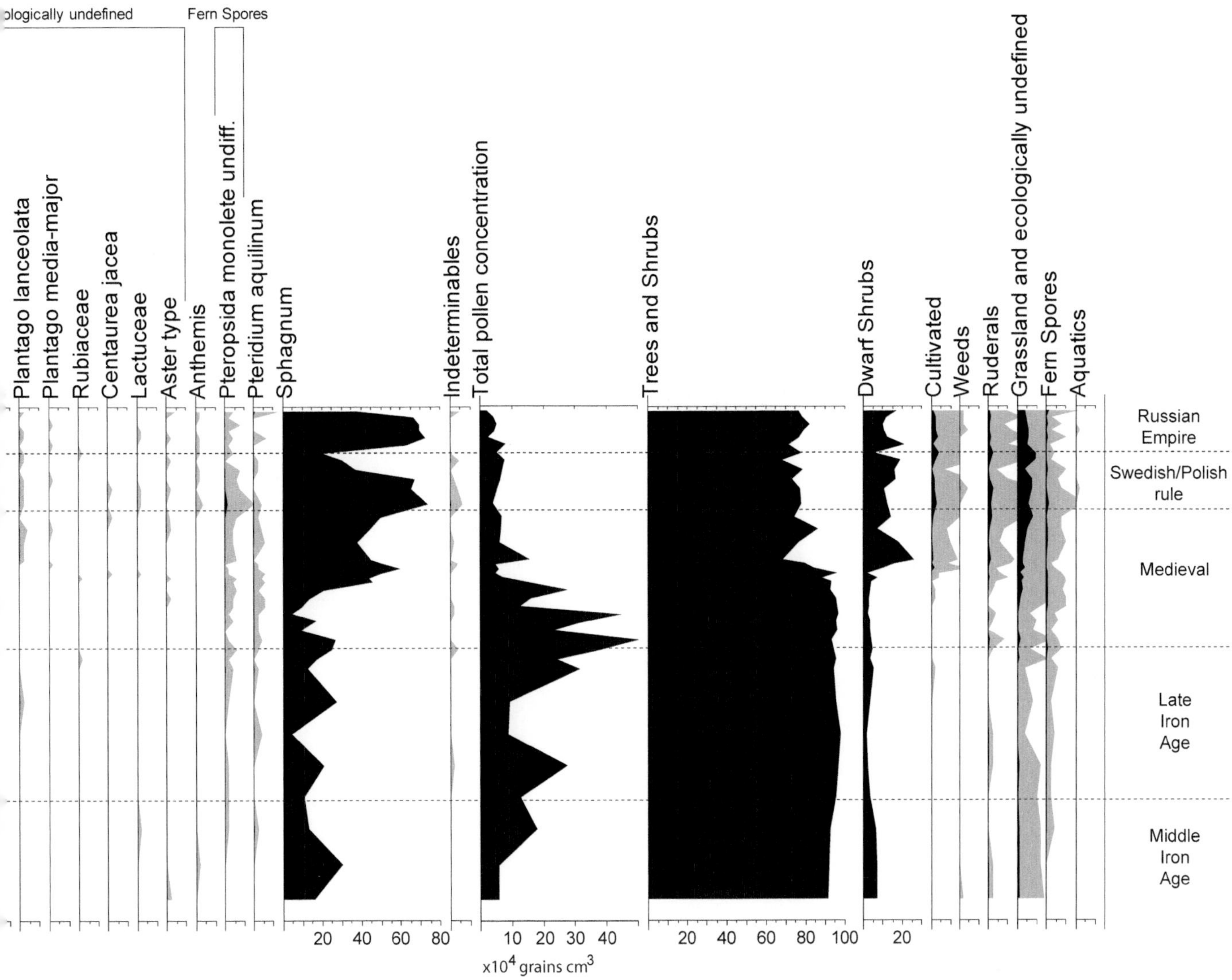

Figure 5.2: Blusu Mire, selected taxa percentage pollen diagram.
Black curve = percentage of taxon, grey curve = values exaggerated × 10.

likely to reflect the activity of wild grazers and browsers within the mire. However, microscopic charcoal levels are highest at the base of the pollen sequence (sixth to early eighth centuries) and may reflect small-scale slash and burn agriculture in the surrounding landscape not visible in the pollen record. There is little significant change in the vegetation until the early/mid-fourteenth century when there is the first appearance of rye (*Secale cereale*) and an increase in grasses and ruderals, intensifying during the fifteenth century and including pollen grains of oat-wheat (*Avena-Triticum*), barley (*Hordeum*) and occasional buckwheat (*Fagopyrum esculentum*). The increase in anthropogenic indicators is accompanied by a decline in arboreal pollen, particularly pine and spruce, although this partly reflects the localised expansion of dwarf shrubs and *Sphagnum* on the bog surface; the increase in *Sphagnum* occurs at the transition from a largely herbaceous to moss peat, reflecting the development of ombrotrophic (rain-water fed) raised bog. The increase in heathers (*Calluna vulgaris*) likely reflects its growth on drier hummocks and ridges in the bog that often forms in-between wetter areas dominated by *Sphagnum*. The reduction in tree cover on and around the bog will have assisted the ingress of pollen from the adjacent dry ground where arable and pastoral land is indicated. However, higher values for cultivated, disturbed, and open ground flora are most apparent after the medieval period, both during the phase of Swedish/Polish and subsequent Russian rule. Nevertheless, there is a decline in cultivated, ruderal, and grass pollen during the mid-seventeenth century accompanied by lower palynological diversity.

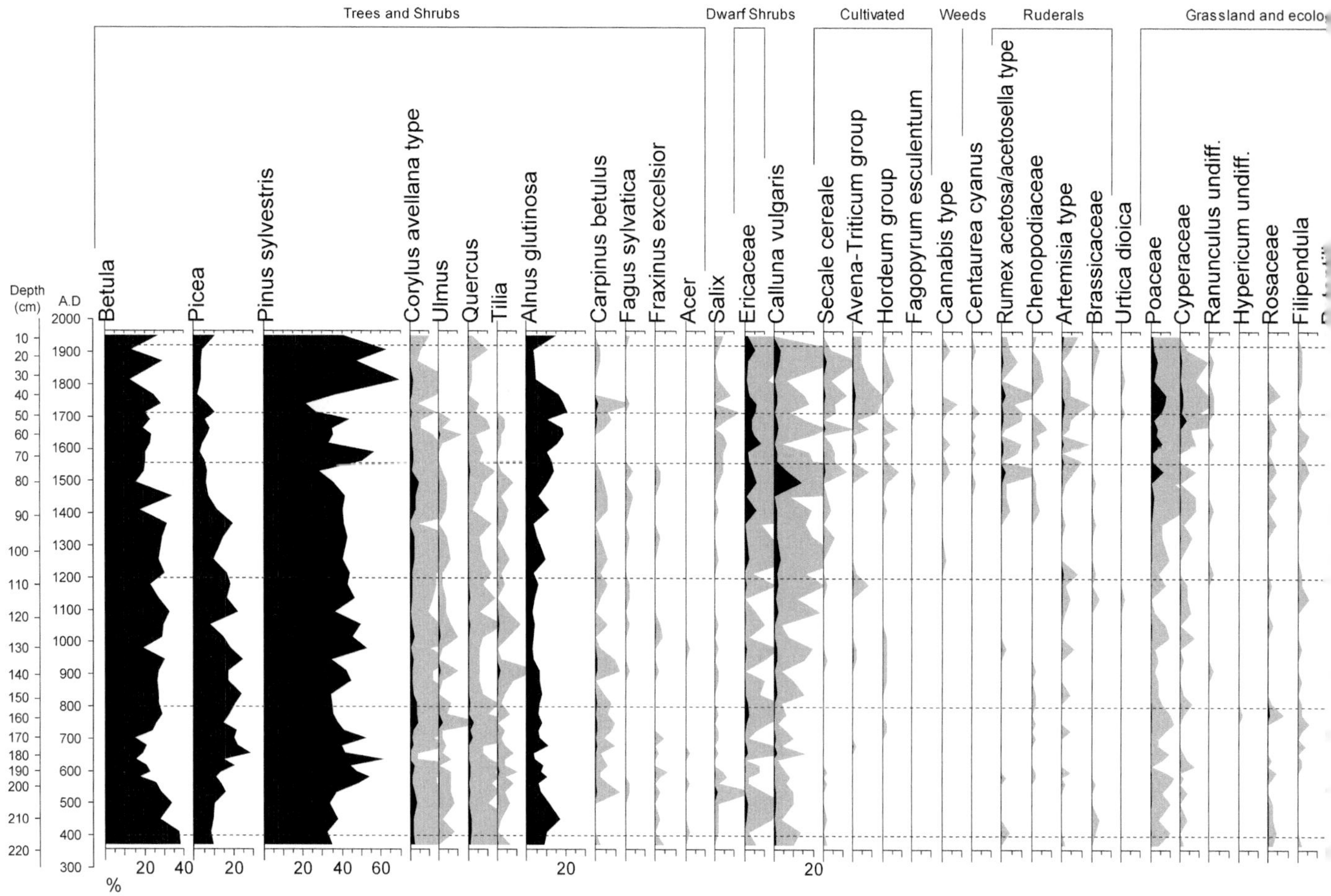
Trees and Shrubs
Dwarf Shrubs
Cultivated
Weeds
Ruderals
Depth (cm)
A.D
Betula
Picea
Pinus sylvestris
Corylus avellana type
Ulmus
Quercus
Tilia
Alnus glutinosa
Carpinus betulus
Fagus sylvatica
Fraxinus excelsior
Acer
Salix
Ericaceae
Calluna vulgaris
Secale cereale
Avena-Triticum group
Hordeum group
Fagopyrum esculentum
Cannabis type
Centaurea cyanus
Rumex acetosa/acetosella type
Chenopodiaceae
Artemisia type
Brassicaceae
Urtica dioica
Poaceae
Cyperaceae
Ranunculus undiff.
Hypericum undiff.
Rosaceae
Filipendula
%

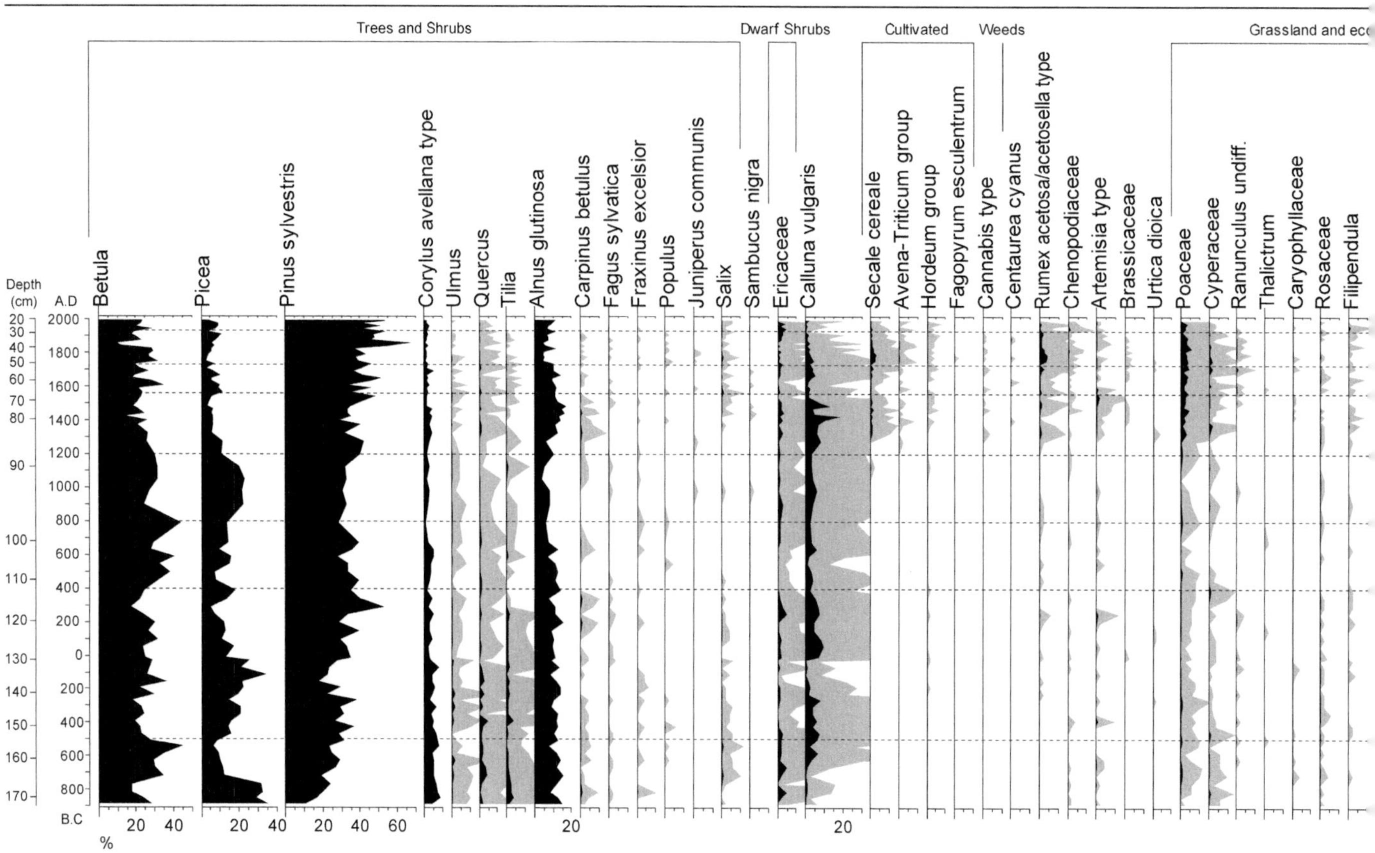
Trees and Shrubs
Dwarf Shrubs
Cultivated
Weeds
Depth (cm)
A.D
B.C
Betula
Picea
Pinus sylvestris
Corylus avellana type
Ulmus
Quercus
Tilia
Alnus glutinosa
Carpinus betulus
Fagus sylvatica
Fraxinus excelsior
Populus
Juniperus communis
Salix
Sambucus nigra
Ericaceae
Calluna vulgaris
Secale cereale
Avena-Triticum group
Hordeum group
Fagopyrum esculentrum
Cannabis type
Centaurea cyanus
Rumex acetosa/acetosella type
Chenopodiaceae
Artemisia type
Brassicaceae
Urtica dioica
Poaceae
Cyperaceae
Ranunculus undiff.
Thalictrum
Caryophyllaceae
Rosaceae
Filipendula
%

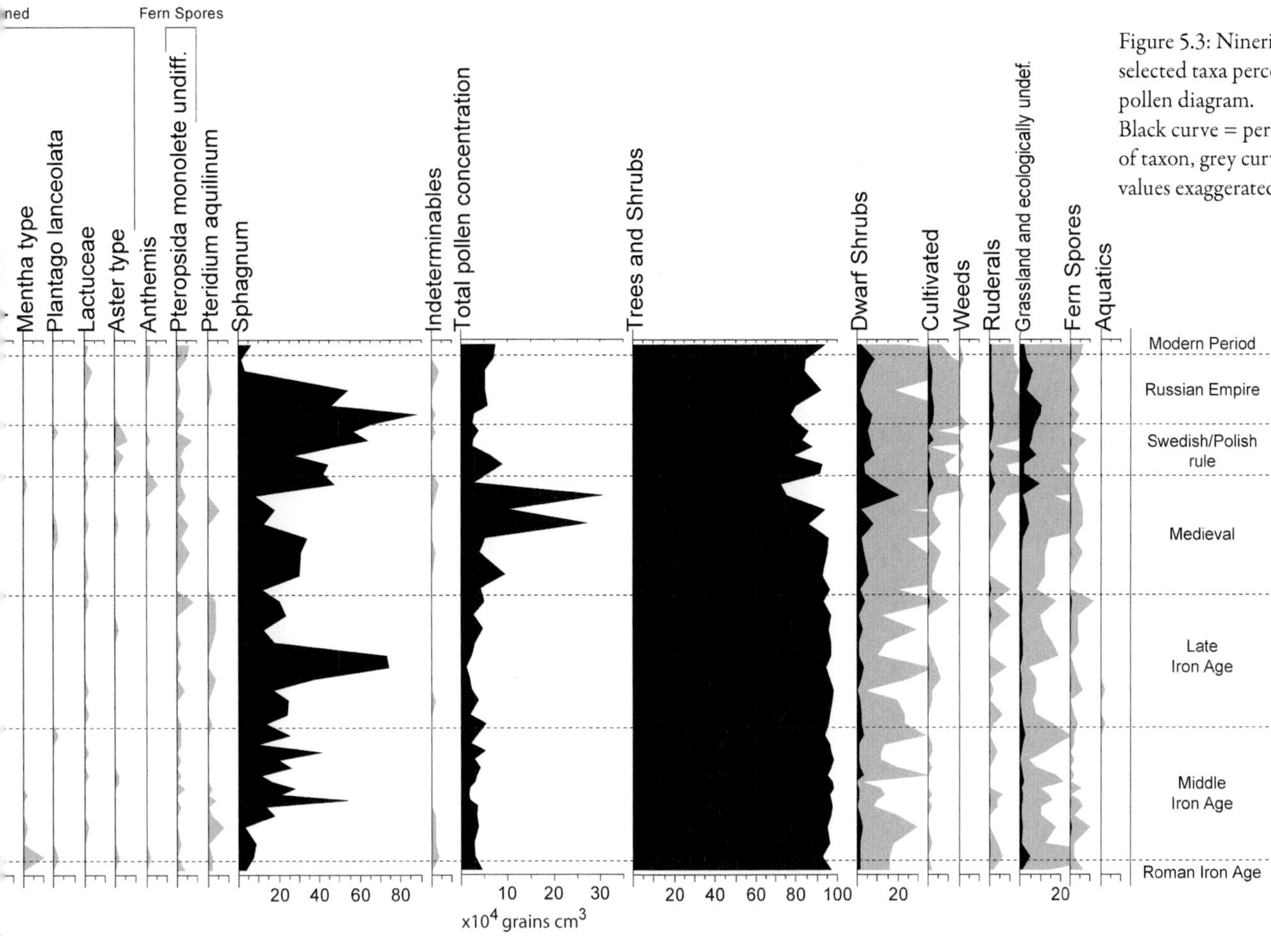

Figure 5.3: Nineris Mire, selected taxa percentage pollen diagram. Black curve = percentage of taxon, grey curve = values exaggerated × 10.

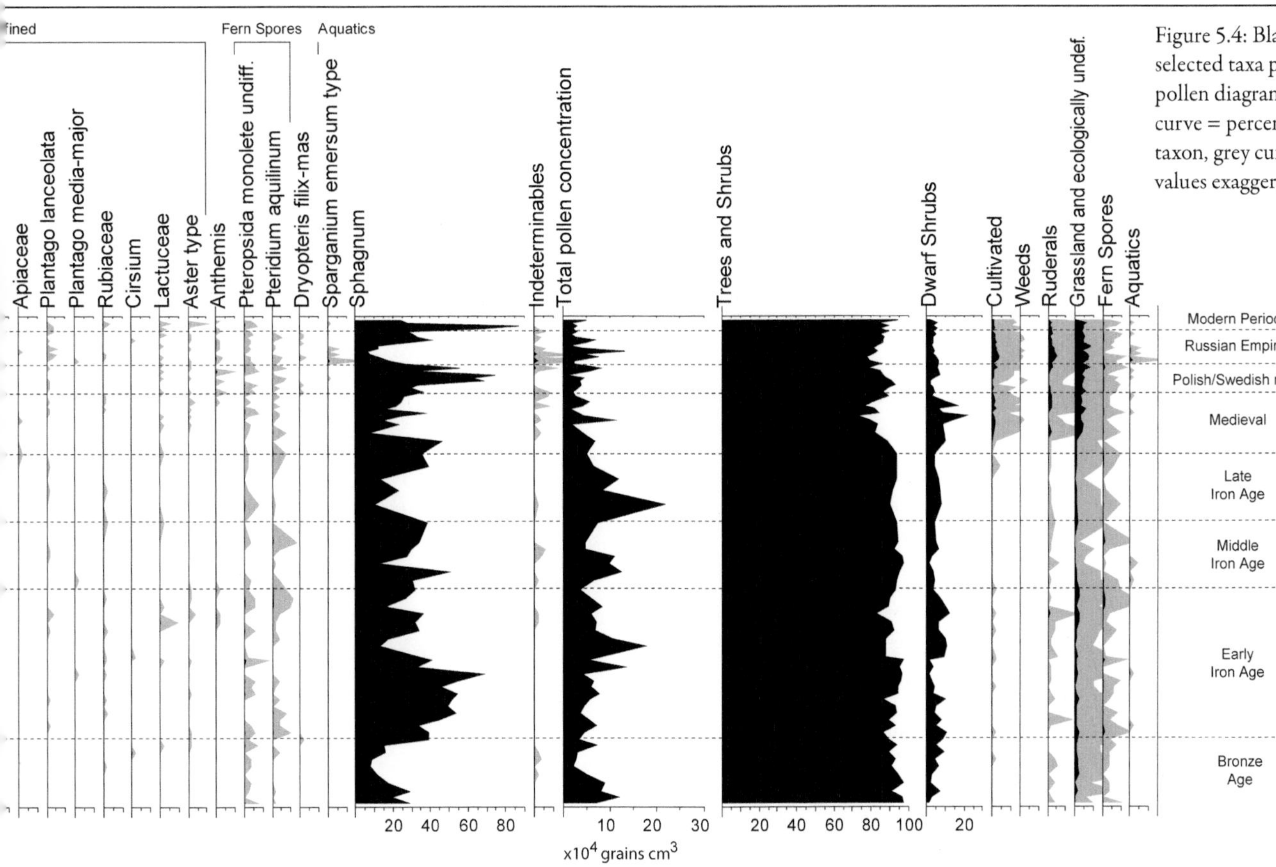

Figure 5.4: Blanku Mire, selected taxa percentage pollen diagram. Black curve = percentage of taxon, grey curve = values exaggerated × 10.

Nineris Mire (Fig. 5.3)

The sequence from Nineris Mire is located 5.5 km north-north-west of Blusu Mire and displays a broadly similar vegetation sequence. The sequence comprises a poorly humified moss peat exhibiting a fairly constant sediment accumulation rate and excellent chronological resolution, with 2.2 m covering the last *c.* 1600 years. Woodland predominates throughout the Iron Age, dominated by pine and birch with spruce increasing in importance from the seventh century. There are occasional cereal pollen grains of the oat-wheat group during the Late Iron Age (tenth century), and at the turn of thirteenth century, that could reflect nearby small-scale agricultural activity. Alternatively, in view of the lack of associated anthropogenic indicators, the grains of oat-wheat could derive from wild grasses that produce equally large pollen grains growing within the bog or along the edges of the nearby lake. Occasional rye pollen grains do occur from the early fourteenth century, with a more noticeable increase in cereals and ruderals only apparent at the end of the medieval period. The post medieval is characterized by an expansion of dwarf shrubs and *Sphagnum* moss on the raised bog, with alder growing on moist soils. The period of Polish/Swedish rule sees a decline in evidence for agricultural activity, although values for cereal and ruderal pollen increase again into the eighteenth century.

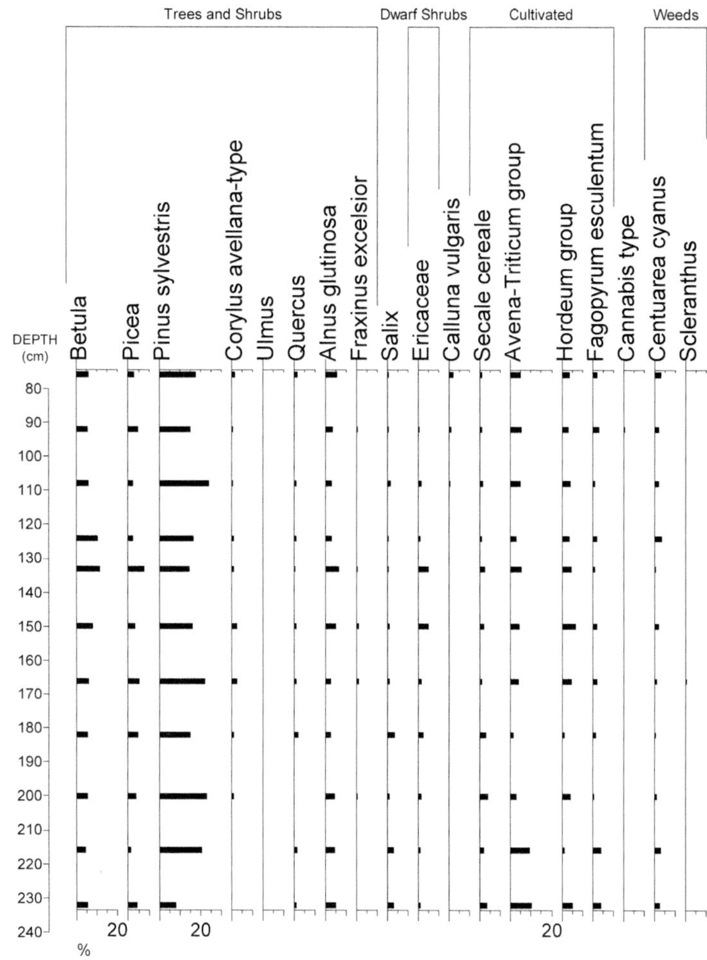

Blanku Mire (Fig. 5.4)

In line with the sequences from Blusu Mire and Nineris, Blanku Mire further demonstrates the regional dominance of pine-birch-spruce woodland during the Iron Age, although spruce declines in importance from the twelfth century. The entire sequence is composed of moss peat, moderately humified until 72 cm and thereafter increasingly well humified, with *Sphagnum* spores and dwarf shrubs pollen characteristic of raised bog. Woodland, mainly pine, is likely to have encroached onto the bog from the adjacent dry ground, forming a component of the surface vegetation much as it does today. There is minimal evidence for human activity during the Iron Age, with only a single rye pollen grain from the early twelfth century suggesting small-scale agricultural land-use within the surrounding landscape. Pollen of cultivated and ruderal species only increase towards the end of the thirteenth century, with wind-pollinated rye most prominent along with increasing frequencies of common sorrel (*Rumex acetosa*-type), mugworts and grasses (Poaceae) suggesting areas of meadow and pastureland. The evidence for increasing agricultural activity during the medieval period, from the thirteenth century, contrasts with the latter half of the sixteenth and seventeenth century during the period of Polish-Swedish rule. Lower values for cereal and ruderal pollen suggest some regression in arable and pastoral land-use at this time, accompanied by an increase in arboreal pollen.

Cēsis Castle Moat (Fig. 5.5)

Pollen from the moat at Cēsis complements the plant macrofossil analysis (Brown and Badura 2017) by providing an indication of the local environmental setting around the castle from the late fourteenth century, and to a lesser extent the range of on-site activities. There is no indication if the moat was ever cleared out, but it is assumed that the sequence reflects the gradual accumulation of sediment following the construction of a timber drain, dendrochronologically dated *c.* 1374/75. In contrast to the off-site sequences, the moat sequence includes significantly less arboreal pollen, with herbaceous pollen types suggesting a locally more open, mosaic environment. There are significant quantities

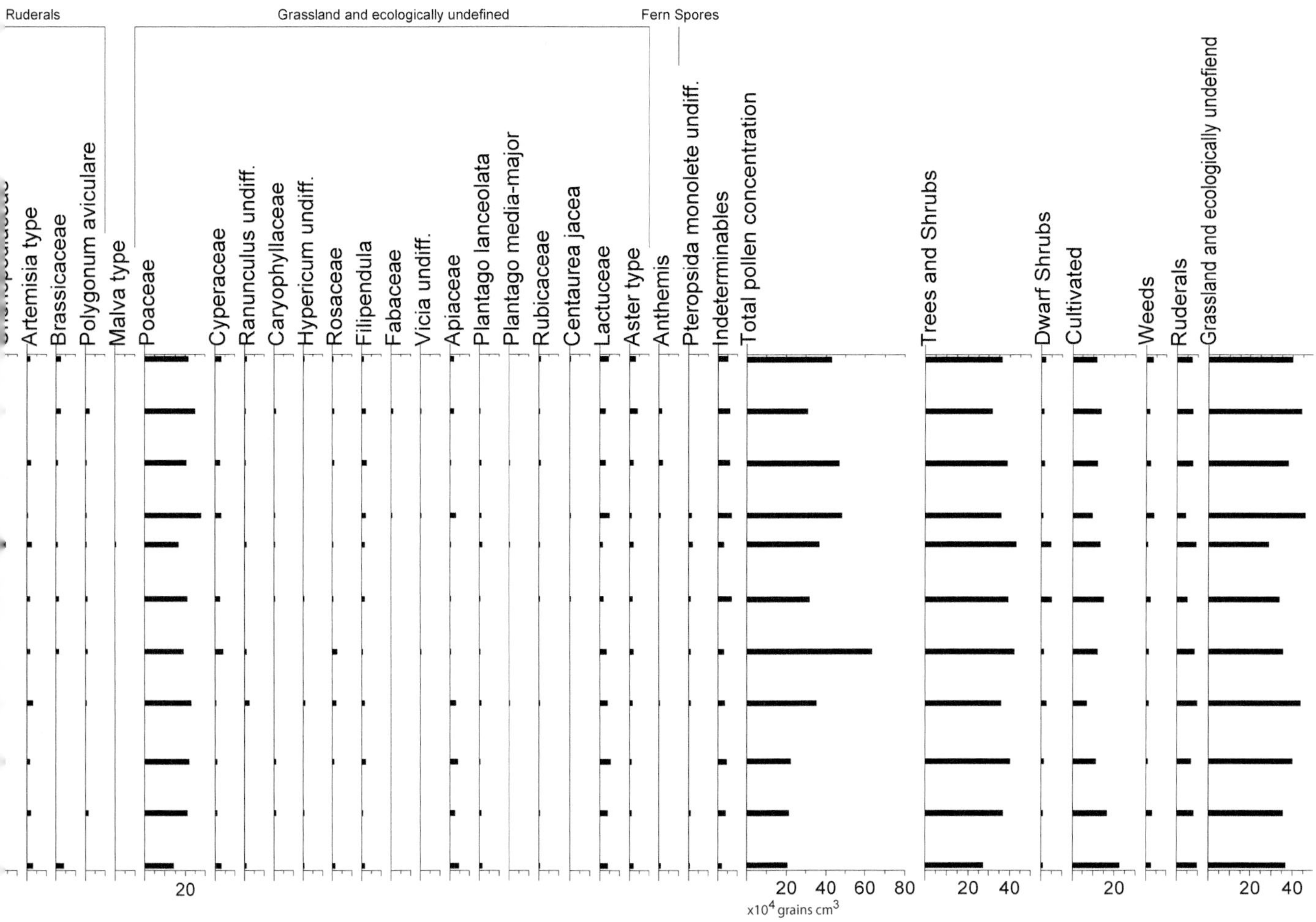

Figure 5.5: Cēsis, moat, selected taxa percentage pollen diagram. Black bars = percentage of taxon.

of cereal pollen, including oat-wheat, barley, rye, and buckwheat, accompanied by pollen of field weeds (cornflower, *Centaurea cyanus*; knawels, *Scleranthus*; and common corn-cockle, *Agrostemma githago*) and ruderals (common sorrel, goosefoots, mustards/cabbages, and mugworts). Many of these cereals, weeds, and ruderals are present in both the macrofossil and pollen record from the moat. In this context it is likely that the moat received pollen from multiple sources, including pollen contained in kitchen and latrine waste flushed into the moat, from grain stores within the outer bailey, and from nearby cultivated fields, meadows and waste ground. The higher proportions of daises (*Aster*-type) and lettuces/dandelion (Lactuceae) may reflect differential preservation of pollen favouring those grains more resistant to decay, although the combination of high values for grasses and entomophilous pollen does strongly indicate meadows, pastures, and disturbed ground in the vicinity.

Rāmuli (Fig. 5.6)

Woodland dominates the sequence, largely comprising spruce with an increasing component of birch and pine. Alder forms an important component of woodland on wetter soils. There is no evidence for human activity in the pollen sequence prior to the thirteenth/fourteenth century when arboreal pollen values begin to decline accompanied by an increase in grasses and intermittent cereal pollen grains of rye, wheat-oat, and barley. Higher values for cereals and grass pollen, along with occasional cornflower, are apparent during the subsequent periods of Swedish/Polish and Russian rule.

Lake Trikāta (Fig. 5.7)

The landscape surrounding Lake Trikāta was heavily wooded over the course of the Iron Age, dominated during the pre-Roman, Roman, and Middle Iron Age by spruce-birch-pine woodland, with alder growing on wetter soils around the lake. Spruce becomes an increasingly dominant component of the woodland canopy during the seventh to tenth centuries, reaching

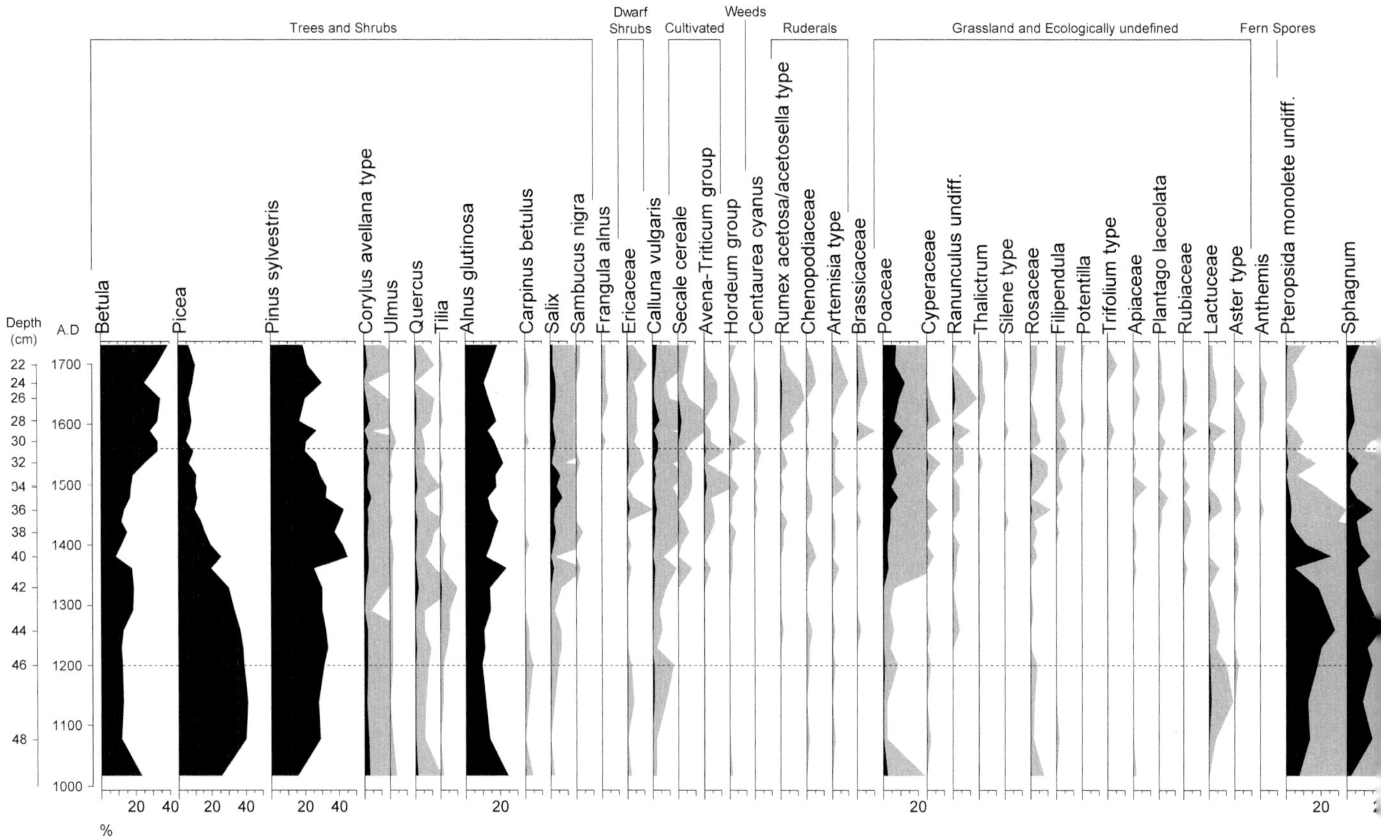
Trees and Shrubs
Dwarf Shrubs
Cultivated
Weeds
Ruderals
Grassland and Ecologically undefined
Fern Spores
Depth (cm)
A.D
Betula
Picea
Pinus sylvestris
Corylus avellana type
Ulmus
Quercus
Tilia
Alnus glutinosa
Carpinus betulus
Salix
Sambucus nigra
Frangula alnus
Ericaceae
Calluna vulgaris
Secale cereale
Avena-Triticum group
Hordeum group
Centaurea cyanus
Rumex acetosa/acetosella type
Chenopodiaceae
Artemisia type
Brassicaceae
Poaceae
Cyperaceae
Ranunculus undiff.
Thalictrum
Silene type
Rosaceae
Filipendula
Potentilla
Trifolium type
Apiaceae
Plantago laceolata
Rubiaceae
Lactuceae
Aster type
Anthemis
Pteropsida monolete undiff.
Sphagnum
%

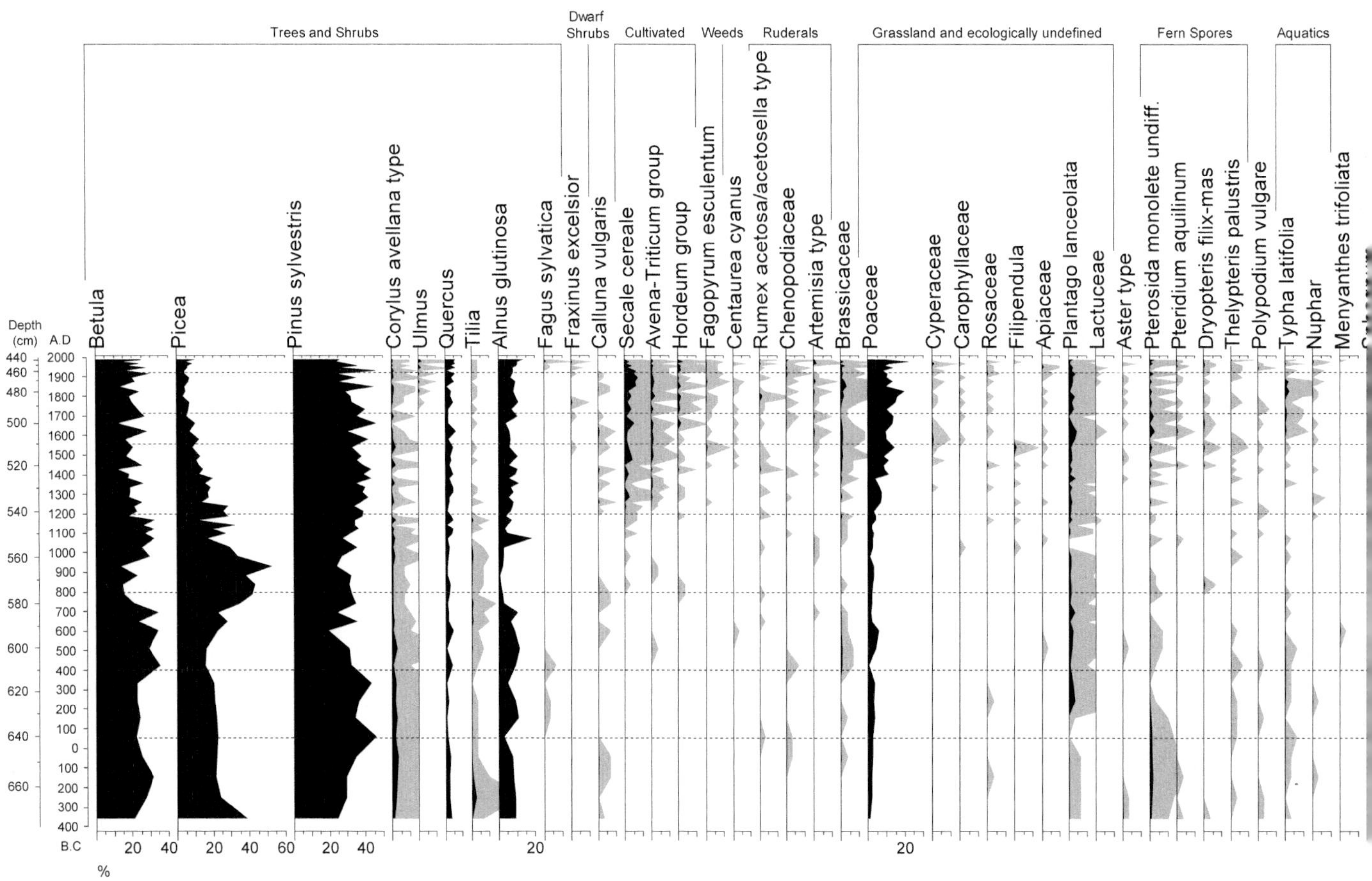
Trees and Shrubs
Dwarf Shrubs
Cultivated
Weeds
Ruderals
Grassland and ecologically undefined
Fern Spores
Aquatics
Depth (cm)
A.D
B.C
Betula
Picea
Pinus sylvestris
Corylus avellana type
Ulmus
Quercus
Tilia
Alnus glutinosa
Fagus sylvatica
Fraxinus excelsior
Calluna vulgaris
Secale cereale
Avena-Triticum group
Hordeum group
Fagopyrum esculentum
Centaurea cyanus
Rumex acetosa/acetosella type
Chenopodiaceae
Artemisia type
Brassicaceae
Poaceae
Cyperaceae
Carophyllaceae
Rosaceae
Filipendula
Apiaceae
Plantago lanceolata
Lactuceae
Aster type
Pterosida monolete undiff.
Pteridium aquilinum
Dryopteris filix-mas
Thelypteris palustris
Polypodium vulgare
Typha latifolia
Nuphar
Menyanthes trifoliata
%

Figure 5.6: Rāmuli, selected taxa percentage pollen diagram. Black curve = percentage of taxon, grey curve = values exaggerated × 10.

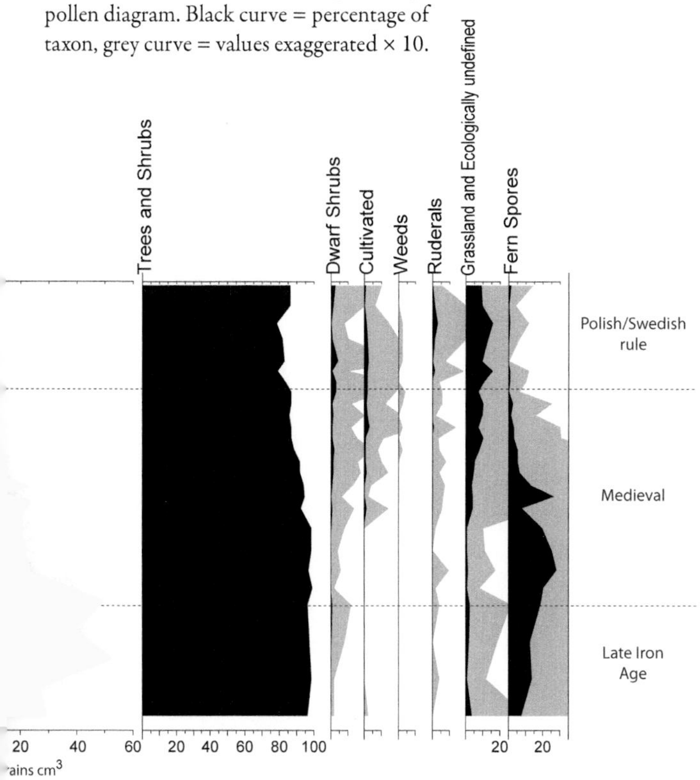

Figure 5.7: Lake Trikāta, selected taxa percentage pollen diagram. Black curve = percentage of taxon, grey curve = values exaggerated × 10.

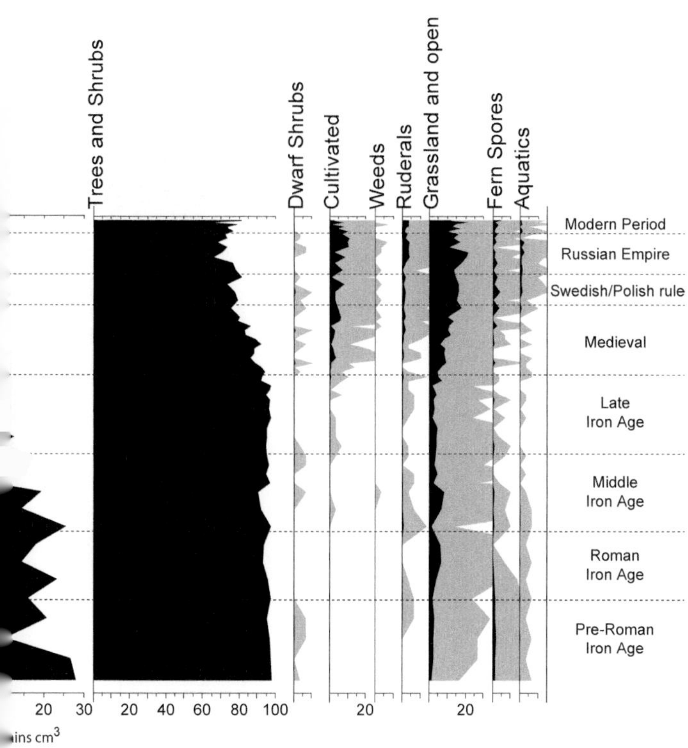

c. 52%, though declining thereafter. There are few signs of nearby human activity in the pollen record until the Middle Iron Age, with very occasional cereal grains of oat-wheat and a slight increase in grasses and ruderal pollen suggesting small-scale intermittent arable/pastoral activity. The picture is similar for the Late Iron Age with only very occasional grains of rye, oat-wheat, and barley recorded; there are small increases in pollen of ruderal and grassland species that may reflect areas of disturbed ground and pastureland. Significant changes in vegetation and land-use are apparent during the medieval period, with increasing values for cereals, ruderals, and grasses. There is a minor increase in cereals during the mid- to late twelfth century, but only a more protracted peak from the late thirteenth century and during the fifteenth century and later, separated by fluctuations in values. Grass pollen increases from the beginning of the medieval period, followed by an increase in ruderal pollen and declining arboreal pollen values, suggesting gradual clearance of the surrounding woodland for arable and pastoral activities. There is a degree of continuity in arable land-use during the post-medieval period, with a further increase in the intensity of arable and pastoral land-use during the subsequent period of Russian rule, particularly during the nineteenth century when arboreal pollen reaches its lowest values.

Fellin Commandery

Lake Viljandi (Fig. 5.8)

Radiocarbon dating of a single gyttja sample from Lake Viljandi showed the effects of a hard water error ($\delta^{13}C$ (‰) -31.7), producing a spuriously early date of 1145±35 BP (cal. AD 780–980). It has been impossible to determine the degree of hard water error, and further dating has proven difficult because of the lack of terrestrial plant macrofossils. However, a probable medieval or later date for the sequence is suggested by single pollen grains of buckwheat at 9.36, 9.20, and 8.96 m. Pollen of buckwheat is generally rare in pollen sequences, but the first archaeobotanical records from Estonia date no earlier than the thirteenth to fourteenth centuries, and are more frequent from the post-medieval period (Sillasoo and Hiie 2007). The pollen sequence suggests some woodland on the slopes dominated by birch and pine with alder and willow growing along the lake edge and rivers. Spruce is locally presence with isolated stands of other trees. Cultivated of rye, oat-wheat, barley, buckwheat, and hemp is apparent, with grasses and a range

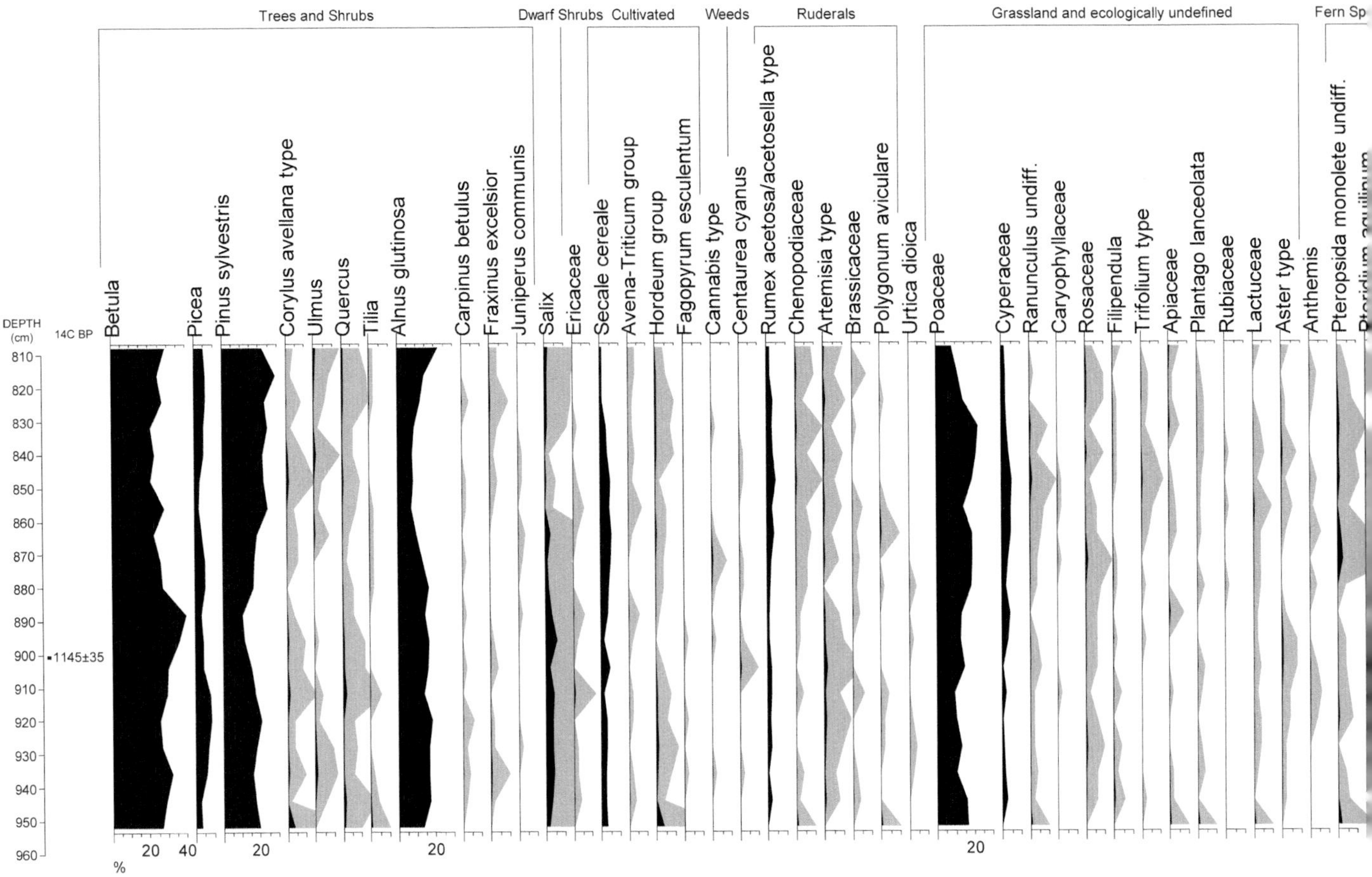
Trees and Shrubs
Dwarf Shrubs
Cultivated
Weeds
Ruderals
Grassland and ecologically undefined
Fern Sp
Betula
Picea
Pinus sylvestris
Corylus avellana type
Ulmus
Quercus
Tilia
Alnus glutinosa
Carpinus betulus
Fraxinus excelsior
Juniperus communis
Salix
Ericaceae
Secale cereale
Avena-Triticum group
Hordeum group
Fagopyrum esculentum
Cannabis type
Centaurea cyanus
Rumex acetosa/acetosella type
Chenopodiaceae
Artemisia type
Brassicaceae
Polygonum aviculare
Urtica dioica
Poaceae
Cyperaceae
Ranunculus undiff.
Caryophyllaceae
Rosaceae
Filipendula
Trifolium type
Apiaceae
Plantago lanceolata
Rubiaceae
Lactuceae
Aster type
Anthemis
Pteropsida monolete undiff.
DEPTH (cm)
14C BP
810
820
830
840
850
860
870
880
890
900
910
920
930
940
950
960
1145±35
20
40
20
20
20
%

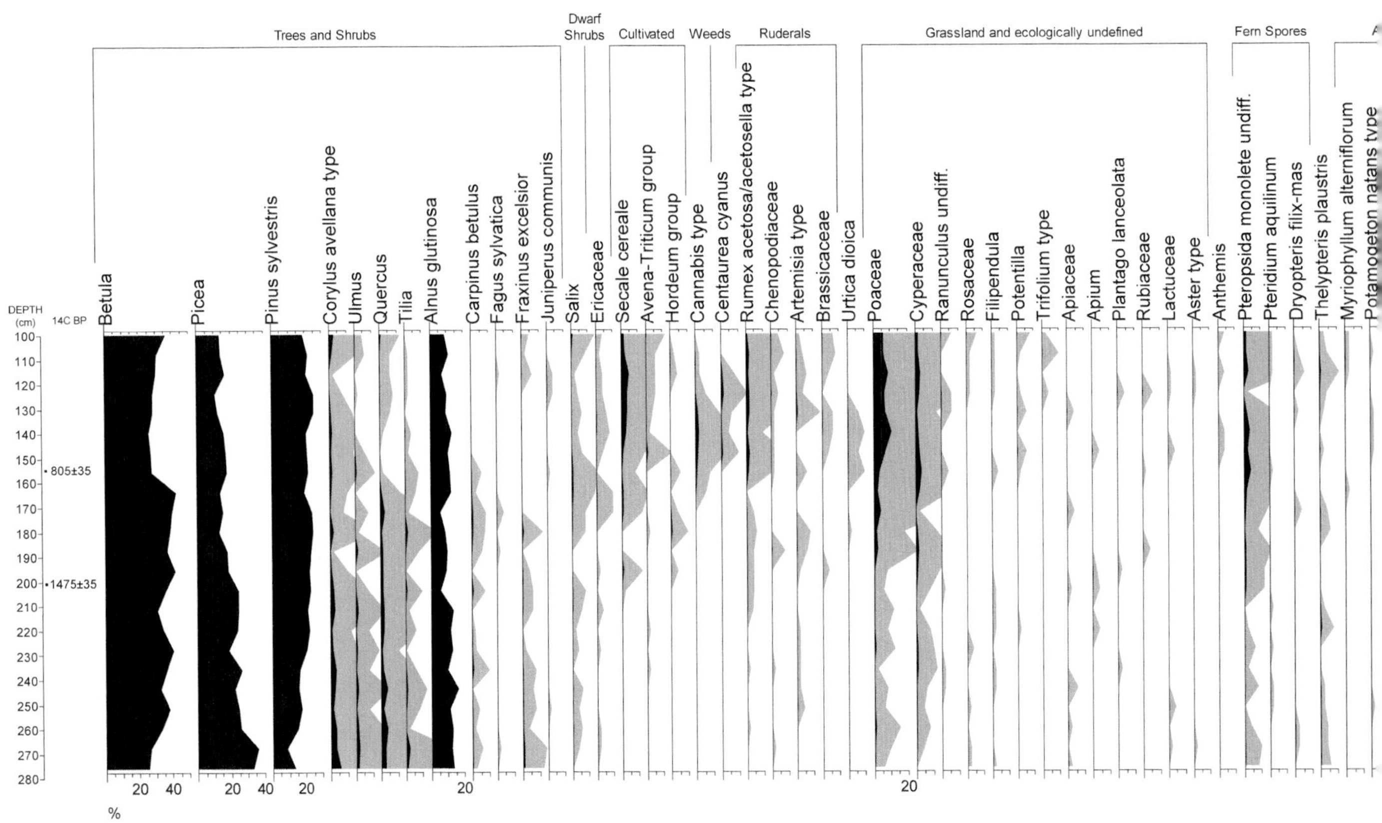
Trees and Shrubs
Dwarf Shrubs
Cultivated
Weeds
Ruderals
Grassland and ecologically undefined
Fern Spores
Betula
Picea
Pinus sylvestris
Corylus avellana type
Ulmus
Quercus
Tilia
Alnus glutinosa
Carpinus betulus
Fagus sylvatica
Fraxinus excelsior
Juniperus communis
Salix
Ericaceae
Secale cereale
Avena-Triticum group
Hordeum group
Cannabis type
Centaurea cyanus
Rumex acetosa/acetosella type
Chenopodiaceae
Artemisia type
Brassicaceae
Urtica dioica
Poaceae
Cyperaceae
Ranunculus undiff.
Rosaceae
Filipendula
Potentilla
Trifolium type
Apiaceae
Apium
Plantago lanceolata
Rubiaceae
Lactuceae
Aster type
Anthemis
Pteropsida monolete undiff.
Pteridium aquilinum
Dryopteris filix-mas
Thelypteris plaustris
Myriophyllum alterniflorum
Potamogeton natans type
DEPTH (cm)
14C BP
100
110
120
130
140
150
160
170
180
190
200
210
220
230
240
250
260
270
280
805±35
1475±35
20
40
20
40
20
20
20
%

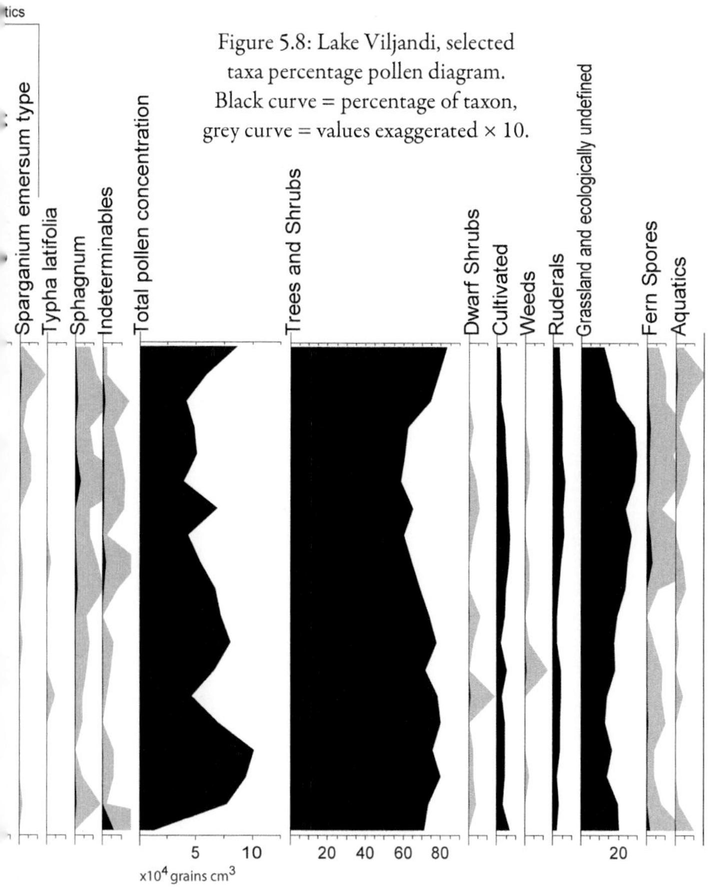

Figure 5.8: Lake Viljandi, selected taxa percentage pollen diagram. Black curve = percentage of taxon, grey curve = values exaggerated × 10.

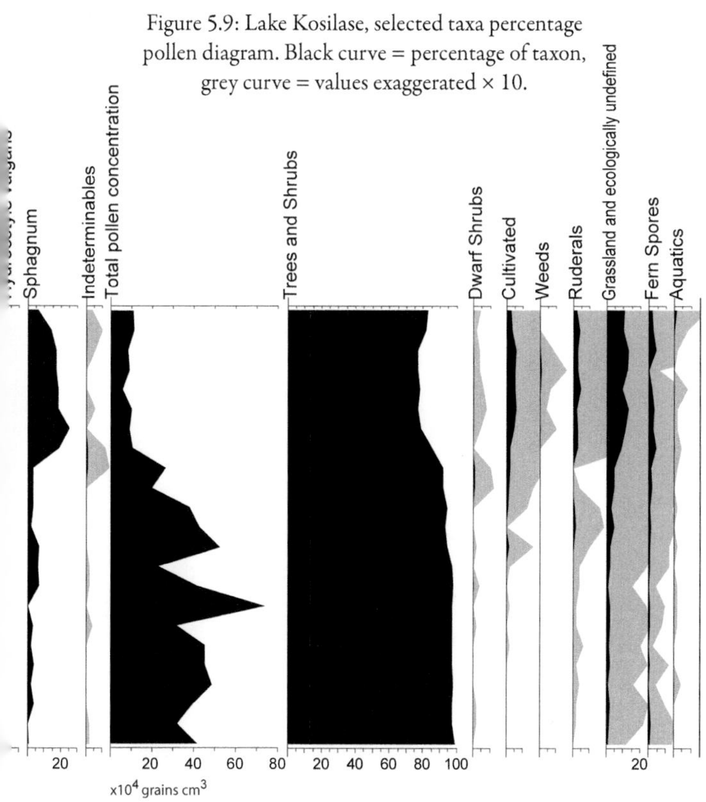

Figure 5.9: Lake Kosilase, selected taxa percentage pollen diagram. Black curve = percentage of taxon, grey curve = values exaggerated × 10.

of ruderal taxa indicating the presence of meadows and pastureland in the surrounding landscape.

Lake Kosilase (Fig. 5.9)

Radiocarbon dating of sediments from Lake Kosilase included high $\delta^{13}C$ values (-31.2 and -31.8 $\delta^{13}C$ (‰)) that indicate the possible effects of a hard water error. The degree of error is impossible to determine as the limited quantities of terrestrial plant macrofossils contained insufficient carbon for dating. The majority of the sequence is characterized by high values for arboreal pollen reflecting woodland dominated by spruce and birch along with alder. Oak, elm, and lime are more prevalent in the base of the sequence but decline by *c.* 1.80 m; spruce values decline gradually from 2.8 m. Rye pollen first appears at 2.04 m, but with more consistent values from 1.72 m and a notable increase in values between 1.40–1.16 m (3.5–4.9%). It is probably that the significant increase in rye pollen from 1.40–1.16 m is post-medieval, consistent with similar increases in rye pollen described here from other dated pollen sequences. The pollen includes evidence for cultivation of rye, wheat/oat, barley, and hemp, with weeds of cornflower and ruderal pollen indicating permanent arable fields, pastures, and meadows. The underlying sediments to *c.* 1.70 m contain consistent although lower values for rye, along with occasional oat-wheat, barley and hemp. Although the chronology of the sequence is highly generalised in the absence of secure dating, bulk gyttja at 1.54 cm dated to 805±35 BP (GU-27670, cal. AD 1170–1275) could suggest a likely medieval date for sediments at and below this horizon (*c.* 1.50–170 m). The underlying date at 2.00 m of 1475±35 BP (GU-27671, cal. AD 535–652) could equally imply a Late Iron Age date associated with the first appearance of rye pollen grains.

Karkus Advocateship

Karksi High Castle (Fig. 5.10)

Pollen was well preserved within context 13, interpreted as a dumped fodder layer. The four samples from the base of this deposit, corresponding to unit 13C, are characterized by higher values for arboreal pollen, particularly pine and spruce. There are numerous fragments of wood, leaves, and bark along with occasional spruce needles within this unit that are suggested to have accumulated naturally (see Chapter 7) and may indicate nearby stands of pine and spruce. The majority of unit 13C comprises large quantities of grass and cereal pollen, particularly oat-wheat; a sizeable quantity of the cereal pollen was heavily crumpled and is identified only generally as cereal-type. No pollen survived in samples from unit 13B, with the majority of samples derived from unit 13A. Grass pollen dominates (48–64%), with large quantities of cereal pollen, mostly oat-wheat and barley with less rye. There are very few weeds apart from single records of common corn-cockle, with similarly small quantities of ruderals. There is a wide range of herbaceous pollen types suggestive of damp and meadow environments in the vicinity. The dominance of both cereals and grass pollen within context 13 is consistent with the interpretation of the deposit as dumped fodder, but also including pollen from vegetation growing within the vicinity of the site. The immediate area around the castle appears to have been largely cleared of trees. Preservation of pollen in the overlying midden (context 12) was generally poor. There are clear differences in the pollen signals however, with less pollen of grasses and cereals in the midden. Instead there are greater quantities of mustards, buttercups, champions (Caryophyllaceae), meadowsweets (*Filipendula*), along with pollen of the rose and a carrot (Apiaceae) families. Pollen of the latter two families, along with mustards, could reflect the cultivation or deliberate collection and use/disposal of herbs and vegetables, though none are apparent in the plant macrofossil record from the midden. Other species are likely to reflect plants growing in grassy and damp areas, perhaps including nearby areas of pastures and meadows. There is a significant peak in birch pollen at the top of the midden, which corresponds to a thin leaf litter layer identified on the top of the midden.

Univere (Fig. 5.11)

The pollen sequence from Univere is the closest off-site sequence to Karksi Castle (1 km). Woodland dominates much of the sequence, characterized by spruce-pine woodland during the Iron Age. Spruce values are highest during the first to fifth centuries, but decline gradually thereafter, particularly from the tenth century with a greater contribution of pine and birch. There is a rapid change in pollen signal corresponding to a sharp transition from wood peat to clay-silt (0.55 m). The sharp change in sediment reflects a significant shift in the depositional environment, involving greater erosion and deposition of minerogenic material from the surrounding environment, although the possibility of an erosional break in the sequence should not be excluded. The sediments overlying the peat are all post-medieval, and whilst there is some evidence for arable and disturbed land in the vicinity, the pollen primarily reflects the

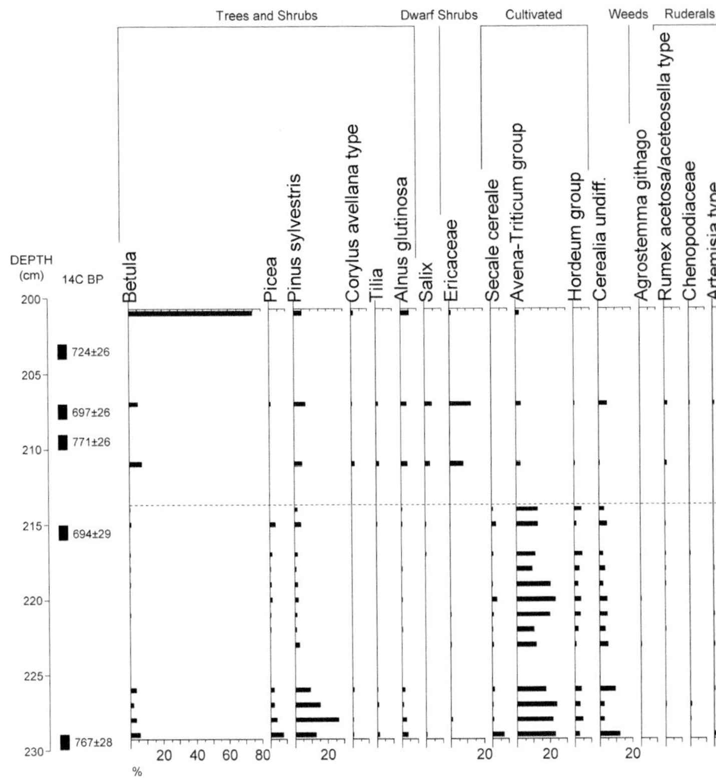

localised development of tall herb swamp communities dominated by sedges and grasses.

Äriküla (Fig. 5.12)

The vegetation during the Iron Age surrounding Äriküla is dominated by spruce-birch-pine woodland with a component of alder and smaller quantities of oak, lime, elm, and hornbeam; pollen of other trees occurs more sporadically. Birch and pine, along with hazel, suggest a relatively open canopy with an understorey of dwarf shrubs and *Sphagnum* mosses reflecting the presence of raised mire communities. There are occasional cereal pollen grains and ruderals that may hint at human activity on the nearby dry ground surrounding the bog, although no certain finds of Late Iron Age date have been found from the area, though there are occasional stray finds. The medieval period is characterized by a significant decline in spruce as a component of woodland during the thirteenth century, with increasing pine. There are visible increases in cereal pollen grains of oat-wheat from the mid-thirteenth century, rye from the early fourteenth century, and buckwheat during the fifteenth century, along with an increase in ruderal and grasses, reflecting localized settlement in the vicinity. The settlement at Äriküla is likely of medieval date where some medieval pottery sherds have been found, with another medi-

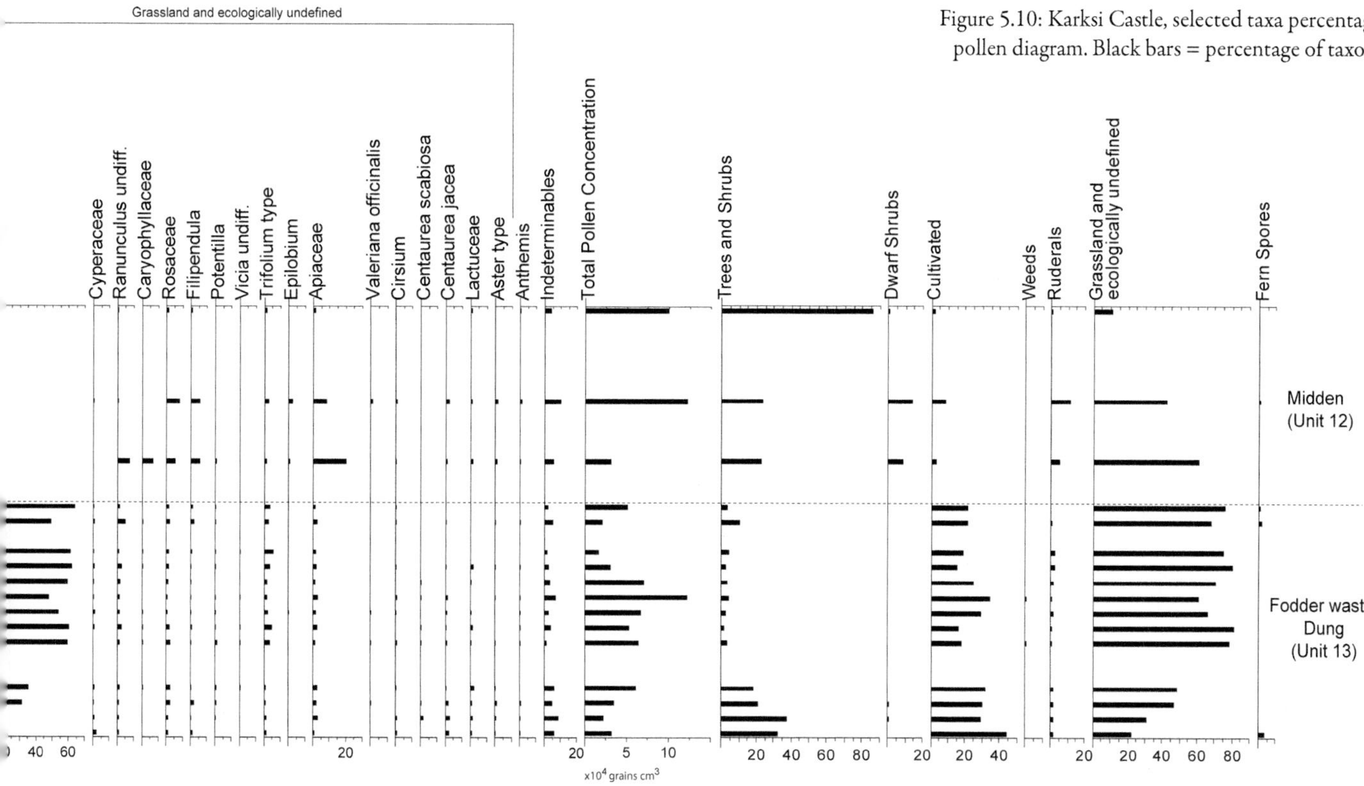

Figure 5.10: Karksi Castle, selected taxa percentage pollen diagram. Black bars = percentage of taxon.

eval settlement, possibly of thirteenth century and later date located some 2.3 km east at Kiisa. *Sphagnum* spores increase significantly from the late thirteenth century, reflecting the local dominance of raised bog, but decline gradually thereafter. Birch increases during the late sixteenth/early seventeenth centuries followed by increasing values for heathers forming drier hummocks in the bog with an open canopy of birch and pine. Oak, elm, lime, and hornbeam almost completely disappear during the post-medieval period, with only isolated stands of oak and hornbeam suggested. There is continued agricultural activity on the surrounding dry ground, with an increase in rye towards the end of the seventeenth century. Cereal cultivation is most apparent, however, during the eighteenth century along with an increase in grass and ruderal pollen, particularly common sorrel, suggesting intensifying arable and pastoral land-use. By the mid-nineteenth century, there is a decline in evidence for agricultural land-use with an increase in dwarf shrubs on the bog surface and regeneration of pine and some spruce in the surrounding woodland.

Discussion

The results of palynological analysis are discussed within three main areas; the territories of Riga, Wenden, and Fellin/Karkus. Data from Riga and Wenden are considered in the context of existing palynological studies from southern Livonia, with those from Fellin/Karkus considered in the context of studies from northern Livonia. In addition to the twelve sequences presented here, a further forty-five pollen sequences with radiocarbon dates ≤ 1500 ^{14}C years BP, extending from the Iron Age, are available from Livonia; eighteen from southern Livonia, and twenty-seven from northern Livonia, with an additional eleven from the Duchy of Estonia (Map. 5.2). Many other sequences were excluded either because they lack deposits of the relevant age or contain Late Iron Age and medieval deposits that were not directly dated and therefore lack a precise chronological context. Of these forty-five sequences, comprising a hundred ^{14}C dates ≤ 1500 ^{14}C years BP (twenty-three from Latvia, sixty-seven from Estonia), nearly half of the forty-five sequences (twenty-one) have only one ^{14}C date. This lack of a precise chronology presents difficulties in understanding the timing of vegetation change and patterns of human impact and land-use at both the local and regional scale, although a growing number of well-dated pollen-based studies have emerged over the last decade (e.g., Niinemets and Saarse 2009; Niinemets 2010; Stivriņš et al. 2015; Veski et al. 2005).

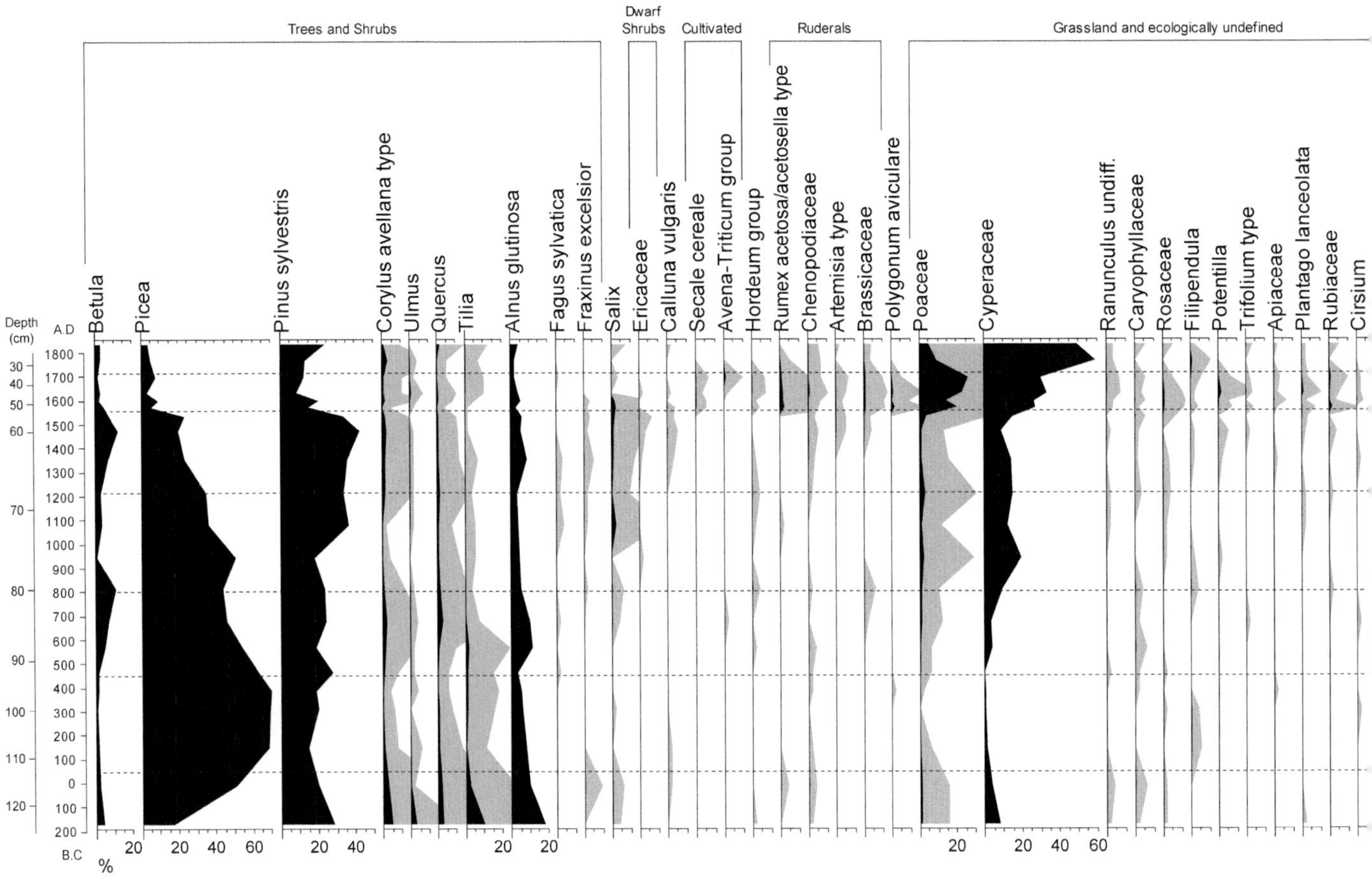
Trees and Shrubs
Dwarf Shrubs
Cultivated
Ruderals
Grassland and ecologically undefined
Depth (cm)
A.D
B.C
%
Betula
Picea
Pinus sylvestris
Corylus avellana type
Ulmus
Quercus
Tilia
Alnus glutinosa
Fagus sylvatica
Fraxinus excelsior
Salix
Ericaceae
Calluna vulgaris
Secale cereale
Avena-Triticum group
Hordeum group
Rumex acetosa/acetosella type
Chenopodiaceae
Artemisia type
Brassicaceae
Polygonum aviculare
Poaceae
Cyperaceae
Ranunculus undiff.
Caryophyllaceae
Rosaceae
Filipendula
Potentilla
Trifolium type
Apiaceae
Plantago lanceolata
Rubiaceae
Cirsium

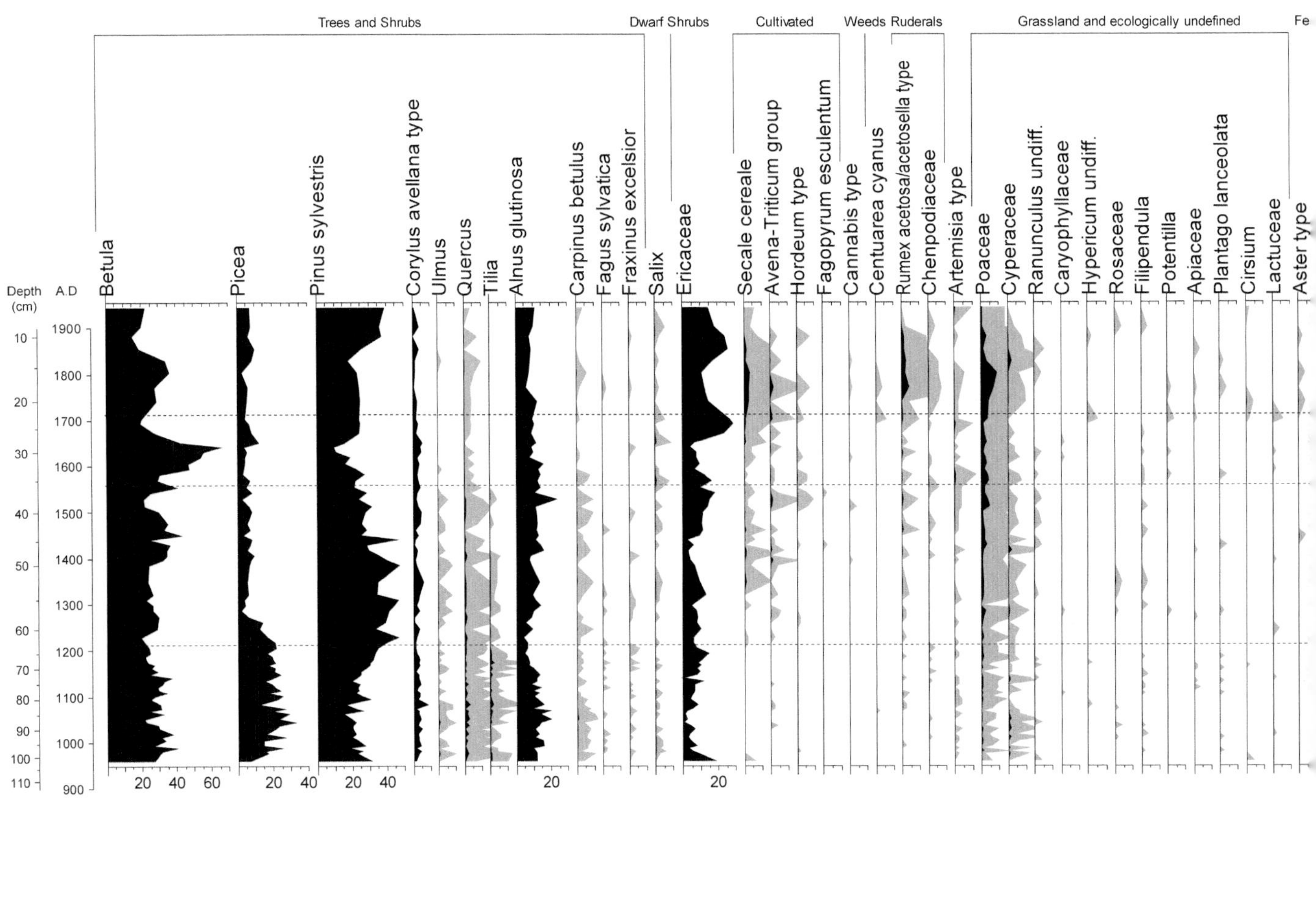
Trees and Shrubs
Dwarf Shrubs
Cultivated
Weeds
Ruderals
Grassland and ecologically undefined
Fe
Depth (cm)
A.D
Betula
Picea
Pinus sylvestris
Corylus avellana type
Ulmus
Quercus
Tilia
Alnus glutinosa
Carpinus betulus
Fagus sylvatica
Fraxinus excelsior
Salix
Ericaceae
Secale cereale
Avena-Triticum group
Hordeum type
Fagopyrum esculentum
Cannabis type
Centuarea cyanus
Rumex acetosa/acetosella type
Chenpodiaceae
Artemisia type
Poaceae
Cyperaceae
Ranunculus undiff.
Caryophyllaceae
Hypericum undiff.
Rosaceae
Filipendula
Potentilla
Apiaceae
Plantago lanceolata
Cirsium
Lactuceae
Aster type

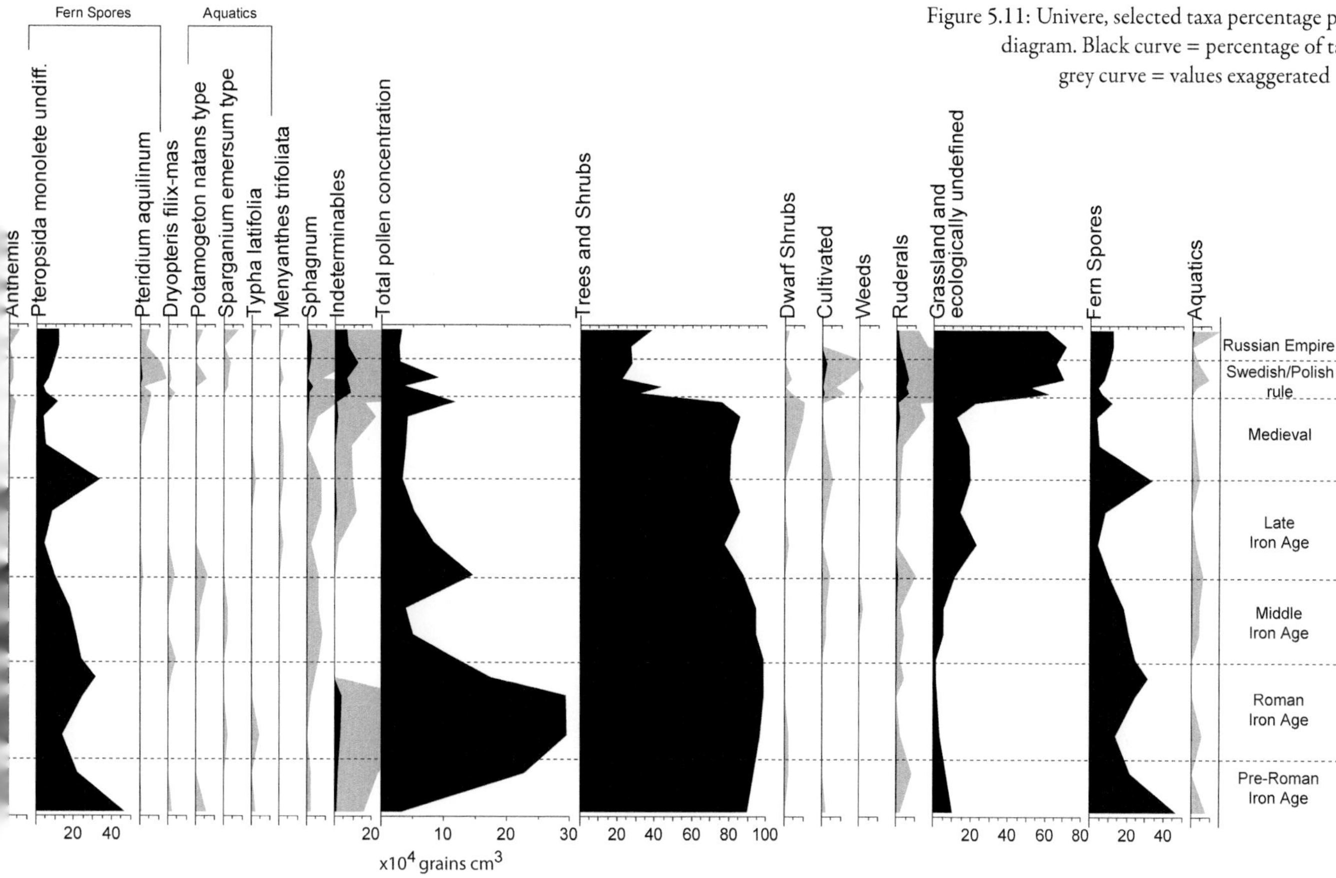

Figure 5.11: Univere, selected taxa percentage pollen diagram. Black curve = percentage of taxon, grey curve = values exaggerated × 10.

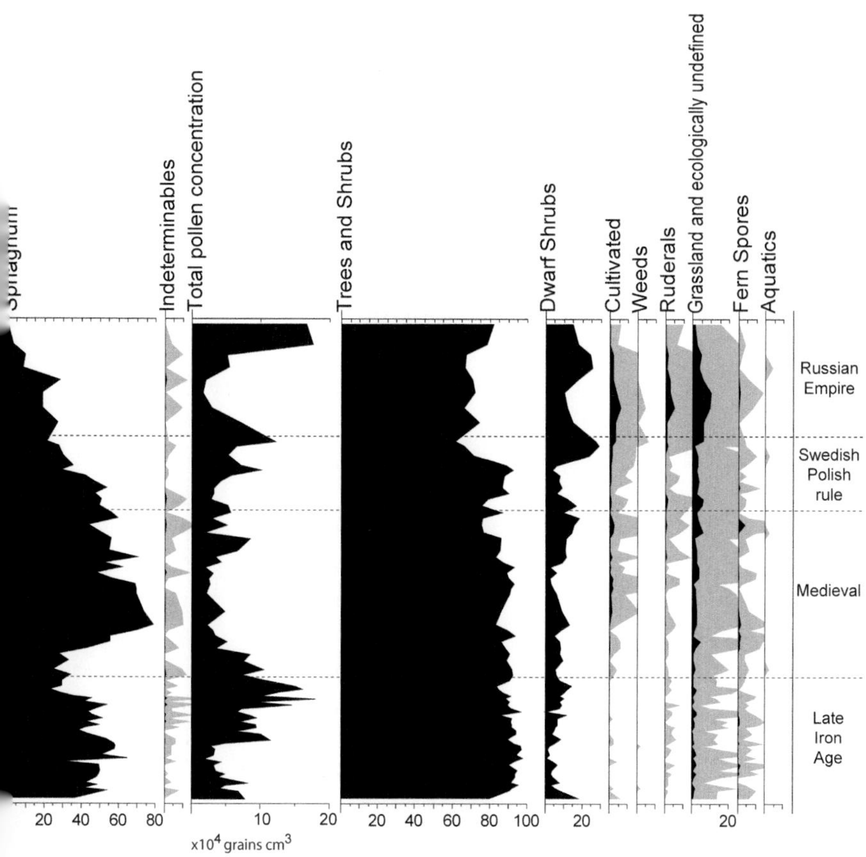

Figure 5.12: Äriküla, selected taxa percentage pollen diagram.
Black curve = percentage of taxon, grey curve = values exaggerated x 10.

1 Āraiši (Stivriņš et al. 2015)
2. Eipurs Mire (Kušķe et al. 2010)
3. Dzelve-Kronis Mire (Kušķe et al. 2010)
4. Lake Ķūži (Kangur et al. 2009)
5. Bazi Mire (Pakalne and Kalniņa 2005)
6. Lake Kurjanovas (Heikkilä and Seppä 2010)
7. Rožu Bog (Kušķe et al. 2011; Kalniņa et al. 2014)
8. Zilais Mire (Kalniņa et al. 2014)
9. Ramava Mire (Kalniņa et al. 2014, Kalniņa et al. 2016)
10. Lake Velneezers (Kalniņa et al. 2014)
11. Zosna peatland (Ozola 2013)
12. Lake Verijärv (Niinemets and Saarse 2009)
13. Lake Lasva (Niinemets and Saarse 2009)
14. Parika Mire (Niinemets, Saarse, and Poska 2002)
15. Lake Paidre (Saarse et al. 1995)
16. Lake Touglärv, Rouge (Poska et al. 2008; Veski et al. 2005)
17. Lake Planni (Niinemets and Saarse 2009)
18. Lake Ala-Pika (Kihno and Valk 1999)
19. Lake Viitna Linajärv (Punning et al. 2007)
20. Lake Kahala (Poska and Saarse 1999)
21. Lake Mardu (Veski and Lang 1996)
22. Surusoo Mire (Poska and Saarse 2002)
23. Vedruka Mire (Poska and Saarse 2002).

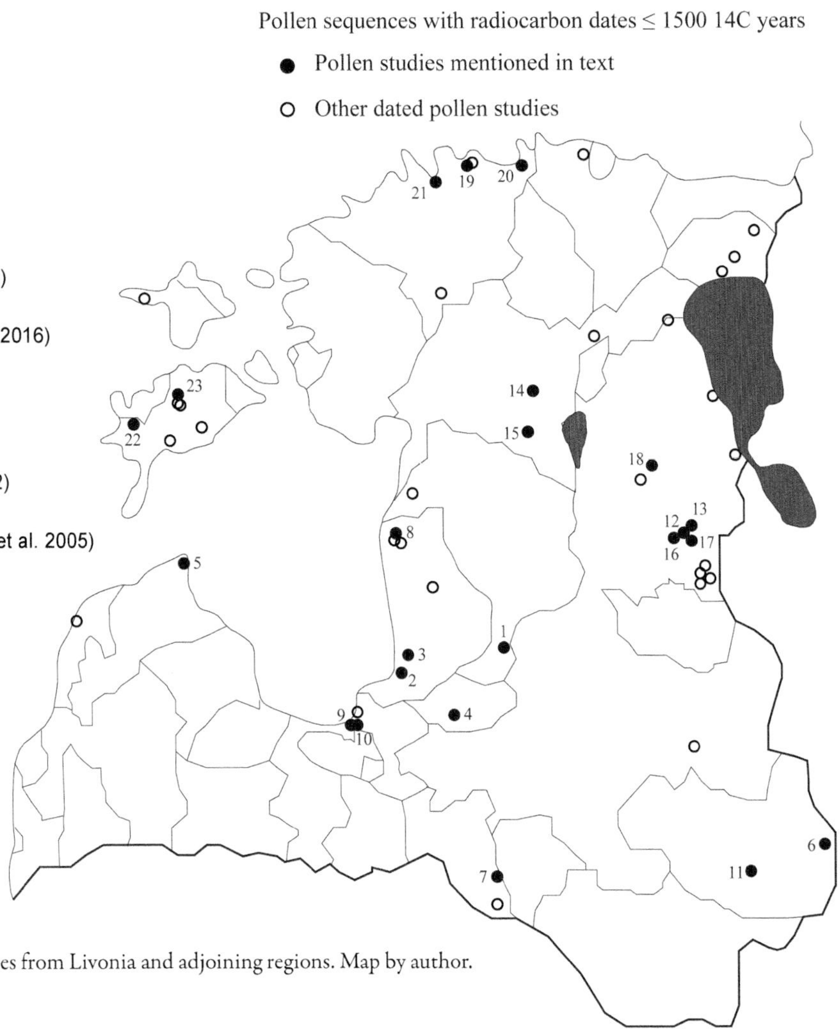

Map 5.2: Location map of existing palynological studies from Livonia and adjoining regions. Map by author.

Riga

The palaeoecological record from the Riga Liv quarter, although predominantly reflecting the activities of the inhabitants of this part of the town (Brown et al. 2017), can also provide useful indications about the vegetation environment surrounding Riga in the fourteenth century. The consistent and complementary nature of the plant macrofossil and pollen records suggest that both, at least in part, reflect the diversity of habitats occurring locally around the city, whilst also recognizing the potentially wide source areas of each proxy from urban contexts, including materials derived both locally and transported over longer distances (e.g., Latałowa et al. 2003; Święta-Musznicka et al. 2013).

The consistently low levels of arboreal pollen demonstrate that the immediate landscape was largely deforested, with significant quantities of non-arboreal taxa indicative of open, disturbed, and grassland habitats, including both wet and dry meadows and pasture. Arboreal pollen occurs in similarly low frequencies in samples from buildings dating to the early to mid-thirteenth century. The exceptions are moss samples 48 and 59 that include higher quantities of birch and pine pollen, although in neither case are the values high enough to suggest a dominance rather than local presence of either species. The moss samples may have been collected from areas of wetland close to woodland, used for waterproofing the gaps between building planks in a similar fashion to caulking on ships (e.g., Deforce et al. 2014). Both moss samples include large quantities of cereal and grass pollen, no doubt incorporated at a later stage, perhaps reflecting the storage and processing of cereals or the presence of large quantities of grass and cereal waste; the moss itself acts as a perfect trap for pollen (Pardoe et al. 2010). The picture of a predominantly deforested landscape in the immediate vicinity of Riga is supported by earlier pollen studies from the Dome Square and from below the Triangular Bastion that contains little arboreal pollen but large quantities of wetland, weed, and ruderal pollen (Kalniņa pers comm.).

Values for lime, however, reach up to 8% in the latrine (Fig. 5.7). This is a significant quantity for a tree that flowers less profusely than other arboreal taxa, is insect pollinated, and has poorly dispersed pollen grains that are typically under-represented in pollen spectra. Lime values from Latvian pollen sequences generally do not exceed *c.* 2% at this time (Pakalne 2008; Brown and Pluskowski 2013; Stivriņš et al. 2015), although this partly reflects the intolerance of lime to waterlogged soils and tendency to be restricted to drier soils along the edges of lakes and bogs located further from pollen profiles. Nonetheless, the elevated levels of lime pollen from the latrine may reflect the highly localized presence of lime trees (or tree) either within the Liv quarter or very close by. Importantly, lime trees were sacred to indigenous pagan societies in the eastern Baltic, associated with the goddess Laima (Vaitkevičius 2004, 20). The persistence of lime within the Liv quarter of Riga into the fourteenth century could suggest the maintenance or survival of a sacred tree(s), cultic site, and/or of pre-Christian practices. This hypothesis cannot be proven for certain, but there is ample documentary evidence that local traditions were respected and sacred forests protected by the conquering forces, at least until the arrival of the Jesuits in the seventeenth century (Kļaviņš 2011). Alternatively, elevated lime pollen within the latrine may reflect human waste derived through the consumption of lime-tree honey, or mead based on such honey. Bee-trees are consistently mentioned in the documentary sources from Livonia, including Riga, forming an important part of the economy, both in food (e.g., as a sweetener) but also wax production (e.g., used to manufacture candles) (see Chapter 8).

Documentary sources relating to Riga indicate that there was a lack of timber resources in the surrounding area by AD 1255, with a subsequent increase in the exploitation of woodland along the River Daugava and its tributaries (Zunde 1998/99); this may refer to the availability of suitable constructional timber rather than the presence/absence of woodland per se. However, mixed coniferous-deciduous woodland, largely comprising pine and birch, dominate the pollen sequence from Ramava Mire, located on an island in the River Daugava 2.5 km south-east of medieval Riga, and similarly in the pollen sequence from Lake Velnezers, *c.* 8.5 km to the north-east (Kalniņa et al. 2014). Although the immediate surroundings of the city were cleared of woodland, it is evident from the pollen record that woodland remained the key component of the vegetation within the nearby surrounding landscape.

Wenden Commandery

Pollen analysis from sites around Cēsis (Figs 5.2–5.7) showed a heavily wooded landscape throughout the Late Iron Age and medieval periods, with limited evidence for significant human impact beyond the immediate surroundings of the key castles and urban centres. Mixed boreal coniferous and deciduous woodlands chiefly comprising pine, spruce, and birch dominate the landscape during the Late Iron Age. Occasional pollen grains of cereal and ruderal taxa in the Nineris Mire sequence (2.6 km north of Cēsis Castle) point towards some small-scale agricultural land-use in the vicinity during the Late Iron Age, perhaps within small clearings along the shores of the adjacent lake. There is currently no palaeoenvironmental data from the immediate landscape of Cēsis Castle contemporary with the occupation of the Riekstu Hillfort, but one can hypothesize that although the wider landscape was heavily wooded, the immediate environs of the hillfort would most likely have been largely cleared of trees.

The generally low level of human activity in the mire pollen sequences is not unsurprising, however; many are located in isolated settings peripheral to areas of human settlement or, as in the case of Nineris, Blusu, and Blanku Mire, are located in areas with largely sandy soils (European Soils Portal) of marginal to poor agricultural potential. Predominantly sandy soils are distributed to the north, north-east and north-west of Cēsis and today these areas retain a heavy woodland cover. However, heavy boulder clay soils more suited to intensive cultivation are distributed to the south-east of Cēsis, where palaeoenvironmental analysis suggests a more mosaic vegetation landscape in places during the Late Iron Age. Pollen and fungal spore analysis from Lake Āraiši, 6.5 km south of Cēsis, revealed significant evidence for human activity during the Late Iron Age (Stivriņš et al. 2015) associated with the lake settlement occupied from *c.* AD 780 to the early/mid-eleventh century (Apals 2002; Meadows and Zunde 2014; Punning et al. 1968). The study at Lake Āraiši is the first detailed multi-proxy palaeoenvironmental investigation associated with a Late Iron Age settlement in Latvia. Although the first appearance of cereals and a gradual opening up of the woodland environment occurred towards the end of the Middle Iron Age (AD 400–800), land-use was most intensive between the mid-eighth to mid-eleventh centuries, resulting in a semi-open landscape surrounding the lake comprising a mosaic of cultivated land, meadows, and pasture. Intensive land-use resulted

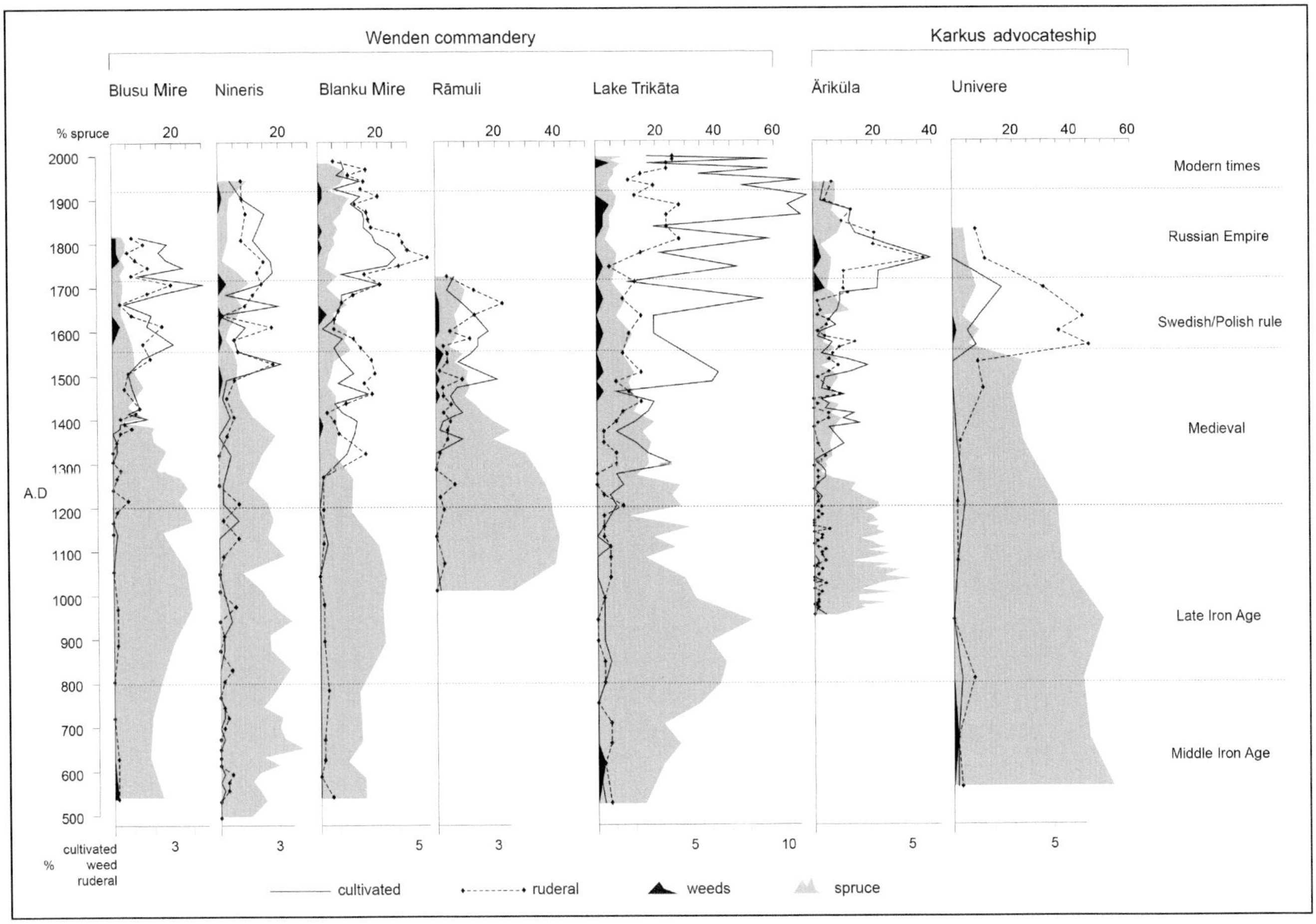

Figure 5.13: Synthesis of cultivated, ruderal, and field weed pollen from selected pollen studies plotted against values for spruce pollen.

in the increasing eutrophication of the lake, reflected by an expansion of cyanobacteria, with high values for coprophilous fungi emphasizing the localised presence of grazing animals. The finds of Arabic dirhams from the lake settlement suggest that this was a relatively prosperous community producing sufficient goods for trade and exchange (Apals 2012). Despite the abandonment of the lake settlement by the mid-eleventh century, intensive agricultural land-use continued in the surrounding landscape for the remainder of the late Iron Age (Stivriņš et al. 2015, Fig. 7); the location was clearly an important focus for settlement over several centuries. By comparison, pollen studies from Lake Trikāta, 37 km north-north-west of Wenden, suggest a heavily wooded landscape with limited evidence for agricultural land-use during the Late Iron Age (Fig. 5.7; Stivriņš et al. 2016), despite the reference in Henry's *Chronicle of Livonia* (*HCL* IV, 12) in 1208 to Trikaten as the location of one of two strongholds ruled over by the local Lettish chieftain. There have been no archaeological investigations to confirm the existence of an earlier stronghold below the Order's castle, but the palaeoenvironmental evidence does not suggest an intensive use of the local landscape during the Late Iron Age, instead involving small-scale slash-and-burn agriculture.

The decline in spruce has been linked to an increase in agricultural activity (Reitalu et al. 2013), with a clear link between decreasing spruce pollen and an increase in cultural indicators in the pollen from Lakes Trikāta and Āraiši. The relationship between spruce and human activity is less clear in other pollen sequences, with declines dating to the twelfth (Blanku Mire), thirteenth (Blusu Mire), fourteenth century (Rāmuli), and tenth, thirteenth, and fifteenth centuries from Nineris Mire (Fig. 5.13). Only in the cases of Rāmuli and Nineris Mire does there seem to be a clear link between increasing human activity and a decrease in spruce that might otherwise be invisible in the sequences from Blanku and Blusu Mire (see below).

The limited number of existing radiocarbon-dated pollen studies from southern Livonia support the picture of a heavily wooded landscape across much of the landscape, at least beyond the immediate locales of settlements, with several sequences from peripheral locations, including Rožu and Zilais Mires, showing little or no evidence for human activity during either the Iron Age or medieval periods (Kalniņa et al. 2014; Ozola 2013). Small increases in pasture and arable land was recorded in the Lake Kūži sequence from 925±30 BP (cal. AD 1029–1180, Kangur et al. 2009), from *c.* 1000 at Zosna peatland (Ozola 2013), with a continuous *Cerealia* pollen curve from *c.* 500 BC and intermittent rye pollen from the tenth century at Lake Kurjanovas (Heikkilä and Seppä 2010). The sparse palynological evidence from Latvia for intensifying land-use during the Late Iron Age clearly reflects the peripheral location of many pollen sequences away from archaeological sites and contrasts with the archaeological and historical record from the eastern Baltic that demonstrates this was a period of demographic and economic expansion (Khino and Valk 1999). Numerous hillforts, settlements, and cemeteries are distributed across Latvia, particularly along the Curonian coast and the course of major rivers including the Daugava and Gauja; excavations of hillforts and settlement sites are characterized by the almost ubiquitous finds of farming tools and emphasize the important role of farming in society. This is clearly reflected in the palaeoenvironmental evidence for intensive agrarian land-use from Lake Āraiši (Stivriņš et al. 2015), and it highlights the need for additional palaeoenvironmental research on Late Iron Age settlements. Palaeoenvironmental analysis from Iron Age Curonian strongholds in north-western Lithuania has shown clear evidence for significant landscape impact during the Late Iron Age. At Impilits, land-use is most intensive between 1050 and 1250, with the regression of woodland habitats, particularly spruce, accompanied by an expansion in pastoral and arable activities (Stančikaitė et al. 2009). The development of an open mixed pastoral/arable landscape is also apparent during the Viking Age/Late Iron Age from the Skomantai settlement prior to its destruction by the Teutonic Order (Stančikaitė et al. 2013). Similarly, there is widespread evidence for intensifying settlement and land-use apparent in the pollen and archaeological record from northern Livonia and particularly north-eastern Estonia and Saaremaa (see below).

The palynological evidence from the Wenden commandery does not suggest a significant change in the intensity of land-use at the beginning of the medieval period (Fig. 5.13). Sustained increases in cereal and ruderal pollen is only apparent from the late thirteenth/early fourteenth century at Blanku Mire, late fourteenth/early fifteenth century at Blusu Mire and Rāmuli, and from the early sixteenth century at Nineris Mire. This pattern is reflected in many radiocarbon dated pollen sequences across southern Livonia (Map 5.2) that show only small increases in pollen indicative of pasture/meadow and arable land during the medieval period; at Eipurs Bog from a level dated 689±50 BP (cal. AD 1228–1398), Dzelve-Kronis Bog after 757±55 BP (cal. AD 1169–1381; Kuške et al. 2010) with intermittent increases in rye pollen at Bazi Mire from 742±90 BP (cal. AD 1046–1408; Pakalne and Kalniņa 2005). Pollen from extensive mires located in peripheral areas of otherwise low agricultural potential is unlikely to produce evidence for significant human impact on the environment, providing potentially false-negative evidence for human activity. However, the consistent although low values for rye and other cereals, and intermittent weeds, at least point to an increase in arable fields and pastoral land in the wider landscape surrounding Cēsis from the late thirteenth century. Although wind-pollinated rye typically occurs in larger quantities in pollen sequences than other self-pollinated cereals, it may have been preferred in areas with nutrient poor sandy soils because of the ability of its fibrous root system to exploit water to a greater depth than other cereals (Vaughan and Geissler 1997).

The reduction in arboreal pollen in sequences surrounding Cēsis, particularly spruce, also points towards increasing human impact within the landscape during the medieval period. Declining spruce values have been linked in parts of Sweden and Finland to expanding agrarian land-use (Segerström 1990) including canopy thinning to create meadows and forest pastures (Segerström and Emanuelsson 2002); such a strategy may have been employed around Cēsis in areas of poorer arable potential. In addition, there is widespread evidence across Scandinavia, and from parts of the eastern Baltic, for the development of heathland as a result of increased grazing activity (Lagerås 2007; Moe et al. 2005). Higher frequencies of heather pollen are apparent in the Blusu Mire, Blanku Mire, Nineris, and Rāmuli sequences (Figs 5.2–5.4, 5.6). The increase in heather at all four sites occur contemporary with declining arboreal pollen and an increase in cereal and ruderal taxa, lending support at least superficially for the increased grazing hypothesis. However, in peat bogs an expansion in heather more likely reflects the gradual development of nutrient depleted raised bog and is reflected at all four sites by the synchronous increase in values for *Sphagnum* spores characteristic of bog environments. Although increased grazing pressure may have occurred along the margins of bogs, the pollen data most likely reflects two contemporary but largely unconnected processes occurring with the wetland and dry ground; the former related to natural development of raised bog and the latter connected to some expansion in arable and pastoral land.

Prior to the fourteenth century, the picture from pollen sequences therefore appears to be one of continuity or even decline in the intensity of agricultural activity during the thirteenth century. However, at Lake Trikāta major changes in vegetation and intensification in land-use are apparent in the pollen data following the crusades (Fig. 5.7). Declining arboreal pollen is accompanied by an increasing range and frequency of cultivated and ruderal taxa from the mid-thirteenth century, suggesting significant clearing of the surrounding woodland and the development of a mosaic landscape of arable, pastoral, and disturbed ground. There are no direct sources to confirm the date of construction of the adjacent castle, but it is considered to predate the formation of the Parish of St John in 1282–87; the significant changes in vegetation surrounding the lake are therefore most likely related to activity following the construction of the castle and associated settlement. Trikaten was located on one of the main north–south routes within a relatively stable part of Livonia and is likely along with other key castle to have played an important provisioning role in the initial decades following the crusades; the pollen evidence for intensified land-use provides some indirect evidence in support of this hypothesis. Documentary sources from the mid-sixteenth century do, however, list Trikaten as one of several of the Order's castles acting as *Kornhausen* (corn houses), emphasizing their role in the collection, storage, and redistribution of agricultural produce (Caune and Ose 2004; Hartmann 2008, 189). The mosaic nature of the landscape suggested by the pollen is further emphasised by documentary sources listing a range of different land-uses. Descriptions of a boundary dispute, dating from 1472 and involving the subjects of Trikaten, contain references to wood-felling sites, meadows, fields, and bee gardens within the wider district of Trikaten, including land to the north around Ērģeme located close to the current border between Latvia and Estonia (LGU I, 487). Documentary sources from across medieval Livonia frequently make reference to fields, meadows, mills, and rights related to pasturing, fishing, cutting of firewood, and bee-keeping (see Chapter 8), the latter no doubt for wax and honey production. Fishing and fish weirs in particular are frequently mentioned in documentary sources, and it is likely that the rivers and lakes adjacent to castles such as Trikaten and Arrasch were actively used for this purpose.

At Lake Āraiši, there are slight increases in rye pollen and coprophilous fungi during the fourteenth century that suggest some intensification in arable and pastoral land-use in the surrounding landscape following the foundation of the castle (Stivriņš et al. 2015). The archaeological evidence is ambiguous, however; large quantities of animal bones recovered during excavations of the western range during the 1980s (Apals 2012) suggest some storage or control of local resources, but the absence of either artefactual or structural remains from the outer bailey and eastern range suggests that the castle performed a rather limited role in this respect.

Alternatively, agricultural produce was transported the short distance (8 km) from Arrasch to Wenden where large stores of food are recorded during a visitation in 1451, including significant quantities of rye, barley, and oats. Apparently, the grain stored in Wenden at this time accounted for almost a third of all stored grain in Livonia, a significant part of which was used to concentrate power with the Grand Master. Pollen from the moat at Cēsis Castle suggests that the immediate environs were largely open by the fourteenth century when the castle underwent a major phase of expansion, although surrounded by the pine-birch, spruce woodland that characterizes the off-site pollen sequences. The significant quantities of cereal pollen in the moat, including oat, wheat, barley, rye, and buckwheat, are likely to reflect inclusion of pollen from multiple sources. The poor production and dispersal rates of most cereals, excluding wind-pollinated rye, suggest surrounding arable fields are unlikely to have contributed the significant amounts of cereal pollen recorded from the moat. Instead, the majority of cereal pollen was most likely contained within grain stored within the outer bailey, used for human consumption or as a component of animal fodder, and also contained with kitchen, animal dung, and latrine waste flushed into the moat: many of the field weeds are likely to have been brought into the castle along with the grain. Experimental work by Greig (1982) demonstrated that cereal pollen was often retained in grain products (e.g., bread), but also in chaff used as animal fodder. In both cases grain products and animal fodder will have resulted in rotting food waste, along with human and animal faeces, that are likely to have attracted insects, themselves transporting pollen from plants growing with the surrounding environment. The presence of dung has been suggested as one of the primary factors in the high percentage of entomophilous taxa on many urban archaeological sites relative to their likely presence (e.g., Święta-Musznicka et al. 2013); this is emphasized by the large quantities of invertebrates recorded from the Riga latrine (King 2019) that no doubt deposited quantities of entomophilous pollen within these deposits. Entomophilous taxa in the Cēsis moat reflect a variety of ecological niches within the castle, urban, and fringing environs, including disturbed waste ground, meadows, and pastures. The high values of grass likely reflect the variety of open ground settings within the castle and town, but may also have been contained within hay used as fodder, perhaps for horses stabled in the castle. The town of Wenden developed together with the castle and was situated on an important Hanseatic trade route. Documentary sources emphasize that during the fifteenth and sixteenth centuries, in addition to goods from Russia and the west, residents of Wenden also sold their own goods, including linen, hemp, flax, hops, leather, honey, and beeswax (Kļaviņš 2012), no doubt grown on farms in the surrounding landscape.

Fellin Commandery/Karkus Advocateship

Palaoenvionmental analysis from southern Estonia focused on the landscapes surrounding the ruins of the castles at Viljandi and Karksi. Pollen sequences were taken from several locations, particularly around Viljandi, but in many cases these either failed to preserve pollen or produced sequences where the medieval and later deposits had been truncated or removed entirely. However, despite the smaller number of sequences analysed, when taken together with existing palaeoenvironmental studies across southern Estonia, a picture of vegetation history and land-use is apparent.

The settlement record for Estonia shows an increase in new sites during the eleventh to thirteenth centuries (Laul and Valk 2007), and in southern Estonia these concentrate mainly in the south-east, where palaeoenvironmental data suggests increasing landscape openness in places during the Late Iron Age (Poska et al. 2008, 2014; Niinemets and Saarse 2009; Kihno and Valk 1999). South-western Estonia by comparison was more sparsely populated, largely as a consequence of the many large bogs and vast tracts of woodland that cover the landscape, with several large bogs also located to the north and north-east of Lake Võrtsjärv. Distribution maps of Migration Period, pre-Viking, and Viking Age sites, for example, show a complete absence of sites of any form in these areas (Tvauri 2012, Figs 2–4), a pattern reflected in the later distribution of castles and medieval settlements. Viljandi and Karksi are both located close to the western limits of the more intensively settled areas of south-eastern Estonia, and it seems plausible that south-western Estonia has always been rather more peripheral in terms of settlement and retains much of its natural character in terms of both vegetation and landforms.

Fellin itself was situated on important river and land routes, and was the location of one of the strongest Late Iron Age hillforts in the region (Rannamäe and Valk 2013). However, the pollen data from around Viljandi is inconclusive with regards to the intensity of land-use, largely as a consequence of the poor chronological resolution of the pollen sequences that make it difficult to distinguish between sediments of Late Iron Age and medieval date. Coring along the margins of Lake Viljandi directly below the castle identified a small hollow with a two-metre sequence of peat, silt, and sand, with dates at 0.90 and 2.04 m of 848±30 BP (Wk-24857, cal. AD 1150–1270 91%) and 2481±30 BP (Wk-24858, 775–483 cal. BC) respectively. Pollen was absent from the majority of the sequence except two samples at 1.08 and 1.16 m, likely of Late Iron Age date, which were both dominated by pine and spruce, and in the case of 1.16 m, by approximately 20% lime. Pine, birch, and spruce are the key woodland components in the sequence from nearby Lake Kosilase, and also from two peat sequences assessed for pollen, but not subsequently analysed, located to the north-west and west of Viljandi. At Pärsti, *c.* 8 km north-west of Viljandi, peat at 0.68 m produced a date of 2126±34 BP (GU-31597, 210–48 cal. BC), with pollen from above this level containing comparable quantities of pine, spruce, and birch. There are small increases in pollen of rye and ruderal taxa from Lake Kosilase that hint at increasing agrarian activities within the surrounding landscape, albeit of uncertain date. It is worth noting that the beginnings of winter rye cultivation in Estonia date to around *c.* AD 800–1000 (Poska and Sarse 1999; Poska et al. 2014), providing a useful context for considering the pollen data from Lake Kosilase. The picture of land-use in the landscape surrounding Viljandi is therefore poorly resolved for the Iron Age, although it is logical to assume that the surroundings of the Late Iron Age stronghold are likely to have included a significantly open component to the vegetation, with pollen from surrounding locations suggesting a greater extent of woodland. The presence of undated plough marks identified during excavations in Viljandi hint at agricultural activity, perhaps related to Viking and Late Iron Age activity. Pollen analysis from Lake Päidre, 6.5 km south of Viljandi, produced limited evidence for human impact (Saarse et al. 1995). The small number of pollen sequences from the Karksi landscape suggests a heavily wooded landscape during the Iron Age similarly dominated by pine, spruce, and birch. Occasional finds of late prehistoric and Roman Iron Age date are known from the vicinity of the pollen sequence from Äriküla, but the pollen data suggests a rather minimal human impact on the landscape, with only very occasional rye pollen grains during the tenth and eleventh centuries (Figs 5.12, 5.13).

The majority of existing pollen studies from southern Estonia is distributed within the south-east, particularly within the Haanja Heights. Here pollen analysis shows an increasing intensity in land-use during the Iron Age, with studies at Lakes Plaani, Verijärv (Niinemets and Saarse 2009), Ala-PIka (Kihno and Valk 1999), and Rõuge Tõugjärv (Poska et al. 2008) all showing an increase in the quantities of cereal, ruderal, and open ground taxa from the seventh/eighth century. Quantitative land-cover reconstructions from Lake Verijärv suggest that although the levels of landscape openness were relatively low around 1000, they increased rapidly after this with the development of a semi-open landscape during the Late Iron Age and medieval periods (Poska et al. 2014). The picture from south-eastern Estonia corresponds well with the palynological data from north-eastern Estonia and the island of Saaremaa that show intensifying land-use during the Late Iron Age. Pollen sequences from the northern coast of Estonia, including Lake Viitna Linajärv (Punning et al. 2007), Lake Kahala (Poska and Saarse 1999), and Lake Mardu (Veski and Lang 1996) show increasing agricultural intensification during the Viking Age and later. Saaremaa was argued to have been the most developed part of Estonia during the Iron Age, in part because of its favourable location on the Viking Baltic-Black Sea trade routes (Mägi 2004), with the *Liber census Daniæ* (Danish Book of Land Taxation) listing three thousand farmsteads on the island by *c.* 1300 (Ligi 1992). The pollen evidence from several sequences show increasing agricultural activity from *c.* 800, with peaks in rye, barley, and cornflower at Surusoo and Vedruka Mires emphasizing the development of permanent field systems during the Viking and subsequent Late Iron Age (Poska and Saarse 2002).

The medieval period in southern Estonia is marked in pollen sequences by contrasting patterns of vegetation change and land-use, characterized in some areas by increasing intensity in agricultural activity and in other areas by stability or even decline in land-use. The disappearance of cereal pollen in the sequence from Parika Mire *c.* 1200–50 was argued to reflect the direct impact of battle and plagues (Niinemets et al. 2002), presumably connected with the conquest. Pollen analysis from Lake Rõuge Tõugjärv showed very low non-arboreal pollen values at the beginning of the medieval period and a decrease in human impact indicators during the thirteenth century (Veski et al. 2005). Spatial reconstructions of land-cover suggest a more open landscape surrounding the lake than suggested by the pollen alone, but nonetheless showing a similar decline in the amount of open land surrounding Lake Rõuge Tõugjärv by the end of the Late Iron Age (Poska et al. 2008). From south-eastern Estonia, within the Dorpat bishopric, a pattern of relative stability in land-use is suggested by the consistent levels of arable and pastoral indicators from lakes Plaani and Verijärv, whilst from Lake Lasva, and a number of other pollen sequences across eastern and northern Estonia, there is an intensification in land-use from the beginning of the medieval period (Niinemets and Saarse 2009; Sillasoo et al. 2009). The shift towards a cooler climate during the thirteenth to mid-fourteenth century appears to have had little effect on cultivation, although the impact of famines and plague is argued to have left their mark in some pollen records. This is most noticeable in the Lake Lasva sequence, with a clear decline in arable and pastoral indicators during the mid-fourteenth century ascribed to the impact of the Black Death (Niinemets and Saarse 2009). There is a clear danger in linking historical events and palaeoecological data without a sufficiently robust chronology, and many pollen sequences are rather ambiguous without clear evidence for the impact of either famine or disease. Moreover, very little is known about the Black Death in the eastern Baltic, apart from Prussia (Benedictow 2012), but the late arrival and relatively sparse population density of Livonia may have largely limited the impact of the plague to the key urban centres with little impact in more isolated rural areas.

The picture of land-use from Viljandi and Karksi is rather patchy for the medieval period. The sequence from Lake Viljandi includes single pollen grains of buckwheat from near the base of the sequence (9.36 m), suggesting the sediments are unlikely to predate the earliest archaeobotanical and historical records of buckwheat in Estonia from the thirteenth/fourteenth centuries (Sillasoo and Hiie 2007). The catchment of Lake Viljandi covers 66.8 km^2, extending over 10 km to the north and no more than 2–3 km to the east (Vassiljev et al. 1995, Fig. 2). It is very likely that a proportion of the pollen will have been transported fluvially into Lake Viljandi

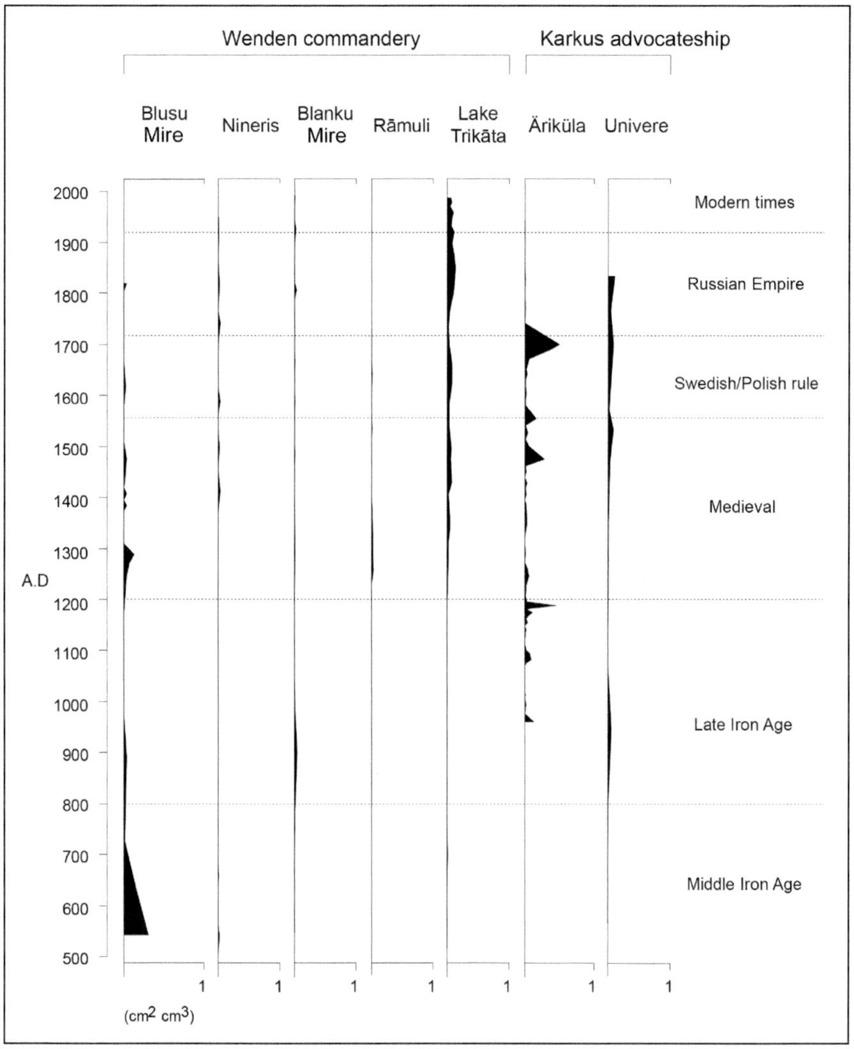

Figure 5.14: Synthesis of microscopic charcoal data from selected pollen studies.

from within the catchment reflecting the vegetation of this broader landscape. The current landscape within the boundaries of the catchment comprises 15–20% woodland, with much of the rest used for agriculture. Arboreal pollen within the top of the sequence accounts for 80% of the pollen sum and serves to emphasise the bias in representation of woodland habitats within an otherwise largely open landscape. Arboreal pollen values decline through the sequence, with consistent quantities of cereal and ruderal pollen suggesting an open mosaic landscape surrounding the castle, town, and lake. It is likely that the plateaus and western slopes overlooking the lakes were largely cleared and used for a mix of arable and pastoral land-use, with more woodland likely surviving on the steeper eastern slopes opposite the castle, with stands of alder growing in places on wetter soils along the lake and river margins. Pollen from Lake Kosilase, although undated, includes intensification in arable and pastoral land-use within the tope *c.* 0.50 m that likely reflects the activities of medieval and later communities.

The base of sediment sequence from Lake Viljandi comprises bands of eroded Devonian sandstone interleaved with gyttja. The construction of the castle entailed significant modification of the plateau, particularly during the phase of brick construction, truncating earlier traces of occupation (Rannamäe and Valk 2013). The cross cutting glacial valleys acted as natural defensive moats and were likely deepened, mobilizing further sediment into the lake. The adjacent town was constructed after the crusades but may also have involved significant modification and construction of the surrounding topography. It is tempting therefore to interpret the origin of the eroded sandstone at the base of the sequence as reflecting the intensive modification and settlement of the plateau following the conquest. The present-day landscape of Viljandi County, including both Viljandi and Karksi, has many areas of nature protection and includes a significant number of individual and groups of protected primeval trees. The majority of groups of trees are represented by stands of lime and oaks, with single stands largely comprising oak along with elm and some spruce and juniper (Estonian Environment Board 2012). Several of these are known as sacrificial sites and sacred groves that along with springs and natural stones may have had pre-Christian origins and continued to be used during medieval and later times (Valk 2009).

The pollen evidence from Äriküla suggests that intensification in agrarian land-use occurred during the fourteenth century, similar to some of the pollen sequences from around Cēsis (Fig. 5.13). There is also a notable decline in spruce from the beginning of the thirteenth century that reflects a shift to more open pine-birch dominated woodland. Studies by Reitalu et al. (2013) suggest that spruce is negatively associated with human impact, and that it tended to dominate in those areas with more limited human impact, with agricultural activity seen as a primary driver in its subsequent reduction (Reitalu et al. 2013; Poska et al. 2004). The reduction in spruce in the Äriküla sequence is not associated with an increase in arable or pastoral indicators, but may nonetheless reflect a degree of woodland clearance within the wider landscape. At Univere, the nearest off-site pollen sequence to Karksi, spruce values decline steadily from the fifth century but most sharply contemporary with an increase in open ground from the sixteenth century. There is little evidence for human activity in the surrounding landscape prior to this, although pollen in mid- to late thirteenth-century organic deposits from the high castle at Karksi suggests that the immediate surroundings of the castle were largely open. The significant quantities of cereal pollen in the dumped deposits could reflect the disposal of agricultural waste or cereal pollen contained within animal fodder, but may reflect crop processing within or near the castle, including threshing and winnowing that release significant quantities of cereal pollen (Bower 1992). Indirectly at least the cereal pollen reflects some control and perhaps storage of local agricultural produce, reflected also in the zooarchaeological remains from these deposits (see Chapter 6).

Conclusion

The palynological data from Livonia suggests that the conquest of Latvia and southern Estonia was accompanied by a complex pattern of both change and continuity in vegetation and land-use, variously characterized in places by intensification, stabilization, and declining agrarian land-use. Intensification was most apparent around the key urban centres and castles where their establishment involved new concentrations of people and animals with increasing provisioning requirements. Colonization was largely limited to towns, while rural landscapes show a degree of continuity in land-use, particularly those areas already densely settled and where an intensification of agrarian activities is apparent from the Late Iron Age. Many settlements and strongholds were attacked and destroyed during the initial conquest (see Lang and Valk 2011), but few of these sites have associated palaeoenvironmental data. There is evidence in some pollen studies for declining agrarian activities that might be associated with a degree of instability within the landscape following the crusades. In southern Livonia and along the Curonian coast, the documented destruction of Iron Age strongholds is accompanied in the palaeoenvironmental record by a regression in human activity (Stančikaitė et al. 2009, 2013): what was an actively settled landscape developed into a frontier zone between Livonia and the Grand Duchy of Lithuania.

One of the noticeable aspects of the palynological sequences presented here is the relatively small quantities recorded of microscope charcoal (Fig. 5.14), suggesting that fire formed a relatively minor aspect of land-management practices. However, Jääts et al. (2010) reviewed the historical evidence for the long-term maintenance of fire cultivation in Estonia up to the nineteenth century. They argue Estonia formed part of a wider zone of fire cultivation, including Latvia, Scandinavia, and parts of northern Russia, where this form of land-use was both more prevalent than previously thought and persisted for far longer than in other parts of Europe. In addition to slash and burn and burnbeating, the latter technique involving the burning of branches covered in soil, small plots of land were also cultivated in areas covered by young trees and bushes, termed 'bushlands'. These bushlands are argued to have made up as much as 25% of all farmlands in southern Estonia by the nineteenth century (Meikar and Uri 2000), replacing earlier techniques of fire cultivation which were argued to have been more widespread across Livonia during the sixteenth and seventeenth centuries (Ligi 1963).

These forms of agricultural land-use therefore seem likely to have persisted in use during the medieval period alongside the development of more intensive cultivation within permanent fields. Fire cultivation is potentially rather difficult to identify in the palynogical record from Estonia and Latvia, limited to small clearings within woodland that will have severely constrained the wider dispersal of microscopic charcoal particles and cereal and herb pollen. Declines in alder and spruce pollen during the Iron Age have been linked with developments in agriculture, such as the use of iron implements, and an increasing requirement for more agricultural land as populations expanded (Saarse et al. 2010). Clearance of spruce often followed the clearance of alder, a pattern that is visible to varying degrees in the pollen sequences from Äriküla, Blusu Mire, Blanku Mire, Rāmuli, Trikāta, and Āraiši during both the Iron Age and medieval period (Fig. 5.13; Stivriņš et al. 2015).

It has been highlighted already that the declines in spruce in several cases precede and are apparently unconnected to increases in human indicators. It seems likely however in these cases that spruce declines are reflecting land-use occurring within the wider landscape that is otherwise palynologically invisible, perhaps of the form discussed by Jääts et al. (2010), and occurring largely within small forest clearings. The increase in cereal, field weed, and ruderal pollen from the fourteenth century reflects intensification in crop cultivation, characterized in part by the expansion of permanent rotation field systems. This occurs against a backdrop of increased political stability, the growth of urban centres, the growing significance of the Hanseatic League and foreign trading networks, the establishment of serfdom, and the development of the manorial system that created an increased demand for agricultural produce. The growing importance of cereals, particularly rye, is demonstrated not only by the pollen evidence but through documentary and archaeobotanical evidence for its growing significant as a consumable, tradable, and taxable commodity (Chapter 13; Sillasoo and Hiie 2007). It seems likely, however, that traditional forms of land-use continued in use, perhaps persisting in more rural areas further from the principal towns. Colonization occurred at a significantly lower rate within Livonia, and the continuance of pre-conquest practices is reflected not just in the use by indigenous communities of Iron Age farming implements on medieval sites (e.g., Mugurēvičs 1990, 2008) but also the survival of traditional burial customs and continued use of sacred natural places (Laul and Valk 2007; Valk 2009). The palynological evidence contributes to this picture of change and continuity.

Works Cited

Primary Sources

HCL Brundage, J. A., trans. 2003. *The Chronicle of Henry of Livonia/Henricus Lettus*, New York: Columbia University Press

LGU Bruiningk, H. von. 1908. *Livländische Güterurkunden (aus den Jahren 1207 bis 1545)*, 2 vols, Riga: Komissionsverlag von Janck und Poliewsky

Secondary Works

Apals, J. 2002. 'Araisi Lake Fortress in Latvia', *Viking Heritage Magazine* 4, 24–25

——, 2012. *Āraišu ezerpils: rakstu izlase un draugu atmiņas*, Rīga: LU Latvijas vēstures institūts

Benedictow, O. J. 2012. *The Black Death 1346–1353: A Complete History*, Woodbridge: Boydell

Bower, M. 1992. 'Cereal Pollen Dispersal: A Pilot Study', *Cambridge Archaeological Journal* 2/2: 236–241

Brown, A. D., and Badura, M. 2017. 'Plant Macrofossil and Pollen Analysis from the Moat and Outer Bailey at Cēsis Castle: Indicators of Local Environment and Plant-Use', in *Cēsu pils raksti* (vol. I), ed. G. Kalniņš, Cēsis: Cēsu pils saglabāšanas fonds, 101–113

Brown, A., and Pluskowski, A. G. 2013. 'Medieval Landscape Transformation in the South-Eastern and Eastern Baltic: Palaeoenvironmental Perspectives on the Colonization of Frontier Landscapes', *Archaeologia Baltica* 20, 24-46

Brown, A. D., Badura, M., King, G., Gos, K., Cerina, A., Kalnina, L., Kalējs, U., and Pluskowski, A. G. 2017. 'Plant Macrofossil, Pollen and Invetebrate Analysis of a Mid-14th Century Cesspit from Medieval Riga, Latvia (Eastern Baltic): Taphonomy and Indicators of Human Diet', *Journal of Archaeological Science Reports* 11, 674–682

Caune A., and Ose I. 2004. *Latvijas 12. gadsimtā beigu - 17. gadsimtā vācu piļu leksikons*, Rīga: Latvijas vēstures institūta apgāds

Deforce, K., Allemeersch, L., Stieperaere, H, and Haneca, K. 2014. 'Tracking Ancient Ship Routes through the Analysis of Caulking Material from Shipwrecks? The Case Study of Two 14th Century Cogs from Doel (northern Belgium)', *Journal of Archaeological Science* 43, 299–314

Estonian Environment Board, 2012. Protected Natural Objects in Viljandi County. Trükikoda: AS Printon European Soils Portal <http://www.eusoils.jrc.ec.europa.eu> [accessed multiple occasions 2014–15]

Forest Europe, UNECE, and FAO, 2011. *State of Europe's Forests 2011: Status and Trends in Sustainable Forest Management in Europe*, Oslo: Forest Europe

Greig, J. 1982. 'The Interpretation of Pollen Spectra from Urban Archaeological Deposits', in *Environmental Archaeology in the Urban Context*, eds A. R. Hall and H. K. Kenward, CBA Research Report 43, York: Council for British Archaeology

Hartmann, S. 2008. *Herzog Albrecht von Preussen und Livland (1560–1564)*, Köln: Böhlau

Heikkilä, M., and Seppä, H. 2010. 'Holocene Climate Dynamics in Latvia, Eastern Baltic Region: A Pollen-Based Summer Temperature Reconstruction and Regional Comparison', *Boreas* 39, 705–19

Jääts, L., Kihno, K., Tomson, P., and Konsa, M., 2010. 'Tracing Fire Cultivation in Estonia', *Forestry Studies* 53, 53–65

Kalniņa, L., Stivriņš, N., Kuske, E., Ozola, I., Pujate, A., Zeimule, S., Grudzinska, I., and Ratniece, V. 2014. 'Peat Stratigraphy and Changes in Peat Formation during the Holocene in Latvia', *Quaternary International 383* (5), 186–95

Kalniņš, G. 2014. *Cēsu Pils (Cēsis Castle)*, Cēsis: Cēsis Culture and Tourism Centre

Kangur, M., Koff, T., Punning, J.-M., Vainu, M., and Vandel, E. 2009. 'Lithology and Biostratigraphy of the Holocene Succession of Lake Kuži, Vidzeme Heights (Central Latvia)' *Geological Quarterly* 53, 199–208

Kihno, K., and Valk, H. 1999. 'Archaeological and Palynological Investigations at Ala–Pika, Southeastern Estonia', *PACT* 57, 221–237.

King, G. 2019. 'A Foul Perspective: Analysis of Invertebrate Remains in a Mid-Fourteenth Century Latrine from Riga, Latvia', in *in Ecologies of Crusading, Colonization, and Religious Conversion in the Medieval Baltic: Terra Sacra II*, ed. A. Pluskowski, Environmental History of the North Atlantic World, 3, Turnhout, Brepols (forthcoming)

Kļaviņš, K., 2011. 'Sacred Forests and Trees in Latvia and in the Latvian Mind', in *Environmental Philosophy and Landscape Thinking*, eds L. Lukas, U. Plath, and K. Tüür, Tallinn: Underi ja Tuglase Kirjanduskeskus, 259–67

——, 2012. *Cēsis — A Symbol of Latvian History*, Cēsis: Municipal Agency, Cēsis Culture and Tourism Centre

Kušķe, E., Silamiķele, I., Kalniņa, L., and Kļaviņš, M. 2010. 'Peat Formation Conditions and Peat Properties: A Study of Two Ombrotrophic Bogs', in *Mires and Peat*, ed. M. Klavins, Rīga: University of Latvia Press, 56–70

Kušķe E., Kalniņa L., Stivriņš N., Diņķīte A., Reča M., and Bigačs R., 2011. 'Kūdras uzkrāsānās un paleoveģetācijas raksturs Rožu purva attīstības gaitā', in *Ģeogrāfija. Ģeoloģija. Vides zinātne*, Rīga: LU Akadēmiskais apgāds, 453–55

Lagerås, P. 2007. *The Ecology of Expansion and Abandonment: Medieval and Post-Medieval Land-Use and Settlement Dynamics in a Landscape Perspective*, Lund: National Heritage Board

Lang, V., and Valk, H. 2011. 'An Archaeological Reading of the Chronicle of Henry of Livonia: Events, Traces, Contexts and Interpretations', in *Crusading and Chronicle Writing on the Medieval Baltic Frontier*, eds M. Tamm, L. Kaljundi, and C. S. Jensen, Aldershot: Ashgate, 291–316

Latałowa, M., Badura, M., and Jarosińska, J. 2003. 'Archaeobotanical Samples from Non-Specific Urban Contexts as a Tool for Reconstructing Environmental Conditions (Examples from Elbląg and Kołobrzeg, Northern Poland', *Vegetation History and Archaeobotany* 12, 93–104

Laul, S., and Valk, H. 2007. *Siksälä: A Community at the Frontiers. Iron Age and Medieval*, Tartu: University of Tartu

Ligi, H., 1963. *Põllumajanduslik maakasutus Eestis XVI–XVII sajandil*, Tallinn: Eesti NSV Teaduste Akadeemia Ajaloo Instituut

Ligi, P. 1992. 'The Prehistory of Saaremaa', *PACT* 37, 163–73

Mägi, M. 2004. 'Ships Are their Main Strength: Harbour Sites, Arable Lands and Chieftains on Saaremaa', *Estonian Journal of Archaeology* 8, 128–55

Meadows, J., and Zunde, M. 2014. 'A Lake Fortress, a Floating Chronology and an Atmospheric Anomaly: The Surprising Results of a Radiocarbon Wiggle-match from Araisi, Latvia', *Geochronometria* 41, 223–33

Meikar, T., and Uri, V. 2000. 'Managing the Bushlands in Estonia', *Akadeemilise Metsaseltsi Toimetised* 11, 103–20

Moe, D., Savukynienė, N., and Stančikaitė, M. 2005. 'A New 14C (AMS) Date from Former Heathland Soil Horizons at Kuršių Nerija, Lithuania', *Baltica* 18, 23–28

Mugurēvičs, Ē. 1990. 'Interactions between Indigenous and Western Culture in Livonia in the 13th to 16th Centuries', in *From the Baltic to the Black Sea*, eds D. Austin and L. Alcock, London: Routledge, 168–78

——, 2008. *Viduslaiku ciems un pils Salaspils novadā*, Rīga: Latvijas vēstures institūta apgāds

Niinemets, E. 2010. *Holocene Vegetation Dynamics in SE–Estonia: Palynological Records from Three Lakes of the Haanja Heights*, Saarbrücken: VDM

Niinemets, E., and Saarse, L. 2009. 'Holocene Vegetation History and Land-Use Dynamics of South–Eastern Estonia', *Quaternary International* 207, 104–16

Niinemets, E., Saarse, L., and Poska, A. 2002. 'Vegetation History and Human Impact in the Parika Area, Central Estonia', *Proceedings of the Estonian Academy of Science Geology* 51, 241–58

Ozola, I. 2013. *Holocēna organogēnie nogulumi un to uzkrāšanās apstākļu izmaiņas purvos ziemeļ Vidzemē* (unpublished doctoral thesis, University of Rīga)

Pakalne, M. 2008. *Mire Conservation and Management in Especially Protected Nature Areas in Latvia*, Jelgava: Latvian Fund for Nature

Pakalne, M., and Kalniņa, L. 2005. 'Mire Ecosystems in Latvia', in *From Siberia to Tierra del Fuego*, ed. G. M. Steiner, Linz: Biologiezentrum der Ober Biologiezentrum der Oberösterreichischen Landesmuseen, 147–74

Pardoe, H. S., Giesecke, T., van der Knaap, W., Scitavska-Svobodová, H., Kvavadze, E. V., Panajiotidis, S., Gerasimidis, A., Pidek, I. A., Zimny, M., Święta-Musznicka, J., Latałowa, M., Noryskiewicz, A. M., Bozilova, E., Tonkov, S., Filipova-Marinova, M. V., van Leeuwen, J. F. N., and Kalniņa, L. 2010. 'Comparing Pollen Spectra from Modified Tauber Traps and Moss Samples: Examples from a Selection of Woodlands across Europe', *Vegetation History and Archaeobotany* 19, 271–83

Poska, A., and Saarse, L. 1999. 'Holocene Vegetation and Land-Use History in the Environs of Lake Kahala, Northern Estonia', *Vegetation History and Archaeobotany* 8, 185–97

——, 2002. 'Biostratigraphy and ^{14}C Dating of a Lake Sediment Sequence on the North-West Estonian Carbonaceous Plateau, Interpreted in terms of Human Impact in the Surroundings', *Vegetation History and Archaeobotany* 11, 191–200

Poska, A., Saarse, L., Koppel, K., Nielsen, A. B., Avel, E., Vassiljev, J., and Väli, V. 2014. 'The Verijärv Area, South Estonia over the Last Millennium: A High Resolution Quantitative Land-Cover Reconstruction Based on Pollen and Historical Data', *Review of Palaeobotany and Palynology* 207, 5–17

Poska, A., Saarse, L., and Veski, S. 2004. 'Reflections of Pre- and Early-Agrarian Human Impact in the Pollen Diagrams of Estonia', *Palaeogeography, Palaeoclimatology, Palaeoecology* 209, 37–50

Poska, A., Seep, E., Veski, S., and Koppel, K. 2008. 'Using Quantitative Pollen-Based Estimations and a Spatial CA_Markov Model to Reconstruct the Development of Cultural Landscape at Rõuge, South Estonia', *Vegetation History and Archaeobotany* 17, 527–41

Punning J-M., Liiva A., and Ilves, E. 1968. 'Tartu radiocarbon dates III', *Radiocarbon* 10, 379–83

Punning, J-M., Koff, T., Sakson, M., and Kangur, M. 2007. 'Holocene Pattern of Organic Carbon Accumulation in a Small Lake in Estonia', *Polish Journal of Ecology* 55, 5–14

Rannamäe, E., and Valk, H., 2013. 'Some Spatial and Temporal Aspects of Animal Utilisation in Viljandi, Medieval Livonia', *Archaeologia Baltica* 20, 47–58

Reintam, L., Rooma, I., Kull, A., and Kõlli, R. 2005. 'Soil Information and its Application in Estonia', in *Soil Resources of Europe*, eds R. J. A. Jones, B. Houskova, P. Bullock, and L. Montanarella, Luxembourg: EUR 20559 EN, Office for Official Publications of the European Communities, 121–32

Reitalu, T., Seppā, H., Sugita, S., Kangur, M., Koff, T., Avel, E., Kihno, K., Vassiljev, J., Renssen, H., Hammarlund, D., Heikkila, M., Saarse, L., Poska, A., and Veski, S. 2013. 'Long-Term Drivers of Forest Composition in a Boreo-Nemoral Region: The Relative Importance of Climate and Human Impact', *Journal of Biogeography* 40, 1524–34

Saarse, L., Niinemets, E., Poska, A., and Veski, S. 2010. 'Is there a Relationship between Crop Farming and the Alnus Decline in the Eastern Baltic Region?', *Vegetation History and Archaeobotany* 19, 17–28

Saarse, L., Veski, S., Heinsalu, A., Rajamäe, R., Kihno, K., and Martma, T. 1995. 'Litho- and Biostratigraphy of Lake Päidre, South Estonia', *Proceedings of the Estonian Academy of Science Geology* 44, 45–59

Seepä, H., and Poska, A. 2004. 'Holocene Annual Mean Temperatures Changes in Estonia and their Relationship to Solar Insolation and Atmospheric Circulation Patterns', *Quaternary Research* 61, 22–31

Segerström, U. 1990. 'The Natural Holocene Vegetation Development and the Introduction of Agriculture in Northern Norrland, Sweden: Studies of Soil, Peat and Especially Varved Lake Sediments' (unpublished doctoral dissertation, University of Umeå)

Segerström, U., and Emanuelsson, M. 2002. 'Extensive Forest Grazing and Hay-Making on Mires — Vegetation Changes in South-Central Sweden due to Land Use since Medieval Times', *Vegetation History and Archaeobotany* 11, 181–90

Sillasoo, Ü., and Hiie, S. 2007. 'An Archaeobotanical Approach to Investigating Food of the Hanseatic Period in Estonia', in *Medieval Food Traditions in Northern Europe*, ed. S. Karg, Copenhagen: National Museum of Denmark, 73–96

Sillasoo, Ü., Poska, A., Seppā, H., Blaauw, M., and Chambers, F. M. 2009. 'Linking Past Cultural Developments to Palaeoenvironmental Changes in Estonia', *Vegetation History and Archaeobotany* 18, 315–27

Stančikaitė, M., Bliujiene, A., Kisielienė, D., Mažeika, J., Taraškevičius, R., Messal, S., Szwarczewski, P., Kusiak, J., and Stakėnienė, R. 2013. 'Population History and Palaeoenvironment in the Skomantai Archaeological Site, West Lithuania: Two Thousand Years', *Quaternary International* 308–09, 190–204

Stančikaitė, M., Šinkūnas, P., Risberg, J, Šeirienė, V., Blažauskas, N., Jarockis, R., Karlsson, S., and Miller, U. 2009. 'Human Activity and the Environment during the Late Iron Age and Middle Ages at the Impiltis Archaeological Site, NW Lithuania', *Quaternary International* 20, 74–90

Stivriņš, N., Brown, A. D., Reitalu, T., Ratniece, V., Austin, J.E., Liiv, M., Cerina, A., and Veski, S. 2016. 'Palaeoenvironmental Evidence for the Impact of the Crusades on the Local and Regional Environment in Northern Latvia: Terrestrial Evidence', *The Holocene* 26, 61 –69.

Stivriņš, N., Brown, A., Reitalu, T., Veski, S., Heinsalu, A., Banerjea, R. Y., and Elmi, K. 2015. 'Landscape Change in Central Latvia since the Iron Age: Multiproxy Analysis of the Vegetation Impact of Conflict, Colonization and Economic Expansion during the Last 2,000 Years', *Vegetation History and Archaeobotany* 24, 377–91

Święta-Musznicka, J., Latałowa, M., Badura, M., and Gołembnik, A. 2013. 'Combined Pollen and Macrofossil Data as a Source of Reconstructing Mosaic Patterns of the Early Medieval Urban Habitats — A Case Study from Gdańsk, N. Poland', *Journal of Archaeological Science* 40, 637–38

Tvauri, A. 2012. *The Migration Period, Pre-Viking Age and Viking Age in Estonia*, Estonian Archaeology, 4, Tartu: Tartu University Press

Vaitkevičius, V. 2004. *Studies into the Balts' Sacred Places*, BAR international Series, 1228, Oxford: Archaeopress

Valk, H. 2009. 'Sacred and Natural Places of Estonia: Regional Aspects', *Folklore* 42, 45–66

Vassiljev, J., Harrison, S., and Haxeltine, A. 1995. 'Recent Lake-Level and Outflow Variations at Lake Viljandi, Estonia: Validation of a Coupled Lake-Catchment Modelling Scheme for Climate Change Studies', *Journal of Hydrology* 170, 63–77

Vaughan, J., and Geissler, C. 1997. *The New Oxford Book of Food Plants*, Oxford: Oxford University Press

Veski, S., and Lang, V., 1996. 'Prehistoric Human Impact in the Vicinity of Lake Maardu, Northern Estonia. A Synthesis of Pollen Analytical and Archaeological Results' *PACT* 51, 189–205

Veski, S., Koppel, K., and Poska, A. 2005. 'Integrated Palaeoecological and Historical Data in the service of Fine Resolution Land Use and Ecological Change Assessment during the Last 1000 Years in Rõuge, Southern Estonia', *Journal of Biogeography* 32, 1473–1488

Zelčs, V., and Markots, A. 2004. 'Deglaciation History of Latvia', in *Quaternary Glaciations: Extent and Chronology of Glaciations*, eds J. Ehlers, L. Gibbard, and P. Hughes, Amsterdam: Elsevier, 225–43

Zunde, M. 1998–1999. 'Timber Export from Old Riga and its Impact on Dendrochronological Dating in Europe', *Dendrochronologia* 16–17, 119–30

6. Farming, Hunting, and Fishing in Medieval Livonia: The Zooarchaeological Data

Mark Maltby, Aleksander Pluskowski, Eve Rannamäe, and Krish Seetah
With contributions by Sheila Hamilton-Dyer, Katherine French, and Richard Madgwick

Introduction

Before the crusades, the indigenous societies of the eastern Baltic had an established pastoral culture, revolving around the raising of cattle, pigs, sheep, goats, and poultry. Regional and local variations in the representation of these species can be linked to the socio-economic characteristics of individual strongholds and their interdependent settlements (Blomkvist 2009, 444). Little is known about the social organization of animal husbandry, and the relationships between strongholds and settlements within their territories are assumed to mirror the framework of stratified power relations alluded to by other types of archaeological evidence (see Chapter 4; Zemītis 2014, 451–55). In order to support the influx of crusaders, merchants, colonists, and their heavy cavalry, a complex provisioning system was essential from the earliest arrival of crusading armies. The establishment of colonies in Üxküll (Latv. Ikšķile) and Riga enabled fields, pastures, and herds to be rapidly procured and reorganized. These were immediately vulnerable; in the summer of 1203 the Rus' leader of Yersika, along with the Lithuanians, stole the cattle from Riga's pasture (*pascuis*) and captured two priests who were cutting trees, together with crusaders at the 'Old Hill' (*HCL* vii, 8). Riga had been founded a few years earlier within the vicinity of existing Liv settlements, and the incoming colonists would have adopted the existing provisioning system. The pastoral economy at this time would have been highly volatile, with raiding accentuating the impact of environmental factors on herds (Goldschmidt 1979).

The seizure of livestock during raids was common. This supplemented the provisioning requirements of the victors, deprived their enemies of vital resources, and also represented a measure of wealth; referring to the aftermath of the slaughter of Lithuanians and Estonians in 1205, Henry describes the Germans and their Semigallian allies seizing 'untold loot, both in horses and flocks' (*HCL* ix, 4). The crusaders' resources were also targeted; in the autumn of 1218 an Öselian fleet came up the Daugava, took captives, and stole many flocks (*HCL* xxii, 8). This continued throughout the thirteenth century, with all the narratives of the Livonian Crusade punctuated by seizures of livestock, particularly horses, and grain. With the onset of the 'eternal crusade' against Lithuania, livestock seizure continues to be documented in the southern Livonian frontier. In February 1367, the marshal of the Order attacked northern Lithuania and captured fifty mares from the Grand Duke's stud farm (*equirriam*); three years later the commander of Goldingen (Latv. Kuldīga) seized 430 cattle alongside horses during an attack across the Lithuanian border (*HWC*, 105, 119). When tensions flared up between the Order and its tenants, animals — as valuable property — were taken. Early in 1454 when the Order's men set fire to some hamlets belonging to the town of Riga near Neuermühlen (Latv. Bukulti) and Depena (Latv. Sarkandaugava), they seized the farmers' cattle and property; conversely, in March 1482 the burghers of Riga broke into the outer bailey of the Order's castle and took horses, oxen, and food (*HHC* ii, 743, 778).

The incremental conquest and seizure of tribal lands was accompanied by the annexation of livestock, pasture, and farmland. Although many indigenous power centres were destroyed during the crusades, a significant part of the population remained intact, along with local political and economic structures that were incorporated into the newly established *Ordensland* (Pluskowski and Valk 2016). At the same time, the provisioning requirements of newly established towns

Table 6.1: Animal bone assemblages included within the study from Livonia, modern Latvia and Estonia.

	Riga Town 1300–1500	**Riga Castle 1 1200–1330**	**Riga Castle 2 1330–1550**	**Vecdole 1220–1290**	**Āraiši Lake 780–1050**	**Āraiši Livonian 1 1300–1394**	**Āraiši Livonian 2 1395–1421**	**Āraiši Polish 1573–1624**	**Ventspils 1300–1450**	**Ropaži**	**Cēsis LO**
Cattle	1751	120	77	452	77	21	138	322	242	18	64
Sheep/Goat	307	41	48	150	20	19	193	394	127	25	25
Sheep	27	4	7	19	5	1	21	30	13	3	0
Goat	25	10	2	2	0	0	6	5	6	0	0
Pig	669	52	30	351	36	15	127	230	70	7	24
Horse	30	1		42	36	1	8	11	1	1	25
Dog	78	1		6	1		3	8	3		2
Cat	14	2		8				3	5		
Aurochs/Bison					4						
Elk	9	2	1	1	42					23	
Red Deer	2	3			1						
Roe Deer	1		1								
Cervid							1				
Bear									1		
Wolf											
Lynx											
Hare			2				29	31	8		1
Beaver	4			1	50						
Marten											
Rat							2	3			
Vole				1							
Hedgehog											
Mole				2				2	1		
Total Mammal	***2865***	***222***	***159***	***1014***	***267***	***56***	***501***	***1004***	***458***	***74***	***36***
Domestic Fowl	13	8	14	51		21	112	79	34	4	2
Goose	3		1	9		1	5	7	16	3	
Mallard	1			1				1	3		
Ducks		1	1	1					5	1	
Teal				1							
Swan											
Crane											
Capercaillie											
Small Galliform											
Owl				1							
Woodpigeon				1							
Eagle											
Hawk											
Gull	1										
Rook/Crow											
Passerine											
Total Bird	***18***	***9***	***16***	***65***	***0***	***22***	***117***	***87***	***58***	***8***	***2***

Cēsis EPM 1550–1600	Cēsis 1996	Karksi Period II 1250–1350	Karksi Period III 1250–1550	Karksi Period IV 1550–1700	Viljandi Town P.II 1250–1350	Viljandi Town P.III 1250–1550	Viljandi Town P.IV 1550–1700	Viljandi Suburb 1250–1550	Viljandi Castle P.II 1250–1350	Viljandi Castle P.III 1250–1550	Viljandi Castle P.IV 1550–1700
21	22648	853	685	497	765	2789	341	4335	2247	3227	976
5	14755	753	357	200	553	1688	95	1576	1154	3678	913
0	37										
0	26										
14	4982	1144	166	92	455	981	47	950	23	1045	311
2	131	6	1		4	42	26	321		7	13
9	405	1	9	2	5	119	5	48	1	28	25
	101		3		10	77	1	22		7	
					6	2		2		7	
											5
	42			2				1		2	4
		24							14		
			1	1				5	6		
	14										
										1	
	198	73	44	15	1	5		14	1	111	52
		1		1							1
	15										
		24		2		8			6		
											30
51	***43291***	***2879***	***1266***	***812***	***1799***	***5711***	***515***	***7274***	***3452***	***8113***	***2330***
		2621	57	42	23	121	24	63	19	659	139
		36	3	5		10		4	4	12	35
								11		209	31
						5					
						2					
						5	2				
		13									
			1								
				1							
			1								
						3		3			9
0		***2670***	***62***	***48***	***23***	***146***	***26***	***81***	***23***	***880***	***214***

Table 6.1: Animal bone assemblages included within the study from Livonia, modern Latvia and Estonia. *(cont.)*

	Riga Town 1300–1500	Riga Castle 1 1200–1330	Riga Castle 2 1330–1550	Vecdole 1220–1290	Āraiši Lake 780–1050	Āraiši Livonian 1 1300–1394	Āraiši Livonian 2 1395–1421	Āraiši Polish 1573–1624	Ventspils 1300–1450	Ropaži	Cēsis LO
Cyprinid				6				24			
Pike	1			3				30			
Perch	5							90			
Zander	3										
Flatfish	1			1				1	14		
Salmonid	1			4							
Trout											
Burbot											
Eel											
Cod								8	3		
Gadoid	7							12	42		
Herring											
Smelt											
Haddock	5										
Sturgeon	7			1			1	1			
Whitefish											
Catfish	1			1							
Total Fish	*31*	*0*	*0*	*16*	*0*	*0*	*1*	*166*	*59*	*0*	*0*
Frog											
Toad						2		2			
Frog/Toad						1	2				
Total Amphibian	***0***	***0***	***0***	***0***	***0***	***3***	***2***	***2***	***0***	***0***	
Total Identified	***2914***	***231***	***175***	***1095***	***267***	***81***	***621***	***1259***	***575***	***82***	***36***
Large Mammal	796	45	1	267	49	64	309	291	105	44	27
Medium Mammal	191	17	1	165	9	91	653	323	98	36	94
Small Mammal		1	1	2		1		3	1		
Unid. Mammal	112	1		80	1	33	328	334	4	5	11
Unid. Bird	6	7	2	15			13	20	20	6	4
Unid. Fish	24			6				119	85	1	
Total Unidentified	1129	71	5	535	59	189	1303	1090	313	92	38
Total	**4043**	**302**	**180**	**1630**	**326**	**270**	**1924**	**2349**	**888**	**174**	**75**

Counts are of numbers of individual specimens (NISP)

Total includes bones in associated bone groups (ABGs) and bones from sieved samples

Karksi and Viljandi data adapted from Rannamäe and Lõugas (2019)

Cēsis 1996 data adapted from Mugurēvičs (unpublished)

Cēsis EPM 1550–1600	Cēsis 1996	Karksi Period II 1250–1350	Karksi Period III 1250–1550	Karksi Period IV 1550–1700	Viljandi Town P.II 1250–1350	Viljandi Town P.III 1250–1550	Viljandi Town P.IV 1550–1700	Viljandi Suburb 1250–1550	Viljandi Castle P.II 1250–1350	Viljandi Castle P.III 1250–1550	Viljandi Castle P.IV 1550–1700
		389	2	6		35					
		220	29	30		13					
		85	1	12		42					
		1	5	4							
						1					
		1									
		3									
		5		1							
		1	5	8							
		1				5					
						8					
						1					
0		*706*	*42*	*61*		*105*					
51	***43291***	***6255***	***1370***	***921***	***1822***	***5962***	***541***	***7355***	***3475***	***8993***	***2544***
39											
25		1715	683	402							
		24	1	4							
		1403	1515	567							
		121	30	28							
64		3263	2229	1001							
115		**9518**	**3599**	**1922**							

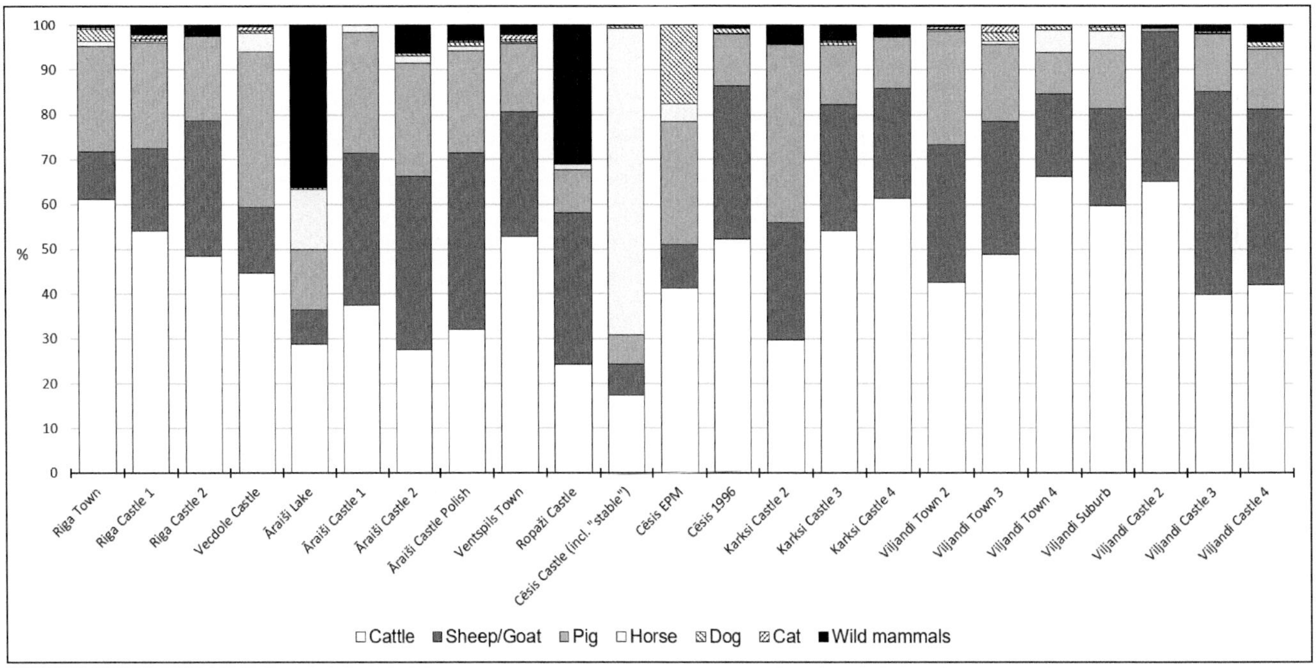

Figure 6.1: Mammal species relative percentages by site and phase. See Table 6.1 for sample sizes. Figure by M. Maltby and A. Pluskowski.

and castle communities, the commercial opportunities provided by the Hanse, and the reorganization of land ownership by the theocratic elites (see Chapter 9) would have necessitated degrees of intensification in animal husbandry and fishing. Shifts in the representation of wild mammals and birds are also evident after the crusades and can be variously synchronized with habitat transformation and newly imposed forms of social control. With the merging of the Sword Brothers and the Teutonic Order in 1237, a standardized dietary regime would have been adopted across all the Livonian convents, one that was clarified in 1264 as a means of distancing the Order from its initial model based on the Templars (Sterns 1969, 70–71). This regime allowed the consumption of more meat: three times each week, with two weekly fast days when fish and other aquatic species such as beaver could be eaten (Mänd 2004). Whether a Christian fasting culture was effectively imposed on the indigenous population and enforced is unclear, but it would have been expected in towns, and perhaps also within episcopal territories, where the Livonian Church exerted the most effort in converting its indigenous subjects (Valk 2019).

Few detailed inventories have survived from the Order's convents in Livonia, in contrast to Prussia. However, one such register from the convent of Goldingen, dated to 28 September 1341, provides a glimpse of the diversity and location of animal resources which amounted to 146 horses, 245 cattle, 300 sheep, and 40 goats (Bauer 1924). The inventory contains separate sections on horses, cattle, sheep, goat, grain reserves, and fish. Horses feature as the most important animal, subdivided by sex, age, and function; seventeen horses are listed as under the charge of the marshal; seven horses were located in the manor of the agricultural *curia*, an unspecified number in the manor of the gardener, a half-brother; the manor of Alsunga had forty young (*juvenes*), middle-aged (*pariter*), and old (*senes*) riding mares and three stallions, as well as twelve colts of both genders, all one-year-old. Fourteen male colts are listed as at the disposal of the convent's brethren, and the commander's stable contains fifty-five riding horses (*equos equitales*), with the total number of horses and colts equalling 146. Within the convent's granaries, nine out of the thirty-six *last* of grain are designated as fodder for the brethren's horses. The sections on livestock are shorter. In the *marstabulo* forty oxen and five cows are listed; in the agricultural manor there are forty oxen; in the half-brother Rovke's manor there are 125 small cattle (*capita pecorum*) of both genders, and in the manor at Alsunga there are thirty-five small cattle. The location of sheep is not specified, only that three hundred sheep and forty goats are listed on the convent's lands. In some districts local communities are also listed as renting 205 cows and seventy-five goats from the Order. The convent's kitchen included a reserve of unspecified meat (*carnium*), with an additional note on 130 slaughtered sheep and forty live cattle to be fattened for the kitchen's needs, but its

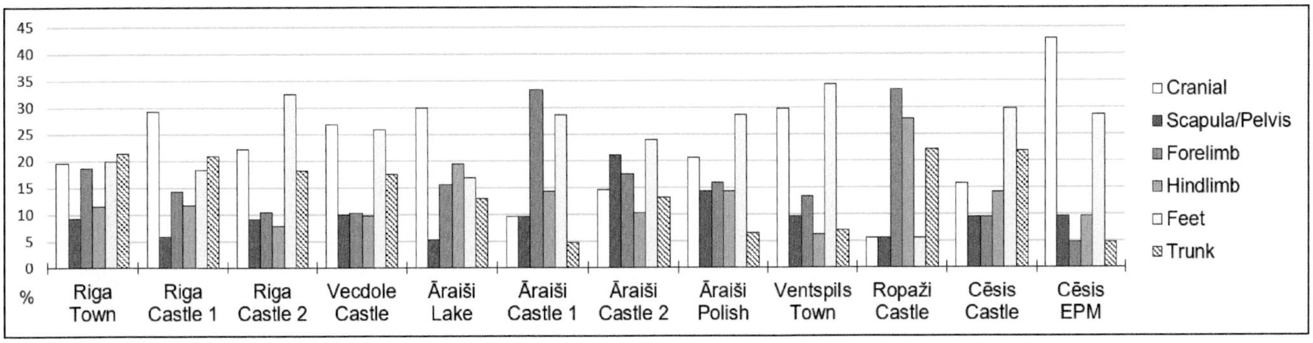

Figure 6.2: Body part representation of cattle from Latvian sites. See Table 6.1 for sample sizes. Figure by M. Maltby and A. Pluskowski.

fish stocks were more diverse and abundant, dominated by freshwater species, particularly pike and vimba.

In comparison to the detailed written sources, the zooarchaeological record for medieval Livonia is equally fragmented, but its resolution is coarse. It is, however, valuable from a comparative perspective, particularly in the general absence of detailed inventories for the majority of the sites. Moreover, diachronic comparisons provide valuable information on the biological profiles of the Order's castle communities (Rannamäe and Lõugas 2019). For the purposes of this study, faunal assemblages were made available for recording and analysis from ten sites in Livonia. With the exception of Viljandi (Ger. Fellin) and Karksi (Ger. Karkus), located in Estonia, all sites are situated in Latvia. Wherever possible, secondary data from published and unpublished reports was included. However, it must be noted that much of this comparative material only consisted of basic NISP, MNI, and anatomical element counts. The most detailed and comprehensive comparative study on long-term trends in faunal exploitation which includes the final centuries of the Iron Age has been published by Liina Maldre for Estonia (2012). This has provided a useful context for the data presented in this chapter. All primary and secondary sites are presented in Table 6.1 and Fig. 6.1 (for Karksi and Viljandi phasing, see Rannamäe and Lõugas 2019). Species are presented below in order of importance, and subdivided into domestic and wild species. There is only space to provide comparative summaries, and more detailed site-based studies will be published in due course.

Livestock

Cattle (*Bos taurus*)

Cattle elements were the most commonly identified in most of the assemblages under consideration. Comparing NISP counts of the main domestic mammal species, cattle outnumbered pig and sheep/goat in nineteen of the twenty-three assemblages (Table 6.1). Cattle dominated the small assemblage from the earlier fortified lake settlement at Āraiši (Ger. Arrasch) and were also prominent in the assemblages from: the town and suburbs from Riga, Ventspils (Ger. Windau), and Viljandi; the castles at Riga and Cēsis (Ger. Wenden); and the earliest phase at Viljandi Castle, where their percentages decreased significantly in later levels. The cattle percentage was inflated in the assemblage from Vecdole Castle (Ger. Alt-Dahlen) by the inclusion of 109 bones from the calf skeletons discussed below.

Although NISP counts are usually biased towards cattle because fragmentation and retrieval biases favour their recovery, there is little doubt that beef constituted the major meat source for all the settlements under consideration (Fig. 6.2), even in cases where cattle elements were outnumbered by sheep and goats. The dominance of cattle in most of the urban and castle assemblages reflects both the favourable environment for keeping herds in Livonia (see Chapter 5) and the need to ensure a good supply of meat to these centres. A full-sized cattle carcass can supply around ten times more meat than a sheep or goat and over four times as much as a pig. It is not surprising that cattle feature in reports of raiding and enforced procurement in Livonia.

Mortality evidence for cattle was obtained from mandibular tooth ageing and epiphyseal fusion data (Figs 6.3–6.4). Only fusion data was obtained from the lake settlement at Āraiši, which indicated that although over half the cattle bones belonged to adults, some immature and subadult cattle were also represented. In the late Viking Age assemblage from Pada (Estonia), nearly a quarter of cattle were slaughtered before reaching two years, whilst just over 30% were butchered before reaching four years (Maldre 2007). The Late Iron Age assemblage from Linnaaluste (site 2) also consisted of mainly adults, with five individuals aged over two years (Maldre

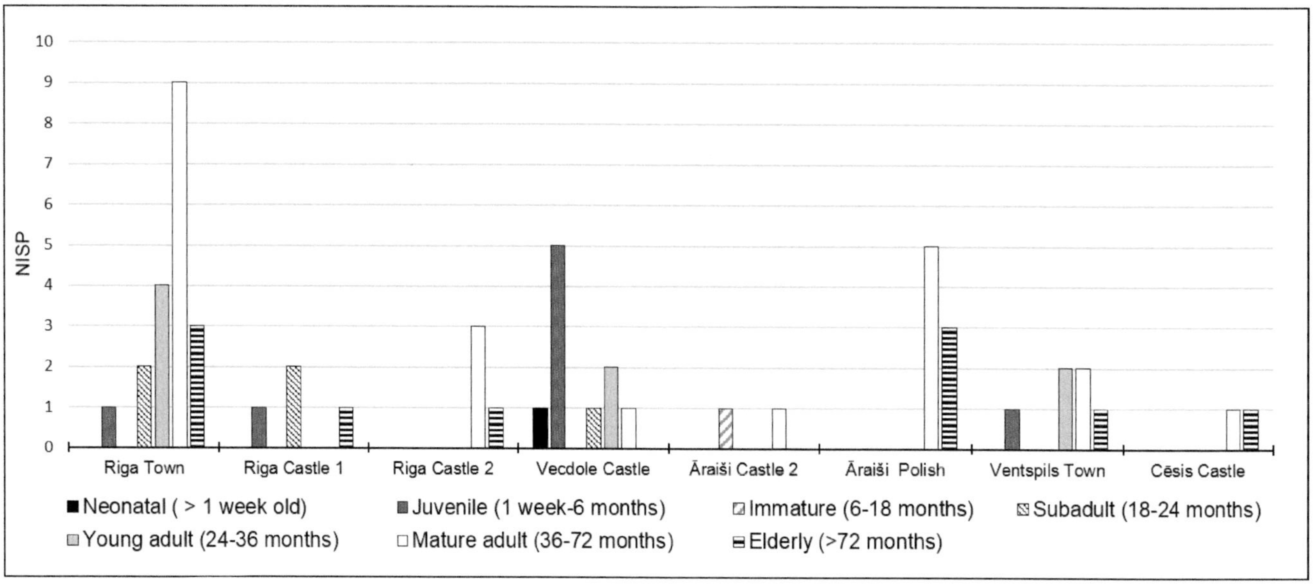

Figure 6.3: Cattle mandibular ageing data from Latvian sites. Figure by M. Maltby and A. Pluskowski.

2012, 113). More adults were represented in the final phase of the Curonian hillfort at Beltes (Padure, Latvia), where only one animal was aged to one month when slaughtered, with the others at least seventeen to twenty-four months through to five to six years old (Vasks et al. 2011, 92). In the later assemblage from Riga town, 60% of the mandibles belonged to cattle over three years old. The dominance of adult cattle was also reflected in the fusion data, with over 61% of the latest-fusing epiphyses being fully fused, indicative of animals over three years old. However, at least 30% of the mandibles belonged to cattle slaughtered between 18 and 36 months old, and only 15% fell within the elderly category (over six years old). This indicates a focus on supplying the town with good quality beef. There was limited evidence from Riga Castle, although there was a higher incidence of sub-adult cattle represented in the thirteenth century deposits than in the later assemblage, which produced similar results to the town. Ventspils produced very low percentages of cattle over three years old, perhaps indicating greater intensity of beef production, although this is based on a small sample. Livonian levels at Āraiši Castle produced little ageing evidence, whilst the early post-medieval Polish assemblage from the site was dominated by adult cattle. Ageing data from both the castle and town assemblages from Viljandi indicated that most cattle were over three years old, but the fusion data indicated a substantial kill-off of cattle between three and five years (Rannamäe 2010, 32; Haak et al. 2012, 303), which echo the results from Riga.

Mandibles and bones of neonatal and juvenile calves were rarely encountered in any of the assemblages discussed above. This is also reflected in the low percentages of porous bones recorded. This suggests that intensive dairy production, which would result in the slaughter

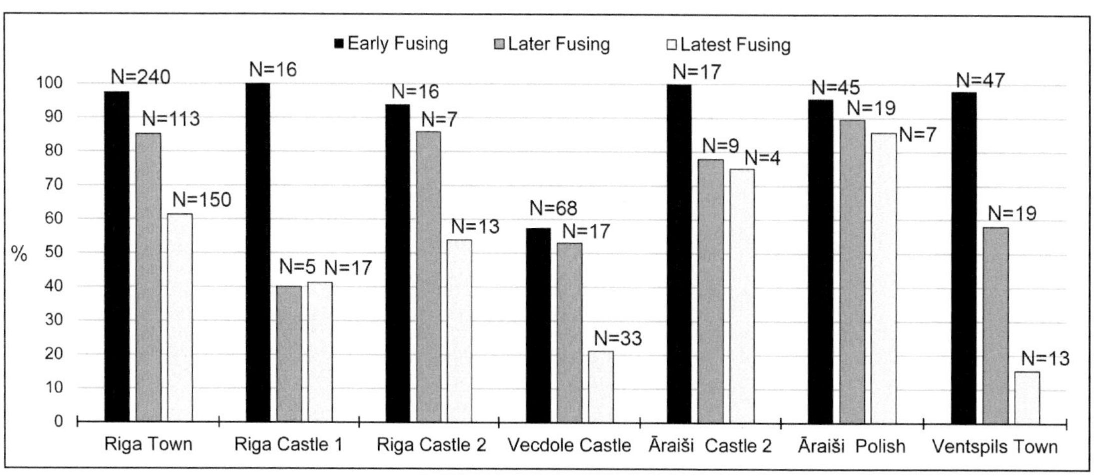

Figure 6.4: Cattle epiphyseal fusion data from Latvian sites (percentages of fused specimens). Figure by M. Maltby and A. Pluskowski.

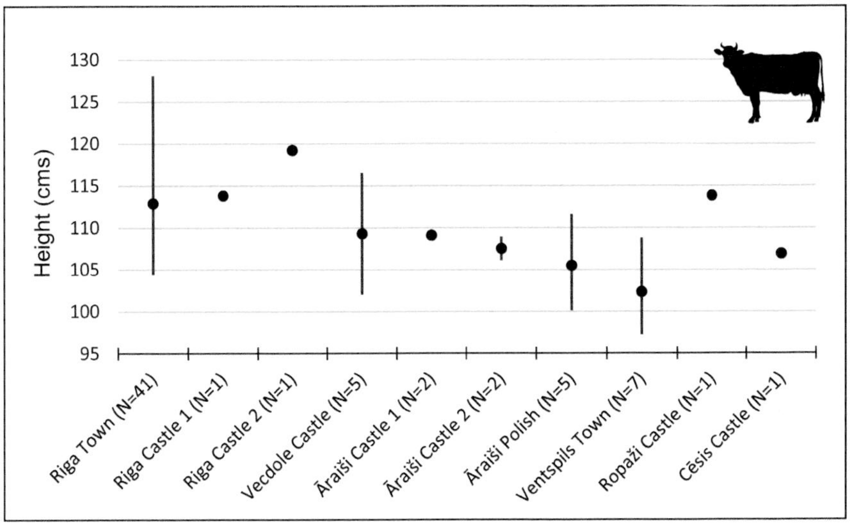

Figure 6.5: Withers height estimates for cattle from Latvian sites.
Figure by M. Maltby and A. Pluskowski.

of large numbers of veal calves, was not practised. This contrasts markedly with the assemblage from Vecdole Castle, where bones of young calves were prominent. Most of these were associated with the skeletons of two young calves, at least one of which appears to have been a foundation deposit (see below), although isolated bones of other calves were also found. However, this does not necessarily mean that dairy production was a prominent feature in cattle husbandry at Vecdole. The animals involved could have been from herds kept and bred in the vicinity of the castle and represent either natural mortalities or calves sacrificed for ritual purposes.

Sexing data from metrical analysis of cattle metacarpals from Viljandi's suburbs indicated that most (*c.* 75%) of the adult cattle were cows (Rannamäe 2010, 32–35). This would be expected, as surplus males are more likely to have been killed off before adulthood. Metrical data from cattle metacarpals was limited in the Latvian assemblages. In the assemblage from Riga town, nineteen (73%) out of twenty-six complete metacarpals were sufficiently slender to be attributed to cows. At Ventspils, there was a more even distribution, with four of the nine metacarpals having dimensions ascribable to oxen or bulls. Oxen are mentioned specifically in documentary records, and many would have been required for ploughing. It is conceivable that some of these oxen, after they had served their working lives, may have been considered too old to be brought for slaughter to the urban and castle consumer sites, where demand seems to have been for beef from rather younger animals. However, as O'Connor (2011) has pointed out in relation to Anglo-Saxon sites, five-year-old oxen could have been used as plough animals for two or three years prior to slaughter. There were very few instances of pathological conditions commonly associated with the prolonged use of cattle for ploughing (Bartosiewicz et al. 1997), echoed by Maldre's (2008) findings in Tallinn (Ger. Reval), Pärnu (Ger. Pernau), and Tartu (Ger. Dorpat), where the majority of pathologies were dental and mandibular.

Withers height estimates (following Matolcsi 1970) showed that most of the cattle from Livonia were of small stature, with average heights from most of the sites averaging <110 cm (Fig. 6.5). Although analysis was limited by the small number of complete limb bones that could be measured from most sites, there were some variations. Most of the cattle from Ventspils were generally very small, whereas the cattle from Riga were generally larger than those from the other sites. There were no clear chronological trends, and no evidence of any diachronic 'improvements'. Variations in cattle sizes are implied by the documentary references to 'small cattle', and it is possible that there were regional variations in cattle herds. Further discussion of stature can be found in the comparisons with Novgorod (Maltby 2019).

Pigs (*Sus scrofa*)

Pig bones have formed a significant percentage of most faunal assemblages from the Livonian sites examined during this project (Table 6.1). This supports evidence from previous studies. In Rositten (Latv. Rēzekne) around a third of the assemblage consisted of pig remains (Mugurēvičs 1985, 69), in Jersika they represented over 50% of domesticates (Vilcāne 2001, 68), and in Asote they comprised 72% of the assemblage (Shnore 1961, 93). At the earlier Āraiši Lake settlement, suid elements ranked second behind cattle, although these include five bones of wild boar. Comparative assemblages from Estonia indicate variability in the representation of pigs at sites, with the species dominating the assemblage from Lõhavere Hillfort, ranking second at Saadjärve Hillfort and third at every other site (Maldre 2012). This prevalence, particularly on Latvian sites, has been interpreted as reflecting the use of surplus animals for socially and politically symbolic feasting (Šnē 2006, 70).

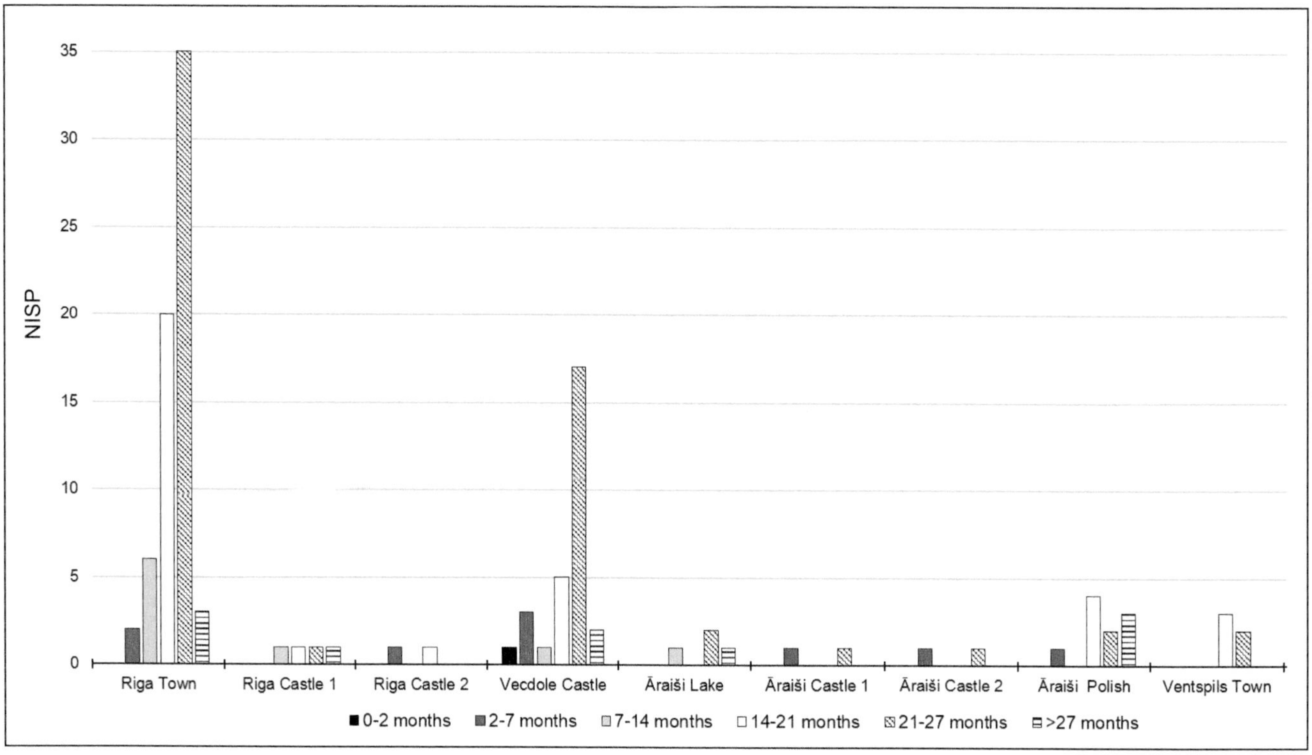

Figure 6.6: Pig mandibular ageing data from Latvian sites.

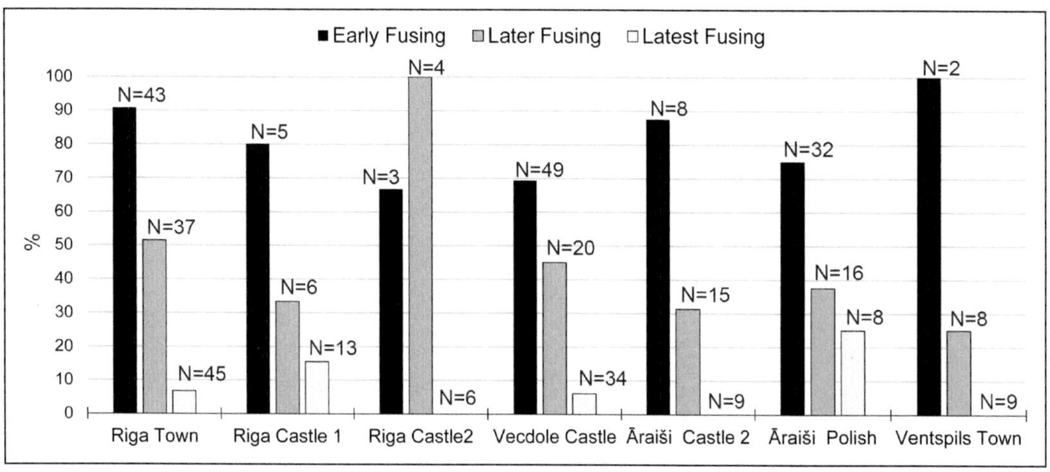

Figure 6.7: Pig epiphyseal fusion data from Latvian sites. (percentages of fused specimens). Figures by M. Maltby and A. Pluskowski.

Excluding the bones from the calf skeletons, pigs were the most commonly identified species in the assemblage from Vecdole Castle. In Livonian and Polish period deposits from Āraiši Castle, pig ranked behind both cattle and sheep/goat but still provided around 24%–27% of the elements identified to these species. They outnumbered sheep/goat in the early Livonian assemblage from Riga Castle (24%) but ranked third behind cattle and sheep/goat in the later phase (19%). Both those samples were much smaller than those from Riga town, in which pig provided 25% of the total cattle, sheep/goat, and pig elements. However, at other urban sites in Ventspils and Viljandi, pig elements ranked a distant third (13%–16%) in cattle-dominated assemblages. They were particularly poorly represented (4%) in the earliest castle deposits at Viljandi but increased to levels comparable with the urban deposits from the town in later levels. The opposite trend is seen at Karksi Castle, where pig elements (42%) dominated over cattle in the early Livonian phase (see also Valk et al. 2012; Rannamäe and Lõugas 2019). Could this reflect a deliberate emphasis of authority and the appropriation of a local surplus in stock, given the Order's comparatively new presence at the site? Their percentages declined spectacularly in the late medieval assemblages to 12%–14% of the total cattle, sheep/goat, and pig.

Various reasons could account for the fluctuations in pig percentages, including regional differences in the

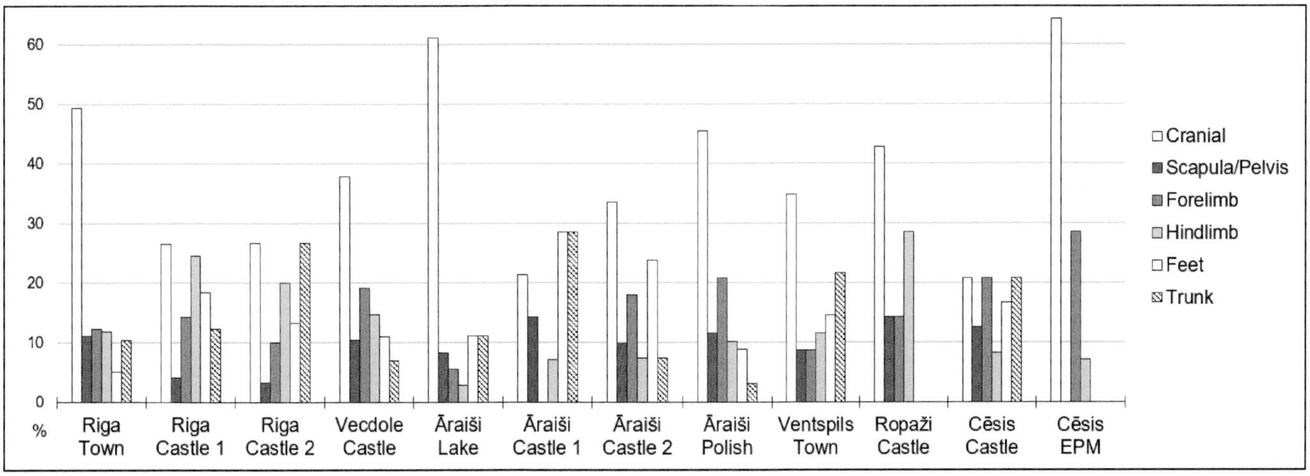

Figure 6.8: Body part representation of pigs from Latvian sites. See Table 6.1 for sample sizes.

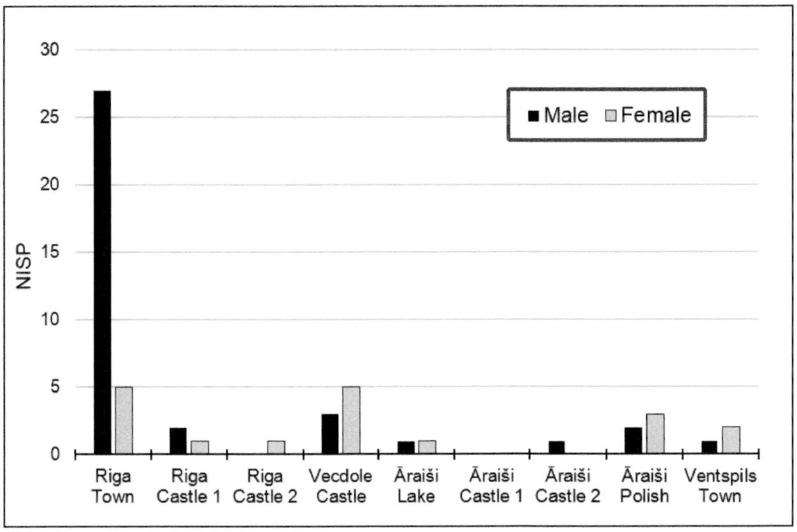

Figure 6.9: Pig sexing data from Latvian sites based on numbers of mandibles and maxillae (N = 55). Figures by M. Maltby and A. Pluskowski.

amount of woodland available for pig pannage, variations in meat-supply patterns, and biases within the archaeological samples. The assemblage from Vecdole included a higher proportion of jaws and bones of neonatal and juvenile pigs than from any of the other Latvian sites (Figs 6.6–6.8), implying that some pigs (as well as calves) might have been housed within the castle. Apart from Vecdole, Riga town, Viljandi, and Karksi, pig assemblages were too small to allow detailed analysis of ageing and sexing data. In Riga, the tooth ageing evidence indicated that there was an emphasis on the acquisition of sub-adult, or young adult pigs, probably mainly slaughtered in their third year. The fusion evidence gave similar results with over half the pigs surviving into their third year (fused later-fusing epiphyses) but very few survived beyond three to four years (fused latest-fusing epiphyses). The same pattern was apparent at Vecdole, for the pigs that survived neonatal mortality. However, one major difference is that most of the pig jaws at Riga came from males, whereas there was a much more even balance of sows and boars in the Vecdole sample. It suggests that boars were preferentially selected to supply Riga with pork (Fig. 6.9). The epiphyseal fusion evidence from the Estonian sites also showed that nearly all the pigs were slaughtered under four years old, with the majority within the first two years of their life (Rannamäe 2010, 44; Maldre 2012). In Karksi, the largest number of pig bones derived from neonatal animals up to six months of age (45%), followed by animals over one year (30%), and juveniles under or around one year (24%) (Rannamäe and Lõugas 2019).

The small size of nearly all the measurable pig bones and teeth confirmed that the vast majority of the suid bones were from domestic animals. For example, the lengths of eight lower third molars from Riga town and all seven specimens from Vecdole Castle fell below 32.7 mm, comfortably within the domestic pig range. There was one larger specimen from Riga measuring 36.2 mm, which could have belonged either to a large domestic pig or a small wild boar. Similarly, withers height estimates of nine bones from the Viljandi suburb ranged between 67 cm and 80 cm, comfortably within the domestic range (Rannamäe 2010, 44). One much larger specimen provided an estimate of 93 cm, which falls within the range of wild boar. Further discussion of pig stature will be considered in comparisons with Novgorod (Maltby 2019).

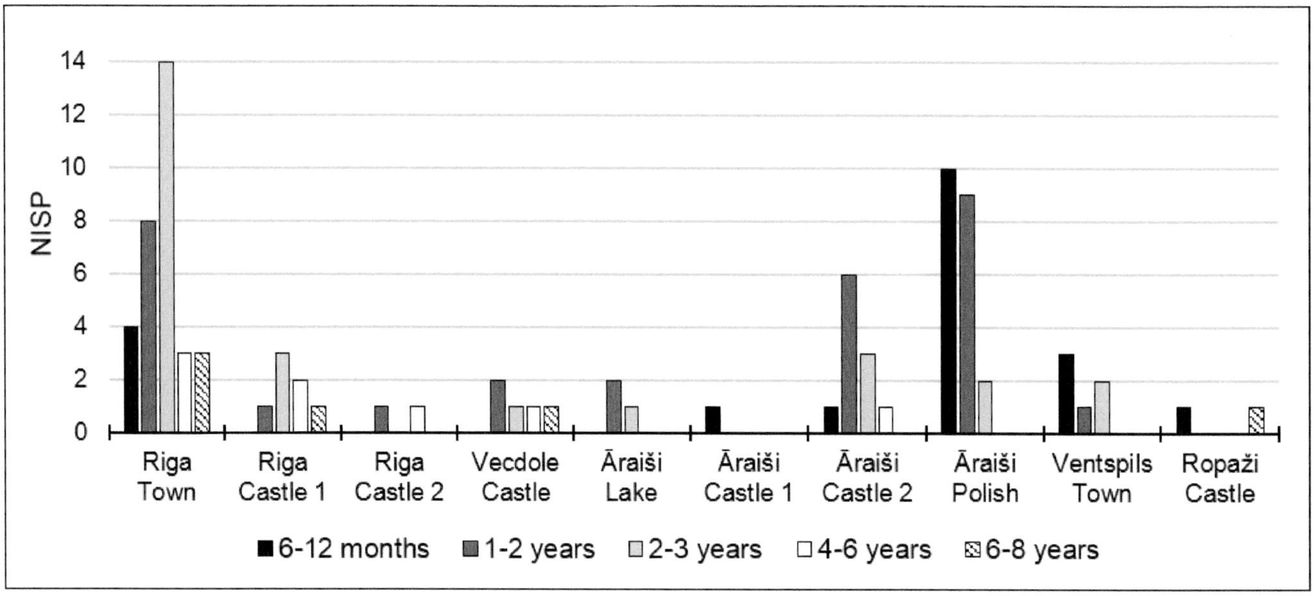

Figure 6.10: Sheep/goat mandibular ageing data from Latvian sites.

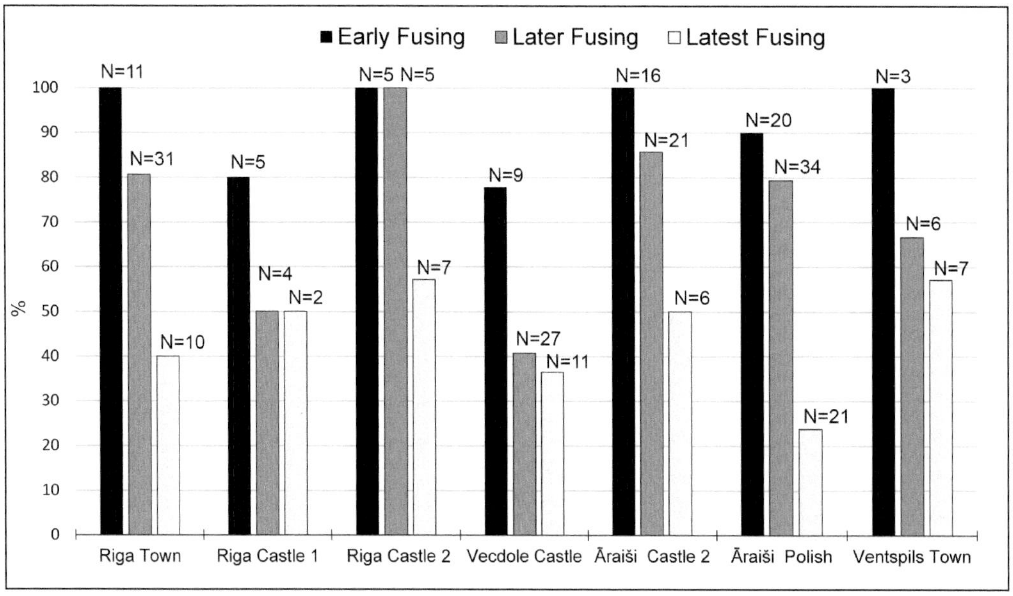

Figure 6.11: Sheep/goat epiphyseal fusion data from Latvian sites. Figures by M. Maltby and A. Pluskowski.

Sheep (*Ovis aries*) and Goat (*Capra hircus*)

Identifications of sheep and goats were based on distinguishing characteristics of some of the more complete diagnostic bones, mainly horn cores, mandibles, metapodials, radii, ulnae, and humeri. Counts from the Latvian sites are provided in Table 6.1. Overall, the assemblage from Riga town produced fairly equal numbers of identifications of sheep and goats. However, eighteen of the twenty-five goat elements were horn cores compared with only six of sheep. In contrast, sheep provided twenty-one of the twenty-eight other elements identified to species. A similar contrast was observed in the Viljandi suburb assemblage, where ninety-seven goat horn cores were recorded compared with only eleven of sheep. However, twenty-one of the thirty-two metacarpals distinguished to species were sheep (Rannamäe 2010, 37). In both cases, there is evidence that the horns of goats, particularly the larger ones of males, were imported for working, and horn cores from cattle were significantly more prevalent on Livonian urban sites, which came to represent major centres of manufacturing (cf. Luik 2015). Two goat horn cores were also identified in the Polish levels at Āraiši Castle, and one was found in the Ventspils assemblage. In most of the assemblages, positive identifications of sheep comfortably outnumbered those of goats (Table 6.1). The thirteenth-century assemblage at Riga Castle (belonging to the pre-castle urban phase) was an exception, where ten goat limb bones were identified compared with only four elements of sheep.

Maldre's (2012, 119) comparative study of sheep/goat representation from the Viking to Final Iron Age

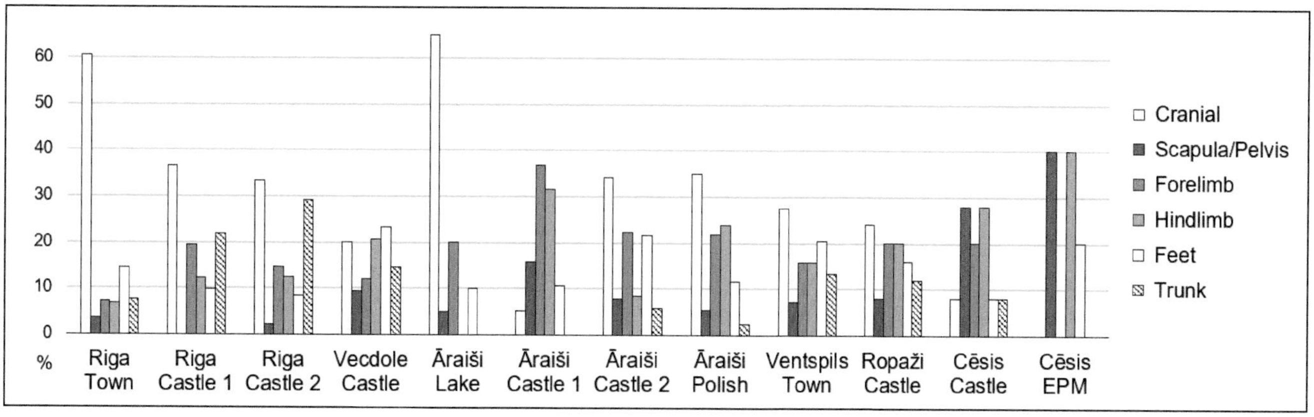

Figure 6.12: Body part representation of sheep from Latvian sites. Figure by M. Maltby and A. Pluskowski.

suggests regional differences; the rearing of these species decreased slightly in northern Estonia and Saaremaa between the two periods, and increased in southern Estonia, although most of the later assemblages derive from hillforts which may have operated distinct husbandry regimes. Percentages of sheep/goat also varied significantly between different assemblages included within the project (Table 6.1). Comparing cattle, sheep/goat, and pig elements only, those of sheep/goat were the least common in the assemblages from the Āraiši Lake settlement, Riga town, the early (pre-castle) phase at Riga Castle, Vecdole, and the early post-medieval phase at Cēsis Castle. Percentages of sheep/goat from both phases of the recent excavations at Cēsis were both lower than in the large assemblage from earlier excavations, in which sheep/goat were much better represented than pig (Mugurēvičs unpublished). It is possible that the small samples from the recent excavations are unrepresentative or there were significant differences in deposition practices in different parts of the castle. Sheep/goat were much better represented at Āraiši Castle and were the best represented species in the later Livonian and Polish phases. Sheep/goat provided 25%–30% of the cattle, sheep/goat, and pig elements in the medieval phases at Karksi Castle and 50% of the tiny sample from Ropaži. At Viljandi, sheep/goat were consistently better represented in the castle than in the town deposits. Apart from Karksi and Cēsis, sheep/goat were more poorly represented in the earliest phases of multi-period sites, suggesting that they may have become more important in the later Livonian and subsequent periods.

Ageing evidence for sheep/goat must take into account the possibility that sheep and goats may have had different mortality profiles. However, all but one of the twenty ageable mandibles that were identified to species from the various Latvian sites in the project belonged to sheep, so it can be confidently assumed that the remaining mandibles were also mainly sheep. With the possible exception of the earlier phase at Riga Castle, the same applies to the fusion data. Only Riga town and Āraiši Castle produced sufficient numbers of mandibles worthy of analysis (Fig. 6.10). At the former, most of the sheep represented were slaughtered as sub-adults and young adults between the ages of eighteen and forty-eight months. In the Livonian phases at Āraiši, most of the sheep were eighteen to thirty months old, whereas in the Polish phase, sheep mandibles of this age were accompanied by almost equal numbers of sheep aged between six and twelve months, and none of the mandibles belonged to mature adults. A very high percentage of the latest-fusing epiphyses from the Polish phase were unfused (Fig. 6.11). A focus on the slaughter of sheep and goat in their second year was evident in the fusion data from Viljandi (Rannamäe 2010, 39). This suggests that meat production was the main focus of sheep exploitation during the Livonian period, and perhaps became more intensive in the Polish phases at Āraiši. All sheep that survived winters would have provided wool, but the ageing evidence from these sites suggests that wool production was not the main focus of sheep exploitation.

The lack of mandibles and the low percentages of unfused early-fusing epiphyses and porous bones belonging to neonatal and juvenile lambs and kids also suggest that milk production was not of high priority and that sheep and goats were not being kept in large numbers within the Latvian towns and castles (Fig. 6.12). One possible exception is again Vecdole Castle, where 12% of the sheep/goat bones were porous. Withers height estimates based on Teichert's (1975) and Schramm's (1967a, 1967b, 1976) conversion factors for sheep and goat metapodials respectively (Fig. 6.13) showed that sheep from Riga were generally slightly larger than those from other

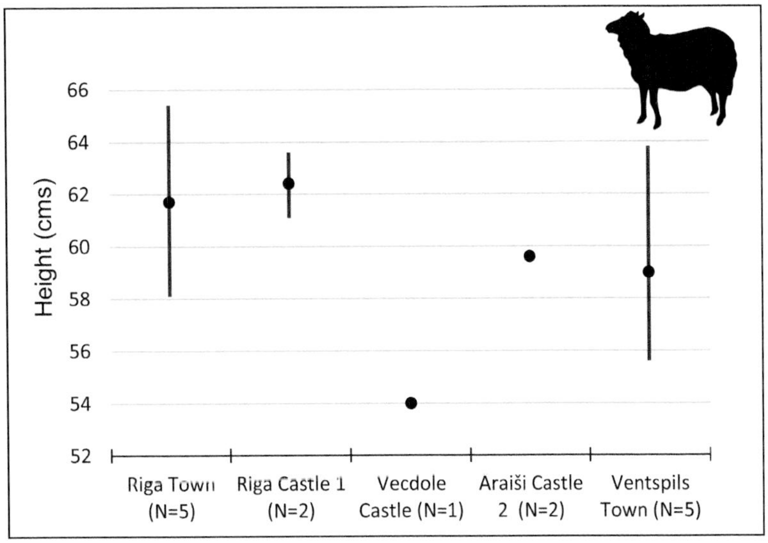

Figure 6.13: Withers heights estimates for sheep from Latvian sites. Figure by M. Maltby and A. Pluskowski.

sites, and that goats from Riga were substantially larger than those from Viljandi (Rannamäe 2010, 41–42). Whether this relates to regional or chronological variations cannot be established from these data. Further discussion of stature of animals from Livonian sites can be found in Maltby (2019).

Horses (*Equus sp.*)

The horse came to define crusading armies in both Livonia and Prussia, with the documented introduction of larger breeds designed to carry heavily armoured knights. This clearly made an impression in battle, where 'the trappings of the horses threw terror into the enemy' (*HCL* xv, 3). As a result, the warhorse, along with the crossbow, has been seen as pivotal to the military successes of crusading armies in the eastern Baltic (Ekdahl 1998). Written sources indicate indigenous military uses of cavalry in the thirteenth century were for travelling to and from battlefields, and that in turn the military training of horses was relatively limited (Nikžentaitis 2009, 404). Professional cavalry would only develop in the eastern Baltic following the crusades, under the management of the new regime. Glimpses of this are evident in obligations of military service to the Order which were maintained into the sixteenth century. In a document issued by the Livonian Master Wolter von Plettenberg to one Draggun on 20 December 1503 (Švābe 1938, 190–91), the recipient was obliged to accompany the commander of Goldingen to war with one horse, and if required to voluntarily accompany him on his travels, also with one horse. References to vassals keeping warhorses continue into the seventeenth century, as in the Swedish Audit of the Wenden district in 1601. Horse management was also organized at both convents and towns. A document issued by the Livonian Master in 1526 concerning a fief near the Riga road referred to the town of Wenden's horse pasture called *kuppel* which was situated in Duppen parish (*LGU* II, 442) (see also Chapter 8).

With respect to this study, horse bones were identified in all but three of the assemblages, albeit usually only forming a small percentage of the mammal elements in most cases (Table 6.1). It should be noted that although it is assumed that all the equid bones belonged to horse, the presence of mules cannot be ruled out. Although there is some evidence that horseflesh was eaten at some of these settlements, its consumption by Christians was discouraged following the decree of Pope Gregory III in AD 732 (Bartosiewicz 2003, 117). The extent to which this decree was subsequently enforced is a matter of debate, as there is variable evidence for horse consumption from certain parts of Christendom, such as the Hungarian Plain. In the case of Livonia, given the animal's role in

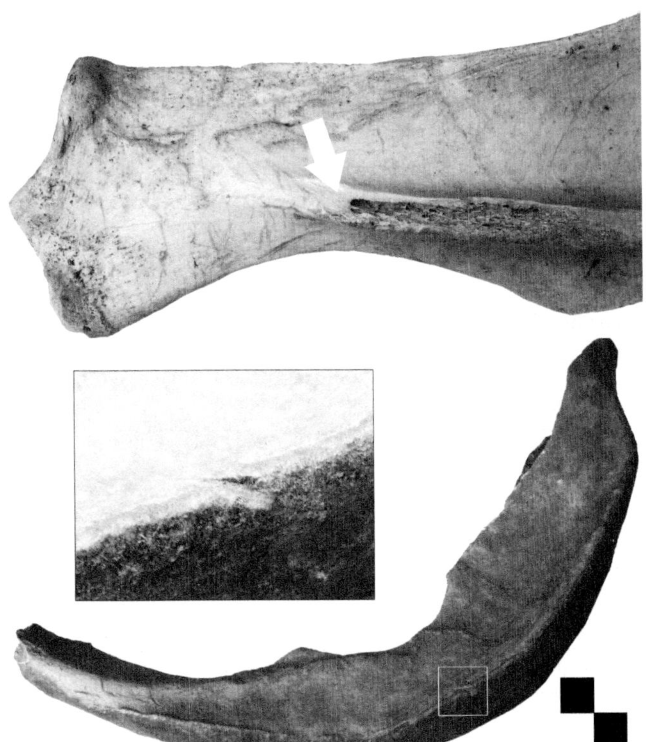

Figure 6.14: A butchered horse scapula and mandible from the Liv quarter in Riga. Photo by K. Seetah.

Table 6.2: Body part representation of horses from Latvian sites.

Element	**RigaTown 1300–1500**	**Riga Castle 1 1200–1330**	**Riga Castle 2 1330–1550**	**Vecdole 1220–1290**	**Āraiši Lake 780–1050**	**Āraiši Livonian 1 1300–1394**	**Āraiši Livonian 2 1395–1421**	**Āraiši Polish 1573–1624**	**Ventspils 1300–1450**	**Ropaži**	**Cēsis LO 1350–1538**	**Cēsis EPM 1550–1600**
Skull fragment				1	1						2	
Mandible	4										3	
Hyoid											1	
Loose Teeth		1		18	7	1	2	4			13	
Scapula	4				1		1	1			3	
Humerus				1	1						2	
Radius				2	1		1	3			3	
Ulna				2							3	
Pelvis					3						4	
Femur	1			1							2	
Patella											1	
Tibia	4			3	2					1	8	
Fibula											1	
Carpals					2						7	
Astragalus	1			2	4						3	
Calcaneus				3	4						2	
Other Tarsals	1										4	1
Metacarpal	4			2							2	
Metatarsal	2			3	1		2				1	
Peripheral Mp	4			2	1			2			6	
Phalanx 1				1	3		1		1		5	1
Phalanx 2				1	2		1				5	
Phalanx 3	1										5	
Sesamoids											6	
Atlas (VC1)	1											
Axis (VC2)					2						1	
Cervical V					1						9	
Thoracic V											44	
Lumbar V											11	
Sacral V								1			3	
Caudal V											4	
Unid. V											1	
Ribs	3										77	
Sternebrae											9	
Total	**30**	**1**	**0**	**42**	**36**	**1**	**8**	**11**	**1**	**1**	**251**	**2**

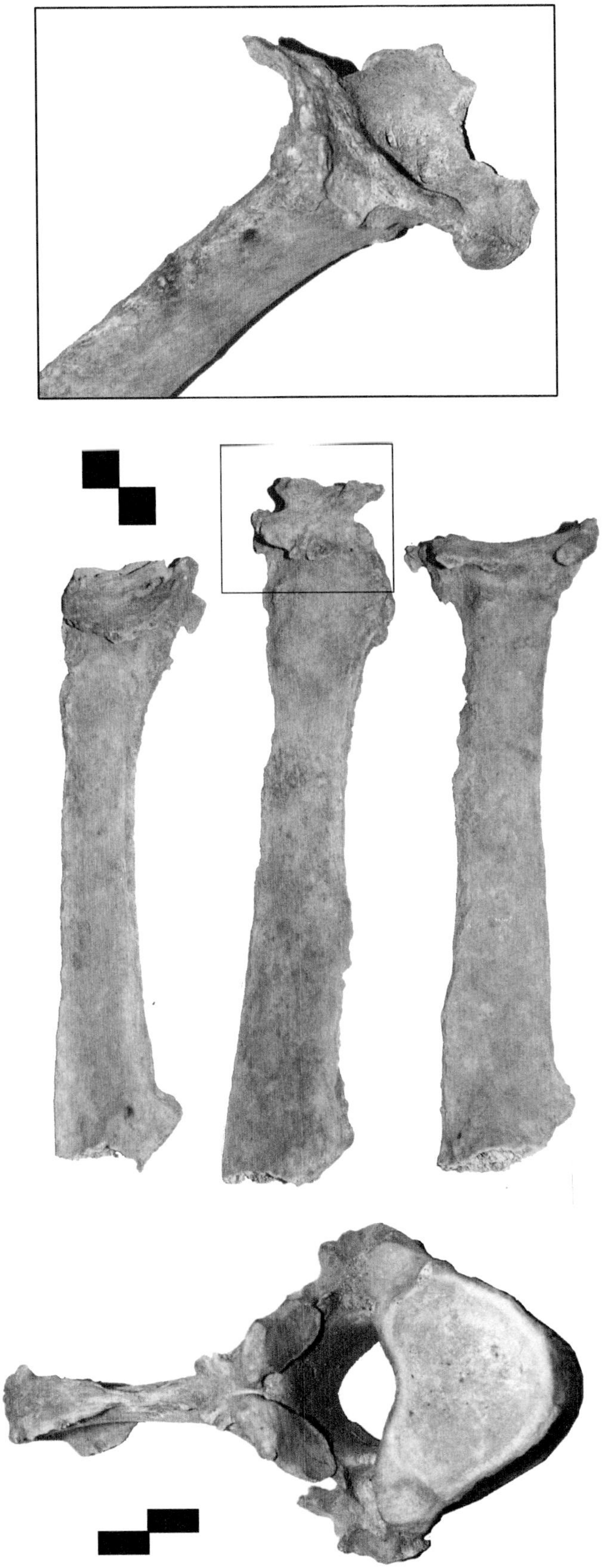

Figure 6.15: Pathological horse vertebrae from the outer bailey trench in Cēsis Castle. Photo by K. Seetah.

pre-Christian Balt society, it is therefore not surprising that horse bones appear only rarely in assemblages dominated by food refuse, with cut marks (Table 6.2). The highest number of horse bones displaying cut marks came from the Liv quarter in Riga, pointing to the continuation of horse consumption amongst the indigenous population (Fig. 6.14). This will require future verification with comparisons of assemblages from medieval rural settlements.

The only assemblages where horse provided over 3% of the mammal assemblage were from the Viljandi suburb, the latest phase in Viljandi town, Vecdole, Āraiši Lake settlement, and, especially Cēsis Castle, where horse provided over 68% of the Livonian period assemblage from the outer bailey. However, in the case of the latter site most of the bones came from a destruction layer and were U-series dated to the final phase of the Order's rule; *c.* 1468–1538 (see Chapter 3). At least three horses were represented, one consisting of a fairly complete skeleton. There were several other associated groups of horse bones (particularly vertebrae) along with other horse bones dispersed in this burnt layer. All the bones belonged to adult horses, and the most complete skeleton belonged to a male. Three of the horse bones bore butchery marks, indicating that some of the flesh was removed for human or dog consumption. However, the horses were not heavily processed. Six of their limb bones were sufficiently complete to estimate withers height using the conversion factors of Vitt (1952). These estimates ranged between 135 cm and 147 cm, averaging at 141 cm. Therefore, these were large horses, significantly larger than most recorded from late medieval Prussia (see Chapter 12) and the prehistoric ranges for Latvia and Estonia. The most complete skeleton had pathology on several of the thoracic vertebrae consisting of overriding spinous processes on the dorsal surfaces, as well as spondylotic spurs of newly formed bone on the ventral and lateral surfaces of their lower bodies (Fig. 6.15). These are relatively minor pathologies, which could have developed through stresses instigated by being subjected to various types of traction such as pulling carts or pulling ploughs. These are certainly documented in fourteenth- through to sixteenth-century Livonian sources. However, given the large size of the horses and their association with high status riding tack, it is highly feasible that these pathologies developed because of the stresses of being ridden by heavily armoured knights. Examples of pathological horse thoracic vertebrae have also been found in medieval suburban contexts in Viljandi (Rannamäe 2010,

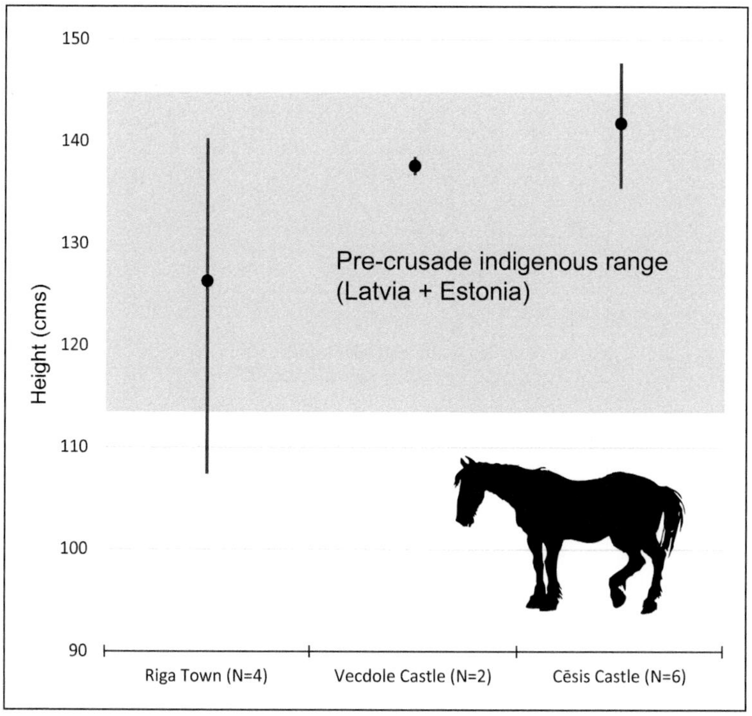

Figure 6.16: Withers height estimates for horses from selected Latvian sites. Figure by M. Maltby and A. Pluskowski.

Appendix 4, 109), and pathological thoracic and lumbar vertebrae have also been observed on horse remains recovered from the outer bailey of the castle in Malbork (Pluskowski et al. 2010).

Some of the horses recovered from the other sites were also of a large size, sometimes over 135 cm in height (Fig. 6.16). Seven complete horse bones from Viljandi produced withers height estimates between 128 cm and 144 cm (Rannamäe 2010, 46). Two horse bones from Riga town, however, were from much smaller animals, with withers heights of only 107 cm and 117 cm. Ekdahl (1998) notes that the warhorses of the Teutonic Knights, which were bred on the Order's estates, were substantially larger than the small indigenous horses which ranged in height from 112–144 cm. The metrical data confirm the presence of both small and large horses in Livonia, suggestive of a dual population of imported animals and indigenous stock comparable to that in Prussia (see Chapter 12).

The discovery of four very worn and shed deciduous teeth at Vecdole Castle indicates that horses were been kept in the castle. Vecdole and the earlier Āraiši Lake settlement were the only sites which produced any unfused horse limb bones including a few very porous bones of young foals, probably indicating that some horses were being kept at these settlements. The limited epiphyseal fusion data from the other sites produced no evidence for the presence of immature horses (Table 6.3). Most horses could expect to have been kept until adulthood because of their value for working and riding. Due to the limited survival of documentation from Livonian commanderies, numbers of horses are difficult to quantify. However, as in the case of Prussia (see Chapter 15), there is evidence these populations fluctuated. In 1341, the convent of Goldigen had fifty-five horses for ploughing (*equos uncales*) and another three in the half-brother gardener's (*ortulanus*) manor simply defined as *equos*. By 1346, the commandery maintained 146 horses, of which fifty-five were for riding (*equos equitales*) (Bauer 1924, 182–85). In 1442 this had dropped to horses for thirty-one men and an additional three in the possession of Goddert von Bachem, and by 1451 only the commander and his deputy had listed horses (*LUB* 11, 160). Indeed, where they were documented the numbers of horses in the Order's Livonian houses were relatively low in that year, with only two listed in Windau, thirteen in Riga, eleven in Segewold (Lat. Sigulda), twelve in the subservient castle of Rositten, seventeen horses in Wenden, two in Soneburg (Est. Maasilinn), five in Narva (Est. Narva) and Overpal (Est. Põltsamaa) (*LUB* XI, 160). A century later, the situation had significantly changed, and by 1555 Wenden's commandery could field three hundred horses (Hartmann, 2008, 190).

The origins of these horses would have been variable, and not solely from the various farms on the Order's estates. Horses could be readily moved within the Order's interregional networks; the Livonian land marshal sent the Grand Master horses in 1463 (*LUB* XII, 133). In January 1501, the Grand Master wrote to the Livonian Master to inform him that he was going to send the advocate of Brandenburg to Livonia and asked to supply him with carts (*fure*), horses, and other things that the advocate may need for his trip (*LUB* II, 10). Horses could also derive from neighbouring regions; in *c.* 1450, the Voivode of Polotsk complained to the town council of Riga that he had received back only one of the two horses that he had left in the care of the town's horse herd (*LUB* XI, 91). The ongoing conflict between the Livonian episcopates and the Order was also characterized by the seizure of valuable animals, as during the crusading period. In June 1383, the Order's brethren attacked members of Riga's cathedral chapter in

Table 6.3: Horse epiphyseal fusion data from Latvian sites.

	Riga Town 1300–1500		**Vecdole 1220–1290**		**Āraiši Lake 780–1050**		**Āraiši Livonian 2 1395–1421**		**Āraiši Polish 1573–1624**		**Ventspils 1300–1450**		**Ropaži**		**Cēsis LO 1350–1538**		**Cēsis EPM 1550–1600**	
Early Fusing	U	F	U	F	U	F	U	F	U	F	U	F	U	F	U	F	U	F
Scapula D		4				1				1						2		
Radius P				2						1						3		
Humerus D				1		1										2		
Acetabulum						3												
1st Phalanx P				1		3		1				1				3		1
2nd Phalanx P				1		2		1								5		
Total	**0**	**4**	**0**	**5**	**0**	**10**	**0**	**2**	**0**	**2**	**0**	**1**	**0**	**0**	**0**	**15**	**0**	**1**
Later Fusing																		
Tibia D		3	2	1		1								1		3		
Metacarpal D		4		1												2		
Metatarsal D				1												1		
Metapodial D																		
Total	**0**	**7**	**2**	**3**	**0**	**1**	**0**	**0**	**0**	**0**	**0**	**0**	**0**	**1**	**0**	**6**	**0**	**0**
Latest Fusing																		
Ulna P																3		
Femur D																1		
Radius D						1		1		2						2		
Humerus P																2		
Femur P		1														2		
Calcaneus P			1			1										2		
Tibia P						1								1		4		
Total	**0**	**1**	**1**	**0**	**0**	**3**	**0**	**1**	**0**	**2**	**0**	**0**	**0**	**1**	**0**	**16**	**0**	**0**
% Porous	0.0		4.2		6.9		0.0		0.0		0.0		0.0		0.0		0.0	
P = proximal; D = distal;																		

Ance and took their eight horses, whilst the Bishop of Courland's retinue also stole horses (and other livestock) from the chapter at various points in the same year (*LUB* III, 1248). Horses were also seized during raids into Lithuania, and in 1492, the Livonian Master forbade peasants within the frontier advocateship of Bauska to sell horses (and weapons) to the Lithuanians, on pain of death (Arbusow 1924–26, 105–08).

Dogs (*Canis familiaris*) and Cats (*Felis catus*)

Dogs provided only a small proportion of the animal bones recovered in most of the assemblages (Table 6.1). This is unsurprising in deposits that were largely derived from butchery and consumption waste. Although it is known that dogs were sometimes eaten in medieval times, particularly in periods of famine, they did not form a regular dietary supplement. Two of the dog humeri from the Polish levels at Āraiši Castle did bear butchery marks associated with meat procurement, but there is no evidence that any of the other dog bones from the Latvian sites had been butchered and many survived intact. However, eighteen of the forty-eight dog bones from the Viljandi suburbs bore cut marks, several of which were located on the main meat bones (Rannamäe 2010, 47). The dog assemblage included greater numbers of partial or complete skeletons than for the main food-bearing species, indicating that their carcasses were deposited without being butchered. Three dog skeletons were found in pits on one site in Viljandi town (Rannamäe 2010, 84). Riga town produced several partial skeletons including: twenty-seven bones from a three- to six-month-old puppy; the skull, mandibles, and all seven cervical vertebrae of an adult dog; eighteen ribs of another adult and six vertebrae, possibly from the same animal. Five of the six dog bones from Vecdole consisted of ribs and vertebrae probably from one individual. All nine bones from the early post-medieval levels at Cēsis Castle could have belonged to one adult dog. This individual was quite large, measuring around 54–55 cm at the shoulder (Harcourt 1974). Three dogs from Riga town were of medium size, having shoulder heights of around 33 cm, 39 cm, and 49 cm, but one of the skulls belonged to a smaller round-headed lapdog. In the Viljandi suburbs assemblage, dogs shoulder heights ranged between 35 cm and 60 cm (Rannamäe 2010, 48). A wide range in dog sizes is to be expected, as some would have been bred for specific purposes. The presence of dogs on all the sites is attested by the frequent occurrence of bones damaged by gnawing. One of the Riga skulls also had evidence for a healed depressed fracture above its eye socket. Such fractures are not unusual and may sometimes have been caused by blows inflicted by humans.

Bones of domestic cat were recorded in several of the assemblages but never provided more than 1% of the mammal assemblages (Table 6.1). The skull, mandible, and atlas from an adult cat were found at Vecdole, and a more complete cat skeleton was found amongst the same pits as the dogs in Viljandi (Rannamäe 2010, 84). A complete skeleton of a sub-adult cat was recovered from excavations in the Liv quarter in Riga, most likely discarded on a midden after expiring. It is unlikely that cats would have often provided meat for human consumption, although cut marks were found on a humerus and femur possibly from the same immature cat from Viljandi suburb (Rannamäe 2010, 48). Moreover, none of the sites produced definite evidence for skinning, which is attested in other parts of Northern Europe. Bones of kittens and sub-adult cats were recorded in Riga town, Vecdole, and in the Polish levels at Āraiši. Several bones, mainly of birds, had evidence of puncture marks from cats' teeth. Cats as well as dogs would have had access to kitchen and butchery waste deposited within the settlements. Although cats occasionally feature on Late Iron Age sites in the eastern Baltic, they are more regularly associated with medieval sites. The process of urbanization, along with the presence of equivalent ecosystems concentrating organic waste in the Order's convents, would have provided suitable habitats for the proliferation of the species (see also Chapter 12).

Poultry

Bird bones were recovered from all the sites apart from the Āraiši Lake settlement (Table 6.1). By far the most commonly recorded was domestic fowl (*Gallus gallus domesticus*). Chicken bones never formed less than half of the total bird bones identified and in most cases they provided over 70% of the bird assemblage (Fig. 6.17). Recovery biases using normal excavation techniques are likely to have resulted in their under-representation in relation to the larger domestic mammals. However, relative comparisons of abundance can still be made. Comparing domestic fowl NISP counts with those of pig show interesting variations. Assemblages from castle sites consistently show high percentages of domestic fowl. The most spectacular results have been obtained from the early Livonian phase at Karksi Castle (Valk et al. 2012, 81; Rannamäe and Lõugas 2019), where domestic fowl provided 78% of the total pig and fowl, and indeed provided nearly as many bones as all the mammals combined (Table 6.1). However, domestic fowl were very well represented in all the castle assemblages, regularly providing over 30% of the total pig and fowl counts. At Riga they were found significantly more frequently in the castle deposits than in the assemblages from the associated town. They also increased significantly in the assemblage from Riga Castle after the Order's convent had been established. They were, however, also well represented in the urban deposits at Ventspils, in the latest phase at Viljandi town and on some of the late prehistoric hilltop sites at Viljandi (see Rannamäe and Lõugas 2019). Domestic fowl were less well represented at Vecdole that in other castle assemblages, although the assemblage from Vecdole is unusual in containing more bones of juvenile animals and may be atypical. The overall impression is that domestic fowl were more common on both high status and urban sites.

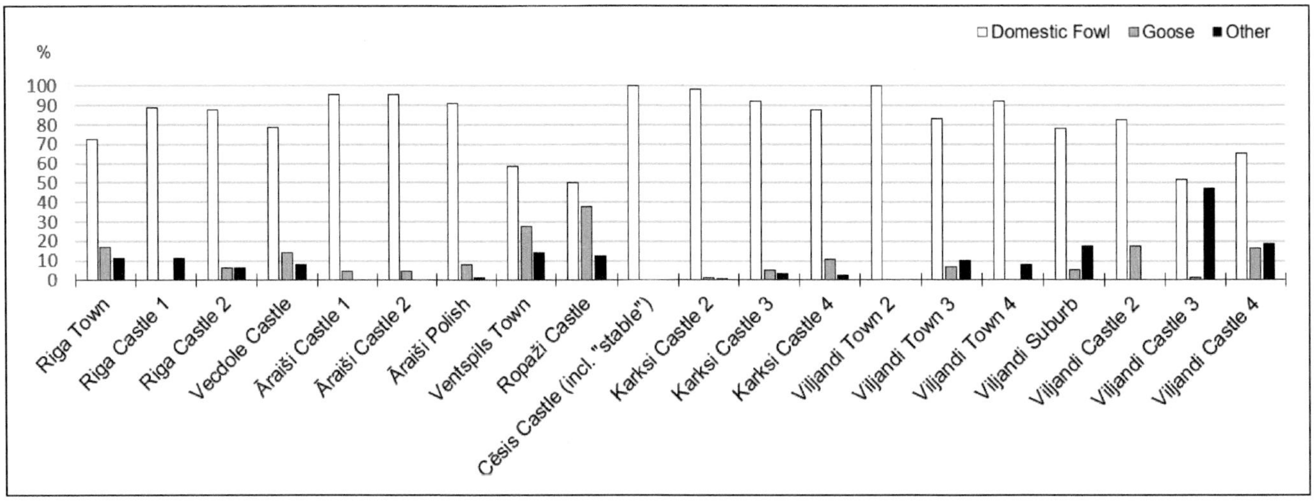

Figure 6.17: Bird bone frequencies from Livonia and Estonia. See Table 6.1 for sample sizes.

The number of domestic fowl bones at Vecdole was inflated by the inclusion of twelve bones from a juvenile chick and six bones from an adult fowl. The presence of the very young chickens probably indicates that they were being bred within the castle, as in the case of Karksi. Other bones of very young chickens were recorded in small numbers at Āraiši Castle and Ventspils. Bones of more developed, but still immature, chickens were common in most Latvian assemblages, outnumbering bones from adults in most cases. The percentages should be regarded as minimum figures, as the recovery and survival of immature bones will be particularly vulnerable to preservation and retrieval biases. Comparisons are difficult to find because ageing data have not been routinely recorded, but on British medieval sites percentages of immature fowl usually range between 20%–40% of the total chicken bones (Serjeantson 2006, 140). The very high percentages of immature chickens on the Livonian sites are likely to represent a focus on the raising, acquisition, and consumption of young chickens for meat. Cut marks were observed on a few bones from several sites. Evidence for the consumption of eggs is less clear. However, eggshells were found in the thirteenth-century midden at Karksi Castle. These have as yet not been

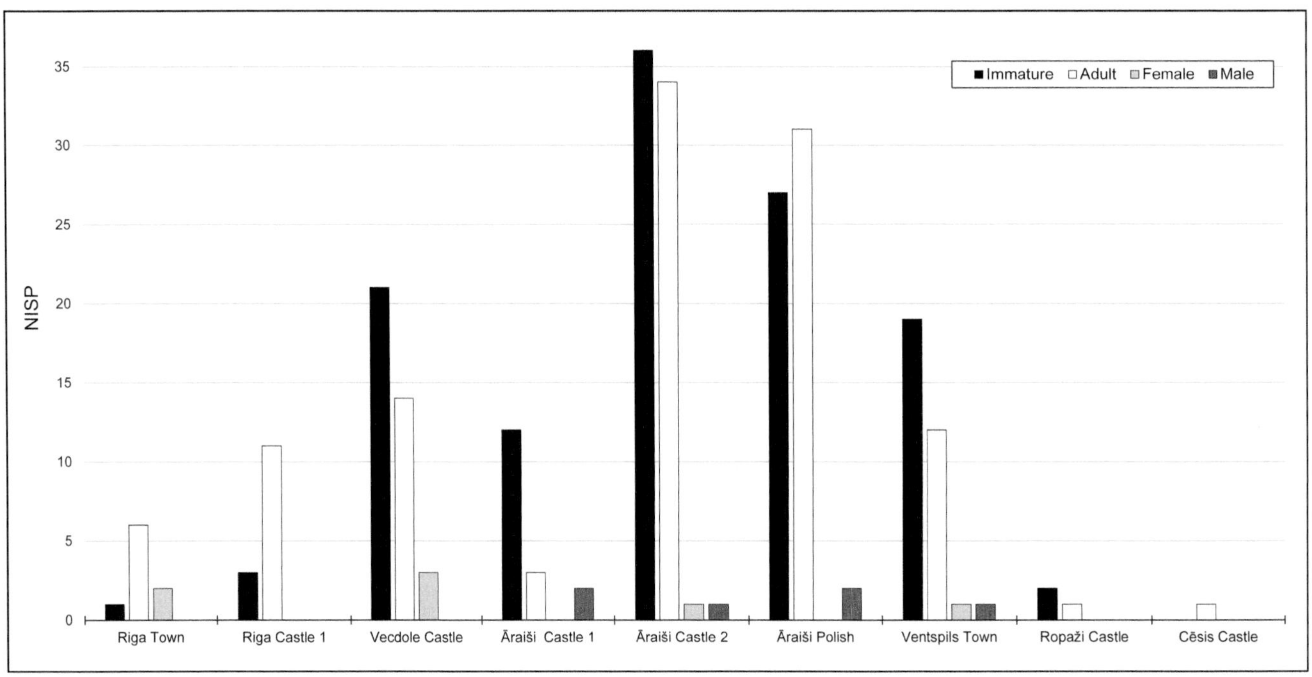

Figure 6.18: Domestic fowl ageing and sexing data from Latvian sites, based on presence or absence of spurs and their scars on tarsometatarsi fragments.

speciated but are likely to have belonged to hens given the predominance of domestic fowl in this assemblage. However, it has not yet been established whether these were eggs of hatched chickens or newly laid eggs that were eaten. The latter is more likely given that the midden contents were substantially food waste. Hens, when they are in lay, form medullary bone in the shafts of some of their bones, and examples of these were recorded from Āraiši Castle, Karksi, and Ropaži (Fig. 6.18). Again, it is not known whether these hens were laying eggs for human consumption or brooding chicks. Several other bones did not have medullary bone, and although some of these may have belonged to cocks or capons, the low frequency of medullary bone plus the high frequency of immature birds suggests that the focus was on meat rather than egg production. The presence or absence of spurs on the tarsometatarsi was recorded, but the samples were too small to determine the relative abundance of male or female fowl. Similarly, there was insufficient metrical data in the assemblages to determine any sexual dimorphism patterns.

Bones of geese were recovered from most of the sites. They were always outnumbered by domestic fowl but formed substantial percentages of the bird bones from Ventspils. Measurable bones were all of a size comparable to grey lag goose (*Anser anser*) and were probably all from domestic birds. In contrast to domestic fowl, nearly all the goose bones belonged to adults, with just single examples from the Polish levels at Āraiši and Ventspils of bones from immature geese. Butchery marks on specimens from Riga town and Livonian levels at Āraiši confirm their flesh was consumed, but they were not exploited as intensively as chickens. Finally, bones of mallard-sized ducks (*Anas platyrhynchos*) were found on four of the Latvian sites in very small numbers. There is no proof that these were domestic birds. The evident role of poultry in medieval Livonian sites contrasts with its general absence in extant inventories, which focus on mammal livestock, fish, and grain. However, clearly some chickens were delivered to castles: the Cathedral Chapter of Riga received three hundred hens from their subjects in Dundagen (Latv. Dundaga) in the 1380s (*LUB* III, 1248); in the late fifteenth century peasants in Bauska were obliged to deliver two hens to the Order's advocate each year (Arbusow 1924–26, 105–08), and dues of hens are specified in the late sixteenth centuries audits of Wenden (Ose 2011, 428, 432). A pasture for geese (*gozeweyde*) is also listed in 1479 in Riga's revenue book (*LR*, 164). In this respect poultry, whilst present and potentially linked to high-status consumption (Rannamäe and Lõugas 2019), do not appear to have been commercially significant within the economic paradigms of Livonian commanderies. They do, however, feature in the ritual praxis of the indigenous population (see below).

The Religious Significance of Domestic Animals

The horse is the most important animal in pre-Christian cult praxis in the eastern Baltic (see papers in Bliujienė 2009; also Chapter 12). Deposits of horses feature in cemetery sites across the region, as far north as central Latvia. There is a general regional trend in deposits of whole horses, largely confined to Prussian and Lithuanian territories, with only seven cemetery sites in Latvia containing individual animals and none reported from Estonia, although horses were present in cremations interred within barrows in southern Estonia from the second half of the first millennium AD. Out of 115 investigated burials, twenty-one contained horse bones, with sporadic examples of a range of anatomical elements suggesting that occasionally whole animals were cremated (Allmäe and Maldre 2005; Aun et al. 2008). The inhumed horses in southern Latvia have been interpreted as representing the presence of Selonians or Selonian influence, in turn reflecting Lithuanian cultural trends. More common in Liv sites are deposits of horse body parts (Brūzis and Spirģis 2009, 290), and the inclusion of equestrian equipment, such as harnesses and spurs, in the graves of high-status males within Selonian and Latgalian cemeteries from the mid-tenth century (Vilcāne 2009). The latest dated (and most unusual) horse deposits which can be linked to a cultic context are from the Liv cemetery at Ogresgala Čabas, with graves dating to the twelfth to thirteenth centuries, and fire pits, which include the remains of horses and other species, dated from the eleventh to fourteenth centuries. The nearby settlement appears to have also been inhabited into the fourteenth century. In the cemetery, the deposit of a single horse in a compact, circular pit and covered in dolomite slabs is unique (Fig. 6.19), with a potential parallel in the cemetery next to Daugmale Hillfort (Brūzis and Spirģis 2009, 288–89). Written sources hint at extra-funerary roles for horses in Liv culture, most famously the white stallion used by the Livs of Treiden (Latv. Turaida) as a divine proxy for determining the fate of the Cistercian missionary Theoderic in *c.* 1190 (*HCL* I, 10; Zemītis 2009).

The most widely reported category of animal deposits is that of *Bauofpern*, or foundation deposits. These consist of the bodies or body parts of domestic animals delib-

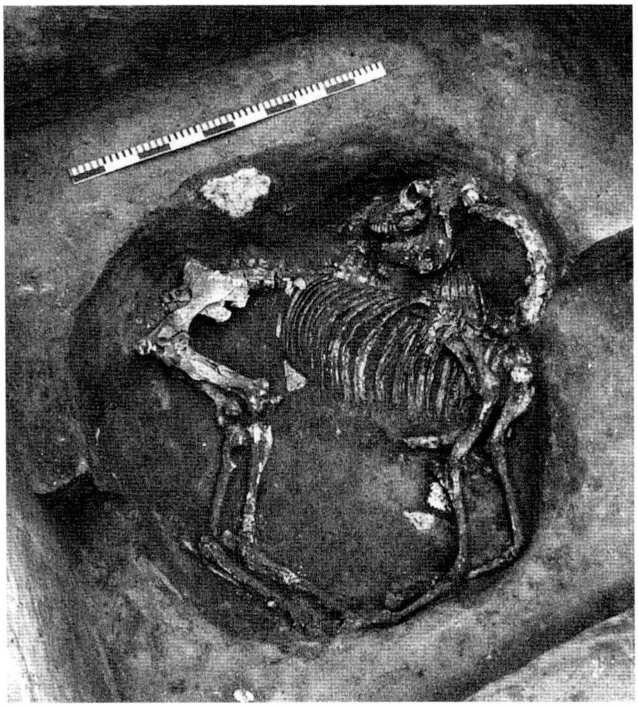

Figure 6.19: The horse deposit in the Liv cemetery at Ogresgala Čabas (after Spirģis 2008).

erately placed under the floors and walls of buildings in both pre-crusade and medieval contexts. Whole animals again are rare, and from the period of active crusading include the deposit of a calf, placed on its back, across the threshold of the main gate of Vecdole Castle (Atgāzis 1999, 337). More common is the deliberate placement of skulls or cranial elements within pre-crusade domestic contexts, where there may be evidence for the continuation of depositional practices into the thirteenth century. The decapitated skull of a fifteen-year old stallion along with its lower jaw was found during excavations in Musumägi (Kissing Hill) within the Viking Age complex in Viljandi, and may have represented a ritual deposit. An inverted cow skull from the twelfth-century Liv settlement preceding Riga is contrasted with the placement of horse and goat skulls under rebuilt sections of Riga's northern wall in the thirteenth century. A pig mandible was also placed in two furnished human graves within the thirteenth/fourteenth century 'grey cemetery' next to the cathedral (see Caune 2007). These types of deposits are typically regarded as a continuation of local customs. However, *Bauopfern* are a feature of twelfth-century Christian societies in central and Western Europe (including cattle deposits in the emerging Hanseatic hub of Lübeck), and can just as easily reflect the depositional practices of German, Slavic, or Scandinavian colonists (Schmidt 2001; Ermischer 2006).

There is certainly some evidence for animal sacrifice. When the Estonians attacked the Livs allied to the crusaders, they drove their cattle and other livestock back to their base, where they slaughtered them and immolated them as offerings to their gods. Henry notes that 'the flesh which they cut off fell on the left side, which indicated that their gods were displeased' (*HCL* xv, 3). War sacrifices are also documented amongst the Lithuanians; the Samogitians determined to burn armour and horses for their gods if they were victorious in battle against the Teutonic Order (*LRC*, 4675), and in the fourteenth century horses and officials of the Teutonic Order captured by the Lithuanians were sacrificed by immolation. Widely reported 'fire pits' in pre-Christian cemeteries across the eastern Baltic, features containing charcoal, ash, and animal bones, may represent such offerings, more broadly linked to feasting at the graveside. Later episcopal edicts sought to end such practices, consistent with Christian theological opposition to animal sacrifice (Ullucci 2012). However, the inclusion of joints of meat within graves is evident in the fourteenth and fifteenth centuries, and public feasting in cemeteries appears to have continued into the post-medieval period across Livonia, Estonia, and Lithuania (Vaitkevičienė 2011, 108–09; Valk 2001b). At the domestic level, expressions of ritual animal use appear to have continued if the Jesuit accounts of the early seventeenth century are to be believed. One report from 1606 refers to the sacrifice of a piglet, chicken, and bull (all black in colour) to a deity associated with the fields and grain, whilst another from 1618 (concerning Dünaburg (Latv. Daugavpils), Ludsen (Latv. Ludza), and Rositten parishes) refers to spilling the blood of slaughtered animals on a set of sacred stones kept in the kitchen, barn, or store (Laurinkienė 2011, 58–59); one of several examples of animal blood libations documented amongst rural communities in Latvia into the nineteenth century (Vaitkevičienė 2011, 111). Whilst these may be authentic ethnographic observations, their connection with the pre-Christian period remains vague, except that similar customs involving the use of animals were documented in the narratives of the Livonian Crusade.

Hunting and Trapping

With the exception of small mammals whose remains are typically the end result of natural expiration, the bones of wild animals reflect hunting and trapping activities conducted within the territories of settlements. They also usually represent less than 2% of the overall animal bone assemblage from Livonian sites included within this study and indicate hunting, as in many other medieval European contexts, was practised infrequently and played a relatively insignificant nutritional role. On Late Viking Age and Iron Age sites the representation of wild species is overall comparatively higher, but varies regionally, from very low proportions in north Estonian sites (e.g., less than 1% in Pada; Maldre 2007, 74), whilst at Rõuge Hillfort in the south-east they form a striking 57% of the assemblage, with a similar proportion in the nearby settlement (Paaver 1965; Maldre 2012, 122). The regional differences may reflect relative biodiversity across the region, and further south, in strongholds such as Asote and Beltes, wild species represent larger proportions of assemblages (Mugurēvičs 2002; Vasks et al. 2011). These have been interpreted as reflecting the beginnings of systematic exploitation of wild species in the Late Viking Age, linked to the political control exerted by elites residing in strongholds. Woodlands in particular would have been an accessible source of pelts for an international market fully established by the thirteenth century (Asaris 2007, 247–50). It is difficult to gauge the impact of the crusading period in Livonia on the resident wildlife; the few sites with thirteenth-century material and the crusading narratives indicate hunting was occasionally practised. Modern warfare and civil strife can certainly result in environmental degradation. In contrast, periods of warfare in the past which resulted in depopulation are typically associated with the regeneration of wildlife and habitats, following the reduction in human activity (Dudley et al. 2002). The general depopulation of the Livonian-Lithuanian borderlands may have enabled wildlife to regenerate or concentrate in *refugia*, but unfortunately no archaeological material was available from such frontier sites for the purposes of this study (for Prussia, the situation is different; see Chapter 12).

The Order's Rule permitted hunting (except with hounds and hawks), and this would have been primarily used as a means of training its regiments in the use of weapons and tactical formations 'in the field'. However, in Livonia the local population remained armed, playing an important role in the provision of military service for the Christian theocracy and clearly engaging in hunting and trapping (Kļaviņš 2009). There is some evidence for the introduction of seigneurial hunting rights. In 1272, when subdividing annexed Semigallian territory, the Dean of Riga and the Livonian Master agreed that any game being pursued which crossed from the territory of one lord to another would belong to the hunter, but the respective lord would be entitled to receive the animal's shoulder (*armum*); a standard cut taken as a feudal due (*LUB* I, 432). Permission to hunt was freely distributed; in 1291, the commander and advocate of Memel (Lith. Klaipėda) when subdividing territory between the Bishop of Courland and the Chapter of Riga designated that the large Forest of Okte (*magna silva de Octen*) remained the shared property of the bishop, chapter, and their subjects for cutting firewood and hunting (*LUB* I, 540).

The introduction of a legal separation between German and non-German in Livonia was accompanied by social restrictions in hunting, but it is difficult to determine how effectively this was enforced. Regional regulations surviving from Curonia indicate that 'free hunting' was legally recognized, but hunting certain species, particularly cervids, had become the privilege of the landowner by at least the fifteenth century (Mugurēvičs 2009, 178). In a reply to the Bishop of Courland in 1465 concerning hunting rights on the territory of the *landknecht* of Skrunda, the commander of Goldingen clarified that it was possible to hunt red deer, roe deer, and elk only after the owners of the land have hunted first (*LUB* XII, 293). However, the series of so-called 'Livonian Peasants' Rights', initially functioning as a *Landrecht* (land law) with documents surviving from the sixteenth and seventeenth centuries, makes no mention of hunting restrictions, and so these almost certainly varied between different regions according to historical precedents and the customs imposed by commanders, advocates, and other officials. By the fifteenth century, the Prussian court of the Grand Master had become increasingly secularized, partly expressed in the trappings of aristocratic hunting culture. The Livonian Masters appear to have participated in this as well; on 10 November 1446, the Grand Master wrote to thank the Livonian Master for sending him falcons (*habichte*) and wind dogs (*winde*) that he regretted had died (*LUB* X, 276).

The remains of wild species can be most usefully subdivided by their broad habitat preferences. Assigning habitats to species as ecological proxies often underestimates their behavioural plasticity and adaptability evident in modern biological studies; however, a high level of resolution is not required, particularly given the limited role of wild species in medieval eastern Baltic sites. What is more significant are: long-term trends in relative biodiversity, the presence of species sharing comparable habitat preferences, and shifts in cultural values assigned to these species. Changes in levels of biodiversity at and across sites (calculated using Simpson's (1949) Diversity Index) indicate a reduction in the measure of diversity between the centres of power in the Late Viking Age and the Middle Ages; at Āraiši this is particularly pronounced with a shift from 2.31 (lake settlement) to 1.84 (castle). Questions concerning the diachronic abundance and integrity of wild populations can only be effectively answered with genetic studies. In some instances, the complete absence of a species from the archaeological record after a certain time may be indicative of its extinction, but the cause and biogeography of that process remains vague. In the case of the eastern Baltic, this concerns only one species: the aurochs (*Bos primigenius*) (see also Chapter 12).

Woodland Refugee Species

Aurochs and bison (*Bison bonasus*) are traditionally assumed to have shared similar wooded habitats, but recent isotopic studies have suggested that early Holocene European bison were not confined to dense woodland but preferred more open landscapes. Aurochs overlapped with bison, ranging from woodland through to open habitats, but with contrasting dietary preferences. Postglacial afforestation and increasing hunting pressure pushed both species into denser wooded areas, which were comparatively sub-optimal habitats and only enabled them to survive at lower densities, resulting in further genetic fragmentation (Bocherens et al. 2015; Benecke 2005). As a result, both became refugee species (Kerley et al. 2012), confined by the crusading period to the belts of woodland which came to form the 'Great Wilderness'. The youngest remains of this species in both Estonia and Latvia are datable to the Iron Age. However, it is possible that bison migrated to Estonia in the Middle Iron Age as their bones have not been found on earlier sites. Tentatively identified fragments of bison bone have been identified at Rõuge (Paaver 1965, 293), Viljandi (where identifiable bones consist of phalanges: Saks and Valk 2002; Rannamäe and Lõugas 2019), and Unipiha (Paaver 1965), but these have been difficult to verify with the exception of Viljandi. A less optimistic survey indicates that bison are generally absent from the pre-crusading period in Estonia and north Livonia. Further south, at the lake settlement in Āraiši, bison or aurochs bones were identified on the basis of their size (the estimated withers height of a complete metatarsal was 133.5 mm, much larger than the domestic cattle, but their preservation was too poor for further analysis).

The most recent survey of aurochs and bison remains indicates a Late Iron Age distribution in southern Livonia (southern Latvia) at sites such as Beltes Hillfort (Vasks et al. 2011), and Paaver (1965) reported the presence of bison at nine medieval sites in Latvia. Aurochs and bison meat is documented in the stores of the Teutonic Order's frontier convents in eastern Prussia, rather than Livonia, and there is no evidence that the species was accessible in the Livonian-Samogitian borderland (Calkin 1961; Mugurēvičs 2009, 178). The extinction of the northernmost population of Baltic aurochs may therefore already have taken place during the Late Iron Age, coinciding with the intensification of hunting associated with strongholds. Both species, rare and almost certainly confined to wooded *refugia*, would have only been encountered during the Order's *Reisen* into Lithuania. It is entirely plausible that the depopulation of the frontier sustained populations of aurochs for longer than if Iron Age trends in hunting pressure had continued uninterrupted, including in western Lithuania, and as a result the last documented individuals — in captivity within the vicinity of Augustów — survived until the seventeenth century (see also Chapter 12).

Woodland Fur-Bearing Mammals

The presence of extensive belts of woodland in Livonia into the modern period would have supported abundant and diverse populations of fur-bearing mammals, particularly the mustelids and squirrels prized by furriers and merchants. However, there is a striking difference between the organization of fur exploitation before and after the crusades. Some Late Iron Age southern Livonian strongholds have a large and diverse proportion of fur-bearers (Mugurēvičs 2009), and the fur-trade was clearly an important economic component from the eleventh century when specialized iron blades for dressing pelts are first evident on Liv sites in the Lower Daugava, including at Daugmale (Vasks 2010). In the equivalent medieval centres of power, the Order's Livonian castles, the remains of these mammals only occur in minute quantities, with the most valuable furs deriving from mustelids least well represented. Marten (*Martes martes*) bones (fifteen) have only been found at Cēsis Castle within the study area, although there are more references to its pelts in trading records. Fragments of badger were also reported from Valmiera Castle (Ger. Wolmar), although the fur of this species rarely features in trade inventories. Fox remains are more common, although the quantities are typically negligible and derive from a minimum of one or two individuals, as at Turaida (seven), whilst the Order's phase at Rakvere (Ger. Wesenberg) Castle, in Estonia, is associated with the highest percentage of fox bones of any site, representing 80% of the wild species assemblage (Paaver 1965, 323). Larger carnivores were hunted, but their furs were of limited commercial interest after the crusading period. Bears (*Ursus arctos*) documented on a number of pre-crusade sites are rarely encountered in medieval contexts within the study area. The popularity of bear fur in the Viking Age, as suggested by the inclusion of bear pelts into graves in southern Scandinavia and the quantities of ursine bones in eastern Baltic sites, contrasts sharply with the post-crusade period. Six bear elements were found at Viljandi Castle. Fragments of bear bone were found in both the Liv and castle contexts at Turaida, and a metapodial from a juvenile was found in the later medieval contexts at Karksi. They feature in two urban assemblages, Ventspils and Viljandi suburbs (Table 6.1), and in the case of the former, the arthritic phalanx of a bear may potentially represent an animal kept within a confined space and perhaps used for baiting or dancing. Alternatively, this may be a fragment of a pelt, and in fact most bear remains from medieval Livonian contexts derive from paws (Rannamäe and Lõugas 2019). Bear claws continue to feature as centre pieces of jewellery deposited in human graves into the fifteenth century (see below).

Of the other wild species which rarely feature on medieval sites, populations of wolf (*Canis l. lupus*) and lynx (*Lynx lynx*) are documented into the twentieth century. The teeth of both species were used as pendants in the Late Iron Age (Jonuks and Rannamäe 2018). However, the lynx, which prefers corridors of unbroken woodland supporting abundant populations of wild ungulates, is arguably the rarest mammal taxa encountered on pre-crusade and medieval sites in Livonia (Mugurēvičs 2009, 179). However, its well-documented rapid recolonization of areas of Latvia following intensive periods of hunting in the nineteenth century — a period of no more than three decades — suggests the species was widely distributed in earlier centuries (Ozoliņš et al. 2007). Like other fur-bearers, it was rarely hunted by the households of the Order's Livonian convents. Wolves, notoriously difficult to identify in fragmentary archaeological assemblages, have been reported in small quantities on a number of sites, including Cēsis Castle (Table 6.1). Given the mobility of wolves, it is difficult to connect these sparse fragments with anything other than opportunistic and infrequent hunting. Wolf fur itself had relatively limited commercial value and rarely features in Hanseatic trade documentation (Delort 1978). Although both lynx and wolf, along with several other wild mammal species, have been identified on prehistoric sites around Viljandi, only a single bone of lynx was recorded at Viljandi Castle (Rannamäe and Lõugas 2019). No further bones of either lynx or wolf were positively identified in assemblages examined during this project. A canid ulna from the later Livonian levels at Āraiši Castle was similar in size to wolf but is more likely to have been from a large hound.

The fur trade was supplied by a network of hunters and trappers largely based in rural settlements. Furs appear to have been removed from carcasses and prepared in these settlements, and then pelts brought to castles which functioned as points of storage and redistribution (along with other agrarian produce). As in the case of Novgorod (see Maltby 2019), the most important centre of the north-east European fur trade in the late medieval period, the storage of pelts has left virtually no archaeological traces. However, inventories of goods brought to Riga for export across the Hanseatic network indicate that tens of thousands of furs were shipped from this town from the fourteenth to sixteenth centuries. In 1406, when English pirates captured three ships from Riga, 394,864 pieces of fur were discovered in their holds, with an estimated 1.5 million pieces of fur exported annually from the eastern Baltic at the time. The majority of these derived from Russian territories, but the source of the trade appears to have shifted at the end of the fifteenth century to Livonia (Delort 1978, II, 1042ff.; Jahnke 2015, 222). Pelts could have been acquired across Livonia and redistributed via the Order's own network. In this respect the majority of pelts would not have remained long in castle stores, and only a small percentage would have been used by the household. Local use may be represented by occasional finds of bones from fur-bearing mammals, sometimes with skinning marks. This accounts for the composition of the assemblage from Cēsis Castle, the largest from any Livonian site (Table 6.1). Despite evidence for extensive woodland within the castle's territory, the brethren and extended household clearly did not engage in hunting very often. Furriers were more likely to have been situated in the town, where significant assemblages of animal bone are lacking. Written sources provide a more accurate measure of the quantities of fur being handled through the networks of the theocratic authorities of Livonia. When the Semigallians concluded a treaty with the Archbishop of Riga and the Livonian Master in 1272, in the absence of sufficient quantities of grain for making payments to the new regime, this would be substituted by two marten pelts and eight squirrel pelts (*LUB* I, 1272). In the later fourteenth century, the Chapter of Riga received from the properties of its castle in Dundagen (Latv. Dundaga) in Curonia, an annual quantity of two thousand furs of weasel, ermine, and marten, as well as other animals worth twelve Riga marks per thousand (*LUB* III, 1248). Future excavations of rural settlements may pinpoint, at least on a regional level, the relative importance of fur acquisition within the Order's individual administrative districts. By way of comparison, in early medieval Masovia, the high representation of fur-bearing species in rural settlements may be interpreted as reflecting the acquisition of fur (and wild meat) for payment as tax or feudal dues (Dulinicz and Iwaszczuk 2008, 165–67), with a similar situation, albeit on a larger scale, visible in the territory of Novgorod (Maltby 2019).

Ungulates within and at the Fringes of Woodland

Two further cervid species represented on Livonian sites are red deer (*Cervus elaphus*) and roe deer (*Capreolus capreolus*). However, with the exception of from roe deer at Cēsis Castle, their remains are found in minute quantities (Table 6.1). Their biogeography in the eastern Baltic is also unclear, as red deer may have been completely absent in Estonia in the Late Iron Age with antler fragments representing imports derived from southern populations (Rannamäe and Lõugas 2019; Haak et al. 2012; Luik 2015, 103–04). Both species favour woodland, with roe deer preferring the more foliage-rich edges of habitats. However, where roe deer appear alongside hare in archaeological assemblages, this presents a compelling argument for the presence of fragmented habitats and agricultural plains (Hewison et al. 2001), with roe deer readily able to colonize these and substitute hedgerows for woodland (Morellet et al. 2011) (see also Chapter 12). This combination is evident in a number of sites: at the castle of Cēsis, Turaida, the Order's phases of the castles in Riga and Viljandi, as well as the suburbs of the associated town of the latter. Both meat and antler from red and roe deer appear to have been exploited, with antler used as a multi-purpose raw material. Evidence for wild boar (*Sus scrofa*) is equally scarce, although five large pig specimens from the earlier Āraiši Lake settlement were probably from wild boar. However, only two boar femur fragments were recorded from Riga Castle, and very few suid bones from other sites were sufficiently large to be considered to be wild boar. All three species appear to have been of limited interest to the Livonian branch of the theocracy (Mugurēvičs 2009, 178).

Wetland Mammals

Two wild species which prefer wetland habitats are found on Livonian sites; elk (*Alces alces*) and beaver (*Castor fiber*). Both dominate the pre-crusade assemblages of wild species in southern Estonia and northwest Latvia, including the late prehistoric sites around Viljandi (Maldre 2007, 75; 2012, 124; Rannamäe and Lõugas 2019), after which they are typically found in small quantities. Elk was exploited within the Lower Daugava Valley, with its remains present in the Order's phases of Riga Castle, and in the Liv quarter of the town, within the animal bone waste from Vecdole Castle (Table 6.1), and at nearby Mārtiņsala. Further up the Daugava, elk is the second most represented wild species at the hillfort at Jersika (Mugurēvičs unpublished). To the east of Riga, elk was also present in the wetlands of the Lielā Jugla Valley at the site of Ropaži Castle (Table 6.1). Further north, within the Gauja river basin, the species was present, but here there is some evidence for its depletion by the crusading period. On the western side of the valley, a large quantity of elk bones was recovered from excavations in 1980–96 at Turaida, where thirty-eight fragments were identified from the earlier Liv settlement and thirteen from the episcopal castle (Renga 1997), with a further 104 fragments from mixed layers (Renga 1999). Likewise, an abundance of elk bones (forty-two) from the Āraiši Lake settlement contrasted with the complete absence of the species in medieval and post-medieval phases from the nearby castle. Elk is also noticeably absent from Cēsis Castle, whilst it was present in small quantities further up the Gauja at Valmiera Castle (Mugurēvičs unpublished). In north Livonia (southern Estonia), elk is widely distributed in the wetlands today, and its bones have been found in both the castle and town of Viljandi (see also Rannamäe and Lõugas 2019). A more detailed analysis of the anatomical element representation does not indicate any site-related preferences for specific joints; whole carcasses were processed at Ropaži and the Liv quarter in Riga. There are also rare examples of the use of elk antler, which features more widely in documentary sources. The largest quantity of elk antler has been found at Viljandi Castle, where the fragments represent working refuse (alongside pieces of red deer antler) that can be related to the production of crossbow parts (Luik 2015).

The distribution of beaver at Livonian sites mirrors that of elk; it is typically found on the same sites, although in smaller quantities, particularly in southern Estonia. At Jersika, the two most abundant wild spe-

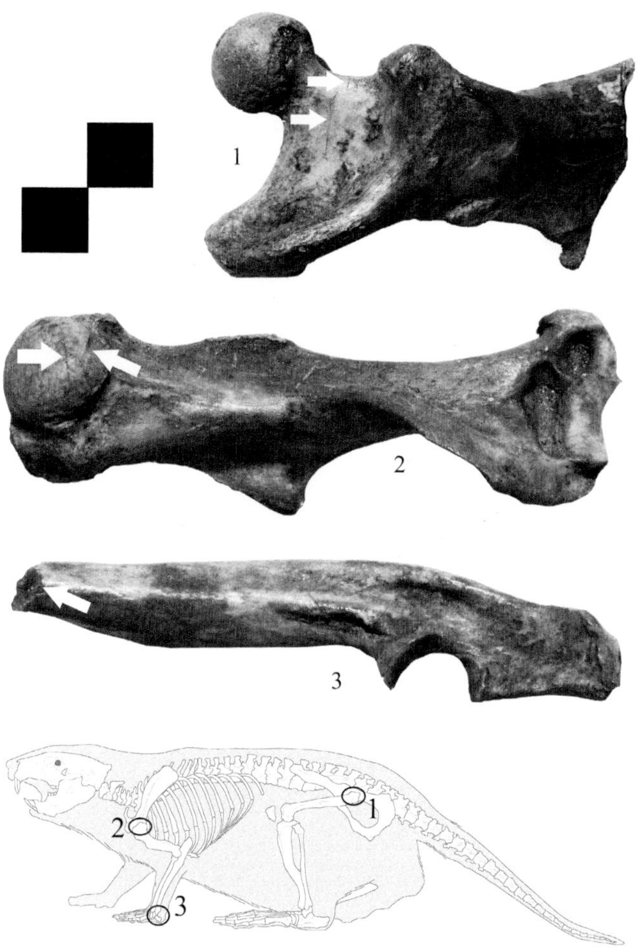

Figure 6.20: Beaver bones (femur, humerus, ulna) from the Liv quarter in Riga (photos by K. Seetah), showing location of cut marks on the skeleton (M. Coutureau, after Pales and Garcia 1981, 169, adapted by A. Pluskowski).

cies are beaver (seventy-nine) and elk (fifty-one); both are the dominant wild species at Āraiši, and at Turaida beaver occurs in much smaller quantities. However, the exploitation of *Castor* in medieval Livonia had a more commercially tangible dimension. Its fur was in high demand in European markets in the fourteenth and early fifteenth century, and it is clear the species was already widely exploited in the Viking Age. Its remains have been found in a number of strongholds within the Daugava Valley, and also feature on some Estonian sites (Maldre 2012). Beaver bones from Livonian sites, however, largely represent food waste, with evidence for skinning present on only a few fragments across the region. Given its association with aquatic environments, beaver was an acceptable Lenten food documented in the medieval eastern Baltic (Mänd 2004). Its presence at castles is more readily linked to consumption rather than fur working. This contrasts with the documented quantities of beaver pelts passing through Riga. The specific geographic origin of these pelts is unknown, but it is likely that they were obtained from indigenous communities exploiting local wetlands and handed to the Order's officials as payments of tenurial dues. For example, in 1466 each merchant entering and leaving Dünaburg (Latv. Daugavpils) was taxed one marten pelt by the Order. It is also likely they were obtained from the traditional fur-producing neighbouring regions (Delort 1978, I, 113; II, 1198). Given the general lack of excavations of medieval Livonian rural settlements, this aspect of the fur trade is largely invisible. However, the distribution of beaver remains indicates that whole animals were processed in the Liv quarter in Riga — very plausibly for export — as well as the lake settlement at Āraiši (Fig. 6.20); the presence of the species in castles may also reflect the activities of the local population. The absence of elk and beaver from key medieval-period sites within the southern part of Wenden's commandery (Cēsis and Āraiši), despite the patchwork of wetlands and rivers, may indicate their populations had already been locally depleted by the fourteenth century. The absence of references to beavers in Livonia from the seventeenth century has been used to infer the collapse of the local population (Mugurēvičs 2009, 179), although this cannot be verified by zooarchaeological studies of post-medieval assemblages which are virtually nonexistent.

Wild Birds

Small numbers of a diverse range of wild birds were recovered from Late Iron Age contexts in Viljandi, where fragments from Anseriformes (various species of geese and ducks), Accipitriformes (hawks, falcons, golden eagle, white-tailed eagle), Laridae (gulls and terns), Strigidae (boreal owl), Galliformes (capercaillie, quail, grouse), Columbidae (woodpigeon), and Corvidae (magpie, raven) were identified (Ehrlich 2016). Talons from ospreys and white-tailed eagles (and perhaps golden eagles) were used in the Late Iron Age as pendants or grave goods (Jonuks and Rannamäe 2018). The representation of wild avian species after the crusades is comparatively limited. Written sources indicate that by the fifteenth century, the earlier prohibition against the use of falcons had long been abandoned. In 1417, the commander of Windau is documented as sending hunting hawks (*falken*) to the Grand Master of the Teutonic Order following an old custom (*nach elder gewonheit*) and asking him to pay for the hawks in the new currency (*LUB* I, 536). However, very few remains

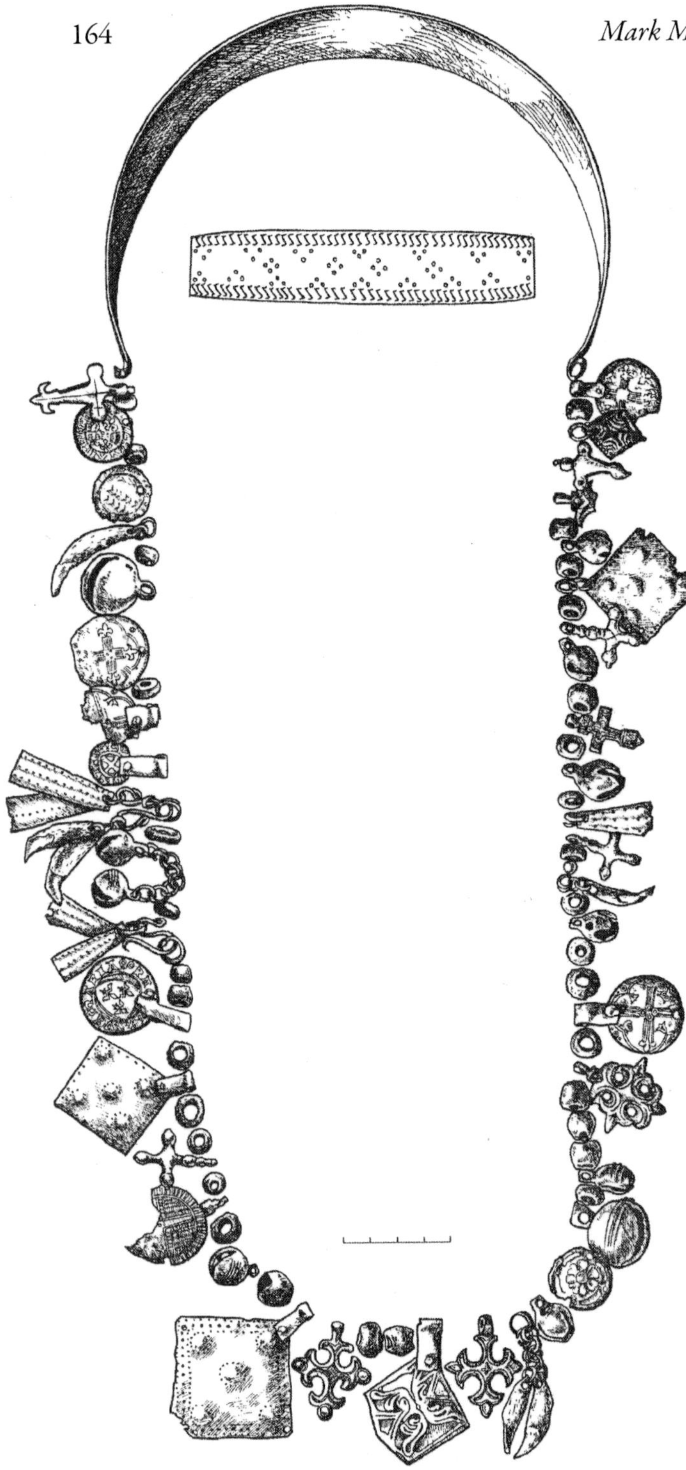

Figure 6.21: A necklace with a number of carnivore teeth included within a grave at the cemetery of Augustinišķu (Augšzeme, south-east Latvia) (Berga 2007, 50).

of wild birds have been recovered from the Order's sites in Livonia investigated in this project (Table 6.1). A few bones of medium-sized and small ducks were recovered from six of the sites. These included two bones closely comparable to a tufted duck (*Aythya fuligula*) from Ventspils. This is a diving duck which is a summer visitor to this coastal region. One bone of a teal (*Anas crecca*) was found at Vecdole. A large number of duck bones were found at Viljandi Castle (Rannamäe 2010). These were not further identified and could have included bones from larger domestic ducks and/or wild mallards. Other water birds and waders have been recorded only occasionally. A gull ulna was recorded from Riga town, and possible bones of swan (*Cygnus* sp.) and crane (*Grus grus*) were found on the Pikk Street site in Viljandi town (Rannamäe 2010, 84; for latest identifications, see Ehrlich 2018). The same town produced seven bones of capercaillie (*Tetrao urogallus*). The only other evidence for potential gamebirds consisted of thirteen bones of small galliforms from Karksi, along with single fragments of hawk- and eagle-sized Accipitridae, and at Vecdole a coracoid of a woodpigeon (*Columba palumbus*). The latter may have been derived from a bird nesting nearby, which could also account for the presence of a tibiotarsus of a Tengmalm's owl (*Aegolius funereus*) from the same site.

Small Commensals

The establishment of castles and towns would have provided new and abundant sources of concentrated organic waste, creating attractive habitats for diverse communities of small mammals as well as birds. Medieval urban food webs, in particular, have been extensively studied, and those in Livonia would be comparable to other parts of Europe. However, the bones of small mammals have been infrequently recovered from Livonian sites due to a lack of routine sieving during excavations. The presence of black rat (*Rattus rattus*) is evident in the later medieval and Polish phases of Āraiši Castle. Hedgehogs (*Erinaceus europaeus*), which can exhibit commensal behaviour, are only represented by a large number of bones from one individual in the late phase of Viljandi Castle, with fragments of voles and moles present on a few sites in negligible quantities and bones of passerines have been recorded on several sites in Viljandi (Table 6.1).

The Religious Significance of Wild Animals in Livonia

The religious significance of wild animals is largely represented by the use of teeth and other bones as pendants (Fig. 6.21). Perforated carnivore teeth derived from foxes, wolves, and bears, are widely reported as grave goods, alongside astragali from beavers and martens used as pendants. Their meaning is unclear, but in the Grand Duchy of Lithuania the use of bear claws within necklaces is represented by 114 examples from

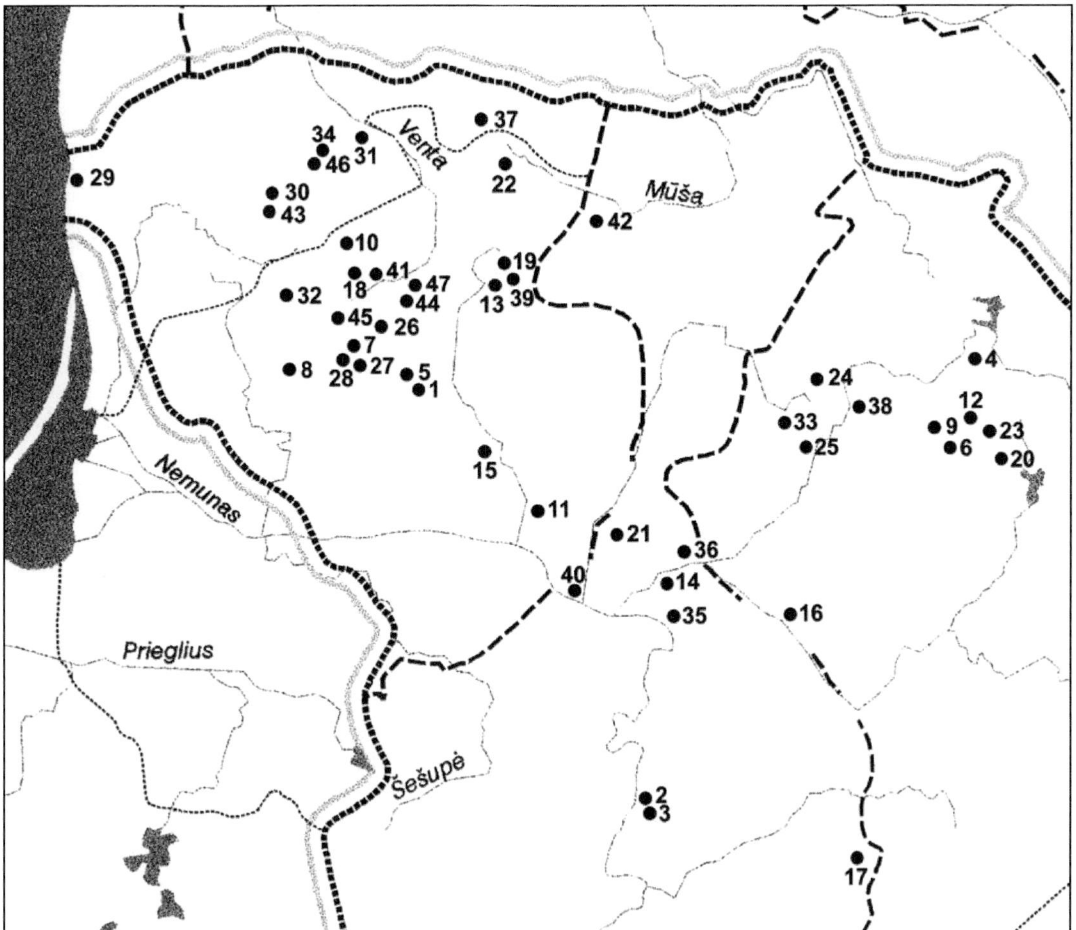

1 Aleknai; **2** Alytus; **3** Alovė; **4** Antandraja; **5** Arglaičiai; **6** Ažugiriai; **7** Budraičiai; **8** Dargaliai; **9** Diktarai; **10** Dvarviečiai; **11** Gėluva; **12** Geniakalnis; **13** Jakštaičiai; **14** Karmėlava; **15** Kejėnai; **16** Kernavė; **17** Kukliai; **18** Kuršai; **19** Lepšiai; **20** Liepiniškės; **21** Masteikiai; **22** Mažeikiai; **23** Narkûnai; **24** Noriûnai; **25** Obeliai; **26** Paalksniai; **27** Paežeris; **28** Payžnys; **29** Palanga; **30** Patumšaliai; **31** Pavirvytė; **32** Pribitka; **33** Raguva; **34** Rūdupiai; **35** Rumšiškės; **36** Skrebinai; **37** Šapnagiai; **38** Šeimyniškėliai; **39** Šiauliai; **40** Šilelis; **41** Šlapgiris; **42** Šukionys; **43** Užgiriai; **44** Užventis; **45** Varniai; **46** Vembūtai; **47** Žalakiai.

Figure 6.22: The distribution of bear claw pendants in the Grand Duchy of Lithuania in the late fourteenth and fifteenth centuries (Svetikas 2008, 175).

forty-seven sites and dating to the late fourteenth and fifteenth centuries (Fig. 6.22), whilst in the Order's territories thirteen amulets from eleven burials in four cemeteries have also been dated from the mid-fourteenth to mid-fifteenth centuries. Two thirds of these artefacts were found deposited with the bodies of elderly women, and they have been interpreted as a hybridized merging of two symbolic worlds, perhaps mediating the process of religious conversion: the Christian ursine symbol of resurrection combined with a powerful, indigenous animal (Svetikas 2008). The significance of the bear is occasionally alluded to in written sources. A document from 1417 outlining a grant of land to the commandery of Windau from the Livonian Master refers to various landmarks, including a stone on which a bear's paw (*bahrenklawe*) is carved (*LUB*, v, 2168). No such diachronic trends are evident in Livonia, where animal amulets continue to be deposited in indigenous burials, including within Riga itself. Beaver and hare astragali (previously identified as marten) in Estonia have been interpreted as potential markers of social status, specifically of fur traders, rather than having religious significance and in burials they tend to be associated with women, rather than claw pendants typically found with men (Luik 2003).

There is very little mention of wild animals in written sources, whilst place names allude to the importance of certain species. Later folklore suggests that certain animals in specific circumstances were perceived as having supernatural agency, which could be linked to individual deities. However, it is difficult to demonstrate the antiquity of such beliefs. Hunting (and gathering) within sacred groves appears to have been prohibited, although the Curonian 'kings' specifically hunted within such designated woodlands, with roe deer and hare specified as targeted species. This activity appears to itself have had a sacral function, and one that may even have emerged as an indigenous response to the Christian theocracy. Whilst it is tempting to see all these late medieval customs as relics of the Iron Age, it is important to acknowledge that beliefs and practices invariably changed, and may have been heightened or merged with understandings of Christianity during and after the crusading period.

Fishing

Fish clearly played a fundamental role in the alimentation of Late Iron Age and medieval societies in Livonia. However, the lack of routine sieving on archaeological sites has meant that fish bones are typically underrepresented, and in some cases completely absent. Exceptions include the Late Iron Age hilltop sites in Viljandi, where over nine hundred fish bones were identified (Rannamäe and Lõugas 2019). These consisted entirely of freshwater species, mainly perch (*Perca perca*), pike (*Esox lucius*), and cyprinids, indicating exploitation of the local rivers and lakes. However, the complete absence of fish from the archived assemblage of the lake settlement at Āraiši can plausibly be interpreted as a result of archaeological recovery techniques, rather than reflecting a lack of exploitation. Fishing equipment has also been found in a number of sites in Estonia and Latvia, including hooks, which would have been used for catching a range of species (e.g., Riekstu Hillfort; Apala and Apals 2014, 128).

Fishing, like other forms of provisioning, was organized from the onset of the crusading period (see Chapter 8). Henry's *Chronicle* (*HCL* v, 4) mentions two of Bishop Albert's fishermen (*piscatores*) caught up in a Lithuanian raid in 1202, and in 1210 fishermen, on both sides of the Daugava, alerted the town of Riga to the approaching Curonian fleet (*HCL* xiv, 5). Fishing rights were subsequently an important feature of the Order's management. Following the reorganization of territory, lakes, and rivers were reassigned to the Order's new tenants, but some became incorporated into the provisioning systems of castles. In 1274, the Livonian Master concluded an agreement with Riga's Cathedral Chapter regarding the building of fish weirs in the Irbe River in northern Curonia (*HWC*, 53; see also Chapter 8). Fishing continued to be practised by indigenous communities, and perhaps fishermen were drawn exclusively from this population. The most striking evidence for this comes from the Liv quarter in Riga, where eggs from fish tapeworm were identified in a mid-fourteenth-century latrine fill, indicative of handling raw fish (Yeh et al. 2014). From the recent excavations in the town, bones of haddock (*Melanogrammus aeglefinus*), gadoids (probably all cod (*Gadus morhua*)), perch, zander (*Sander lucioperca*), salmon (*Salmo* sp.), pike, and sturgeon (Acipenseridae) were also found in waste deposits, indicating access to marine and migratory species as well as freshwater species (Table 6.1).

Access to fishing sites in the Daugava was regulated and contested into the sixteenth century. In 1394 there is reference to a *piscinam* near the Sand Gate (*LR*, 91), and specialized fishing sites include references to an eel weir in the Engure Rivulet in 1353 (Bauer 2003, 184), a lamprey weir (*negenoughe weer*) in the Gauja River in 1435 (*LUB* viii, 1019), and a salmon weir in the Daugava near Ikšķile is mentioned in 1439 (*LUB*, ix, 454). Salmon, in particular, were the focus of disputes in 1469 perhaps because their numbers were being depleted. The Archbishop of Riga demanded the townspeople stopped catching them and demolish the weirs they had built in the Daugava (*LUB* xii, 653; *LGU* i, 453). Fishing equipment does not appear to have changed following the crusades, and hooks used in the medieval period are identical to those used in the Late Iron Age. However, the documents specify the regulation of types of equipment, from a fishing-rod (*stupent*) and small nets (*settenetten*), through to large drag nets (*dryffnetten*) of varying lengths. In 1503, the Dean of Riga and the *Kumpan* of the Order in Kircholm (Latv. Salaspils) clarified fishing rights for vassals in the Daugava in some detail. Peasants who lived in Memorgen (Latv. Memurga), Kekau (Latv. Ķekava), Koven, and Dole Island could keep forty ells-long drag-nets (*dryffnetten*). Peasants who lived downstream from Dole could keep sixty ells-long drag-nets and fish from both banks of the Daugava. Peasants living on either bank near Dole and Kirchholm could keep forty ells-long nets and fish in the stretch from Rumbula to Jungfernhof (Latv. Jumpravmuiža). Peasants from Üxküll (Latv. Ikšķile), Lennewarden (Latv. Lielvārde), Ascheraden (Latv. Aizkraukle), and Jungfernhof who had not had the right to fish from former times could henceforth use a small net (*nette*), a boat, and a fishing-rod. The agreement continued with specifying during which periods fishing could take place, the installation of various weirs, and dates of future inspections (*LUB* ii.2, 427).

What did change was a gradual increase in the exploitation of local coastal waters. Isotopic provenancing indicates that stockfish imported from Scandinavian waters during the eleventh and twelfth centuries were supplemented by locally caught cod by the fifteenth century (Orton et al. 2019). All seven gadoid bones recovered from Riga town were precaudal vertebrae, probably all from large cod. These most likely derived from fish preserved by salting and drying that had been transported to the town once their heads had been removed. All five haddock bones were cleithra from quite large fish. These again are likely to have been bones attached to preserved portions of filletted fish imported to the town. The bones from large sturgeon and salmon are also likely to have been from imported fish. The central role of Riga in the Baltic Hanseatic network also meant the town would have been a major recipient of the trade in herring. However, no bones of herring (*Clupea harengus*) were found in any of the recent excavations in Riga, although this may be because of recovery bias. They have been found in small numbers at Karksi and Viljandi town (Rannamäe and Lõugas 2019). In the earlier thirteenth century deposits at the Order's castle at Vecdole, there is evidence that the community was already supplementing its locally available cyprinids and pike with possible imports of salmon, catfish, sturgeon, and plaice (*Pleuronectes platessa*). However, in the late thirteenth century deposits at the inland site at Karksi Castle, the community seems to have continued to rely initially entirely on local cyprinids, perch, and pike for its fish supplies (Valk et al. 2012). It was not until the later medieval period that these supplies were supplemented by imports of cod and possibly zander (Table 6.1). A similar picture can be found in the nearly town of Viljandi, where most of the fish eaten were cyprinids, perch, and pike, occasionally supplemented by imports of cod, salmon, and sturgeon.

The identifiable fish bones recovered from Livonian period levels at Āraiši Castle consisted of a sturgeon fragment, three cod bones, two cyprinids, and a flat fish. During the period of Polish occupation in the early post-medieval period, there was a mix of freshwater (perch, pike, and sturgeon) and marine fish (gadoids and flatfish). Most of the perch and pike remains from here consisted of scales indicating the processing of fish caught locally. The gadoid bones probably mainly consisted of small or medium-sized cod, although whiting (*Merlangius merlangus*) may also be represented. Most of the gadoid bones came from the head. This could indicate on-site processing of marine fish despite being over 80 km from the coast. It is perhaps more likely that these smaller gadoids were preserved and transported as complete fish. The flatfish were probably all plaice. Less surprisingly, all the fish bones recovered from the coastal town of Ventspils were gadoids and flatfish, all of which could have been caught locally. Most of the gadoids were small cod and consisted mainly of bones from the head, indicative of local processing.

Did the importance of fish change with the introduction of a Christian fasting culture? There was certainly an emphasis on fasting during and after the crusades, but it is difficult to quantify fish consumption in both the Iron Age and medieval period in the eastern Baltic. Earlier studies which tackled this question on a much larger scale hesitated to link the rise of commercial fishing in the eleventh century with the popularization of Lenten fish consumption (Barrett et al. 2011). Given the limitations in introducing even the basic tenets of Christianity amongst the indigenous population, it seems likely that a fasting culture was most prevalent if not confined to castle and town communities. A fasting ideology persisted, at least within the higher ranks of the Order, into the fifteenth century. On 9 April 1450, the Grand Master wrote to the Order's representative at the papal court instructing him to obtain a permit from the pope for the Livonian Master, who was old and sick and thus, unlike the others, unable to sustain himself on fish during fast days, and to be permitted to eat dairy products and eggs during fasts (*LUB* XI, 25).

Conclusion

To date, trends in faunal exploitation before the crusades in the eastern Baltic are best known from Estonia, where broad regional differences subdivide northern and southern sites in terms of animal husbandry practices (Maldre 2012). In terms of diachronic observations, there were no fundamental changes in the exploitation strategies of cattle (mostly slaughtered as adults) and pigs (slaughtered during the first two years of their life, more prevalent in areas of woodland) between the Viking Age and Final Iron Age. The most evident change is in the husbandry of sheep and goats, with older animals clearly kept for secondary products (milk and wool) being more prevalent in the Final Iron Age. The trends evident on Late Iron Age sites in Latvia broadly correspond to this, although there is evidence for local variation in the relative representation of certain species, as in the case of Estonia (Fig. 6.23a, b) which can be attributed to variability in local environmental and social contexts, as well as sample size, recovery techniques, and chronological resolution. Horse represen-

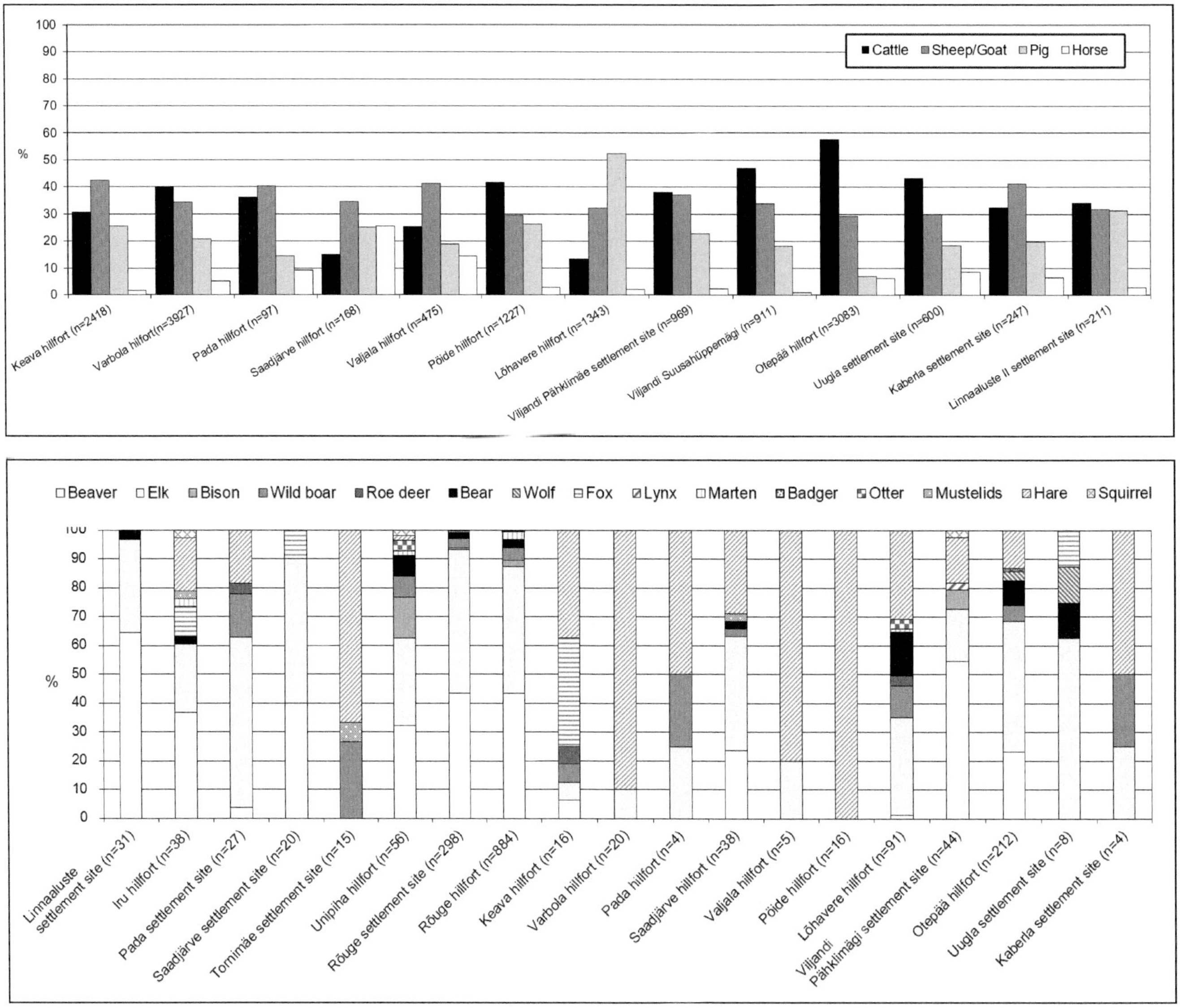

Figure 6.23: Percentages of (a) domestic and (b) wild mammals at pre-crusade Estonian sites (for dates and details, see Maldre 2012; data provided by Liina Maldre).

tation remained low on Estonian sites, which parallels the general absence of the species in the ritual sphere of pre-Christian Estonian society. This contrasts with their distinctive role in the mortuary culture of Prussians and Lithuanians, and to a lesser extent amongst the southern Livs and Semigallians.

This study has contributed new material from a range of Livonian sites, although in quantity it does not match the significantly more abundant dataset from Prussia (Chapter 12). The evidence from medieval Livonian assemblages indicates that the incoming colonists adopted the indigenous livestock husbandry culture, and as a result the absence of any regional diachronic variation is not so surprising. The provisioning of meat to castles and towns was reliant on existing pastoral systems in the countryside, run by indigenous communities, with the only significant change consisting of the reorientation of higher purchasing power communities from strongholds to castles and towns. There were also changes in ownership, with the Order and episcopate in Livonia representing new managers of substantial animal resources. Did numbers of livestock increase? Both zooarchaeological and written datasets are too fragmentary to be sure, but the appearance of urban centres certainly promoted a new demand for beef above all other meat, also a feature of medieval Prussia (Fig. 6.24, see also Chapter 12). The distinctiveness of the Order's sites is emphasized by access to good-quality meat and the introduction of new meat-processing technology, linked to the use of heavy cleavers, as well as diachronic

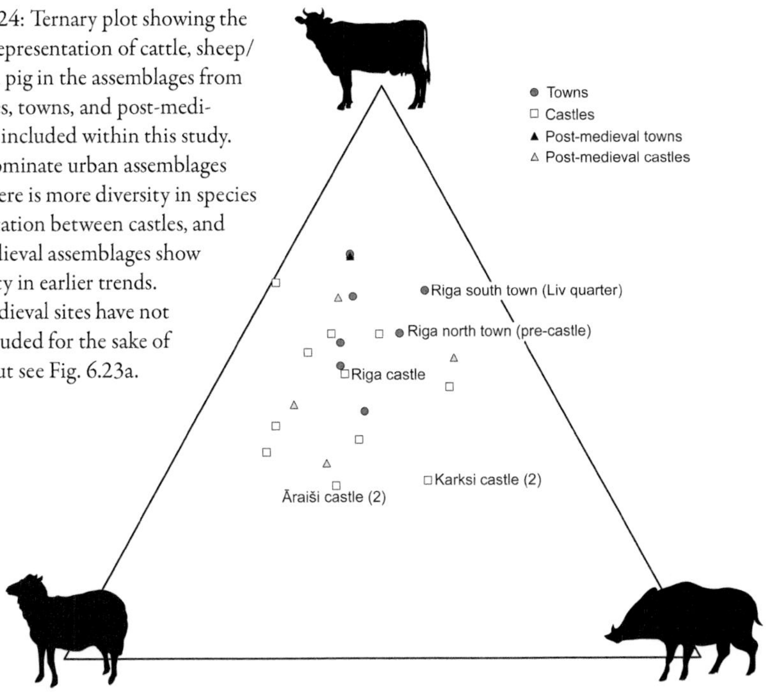

Figure 6.24: Ternary plot showing the relative representation of cattle, sheep/goat, and pig in the assemblages from the castles, towns, and post-medieval sites included within this study. Cattle dominate urban assemblages whilst there is more diversity in species representation between castles, and post-medieval assemblages show continuity in earlier trends. Early medieval sites have not been included for the sake of clarity, but see Fig. 6.23a.

shifts in husbandry strategies at individual sites reflecting local adaptations and the purchasing power of individual castle communities (e.g., Viljandi and Karksi; see Rannamäe and Lõugas 2019). The networks created by the Order in Livonia, which mapped onto those of the Hanse, facilitated the movement of raw materials (e.g., antler), meat (whether preserved wild or on the hoof domestic) and fish around the *Ordensland*, framed by the communication routes linking the Order's castles with Reval and Riga. At the same time, it can be assumed that the indigenous elites who were incorporated into the new power structures retained the same access to good quality produce, although this can only be tested with future excavations of medieval rural settlement sites in Livonia which take into account socio-topography as a clear research aim.

The exploitation of wild species does appear to have changed, at least in southern Livonia and southern Estonia, where pre-crusade levels of biodiversity are comparatively higher than in the medieval period (Paaver 1965; Maldre 2012). All of the case study micro-regions are within the heartlands of the Livonian *Ordensland*. However, the wild mammal and bird assemblages from the sites included within this study point to the presence of heterogeneous habitats within Livonian commanderies. Woodlands of varying character, wetlands, and meadows were clearly exploited. Levels of biodiversity appear to have been relatively higher in the Late Iron Age, and there is a marked decline in the relative abundance and diversity of wild species after the crusades. This could either reflect shifts in local biodiversity or be the result of changes in hunting and trapping practices. It could of course reflect both, although the only species which vanish completely from the Livonian landscape by the medieval period are the woodland refugees-aurochs. At the local scale this varied from one commandery to the next, with relatively higher levels of biodiversity present at the eastern Prussian frontier of the 'Great Wilderness', as evident from the faunal assemblages of the castles of Klaipėda and Ełk (Lyck) dating to the fifteenth and sixteenth centuries (see Chapter 12). In contrast to the impact of the Norman Conquest (Sykes 2007), the control of hunting in the eastern Baltic was not used by the Order in Livonia to assert its authority until the later fourteenth century, and the presence of wild species in the Liv quarter in Riga and the village in Mārtiņsala reinforces this.

Given the documentary evidence for the supply of furs, the most convincing scenario is that following the crusades, the theocratic rulers of Livonia fostered a hunting culture prioritizing the acquisition of fur for the international market. This was paralleled in Prussia; in *c.* 1400, fur represented 6.3% of the net profits derived from exports managed by the Order's *Großschäffer*, in fourth place after amber, rye, and wax (Sarnowsky 1993, 295, tab. 29). The depiction of squirrel hunting in the Livonian Master's chamber in Cēsis Castle is the only known representation of the exploitation of wild species by the Order and underlines the importance placed on the fur trade over the pageantry of the aristocratic hunt. At the same time, the Order's lordship in Livonia was partly articulated through the control of wild resources, echoing the better documented evidence from Prussia where this relationship is emphasized in town regulations set out during the period of active crusading in the 1230s (see Chapter 12). This is only evident in association with the Grand Master's court in Marienburg (Pol. Malbork), which contains the sole-documented (and short-lived) hunting 'park' of the Order in the fifteenth century. In late fifteenth- and sixteenth-century Livonia, there are instances where wild animals, including beavers and in one case a stag (possibly an elk rather than a red deer, which would have had to have been imported), alongside horses, are documented as gifts presented to and from officials of the Order, combining commercial value, symbolism, and exoticism, and emulating common practices in other parts of Europe (Mänd 2016).

Although hunting was carried out by members of the Order both in Prussia and Livonia, and perhaps most of all by their guests participating in the fourteenth-century *Reisen*, it is generally absent from the iconographic programmes of the Order's convents. This is in contrast with the glorification of hunting by the secular aristocracy in other parts of Europe in the fourteenth and fifteenth centuries. Indeed, the commander of the Windau convent permitted his leading vassals, the local Curonian 'kings', to hunt deer and hare in their sacred forests — an activity that resonates with the contemporary secular emphasis on performance and display. Even a prominent room in the papal palace in Avignon was famously frescoed with hunting scenes, commissioned by Clement VI in 1343 (Anheim 2008). This puts the relationship between the Teutonic Order and hunting, an activity it engaged with relatively infrequently in the eastern Baltic, into perspective. The crusading period is poorly represented in this respect. Henry's *Chronicle* suggests hunting was practised intermittently, perhaps as a meat supplement (Mugurēvičs 2009, 180), and the few sites dating to the thirteenth century (e.g., Karksi, Vecdole, Mārtiņsala) indicate it was of relatively minor importance. With the suppression of Baltic tribal resistance, the Order's efforts and those of its seneschals focused on animal husbandry, agricultural production, and forestry. However, the fragmentation and reduction of habitat associated with this agrarian regime would have had a profound impact on biodiversity. In this respect the Order contributed to the depletion of wild populations that had already been heavily exploited. The study of the Livonian borderlands will be a useful avenue for future understanding of the impact of warfare and depopulation on levels of biodiversity, for this is one region of Europe with a clear reduction in human settlement for a period of several centuries, followed by an increase that continued into the twentieth century. In this respect the period of the 'Great Wilderness' in the eastern Baltic, much like the modern reserve in Białowieża Forest or the unintended *refugium* of the post-Chernobyl landscape, represents a valuable laboratory for investigating the relationship between human activity, the dynamism of trophic cascades, and levels of biodiversity.

Works Cited

Primary Sources

Arbusow, L. 1924–26. *Die altlivländischen Bauerrechte: Mitteilungen aus der livländischen Geschichte*, Rīga: Gesellschaft für Geschichte und Altertumskunde zu Riga

Bauer, A. 1924. *Die Wartgutsteuresliste der Komturei Goldingen: Mitteilungen aus der livländischen Geschichte*, Rīga: Gesellschaft für Geschichte und Altertumskunde zu Riga

Bauer, A. 2003. Transcripts from the Sources of the History of Manors in Courland collected by Albert Bauer, digitized by the Herder Institute in Marburg (the date refers to the completion of the digitization project)

Hartmann, S. 2008. *Herzog Albrecht von Preussen und Livland (1560–1564)*, Köln: Böhlau

HCL Brundage, J. A., trans. 2003. *The Chronicle of Henry of Livonia/Henricus Lettus*, New York: Columbia University Press

HHC Fuchs, M., ed. 1853. *Das rothe Buch inter Archiepiscopalia/ Scriptores Rerum Livonicarum*, 2 vols) Rīga, Leipzig

HWC Mugurēvičs, Ē., ed. 2005. *Vartberges Hermaņa hronika*, Rīga: Latvijas vēstures institūta apgāds

LGU Bruiningk, H. von. 1908. *Livländische Güterurkunden (aus den Jahren 1207 bis 1545)*, 2 vols, Riga: Komissionsverlag von Janck und Poliewsky

LR Napiersky, J. G. L. 1881. *Die Libri Redituum der Stadt Riga*, Leipzig: Vjeelag Von Duncker and Humblot

LRC Smith, J. C., and Urban, W., trans. 2006. *The Livonian Rhymed Chronicle*, London: Routledge Curzon

LUB Bunge, F. G. von, ed. 1853–57. *Liv-, Est- und Kurländisches Urkundenbuch nebst regesten* (vols I–III), 13 vols to date, Reval: Kluge and Ströhm

Ose, I., ed. 2011. 'Cēsu pils 1582. gada revīzija', *Latvijas viduslaiku pilis* 7, 429–32

Secondary Works

Allmäe, R., and Maldre, L. 2005. 'Rõsna-Saare I kääbaskalmistu – esialgseid osteoloogilisi andmeid', *Setumaa kogumik* 3, 121–37

Anheim, E. 2008. 'La *Chambre du Cerf*: image, savoir et nature à Avignon au milieu du XIVe siècle', *Micrologus* 16, 57–124

Apala, Z., and Apals, J. 2014. 'The Vendic Hill Fort on Riekstu Kalns in Cēsis', in *Strongholds and Power Centres East of the Baltic Sea in the 11th–13th Centuries*, ed. H. Valk, Tartu: Institute of History and Archaeology at the University of Tartu, Department of Archaeology at the Institute of History of Tallinn University, 115–38

Asaris, J. 2007. 'Valgale and Pāce Hillforts in Northern Curonia', in *The Significant Detail: Europeanization at the Base of Society. The Case of the Baltic Rim 1100–1400*, eds N. Blomkvist and T. Lindström, Visby: Gotland University, 245–48

Atgāzis, M. 1999. 'Vecdoles pils nocietinājumi un apbūve', *Latvijas Viduslaiku Pilis* 1, 312–48

Aun, M., Allmäe, R., and Maldre, L. 2008. 'Pikk-kääbaste tähendusest (Rõsna küla kääbaskalmistuste materjali põhjal)', *Setumaa Kogumik* 4, 269–90

Barrett, J., Orton, D., Johnstone, C., Harland, J., Van Neer, W., Ervynck, A., Roberts, C., Locker, A., Amundsen, C., Enghoff, I. B., Hamilton-Dyer, S., Heinrich, D., Hufthammer, A., Jones, A. K. G., Jonsson, L., Makowiecki, D., Pope., P., O'Connell, T. C., de Roo, T., and Richards, M. 2011. 'Interpreting the Expansion of Sea Fishing in Medieval Europe Using Stable Isotope Analysis of Archaeological Cod Bones', *Journal of Archaeological Science* 38/7, 1516–24

Bartosiewicz, L. 2003. 'A Millennium of Migrations: Proto-Historic Mobile Pastoralism in Hungary', *Bulletin of the Florida Museum of Natural History* 44, 101–30

Bartosiewicz, L., Van Neer, W., and Lentacker, A. 1997. *Draught Cattle: Their Osteological Identification and History*, Tervuren: Annales du Musée Royal de l'Afrique Centrale, Sciences Zoologiques 281

Benecke, N. 2005. 'The Holocene Distribution of European Bison: The Archaeozoological Record', *Antropologia-Arkeologia* 57, 421–28

Berga, T. 2007. *Augšdaugavas 14.–17. gadsimtā senvietas: no Krāslavas līdz Slutišķiem*, Rīga: Latvijas Vēstures Inst. Apgāds

Bliujienė, A., ed. 2009. *The Horse and Man in European Antiquity* (= *Archaeologia Baltica* 11)

Blomkvist, N. 2009. '12th–14th century European Expansion and its Reception in the Baltic North', in *The Reception of Medieval Europe in the Baltic Sea Region*, ed. J. Staecker, Visby: Gotland University Press, 431–84

Bocherens, H., Hofman-Kamińska, E., Drucker, D. G., Schmölcke, U., and Kowalczyk, R. 2015. 'European Bison as a Refugee Species? Evidence from Isotopic Data on Early Holocene Bison and Other Large Herbivores in Northern Europe', *PLOS ONE* 10/2, e0115090 DOI: <10.1371/journal.pone.0115090>

Brūzis, R., and Spirģis, R. 2009. 'The Ogresgala Čabas Cemetery and Horse Sacrifice', *Archaeologia Baltica* 11, 283–94

Calkin, V. I. 1961. *Fauna iz raskopok srednevekovyh gorodiŝ na territorii Latvii*, Rīga: Izdatelstvo Akademij Nauk Latvijskoj SSR

Caune, A. 2007. *Pētījumi Rīgas archeoloģijā*, Rīga: LVIA

Delort, R. 1978. *Le commerce des fourrures en occident á la fin du Moyen âge: (vers 1300–vers 1450)*, 2 vols, Rome: École Française de Rome

Dudley, J. P., Ginsberg, J. R., Plumptre, A. J., Hart, J. A., and Campos, L. C. 2002. 'Effects of War and Civil Strife on Wildlife and Wildlife Habitats', *Conservation Biology* 16/2, 319–29

Dulinicz, M., and Iwaszczuk, U. 2008. 'Interdyscplinarne badania wybranych wczesnośredniowiecznych stanowisk z Mazowsza', in *Człowiek i środowisko przyrodnicze we wczesnym średniowieczu w świetle badań interdyscyplinarnych*, ed. W. Chudziak, Toruń: Wydawnictwo Naukowe Uniwersytetu Mikołaja Kopernika, 149–70

Ehrlich, F. 2016. 'Kana ja teised linnud muinasaegse Viljandi Lossimägedes. Zooarheoloogiline uurimus' (unpublished bachelor's thesis, Archaeology Archive, University of Tartu)

Ehrlich, F. 2018. 'Linnuluud Viljandi keskaegses ja varauusaegses zooarheoloogilises materjalis' (unpublished master's thesis, University of Tartu)

Ekdahl, S. 1998. 'Horses and Crossbows. Two Important Warfare Advantages of the Teutonic Order in Prussia', in *The Military Orders*, II: *Welfare and Warfare*, ed. H. Nicholson, Farnham: Ashgate, 119–51

Ermischer, G. 2006. 'Hier liegt der Hund begraben', in *Die Ketzelburg in Haibach. Eine archäologisch-historische Suprensuche*, ed. H. Rosmanitz, Neustadt a. d. Aisch: Heimat- und Geschichtsverein, 99–102

Goldschmidt, W. 1979. 'A General Model for Pastoral Social Systems', in *Pastoral Production and Society*, ed. L'Equipe écologie et anthropologie des sociétés pastorals, Cambridge: Cambridge University Press, 15–27

Haak, A., Rannamäe, E., Luik, H., and Maldre, L. 2012. 'Worked and Unworked Bone from the Viljandi Castle of the Livonian Order (13th–16th centuries)', *Lietuvos Archaeologija* 38, 295–338

Harcourt, R. 1974. 'The Dog in Prehistoric and Early Historic Britain', *Journal of Archaeological Science* 1, 151–75

Hewison A. J. M., Vincent, J.-P., Joachim, J., Angibault, J.-M., Cargnelutti, B., and Cibien, C. 2001. 'The Effects of Woodland Fragmentation and Human Activity on Roe Deer Distribution in Agricultural Landscapes', *Canadian Journal of Zoology* 79, 679–89

Jahnke, C. 2015. 'The Baltic Trade', in *A Companion to the Hanseatic League*, ed. D. J. Harreld, Leiden: Brill, 194–240

Jonuks, T., and Rannamäe, E. 2018. 'Animals and Worldviews: A Diachronic Approach to Tooth and Bone Pendants from the Mesolithic to the Medieval Period in Estonia', in *Bioarchaeology of Ritual and Religion*, ed. A. Livarda, R. Madgwick, and S. R. Mora, Oxford: Oxbow, 162–78

Kerley, G. I. H., Kowalczyk, R., and Cromsigt, J. P. G. M. 2012. 'Conservation Implications of the Refugee Species Concept and the European Bison: King of the Forest or Refugee in a Marginal Habitat?', *Ecography* 35, 519–29

Kļaviņš, K. 2009. 'The Significance of the Local Baltic People in the Defence of Livonia (Late Thirteenth–Sixteenth Century)', in *The Clash of Cultures on the Medieval Baltic Frontier*, ed. A. V. Murray, Farnham: Ashgate, 321–40

Laurinkienė, N. 2011. 'Stone-Goddesses in Granaries', *Archaeologia Baltica* 15, 56–60

Luik, H. 2003. '*Luuesemed* hauapanustena rauaaja Eestis', in *Arheoloogiaga Läänemeremaades: Uurimusi Jüri Seliranna auks*, eds V. Lang and Ü. Tamla, Tallinn: Eesti Teaduste Akadeemia Kirjastus, 153–72

——, 2015. 'Bone-Workers in Medieval Viljandi, Estonia: Comparison of Finds from Downtown and the Order's Castle', in *Everyday Products in the Middle Ages: Crafts, Consumption and the Individual in Northern Europe, c. AD 800–1600*, eds G. Hansen, S. P. Ashby, and I. Baug, Oxford: Oxbow, 91–109

Maldre, L. 2008. 'Pathological Bones amongst the Archaeozoological Material from Estonian Towns', *Veterinarija ir Zootechnika* 42/64, 51–57

——, 2012. 'Archaeozoological Material from the Keava Hill Fort and Linnaaluste Settlement Sites', *Estonian Journal of Archaeology* suppl. 1, 106–26

Maltby, M. 2019. 'Animal Exploitation in Medieval Novgorod', in *Ecologies of Crusading, Colonization, and Religious Conversion in the Medieval Baltic: Terra Sacra II*, ed. A. Pluskowski, Environmental History of the North Atlantic World, 3, Turnhout: Brepols (forthcoming)

Mänd, A. 2004. 'Beaver Tails and Roasted Herring Heads: Fast as Feast in Late-Medieval Livonia', *Medium Aevum Quotidianum* 50, 5–12

——, 2016. 'Horses, Stags and Beavers: Animals as Presents in Late-Medieval Livonia', *Acta Historica Tallinnensia* 22, 3–17

Matolcsi, J. 1970. 'Historische Erforschung der Körpergröße des Rindes auf Grund von ungarischem Knochenmaterial', *Zeitschrift Tierzüchtung und Züchtungsbiologie* 87, 89–137

Morellet, N., Van Moorter, B., Cargnelutti, B., Angibault, J.-M., Lourtet, B., Merlet, J., Ladet, S., and Mark Hewison, A. J. 2011. 'Landscape Composition Influences Roe Deer Habitat Selection at Both Home Range and Landscape Scales', *Landscape Ecology* 26, 999–1010

Mugurēvičs, A. (unpublished). 'Medieval Zooarchaeology in Latvia'

Mugurēvičs, Ē. 1985. 'Rēzeknes pils un tās apkārtnes novads 9.-17.gs.', *LZAV* 3, 48–71

——, 2002. 'Forest Animals and Hunting in Medieval Livonia', in *Core-Centre-Periphery: Medieval Europe II*, eds G. Helmig, B. Scholkmann, and M. Untermann, Basel: Folio Verlag Dr G. Wesselkamp, Archäologische Bodenforschung Basel-Stadt, 177–81

Nikžentaitis, A. 2009. 'Die Möglichkeiten der alternativen Geschichte: Das Alltagsleben im Baltikum des 13. und 14. Jahrhunderts', in *The Reception of Medieval Europe in the Baltic Sea Region*, ed. J. Staecker, Visby: Gotland University Press, 397–419

O'Connor, T. 2011. 'Animal Husbandry', in *The Oxford Handbook of Anglo-Saxon Archaeology*, eds H. Hamerow, D. Hinton, and S. Crawford, Oxford: Oxford University Press, 363–78

Orton, D. C., Makowiecki, D., de Roo, T., Johnstone, C., Harland, J., Jonsson, L., Heinrich, D., Enghoff, I. B., Lõugas, L., Van Neer, W., Ervynck, A., Hufthammer, A. K., Amundsen, C., Jones, A. K. G., Locker, A., Hamilton-Dyer, S., Pope, P., MacKenzie, B. R., Richards, M., O'Connell, T. C., and Barrett, J. H. 2011. 'Stable Isotope Evidence for Late Medieval (14th–15th C) Origins of the Eastern Baltic Cod (*Gadus morhua*) Fishery', *PLoS ONE* 6, e27568

Orton, D., Rannamäe, E., Lõugas, L., Makowiecki, D., Hamilton-Dyer, S., Pluskowski, A., O'Connell, T., and Barrett, J. 2019. 'The Teutonic Order's Role in the Development of a Medieval Eastern Baltic Cod Fishery', in *Ecologies of Crusading, Colonization, and Religious Conversion in the Medieval Baltic: Terra Sacra II*, ed. by A. Pluskowski, Environmental History of the North Atlantic World, 3, Turnhout: Brepols (forthcoming)

Ozoliņš, J., Bagrade, G., Ornicāns, A., Pupila, A., and Vaiders, A. 2007. *Action Plan for the Conservation of Eurasian Lynx (Lynx lynx) in Latvia*, Salaspils: Latvian State Forestry Research Institute 'Silava'

Paaver, K. 1965. *Forminovanie teriofauny i izmencivost' mlekopitajusich Pribaltiki v golocene*, Tartu: Tartu: Akademija Nauk Estonskoj SSR

Pales, L., and Garcia, M. A. 1981. *Atlas ostéologique pour servir à l'identification des Mammifères du Quaternaire - Herbivores*, Paris: CNRS

Pluskowski, A. G., Maltby, M., and Seetah, K. 2010. 'Potential Osteoarchaeological Evidence for Riding and the Military Use of Horses at the Castle in Malbork, Poland', *International Journal of Osteoarchaeology* 20/3, 335–43

Pluskowski, A. G., and Valk, H. 2016. 'Conquest and Europeanisation: The Archaeology of the Crusades in Livonia, Prussia and Lithuania', in *The Crusader World*, ed. A. Boas, London: Routledge, 568–92

Rannamäe, E. 2010. *A Zooarchaeological Study of Animal Consumption in Medieval Viljandi* (unpublished MA thesis, University of Tartu)

Rannamäe, E., and Lõugas, L. 2019. 'Animal Exploitation in Karksi and Viljandi (Estonia) in the Late Iron Age and Medieval Period', in *Ecologies of Crusading, Colonization, and Religious Conversion in the Medieval Baltic: Terra Sacra II*, ed. A. Pluskowski, Environmental History of the North Atlantic World, 3, Turnhout: Brepols (forthcoming)

Renga, I. 1997. *Turaidas pils osteoloģiskā materiāla anlize veikta 1997. gadā*, Turaida: Turaidas muzejrezervāts (unpublished)

——, 1999. *1998. Gadā analizētais Turaidas pils osteoloģiskā materiāla*, Turaida: Turaidas muzejrezervāts (unpublished)

Saks, P., and Valk, H. 2002. 'Loomaluude leiud Viljandi Pähklimäelt', *Viljandi Muuseumi Aastaraamat 2001*, 52–58

Sarnowsky, J. 1993. *Die Wirtschaftsführung des Deutschen Ordens in Preußen (1382–1454)*, Cologne: Böhlau

Schmidt, V. 2001. 'Tierische und menschliche Bauopfer bei den Nordwestslawen', *Studia Mythologica Slavica* 4, 25–34

Schramm, Z. 1967a. 'Kości długie a wysokość w kłębie u kozy', *Roczniki Wyzszej Szkoly Rolniczej w Poznaniu* 36, 89–105

——, 1967b. 'Szczątki kostne wczesnośredniowiecznej owcy i kozy z wykopalisk północno-zachodniej Polski', *Roczniki Wyzszej Szkoly Rolniczej w Poznaniu* 36, 135–74

——, 1976. *Bydło wczesnośredniowieczne na ziemiach Wielkopolski i Kujaw. Studium morfologiczne na podstawie wykopaliskowego materiału kostnego z Kruszwicy, Międzyrzecza i Santoka*, Poznań: Akademia Rolnicza w Poznaniu

Serjeantson, D. 2006. 'Birds: Food and a Mark of Status', in *Food in Medieval England*, eds C. M. Woolgar, D. Serjeantson, and T. Waldron, Oxford, Oxford University Press, 131–47

Shnore, E. D. 1961. *Asotskoje gorodishche*, Rīga: Izdatelstvo Akademij Nauk Latvijskoj SSR

Simpson E. H. 1949. 'Measurement of Diversity', *Nature* 163, 688

Šnē, A. 2006. 'The Economy and Social Power in the Late Prehistoric Chiefdoms in Eastern Lativa', *Archaeologia Baltica* 6, 68–78

Sterns, I. 1969. 'The Statutes of the Teutonic Knights: A Study of Religious Chivalry' (unpublished doctoral dissertation, University of Pennsylvania)

Švābe, A. 1938. *Kuršu ķoniņi un novadnieki*, Pielikumi: Straumes un avoti

Svetikas, E. 2008. 'XIV a. Pabaigos-XV a. amuletai iš apkaustyto okio nago lietuvos didžiojoje kunigaikštystėje ir kaimyniniuose kraštuose', *Lietuvos Archeologija* 34, 171–210

Sykes, N. 2007. *The Norman Conquest: A Zooarchaeological Perspective*, Oxford: Archaeopress

Teichert, M. 1975. 'Osteometrische Unt ersuchungen zur Berechnung der Widerristhohe bei Schafen', in *Zooarchaeological Studies*, ed. A. Clason, Amsterdam: New Holland, 51–69

Ullucci, D. C. 2012. *The Christian Rejection of Animal Sacrifice*, Oxford: Oxford University Press

Vaitkevičienė, D. 2011. 'Libation in Baltic Religious Practices', *Archaeologia Baltica* 11, 105–19

Valk, H. 2001. 'Besieging Constructions from 1223 in Viljandi', *Arheoloogilised välitööd Eestis/Archaeological Field Works in Estonia*, 2000, 65–79

——, 2019. 'Sacred Landscapes of Medieval Livonia: Lands of the Teutonic Order in Southern Estonia', in *Ecologies of Crusading, Colonization, and Religious Conversion in the Medieval Baltic: Terra Sacra II*, ed. A. Pluskowski, Environmental History of the North Atlantic World, 3, Turnhout: Brepols (forthcoming)

Valk, H., Pluskowski, A., Brown, A., Rannamäe, E., Malve, M., and Varul, L. 2012. 'The Castle of the Teutonic Order in Karksi: Preliminary Excavation Results', *Arheoloogilised välitööd Eestis/Archaeological Field Work in Estonia*, 2011, 47–56

Vasks, A. 2010. 'Slīmests vai ādu apstrādes rīks?', *Latvijas Vēstures Institūta Žurnāls* 2, 5–17

Vasks, A., Kalniņa, L., and Daugnora, L. 2011. 'Beltu pilskalns', *Arheoloģija un etnogrāfija* 25, 73–99

Vilcāne, A. 2001. 'Par sēļu iespējamību Dubnas baseinā', *LZAV* 5/6, 63–71

——, 2009. 'Findings of Harness Items and the Cult of the Horse in Latgallian and Selonian Territories', *Archaeologia Baltica* 11, 254–69

Vitt, V. O. 1952. 'Horses of the Pazyryk Kurgans [Lošadi Pazyrykskikh Kurganov]', *Sovetskaja Archeologija* 16, 163–205

Yeh, H. Y., Pluskowski, A., Kalējs, U., and Mitchell, P. D. 2014. 'Intestinal Parasites in a Mid-14th-Century Latrine from Riga, Latvia: Fish Tapeworm and the Consumption of Uncooked Fish in the Medieval Eastern Baltic Region', *Journal of Archaeological Science* 49, 83–89

Zemītis, G. 2009. 'The "Divine Trial" in 12th-Century Turaida in Historical Context', *Archaeologia Baltica* 11, 270–74

——, 2014. 'The Hill Fort of Satesele — A Power Centre in the Territory of the Livs in the 11th–13th Centuries', in *Strongholds and Power Centres East of the Baltic Sea in the 11th–13th Centuries*, ed. H. Valk, Tartu: Institute of History and Archaeology at the University of Tartu, Department of Archaeology at the Institute of History of Tallinn University, 445–61

7. Settlement Life in Livonia and the Impact on its Territories: The Geoarchaeological and Archaeobotanical Evidence

Rowena Banerjea and Monika Badura

A geoarchaeological approach has enabled the extent of the impact of the early colonizers in Livonia to be examined at high resolution, with specific reference to the same themes as for Prussia (see Chapter 14): the development of occupation and settlement use, moats, and the flow of effluent on sites within their immediate vicinities. High resolution analyses of sediments focused on four Livonian sites. Two sites in Latvia, the excavation of the Liv quarter in the Old Town of Riga, and the stronghold on top of Riekstu Hill in Cēsis (Ger. Wenden), present the opportunity to examine and compare aspects of co-existence between indigenous communities and the incoming colonists within the newly planned town, alongside the nature of the pre-crusade stronghold occupation. At Riekstu Hill, micromorphological analysis of organic layers adjacent to the burnt timber fortifications was conducted to understand the destruction process of the timber fortifications, and on the 'cultural layer' from the centre of the stronghold to understand the nature of the occupation. Geochemical analysis of occupation deposits from the outer bailey and moat within the adjacent castle at Cēsis contribute to our understanding of the activities and development of occupation at the site. A sediment sequence from the lake adjacent to the castle at Āraiši (Ger. Arrasch) was also analysed using palynological and geochemical techniques (Stivriņš et al. 2015). In Estonia, analysis focused on Karksi Castle (Ger. Karkus): here, the geoarchaeological investigation of a midden in Trench 1, located in the centre of the high castle, presented a unique opportunity to study the early colonizing activities of the Teutonic Order in this key administrative centre, as well as the development and diachronic use of this area within the early castle. These microscopic insights into aspects of daily life and soil formation complement the palynological (see Chapter 5), zooarchaeological (see Chapter 6), and written sources (see Chapter 8) for the Livonian sites included within this study.

The Development of Occupation and Settlement Use

The Development of Urbanism in Medieval Riga

Riga was arguably the most important Hanseatic city in the eastern Baltic. The results of geochemical data from internal and external occupation deposits, within and around several buildings in the Liv quarter in the medieval town, contribute important information about the activities and daily life of the town's inhabitants. They also provide a diachronic perspective on the development of occupation within the town, specifically during the incorporation of the pre-crusade Liv village into the medieval urban area. Founded in 1201, Riga quickly developed as a port, a base for further crusading, and a major power centre in the eastern Baltic, and by 1225, a circuit of walls enclosed the city and incorporated two pre-existing indigenous settlements within its boundaries (see Chapter 4). Excavations within the south-eastern corner of the historic centre of the Old Town revealed part of an Liv quarter of Riga, dating to the thirteenth to sixteenth centuries (see Chapter 4). The multiphase occupation was exceptionally well preserved and comprises traces of seventeen timber buildings, timber planked streets, and latrines. Samples were collected for a range of environmental analyses: geochemical, macrobotanical, pollen (see Chapter 5), and zooarchaeological (see Chapter 6) datasets to complement each other to characterize the occupation deposits and activities, and to interpret the spatial layouts of buildings (Banerjea et al. 2017).

Table 7.1: Element concentrations (PPM) and enrichment factors (EF) for selected elements from Buildings 4, 5, 15, 16, and 17, and the latrine in Riga.

Sample	Building	Layer	Ca (PPM)	Ca EF	P (PPM)	P EF	Cu (PPM)	Cu EF	Pb (PPM)	Pb EF	Sr (PPM)	Sr EF	Zn (PPM)	Zn EF
Riga 2	4	4	63822.7	23.6	14585.2	60.4	158.8	40.6	424.4	197.4	189.3	41.6	220.4	22.8
Riga 4	4	4	95555.4	35.3	16684.3	69.0	155.4	39.7	2366.9	1100.9	279.7	61.5	300.5	31.0
Riga 14	4	5	44239.9	16.3	13083.9	54.1	74.1	18.9	20.4	9.5	104.3	22.9	172.8	17.9
Riga 15	4	5	34591.5	12.8	10469.7	43.3	75.2	19.2	23.1	10.7	102.0	22.4	219.3	22.7
Riga 23	4	4	31160.9	11.5	7096.2	29.4	77.2	19.7	18.2	8.5	74.3	16.3	191.7	19.8
Riga 24	4	4	248143.8	91.6	26896.6	111.3	168.1	43.0	160.4	74.6	944.5	207.6	578.1	59.7
Riga 25	4	4	144512.3	53.3	25718.2	106.4	177.4	45.4	54.6	25.4	482.7	106.1	464.1	47.9
Riga 26	4	4	58962.8	21.8	8409.8	34.8	27.9	7.1	15.2	7.1	154.0	33.8	124.4	12.8
Riga 10	5	4	35377.7	13.1	9518.3	39.4	41.4	10.6	31.6	14.7	106.2	23.3	112.8	11.6
Riga 11	5	4	51672.8	19.1	17417.5	72.1	94.3	24.1	97.0	45.1	155.9	34.3	140.5	14.5
Riga 79	15	7	64751.8	23.9	13306.4	55.1	207.2	53.0	435.5	202.6	216.1	47.5	227.3	23.5
Riga 80	15	7	35234.7	13.0	3356.1	13.9	18.4	4.7	21.2	9.9	94.0	20.7	67.7	7.0
Riga 81	15	7	59677.5	22.0	16247.9	67.2	257.0	65.7	6788.0	3157.2	206.8	45.5	185.6	19.2
Riga 82	15	7	35592.1	13.1	2330.5	9.6	14.4	3.7	11.7	5.4	94.4	20.7	71.1	7.3
Riga 83	15	7	31446.8	11.6	6306.3	26.1	26.9	6.9	106.5	49.5	103.2	22.7	119.4	12.3
Riga 69	16	7	60392.2	22.3	10098.8	41.8	168.6	43.1	759.2	353.1	159.1	35.0	123.6	12.8
Riga 70	16	7	46384.0	17.1	4918.5	20.4	19.6	5.0	19.3	9.0	99.7	21.9	61.0	6.3
Riga 71	16	7	33019.1	12.2	5682.2	23.5	23.0	5.9	24.5	11.4	96.9	21.3	72.1	7.4
Riga 75	17		42953.5	15.9	6140.4	25.4	28.1	7.2	22.7	10.6	103.8	22.8	92.8	9.6
Riga 76	17	6	36521.2	13.5	5149.8	21.3	15.2	3.9	20.9	9.7	92.6	20.4	69.9	7.2
Riga 77	17	6	31160.9	11.5	3543.7	14.7	17.5	4.5	22.3	10.4	94.4	20.7	78.4	8.1
Riga 78	17	7	32090.0	11.8	5874.2	24.3	18.4	4.7	17.7	8.2	94.0	20.7	83.8	8.7
Riga 19	Latrine	2	54674.6	20.2	57031.4	236.0	284.5	72.8	65.0	30.2	161.8	35.6	887.0	91.6
Riga 28	Latrine	2	34091.2	12.6	52627.9	217.8	230.6	59.0	53.5	24.9	111.5	24.5	670.2	69.2
Riga 29	Latrine	2	36306.8	13.4	64708.0	267.8	206.4	52.8	35.0	16.3	88.4	19.4	612.8	63.3
Site average			**47952.9**	**17.7**	**12116.3**	**50.1**	**91.7**	**23.4**	**257.5**	**119.8**	**141.6**	**31.1**	**185.9**	**19.2**

The geochemical data (Table 7.1) show that certain buildings were less regularly maintained than others, and those that were 'clean' houses in terms of the absence of notable geochemical enrichments and build-up of macrobotanical remains (Banerjea et al. 2017). Distinctive activities relating to food preparation, storage, and metalworking have also been identified as a result of specific geochemical enrichments. Geochemical enrichments have been calculated as a factor against the surrounding surface geology (see Chapter 2), and due to particularly high factors of enrichments (consistently well above the 2.95 factor that is considered high), they are also compared to the average enrichments for the site (Table 7.1).

Building Maintenance

In comparison with Buildings 4 and 15, Buildings 5 and 17 have an absence of notable geochemical enrichments, specifically Ca (calcium), P (phosphorous), Sr (strontium), Cu (copper), Zn (zinc), Pb (lead), Sn (tin), and Ag (silver). In addition, plant macrobotanical assemblages were also significantly less abundant and diverse in Buildings 5 and 17, which also suggests that these structures were more frequently maintained than Buildings 4 and 15, and that there was a different use of space. Zooarchaeological assemblages were predominantly recovered from deposits outside the 'dirtier'

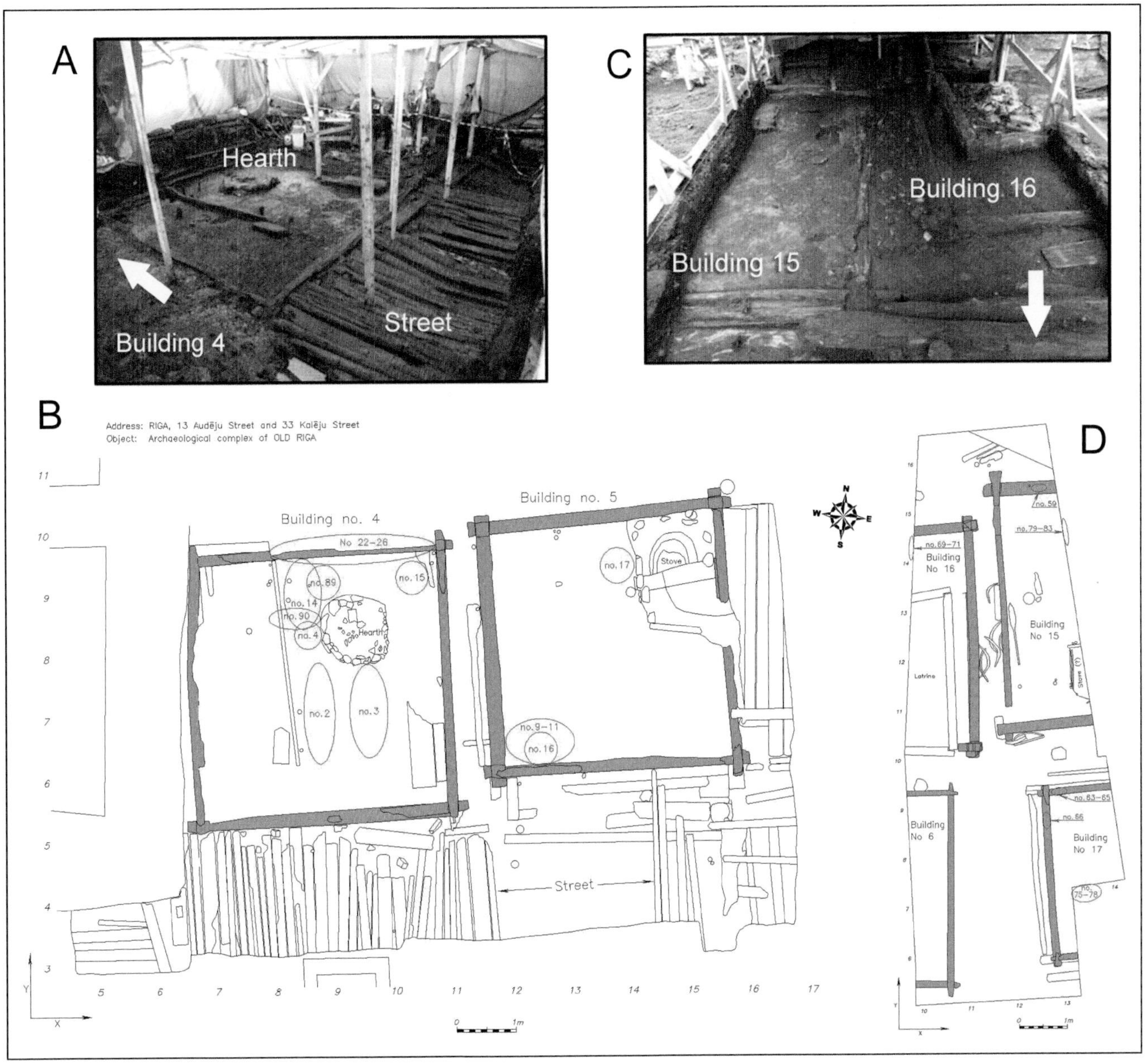

Figure 7.1: Riga excavations: (a) photograph of Building 4 during excavation; (b) plan of Buildings 4 and 5 and the surrounding external space with sample locations; (c) photograph of Buildings 15 and 16 during excavation; (d) plan of Buildings 15, 16, and 17 and the surrounding external space with sample locations. Photos and plans by U. Kalējs, modified by R. Banerjea.

or less well-maintained buildings (Banerjea et al. 2017). Buildings 4 and 5 (Fig. 7.1a, b) are situated adjacent and in close proximity to each other, and are later in date than Buildings 15, 16, and 17 (Fig. 7.1c, d), which are also in close proximity to each other, with Building 17 dated by dendrochronology to *c.* 1209 and as such being one of the earliest buildings in Riga (see Chapter 4).

Evidence for Specific Activities at Riga

As well as comparing household maintenance practices between the buildings, the integrated analysis of geochemistry and plant macro-remains identified traces of specific activities within Buildings 4 and 15, the less well-maintained buildings. Geochemical analysis is a particularly useful analytical tool for identifying in situ metalworking activities within buildings as, due to the maintenance of buildings while they are in use, larger artefacts such as broken tools and industrial by-products may be discarded into nearby rubbish pits, or into nearby abandoned buildings (LaMotta and Schiffer 1999). Geochemical analysis of sediment samples of occupation deposits within these buildings produced significantly elevated enrichments of Pb, Sn, Cu, and Ag,

Table 7.2: Key descriptive classification attributes of microstratigraphic units from Riekstu Hill, Cēsis.

Location	Slide	Unit	Deposit Classification	Formation Features	Notable Inclusions
Cultural layer, central trench	1	1	Post-depositional soil formation	*c.* 5–7.5 cm thick (*c.* 20–40 cm in the field section) Post-depositional soil formation of midden-like material and minerogenic sediment	Possible roundworm parasite egg (may indicate the presence of herbivore dung); Fragment of omnivore coprolite; Elongate smooth grass phytolith; Fragment of calcined bone; Fragment of decayed bone; Degraded fragment of brick/tile/ceramic
Cultural layer, central trench	1	2	Reworked sand substrate	Reworked blown sand substrate with evidence of bimodal sorting Soil development — well developed ped structure Very abundant root bioturbation	Mainly mineral Phytoliths include elongate, smooth cells, and rondels
Trench B	2	3	Reworked sand deposit	Unsorted, which may indicate redeposited sand rather than a blown sand origin Some soil development — weakly developed ped structure Very abundant root bioturbation	Mainly mineral Some organic inclusions including charred wood fragments, 10%, charred plant remains, <5%, amorphous charred material, 10%, and phytoliths, <2% Phytoliths include rondels, hair, dendritic, amongst the amorphous charred material
Trench B	2	4	Post-depositional soil formation of burnt material	Some soil development — weakly developed ped structure Very abundant root bioturbation and organic staining	Dominant sand component Organic inclusions are burnt and include charred wood, 20%, and amorphous charred material, 10%
Trench B	2	5	Reworked sand substrate	Reworked blown sand substrate with evidence of bimodal sorting Soil development — well-developed ped structure Very abundant root bioturbation	Occasional organic inclusions including charred wood, <5%, amorphous charred remains, <5%, plant tissue with cellulose and ferruginous, 5%, phytoliths, <2% Phytoliths include long smooth cells multi-cell formation, and single cells amongst amorphous charred remains that have been reworked in from above unit 4
Trench E	3	6	Reworked accumulations of charred organics	Soil development — well-developed ped structure Abundant root bioturbation Silty clay coatings are 'broken' indicating that they have been reworked	Dominant organic component including charred wood, 20%, charred seeds, <5%, amorphous charred remains, 20%, and phytoliths <2% Phytoliths include bilobates
Trench E	3/4	7/8	Reworked blown sands	Reworked lenses of sand, loamy sand, and sand loam. Bimodal sorting — larger quartz mineral aligned with basal boundary Limited soil development Calcite precipitation around roots	Mainly mineral Some organic inclusions including charred wood fragments, 10%, amorphous charred material, <5%
Trench E	4	9	Trampled charred remains	Strongly developed platey peds Larger charcoal frags, >0.5 mm, are moderately oriented parallel, aligned to basal boundary. Other inclusions are unoriented, unrelated, random, and unreferred Traces of microlaminations	Dominant organic component including charred wood, 40%, and amorphous charred remains, 20%

continued opposite

Location	Slide	Unit	Deposit Classification	Formation Features	Notable Inclusions
Trench E	4	10	Blown sands	Bimodal sorting Sand particles are locally oriented and clustered referred to each other. Elogated mica particles have parallel orientation strongly aligned with basal boundary Limited soil development Abundant root bioturbation	
Trench E	4	11/12	Eroded sands	Unsorted Sand particles are locally oriented and clustered referred to each other Limited soil development	Mostly quartz mineral, 70%. Unit 12 has very few fragments of charred wood, <5%, and amorphous charred remains, <5%
Trench E	5	13	Blown/ eroded sands	Mixture of two microfabric types: type 1 similar to unit 10 in formation; type 2 similar to units 11 and 12	Mostly quartz mineral, 70%
Trench E	5	14	Trampled charred remains	Strongly developed platey peds Charcoal fragments, are moderately oriented parallel, aligned to basal boundary. Other inclusions are unoriented, unrelated, random, and unreferred Traces of microlaminations	Dominant organic component including charred wood, 20%, and amorphous charred remains, 20%
Trench E	5	15	Blown/ eroded sands	Bimodal sorting Sand particles are locally oriented and clustered referred to each other. Elogated mica particles have parallel orientation strongly aligned with basal boundary	Mostly quartz mineral, 70%
Trench E	5	16	Blown/eroded material	Unsorted Inclusions are mostly unoriented, unrelated, random, and unreferred. Some charred plant remains are moderately oriented aligned parallel with basal boundary	Quartz, 30% Other rock and minerals, 40% Charred wood, 10% Amorphous charred remains, 20%
Trench E	5	17	In situ charred remains	Unsorted Charred remains are moderately oriented parallel to basal boundary. Minerals are locally oriented and clustered referred to each other Traces of microlaminations	Dominant organic component comprising charred wood, 20%, amorphous charred remains, 30%, and phytoliths, <2% Phytoliths included elongate smooth cells Clay-rich sediment aggregate with calcite, possibly precipitated ash that forms large (non faecal) spherulites at times
Trench E	5	18	Reworked poorly developed soil	Reworked blown sands Soil development Abundant root activity Traces of microlaminations, possibly originally like unit 13 but now disturbed	Mainly quartz mineral, 50% Fragments of roots, 2%
Trench E	6	19	Podzolic horizon	Soil development Abundant root bioturbation Colour: orange/dark brown (PPL); orange/greyish brown (XPL)	Mainly quartz mineral, 40%, with sediment aggregates, 20%
Trench E	6	20	Poorly developed soil	Poorly developed soil Abundant root bioturbation	Mainly quartz mineral, 70% Very few charred wood fragments, 5%, some are ferruginous Phytoliths, <2%, including elongate smooth cells

and are considered to relate to in situ non-ferrous metalworking (Banerjea et al. 2017). Spatial geochemical surveys of experimental buildings, in which non-ferrous metalworking took place, provide interesting comparative data for identifying in situ metalworking activities within Buildings 4 and 15 through the spatial distribution of enrichments (Banerjea 2008; Carey et al., 2014). Within Building 4, the greatest enrichments of lead and tin were recorded in sample 4, collected from the area directly adjacent to the hearth, suggesting that this area was a location for metalworking activities such as smelting or casting. In Building 15, a casting area may have been located away from the actual hearth. Although no tools or production by-products from non-ferrous metalworking were recovered from these buildings, previous excavations within Riga, including areas of the Liv village, have recovered wooden and stone moulds (thirteenth and fourteenth centuries) made in accordance with indigenous Late Iron Age traditions, which, as attested by experimental reconstruction and use, are only suitable for Pb or Sn alloy working (Svarāne 2002).

The area against the back-wall of Building 4 contained a deposit described during excavation as a 'limey layer' and is represented in sample 24 (Table 7.1). Samples 23–26 were collected from the northern end of Building 4, adjacent to the back wall (Fig. 7.1b). Sample 24, collected from the 'limey layer', has above average enrichment factors in Ca (x 92), P (x 111), Sr (x 208), and also the trace elements Cu, Zn, and Pb, of which Ca, P and Sr enrichments have been observed from the soils within fish processing areas and smokehouses (Knudson et al. 2004). Samples from Building 4 produced the highest amount of fish remains (N = 129), in comparison with Buildings 5, 15, 16, and 17, and sample 24 from the 'limey layer' was particularly rich (Banerjea et al. 2017).

The archaeobotanical assemblages within Buildings 4, 5, 15, 16, and 17 provide details of plant use within the Liv district of the town, specifically activities relating to the cultivation of cereals and vegetables, gathered foodstuffs and medicines, plants used in manufacture and construction, and plants used as fodder or bedding for livestock. Most of the useful plants relate to food, representing both cultivated and collected plants. The following cereals species have been identified at 33 Kalēju iela site: oat (*Avena* sp.), barley (*Hordeum vulgare*, including *Hordeum* sp.), rye (*Secale cereale)*, wheat (*Triticum aestivum*, incluing *Triticum* sp.), millet (*Panicum miliaceum)*, and foxtail millet (*Setaria italica)*. Almost all of the cereal remains were recovered from Building 4 (Banerjea et al. 2017) which, as the faunal remains and geochemical enrichments suggest, was an intensively used building, and typical of those used across medieval Europe, and particularly in Hanseatic towns, as well as in the territories ruled by the Teutonic Order (Wasylikowa et al. 1991; Karg 2007; Valk et al. 2013). These cereals, with the exception of foxtail millet, are previously known from medieval Riga as well as other sites from Latvia, dating from the eleventh/twelfth to the sixteenth centuries (Sillasoo and Hiie 2007).

Knowledge of vegetable cultivation in the area of Livonia was spread with German colonizers and Hanseatic trade (Sillasoo and Hiie 2007), and the plant macroremains indicate that it is possible that in the vicinity of the buildings there were places where the cultivation of vegetables could have been carried out; although, few remains of vegetables were recovered from the buildings, and mostly within Building 4, and comprise parsnip (*Pastinaca satvia*), cabbage (*Brassica oleracea*), and turnip (*Brassica rapa*). Fennel (*Foeniculum vulgare*) and black mustard (*Brassica nigra*) represent cultivated spices (Renfrew and Sanderson 2005). The latter was preserved in Buildings 4 and 15 in the form of crushed seeds, which may indicate food preparation. The fruits of hop (*Humulus lupulus*) were recorded only in Buildings 4 and 15, and may represent the production of beer (Behre 1999; Miśkiewicz 2008), but it is unclear if these remains belonged to the cultivated or collected plant.

Gathered foodstuffs are represented by the occurrence of 'wild' spices, the inclusion of which could be accidental, and fruits and nuts, which were preserved in the buildings in very low quantities and include fruits such as wild strawberry (*Fragaria vesca*), raspberry (*Rubus idaeus*), dewberry (*Rubus caesius*), bilberry (*Vaccinium myrtillus*), and hazel (*Corylus avellana*). No exotic fruits were noted, and this reinforces the characterization of the Liv district as being inhabited by a community of low socio-economic status. Plant materials were also gathered for other purposes such as for use as animal fodder, medicines, and construction. The flowers of white clover (*Trifolium repens)* and red clover (*Trifolium pratense)* occur in Building 4 and could indicate places where people stored fodder (Rutkowska 1984). Needles of Norway spruce *Picea abies* occur and could represent timber collected for the construction of buildings or pieces of furniture and other equipment. Rushes (*Juncus* sp.) and bulrush (*Typha* sp.) seeds occur relatively frequently in the samples and may represent roofing material or matting materials (Rutkowska 1984). Species such as St John's wort (*Hypericum perforatum)* or

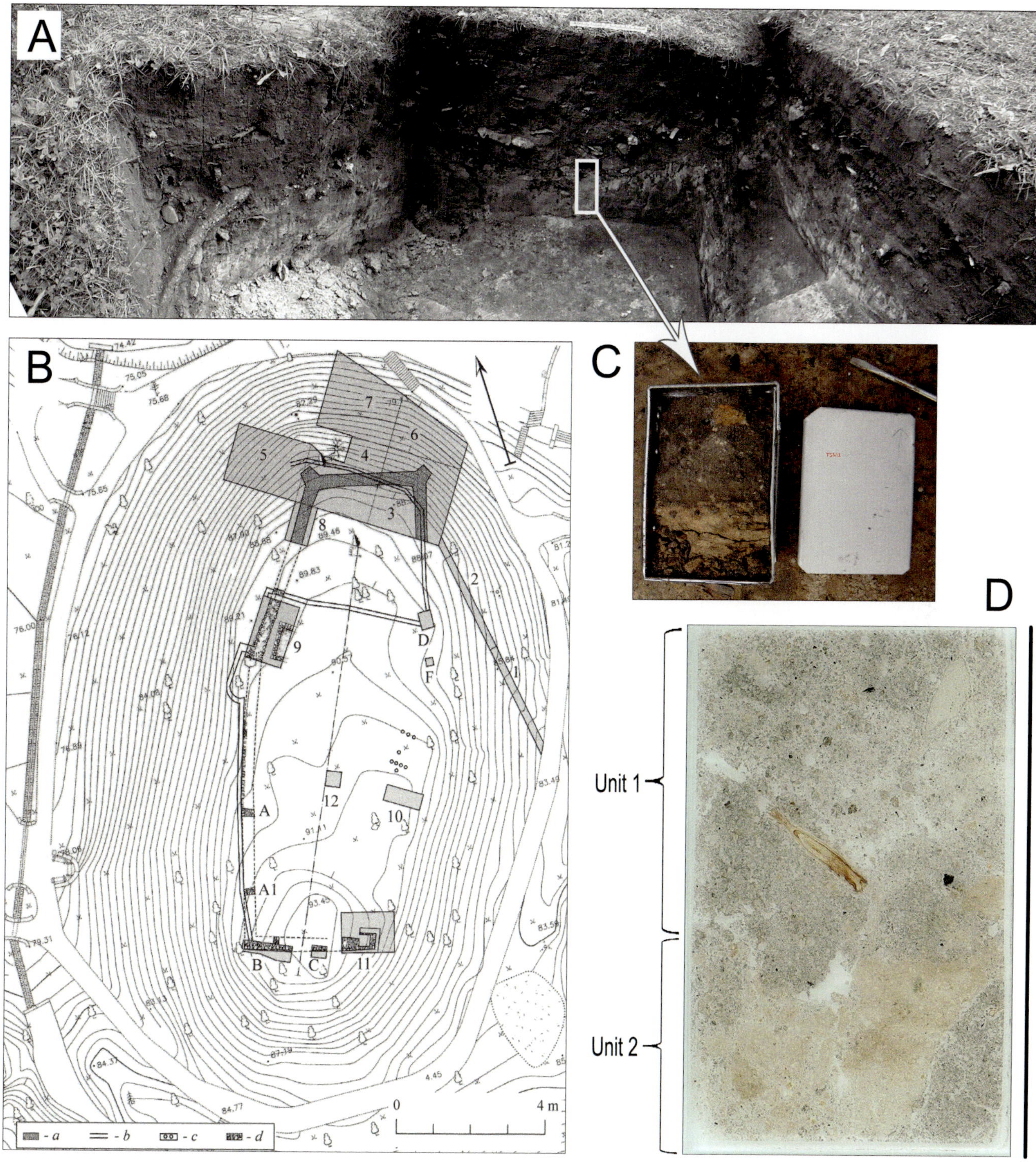

Figure 7.2: Riekstu Hillfort, Cēsis: (a) photograph of the profile through the 'cultural layer'; (b) plan of the hillfort with the location of excavation trench (Trench 12); (c) micromorphology sample collected from the 'cultural layer' and the underlying substrate; (d) scan of the micromorphology sample with units 1 and 2 marked. Photos by R. Banerjea, plan by Z. Apala.

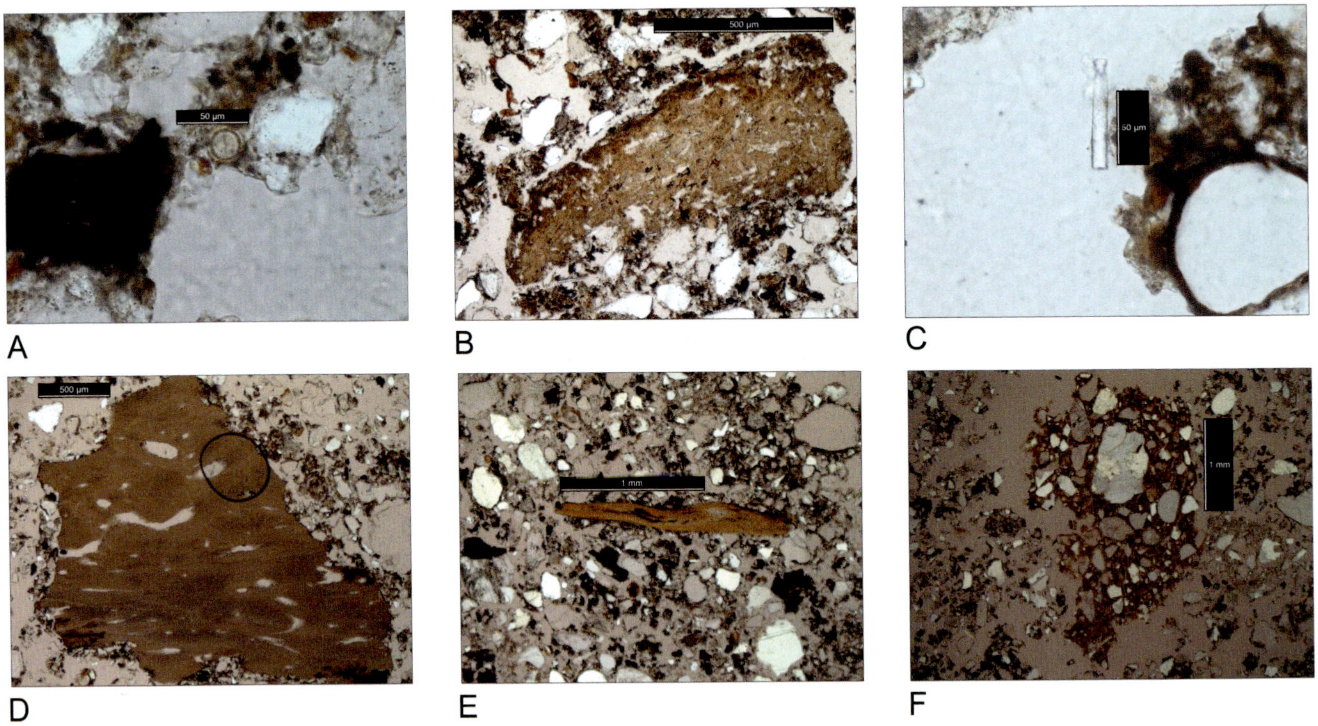

Figure 7.3: Photomicrographs of inclusions within the 'cultural layer', Riekstu Hillfort, Cēsis:(a) a possible roundworm parasite egg (that may indicate the presence of omnivore/herbivore dung); (b) a fragment of coprolite; (c) elongate smooth grass phytolith; (d) fragment of calcined bone; (e) fragment of decayed bone; (f) a degraded fragment of brick/tile/ceramic. Photos by R. Banerjea, plan by Z. Apala.

bilberry could be used as natural dyes. Common bearberry (*Arctostaphylos uva-ursi)* contains a significant content of tannins within its leaves and could have been used in tanning; however, it is also a species that has been used for centuries in folk medicine as a mild diuretic and astringent (Grieve 1971; Podbielkowski 1992; Podbielkowski and Sudnik-Wójcikowska 2003). The presence of moss remains is interesting. Their composition indicates that they were gathered from a variety of locations. Most of them represent epiphites, which grow on trees. Mosses have multiple uses; they could be used as matting, woven and used to make baskets, seals, fillings, for latrine hygiene, or dressings for wounds.

Despite Riga being well established within long-distance trade networks during the twelfth and thirteenth centuries, and the apparent existence of a group of wealthy people with access to variety and expensive food within the town, the Liv district was inhabited by poorer people of a low socio-economic status. The archaeobotanical results show that the local diet within the Liv district was based on traditional products that were cultivated or gathered in the vicinity of the town, rather than reflecting the consumption of exotic produce. Non-ferrous Pb and Sn alloy metalworking took place in Buildings 4 and 15. These metalworking activities also seem to have been organized at the household level alongside other activities such as fish processing, the storage of cultivated crops, and animal fodder or bedding, alongside archaeobotanical evidence for kitchen waste.

The Development of Occupation at Cēsis: A Place of Co-Existence or Conflict?

Henry's *Chronicle of Livonia* states that the stronghold at Riekstu Hill was occupied by the Vends who voluntarily converted to Christianity in 1206. The Sword Brothers established a base here and co-existed alongside the Vends from 1208 until, in 1218, they constructed their stone castle alongside the timber fort on Riekstu Hill (Apala and Apals 2014; Kļaviņš 2012, 10; see Chapter 4). Micromorphological analyses of occupation and potential destruction deposits associated with the timber fortifications on Riekstu Hill contribute to examining the validity of the suggestion of religious assimilation and peaceful co-existence between the Vends and the Sword Brothers at Wenden during the thirteenth century. Geochemical and archaeobotanical analyses of occupation deposits from a structure in the outer bailey of the stone castle, considered during excavation to be a

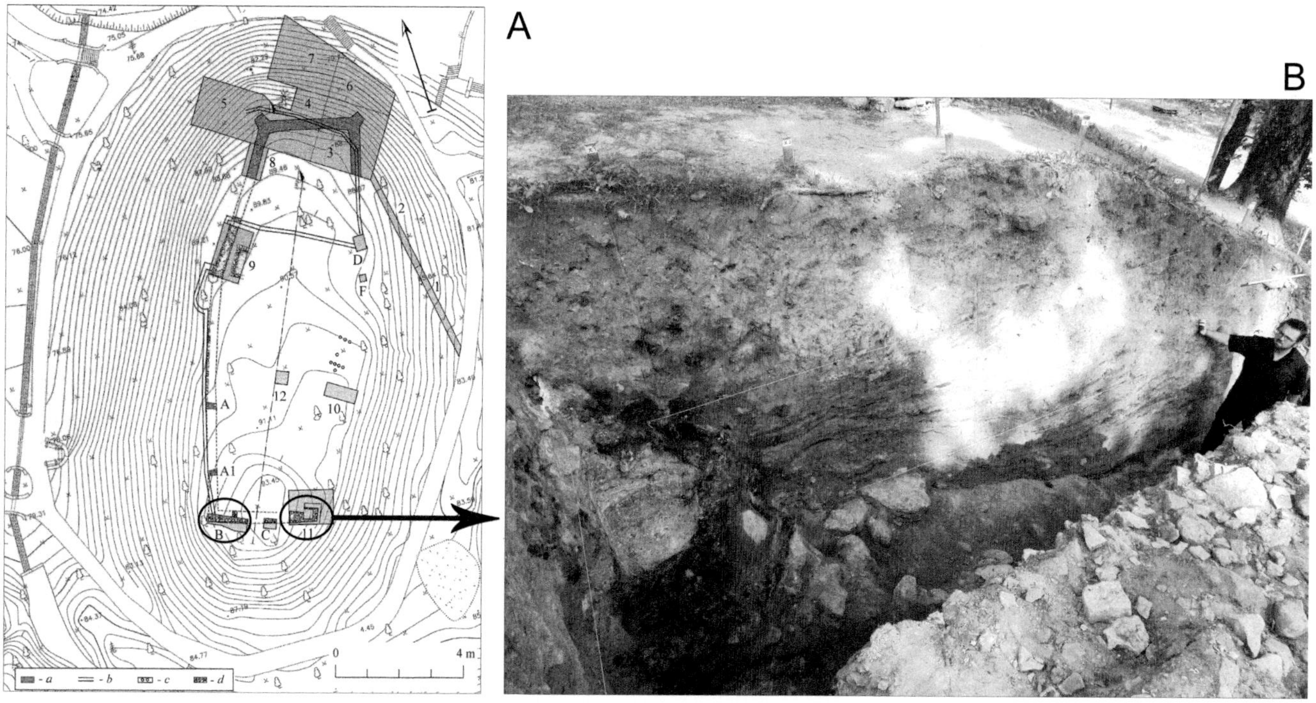

Figure 7.4: Riekstu Hillfort, Cēsis: (a) plan of the hillfort with the location of excavation Trenches B and E (circled); (b) photograph of the profile through Trench E. Photo by A. Pluskowski, plan by Z. Apala.

stable, contribute to the understanding of the use of the service buildings within the outer bailey following the expansion of the castle by the Teutonic Order in the later fourteenth century (Kalniņš 2014, 4).

The Activities and Occupation on Riekstu Hill

If the settlement on Riekstu Hill functioned as a typical stronghold, there should have been an associated settlement below, but none of that has survived or been documented archaeologically. During excavations in 2009 by Zigrīda Apala, Trenches B and E were cut through the fortifications to ascertain the extent, character, and depth of the stone wall, and to recover data associated with its construction. The stone wall is thought to have been constructed by the Sword Brothers or during their occupation, replacing the earlier timber fortification (Apala and Apals 2014; Kalniņš 2014, 2). A rubbish spread or 'cultural layer' was excavated in the middle of the stronghold. The U-series date from animal bone in the 'cultural layer' produced dates from the mid-thirteenth century (see Chapter 4; Appendix B), and the cut marks on the bones indicate a typical indigenous technology pattern (Seetah et al. 2013). A micromorphology sample, slide 1 (Table 7.2), was collected from the 'cultural layer' in Trench 12 (Fig. 7.2a, b), which resembled a 'dark earth', to determine the nature of the rubbish spread, specifically its composition, if it was deposited over time, or if there was one depositional event representing a single period of occupation. Micrormorphological analysis shows that slide 1 comprises two units (Table 7.2; Fig. 7.2c, d): unit 1 is a soil horizon that has formed from the decomposition and biological reworking of midden-like material, by similar processes to those ascribed to a 'dark earth', *sensu lato* (Table 7.2; Fig. 7.3); and minerogenic sediment from the aeolian sand substrate, unit 2. The study and definition of 'dark earth' has been a much debated issue during the last decades and has seen its definition explored beyond characterizing thick, dark, homogenous units within urban stratigraphy that mark the transition between the late Roman and early medieval periods in the United Kingdom (Macphail et al. 2003; Yule 1990) to understand its origin and formation (Borderie et al. 2015; Nicosia and Devos 2014), and to understand the differences and similarities between European and Amazonian 'dark earth' (Arroyo-Kalin et al. 2003; Nicosia and Devos 2014). 'Dark earth' is not a scientific soil classification term but as 'archaeological jargon' it is useful for identifying these diverse deposits on account of their common features (Nicosia and Devos 2014). Micromorphology has played an integral role in numerous interdisciplinary scientific studies that have examined the genesis and formation of 'dark earth' (Borderie et al. 2015; Devos et al. 2009). 'Dark earths' develop as a result of pedogenesis, which reworks the underlying parent material of occupation materials and archaeological strata, and provide a lens through which the former activities and land-use within an urban area can be ascertained (Borderie et al. 2015; Nicosia and Devos 2014); however, 'dark earths' need to be studied on a case-by-case basis as there is not one single model that explains them all, and importantly, interpretations must be made on the basis of context (Goldberg and Macphail 2006, 271–73; Nicosia and Devos 2014). Experimental archaeology has demonstrated archaeological deposits in external or unroofed spaces in temperate environments can be transformed into soil within 2.5 years (Banerjea et al. 2015). The dark soil, unit 1, that forms the 'cultural layer', as at Biała Góra in north Poland (see Chapter 14), formed on sand and most probably represents a dark earth forming under short grassland (Pluskowski et al. 2014; Macphail and Crowther 2009), reflecting a short period of intense occupation. This was potentially formed within 2.5 years and, as a buried soil, will have been continuously reworked by pedogenesis (French 2003; Retallack 2001).

Five micromorphology samples, slides 2–6 (Table 7.2), were collected from Trenches B and E to determine the formation, destruction, and composition of units that, during excavation, were speculatively described as 'wood lenses'. More specifically, the samples were collected to ascertain:

- Did the associated timber structure (the earlier fortifications) consist of multiple structures that were destroyed?
- Were the timber fortifications rebuilt or was this was one event?
- What is the composition of the 'wood lenses'?
- Is there any evidence that grass/turf and soil coverings may have been included alongside the timbers?

Both trenches show an Aeolian sand substrate, which has undergone podzolization, unit 5, Trench B, and soil development in Trench E, units 18–20. It is overlaid by a series of charred organic units, dumped sand, blown sand, and trampled materials (Table 7.2), and there is evidence of soil development in all units. Slides 3–6, Trench E, show a much more complex microstratigraphic sequence than slide 2, from Trench B, which most probably results from material being eroded, blown, and trampled down the steeper slope on the side of Trench E (Fig. 7.4).

Unit 17, Trench E, has been classified as an in situ formation of charred organic remains, where the charred remains have landed on the ground in close proximity to the burning of the fortifications (Table 7.2), that is, a deposit that has not been removed from its initial or primary place of deposition (*sensu stricto* Schiffer 1987). The parallel orientation of charred remains, aligned to the basal boundary, and microlaminations, indicative of period accumulation over time, are sediment characteristics that occur in primary contexts within hearth areas (Banerjea et al. 2015, 95–96, 102; Goldberg and Macphail 2006; Matthews 1995). This is the only unit with a dominant charred organic component, >50%, within the microstratigraphic sequence that shows attributes associated with primary deposition, and so suggests that unit 17 represents a burning event associated with the initial destruction of the timber fortifications. Unit 17 also contains calcitic ashes (Table 7.2), which have been re-precipitated within the unit as a result of chemical weathering (Weiner 2010). All later units that contain charred wood and organics show evidence of being blown or trampled into their place of deposition, and contain lower frequencies of fragments of charred wood and amorphous charred organics (Table 7.2).

The organic component of units 4, 6, 9, 14, and 17, which were described during excavation as 'wood lenses', consists of varying frequencies of charred wood, amorphous charred material, and charred seeds (unit 6) and phytoliths (units 6 and 17). The phytoliths within these units represent grass species. Bilobate forms (unit 6) are short-cell phytoliths that occur within grasses that prefer a damp habitat (Twiss 1992; Barboni et al. 1999). Elongated, smooth cell forms occur within the leaf or stem of taller, cereal-type wild and/or domesticated grasses (Twiss 1992). Consequently, it is plausible that grass/turf and soil coverings may have been included alongside the timbers during construction.

Phytoliths also occur in units 3, 5, and 20 (Table 7.2), which are predominantly minerogenic and show attributes of pedogenesis, and represent the taller, cereal-type wild and/or domesticated grasses (Twiss 1992). Phytoliths within these may result from short periods of grass growth when these soil units stabilized at the surface and briefly supported vegetation, with the caveat that some of these phytoliths may have been reworked into the minerogenic units from the organic lenses. Stability horizons could represent intermittent periods of low activity around the hillfort, or simply that soils did not develop well on the sandy substrate, particularly in conjunction with erosion processes on the slope.

Table 7.3: List of taxa — Cēsis, outer bailey; abundance (determined separately for each of the site): + - single. Sample no. 307/cont. 044, in which no diasporas were found was not included.

TAXA	152 cont. 020	312 cont. 044	292 cont. 053	297 cont. 053	298 cont. 053	303 cont. 055
USEFUL PLANTS						
Cereals						
Avena sp.; oat	-	-	-	-	-	+
Cerealia indet.; cereals	-	+	+	-	+	+
Hordeum vulgare L.; barley	-	+	-	-	-	+
Secale cereale L.; rye	-	+	-	-	-	-
Triticum sp.; wheat	+	-	-	-	-	-
Oil and Fibre Plants						
cf. *Canabis sativa* L.; hemp	-	+	-	-	-	-
Fruits and Berries						
Rubus idaeus L.; raspberry	-	-	-	-	+	-
NON-USEFUL PLANTS						
Galium spurium L.; false cleavers	-	-	+	-	+	+
Fumaria officinalis L.; common fumitory	-	-	+	-	+	-
Carex sp.; sedge	+	-	-	-	-	-
Galeopsis sp.; hemp nettle	-	-	-	+	-	-

Table 7.4: Geochemical enrichment factors for selected major and trace elements, outer bailey, Cēsis.

Sample Information			Major Elements					Trace Elements			
Sample no.	Context no.	Notes	Mg EF	P EF	Ca EF	Mn EF	Fe EF	Cu EF	Zn EF	Sr EF	Pb EF
CES 153	20/22	Red clay, AD 1599, post-medieval	6.43	3.20	13.12	1.50	1.69	2.00	3.67	1.40	2.25
CES 267	34	Charcoal-rich, sixteenth century, post-medieval	9.78	15.70	28.30	5.00	1.34	18.53	16.20	1.91	6.27
CES 288	33	Mixed rubble layers	9.25	12.30	24.14	2.50	1.59	4.80	14.37	1.68	2.16
CES 294	53	Charcoal-rich, organic, fish scales and bone	15.25	14.40	46.38	3.50	2.29	9.47	8.61	2.09	4.90
CES 296	53	Charcoal-rich, organic, fish scales and bone	8.68	34.40	25.82	5.50	2.79	17.73	21.65	2.45	9.12
CES 302	53/55		10.28	34.00	45.14	8.75	2.77	30.93	29.37	3.08	7.45
CES 306	55	Organic, charcoal-rich	6.20	8.90	13.94	1.75	1.27	7.07	3.35	1.49	1.47

The Use of the Outer Bailey, Cēsis Castle: Integrated Plant Macrobotanical and Geochemistry Evidence

Samples for plant macrobotanical and geochemical analysis were collected from the 'first bailey' of the stone castle at Cēsis, which is located on the southern side of the main castle building (Kalniņš 2014, 44; see Chapter 4). There are two additional baileys or wards relating to castle complex: the second bailey on the north-eastern corner of the complex, for which the earliest written evidence dates to 1590; and a third bailey on the northern side of the main castle building dating to the sixteenth century (Kalniņš 2014, 48–49). During the medieval period the outer bailey was the castle's area with the high concentration of human activity and included, for example, workshops, stables, or storage facilities. This ward was intensively used and transformed, and so was regularly disturbed, often containing wet areas. Consequently, it would be a place where mainly ruderal

Figure 7.5: Section through the middle of Trench 1, Karksi. Sample 242 was collected from deposits above sample 241, and was sub-sampled (right) for micromorphology and phytolith analysis. Photos by R. Banerjea.

vegetation could grow. If the outer bailey at Cēsis was an area of intense human occupation, the expectation would be that it would contain abundant quantities of plant remains, such as the fruits and seeds of segetal weeds brought into the ward mixed in with cultivated plants, alongside meadow plants transported as sources of fodder. However, both the plant macrobotanical and geochemical data indicate discrete middening relating to the disposal of kitchen waste in the area that was sampled, rather than this being a place of intensive occupation.

Within the outer bailey, seven archaeobotanical samples, which represent four contexts dated to the fourteenth to sixteenth centuries, were collected. Waterlogged timbers, one of which was dated to *c.* 1418 by dendrochronology, were preserved at the base of the sequence but with no organic layers. The above contexts that were sampled contained well-preserved organic material with visible charcoal fragments as well as animal bones and fish scales. However, they were surprisingly sparse in terms of plant macrofossils (Table 7.3). The dominant taxa represent cultivated plants. The remains of rye (*Secale cereale*), barley (*Hordeum vulgare*), wheat (*Triticum* sp.), oat (*Avena* sp.), and cereals that could not be identified to the level of species are preserved as single charred grains, whereas the fruit of hemp (*Cannabis sativa*) – oil and fibre plants – were preserved in waterlogged conditions. Most of these remains occur in context 44, which is rich in charcoal fragments. In addition, two stones of raspberry (*Rubus idaeus*), one of the most commonly collected plants in medieval Europe, were found in context 53. Other taxa belong to the group of weeds, specifically false cleavers (*Galium spurium*), which is associated with winter crops, while common furmitory (*Fumaria officinalis*) can be found in gardens and rarely in fields.

The limitations of the archaeobotanical material make it difficult to describe the function of the area within the castle's outer bailey in any detail. It is possible that elements of the castle's waste were discarded nearby, including the remains of any fires or hearths: unburnt plant diasporas could have been transported to the castle along with agricultural products. This hypothesis is supported by the geochemical analysis from the outer bailey.

Geochemical enrichment factors that are more than 2.95 times the background mean value were considered high; enrichment factors between 2.50–2.95 were considered medium level; and enrichment factors between 1.5–2.5 were considered low level (see Chapter 2). All samples from the outer bailey show particularly high geochemical enrichment factors of Mg (magnesium), P (phosphorus), Ca (calcium), Mn (manganese), Fe (iron), Cu (copper), Zn (zinc), Sr (strontium), and Pb (lead), with enrichment levels well above the 2.95 enrichment factor that is considered to be high (Table 7.4). Samples from contexts 53 and 55 show the highest geochemical enrichments, particularly samples 296 and 302, which have P enrichments thirty-four times, Ca up to forty-five times, and Cu and Zn up to thirty-one times the background levels (Table 7.4). The samples from the post-medieval layers are again enriched in these elements, but not as high as samples 296 and 302 (Table 7.4). Contexts 53 and 55 were described during excavation as organic, charcoal-rich deposits that contained bone and fish scales. The geochemical enrichments of P, Mn, Ca, Cu, Zn, Sr, and Pb reflect the deposition of midden material containing organic refuse, bone, and fish remains (Knudson et al. 2004; Wilson et al. 2005), and the range of elements that is enriched compares favourably to the fish processing waste in Building 4 ('limey layer', sample 24), Riga (Table 7.1).

During the Teutonic Order's use of the castle from the mid-thirteenth to mid-sixteenth centuries, the outer baileys or wards had various functions, which included the location of workshops, stables and storage facilities, comparable to the use of the Order's conventual castle wards in Prussia. These areas were intensively used and transformed. Ruderal vegetation dominated wet and disturbed soils, enriched in nitrogen, within these wards, whilst fruits and the seeds of segetal weeds were brought in along with cultivated plants and stored for local consumption and export. Meadow plants were brought in as sources of fodder. The plant macroremains and geochemical results from the first bailey at Cēsis indicate that the excavated area was used as a midden where organic refuse, bone and fish remains were deposited.

The Early Activity and Development of Occupation at Karksi Castle

Micromorphology has been instrumental in interpreting how the early colonists used and developed the area that later became the high castle. Micromorphology samples were collected from Trench 1 (Fig. 7.5a) during the 2012 excavation (Valk et al. 2012). The excavation of Trench 1 began in 2011 as a 20 m^2 plot, which focussed on the southern half (8.5 m^2) during the 2012 season as it contained a substantial quantity of cultural layers (Fig. 7.6a). Monoliths 241 and 242 (Fig. 7.5a, 7.6b) were sub-sampled for micromorphological analysis. Monolith 241 was collected from the earliest layers, which represent the activity of the earliest castle community, through the infilling of the shallow, basin-like depression prior to the construction of the high castle, from which deposits occur in monolith 242. Samples for radiocarbon dating were collected from monolith 241, and the modelled dates suggest that the deposits, which include the deposition of midden material, formed between *c*. AD 1272 and 1290 (Valk et al. 2012, 79).

Micromorphology has been able to examine the following hypotheses that were proposed about the formation of the deposits in this area following excavation and analysis of the zooarchaeological, pollen, and plant macro-remains (Valk et al. 2012, 88–90):

- Prior to occupation, a natural, spring-fed, shallow (0.7 m) basin-like depression (diameter 7–8 m) existed. This may have formed a small pool, and during excavation natural spring water entered the western part of the trench.
- The basin-like depression was filled with redeposited clay, which contained residual building materials, between 1272 and 1280 AD. The area continued to be wet.
- An anthropogenic peat deposit formed (context 13) from successive rapid dumping events of organic materials.

Table 7.5: List of taxa — Karksi, Trench 1; abundance: + - single, ++ - numerous, +++ - abundant.

Taxon	106 cont. 5	107 cont. 6	108 cont. 7	139 cont. 8	226 cont. 10	225 cont. 1
USEFUL PLANTS						
Cereals						
Avena sp.; Oat	+	–	–	–	–	–
Cerealia indet.; cereals	+	–	+	–	–	–
Hordeum vulgare L.; barley	+	–	–	–	–	–
Secale cereale L.; rye	+	–	+	–	–	–
Triticum aestivum L.; wheat	–	–	–	–	–	–
Vegetables						
Brassica rapa L.; turnip	–	–	–	–	–	–
Oil and Fibre Plants						
Cannabis sativa L.; hemp	–	–	–	–	–	–
Fruits, Berries and Nuts						
Cerasus avium (L.) Moench/ *vulgaris* Mill.; sweet cherry/sour cherry	–	–	–	–	–	–
Corylus avellana L.; hazel	–	–	–	–	+	+
Fragaria vesca L.; wild strawberry	–	–	–	–	–	–
Rubus idaeus L.; raspberry	–	–	++	–	–	+
NON-USEFUL PLANTS						
Segetal and Ruderal Weeds						
Chenopodium album (L.) s. str; white goosefoot	–	+	–	+	–	–
Epilobium cf. *roseum* Schreb.; pale willowherb	–	–	–	–	–	–
Fallopia convolvulus (L.) Á. Löve; black-bindweed	–	–	–	–	–	–
Galeopsis tetrahit L.; common hemp-nettle	–	–	–	–	–	–
Galeopsis cf. *tetrahit* L.; common hemp-nettle	–	–	–	–	–	–
Lapsana communis L.; nipplewort	–	–	–	–	–	–
Melandrium album (Mill.) Garcke; white campion	–	–	–	–	–	–
Polygonum aviculare L.; knotgrass	–	–	–	–	+	–
Polygonum hydropiper L.; water pepper	–	–	–	–	–	–
Polygonum lapathifolium L.; pale persicaria	–	–	–	–	–	–
Plantago major L.; greater plantain	–	–	–	–	–	–
Rumex acetosella L.; sheep's sorrel	–	–	–	–	+	–
Rumex crispus L.; curled dock	–	–	–	–	–	–
Spergula arvensis L.; subsp. *sativa* (Boenn.) Čelak; corn spurry	–	–	–	–	–	–
Spergula arvensis L.; subsp. *maxima* (Weihe) O. Schwarz; corn spurry	–	–	–	–	–	–
Urtica dioica L.; common nettle	–	–	–	–	–	–
Viola arvensis Murray/tricolor L.; field/wild pansy	–	–	++	–	+	–
Achillea millefolium L.; s. str; yarrow	–	–	–	–	–	–
Filipendula ulmaria (L.) Maxim.; meadowsweet	–	–	–	–	–	–
Linum catharticum L.; fairy flax	–	–	–	–	–	–
Lychnis flos-cuculi L.; ragged robin	–	–	–	–	–	–

228 ·nt. 12	230 cont. 12	232 cont. 12	236 cont. 12	240 cont. 14	233 cont. 15	234 cont. 15	126 no cont.	134 no cont.	149 no cont.	159 no cont.	183 no cont.	198 no cont.
−	−	−	−	−	−	−	−	−	−	−	−	−
−	−	+	−	−	−	−	−	−	−	−	+	−
−	−	−	−	−	−	−	−	+	−	+	+	−
−	−	−	−	−	−	−	−	++	−	−	+++	−
−	−	+	−	−	−	−	−	−	−	−	+	+
−	−	+	−	−	−	−	−	−	−	−	−	−
+	−	−	−	−	−	−	−	−	−	−	−	−
−	−	−	+	−	−	−	−	−	−	−	−	−
+	+	+	+	−	−	+	−	−	−	+++	−	++
−	−	−	−	−	−	−	−	−	+	+	−	−
+	−	−	−	−	−	−	−	+	−	−	−	−
−	−	+	+	−	−	−	−	−	+	+	−	−
−	−	−	−	−	−	−	−	−	−	−	+	−
−	−	−	+	−	−	−	−	−	−	−	−	−
−	−	−	+	−	−	−	−	−	−	−	−	−
−	−	−	+	−	−	−	−	−	−	−	−	−
−	−	+	−	−	−	−	−	−	−	−	−	−
−	−	−	−	−	−	−	−	−	−	+	−	−
−	−	−	−	−	−	−	−	+	−	−	−	−
−	−	+	−	−	−	−	−	−	−	−	−	−
−	−	+	−	−	−	−	+	−	−	+	−	−
−	−	−	+	−	−	−	−	−	+	−	−	−
−	−	−	−	−	−	−	−	−	−	−	−	−
−	−	+	−	−	−	−	−	−	−	−	−	−
−	−	+	−	−	−	−	−	−	−	−	−	−
−	−	+	−	+	−	−	−	−	−	−	−	−
−	−	−	−	−	−	−	−	−	−	−	+	−
−	−	−	+	−	−	−	−	−	−	−	−	−
−	−	−	−	−	−	−	−	−	−	+	−	−
−	−	−	+	−	−	−	−	−	−	−	−	−
−	−	+	−	−	−	−	−	−	−	−	−	−
−	−	−	−	−	−	−	−	−	−	++	−	−

continued overleaf

Taxon	106 cont. 5	107 cont. 6	108 cont. 7	139 cont. 8	226 cont. 10	225 cont.
Prunella vulgaris L.; self-heal	–	–	–	–	–	–
Potentilla erecta (L.) Raeusch.; tormentil	–	–	+	–	–	+
Ranunculus acris L.; meadow buttercup	–	–	–	–	+	–
Ranunculus repens L.; creeping buttercup	–	–	–	–	–	–
Scirpus sylvaticus L.; wood club-rush	–	–	–	–	–	–
Silene vulgaris (Moench) Garcke.; bladder campion	–	–	–	–	–	–
Valeriana officinalis L.; valerian	–	–	–	–	–	–
Plants from Woodlands and Forest Edges						
Betula sect. *Albae*; birch	–	–	–	–	+	–
Picea abies (L.) H. Karst; Norway spruce	–	–	–	–	–	–
Pteridium aquilinum L.; bracken	–	–	–	–	–	–
Quercus sp.; oak	–	–	–	–	–	–
Salix sp.; willow	–	–	–	–	–	–
Plants from Wet and Moist Habitats						
Carex elata All.; tufted sedge	–	–	–	–	–	–
Carex elata All./*nigra* Reichard; tufted sedge/common sedge	–	–	–	–	+	–
Carex nigra Reichard; common sedge	–	–	–	–	–	+
Carex ovalis Gooden.; oval sedge	–	–	–	–	–	–
Carex sp.; sedge	–	–	+	+	+	–
Juncus sp.; rush	–	++	++	–	–	–
Mentha cf. *aquatica* L.; water mint	–	–	–	–	–	–
Schoenoplectus lacustris (L.) Palla; common club-rush	–	–	–	–	–	–
Sphagnum sp.; peat moss	–	–	–	–	–	–
Typha sp.; bulrush	–	–	–	–	–	–
Other Plants						
Astragalus sp.; milkvetch	–	–	–	–	–	–
Caryophyllaceae indet.; pink family	–	–	–	+	+	–
Chara sp.; stonewort	–	–	–	–	–	–
Chenopodium sp.; goosefoot	–	+	++	+	–	+
Chenopodiaceae indet.; goosefoot family	–	+	–	–	–	–
Hypericum sp.; St John's wort	–	–	–	–	–	–
Lamiaceae indet.; mint family	–	–	–	–	–	–
Mentha arvensis L./*aquatica* L.; wild mint/water mint	-	-	-	-	-	-
Poaceae indet.; grass family	-	+	-	-	-	-
Potentilla sp.; cinquefoil	-	-	-	-	-	-
Ranunculus sp.; buttercup	-	-	-	-	-	-
Rosaceae indet.; rose family	-	-	-	-	-	-
Silene vulgaris (Moench) Garcke; bladder campion	-	-	-	-	-	-
Viola sp.; pansy	-	-	-	-	-	-
Varia	-	-	-	-	-	+

228 nt. 12	230 cont. 12	232 cont. 12	236 cont. 12	240 cont. 14	233 cont. 15	234 cont. 15	126 no cont.	134 no cont.	149 no cont.	159 no cont.	183 no cont.	198 no cont.
–	–	++	+	–	–	–	–	–	–	–	–	–
–	–	+	–	–	+	–	–	–	–	+	–	–
–	–	–	+	–	–	–	–	–	–	+	–	–
–	–	+	–	–	–	–	–	–	–	+	–	–
+	–	++	+	–	–	–	–	–	+++	+++	–	–
–	–	–	+	–	–	–	–	–	–	–	–	–
–	–	+	–	–	–	–	–	–	+	+	–	–
–	+	–	–	+	–	–	–	–	–	–	+	–
–	–	–	–	–	+	–	–	–	–	–	–	–
–	–	+	–	–	–	–	–	–	–	–	–	–
–	–	+	–	+	+	–	–	–	–	–	–	–
–	–	–	–	+	–	–	–	–	–	–	–	–
–	–	+	–	–	–	–	–	–	–	+	–	–
+	–	–	–	–	–	–	–	–	–	+	–	–
–	–	–	–	–	–	–	–	–	–	–	–	–
–	–	+	+	–	–	–	–	–	+	+	–	–
–	–	+	+	+	–	–	–	+	–	++	–	–
–	–	+	–	–	–	–	–	–	+	+++	++	–
–	–	–	–	–	–	–	–	–	+	–	–	–
–	–	–	–	–	–	–	–	–	–	+	–	–
–	–	–	–	–	–	–	–	–	–	+	–	–
–	–	–	–	–	–	–	–	+	–	–	+	–
–	–	–	–	–	–	–	–	–	–	–	+	–
–	–	–	–	–	–	–	–	–	–	–	–	–
–	–	–	–	+	–	–	–	–	–	++	–	–
+	+	+	+	–	–	–	+	+	+	+	–	–
–	–	–	–	–	–	–	–	–	–	–	–	–
–	–	–	–	–	–	–	–	+	–	++	–	–
–	–	+	–	–	–	–	–	–	–	–	–	–
-	-	+	-	-	-	-	-	-	-	-	-	-
-	+	+	+	-	-	-	+	-	-	+++	+	-
-	-	-	-	-	-	-	-	-	+	++	-	-
-	-	+	_	-	-	-	-	-	-	-	-	-
-	-	+	-	-	-	-	-	-	-	-	-	-
-	-	-	+	-	-	-	-	-	-	-	-	-
-	+	-	-	-	-	-	-	+	-	-	-	-
+	-	-	-	-	-	+	-	-	-	+	-	-

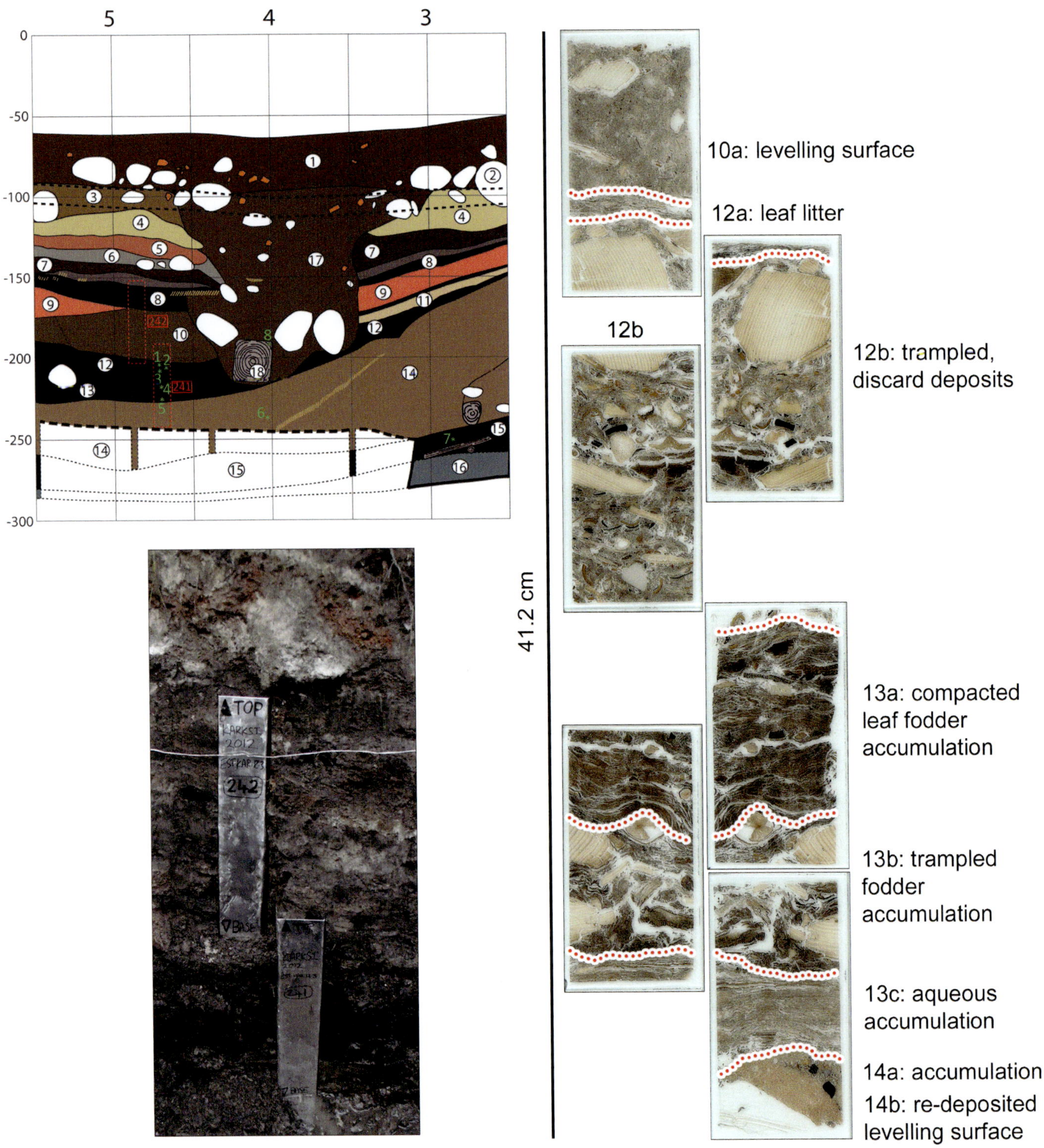

Figure 7.6: Trench 1 Karksi: (a) Section through the middle of Trench 1. C14 model indicates that 'midden' deposits (14, 13, 12, and 10) in sample 241 formed between AD 1270–95; (b) photograph of monoliths 241 (lower) and 242 (upper) in situ; (c) scans of micromorphology slides from monolith 241. Section drawing by E. Rannamäe; photos by R. Banerjea.

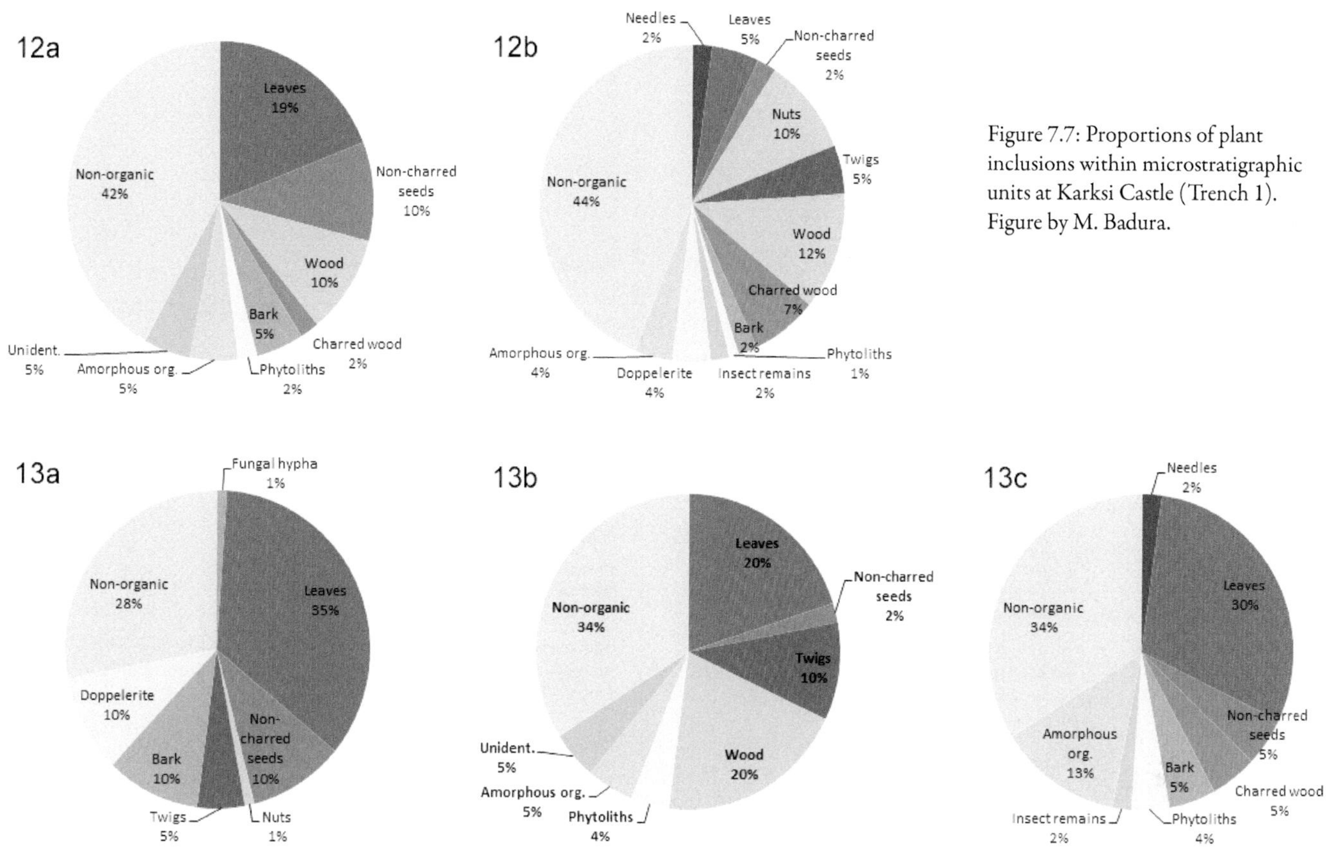

Figure 7.7: Proportions of plant inclusions within microstratigraphic units at Karksi Castle (Trench 1). Figure by M. Badura.

- Refuse material, forming a midden (context 12), was deposited over the peat deposit, which had not formed as a peat prior to the deposition of the midden, between 1272 and 1279 AD. Context 13 is not part of the midden due to the absence of anthropogenic debris.

Microstratigraphy and Formation Processes in Trench 1

Micromorphology has refined the stratigraphic sequence in Trench 1 through the identification of new microstratigraphic units (Fig. 7.6c), and by interpreting the formation processes within these new microstratigraphic units. The microstratigraphic sequence is described starting at the base in order to outline clearly the sequence of depositional events.

Context 14 at the base of the profile (Fig. 7.6a) has been further divided into two microstratigraphic units (Fig. 7.6c) (henceforth MU): 14b and 14a. MU14b has been interpreted as redeposited sediment comprising predominantly of mineral (90%), which has been used to dam the spring waters to create a pond, and is further supported by the identification of *Chara* sp. — freshwater algae — in the plant macroremains (Table 7.5). MU14a is a thin accumulation of anthropogenic residues, including charcoal, brick, and mortar, which has collected at the bottom of the pond, possibly eroded out of MU14b.

Context 13, the 'peat deposit', comprises three separate depositional events: MU13a, MU13b, and MU13c (Fig. 7.6c). The lowest, MU13c, has been interpreted as pond sediment. Pond sediment tends to be fine textured, with at least 20–30% clay content, and contain a higher proportion of organic matter than terrestrial soils (Boyd 1995, 4–5). MU13c has a loamy sand particle size and consists mainly of organic components, 66%, with some mineral inclusions, 34% (Fig. 7.7). Plant fragments are strongly oriented, linear and parallel to basal boundary, and interspersed with lenticular platy peds of sediment with locally oriented and clustered minerals, which may have formed due to the freezing of the pond (Van Vliet-Lanoë 2010, 83). MU13b is a peat consists mainly of organic components, 66%, with some mineral inclusions, 34%; wood and twigs are the dominant components of the organic remains (Fig. 7.7). There are no microlaminations and inclusions are unoriented, unrelated, random, and unreferred, both of these features are different to MU13a above, which is also a peat, but has a microlaminated bedding structure and plant fragments are strongly oriented, aligned parallel to basal boundary. The material within MU13b has been trampled leading to the organic

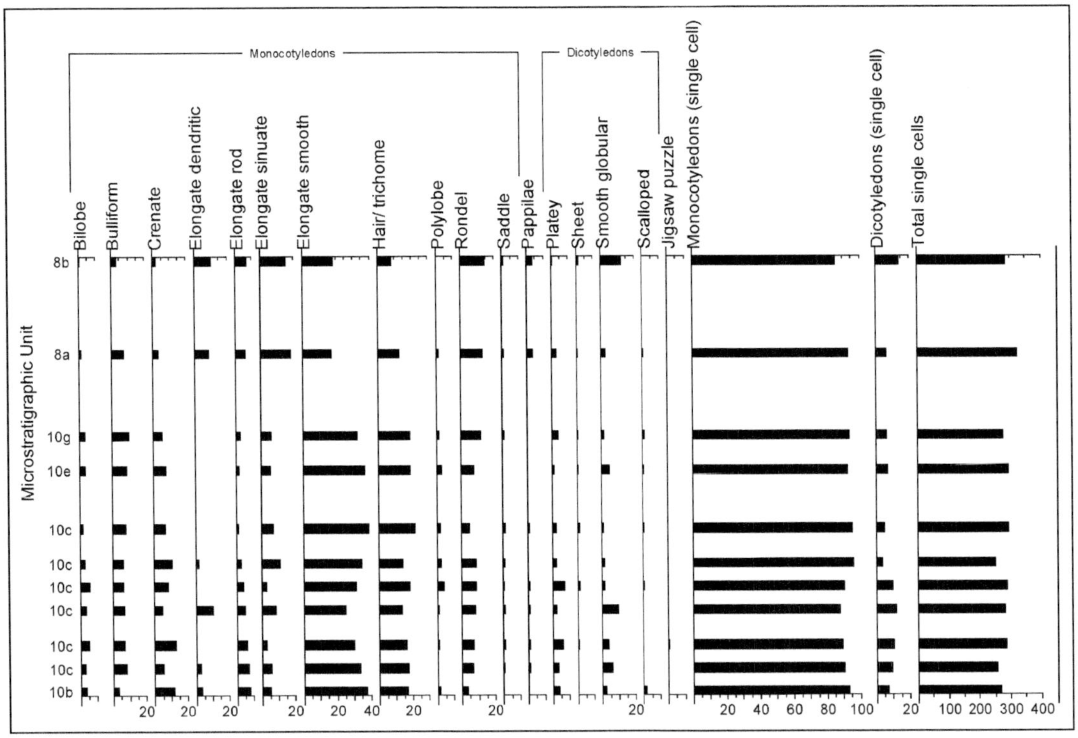

Figure 7.8: Proportions of monocotyledon and dicotyledon phytoliths within microstratigraphic units in monolith 242, Karksi Castle (Trench 1). Figure by R. Banerjea.

remains being broken, which is clearly visible macroscopically in thin-section (Fig. 7.6c). The pond was infilled by at least two events relating to the deposition of foliage: MU13b predominantly wood and twigs; and MU13a predominantly tree leaves. These 'peat' deposits are better classified as anthropogenic peats. The formation process shows similar characteristics to natural peat forming processes, but where organic materials from around the settlement have rapidly accumulated where there is a high groundwater table, which has preserved the organic matter (Ismail-Meyer et al. 2013, 331).

Context 12, the 'midden', comprises two separate depositional events: MU12a and MU12b (Fig. 7.6c). MU12a is a thin, 0.6–1.5 cm, accumulation of leaf litter, which overlies a more substantial horizon formed from midden material, MU12b. MU12b has a microlaminated bedding structure and the organic and anthropogenic inclusions are moderately to strongly oriented, linear and parallel to basal boundary, indicating episodic accumulation overtime (Goldberg and Macphail 2006), and are interspersed with platey peds of sediment with locally oriented and clustered minerals. A platey ped microstructure can indicate that sediment froze due to frost action (Van Vliet-Lanoë 2010, 83), or it can form as a result of trampling in external areas (Milek 2012; Shillito and Ryan 2013).

Within monolith 242 (Fig. 7.5a), a series of external deposits (Fig. 7.5b, c) overlie the midden, MU12b, starting with a leaf litter horizon, MU12a, and followed by horizons where soils have started to form (MU10a, 10c, 10f, 10g, 8a), trampled/frost-affected layers (10b, 8b), trampled soils (10d, 10e), and a levelling surface that has been reworked by biological activity (MU9). With the exception of MU8a and 8b, anthropogenic inclusions are scarce in the microstratigraphic units within monolith 242. MU8a and 8b contain fragments of brick, unburnt bone, charred wood, charred seeds, and charred plant remains, and overlie the levelling surface (MU9), which may indicate a change in use for the area, possibly a point when there is an intensification of activity from the newly constructed high castle. The phytolith assemblages through monolith 242 also show a difference between MU8a and 8b and10b, 10c, 10e, and 10g (Fig. 7.8). Phytoliths from monocotyledons, particularly from the leaves and stems of grasses (elongate sinuate, elongate smooth, crenates) decline in MU8a and 8b; whereas dicotyledons, particularly smooth globular cells from foliage, increase (Fig. 7.8), which is consistent with the plant remains that were identified in thin-section as fragments of wood, charred wood, and bark are more abundant, and amorphous plant tissue is less abundant in MU8a and 8b than the units further down the profile. This may represent activities associated with the high castle such as the collection of wood for fuel and construction, rather than the accumulation of grass and leaves in trampled soil prior to the construction of the high castle.

Early Activities and Refuse Disposal

The radiocarbon dating model indicates that the sediments within monolith 241 were deposited and formed rapidly in a period of 25 years between AD 1270–95 (Valk et al. 2012), which suggests that the episodic midden depositions, MU12b, formed in a very short period of time. Anthropogenic inclusions within the midden such as charred wood, unburnt and burnt bone, egg shell, hazelnut shells, strawberry seeds, also recorded in the plant macroremains (Table 7.5), and herbivore dung represent activities relating to food preparation and animal husbandry. In addition, fragments of brick and omnivore coprolite fragments are also present in this unit. No refuse associated with industrial activities were identified within MU12b.

A

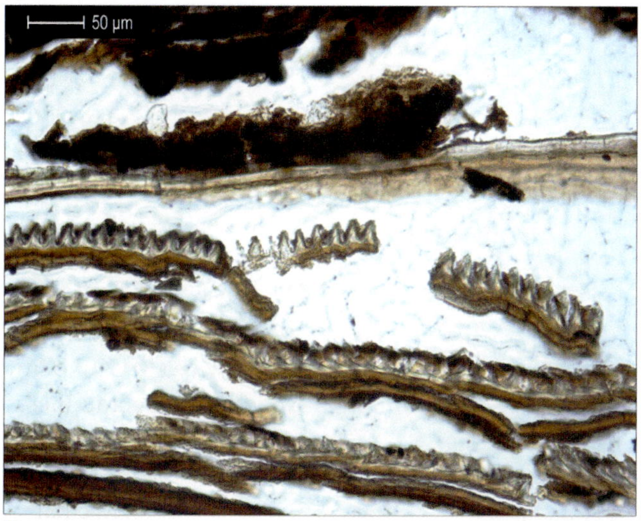

B

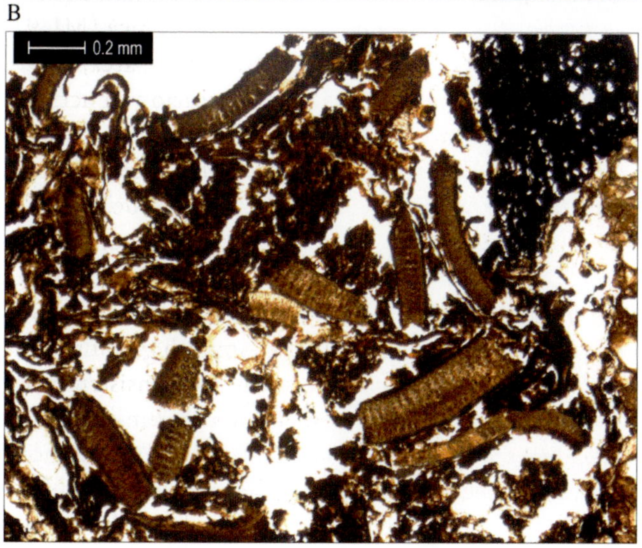

Figure 7.9: Photomicrographs of inclusions within monolith 241, Karksi Castle (Trench 1): (a) fish scales within microstratigraphic unit 13a; (b) millet embedded within a fragment of caprid coprolite within microstratigraphic unit 12b. Photos by R. Banerjea.

Early Activities and Livestock Alimentation: Micromorphology, Trench 1

Micromorphological analysis produced important data pertaining to the management of livestock by the early castle community. The creation of a pond provided a water source for livestock and, as attested by the presence of fish scales (Fig. 7.9), presumably enabled some management of fish as a food resource. The consumption and management of fish by religious orders, including the Teutonic Order, during the medieval period is well attested (Aston 1988; Pluskowski 2013, 328–30; Müldner et al. 2014; see also Chapter 6); however, the establishment of a pond by the initial occupants on their arrival, where fish were potentially present, demonstrates the importance of this resource to their diet. Fish bones were identified in the midden that infills the pond, in both thin-section and in the zooarchaeological assemblage (Valk et al. 2012), which suggests that at this point that fish were being caught from the river at the valley bottom.

The pond, which could have also served as a water source for livestock, was infilled by at least two events relating to the deposition of foliage, which may represent the use of leaf fodder for livestock; branches and twigs have been demonstrated to be an important addition to leaf fodder, particularly in times of food scarcity such as winter (Rasmussen 1993). The first, MU13b, in comparison with MU13a above, comprises predominantly larger wood fragments (20%), twigs (10%), but fewer leaves (Fig. 7.7). Arguably, on the basis of their large size, >1 cm, the wood and twig fragments within MU13b could represent uneaten fodder remains, which have been made available to the livestock adjacent to their water source; twigs and leaves with a diameter <5 mm have been recorded within cow pats and sheep/goat dung pellets (Akeret and Jacomet 1997; Akeret and Rentzel 2001; Akeret et al. 1999; Favre and Jacomet 1998). MU13a, in comparison with MU13b below, contains more leaves and smaller fragments of twigs and may represent the dumping of leaf fodder. The use of leaf fodder by the initial colonizers may represent a period before cereal crops were widely available as animal feed as the pollen evidence shows a gradual move to more agrarian landscape by the post-medieval period; around Karksi in the tenth century there is birch woodland, followed by signification sediment erosion in the medieval period (Chapter 5).

Within the midden deposit, MU12b, ruminant coprolites have fragments of millet embedded within them (Fig. 7.9), indicating the use of millet as fodder.

Table 7.6: Geochemical enrichment factors for selected major and trace elements in the moat at Cēsis.

Sample/ context number	Notes	Major Elements							
		Mg (ppm)	Mg EF	P (ppm)	P EF	Ca (ppm)	Ca EF	Ti (ppm)	Ti EF
CES 102/ 005	60-70cm	15859.61	6.57	3655.86	8.38	56820.75	15.9	2756.94	1.53
CES 106/ 005	90-100cm	15437.49	6.4	3360.37	7.7	62109.94	17.38	2277.47	1.27
CES 128/ 005	120-140cm	16342.03	6.77	2269.34	5.2	49745.13	13.92	2936.74	1.63
CES 133/ 005	140-150cm	18814.44	7.8	2574.83	5.9	62467.31	17.48	3176.47	1.77
CES 138/ 005	160-170cm	15859.61	6.57	2182.06	5	47172.11	13.2	2876.8	1.6
CES 143/ 005	180-190cm	14653.55	6.07	1789.29	4.1	42740.79	11.96	2517.2	1.4
CES 148/ 005	200-210cm	17548.08	7.27	2007.49	4.6	53461.72	14.96	3236.4	1.8
CES 159/ 013	Brown clay under 005	16583.24	6.87	2662.11	6.1	64111.18	17.94	3116.54	1.73
CES 164/ 011	Grey sand pocket	7296.63	3.02	1091.03	2.5	16939.08	4.74	1678.13	0.93
CES 169/ 014	Red silty sand	9407.22	3.9	698.26	1.6	9505.89	2.66	2217.54	1.23
CES 174/ Ft1	Possible fill of drain	10372.06	4.3	2400.26	5.5	27374.12	7.66	1917.87	1.07
CES 208	Red silty sand below timbers	9346.92	3.87	610.98	1.4	7719.07	2.16	2217.54	1.23
CES 213	Red silty sand below timbers	9588.13	3.97	610.98	1.4	9077.06	2.54	2277.47	1.27
CES 218	Red silty sand below timbers	10794.18	4.47	654.62	1.5	11935.97	3.34	2277.47	1.27
CES 223	Sediment under Ft1, drk organic	9045.4	3.75	872.82	2	10720.93	3	2217.54	1.23

This is particularly useful information as the plant macro assemblage from with the midden was noticeably poor, with the remains that were present mostly representing the local damp environment (Valk et al. 2012). Millet may well have been imported by the inhabitants as this is a crop that has productivity and a short growing season under dry, high-temperature conditions such as those in semi-arid regions.

Early Activities and Livestock Alimentation: Archaebotanical Results, Trench 1

Archaeobotanical samples were obtained from separate contexts in Trench 1 relating to the thirteenth-century midden (225, 226, 228, 230, 232, 236) and other layers with a different composition of organic and mineral elements (106, 107, 108, 126, 134, 149, 159, 183, 198, 233, 234). Monolith 241 contained samples 225, 228, 230, 232, 236; and Monolith 242, samples 139 and 226 (Fig. 7.6a, b).

Plant macrofossils from the midden (field context 12, micostratigraphic units 12a and 12b) represent a range of habitats. Context 12 contained mostly remains of species from wet/damp areas. The presence of pastures and meadows is suggested by fruits of self-heal (*Prunella vulgaris*), wood club-rush (*Scirpus sylvaticus*), tormentil (*Potentilla erecta*), and fairy flax (*Linum catharticum*). Damp places are represented further by the remains of sedge (*Carex* sp.) and rush (*Juncus* sp.). Hazel (*Corylus avellana*), raspberry (*Rubus idaeus*), wild strawberry (*Fragaria vesca*); plants characteristic of woodlands and wooded fringes are probably represented by collected edible fruits. Bracken (*Pteridium aquilinum*), a fern which grows in the forest, but also can colonize different types of habitats, could have been collected as an ornamental plant or used as bedding. Although Trench 1 was located in a disturbed area, the number of ruderal plants is relatively low. Single fruits were represented by pale persicaria (*Polygonum lapathifolium*), water pepper (*P. hydropiper*), white goosefoot (*Chenopodium album*), common hemp-nettle (*Galeopsis tetrahit*), and nipplewort (*Lapsana communis*). These species are typical for nitrophilous, waste, and wet places, but some could have derived from gardens and root crops. In the midden there are only single fruits of black-bindweed (*Fallopia convolvulus*) and sheep's sorrel (*Rumex acetosella*), which could be traces of winter crops. The remains of cultivated plans were sporadically present in the midden, but in the contexts above and adjacent to the midden their abundance was quite high. Cereals were mostly preserved within mineral contexts in Trench 1. Many charred grains of rye (*Secale cereale*) (e.g., sample no.

				Trace Elements							
(ppm)	Mn EF	Fe (ppm)	Fe EF	Cu (ppm)	Cu EF	Zn (ppm)	Zn EF	Sr (ppm)	Sr EF	Pb (ppm)	Pb EF
54.67	1.5	16945.1	1.45	45	6	104	6.58	92	1.4	18	1.76
37.23	1.25	15623.7	1.34	60	8	141	8.92	90	1.37	23	2.25
99.78	1	16867.37	1.45	36	4.8	95	6.01	81	1.23	16	1.57
54.67	1.5	20676.14	1.77	44	5.87	118	7.47	90	1.37	17	1.67
99.78	1	17256.02	1.48	35	4.67	91	5.76	76	1.16	27	2.65
32.34	0.75	13991.37	1.2	27	3.6	72	4.56	64	0.98	12	1.18
37.23	1.25	21997.54	1.89	40	5.33	106	6.71	76	1.16	17	1.67
37.23	1.25	18344.24	1.57	62	8.27	169	10.7	95	1.45	26	2.55
7.45	0.25	6607.04	0.57	25	3.33	38	2.41	41	0.63	9	0.88
7.45	0.25	11970.39	1.03	8	1.07	28	1.77	50	0.76	8	0.78
54.89	0.5	11193.1	0.96	43	5.73	50	3.16	56	0.85	13	1.27
7.45	0.25	12203.58	1.05	5	0.67	21	1.33	42	0.64	5	0.49
7.45	0.25	12203.58	1.05	5	0.67	21	1.33	45	0.69	9	0.88
7.45	0.25	11737.21	1.01	6	0.8	19	1.2	43	0.66	7	0.69
4.89	0.5	12592.23	1.08	14	1.87	32	2.03	49	0.75	9	0.88

183–32/5 l.), charred diasporas of barley (*Hordeum vulgare*), and wheat (*Triticum aestivum*) were recovered. This result is similar to the palynological analysis of sediments from Trench 1, which indicated a high value for cereals, particularly in MU13a directly below the midden (see Brown this volume). Cereal pollen grains may have been transported into the material infilling the pond as a result of nearby dumps of fodder, cereal processing activities, or even on the fur and hooves of livestock.

Contexts 8 and 10 overlying the midden, which have been examined using micromorphology (Fig. 7.5), are not rich in plant macroremains (Table 7.5), which is probably linked to soil development. This is indicative of aeration and biological reworking, as well as destruction of plant remains by trampling. The microstratigraphic units comprising contexts 8 and 10 represent external deposits above the midden. Context 8, which comprises a soil (MU8a) and overlaying trampled material (MU8b) contains single remains of ruderal weeds, sedge, and Caryophyllaceae indet. Context 10 is formed from seven microstratigraphic units (10a–10g) comprising soils and trampled horizons again contains single grains of the aforementioned species, as well as additional remains of hazel, ruderal weeds, meadow buttercup (*Ranunculus acris*), and birch (*Betula* sect. *albae*). A high number of raspberry fruit stones is present in context 7 (Fig. 7.6a), loosely described as 'black cultural layers' during excavation.

The distinctive features of non-described contexts from Trench 1 were present in plant diasporas derived from fresh meadows and pastures as well as other wet habitats. Numerous seeds of rush, single fruit of common club-rush (*Schoenoplectus lacustris*), seeds of bulrush (*Typha* sp.), and leaves of peat moss (*Sphagnum* sp.) indicate vegetation which favours riverbanks. It is possible that plants were brought to the castle for use as bedding or fodder, in particular ragged robin (*Lychnis flos-cuculi*), yarrow (*Achillea millefolium*), and the aforementioned tormentil and wood club-rush. Particularly interesting in context 14, at the very base of the dammed pond (Fig. 7.6) is the presence of stonewort (*Chara* sp.) oospores; typical aquatic green algae, characteristic of fresh water. Oospores of *Chara* sp. also appeared together with seeds of rush in sample 159, which is without context.

In summary, the macrobotanical results from the midden and other organic layers at Karksi Castle indicate that plants did not grow in this area, but reflect a wide range of habitats exploited by the castle community. The remains of cultivated or collected plants represent household and table waste.

Table 7.7: List of taxa — Cēsis, moat; abundance: + - single, ++ - numerous, +++ - abundant.

TAXA	163 cont. 011	158 cont. 013	168 cont. 014	207 below timbers 265–75	212 below timbers 285–95	217 below timbers 305–15	173 feature 1	222 under feature 1
USEFUL PLANTS								
Farinaceous Plants								
Avena sp.; oat	+	-	-	-	-	-	-	-
Fagopyrum esculentum L.; buckwheat	+	+	-	-	-	-	-	-
Secale cereale L.; rye	-	-	-	-	-	-	+	+
cf. *Secale cereale* L.; rye	+	-	-	-	-	-	-	-
Triticum sp.; wheat	-	-	-	-	-	-	+	-
Vegetables and Spices								
Brassica nigra L.; black mustard	+	-	-	-	-	+	+	-
Brassica rapa L.; turnip	+	-	-	-	-	-	-	-
Oil and Fibre Plants								
Cannabis sativa L.; hemp	-	-	-	-	-	+	+	-
Fruits, Berries, and Hop								
Ficus carica L.; fig	-	-	-	-	-	-	+	-
Fragaria vesca L.; wild strawberry	+	+	-	-	-	-	+	+
Humulus lupulus L.; hop	+	-	-	-	-	-	+	-
NON-USEFUL PLANTS								
Segetal and Ruderal Weeds								
Anthriscus sylvestris (L.) Hoffm.; cow parsley	+	+	-	-	-	-	-	-
Centaurea cyanus L.; cornflower	+	-	-	-	-	-	-	-
Chelidonium majus L.; greater celandine	-	+	-	-	-	-	-	-
Chenopodium album L.; white goosefoot	+	+	+	-	+	+	+	+
Chenopodium glaucum L./; *urbicum* L.; oakleaf/city goosefoot	-	+	-	-	-	-	-	-
Chenopodium sp.; goosefoot	++	+	+	+	+	-	+	-
Conium maculatum L.; hemlock	-	-	-	+	+	+	-	-
Descurainia sophia (L.) Webb ex Prantl; flixweed	+	-	-	+	-	+	+	+
Fallopia convolvulus (L.) Á. Löve; black-bindweed	+	+	-	-	-	-	+	-
Fumaria officinalis L.; common fumitory	+	-	-	-	-	-	-	-
Galeopsis tetrahit L.; hemp-nettle	+++	+	-	-	-	+	+	-
Galium spurium L.; false cleavers	+	+	+	-	-	-	-	-
Hyoscyamus niger L.; henbane	+	-	+	-	-	-	-	-
Lamium album L.; white nettle	-	+	-	-	-	-	-	-
Leonurus cardiaca L.; motherwort	+	-	+	-	-	-	+	-
Polygonum aviculare L.; knotweed	+	+	-	-	-	-	+	-
Polygonum hydropiper L.; water pepper	++	-	+	-	-	-	-	-
Polygonum lapathifolium L.; pale persicaria	+++	+	-	-	-	-	+	-
Raphanus raphanistrum L.; wild radish	+	-	-	-	-	-	-	-
Rumex acetosella L.; sheep's sorrel	+	+	-	-	-	-	+	-

TAXA	163 cont. 011	158 cont. 013	168 cont. 014	207 below timbers 265–75	212 below timbers 285–95	217 below timbers 305–15	173 feature 1	222 under feature 1
Scleranthus annuus L.; annual knawel	+	-	-	-	-	-	+	-
Solanum nigrum L. emend. Mill.; black nightshade	+	-	+	-	-	-	+	-
Spergula arvensis L. subsp. *maxima* (Weihe) O. Schwarz; corn spurry	+	-	-	-	-	-	-	+
Spergula arvensis L. subsp. *sativa* (Boenn.) Čelak; corn spurry	+	-	-	-	-	-	-	-
Spergula arvensis L. subsp. *vulgaris*; corn spurry	+	+	-	-	-	-	+	-
Stellaria media L.; common chickweed	+++	+	+	-	-	-	+	+
Thlaspi arvense L.; field penny-cress	+	-	-	-	-	-	-	-
Urtica dioica L.; common nettle	+	++	++	+	++	++	++	++
Viola arvensis Murray/*tricolor* L. s. str.; field pansy/wild pansy	+	-	-	-	-	+	-	-
Fresh Meadows and Pastures								
Potentilla erecta (L.) Raeusch.; tormentil	-	+	-	-	-	-	-	-
Potentilla cf. *erecta* (L.) Raeusch.; tormentil	-	-	-	-	-	-	+	-
Prunella vulgaris L.; self-heal	+	-	-	-	-	-	-	-
Ranunculus acris L.; meadow buttercup	+	-	-	-	-	-	-	-
Ranunculus repens L.; creeping buttercup	+	-	-	-	-	-	-	-
Plants from Wet and Moist Habitats								
Lycopus europaeus L.; gypsywort	+	-	-	-	-	-	-	-
Carex elata All.; tufted sedge	-	+	-	-	-	-	+	-
Carex ovalis Gooden.; oval sedge	+	-	-	-	-	-	-	-
Carex elata All./*nigra* L.; tufted/common sedge	-	+	-	-	-	-	-	-
Ranunculus flammula L.; lesser spearwort	+	-	-	-	-	-	-	-
Ranunculus sceleratus L.; celery-leaved buttercup	-	+	-	-	-	-	+	-
Sphagnum sp.; peat moss	-	+	-	+	-	-	+	-
Other Plants								
Arctium sp.; burdock	+	-	-	-	-	-	-	-
Asteraceae indet.; aster family	-	-	-	-	-	+	+	-
Atriplex sp.; orache	+	-	-	-	-	-	-	-
Caryophyllaceae indet.; pink family	-	-	-	-	-	-	-	+
Carex sp.; sedge	+	-	-	-	-	-	-	-
Galeopsis sp.; hemp-nettle	++	-	-	-	-	-	-	-
Lamiaceae indet.; mint family	-	-	-	-	-	-	-	+
Poaceae indet.; grass family	+	-	-	-	-	-	-	-
Ranunculus sp.; buttercup	+	-	-	-	-	-	+	-
Stachys sp.; hedgenettle	+	-	-	-	-	-	-	-
Viola cf. *canina* L. s. str.; heath dog-violet	+	-	-	-	-	-	-	-
Viola sp.; pansy	+	-	-	-	-	-	-	-
Luzula multiflora (Retz.) Lej.; / *campestris* (L.) Dc.; wood/field wood rush	+	-	-	-	-	-	-	-

Moats and the Flow of Effluent

The stone castle at Cēsis was built on elevated ground with steep slopes on the northern and western sides, and the area of the castle was enclosed on the southern and eastern sides with an artificial moat (see Chapter 4). The deep moat had a defensive role, as well as being used as a dolomite stone quarry during its construction and was not thought to have been filled with water (Kalniņš 2014, 3, 42), but appears to have collected and drained water from other areas of the castle.

The Geochemistry of the Moat, Cēsis Castle

Geochemical samples were collected from Trench N, Level S, Profile 1, at intervals through context 005, a 'black clay' deposit, and contexts 011, 013, and 014 below (Fig. 7.10; Table 7.6). Contexts 011, 013, and 014 overlay what was considered to be an infilled timber drain, Feature 1, and sediments infilling this feature and below it were also sampled (Fig. 7.10). The sediments beneath/abutting timbers (context 014 and samples 208, 213, 218, and 223) are the most sterile with the lowest geochemical enrichment factors overall. The samples from contexts 005 and 013 show the highest enrichment factors (Table 7.6), particularly for Mg, P, Ca, Mn, Cu, Zn, and Pb, which may represent the inclusion of waste materials from the castle (Wilson et al. 2005). The sediment within Feature 1, considered to be the fill of the drain, also showed enrichments of these elements, but not as high as contexts 005 and 013, which suggests that some waste materials and effluent may be included in the sediment. The sediment samples from a timber latrine within the Liv district, Riga, dated using dendrochronology to *c.* AD 1356, showed absolute concentrations (ppm) and enrichment factors for P and trace elements that are significantly higher than those within the moat at Cēsis (Tables 7.1, 7.6), for example: the enrichment factor for P from the latrine ranges between x 217–278 the background geology, but in contexts 005 and 013 in the moat at Cēsis is x 4–8.5 the background geology; however, for Ca from the latrine ranges between x 12–20 the background geology, and similar in contexts 005 and 013 in the moat x 11–18 the background geology. The higher enrichment factors of P and trace elements from the latrine at Riga (Table 7.1) than from the moat at Cēsis (Table 7.6) may be due to factors such as more intensive and longer in an urban site (Riga), as well as the primary function of the feature was as a latrine; whereas both human effluent and kitchen waste may have occurred in moat at Cēsis, as identified in the plant macroremains below, and due to less intensive occupation at the castle than on an urban site such as Riga. The similarity between the enrichment factors of Ca at the two sites may be due to the calcareous dolomite geology of the Cēsis moat, which has raised the levels of Ca. Given the low levels of phosphorous throughout the profile (Fig. 7.10, Table 7.6), translocation as a result of temporary water saturation is not thought to the cause of the low levels. Phosphorous does not seem to be washing through the profile from the deposits higher up, but without micromorphological analysis, it cannot be discounted completely.

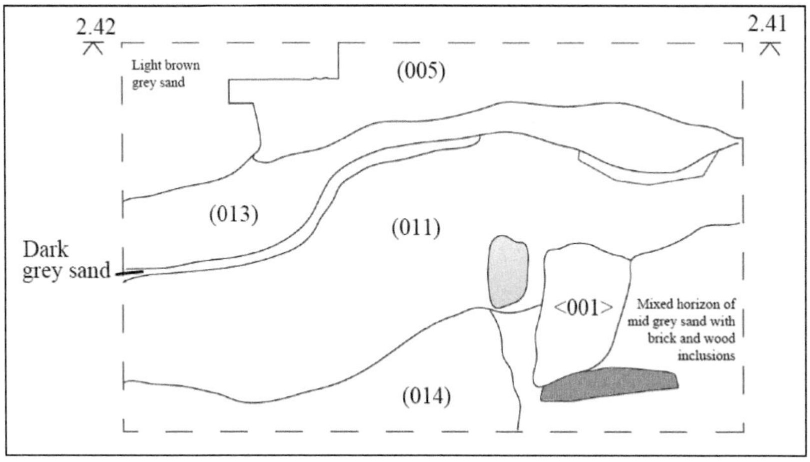

Figure 7.10: Trench N, Level S, Profile 1, Cēsis Castle. Plan redrawn by R. Banerjea.

Occupation Waste and Moat Ecology: Archaeobotanical Results

Eight samples from the moat of Cēsis Castle, dated to the second half of the fourteenth century, were analysed for archaeobotanical remains (Table 7.7). They represented individual contexts and sediments under the timber channel at the base of the moat, as well as Feature 1, the fill of the timber drain.

All materials contained mineral parts (sand, clay) with organic components. Plant remains were found mostly in samples from the moat's contexts and Feature 1 (Fig. 7.10). All of them reflect different types of activities around the castle. The archaeobotanical analysis indicates a mixed composition of useful plants and plants from anthropogenic vegetation. Most samples are characterized by synanthrope vegetation (ruderal plants) from disturbed soils areas used for waste disposal. In the vicinity of the moat a plant community with hemp-nettle (*Galeopsis tetrahit*), pale persicaria (*Polygonum lapathifolium*) and common chickweed (*Stellaria media*) grew. All three taxa are nitrophilous plants, which prefer moist soil. It is possible that they derived from an adjacent small garden plot, as well as black nightshade (*Solanum nigrum*) and common fumitory (*Fumaria officinalis*). Wetter areas near the moat, but also with fertile soils, were occupied by thickets of common nettle (*Urtica dioica*). The most significant taxa were segetal weeds, which are found today amongst winter crops on light and sandy soils, such as: cornflower (*Centaurea cyanus*), black-bindweed (*Fallopia convolvulus*), sheep's sorrel (*Rumex acetosella*), and annual knawel (*Scleranthus annuus*). Other species with a broader distribution were: wild radish (*Raphanus raphanistrum*), field pennycress (*Thlaspi arvense*), and water pepper (*Polygonum hydropiper*). The other group of uncultivated plants is associated with meadows and they reached the moat as the remains of fodder or waste: meadow buttercup (*Ranunculus acris*), creeping buttercup (*R. repens*), self-heal (*Prunella vulgaris*), tormentil (*Potentilla erecta*), and derived from wetlands: for example, tufted sedge (*Carex elata*), lesser spearwort (*Ranunculus flammula*), and celery-leaved buttercup (*R. sceleratus*).

On the basis of the plant macroremains and geochemical enrichments, it is likely that waste from the castle's kitchen and drains ultimately reached the moat. The timber channel was most probably constructed to improve the flow of effluent with the expansion of the castle in the later fourteenth century. This is reflected by the presence of edible plants, which are not exceptionally abundant in the samples but represent various elements of food. The diet of the castle community was based on farinaceous products, consisting of rye (*Secale cereale*), wheat (*Triticum* sp.), oat (*Avena* sp.) and buckwheat (*Fagopyrum esculentum*), with rye and wheat representing the principal bread cereals. Oat was used in the production of coarse and cheap bread, porridges and may also have served as animal fodder. Buckwheat was considered an important component of a low socio-economic status diet, but its consumption could be associated with any social group.

The archaeobotanical results from a site located near to Cēsis also indicate the presence of buckwheat (Sillasoo 2013. The resistant fruitscales of buckwheat, as well as charred grains of oat, were found in Feature 1, whereas well-preserved rye was more abundant in the fill of the timber drain. What is interesting is that within these samples there were no traces of millet (*Panicum miliaceum*), a cereal which was very popular in the Middle Ages (Karg 2007). The seeds of turnip (*Brassica rapa*) indicate the presence of vegetables, while seeds of black mustard (*B. nigra*) represent the only traces of spices from Cēsis Castle. Black mustard was known in many towns and castles along the Baltic coast, both in Prussia and Livonia (Latałowa et al. 2007; Sillasoo and Hiie 2007; Badura et al. 2015; Chapter 13). The seeds were also a source of medicinal oil, and the plant's young leaves could be eaten in the form of a salad (Grieve 1971; Renfrew and Sanderson 2005). In the materials from the moat in Cēsis, single remains of hemp (Cannabis sativa) were also found. In the Middle Ages hemp could have been used for oil, fibres, food, or medicine. Although fruits and berries, both cultivated and collected, were a very common feature of medieval diets, this group is very poorly represented in the samples from the moat at the castle. The remains included only gathered species (wild strawberry, *Fragaria vesca*) and exotics (fig, *Ficus carica*). The latter is undoubtedly evidence of the long-distance trade organized by the Teutonic Order in Livonia and indicative of the high status of the castle within the hierarchy of the *Ordensland*. Amongst the useful plants found at Cēsis, a few fruits of hop (*Humulus lupulus*) were found. The remains of this plant were largely found in the fill of the timber drain. Hop is commonly found in Prussia and Livonia. In the Middle Ages it provided important flavouring in beer and also functioned as a beer preservative, so it can be plausibly connected with brewing within the castle. At the same time, it is also possible that it could have occurred naturally in local shrubbery and at the edges of woodlands in close proximity to water. In summary, the castle moat must have been a muddy environment, developed as a result of the presence of standing or slow-moving water with impurities. The fertile and unprotected banks of the moat could have been readily colonized by nitrophilous plants, already present in the disturbed areas of the castle.

Summary of Results

The Initial Phase of Conquest and Colonization

Integrated palynological and geochemical analysis of a lacustrine sediment core from Lake Āraiši (Stivriņš et al. 2015) suggest that the relative impact of the crusades in the eastern Baltic (Livonia) differs significantly to the crusades in Prussia (present day north-east Poland), where the conquest by the Teutonic Order and its allies was accompanied by significant colonization, evident in the form of both urban foundations and numerous rural settlements, secured with fortified *castra* (Pluskowski 2013). Colonization in the territory designated as Livonia was limited largely to the major urban centres, such as Riga, with very limited colonization of the countryside by external migrants. As a result, indigenous populations continued to perpetuate existing patterns of land-use and this is reflected in the data from Lake Āraiši, where slight increases in cereal cultivation in the post-medieval period suggest that human impact on the environment during the period of the Order's rule was fairly limited within the vicinity of the castle (Stivriņš et al. 2015). Furthermore, the archaeological evidence suggests relatively low levels of activity at the castle, which was constructed during the first half of the fourteenth century (see Chapter 4). The intensification of crop cultivation and possibly a more open landscape is evident from *c*. AD 1750 to 1900, as indicated by the highest values of magnetic susceptibility from the input of minerogenic sediment, probably as a result of soil erosion (Stivriņš et al. 2015).

The geoarchaeological and archaeobotanical evidence presented here from Riga, a major urban centre, Cēsis Castle and its adjacent indigenous hillfort, and Karksi Castle, a new administrative centre with no previous occupation, supports the palynological evidence for low-level impact as a result of the crusades in Livonia outside of urban centres. In Riga, the occupation within the indigenous Liv district seems to have been unchanged during the time of the crusades in the early decades of the thirteenth century. Subsequently, the nature of the occupation in this district continued largely unchanged, with the same division of space (poorly maintained buildings adjacent to those that were well maintained) and similar activities after the imposition of Christian theocratic rule. This community seems to have been fairly self-contained despite being located within a major Hanseatic trading port.

The geoarchaeological evidence supports a hypothesis that Cēsis was a place of co-existence between the indigenous Vends and the Sword Brothers rather than one of tensions and conflict. Micromorphology shows that the timber fortification of the hillfort was only destroyed once, and that subsequent thin horizons that contain charred material were formed as a result of erosion and trampling processes, rather than due to several destruction or burning episodes. The erosion and trampling events were interspersed with periods of surface stabilization where pedogenesis began, which indicates periodicity in their formation. This shows that the timber structures were not repeatedly destroyed and rebuilt prior to the construction of the masonry fortifications. Additionally, the Vends occupied the hillfort at Cēsis at the same time as the stone castle was constructed, as evident from animal bone with cut marks consistent with indigenous technology (Seetah et al. 2013), U-series dated to the mid-thirteenth century (see Chapter 4; Appendix B) in the 'cultural layer', which is a unit of post-depositional soil formation consisting of midden material on the hillfort. The excavated contexts in the outer bailey of the stone castle are dated to the fourteenth to sixteenth centuries, with archaebotanical and geochemical evidence indicative of discrete middening relating to the disposal of kitchen waste in the area sampled rather than of intensive occupation. The sediments in the moat do not show particularly high geochemical enrichments, such as those within the latrine sediments at Riga. On this basis, it is possible that Cēsis Castle was not intensively occupied at this stage of its use.

Karksi Castle was a key administrative centre governed by an advocate. The geoarchaeological and archaeobotanical analysis of the stratigraphic sequence through Trench 1 provides detailed evidence relating to the activities of the early occupants, and how they controlled and organized their food, livestock, and settlement space. This analysis of this sequence shows that one of the first activities that took place in the area which later became the high castle was the construction of a pond by damning a natural stream, which suggests that the control and supply of fish was particularly important. The pond also served as a watering hole for livestock. As also observed at the outer bailey of the castle at Elbląg (Ger. Elbing) in Prussia, the livestock of the earliest occupants appear to have been fed using leaf fodder, which indicates that grain for use as animal fodder may not yet have been readily available, and that livestock was initially kept within the secure confines of the castle rather than put out to pasture. As the pond became disused it was reused as a midden in which herbivore coprolites containing millet were recorded, and the millet was most probably a crop that was imported by the colonists.

The Use of Settlements and Castles

The data presented above show the range of activities that took place on the sites and their spatial organization. In particular, the results have highlighted the alimentary importance of fish, both within the indigenous Liv district of Riga, and the Order's castles at Cēsis and

Karksi. A substantial fish processing area was identified within Building 4 at Riga, and the middens at Cēsis and Karksi Castles also produced a significant amount of fish processing remains (see also Chapter 6). The construction of what may be interpreted as a pond at Karksi by the early occupants underlies the importance of accessibility to fish.

At Riga, processing and craft, specifically non-ferrous metalworking activities within buildings, are located within a single building with an adjacent property used as a 'cleaner', perhaps living space, which could indicate a separation of domestic and commercial space at the household level. The prevalence of indigenous technologies and the insular nature of the production and consumption economy within the Liv district suggests that indigenous artisans may have deliberately clustered in this area of the town during the foundation of Riga. These later developed into craft guilds that reflected the social diversity of the medieval town — already known from Riga during the crusading period in the thirteenth century — into the Hanseatic period (Czaja 2005).

At the hillfort and the outer bailey of the Order's castle in Cēsis, and at Karksi Castle, which are fortified sites, refuse disposal took the form of midden spreads in external areas. A midden is defined as an occupation deposit relatively rich in refuse, including less tangible elements such as decayed organics and comminuted structural clay, and with evidence for the deliberate and sequential accumulation of refuse at one location (Needham and Spence 1997, 80). If the deposit was not due to deliberate and persistent acts over a period of time, then it is not a midden, but a unitary dump (Needham and Spence 1997, 80). The sequence from Trench 1 at Karksi very clearly shows the deliberate and sequential deliberate and sequential accumulation of alternating microstratigraphic lenses containing food and animal waste overlying the now disused fishpond, which relate to the maintenance of nearby 'use zones' within and around structures (Needham and Spence 1997). Unlike at Karksi, where the midden was rapidly buried by naturally accumulated and trampled sediment and soils, at Riekstu Hillfort in Cēsis the external refuse (or 'cultural layer') most probably remained exposed on the surface, as it has been reworked and continuously reworked into a soil.

Works Cited

Akeret, O., Haas, J. N., Leuzinger, U., and Jacomet, S. 1999. 'Plant Macrofossils and Pollen in Goat/Sheep Faeces from the Neolithic Lake-Shore Settlement Arbon Bleiche 3, Switzerland', *The Holocene* 9/2, 175–82

Akeret, O., and Jacomet, S. 1997. 'Analysis of Plant Macrofossils in Goat/Sheep Faeces from the Neolithic Lake Shore Settlement of Horgen Scheller: An Indication of Prehistoric Transhumance?', *Vegetation History and Archaeobotany* 6, 235–39

Akeret, O., and Rentzel, P. 2001. 'Micromorphology and Plant Macrofossil Analysis of Cattle Dung from the Neolithic Lake Shore Settlement of Arbon Bleiche 3', *Geoarchaeology* 16/6, 687–700

Apala, Z., and Apals, J. 2014. 'The Vendic Hill Fort on Riekstu Kalns in Cēsis', in *Strongholds and Power Centres East of the Baltic Sea in the 11th–13th Centuries*, ed. H. Valk, Tartu: Institute of History and Archaeology at the University of Tartu, Department of Archaeology at the Institute of History of Tallinn University, 115–38

Arroyo-Kalin, M., Neves, E. G., Petersen, J., and Bartone, R. 2003. 'Geoarchaeological Approaches to the Study of Amazonian Anthropic Dark Earths', in *Second International Conference on Soils and Archaeology Pisa, 12th–15th May, 2003, Extended Abstracts*, ed. G. Boschian, Pisa: Felici, 7–9

Aston, M. 1988. *Medieval Fish, Fisheries, and Fishponds in England, Part i*, British Archaeological Reports, Oxford: Archaeopress

Badura M., Możejko B., Święta Musznicka J., and Latałowa M. 2015. 'The Comparison of Archaeobotanical Data and the Oldest Documentary Records (14th–15th Century) of Useful Plants in Medieval Gdańsk, Northern Poland', *Vegetation History and Archaeobotany* 24/3, 441–54

Banerjea, R. Y. 2008. 'Experimental Geochemistry: A Multi-Elemental Characterisation of Known Activity Areas', *Antiquity* 82, 318 Project Gallery

Banerjea, R. Y., Bell, M., Matthews, W., and Brown, A. 2015. 'Applications of Micromorphology to Understanding Activity Areas and Site Formation Processes in Experimental Hut Floors', *Journal of Archaeological and Anthropological Sciences* 7, 89–112

Banerjea, R. Y., Badura, M., Kalējs, U., Cerina, A., Gos, K., Hamilton-Dyer, S., Maltby, M., Seetah, K., and Pluskowski, A. 2017. 'A Multi-Proxy, Diachronic and Spatial Perspective on the Urban Activities Within an Indigenous Community in Medieval Riga, Latvia', *Quaternary International* 460, 3–21

Barboni, D., Bonnefille, R. A. A., and Meunier, J. D. 1999. 'Phytoliths as Paleoenvironmental Indicators, West Side Awash Valley, Ethiopia', *Palaeogeography, Palaeoclimatology, Palaeoecology* 152, 87–100

Behre K.-E. 1999. 'The History of Beer Additives in Europe: A Review', *Vegetation History and Archaeobotany* 8, 35–48

Borderie, Q., Devos, Y., Nicosia, C., Cammas, C., and Macphail, R. I. 2015. 'Les terres noires dans l'approche géoarchéologique des contextes urbains', in *La géoarchéologie française au xxie siècle*, eds N. Carcaud and G. Arnaud-Fassetta, Paris: CNRS, 217–26.

Boyd, C. E. 1995. *Bottom Soils, Sediment, and Pond Aquaculture*, New York: Chapman & Hall

Carey, C. J., Wickstead, H. J., Juleff, G., Anderson, J. C., and Barber, M. J. 2014. 'Geochemical Survey and Metalworking: Analysis of Chemical Residues Derived from Experimental Non-Ferrous Metallurgical Processes in a Reconstructed Roundhouse', *Journal of Archaeological Science* 49, 383–97

Czaja, R. 2005. 'Patrician Guilds in Medieval Towns on the Baltic Coast', *Acta Poloniae Historica* 92, 31–51

Devos, Y., Vrydaghs, L., Degraeve, A., and Fechner, K. 2009. 'An Archaeopedological and Phytolitarian Study of the "Dark Earth" on the Site of Rue de Dinant (Brussels, Belgium)', *Catena* 78, 270–84

Favre, P., and Jacomet, S. 1998. 'Branch Wood from the Lake Shore Settlements of Horgen Scheller, Switzerland: Evidence for Economic Specialization in the Late Neolithic Period', *Vegetation History and Archaeobotany* 7, 167–78

French, C. 2003. *Geoarchaeology in Action*, London: Routledge

Goldberg, P., and Macphail, R. I. 2006. *Practical and Theoretical Geoarchaeology*, Oxford: Blackwell

Grieve, M. 1971. *A Modern Herbal*, 2 vols, New York: Dover

Ismail-Meyer, K., Rentzel, P., Wiemann, P. 2013. 'Neolithic Lakeshore Settlements in Switzerland: New Insights on Site Formation Processes from Micromorphology', *Geoarchaeology* 28, 317–39

Kalniņš, G. 2014. Cēsu Pils (Cēsis Castle), Cēsis: Cēsis Culture and Tourism Centre

Karg, S., ed. 2007. *Medieval Food Traditions in Northern Europe*, Copenhagen: National Museum

Kļaviņš, K. 2012. *Cēsis — A Symbol of Latvian History*, Cēsis: Municipal Agency, Cēsis Culture and Tourism Centre

Knudson, K., Frink, L., Hoffman, B. W., and Price, T. D. 2004. 'Chemical Characterization of Arctic Soils: Activity Area Analysis in Contemporary Yuk'ik Fish Camps Using ICP-AES', *Journal of Archaeological Science* 31, 443–56

LaMotta, V. M., and Schiffer, M. B. 1999. 'Formation Processes of House Floor Assemblages', in *The Archaeology of Household Activities*, ed. P. M. Allison, New York: Routledge, 19–29

Latałowa, M., Badura, M., Jarosińska, J., and Święta-Musznicka, J. 2007. 'Useful Plants in Medieval and Post-Medieval Archaeobotanical Material from the Hanseatic Towns of Northern Poland (Kołobrzeg, Gdańsk and Elbląg)', *Studies in Archaeology and History (National Museum (PNM), Copenhagen)* 12, 39–72

Macphail, R. I., and Crowther, J. 2009. 'Soil Micromorphology, Chemistry and Magnetic Susceptibility', in *Living and Working in Roman and Later London: Excavations at 60–63 Fenchurch St*, eds V. Birkbeck and J. Schuster, Salisbury: Wessex Archaeology, 113–20

Macphail, R. I., Galinié, H., and Verhaeghe, F. 2003. 'A Future for Dark Earth?', *Antiquity* 77, 349–58

Matthews, W. 1995. 'Micromorphological Characteristics of Occupation Deposits and Microstratigraphic Sequences at Abu Salabikh, Southern Iraq', in *Archaeological Sediments and Soils: Analysis, Interpretation and Management*, eds A. J. Barnham and R. I. Macphail, London: Institute of Archaeology, University College, 41–76

Milek, K. B. 2012. 'Floor Formation Processes and the Interpretation of Site Activity Areas: An Ethnoarchaeological Study of Turf Buildings at Thverá, Northeast Iceland', *Journal of Anthropological Archaeology* 31, 119–37

Miśkiewicz, M. 2008. *Europa wczesnego średniowiecza v–xiii w*, Warsaw: Wydawnictwo TRIO

Müldner, G., Britton, K., and Ervynck, A. 2014. 'Inferring Animal Husbandry Strategies in Coastal Zones through Stable Isotope Analysis: New Evidence from the Flemish Coastal Plain (Belgium, 1st–15th Century AD)', *Journal of Archaeological Science* 41, 322–32

Needham, S., and Spence, T. 1997. 'Refuse and Formation of Middens', *Antiquity* 71, 77–90

Nicosia, C., and Devos, Y. 2014. 'Urban Dark Earth', in *Encyclopedia of Global Archaeology xi*, ed. C. Smith, New York: Springer, 7532–40.

Pluskowski, A. G. 2013. *The Archaeology of the Prussian Crusade: Holy War and Colonisation*, London: Routledge

Pluskowski, A. G., Sawicki, Z., Shillito, L-M., Badura, M., Makowiecki, D., Zabilska-Kunek, M., Seetah, K., and Brown, A. 2014. 'Biała Góra: The Forgotten Colony in the Medieval Pomeranian-Prussian Borderlands', *Antiquity* 88.341, 863–82

Podbielkowski, Z. 1992. Rośliny użytkowe, Warsaw: Wydawnictwo Szkolne i Pedagogiczne

Podbielkowski, Z., and Sudnik-Wójcikowska, B. 2003. *Słownik roślin użytkowych. Państwowe Wydawnictwo Rolnicze i Leśne*, Warsaw: Państwowe Wydawnictwo Rolnicze i Leśne

Rasmussen, P. 1993. 'Analysis of Sheep/Goat Faeces from Egolzwil 3, Switzerland: Evidence for Branch and Twig Foddering of Livestock in the Neolithic', *Journal of Archaeological Science* 20, 479–502

Renfrew J., and Sanderson, S. 2005. 'Herbs and Vegetables', in *The Cultural History of Plants*, eds G. Prance and M. Nesbitt, London: Routledge, 93–132

Retallack, G. J. 2001. *Soils of the Past: An Introduction to Paleopedology*, Oxford: Blackwell Science

Rutkowska, B. 1984. *Atlas roślin łąkowych i pastwiskowych*, Warsaw: Państwowe Wydawnictwo Rolnicze i Leśne

Schiffer, M.B. 1987. *Formation Processes of the Archaeological Record*, Albuquerque: University of New Mexico Press

Seetah, K., Pluskowski, A. G., Makowiecki, D., and Daugnora, L. 2013. 'New Technology or Adaptation at the Frontier? Butchery as a Signifier of Cultural Transitions in the Medieval Eastern Baltic', *Archaeologia Baltica* 20, 84–101

Shillito, L.-M., and Ryan, P., 2013. 'Surfaces and Streets: Phytoliths, Micromorphology and Changing Use of Space at Neolithic Çatalhöyük (Turkey)', *Antiquity* 87, 684–700

Sillasoo, Ü. 2013. 'A Cultural History of Food Consumption in Medieval Livonian Towns', in *Landscapes and Societies in Medieval Europe East of the Elbe: Interactions between Environmental Settings and Cultural Transformations*, eds S. Kleingärtner, T. P. Newfield, S. Rossignol, and D. Wehner, Toronto: Pontifcal Institute of Mediaeval Studies, 316–28

Sillasoo, Ü., and Hiie, S. 2007. 'An Archaeobotanical Approach to Investigating Food of the Hanseatic Period in Estonia', in *Medieval Food Traditions in Northern Europe*, ed. S. Karg, Copenhagen: National Museum of Denmark, 73–96

Stivriņš, N., Brown, A., Reitalu, T., Veski, S., Heinsalu, A., Banerjea, R. Y., and Elmi, K. 2015. 'Landscape Change in Central Latvia since the Iron Age: Multiproxy Analysis of the Vegetation Impact of Conflict, Colonization and Economic Expansion during the Last 2,000 Years', *Vegetation History and Archaeobotany* 24, 377–91

Svarāne, D. 2002. 'Archaeological Evidence of Metalworking from Medieval Riga', in *The Medieval Town in the Baltic: Hanseatic History and Archaeology ii*, Tartu: Tartu City Museum, 99–108

Twiss, P. C. 1992. 'World Distribution of C3 and C4 Grass Phytoliths', in *Phytolith Systematics: Emerging Issues*, eds G. Rapp Jr and S. C. Mulholland, New York: Plenum, 113–28

Valk, H., Pluskowski, A., Brown, A., Rannamäe, E., Malve, M., and Varul, L. 2012. 'The Castle of the Teutonic Order in Karksi: Preliminary Excavation Results', *Arheoloogilised välitööd Eestis/Archaeological Field Work in Estonia*, 2011, 47–56

Valk, H., Rannamäe, E., Brown, A., Pluskowski, A., Badura, M., and Lõugas, L. 2013. 'Thirteenth-Century Cultural Deposits at the Castle of the Teutonic Order in Karksi', *Arheoloogilised välitööd Eestis/Archaeological Field Work in Estonia*, 2012, 73–92

Van Vliet-Lanoë, B. 2010. 'Frost Action', in *Interpretation of Micromorphological Features of Soils and Regoliths*, eds G. Stoops, V. Marcelino, and F. Mees, Amsterdam: Elsevier: 81–108

Wasylikowa, K., Cârciumaru, M., Hajnalová, E., Hartyányi, B. P., Pashkevich, G. A., and Yanushevich, Z. V. 1991. 'East-Central Europe', in *Progress in Old World Palaeoethnobotany: A Retrospective View on the Occasion of Twenty Years of the International Work Group of Palaeoethnobotany*, eds W. Van Zeist, K. Wasylikowa, and K-E. Behre, Rotterdam: Balkema, 207–39

Weiner, S. 2010. *Microarchaeology: Beyond the Visible Archaeological Record*, Cambridge: Cambridge University Press

Wilson, C. A., Davidson, D. A., and Cresser, M. S. 2005. 'An Evaluation of Multi-Element Analysis of Historic Soil Contamination to Differentiate Space Use and Former Function in and around Abandoned Farms', *The Holocene* 15, 1094–1099

Yule, B. 1990. 'The 'Dark Earth' and Late Roman London', *Antiquity* 64, 620–28

8. Reorganizing the Livonian Landscape

Aleksander Pluskowski, Kaspars Kļaviņš, Eva Eihmane, Agris Dzenis, and Juhan Kreem

The organization of the landscapes of Livonia changed following the rapid imposition of a new political regime over the course of the thirteenth century. The process of military conquest was initially relatively quick, compared to the crusades in Prussia, and after the Teutonic Order became involved from 1237, the manning of its commanderies took place at roughly the same time as Prussia. The founding of new commanderies would continue until 1290 in southern Livonia, and in the mid-fourteenth century the Order organized a new commandery and advocateships in north Estonia following its acquisition from the Danish crown. The aim of this chapter is to characterize changes in the conceptual organization of space within the Order's territories in Livonia following the crusades. Episcopal territories are touched upon in relation to boundary agreements and disputes, but it is important to note they were organized differently, and a comparative assessment is beyond the remit of this study. In contrast to the case studies presented in Chapter 15, which reflect the significantly better preservation of Prussian castle inventories, this chapter is structured according to broad diachronic themes based on a heavily fragmented dataset.

The Order's territories in Livonia are unevenly represented in the extant written sources, with the Curonian (Ger. Courland) commanderies best represented, and in particular the territory of Goldingen (Latv. Kuldīga). However, the survival of relict political structures in Curonia, particularly the so-called 'kings' (*regis*, *Könige*, *ķoniņi*, and also *tyranni*) who were granted extensive privileges under the Christian theocracy (Mickevičius 1997; Baranauskas 2004, 71, 72, 78; Laime 2009), appear to have developed into an exceptional regional phenomenon in Livonia. Six Curonian families — the Tontegode, Peniķis, Dragūns (later changed to Vidiņš), Kalējs (in the first half of the nineteenth century Germanized as Schmidt), and Bartolts (in the first half of the nineteenth century Germanized as Bergholtz) — were the only indigenous fief-holder families in western Latvia who from the seventeenth to nineteenth centuries formed a clan, defended and restored their rights, and created economically and politically autonomous units — free villages, which existed until 1929. The main privileges of the Curonian kings were personal and hereditary exemptions from *corvée*, taxes, and other imposed liabilities. The social and economic peculiarities of the Curonian kings are in fact analogous to other European free communities (Dzenis 2014, 16, 239–40).

The other territory which is comparatively well represented is Riga. This has resulted in extremely detailed insights into the control of the Lower Daugava from the thirteenth to the sixteenth centuries, the region which saw the most intensive crusading activities in the first decade of the thirteenth century, and was subsequently the focus of disputes between the major political factions of Livonia. The commandery of Wenden (Latv. Cēsis), however, is poorly represented in the extant written record. The most detailed accounts relate to Russian sieges during the Livonian War and inventories dating to the period of Polish and Swedish rule in the later sixteenth and seventeenth centuries. Further north, the written record is equally fragmented, with Reval (Est. Tallinn) and Fellin (Est. Viljandi) best represented from the later fifteenth century (Kreem 2002, 2019).

Territorial organization before the crusades is considered first, followed by how the crusades reorganized and assigned value to the landscape, how this was developed under the rule of the Order in Livonia, contested in the case of Riga, and mediated in relations with the indigenous population into the sixteenth century. The full range of data relating to the ecological changes initiated by the crusades, defined as the cultural landscape, are considered together in Chapter 8 and comparatively in Chapter 17.

Territories before the Crusades

The organization of the landscape before the crusades, in the late twelfth century, is largely known from archaeological sources, and there has been a tendency in the scholarship to map descriptions in the earliest written sources onto these. Territories were centred on strongholds, but their precise extent is difficult to map. At the regional scale, broad cultural zones have been linked to historically documented tribal groupings such as the various regional groups of Estonians, Livs, Latgalians (or Letts), Kurs (or Curonians), and Semigallians, separated into two broad linguistic groups: Finno-Ugrians and Transitional Balts (Map 8.1). These groupings, traditionally understood as ethnically-based tribal communities, have been assigned geographically defined territories on the basis of distinctive material culture, largely recovered from cemeteries and strongholds. The Latgalians, for example, appear as a materially definable group in eastern Latvia from the seventh century AD, whilst the Livs are associated with settling the lower Daugava and Gauja valleys from the tenth century (Šnē 2005; 2006, 68). In fact, there is no consensus on the exact division and subdivision of the historical Livonian territories (Sutrop 2014), and direct correlations between material culture and ethnicity are problematic. Within these broad ethnically defined regions, Late Iron Age settlement patterns have been interpreted in terms of hierarchical power structures (Šnē 2006, 74; Vītola and Urtāns 2014). At the top were strongholds associated with the highest levels of social and economic complexity; centres of socially stratified feasting, specialized manufacturing, and markets for trade and exchange. They consisted of fortified hilltop settlements and adjacent settlements located at or near their base, with associated cemeteries. These independent centres, controlled by noble families headed by leaders sometimes referred to in external written sources as *seniors*, governed a range of territories which varied spatially and temporally.

The most detailed studies of these territories have been conducted in Estonia, where systematic analyses of strongholds has demonstrated how neighbouring power centres rose and fell within close proximity to each other (Valk 2014), and that by the turn of the thirteenth century there were six regional overlordships, consisting of formerly independent lordships which had merged into larger political units through incremental conquest and

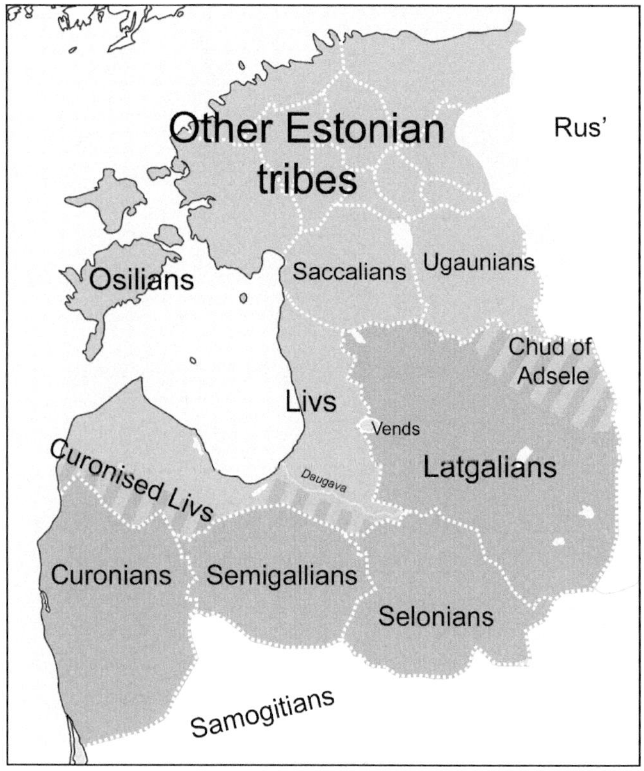

Map 8.1: Map showing the commonly accepted tribal regions in Livonia, although the boundaries should not be considered as clearly defined (after Laakmann, reproduced in Jähnig 2011, 261). Maps by A. Pluskowski.

by forging alliances (Šnē 2007; Oad 2014). In southern Estonia, the regional overlordships were centred on the strongholds at Viljandi (in Saccala, Ger. Fellin) and Otepää (in Ugaunia) (Map 8.2). Whilst the political and cultural boundaries of Viljandi's territory roughly matched, this was not always the case, and the populations within these territories were not necessarily culturally homogeneous. The smaller strongholds within these territories were nominally in charge of districts termed *vakuses*, with each *vakus* containing a number of nucleated settlements associated with units of ploughland. This formed the basis of a system of taxation which was subsequently appropriated by the crusaders. The centres of overlordships drew their resources from their interregional networks, as much as from a local territory defined by agricultural settlements (Lang 2012, 216). When the crusading army attacked Viljandi in 1211, they sent their Liv and Lett allies to despoil the region, and to procure food and grain from its villages (*HCL* xv, 1). In southern Estonia, the strongholds of overlordships were more or less centrally located within their projected territories, close to major communication routes (Oad 2014). This communication network would, in turn, be

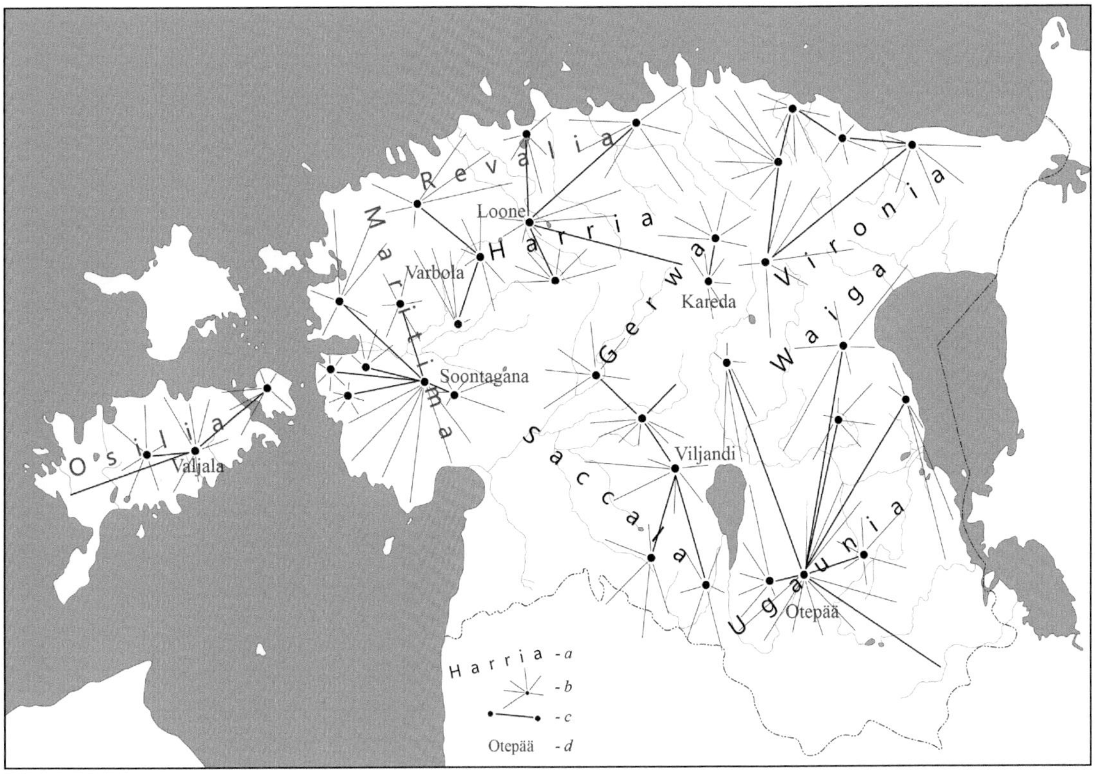

Map 8.2: Regional overlordships in Estonia in *c*. 1200 (Oad 2014, Figure 1): (a) cultural or tribal regions; (b) centre of a formerly independent local lordship and the bonds showing the power of respective kinship groups; (c) bond making an overlordship; (d) centre of an overlordship *c*. 1200.

used by crusading armies and then appropriated by the Christian regime.

A similar settlement hierarchy appears to have operated in southern Livonia, where discrete regions were dominated by strongholds. These varied dramatically in size and extent. The geographic extent of the neighbouring Vendish group, located in the environs of Cēsis, has been linked solely to the Riekstu Hillfort and the cemetery uncovered at the train station in Cēsis; however, this reflects the state of current research, and more Vendish sites may be found in the future (Apala and Apals 2014). Within the territory ascribed to Livs in the northern part of modern Latvia (Latv. Vidzeme), eight strongholds have been identified as functioning in the Late Iron Age, of which four are mentioned in Henry's *Chronicle*. This is from a total of forty-one monuments, of which only thirteen have been investigated archaeologically; in this respect, our understanding of settlement and territorial units remains largely incomplete (Vītola and Urtāns 2014, 419). In the southernmost part of this territory, the stronghold of Daugmale, situated on the southern bank of the Lower Daugava, has been compared to a Scandinavian Viking Age emporium, and with its role as a centre of political authority, trade, and manufacturing, the complex has been considered as the regional precursor to Riga (Radiņš 2001; Šnē 2006, 75) (Map 8.3). The associated settlement pattern can be connected with the Lower Daugava as a major trade artery where, from the mid-eleventh century, locally produced copies of imports appear in a range of settlements, including villages (Radiņš 1998, 186). These villages were situated a few kilometres from each other and varied in size from two to five hectares, with rows of buildings eight to ten metres apart; each settlement may have had around one to two hundred inhabitants, many of whom were clearly engaged in trade as well as agricultural and pastoral activities (Mugurēvičs 1983). The settlement cluster has been broadly defined as 'Liv', although some sites have been associated with multiple ethnic groups, such as the final phase of Aizkraukle (Radiņš, 1998, 182). Daugmale itself appears to have been occupied by Livs and Semigallians, as well as a community of Osilians (from the island of Ösel; Est. Saaremma) whose political interests in the Lower Daugava ultimately prompted attacks on Riga, established after Daugmale was abandoned. The region was also within the political sphere of the princes of Polotsk (Mägi 2005; 2011, 327). South of the Lower Daugava region, there is evidence for further variation. Some of the largest Semigallian centres consisted of double strongholds and were virtually all established on natural elevations overlooking rivers (Jarockis 1998, 48). Here, strongholds appear to have been located at the edges of territories, resulting in fortified frontiers (Stašulāne and Slušinskaitė 2003, 84). This may

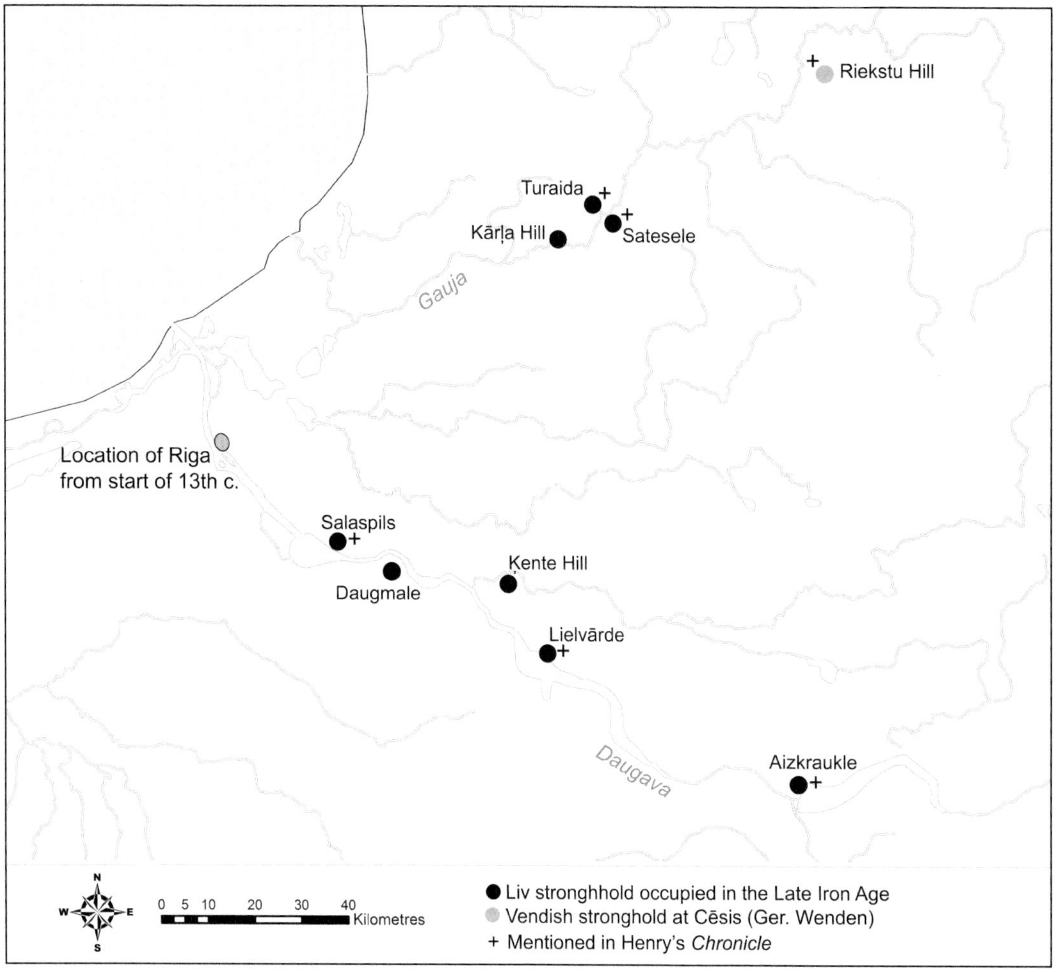

Map 8.3: Strongholds in the southern Liv area inhabited in the Late Iron Age (after Vītola and Urtāns 2014, 419). Map by A. Pluskowski

have reflected the dynamism of settlement networks in adjusting to new threats, in this case as a response to crusading armies and their allies (Nowakiewicz 2010, 498).

It is clear that the arrangement of power relations varied across the eastern Baltic in the twelfth and early thirteenth centuries, expressed in burial rites and the projected geographic influence of regional centres, rather than in the internal spatial configuration of settlements (Šnē 2009, 59; Mägi 2013). However, it is also possible to make some general observations. It is reasonable to assume that the authority of these regional centres, which would have required significant provisioning systems, was linked to the control of natural resources, such as grain and livestock. There is not enough data from the full range of settlements to indicate whether strongholds operated regional command economies (cf. O'Connor 2011), although high-status patterns of consumption involving a prevalence of pigs or chickens are a distinctive feature of some (see Chapter 6). The economic power of the families ruling from strongholds was managed through a system of taxation of some form, most clearly seen in the case of the Estonian *vakus*. However, access to commodities and trade was not exclusively controlled by Late Iron Age *seniors* but can also be associated with a vaguely defined mercantile class, identified archaeologically in Late Iron Age Latvian cemeteries (Šnē 2006, 74), and in Estonia (and elsewhere in the eastern Baltic) potentially in association with commercial fur exploitation (Luik 2003; cf. Loveluck 2013, 302 ff.). In the case of some Latvian regions, wealth appears to have been distributed more evenly in the twelfth century compared to the eighth and ninth centuries. This shift between the Middle and Late Iron Age has been interpreted as an increasingly greater proportion of the population engaging with specialized economic activities and trade (Šnē 2006). Moreover, some of the earliest grants of land to indigenous vassals suggest that a gradual transition from collective to private property had already begun before the crusades, with evidence for family ownership of parcels of land (Blūzma and Lazdiņš 1998; see below).

The creation of the Livonian *Ordensland*, which had been preceded by the formation of an episcopal territory centred in Üxküll (Latv. Ikšķile) and as of 1201, in Riga, contributed to the redefinition of the ownership of land, an ongoing dynamic process articulating the power relations between the new major political factions: the Order, episcopates, and towns. The ability to make decisions concerning territorial organization was removed from the indigenous population, but in some cases restored at the level of individual fiefs through the system of vassalage. From an archaeological perspective, where a few indigenous strongholds continued to be occupied, as in the case of Jersika, this most likely represents the presence of vassals which permitted the preservation of local power structures into the fourteenth century. This is particularly evident in Curonia, where the local *Könige* were designated freeholders. In the case of the strongholds at Cesvaine (Ger. Seßwegen), Asote (near Latv. Jēkabpils), and Rauna (Ger. Ronneburg), there is even evidence for occupation continuing into the fifteenth century. Beyond the destruction of strongholds, the indigenous settlement pattern (where it has been studied) remained intact after the conquest and relatively stable until the sixteenth century, experiencing evident growth in some commanderies such as Wenden. In these regions, the organizational structure of the Late Iron Age landscape was not completely erased. In contrast, the settlement pattern in southern Semigallia, the last region to be incorporated into Christian Livonia, collapsed by the end of the thirteenth century as a result of the military campaigns led by the Order, and this frontier was not repopulated until the second half of the fifteenth century and sixteenth century (Jarockis 1998, 52; Šnē 2008, 92–92). A similar situation is evident in south Curonia, where an absence of indigenous place names from the thirteenth century points to depopulation in what became the frontier with Samogitia (Auns 1998, 265).

Re-defining the Conquered Landscape during the Rule of the Sword Brothers (1202–36)

The acquisition of territory was a clear incentive from the earliest years of crusading in Livonia. Bishop Albert motivated crusaders by enfeoffing them with the lands and strongholds they had captured, whilst at the same time providing security for converts and colonists (Tamm 2013, 454). The motivation behind the foundation of the Sword Brothers in 1202 may have been the defence of the Christian colony in Riga and its allies, which were otherwise reliant on the seasonal presence of crusaders (Jensen 2001, 86), but the Brethren subsequently led the crusading enterprises in Livonia until 1236. Recruited from the families of German nobility and merchants in Riga and maintaining a presence in Lübeck and Visby, they typified the inseparable association between crusading and commerce in the eastern Baltic, the alliance of *ecclesia* and *mercatura* (Blomkvist 2005, 685). The written records of the Brethren's rise to power before their merger with the Teutonic Order in 1237 indicates an overwhelming concern with access to natural resources, linked to the pragmatism of sustaining their garrisons. Military conquests, involving the destruction or appropriation of indigenous strongholds, were accompanied by the immediate annexation of their former territory (or at least physically accessible territory), usually with a resident population situated in villages who were expected to continue working the land and contribute to new provisioning requirements.

When founded, the Sword Brothers were not intended to be territorial lords but comparable in status to vassals. Bishop Albert had received Livonia as a fief from the Holy Roman Emperor in 1207, thus acquiring the sole right to exercise secular power in those territories which his vassals had conquered. At the same time, a third of all conquered territories were granted to the Brethren following their insistence (*HCL* XI, 3). In 1210, Pope Innocent III confirmed the division of land, stipulating the bishop's ecclesiastical and secular overlordship over the military order. When the Brethren subsequently received land within the bishoprics of Dorpat and Ösel-Wiek, they also became subjugated to their bishops. However, conflict erupted between the Brethren and the Danish crown over the subdivision of Estonia, a situation that was not resolved until the Treaty of Stensby in 1238. Curonia was also at the centre of ongoing disputes between the indigenous population and the various political factions of the new regime in the 1230s. The Bishop of Riga, Sword Brothers, and the burghers of Riga all claimed territory there, with the Curonians attempting direct treaties with the papacy. This reflected papal policy at the time, aimed at preserving the freedom of converts in Livonia and even governing the territory. In 1234, the papal emissary renewed fiefs granted to fifty-six burghers of Riga in Curonia, and granted twenty-five *Haken* of land to the priests of St Peter's in Riga (*LUB* I, 135; LUB II, 136), confirming the earlier territorial division. In 1245, William of Modena revoked the earlier treaties and subdivided the region again, with two thirds being granted to the Teutonic Order (who had inherited the Brethren's territorial 'portfolio') on account of the effort the Knights had expended in subjugating the Curonians.

Two new territorial paradigms were introduced during the crusading period. The first concerned property and was defined by the hereditary fief (*feudum*, Ger. *Lehen*), alongside the subdivisions of *Hufe* (hide) and *Hake* (ploughed land). The selective granting of fiefs was used to reward loyal vassals, and resulted in the creation of linear boundaries. The process was initiated by Bishop Albert in 1201, who granted fiefs to a local *senior* and a German knight which centred on the *castra* at Lennewarden (Latv. Lielvārde) and Üxküll (Latv. Ikšķile) (*HCL* v, 2). The granting of fiefs on the Sword Brothers' (and later Order's) territories would be different, typically with the designation of much smaller parcels of land. The difference between the two reflected the ability of the latter to mobilize their own military and administrative resources, whilst the Livonian espicopates were far more reliant on vassals for managing their territories. Whilst the *feudum* as a form of land tenure with specific obligations was new, it appears to have mapped onto existing traditions of private ownership in some parts of Livonia. From late 1207 through to the middle of 1209, the Master of the Sword Brothers granted to Managinte and his brothers (indigenous Livs) the following land described as belonging to their fathers (*heredibus*) as a fief: two *Haken* of land by the old *castrum* of Ydoven, the fields of Pondekeren, the island of Grebasze and five other islands by the bog, and the rivulet of Sloka (which lies in the vicinity of modern Mālpils, south of Sigulda) (*LGU* I, 1). This confirmation of land that already belonged to this Liv family typified the seizure of territory and its return to a disenfranchised population, redefined as vassals. It cemented the new relationship of power between the conquering regime and the indigenous population. The second paradigm was that of the diocese which also introduced the territorial division of the parish, supported by a distinctive form of taxation-tithes (Kala 2001, 10–12). Again, parishes may have been mapped onto earlier administrative structures, as in Semigallia (Jarockis 1998, 52) and on Ösel, where the centres of many (but not all) medieval parishes were related to Late Iron Age territorial units (Mägi 1998). These new territorial paradigms were managed by a new administration, dominated by bishops, officials of the Sword Brothers and their appointees, and at the local level by judges and priests.

Within the framework of broader territorial disputes between the new Christian powers, claims to the ownership of land focused on access to specific resources. In 1211, Bishop Albert and the Sword Brothers settled their dispute over the division of the territories of the Livs and Latgalians, resulting in the Brethren acquiring one-third of the fish weirs located within these lands, as well as a third of the so-called 'King's Island' (*LUB* I, 18). Not long after, in October 1213, the Sword Brothers filed a complaint with the papal curia claiming that the Bishop of Riga did not allow them to build a church in Holm (Latv. Mārtiņsala). In this complaint, the Brethren demanded a third of the churches belonging to the town of Riga, but also a third of the tithe, the judicial authority (*advoctiam*), the right to mint coins (*monetam*), fish sites (*piscationes*), and ultimately one-third of the entire territory (*et ipsus civitatis tertiam partiem*). Pope Innocent III, keen to support the Brethren following their successful crusading campaigns, ordered the abbot of the Cistercian monastery at Dünamünde (Latv. Daugavgrīva) to intervene and shield them from the bishop's claims (*LUB* I, 27). Colonists who settled in Riga soon represented a cohesive political unit, and the town began to feature regularly in territorial negotiations. In December 1225, the Bishop of Semigallia and the town of Riga concluded an agreement, mediated by the papal legate, concerning the castle of Babbit or St Mary (Latv. Babīte) and the surrounding territory, west of Riga across the Daugava. The bishop was to hold half of the extent of the 'Semigallian River' (Latv. Lielupe) and its left bank. Within his share he was entitled to build fish weirs, whilst the citizens of Riga could harvest hay and cut firewood on these lands at any time. It was also noted that those who already had fields or cultivated trees on either side of the river did not suffer any injustice (*LUB* I, 76). The following year, the papal legate William of Modena guaranteed the town an equal share (i.e. a third) of conquered territory, alongside the Sword Brothers and Bishop. The reason for this was almost certainly the participation of Riga's burghers in military actions alongside the Brethren and other crusaders (Jensen 2001, 86). In the same year the town council of Riga was mentioned for the first time. Riga continued to function as the most important assembly point for the various campaigns that made up the Livonian Crusade, where, as the *Livonian Rhymed Chronicle* (*LRC* 11,002–13) notes, the army could supply itself with food and fodder better than anywhere else.

The few monasteries that were established also became important landowners. On 17 March 1226, arbitrators appointed by the papal legate determined the border between the patrimony of the town of Riga and the land of the Cistercians of Dünamünde. In the description of the border there are references to the monastery's properties by the rivulet which issues from Lake Rodenpois (Latv. Ropaži; the modern Lake Juglas located just to the north-east of Riga) and discharges into the Daugava; on this rivulet, the document stated, the monks have a mill and shall also keep a measured plot of land (*spatium*) between this rivulet and the 'new canal' that connected the lake with the river. The 'new canal' can be identified as Mīlgrāvis, a lasting monument to the hydrological investment of the Livonian Cistercians, and the rivulet is now called Dūņupe, with the respective plot of land located on Sūnu Island. Both parties were granted the right to fish in these waters, including in the Laghena (the Langas rivulet that discharges into Lake Ķīšezers), as well as the Gauja, the Daugava, and the sea, except for the weirs which belonged exclusively to those who made them. Within the mill's rivulet and the waters that passed directly by the monastery only the monks were granted permission to fish and make weirs (*LGU* I, 8). The island of Ramesholm (Latv. Mangaļsala) was granted to the Sword Brothers, and the borders of their territory carefully defined. On the other side of the Daugava River at a place called 'St Nicholas's Tree', a canal marked the boundary between the lands of the Cistercians and the Sword Brothers; another canal is mentioned as discharging into this canal, with further border signifiers in the landscape defined by rivulets, an oak tree with a beehive (*mellifluam quercum*), a pine tree located between two hills, the 'Semigallian Road', the 'Semigallian River', and the Pele rivulet (Latv. Spilve) (*LGU* I, 8). This grant contains many of the key features which would continue to define the value of land into the sixteenth century: fishing rights, mills, and beehives.

The outcomes of what must have been complex and at times strained negotiations, requiring the intervention of papal emissaries, indicate the importance of honouring existing land ownership. This notion of land 'custom' defined the understanding of land ownership and related privileges into the sixteenth century. It reflected the recognized importance of satisfying multiple political agents within the power block that emerged after the crusades, which included not just the military orders and Livonian episcopacy, but also towns, monasteries, and, at least on episcopal territories, loyal vassals. Vassals on the Order's lands had no effective political agency. Riga became increasingly powerful as a political entity, and disputes over territory came to a head in March 1226, midway through a series of crusades against Estonia. The papal legate mediated a dispute between the theocratic authorities based in Riga — the Bishop and Cathedral Chapter and the Master of the Sword Brothers — and the citizens of Riga concerning the boundaries of the *Stadtmark* or patrimony of the town. The details of this are worth citing:

On this side of the Daugava River the boundary runs from the Rumula (Latv. Rumbula) rapids in a straight line to Lake Rodenpois until it reaches the bridge across the Pitcorga rivulet (Latv. Piķurga), and from there it proceeds to the end of the lake called the tongue (*lingua*), which on one side is fenced in by the lake, but on the other by the waters of Kilaguais (Latv. Kilagvaja) and further downstream to the border of the properties of Dünamünde monastery on this side of the lake. Within these boundaries there are meadows between Rodenpois bridge and the two hills which are located on the lake shore opposite each other: the one on this side is called Naba (after the rivulet that flows past it), but the one on the other side is called the Eagles' Hill because eagles nest there. The Sword Brothers may keep their fish weirs which they have there on condition that a free corridor is left on one side for ships to pass, and fishing is done only when fish swim downstream and not upstream. On the other side of the Daugava the boundary proceeds in a straight line from Rumbula to the Meisse River (Latv. Misa) and from it to the spot where the Semigallian River joins Babat, and from there further on along the middle of the river to the sea, while on land the boundary line stretches to the border of Dünamünde monastery. The Mark is for the shared use of the burghers of Riga, crusaders or pilgrims (*peregrinorum*; see Blomkvist 2005, 672) and merchants, but may not be used by the Bishop, the Dean of Riga's Cathedral Chapter and the Master who shall keep the share that lies between Babat and the Semigallian River, as well as all the meadows on both sides of the river, because these have again been purchased from the Bishop of Semigallia; likewise all the meadows, that have or may be made in Lake Rodenpois and on this side of the lake as far as the Pitcorga Rivulet, as well as cultivated and uncultivated meadows on both sides of the rivulet. However, fishing rights in Lake Rodenpois shall be shared by all, except for the Sword Brothers' fish weirs.

The rest of the Mark shall be for the common use of everybody: the clergy and lay-people — both for fishing, as well as for pasture, wood-cutting, digging clay, baking bricks and burning lime, and for building kilns for this purpose. The creation of meadows within grasslands and groves that grow in shrubs shall be shared. The fields, meadows, trees, mills, and rivulets that run through them, as well as the old fish weirs that have been made in the Mark must remain the unhindered property of those who have made them — whether clergy or lay-people or the monks of Dünamünde monastery. However, new fish weirs and mills shall only be constructed subject to the approval of all parties. Even if the Selonians or others who have settled in the Mark pay the lord's share (*censum*) to the Master (of the Sword Brothers) or somebody else, they still have to pay taxes to the town as well. The entire island of Osmesara shall remain the property of St Mary's church. Everybody — the Bishop, the Dean of the Cathedral Chapter, the Master, and all

> the burghers of Riga shall have the right to cultivate the uncultivated areas in the Mark of the town in the following way: to use the fruit of the land for eight years and after that, if the respective plot is suitable for a good meadow or pasture, it shall pass over for common use, but if it is suitable for fields, the town shall receive tax for it. If a dispute arises over something in the Mark — cultivated or uncultivated sites, trees, fields of cultivated forests (*agri vel sylvae incultae*), old or new fish weirs — the dispute shall be settled by three sworn citizens of the town appointed together by the Bishop, the Dean of the Cathedral Chapter and the Master. If the arbitrators are unable to come to an agreement, a simple majority vote (two votes against one) will decide the case; if each of the arbitrators holds a different opinion, they shall cast lots. The arbitrators shall be remunerated for their efforts by the conflicting parties. They — the arbitrators — shall also allot one whole *hufen* to the Hospital of the Holy Spirit and half a *hufen* to the collector of dues for the use of Rodenpois bridge within the Mark. (*LUB* I, 78)

The meticulous attention to detail is striking. Resource allocation was carefully defined and included access to all major types of landuse: fields, meadows, and woods. Rivers and lakes feature throughout as both resources and boundary markers, the two functions consistently mapped onto each other in both Livonia and Prussia. There is also evidence for management and forward planning, with the regulation of fishing regularly featuring in such negotiations, in this case banning the harvesting of fish swimming upstream, presumably during the spawning season. Two days after the papal mediators oversaw the signing of the agreement, three arbitrators negotiated the boundaries of the town and the monastery, which highlighted restricted access to fishing, pasture, and woodland (*LUB* I, 80).

The territorial disputes, which defined the economic basis of the political configuration of the eastern Baltic in the first three decades of the thirteenth century, took place at a time when crusading in Livonia was at its most intensive. By 1230, the new administrative system was in place in the new political cores of Livonia. It generated, for the first time in the eastern Baltic, a bureaucracy that enables the control of territory to be scrutinized at a high resolution, although consecutive periods of time are very erratically represented. The indigenous organization of the landscape was partially retained in three ways: through the reuse of earlier power centres, the adoption of the indigenous network of communications, and the redefinition of indigenous vassals' property as fiefs. The granting of fiefs began almost immediately after the annexation of conquered territories in the early decades of the thirteenth century, with the term *heredibus* used in recognition of existing claims to property. However, enfeoffment was highly selective and far from the norm in Livonia, with further differences between the vassals of the episcopate and the Order.

In the territory of Viljandi, the former centre of power was adopted as the location for the regional convent, whilst the medieval road system appears to have been largely based on the prehistoric network of communications. Here, the choice not to reuse certain strongholds reflected the importance of the central routes which came to link the territories of the Order, and those that lost their importance (such as Lõhavere) were located away from the main routes (Veldi 2014, 411–12). Following the destruction of the local nobility, no grants of fiefs were issued around Fellin where the Order became the sole landowner. The Vendish stronghold at what became known as Wenden was also initially utilized by the Sword Brothers, and then superseded by the adjacent castle. Again, there is no evidence of enfeoffment here before the sixteenth century, with the earliest surviving example dating to 1517 (*LGU* II, 130). In contrast to the former organization of indigenous territories, Wenden and Fellin became connected with the main convent in Riga in a way that reflected the imposition of a new supra-regional authority. Even in the Lower Daugava region, which has the earliest examples of enfeoffment, these privileges did not extend to larger indigenous clusters. In 1229, an agreement was concluded between the Prince of Smolensk and German merchants which came to define the future regulation of trade along the Daugava. Noticeably the local indigenous population, who had dominated the organization of trade in the valley from the tenth century, was completely excluded from this agreement (Radiņš 1998, 190).

In 1236, the destruction of most of the Sword Brothers at Saule heralded a new phase in the political development of the Christian governance of Livonia. The bishops of Riga would gradually lose their earlier hegemony, as the remnants of the military order and its lands were seamlessly absorbed into the Teutonic Order. Again, this process was facilitated by the papacy, now with the Order as its sole military arm in the eastern Baltic. In Prussia, the Order outmanoeuvred its episcopal competitors; however, in Livonia the political configuration of the early thirteenth century — represented by the relationship with the episcopates and those territorial negotiations — was already fixed and would determine the character of the Order's rule here. Here, the Order inherited the Brethren's subjugative relation-

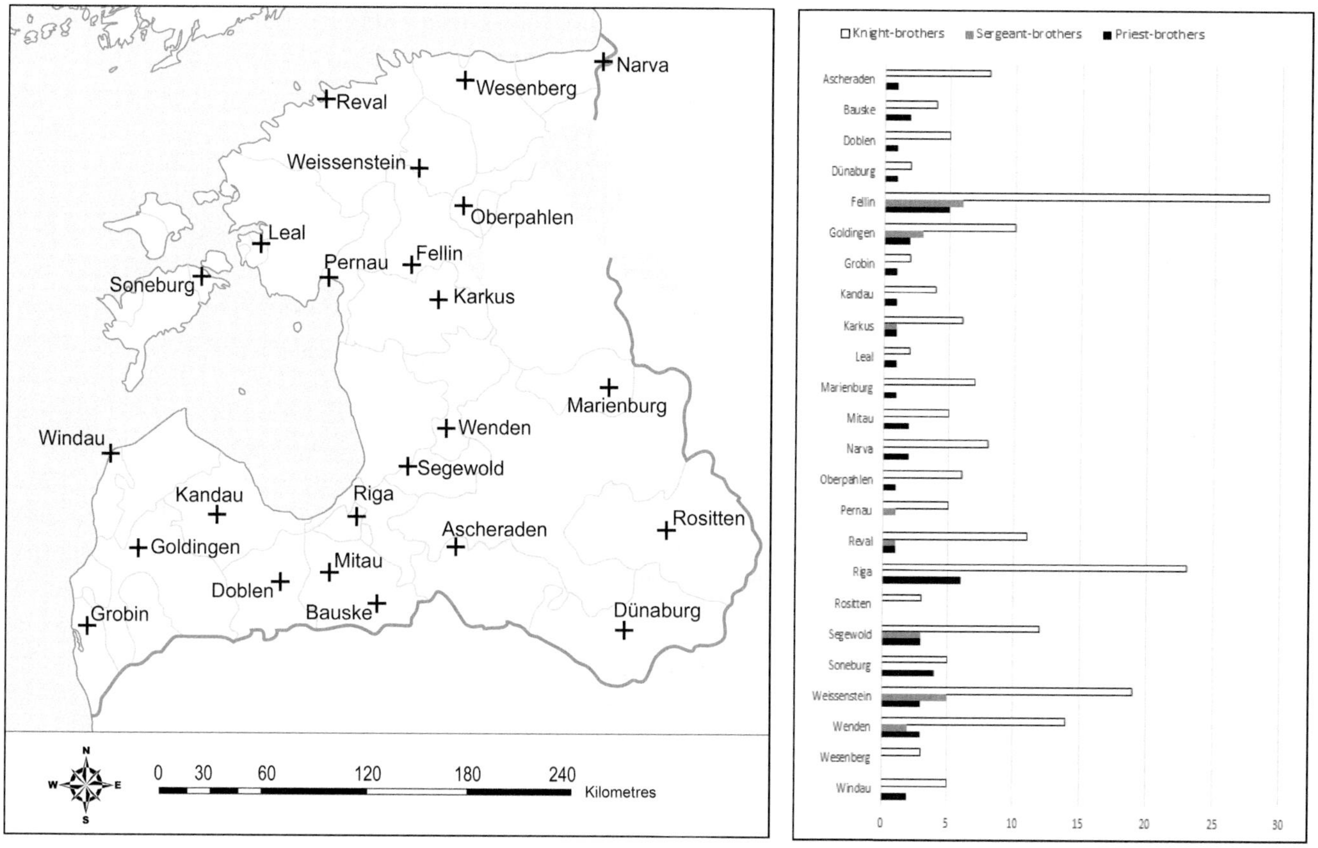

Map 8.4: The major territorial boundaries of medieval Livonia showing numbers of brethren in the Order's houses in 1451 (after Benninghoven 1963).

ship with the archepiscopate of Riga and the episcopate of Dorpat and Ösel-Wiek, with only the episcopate of Courland becoming incorporated into the Order in the thirteenth century. Until the last quarter of the fourteenth century, the Archbishop of Riga and at least one Bishop of Dorpat demanded the Order to swear allegiance to them in the manner of a vassal.

Design and Regulation in the Territories of the Teutonic Order in Livonia (1237–1561)

With the incorporation of the Sword Brothers into the Teutonic Order, a new supra-regional management structure was gradually consolidated on the conquered territories that would come to define medieval Livonia. The Brethren's subservience to the Livonian episcopate was accepted by the Order until the beginning of the fourteenth century, when it became a major point of contention with the Archbishop of Riga. Between 1237 and 1290, when the Order forced a truce on the conquered Semigallians, its armies had been fighting on three major fronts: Prussia, Livonia, and the Holy Land. During this time, its headquarters were relocated twice: from Montfort to Acre in 1271, and from Acre to its *domus* in Venice in 1291. Its provincial headquarters in Elbing (Pol. Elbląg) and Riga were simultaneously concerned with the military logistics of a sustained holy war, the reorganization of conquered territories, and the exploitation of the landscape to sustain its colonies. Over this half a century, the Livonian branch of the Teutonic Order focused on transforming its spheres of political influence into de facto control of the landscape. As new territories were acquired, administrative centres with military garrisons were established to manage their inhabitants and resources. This intent is explicitly stated in the *Livonian Rhymed Chronicle* (*LRC* 6687–90) when referring to Estonia after the conquest, where the 'division of the lands had been carefully thought out and established by charters. There was nothing in Estonia, no clovers (Middle High German *clê*, translated also as "meadows", see Kļaviņš forthcoming), heath, lake, forest, or field that was not duly described and recorded'. Over the course of the Livonian Crusade, the blueprint was established using the fortified convent as the principal tool of the Order's administration. In 1242–44, the Order reoccupied Curonia and established the convent of Goldingen. Following renewed conflict this was consolidated again in 1267. In 1264 the Order had been granted part of the territory of the Latgalians by the Prince of Polotsk, and this was consolidated in 1273 with the establishment of a convent at Dünaburg (Latv. Daugavpils) (Auns 1998, 260–62). In August 1290, at the meeting of the general chapter of the Livonian branch of the Order, the details of the organization of the Curonian commanderies of Windau (Latv. Venstpils) and Goldingen were clarified. They specified how supplies for these convents, which were also dependent on each other, would be drawn from a range of territories in Livonia, including the island of Ösel. The document explicitly stated that the convents had to be self-sufficient and separated from each other, each with its own food and lands, whilst acknowledging their role within the Order's hierarchy under the leadership of the Master (*LUB* I, 586). The conquest of Semigallia was far more protracted, and after hostilities had officially ceased in 1290 with the surrender of Sidrabe and the depopulation of the region, the Order would only establish new centres in the fourteenth century, starting with the reoccupation of Mesothen (Latv. Mežotne) in 1321 (Auns 1998, 264). The primary case study regions considered within this project (Riga, Wenden, and Fellin) were early centres for the Sword Brothers, and despite the lack of data regarding the definition of their territories, it is reasonable to assume that a similar process of designating resources was specified by the Order. The territorial boundaries of the various political units in medieval Livonia are widely replicated in the literature (Map 8.4), but they are largely based on late seventeenth-century Swedish cadastral maps.

In addition to the larger castles, a series of smaller *castra* were built in the thirteenth century, some initially as crusading outposts (e.g., Alt-Dahlen, Latv. Vecdole), others with a more defined administrative role. Those included within this study were Karkus (Est. Karksi) and perhaps even the earliest phase of Arrasch (Latv. Āraiši) given the radiocarbon range of the faunal assemblage recovered from its deepest occupation layers, although the castle is traditionally dated to the fourteenth century (see Chapter 4). These castles, typically associated with the office of advocate (*Vogt*, although the type of official is unclear in the case of Arrasch) where they are first mentioned in written sources, were important centres of governance within the Order's territories. The delineations of their territories are virtually impossible to reconstruct for the later thirteenth and fourteenth century, and in the absence of an earlier centre at the site of Karkus Castle its territory is more likely to have mapped onto one of the subdivisions of Viljandi's overlordship. Within this territory landuse was mapped according to different resources. A visitation report from 1442 stated that Karkus Castle had 1407 and a quarter *Haken* within its territory of which around a third were unoccupied (Biskup and Janosz-Biskupowa 2002, 186). The plough lands for the convents in Fellin and Riga are unknown, but they must have been more substantial. In the Polish audit of 1582, Wenden's associated districts included 185 and a half *Haken*, but this number would have been significantly higher when the territorial extents of the commandery included its subservient advocateships and later the Master's district. The period associated with the expansion of the Order's houses in the fourteenth century is, unfortunately, poorly represented in the documentary record, although it is clear that the Order consistently granted small fiefs. In this respect, environmental changes can be used as proxies for how the territory was being actively transformed by indigenous communities under the managerial scrutiny of the Order. The exception is Riga, where the treaty from 1330 clarified how local resources were to be subdivided between the town and the Order. This treaty arose from the town's surrender to the Order after a bitter conflict and its conditions were purposefully harsh.

The town would allocate a site next to the town for the construction of the Order's castle, although they would share pastures. According to Wartberge, this site was the location of the town's stable, a horse-mill, and a lime-kiln (*HWC*, 74–75). The Order would be free to build dams, ponds, and mills in any place on the mill run next to (St George's) hospital. The town would keep two windmills, but all the fields that the burghers owned on the islands of Rigeholme and Lokeshare, as well as the manor (*curia*), which belonged to Gerlach Rose with everything that goes with it, specifying fields, pastures, and fisheries, would be handed over to the Order to serve the needs of its castle. Five months later, in a gesture of goodwill towards the town and with the stated aim of a lasting peace, the Livonian Master granted its people all the gardens which the Order owned outside Riga (except those owned by the Dominicans), Kalverholm (Latv. Klīversala), several plots of land, the right to fish freely in all the Order's waters (as had been the case formerly), and the right to freely cut firewood on both sides of the Lielupe River, except for construction timber, which remained under the Master's control. The residents of Riga were also given the right to collect birch bark and bast half a mile from Lake Canneierwen (Latv. Kaņiera ezers) (*LUB* II, 741).

The relationship between the Order and the town would continue to fluctuate. In May 1366, the Grand

Master of the Teutonic Order himself settled a dispute between the Archbishop of Riga and the Livonian Master over the rule of Riga. The terms are instructive insofar as they provide details of the Order's holdings, but reiterate the importance of real estate, territorial control, and, above all, access to resources. The Livonian Master was to retain the Order's castle in Riga together with its outer bailey (*suburbium*) and all the buildings located within or outside it, the mill in front of the castle, 'Andrew's Island' (Latv. Andrejsala) in the Daugava, the Order's old buildings in the town of Riga and outside it, as well as the buildings which belonged to the commander of Segewold (Latv. Sigulda) and the advocate of Wenden, along with four chapels in the castle of Riga, while the Archbishop would keep the fifth one (*LUB* II, 1033). The Archbishop would also obtain Brother Bertold's Mill, located outside the town and which until then had belonged to the Order. The newly mortgaged properties outside the town that the burghers had been forced to give to the Order would be returned to them. The Archbishop would also withdraw all of his claims to the castles, buildings, and fortifications at Lake Burtneck (Latv. Burtnieku ezers, also called Astiyerwe), and all the lands that the Teutonic Order had taken over from the Sword Brothers (*LUB* II, 1033). This treaty, which was intended to release the Order from its status as a vassal of the Archbishops of Riga, was not enforced as it was never ratified by the papacy.

In the fifteenth and sixteenth centuries, there is more evidence for the granting and selling of fiefs to vassals with German names on the Order's lands. For example, on 20 June 1434 the Livonian Master granted Wedige von Sacken a plot of land, which the latter had bought in the castle district of Hasenpoth (Latv. Aizpute, in Curonia), as a fief. The borders of this fief are described as follows: 'start from the spot where the Mellupe River flows into the Tebber River (Latv. Tebra), from there go to the stone with a cross carved on it, from this go to another stone with a cross carved on it, from it follow along the edge of the holy woods (*hilgen busch*) until you reach the brook named Absewalke (Latv. Apšuvalks), then proceed along the brook to the Zintere rivulet (Latv. Dzintare), along the Zintere go to the border of the territory of the Bishop of Courland' (Bauer 2003). The site can be localized within the vicinity of Dzintare Hillfort on the right bank of the Tebra River, in the present-day Lažas *pagasts* in Aizpute district. The documents of *feuda* demonstrate how land was defined on the basis of visible natural and artificial landscape features, although this definition was not precise or coherent enough to prevent regular disputes over boundaries. Potentially any landscape feature could be used as a marker, and the Order's grants of land are peppered with references to inscribed trees, where specified usually oaks and pines. These were often marked with crosses (see below), but other symbols are also mentioned: a description of a fief within the advocateship of Rositten (Latv. Rēzekne) in 1475 refers to a pine into which a sword had been carved, almost certainly reflecting a specific type of ownership (*LGU* I, 510). It is also not surprising to find references to bee-trees, containing hives, serving as distinctive markers of fiefs (*LGU* I, 415). Poorly defined boundaries led to disputes, with evidence for falsely marking borders. In 1472, the commander of Fellin, the advocate of Karkus, and the *Kumpan* of Wenden ('comppaen to Wenden') collected evidence regarding a border conflict between vassals in the district of Trikaten (Latv. Trikāta). They noted both discrepancies in the history of the boundary's records, and how one of the fief-holders had secretly chiselled boundary crosses and placed stone markers (*LGU* I, 487). The use of pits filled with charcoal as boundary markers was most likely because these were difficult to remove and destroy without leaving any traces.

The written evidence for more centralized regulation is relatively late, although earlier visitations were most probably conducted orally (Jähnig 2011). In 1441, the Grand Master issued instructions to the Order's inspectors, requiring each commander and advocate to note down how many *Hufen* and *Haken* of land, cultivated and uncultivated, taken and vacant (*czinshaftig adir wuste*), there were in his district, as well as to take note of the teneurial dues that towns, mills, and taverns in each district were obliged to pay. The official of each district would also have to note how many and what kind of brethren were present in each castle, along with quantities of horses and units of armour, and how well the castle was supplied with food. The district officials were forbidden from selling or renting any land, whether inhabited or vacant (*LUB* IX, 794). Shortly after, a set of regulations issued from the Order's headquarters in Marienburg forbade all officials of the Order in Livonia to sell off any assets, aimed at ensuring the integrity of the Order's castles, offices, and territories (*LUB* IX, 716). These regulations came at a time of financial desperation for the Order in Prussia and did not prevent the gradual de facto separation of both branches by the end of the fifteenth century.

By the fifteenth century, Livonian representative assemblies (*Landtags*) were involved in mediating territorial disputes. In December 1435, the *Landtag* in

Walk (Latv. Valka) settled a dispute between the Dean of Riga's Cathedral Chapter and the Order in Livonia. The details reveal an ongoing concern with access to a range of resources between the theocratic authorities in the principal valleys of central Livonia. Both parties agreed that access to the lamprey weir in the Gauja River would alternate between them on an annual basis. Firewood and timber for construction for the requirements of the Archbishop's castles in Lennewarden (Latv. Lielvārde) and Kokenhusen (Latv. Koknese) could be sourced on the other side of the Daugava in woods that were not suitable for fields and bee-trees. The Archbishop's vassals living in the territory of Lennewarden castle were permitted to hunt with traps and beaters on the other side of the Daugava, as they had always done, with the same privilege granted to the subjects of the Order (*LUB* VIII, 1019). The same *Landtag* considered a dispute over the ownership of manors along the Daugava opposite Kirchholm (Latv. Salaspils) and Alt-Dahlen, which throws light on the complexity of shared access to the same stretches of water and even the same processing sites. The settlement included the transference of ownership, in perpetuity, of part of the weir in the Daugava called Zeelsche, from the Dean to the Order. In return, the Order gave the Dean and his successors a part of the weir called Ostingholmische, whilst the Dean was granted the manors of Memorgen (Latv. Memurga) and Keckowe (Latv. Ķekava), along with 'their villages, lands and people, cultivated and uncultivated fields, meadows, fishing sites, lakes, flood-lands, groves, forests and bogs, bee-gardens, wood-felling sites, hunting rights, rights and shares and any other appendixes and benefits' (*LUB* VIII, 1016). The manor, in this respect, was conceptualized above all as the sum of its natural resources, alongside the tenants. The settlement ensured the Dean would also acquire ownership of Bersenbeken Mill (Latv. Bērzes), and whilst the Order would control access to the quern of the mill (and thus the production of flour), the Dean would have ownership of the fishing sites in the adjacent millpond (*LUB* VIII, 1016). The above example also underlines that certain aspects of the landscape were emphasized in the Order's value assigned to territorial units, specifically the control of woodland resources, especially beehives, and access to riverine, lacustrine, and coastal resources.

The Value of Woodland: Timber, Honey, and Wax

During the period of the Order's rule, there is evidence for the maintenance of woodland in each primary case study area (see Chapter 5). Some woods were clearly substantial, and along with marshes are regularly described as places of refuge in the crusading narratives. For example, in 1210, when the Estonians were defeated by the crusaders they did not look for bridges or paths, but ran away through woods and bogs (*viltnisse und bruch*) (*LRC*, 66 ff.; Nielsen 2011); similar tactics were used during the Estonian uprising of 1343–45 according to the *Younger Livonian Rhymed Chronicle*. In the Late Iron Age, substantial woods and wetlands may have functioned as territorial boundaries, and subsequently the boundaries of medieval fiefs were regularly defined by these topographic features. But if Henry's *Chronicle* conceptualized the Livonian forests as something distinctly 'pagan', woodland was valued pragmatically as an important resource, with access to trees for fuel and building material featuring regularly in disputes and treaties concerning territory. Part of the agreement between the Order and Riga in 1330 included defining access to local timber resources, specifically the town's right to freely cut firewood on both sides of the Lielupe River, except for construction timber, which remained under the Master's control. The residents of Riga could also collect birch bark and bast half a mile from Lake Canneierwen (*LUB* II, 741; see also below). Segregating access to woodland between the Order and towns remained a concern into the sixteenth century; the town laws of Windau in *c*. 1550 prohibited cutting trees in the Order's woods, punishable with a fine (*LUB* II, 901). Of the 113 fiefs sampled from across southern Livonia in this study, issued between 1207 and 1561 (with a large proportion from the territory of Riga and Curonia), only twenty-five (around 22%) explicitly included access to woodland, trees, or firewood, although many more refer to woods and trees as boundary markers. In regions with abundant woodland access was most likely implicit and not stated in enffeofment documents.

Some of the documents also hint at a more widespread culture of woodland management. The granting of a fief by the Livonian Master Johann von Mengede to a probable ancestor of the Curonian *König* Andreas Penneck on 24 August 1456, specified a diverse range of natural resources: cultivated and uncultivated fields, meadows and pastures, clearances, wood-felling sites, woods, groves, rivers, waters, brooks, ponds, bee-trees, bee-gardens, fishing, and bird hunting sites (Švābe 1938, 240). A later grant issued by the Master in 1470 included most of these types of landuse, along with young wood stands, sandbanks, and lakes (*LVVA*, 554. f., 3. description, case 1113, sheet 5). This specification of different types of woodland, with young stands, for felling, as well as references to firewood and construction timber, is suggestive of woodland management comparable to other regions of Europe. Important by-products of woodland exploitation were charcoal, ash, pitch, and tar. The demand for the latter two was particularly high with the growth in shipbuilding (Jahnke 2015, 226), and the Order maintained separate ash, pitch, and tar yards in Riga. The other key features of the Livonian (and indeed Prussian) woodland economy was honey and wax production.

Bee-trees (Lat. *arbores apum*, MLG *honnichbomen*) are one of the recurring features of the Order's estate documentation in the eastern Baltic (see also Chapter 15). They referred to trees which contained hives that were protected and maintained, and as suggested by the narratives of the crusades, were already an established use of woodland in Late Iron Age society (Williams et al. 1995). Honey itself may even have formed the basis of a sacred beverage (Kļaviņš 2019). In 1206 the crusaders destroyed the bee-trees (and fields) belonging to Caupo, the *senior* of the Treiden Livs (*HCL* x, 13), and six years later the dispute between the Sword Brothers of Wenden and the Latgalians of Autine also concerned fields and bee-trees (*HCL* xvi, 3), whilst the agreement between the Bishop of Semigallia and the town of Riga in 1225 specified that no harm should come to existing 'cultivated trees' (*arbores cultas*) on either side of the Lielupe River, perhaps referring to hives (*LUB* i, 76). In 1282, during a transfer of land from the monastery of Dünamünde to the Order, it was specified that Curonians could keep their existing bee-trees in compliance with the general 'Law of the Land' (*LUB* i, 447). Following sporadic references to hives in the fourteenth century, they continued to feature in territorial disputes in the fifteenth and sixteenth century. Splitting the ownership of groups of bee-trees between various parties became the subject of precisely defined agreements (e.g., *LGU* ii, 642), although as indicated by various entries in Riga's revenue book, trees were carefully marked to designate ownership (Napiersky 1881).

Hives could also be kept in apiaries referred to as 'bee-gardens' (*honychweyde*) (*LUB* xi, 373); in 1426, the Livonian Master granted a fief to the freemen named Pytkeyanne, the old Jacob, Towte and all of their lawful heirs, which consisted of bee-gardens located in the forest on the other side of the Daugava in the direction of Lithuania. The recipients could use the bee-gardens freely and unimpeded as well as cut trees there for their needs, except for the Order's bee-trees that belonged to the castle of Holm (*LUB* vii, 491). In some cases, space for apiaries was prioritized. In 1435, a settlement of disputes between the Dean of Riga and the Order agreed on during the *Landtag* in Walk (Latv. Valka), specified access to firewood and construction timber on the other side of the Daugava in woods that were not suitable for fields and bee-trees for the episcopal castles in Lennewarden and Kokenhusen (*LUB* viii, 1019). Destroying such valuable resources was considered criminal. In April 1446, the envoys of the Livonian Master complained to the Grand Duke of Lithuania that the residents of Pskov had invaded the territory of Rositten twice, completely demolished the bee-trees, and had stolen the honey (*LUB* x, 214). The sixteenth-century 'Peasants' Rights' in the Archdiocese of Riga specified fines for the destruction of hives, and in the case of Curonia execution by hanging if their value exceeded eight Livonian Artigs (Arbusow 1924–26, 40).

Why were beehives so important in medieval Livonia? Beeswax constituted more than just a major export, it was 'one of the foundation pillars of Hanseatic trade' (Jahnke 2015, 220). The best quality wax was produced in the eastern and south-eastern Baltic region, with Riga mentioned by the Florentine merchant Pegolotti in his *La Pratica della Mercatura*, written in the first half of the fourteenth century, as one of seven international sources of wax (Evans 1936, 293). Wax was readily available on the open market from the lands of the Order and Livonian episcopates, not to mention entrepreneurial vassals, but the town could also benefit from its patrimony. In the mid-fourteenth century, those keeping bee-trees within Riga's territory had to give at least half of the honey they collected, and in some cases also half of the wax, to the town (*LR*, 13, 58, 164). In September 1349, the Livonian Master approved the agreement reached between the town of Riga and the Livs of Kirchholm, who were the Order's vassals and lived within the town's patrimony. This specified a boundary line running from Rumula to Neuermühlen (Latv. Ādaži), where the Livs could keep the existing bee-trees on the other side of the boundary as inheritable property, but from any newly installed bee-trees they had to give a third of honey and wax to the town of Riga (*LGU* i, 71). The Order's officials procured significant quantities from their subjects as dues and from their own bee-trees, for the immediate needs of the households and with a surplus that could be moved to other convents, or to the two major international outlets — Riga and Reval. Honey was a popular sweetener and the basis of mead, which features in the inventories of the Order's castles in both Prussia and Livonia. Wax, however, was directly linked to two characteristic features of the *Ordensland*: bureaucracy and religion. Mixed with turpentine, its use as sealing wax was essential for the exponentially increasing level of documentation generated by towns, convents, and ecclesiastical households. The sacral significance of wax for illuminating religious spaces, and particularly for liturgical needs, would have played a fundamental role from the earliest Livonian church foundations. Indeed, the burgeoning market for wax along with Lenten foods — a demand met by the expansive trade in herring — have been connected with the process of Christianization in the Baltic (Jahnke 2004, 131). The thirteenth century saw the proliferation of the wax candle as a source of illumination in Europe: it was an expensive commodity requiring specialist skills to produce that became closely tied to the emerging cotton industry (Mazzaoui 1981, 102). A document from 1526 issued by the Livonian Master refers to the obligation of the recipient of a fief, to provide each year on St Michael's Day, two pots of wax candles to the castellan of Wenden castle for the constant lightening of the Holy Cross (*LGU* II, 442). But by the 1560s, Riga's importance as an outlet for east Baltic wax was almost completely superseded by Danzig (Pol. Gdańsk); the former accounted for 4.4% of exported wax and the latter 68.9% (Jahnke 2015, 220).

The Value of Water: Fishing and Milling

Access to water was valued for many reasons: rivers were the key arteries of trade linking the Order's network with the Baltic; streams, rivulets, and canals presented opportunities for constructing mills; and along with lakes they were abundant sources of fresh fish and water for brewing and drinking. For all these reasons, the control of waterways and lakes is emphasized from the earliest regulations of the theocratic authorities in Livonia (see above). The documentary record gives the impression of intensive regulation from the later thirteenth through to the fifteenth centuries (see also Chapter 6, and Orton et al. 2019). Fishing rights were included within 28% of the fiefs sampled for this study, although rivers, lakes, rivulets, and canals feature far more frequently as boundary markers in these documents. The conditions of fiefs often specified shared access for fishing. In October 1437, the Livonian Master granted a fief to one Enwold von dem Velde. The fief consisted of a plot of land by Lake Kallen and the Ederitze Rivulet and was granted on condition that the Landmarshal's vassals from Mergenborch (Marienburg, Latv. Alūksne) could fish freely in the lake (*LGU* I, 291). In September 1454, the Livonian Master's grant of a fief in the district of Adsel Castle (Latv. Gaujiena), within the commandery of Wenden, included the freedom and right to maintain a weir (*were*) in the Palsse River (Latv. Palsa) that previously belonged to the Order (*LGU* I, 366). At the same time, the Order maintained its own access. In March 1533, the Livonian Master granted a fief consisting of plots of land in the districts of Ludsen (Latv. Ludza) and Rositten on condition that the recipient of the fief would not interfere with the advocate of Rositten's fishing rights (*LGU* II, 581).

Many disputes in Riga revolved around access to fishing resources, which were numerous and varied within the town's patrimony (also Chapter 6; Orton et al. 2019). These had begun during the time of the Sword Brothers (see above) and continued under the rule of the Teutonic Order. In July 1281, the Livonian Master came to an agreement with the town that a fish weir planned across the Lielupe would be jointly built by the town and the Order, and that catches would be divided into equal shares (*LUB* VI, 3047). Joint-access and maintenance of the weirs was reaffirmed in 1304 (*LUB* II, 610), but after the renegotiation of access to local resources in 1330 after the Order became the overlord of Riga, the fishermen of Riga were henceforth required to give a tithe to the Order for the needs of its castle, and the castle would likewise hand over all the rights to fish in the weirs in Mitau (Latv. Jelgava) (*LUB* II, 741). Tensions over fishing rights with the town continued into the fifteenth century. In April 1400, the commander of Mitau asked the Town Council of Riga to prohibit its fishermen from fishing in the Lielupe River, claiming they had no right to fish there. Indeed, the Order's usual response was a show of force followed by confiscations, and the fishermen of Riga had been divested of their nets and together with their ship escorted back to Bullen (Latv. Buļļi) (*LUB* IV, 1528). Riga's illicit fishing activities ranged across the northern Curonian coast. In August 1454, the Town Council of Riga complained to the commanders of Goldingen and Windau, as well as the advocate of Kandau (Latv. Kandava) about a recent clash between the citizens of Riga and the Order's men on the Curonian coastline of the Daugava over fishing sites. The incident had ended badly for the townspeople, and the Council demanded compensation for their property which had been confiscated by the Order (*LUB* XI, 355). The town itself came to impose regulations on the sale of produce; in 1450, the regulations (*Burspraken*) of Riga prohibited taking fish, meat, and other products into the town if they were not to be sold, whilst at the same time, banning the sale of any meat, dried fish, or other products on the shores of the Daugava.

Figure 8.1: Part of the timber channel at the base of the moat in Cēsis (Ger. Wenden), uncovered during excavations in 2011. Photo by G. Kalniņš.

Alongside fishing, the construction of mills — and the control of the waterways which powered them — features regularly in the regulatory culture of medieval Livonia, and along with wind-powered mills became familiar installations of the conquered landscape. Some were clearly substantial structures; in 1373, the Livonian Master built a mill with six wings or wheels (*molendinum cum sex rotis*) on the Sand Hill near Riga (*HWC*, 130–31). Mills were primarily used for grinding grain, but saw- and oil-pressing mills are also documented in mid-fourteenth-century Riga. Related hydraulic works involving canalization were also extensive, as evident from the numerous references to canals in the documents. The construction of a dam more than one kilometre in length with a bridge over the Jugla River had been planned by Bishop Albert as early as 1221. The river was connected to Lake Baltezers near Bukulti by the 'new canal' or *Mühlenbach* (*LGU* I, 28), which today is paved over by Brīvības Street in Riga. Several of the Order's castles in Livonia utilized local topography to create naturally defensive barriers, but in some cases artificial moat systems were also excavated, connected with ponds, lakes, or rivers. The convent at Riga was segregated from the town by a moat system connected with the Daugava. In the case of Wenden and Arrasch, smaller moats were dug to protect and/or flush single ranges, and excavations revealed that both were lined with timber, in the former example dating to the later fourteenth century (Fig. 8.1). This modification to the channel may have been required to flush higher levels of effluent as household size, food preparation, and/or industrial activities increased.

Skilled hydraulic engineers were not always easy to find, although little is known about this group in Livonia before the fifteenth century. In a letter of 8 September 1417, the Livonian Master asked the Town Council of Reval to send over one of their burghers who was known for his skills in canalizing water. His services were requested for six days in order to channel water to the Order's mill in Riga, as there were no craftsmen at the Master's disposal capable of building sluices and waterfalls (*LUB* V, 2161). In correspondence dated to 11 September 1447, the Grand Master wrote to the Livonian Master with concern that a valued craftsman-carpenter (*werckmeyster uff zymmerwerck*) who built mills in Prussia might start constructing mills for the larger Prussian towns, to the detriment of the Order. The Grand Master offered to send him to Livonia, providing the Livonian Master was willing to keep the craftsman and prevent him from returning to Prussia (*LUB* X, 379).

There was clearly a pronounced political dimension to the ownership of mills, which was carefully regulated by the Order. This ranged from the definition of ownership of different parts of milling complexes, ensuring that the operation of its mills always took priority, to more symbolic statements of authority. Whilst the upper mill in Reval on St John's Brook was rented and pawned to the town by the Order in 1456, the latter retained ownership until secularization (Kreem 2002, 53). This mill, known as the King's Mill during the Danish period, can therefore be regarded as a symbol of the town's overlord. In April 1454, in return for having given up some territory in Semigallia, the Livonian Master had granted the townspeople the two sand mills (*santmolen*) at Riga's Sand Gate along with their associated waters and canals (*watern und reveren*), on the condition that should the level of water powering the mills drop, only the small mill could be operated so that the Order's mill, located next to the castle, could continue to function (*LUB* XI, 335). The construction of new mills was also carefully regulated. In November 1454, one of the concessions offered to Riga by the Livonian Master was permission to build as many mills as the Order possessed, in addition to two new windmills (*LUB* XI, 373). In 1504, a Curonian *König*, Andreas Penneck (and his heirs), was granted the right to build a water mill on a river that flowed through his family's land, on condition that the mill did not damage the peasants' meadows in the Goldingen district, and that the peasants would not be affected by flooding (*LUB* XI, 373). In 1516, the Livonian Master granted a fief to one Johann Anrep which included a mill in Valka, on condition that he must maintain the mill and its buildings in good order from his personal funds and each year grind 6–8 *last* of rye for the needs of Ermes Castle (Latv. Ērģeme) (*LGU* II, 219). As with bee-trees, mills represented viable targets for disaffected groups. In 1481, when the burghers attacked the Order's property, they set fire to the oil-pressing mill (*HHC* II, 777). In sixteenth-century Curonia, stealing from a mill was punishable with racking (Arbusow 1924–26, 41).

Territorial Paradigms and the Indigenous Population

Recent historiography has emphasized the positive relations that developed between the indigenous population and its Christian rulers (Kļaviņš 2009), whilst archaeologists continue to emphasize a visible degree of cultural segregation based on differences in material culture (see Chapter 1). During the period of active crusading, these relations dependend on a shifting mosaic of alliances between indigenous groups and crusaders, in turn reflected in the redistribution of territory, specifically the granting of parcels of land to loyal native vassals. This varied between the various territories of Livonia, with only small parcels of land granted by the Order to its own vassals. In Danish Estonia, most of the land was enfeoffed to major landholders rather than small freeholders. The majority of these were German with smaller numbers of Danes and indigenous vassals. A similar situation was evident in Ugaunia, which was enfeoffed by parishes rather than individual plots of land centred on homsteads, subdivided amongst only four vassals. There is insufficient data from Saccala to indicate how land was subdivided by the Order, but here the elite had been largely destroyed by the crusaders. In the parish of Helme, itself controlled by the Order's castle, there is evidence for settlement proliferation in the Middle Ages, with the first manor — representing a new system of land organization — documented in 1516 (Palang et al. 2005, Fig. 4).

In southern Livonia, there is evidence for more small holders represented by isolated farmsteads, which have rarely been identified in the Late Iron Age. This observation is noted in the *Livonian Rhymed Chronicle* (*LRC* 48 ff.), which states that '[Next to the Selonians' land] there is the Letts' Land (*Lettenlandt*) and the pagans living there have a strange custom: they do not live together, but build their homesteads separate from each other in woods (*buwen besunder in manchen walt*)'. The chronicle was written at the end of the thirteenth century at a time when the redistribution of land was already underway. In this respect the chronicler may have captured a 'snapshot' of the process of new land divisions (see Kļaviņš 2019). But it is also possible this was happening already before the crusading period (Šnē 2006, 70), and prehistoric subdivisions of space based on central places were preserved to a degree, where there is evidence for the long-term continuity of settlement areas (not necessarily direct continuity in individual settlements; Palang et al. 2005).

Many fiefs issued in southern Livonia by the Livonian Master in the fourteenth and fifteenth centuries typically concerned small parcels of land. For example, a document issued in Segewold by the Livonian Master Everhard von Monheim to one Tontegode on 13 May 1333, which specified two *Haken* of land in Syallen parish along with its fields, meadows, and all other accessories, included the stipulation that Tontegode and his heirs should cultivate, use, and govern the plot in perpetuity and were released from delivering the lord's share and from *corvée* (unpaid labour), that is, on the same conditions as the other neophytes in Curonia held their fiefs in tenure (Švābe 1938, 158). In other instances, fiefs were more expansive and included the full suite of local resources. For example, in 1396, the Livonian Master confirmed the rights of two brothers, Onghute and Cure, to freely govern their fief in perpetuity with all the revenues of cultivated and uncultivated fields, pastures, meadows, woods, timber-felling sites, waters, swamps, bee-trees, and hunting and fishing sites (Keyserling 1848, 20–21). A number of documents specify inheritance according to former customs, in particular communities of Livs who were granted privileges based on old customs defined as the Livs' Law (Ger. *Liveschen Recht*). In 1380, the deputy commander of Riga's convent restored the Livs of Rodenpois their former access to woodland, although this was exchanged for one field, and included the Order's additional rights to fields, meadows, and bee-trees within the woods (*LGU* I, 109); a similar expression of inheritance referred to a woodland on the banks of the Jugla in a grant of 1388 (*LGU* I, 124), and in subsequent grants of land in the fifteenth and early sixteenth centuries.

Two broad diachronic trends can be noted within the history of landholding in medieval Livonia relating to the status of the indigenous population. First, there was a gradual shift in vassalage. By the fifteenth century, the majority of fiefs in Livonia in the territories of the Archbishops of Riga were held by the descendants of crusaders and German-speaking immigrants, whilst on the Order's Livonian lands this transition took place later. Second, a system of serfdom was effectively imposed on the peasantry by the sixteenth century, a process already marked by earlier regulations such as the *Landtag*'s ruling in 1422 that peasants unable to repay their debts had to be returned to their creditors. Serfdom became inextricably tied to the emerging, export-driven manorial system (Šnē 2008, 94–95).

Christian and Non-Christian Sacred Landscapes

It has been argued that elements of the Estonian (and more broadly Livonian) nobility in the early thirteenth century would have thought of themselves as at least partially Christian but were confronted with an uncompromising vision of Christianity imported from the West with the crusades (Kurisoo and Jonuks 2013, 92). A standardized Christianity is certainly visible in the new landscape of religious organization mapped onto the conquered territories, defined by the introduction of new cultural symbols: churches, which completely changed the experience of organized religion with monumental buildings and their belfries (Murray 2001). Towns concentrated churches, and Riga was itself envisaged in Henry's *Chronicle* (*HCL* IV, 5) as a beacon of salvation, surrounded by baptismal waters. Riga was also the main repository for relics according to Henry's *Chronicle* (*HCL* VII, 4), where bishops Meinhard and Berthold appear to have been treated as local saints, whilst the need for relics in furthering the cause of the Livonian Crusade was clearly recognized (Tamm 2013, 445, 452), supported by tentative evidence for cult statues containing relics in Livonia during the crusading period (Andreson 2014). However, in contrast to Prussia, the use of religious names was relatively limited. The early convent founded in Üxküll and relocated to Riga had taken the name of the Virgin Mary (*HCL* VI, 4), but only one castle established during the period of active crusading had been given a sacred name (St Mary, or Babbit), with two castles in the late thirteenth century associated with sacred toponyms (Kreutzburg, Latv. Krustpils; Heiligenberg, Latv. Svētkalns) and only four other examples in the fourteenth and fifteenth centuries (Marienhausen, Latv. Viļaka; Marienburg, Latv. Alūksne; Jürgensburg, Latv. Zaube; Frauenburg, Latv. Saldus).

This suggests different conceptualizations of holy war in both regions (see Chapter 15). Perhaps the forbidding 'pagan' landscape of Henry's *Chronicle* was not perceived as such by crusading authorities in Livonia. Moreover, if the crusading narratives created a sense of Christian cores and pagan peripheries (Jensen 2009), episodes of neutralizing the 'pagan landscape' through the deliberate destruction of cult sites are limited to one case: the grove dedicated to 'Therapita' (*HCL* XXIV, 5). The parochial infrastructure was defined by the presence of churches and chapels, which — at least in some regions, such as south-eastern Estonia — were less intensively constructed on the Order's Livonian lands than on those of its episcopal neighbours (Valk 2019). Less

Figure 8.2: A boulder carved with a cross near Āraiši. 1928. O. Erdmanis, National History Museum of Latvia, Ref. 12499.

durable fixtures are noted in fragmentary written sources and would have formed part of the visible landscape of Christianity in Livonia. The clearest expression of this was the use of crosses as boundary markers (Tarkiainen 2012). In 1320, the description of the boundaries of a fief granted by the Livonian Master to Tontegode referred to trees with crosses carved on them as distinct markers; in some cases, particular species were singled out, most often oaks. A document from 1439 specifies oak trees carved with crosses as defining the boundary of a fief, as well as several border stones carved with crosses (Švābe 1938, 154, 178, 244–45). A stone cross is mentioned in an episcopal grant of land to one Johann van der Gaden in 1380, and the same boundary incorporated a sacred wood (*Hilge Holt*, see below) (*LGU* I, 108). One of the most detailed descriptions of such markers dates from a document issued in June 1417, which was a grant of a fief next to Lake Burtnieki from the Livonian Master to one Peter van Duren. This referred to several crosses which marked the boundary of the plot, which began at a wooden cross and subsequently included a stone cross, two further wooden crosses, a further wooden and stone cross, with the boundary of the plot concluding with the initial wooden cross (*LGU* I, 195). Some stone crosses, or stones marked with crosses, have survived in Latvia and Estonia, although they are difficult to date. One example from the commandery of Wenden is located not far from the castle in Arrasch (Fig. 8.2). It is not possible to say whether the cross represented the Order's own emblematic identity or whether it was treated as a general Christian symbol. In all probability, the line between the two was blurred, and the use of crosses as boundary markers in other parts of Northern Europe was not uncommon from the thirteenth century (Schneider 1993; Myśliwski 2002; Altenberg 2003). In Livonia, however, the backdrop to the public use of the cross was that of ongoing Christianization (cf. Kurisoo and Jonuks 2013), of a re-sacralization of the landscape. By the late medieval period, many village cemeteries in Livonia were most likely furnished with a visible Christian symbol — a chapel or a large timber or stone cross, intended to mark and designate the burial ground (Valk 2019). At the same time, cross-marked trees could also have developed localized magical or offering functions as noted, for example, in later Estonian folklore (Tarkiainen 2012; Valk 2019).

Whilst it is plausible to assume the presence of cross-marked trees in the Livonian landscape was ubiquitous before the seventeenth century, when stone markers became the norm, there is also evidence for the recognition of contemporary sites linked to indigenous spirituality within territorial definitions. The published written sources that cover the period before the disintegration of medieval Livonia do not contain any information from the regions of Vidzeme or Latgale. Here, information about sacred sites dates only from the seventeenth century, and as such their antiquity is difficult to verify (see also Chapter 4 for summaries of sacred sites in the case study regions). Most of the medieval documents pertain to fiefs and territorial boundaries in Curonia. Outside of the crusading narratives, sacred woods in Livonia are mentioned for the first time in the document issued jointly by the Bishop of Courland Hermann and the representative of the Grand Master, Eberhard von Seyne, on 18 October 1252. The document concerns the foundation of parishes and the supply of priests in the diocese. Amongst other things, it states that the priests of the newly founded parishes would own wood-felling sites jointly with their parishes, except for those woods which had previously been referred to as holy (*heilich*) (*LUB* I, 240). This can be interpreted as a deliberate agreement, presumably in the interests of the indigenous Curonian vassals, to protect sacred woods from felling. The following year, a record of the division of lordship over Curonia between the Bishop of Courland and the Order mentions that the Curonians shall keep their property rights over the fields, fishing sites, and woods, which are not holy (*in silvis non sanctis*). Moreover, a lake, which is called holy, in the land of Dovzare would not be divided amongst the parties (*LUB* I, 248): this lake can be identified as Lake Pape in the present-day Nīcas district. It follows from the text that holy sites were not supposed to be owned or treated as objects of economic transactions. This regulation would be later relaxed. In August 1476, one Hans Schenke sold the holy woods of a Lettish hamlet (*hilgen busch to Lettendorpe*) located on the Growen side of the road leading to Riga to an individual called Werner Butler, certifying that the buyer had paid for the purchase in full (Bauer 2003, 518). The plot can be identified as the place occupied by 'Gravu' homestead and the vicinity of the Doru or Galtenes 'cult spring', in the present-day Balgales parish in Talsu district.

Sacred woods and individual sacred trees also formed components of territorial boundaries. The agreement of the borders between the lands of the Bishop of Courland and his cathedral chapter in January 1291, shortly after the suppression of Semigallian resistance, when referring to one-third of the land of Pilsaten (today in the Lith. Kretinga district) contains the following text: 'follow the road, which leads from Poys to Sarden, cross the Sardene River, then go to the holy tree (*lignum sanctum*) called Ouse Warpe, from this holy tree to the hill called Galmene, then turn back and go to the Sardene River' (*LUB* I, 540). A decade later, the Bishop of Courland delineated the border between episcopal territory and that of the Order in the land of Zerende. The description of the border contains the following sentence: 'the old road leading from Goldingen to the holy woods of Aralden [shall be the border] and next to these woods there stand crosses and [border] marks, which shall be followed' (*LUB* II, 604). The respective site is located in the vicinity of the present-day hamlet Biržu (Latv. Basu) within Gudenieku parish, in the Alsunga district. In 1422, the description of the layout of the border between the lands of the Bishop of Courland and the Cathedral Chapter of Riga, specified that 'between Targulen and Kolen there stands a large stone; from this stone the border shall go to the holy woods (*hilgen busch*), where there lies a large stone with a lily carved on it; from there along a pile of stones in a straight line to the seashore, called *Krunkelan*.' (*LUB* V, 2575). The lily in this case (and in other instances twinned with crosiers) marked out episcopal ownership (Tarkiainen 2012, 53). Two centuries later, a grant of land in Durben (Latv. Durbe) issued by the Livonian Master in 1516 refers to a honey-bee oak tree with two crosses carved on it; the same boundary included a sacred wood (Bauer 2003, 1746). The location of crosses alongside sacred woods, the mixture of two contrasting symbolic worlds, can also be interpreted as an attempt to mitigate a landscape that would clearly be defined as 'other' within the Christian mindset, whilst at the same time maintaining good relations with indigenous vassals by keeping such woods physically intact.

There was surely no greater sense of 'otherness' embodied in the landscape than that of the sacred sites encountered by the Order's armies as they crossed into Lithuanian territory. The descriptions of roads provided by the Order's guides mention sacred woods as defining landmarks. In 1385, instructions specified going from Milsowe [in the vicinity of Lith. Kelmė] two miles until Santen the holy woods (*heiligenwalde*), where the troops could set up camp. In 1386 Eykint, his brother, and Mase guided the troops from Jensetilte to the holy woods (*heilgin walde*) called Asswyiote (Latv. Užventa in the upper reaches of the Venta River). Holy woods are also mentioned in 1386 in the vicinity of Niemokszty, in 1395 in relation to Jonawdo land and at Rambyn (on the northern shore of the Memel River near Ragnit) (Letto 1936, 126–27). It is difficult to know how these were perceived by participants of the *Reisen*, but it is striking that in contrast to some landscape features that were clearly rebranded as diabolical, that indigenous sacred toponyms were readily incorporated into the Livonian theocracy's geographic paradigm.

In the early sixteenth century, the integrity of some sacred woods continued to be respected. In December 1503, the Livonian Master Wolter von Plettenberg granted a fief of two *Haken* to a person named Draggun. The description of the layout of the plot's borders begins with 'start from the holy woods called Elkewalke' (Švābe 1938, 242). The place name consists of *elks*, meaning a pagan object of worship (literally 'idol') and *valks*, a slow-running brook. The plot granted as a fief to Draggun is now part of Rumba *pagasts* in Kuldīga district, within the vicinity of the former free village of Curonian *Könige* under the name of Dragūnciems (Draggun's village). In the 1960s Edgars Dunsdorfs found amongst the Order's documents in the State Archives of Sweden in Stockholm a sketch of the plan of Draggun's holy wood, drafted in the late fifteenth or early sixteenth century (Dunsdorfs 1971, 42–45; see also Laime 2009). The plan was probably made in order to ensure that the borders of the holy woods were firmly respected and that nobody unintentionally did any harm to the site. However, the description of a fief granted to one Friedrich von Benthem in 1511 within the territory of the Order's castle at Sabell (Latv. Sabile) suggests a different attitude. This specified 'one eighth of the sacred woods and one plot located in Veytzenn (Latv. Viecīši) on the other side of the Abava River' (Bauer 2003, 1710), a subdivision suggesting the wood's spiritual function was at least partially redundant, whilst the toponym had survived.

Two further examples from Curonia provide direct, detailed evidence of woods being treated as places of ritual significance by indigenous communities in the fifteenth and sixteenth centuries. The first was an account by the French knight Guillebert de Lannoy, who journeyed around western Livonia in 1413 and 1414. It includes a description of a sect that cremated the dead in a grove or forest, dressed and adorned with the finest ornaments, and if the smoke rose straight up to the sky, they believed the soul was saved (Spekke 1995, 94–97). The second dates to 1585. An apothecary from Königsberg named Reinhold Lubenau visited Curonia and witnessed an unusual hunt in the Elka grove, next to the village of Ķoniņciems (Latv. Turlava parish). His description is worth citing in full:

> Since it was the Christmas Day, they went to the sacred woods, where on no other day of the entire year any animal is killed or a stick is cut; the killed roe, deer, and hares were skinned, roasted, and placed on a long table, around which a great many wax candles were placed, each for a soul of their parent, ancestor, child, or a relative, and then they ate and drank, standing or walking to and fro and offered the food and drinks to us, too; later they brought an empty beer barrel and drummed on it with two clubs and the women and men danced around the table and children followed the suit; it lasted all night. When they all finally departed for bed, they urged us to eat and to take with us any food we wanted; because they did not eat the leftovers, instead the dogs devoured them; they also refused to take any money from us for the food that we had eaten. (Švābe 1938, 182–83).

This is the most detailed example of practices that must have continued in different regions of Livonia at the local level, reflecting the specific relations between local Christian authorities and indigenous communities (Jonuks 2007, 28). During the period of active crusading, the pope had called for the tolerance of local practices (Švābe 1939, 29); when the papal legate visited the Sword Brothers' *castrum* at Fellin in 1225, he warned them to treat their indigenous subjects fairly, including regarding the extraction of tithes, lest they return to paganism. He then proceeded to do the same at Wenden and Segewold (*HCL* XXIX, 3). It is important to note that the aforementioned examples of tolerating indigenous sacred sites, though seemingly widespread across Livonia as a result of the cultural segregation between convents, towns, and the countryside, are focused on specific regions and particularly associated with the Order's lordships. Episcopal authorities were perhaps less tolerant and keener on evangelizing their subjects (Valk 2019). In 1428, the synod of Riga declared that pagans had to be forced to 'leave the phenomena of nature and the lowly creatures behind, such as thunder which they pray to as their god, snakes, reptiles and insects, and trees that they believe in' (Mannhardt 1936, 156).

Judging by the observations of non-Christian indigenous practices in Livonia recorded by Jesuits in the seventeenth century, it is clear that the reorganization of the physical landscape did not obliterate indigenous spirituality. Expressions of non-Christian religion survived or were reconstituted from the thirteenth to sixteenth century. The records, if accurate, indicate that multiple religious meanings were attached to specific natural places, plants, and animals which contradicted official Christian doctrine. It is perhaps surprising that sacred sites were not deliberately expunged by the Order, but such places would have only constituted a threat if they were politicized in a way that threatened its sovereign rule. If anything, the privileges extended to indigenous elites, particularly to certain Curonian vassals, indicate that working relationships between the colonists and colonized could be sustainable into the sixteenth century, despite the evident apartheid from the mid-fourteenth century, and the increasing Germanization of land ownership. The emphasis on conversion had traditionally come from the papacy, and this period coincides with limited papal influence in promoting Christianity in Livonia and intensive lobbying by the Order and Livonian episcopate in the curia to further political ends, particularly in the fifteenth century (Eihmane 2012).

Conclusion: Landscapes of Regulation

This chapter has considered the transformation and development of an integral component of Livonia's cultural landscape. It is clear that by the end of the twelfth century the indigenous societies of the eastern Baltic were organized by regionally dominant centres of power inhabited by aristocratic families. The crusades resulted in the destruction of a number of these key centres, and the subordination of the existing cultural landscape to the new regime. From the very onset there was a concern with access to resources. The development of political organization during the crusading period was inextricably linked with the control of local resources, defined by ownership of space. These represented a vital infrastructure for the conquering theocratic authorities, their armies, merchants, and other categories of colonists who nonetheless made up a small fraction of the population of Livonia. A culture of regulation was introduced within decades, focusing on arable land, woods, and wetlands. It defined territory with a new level of precision, with clear reference points in the landscape. A process of assigning new toponyms, sometimes renderings of indigenous names, accompanied the bureaucratization of the landscape. The majority of precisely defined fiefs were concerned with land that could be ploughed, the staple landuse of the developing Livonian economy. However, the resources of woods, lakes, and rivers were also highly valued. Given the limited grants of such resources to vassals in southern Livonia, it appears that they largely remained in the general possession of the Order, in contrast to the situation in Danish Estonia or episcopal lands. The gradual industrialization of such landscapes was also carefully regulated, most evidently in the construction and ownership of mills and weirs. The documents indicate that in addition to a direct connection between castles and their administrative districts — commanderies and advocateships — the Order's networks ensured that supplies and commodities could be sourced at regional, interregional, and international scales.

The culture of territorial regulation introduced during the crusading period invariably became a flashpoint for conflict between the major political factions in Livonia, and much of the surviving documentation relates to this. The Order was gradually able to secure the upper hand in these negotiations, although political tensions with Riga, which partially manifested in access to the resources of the new cultural landscape, continued into the sixteenth century. Alongside Riga, the Curonian commanderies are better represented than other regions of Livonia in the written record, where the culture of regulation was counterbalanced by comparatively generous grants of land handed down to vassals in relation to the rest of the Order's Livonian terriories, as well as networks of interdependence and a pronounced tolerance of indigenous sacred sites. In the case of the commanderies of Wenden, Fellin, and the advocateship of Karkus, there is little evidence for enfeoffment: on the one hand, this reflects the uneven preservation of the written sources; on the other, the indigenous experiences of the crusading period were markedly different here, where tribal centres were quickly replaced and the elites removed (as in the case of Fellin) or assimilated (as in the case of Wenden). Here, the palaeoenvironmental record represents an important proxy of the indigenous colonization of the landscape in the Middle Ages, which, within the context of a regulatory society must have been permitted, organized, or encouraged by the Order.

Works Cited

Primary Sources

Arbusow, L. 1924–26. *Die altlivlaendischen Bauerrechte. Mitteilungen aus den livlaendischen Geschichte*, Rīga: Nicolai Kymmel.

Bauer, A. 2003. Transcripts from the Sources of the History of Manors in Courland collected by Albert Bauer, digitized by the Herder Institute in Marburg (the date refers to the completion of the digitization project)

Biskup, M., and Janosz-Biskupowa, I., eds. 2002. *Visitationen im Deutschen Orden im Mittelalter*, I: *1236–1449*, Quellen und Studien zur Geschichte des Deutschen Ordens, 50, Marbug: Elwert

Evans, A., ed. 1936. *Francesco Balducci Pegolotti, La Pratica della Mercatura*, New York: Kraus Reprint (repr. 1970)

HCL Brundage, J. A., trans. 2003. *The Chronicle of Henry of Livonia/Henricus Lettus*, New York: Columbia University Press

HHC Fuchs, M., ed. 1853. *Das rothe Buch inter Archiepiscopalia/Scriptores Rerum Livonicarum*, 2 vols, Rīga

HWC Mugurēvičs, Ē., ed. 2005. *Vartberges Hermaņa hronika*, Rīga: Latvijas vēstures institūta apgāds

Keyserling, H. 1848. '*Beitrage und Nachrichten zur Geschihte der Freibauern in Kurland*', Arbeiten der kurländischen Gesellschaft für Literatur und Kunst, Heft 5, Mitau, 10–34

LGU Bruiningk, H. von. 1908. *Livländische Güterurkunden (aus den Jahren 1207 bis 1545)*, 2 vols, Rīga: Komissionsverlag von Janck und Poliewsky

LR Napiersky, J. G. L. 1881. *Die Libri Redituum der Stadt Riga*, Leipzig: Vjeelag Von Duncker and Humblot

LRC Smith, J. C. and Urban, W., trans. 2006. *The Livonian Rhymed Chronicle*, London: Routledge Curzon

LUB Bunge, F. G. von, ed. 1853–57. *Liv-, Est- und Kurländisches Urkundenbuch nebst regesten* (vols I–III), 13 vols to date, Reval: Kluge and Ströhm (repr. Aalen: Scientia)

Mannhardt, W. 1936. *Letto-preussische Götterlehre*, Rīga: repr. (Hannover-Döhren: Verlag Harro v. Hirschheydt, 1971)

Švābe, A. 1939. *Senās Latvijas vēstures avoti*, Rīga: Latvijas Vēstures institūta apgādiens

Secondary Works

Altenberg, K. 2003. *Experiencing Landscapes: A Study of Space and Identity in Three Marginal Areas of Medieval Britain and Scandinavia*, Stockholm: Almqvist and Wiksell

Andreson, K. 2014. 'The Presence of the Sacred: A 13th-Century Cult Image from Saaremaa (Estonia)', *Baltic Journal of Art History* 8, 7–43

Apala, Z., and Apals, J. 2014. 'The Vendic Hill Fort on Riekstu Kalns in Cēsis', in *Strongholds and Power Centres East of the Baltic Sea in the 11th–13th Centuries*, ed. H. Valk, Tartu: Institute of History and Archaeology at the University of Tartu, Department of Archaeology at the Institute of History of Tallinn University, 115–38

Auns, M. 1998. 'Acquisition of the Acquired: The Establishing of a Real Administration in Livonia', in *Culture Clash or Compromise? The Europeanisation of the Baltic Sea Area 1100–1400* AD, ed. N. Blomkvist, Visby: Gotland Centre of Baltic Studies, Gotland University College, 259–67

Baranauskas, T. 2004. 'Saxo Grammaticus on the Balts', in *Saxo and the Balic Region: A Symposium*, ed. T. Nyberg, Odense: University Press of Southern Denmark, 63–79

Benninghoven, F. 1963. 'Probleme der Zahl und Standortverteilung der livländischen Streitkräfte im ausgehenden Mittelalter', *Zeitschrift für Ostforschung* 12, 601–22

Blomkvist, N. 2005. *The Discovery of the Baltic: The Reception of a Catholic World-System in the European North (AD 1075–1225)*, Leiden: Brill

Blūzma, V., and Lazdiņš, J. 1998. 'Paražu tiesības Latvijas teritorijā līdz XIII gs', in *Latvijas tiesību avoti. Teksti un komentāri. 1.sēj. Seno paražu un Livonijas tiesību avoti 10.gs.–16.gs*, eds J. Lazdiņš, V. Blūzma, and S. Osipova, Rīga: LU žurnāla 'Latvijas Vēsture' fonds, 19–25

Dunsdorfs, E. 1971. 'Ķoniņu svētie meži', *Archīvs: Raksti par latviskām problēmām* 2: 43–46

Dzenis, A. 2014. *Kuršu ķoniņi un citi lēņavīri Rietumlatvijā: Septiņsimt gadus ilga brīvības saglabāšanas pieredze*, Rīga: Domas speks

Eihmane, E. 2012. 'Livonia and the Holy See in 13th–Early 15th Centuries: Centre and Periphery of Christendom', in *The Image of the Baltic: A Festschrift for Nils Blomkvist*, eds M. F. Scholz, R. Bohn, and C. Johansson, Visby: Gotland University Press, 23–40

Jähnig, B. 2011. *Verfassung und Verwaltung des Deutschen Ordens und seiner Herrschaft in Livland*, Berlin: Lit

Jahnke, C. 2004. 'Wege und Absatzmärket im Handel mit Ostseehering (1100–1600), Kontinuität und Wandel', in *Der Ostseeraum und Kontinentaleuropa (1100–1600)*, eds D. Kattinger et al., Schwerin: T. Helms, 131–36

——, 2015. 'The Baltic Trade', in *A Companion to the Hanseatic League*, ed. D. J. Harreld, Leiden: Brill, 194–240

Jarockis, R. 1998. 'Semigallia 1100–1400: A Review of Archaeological and Historical Sources', in *Culture, Clash or Compromise? The Europeanisation of the Baltic Sea Area 1100–1400* AD, ed. N. Blomkvist, Visby: Gotland Centre for Baltic Studies, 45–53

Jensen, C. S. 2001. 'Urban Life and the Crusades in North Germany and the Baltic Lands in the Early Thirteenth Century', in *Crusade and Conversion on the Baltic Frontier, 1150–1500*, ed. A. V. Murray, Aldershot: Ashgate, 75–94

Jonuks, T. 2007. 'Holy Groves in Estonian Religion', *Estonian Journal of Archaeology* 11/1, 3–35

Kala, T. 2001. 'The Incorporation of the Northern Baltic Lands into the Western Christian World', in *Crusade and Conversion on the Baltic Frontier, 1150–1500*, ed. A. V. Murray, Aldershot: Ashgate, 3–20

Kļaviņš, K. 2009. 'The Significance of the Local Baltic People in the Defence of Livonia (Late Thirteenth–Sixteenth Century', in *The Clash of Cultures on the Medieval Baltic Frontier*, ed. A. V. Murray, Farnham: Ashgate, 321–40

——, 2019. 'Reorganising the Livonian Landscape: Some Issues and Research Perspectives', in *Ecologies of Crusading, Colonization, and Religious Conversion in the Medieval Baltic: Terra Sacra II*, ed. A. Pluskowski, Environmental History of the North Atlantic World, 3, Turnhout, Brepols (forthcoming)

Kreem, J. 2002. *The Town and its Lord: Reval and the Teutonic Order (in the Fifteenth Century)*, Tallinn: Tallinna Linnaarhiiv

——, 2019. 'Written Evidence Concerning the Resources of the Teutonic Order in Livonia: The Cases of Karkus and Fellin' in *Ecologies of Crusading, Colonization, and Religious Conversion in the Medieval Baltic: Terra Sacra II*, ed. A. Pluskowski, Environmental History of the North Atlantic World, 3, Turnhout, Brepols (forthcoming)

Kurisoo, T., and Jonuks, T. 2013. 'To be or not to be... a Christian: Some New Perspectives on Understanding the Christianisation of Estonia', *Folklore* 55, 69–98

Laime, S. 2009. 'The Sacred Groves of the Curonian Ķoniņi: Past and Present', *Folklore* 42, 67–80

Lang, V. 2012. 'Settlement Development and Power Structures in the Late Iron Age Harju District', *Estonian Journal of Archaeology* 2012, 201–25

Letto, W. 1936. *Preussische Goetterlehre*, Rīga

Loveluck, C. 2013. *Northwest Europe in the Early Middle Ages, c. AD 600–1150: A Comparative Archaeology*, Cambridge: Cambridge University Press

Luik, H. 2003. '*Luuesemed* hauapanustena rauaaja Eestis', in *Arheoloogiga Läänemeremaades. Uurimusi Jüri Seliranna auks*, eds V. Lang and Ü. Tamla, Tallinn: Institute of History and University of Tartu, 153–72

Mägi, M. 1998. 'Districts and Centres in Saaremaa: 1100–1400', in *Culture, Clash or Compromise? The Europeanisation of the Baltic Sea Area 1100–1400 AD*, ed. N. Blomkvist, Visby: Gotland Centre for Baltic Studies, 147–57

——, 2005. 'On the Mutual Relationships between Late Prehistoric Saaremaa and the Livs', in *Rituals and Relations: Studies on the Society and Material Culture of the Baltic Finns*, ed. S. Mäntylä

——, 2011. 'Ösel and the Danish Kingdom: Revisiting Henry's Chronicle and the Archaeological Evidence', in *Crusading and Chronicle Writing on the Medieval Baltic Frontier*, eds M. Tamm, L. Kaljundi, and C. S. Jensen, Farnham: Ashgate, 317–44

——, 2013. 'Late Prehistoric Societies and Burials in the Eastern Baltic', *Archaeologia Baltica* 19, 177–94

Mazzaoui, F. M. 1981. *The Italian Cotton Industry in the Later Middle Ages, 1100–1600*, Cambridge: Cambridge University Press

Mickevičius, A. 1997. 'Curonian "Kings" and "Kingdoms" of the Viking Age', *Lithuanian Historical Studies* 2, 7–14

Mugurēvičs, Ē. 1983. 'Mittelalterliche Siedlungen und Veränderungen der Siedlungsstruktur am Unterlauf der Daugava im 12. bis 13. Jahrhundert', *Lübecker Schriften zur Archäologie und Kulturgeschichte* 7, 171–78

Murray, A. V. 2001. 'Music and Cultural Conflict in the Christianization of Livonia, 1190–1290', in *Crusade and Conversion on the Medieval Baltic Frontier*, ed. A. V. Murray, Aldershot: Ashgate, 293–305

Myśliwski, G. 2002. 'Boundaries and Men in Poland from the Twelfth to the Sixteenth Century: The Case of Masovia', in *Medieval Frontiers: Concepts and Practices*, eds D. Abulafia and N. Berend, Cambridge: University of Cambridge Press, 217–37

Nowakiewicz, T. 2010. 'Some Remarks on Settlement Systems of Early Medieval Prussians: The Case of Northern Galindia', in *Worlds Apart? Contacts across the Baltic Sea in the Iron Age*, eds U. L. Hansen and A. Bitner-Wróblewska, Copenhagen: Det kgl. Nordiske Oldskriftselskab, 487–504

Oad, K. 2014. 'Structures and Centres of Power in Estonia in 1200 AD: Some Alternative Interpretations', in *Strongholds and Power Centres East of the Baltic Sea in the 11th–13th Centuries*, ed. H. Valk, Tartu: Institute of History and Archaeology at the University of Tartu, Department of Archaeology at the Institute of History of Tallinn University, 239–55

O'Connor, T. 2011. 'Animal Husbandry', in *The Oxford Handbook of Anglo-Saxon Archaeology*, eds H. Hamerow, D. Hinton, and S. Crawford, Oxford: Oxford University Press, 363–78

Orton, D., Rannamäe, E., Lõugas, L., Makowiecki, D., Hamilton-Dyer, S., Pluskowski, A., O'Connell, T., and Barrett, J. 2019. 'The Teutonic Order's Role in the Development of a Medieval Eastern Baltic Cod Fishery', in *Ecologies of Crusading, Colonization, and Religious Conversion in the Medieval Baltic: Terra Sacra II*, ed. A. Pluskowski, Environmental History of the North Atlantic World, 3, Turnhout, Brepols (forthcoming)

Palang, H., Printsmann, A., Konsa, M., and Lang, V. 2005. 'Ideology and Tradition in Landscape Change: A Case of the Helme Parish, Estonia', in *Landscape, Law and Justice*, eds T. Peil and M. Jones, Oslo: Novus, 288–97

Radiņš, A. 1998. 'Some Notes on the Daugava Way: The End of 12th-Beginning of 13th Century. Time of Changes on the Lower Daugava', in *Culture, Clash or Compromise? The Europeanisation of the Baltic Sea Area 1100–1400 AD*, ed. N. Blomkvist, Visby: Gotland Centre for Baltic Studies, 178–91

——, 2001. 'Daugmale, Jersika, Riga: The Development of Economic and Political Centres along the Lower Reaches of the Daugava', in *Lübeck Style? Novgorod Style? Baltic Rim Central Places as Arenas for Cultural Encounters and Urbanisation, 1100–1400 AD*, ed. M. Auns, Rīga: Nordic, 89–94

Schneider, R. 1993. 'Lineare Grenzen vom frühen bis zum späten Mittelalter', in *Grenzen und Grenzregionen*, eds W. Haubrichs and R. Schneider, Saarbrücken: Saarbrücker Druckerei und Verlag, 51–68

Šnē, A. 2005. 'Archaeological Evidence of Social Relations and Power in the Late Prehistoric Societies of the Livs', *Suomalaisen tiedeakatemian toimituksia: Sarja Humaniora* 336, 169–85

——, 2006. 'The Economy and Social Power in the Late Prehistoric Chiefdoms in Eastern Lativa', *Archaeologia Baltica* 6, 68–78

——, 2007. 'Warfare and Power in Late Prehistoric Societies in Territory of Latvia (Tenth to 12th Centuries)', *Archeologia Baltica* 8, 254–62

——, 2008. 'The Medieval Peasantry: On the Social and Religious Position of the Rural Natives in Southern Livonia (13th–15th Centuries)', *Ajalooline Ajakiri* 1/2 (123/24), 89–100

——, 2009. 'The Emergence of Livonia: The Transformation of Social and Political Structures in the Territory of Latvia during the Twelfth and Thirteenth Centuries', in *The Clash of Cultures on the Medieval Baltic Frontier*, ed. A. V. Murray, Farnham: Ashgate, 53–71

Spekke, A. 1995. *Latvieši un Livonija 16. g.s.*, Rīga: A. Gulbis

Stašulāne, I., and Slušinskaitė, M., eds. 2003. *Zemgaļi senatnē-Žiemgaliai senovėje*, Rīga: Latvijas Vēstures muzejs

Sutrop, U. 2014. 'Livonian Landscapes in the Historical Geography of Livonia and the Division of the Livonian Tribes', *ESUKA-JEFUL* 5/1, 27–36

Švābe, A. 1938. *Kuršu ķoniņi un novadnieki*, Rīga: Gulbis

Tamm, M. 2013. 'How to Justify a Crusade? The Conquest of Livonia and New Crusade Rhetoric in the Early Thirteenth Century', *Journal of Medieval History* 39/4, 431–55

Tarkiainen, U. 2012. 'Trees for Marking Boundaries of Landed Property in Premodern Estonia', *Estonian Journal of Ecology* 61/1, 51–63

Valk, H. 2014. 'The Fate of Final Iron Age Strongholds of Estonia', in *Strongholds and Power Centres East of the Baltic Sea in the 11th–13th Centuries*, ed. H. Valk, Tartu: Institute of History and Archaeology at the University of Tartu, Department of Archaeology at the Institute of History of Tallinn University, 333–84

——, 2019. 'Sacred Landscapes of Medieval Livonia: Lands of the Teutonic Order in Southern Estonia', in *Ecologies of Crusading, Colonization, and Religious Conversion in the Medieval Baltic: Terra Sacra II*, ed. A. Pluskowski, Environmental History of the North Atlantic World, 3, Turnhout, Brepols (forthcoming)

Veldi, M. 2014. 'Roads and Hill Forts in Southern Estonia during the German Conquest in Henry's *Chronicle of Livonia*', in *Strongholds and Power Centres East of the Baltic Sea in the 11th–13th Centuries*, ed. H. Valk, Tartu: Institute of History and Archaeology at the University of Tartu, Department of Archaeology at the Institute of History of Tallinn University, 385–416

Vītola, I., and Urtāns, J. 2014. 'Oral Tradition about Late Iron Age Liv Hill Forts in Latvia', in *Strongholds and Power Centres East of the Baltic Sea in the 11th–13th Centuries*, ed. H. Valk, Tartu: Institute of History and Archaeology at the University of Tartu, Department of Archaeology at the Institute of History of Tallinn University, 417–28

Williams, I. H., Padjus, P., Parn, E., Rootslane, M., and Susl, J. 1995. 'A History of Beekeeping in Estonia', *Bee World* 76/1, 18–30

9. The Environmental Impact of the Conquest of Livonia

Aleksander Pluskowski

'For this land (Livonia) is fertile in fields, plentiful in pastures, irrigated by rivers, also sufficiently rich in fish and forested with trees.'

(Arnold von Lübeck, *Chronica Slavorum*, v.30, 213–14; after Tamm 2009, 23–24)

Thirteenth-century texts, such as Arnold von Lübeck's *Chronicle*, described Livonia as a land of plenty, an imagined paradise ready to be conquered (Tamm 2009, 25). To what extent did this reflect the reality encountered by crusaders, and was this followed by a period of intensive exploitation instigated by the conquering Christian theocracy? Moreover, was there subsequently a 'Europeanized paradigm of exploitation', as this passage may suggest? This chapter is a synthesis of all the data for the Livonian case studies presented in the previous chapters; pollen (Chapter 5), animal bones (Chapter 6), sediments (Chapter 7), and written sources (Chapter 8). Additional data has been included from the range of specialist studies on parasites, insects, human remains, and plant macrofossils where appropriate, all contextualized within the current state of knowledge. These are integrated here to focus on two key questions: how did patterns of exploitation change, and why?

The broadly accepted economic history of medieval Livonia can be summarized as follows. The regional economies of indigenous strongholds at the end of the twelfth century were replaced by much larger, interregional economies as a result of the crusades. The establishment of Riga at the start of the thirteenth century resulted in the incorporation of the Lower Daugava trading zone into the European Catholic world system, a process that contributed to the formation of the Baltic mercantile network which came to define the Hanse (Blomkvist 2004, 2005). The crusades resulted in the creation of semi-autonomous, interdependent territories that eventually produced a surplus of grain from their rural estates for export by the late fifteenth century, with wax and fur representing the other major exports (Kreem 2002, 101; Jahnke 2015). The theocracy's new subjects, the indigenous peasantry, remained free people, and their relations with their lords were purely economic and judicial (Šnē 2008, 95). The Teutonic Order's relationship with the principal urban centres of export where it maintained convents — Riga and, from the mid-fourteenth century, Reval — was also technically one of overlord and subject, but in reality it depended on the interconnectedness of interests from all the major political factions, including the episcopate, whilst coercion and conflict typically favoured the Order in the short term. Trade in agricultural produce facilitated the development of private estates and the adoption of a monetary economy at a regional level by the late fifteenth century, with coins widely used in small denominations for commerce and taxation. At the heart of each newly created estate was the elite group which controlled it, whose residences began to be called manors from the sixteenth century (Jarockis 2009, 255). This accompanied a shift in economic thinking, aimed at maximizing profit from estate revenues. By this time the majority of manors were held by German migrants or their descendants, and they represented an important political force within Livonia, collectively embodied in the *Landtag*. The transition to a manorial economy diluted the theocracy's top-down regulation of territory and provided the basis for legally imposed Livonian serfdom amongst the native peasantry (Kreem 2002, 132; Šnē 2008). By the time the Order's Livonian branch was secularized in 1561, noble vassals held as much as half of the land in Livonia. The manorial system that developed after this has traditionally been called *Gutsherrschaft* in German historiography, and is regarded as the economic transition from feudalism to capitalism, from the medieval to the modern (Melton 1988, 319).

This model, which is preoccupied with the Livonian periphery in terms of its commercial subservience to the European centre, needs to be qualified further with respect to the Teutonic Order's role and the characterization of medieval Livonia as a set of 'crusader states' (see Chapter 1). What must be questioned from the very outset is the 'economic mentality' of the Order in Livonia (see also Chapter 16). Within the context of its supra-national role as a crusading institution, the most convincing interpretation of the economic benefits gained from expansion in the Baltic was to ultimately bring greater force to bear in securing the Holy Land. A significant recommitment of the Order's resources to the Levant was already evident from the 1250s, but the situation changed dramatically with the fall of Acre in 1291 (Morton 2009, 127–30). From this time, but particularly from the second decade of the fourteenth century, the Order was committed to maintaining the holy war against Lithuania as its distinctive raison d'être, as a means of securing the north-eastern frontier of Latin Christendom. The Order's superstructure in Livonia was therefore geared towards maintaining this security, as it had previously been in the Holy Land. First and foremost, this involved the maintenance of convents, with regular numbers of recruits from the Holy Roman Empire encouraged to join not through the path of aristocratic *Herrschaft*, which could result in tensions between ecclesiastical lords and lay officials, as it had done in Germany (Arnold 1989, 77f), but through the template of a military order. This stressed corporate obedience and a communal lifestyle which, like the Templars and Hospitallers, can be defined as professed religious (cf. Riley-Smith 2010).

Knights, their supporting households, the buildings they occupied, alongside their equipment, horses, livestock, and everything within the penumbra of conventual life, including the provision of medical support, required a regular income to maintain. In the fifteenth and sixteenth centuries the Order can be increasingly characterized as a fusion of a religious military order and a secular chivalric order, with a conscious retention of its crusading heritage and status (Kreem 2013, 2014). Throughout this period the Livonian branch of the Order needed to maintain its military resources. It was engaged with intermittent internal and external conflicts with other Catholic powers, but more importantly from its ideological perspective, with Russian schismatics; from the end of the fifteenth century, the Turks and Russians held parallel positions as Christendom's worst enemies, with the Order's defence of Livonia mirrored in the Hospitallers' defence of the eastern Mediterranean (Maasing 2015). This moves us away from an understanding of the environmental changes in Livonia on the Order's lands as the result of Europeanization in the sense of economic proto-colonialism where the periphery served the centre, to one of niche construction and adaptation, of the Order prioritizing the maintenance of its houses, networks, and military resources, of crusader ideology, all of which can be linked back to its recast role after 1291 as the defender of Latin Christendom in north-eastern Europe. At the same time, it is important to note the Order — the focus of this study — was only one of the political and commercial actors in Livonia, and the agendas of episcopates, towns and merchants were also important drivers.

With this adjusted framework in mind, the results of each of the primary case studies are now summarized in turn, followed by a discussion characterizing the Order's ecological impact on Livonia subdivided according to periods of 'social time' (see Chapter 3). Detailed references can be found in earlier chapters. The synthesis for Prussia is presented in Chapter 16, and both regions are compared in Chapter 17, where the study's initial objectives outlined in Chapter 1 are addressed. The various caveats and critiques of the individual datasets have been presented in the previous thematic chapters and are not repeated here. What these summaries present are necessarily abbreviations, but the aim is to present a general model which can be challenged and developed with future research. For the sake of comparative clarity, modern names are used to refer to pre-crusade territories.

From Overlordship to Commandery: Viljandi-Fellin

The overlordship of Viljandi can be associated with an intensive impact on its surrounding landscape, a territorial heartland that resulted in the creation of a new cultural landscape, and new ecological niches. Whilst the pollen data from the vicinity of Viljandi is inconclusive with regards to the intensity of land-use, largely as a consequence of the poor chronological resolution (Chapter 5), there is some tentative evidence of cultivation, most likely corresponding to broader regional trends in the adoption of winter rye cultivation in Estonia *c.* AD 800–1000 (Poska and Sarse 1999; Poska et al. 2014). The presence of woodlands dominated by pine, birch, and spruce is more pronounced. This woodland was also found at the periphery of the overlordship's projected territory, as demonstrated by pollen sequences with very limited evidence for cultivation within the vicinity of Karksi (see below). The woodlands, so clearly defined in both areas of the overlordship surveyed in this study, almost certainly served as isolated *refugia* for northernmost remnants of the bison population, which appears to have been extirpated before the convent and town were established. However, there is no significant deforestation until the later medieval period, suggesting this type of exploitation was not directly linked with the destruction of suitable habitats. Zooarchaeological data from the Iron Age settlement spread across the hills surrounding the stronghold at Viljandi provides better evidence of the exploitation of multiple habitats (Chapter 6; Rannamäe and Lõugas 2019). Wetlands were extensively utilised for freshwater fish, as well as for elk and beaver. The presence of beaver astragali pendants, alongside beaver bones, can be linked to the type of fur exploitation developing in the Viking Age which characterized the mercantile function of select central places. This is complemented by evidence for the exploitation of a broad range of fur-bearers; lynx, wolf, fox and mustelids. Horse was also consumed, forming almost 2% of the total assemblage. By the early thirteenth century, and perhaps resulting from the relocation of the local nobility after the crusaders occupied the stronghold in 1215 (Valk 2015, 17), the high-status character of occupation on one of the hills is evident from the consumption of juvenile animals, poultry (particularly chickens), and a greater diversity of freshwater fish. The area of the stronghold was briefly a military frontier during the early thirteenth century, but its central status was soon re-established by the conquering Christian theocracy.

The recapture of the stronghold in 1223 by crusading armies and the subsequent establishment of the convent and town appears to have left the provisioning systems of the overlordship intact enough to support the community of incoming colonists. Henry's *Chronicle* (*HCL* XXVII, 2) describes how the horses and other livestock within the stronghold were appropriated by the crusaders; the same occurred after the seizure of the nearby stronghold at Pala, and more generally within the region. When the Sword Brothers returned to Fellin in 1224, they established churches and provision for priests 'assigned them sufficient revenue both in grain and fields' having extracted a tithe from the indigenous population (*HCL* XXVIII, 9). By the early decades of the fourteenth century, there was significant investment in the expansion of the castle and town, utilizing locally quarried limestone, including dolomite and tufa, imported stone and at least two different sources of clay for producing bricks; a third, distinct type of clay used in later bricks may have derived from the adjacent lake (Hayward 2019). Both convent and town were supplied with a range of livestock on the hoof, with evidence for keeping live animals within or close to the town, alongside the proliferation of manufacturing activities utilizing deadstock (Haak et al. 2012). This included an emphasis on young sheep and goat, where 40% were slaughtered before they had aged ten months with clear skinning marks on the cranium and extremities. Such animals would have been ideal for good quality meat and, equally important, parchment, which would have been essential for the Order's bureaucracy before it became superseded by paper.

In the mid-fifteenth century, the convent contained the largest community of brethren in Livonia, with 29 knights (as well as five priests and six sergeants) (Benninghoven 1963) supported by an extensive household which may have numbered as much as 200. In 1554, the range of personnel recorded at the castle indicates the presence of many different offices, specialist and artizanal staff, including one responsible for hunting and another for fishing (Kreem 2015, 70–72). Its faunal profile reflects a high-status diet, particularly poultry, small amounts of game, and, for the first time, imported marine fish consisting of cod, herring, and smelt (Orton et al. 2019; Rannamäe and Lõugas 2019). Herring, more than any other species, became a dietary staple; in 1451, the convent had 200,000 Baltic herring, two *last* of Skåne herring (a *last* being the largest defined unit of grain which could be carried in a vessel), and 14 tuns (barrels) of salted herring within its stores. The complex is also associated with the intensive processing of deadstock, represented largely by sheep and cattle, as well as red deer antler, most likely imported from the south. Slaughter, butchery, and processing were most likely confined to the outer baileys of the castle by the time the conventual buildings were established, with evidence for prepared cuts of meat reaching the kitchens (Haak 2015, 47). Stored meat is also noted in the 1451 visitation report. The consumption of birds was less prevalent in the adjacent town, which is dominated by meat supplied from slightly older animals, with negligible amounts of wild fauna and at the suburban site in Vaksali Street, a substantial assemblage of horse bones, perhaps reflecting knackering. Imported marine fish only appear in the town from the fifteenth century (Rannamäe and Lõugas 2019).

Significant expanses of pasture land and fodder would have been required for the quantities of livestock brought in on the hoof into the castle and town, and following the imposition of taxes and tithes, as well as the presence of a new urban market, the generation of an agrarian surplus is visible in the documented quantities of grain within the convent's stores. In 1451, the convent had 20 *last* of grain, 42 *last* of meal, 35 *last* of malt, and *c.* 160 kg of hops within its stores, most likely derived

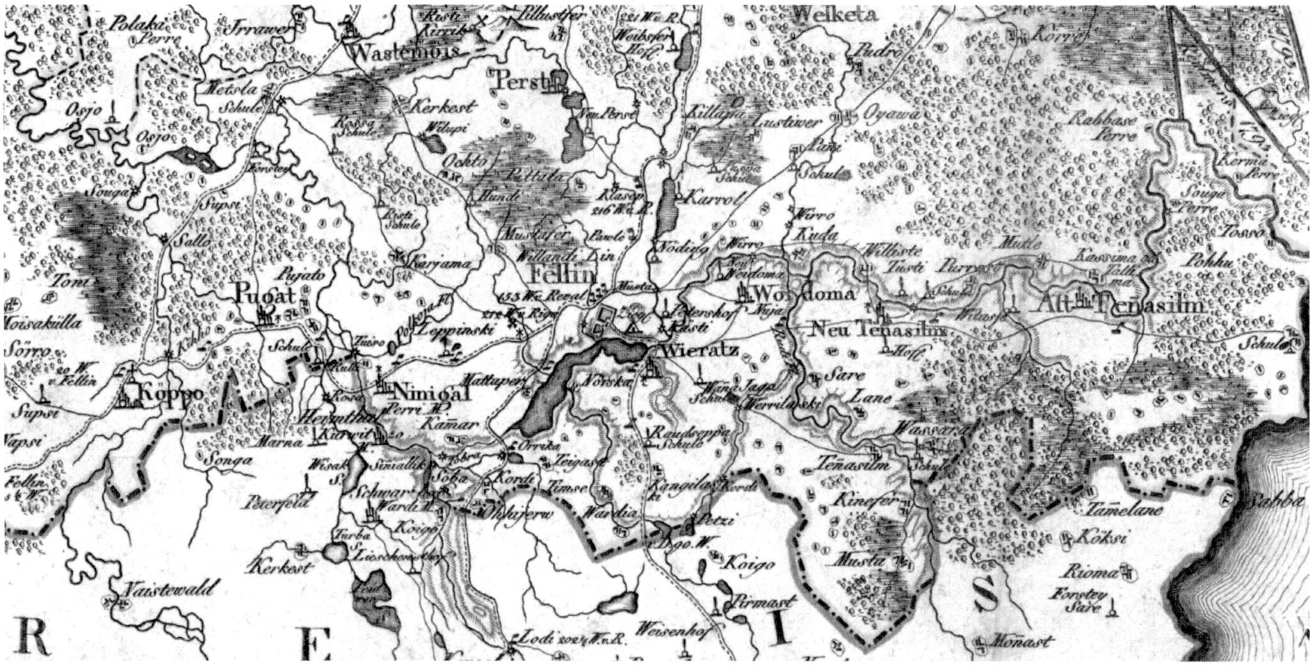

Map 9.1: Ludwig August Mellin's map showing the landscape around Fellin (Est. Viljandi). Map from Mellin et al, 1798, *Atlas von Liefland*. Retrieved from the Library of Congress, <https://www.loc.gov/item/75572471/>.

from its commandery. Some provisions could have been readily sourced from the broader network, as already suggested by the imports of marine fish; in 1554 the commander of Fellin stored 10 *last* of his grain in Karkus, and a report from 1554 indicated annual purchases of oxen from Rujen (Latv. Rūjiena), 68 km away (Kreem 2019). However, it is clear that the proportion of cultivated land expanded with the establishment of the commandery, evident within the vicinity of the convent and town. Arboreal pollen from the core in Lake Viljandi decline through the sequence, with consistent quantities of cereal (including buckwheat which is otherwise recorded in Estonia only from the thirteenth/fourteenth centuries; Sillasoo and Hiie 2007) and ruderal pollen suggesting an open mosaic landscape surrounding the castle, town, and lake (see Chapter 5). Pollen from Lake Kosilase, east of Viljandi, is also suggestive of an intensification in arable and pastoral land-use. This general picture of a mosaic landscape with increasingly fragmented habitats is reinforced by the dominant presence of hare (and to a lesser extent roe deer) in the faunal assemblages from the castle and town, representing a marked drop in biodiversity from the pre-crusader period. Future work could focus on the more peripheral regions of the commandery, to ascertain just how much of the Order's regional 'granary' of Saccala was being utilised in this way (Tuulse 1942, 76). On Mellin's map from 1789 the landscape surrounding Fellin is largely clear, but substantial amounts of woodland are evident at the fringes of its territory and would have almost certainly been present in the Middle Ages (Map 9.1).

Within the Order's corporate hierarchy, it is clear that the convent of Fellin was considered as a particularly important centre of power within Livonia. Judging by the degree of imported tablewares and the quality of its architecture, it was also one of the wealthiest. Investment in the convent may have been related to its geopolitical location before the mid-fourteenth century, when its territories bordered to the north with those of the Danish crown and to the east with the Dorpat episcopate. Its documented territory in the 1580s encompassed much of Saarde parish enabling control of the road between Fellin and Pernau (Est. Pärnu), as well as two other key routes linking the convent with Reval and Dorpat (Valk 2019). The castle was itself the centre of arms production; elements of deadstock were used for making crossbows, and there is evidence for the presence of an armourer (Haak 2015, 36–39). By the mid-fifteenth century, the concentration of military resources at the convent can be linked to the ideology of the defence of the Latin north-east against Russian principalities, with Fellin forming an important (if not the most important) hub connecting the Order's main strategic Livonian houses. The convent housed the Order's 'treasure chest' in Livonia, most likely an emergency war fund, and in 1554, its commandery was to contribute three hundred horses during times of war, whilst the presence of thirty-six 'stable brethren' (*Stallbrüder*, sometimes called 'Black Heads', whose name was subsequently adopted by the urban merchant confraternities) formed part of the convent's ideological heritage (Kreem 2015, 68, 72–74). Both groups of 'Black Heads' venerated St Maurice, whose cult was centred on Magdeburg and had clear associations with crusading, featuring also on the banner of the Livonian Master and promoted with the Order's territory in Livonia (Mänd 2009, 204–05; 2014). However, this was also a time when the Order's corporate structure had broken down into a confederation of officials, and it is unlikely that a conventual lifestyle was still the norm.

From a Periphery to a New Lordship: Karksi-Karkus

The pollen sequences from the vicinity of Karksi point to a heavily wooded landscape during the Late Iron Age similarly dominated by pine, spruce, and birch, with only very occasional evidence of rye cultivation during the tenth and eleventh centuries. This corresponds to the peripheral location of this landscape within the Saccalian overlordship. After the establishment of the convent at Fellin, the area around Karksi was transformed from a frontier into a central place, and one that had a significant degree of autonomy represented by the status of the castle's advocate (*Vogt*). The castle had been constructed to manage the conquered territory, and already during its earliest phase of occupation in the latter decades of the thirteenth century, there is evidence for processing and storing agrarian produce, with tally sticks used to record the quantities (Valk et al. 2013). This appears to have derived from the immediate landscape around the castle, which palynological data suggests was open at this time (see Chapter 5). The early community of castle occupants was defined by its consumption of quality meat derived from lambs, kids, piglets and chickens, with almost negligible quantities of game represented by hare, wild geese and galliforms (Rannamäe and Lõugas 2019).

After the complex was expanded, the castle's community continued to be defined by high-status foods, particularly imported marine fish such as cod. Despite the limited biodiversity within the faunal assemblage from this period, the immediate landscape retained significant areas of woodland and wetland. The remains of bear found at the castle may have derived from these habitats at the fringes of the cultivated area or, given the absence of evidence for processing whole carcasses, these may have been skins imported via the Order's network. The transformation of the landscape within the castle's territory at this time was gradual. The pollen evidence from Äriküla indicates that intensification in agrarian land-use occurred during the fourteenth century, and visitation records from the fifteenth century confirm a substantial amount of cultivated land was controlled by the castle. At this time there was significantly more grain stored in Karkus than in Fellin (although not as much as in Wenden; see below), and it appears this use of the castle as a granary was consistent at least in the middle decades of the fifteenth century (Alttoa 2006, 272; Kreem 2019). The 1451 visitation record also indicates there was a higher quantity of cured meat stored in Karkus than Fellin or at the similarly sized castle at Windau (Latv. Venstpils). These stores were defined by their capacity to supply castles for seasons or a whole year, and a convent such as Fellin and Windau would have had access to supplementary stores beyond its walls. At the same time, Karkus may have functioned as a redistribution centre for meat as well as grain.

The continued expansion of the cultivated area in the sixteenth century indicates that the medieval system of environmental exploitation had laid the foundations for and then been superseded by the manorial system. This was accompanied by a shift in the organization of territorial management and exploitation, from the castle to multiple estates centres. The advocate of Karkus ruled over a number of parishes, including that of Helmet (Est. Helme). The growth of centres in this parish documented by the end of the Middle Ages can ultimately be situated within the broader regional intensification of cultivation. At the same time, it points to a strong continuity in the settlement pattern from the Late Iron Age to the Middle Ages (Palang et al. 2005).

From a Small Tribal Centre to a Major Commandery: Cēsis-Wenden

In contrast to Viljandi, it is inappropriate to define the Late Iron Age settlement complex at Cēsis as a comparable, regional lordship. It was certainly the heartland of the local Vends, but its territorial extent was comparatively limited, and it was situated within a landscape that represented the frontier between other tribal polities. There is evidence for some very small-scale agricultural land-use around Lake Nineris, situated to the north, and the hillfort community kept livestock. From an interregional perspective, its ecological profile is comparable to that of the peripheral Karksi landscape in the Late Iron Age. The story of Wenden during the crusading period

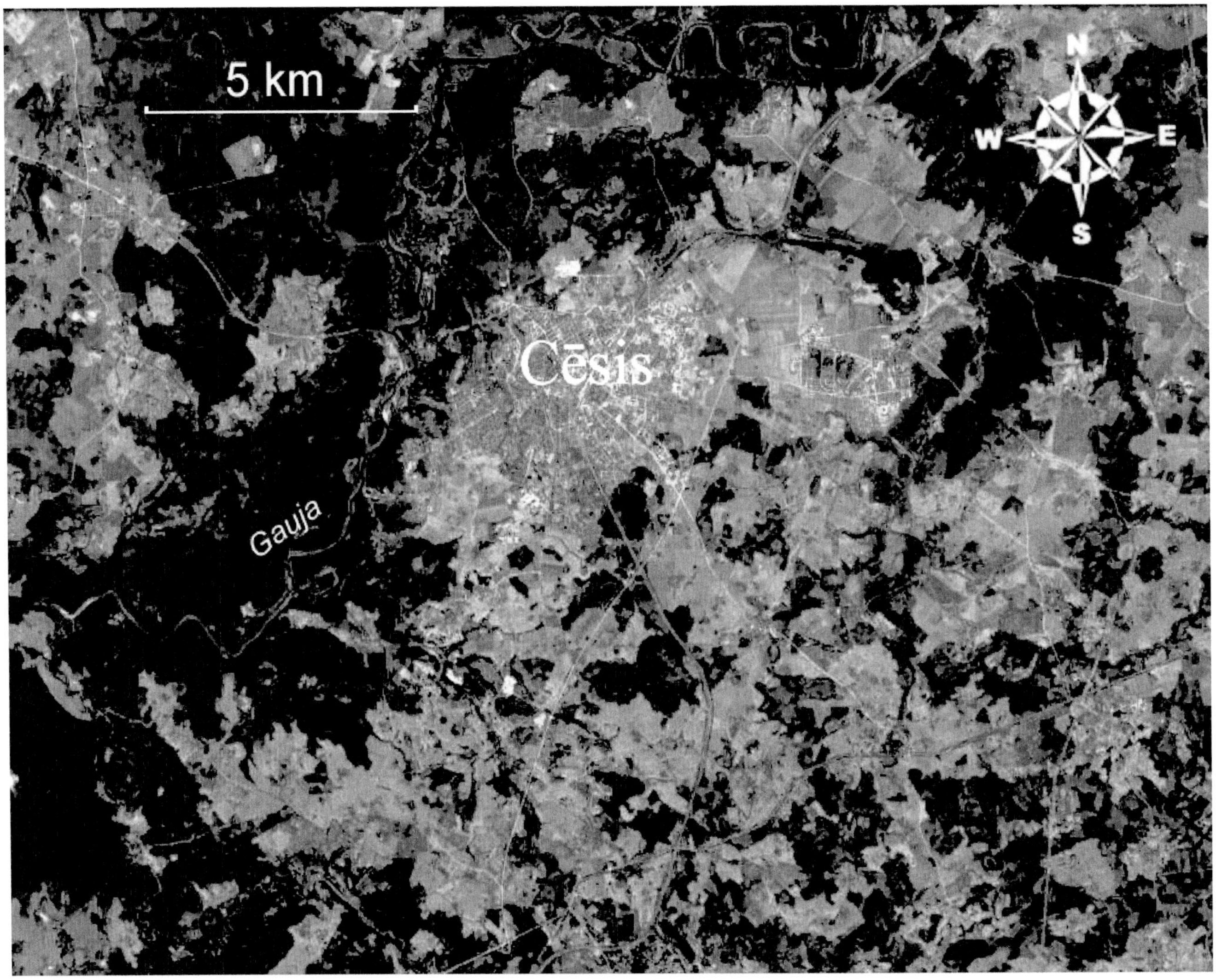

Map 9.2: The landscape within the vicinity of Cēsis today. The dense woodlands in the north-west contrast with the fragmented landscape in the south. Google Earth: 2015. CNES/Astrium.

is one of alliance, Christianization, and coexistence with the Sword Brothers. With the construction of the convent, there was a seamless takeover of the existing centre of power, along with its provisioning systems. These soon developed and became far more expansive as the territory of the commandery expanded to the north and south, incorporating other tribal districts. The Vends themselves would abandon their distinctive identity, although there is virtually no evidence (to date) of how the local indigenous population adapted to the new regime. The development of the convent's supporting network — which consolidated the new taxation system amongst the local population — took place in the latter decades of the thirteenth century, with the castles at Wolmar (Latv. Valmiera), Trikaten (Latv. Trikāta), and Burtneck (Latv. Burtnieki), whilst the southern district of Arrasch may have been functioning by the end of the thirteenth century or start of the fourteenth century at the latest, and Ermes (Latv. Ērģeme) also established in the early fourteenth century.

The Order invested in the expansion of the castle complex, which became an important centre on the route linking Riga and Fellin. This expansion coincided with new building works drawing on quarries located close to the convent, and the abundant woodlands were exploited for constructing parts of the castle complex and the adjacent urban colony; the oldest vernacular buildings in the town square have been dated by dendrochronology to the 1280s (see Chapter 4; Zunde 2019). In the earliest surviving part of the conventual buildings at Wenden, tufa was used in the thirteenth-century chapel, almost certainly derived from the extensively quarried tufa cone at Libani on the River Triecupite, *c.* 4 km north-east of the castle. Dolostone was also used, most likely quar-

Figure 9.1: Reconstructions of the castle at Cēsis (Ger. Wenden) in (a) *c.* 1250; (b) *c.* 1550 (Kalniņš 2014, 3, 8).

ried onsite from the outcrop or spur on which the castle complex was constructed. During the second phase of expansion in the latter half of the fourteenth century, there is evidence for a variety of imported Baltic limestones used in the castle's fabric, as well as in the town's church and for tomb slabs (Hayward 2019) (Map 9.2). The expansion of the castle at this time coincides with an intensification in agrarian land-use within the immediate landscape, but its diachronic spatial distribution is uneven. Sustained increases in cereal and ruderal pollen appear in the late thirteenth/early fourteenth centuries at Blanku Mire, in the late fourteenth/early fifteenth centuries at Blusu Mire and Rāmuli, and from the early sixteenth century to the north of the convent. It is more plausible to link this incremental change in local land-use with the needs of the growing convent and its extended household, as well as the adjacent town, rather than the top-down implementation of a surplus-driven economy. This is supported by the macrofossil and pollen record of cereals, weeds, and ruderals within the moat, with the increase in kitchen waste and effluent from other parts of the castle prompting the installation of a timber-lined channel. In the mid-fifteenth century, the convent contained the largest stores of grain of any of the Order's castles in Livonia, the central point of the *Innere Linie* (see above); however, much of this surplus must have derived from the districts that were subservient to the convent, and in the fifteenth century from the expanded Livonian Master's district (*Kammergebiet*).

Substantial areas of woodland remained within the core of the commandery, and this can be linked to the sandy soils found in the north, north-east, and north-west of Cēsis, which today remain heavily wooded (Fig. 9.1). At the same time, levels of biodiversity appear to have been as low as in the landscape further south (see below), with a negligible amount and range of wild species in the faunal assemblages of the castle. As in the case of the other Livonian castles, the convent at Wenden was able to process large quantities of good quality meat from livestock brought in on the hoof, which may have derived from meadows within the convent's immediate districts, from some of its subservient advocateships or perhaps brought over greater distances. The contribution of Arrasch to the commandery is considered separately below, whilst pollen studies from Lake Trikāta, 37 km north-north-west of Wenden and situated next to the ruins of Trikaten, reinforce the changing character of this landscape. In the Late Iron Age this was a relative periphery, comparable to the Vendish territory around Cēsis, with a heavily wooded landscape and limited evidence for agriculture. In the medieval period there is a clear link between decreasing spruce pollen and an increase in cultural indicators, corresponding to the documented role of the castle in the fifteenth century as one of the Order's 'corn houses', a function it may have already begun to serve from the later thirteenth century (see Chapter 5).

The conventual complex at Wenden underwent a further upgrade when it became assigned as the permanent residence of the Livonian Master, distinguished by a new type of mortar that changed the ingredients that had been used from the thirteenth century (Hayward 2019). Like all other convents, Wenden had multiple service buildings within its outer baileys, including stables and the capacity to mobilize significant cavalry, which played an important role in Plettenberg's wars with neighbouring Russian principalities in the early sixteenth century. The data from the castle's outer bailey provides direct evidence that larger warhorses remained an important component of the Order's armies at this time. It is possible to draw on the details provided in the early seventeenth century to reconstruct components of the immediate landscape of Wenden in during this final phase of the Order's rule. Given the demands of the convent and town, and following analogous medieval sources for other castles in Livonia, the range of features was certainly typical of the post-crusade industrialized landscape and must have gradually emerged over the course of the fourteenth and fifteenth centuries. Unfortunately, the lack of data prevents a more specific diachronic understanding. The Swedish Audit of 1638 referred to an unfinished mill owned by the castle, but also to the former locations of three mills, with a number of ruined mills noted in other associated manors, although none are specified at Arrasch. The two castles also owned a dozen large and small lakes (Dunsdorfs 1939, 454–96). Two maps from 1690 and 1691 record a town mill and associated rivulet (also known as the 'Herring Mill' or *Silckensche Muhle*). Around a kilometre to the north of the town, near the tributary of the Melnupīte Rivulet, was the so-called 'Castle's Brick Hill' (*Schloss Stegelberch*), which may have been a site for extracting clay and preparing bricks. A water-powered mill is also marked on the Dzirnavupīte (a tributary of the Gauja), with a mill pond on the other side of the rivulet. An old lime kiln (*Alte Kalckofen*) is marked to the north-east of the town (Caune 2007, 160). The Swedish Audit also listed the dues paid from the manors, which fall into categories familiar from medieval documents for other commanderies: grain (rye, wheat), fodder (hay, oats, specified as 'warhorses' oats'), useful plant produce (peas, hemp, flax, hop), woodland products (honey, firewood) and animals (hens, sheep, pigs, as well as joints of meat), and coinage, which would have been a late development with the proliferation of monetarization within rural Livonia. In summary, the impact of establishing a new centre of power at Wenden exemplifies the effectiveness of the Order's broader network; changes in the local landscape were not dramatic, whilst the convent itself was an important consumer of natural resources, and the most significant centre in the final period of the Order's rule.

From a Major Centre to a Peripheral Centre: Āraiši-Arrasch

The detailed multi-proxy study of the lake sediments at Lake Āraiši provided clear evidence of the impact of the fortified settlement during the Viking Age (Stivriņš et al. 2015). The first appearance of cereals and a gradual opening up of the woodland environment occurred towards the end of the Middle Iron Age (AD 400–800), whilst the use of the landscape was most intensive between the mid-eighth to mid-eleventh centuries, resulting in a semi-open landscape surrounding the lake comprising a mosaic of cultivated land, meadows, and pasture. Livestock was grazed on the peninsula, as suggested by the high values for coprophilous fungi in the lake, then slaughtered and butchered within the settlement. The presence of calves and foals at the site indicates that the settlement maintained its own herds (see Chapter 6). As with other Late Iron Age central places, the community was engaged in exploiting all the surrounding habitats and contributed to a significant reduction in biodiversity as a result. Elk and beaver were hunted in the nearby wetlands, and aurochs or perhaps bison and wild boar were hunted in increasingly fragmented woodlands. The first phase of the fortified settlement required around 1000 m^3 of timber (Apals 1996a, 71), and there were a further five phases of rebuilding. The first phase was primarily from spruce, with an increasing use of pine in the subsequent phases (Zunde 2019).

The abandonment of the settlement, and presumably the end of the polity it represented, did not result in a collapse of the regional settlement pattern, for intensive agricultural land-use continued in the surrounding landscape for the remainder of the Late Iron Age. When the Order took over this territory and established a castle on the peninsula, perhaps in the late thirteenth century (Tuulse 1942, 241), the location may have been chosen in relation to the existing church (rather than to provide military support to Wenden; Apals 1996b, 35), which would have represented the main central place within the newly defined parish. The new dual-authority, articulated through the collection of taxes and tithes, inherited a mosaic landscape, albeit one with significantly less biodiversity. Local wildlife had not recovered from the impact of Late Iron Age exploitation; both here and in Wenden, there is very little evidence for the exploitation of wild species, and those that feature, particularly hare and roe deer, point to open and fragmented habitats. The castle, dominated by a single range enclosing a rectilinear precinct, was constructed primarily from field stones in a naturally defensive location, surrounded on three sides by the lake and a moat cut into the front of its western range (Fig. 9.2). It was staffed by a garrison but, as with the other castles belonging to the Order, supported by a household drawn from the indigenous population.

Parts of the peninsula may have been used for housing and even grazing livestock, but little evidence of this has been found in the geochemical analyses of sediments from the site, and the cattle, sheep, goats, and pigs slaughtered and butchered within the castle precinct may have been brought in from nearby settlements. Most sheep were slaughtered aged eighteen to thirty months for mutton and leather, with wool as a secondary product of lesser importance. The advocate's household consumed fish from the nearby lake and rivers, but also imported marine species. The presence of small numbers of young chickens makes the castle comparable to the other houses of the Order; however, Arrasch clearly did not manage its territory in the same intensive way as the more powerful and autonomous advocate of Karkus. There are slight increases in rye pollen and coprophilous fungi during the fourteenth century, suggesting some intensification in arable and pastoral land-use in the surrounding landscape following the establishment of the district. Livestock, but particularly grain, would have been sent on to Wenden, to which the officials of Arrasch were directly subservient.

Developing a Riverine Heartland: Riga and the Lower Daugava

Riga was the primary entry point for crusading armies into Livonia, and as a composite social entity the town reorganized and monopolized the trade of the Lower Daugava previously controlled by a string of settlements and strongholds up the river, dominated by the regional multi-ethnic centre at Daugmale. In the Late Iron Age, the mosaic landscape within the vicinity of Riga and the broader territory of Daugmale can be clearly linked to small-scale agriculture and animal husbandry, alongside

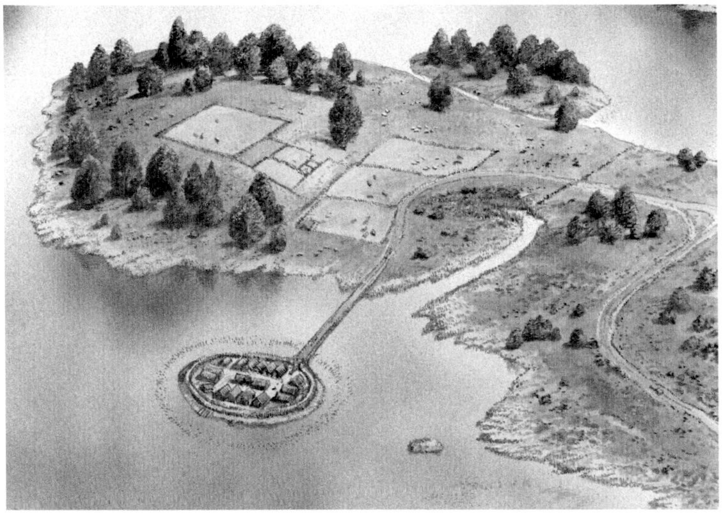

Figure 9.2: Reconstructions of the Viking Age fortified lake settlement (a, b) and the later Teutonic Order's castle (c) at Āraiši (Ger. Arrasch). Figure by P. Dennis.

fluctuating woodlands, with decreases in some broad-leaved tree species (Kalniņa et al. 2019). From the mid-tenth/eleventh centuries the Liv settlements in the Lower Daugava are particularly associated with working and trading furs (Vasks 2010). At least six of these communities appear to have adopted elements of the ritualistic treatment of horses evident in neighbouring Selonian and Latgalian funerary practices, or individuals from the latter ethnic groups had been buried in these predominantly Liv cemeteries. The cemetery associated with Daugmale contained the deposit of a single horse, and a few kilometres up the river at Ogresgala Čabas, a whole horse had been placed in a circular pit and covered in dolomite slabs in the twelfth-/thirteenth-century phase of the cemetery (Brūzis and Spirģis 2009, 289). The Lower Daugava region was a frontier during the crusading period, with Riga as the principal centre for organizing military campaigns, and authority gradually consolidated through the construction of *castra*. This was accompanied by the appropriation of local provisioning networks. One such castle, at Alt-Dahlen (Latv. Vecdole), included the deposit of a calf across the threshold of its main entrance reflecting either local practice or a *Bauopfer* introduced by the colonists. The zooarchaeological assemblage from the castle is unusual for its large quantity of juvenile animals, and the castle appears to have maintained its own herd of pigs, sheep, goats, horses, chickens, and perhaps cattle, with all the associated implications for provisioning. The community also hunted elk and wild birds, fished pike, catfish, migratory salmon and sturgeon, and imported plaice.

The landscape did not *substantially* change after the crusades, presumably because the settlement pattern remained largely intact. The documents attest to the granting of land back to indigenous Liv vassals from the thirteenth to fifteenth centuries, where the so-called Liv's Law (*Liveschen Recht*) was consistently recognized. In the borderland between the fluctuating territories of Riga, the Order, represented by the advocate's castle at Kirchholm from the end of the thirteenth century (Latv. Salaspils), and the Dorpat episcopate, represented by Üxküll (Latv. Ikšķile, and also the early missionary colony), the post-crusade period saw the gradual removal of commercially valuable broad-leaf species, a decrease in pine and the development of shrub, although substantial areas of woodland remained. It is difficult to ascertain whether climate change played a role in these fluctuations (see Zunde 2019), and levels of cultivation remained the same as in the Late Iron Age (Kalniņa et al. 2019), again suggesting a degree of stability in the settlement pattern and an absence of the levels of harvesting. Levels of local biodiversity remained high enough that the castle community of Kirchholm was able to access a range of woodland and wetland mammals.

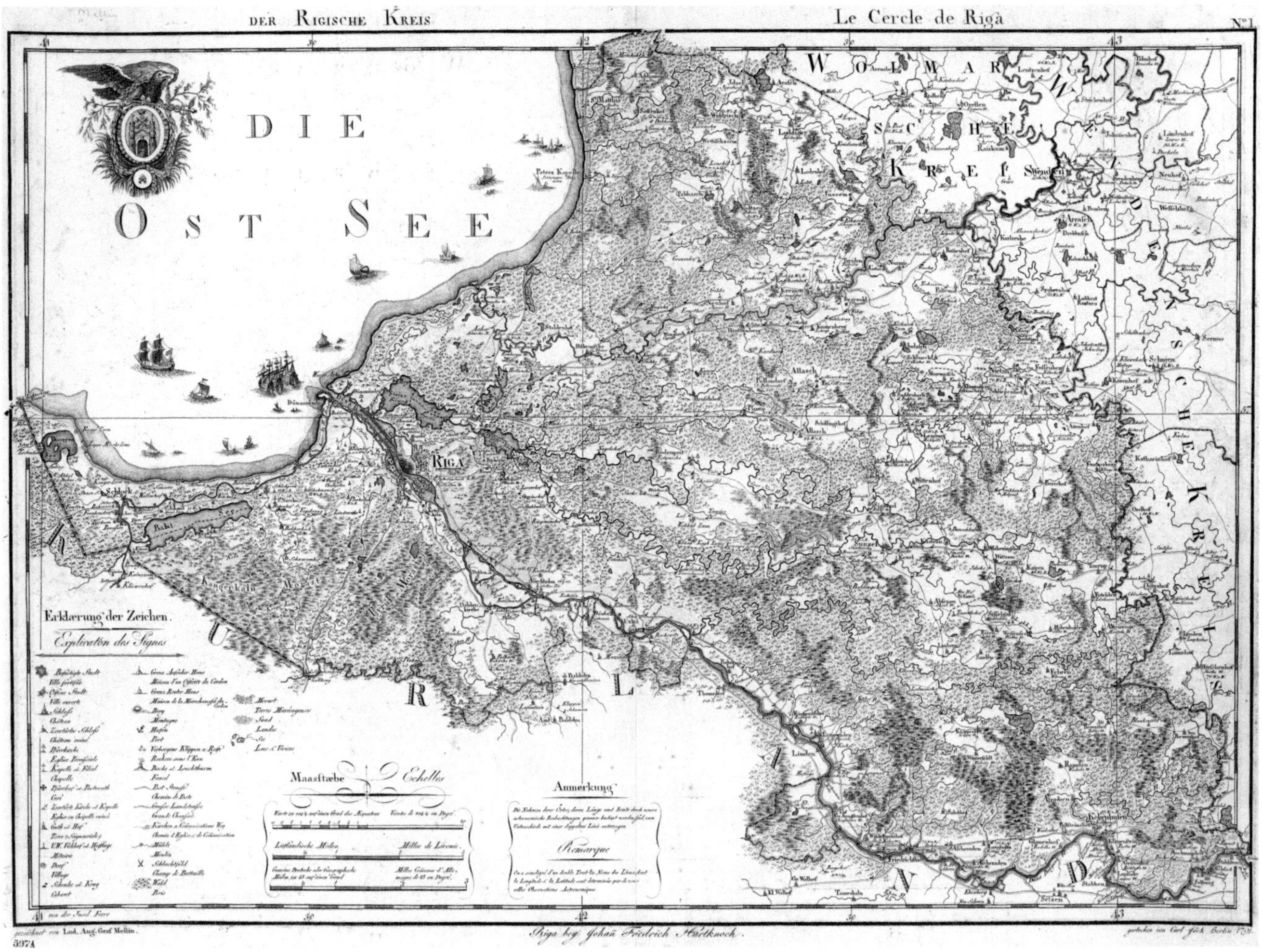

Map 9.3: Ludwig August Mellin's map of the vicinity of Riga showing the woods next to Dünamünde.
Map from Mellin et al. 1798, *Atlas von Liefland*. Riga. Image courtesy of the National Library of Latvia, digital collection.

Access to the landscape around Riga itself was quickly redefined by 1226 and again in 1330, by which time the town's patrimony had dramatically shrunk. The immediate vicinity of the town became deforested, as the rapidly increasing demands for fuel, building material (for housing, urban structures and shipping), and land clearance took their toll (see Chapter 5). Arboreal pollen occurs in low frequencies in samples from buildings in the Liv quarter dating to the early to mid-thirteenth century, and by the mid-fourteenth century palynological data points towards a mosaic landscape of cultivated fields, open, disturbed, and grassland habitats, including both wet and dry meadows and pasture. The town, including the Order's convent, brought in adult cattle on the hoof alongside sub-adult pigs and sheep to provide good quality meat. These animals were larger than found on other Livonian sites, but still within the size range of indigenous stocks and presumably pastured locally; specific pastures for cattle, horses, and geese are documented within the vicinity of the town. As in other urban centres, a significant industry based on the processing of deadstock emerged within Riga, particularly harvesting horn from goats, sheep, and cattle. Commercially valuable *smaschen* (lamb felt) was shipped out from the town (Delort 1978, 1, 234). Riga's elites were also able to supplement their diets with cultivated fruits grown in gardens within and around the town, alongside imports, and the convent appears to have been able to afford chickens, a relative luxury in both Late Iron Age and medieval Livonia.

Beyond the immediate vicinity of the town, commercially valuable species — principally oak — were quickly denuded from Riga's patrimony, partly to construct revetments reinforcing the bank of the Rīdzene River. Timber from oak, birch, ash, elm, and lime was also sometimes used in the town's structures in the thirteenth and fourteenth centuries. By the fifteenth century this commercially valuable timber was being floated to the town from the northern parts of present-day Lithuania and Belarus, both for local constructions and for export (Zunde 2019). Pine was more typically used for the town's buildings, and the palynological data suggests that pine- and birch-dominated woodlands within the patrimony certainly remained substantial, with the wax and honey from their beehives attached to protected oak trees particularly valued alongside firewood. The documentary records relating to the various territorial claims in Riga from the thirteenth to sixteenth centuries indicate a constant emphasis on segregating key resources for the use of the Order's principal Livonian convent (see Chapter 8). A certain proportion of the landscape within the vicinity of Riga consisted of wetlands which were marked on the earliest detailed maps of the region, and which are still present today. But there is also evidence for the long-term maintenance of nearby areas of woodland. To the north of Riga, the commander of Dünamünde (Latv. Daugavgrīva) reserved the woods of 'Urgenumme' for the Order's use when granting land to vassals, and on Mellin's map from 1789, a substantial area of woodland is marked between Dünamünde and Neuermühlen (Latv. Ādaži) which remains today (Map 9.3).

There is limited evidence for fur processing following the crusades. This became largely confined to the opportunistic exploitation of local beavers, whose remains have been found in very small quantities in the Liv quarter in Riga, and the Daugava island sites. It appears that the organization of the fur-trade in the Lower Daugava changed from local supply to long-distance procurement from other parts of Livonia, such as the castle district of Dundagen (Latv. Dundaga) in Curonia which belonged to the Cathedral Chapter of Riga (see Chapter 6), as well as from neighbouring Russian principalities. Hunting otherwise was infrequently practised, rarely featuring in the documentation of territorial disputes and fiefs, or in faunal assemblages from medieval contexts at sites in the Lower Daugava. Elk were occasionally procured from local wetlands, and roe deer from the gradually fragmented woodlands.

More emphasis was placed on the construction and ownership of mills and fishing weirs, which remained the objects of territorial disputes and economic transactions between the Order, episcopate, and town into the fifteenth century, suggesting it was one of the more intensive forms of local resource exploitation (see Chapter 8); the Liv community in Riga was certainly handling raw fish in the mid-fourteenth century (Yeh et al. 2014). Riga's multiculturalism at this time was qualified by socio-economic differences, emphasized by contrasts in language, customs, and material culture, although the Liv quarter appears to have had at least one wealthier household which could afford more expensive fruit, imported fish (although not the Hanseatic staple of herring), better cuts of meats, and perhaps even the latest steel-tipped cleavers brought over by the colonists. At the same time, the indigenous custom of consuming horse meat appears to have continued, and it is possible the horse skulls placed under the rebuilt north wall of Riga were also expressions of local ritual praxis. Future excavations in Riga, and within the settlements of the Lower Daugava, will enable the socio-topography of environmental exploitation following the crusades to be understood at a much higher resolution.

Discussion: Contextualizing and Defining the Impact of the Teutonic Order in Livonia

The Late Iron Age

The impact of the crusades on the tribal territories which came to be referred to as Livonia is only evident when situated within long-term trends. Social organization from the mid-eleventh through to the thirteenth centuries is defined by a settlement hierarchy dominated by multifunctional strongholds, the central places of an increasingly visible stratified society. More strongholds are occupied at this time within Estonia, Latvia, and Lithuania than in any other period; they are also typically larger and associated with more complex fortifications (Valk 2014c). Is it not surprising to find that this coincides with a period of unprecedented niche construction, associated with the intensification of agriculture, animal husbandry, deforestation, and fur exploitation, linked to extensive provisioning systems and the increasingly visible presence of a mercantile stratum of society participating in long-distance trade.

However, this impact was unevenly spread across the region. Significant woodland clearance and the expansion of cultivated land is evident much earlier, already in the Late Viking Age, in north-western Estonia, particularly on the islands, as well as in western Curonia. The emerging trading corridor of the Lower Daugava, focused on Daugmale, saw a gradual opening up of the landscape, with local woods and wetlands exploited for fur. Unfortunately, relatively little data exists from this region, but the landscape north of Daugmale on the other side of the Daugava was a mosaic of cultivated fields, meadows, and woods in the Late Iron Age. Further north, at the western edges of Lettish territory within a broader frontier zone between Livs, Latgalians, and eventually Vends, landscape transformation coincides with the functioning of individual fortified central places, most visibly at Āraiši in the tenth and eleventh centuries. The impact of the small-scale Vendish centre at Cēsis which functioned into the thirteenth century, where there is evidence for livestock husbandry, practised small-scale agriculture but far from comparable to Āraiši. The landscape around Karksi, which had no discernible Late Iron Age stronghold and can be regarded as peripheral to Viljandi's territory, was not developed until the construction of a new centre of power by the Order, and the chronology of landscape transformation here fits within the broader regional trend of south-eastern Estonia and north Latvia. In the case of Viljandi, the stronghold was less reliant on intensifying agricultural production within its immediate landscape, and more dependent on its regional networks of subservient centres. It was also involved in fur production. In this respect, future studies of Late Iron Age stronghold landscapes should find a correlation between their regional status and degrees of environmental transformation. The majority of strongholds which have been excavated, but have not been examined from a landscape perspective, contained assemblages representing a high degree of biodiversity, alongside agricultural tools and in some cases plant remains, all providing complimentary data which connect these sites with substantial ecological transformation within their territories.

All of these central places were engaged in animal husbandry, and in Estonia there is enough data to indicate slight regional and diachronic variations in livestock preferences (Maldre 2012). The hunting activities of a number of strongholds, perhaps reflecting an aristocratic hunting culture, resulted in the depletion of some species and the extirpation of aurochs and bison, although more detailed genetic studies are required to ascertain the specific characteristics of these depletions, and populations may have survived in the corridors of wilderness in southern/south-eastern Livonia. The procurement of fur represents an important exploitation of woodlands at this time, linked directly to the activities of strongholds where processing and redistribution took place. This can also be situated within an emerging mercantile class, with a larger proportion of the population engaged in trade. It is possible to speculate that such a formative period of niche construction resulted in specific and regionally variable sacred landscapes, although these are notoriously difficult to date, and regional clusters most likely represent the uneven survival of data (Vaitkevičius 2004), which in turn may reflect population changes during the crusading period or the Livonian War. What is clear is that very discrete landscape features — individual trees, groves, small woods, hills, lakes, mires, and rivers — designated as sacred with varying functions were widely distributed in relation to settlements. More reliable indicators of religious diversity are regionally specific ritual practices, such as the funerary role of the horse amongst the Daugava Livs, alongside the isolated example of the prophetic horse amongst the Treiden Livs, which were probably influenced by the animal's role amongst the Latgalians and Selonians. Within this cultural zone, the consumption of horse meat, in particular, may have played an important role within the context of funerary feasting. The cult of the horse, as expressed in the culture of the southern Livs, does not appear in the sites of the Vends or Saccalians, although the sacrifice of livestock was practised by both Baltic and Finno-Ugric groups.

The Middle Ages

The crusades resulted in the reorganization of land ownership and resource exploitation, and a gradual move towards centralization encouraged by the hierarchical structure of the Teutonic Order and the commercial opportunities presented by the Hanse. In contrast to the rule of indigenous *seniors* and their families, even those dominating the mighty Estonian overlordships, the new regime was a unified pan-regional corporation. Like any corporation its ranks included opportunistic careerists, but the hierarchy was adhered to, and the mechanics of its lordship were consistent until the sixteenth century, when there is evidence for increasing fragmentation within the upper echelons of the Order's Livonian branch. The Order's ecological impact can be summarized under four themes.

Redefinition and Regulation

As territories were gradually appropriated and subdivided between the conquerors, at times controlled by proxy through loyal vassals, the existing provisioning networks supporting strongholds were realigned, for example, from Daugmale to Riga, from Viljandi to the expanded castle and urban centre of Fellin. By the mid-thirteenth century, the organization of territory in Curonia exemplifies the Order's subsequent management of resources, in relation to other political factions and its indigenous vassals. In April 1253, the Order's representative in Livonia concluded an agreement with the Bishop of Courland concerning the division of the lands of the so-called *Vredecuronia*, which specified access to lakes, meadows, and stone quarries for the Order and episcopate. The sacred lake in Duvzare was to remain undivided, whilst Durbe Lake would be shared between the bishop, the Order, and those Curonians who claimed existing rights to it. The Curonians would also maintain existing ownership of the rivers which flowed from these lakes, along with their hereditary properties which included fields, meadows, fishing sites, woods, and bee-trees (*LUB* I, 248). This pragmatic reorganization of the landscape within the context of redefining lordship in Livonia took place as the Order was reprioritizing its involvement in the defence of the Holy Land. In 1290, the year before Acre fell, the Livonian branch of the Order outlined the relationships between the commanderies of Windau and Goldingen, specifying the importance of interregional networks (Chapter 8). By this time, the major convents were supported by a series of districts run by advocates, and further *Ämter* would be established in the fourteenth century, particularly after the Order's purchase of north Estonia.

The religious landscape of Livonia was a composite, a palimpsest that never erased the indigenous attachment of spiritual values to natural places, whether those enduring after the crusades or those that were subsequently designated. Henry's *Chronicle of Livonia* and *The Livonian Rhymed Chronicle* conceptualize the landscape beyond the Christian hubs of the thirteenth-century crusader landscape as definitively 'pagan' and 'other' (Nielsen 2011; Lazda-Cazers 2009). Churches and the widespread use of crosses as boundary markers were constant reminders of the new religion, but the Order's pragmatism also prevailed when dealing with its indigenous subjects. In Curonia, which has the best examples of sacred sites referred to in medieval documents already from the thirteenth century (see Chapter 8), the preservation or discontinuity of cult sites clearly happened on an ad hoc basis. Agreements defining the rights of Curonian subjects in 1252 and 1253 refer to sacred woods and lakes, although archaeologically visible sacred sites are typically associated with strongholds and settlements. At the Curonian hillfort complex at Strazde, which lay within the advocateship of Kandau (Latv. Kandava) from the mid-thirteenth century, excavations on Baznīckalns hill uncovered ten pits which included the remains of regular fires, human and animal bone, alongside small fragments of ornaments, dating from the eleventh through to the thirteenth/fourteenth centuries (Urtāns 2009, 263). At the southernmost edge of Curonian territory, the site of Palanga continued to function into the fourteenth century, despite the destruction of the centre of power here (Žulkus 2007), perhaps because it was at the fringes of the territories managed by the castles at Memel (Lith. Klaipėda) and Wartha (Latv. Vārtāja). Interestingly this micro-region is associated with a number of sacred toponyms, which perhaps reflect the enduring rural settlement pattern. Following the crusades, the management of the landscape was fully redefined by the theocracy, but a sense of otherness (on both sides) would persist in the separation of urban and rural cultures. Those who went from the countryside to live in towns, or whose settlements were absorbed by urban development, occupied a hybrid status, neither fully belonging to one world or the other. Convents and even the smaller castles were also points of contact between colonists and the indigenous population who staffed their households.

The documentation of territorial rights, overwhelmingly pragmatic in their concern, indicate that the theocratic authorities in Livonia connected their castles directly with the resources of their associated territories, and therefore their principal roles were administrative. In 1387, within the context of a complaint concerning the Chapter of Riga's castle at Dundagen which was being raided by the Order and Bishop of Courland, the castle's district was concisely described as having 'fixed boundaries with cultivated and uncultivated land, fields, forests, bee-trees, various forest animals, including fur-bearing animals that are hunted by various means in forests and forest-covered hills, meadows, waters and fishing sites, coastal stone quarries and amber (in the vulgar tongue called *bernesteyn*) deposits' (*LUB* III, 1248). The castle, administered by an advocate, therefore served as the centre of an estate providing the Chapter with resources which it could sell, including honey, fish, livestock, and their secondary products such as wool, hay, furs, and additional service, altogether providing an annual revenue of 313 Riga marks (*LUB* III, 1248). The castles themselves were clearly intended to be self-sufficient, supported by taxes, tithes, and *corvée* from their vassals, whilst clearly functioning within a hierarchy headed by the major castles of the theocracy, the residences of its leadership. The fortifications and garrison of Dundagen did not prevent the Bishop of Courland from taking the castle by force in 1383 and seizing its goods. This episode exemplifies that the administrative centres of the theocracy were also constructed with security in mind. As in the case of Prussia, all castles which played an important role in the heyday of Livonia were rebuilt or expanded *after* the period of active crusading. They were all fortified and protected with ditches, moats, or natural sheets of water, and their inventories consistently focused on two categories of resources alongside household size: food products and arms.

The Nature of Agricultural Intensification

Trends in the intensification of agriculture can, in the first instance, be generally linked to increases in relative settlement density between the Late Iron Age and the end of the Middle Ages, building upon the existing settlement pattern. The redivision of territory and the requirements of the castles to sustain themselves, as well as the maintenance of a new spiritual infrastructure through religious taxation, promoted a clear agenda of environmental exploitation that in turn could only be facilitated by a denser settlement pattern. In Wenden, the small-scale Vendish centre saw continuity and expansion in terms of the immediate settlement area represented by the castle and town, as was the case in Arrasch after a short hiatus in occupation, but their associated territories saw the proliferation of settlement. The Swedish land audit of 1601 specified that in the time of the Livonian masters, the castle of Wenden had four parishes with at least ninety-five farmsteads and Arrasch had six parishes with at least 116 farmsteads, with a further 122 listed between the two castles (Švābe 1933, 489). The earlier Polish audit from 1582 documented 290 farmsteads within the ten parishes associated with Wenden at the time (Ose 2011). In the parish of Helmet, contributing to the resources of Karkus, the number of centres increased from seven in the Late Iron Age to over sixty by the sixteenth century (Palang et al. 2005). Overall, the population of Livonia is estimated to have doubled during this period (see Chapter 4). Agrarian productivity then can be linked to the activities of the rural population — the suite of paleoenvironmental data considered within this project for Livonia largely reflects the agency of indigenous subjects — but it is possible to qualify the conclusion of Palang et al. (2005, 296) that 'the wider and longer-lasting a change in the appearance of a settlement system is, the more probable that this change is driven not by some interest group, but agents and individuals following their everyday practice'. This recognition of indigenous agency in the longue durée of settlement ignores the multiple roles of the Christian theocracy in amplifying and regulating this trend.

A castle was required to secure the resources from its territory to sustain its community, but the productivity of this territory — if measured in grain represented by store records and the percentage of cultivated land in palynological profiles — varied. The reasons for this variation depended on the quality of local soils and relations with the local population, articulated through systems of taxation, vassalage, and social stability. The Order managed this variation as a corporation. Should stocks run dry, the shortfall could be accommodated by the Order's unique regional and supra-regional networks. Benninghoven (1963, 609–10) defined the so-called *Innere Linie* (inner line) of Livonian castles, where documented large quantities of grain were interpreted as a surplus for export, rather than simply for local consumption. On the basis of the visitation records, this line appears to have concentrated the Order's military and agrarian resources in Livonia. However, grain and other commodities could also be stored in a variety of locations owned, rented, or obliged to the convent's officials (Kreem 2002, 103). The interregional provisioning structure of the commanderies of Goldingen and Windau defined by the Livonian chapter in 1290 (see Chapter 8) was still functioning in the mid-fourteenth century. In 1349, the convent of Goldingen had access to granaries in Alschwangen (Latv. Alsunga) (with 24 *last* of grain), Hasenpoth (Latv. Aizpute) (20 *last*), Altenburg (Latv. Vecpils) (16 *last*), on the premises of the advocate of Kandau (Latv. Kandava) (30 *last* collected *de censu derivaturo* or 'lord's duty'), in Ösel (Est. Saaremaa) (60 *last* collected *de censu derivaturo*), in Pernau (Est. Pärnu)

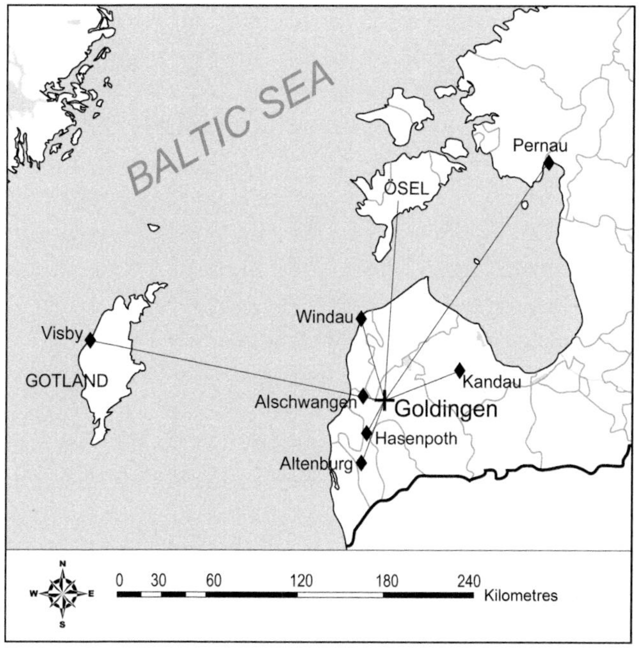

Map 9.4: Map showing the location of the grain stores belonging to the convent of Goldingen in the mid-fourteenth century. Map by author.

(25 *last*), in Gotland (55 *last*), and finally in the Order's granaries in Windau, and at the convent in Goldingen a combined 400 *last* (Map 9.4). This network was partly facilitated by an internal system of taxation; Goldingen's *wartgud* or 'landguards' tax' as specified in an earlier inventory consisted of payments from the commander of Windau, the Bishop of Courland, and the advocate of Dundangen (Latv. Dundaga) (Bauer 1924, 183–84). A century later, the convent's stored provisions consisted of only 28 *last* of grain, six *last* of flour, 15 *last* of barley, six *last* of oats and two *last* of malt (*LUB* XI, 160).

At this time Wenden stands out as the main grain store in Livonia. The core community at Wenden was medium-sized, consisting of an advocate and thirteen brothers, three priest-brothers and two sergeants (*LUB* XI, 160), but the inventory specifies that *half* of the 750 *last* were intended for the consumption of the convent, and the other half was treated as an emergency surplus intended to sustain pressing journeys from the castle. This subdivision of grain stores was not unusual; the inventories of the convent of Goldingen also specified that reserves of grain could be lent to neophytes (Bauer 1924, 184–85); however, there is no mention of grain intended for export through Reval or Riga. It is in fact doubtful whether any significant quantities could be traded from the Order's estates in Livonia. In 1421, the Livonian Master, when making excuses for not providing monetary aid to the Grand Master, wrote that the Order's convent in Riga was dependent on the revenues produced by Wenden and Karkus, which were the two best districts in the land (*das die besten zwe gebitte sein im lande*, perhaps referring to the territories directly controlled by the Master), and that it would require more than a year to collect only half of the requested amount (*LUB* V, 716). A visitation in 1488 also confirmed that the castle of Wenden was well stocked, including with food and grain (Biskup and Janosz-Biskupowa 2004). However, the palynological data from Wenden's immediate landscape does not point to an intensification of cultivation over the fourteenth and fifteenth centuries: instead, substantial expanses of woodland were maintained alongside pasture for the various herds of livestock owned by the castle, town, and rural communities. In contrast, in the northernmost part of Wenden's territory at Trikaten, there is clear evidence for an increase in farming at this time, a dramatic shift from a largely wooded Late Iron Age landscape.

In the case of Karkus, the documented storage of grain and access to arable land synchronizes with the intensification of cultivation over the course of the fourteenth century. The 1442 inventory of the castle states there were sufficient stocks of flour, malt, and smoked and salted fish to last from Easter until Christmas. There were also 6 *last* of salt, two barrels of butter, two barrels of lard, and 100 *last* of rye and barley, enough for a year (*LUB* IX, 834). At this time the castle at Karkus had access to over a thousand *Haken* of actively cultivated land. In contrast, the advocate's castle at Nyenslot (Est. Vasknarva), located right on the frontier with Novgorod's territory, had access to only fifty *Haken* of actively cultivated land. Its meagre stores contained only 3 *last* of flour and 4 *last* of malt. There was no more butter, and all the cows and calves had died, although aside from small quantities of preserved meat, there were enough smoked fish stocks to last from March until the New Year. Over a century later the castle was inevitably reduced to rubble during the Livonian War.

How much grain was kept, and how much was traded? Sustainable storage was a concern for the Order into the fifteenth century, and it is difficult to say how much of any given surplus was intended for sale, or whether this had already been shipped when an inventory was formulated. The provisioning requirements of any given castle were diverse and substantial. The brethren were not the only consumers: provisions were needed for the extended household, and, in far greater quantities, for the herds of cattle, sheep, goats, pigs, poultry, and, most importantly (symbolically if not pragmatically), horses. Whilst the sizes of local herds clearly increased to meet the demands of the growing populations of convents, towns, and villages, this in turn would have put pressure on supplies of fodder; the expansion of cultivated land, as evidence for example in the landscape around Karksi in the fourteenth century, would have been partly driven by this demand. Pathologies evident amongst the livestock from Viljandi and Karksi point to episodes of limited fodder availability, most likely in the winter (see Rannamäe and Lõugas 2019), and in the fifteenth century whole swathes of ploughed fields within the castle's territory lay fallow, perhaps as a result of a localized population crisis. If an emphasis on grain production of trade only appears from the late fifteenth and sixteenth centuries, it can perhaps be synchronized with the gradual disintegration of the unity of the Order's network and the privatization of the resources of individual castles (Kreem 2002, 105), alongside the increase in productivity stimulated by the manorial economy. Despite Plettenberg's emphasis on the Livonian Master's sovereignty, this final phase of the Order's rule was characterized by the agendas of individual commanders. Moreover, the emphasis

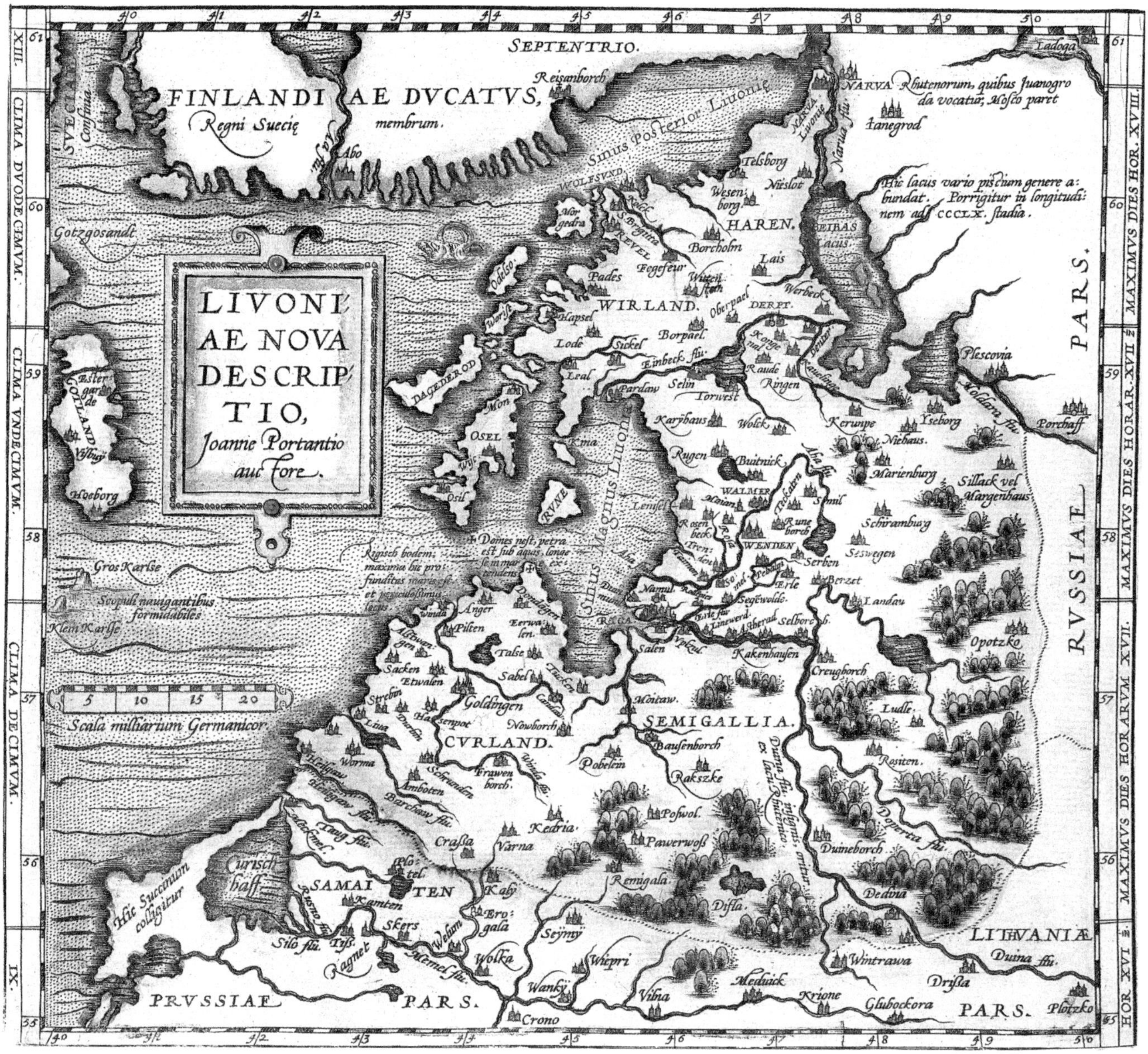

Map 9.5: Abraham Ortelius's map of Livonia, *Theatrum orbis terrarum*. 1573 (VN 1191).
Image reproduced courtesy of the National Library of Latvia, digital collection.

on grain production for trade depended on the interconnected interests of Order, landed nobility, and towns. In 1502, when the *Landtag* in Wenden banned the export of grain due to a famine, the merchants of Riga and Reval complained that this would interfere with their cycles of trade, and with the intervention of the Livonian Master Reval was granted exemption from the ban (Kreem 2002, 125). Trends in environmental exploitation can therefore be regarded as gradually separating from those of the Middle Ages. They were partially interrupted by the Livonian War, after which they can be linked with the manorial economy (see below). At the same time, elements of the medieval legacy continued into the nineteenth century, when, for example, changes in the size of livestock breeds mark a complete break with earlier indigenous forms that had persisted since the Late Iron Age and had been maintained under the Order's rule.

Managing Woodland

The tendency to preserve and manage woodland, and the exploitation of hardwoods for export, is evident across the Order's commanderies. However, in the thirteenth century the conquering Christian elites did not encounter a homogeneously forested landscape. The first significant phase of clearing and cultivation had already begun in the Late Iron Age, associated with the economies of strongholds. One of the earliest maps of Livonia compiled by Abraham Ortelius and published in 1570 depicts the major towns, lakes, rivers, and stretches of woodland in the region (Map 9.5). Although this is a document of relational geography, it is removed from the cartographic paradigm of Olaus Magnus which continued the medieval tradition of symbolizing space with markers such as animals and monstrous hybrids. Ortelius' map shows the former territory of the Livs, in western Latvia, as deforested and relatively densely settled compared to the south-eastern corner of Livonia and Semigallia. The woods are thickest where settlements are few and dispersed, in the borderlands with Russia and Lithuania. This map represents one type of reality at the end of the Middle Ages, not long after the secularization of the Order's Livonian branch. It shows a central Livonian heartland focused on the Lower Daugava and Gauja valleys, as well as coastal Curonia, with a topographically defined border.

At the local scale, trends in deforestation after the crusades varied across the region, and substantial woods remained in the Livonian heartlands of Wenden, Fellin, and Riga: more precise regional maps from the eighteenth century confirm the presence of these woods. In some instances, such as the belt of woodland to the north of Wenden's convent, the sandy soils were not suited to cultivation. Woodland had its uses, not least for producing wax from highly valued apiaries, but in a landscape increasingly dominated by pine, spruce, birch, and alder, it was of increasingly limited interest to Hanseatic timber merchants. The commercially valuable species, principally oak, were heavily depleted in the woodlands of the heartland territories, and as Livonia's networks developed such timbers were increasingly being sourced from areas close to rivers where lumber could be floated (Zunde 2019). But the regular value placed on woodland in territorial documentation suggests it was considered of value to the Order's castles; an essential supply of fuel, building materials, and artefact materials (particularly birch). Hunting rights are sometimes mentioned in grants of fiefs and territorial agreements, however the representation of wild species dropped, and this almost certainly reflected localized reductions in biodiversity by the thirteenth century, rather than the Order's own prescriptions concerning hunting. Livonia is often associated with the medieval fur trade, but this is largely invisible due to the absence of relevant documentation and the lack of faunal assemblages from indigenous villages. If the Order's castles were used to store fur (as in Prussia), they were rarely places where pelts were processed. This also reflected a shift in the function of high-status sites, whereas the communities of strongholds were readily engaged in processing furs, in medieval Livonia this would have been done by the indigenous peasantry in their villages, comparable to how the principality of Novgorod — the greatest fur procurement network in Northern Europe at the time — connected its rural suppliers with its urban boyars and German merchants (Maltby 2019). The main Livonian fur outlets were Riga and Reval, but it is extremely difficult to specify where they sourced their pelts from (Delort 1978, II, 1046), and the majority before the last quarter of the fifteenth century would have derived from Rus' principalities (Jahnke 2015, 222). Some woods remained untouched because they were considered sacred by indigenous vassals or perhaps entire communities; it is difficult to tell who else benefited besides the fief holders, but later documented customs and practices suggest widespread participation in the maintenance or revival of non-Christian indigenous spiritual investment in trees, groves, lakes, and rivers.

One area where the extent of woodland appears to have increased as a result of the crusades was the border with Lithuania. The notion of depopulated, wild landscapes, situated between the southernmost part of the tribal territories of Curonia and Semigallia, and northern Samogitia and Lithuania, is well established in Baltic historiography. This is based on references in thirteenth- and fourteenth-century written sources referring to devastated settlements and forced migrations, the subsequent lack of sources concerning the borderlands, or to famines and epidemics affecting the local population, as well as the absence of fourteenth-and fifteenth-century cemeteries. Written sources relating to the Order's incursions into these borderlands illustrate a wild and depopulated landscape, already during the last decades of the thirteenth century. In *c.* 1280 the advocate of Goldingen led an army to the Semigallian stronghold of Dobele. They travelled along poor roads and through a thick forest, riding and walking, and at times having to construct bridges. Likewise, on the journey from Goldingen to Gresen (near Amboten), the crusading army encountered many bad paths, swamps, and difficult streams, which even an unmounted horse could barely negotiate (*LRC*, 7240–43). These interpretations have been challenged recently, and the reality of the depopulated southern Livonian borderland has been renewed as an object of scholarly inquiry. The problem of dating archaeological data to the fourteenth and fifteenth centuries has been highlighted, whilst continuity in the location of settlements and cemeteries is evident before and after this period, with a likely decrease in the population already in the Late Iron Age (Jarockis 2009, 253–57).

Changes in the environment have not been generally considered as a proxy of long-term human activity in this particular regional frontier. Exceptions include the Curonian stronghold at Impiltis, associated with a significant intensive use of the landscape from the eleventh to thirteenth centuries, which saw significant regrowth of woodland and a noticeable decrease in agriculture from the fourteenth to sixteenth century (Stančikaitė et al. 2009). But depopulation was only partial along the borderland, with a few written sources attesting the presence of populations in southern Semigallia in fourteenth- and fifteenth-century Lithuanian documents (Jarockis 2009, 254). The territories controlled by frontier castles were also precarious and prone to raiding. In September 1374, the Dukes of Polotsk and Odrisk with their armed retinues seized all the livestock (cows, sheep, pigs, and horses) belonging to the convent and local peasants that had pastured next to the castle of Dünaburg (Latv. Daugavpils). Conversely in 1447, during negotiations with the King of Poland in Vilnius, the *hauptmann* of Pskov had reported that Livonians had crossed their borders, cut trees, fished, hunted beavers, and murdered people (*LUB* x, 401).

Relationships with Indigenous Lifestyles

The German colonists in Livonia appropriated existing provisioning systems, and it is clear that the demands of convents and towns, combined with rural population growth, resulted in agricultural intensification. This took place without 'improvements' in agrarian technology: permanent-rotation field systems were in use before the crusades, and whilst a new emphasis was placed on rye, traditional modes of farming prevailed, including fire cultivation. A similar lack of innovation is evident in livestock husbandry. The famous warhorse, imported with the crusades, remained a small component of the equine population which was otherwise drawn from indigenous stock. Whilst the size of cattle and sheep in Fellin and Karkus initially decreased in the period immediately following the crusades, and then increased until this exceeded the prehistoric mean, local livestock types prevailed until the intensive breeding regimes of the nineteenth century. The reason for this is almost certainly the dominating involvement of the indigenous population in stockbreeding. The Order's officials may have quantified herds for the purposes of taxation and provisioning estimates, but they did not promote new husbandry regimes. In this respect, the first significant change in the lifestyle of the Livonian peasantry came with the manorial system and the redefinition of the relationship between lords and their subjects.

Some cultural elements that were introduced by the German colonists remained confined to their enclaves, and it is difficult to determine whether this was the result of economic or social differences between the two populations. Where the cleaver was used to process meat as it was in the Order's castles, it does not appear to have been adopted by the indigenous population, perhaps because native butchers — hardly landed vassals with significant disposable income — could not afford such tools, or because this avoidance was a deliberate act of cultural resistance. Both may have been combined, for butchery in the Late Iron Age, carried out with small blades, was not a highly specialized activity and was probably widely practised. Degrees of standardization are evident in the Liv quarter of medieval Riga, but arguably driven by the demands of an urban market, itself the subject of regulation. The sporadic examples of cleaved bone fragments suggest that at least one affluent Liv burgher embraced the new technology. In a society where food and culinary practices were loaded with social and religious meaning, provisioning was never just about subsistence. The Order's castles maintained a distinctively elite lifestyle, represented by access to good quality meat, an abundance of fish (the staple Lenten foods), alongside imports, including exotic fruits and marine fish. This points to the introduction of a wholly new and exclusive culinary culture, although there was a degree of continuity in the consumption of juvenile animals and poultry at both Late Iron Age strongholds and later castles.

The Post-Medieval Period

Although the focus of this study has been on the transition from the Late Iron Age to the Middle Ages, the long-term impact of the crusades in Livonia can be contextualized further by considering the degree of continuity in trends in environmental exploitation during the period of Polish (1561–1629) and Swedish (1629–1721) rule, which forms a convenient block of 'social time' following the earlier chronological divisions. There is no space here to consider the wealth of written data relating to the organizational changes in the administration of territory or shifts in settlement patterns following the Livonian War, but a useful comparison can be made with unbroken pollen sequences from the primary case study regions, of which seven from Livonia extend into the post medieval period. The secularization of the Order did not result in a significant break in continuity with land ownership or the retention of privileges of lordship. In August 1561, the Livonian Master released the deputy commander of Ascheraden (Latv. Aizkraukle) from his vows to the Order and granted him and his heirs premises in the castle, as well as: property; the right to impose capital punishment on the peasants; homesteads with access to fields, woods, bee-trees, lakes, and rivers; along with hunting and fishing rights (Bauer 2013, 1263). The impact of the Livonian War was devastating, and the Polish audit indicates that over 60% of farmsteads within Wenden's parishes were vacant by 1582 (Ose 2011). The Swedish Audit of 1601 also noted that no cattle had survived on any of the manors (Švābe 1933, 489). Livestock herds were eventually replenished, and the zooarchaeological assemblage from the town of Viljandi dating from the mid-sixteenth to seventeenth centuries indicates no significant changes in husbandry regimes and urban provisioning (Rannamäe and Lõugas 2019).

The pollen data (Fig. 5.13) suggests a decline in agricultural land-use in some pollen sequences during the late fifteenth and early sixteenth centuries, but with indications of increasing intensity during the late seventeenth and eighteenth centuries. This can be plausibly interpreted as reflecting the impact of the Livonian War on the fledging manorial system which only recovered during the period of Swedish rule. At both Blanku Mire (Fig. 5.5) and Äriküla (Fig. 5.12) the pollen shows a peak in arable land-use during the mid-eighteenth century, but with generally declining levels of arable and pastoral indicators from the mid-eighteenth century onwards. Two pollen sequences from Trikāta (Ger. Trikaten) and Āraiši, both located in proximity to medieval castles and later post-medieval manors, both clearly show a pattern of intensive agrarian land-use peaking from the mid-seventeenth to early twentieth century (Stivriņš et al. 2015, 2016), reflected to a lesser degree in the pollen sequence presented in Chapter 5. The castle at Trikaten continued to be used until the Great Northern War (1700–21), when it was destroyed. Here, the conflict does not appear to have interrupted the intensified cultivation of the surrounding landscape which took place under the rule of the *Landtag*, when Livonia was under Russian control. When the castle at Āraiši was being used as a *folwark* or specialised farm during the period of Polish rule, medieval trends in animal husbandry continued, with the same types of livestock and virtually no evidence of hunting or trapping. The farm continued to be used for slaughtering, butchering, and processing livestock, dominated by at this time by adult cattle. The medieval trade routes were still operational, as marine fish species were still being consumed at the site. At the same time, the surrounding settlements and fields had been badly affected by the Livonian War, as within the vicinity of Wenden. After the *folwark* was abandoned, it was only during the period of Swedish rule that the region appears to have been gradually repopulated, resulting in an intensification of farming. In the vicinity of Riga, the period from the mid-seventeenth century is also associated with deforestation and an expansion in cultivated land, in a significant break from Late Iron Age and medieval trends (Kalniņa et al. 2019). In summary, the post-medieval trends in environmental exploitation reflect to varying degrees the impact of the Livonian War, which devastated the local population. However, subsequently the intensification in cultivation evident within all the macro-regions can be linked to the dominance of the manorial economy. Riga remained a major international port, and it became the main outlet for shipping pine mast-trees from increasingly distant sources (Zunde 2019).

Conclusion

Both the level of regulating environmental exploitation and the interregional networks created in the process were completely novel in Livonia and ensured the new centres of theocratic power could be maintained. For much of the thirteenth century, the purpose of this organizational superstructure which connected the conquered landscape with the Order's houses, in turn with its supra-regional and commercial networks, was to sustain the defence of the Holy Land, as well as to secure the newly annexed domains. As in the Latin East, the new Christian authorities in the eastern Baltic worked with the system they had annexed, which resulted in the retention of existing settlement patterns, the tolerance of non-Christian sacred sites, and the granting of privileges to loyal native vassals. What eventually changed in the fifteenth and sixteenth centuries was the character of ownership of these estates, which reflected a shift in power relations within the context of locally developing German identity. However, on the Order's lands the proliferation of manors was relatively late, no doubt reflecting the Knights' concern with self-sufficiency and sustainability. The Order — as an organization — consistently refused to dilute its control over the landscape and remained the principal landowner in Livonia until the point of secularization in 1561.

The Order's fiery disputes with the town and Archbishop of Riga which eventually concluded with the relocation of the preferred residence of the Livonian Master to Wenden, encapsulate this paradigm of control, as do the edicts of the Grand Master in the fifteenth century: these serve as a reminder of the supra-regional organization of the Teutonic Order, a structure that was only loosened with the ascendancy of Wolter von Plettenberg, and not long before the Prussian branch was secularized and redefined as a vassal of the Polish crown. Within the framework of the long-term ecological history of north-eastern Europe, the impact of the crusades on Livonia and the imposition of a theocratic regime was comparatively less significant than the preceding period of stronghold expansion in the Late Iron Age. The crusades broke the power of the strongholds, and in the process halted the growth of their ecological niches which had seen localized reductions in biodiversity. These were then reconfigured to meet the demands of convents and new urban markets, now connected at regional and supra-regional levels. Whilst some Late Iron Age heartlands were maintained, new ones were also created with corresponding ecological profiles. In some regions this resulted in the gradual expansion of cultivated land. This was briefly interrupted by the Livonian War and continued under the system of *Gutsherrschaft*, itself a legacy of the crusades in the eastern Baltic. In the southernmost borderlands of Livonia, our current understanding of the impact of the crusades is significant depopulation and the expansion of wilderness, although this may yet be challenged by future research.

Works Cited

Primary Sources

Bauer, A. 1924. *Die Wartgutsteuresliste der Komturei Goldingen: Mitteilungen us der livlaendischen Geschichte*, Rīga: Gesellschaft für Geschichte und Altertumskunde zu Riga

——, 2003. Transcripts from the Sources of the History of Manors in Courland collected by Albert Bauer, digitized by the Herder Institute in Marburg (date refers to the completion of the digitization project)

Biskup, M., and Janosz-Biskupowa, L. eds. 2004. *Visitationen im Deutschen Orden im Mittelalter, II: 1450–1519*, Quellen und Studien zur Geschichte des Deutschen Ordens, 50, Marburg: Elwert

Dunsdorfs, E., ed. 1939. *Vidzemes 1638. gada arklu revīzija*, Izdeva: Latvijas Vēstures institūta apgādiens

HCL Brundage, J. A., trans. 2003. *The Chronicle of Henry of Livonia/Henricus Lettus*, New York: Columbia University Press

LRC Smith, J. C. and Urban, W., trans. 2006. *The Livonian Rhymed Chronicle*, London: Routledge Curzon

LUB Bunge, F. G. von, ed. 1853–57. *Liv-, Est- und Kurländisches Urkundenbuch nebst regesten* (vols I–III), 13 vols to date, Reval: Kluge and Ströhm

Mellin, L. A, et al., 1798. *Atlas von Liefland, oder von den beyden Gouvernementern u. Herzogthümern Lief- und Ehstland, und der Provinz Oesel: entworfen nach geometrischen Vermessungen, den neusten astronomischen Beobachtungen und nach sorgfaeltiger Untersuchung und Kentnis der Gegenden: das Ganze besteht aus einer Generalkarte und vierzehn Kreiskarten.* Riga. Available online at [http://www.loc.gov/item/75572471/].

Ose, I. ed. 2011. 'Cēsu pils 1582. gada revīzija', *Latvijas viduslaiku pilis* 7, 429–32

Secondary Works

Alttoa, K. 2006. 'Karksi keskajal. Mõningaid probleeme', *Viljandi Muuseumi Aastaraamat* 2008, 266–74

Apals, J. 1996a. *Senie mājokļi Latvijā: No vissenākajiem laikiem līdz 13. gs*, Rīga: RaKa

——, 1996b. '*Vēstures avotu ziņas* par *Āraišu mūra pili*', *Arheoloģija un etnogrāfija* 18, 27–37

Arnold, U. 1989. *Zur Wirtschaftsentwicklung des Deutschen Ordens im Mittelalter*, Marburg: Elwert

Benninghoven, F. 1963. 'Probleme der Zahl und Standortverteilung der livländischen Streitkräfte im ausgehenden Mittelalter', *Zeitschrift für Ostforschung* 12, 601–22

Blomkvist, N. 2004. 'The Medieval Catholic World-System and the Making of Europe', in *The European Frontier: Clashes and Compromises in the Middle Ages*, ed. J. Staecker, Lund: Almqvist and Wiksell, 15–34

——, 2005. *The Discovery of the Baltic: The Reception of a Catholic World-System in the European North (ad 1075–1225)*, Leiden: Brill

Brūzis, R., and Spirģis, R. 2009. 'The Ogresgala Čabas Cemetery and Horse Sacrifice', *Archaeologia Baltica* 11, 283–94

Caune, A. 2007. 'Cēsis 14.–18. gadsimtā: plānojums, apbūve un iedzīvotāji', in *Cēsīm 800. Quo vadis, Cēsis? Vēsture un mūsdienu nosacījumi pilsētas attīstībai*, ed. V. Rozentāle et al., Cēsis: Cēsu pašvaldības aģentūra, Vidzemes Vēstures un tūrisma centrs, 155–66

Delort, R. 1978. *Le commerce des fourrures en occident á la fin du Moyen âge: (vers 1300 - vers 1450)* (2 volumes), Rome: École Française de Rome

Haak, A. 2015. 'Arheoloogia andmed Viljandi ordulinnuse kesk- ja uusaja kohta', in *Viljandi ordulinnus ja Lossimäed läbi aegade*, Viljandi: Viljandi Muuseumi Toimetised, 27–64

Haak, A., Rannamäe, E., Luik, H., and Maldre, L. 2012. 'Worked and Unworked Bone from the Viljandi Castle of the Livonian Order (13th-16th centuries)', *Lietuvos Archaeologija* 38, 295–338

Hayward, K. M. J. 'Resources for Castle Building in Prussia and Livonia', in *Ecologies of Crusading, Colonization, and Religious Conversion in the Medieval Baltic: Terra Sacra II*, ed. by A. Pluskowski, Environmental History of the North Atlantic World, 3, Turnhout, Brepols (forthcoming)

Jahnke, C. 2015. 'The Baltic Trade', in *A Companion to the Hanseatic League*, ed. D. J. Harreld, Leiden: Brill, 194–240

Jarockis, R. 2009. 'Graves, Plains and Settlement Development. Semigallia 600–1600', in *The Reception of Medieval Europe in the Baltic Sea Region*, ed. J. Staecker, Visby: Gotland University Press, 243–62

Kalniņa, L., Ceriņa, A., Kiziks, K., Stankeviča, K., Pujāte, A., and Dručka, A. 2019. 'Evidence of Human Impact and Vegetation Change during the Late Iron Age and the Medieval Livonian Period at Some Sites along the Lower Course of the River Daugava', in *Ecologies of Crusading, Colonization, and Religious Conversion in the Medieval Baltic: Terra Sacra II*, ed. by A. Pluskowski, Environmental History of the North Atlantic World, 3, Turnhout, Brepols (forthcoming)

Kalniņš, G. 2014. 'Cēsis Castle, Latvia: Landscape Context of the Teutonic Order's Castle', unpublished MA dissertation, Leicester: University of Leicester.

Kreem, J. 2002. *The Town and its Lord: Reval and the Teutonic Order (in the Fifteenth Century)*, Tallinn: Tallinna Linnaarhiiv

——, 2013. 'Crusading Traditions and Chivalric Ideals: The Mentality of the Teutonic Order in Livonia at the Beginning of the Sixteenth Century', *Crusades* 12/1, 233–50

——, 2014. 'Der Deutsche Orden im 16. Jahrhundert. Die Spätzeit einer geistlichen Adelskorporation in Livland', in *Leonid Arbusow (1882–1951) und die Erforschung des mittelalterlichen Livland*, eds I. Misāns and K. Neitmann, Köln: Böhlau, 287–96

——, 2015. 'Saksa ordu ja Viljandi linnus', in *Viljandi ordulinnus ja Lossimäed läbi aegade,* Viljandi: Viljandi Muuseumi Toimetised, 65–86

——, 2019. 'Written Evidence Concerning the Resources of the Teutonic Order in Livonia: The Cases of Karkus and Fellin' in *Ecologies of Crusading, Colonization, and Religious Conversion in the Medieval Baltic: Terra Sacra II*, ed. by A. Pluskowski, Environmental History of the North Atlantic World, 3, Turnhout, Brepols (forthcoming)

Lazda-Cazers, R. 2009. 'Landscape as Other in the *Livländische Reimchronik*', *Amsterdamer Beiträge zur älteren Germanistik* 65, 183–209

Maasing, M. 2015. 'Infidel Turks and Schismatic Russians in Late Medieval Livonia', in *Fear and Loathing in the North: Jews and Muslims in Medieval Scandinavia and the Baltic Region*, eds C. Heß and J. Adams, Berlin: De Gruyter, 347–88

Maldre, L. 2012. 'Archaeozoological Material from the Keava Hill Fort and Linnaaluste Settlement Sites', *Estonian Journal of Archaeology* suppl. 1, 106–26

Maltby, M. 2019. 'Animal Exploitation in Medieval Novgorod', in *Ecologies of Crusading, Colonization, and Religious Conversion in the Medieval Baltic: Terra Sacra II*, ed. by A. Pluskowski, Environmental History of the North Atlantic World, 3, Turnhout, Brepols (forthcoming)

Mänd, A. 2009. 'Saints' Cults in Medieval Livonia', in *The Clash of Cultures on the Medieval Baltic Frontier*, ed. A. V. Murray, Farnham: Ashgate, 191–223

——, 2014. 'Black Soldier - Patron Saint: St Maurice and the Livonian Merchants', *Iconographisk Post* 1, 56–75

Melton, E. 1988. '*Gutsherrschaft* in East Elbian Germany and Livonia, 1500–1800: A Critique of the Model', *Central European History* 21/4, 315–49

Morton, N. E. 2009. *The Teutonic Knights in the Holy Land: 1190–1291*, Woodbridge: Boydell

Nielsen, T. K. 2011. 'Henry of Livonia on Woods and Wilderness', in *Crusading and Chronicle Writing on the Medieval Baltic Frontier*, eds M. Tamm, L. Kaljundi and C. S. Jensen, Farnham: Ashgate, 157–78

Orton, D., Rannamäe, E., Lõugas, L., Makowiecki, D., Hamilton-Dyer, S., Pluskowski, A. G., O'Connell, T., and Barrett, J. 2019. 'The Teutonic Order's Role in the Development of a Medieval Eastern Baltic Cod Fishery', in *Ecologies of Crusading, Colonization, and Religious Conversion in the Medieval Baltic: Terra Sacra II*, ed. by A. Pluskowski, Environmental History of the North Atlantic World, 3, Turnhout, Brepols (forthcoming)

Palang, H., Printsmann, A., Konsa, M., and Lang, V. 2005. 'Ideology and Tradition in Landscape Change: A Case of the Helme Parish, Estonia', in *Landscape, Law and Justice*, eds T. Peil and M. Jones, Oslo: Novus, 288–97

Poska, A., and Saarse, L. 1999. 'Holocene Vegetation and Land-Use History in the Environs of Lake Kahala, northern Estonia', *Vegetation History and Archaeobotany* 8, 185–97

Poska, A., Saarse, L., Koppel, K., Nielsen, A. B., Avel, E., Vassiljev, J., and Väli, V. 2014. 'The Verijärv area, South Estonia over the last Millennium: A high Resolution Quantitative Land-Cover Reconstruction based on Pollen and Historical Data', *Review of Palaeobotany and Palynology* 207, 5–17

Rannamäe, E., and Lõugas, L. 2019. 'Animal Exploitation in Karksi and Viljandi (Estonia) in the Late Iron Age and Medieval Period', in *Ecologies of Crusading, Colonization, and Religious Conversion in the Medieval Baltic: Terra Sacra II*, ed. by A. Pluskowski, Environmental History of the North Atlantic World, 3, Turnhout, Brepols (forthcoming)

Riley-Smith, J. 2010. *Templars and Hospitallers as Professed Religious in the Holy Land*. Notre Dame, Ind.: University of Notre Dame Press

Sillasoo, Ü., and Hiie, S. 2007. 'An Archaeobotanical Approach to Investigating Food of the Hanseatic Period in Estonia', in *Medieval Food Traditions in Northern Europe*, ed. S. Karg, Copenhagen: National Museum of Denmark, 73–96

Šnē, A. 2008. 'The Medieval Peasantry: On the Social and Religious Position of the Rural Natives in southern Livonia (13th–15th centuries)', *Ajalooline Ajakiri* 1/2 (123/24), 89–100

Stančikaitė, M., Šinkūnas, P., Risberg, J, Šeirienė, V., Blažauskas, N., Jarockis, R., Karlsson, S., and Miller, U. 2009. 'Human Activity and the Environment during the Late Iron Age and Middle Ages at the Impiltis Archaeological Site, NW Lithuania', *Quaternary International* 20, 74–90

Stivriņš, N., Brown, A., Reitalu, T., Veski, S., Heinsalu, A., Banerjea, R. Y., and Elmi, K. 2015. 'Landscape Change in Central Latvia since the Iron Age: Multiproxy Analysis of the Vegetation Impact of Conflict, Colonization and Economic Expansion during the Last 2,000 Years', *Vegetation History and Archaeobotany* 24, 377–91

Stivriņš, N., Brown, A.D., Reitalu, T., Ratniece, V., Austin, J.E., Liiv, M., Cerina, A., and Veski, S. 2016. 'Palaeoenvironmental Evidence for the Impact of the Crusades on the Local and Regional Environment in Northern Latvia: Terrestrial Evidence', *The Holocene* 26, 61–69

Švābe A. 1938. *Kuršu ķoniņi un novadnieki*. Pielikumi. Straumes un avoti

Tamm, M. 2009. 'A New World into Old Words: The Eastern Baltic Region and the Cultural Geography of Medieval Europe', in *The Clash of Cultures on the Medieval Baltic Frontier*, ed. A. V. Murray, Farnham: Ashgate, 11–36.

Tuulse, A. 1942. *Die Burgen in Estland und Lettland*, Dorpat: Aasta

Urtāns, J. 2009. 'Semigallian Church Sites in Folklore and the Archaeological Evidence', in *The Reception of Medieval Europe in the Baltic Sea Region*, ed. J. Staecker, Visby: Gotland University Press, 263–71

Vaitkevičius, V. 2004. *Studies into the Balts' Sacred Places*, BAR international Series 1228, Oxford: Archaeopress

Valk, H. ed. 2014. *Strongholds and Power Centres East of the Baltic Sea in the 11th–13th Centuries*, Tartu: Institute of History and Archaeology at the University of Tartu, Department of Archaeology at the Institute of History of Tallinn University

——, 2015. 'Viljandi muinasaeg', in *Viljandi ordulinnus ja Lossimäed läbi aegade*, Viljandi: Viljandi Muuseumi Toimetised, 7–26

——, 2019. 'Sacred Landscapes of Medieval Livonia: Lands of the Teutonic Order in southern Estonia', in *Ecologies of Crusading, Colonization, and Religious Conversion in the Medieval Baltic: Terra Sacra II*, ed. by A. Pluskowski, Environmental History of the North Atlantic World, 3, Turnhout, Brepols (forthcoming)

Valk, H., Rannamäe, E., Brown, A., Pluskowski, A., Badura, M., and Lõugas, L. 2013. 'Thirteenth Century Cultural Deposits at the Castle of the Teutonic Order in Karksi', *Arheoloogilised välitööd Eestis/Archaeological Field Work in Estonia*, 2012, 73–92

Vasks, A. 2010. 'Slīmests vai ādu apstrādes rīks?', *Latvijas Vēstures Institūta Žurnāls* 2, 5–17

Yeh, H. Y., Pluskowski, A., Kalējs, U., and Mitchell, P. D. 2014. 'Intestinal Parasites in a mid-14th Century Latrine from Riga, Latvia: Fish Tapeworm and the Consumption of Uncooked Fish in the medieval eastern Baltic Region', *Journal of Archaeological Science* 49, 83–89

Žulkus, V. 2007. *Palanga in the Middle Ages: Ancient Settlements*, Vilnius: Versus Aureus

Zunde, M. 2019. 'Timber and its Use from the Late Iron Age to the end of the Medieval Period in Latvia', in *Ecologies of Crusading, Colonization, and Religious Conversion in the Medieval Baltic: Terra Sacra II*, ed. by A. Pluskowski, Environmental History of the North Atlantic World, 3, Turnhout, Brepols (forthcoming)

PRUSSIA

Bison in Białowieża Forest, Poland.
Photo by M. Elander.

10. Sites in Prussia: The Historical and Archaeological Background

Aleksander Pluskowski, Marc Jarzebowski, Małgorzata Karczewska, and Maciej Karczewski

The term 'Prussia', as it is understood in terms of historical geography, was ultimately a creation of the Teutonic Order. Peter von Dusburg defined the region by its borders: the Vistula to the west, the Baltic Sea to the north, the Memel to the north-east, and the territories of Rus' principalities, Masovia and Dobrin, to the south and south-east. Prussia itself was then subdivided into eleven territories: Kulm (referred to throughout this study as the Kulmerland) and Lubovia, Pomesania, Pogesania, Warmia, Natangia, Sambia, Nadruvia, Scalovia, Sudovia, Galindia, and Barta (*KP* III, 3; see Bojtár 1999, 156). Other sources mention Sasna, situated between Lubovia and Galindia, subdivided between the Teutonic Order and Polish principalities in 1240. The Kulmerland had been a Polish territory: western Pomesania had been partially colonized by Pomeranians and also by settlers from the Kulmerland in its southern stretches, before Prussian incursions and settlements provided the justifications for crusading campaigns from the 1230s. Within each area there were several sub-groupings, also referred to as 'tribes' in the literature (Nowakowski 2003; for detailed etymological discussions of the Prussian tribes, see Bojtár, 1999; Powierski, 2003). The Sudovians of German sources are synonymous with the Jatvingians of Rus' and Polish sources, although opinion remains divided on whether this easternmost group spoke a Prussian dialect or had its own separate West Baltic language (Bojtár 1999, 158). The Prussian Crusade, a whole series of individual campaigns beginning with the reclamation of Christian control of the Kulmerland in 1230, concluded with the conquest of Sudovia in 1283. Whilst the Teutonic Order's chroniclers described peoples associated with specific territories, they also merged these into a single, derogatory stereotype: the Prussians were in many respects the 'other'. But over the course of the fourteenth century, the term 'Prussia' was adopted by the Order to refer to its own territory, including the Kulmerland and annexed Pomerania, and its leaders began to describe themselves as the 'German Order in Prussia', a label also used by the papal court (Biskup 2002). In Polish, *Prusowie* came to denote the pre-Christian inhabitants of the region and *Prusacy* the German colonists (Okulicz-Kozaryn 1997, 11). The name of the

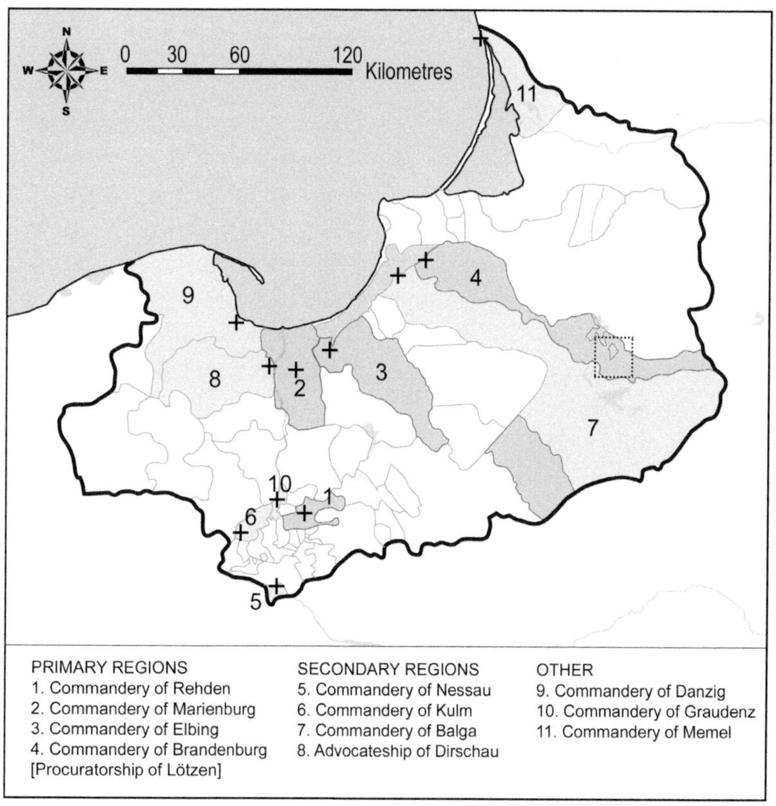

Map 10.1: Primary and secondary sites in Prussia. Map by A. Pluskowski.

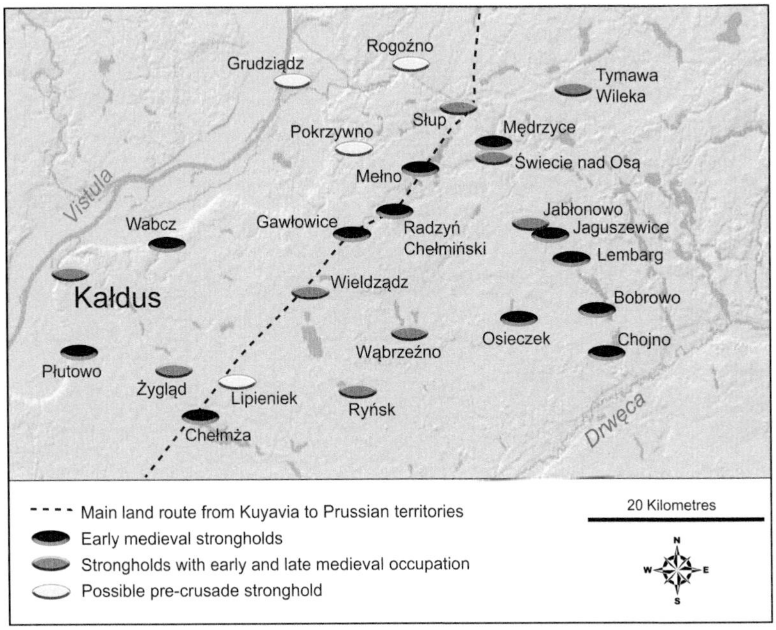

Map 10.2: The distribution of strongholds in the early medieval Kulmerland. Map by A. Pluskowski.

Order's territories was thus conferred onto its ethnically diverse population. Following the annexation of Pomerelia, the Order's territories encompassed an area of 48,000 km^2, expanding by an additional 32,100 km^2 by the early fifteenth century, although with the exception of the Neumark these later gains were only held for a short period of time (Tandecki 2009, 109).

Although the archaeology of the crusading period indicates a hiatus between early and later medieval culture in the political centres of Prussia (Pluskowski 2013), the indigenous population was not obliterated by the crusades. Labuda and Biskup (1986, 177–78) estimated that at the end of the twelfth/start of the thirteenth century the population of Prussia numbered around 170,000. Marceli Kosman (1976, 12) estimated that in the thirteenth century there were an estimated 120,000 Prussians and only around 12,000–15,000 Germans. In 1309, the Prussian population has been estimated at around 90,000 Prussians, with 15,000 Germans and 30,000 Poles (Biskup 2002, 133); the significant change in the ethnic composition of Prussia would take place over the course of the fourteenth and fifteenth centuries with the onset of colonization. The ethnic diversity of the Order's state in the fourteenth and fifteenth centuries is not easy to identify archaeologically, especially as no villages occupied by the indigenous population and governed under 'Prussian Law' have been properly excavated, the largest number of which are documented in the Sambian diocese (Długokęcki 2009d, 469–74). Prussian names recorded by the Order's copyists were adapted to German, and they were also subject to 'Polonization' as a result of the high number of incoming Polish settlers in the southern part of Prussia from the fourteenth and fifteenth centuries; names of lakes were frequently transferred to settlements created in their vicinity, and sometimes the same name would be given to more than one place. The loss of indigenous identity was partially associated with these linguistic changes (Ekdahl 2004, 7). However, traces of Old Prussian names did survive in some places up to the seventeenth century (Pospiszylowa 1986). The Duchy of Prussia created from the secularized remnants of the Order's territories in 1525 ultimately developed into the Kingdom of Prussia. Following the incorporation of East Prussia into the USSR in 1945, German toponyms were systematically replaced by Russian ones (Krickus 2002, 39). Place names also changed in former eastern Prussian territories assigned to Lithuania and Poland (Chojnacki 1946).

The purpose of this chapter is to present the historical and archaeological background to primary and secondary sites (see Chapter 2) included within this study. As in the case of Livonia (Chapter 4), each site is presented in turn with an outline of its known historical development, a summary of its territory, which includes what is presently known about the distribution of sacred sites, followed by a summary of the archaeology and how this relates to the present study, concluding with a characterization of the site within the context of a frontier (see Chapter 2).

Primary Sites and Regions

The primary sites in Prussia included within the study are situated in modern north-eastern Poland. Their details are outlined below with sites, and their territories listed from south-west to north-east (Map 10.1).

1. The Commandery of Rehden (Radzyń Chełmiński)

The study site at Radzyń Chełmiński is located in the Kulmerland (Pol. Ziemia Chełmińska), a region in present day northern-central Poland (today within the Kuyavian-Pomeranian Voivodeship). The Kulmerland is one of the most intensively studied regions associated

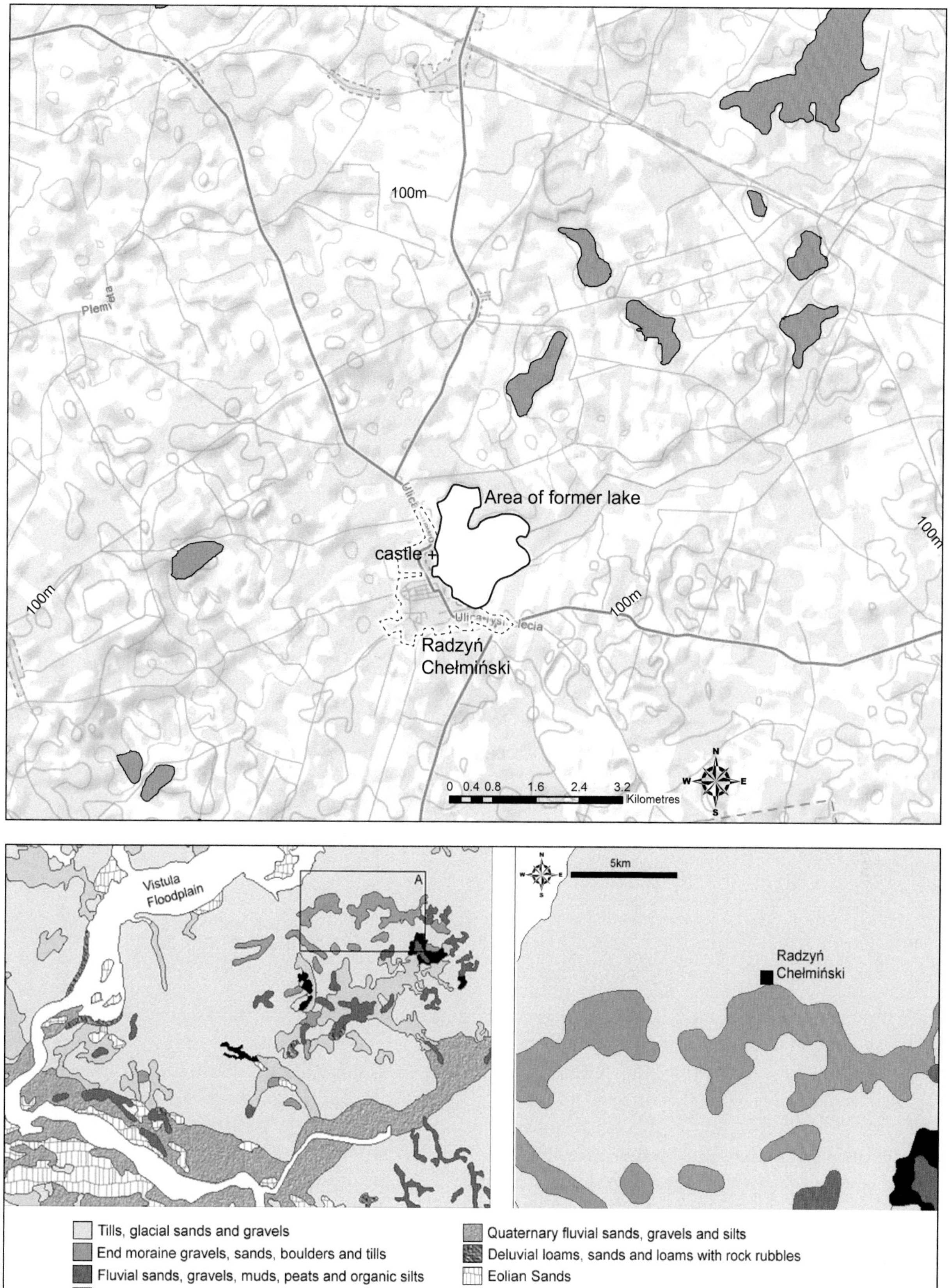

Map 10.3: (a) The landscape within the immediate vicinity of the convent and town of Rehden (Pol. Radzyń Chełmiński); (b) a map of the superficial geology (1:500,000, after Central Geological Database, Polish Geological Institute). Map by A. Pluskowski.

Figure 10.1: (a) The ruins of the conventual castle at Radzyń Chełmiński, view from the south-east (photo by Magnus Elander); (b) view of the interior (photo by A. Pluskowski); (c) plan of archaeological site (Wiewióra 2013a); (d) reconstruction of castle (after Arszyński 2010, 14).

with the frontier of the early medieval Polish state, and one of the heartlands of the later *Ordensland* (Chudziak 1996, 2000; Poliński 2003). From the eighth century the region saw increasing Slavic colonization and subsequently came under Polish control from the second half of the tenth century, becoming a frontier with Prussian territories. A network of fortified strongholds was constructed around the margins of the Kulmerland, interpreted as reflecting the threat from neighbouring Prussian tribes (Chudziak 1996, 167–70); at its core was the complex at Kałdus (see below), interpreted as a castellany, which may have even functioned as a missionary centre, with its stone church representing the furthermost from the core of the Polish state (Chudziak 2003). Increasingly intensive colonization was accompanied by unsuccessful attempts to Christianize the neighbouring Prussians.

The stronghold at Radzyń, part of this defensive ring, was situated in the north-eastern part of the territory, at the very edge of the borderland (Map 10.2). It was one of a number of central places defining a 'meso-regional' cluster of settlement in the north-east of the Kulmerland focusing on valley systems (Chudziak 1996, 174, Fig. 33). The site itself was located on the edge of a bedrock peninsula next to a lake within a landscape of wetlands, punctuated by outcrops of high ground. It was occupied from the late eleventh through to the twelfth or possibly early thirteenth century; its final phase associated with a burned timber layer, potentially the result

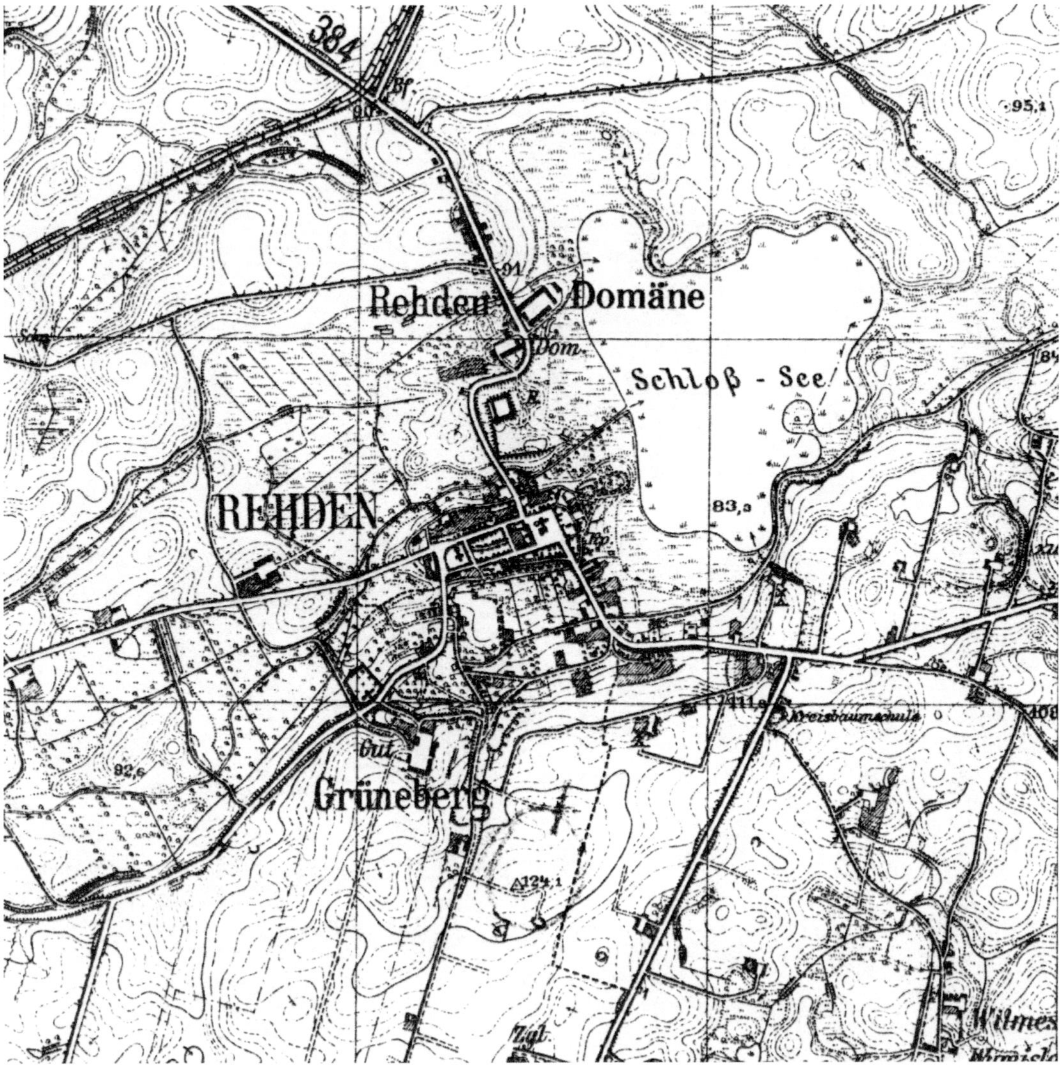

Map 10.4: Map of Rehden (Pol. Radzyń Chełmiński) showing the shrunken extent of the adjacent lake in 1942 (1:25,000; Reichsamt für Landesaufnahme, Wydział Geografii i Studiów Regionalnych Uniwersytetu Warszawskiego).

of a Prussian attack (Chudziak 1994). Following the temporary occupation of the Kulmerland by Prussians in the early thirteenth century, in 1225 the Teutonic Order was granted territory in the region by Duke Konrad of Masovia in return for reclaiming the region and protecting Christian missionaries and colonists. This was confirmed by the Holy Roman Emperor. Five years later the Order initiated its military campaigns which eventually stabilized the region, forming a base for further crusades into Prussia.

The Order's presence in the Kulmerland was secured with a series of *castra*, a number reusing or located near existing strongholds. The territory of Rehden came into the possession of Christian, the Bishop of Prussia, in 1218, and was acquired by the Order in 1231 (Wiewióra 2013b, 104). The first *castrum* was constructed in 1234, at the same time as the adjacent town was granted privileges, located directly to the south of the later castle. The commandery itself was founded in *c.* 1251, one of four established by the Order in the region (Poliński 2003, 23). The surviving brick-built castle, with its construction variously dated to *c.* 1270–1330 (Wiewióra 2013b, 104), is situated 500 m north-west of the early medieval stronghold, located on an elevation at the eastern end of the valley which ended with the lake (Map 10.3). Even in the early twentieth century, the surrounding landscape consisted of wetlands and streams (Map 10.4), now largely drained and reclaimed, and it is assumed a similar landscape existed within the valley system in the fourteenth century. The castle complex consisted of three separate elevations (effectively islands) connected by bridges, one of which was squared off, and contained the conventual castle (with dimensions of 49.4 × 49.6 m) surrounded on all sides by an outer bailey that was raised 2.5–3 m, with the two flanking outcrops serving as outer baileys (Fig. 10.1). The banks of all three islands were reinforced with brick walls, and the channels separating them functioned as a moat system connected with the lake, parts of

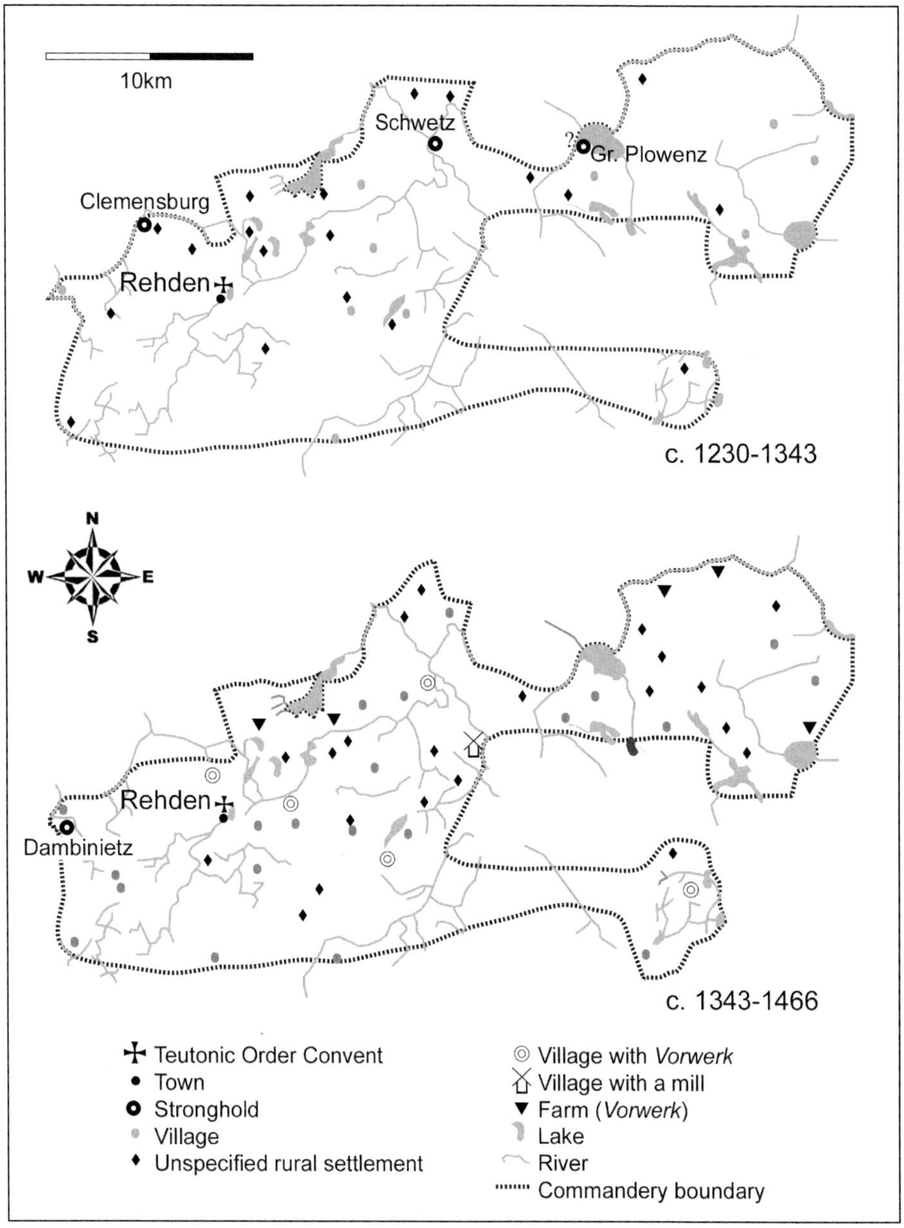

Map 10.5: The commandery of Rehden (Pol. Radzyń Chełmiński) (after Poliński 2003). Map by A. Pluskowski.

within the commandery. Until the early fourteenth century, the castle was served by the *Vorwerk* (and later village) at Dorf Rehden (Pol. Radzyń Wieś) situated to the south-west, which also had seen earlier occupation from the eleventh to twelfth centuries (Chudziak 1996, 246, cat. 656–58), and thirty-eight other settlements are documented within its administrative region. From 1343 to 1466, the settlement of the commandery intensified, with forty-eight settlements (including smaller fortified sites) and nine *Vorwerke* documented (Poliński 2003, 75, 104, Table 13). The majority of settlements were located within the north-eastern section of the commandery, with relatively few near the town and castle; the two nearest *Vorwerke* were located 2.5 km to the north (Königlich Neuhof, Pol. Nowy Dwór) and 5 km to the south-west (Dorf Rehden). The colonists were German and Polish, and of the villages established in the commandery, the largest fourteen belonged to the Order, nine to knightly families, and four were under joint-ownership.

Inventory records from the late fourteenth century show significant provisioning of cereals, horses, and livestock in the castle from across the commandery.

which were contained by walled embankments (Torbus 1998, 597–98; Wiewióra 2013b).

The territory governed by Rehden was gradually shaped from the mid-thirteenth century as the administrative divisions of the Kulmerland became increasingly complex (Map 10.5). By 1343 the delineation of the commandery can be ascertained (Poliński 2003, Map 1), encircled by a series of commanderies and advocate's territories and in the north-east the lands of the episcopate of Kulm. From 1283 until *c.* 1409, there was a period of relative stability free of major conflicts, corresponding to an increase in documented settlements

However, between 1409 and 1411, before and after the defeat of the Teutonic Order at the Battle of Grunwald (Ger. Tannenberg), there is a dramatic drop in the provisioning of cereals, horses, and livestock; the role of the castle significantly diminished after 1410 and the ongoing conflict with Poland-Lithuania. In 1416, the convent obtained part of the territory formerly within the commandery of Engelsberg (Pol. Pokrzywno) (Poliński 2003, 25). However, a few decades later in 1454 the commandery of Rehden was itself dissolved, becoming the centre of an eldership; a component of the late medieval administration of the Polish-Lithuanian state. By

Map 10.6: Map from 1909 showing the hummocky landscape and wetlands around Kgl. Rehwalde (Pol. Rywałd) (1:25,000; Geognostisch-Agronomische Ausgabe, Wydziałowe Archiwum Kartograficzne, Wydział Nauk Geograficznych Uniwersytetu im. Adama Mickiewicza, Poznań).

the sixteenth century, much of the western wing of the castle was a disused ruin, and the structure was damaged further during the Swedish 'Deluge' (1626–29). The castle was partly dismantled up to the nineteenth century, after which the first major restoration work began under the direction of Conrad Steinbrecht, as with a number of other Prussian castles. The process of restoration has continued into the present day.

Archaeology

There have been several excavations within the castle grounds (at the edges of the outer bailey, the moat, and inside the castle). In 2008–09 excavations directed by Daniel Gazda and Marcin Wiewióra did not uncover undisturbed medieval contexts within the castle itself but did identify earlier construction phases in the ward surrounding the conventual building (Wiewióra 2013a). Small-scale excavations and conservation work has continued under Gazda's direction until today. Due to the lack of suitably dated deposits, it was not possible to

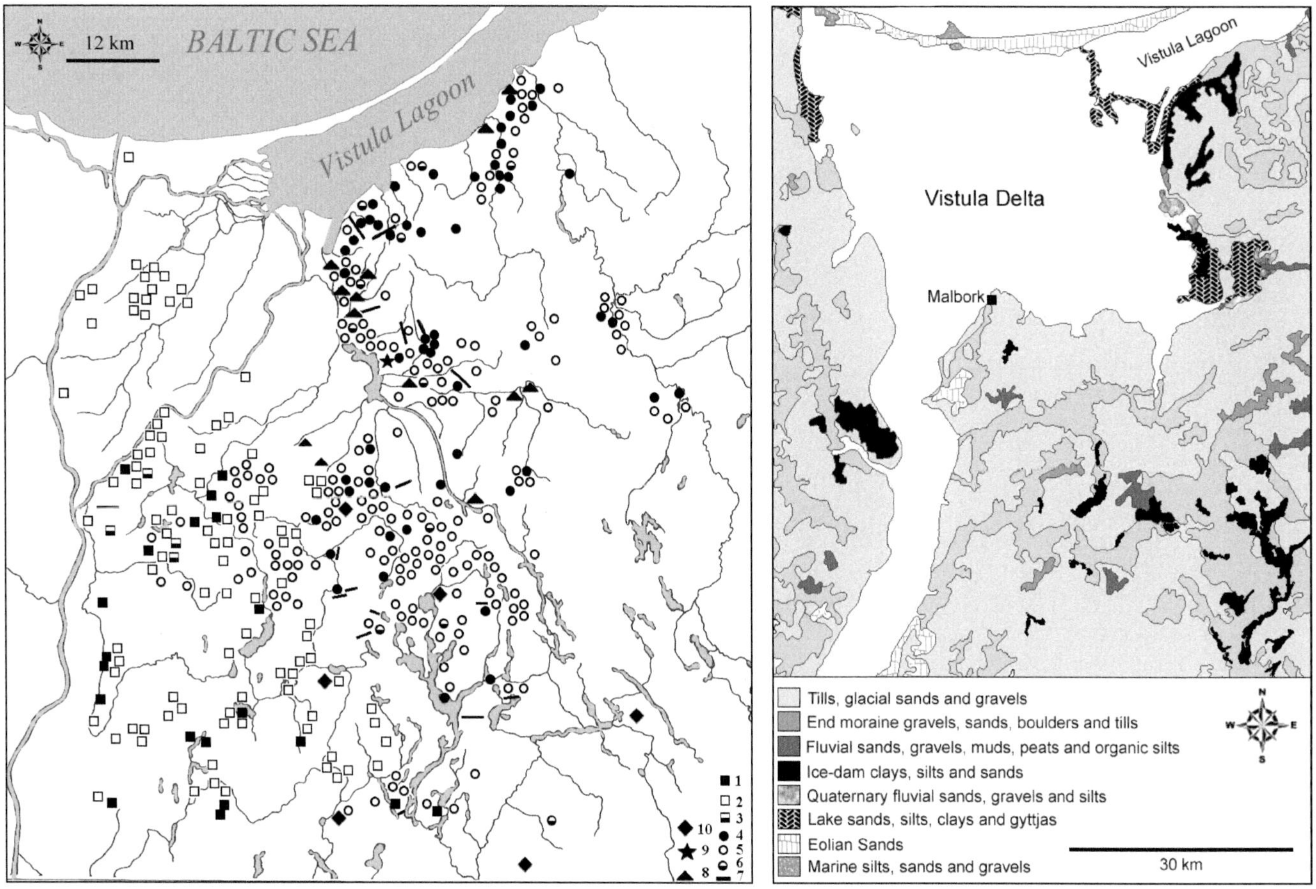

Map 10.7: (a) The pre-crusade settlement pattern between the Vistula and Dzierzgoń rivers (Jagodziński 2004b, 38; with updated records from Szczepański 2016) 1 = Slavic strongholds; 2 = Slavic settlements; 3 = Slavic cemeteries; 4 = Prussian strongholds; 5 = Prussian settlements; 6 = Prussian cemeteries; 7 = Substantial defensive embankments; 8 = Traces of Scandinavian settlement; 9 = Truso; 10 = Early medieval hoards (b) a map of the superficial geology (1:500,000, after Central Geological Database, Polish Geological Institute).

analyse any faunal, botanical, or occupation-related sediments from the site. Instead, the data for this case study derives from three cores taken from the lake adjacent to the castle, and from a peat core obtained near the village of Rywałd Królewski (Ger. Königlich Rehwalde), *c.* 7 km east of Radzyń Chełmiński (Map 10.6). The settlement was founded in 1312 by the commander of Rehden with the privileges of the Kulm Law (Poliński 2003, 57, cat. no. 414). The constraints of time did not allow more cores to be taken and analysed, but given the extensive distribution of wetlands within this region future palynological and geoarchaeological research may enhance the resolution of the landscape changes presented in this study. Some samples of medieval brick debris resulting from recent restoration work were also taken for analysis (Hayward 2019). The convent is also associated with twenty-two separate inventory records between 1377 and 1449. They reflect its role as a centre for storage and provisioning, particularly in preparation for war from the end of the fourteenth century.

Regional Status

The landscape associated with the commandery of Rehden is an example of a frontier (Polish-Prussian; temporarily occupied by Prussians, then a short-lived borderland during the Prussian Crusade) that was transformed into a heartland within the *Ordensland*. It then became a short-lived frontier between the polities of the Teutonic Order and Poland, before eventually being incorporated into the latter.

2. *The Commandery of Marienburg (Malbork)*

The commandery of Marienburg is associated with the largest castle constructed by the Teutonic Order, today named Malbork (Pomeranian Voivodeship), and one of the most important historical monuments in Poland (Figs 10.2–3). The territory was carved out across the delta of the Lower Vistula, a landscape of fens and floodplains, flanked by major escarpments.

Figure 10.2: The western side of the castle at Malbork (Ger. Marienburg) today, with the Nogat in the foreground. Photo by M. Elander.

The earliest written sources describe this region as the borderland between eastern Pomerania, also known as Pomerelia, and the adjacent territories of Pomesania and Pogesania. The Nogat itself did not mark this boundary, as Pomeranian settlements are noted east of the tributary in the eleventh and twelfth centuries, along with Slavic toponyms. In the Viking Age, Pomeranian settlements spreading across the Vistula floodplain and its eastern tributary, the Nogat, eventually gave way to villages and strongholds associated with Prussians, largely on the eastern side of the belt of woodland stretching from Kwidzyn (Ger. Marienwerder) to Lake Drużno, and clustered within the vicinity of Dzierzgoń (Haftka 1987; Jagodziński 2004b; Szczepański 2016) (Map 10.7). Scandinavians also formed an important component in this multi-ethnic landscape, particularly associated with the emporium of Truso (Janów Pomorski) (Jagodziński 2010). By the end of the tenth century, the political reach of the Polish Christian state had extended to this region (Buko 2005, 183–86), with Gdańsk superseding Truso as the main regional emporium, and its dukes, allied with the Polish crown, increasingly flexed their political control over this borderland through strongholds such as Węgry (Ger. Wengern), occupied from the eleventh through to the late twelfth century. Węgry has been interpreted as the centre of a castellany, the political focal point for Pomeranian colonization in much the

Figure 10.3: (a) The eastern side of the castle at Malbork today, view of the high castle, middle castle, and moat from the outer bailey; (b) view of the middle castle and outer bailey from the high castle looking north. Photos by M. Elander.

Figure 10.4: The site at Biała Góra 3 in relation to the village of Biała Góra (Ger. Weißenberg), and the plan showing the location of trenches from 2007 to 2011. Figure by Z. Sawicki.

same way as settlements clustered in the Kulmerland. The stronghold had been destroyed in the 1160s, and its associated settlements had been abandoned, as well as the settlement complex to the north at Wielbark-Malbork (site 2) (Haftka 1987, 37–41; Haftka and Wadyl 2015, 180–81), perhaps as a result of Prussian incursions.

This process of colonization was accompanied by the development of religious infrastructure. Abbeys in the heartlands of the Polish state began to play important roles in the management of the Vistulan trade route, linking the Pomeranian strongholds with Gdańsk and the Baltic coast (Powierski 2001, 161). Some became actively involved in missionary activity targeting the neighbouring Prussian tribes. During the twelfth century, borderland tensions increased as Polish magnates raided Prussian lands, whilst Prussian incursions against Pomeranian centres reached as far as the Nogat and occasionally into Gdańsk. Whilst Pomeranian sites east of the Nogat appear to have been abandoned before the crusades (Jagodziński, 2004b), with a visible, albeit in some cases short, break between the early and later medieval settlement pattern (Haftka 1987, 41), a site occupied during this transitional period has been identified to the north of the village of Biała Góra (Ger. Weißenberg), *c.* 18 km southwest of Malbork (Fig. 10.4). It represents a unique example of an excavated medieval rural colony in the Vistula borderlands, and from 1309 situated within the heartland of the Prussian *Ordensland*. The site at Biała Góra (3), first discovered during a field survey in 1972 has been the subject of excavations in 2007 and 2008 directed by Zbigniew Sawicki, and in 2011 and 2013 within the framework of this study. The site has been tentatively associated with 'Zantyr' (or Zantir), which began as a Pomeranian, Cistercian missionary outpost functioning from 1235 to 1236. The location was strategic, as the site was situated next to the confluence of the Vistula and Nogat. By 1233 crusading armies had reached the Lower Vistula, and Zantyr was obtained by the Teutonic Order in 1236. Its acquisition by the Order and tensions over the control of the Vistula led to a protracted war with Duke Świętopełk of Gdańsk in 1242–53, during which the stronghold was briefly recovered. It was subsequently returned to the Prussian Church and ceded to the Order by 1251. The settlement at Biała Góra was certainly refounded or further developed by a new wave of settlers who introduced a different type of firing technology, represented by significant quantities of greyware ceramics (Pluskowski et al. 2014; Sawicki et al. 2015).

As the political situation stabilized in the early 1250s, the Order established the commandery of Zantyr; the rest of Pomesania was governed by the convent established in Christburg (Pol. Dzierzgoń). The convent at Zantyr was officially relocated to Marienburg *c.* 1280, four years after both are mentioned as the responsibility of the same commander. The exact dates remain disputed: Jan Powierski (1979) argued the Order had a

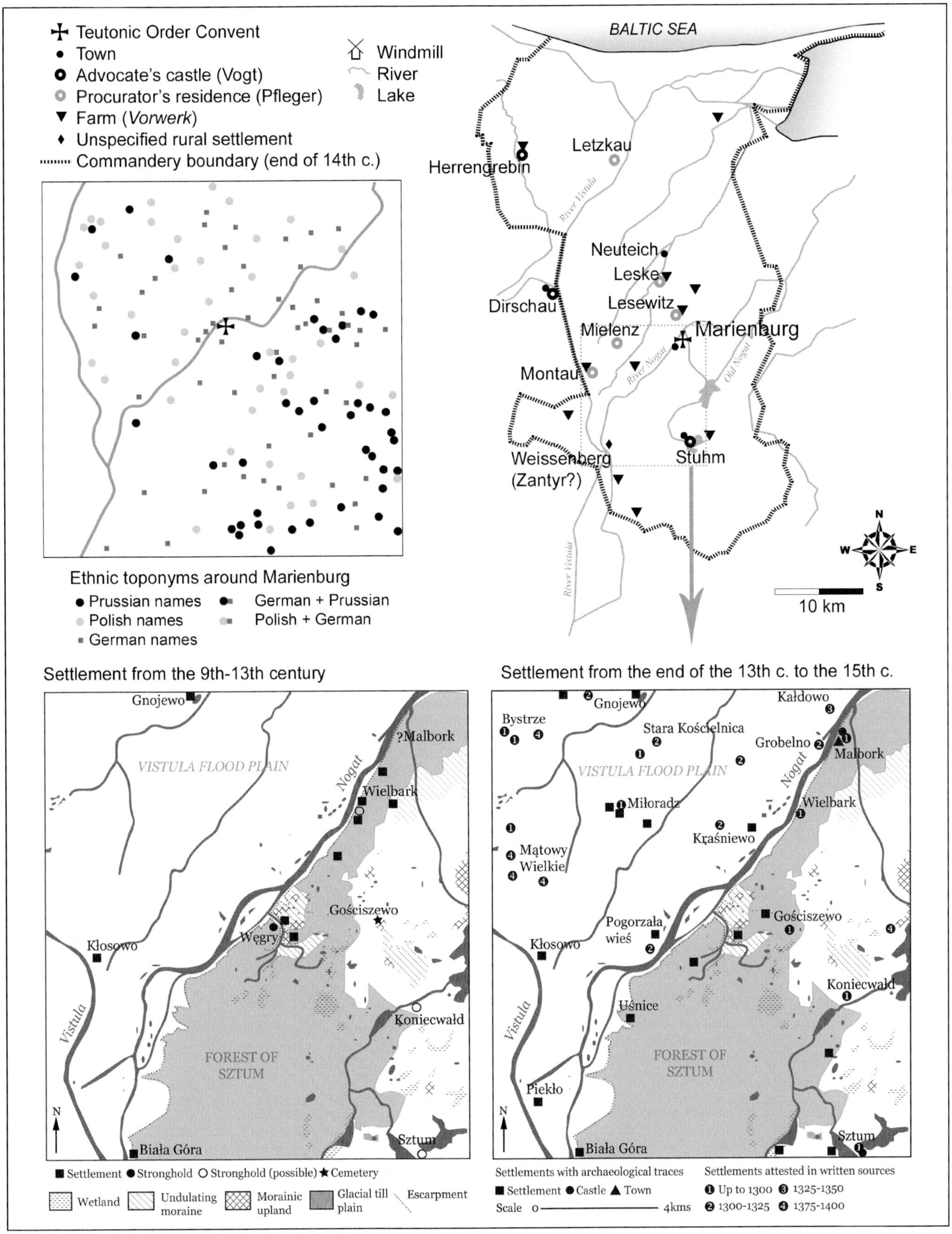

Map 10.8: The commandery of Marienburg (Pol. Malbork); Polish, German, and Prussian settlement place names in the southern part of Marienburg's (Pol. Malbork) commandery (Górnowicz 1980; settlement maps after Haftka 1987.).

Map 10.9: Map of the vicinity of Neuteich (Pol. Nowy Staw) in 1925 showing a completely poldered landscape, a process which began in the fourteenth century (1:25,000; Archiwum Map Zachodniej Polski).

presence at Marienburg from 1272, and the convent was formally established there in 1276. Jóźwiak and Trupinda (2007, 53) have argued for a date of 1278 for the construction of the castle and 1279 for the relocation of the convent. Peter von Dusburg's *Chronicle* (*CTP* III, 208) states the date for relocation was 1280, whilst various other documents refer to relocation dates from 1279 to 1282 (Pollakówna 1997, 416–17). The last decades of the thirteenth century saw the consolidation of Marienburg's commandery, following the Order's acquisition of land on the western side of the Vistula in 1282 (Powierski 2003, 80), although full expansion would only take place with the annexation of Pomerelia in 1309, with the western edge of Marienburg's commandery partly delineated by the Lower Vistula valley (Map 10.8).

The castle was constructed in three phases (Jóźwiak and Trupinda 2007, 53–71). The earliest, the high castle, which was situated on the highest elevation overlooking the Nogat, consisted of a northern range with an enclosure wall forming a quadrangle of 61 × 52 m, to which a western range was added in *c.* 1300. In 1309 when the Order annexed Gdańsk and Pomerelia, the convent of Marienburg was designated as the headquarters of the Order by Siegfried of Feuchtwangen, prompting a programme of expansion which lasted into the fifteenth century. From 1310 to 1350, additional ranges were added to the high castle to create the classic conventual form, along with an extension of the chapel in the north range, and the construction of three ranges of buildings to the north (in an area that probably functioned as the earlier *Vorburg*), which came to be referred to as the middle castle and came to house the largest space in the castle, 'the Great Refectory', as well as the 'Grand Master's Palace'. To the north of the middle castle, a substantial outer bailey was constructed, which included industrial buildings, stables, and barns. From the second quarter of the fourteenth century, an outer wall (the third defensive wall) was built in stages to enclose all three areas of the castle complex. A bridge was also constructed across the Nogat. In 1411–14, embankments were constructed beyond the outer wall, and between 1417 and 1448 new defensive structures were installed in the eastern and southern side of the castle complex, encompassing by then an area of around 21 hectares. This was the last phase of castle building, and all subsequent modifications have involved refining the existing structure. The castle remained the property of the Order until it was sold to the Polish Crown in 1457, and from the late nineteenth century it was the object of restoration and conservation under the direction of Conrad Steinbrecht. After suffering damage in 1945, the process of restoration began again and continues into the present day. In 1997 the castle was designated a UNESCO World Heritage Site.

The landscape around Marienburg was recolonized and developed following the establishment of the commandery. The town of Marienburg, established directly to the south of the castle shortly after the foundation of the convent, is largely known from fragmentary written sources. It was fortified and contained civic buildings that were typical of medieval Prussian towns, such as a town hall, church (dedicated to St John), school, and Artus court. It was almost completely destroyed during WWII, and subsequent urban renewal has not included systematic archaeological investigations. Whilst the castle towns of Marienburg and Dirschau were situated on morainic uplands overlooking the floodplain between the Nogat and Vistula, the Delta region (referred to as *Werder* in the Order's sources) saw extensive colonization in the fourteenth and fifteenth centuries, a process that had already begun before the annexation of Pomerelia. Settlements were founded or reorganized under the Kulm Law, with twenty-five villages documented in the Żuławy Gdańskie, over forty in the Żuławy Wielkie, and around thirteen in the Małe Żuławy Malborskie (Długokęcki 1992). These villages are poorly known archaeologically: only the red brick churches have survived from this period of settlement, although a number have been rebuilt. In some cases, it is clear these were founded on islets of higher ground that had previously seen settlement, as with the town of Neuteich (Pol. Nowy Staw) (Map 10.9). Here, the Order founded a town adjacent to this earlier Pomeranian village, which became an important centre for cattle and sheep breeding in the commandery. Just to the south of the town at Leske (Pol. Laski) was an advocate's complex specializing in, amongst other things, horse rearing. The town's wealth is attested in the early fifteenth century by the significant investment in its largest church (dedicated to St Matthew) (Fig. 10.5), a town hall, a hospital, and chapel dedicated to St George.

Figure 10.5: The church of St Matthew at Nowy Staw (Ger. Neuteich). Photo by M. Elander.

In the southern part of the commandery, with the exception of Biała Góra there is no convincing evidence to date of continuity between the pre- and post-crusade settlement pattern (Haftka 1987), although across the

Figure 10.6: The so-called 'Baker's Mill' (Ger. Obermühle) at ul. Konopnickiej 2 in Malbork today, first documented in *c.* 1401. Photo by A. Pluskowski.

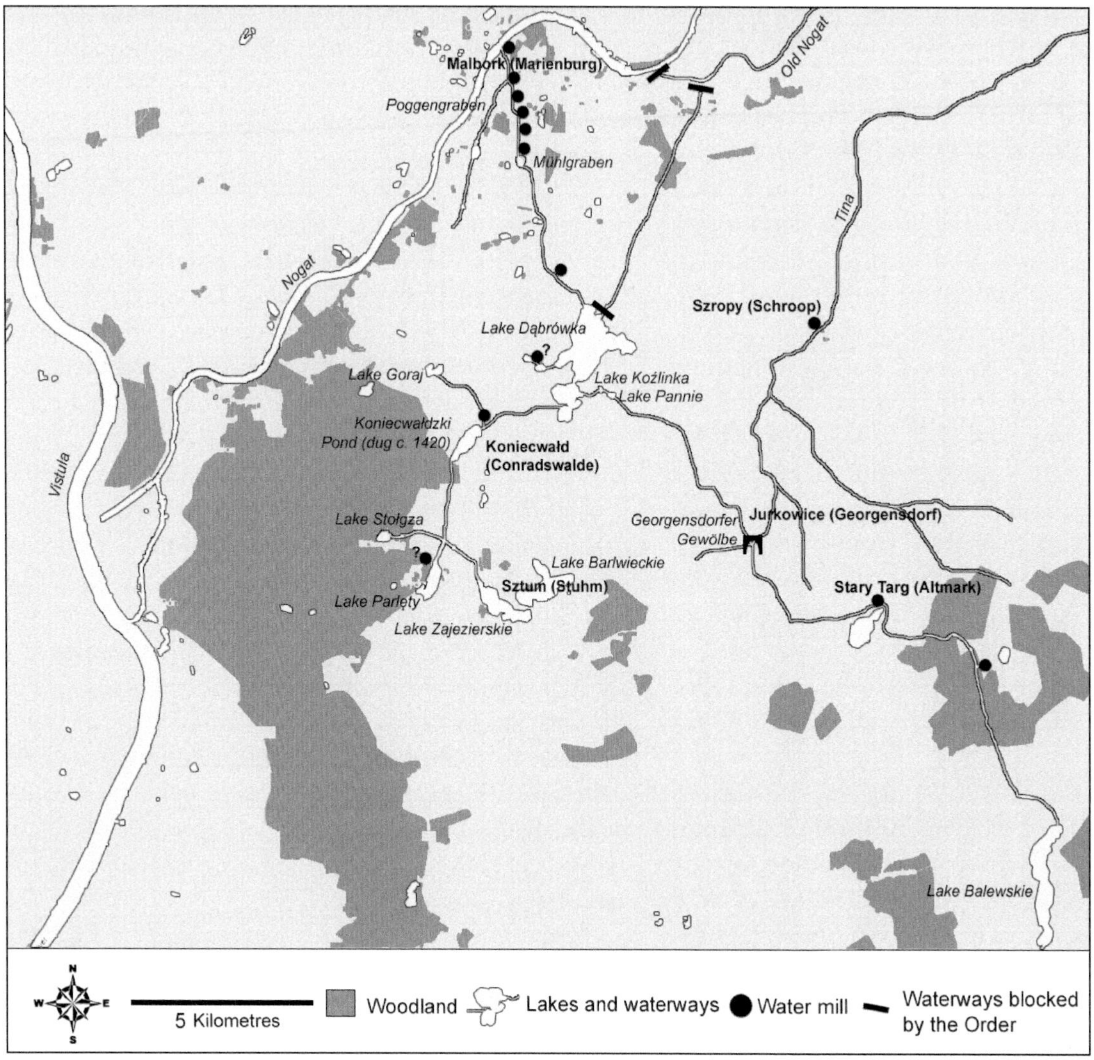

Map 10.10: The hydrological network linked to the convent of Marienburg's (Pol. Malbork) moat system in the first half of the fifteenth century (after Długokęcki et al. 2004), set against the modern distribution of woodland and the present course of the Vistula and Nogat.

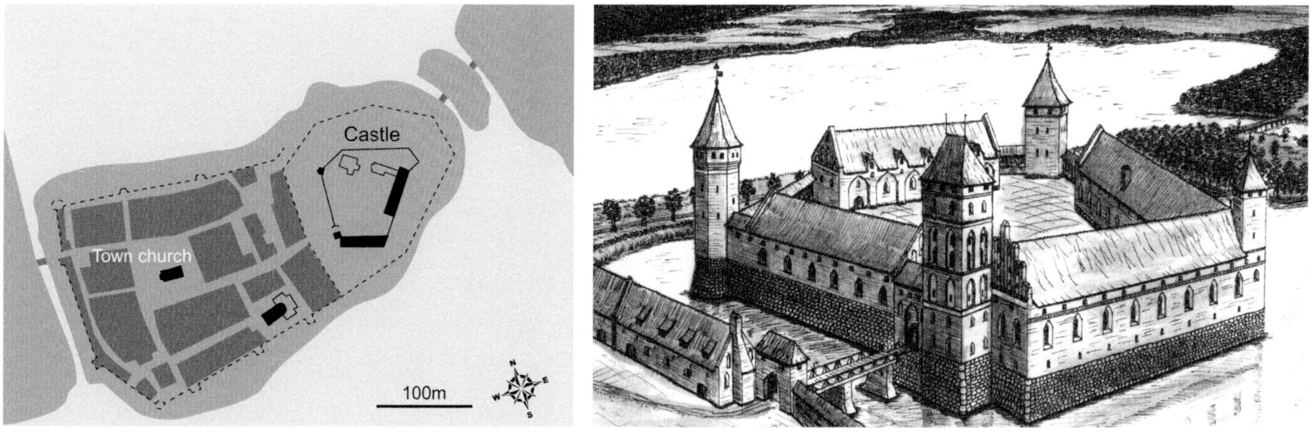

Figure 10.7: (a) Plan of the castle and town of Stuhm (Pol. Sztum) (after Schmid, in Haftka 1999, 305); (b) reconstruction of the castle at Stuhm (Klat and Mykowski 2002, 88).

commandery the range of toponyms recorded in the fourteenth century indicates a mixture of Polish, German, and even Prussian settlements (Górnowicz 1980, 13) (Map 10.8). The integrity of the largest stretch of woodland in the convent's territory was respected, suggesting a degree of management by the Order (see Chapter 15). Hydrological modifications included the construction of a canal (Map 10.10), linking Lake Dąbrówka (in turn fed by a number of smaller lakes) to the moat complex at Marienburg, 12 km northwest. The canal (Ger. Marienburger Mühlgraben), which is still visible today, is a major hydrological feature in the landscape and was constructed to channel water through several mills located along its length and feed the outer moat of the castle complex (Fig. 10.17), finally draining into the Nogat (Długokęcki et al. 2004, 17; see also Chapter 13).

The administration of the commandery was delegated to a series of officials, including districts run by bailiffs (sing. *Kammeramt*), at least two advocates (*Vogtei*), and five procurators (*Pflegeramt*). To the south-east of Marienburg, an irregular-sided castle was constructed in 1326–35 at Stuhm (Pol. Sztum), with its outer bailey reworked into a planned town in 1416 (Haftka 1999, 305) (Fig 10.7a,b). From 1333, the castle is associated with an advocate, and in the fourteenth and fifteenth century it was used as a summer residence for the grand masters, with a park containing predominantly deer documented in close proximity (Torbus 1998, 172; see Chapter 15). This functioned as one of nineteen *Vorwerke* documented in the commandery of Marienburg by the end of the fourteenth century, increasing to at least twenty five in the fifteenth century, supplying the convent with woodland, arable, and wetland resources (Gancewski 2013). Excavations of these have been very limited, but Bönhof (Pol. Benowo) consisted of a self-contained complex which included a residential building, chapel, and service buildings, with no evidence of fortifications or moats (Chęć 2004). The convent at Marienburg was also the recipient of goods and income from the other territories held by the Order, following the centralization of its administration and bureaucracy (Józwiak 2009a, 131–32).

Sacred Sites

Given the intensity of Pomeranian colonization in the Vistula Delta region, it is not surprising that very few pre-Christian sites are recorded here in the medieval period. The Prussian sacred sites of Pomesania have been surveyed by Kawiński (2011, 92–102), and the nearest are located in the vicinity of Dzierzgoń, within the commandery of Christburg. In his reconstructions of medieval settlement patterns in the landscape south of Malbork, Haftka (1987) speculated about the presence of a cult site within the Forest of Sztum characterized by a circle of stones, a place later referred to as 'Hexentanzplatz'. However, it has not been possible to verify this. As in many other respects, this landscape was one of transition, with evidence for the creation of a Christian infrastructure already associated with Pomeranian missionary activity. When the conditions of the Treaty of Christburg (1249) required the defeated Prussians to build thirteen churches, of which six have been identified by Szczepański (2013) as being located east of the Prussian centre at Dzierzgoń, it is clear the landscape of northern Pomesania had already been the focus of intensive Christianization.

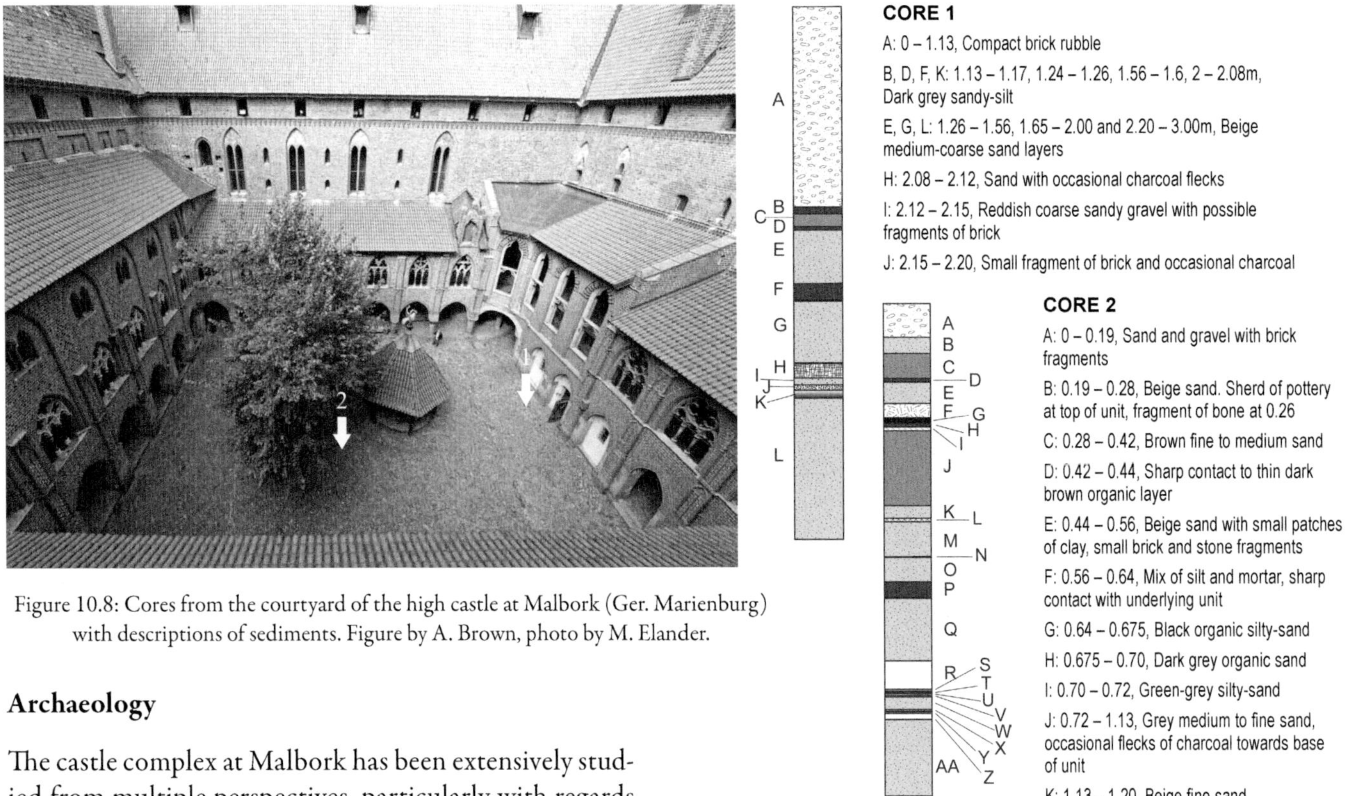

Figure 10.8: Cores from the courtyard of the high castle at Malbork (Ger. Marienburg) with descriptions of sediments. Figure by A. Brown, photo by M. Elander.

Archaeology

The castle complex at Malbork has been extensively studied from multiple perspectives, particularly with regards to the conservation and restoration of its architecture. There have also been numerous excavations within the complex from the late nineteenth century, and most recently under the direction of Mieczysław Haftka and Zbigniew Sawicki, and research-focused excavations led by Maria Dąbrowska (2007) in various parts of the outer bailey from 1998. As a result of the continuous occupation of the site, and the multiple phases of building and restoration, archaeological contexts that can be associated with the Teutonic Order's period are typically small and truncated. Geophysical survey was carried out in parts of the outer bailey and high castle between 2011 and 2013. Two cores were obtained from the courtyard of the high castle which provided a stratigraphic sequence (Fig. 10.8): one included an organic layer that was radiocarbon-dated to the earliest phase of the complex (*c.* AD 1267–1311, 68.8% probability), whilst a core from the fishpond located to the north of the outer curtain wall of the *Vorburg* did not contain any material suitable for further analysis (Fig. 10.9). The faunal assemblages from Dąbrowska's excavations had been made available at the pilot stage of the project (Maltby et al. 2009; Pluskowski et al. 2010) and in 2011, a trench was opened in the outer moat to obtain a suite of palaeoenvironmental data (Fig. 10.10), with a fragment of bark from the lowest stratigraphic unit radiocarbon dated to *c.* AD 1410–1520 (88.2% probability). A core taken south-east of Malbork in the pilot stages of the project (Brown and Pluskowski 2014) was supplemented with a number of cores taken from the south-eastern section of the commandery (see Chapter 11). A series of brick and mortar samples were also obtained from the earliest phases of the high castle and compared with those excavated at Biała Góra (Hayward 2018). Unfortunately, post-medieval hydrological modifications, along with damage to flood defences in 1945, had destroyed the palaeoenvironmental record in much of the landscape in the north-western part of the commandery, that is, in the floodplain of the Nogat and Vistula. Attempts were made to take cores from peat deposits within this area, but all failed to obtain suitable material. It is possible that future surveys of smaller pockets of peat in other parts of the fens will locate isolated, undisturbed deposits, enabling the long-term environmental changes associated with the colonization of the Delta to be investigated in more detail. However, it is expected this will follow the trends presented in the

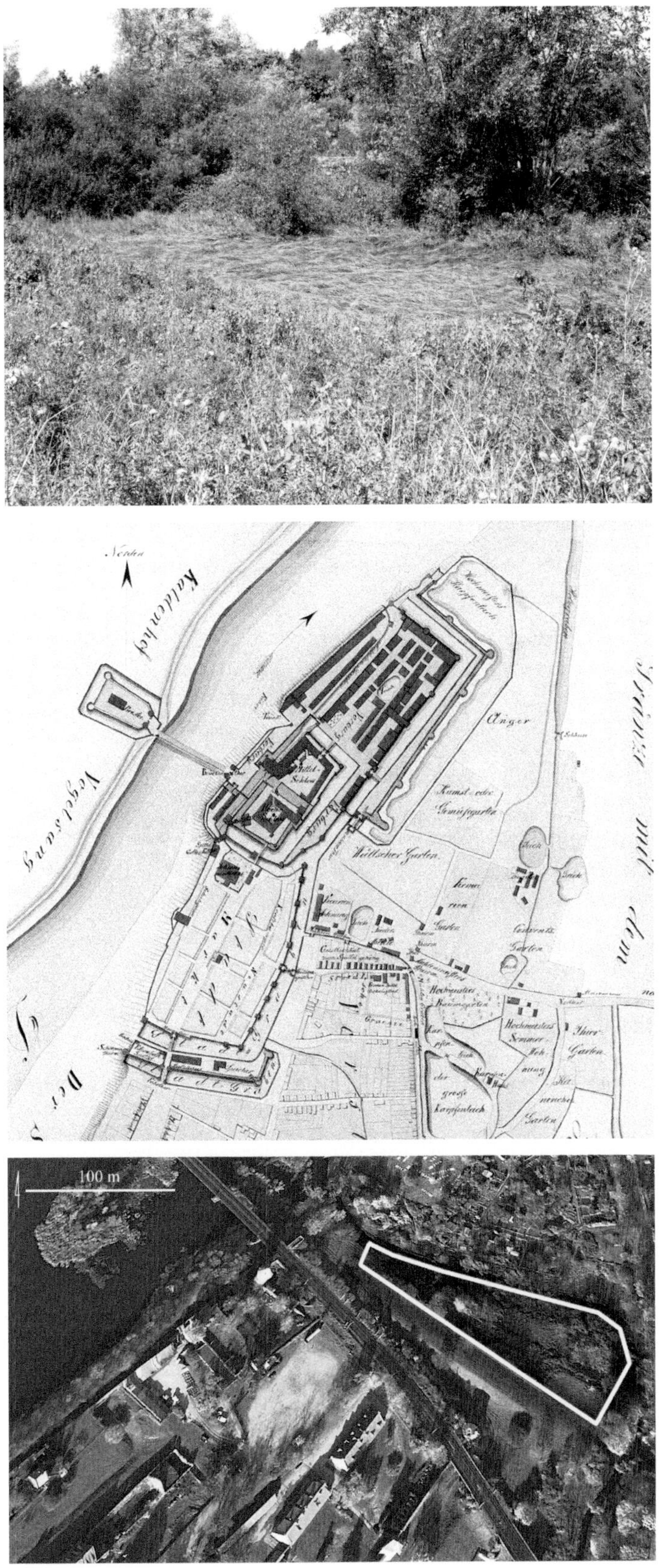

Figure 10.9: (a) The fishpond to the north of the castle complex at Malbork (Ger. Marienburg) (photo by A. Pluskowski); (b) as represented on a map of 1855 (1:5100) (Czaja and Czacharowski 2003); (c) the area today (Google Earth 2015, MGGP Aero).

model for the rest of the commandery, reflecting the relative impact of the two phases of colonization in the early and later medieval periods.

The castle has an impressive archive of written sources, in part reflecting its role as the chancellery of the *Ordensland* and the seat of the Grand Master. Its impact on the environment, however, has often only been referred to indirectly from a predominantly economic perspective in discussions of provisioning, hunting, water management, construction materials, and fuel. These discussions have been almost exclusively informed by written sources, which predominantly date from the latter half of the fourteenth and fifteenth centuries (Jóźwiak and Trupinda 2007, 2012; Chęć 2005; Chęć and Gancewski 2009; Długokęcka and Długokęcki, 2010). The castle in Stuhm is also represented by a number of records.

Regional Status

The changing status of the landscape around Malbork is one of the most significant in the eastern Baltic region. The volatile early medieval frontier is transformed into a heartland with the annexation of Pomerelia and the relocation of the Order's headquarters to Marienburg. This is associated with the reorganization of administration and settlement and the beginning of intensive colonization of the Delta region.

3. *The Commandery of Elbing (Elbląg)*

The abandonment of Truso at the start of the eleventh century did not result in the collapse of the settlement pattern within the vicinity of Lake Drużno. Indeed, this remained the westernmost extent of permanent clusters of Prussian settlement in the twelfth century. Further south, around the vicinity of Morąg, the population density appears to have been much smaller (Jagodziński 1996; Długokęcki 2009c, 37). The region, as elsewhere in Prussia, was defined by strongholds functioning as political centres, particularly in the Elbląg uplands. The largest fortified complex was located north of the village of Weklice (Ger. Wöklitz), consisting of four strongholds located in close proximity to each other. Occupied from the twelfth to mid-thirteenth century, these represented a unified defensive system focused on the most extensively fortified stronghold (site 2), situated on the highest ground and almost certainly the central point of political organization within the

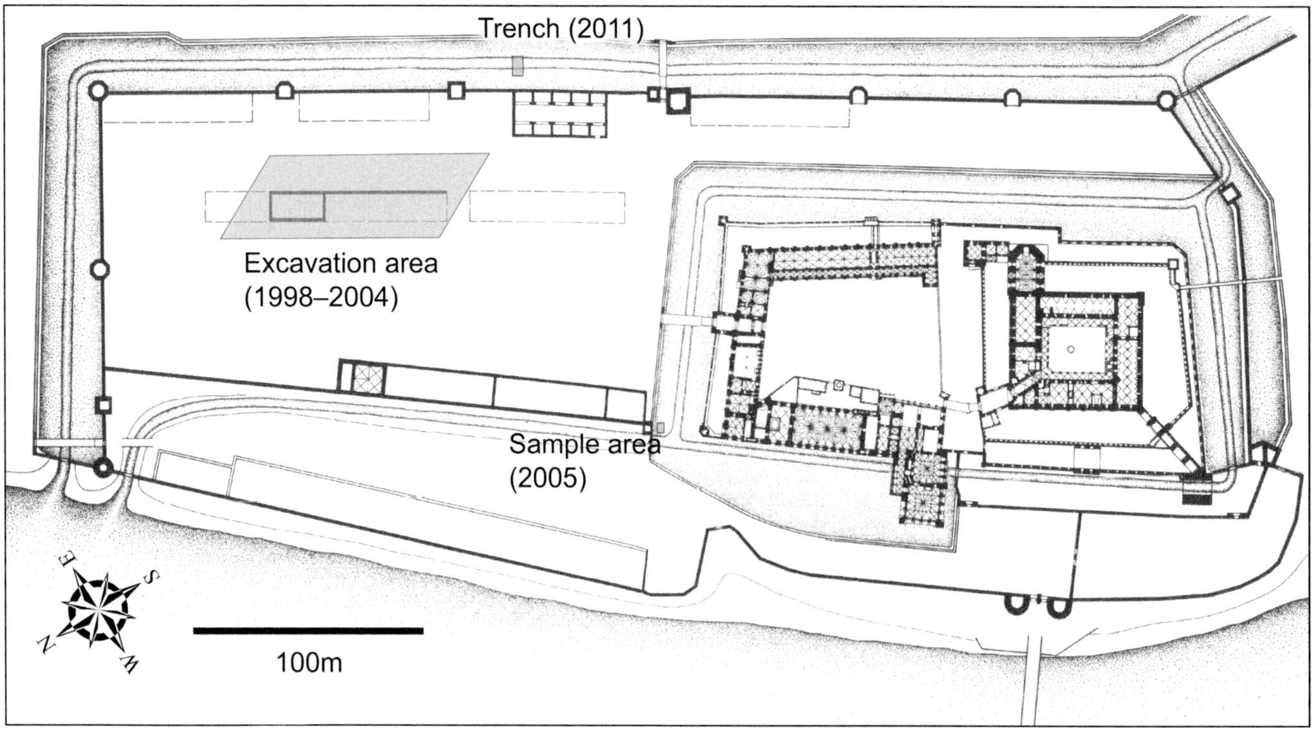

Figure 10.10: Location of excavations, trench in the outer moat, and sampling area at Malbork Castle (Ger. Marienburg) (simplified plan after Arszyński 2010, 29; Dąbrowska 2007, 305).

macro-region. These were not exclusively military structures, as there was evidence for provisioning and industrial and trade activities (Pawłowski 1991).

After the conquest of Pomesania, Peter von Dusburg described how the Order constructed a fort on an island in the River Elbing in 1237, from material prepared beforehand at Marienwerder (Kwidzyn) (*CTP* III, 16). The crusading contingent of Henry III, the Burgrave of Meissen, founded the nearby settlement which was defined as a *civitas*, and saw the rapid construction of a Dominican friary (Długokęcki 1992, 137, 167; Domagała 2004, 89). There has been some debate as to whether there was any continuity between the thirteenth-century planned town and the early medieval emporium at Truso (Janów Pomorski), but there is no evidence for twelfth-century occupation at either site (Jagodziński, 2004a). In 1239/40, the Order started to build the castle from stone and brick in place of the earlier timber fortress. There were also plans to build a substantial church, fitting for the planned headquarters of the Prussian Master. At the time, the location was well connected via the waterways of the Vistula Delta and facilitated effective communications with Lübeck — the source of the first documented wave of colonists to settle in the vicinity of the Order's fort after it was stormed by the Prussians (Haftka 1999, 80–81).

Elbing subsequently functioned as the main base for launching crusades into Pogesania and Warmia (Ger. Ermland), and also as the political centre of the Teutonic Order in Prussia from 1251 to 1309 (Jóźwiak 2001, 417). The first commander is documented in Elbing in 1246, and the convent hosted the meeting of the Order's general chapter from 1251. The castle, situated to the south of the town, formed its own quarter consisting of the main conventual building and three outer baileys, all divided by moats linked to the river (Torbus 1998, 387–99) (Fig. 10.11). Although the convent became of secondary importance after the headquarters were established in Marienburg, it ranked amongst the most important in Prussia and included the residence of the Order's Hospitaller (*Spittler*) who also served as the convent's commander from 1312. Its chapel dedicated to St Andrew, containing the relics of the True Cross, remained an important centre of pilgrimage (Haftka 1999, 81). It was also one of the largest convents in the Order's territories; in 1451 the community included forty-nine knight-brothers and four chaplains.

The town's mercantile community developed its own civic identity within a few decades. The town acquired the Lübeck Law, probably in 1240 (grudgingly confirmed by the Order in 1246), with its rights listed in eighty-two articles; this number had virtually doubled

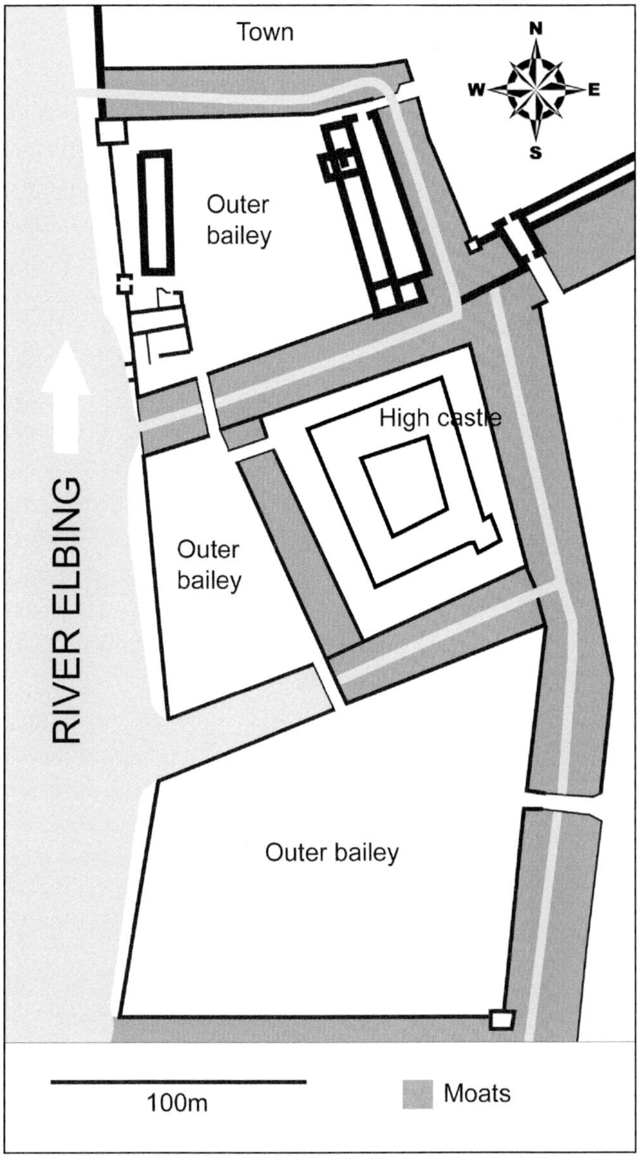

Figure 10.11: Plan of the castle of Elbing (Pol. Elbląg) (after Hauke, Stobbe, and Nawrolski, in Haftka 1999, 81). Given the lack of excavations in the castle area, much of this has yet to be verified, and the detailed plan of the medieval castle is relatively unknown (for comparison, see the plan in Torbus 1998).

by 1275, and by 1295 there were 256 articles listed in the town's codex (Domagała 2004, 92). Archaeologically, the early phase of the town is well known from the mid-thirteenth century onwards (Fig. 10.12). The town was laid out over 15 hectares, with the built-up area subsequently enlarged to 24 hectares (Nawrolska 1997, 21). A central area developed around St Nicholas' Church, and the majority of early townhouses were timber-framed with clay or brick walls, although stone foundations of four residential towers were also discovered during excavations dating to the 1260s and 1280s and, complementing the appearance of imported luxuries, they represented the wealthiest inhabitants of the early town and reflect the rapid development of social stratification (Nawrolska 2008, 512). This also reflects how quickly Elbing developed into a trading centre. The town seal from 1242 — the oldest known Hanseatic seal — depicts a sea-going ship, and most trade connections were linked to the Baltic, although the town was also situated on the overland route from Lübeck through to Königsberg. By 1293, Elbing was trading with England (Nawrolska 1999, 373). The town resisted Pomeranian attacks in the 1260s only to suffer destruction by fire in 1288. The period after, from the 1290s to 1340s, saw a phase of intensive rebuilding, with a slightly different urban layout and the increasing use of brick; between 1330 and 1360 there is evidence for 116 brick houses (Nawrolska 1997). The town is estimated to have had five thousand people in its heyday, with most of the original inhabitants coming from Westphalia, the Rhineland, and Thuringia, but including burghers from Prussia, Pomerelia, and amongst the Patrician class, from northern and central Germany (Nawrolska 2008, 510). In 1326, the border of the town was moved in the direction of the river, whilst the number of town plots and households dramatically increased over the fourteenth century (Nawrolska 2008, 520). This resulted in a second town or suburb built between 1335 and 1337, and later granted privileges under the Lübeck Law. The New Town of Elbing functioned as a separate unit until 1478. By the fifteenth century, the Old Town occupied an area of around 24 hectares containing 658 residential properties, and together with its satellite settlement was one of the largest urban complexes in the *Ordensland*. It

Figure 10.12: Remains of a thirteenth-century building excavated in the Old Town of Elbląg (Ger. Elbing) (after Fonferek et al. 2012, 27).

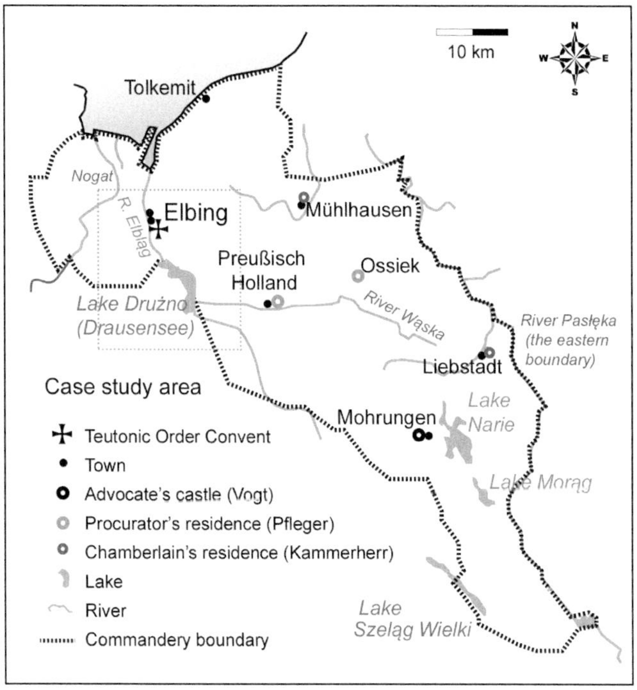

Map 10.11: The northern sector of the commandery of Elbing (Pol. Elbląg). Map by A. Pluskowski.

remained an important trading centre until the gradual shrinkage of Lake Drużno resulted in the reduction of waterways linking it with the Vistula Lagoon. Elbing's burghers were amongst a number of towns which rebelled against the rule of the Order and resulted in the destruction of the castle in 1454 (Czaja 1997).

The commandery of Elbing was separated into two parts, split by the territories of the episcopate of Ermland (Map 10.11). To the west, the convent's territory bordered with the commanderies of Marienburg and Christburg; in the south with Osterode, Balga, and the northern frontier of Masovia. In its northernmost segment, the commandery encompassed the eastern extent of the Vistula Delta region, the uplands to the north-east which remain heavily wooded today, and the valley of the Pasłęka, which flows through Warmia. As with neighbouring commanderies, there was a process of intensive colonization in the fourteenth century. By the fifteenth century, the commandery was supported by three procuratorships and one district run by an advocate based at Mohrungen (Pol. Morąg). The territory within the vicinity of Elbing was split between the Order and the town, although the patrimonium of the latter was the largest out of all the Prussian towns, covering an area of *c.* 200 km^2 and located predominantly west of the town across the Vistula Delta as far as Nowy Dwór (Ger. Tiegenhof) (Nawrolska 2008, 518, Fig. 22).

Sacred Sites

The surviving evidence for pre-Christian sacred sites in Pogesania is limited and has been surveyed by Kawiński (2011), who observed some trends in their location in relation to settlement clusters. One category of sites was typically found on boundaries, perhaps defining them and even bringing together communities from adjacent territories. The other type was more centrally located and can be linked to the reinforcement of political authority. The range of features is comparable to other parts of Prussia and the eastern Baltic, and they are equally difficult to date. The famous site of Nowinka with its horse deposits was used much earlier by the Elbląg group in the Late Migration Period (Kontny et al. 2011). Later cemeteries are perhaps the most easily identifiable sites of cult activity, and the Prussian toponyms associated with fields and forests with the element 'Kaukai' (Prussian mythological psychopomps whose responsibility is to escort souls from Earth to the afterlife) have been linked to burial grounds with an emphasis on their sacral function. In the case of the site at Żabi Róg, a large glacial erratic was surrounded by smaller stones, underneath which a charred archaeological layer was uncovered. Erratics with early medieval cultural associations are otherwise rare; a sword dating to the tenth and eleventh centuries was found under the so-called 'offering stone'

Figure 10.13: The oaks from excavations in the courtyard of the Archaeology and History Museum of Elbląg (Ger. Elbing). Photo by M. Marcinkowski.

Map 10.12: Heiligenwalde (Pol. Święty Gaj) on Schroeder's map, published 1801 (Staatsbibliothek zu Berlin - Preußischer Kulturbesitz; Kart. N 1020; image reproduced with permission).

near the village of Sambród in the 1940s. More certain is the likely cult site within the vicinity of Świątki and Garzewo, east of Morąg, consisting of a sacred grove, field (cemetery), and valley. Conversely, the name Diabla Góra (formerly Teufelsberg, 'Devil's Hill') near Książnik may refer to a perceived pre-Christian site. It is difficult to ascertain the antiquity of the toponym Święta Góra (Sacred Hill), which features at two places near the villages of Kiełkuty and Słonecznik, as this name does not appear in medieval documents. In a few instances churches were constructed within the vicinity of these toponyms, and perhaps deliberately for this reason. The relationship between church construction and earlier pre-Christian sites, the memory of which may have been retained for generations by the surviving Prussian population, is one that requires more investigation. Where settlements contained a mixture of German and Prussian settlers (as on the territories of the Sambian and Warmian episcopates), the intention may have been to promote Christianity amongst the indigenous population (Długokęcki 2009a, 204).

The largest timbers uncovered during excavations in the courtyard of the castle in Elbląg were whole oaks measuring almost a metre in diameter and estimated to have been *c.* two hundred years old when felled (Fig. 10.13). Given that individual oaks of this age would have formed striking components of woodland and were most likely to attract devotion, it is not impossible that such trees derived from a sacred grove, and the act of incorporating them into the superstructure of the castle during the period of active crusading would have made a very clear religious and political statement. The nearest documented toponym relating to 'sacred woods' would have been at Heylgewald, later rendered as Heiligenwalde (first mentioned in 1324 when a church was most likely built at the village, today

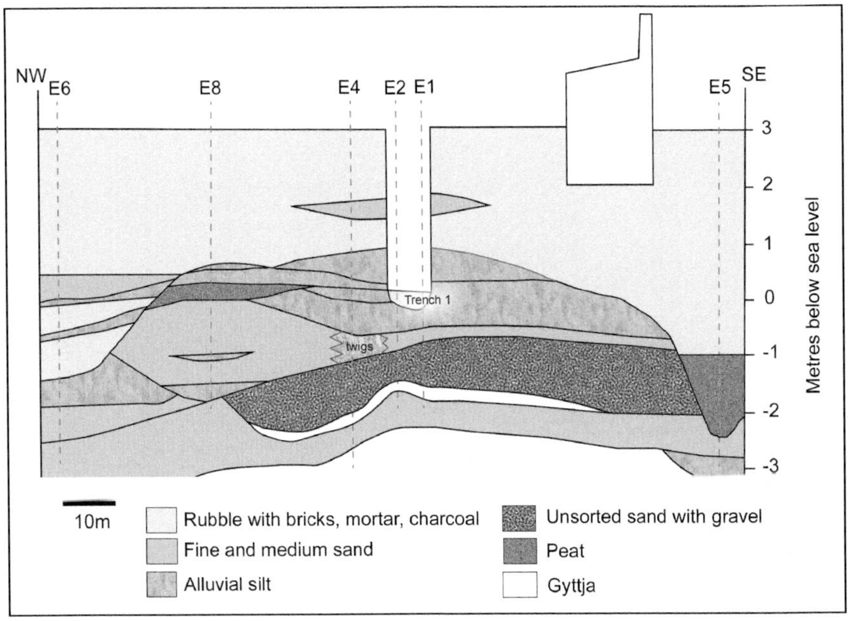

Figure 10.14: Stratigraphic reconstruction based on a coring transect (E numbers) and excavation (Trench 1) in the courtyard of the Archaeology and History Museum of Elbląg (Ger. Elbing) (after Nitychoruk et al. 2012).

Święty Gaj), located *c.* 20 km south of the convent and situated within the midst of an earlier Prussian settlement cluster (Długokęcki 2009a, 95–96) (Map 10.12). The Order's early *castrum* was itself imbued with spiritual power. In 1232, the Holy Roman Emperor had presented Grand Master Hermann von Salza with a relic of the true cross, which was then subsequently deposited at the Order's church in Elbing (Józefczyk 1995, 36). The following year, Pope Gregory granted a ten days' indulgence to crusaders who honoured the relic, echoing not only its importance in the religious and military life of the Kingdom of Jerusalem (Fonnesberg-Schmidt 2007, 195–96) but also the significance of this Christian beacon in a spiritually contested landscape.

Archaeology

Elbląg (today in the Warmian-Masurian Voivodeship) is one of the best excavated towns in the southern Baltic. Around 98% of the buildings in the Old Town were estimated to have been destroyed in WWII, although the medieval layout survived and began to be documented in the 1950s. The first major excavations took place at the site of the Dominican monastery in 1959, continuing in 1975 and 1977, as well as in the castle in 1977. The first systematic excavations did not begin until 1980 under the direction of Tadeusz Nawrolski and Grażyna Nawrolska (Lubocka-Hoffmann 1992; see Nawrolska references). Subsequently, Mirosław Marcinkowski became involved in excavating and analysing fabric from the Old Town, particularly the ceramic assemblages, and most recently in the castle. Excavations in the Old Town have yielded (and continue to yield) some of the most impressive medieval urban archaeology in Europe, with the waterlogged soil preserving an incredible array of organic material, ranging from wooden structures through to musical instruments, textiles, and wax tablets. As a result, the town also has one of the best dendrochronological sequences in Europe, with over a thousand dated samples (Ważny 2004).

In comparison to the town, the castle is poorly known. Excavations at the castle in 1914 and 1919 uncovered various architectural fragments, indicating a significant level of investment had gone into the early castle before the relocation of the headquarters of the Order to Marienburg. Nawrolski's reconstruction of the high castle estimated its dimensions as 65 × 78 m (Nawrolski 1986, see also Torbus 1998, 389). In 2012, excavations in the courtyard of the outer bailey of the castle (in the grounds of the Archaeology-History Museum), uncovered multiple timber structures, with at least three phases of construction and one clear horizon of destruction by fire in *c.* 1305–08, when the brick castle already existed (see Fonferek forthcoming). The earliest structure was dated by dendrochronology to *c.* 1240–42, with subsequent phases from the latter half of the thirteenth century and early fourteenth century. A coring transect across the courtyard enabled the underlying stratigraphy to be mapped, revealing the presence of an outcrop of sediment under the castle corresponding to an island (Nitychoruk et al. 2012). This had been filled and levelled with the expansion of the conventual buildings (Fig. 10.14). The brick castle appears to have been built from the 1240s, with the timber structures representing the remains of the earliest *Vorburg*. Following magnetometry and GPR surveys in select parts of the courtyard, excavations in 2013 uncovered a bank supported by timber which was interpreted as part of the earliest *castrum* built by the Order. The excavations provided botanical, faunal, and geoarchaeological materials, linked to the early phases of the convent. A core was also taken from Lake Drużno nearby, with the aim of capturing the timeline that encompassed both Truso and Elbing. Coring was also attempted in other parts of the northern district of the commandery, as well as in the Elbląg uplands, but failed to obtain any suitable sequences. The commandery of Elbing remains to be investigated in more detail, and within the framework of the project the focus was very much on the early phases of the convent.

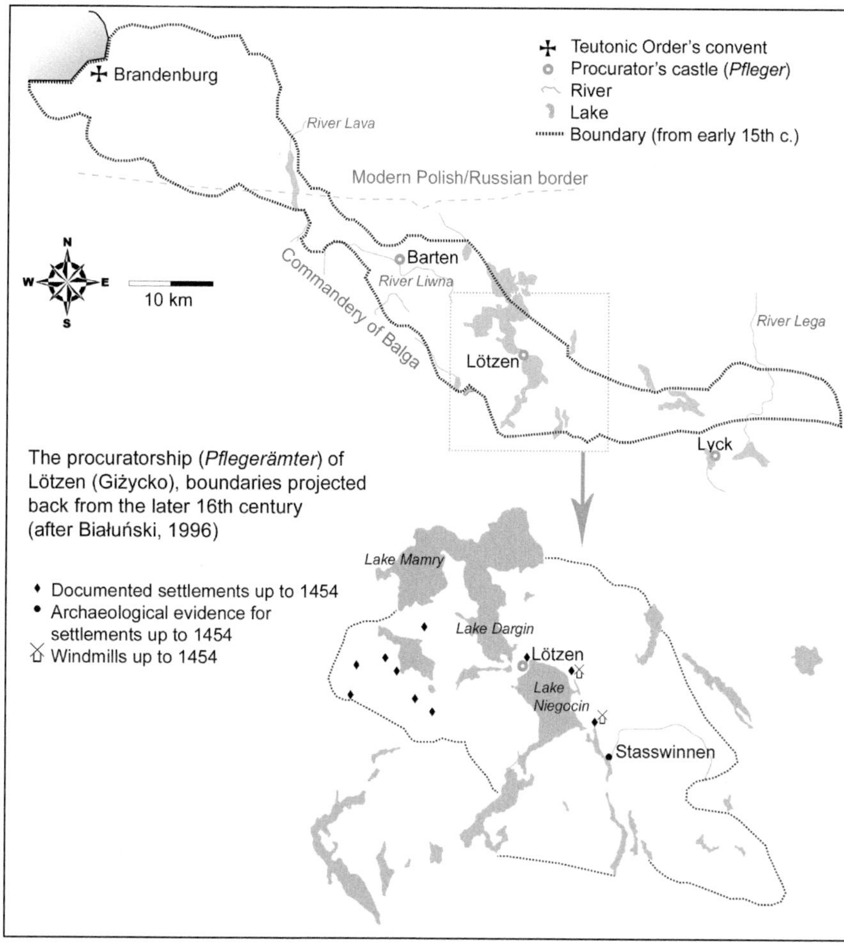

Map 10.13: The settlement pattern according to written sources within the vicinity of Lötzen (Pol. Giżycko) (after Białuński 1996). Map by A. Pluskowski.

Regional Status

This case study can be considered as an indigenous Prussian heartland, albeit within the broad frontier landscape defined by Pogesania, and one that functioned as a military border during the early decades of the Prussian Crusade. Subsequently, the commandery became one of the heartlands of the *Ordensland*.

4. The Procuratorship of Lötzen (Giżycko) (Commandery of Brandenburg)

Lötzen (Giżycko)

The full extent of the commandery of Brandenburg cut through four documented tribal territories: Natangia, Barta, Galindia, and Sudovia. The convent was established during the crusading period, with the first reference in 1266. Excavations conducted at the castle of Brandenburg (Pol. Pokarmin; Rus. Ushakovo) in the late nineteenth century focused on the architectural history of the main conventual building, suggesting at least two phases of construction; the preliminary phase consisting of a three-range structure and a second phase which added a further range to create the classic quadrangular *Konventsburg* of 65.4 × 52 m, with an attached outer bailey (Torbus 1998, 370–74). Today the ruins of the castle lie within the Russian Kaliningrad Oblast, and it was not possible to include further investigations at the site within the framework of the project. Instead, the focus was on a sub-district of the commandery situated to the south-east within the Great Masurian Lake District, today in Poland; the procuratorship of Lötzen (Pol. Giżycko) which encompassed part of the Prussian territory of Galindia.

Throughout its early history, the castle and its adjacent village of Lötzen (also called Neuendorf) are characterized by their peripheral location within the territory of the Teutonic Order, on the fringes of the 'Great Wilderness' that marked the frontier with Lithuania. Before its first mention in the written sources in 1340, the early history of the castle (and the Order's activities in this region) has been the topic of extensive speculation. Caspar Hennenberger (1595) and Christian Hartknoch (1684) assumed that the Order had constructed an outpost at the strategically important pass between Lake Löwentin (Pol. Niegocin) and Lake Kissein (Pol. Kisajno), following the conquest of Sudovia in 1283. Later historians adopted this opinion. Archaeological finds near the later castle in 1969 were regarded as relics from this early construction, but more likely derived from a former Prussian stronghold. Today Piękna Góra (Ger. Schönberg), a place situated 3 km west of Giżycko, which is surrounded by water, is seen as a more likely location for the Order's first outpost.

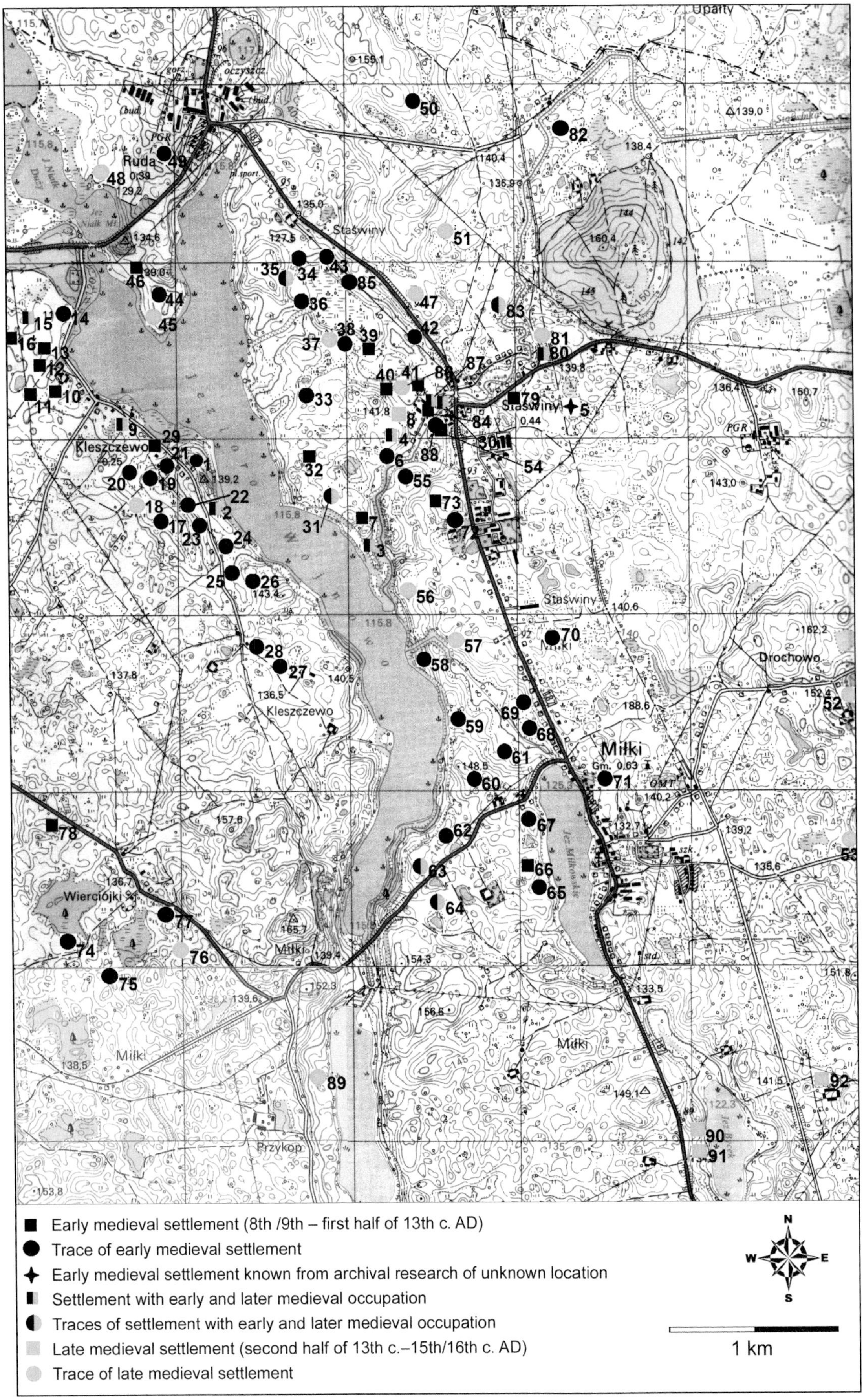

Map 10.14: Archaeological evidence for pre- and post-crusade settlement within the vicinity of Lake Wojnowo. Map by M. Karczewski.

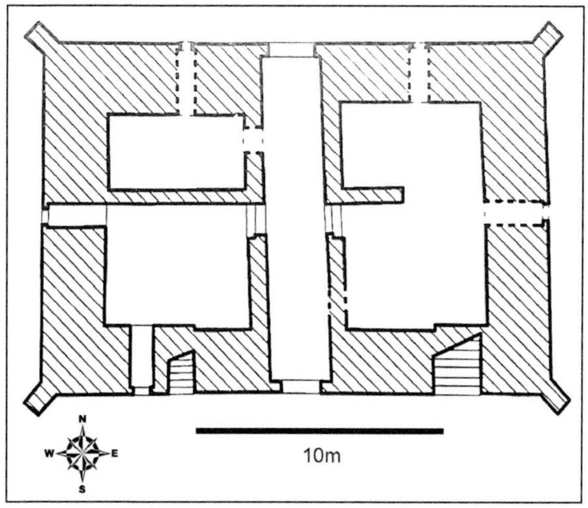

Figure 10.15: (a) The castle of Lötzen (Pol. Giżycko) in the 1920s before its recent rebuilding; (b) its former appearance (after Herrmann 2007, 577, Figure 777).

While Dietrich von Altenburg (1335–41) was Grand Master, he was engaged with developing the administrative structure of this region, and the founding charter for the village of Groß Kämlack (Pol. Kiemławki Wielkie), about 40 km northwest of Giżycko, from 1340 mentions a procurator in Lötzen for the first time. The chronicler Wigand von Marburg refers to the significance of the castle in the campaigns against the Lithuanians in the 1360s, during which Grand Duke Kynstut was captured and brought to Marienburg. The castles of this contested region, including Lötzen, Barten (Pol. Barciany), Angerburg (Pol. Węgorzewo), Rastenburg (Pol. Kętrzyn), Rhein (Pol. Ryn), Seehesten (Pol. Szestno), Lyck (Pol. Ełk), and Johannisburg (Pol. Pisz), fulfilled clear military functions in the conflict with Lithuanian and Polish forces, with administrative and economic roles seemingly of lesser importance. Inventories from all these castles only exist after the battle of Grunwald in 1410 and the first Peace of Thorn in 1411, which reflected a period of easing in the conflict. While Lötzen usually belonged to the commandery of Brandenburg, it was temporarily under the control of the commander of Rhein. Until 1466, the nine documented colonies within the procuratorship were limited to the west of Lötzen (Map 10.13). After the end of the Thirteen Years' War, colonists from Masovia were invited to settle within a number of the south-eastern territories of the Order, including the district of Lötzen. These were initially knightly families, followed from the end of the fifteenth century by peasant communities (Białuński 1996, 2005; Długokęcki 2009b, 380). A component of the indigenous Prussian population remained within this region (Map 10.14). In the commandery of Brandenburg, 1598 Prussian families were listed at the end of the fourteenth/start of the fifteenth century, amounting to around eight thousand people. The participation of Prussian settlers reflects the high number of toponyms containing Prussian elements, one of the arguments put forward for the survival of an indigenous population in Galindia (Karczewska et al. 2005, 55–57).

Archaeology

The castle in Giżycko (today in the Warmian-Masurian Voivodeship) has been the focus of two recent excavations: the first targeting the moat under the 'dansker' tower from which a very small assemblage of animal bones was made available to the project; and the second which saw the castle refitted and effectively rebuilt as a hotel (Fig. 10.15). Unfortunately, no material remains were recovered during the latter works, and the entire archaeological context of the castle was lost in the process. There is a body of documentary evidence associated with the final phase of occupation of the castle during the Teutonic Order's period of rule. The earliest inventory dates from 16 October 1420, when the property of the *Kammerämter* of Barten, Leunenburg, Rastenburg, Rhein, Lötzen, Szestno, and Lyck was recorded following the resignation of the commander of the Rhein convent, Luppolt Reitenbach. The remainder of the inventories and accounts are from the sixteenth century.

The lauks of Staświny

Within the central part of the procuratorship the primary sites included within the project were situated within a micro-region tentatively defined as the *lauks* of Staświny

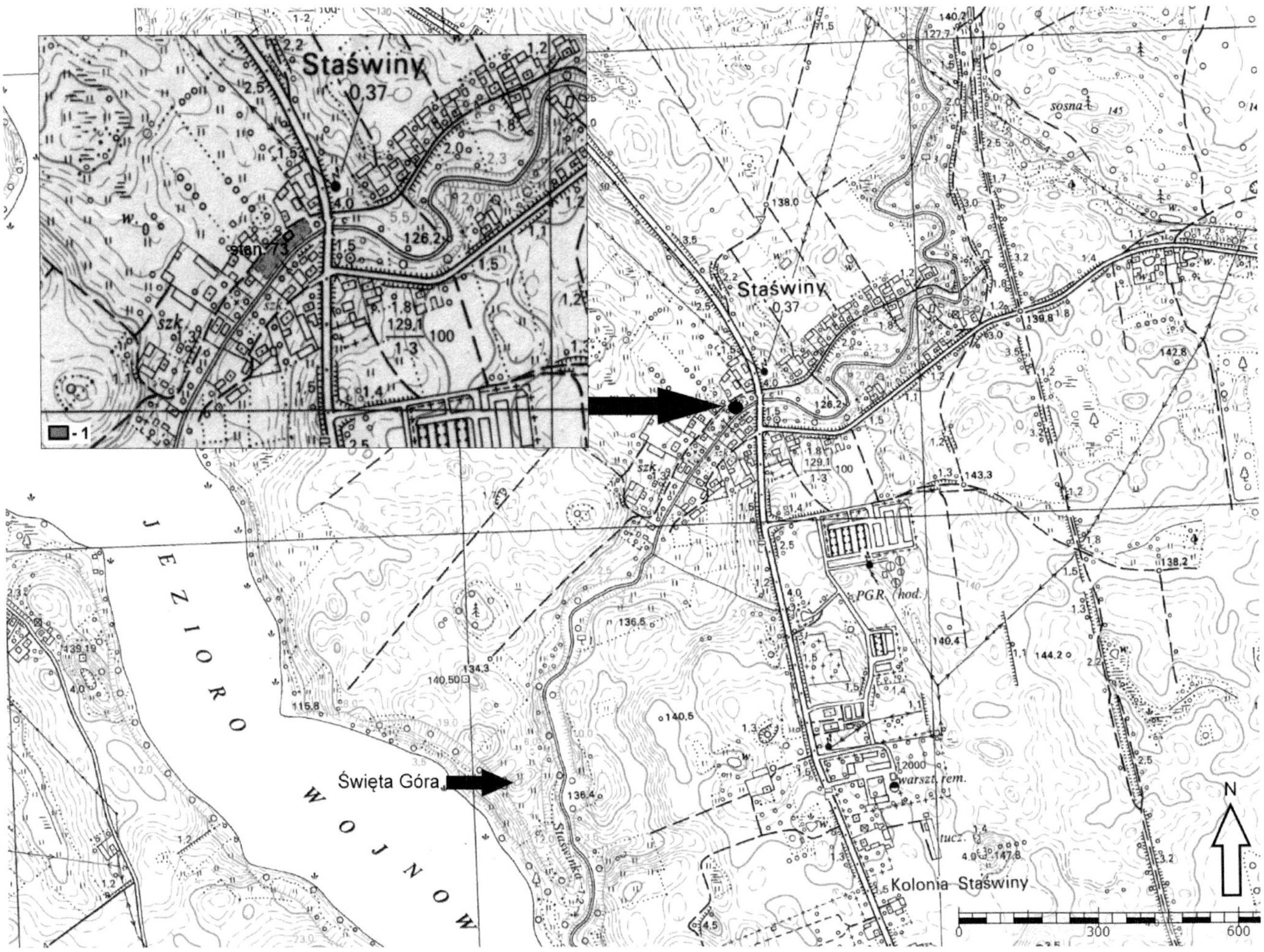

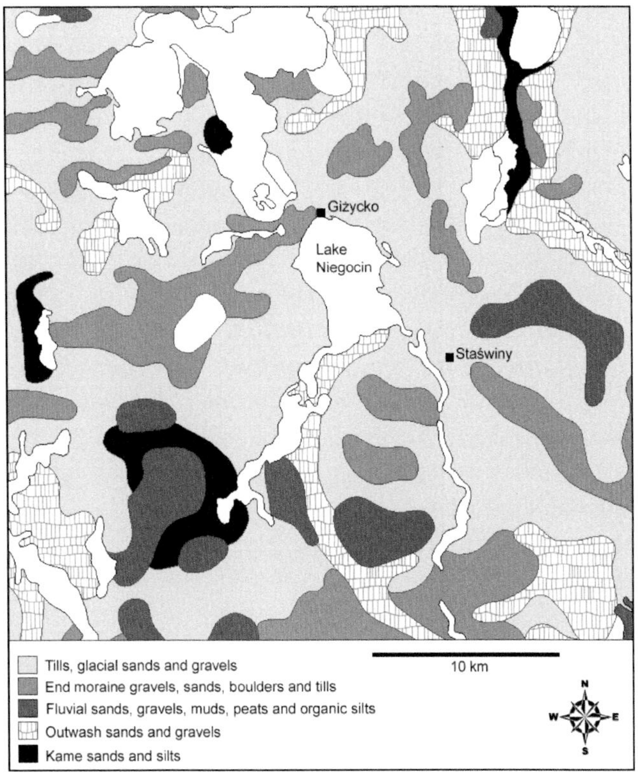

Map 10.15: (a) The village of Staświny (site 73) and the nearby hill of Święta Góra (Staświny site 1) in their landscape setting; (b) a map of the superficial geology (1:500,000, after Central Geological Database, Polish Geological Institute).

(Map 10.15). A *lauks* was the smallest unit of medieval Prussian settlement and reflected a territorial community consisting of a few families (Okulicz-Kozaryn 1997, 175–77; see Chapter 15). Several names of Prussian *lauks* are still visible in the toponyms of eastern Masuria, which had been rendered into German as *lauken* and Polish as *ławki*, such as Bezlauken (Bezławki), Lauken (Ławki), and Stürlack (Sterławki). In 1998 and during the last few years the territory around Staświny was the focus of archaeological surveys, carried out as a part of the Polish Archaeological Record programme (AZP, *Archeologiczne Zdjęcie Polski*). As a result, several surface traces of early medieval settlements were uncovered, most probably the remains of two Prussian settlement micro-regions situated around Lake Wojnowo. The micro-region defined as *lauks* Staświny encompassed the area along the eastern shore of Lake Wojnowo, the lower stretch of the River

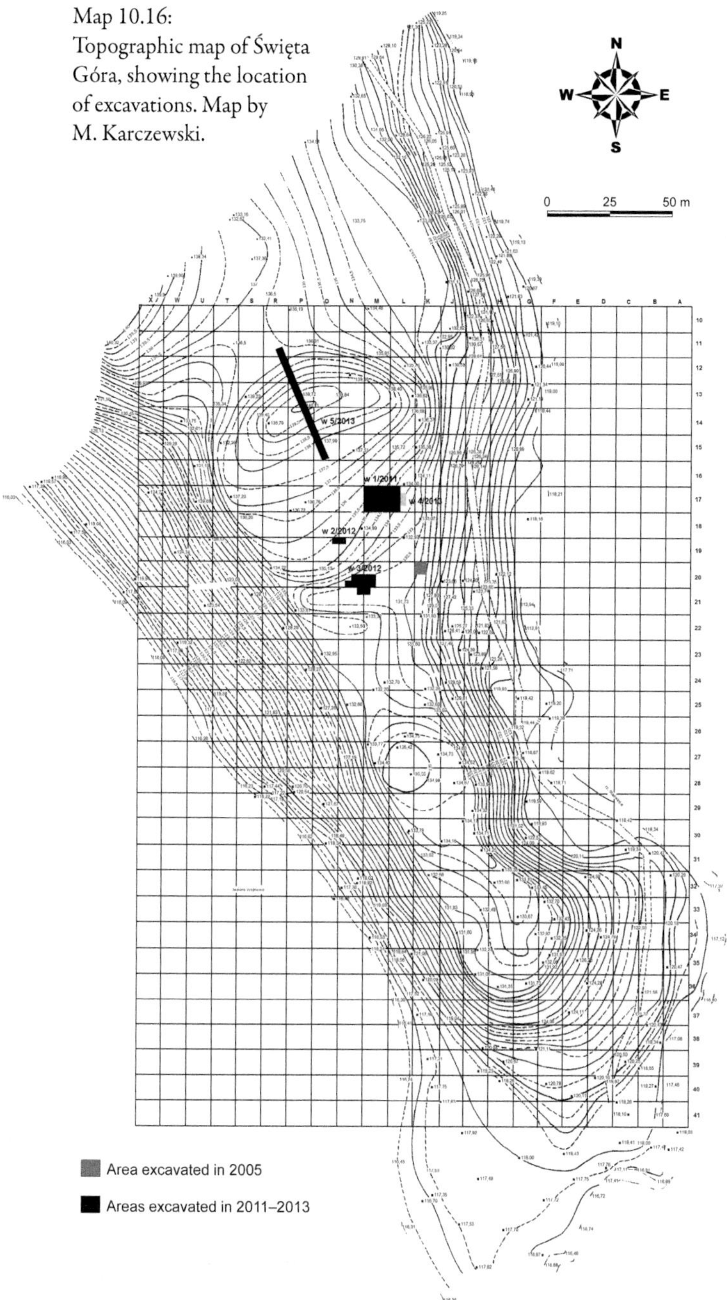

Map 10.16: Topographic map of Święta Góra, showing the location of excavations. Map by M. Karczewski.

Staświnka and two small nameless lakes situated to the north of the river. It included two principal sites: Staświny site 1, a morainic elevation with the toponym Święta Góra ('Sacred Hill'), and site 73 (the village of Staświny). In 2005, Piotr Banaszuk conducted geological and geomorphological analyses of Sacred Hill and the shore of Lake Wojnowo, demonstrating that the water level in the lake was around a metre higher in the past, and that the estuary of the Staświnka River had been located at the southern edge of the hill. Today it is situated *c.* 100 m to the south. Cores taken from Lake Wojnowo in the vicinity of Staświny village also provided important palynological profiles (Wacnik 2009, see also Chapter 11).

Sacred Sites

German and Polish place names incorporating the Prussian word *svin*, meaning 'sacred', can be used to reconstruct elements of the sacred landscape. This element is present in the name of Lake Swinteseyte (Pol. Święcajty), the village of Stasswinen (Pol. Staświny), a meadow located near this village which until the nineteenth century was called Lake Stasswin/Stasswienen See (Pol. Łąki Staświńskie) and a small river called Stasswin (Pol. Staświnka) connecting this meadow with Lake Wojnowo. In the sixteenth century, the Prussian linguistic appellation *Stas Swins* meant 'this sacred [place]' (Klimek 2014, 64–65). On this basis one of the former Prussian cult sites or sacred areas can be identified with the present-day village of Staświny and its vicinity. Nearby, the toponym Święta Góra is suggestive of a sacred function, but excavations to date have revealed no traces of cult-related activities, and it is possible this was a much later designation. The provision of a Christian pastoral infrastructure within the micro-region of the *lauks* of Staświny is not evident until the foundation of the church at Milken (Pol. Miłki) in the last decades of the fifteenth century (most likely in *c.* 1481). The building was converted to an Evangelical church shortly after 1525 following the secularization of the Teutonic Order and rebuilt after being burned in 1656 (Karczewska et al. 2005, 108). It is possible it was preceded by an earlier church, but the alternative scenario is of Galindian communities maintaining their earlier pre-Christian beliefs and living in the shadow of the castle at Lötzen.

Święta Góra

Site 1 or 'Sacred Hill' also called *Swienta Gora*, *Heiliger Berg*, or *Pruzzenhöhe*, is situated on the extended morainic elevation to the north of the village of Staświny. The site is naturally well protected. The eastern and southern sides of the hill drop steeply down to the valley of the River Staświnka, which drains into Lake Wojnowo directly to the south of Sacred Hill. The western hillside reaches the shore of the lake and the only point of easy access to the hill is from the north, which is protected by an embankment constructed from clay, sand, and stones. This encloses an inner occupation area, triangular in shape, which is significantly higher in its southern part (by *c.* 2 m) (Map 10.16). From the nineteenth century, Sacred Hill was regarded as a Prussian stronghold and place of worship (Beckherrn 1895, 367; Engel 1930, 7), and was the focus of five excavations; in 1881 by Scherbing, the teacher from the primary school

Figure 10.16: Excavations at Święta Góra in 2011. Photo by A. Brown.

in Staświny; in 1882 by the antiquarian H. Lorek; and in 1922 and 1923 by Hans Kurt Hess von Wichdorff, a geologist and archaeologist who speculated that the southern part of the hill was the location of the stronghold of Ysegubs/Jezegub, the legendary 'king' of Galindia (Hess von Wichdorff 1925, 2–3). Both Scherbing and Lorek searched for the Prussian cremation cemetery on Sacred Hill (Lorek 1882/83, 45–47). In fact, the cemetery was located on the elevation at the north-western edge of the village of Staświny. There, in the nineteenth century, the villagers found on several occasions spearheads, rings, and other ornaments originating from cremation graves, but Lorek did not speak the local language (Masurian) and failed to understand the location of these finds. In 2005 excavations by the archaeological expedition of the University in Białystok, led by Maciej Karczewski and Małgorzata Karczewska, identified two cultural layers in the eastern part of the inner occupation area, dated to the late phase of the Migration Period and the early Middle Ages (Karczewska and Karczewski 2007, 134–37, 143–52.).

Within the framework of this study, the Sacred Hill site was excavated in 2011, 2012, and 2013, preceded by geophysical survey. One season of excavations took place at site 73 in 2012. The excavations on Sacred Hill focused on the middle and northern part of the inner occupation area and the embankment (Fig. 10.16). A total of fifty-two features were uncovered, consisting largely of agrarian pits which had been used for waste disposal in the last stages of their use. Two features (18 and 24) were also used as cesspits. A series of postholes were also uncovered (see Chapter 14), alongside traces from the First World War consisting of military trenches and pits caused by shelling. The features contained an abundant range of artefacts and ecofacts which formed the principal datasets used within the study. The chronology of the features uncovered during these excavations was based on a combination of typological, morphological, and technological analysis of ceramics, the stratigraphy of feature fills, and some AMS dates of samples from feature 40. Interpretation of the early medieval ceramics was compounded by its heterogeneous styles. The so-called 'Prussian' style was represented by vessels used as a cooking pots, the 'Germanic' style was represented by table vessels, whilst the 'Slavic' style of pottery was identified on the basis of Masovian analogues. Found separately, each stylistic group should be dated differently: the 'Germanic' to the sixth to seventh century AD, 'Prussian' pottery and the 'Slavic' to the tenth and eleventh/twelfth century. However, all these styles of ceramics were found in feature 40, where the fill had formed in a very short time. The radiocarbon dates from this feature indicated a short chronology from the end of the ninth until the end of the tenth century. On this basis, all three styles can be regarded as contemporary and characteristic of Prussian pottery in the early phase of the early Middle Ages.

In summary, the site of Sacred Hill was settled four times: in the Early Iron Age (fifth to first century BC), in the late phase of the Migration Period (second half of the fifth century–seventh / start of eighth century AD), in early and late phases of the Middle Ages (from the end of the ninth century until the end of the tenth / start of the eleventh century, and from the end of the tenth / start of eleventh until the thirteenth or fourteenth century). The final stage of the Prussian settlement was

Figure 10.17: Excavations in the village of Staświny (site 73) in 2012. Photo by M. Karczewski.

radiocarbon-dated to the first half of the thirteenth century (with material from feature 32, a posthole), and confirmed by the discovery of 'greyware' pottery, typical of the ceramic technology associated with the process of colonization accompanying the Teutonic Order's conquest. The site subsequently functioned as a military camp, most probably in the seventeenth century, at which point the embankment located on the north edge was constructed. A substantial range of artefacts and ecofacts, particular plant macrofossils, animal bones and sediment samples, was recovered from the site and included within this study.

Staświny

The village of Staświny is documented as being established in 1475 by Bernhard von Balzhofen, the commander of the Brandenburg convent. The medieval name of the village — Eisermühl — suggests that at its foundation there was an ironworks within the village or its vicinity (Białuński 1996, 56–57). The aim of the excavation of site 73, situated in the central part of the area of the historical village, on the north bank of the River Staświnka (Fig. 10.17), was to ascertain if there was any continuation in occupation between the Prussian open settlement on Sacred Hill and the later village of Staświny. The excavation indicated that the settlement began there in the thirteenth century and continued into the period of the historically attested foundation. Feature 5 consisted of the remains of a building connected with iron and glass production (providing a link with the medieval name 'Eisermühl'), with layer 5 containing artefacts, mainly potsherds, from the late phase of the early Middle Ages and the late Middle Ages, analogous to the pottery found at the Sacred Hill site, whilst a fragment of wood from layer 12 was radiocarbon dated to *c.* 1414–58. It is therefore highly possible that in the thirteenth century both settlements existed on the territory of the *lauks*. The establishment of the village in 1475 was not a new foundation but rather the commander's reorganization of the existing Prussian settlement, integrating it within the Order's economic, political, cultural, and ideological framework. This may have resulted in significant changes in the layout and appearance of the village and its economic area. Although no plant macrofossils were preserved in samples taken from site 73, a faunal assemblage was recovered.

Regional Status

Lauks Staświny can be considered a localized core area in the early medieval period, despite the collapse of settlement elsewhere in Galindia, after which it became incorporated into a frontier commandery. This region remained a frontier until the cessation of hostilities with Lithuania but remained within the truncated territories of the Teutonic Order after 1466, redefined as the Duchy of Prussia in 1525. The foundation of the settlement of Eisermühl can be situated within the general trend of increasing colonization, following the stabilization of the political and military situation.

Secondary Sites

The number of secondary sites from the area of the historical Prussian *Ordensland* included within this study is quite substantial, largely due to the number of faunal assemblages analysed by Makowiecki and isotopically by Müldner and Scull (see Chapter 12; Scull et al. 2019). Where two or more categories of data were made available from the same site, or where samples were taken from excavations specifically for inclusion within this study, the details are outlined briefly below. In a few cases the excavations reports have yet to be published, and so the details are summarized below along with relevant chronologies. Other sites are referred to in the relevant chapters.

The Commandery of Nessau (Mała Nieszawka)

The history of Nessau was characterized by its situation at the southernmost edge of the Prussian *Ordensland*, being the starting point of the Prussian Crusade, the bridge from Kuyavia across the Vistula into the contested Kulmerland, and subsequently a point of contention between the Order and Polish crown following increasingly hostile relations (Märcker 1899, 395–98). The first fortification in this area, named Vogelsang, was subsequently replaced by Nessau, although both appear to have been constructed in different locations. In 1230, Konrad of Masovia gave the Order the small fortress of Nissowe together with four villages, providing the first territorial base for organizing military campaigns across the Vistula (*PrUB* I.1, 76). The commander of Nessau is first mentioned in 1255 (Dygo 2009a, 63, 74), but the castle is only referred to from 1348 (*PrUB* IV, 344). At this time the convent was managing a small commandery within the region of north Kuyavia, separated from Thorn by the Vistula. The entire district was oriented to the river, where the key villages of Kostbar, Kostrin, Ober-Nessau, Nieder-Nessau, Orlau, Rudak, Groß-Stab and Klein-Stab were located (Märcker 1899, 322f). Some eyots (river islands) also belonged to the convent, mentioned in charters between 1294 and 1404 (Märcker 1899, 10, 619, 629, 640f). Excavations between 1975 and 1990 at the site of Mała Nieszawka (Kuyavian-Pomeranian Voivodeship) did not find any traces of an earlier fortification (Poliński 2007, 43, 50–51; Dygo 2009, 64) but uncovered the remains of a regular quadrangular *Konventsburg* constructed from brick on top of stone foundations, with dimensions of 35 × 35.75 m (Torbus 1998, 556). This was not completely enclosed, and the north-eastern area by the main gate was integrated with the paved courtyard, where a well was also found. The main residential part of the castle was identified with the southern wing, which contained cellars interpreted as storage areas. The main complex was separated from its outer bailey to the north and east by a moat, linked with a single bridge (Haftka 1999, 195–96). Changing political relations with Poland saw the convent on the geographic frontline of the conflict, ultimately resulting in the complete dismantlement of the castle in *c.* 1424. When the Order took the nearby Polish castle at Dybów, they named it Neu Nessau, although in the treaty of Brest in 1435 the Knights relinquished their claims to both the old castle site and Dybów.

The excavations at Mała Nieszawka recovered a substantial faunal assemblage, elements of which had been published earlier (Iwaszkiewicz 1991; see also Chapter 12) and which was included within the project. The faunal assemblage was associated with one phase of occupation which has been variously dated to the fourteenth century, and most recently 1320/40–1424; no thirteenth-century phases were identified during the excavations (Torbus 1998, 557–58). There is also a body of inventories and tax records relating to the convent that was included within the project, which reflect the changing political situation in the early fifteenth century. The convent was relatively wealthy in the late fourteenth century, reflected in part by significant construction work in the 1380s, when the listed numbers of bricks kept at Nessau decreased from 89,000 to 18,000 between 1382 and 1386 (*GAD*, 476–77). There was a clear break in the convent's status after 1410 (as in the case of Rehden), with reduced quantities of provisions until 1421 and a heterogenous picture of supply in the few years between 1431 and 1435 at the castle at Dybów.

Kałdus (In Culmine)

The western Slavic (Łużycka culture) settlement of Kałdus (Lat. In Culimne), which the most recent excavations dated to the mid-seventh century, came under Polish dominion in the second half of the tenth century (Chudziak 2003). The range of finds and international contacts with Rus' lands, Scandinavia, and Central Europe, indicates this was the most important early medieval centre in the southern Polish–Prussian borderlands, situated in the far west of the Kulmerland. The presence of a fortified stronghold and the church reflects political aspirations and most likely the creation of a base for missionary activity; part of the church foundation overlay a stone construction interpreted as a pre-Christian altar. In the twelfth century, the settlement

Figure 10.18: The brick structure at ul. Zamkowa in Tczew (Ger. Dirschau) uncovered by excavations in 2009. Photo by Z. Sawicki.

was the centre for the regional Polish castellan (Buko 2005, 283–84). The final phase of occupation at Kałdus is datable to the second half of the thirteenth century, as suggested by ceramic fragments and a Teutonic Order bracteate (Chudziak, 2003). Historical sources point to Kałdus as the most likely site of the castellan's stronghold of Colmen attached to the settlement In Culmine, which suffered Prussian incursions in the decades leading up to the Prussian Crusade and throughout the thirteenth century. This final phase has been identified with the Order's short-lived stronghold of Potterberg which was dismantled and the materials used to build the fortress at Mewe in 1282 (Poliński 2003, 179). The *castrum* at Starogród, in close proximity to Kałdus and Chełmno (Ger. Kulm), is mentioned in 1232. It quickly became an important administrative centre in the Kulmerland and ultimately the focus of a commandery. Ongoing excavations at the site have yielded significant faunal assemblages enabling diachronic socio-topographic trends to be investigated (Makowiecki 2007) and were included in this study for comparative purposes, as well as being sampled for isotopic analyses. The site has also been studied from a range of palaeoenvironmental perspectives which provided important baselines for the Kulmerland as a macro-region.

The Advocateship of Dirschau (Tczew)

At the end of the twelfth century, Tczew was a poly-focal settlement which came to function as an important trading centre within the territories of the dukes of Gdańsk. In 1252/53 it became a ducal residence, and in 1260 it was granted town privileges under the Lübeck Law. In 1289, the duke of Gdańsk encouraged the Dominicans to establish a friary in the town. The Order's annexation of Pomerelia resulted in the partial destruction of Tczew, and the town was subsequently reorganized under the Kulm Law (Czaja 2009, 190–91). The urbanization of the Vistula Delta region, completed by the mid-fourteenth century, consolidated the Vistula trade network which connected Gdańsk and the Baltic with the Order's core Prussian commanderies and the Kingdom of Poland. Dirschau, situated on the border with the commandery of Marienburg, was an important point along this route. It was assigned to an advocate who governed an extensive territory that stretched out to the west, bordering to the north with the newly formed commandery of Danzig and to the south with the commanderies of Mewe (Pol. Gniew) and Tuchel (Pol. Tuchola). The town changed ownership several times during the conflicts with Poland in the fifteenth century, but remained in the Order's possession until 1466.

In 2009, excavations at ul. Zamkowa on the eastern edge of the town (today in the Pomeranian Voivodeship) in close proximity to the Vistula uncovered the remains of a substantial brick structure (Fig. 10.18), of which the lower levels had been destroyed by fire (Jonakowski 2009/2010). The thick charred horizon, which included large quantities of burnt seeds, was radiocarbon-dated to *c.* 1395–1475 (93.6% probability). The macrobotanical material from this horizon was sampled and included within the framework of the study. The space was interpreted as a storage area, connected with the movement of grain up the Vistula in the fifteenth century. Further details and contextual discussion can be found in Chapter 13.

The Procuratorship of Lyck (Ełk) (Commandery of Balga)

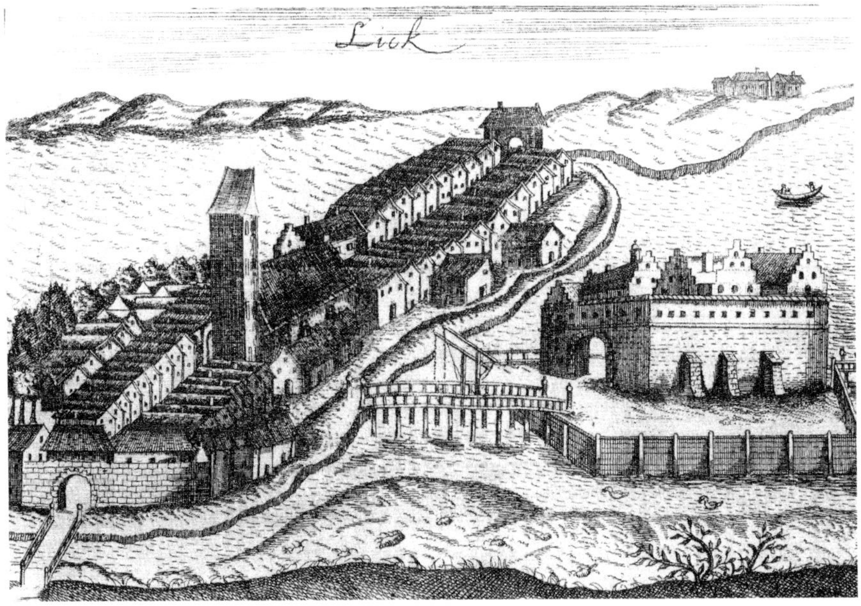

Figure 10.19: Hartknoch's drawing of the castle and town of Lyck (Pol. Ełk) in 1684 (after Haftka 1999, 87).

Konrad von Jungingen initiated the construction of the castle at Lyck, deep within the 'Great Wilderness', reflecting increasing confidence in colonizing the fringes of the eastern border of Prussia. Situated at the south-eastern edge of the commandery of Balga, its foundation was partly intended to strengthen the frontier with Lithuania and to develop the exploitation of natural resources in the locally abundant woods; the castle is initially referred to as a *Wildhaus*. The castle is first mentioned in 1398, but appears to have been rebuilt or finished in 1408, by which time its procurator was already demanding rents from bee-keepers in Goniondz (Pol. Goniądz) for access to the woodland hives on the right bank of the Biebrza (Długokęcki 2009d, 490; Białuński 2002, 76). Recent excavations at the site in 2011–14 (today in the Warmian-Masurian Voivodeship), directed by Radosław Herman (see Herman and Dudak 2015), revealed the earliest phase consisted of a residential-defensive tower situated on the southern edge of the island, subsequently incorporated into the larger brick building. The dendrochronology of the oak timbers indicated the first archaeological phase was also dated to the start of the fifteenth century. Locating the castle on a small island between two lakes provided it with excellent defensive capability (Fig. 10.19). In addition to the evident natural barrier encircling the site, the tower itself was also surrounded by a moat on three sides cutting into the island. By the eastern side of the island, the remains of a timber bridge and three sets of palisades were discovered, which were dated to the later decades of the fifteenth century, and the excavations revealed the island had been artificially shaped during the process of construction and the banks stabilized with posts made from pine.

The expansion of the complex however points to a change in function, most likely reflecting the frontier outpost transforming into a centre for managing local woodland resources (Haftka 1999, 86–88). Excavations recovered significant quantities of ceramic, brick, and animal bone fragments from late medieval and predominantly post-medieval contexts; the range of dendrochonological dates indicates regular upgrades to the infrastructure of the site throughout the fifteenth century. Shortly after its foundation the procurator's residence was associated with a settlement that was granted privileges in 1425 (and redefined as a town in 1440), with the process of more intensive colonization of the procuratorship lasting until 1525 (Białuński 2002; Długokęcki 2009b, 380). At the same time as settlement was being developed at the edges of the eastern frontier, administrative and fiscal changes within the Order's territories prompted the gradual dissolution of smaller estate centres from the early fifteenth century (Jóźwiak 2009b, 137). The recent excavations at the castle and island in Ełk (ul. Zamkowa) produced an important faunal assemblage and significant quantities of waterlogged timbers. Unfortunately, samples taken for plant macrofossil analysis produced no results. A series of cores taken from peat deposits within the territory of the procuratorship also produced no suitable material; however, existing studies from nearby regions have provided suitable palynological profiles (see Chapter 11).

Conclusion

The primary Prussian sites included within the study represent a range of frontier types, all of which can be broadly defined as transitional in character. The only sustained heartlands are represented by Radzyń Chełmiński and Elbląg, although both were located within broader frontier zones which became cores of the *Ordensland*. Future work would particularly benefit from investigating sites and their territories in the Sambian Peninsula.

Works Cited

Primary Sources

CTP Scholz, K. and Wojtecki, D., trans. 1984. *Chronik des Preussenlandes/Peter von Dusburg*, Darmstadt: Wissenschaftliche Buchgesellschaft

GAD Ziesemer, W., ed. 1968. *Das große Ämterbuch des Deutschen Ordens: Mit Unterstützung des Vereins für die Herstellung und Ausschmückung der Marienburg*, Wiesbaden: M. Sändig (repr. 2009)

Hartknoch, C. 1684. *Alt- und Neues Preußen*, Leipzig, Frankfurt, Königsberg: Hallervorden

Hennenberger, Caspar. 1595. *Erklärung der Preüßischen größern Landtaffel oder Mappen, Königsberg* (München, Bayerische Staatsbibliothek, Hbks / F 63 vd)

KP Fischer, M. 2010. *The Chronicle of Prussia by Nicolaus von Jeroschin: A History of the Teutonic Knights in Prussia, 1190–1331*, Farnham: Ashgate

PrUB *Preußisches Urkundenbuch* (*Prussian Records Book*) <http://www.spaetmittelalter.uni-hamburg.de/Urkundenbuch/ [accessed December 2017]

Secondary Works

Arszyński, M. 2010. 'Architektura warowna Zakonu Krzyżackiego w Prusach', in *Fundacje artystyczne na terenie państwa krzyżackiego w Prusach*, ed. B. Pospieszna, Pelplin: Wydawnictwo Bernardinum, 7–45

Beckherrn C. 1895. 'Über die Benennungen der ostpreußischen "Burgwalle" und die Pillberge in Samlande', *Altpreußische Monatsschrift* 32, 353–410

Białuński, G. 1996. *Osadnictwo regionu Wielkich Jezior Mazurskich od XIV do początku XVIII wieku — starostwo leckie (giżyckie) i ryńskie*, Olsztyn: Ośrodek Badań Nauk im. Wojciecha Kętrzyńskiego

——, 2002. *Kolonizacja 'wielkiej Puszczy' (do 1568 roku) — starostwa piskie, ełckie, starduńskie, zelkowskie i węgoborskie (węgorzeweskie)*, Olsztyn: Ośrodek Badań Nauk. im. Wojciecha Kętrzyńskiego

——, 2005. *Siedlungswesen im Bereich der Großen Masurischen Seen vom 14. bis zum 18. Jahrhundert — Ämter Lötzen und Rhein*, Hamburg: Selbstverlag des Vereins für Familienforschung

Biskup, M. 2002. 'Etniczno-demograficzne przemiany Prus Krzyżackich w rozwoju osadnictwa w średniowieczu', in *Opera Minora: Studia z dziejów zakonu Krzyżackiego* (collected essays), Toruń: Towarzystwo Naukowe w Toruniu, 129–50

Bojtár, E. 1999. *Foreword to the Past: A Cultural History of the Baltic People*, Budapest: Central European University Press

Brown, A. D., and Pluskowski, A. G. 2014. 'Medieval Landscape Transformation in the South-Eastern and Eastern Baltic: Palaeoenvironmental Perspectives on the Colonisation of Frontier Landscapes', *Archaeologia Baltica* 20, 24–46

Buko, A. 2005. *Archeologia Polski wczesnośredniowiecznej*, Warsaw: Trio (repr. in 2008)

Chęć, A. 2004. 'Wystrój wnętrz w folwarkach krzyżackich na podstawie źródeł historycznych (na przykładzie Mątów Małych i Benowa)', *Archaeologia Historica Polona* 14, 245–49

——, 2005. 'Hodowla koni w komturii malborskiej', *Archaeologia Historica Polona* 15/2, 359–67

Chęć, A., and Gancewski, J., 2009. 'Urząd leśniczego w państwie zakonu krzyżackiego —funkcjonowanie, znaczenie dla gospodarki', in *Las w kulturze polskiej, t. VI*, ed. W. Łysiak, Poznań: Eco, 47–56

Chojnacki, W. 1946. *Słownik polskich nazw miejscowości w b. Prusach Wschodnich i na obszarze b. Wolnego Miasta Gdańska według stanu z 1941 g*, Poznań: Wydawnictwo Instytutu Zachodniego

Chudziak, W. 1994. 'Podstawy klasyfikacji chronologicznej grodzisk', in *Wczesnośredniowieczne grodziska Ziemi Chełmińskiej*, ed. J. Chudziakowa, Toruń: Universytetu Mikołaja Kopernika, 24–280

——, 1996. *Zasiedlenie strefy chełmińsko-dobrzyńskiej we wczesnym średniowieczu (VII–XI wiek)*, Toruń: Wydawnictwo Naukowe Uniwersytetu Mikołaja Kopernika

——, 2000. 'Wyniki badań weryfikacyjnych i rozpoznawczych na Ziemi Chełmińskiej', in *Osadnictwo i architektura ziem polskich w dobie Zjazdu Gnieznienskiego*, eds A. Buko and Z. Świechowski, Warsaw: Instytut Archeologii i Etnologii PAN, 85–100

——, 2003. *Wczesnośredniowieczna przestrzeń sakralna in Culmine na Pomorzu Nadwiślańskim*, Toruń: Wydawnictwo Naukowe Uniwersytetu Mikołaja Kopernika

Czaja, R. 1997. 'Uwagi nad socjotopografią Starego Miasta Elbląga w średniowieczu i w czasach nowożytnych', *Archaeologia Elbingensis* 2, 91–98

——, 2009. 'Urbanizacja kraju', in *Państwo zakonu krzyżackiego w Prusach: Władza i społeczeństwo*, eds M. Biskup et al., Warsaw: Państwowe Wydawnictwo Naukowe, 177–230

Czaja, R., and Czacharowski, A. 2003. *Malbork: Atlas historyczny miast polskich*, Toruń: Wydawnictwo Uniwersytetu Mikołaja Kopernika

Dąbrowska, M. 2007. 'Badania archeologiczno-architektoniczne na terenie zamku niskiego w Malborku w latach 1998–2004', in *XV Sesja Pomorzoznawcza*, ed. G. Nawrolska, Elbląg: Muzeum Archeologizno-Historyczne w Elblągu, 303–16

Długokęcka, O., and Długokęcki, W. 2010. 'The Sumptuous Use of Food at Castle Marienburg (Malbork) at the Start of the Fifteenth Century', *Acta Poloniae Historica* 102, 101–26

Długokęcki, W. 1992. *Osadnictwo na Żuławach w XIII i na początku XIV wieku*, Malbork: Muzeum Zamkowe w Malborku

——, 2009a. 'Kolonizacja ziemi chełmińskiej, Prus i Pomorza Gdańskiego do 1410 r.', in *Państwo zakonu krzyżackiego w Prusach: Władza i społeczeństwo*, ed. M. Biskup et al., Warsaw: Państwowe Wydawnictwo Naukowe, 200–17

——, 2009b. 'Odbudowa osadnictwa, kolonizacja i przemiany wsi w państwie krzyżackim (1411–1525)', in *Państwo zakonu krzyżackiego w Prusach: Władza i społeczeństwo*, ed. M. Biskup et al., Warsaw: Państwowe Wydawnictwo Naukowe, 375–83

——, 2009c. 'Prusy w starożytności i we wczesnym średniowieczu', in *Państwo zakonu krzyżackiego w Prusach: Władza i społeczeństwo*, ed. M. Biskup et al., Warsaw: Państwowe Wydawnictwo Naukowe, 25–50

——, 2009d. 'Społeczeństwo wiejskie', in *Państwo zakonu krzyżackiego w Prusach. Władza i społeczeństwo*, ed. M. Biskup et al., Warsaw: Państwowe Wydawnictwo Naukowe, 460–94

——, 2004. 'Młynarstwo w komturstwie malborskim w XIII–XV w.', in *Młyny w Malborku i okolicy od XIII do XIX wieku*, eds W. Długokęcki, J. Kuczyński, and B. Pospieszna, Malbork: Muzeum Zamkowe w Malborku, 7–36

Domagała, T. 2004. 'Zagubiony Elbląski kodeks prawa lubeckiego z 1240 roku: Próba identyfikacji i rekonstrukcji tekstu', in *Archaeologia et historia urbana*, eds R. Czaja, G. Nawrolska, M. Rębkowski, and J. Tandecki, Elbląg: Muzeum w Elblągu, 89–93

Dygo, M. 2009. 'Ekspansja terytorialna i władztwo terytorialne (1249–1309)', in *Państwo zakonu krzyżackiego w Prusach: Władza i społeczeństwo*, ed. M. Biskup et al., Warsaw: Państwowe Wydawnictwo Naukowe, 79–104

Ekdahl, S. 2004. 'Crusades and Colonisation in the Baltic: A Historiographic Analysis', *Rocznik Instytutu Polsko-skandynawskiego* 19, 1–42

Engel, C. 1930. 'Tätigkeitsbericht des Prussia-Museum in Königsberg (Ostpreußen) für das Jahr 1929', *Nachrichtenblatt für deutsche Vorzeit* 6, 4–8

Fonferek, J. 2019. 'The Environmental Setting of the Earliest Teutonic Order Stronghold in Elbląg', in *Ecologies of Crusading, Colonization, and Religious Conversion in the Medieval Baltic: Terra Sacra II*, ed. by A. Pluskowski, Environmental History of the North Atlantic World, 3, Turnhout: Brepols (forthcoming)

Fonferek, J., Marinkowski, M., and Sieńkowska, U. 2012. *Elbląg — życie codzienne w porcie hanzeatyckim*, Elbląg: Muzeum Archeologiczno-Historyczne w Elblągu

Fonnesberg-Schmid, I. 2007. *The Popes and the Baltic Crusades: 1147–1254*, Leiden: Brill

Gancewski, J. 2013. 'Początki gospodarki folwarcznej w kompleksie zamkowym w Malborku i na terenie komturstwa malborskiego', *Echa Przeszłości* 14, 29–38

Górnowicz, H. 1980. *Toponimia Powiśla Gdańskiego*, Wrocław: Zakład Narodowy im. Ossolińskich

Haftka, M. 1987. 'Mikoreregion osadniczy Węgry-Gościszewo-Malbork w świetle kilkunastoletnich obserwacji terenowych', in *Badanaia archeologiczne w woj, elblaskim w latach 1980–83*, ed. A. Pawłowski, Malbork: Muzeum Zamkowe w Malborku, 27–42

——, 1999. *Zamki krzyżackie w Polsce: Szkice z dziejów*, Malbork: Wydawnictwo CONSORT i Muzeum Zamkowe w Malborku

Haftka, M., and Wadyl, S. 2015. *Węgry. Zespół osadniczy na pograniczu Pomorsko-Pruskim w XI–XII w.*, Malbork: Muzeum Zamkowe w Malborku

Hayward, K. M. J. 2019. 'Resources for Castle Building in Prussia and Livonia', in *Ecologies of Crusading, Colonization, and Religious Conversion in the Medieval Baltic: Terra Sacra II*, ed. by A. Pluskowski, Environmental History of the North Atlantic World, 3, Turnhout: Brepols (forthcoming)

Herman, R., and Dudak, W. 2015. 'Wyniki badań naukowych a projekt adaptacji zamku. Wzajemne uwarunkowania na przykładach zamków w Uniejowie, Lidzbarku i Ełku', in *Renovatio et restitutio. Materiały do badań i ochrony założęń rezydencjonalnych i obronnych*, eds P. Lasek and P. Sypczuk, Warsaw: IS PAN, 27–57

Herrmann, C. 2007. *Mittelalterliche Architektur in Preussenland: Untersuchungen zur Frage der Kunstlandschaft und -geographie*, Petersberg: Michael Imhof Verlag

Hess von Wichdorff, H. K. 1925. 'Auf der Burg des Königs Ysegubs von Galindien', *Nachrichtenblatt für deutsche Vorzeit* 1, 2–4

Iwaszkiewicz, M. 1991. 'Szczątki ryb z zamku krzyżackiego w Małej Nieszawce (woj. Toruńskie)', *Roczniki Akademii Rolniczej w Poznaniu* 227, 3–5

Jagodziński, M. F. 1996, 'Osadnictwo pradziejowe i wczesnośredniowieczne okolic Elbląga. Pograniczny character stosunkow etnicznych', in *750 lat praw miejskich Elbląga. Księga pamiątkowa*, ed. A. Groth, Gdańsk: Marpress, 11–20

——, 2004a. 'Czy Elbląg był kontynuacją Truso? Jeszcze jeden głos w dyskusji', in *Archaeologia et historia urbana*, eds R. Czaja, G. Nawrolska, M. Rębkowski and J. Tandecki, Elbląg: Muzeum w Elblągu, 95–105

——, 2004b. 'Podstawy źródłowe – analiza. Przekazy pisane – odkrycia archeologiczne', in *Pacifica Terra: Prusowie-Słowianie-Wikingowie u ujścia Wisły*, ed. J. Trupinda, Malbork: Muzeum Zamkowe w Malborku, 21–40

——, 2010. *Truso: Między Weonodlandem a Witlandem*, Elbląg: Muzeum Archeologiczno-Historyczne w Elblągu

Jonakowski, M. 2009/10. *Tczew, Ul. Zamkowa, działka nr 747/2: Sprawozdanie archeologiczne z przeprowadzonych prac*, Elbląg (unpublished report)

Józefczyk, M. 1995. *Sredniowiecze Elbląga*, Pelplin: Wydawnictwo Diecezjlane w Pelplinie.

Jóźwiak, S. 2001. *Centralne i terytorialne organy władzy zakonu krzyżackiego w Prusach w latach 1228–1410: rozwój, przekształcenia, kompetencje*, Toruń: Wydawnictwo Uniwersytetu Mikołaja Kopernika

——, 2009a. 'Centralne organy władzy', in *Państwo zakonu krzyżackiego w Prusach: Władza i społeczeństwo*, ed. M. Biskup et al., Warsaw: Państwowe Wydawnictwo Naukowe, 123–32

——, 2009b. 'Podziały administracyjne', in *Państwo zakonu krzyżackiego w Prusach: Władza i społeczeństwo*, ed. M. Biskup et al., Warsaw: Państwowe Wydawnictwo Naukowe, 132–37

Jóźwiak, S., and Trupinda, J. 2007. *Organizacja życia na zamku krzyżackim w Malborku w czasach wielkich mistrzów (1309–1457)*, Malbork: Muzeum Zamkowe w Malborku

——, 2012. *Krzyżackie zamki komturskie w Prusach*, Toruń: Wydawnictwo Naukowe Uniwersytetu Mikołaja Kopernika

Karczewska, M., and Karczewski, M. 2007. 'Grodzisko Święta Góra w Staświnach w Krainie Wielkich Jezior Mazurskich: Archeologia archiwalna i nowa', *Komunikaty Mazursko-Warmińskie* 2/256, 131–63

Karczewska, M. et al. 2005. *Miłki: Monografia krajoznawcza gmniy mazurskiej*, Białystok/Miłki: Wydawnictwo Kwadrat

Kawiński, P. 2011. 'Organizacja pogańskiej przestrzeni sakralnej Prusów na tle osadnictwa w okresie plemiennym - przykład Pomezanii, Pogezanii i Warmii', *Pruthenia* 6, 89–128

Klat, M., and Mykowski, J. 2002. *Krzyżackie tajemnice: zamki, skarby, odkrycia*, Gdańsk: Oficyna Pomorska

Klimek R. 2014. 'Pogańskie kamienne ołtarze Prusów', in *Zjawiska magiczno-demoniczne na tle dawnych ziem pruskich na tle porównawczym*, ed. K. Grążawski and J. Gancewski, Olsztyn: Wydawnictwo Uniwersytetu Warmińsko-Mazurskiego, 47–66

Kontny, B., Okulicz-Kozaryn, J., and Pietrzak, M. 2011. *Nowinka, Site 1: The Cemetery from the Late Migration Period in the Northern Poland*. Gdańsk: Archaeological Museum in Gdańsk and Warszawa University

Kosman, M. 1976. 'Zanik pogaństwa w Prusach', *Komunikaty Mazursko-Warmińskie* 1 /131, 3–29

Krickus, R. J. 2002. *The Kaliningrad Question*, Lanham: Rowman & Littlefield

Labuda, G., and Biskup, M. 1986. *Dzieje Zakonu Krzyzackiego w Prusach*, Gdańsk: Wydawnictwo Morskie

Lorek H. 1882/83. 'Der Swinta Gura der heilige Berg auf der Feldmark Stasswinnen, Kr. Lötzen', *Prussia* 9, 45–47

Lubocka-Hoffmann, M. 1992. 'The Old Town in Elbląg – Research and Purpose', *Archaeologia Elbingensis* 1, 35–37

Makowiecki, D. 2007. 'Gospodarka zwierzętami w strefie chełmińskiego pogranicza słowiańsko-pruskiego i krzyżackiego w średniowieczu', in *Pogranicze polsko-pruskie i krzyżackie*, ed. K. Grążawski, Włocławek-Brodnica: Lega, 23–37

Maltby, M., Pluskowski, A. G., and Seetah, K. 2009. 'Animal Bones from an Industrial Quarter at Malbork, Poland: Towards an Ecology of a Castle Built in Prussia by the Teutonic Order', *Crusades* 8, 191–212

Märcker, H. 1899. *Geschichte der ländlichen Ortschaften und der drei kleineren Städte des Kreises Thorn, Quellen und Darstellungen zur Geschichte Westpreußens 2*, Danzig: Th. Bartling

Nawrolska, G. 1997. 'Stand, Aufgaben und Perspektiven der Archäeologie in Elbing (Elbląg)', in *Lübecker Kolloquium zur Stadtarchäologie im Hanseraum I: Stand, Aufgaben und Perspektiven*, ed. D. Mührenberg, Lübeck: Schmidt-Römhild, 291–303

——, 1999. 'Archaeological Evidence for Trade in Elbląg from the 13th to the 17th Centuries', in *Lübecker Kolloquium zur Stadtarchäologie im Hanseraum*, II: *Der Handel*, eds. R. Dunckel, M. Gläser, and U. Oltmanns, Lübeck: Schmidt-Römhild, 373–85

——, 2008. 'A Way of Life. Luxury in a Medieval Town', in *Lübecker Kolloquium zur Stadtarchäologie im Hanseraum*, VI: *Luxus und Lifestyle*, eds C. Kimminus-Schneider, D. Mührenberg, M. Schneider, and C. Sheehan, Lübeck: Schmidt-Römhild, 509–27

Nawrolski, T. 1986. 'Zamek Elbląski w świetle nowych badań archaeologicznych', *Kwartalnik Architektury i Urbanistyki* 31/3, 91–107

Nitychoruk, J., Bińka, K., and Szymanek, M. 2012. *Sytuacja geologiczna i interpretacja paleośrodowiskowa obszaru dziedzińca Muzeum Arhceologiczno — Historycznego w Elblągu, Bulwar Zygmunta Augusta 11, 82–300 Elbląg*. Warszawa (unpublished report)

Nowakowski, W. 2003. 'Retrospekcja w archeologii: Galindai/Galinditae oraz Soudinoi/ Sudavitae w świetle źródeł historycznych i znalezisk archeologicznych', in *Studia Galindzike* (vol. I), ed. W. Wróbelski, Warsaw: Instytut Archeologii Uniwersytetu Warszawskiego, 7–12

Okulicz-Kozaryn, Ł. 1997. *Dzieje Prusów*, Wrocław: Fundacja na Rzecz Nauki Polskiej

Pawłowski, A. J. 1991. 'Z badań wczesnośredniowiecznych grodzisk Pomezanii i Pogezanii', in *Archeologia baltyjska. Materiały z konferencji Olsztyn, 24–25 kwietnia 1988 roku*, ed. A. J. Pawłowski, Olsztyn: Ośrodek Badań Naukowych im. Wojciecha Kętrzyńskiego w Olsztynie, 156–73

Pluskowski, A. G. 2013. *The Archaeology of the Prussian Crusade: Holy War and Colonisation*, London: Routledge

Pluskowski, A. G., Maltby, M., and Seetah, K. 2010. 'Potential Osteoarchaeological Evidence for Riding and the Military Use of Horses at the Castle in Malbork, Poland', *International Journal of Osteoarchaeology* 20/3, 335–43

Pluskowski, A. G., Sawicki, Z., Shillito, L.–M., Badura, M., Makowiecki, D., Zabilska-Kunek, M., Seetah, K., and Brown, A. 2014. 'Biała Góra: The Forgotten Colony in the Medieval Pomeranian-Prussian Borderlands', *Antiquity* 88/341, 863–82

Pollakówna, M. 1997. 'Zantyr', *Komunikaty Mazursko-Warmińskie* 3, 414–25

Poliński, D. 2003. *Późnośredniowieczne osadnictwo wiejskie w ziemi chełmińskiej*, Toruń: UMK

——, 2007. 'Wczesne warownie krzyżackie w Prusach w kontekście miejscowych obiektów obronnych w umocnieniach drewniano-ziemnych', *Archaeologia Historica Polona* 16, 41–61

Pospiszylowa, A. 1986. 'Polonization of Old Prussian Geographical Names in the Region of South Warmia', *Lingua Posnaniensis* 29, 61–75

Powierski, J. 1979. 'Chronologia początków Malborka', *Zapiski Historyczne* 44/2, 5–31
——, 2001. *Prusowie, Mazowsze i sprowadzenie krzyżaków do Polski* (vol. II.1), Malbork: Muzeum Zamkowe w Malborku
——, 2003. *Prussica* (vol. I), Malbork: Muzeum Zamkowe w Malborku
Sawicki, Z., Pluskowski, A. G., Brown, A., Badura, M., Makowiecki, D., Shillito, L.-M., Zabilska-Kunek, M., and Seetah, K. 2015. 'Survival at the Frontier of Holy War: Political Expansion, Crusading, Environmental Exploitation and the Medieval Colonising Settlement at Biała Góra, North Poland', *European Journal of Archaeology*, 18/2, 282–311
Scull, C., Müldner, G., and Makowiecki, D. 2019. 'Carbon and Nitrogen Stable Isotope Evidence for Animal Husbandry and Environmental Change in the Medieval Kulmerland', in *Ecologies of Crusading, Colonization, and Religious Conversion in the Medieval Baltic: Terra Sacra II*, ed. A. Pluskowski, Environmental History of the North Atlantic World, 3, Turnhout: Brepols (forthcoming)
Szczepański, S. 2013. 'Chomor Sancti Adalberti (1249): a możliwości lokalizacji terenowej wybranych kościołów pomezanii', *Komunikaty Mazursko-Warmińskie* 1/279, 19–45
——, 2016. *Pomezania pruska: Dzieje osadnictwa w XIII–XV wieku*, Olsztyn: Rozprawy i Materiały Ośrodka Badań Naukowych im. Wojciecha Kętrzyńskiego w Olsztynie
Tandecki, J. 2009. 'Rozwój terytorialny państwa zakonnego w Prusach', in *Państwo zakonu krzyżackiego w Prusach: Władza i społeczeństwo*, ed. M. Biskup et al., Warsaw: Państwowe Wydawnictwo Naukowe, 105–09
Torbus, T. 1998. *Die Konwentsburgen im Deutschordensland Preussen*, Munich: Oldenbourg
Wacnik, A. 2009. 'Galindowie i Krzyżacy — oddziaływania na lokalną roślinność w rejonie Miłek i Staświn (Kraina Wielkich Jezior Mazurskich, północno-wschodnia Polska', *Wiadomości Botaniczne* 53, 21–34
Ważny, T. 2004. 'Zastosowanie dendrochronologii w badaniach starego miasta w Elblągu', in *Archaeologia et Historia Urbana*, eds R. Czaja, G. Nawrolska, M. Rębkowski, and J. Tandecki, Elbląg: Muzeum w Elblągu, 377–82
Wiewióra, M. 2013a. 'Badania archeologiczno-architektoniczne zamku głównego w Radzyniu Chełmińskim w latach 2008–2009', in *XVII Sesja Pomorzoznawcza*, II: *Od późnego średniowiecza do czasów nowożytnych*, eds H. Paner and M. Fudziński, Gdańsk: Muzeum Archeologiczne w Gdańsku, 169–78
——, 2013b. 'Zamki krzyżackie na ziemi chełmińskiej w świetle najnowszych badań wybranych obiektów', *Kwartalnik Architektury i Urbanistyki* 58/1, 97–123

11. Vegetation Change in Prussia: The Palynological Data

Alexander Brown
With contributions by Anneli Poska and Amanda Dawn-Wynne

Introduction

The commanderies of Rehden, Marienburg, Elbing, and Brandenburg (Lötzen procturatorship) form the focus for palynological analysis from Prussia, with data presented from four peat deposits, five lakes, and samples from archaeological excavations at Malbork and Elbląg (Map 11.1). The palynological studies reflect a variety of geographical and archaeological contexts, from deposits located either adjacent to or directly associated with castles, towns, and earlier Slavic and Prussian strongholds, to sequences located in rural locations or more isolated setting within existing areas of woodland. The palynological data from Prussia, as with those from Livonia (Chapter 5), are presented geographically by commandery according to the broad chronological order of conquest. These studies provide an important context for understanding how local environments were transformed during the late medieval period, considered within the context of evidence for land-use change associated within earlier documented phases of colonization occurring from the seventh century onwards. Moreover, the wide geographical focus of research allows comparison of the environmental histories of contrasting regions across Prussia. These areas have distinctly different pre-crusading histories, were conquered at different stages during the crusades, and subsequently saw varying histories and intensities of rural colonization and urban development during the late medieval period. Furthermore, these regions cover a diverse range of geographical settings with varying flora and fauna, from the marshlands of the Vistula Delta to the fertile gently undulating landscapes of the Lower Vistula and the numerous lakes, marshes, and woodlands that characterize the Great Masurian Lake District. The palynological records provide an important source of data for identifying variations and trends in environmental change and land-use across Prussia since the end of the Migration Period, and are considered in the context of existing palynological studies from across north-eastern Poland. Together these data provide an important lens for considering the impact of the Crusades on the physical landscapes of the south-eastern Baltic.

Present-Day Biogeography, Relief, and Climate

The vegetation of Poland belongs to the Central European Upland-Lowland region, but within the north-east has a transitional character located at the range limits of several key European taxa, belonging to the hemiboreal vegetation group (European Environment Agency 2007) with coniferous and mixed broad-leaved-coniferous forests. Woodland currently constitutes *c.* 30% of the Polish landmass (Forest Europe, UNECE and FAO 2011, table A1.1), the majority of which in north-eastern Poland occur across the Lake District, an area comprising several Lake Districts and National Parks. The vast majority of woodlands are planted, or else heavily influenced by human intervention, dominated by pine and mixed pine woodlands. The potential natural woodland cover (i.e., in the absence of human intervention) would be characterized by sub-boreal mixed oak-lime-hornbeam dominated woodlands, not too dissimilar to those present within the Białowieża National Park (Falinski 1986), considered by many (e.g., Rackham 2003; Peterken 1981; Falinski 1986; Bernadzki et al. 1998) to represent the last stands of old growth woodland in Europe. The relief of northern Poland is heavily influenced by the last glaciation (Weichselian) and deglaciation, with numerous lakes and peatlands, particularly within the lake districts of north-eastern Poland. The climate of northern Poland is transitional in character, with the coldest mean annual temperatures within the north-eastern Lake Region, with mean January and July temperatures of -5°C and +17°C (Gałka et al. 2014; Szal et al. 2014; Wacnik et al. 2014).

Rehden Commandery

Radzyń Chełminski *(53°23′ N, 18°56′ E)*: three sequences were sampled from locations along the western edge of the Castle Lake, a large (75 hectares) and now predominantly infilled basin, adjacent to the Slavic stronghold and Teutonic Order's castle and town. The lake is considered to have formed part of a much larger wetland system surrounding the late medieval castle and town and extending to the east, now largely under cultivation with the lake covered in sedge fen. The surrounding landscape is dominated by agricultural fields with only small isolated patches of woodland; the largest tracts of woodland are *c.* 10 km to the west around Grudziądz. The Slavic stronghold is situated on a bedrock peninsula on the eastern edge of Radzyń Chełminski, with the castle situated 500 m to the north. The first Radzyń sequence (Radzyń1) was sampled in proximity to the castle 50 m from the edge of the lake adjacent to the outer bailey. The sequence comprises a well-humified herbaceous peat (0–0.23 m), underlain by silt peat (0.23–0.69 m) and dark grey gyttja. The lake was cored to a depth of 7.33 m, with the top 4 m sampled but with analysis focused on the top 1.35 m dating from the first century BC. The second Radzyń sequence (Radzyń2) is located 225 m south-east of Radzyń1, close to the Slavic stronghold, and comprises a sequence of herbaceous peat (0–0.12 m) and gyttja of dark-green (0.12–0.48 m) and reddish grey (0.48–1.02 m) hue. Pollen was poorly preserved in Radzyń1 and 2 after *c.* 1500, so a third Radzyń sequence (Radzyń3) was sampled 150 m south-east of Radzyń2. The sediments were sampled to a depth of 1.25 m, again comprising herbaceous peat (0–0.1 m) and gyttja of dark grey (0.1–0.82 and 1.17–1.25 m) and dark reddish grey (0.82–1.17 m) hues.

Rywałd *(53°22′27.4″ N, 19°03′13.7″ E)*: large lake (38 hectares) 7.5 km east of Radzyń, also now largely infilled and cover by sedges; a second extant lake is located immediately to the north. The lake is located 1.2 km west of the medieval village of Rywałd (Ger. Rehwalde) within a dominantly open agricultural landscape. The top 2.25 m of sediment was sampled with analysis focused on the top 1.5 dating from the second century AD. The sediments comprise a humified herbaceous peat with silt (0–0.52 m) and detrital organic lake muds of varying hue from very dark greenish brown (0.52–0.70 m), olive brown (0.70–0.77 and 0.88–1.40 m), to dark olive brown (0.77–0.88 and 1.40–1.50 m).

Marienburg Commandery

Malbork high castle *(52°02′22″ N, 19°01′39″ E)*: coring identified a complex series of sediments under the cobbled surface of the high castle, including several thin organic rich horizons (1.13–1.17, 1.24–1.26, and 1.56–1.65 m) separated by more substantial sandy deposits considered to represent potential garden layers and levelling/construction horizons. The top 1.13 m underlying the coble surface of the courtyard was composed of brick rubble. Pollen was largely absent from the organic horizons but was preserved in a single deposit at 1.14 m. An organic layer from an adjacent core at 0.66–0.74 m was ^{14}C dated 690±29 BP (GU-301089, cal. AD 1267–1388), suggesting these deposits are likely part of the early history of the castle (see Chapter 11).

Malbork Castle moat *(52°02′28″ N, 19°01′55″ E)*: a single core was taken from the eastern section of the outer moat adjacent to the inner curtain wall of the castle. The sequence was relatively shallow (1.5 m) but included alternating layers of sand (0.58–0.71, 0.73.5–0.74.5, 0.83–0.88, 1.17–1.50 m) and organic-rich silts (0.50–0.58, 0.71–0.73.5, 0.88–1.77 m). Pollen was well preserved in the organic layers, with the basal layer also containing quantities of plant macrofossils and animal bones.

Parpary *(53°58′11.3″ N, 18°56′27.9″ E)*: small basin infilling a shallow wooded depression (0.6 hectares) located 150 m from the current northern edge of the Forest of Sztum and just 400 m from the edge of the late medieval colonizing settlement at Parpary (Ger. Parpahren). The settlement at Parpary is currently located within a large clearing (94 hectares) within the woodland, surrounded by coniferous plantation. The sediments, sampled to 1.5 m, were heavily compacted with the top 0.5 m covering the last two thousand years and comprising a well-humified friable structureless peat; the top 0.22 m comprised a more moderately humified moss peat with occasional twigs and leaf fragments towards the top.

Benowo *(53°54′30″ N, 18°56′24.8″ E)*: small peat bog (4 hectares) located within the northern part of the Forest of Sztum, 1.5 km north-east of the late medieval colonizing settlement at Benowo (Ger. Bönhof) located on the edge of woodland. The peat bog is partly covered in planted conifers and some birch, with a ground flora dominated by bryophytes. The peat, comprising a poorly humified fibrous moss peat, was sampled to a depth of 2.5 m but only analysed to 1 m covering the last 1500 years.

Map 11.1: Location map of pollen studies from Prussia within the Rehden, Marienburg, Elbing, and Brandenburg commanderies. Map by author.

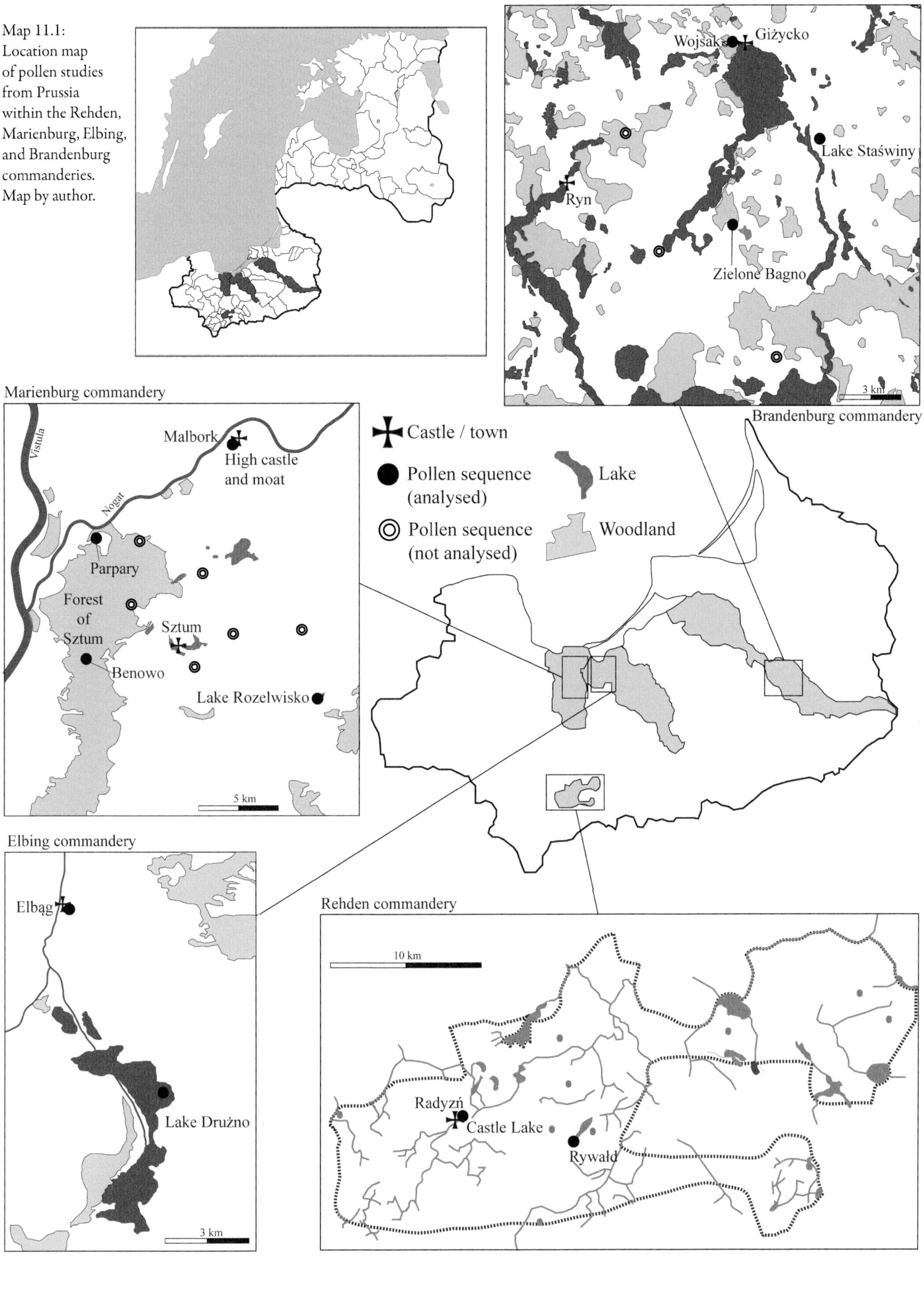

Lake Rozlewisko *(53°54'46.9" N, 19°10'00.5" E)*: large lake (51 hectares) located 1.5 km south of the late medieval colonizing settlement at Stary Targ (Ger. Altmark). The pollen core was taken within an area of semi-terrestrial reed swamp at the south-west end of the lake. Sediments were sampled to a depth of 1.5 m and comprised a herbaceous peat with silt (0–1.04 m) including abundant *Phragmites* roots within the top 0.6 m, underlain by olive-green gyttja. Analysis was focused on the top metre, although pollen was absent above 0.52 m.

Elbing Commandery

Elbląg Castle *(54°09'24" N, 19°23'39" E)*: excavations within the area of the former high castle in advance of improvement works revealed well-preserved and in cases waterlogged occupation deposits. Samples were taken primarily for the purpose of micromorphological analysis within what is considered to have been the courtyard of the high castle, with pollen samples taken from these blocks prior to oven-drying. The castle is located adjacent to the River Elbląg on the southern edge of the old town (now rebuilt following its destruction during World War Two).

Lake Drużno *(54°06'32.2" N, 19°38'03.9" E)*: large although shallow lake located to the south of Elbląg. Figures for the extent of the lake vary between 12.6 to 29.8 km² dependent on the water level. Currently an inflow-outflow lake, Drużno is connected to the Vistula Lagoon via the Elbląg River, but has several inflowing streams and canals with a large catchment of over 1000 km². The edges of the lake, which are highly indented, are surrounded by reed swamps and communities of floating and submerged aquatics, with alder carr woodland along the southern shores of the lake (Zachowicz and Kępińska 1987). Two cores were sampled from the northern area of Lake Drużno, 7.2 km south-south-east from Elbląg Castle but only *c.* 1.2 km from Truso, a Viking Age trading Emporium situated along the former shores of the lake (Bogucki and Jagodziński 2012, Figs 2–4; Nitychoruk 2016). Analysis was focused on the second of the two cores, located 330 m from the current lake edge. The water depth was just 0.9 m, with sediments sampled to a depth of 2.65 m below the frozen surface of the lake. The sediments comprise dark olive grey to greenish-grey gyttja with abundant mollusc shells visible between 1.09–1.12 and 1.29–1.33 m. Two previous cores were analysed by

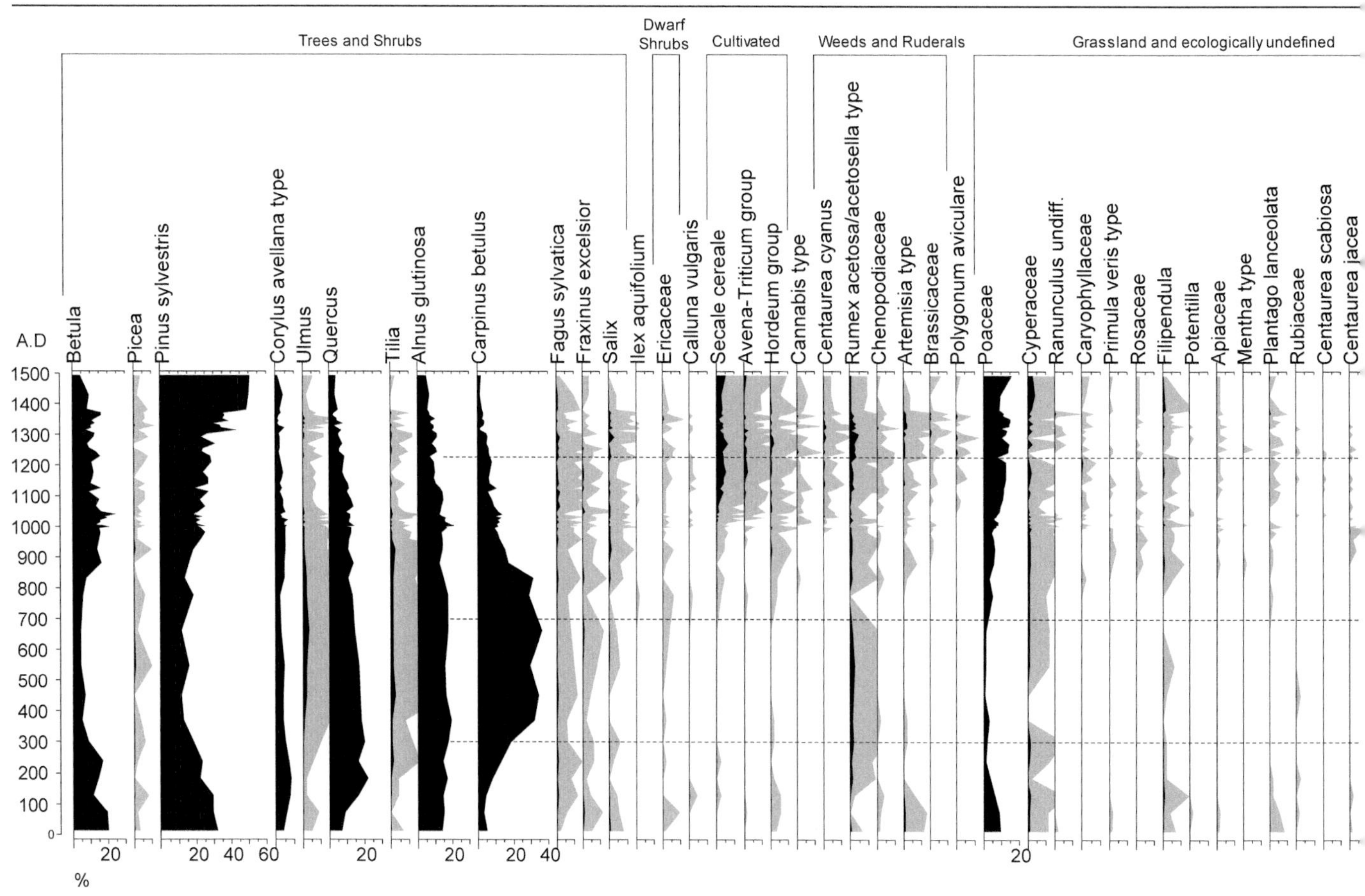

Zachowicz and Kępińska (1982, 1987) from the same portion of the lake, but analysis focused on the entire Late Glacial/Holocene sequence with the only radiocarbon dates of Mesolithic Age.

Brandenburg Commandery (Lötzen Procuratorship

Zielone Bagno *(53°54'18.6' N 21°44'9.8' E)*: large mire complex (90 hectares) located to the east of Lake Jagodne. The pollen sequence was taken from the southern edge of the largest section of the mire, 50 m from adjacent agricultural fields and 1.4 km south of the Prussian and later settlement at Jagodne Małe. Peat with a depth of two metres was retrieved resting on sands. The top 0.25 m was not sampled; the high water content and unconsolidated nature of the sediments made it difficult to retain undisturbed material in the chamber of the auger. Pollen was analysed to a depth of 0.9 m, comprising a poorly humified peat with significant quantities of matted plant material (0.25–0.41 m), overlying a moderately well-humified peat with silt (0.41–1.61 m). The vegetation surrounding the sampling site is dominated by birch carr woodland, with a swampy semi-terrestrial understorey comprising sedges, often forming tussocks, and wetland grasses.

Lake Staświny *(53°58'10" N, 21°51'01" E)*: two cores were taken from a small lake (7 hectares) 0.5 km north-west of the village of Staświny. The lake is located within a small north-west to south-east oriented depression, currently surrounded by agricultural land and with no obvious inflowing or outflowing streams, although a later channel has been dug that connects the lake with the much larger Lake Wojnowo. Pollen was assessed from both cores but only analysed from the second core due to the close proximity and similarity of the pollen signals. The sediments were sampled to a depth of 3.15 m (including water to 1.48 m), comprising gyttja varying in hue from dark grey (1.48–1.69, 1.89–2.0 m) to olive grey (1.69–1.89, 2.09–3.15 m), with a highly organic band of black gyttja from 2.0–2.09 m.

Wojsak *(54°02'49.5" N, 21°44'17.8" E)*: located on the north-western edge of present-day Giżycko, 1.8 km north-west of the procurators Castle, Wojsak comprises a complex of small isolated mires, dispersed over an area of 21 hectares. The area is heavily wooded with a mixed of wet carr woodland and mixed broad-leaved and deciduous woodland (largely pine) on drier soils. The sequence was taken from the southern edge of the mire and comprises a shallow depth of peat totalling 0.8 m, overlying

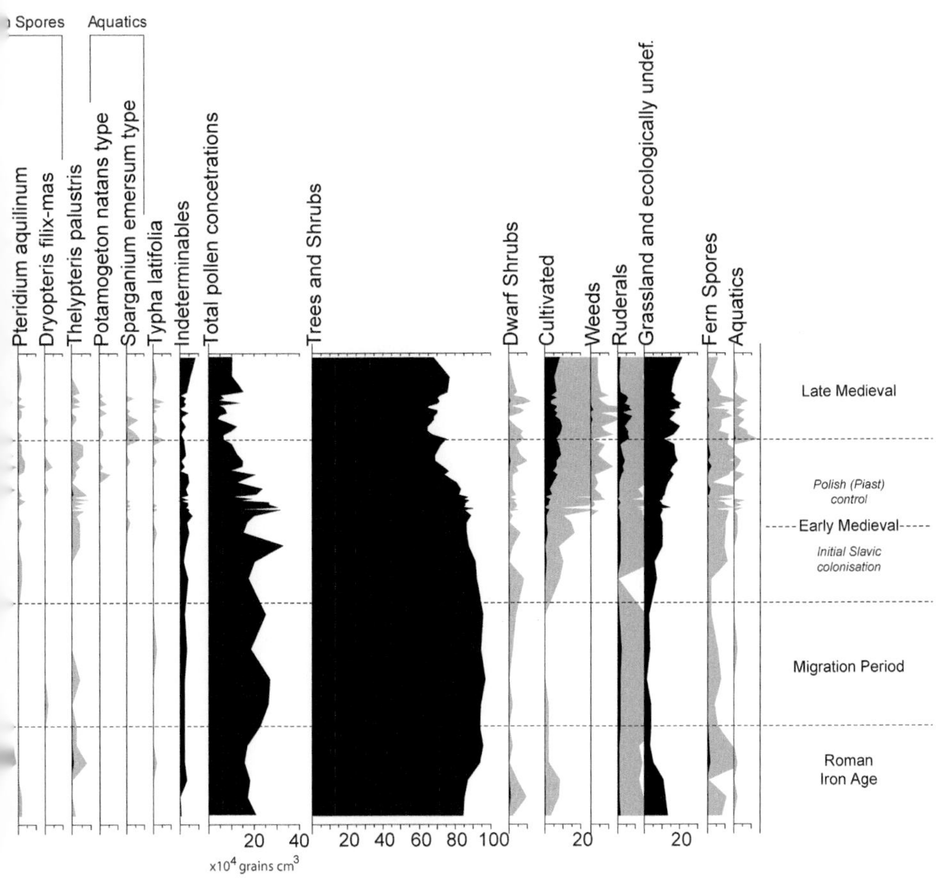

Figure 11.1: Radzyń 1, selected taxa percentage pollen diagram. Black curve = percentage of taxon, grey curve = values exaggerated x 10. Figure by author.

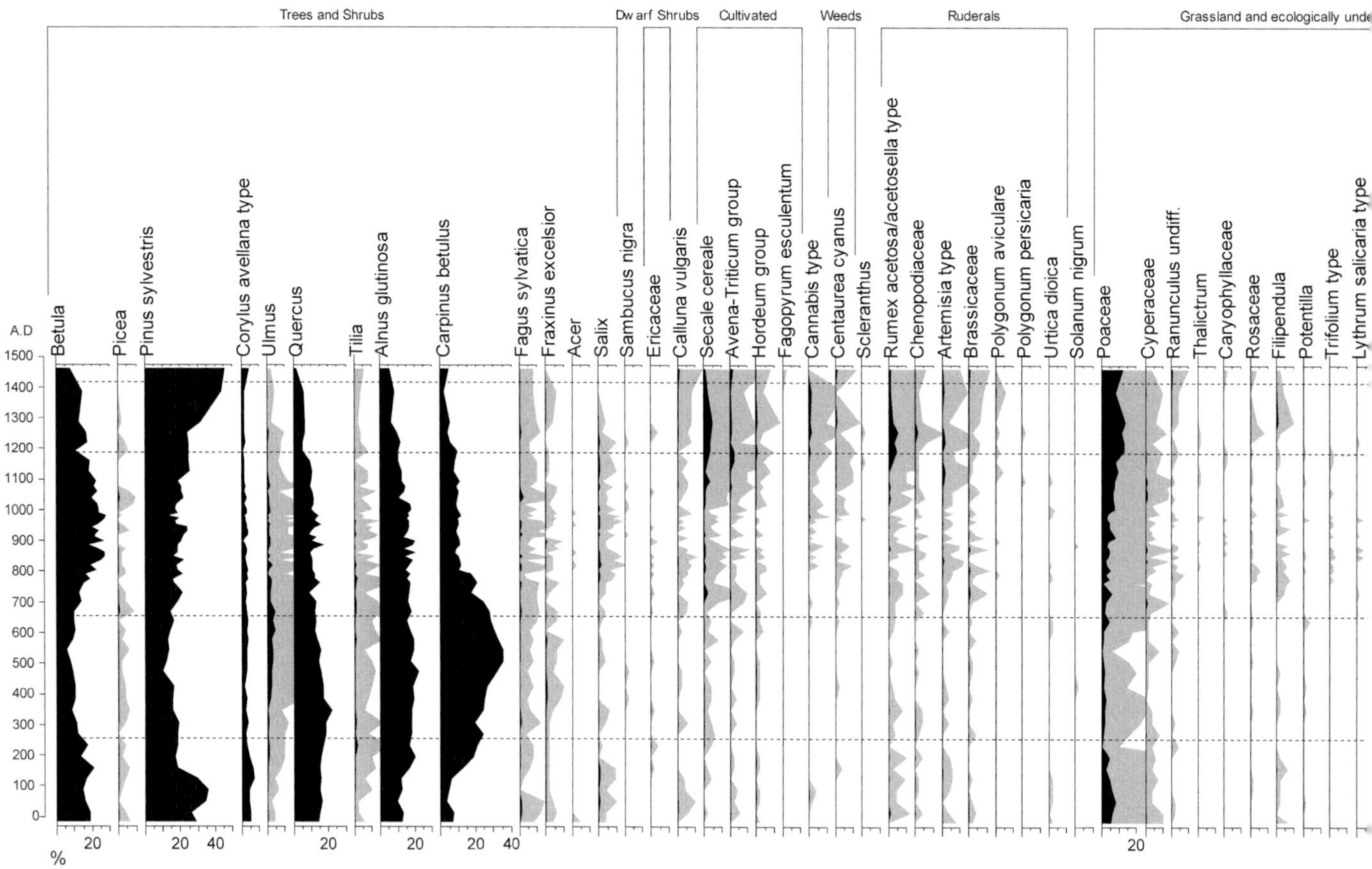
Trees and Shrubs
Dwarf Shrubs
Cultivated
Weeds
Ruderals
Grassland and ecologically unde
A.D
Betula
Picea
Pinus sylvestris
Corylus avellana type
Ulmus
Quercus
Tilia
Alnus glutinosa
Carpinus betulus
Fagus sylvatica
Fraxinus excelsior
Acer
Salix
Sambucus nigra
Ericaceae
Calluna vulgaris
Secale cereale
Avena-Triticum group
Hordeum group
Fagopyrum esculentum
Cannabis type
Centaurea cyanus
Scleranthus
Rumex acetosa/acetosella type
Chenopodiaceae
Artemisia type
Brassicaceae
Polygonum aviculare
Polygonum persicaria
Urtica dioica
Solanum nigrum
Poaceae
Cyperaceae
Ranunculus undiff.
Thalictrum
Caryophyllaceae
Rosaceae
Filipendula
Potentilla
Trifolium type
Lythrum salicaria type
1500
1400
1300
1200
1100
1000
900
800
700
600
500
400
300
200
100
0
20
20 40
20
20
20 40
20
%

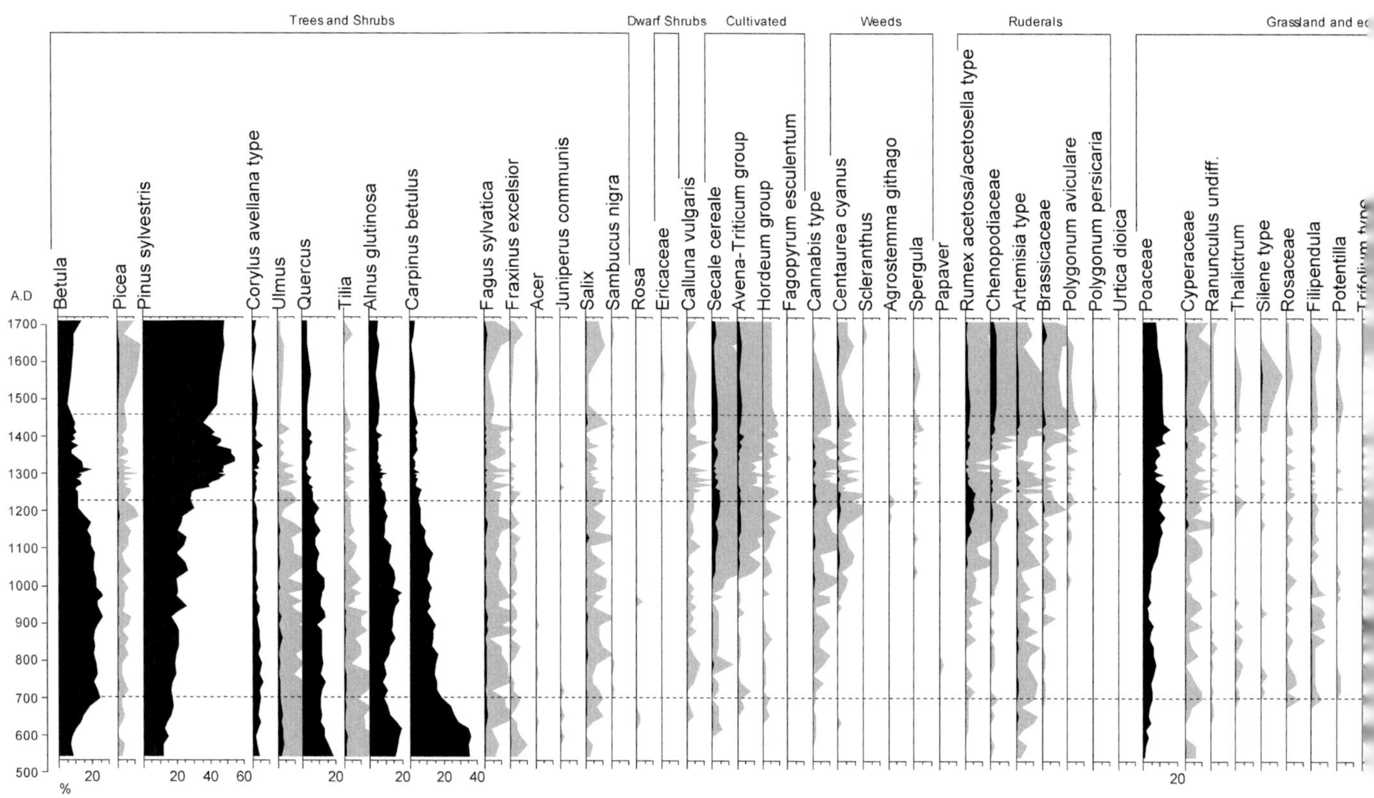
Trees and Shrubs
Dwarf Shrubs
Cultivated
Weeds
Ruderals
Grassland and ec
A.D
Betula
Picea
Pinus sylvestris
Corylus avellana type
Ulmus
Quercus
Tilia
Alnus glutinosa
Carpinus betulus
Fagus sylvatica
Fraxinus excelsior
Acer
Juniperus communis
Salix
Sambucus nigra
Rosa
Ericaceae
Calluna vulgaris
Secale cereale
Avena-Triticum group
Hordeum group
Fagopyrum esculentum
Cannabis type
Centaurea cyanus
Scleranthus
Agrostemma githago
Spergula
Papaver
Rumex acetosa/acetosella type
Chenopodiaceae
Artemisia type
Brassicaceae
Polygonum aviculare
Polygonum persicaria
Urtica dioica
Poaceae
Cyperaceae
Ranunculus undiff.
Thalictrum
Silene type
Rosaceae
Filipendula
Potentilla
1700
1600
1500
1400
1300
1200
1100
1000
900
800
700
600
500
20
20 40 60
20
20
20 40
20
%

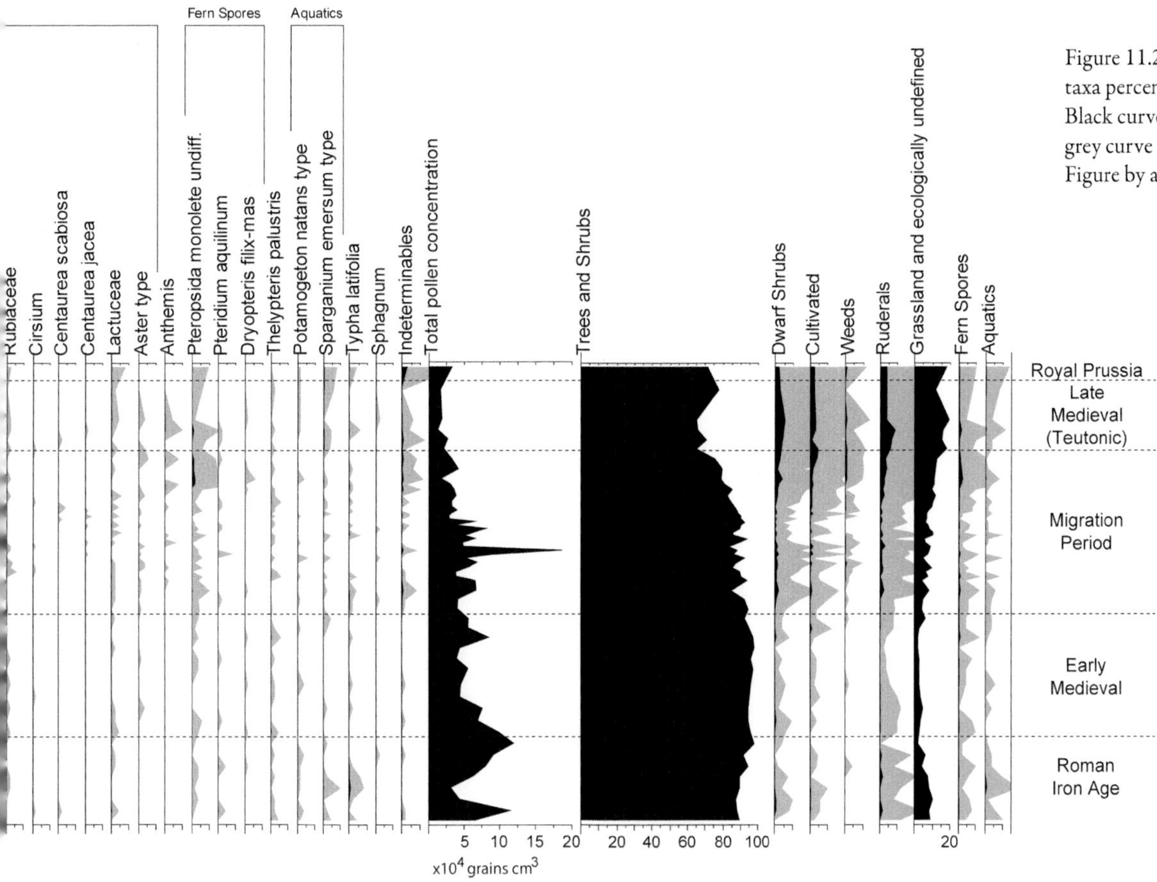

Figure 11.2: Radzyń 2, selected taxa percentage pollen diagram. Black curve = percentage of taxon, grey curve = values exaggerated x 10. Figure by author.

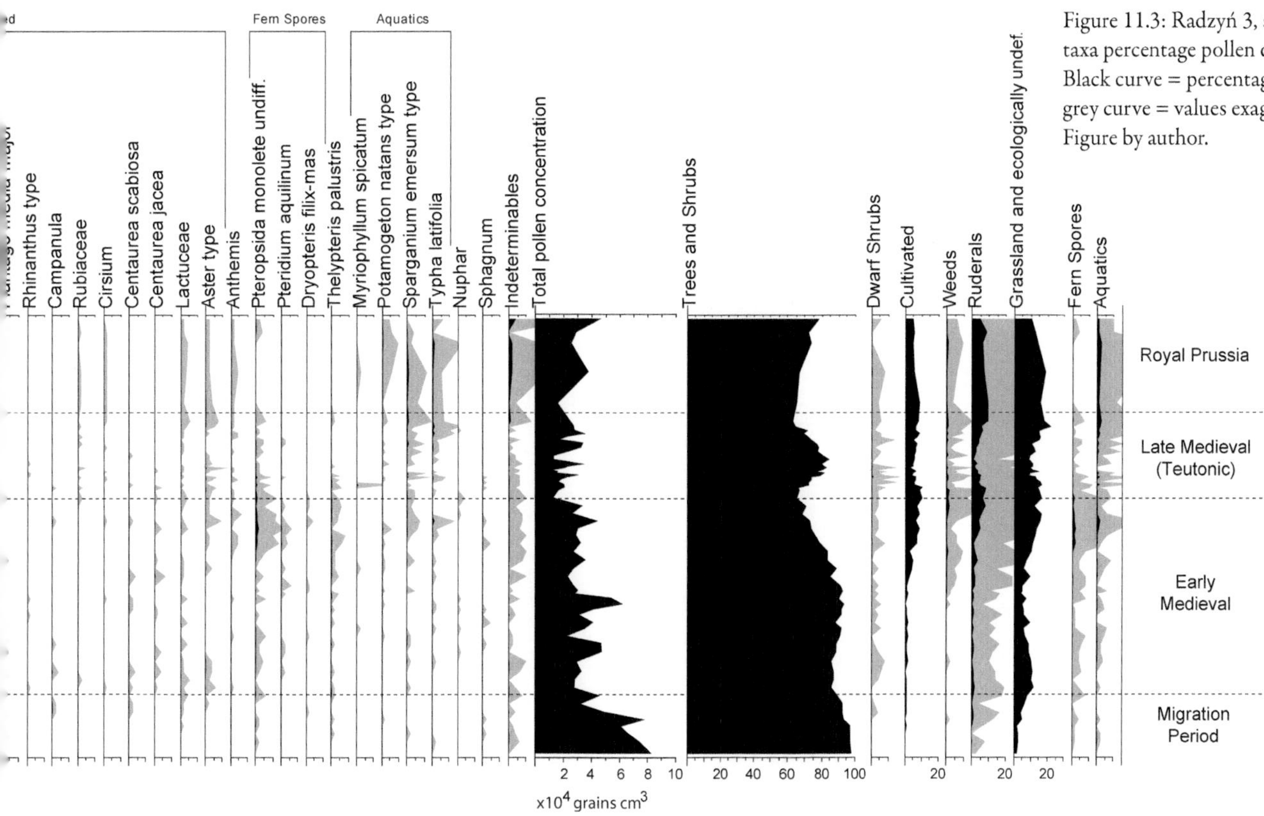

Figure 11.3: Radzyń 3, selected taxa percentage pollen diagram. Black curve = percentage of taxon, grey curve = values exaggerated x 10. Figure by author.

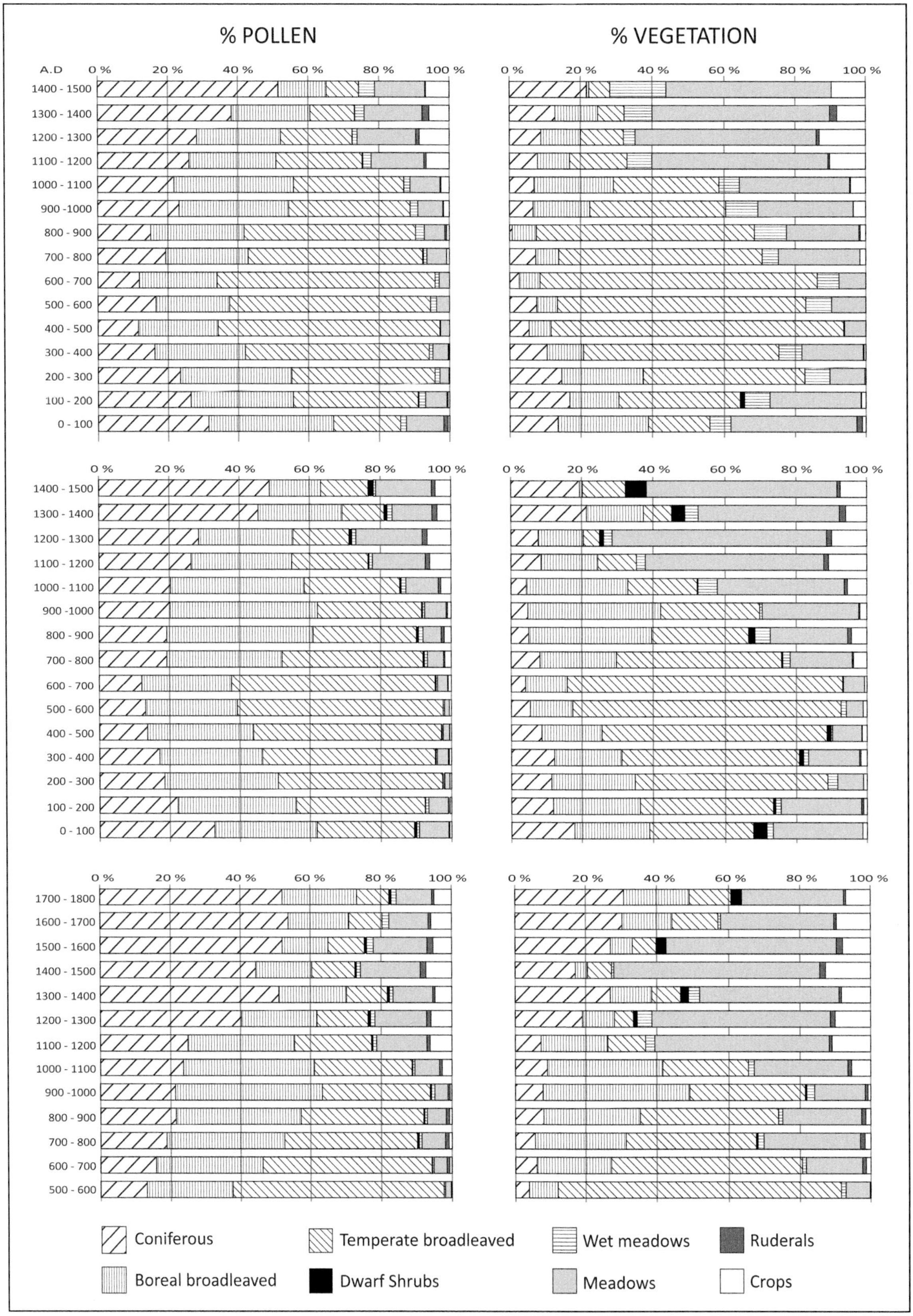

Figure 11.4: Radzyń, proportions of pollen taxa modelled using the Landscape Reconstruction Algorithm (LRA) (LOVE), summarized within eight groups: (1) coniferous trees; (2) boreal broad-leaved trees; (3) temperate trees; (4) dwarf shrubs; (5) wet meadows; (6) meadows; (7) ruderal; (8) crops, and presented in hundred-year time slices. Regional Vegetation Reconstructions are based on pollen data from the Castle Lake and Lake Rozlewisko. Model by A. Poska.

silts resting on gyttja. The peat comprises a sequence of well-humified herbaceous peats with silt (0–0.23, 0.26–0.41 m) and herb/moss peats (0.23–0.26, 0.41–0.8). Pollen was only preserved in the top 0.64 m.

Sequences Assessed but not Analysed

Fourteen additional cores were sampled, assessed but not analysed (Map 11.1), in the majority of cases due to poor pollen preservation. The majority of these cores (eight) were located within the Marienburg commandery. Pollen was either absent or poorly preserved in two cores taken from a peat deposited located 1.3 km south-east of the procurators castle at Sztum, and likewise from a peat deposit on the southern edge of the medieval colonizing settlement at Pietrzwałd. Another core was taken from a peat bog midway between the colonizing settlements at Kalwa and Stary Targ. Assessment demonstrated that the peat had been truncated despite excellent pollen preservation. The lack of a peak in hornbeam so characteristic of pollen sequences for the Migration Period in the Vistula Basin, and absence of hornbeam below 0.50 m suggests the sequence extends no later than *c.* 1500 BC (see Ralska-Jasiewiczowa et al. 2004, Fig. 36; Noryśkiewicz 2013). Three additional cores were also taken within and along the margins of the Forest of Sztum; the first on the northern edge close to the Slavic Pomeranian stronghold at Węgry was heavily truncated and dated to 8699±28 BP (GU-31229, 7784–7601. cal BC), whilst pollen was poorly preserved in a sequence from Sztumskie Pole. The sequence from Koniecwałd preserved pollen but dates at 0.52 m (1070±29 BP, cal. AD 896–1021) and 0.73 m (3222±36 BP, 1424–1609 cal. BC) suggest either a significant hiatus in sediment accumulation or some mixing of sediment. Although rye and other cereal pollen are present (rye from 0.52 m), the sequence lacks an associated pollen signature characteristic of the hornbeam-dominated woodland apparent from nearby and surrounding pollen sequences of the same date; this raises serious doubts over the integrity of the sequence. Pollen was likewise absent from the second of two cores retrieved from the courtyard of the high castle at Malbork that revealed a sequence of thin organo-mineral deposits separated by sterile sands. Three additional cores were sampled from within the Lötzen procuratorship. Pollen was poorly preserved in a shallow depth (1 m) of peat from near Szymonka, located midway between the southern end of Lake Jagodne and Lake Szymonka. Pollen was likewise poorly preserved in a sequence adjacent to the Prussian stronghold at Jeziorko, and absent entirely from a small peat bog at Tuchlin, located *c.* 4.5 km west of the procurators castles in Okartowo (Ger. Eckersberg). Although pollen was well preserved in a sequence from Matule, a forested peat bog 10 km south of Dzierzgoń (Christburg commandery), the high values for hornbeam at the very top of the sequence suggest the medieval deposits may have been truncated. Pollen was absent entirely, however, from an initial core taken from the north-western edge of former Lake Rywałd. A single core from Zaine within the Königsberg commandery was likewise absent of pollen.

Results

Rehden Commandery

Castle Lake, Radzyń Chełmiński (Figs 11.1–11.4)

The three sequences from the Castle Lake at Radzyń are discussed in parallel due to the close proximity of the sequences and similarity in pollen signals. The timespans of the pollen profiles vary, with Radzyń1 and 2 covering the first to fifteenth centuries and Radzyń3 covering the sixth to seventeenth centuries. The base of the Radzyń1 and 2 sequences correspond to the Roman Iron Age (50 BC–AD 300, also called the period of Roman influence); low values for cereal pollen of rye, wheat/oat and barley, grass pollen values of *c.* 10%, and ruderal pollen taxa, particularly common sorrel, indicate the presence of small-scale cultivation along with grassland and meadow environments. Pine and birch dominate arboreal pollen values, increasing in frequency as a consequence of intensifying settlement and land-use during the Roman Iron Age. Pine and birch are both colonizing species, encroaching into areas of cleared and disturbed woodland, formerly dominated by hornbeam, with their profuse pollen being widely distributed across the landscape. Despite increasing anthropogenic activity, oak maintains values of *c.* 10%, suggesting it formed an important component of existing woodland. LOVE models for Radzyń1 and 2 show landscape openness varying between 25–40% during the first and second centuries AD, suggesting a more mosaic landscape surrounding the lake during the Roman Iron Age, with a distinct focus on meadow rather than arable land. There is the start of a gradual regeneration of woodland from the third century towards the end of the Roman Iron Age, but it is most apparent from the fourth century, corresponding to the Migration Period (AD 300–700), with increasing values for oak and hornbeam. Hornbeam

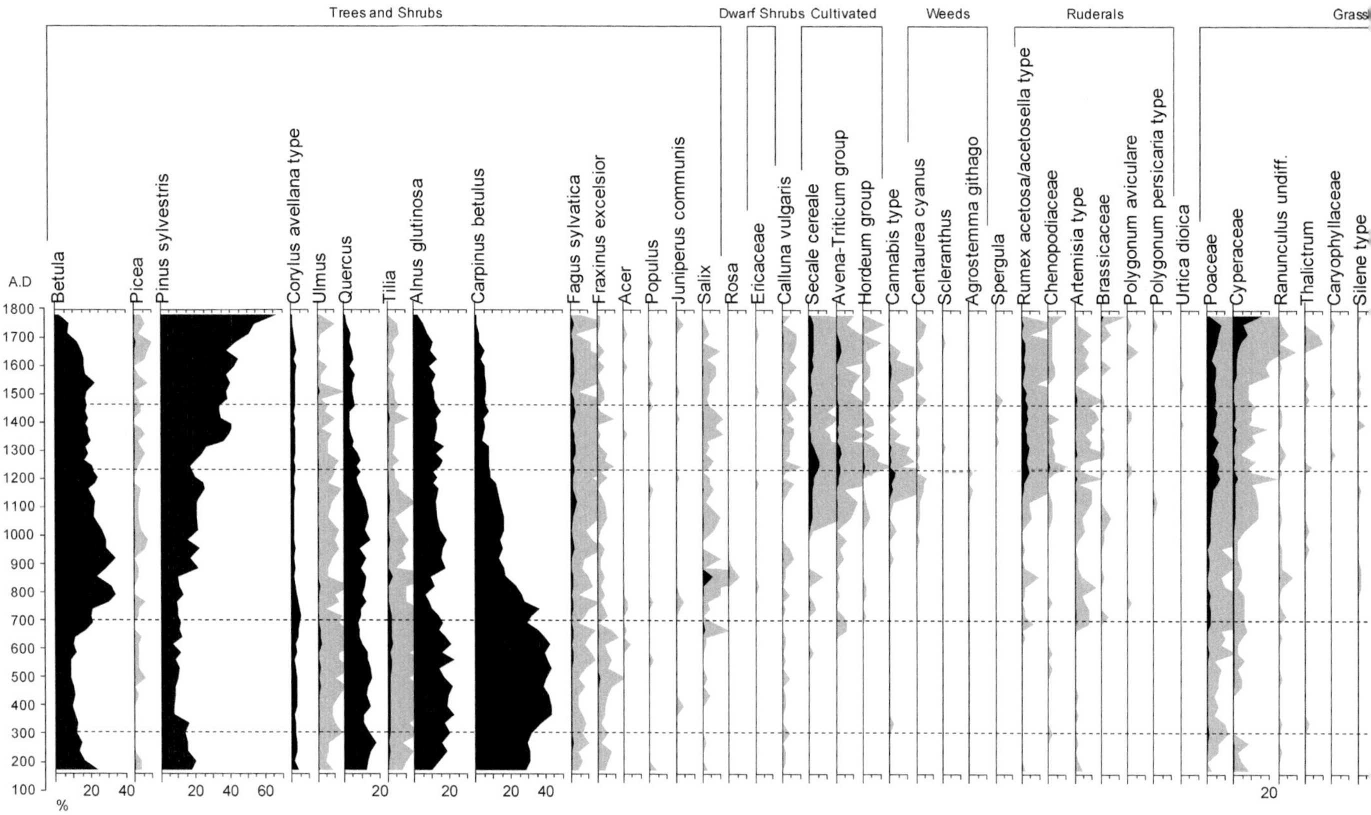

Figure 11.5: Rywałd, selected taxa percentage pollen diagram. Black curve = percentage of taxon, grey curve = values exaggerated x 10. Figure by author.

would have formed the lowest level of the woodland canopy, throwing dense shade with a relatively species poor ground flora, with the secondary canopy layer including oak, elm, maple, and lime. Intermittent values for pollen of ruderals, particularly common sorrel (Radzyń1) and cereal-type pollen, do suggest continued albeit small-scale mixed agriculture in the vicinity, perhaps within some grazing around the lake edge.

One of the characteristic features of the Radzyń pollen sequences and LOVE models is the marked decline in broad-leaved woodland species, particularly hornbeam, towards the end of the Migration Period (Figs 11.1–11.4). Hornbeam values decline significantly in Radzyń3 over the course of the seventh century (*c*. 36–19%), but more gradually in Radzyń2 over the course of the seventh to eighth centuries (*c*. 30–17%). Hornbeam values in Radzyń1 decline marginally during the seventh century (*c*. 36–28%) but not more significantly until the early ninth century (*c*. 32–17%). The decline in hornbeam-/oak-dominated woodland during the seventh to eighth centuries is accompanied by increases in cultivated (*c*. 1–1.5%), ruderal (*c*. 2%), and open/grassland (*c*. 8–10%) pollen taxa. The increase in birch values again reflects its ability to encroach on areas disturbed by human activity, particularly along the edges of woodland clearings. The pattern of small-scale mixed arable/pastoral land-use continues through to the mid-eleventh century when there is a marked increase in values for cultivated, weed, ruderal, and grassland pollen taxa. This is most apparent in the Radzyń3 sequence with increasing quantities of rye and oat-wheat, accompanied in particular by greater quantities of segetal plants, most notably cornflower. Small quantities of cornflower are apparent prior to the mid-eleventh century, but the increase from this point clearly reflects an overall intensification in agricultural activity. An increase in Cannabis-type pollen (up to 2%) may also reflect the cultivation of hemp. There is also a marked increase in common sorrel along with other ruderal taxa and grasses, emphasizing the increasing importance of meadows and pastures during the eleventh and twelfth centuries. The increasing openness of the surrounding landscape from the mid-eleventh century is reflected in declining arboreal pollen values, particularly for broad-leaved species. Hornbeam values decline to ≤5% by the early thirteenth century, similar to oak, suggesting a local but diminished presence. The growing pressure on woodlands is apparent also in the declining values for birch that although initially able to encroach around clearings from the seventh century, was increas-

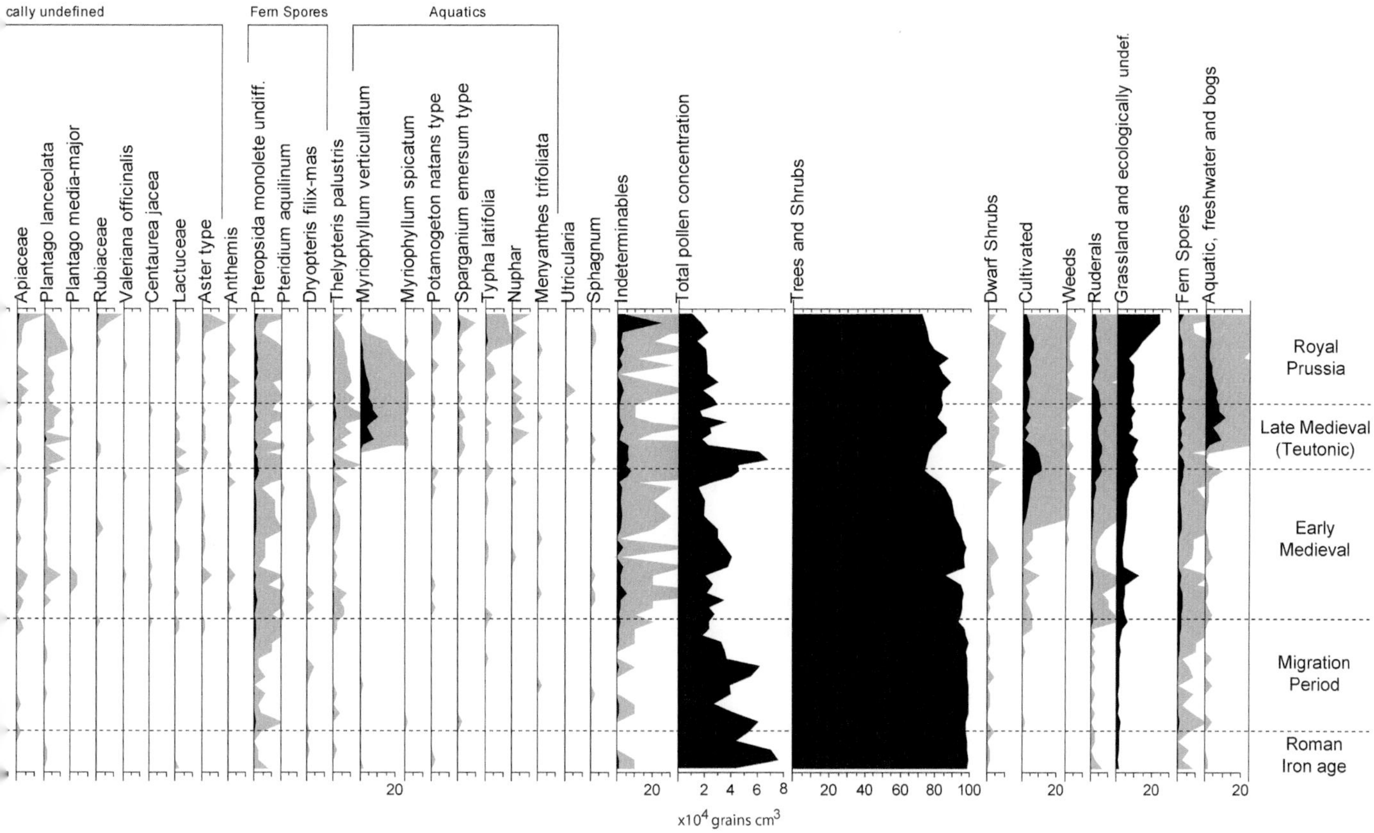

ingly cleared along with other broad-leaved trees as the requirement for agricultural land grew. The progressively open nature of the landscape surrounding the Castle Lake is clearly demonstrated by the LOVE models that show a significant increase in landscape openness to 60% by the twelfth century, with an overwhelming focus on meadows and pastures (Fig. 11.4).

The late medieval period presents a more complicated picture. In all three sequences values for cultivated plants, ruderals, and open-ground taxa remain high during the first half of the thirteenth century, reflected in the LOVE models, suggesting a picture of relative stability despite the possible unstable nature of the landscape during the initial decades of the crusades. The pollen indicates the presence of arable fields, pastures, and meadows surrounding the Castle Lake. The town was captured by the Prussians during the Great Prussian Uprising in the 1260s, although the castle remained in the hands of the Order. In all three Radzyń sequences there is a decline in pollen of cultivated taxa from the mid-/late thirteenth century whilst values for ruderal taxa remain stable but decline marginally during the fourteenth century, implying a reduction in the intensity of arable land-use despite the documented increase in settlements within the wider administrative region of Rehden. Increasing pine values may reflect a measure of colonization of open/disturbed environments, but also a growing dominance of pine within those remaining woodlands increasingly cleared of broad-leaved trees. However, the lack of pine stomata suggest few if any nearby stands of pine, at least around the western side of the lake near the castle and town. The presence of oat-wheat pollen does, however, suggest the presence of nearby arable fields, increasing in Radzyń3 during the fifteenth century.

Only the Radzyń3 sequence extends beyond the late medieval period. Cereal pollen values decline marginally whilst ruderal pollen increases, particularly for goosefoot, species of which are equally commonplace on cultivated as well as disturbed ground. The pollen signal clearly reflects a measure of continued stability in agricultural land-use during the late fifteenth to seventeenth centuries.

Rywałd (Figs 11.5–11.6)

In contrast to the evidence from Radzyń for human activity during the Roman Iron Age, pollen from the former lake at Rywałd is composed almost entirely of arboreal taxa (up to 99%), reflecting the local dominance of woodland at this time. Hornbeam and oak form the key components of the woodland with scattered lime, elm, ash and beech, and alder growing on wetter soils

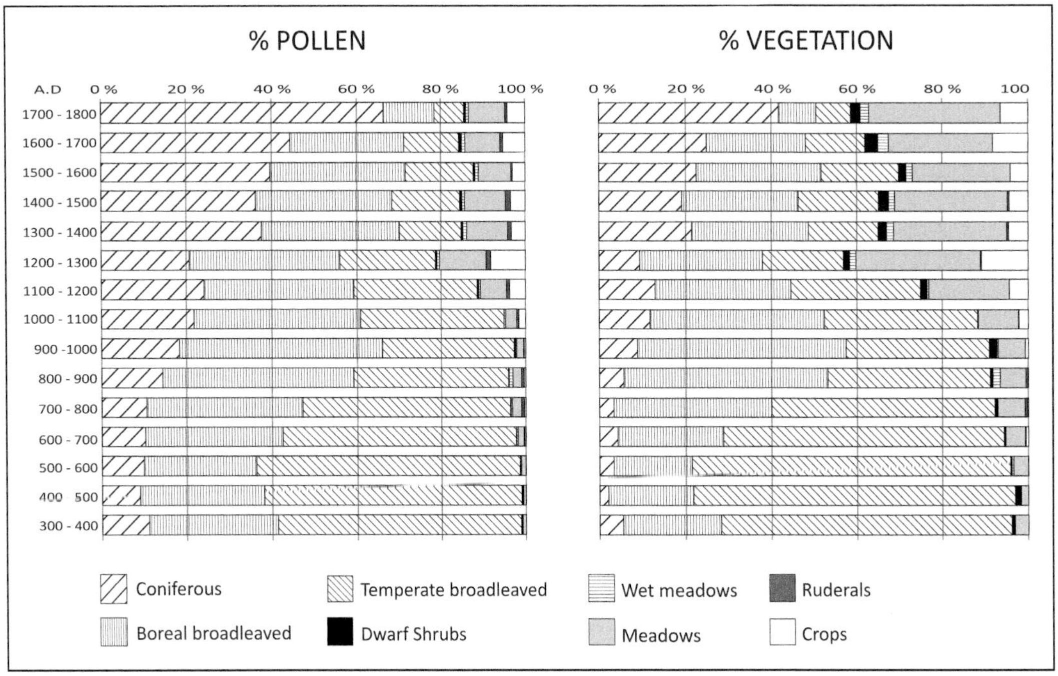

Figure 11.6: Rywałd, proportions of pollen taxa modelled using the Landscape Reconstruction Algorithm (LRA) (LOVE), summarized within eight groups: (1) coniferous trees; (2) boreal broadleaved trees; (3) temperate trees; (4) dwarf shrubs; (5) wet meadows; (6) meadows; (7) ruderal; (8) crops, and presented in hundred-year time slices. Regional Vegetation Reconstructions are based on pollen data from the Castle Lake and Lake Rozlewisko (Figure 2.5). Model by A. Poska.

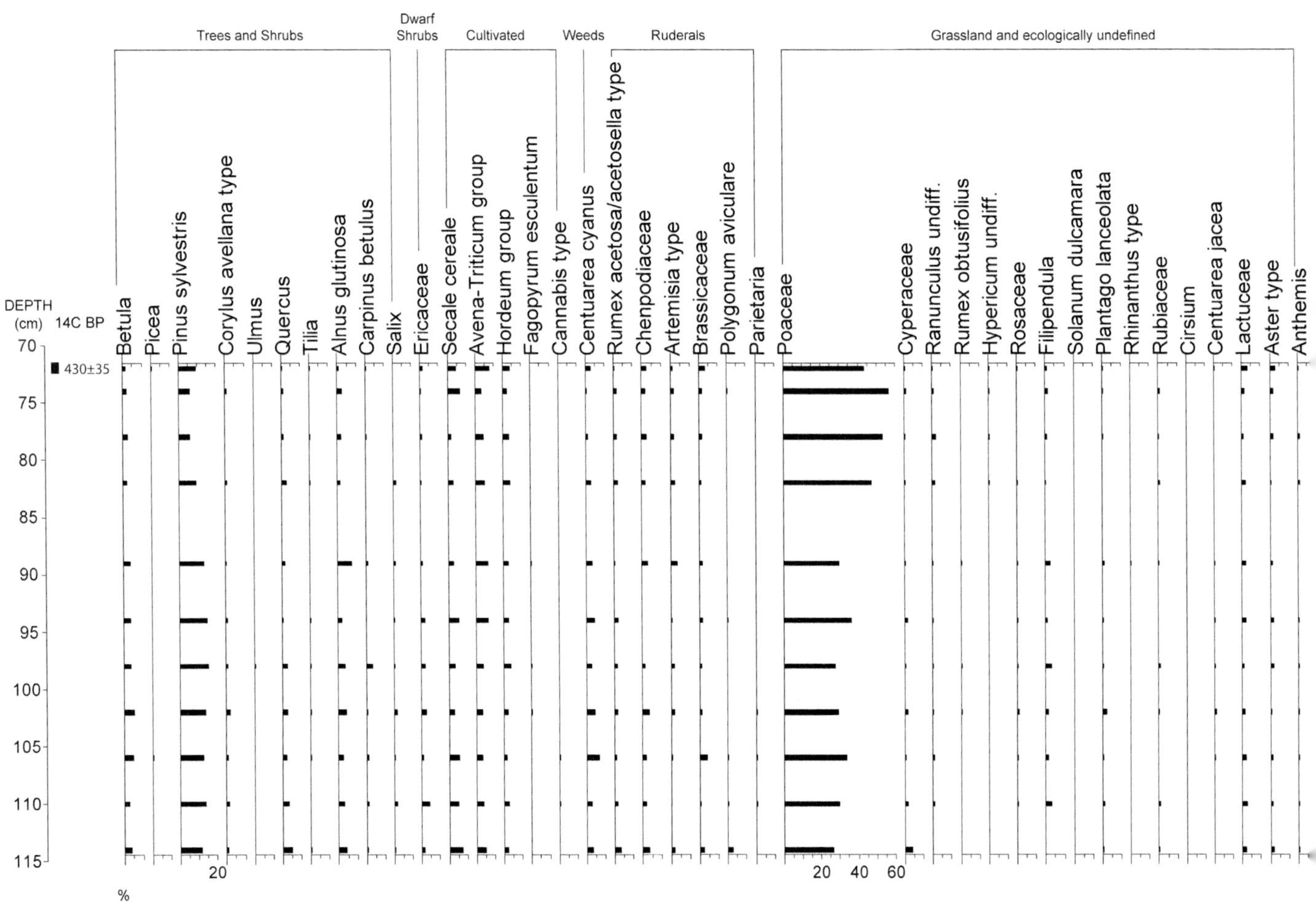

Figure 11.7: Malbork Castle moat, selected taxa percentage pollen diagram. Black bars = percentage of taxon.

around the lake. There is little evidence for human activity in the vicinity, although the presence more widely in the landscape of open/disturbed ground is suggested by high values for the profuse and well-dispersed pollen of light-demanding trees such as pine and birch. Values for pine and birch decline, however, with the transition to the Migration Period. High values for hornbeam (up to 43%) further emphasize its dominance of the woodland environment at this time. LOVE modelling reinforces the picture of a heavily wooded landscape dominated by temperate mixed broad-leaved woodland, but with some indications for small increases in landscape openness (<5%) from the sixth century (Fig. 11.6). Hornbeam and oak decline from the seventh century with an increase in birch reflecting renewed clearance within the surrounding woodland landscape. However, the relatively low quantities of cereals, ruderals and grassland pollen suggest relatively small-scale agricultural activity in the vicinity with the major changes in woodland composition reflective of broader processed occurring across the wider landscape. This is supported by the LOVE model which shows only a small increase in levels of landscape openness (<10%) with the surrounding landscape still dominated by woodland. The principal phase of woodland clearance occurs during the course of the seventh–ninth centuries, after which values for hornbeam stabilise and oak even increases. Pine increases gradually from the tenth century, again suggesting some colonization of open/disturbed habitats, though the absence of pine stomata in the Rywałd sediments does not point towards a significant local presence. Human activity is clearest from the mid-eleventh century, with a distinct increase in pollen of cereals, particularly rye and wheat/oat (up to 2.6% an 2% respectively during the eleventh and twelfth centuries), accompanied by increasing quantities of ruderal, weed, and grassland taxa. Landscape openness increases noticeably during the twelfth century, both in the quantities of arable and particularly meadowland. Cultivation of hemp is also suggested by the distinct increase in Cannabis-type pollen (max 3.5%) during the late twelfth/early thirteenth centuries. There is a notable increase in cereal pollen during the early thirteenth century, with rye increasing to 6% and oat-wheat to over 3%; values for the latter clearly reflect the presence of nearby agricultural fields since wheat and oat only produce very small quantities of large pollen grains which are transported only short distances from source. The combination of cereals emphasizes the development of an intensive system of permanent rotational field systems involving the cultivation of both winter and summer crops. However, values for cereals decline by the late thirteenth century, although ruderal and grassland taxa feature prominently, with LOVE modelling emphasizing the sustained importance of meadows and pastures during and after the late medieval period. Broad-leaved woodland was cleared further during the fourteenth century, although pollen values for hornbeam and oak and the LOVE model suggest they still formed an important component of remnant woodland surviving around the lake. The increasing quantities of pine again reflect a measure of both colonization of open/disturbed ground and an increasing contribution to remnant woodland. There is a considerable measure of continuity in land-use during the fourteenth to sixteenth centuries with little fluctuation in the values for cultivated, ruderal, weed, and grassland taxa, with landscape openness of around 30%. There is evidence for renewed clearance of broad-leaved woodland during the seventeenth and eighteenth centuries, but with no overall increase in landscape openness and instead a large increase in coniferous woodland, primarily represented by pine.

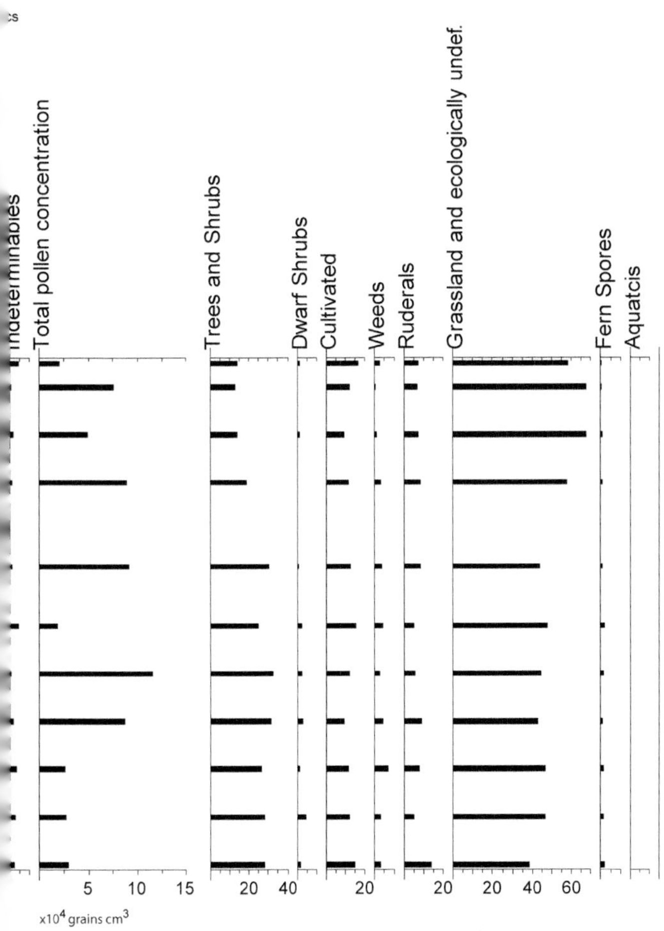

Marienburg Commandery

Malbork Castle Moat (Fig. 11.7)

Pollen from the outer moat adjacent to the inner curtain wall focused on a series of organic rich silt layers between 1.17–0.72 m. These are considered to reflect the ponding of fine sediment within a low energy environment, separated by sand bands deposited under higher energy environments, perhaps related to periodic flushing of the moat. The date of moat construction has been well established to *c.* 1320–40; a sample from 0.715–0.72 m dated 430±30 BP (cal. AD 1410–1520, 88.2%) is consistent with the historical date. However, an earlier date on seeds from 0.94–1.00 m of 899±29 BP (cal. AD 1040–1212) predates the construction of the castle, suggesting the likely erosion and inclusion of earlier material into the moat. Pollen within the moat is characterized by high non-arboreal pollen values, particularly grasses, but also a range of cultivated plants and members of the mustard and carrot family perhaps reflecting herbs and vegetables. Areas of pasture and disturbed ground are suggested by pollen of grasses, ribwort plantain, buttercups, and the daisy and lettuce family. Arboreal pollen is represented mainly by pine growing within a predominantly open landscape. Stands of alder, willow, and birch are likely to be growing in wetter areas.

Malbork High Castle (Table 11.1)

Pollen was only preserved in one sample from core MHC1 from organic sandy sediments containing charcoal, sealed by brick rubble. Significant quantities of pollen were preserved, dominated in particular by pollen of mustards, celery, and the carrot and rose family. The atypical nature of the pollen assemblage suggests that the organic layer, located in the high castle, may have partially functioned as an herb/vegetable garden. There are high values for willow (9%), most probably growing on wet ground along the Nogat, with indications of isolated stands of oak, maple, lime, birch, and pine in the surrounding landscape. The wide range of herbaceous pollen types also suggests pastures/meadows in the vicinity of the castle.

Parpary (Figs 11.8–11.9)

The Migration and early medieval periods are characterised by the dominance of hornbeam-oak woodland with little evidence for clearance until the fourteenth century. Pine, hazel, elm, and lime form consistent components of the woodland, with alder on wetter soils. There are occasional cereal, weed, and ruderal pollen that suggest

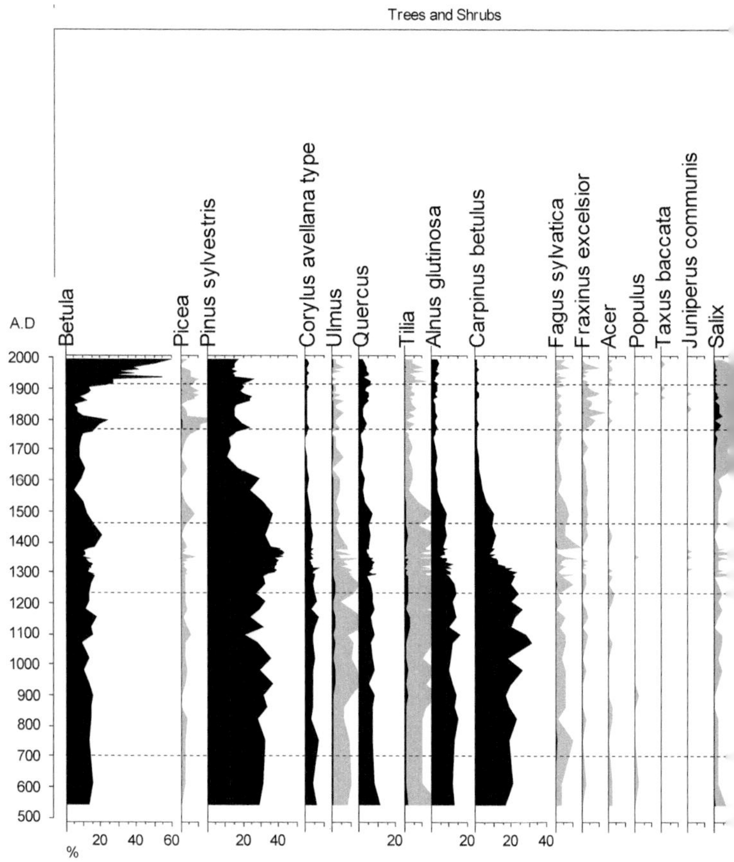

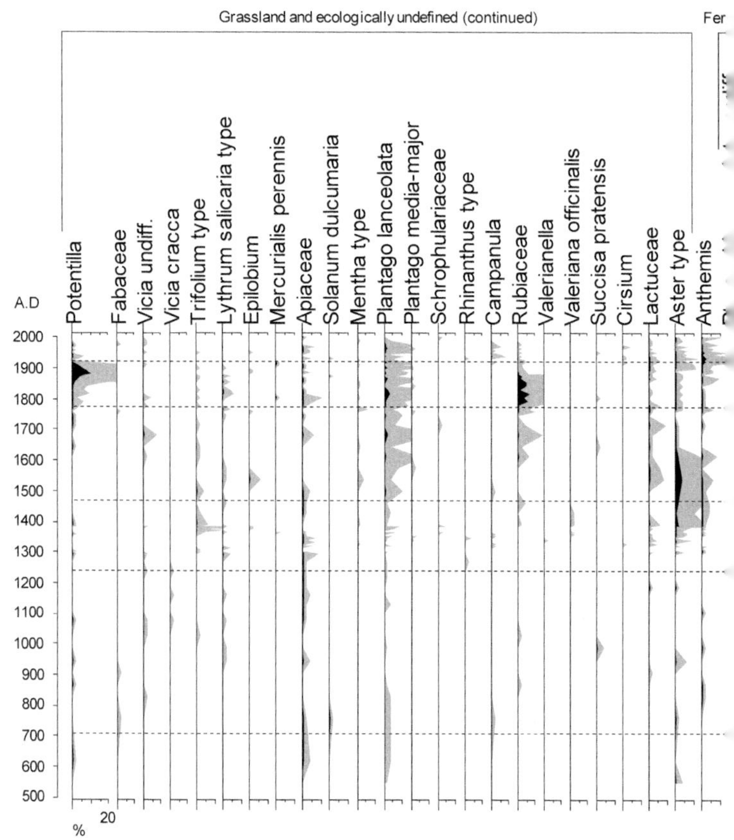

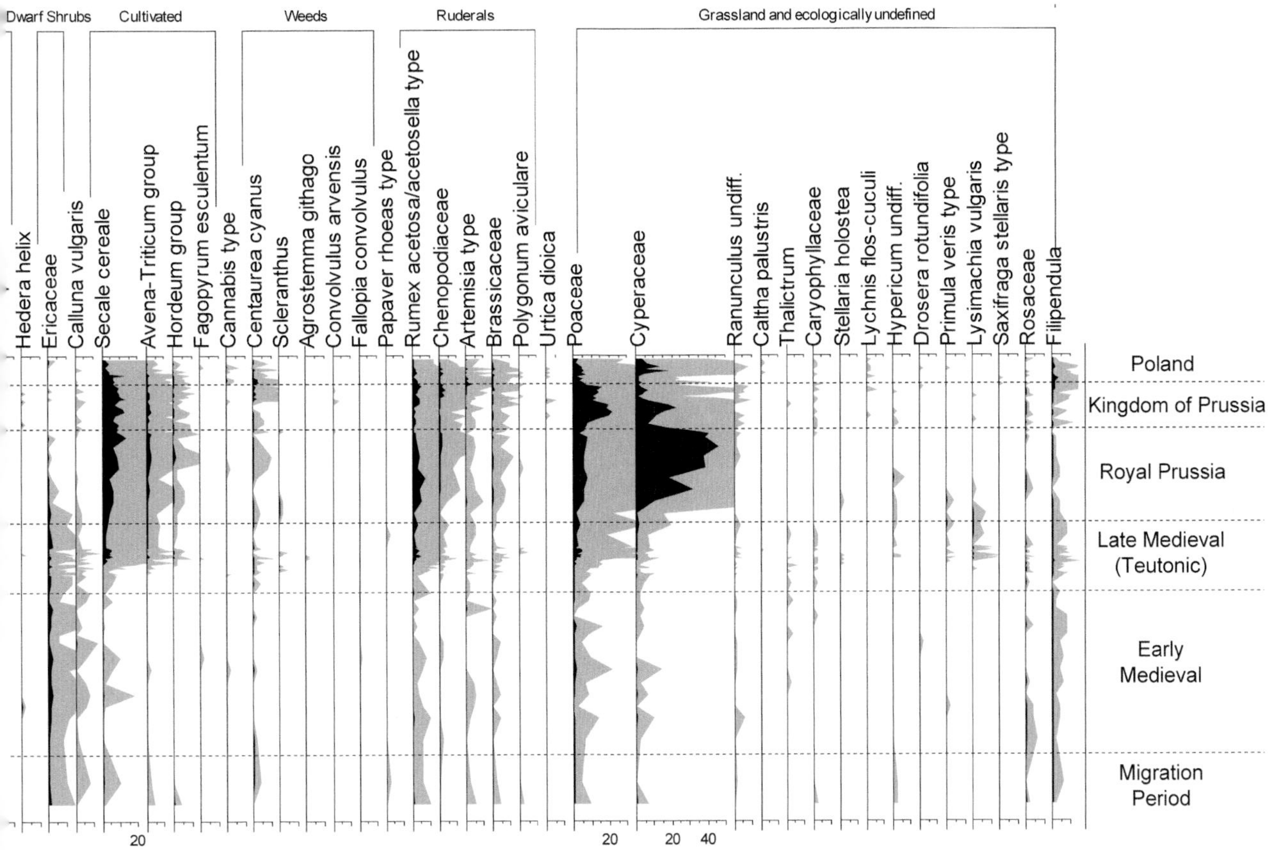

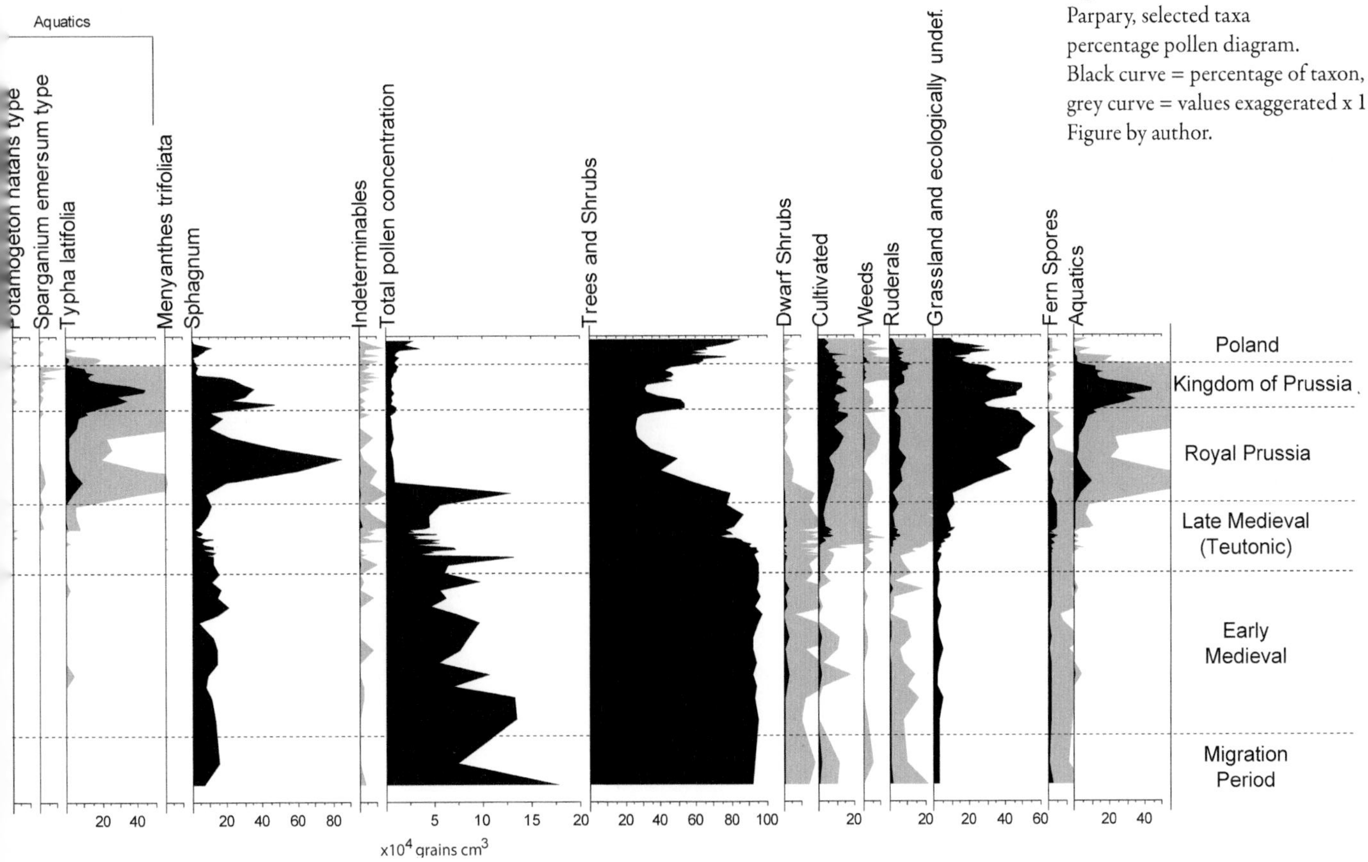

Figure 11.8 (above and below): Parpary, selected taxa percentage pollen diagram. Black curve = percentage of taxon, grey curve = values exaggerated x 10. Figure by author.

small-scale, intermittent land-use in the surrounding landscape during the early medieval period. More substantial changes to the vegetation only occur from the early fourteenth century with a decrease in hornbeam and notable increases in cereal, weed, ruderal, and grassland pollen. LOVE modelling of the Parpary sequence shows landscape openness increasing to *c.* 15% in the fourteenth century with a significant proportion of this used for crop cultivation; permanent field-systems of winter and summer crops are suggested along with some meadows and pastures. Values for hornbeam and oak indicate they remained important components of a surrounding landscape still heavily wooded. The relatively stable oak values could suggest selective clearance and active management of woodland focusing on the removal of hornbeam within the lowest layers of the canopy. The most extensive phase of woodland clearance occurs from the sixteenth century with a reduction in pollen of all arboreal taxa and a significant increase in pollen of rye and higher values for grassland and ruderal communities. LOVE modelling shows only modest increases in landscape openness to *c.* 20%, most of this reflected in an increase in meadowland. There is a massive increase in wet-meadow during the sixteenth to eighteenth centuries, with an increase of mosses, reeds, and bulrushes reflecting the highly localized vegetation around the sampling location; this is most probably a consequence of increasing woodland clearance and run-off of water and mineral sediment into the basin favouring the development of tall herb swamp. Cartographic data from 1913 demonstrate Parpary was located on the edge of the woodland, though now surrounded by conifer plantation with birch growing around the sampling location and increasing in importance during the twentieth century.

Benowo (Figs 11.10–11.11)

Located 1.2 km from the current western edge of the Forest of Sztum, the pollen sequence from Benowo demonstrates a pattern of woodland continuity. Similar to Parpary, hornbeam forms an important component of the woodland canopy along with oak and pine and more scattered elm, lime, beech, ash, and birch. The presence of light-demanding species such as pine, hazel, and oak indicates a mosaic of woodland habitats including areas of more open and closed canopy. The high values for *Sphagnum* moss, and increasing values for heather from the eighth century, suggest areas of more open woodland particularly around boggy areas, but also around the margins of the Forest. Hornbeam declines gradually over the course of the eighth to eleventh centuries, although values remain high and suggest a considerable presence within the woodland. Indications of arable and pastoral activity during the early medieval period are provided by occasional cereal pollen grains of rye, wheat/oat, and barley along with intermittent weed and ruderal pollen taxa. Changes in woodland structure from the eleventh/twelfth centuries are apparent in the decline in hornbeam and an increase in wet mire communities, with alder growing on wetter soils and an increase in

Table 11.1: Malbork (Ger. Marienburg) high castle, selected taxa pollen percentages.

Taxa	%
Poaceae	1.8
Thalictrum	1.6
Silene type	1.0
Rosaceae	8.1
Trifolium type	2.4
Lythrum salicaria type	2.0
Apiaceae	11.3
Rubiaceae	2.0
Centaurea scabiosa	1.0
Centaurea jacea	2.0
Anthemis	1.4
Pinus sylvestris	1.8
Quercus	4.5
Salix	9.1
Rosa	1.2
Calluna vulgaris	2.0
Avena-Triticum group	0.6
Hordeum group	0.2
Centaurea cyanus	0.4
Rumex acetosa/acetosella type	0.8
Brassicaceae	25.1
Apium	9.1

	%
Trees and shrubs	**20.9**
Dwarf shrubs	**2.6**
Cultivated	**0.8**
Field weeds	**1.2**
Ruderal (incl. possible herbs/vegetables)	**34.6**
Grassland and ecologically undefined	**39.9**

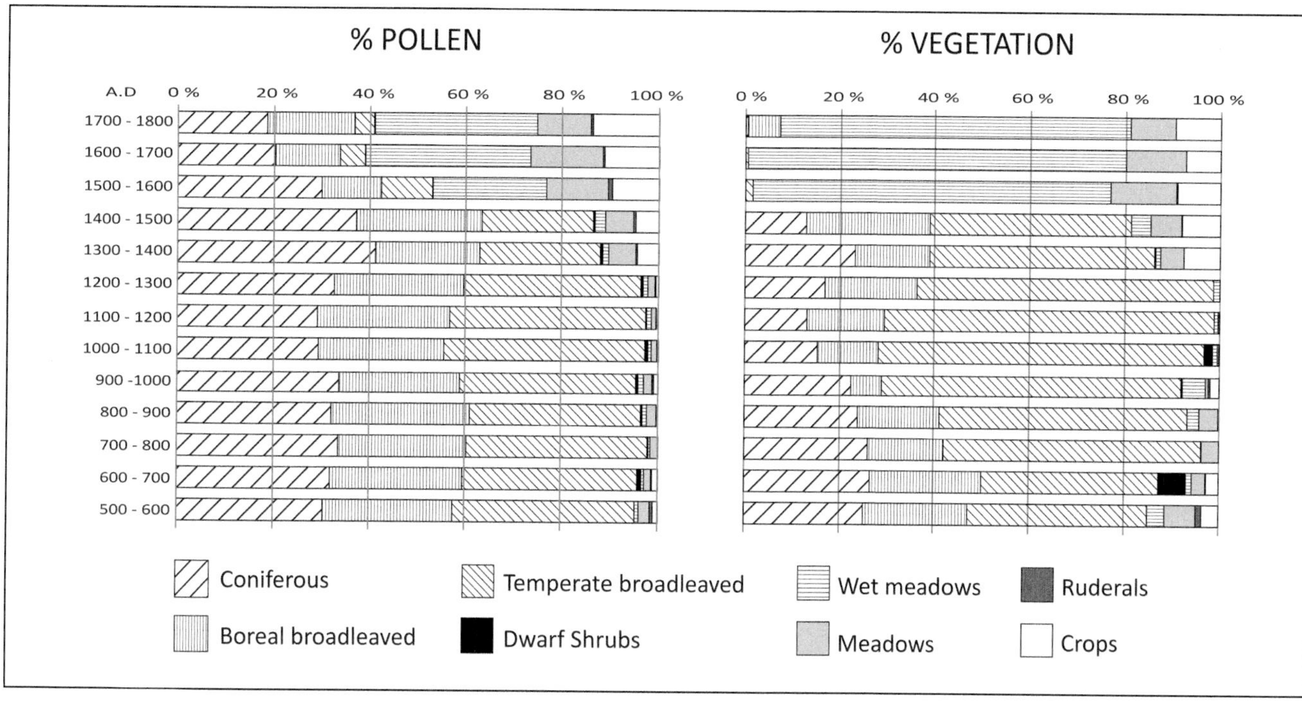

Figure 11.9: Parpary, proportions of pollen taxa modelled using the Landscape Reconstruction Algorithm (LRA) (LOVE), summarized within eight groups: (1) coniferous trees; (2) boreal broad-leaved trees; (3) temperate trees; (4) dwarf shrubs; (5) wet meadows; (6) meadows; (7) ruderal; (8) crops, and presented in hundred-year time slices. Regional Vegetation Reconstructions are based on pollen data from the Castle Lake and Lake Rozlewisko (Figure 2.5). Model by A. Poska.

Sphagnum moss and development of heather, probably forming drier hummocks in the mire. LOVE modelling emphasizes the importance of dwarf shrub communities over the course of the eighth to fifteenth centuries with little or no evidence for human activity apart from a small increase in meadowland during the eleventh century (Fig. 11.11). Woodland still remains the dominant vegetation, although increasingly comprising those more light-demanding species. Oak values remain stable and may reflect selected preservation or management of oak, or perhaps the presence of wood pasture. There are indications of arable and pasture/meadows in the surrounding landscape during the late medieval period. The peak in cornflower during the mid- to late fourteenth century (up to 3.7%) (Fig. 11.10) is strongly suggestive of nearby arable fields, most probably surrounding colonizing settlement located along the edges of the Forest of Sztum; the nearest is located 1.5 km south-south-west at Benowo. Arable and pastoral land-use increases during the sixteenth to eighteenth centuries, with a further decline in broad-leaved trees at the end of the seventeenth century. Pine increases during the eighteenth century, with a large increase in birch during the nineteenth century likely to reflect some colonization of open areas within the woodland.

Lake Rozlewisko (Figs 11.12–11.13)

The pollen sequence covers the fifth to mid-fifteenth centuries and shows a similar regional woodland environment during the Migration Period in which hornbeam and oak again formed the principal components. Clearance of woodland is apparent from the late seventh/early eighth centuries with the characteristic decline in hornbeam pollen, accompanied by a continuous curve for cereal pollen and a range of ruderal and grassland taxa. The initial decline in woodland is followed by consistent values for arboreal, cultivated, ruderal, and grassland pollen over the course of the eighth to twelfth centuries, suggesting some stability in land-use in the surrounding landscape during the early medieval period. LOVE modelling shows meadowland forming an important component of an increasingly open landscape from the tenth century, with landscape openness of *c.* 40%. There is evidence for cultivation of both winter and summer crops, along with hemp, with a distinct peak in rye pollen during the early thirteenth century (up to 3.5%) at the transition between the early and late medieval periods. Rye pollen values subsequently decline and are generally around 2% during the thirteenth to fifteenth centuries. Hornbeam and oak are almost com-

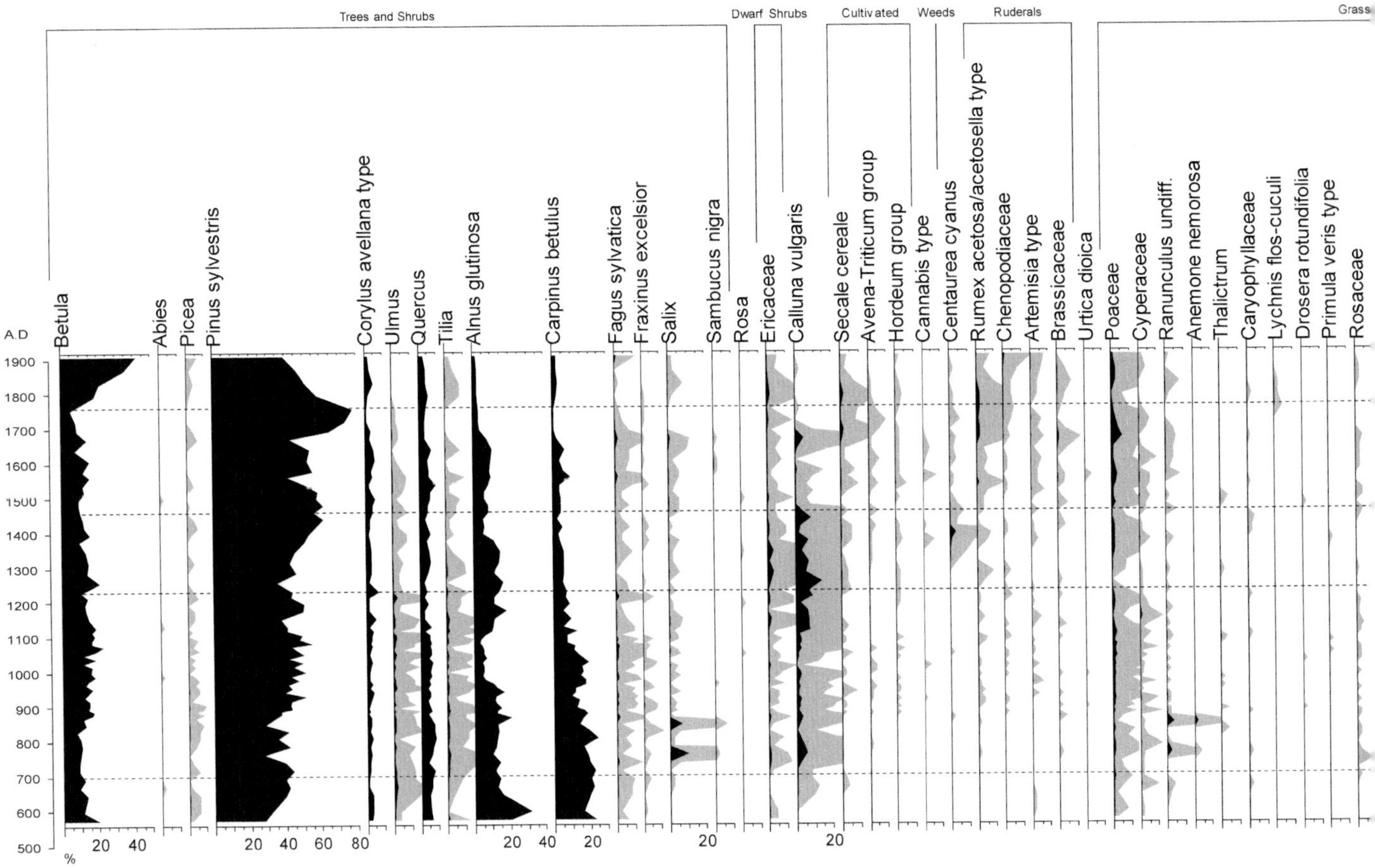

pletely cleared during the fourteenth century along with other broad-leaved trees, suggesting an increasingly open agricultural landscape. Stands of alder and willow persist around the lake, whilst pine increases as a component of the wider landscape, colonizing within open/disturbed areas. LOVE modelling shows that wet meadow communities formed an important component of the surrounding environment, particularly from the fifteenth century, most likely reflecting localized stands of tall herb swamp ringing the lake.

Elbing Commandery

Elbląg Castle (Fig. 11.14)

Thirteenth-century deposits from within the courtyard of the outer bailey at Elbląg Castle contained significant quantities of well-preserved pollen. The samples within the top 0.12 m of the sequence show a high degree of uniformity in the pollen sequence with large quantities of grass pollen up to *c.* 60% and significant quantities of wheat/oat pollen (up to 27.6%). There are large quantities of pollen of weed taxa, particularly cornflower (up to 29.2%), along with corncockle, knawels, and corn spurrey. Pollen of buttercups and chamomiles also occur in large quantities, particularly within the occupation layer. The occupation deposits contain large quantities of animal dung and are likely to reflect a combination of fodder comprising hay and agricultural waste, but perhaps also reflecting hay used in flooring of the courtyard or nearby storage of agricultural produce. There are slightly higher pine values at 1.82 m within the alluvial deposits; pine is typically over represented in alluvial samples because of the profuse production, long-distance transport and buoyancy of the pollen grains (Bennett 1984), and is unlikely to reflect a significant local presence. Pollen samples from 2.05–2.09 m contain significantly higher quantities of wheat/oat (67–80%), with a smaller component of weed, ruderal, grass, or other herb pollen. Cellular material from cereal husks is visible on the pollen slides along with clumps of cereal pollen grains. The samples derive from a trampled surface including dung and probably include material again reflecting agricultural waste used as animal feed, but also perhaps nearby storage or processing (e.g., threshing) of agricultural products within associated out-buildings. The samples towards the base of the sequence occur within probable floor deposits and likewise include high values

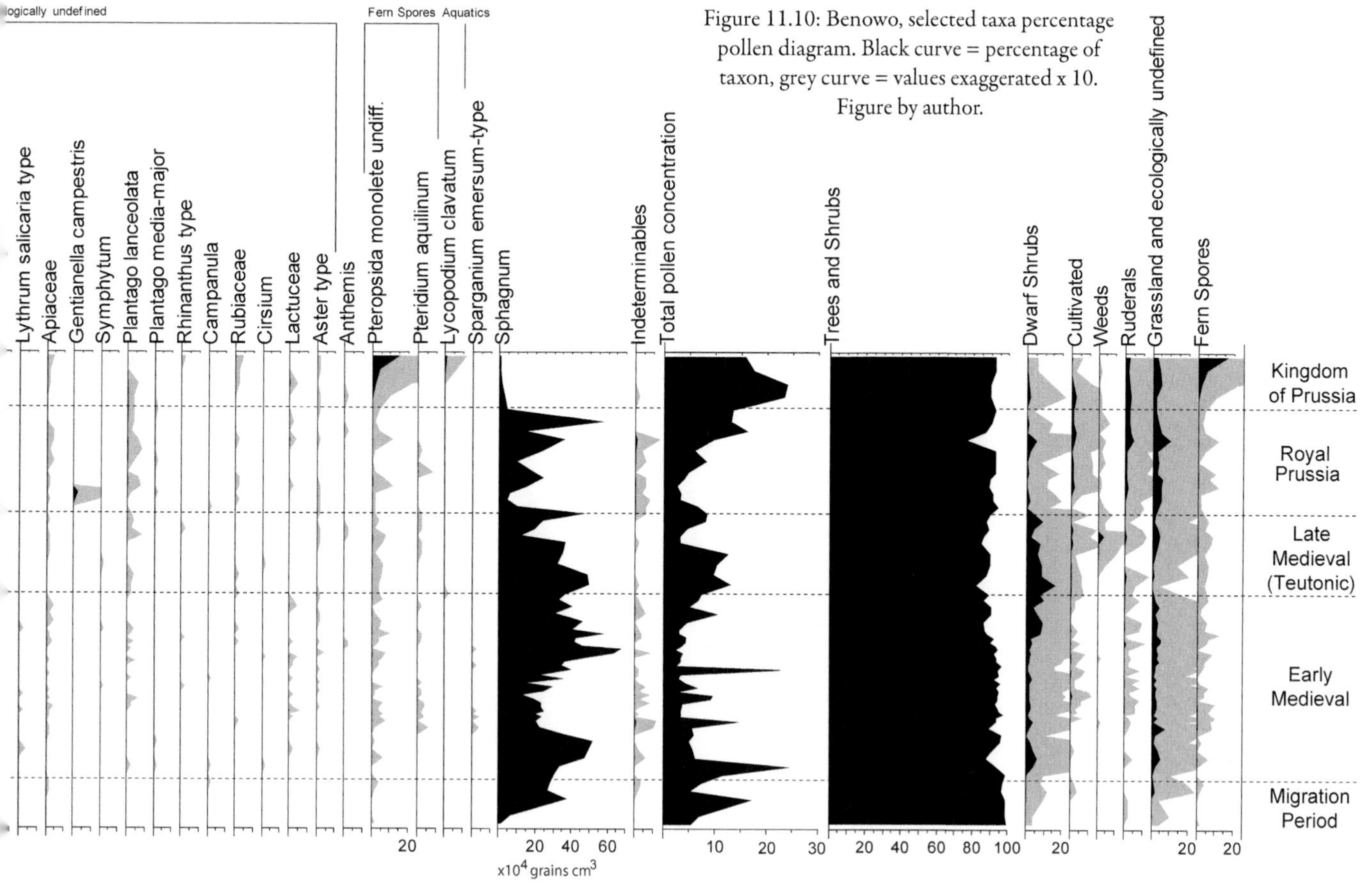

Figure 11.10: Benowo, selected taxa percentage pollen diagram. Black curve = percentage of taxon, grey curve = values exaggerated x 10. Figure by author.

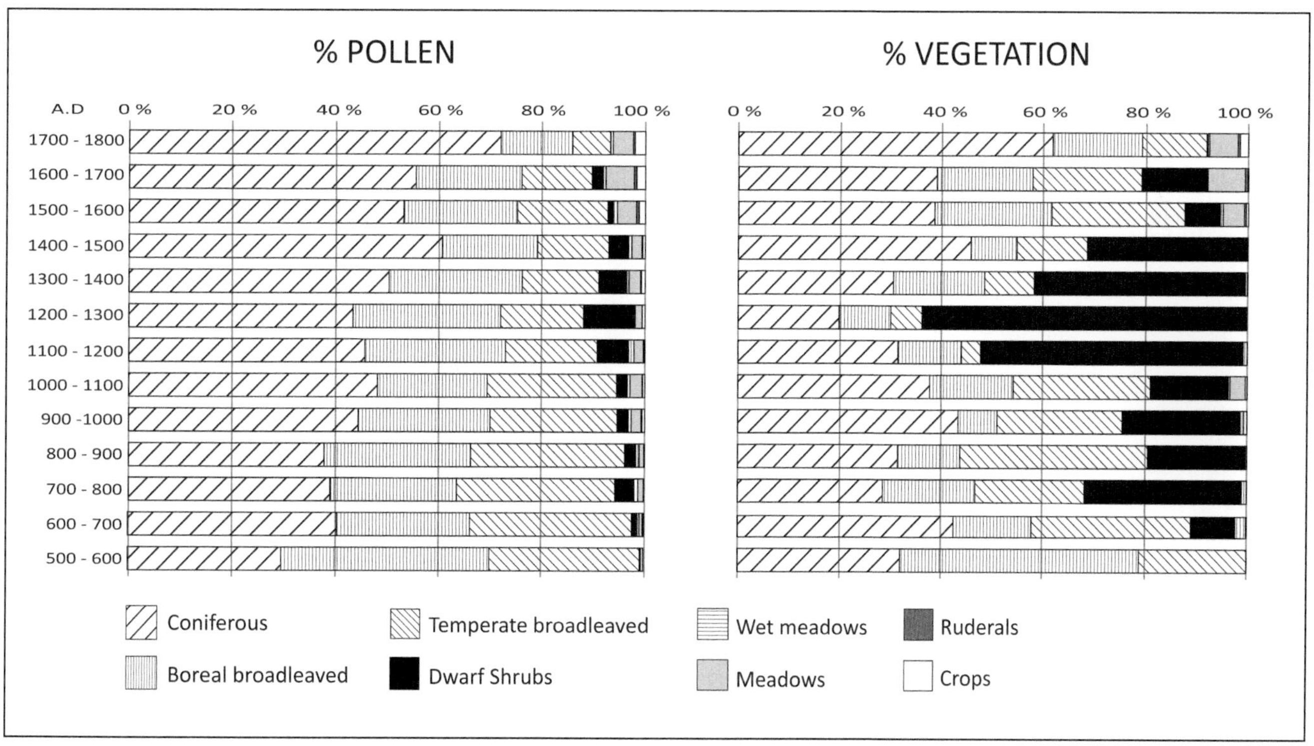

Figure 11.11: Benowo, proportions of pollen taxa modelled using the Landscape Reconstruction Algorithm (LRA) (LOVE), summarized within eight groups: (1) coniferous trees; (2) boreal broad-leaved trees; (3) temperate trees; (4) dwarf shrubs; (5) wet meadows; (6) meadows; (7) ruderal; (8) crops, and presented in hundred-year time slices. Regional Vegetation Reconstructions are based on pollen data from the Castle Lake and Lake Rozlewisko (Figure 2.5). Model by A. Poska.

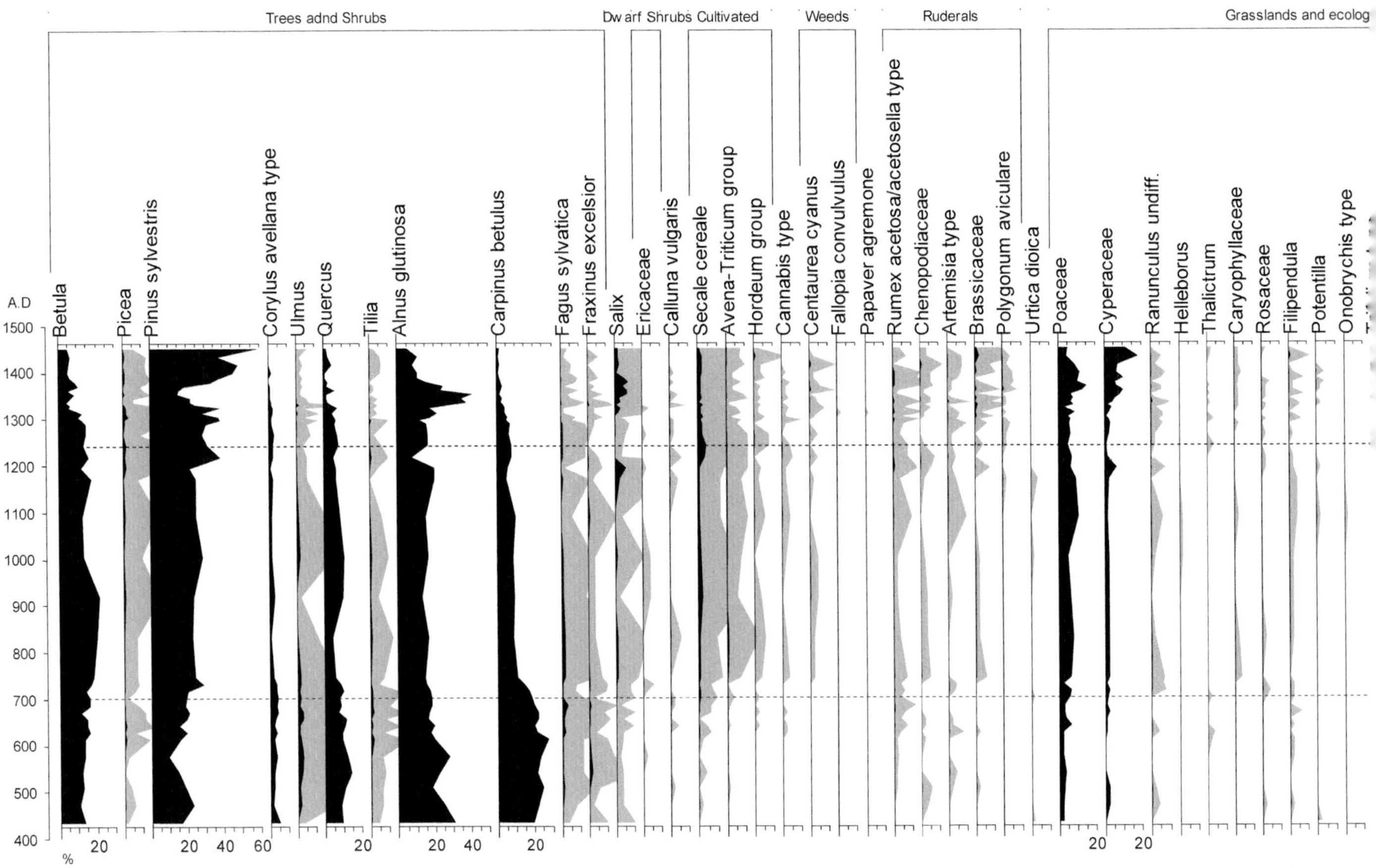

for wheat/oat pollen, with lesser values of grasses and only very small quantities of weeds, ruderals, and other herbaceous plants. The higher quantities of pine within the top of the floor deposit may indicate some mixing of alluvial sediment from the overlying layer where, as previously indicated, pine is often over represented.

Lake Drużno (Fig. 11.15)

Palynomorphs are well preserved, but the high mineral content of the sediments and inclusion of mollusc shells presented a significant challenge for dating. Pollen concentrates from 2.00 m produced an erroneously old date of 2995±34 BP (GU-30864, 1337–1124 cal. BC), reflecting the influence of hard and freshwater reservoir effects on the lake sediments. Subsequent attempts to U-series date mollusc shells also proved unsuccessful, with samples from five levels producing consistently young dates with large uncertainties (Appendix B, Table B.6). However, it is possible to make some comment on the surrounding vegetation, although the large size and pollen catchment of the lake basin will reflect the regional vegetation environment. Arboreal pollen dominates with large quantities of pine along with alder, birch oak and hornbeam; pine is typically over-represented, particularly in lake and estuarine settings, due to the profuse production, dispersal and buoyancy of its pollen grains. Alder would have been growing on wetter soils along the lake margins with oak and hornbeam forming important components of the surrounding woodland on the dry ground. There are two distinct peaks in hornbeam, the latter increase (2.05–1.65 m) dating to the Migration Period; the subsequent decline in hornbeam is well dated to the seventh century onwards and is accompanied by an increase in pollen of cultivated plans, including rye and cannabis-type, and a range of ruderal pollen taxa reflecting increasing agrarian use of the surrounding landscape. Values for cultivated, ruderal, and open habitats increase towards the top of the sequence (*c.* 1.30–1.15 m), probably dating to the post-medieval period.

Brandenburg Commandery/Lötzen Procuratorship

Zielone Bagno (Fig. 11.16)

Pollen analysis indicates a heavily wooded landscape throughout the majority of the Zielone Bagno sequence. During the Early Iron Age and subsequent Roman Iron Age, woodland comprises pine and birch, with

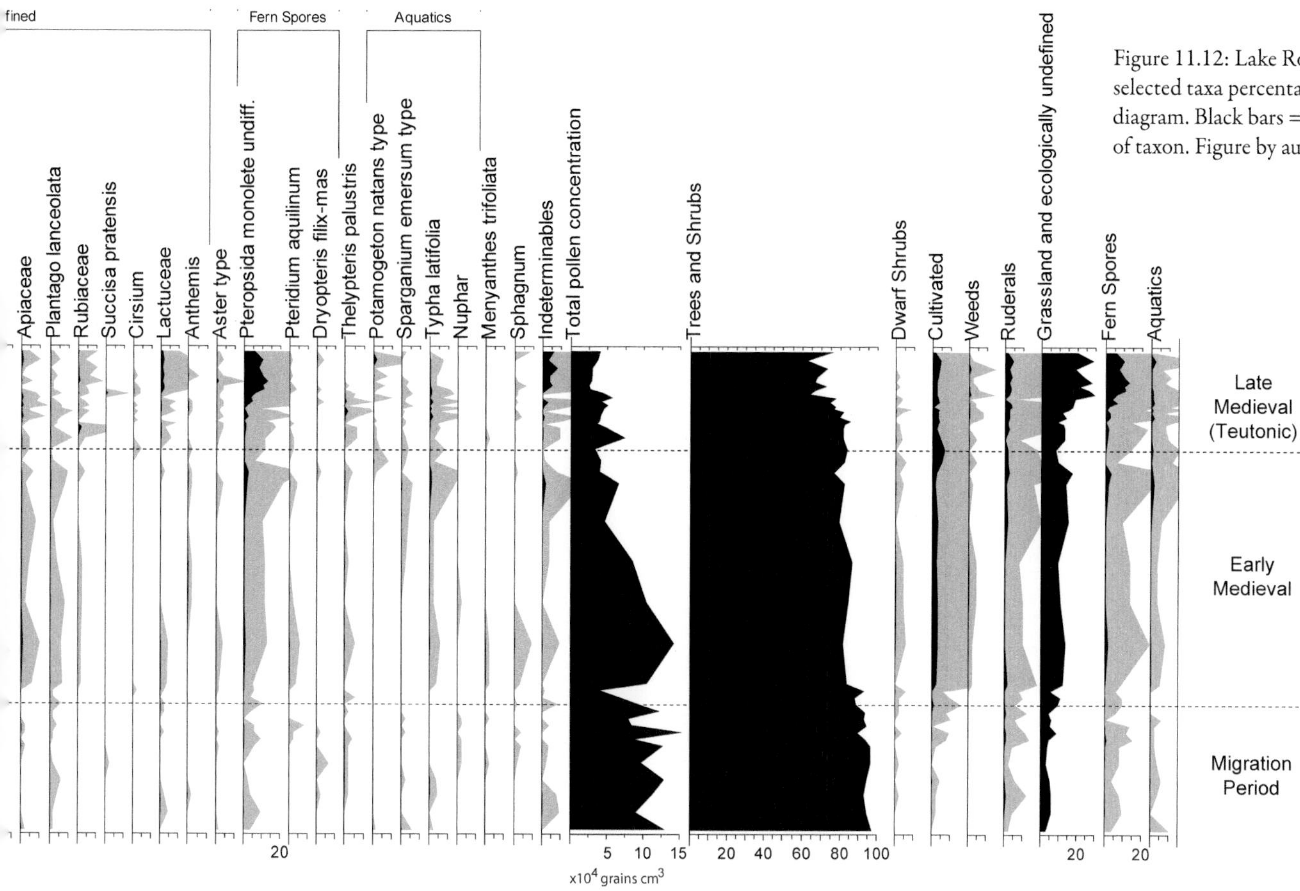

Figure 11.12: Lake Rozlewisko, selected taxa percentage pollen diagram. Black bars = percentage of taxon. Figure by author.

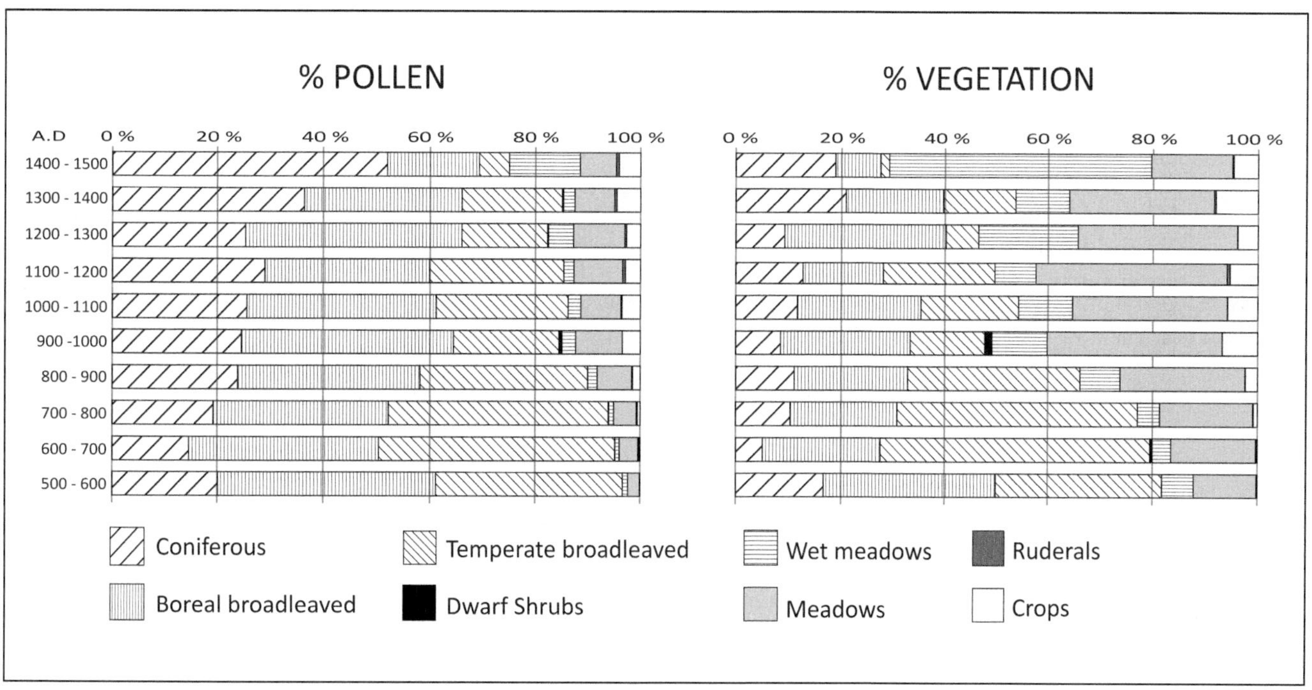

Figure 11.13: Lake Rozlewisko, proportions of pollen taxa modelled using the Landscape Reconstruction Algorithm (LRA) (LOVE), summarized within eight groups: (1) coniferous trees; (2) boreal broad-leaved trees; (3) temperate trees; (4) dwarf shrubs; (5) wet meadows; (6) meadows; (7) ruderal; (8) crops, and presented in hundred-year time slices. Regional Vegetation Reconstructions are based on pollen data from the Castle Lake and Lake Rozlewisko (Figure 2.5). Model by A. Poska.

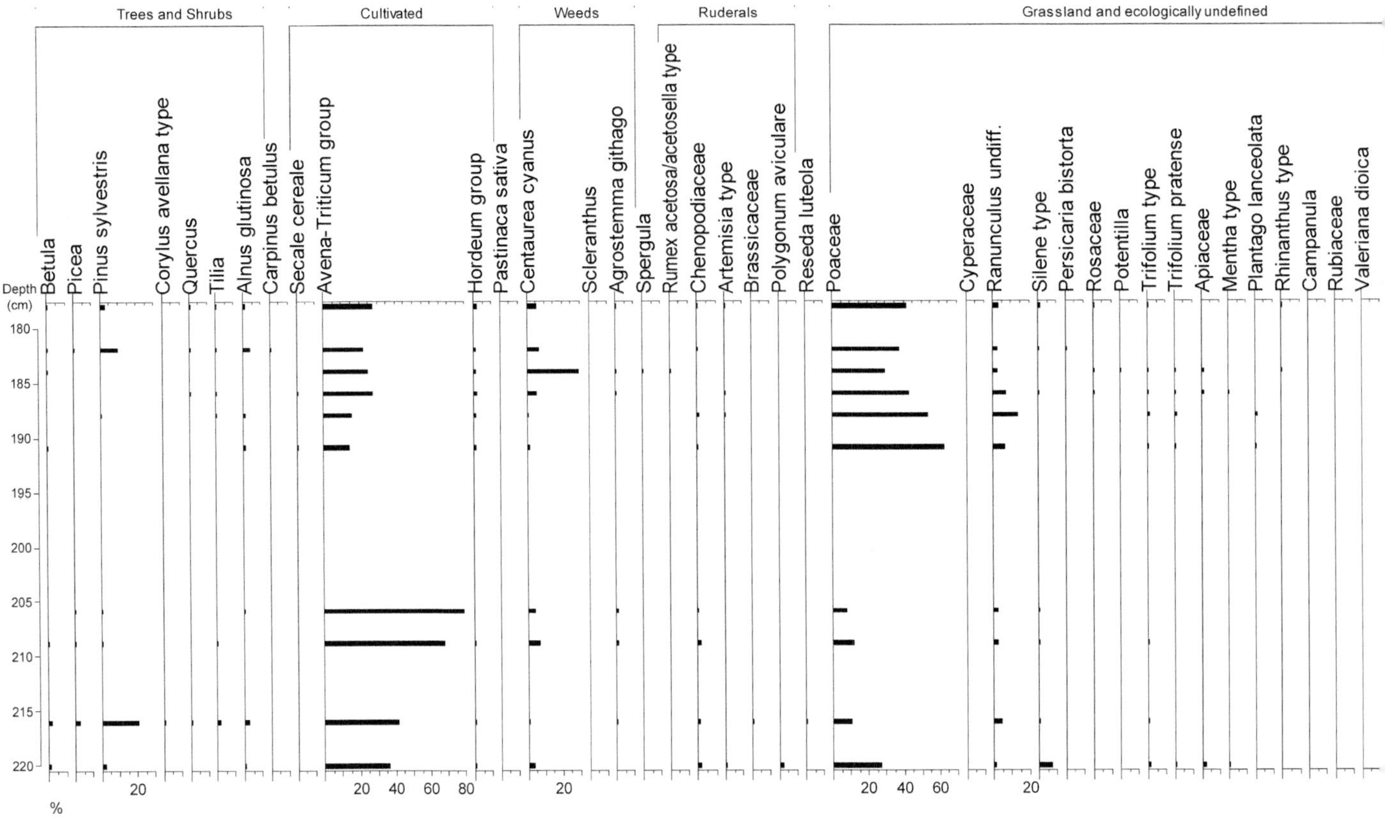

Trees and Shrubs
Cultivated
Weeds
Ruderals
Grassland and ecologically undefined
Depth (cm)
Betula
Picea
Pinus sylvestris
Corylus avellana type
Quercus
Tilia
Alnus glutinosa
Carpinus betulus
Secale cereale
Avena-Triticum group
Hordeum group
Pastinaca sativa
Centaurea cyanus
Scleranthus
Agrostemma githago
Spergula
Rumex acetosa/acetosella type
Chenopodiaceae
Artemisia type
Brassicaceae
Polygonum aviculare
Reseda luteola
Poaceae
Cyperaceae
Ranunculus undiff.
Silene type
Persicaria bistorta
Rosaceae
Potentilla
Trifolium type
Trifolium pratense
Apiaceae
Mentha type
Plantago lanceolata
Rhinanthus type
Campanula
Rubiaceae
Valeriana dioica
180
185
190
195
200
205
210
215
220
%
20
40
60
80

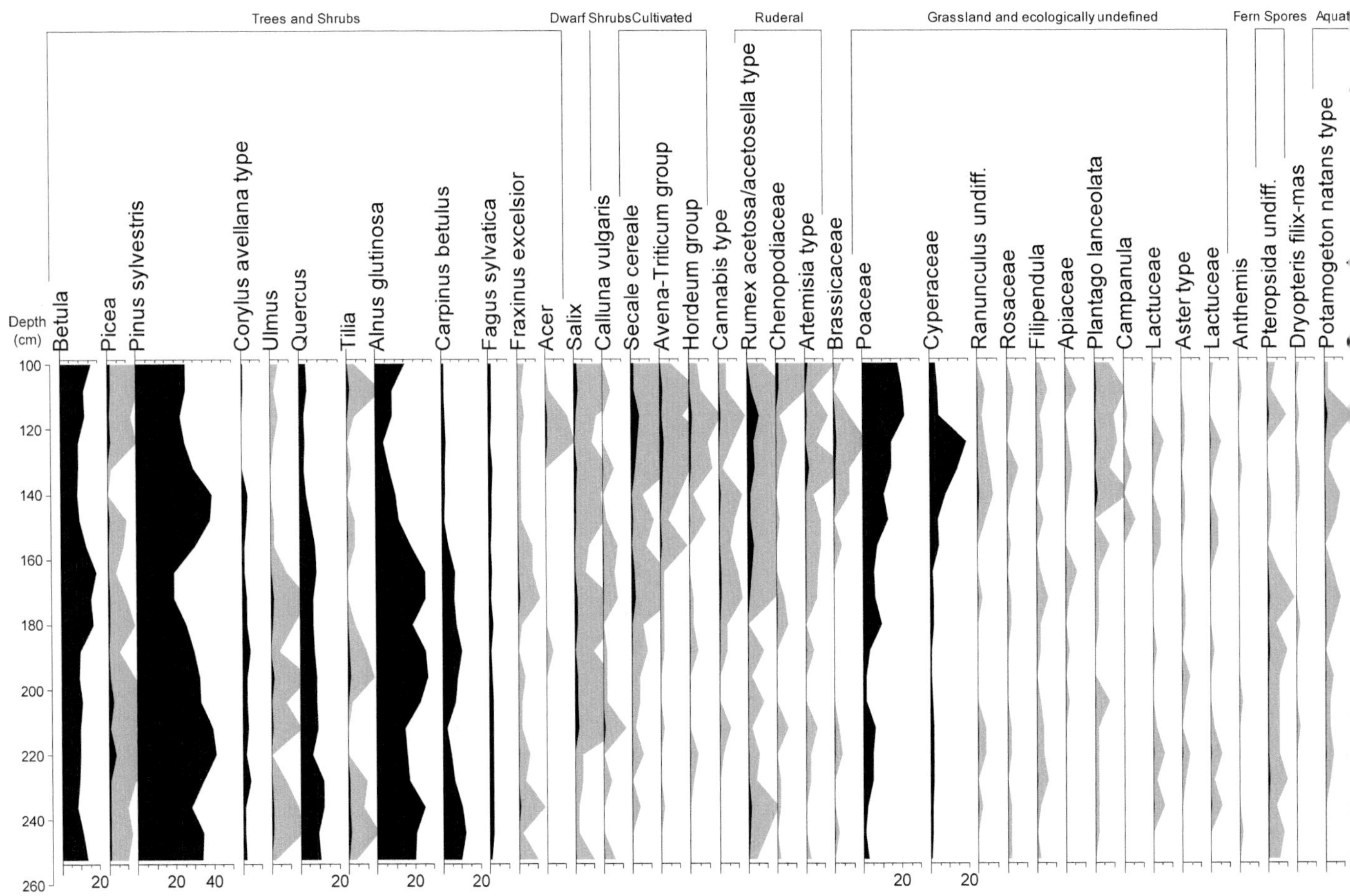

Trees and Shrubs
Dwarf Shrubs
Cultivated
Ruderal
Grassland and ecologically undefined
Fern Spores
Aquat
Depth (cm)
Betula
Picea
Pinus sylvestris
Corylus avellana type
Ulmus
Quercus
Tilia
Alnus glutinosa
Carpinus betulus
Fagus sylvatica
Fraxinus excelsior
Acer
Salix
Calluna vulgaris
Secale cereale
Avena-Triticum group
Hordeum group
Cannabis type
Rumex acetosa/acetosella type
Chenopodiaceae
Artemisia type
Brassicaceae
Poaceae
Cyperaceae
Ranunculus undiff.
Rosaceae
Filipendula
Apiaceae
Plantago lanceolata
Campanula
Lactuceae
Aster type
Lactuceae
Anthemis
Pteropsida undiff.
Dryopteris filix-mas
Potamogeton natans type
100
120
140
160
180
200
220
240
260
20
40

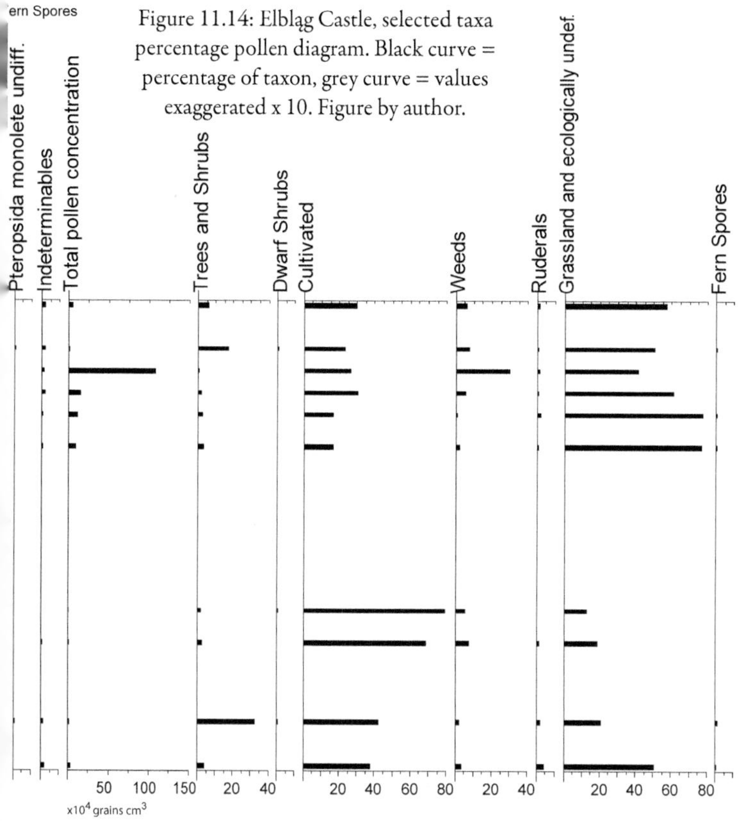

Figure 11.14: Elbląg Castle, selected taxa percentage pollen diagram. Black curve = percentage of taxon, grey curve = values exaggerated x 10. Figure by author.

Figure 11.15: Lake Drużno, selected taxa percentage pollen diagram. Black curve = percentage of taxon, grey curve = values exaggerated x 10. Figure by author.

alder growing in wetland habitats. Oak and hornbeam formed a lesser component along with hazel, lime and spruce; other trees occurred more sporadically. There is an increase in grasses and ruderal pollen during the Roman Iron Age, along with intermittent cereal type pollen grains of rye and wheat/oat and occasional weed taxa (e.g., cornflower) that suggest nearby human activity. The Migration Period is characterized by continued dominance of woodland but with an increasing component of pine. There are very occasional cereal pollen grains, ruderal, and weed taxa that suggest some agricultural activity, although not necessarily intensive or close to the sample site. There is little change in vegetation during the succeeding Viking Age/Early medieval period. Woodland composition remains consistent through to the seventeenth century, but there are indications of increasing human activity during the Prussian phase from *c.* 1000. There are consistent, albeit small, values for rye pollen along with oat/wheat and barley, cornflower, and ruderal pollen, indicating the nearby presence of permanent rotational field systems. It is not until the seventeenth century that values for cultivated, ruderal, and grassland taxa begin to rise associated with a significant decrease in arboreal pollen. This is most apparent in the decline in pine (4–17%), with a corresponding increase in birch, suggesting opening up of the woodland canopy and some colonization by birch along the edges of clearings. Cereal pollen values increase notably in the early eighteenth century, particularly rye (up to 7%) and wheat/oats (1–2%), although an increase in pine and oak suggests some subsequent regeneration of woodland from the mid-eighteenth century/early nineteenth century. Lower cereal pollen levels but increasing values for grasses and ruderals could suggest a greater focus on pasture/meadows.

Lake Staświny (Fig. 11.17)

Due to the lack of terrestrial plant macrofossils for dating, and high carbonate content of the sediments (emphasised by the presence of mollusc shells), it was decided to radiocarbon date terrestrial pollen concentrates. The results demonstrate that the 2.1 m depth of sediments recovered from Lake Staświny cover *c.* five thousand years. The vegetation sequence is comparable to nearby pollen sequences from Lake Wojnowo (1 km west) and Lake Miłkowski (2 km south) (Wacnik et al. 2012), although exhibiting a significantly slower sediment accumulation rate, and suggests that the radiocarbon-dated pollen concentrates are broadly reliable. The majority of the sequence therefore significantly predates the main archaeological periods of interest. The pollen assemblage is dominated by arboreal pollen reflecting mixed deciduous woodland initially comprising oak, hazel, elm, lime, and ash, but with a shift after 1500 BC towards larger quantities of birch, pine, and hornbeam with values for oak remaining consistent. There are few indications of human activity before the Roman Iron-Age, although short-term impacts may be missed due to the coarse chronological interval between samples; settlement traces from the Roman Iron Age attest to activity around the lake, with an increase in rye, cannabis, ruderal, and open ground taxa pointing towards arable and pastoral land-use. The Migration Period is characterized by woodland regeneration, with an increase particularly in hornbeam, oak, and birch. More significant changes in the vegetation composition are apparent from *c.* AD 700 with a decrease in arboreal pollen, particularly birch, and an increase in rye and ruderal pollen, intensifying from *c.* 1000; rye values increase along with oat-wheat and pollen of cornflower, suggesting permanent arable fields along with meadows and pastures, declining from the seventeenth century. Deciduous oak-hornbeam woodland was significantly cleared during the Prussian and medieval periods with pine forming the primary component of surviving woodlands.

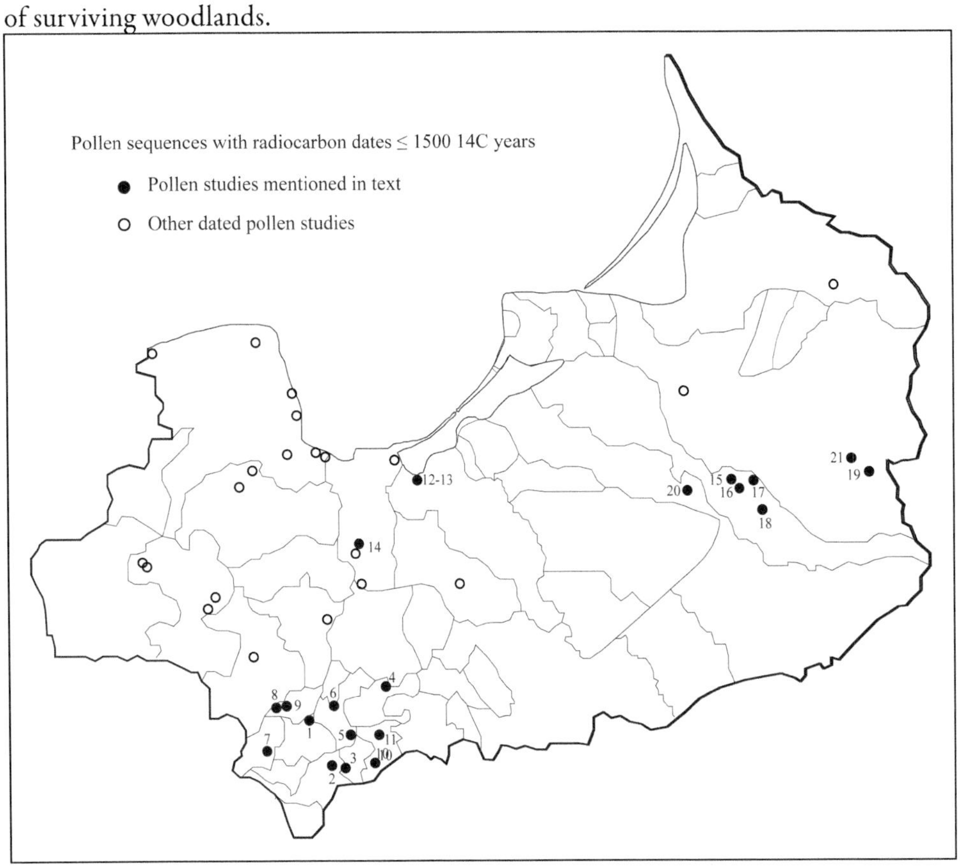

Map 11.2: Location map of existing palynological studies from Prussia and adjoining regions. Sites: 1 — Lake Czyste, 2 — Lake Kamionkowskie, 3 — Gronowo, 4 — Lake Mełno, 5 — Lake Wieczno, 6 — Lake Robakowski, 7 — Linje (Noryskiewicz 2013), 8 — Chełmno-Rybaki (Noryskiewicz 2004a), 9 — Uść (Noryskiewicz 2004b), 10 — Napole, 11 — Czstochleb (Filbrandt-Czaja and Noryśkiewicz 2003), 12 — Lake Drużo (Zachowicz, J and Kępińska 1982), 13 — Elbląg Castle (Nitychoruk et al. 2012), 14 — Gościszewo (Brown and Pluskowski 2011), 15 — Szczepanki, 16 — Lake Wojnowo, 17 — Lake Miłkowski, 18 — Lake Łazduny (Wacnik et al. 2012), 19 — Lake Wigry (Kupryjanowicz 2007), 20 — Lake Salęt (Szal et al. 2014), 21 — Lake Linówek (Gałka et al. 2014). Map by author.

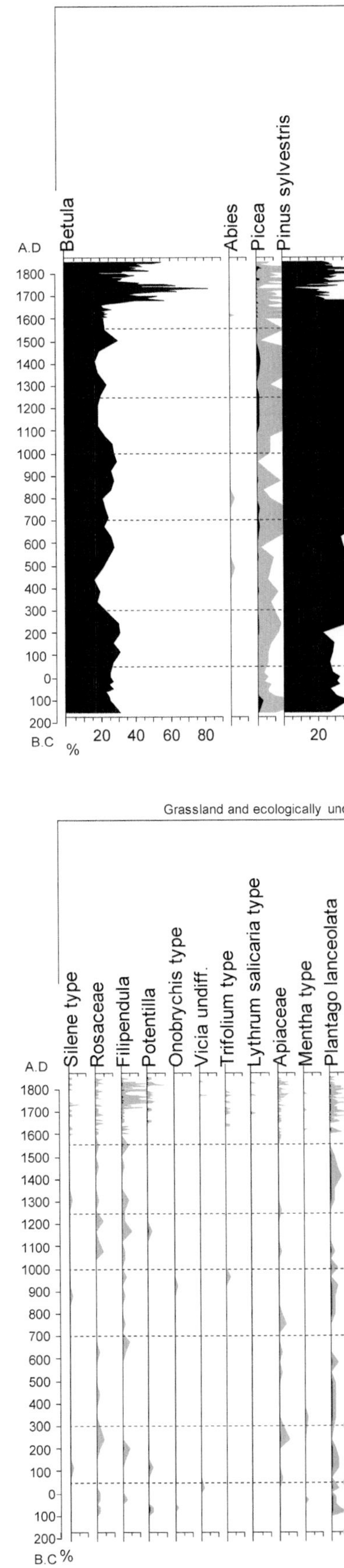

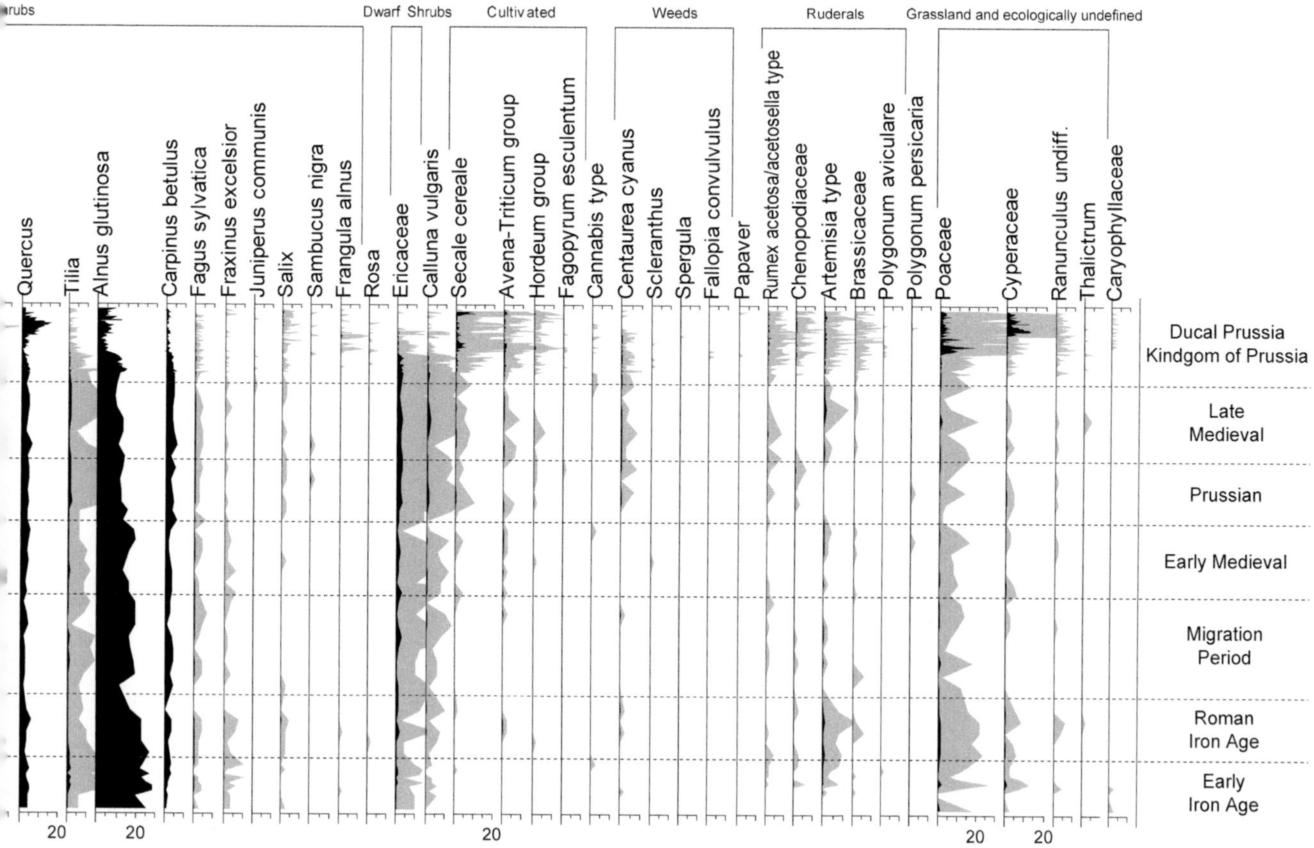

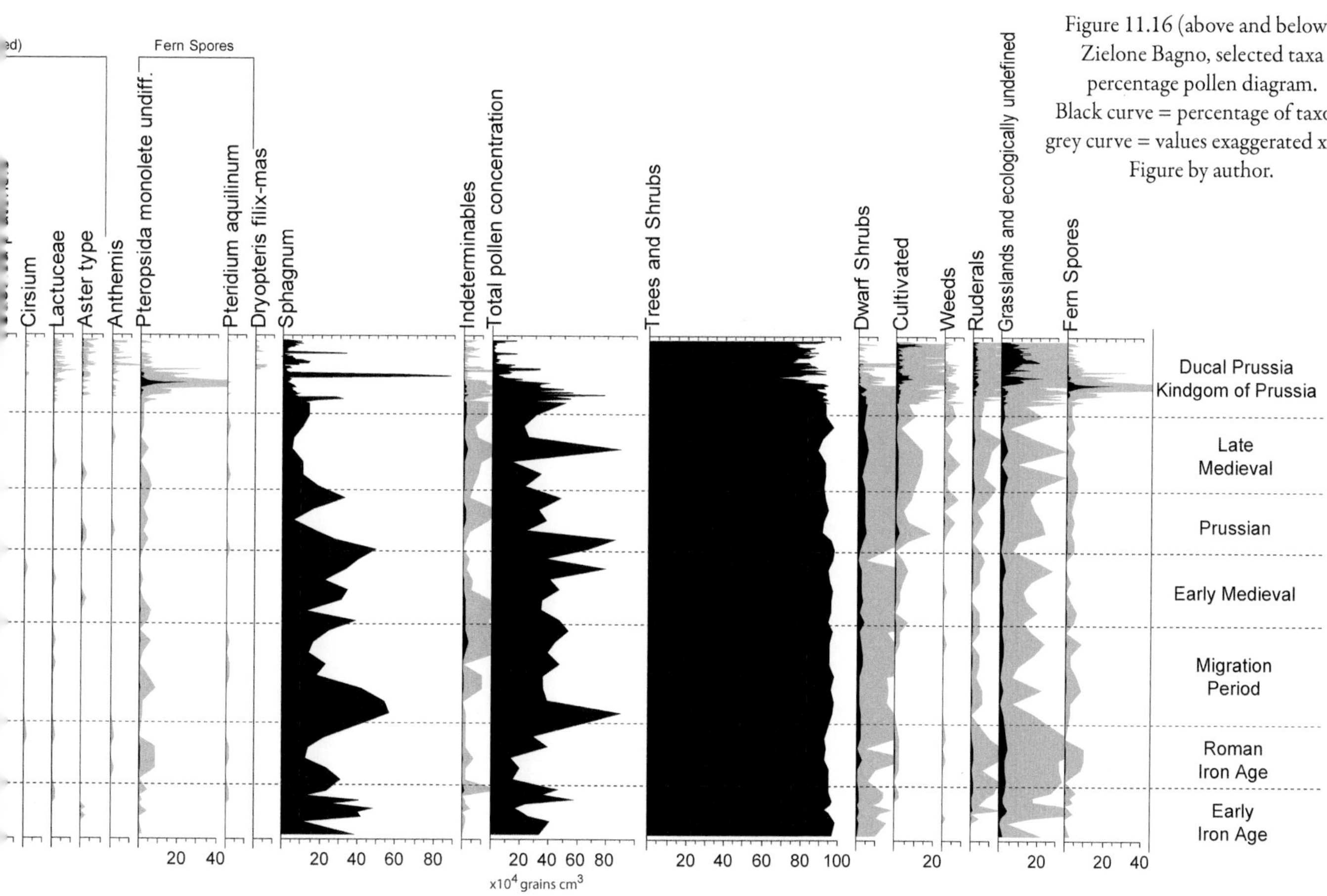

Figure 11.16 (above and below): Zielone Bagno, selected taxa percentage pollen diagram. Black curve = percentage of taxon, grey curve = values exaggerated x 10. Figure by author.

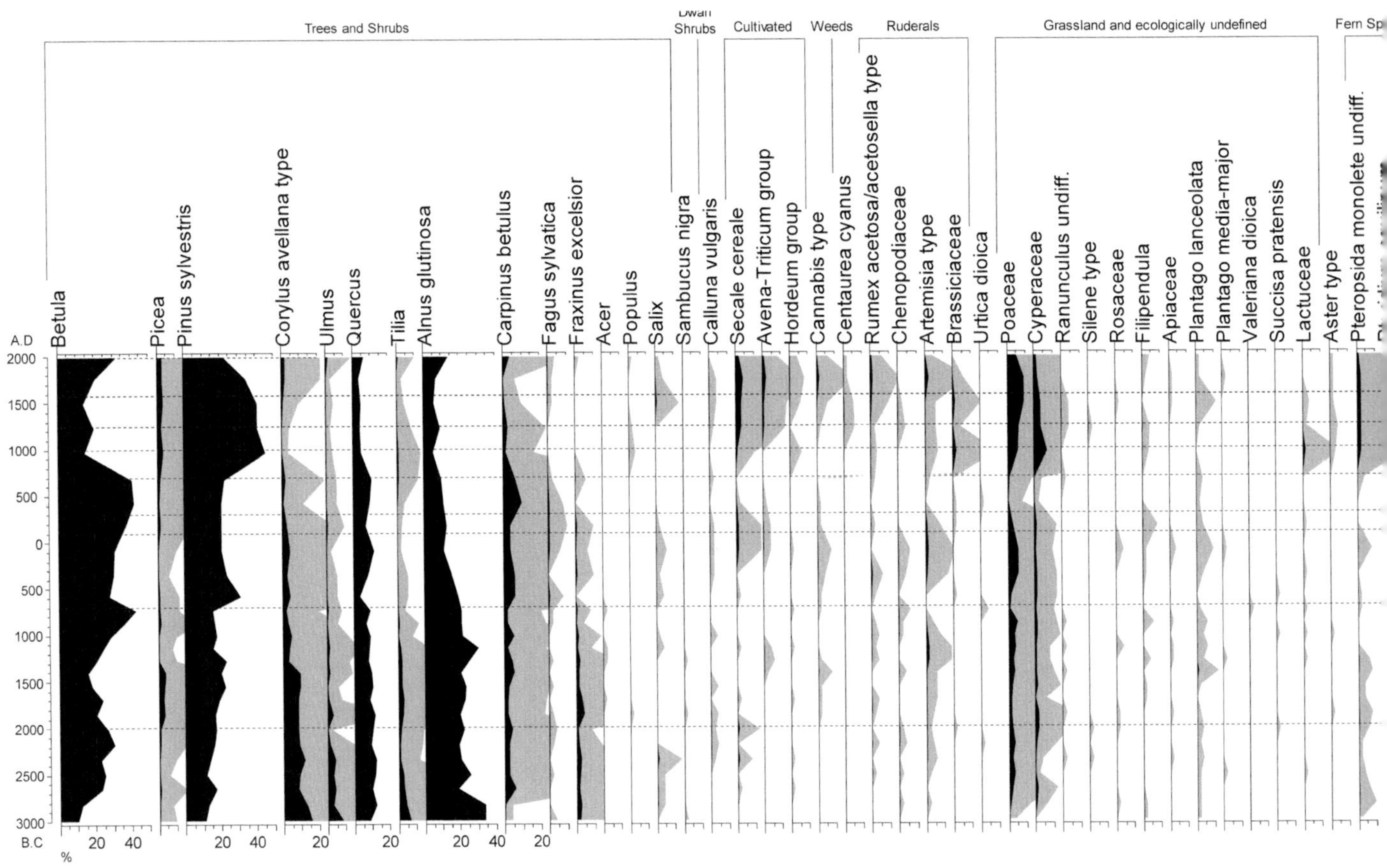

Trees and Shrubs
Shrubs
Cultivated
Weeds
Ruderals
Grassland and ecologically undefined
Fern Sp
Betula
Picea
Pinus sylvestris
Corylus avellana type
Ulmus
Quercus
Tilia
Alnus glutinosa
Carpinus betulus
Fagus sylvatica
Fraxinus excelsior
Acer
Populus
Salix
Sambucus nigra
Calluna vulgaris
Secale cereale
Avena-Triticum group
Hordeum group
Cannabis type
Centaurea cyanus
Rumex acetosa/acetosella type
Chenopodiaceae
Artemisia type
Brassiciaceae
Urtica dioica
Poaceae
Cyperaceae
Ranunculus undiff.
Silene type
Rosaceae
Filipendula
Apiaceae
Plantago lanceolata
Plantago media-major
Valeriana dioica
Succisa pratensis
Lactuceae
Aster type
Pteropsida monolete undiff.
A.D
2000
1500
1000
500
0
500
1000
1500
2000
2500
3000
B.C
20 40
%
20 40
20
20
20 40
20

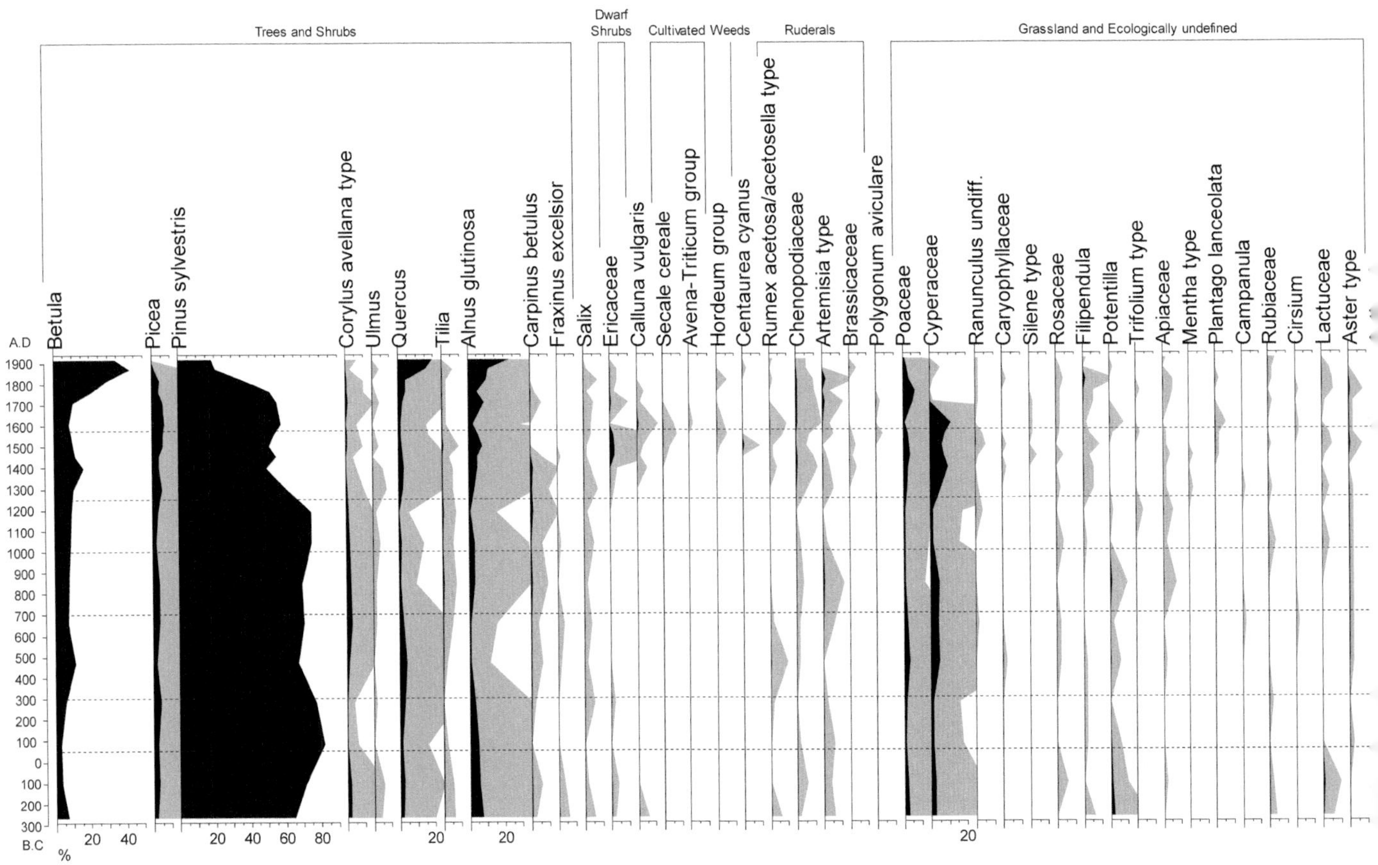

Trees and Shrubs
Dwarf Shrubs
Cultivated Weeds
Ruderals
Grassland and Ecologically undefined
Betula
Picea
Pinus sylvestris
Corylus avellana type
Ulmus
Quercus
Tilia
Alnus glutinosa
Carpinus betulus
Fraxinus excelsior
Salix
Ericaceae
Calluna vulgaris
Secale cereale
Avena-Triticum group
Hordeum group
Centaurea cyanus
Rumex acetosa/acetosella type
Chenopodiaceae
Artemisia type
Brassicaceae
Polygonum aviculare
Poaceae
Cyperaceae
Ranunculus undiff.
Caryophyllaceae
Silene type
Rosaceae
Filipendula
Potentilla
Trifolium type
Apiaceae
Mentha type
Plantago lanceolata
Campanula
Rubiaceae
Cirsium
Lactuceae
Aster type
A.D
1900
1800
1700
1600
1500
1400
1300
1200
1100
1000
900
800
700
600
500
400
300
200
100
0
100
200
300
B.C
20 40
%
20 40 60 80
20
20
20

Figure 11.17: Lake Staświny, selected taxa percentage pollen diagram. Black curve = percentage of taxon, grey curve = values exaggerated x 10. Figure by author.

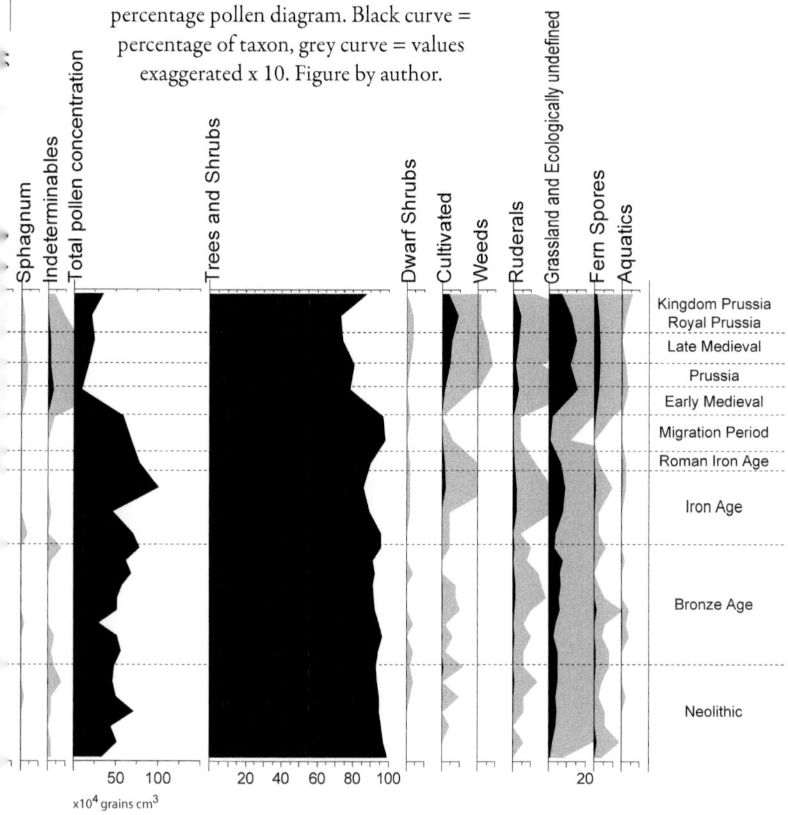

Figure 11.18: Wojsak, selected taxa percentage pollen diagram. Black curve = percentage of taxon, grey curve = values exaggerated x 10. Figure by author.

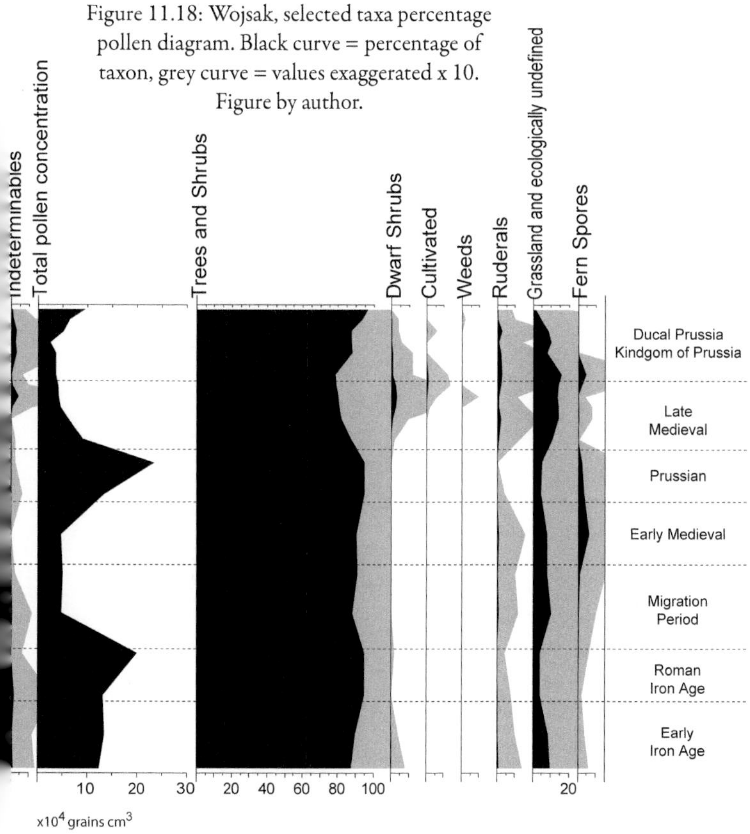

Wojsak (Fig. 11.18)

Despite the shallow depth of peat, the sequence from Wojsak covers a significant period of time from the second century BC; however, the restricted size of the basin, forming part of a larger complex of smaller wetlands, suggests that the pollen reflects the highly localized vegetation around the sample site, with only a minor component derived from further afield. The majority of the sequence is characterised by woodland dominated by pine with only limited quantities of other arboreal species, and no evidence for human activity. Occasional ruderal pollen grains are more likely to reflect natural processes of disturbance within the woodland, for example, resulting from tree throws or the actions of grazing animals. Woodland composition remains stable until the thirteenth century when pine begins to decline, although still remaining the dominant component of the canopy. Ruderal pollen increases from the fourteenth century and may reflect human impact in the vicinity; occasional rye and cornflower pollen during the fifteenth century suggest nearby arable fields. The site is located 1.8 km from the castle at Giżycko, established as the seat of the procurator of Lötzen *c.* 1341, although more sustained colonization did not begin until the fifteenth century.

Discussion

The results of analysis on peat and lake sediments are discussed from the three main areas of study (Kulmerland, Marienburg/Elbing commanderies, and the Great Masurian Lake District) and considered in the context of existing palynological data from these and adjoining regions of former east Prussia (Map 11.2). The pollen evidence supports a pattern of significant impacts on the landscape that reflect the process of territorial expansion and colonization occurring across the southern Baltic from the eighth to ninth century, with important variations apparent between the three principal areas. Syntheses of the pollen and charcoal data are presented in Figs 11.19 and 11.20, although in the latter case the quantities recorded of microscopic charcoal area were minimal.

Kulmerland

The pollen studies from Radzyń and Rywałd provide an important wider environmental context for what is archaeologically one of the most intensively studied regions of the early Slavic and subsequent Teutonic Order State (Poliński 2003). Although there is a significant number of existing pollen studies across the Kulmerland (forty sequences; see Noryśkiewicz 2013, Table 1), the majority lack sufficient sample or chronological resolution to allow for any detailed comparison with the cultural record. However, recently published research by Noryśkiewicz (2004a, 2004b, 2005, 2013) include several detailed and well-dated palynological studies: the most important of these include sequences from Uść and Chełmno-Rybaki closely associated with excavations of the important Slavic settlement at Kałdus (Chudziak 1994). Further important palynological studies were undertaken by Noryśkiewicz from peat mires at Gronowo and Linje and lake sediments at Czyste, Robakowskie, Wieczno, Kamionkowskie, and Mełno (Noryśkiewicz 2013); these studies were focused on tracing changes in settlement intensity in light of vegetation and environmental history for the last five thousand years and provide an important wider landscape context for the more focused studies at Radzyń and Rywałd. Bayesian chronological modelling using Bacon (Blaauw and Christen 2011) was undertaken on existing well-dated, but currently un-modelled, sequences to enable comparison with the pollen sequences presented here.

The pollen studies from Radzyń1 and 2 and Rywałd cover the Roman Iron Age during which time the Wielbark culture dominated the Lower Vistula region east of the Vistula to the Pasłęka River (Barford 2001); the Roman Iron Age is considered a time of intensive settlement development within present-day Poland and provides a useful comparison for later human impact. The Wielbark culture is generally characterized by open settlements with a subsistence economy including cultivation of wheat, millet, rye, and barley with cattle the most common domesticated animal (Prygała 1975; Makowiecki 2006). The pollen and LOVE models emphasize that the landscape surrounding the Castle Lake at Radzyń included a significant open component, particularly during the first and second centuries, with a largely pastoral focus to land-use. This compares with Rywałd, where there is little evidence for contemporary land-use and suggests the lake and wetlands at Radzyń were a particular focus for activity at this time. The pollen studies of Noryśkiewicz generally include small quantities of cereals, and few if any segetal taxa, likewise suggesting small-scale cultivation with perhaps a greater focus on pastoral land-use. Iso-pollen mapping show higher cereal and non-arboreal pollen values in the west of the Kulmerland around Chełmno with lower values in the east of the region (Noryśkiewicz 2013, pls 11–13). Despite the evidence for increasing landscape openness during the Roman Iron Age, increasing oak values are apparent in a number of sequence across the Kulmerland (e.g., Filbrandt-Czaja and Noryśkiewicz 2003; Noryśkiewicz 2013) including Radzyń1, perhaps encouraged either by warmer temperatures, selective removal of trees avoiding oak, or as a component of grazed wood-pasture as noted elsewhere in Europe (Vera 2000, 300–01; Rackham 2003; Makohonienko 2004). However, there is a progressive regeneration of woodland at the end of the Roman Iron Age, most apparent from the fourth century in increasing values for oak and hornbeam, perhaps related to the southern migration of communities associated with the Wielbark culture.

Hornbeam-dominated woodlands reach their greatest extent and highest Holocene pollen values during the Migration Period, reaching up to 43% at Rywałd and 36% in the Radzyń1 and 3 sequences (Fig. 11.19), with consistent values of between 20–40% in pollen sequences across the Kulmerland, but with the highest values in sequences to the north-east and south-east (Noryśkiewicz 2013, pl. 3). Regeneration of hornbeam was probably aided by the open conditions at the end of the Roman Iron Age, as suggested by the LOVE models, favouring a greater survival and spread of seedlings combined with the shorter juvenile stage prior to maturity; under conditions of dense shade, the juvenile stage is longer, and a greater number of seedlings die (Rackham 2003; Ralska-Jasiewiczowa et al. 2004). Within the strict reserve of the Białowieża National Park in north-east Poland, hornbeam typically dominates the lower canopy of a mixed broad-leaved woodland that is considered by many to reflect the primeval vegetation of Central Europe (e.g., Rackham 2003; Peterken 1981; Falinski 1986; Bernadzki et al. 1998) and our best analogue for the woodlands that came to dominate the Kulmerland and surrounding regions during the Migration Period.

Renewed woodland clearance is apparent in pollen sequences across the Kulmerland from the seventh to eighth century (Fig. 11.19), with sharp declines particularly in hornbeam and oak associated with a rise in pollen taxa indicative of human impact (Filbrandt-Czaja and Noryśkiewicz 2003; Noryśkiewicz 2004a, 2004b, 2005, 2013). Chronological modelling on existing published sequences from the Kulmerland (Noryśkiewicz 2013) demonstrates a broad consistency in the timing of renewed woodland clearance, beginning in the seventh century at Lake Czyste, Wieczno, and Chełmno-Rybaki, with slightly later dates in the seventh to eighth centuries at Lake Robakowskie and the eighth to ninth centuries at Linje Mire and Uść. The sequence from Gronowo includes an initial phase of woodland clearance in the seventh century, followed by a return to hornbeam-dominated woodland during the eighth century with renewed woodland clearance during the ninth century. Single radiocarbon dates are only available on the sequences from Czystochleb and Napole (Filbrandt-Czaja and Noryśkiewicz 2003), with hard water errors affecting the dates from Lakes Mełno and Kamionkowskie (Noryśkiewicz 2013), although similar patterns of vegetation change are apparent at these sites. Despite the slight variation in dates for woodland clearance between pollen sequences, the broad consistency in pollen signals argues for a synchronous pattern of land clearance occurring across the Kulmerland, beginning in the seventh century and expanding during the following centuries.

The deforestation of the Kulmerland has been linked with an intensification of settlement and movement of populations from Kuyavia and Greater Poland and corresponds with the appearance of strongholds, all relatively small in size and dating from the seventh through to the ninth century (Chudziak 1996; Buko 2005). The Slavic stronghold at Radzyń dates from the eleventh–early thirteenth century (Chudziak 1994), but it is clear that the land surrounding the Castle Lake was a focus for renewed settlement from the seventh century. Declining arboreal pollen is typically associated with increases in pollen of grasses, but with generally lower values for cereals and small quantities of segetal plants, suggesting a greater focus on pastoral rather than arable land-use. This is emphasized by the LOVE models from Radzyń and Rywałd that show larger quantities of meadowland with only small increases in cereals (Figs 11.4 and 11.6) However, the significant levels of early medieval woodland clearance yet relatively small increases in cereals compared to successive periods have led to suggestions that communities may have been cultivating more millet in the early medieval period (Latałowa pers. comm.), with rye only becoming more popular in the later medieval period. Millet pollen is indistinguishable from grass, yet archaeobotanical samples from some medieval urban contexts, including Gdańsk (Badura 2011), do show a greater emphasis on millet in the early medieval. It is possible therefore that some of the grass pollen included in the meadow group of the LOVE models could reflect cultivation of millet.

Intensification in agrarian land-use is most apparent in the Radzyń sequences from the eleventh century, with landscape openness increasing significantly up to 60% (Fig. 11.4), dominated by meadows and pasture, but with more areas under cultivation. Values for rye pollen increase noticeably along with pollen of taxa such as cornflower, a field weed typically associated with cultivation and which has been argued to reflect the presence of permanent rye fields (Vuorela 1986). Rye would have been largely grown as a winter crop, sewn in autumn, most probably along with winter wheat, with oats and barley sewn in the spring and harvested in summer (e.g., Scott 2002). The increase in cereals including rye reflects both an intensification and diversification in crop cultivation, most likely as part of a three-field system of crop rotation. There is a noticeable increase in microscopic charcoal during the eleventh and twelfth centuries in the Radzyń1 sequence, perhaps related to burning of agricultural fields, although charcoal values are negligible in the other sequences from the Kulmerland (Fig. 11.20). It is probably therefore that some of the expansion in meadowland in the pollen and LOVE models reflects an increase in fallow land as part of the three-field system. Rye cultivation also increases in the Rywałd sequence from the eleventh century, but with a greater survival of broad-leaved woodland, suggesting less settlement within the immediate landscape. Slightly earlier intensifications in agrarian land-use are apparent from the tenth century at Lakes Czyste, Robakowskie, and Uść, with an eleventh century intensification in land-use apparent at Linje Mire (Noryśkiewicz 2013). The increasing intensity in land-use apparent in sequences from the tenth century most likely reflects growing Polish control and the development of the Polish regional castellany centred on the important settlement at Kałdus, and which saw increasing settlement of the Kulmerland from the mid-tenth century (Chudziak 1996; Buko 2008; Makowiecki 2007). Iso-pollen mapping by Noryśkiewicz show higher values for non-arboreal and cereal pollen during the early medieval period in the north-west of the Kulmerland, an area including Kałdus and extending east towards Radzyń (Noryśkiewicz 2013, plates 9 and 11). The results from Radzyń and Rywałd fit well within this pattern and correspond with the distribution of settlements from the end of the early medieval period that show a similar concentration towards the north-west of the Kulmerland (Poliński 2003, Map 1a).

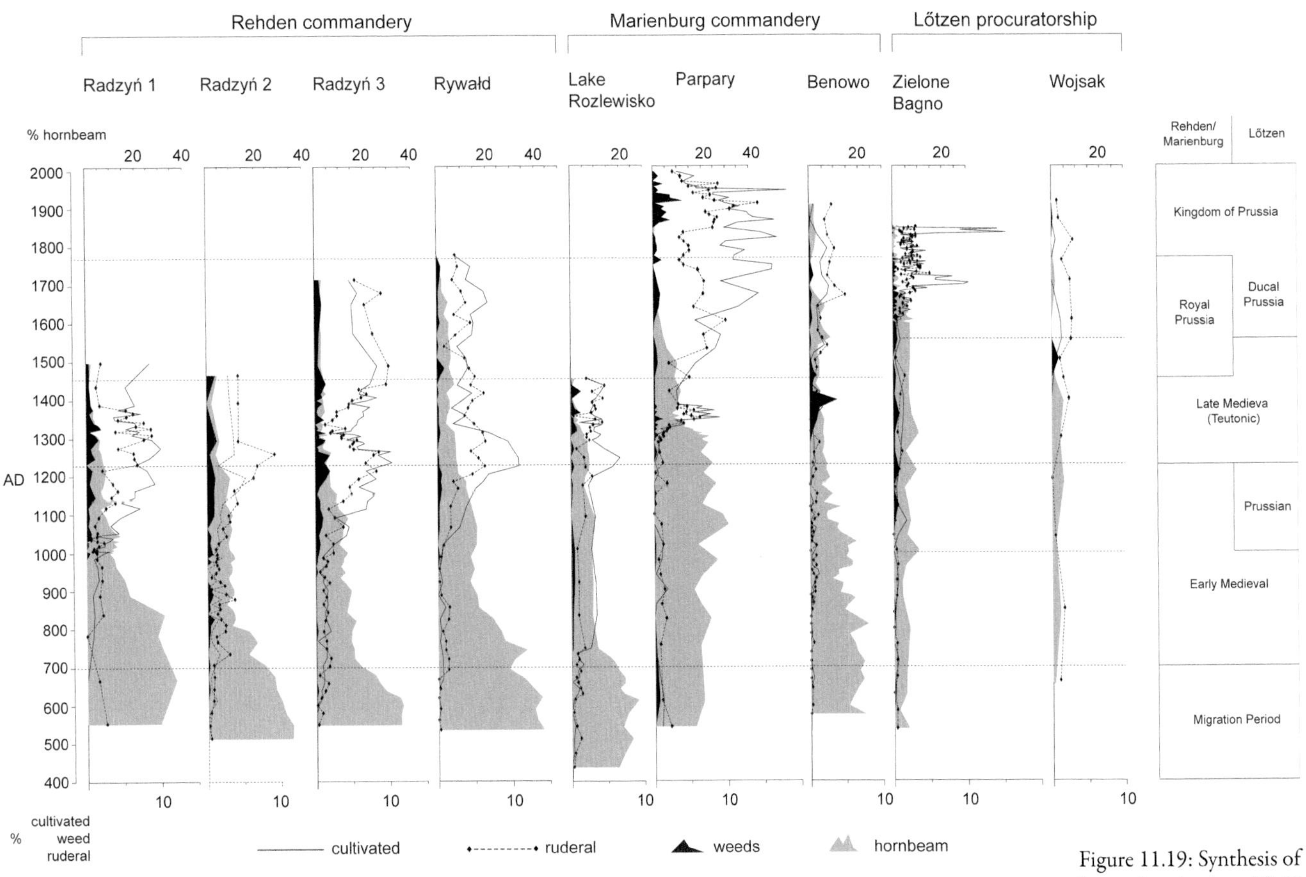

Figure 11.19: Synthesis of cultivated, ruderal, and field weed pollen from selected pollen studies plotted against values for hornbeam pollen. Figure by author.

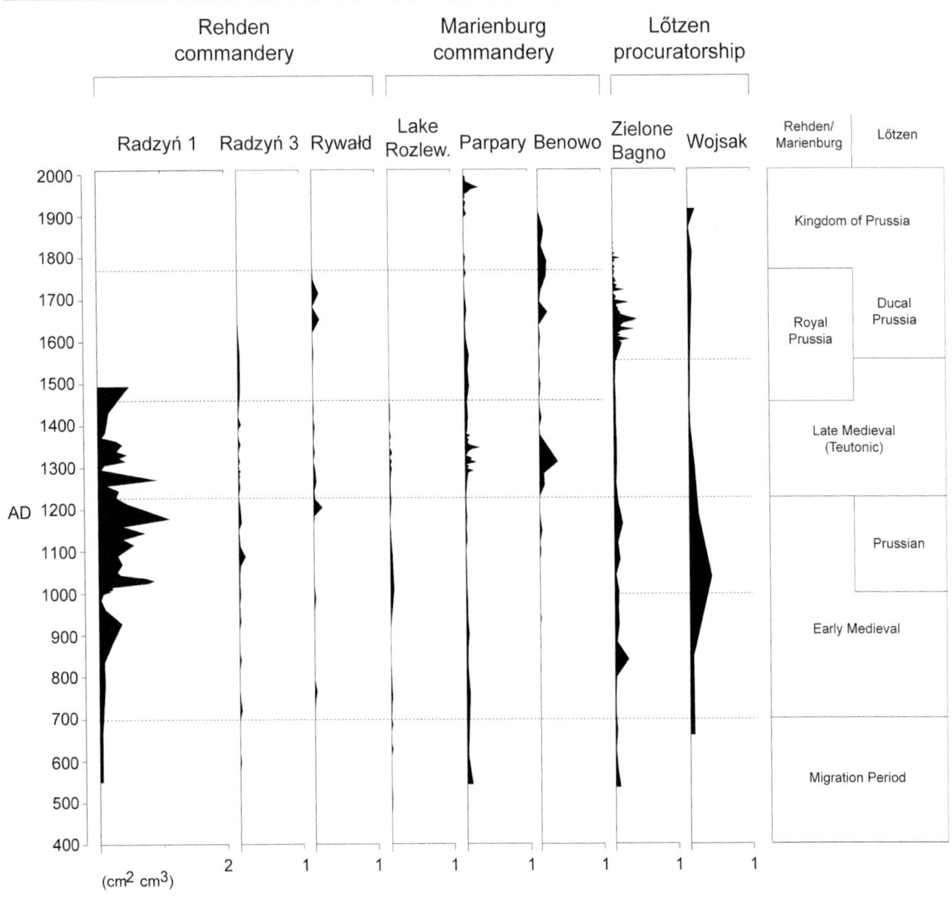

Figure 11.20: Synthesis of microscopic charcoal data from selected pollen studies. Figure by author.

The Kulmerland became an important frontier with Pomeranian lands to the west, Prussian tribal lands to the north-east, and Masovian lands to the south-east. The Slavic stronghold at Radzyń was located in a particularly vulnerable location in proximity to Prussian Pomesania, yet the archaeological and palynological data, both from Radzyń and across the Kulmerland, suggests the region was stable enough during the tenth and eleventh centuries for sustained agriculture to develop. However, during the second half of the twelfth century sites began to be abandoned, coinciding with a documented period of Prussian incursions, culminating in Konrad of Masovia's attempts to regain this territory. This phase of settlement instability is not obviously apparent in pollen records, perhaps because of the intermittent nature of Prussian incursions. Indeed the pollen and geochemical data from Radzyń shows a measure of stability, despite the destruction of the stronghold in the mid-twelfth century.

The landscape of the Kulmerland had therefore already experienced significant clearance of woodland and intensive agrarian modification by the time of the crusades. However, much of the surrounding landscape would have been an unstable frontier zone during the initial decades of the crusades, with Prussian armies and warbands periodically laying waste to the countryside. Many of the castles and associated urban foundations in the Kulmerland were periodically attacked and particularly badly damaged during the Great Prussian Uprising in the 1260s; the town at Radzyń was captured by the Prussians, although the castle remained in the hands of the Order. Poliński (2003) has argued that settlement in the region was largely hindered until the 1280s when military action against the Prussians ceased. However, the main colonies and towns were not only founded but prospered during the 1230s to 1280s, despite intermittent attack. These towns required provisioning to support their populations, suggesting sufficient stability existed in the surrounding landscape for their maintenance.

The palynological data lend some support to this picture. The pollen and LOVE modelling of the Radzyń and Rywałd sequences all show increasing arable and pastoral land-use from the twelfth into the thirteenth centuries, with landscape openness of *c*. 50–60% (Figs 11.4 and 11.6). However, the pattern is more complex, with the late twelfth to mid/late thirteenth centuries representing a peak in cereal pollen levels during the late medieval period. Both the percentage and modelled pollen show a decline in the proportion of arable land from the late thirteenth century, but with no apparent decline in ruderal or meadowland. Similar fluctuations in pollen curves for cereals and segetal plants are apparent at some, but not all, sites across the Kulmerland. Fluctuations are most apparent in those sites with higher sample intervals, including Lakes Czyste, Robakowskie, and Mełno, whilst at other sites, such as Linje Mire, values for key taxa appear more stable; sediments of late medieval date were either absent or poorly dated in the Chełmno-Rybaki, Gronowo, and Wieczno sequences (Noryśkiewicz 2013). The pollen evidence for a possible decline in arable activity in the Radzyń and Rywałd sequences occurs against a background of increasing colonization of the Kulmerland followed the end of the crusades in Prussia in 1283. This represents a period of relative stability free of major conflict that lasted until 1409, with an increase in documented settlements within the commandery of Radzyń. Until the early fourteenth century, the castle was served by the Vorwerk, and later village, at Radzyń Wieś, situated to the south-west, and thirty-eight other documented settlements within the commandery. From *c*. 1343 to 1466, the population of the commandery increased, with forty-eight settlements and nine *Vorwerks* documented (Poliński 2003, 75, 104). The lack of nearby settlements around Radzyń (Poliński 2003, Figs 1 and 2) could suggest that the focus of agricultural activity moved further away from the castle as the wider landscape was increasingly colonized from the late thirteenth century onwards, but this hypothesis is not borne out by the evidence from Rywałd, or Lake Mełno located in the northern edge of the commandery (Noryśkiewicz 2013, Fig. 13), that show a similar pattern to Radzyń. The more likely hypothesis is that there was a greater focus on stock rearing, emphasized by the significant proportions of meadowland suggested by the pollen modelling. Further analysis and dating of suitable pollen-bearing deposits is necessary to investigate the variability or otherwise of this pattern more widely across the Kulmerland.

The relative stability of the thirteenth and fourteenth centuries was followed by a series of conflicts that lasted from 1409 until 1466 and which resulted in the destruction and abandonment of many villages, with rural communities fleeing to the shelter of the fortified towns and an overall decrease in the population of the Kulmerland from the mid-fifteenth to the start of the sixteenth century (Poliński 2003). The pollen record, however, indicates continuity in levels of agrarian land-use with little convincing evidence for decline or abandonment. Inventory records from Rehden (see Chapter 15) show the dramatic drop in provisioning of cereals, horses, and livestock after 1410, emphasizing the diminishing role of the castle thereafter. The town did not develop significantly like other urban centres with castles such as Thorn and Graudenz, so it is perhaps not surprising that the levels of land-use remain relatively unchanged into the post-medieval period.

Marienburg and Elbing Commanderies

The pollen studies from the Marienburg commandery provide the first detailed palynological studies from this part of Poland; these are considered along with the nearby pollen studies from Lake Drużno and Elbląg Castle. Attention was focused on land to the south-east of Malbork because of the lack of suitable coring locations within the Vistula fenlands. The vast floodplains of the Vistula and its tributaries were progressively embanked from the fourteenth century for agriculture and settlement, but particularly between the sixteenth to nineteenth centuries (Kowalik 2008). During the medieval period this would have been a dynamic environment of semi-terrestrial marshland, periodically flooded and flanked in places by inland river dunes, perhaps with occasional dry land islands scattered across the wetlands. South-east of the River Nogat, occasional small mires and lakes provide an opportunity to reconstruct local vegetation environments and land-use history; pollen studies were largely focused on suitable deposits south of Malbork around the Forest of Sztum (Map 11.1). The Forest forms a significant natural feature of the present landscape, extending 20 km to the south near Kwidzyn, representing the remnants of a previously more extensive woodland bordering the Vistula and Nogat.

The pollen studies from Parpary, Benowo, Lake Rozlewisko, and Lake Drużno (Figs 11.8, 11.10, 11. 12, 11.15) to the east all show a heavily wooded environment during the Migration Period, with occasional cereal and ruderal pollen grains pointing towards small-scale intermittent human activity. Hornbeam and oak form important components of this woodland, along with smaller quantities of beech, ash, lime, elm, and other broad-leaved trees, but with a greater contribution of pine most likely growing on poorer sandy soils. Values for beech are slightly higher in the Lake Drużno sequence and may reflect a greater contribution of beech to woodlands on the adjacent Elbąg upland where today beech remains an important component (Zachowicz and Kępińska 1982). The chronological precision of the Gościszewo sequence is poor (Brown and Pluskowski 2011), but increasing landscape openness and agrarian land-use is apparent at Lake Rozlewisko from the eighth/ninth centuries at a time when the first Pomeranian strongholds appear in the landscape. The remains of possible Pomeranian strongholds have been suggested near Lake Rozlewisko at Stary Targ, and 3 km north at Kalva; excavations at the latter site revealed two phases of activity in the ninth to twelfth and first half of the thirteenth centuries respectively (Długokęcki and Haftka 2000). Clearance of woodland is likewise apparent in the Lake Drużno sequence (Fig. 11.15), with declining values for hornbeam and increasing values for grasses, ruderal, and cereals. Despite the lack of radiocarbon dates, it is likely that increasing land-use reflects a process of intensifying settlement in the surrounding landscape from the sixth century. Settlement sites proliferate on the Elbląg uplands and particularly in the area of Janów Pomorski (Truso) that would develop as one of series of coastal trading centres from the ninth to eleventh centuries (Pluskowski 2013, Fig. 2.4; cf. Jagodziński 2004). The site of Janów Pomorski is located only 1.2 km east of the pollen core, but the large pollen source area of the lake will reflect the vegetation history, settlement, and land-use of the wider regional landscape. This is particularly the case with rye, which is wind pollinated and may be transported over longer distances; it is also possible that rivers flowing into the southern end of Lake Drużno introduced a background pollen component from within the wider lake catchment. There is a decline in rye pollen from *c.* 1.65 m, but it is impossible without any associated chronology to link this with changes in settlement intensity, particularly the decline of Truso in the eleventh century.

The pollen sequences from Parpary and Benowo paint a very different picture, with woodland remaining largely intact in the area of the Forest of Sztum during the early medieval period. There are indications of limited agrarian activity in the pollen and LOVE models, particularly from Parpary between the sixth to tenth centuries, but with a noticeable absence of human activity in both the Benowo and Parpary sequence during the eleventh to mid-thirteenth centuries (Figs 11.9 and 11.11). Securely dated archaeological traces of early medieval settlement in the area are few and support the palynological picture of a rather sparse settlement pattern in the surrounding landscape at this time. Only the nearby Pomeranian stronghold and settlement complex at Węgry, *c.* 1.8 km north of Parpary, has been confidentially assigned to the early medieval period, occupied from *c.* 1035 to 1170/80, but it appears to have had a relatively limited impact on the surrounding landscape. It is possible, however, that Parpary is largely reflecting the persistence of more localized woodland habitats, as the pollen source area of similar small basins (*c*. 25–75 m radii) has been argued to vary between 1200 to 2300 m (Hellman et al. 2009). The palynological evidence from Gościszewo located *c*. 3 km north of Węgry, and close to the early medieval settlement complex at Wielbark, shows significant clearance of woodland from the eleventh century (Brown and Pluskowski 2011), suggesting the development of a more mosaic landscape to the north of the Forest of Sztum.

Palynological sequences west of the Vistula within the Kashubian Lake District and Tuchola region to the south show similar evidence for increasing human activity at stages during the early medieval period, particularly the eighth/ninth and tenth centuries (Lamentowicz et al. 2008; Pidek et al. 2009; Hjelmroos 1981); this occurs against the backdrop of increasing control of Pomeralia by the Polish Piast state from the 960s. Pollen from the territory of Gdańsk, the dominant economic and political power in the region, show large-scale woodland clearance and cereal cultivation from the tenth century (Latałowa et al. 2009), with pollen and macrofossils from urban excavations suggesting the immediate environs were almost totally deforested by the beginning of the twelfth century (Święta-Musznicka et al. 2013).

The lower Vistula remained a permeable frontier between Pomeralia and Pomesania into the thirteenth century, with pollen evidence from Benowo and Parpary emphasizing continuity of woodland. The initial decades of the crusades saw little impact on the environment surrounding these sites, with the Forest of Sztum remaining a significant landscape feature. This remains the case at Benowo to the present day, although there is some evidence for development of meadows from the sixteenth century, perhaps involving some forest grazing of animals. At Parpary, however, there is clear evidence for intensifying land-use from the early fourteenth century, although LOVE models suggest landscape openness of only around 15% with significant remaining woodland dominated by hornbeam and oak (Figs 11.8–11.9). However, there are strong indications of arable fields, with increasing values for cereal pollen accompanied by pollen of segetal plants, most notably cornflower. There is a defined increase in microscopic charcoal particles during the late thirteenth to early fourteenth centuries in both the Parpary and Benowo sequence (Fig. 11.20); the values are rather low and suggest the limited use of fire, perhaps involved in some clearance or management of woodland or within agricultural regimes. Intensifying land-use in the Parpary sequence clearly reflects the development of the nearby colonizing settlement, itself one of several settlements that developed around the Forest of Sztum from the early fourteenth century. The relocation of the headquarters of the Order to Malbork in AD 1309 clearly provided an impetus for increasing colonization and landscape reorganization from the fourteenth century. There is a notable increase in rye pollen in the Gościszewo sequence from the fourteenth century (Brown and Pluskowski 2011), with pollen from the moat at Malbork, although dating from the fifteenth century, showing a significantly open agricultural landscape within the vicinity of the castle. The Lake Rozlewisko sequence is more equivocal; despite an increase in rye pollen during the early to mid-thirteenth century, LOVE models show declines in arable and pastoral land during the thirteenth and particularly fifteenth centuries (Figs 11.12–11.13). Although the proportion of wet meadow increases, this seems likely to reflect the localized expansion in alder-willow carr-woodland habitats along the lake margin. The dense intervening vegetation will have filtered out a proportion of pollen from surrounding agricultural land, particularly those cereals, ruderal, and other herbaceous plants with poor dispersal mechanisms, skewing the LOVE model and giving a false impression of agricultural decline. The surrounding landscape was heavily colonized, mainly from the fourteenth century, with nearby settlements at Stary Targ (1.5 km north) and Nowy Targ (1.5 km south-east). The almost complete disappearance of oak and hornbeam over the course of the fourteenth and fifteenth centuries (Fig. 11.12) serves to emphasize the increasing pressure on surviving woodland.

Woodland decreases significantly in the Lake Drużno sequence during the medieval period, with pollen from thirteenth- and fourteenth-century deposits within the courtyard at Elbląg Castle suggesting a predominantly open environment in the vicinity (Figs 11.14–11.15). At this time the castle was located on an island at the northern end of Lake Drużno that was in all likelihood still joined to the Vistula Lagoon (Zachowicz and Kępińska 1987). The deposits within the courtyard are dominated by cereal, segetal, and ruderal pollen, considered to reflect agricultural waste used as animal feed and the nearby storage of agricultural produce in out-building. Earlier analysis of a core located at the north-western edge of the courtyard, sampled prior to excavations in 2012, includes sediments from the earliest thirteenth century deposits of the castle and likewise contains significant quantities of cereal pollen and only small quantities of arboreal pollen (Nitychoruk et al. 2012). Although the pollen from the castle may reflect the highly localized depositional context, it is likely that the construction of the castle and adjacent town will have necessitated the removal of much of the surrounding woodland. Pollen in stratigraphically earlier deposits include greater quantities of arboreal pollen, particularly pine, birch, and alder along with oak and hornbeam likely growing on the nearby dry ground. There are small quantities of cereal pollen in these sediments (<1%) that hint at nearby agricultural activity. Although the chronology of these deposits has not been firmly established, the quantities of hornbeam (5%) and oak (10–13%) pollen are broadly comparable with values for these taxa in off-site pollen sequences dating to the early medieval period.

The requirement for wood for building, fuel, and industrial processes (including charcoal for firing brick) may have required careful management of existing woodland resources, particularly in the context of overall declining levels of arboreal pollen during the medieval period. Indeed, documentary sources describe the Forest of Sztum as an actively managed woodland during the fourteenth and fifteenth centuries (Chęć and Gancewski 2009). Settlements such as Parpary and *Vorwerks*, including Benowo, would have played an important role in exploiting and managing both the floral and faunal resources of the woodland, subservient to the castle as Malbork. Specific management practices are difficult to identify in the palynological record and are not described in documentary sources. However, the relatively stable values of oak and hornbeam in the Benowo and Parpary sequences during the fourteenth and fifteenth centuries might point towards some selective felling and/or maintenance of hardwood species. There may also have been a greater survival of mixed deciduous woodland under conditions of moderate anthropogenic pressure, enduring deep within the forest and on those more nutrient poor soils with less agricultural potential. Management of the woodland for charcoal production seems unlikely, at least on an industrial scale, because of the consistently low values of microscopic charcoal in pollen sequences. No charcoal-burning platforms or saw pits are known from the Forest of Sztum, although there has been a lack of detailed archaeological surveys within the woodland itself, and the quantities of microscopic charcoal recorded from pollen sequences was uniformly low (Fig. 11.20). The occasional fruits and nuts recorded in macrofossil samples at Biała Góra may reflect some exploitation of local resources, but only of the woodland margin (Chapter 14). Documentary sources suggest parts of the woodland functioned as a hunting reserve, with a park containing wild animals documented near the castle in Sztum, that by the end of the fourteenth century was used as a hunting lodge for the grand masters (Haftka 2010). Moreover, it is probable that the woodland also provided readily available seasonal pannage for pigs and pasturing of cattle (see Chapter 12). In reality, medieval woodlands were often managed for multiple uses, including timber, pasture, and recreation. The Forest of Sztum survived because it had an economic value, but it was also located on largely nutrient-poor sandy soils of limited agricultural potential, whilst surrounding areas with greater agricultural potential witnessed more significant agricultural use.

Lötzen Procuratorship and the Great Masurian Lake District

The results of recently published pollen analysis from Lakes Wojnowo, Miłkowski, Łazduny, and Szczepanki (Wacnik et al. 2012, 2014) provided the first detailed evidence for vegetation and land-use within the Great Masurian Lake District. These studies represent an important point of departure for considering the pollen evidence from Zielone Bagno, Staświny, and Wojsak (Figs 11.16–11.18), and they provide an important wider landscape context for documented medieval settlement within the Great Masurian Lakelands and adjoining lake regions of north-east Poland. They demonstrate that significant changes in vegetation were largely the result of human activity but with distinct differences visible between sites dependent on the local topographical context and settlement histories. Wacnik et al. (2012) identify four main stages of anthropogenic activity most clearly resolved in the Wojnowo and Miłkowski sequences, the last two of most relevance to the medieval period. The third stage, dating to AD 50–600, was linked with settlement activities of the Bogaczewo culture and Prussian tribes, involving an increase in agricultural activity, but within a still largely wooded landscape The fourth stage is characterized by extensive clearance of woodland and the appearance of intensive agricultural activity from *c.* 1100, and involved the almost complete removal of woodland surrounding the lakes. The pollen sequence from Lake Staświny (Fig. 11.17), although lacking the sample and chronological resolution of the nearby Wojnowo and Miłkowski studies, shows a similar vegetation sequence with increasing impacts on the woodland and intensifying agricultural land-use from *c.* 1000. Distribution maps show several settlements, and traces of settlement of Roman Iron Age and Migration Period date focused around Lakes Wojnowo and Staświny, with a cemetery of first- to sixth-/seventh-century date close to the village of Staświny. The number of recorded settlement sites increases for the early and late medieval periods (Karczewski 2008) and supports the palynological picture for intensifying land-use within the surrounding landscape focused on the lakes. Moreover, there is significant continuity in the intensity of agricultural land-use in both the Lake Wojnowo, Miłkowski, and Staświny sequences through the early and late medieval periods. Continuity in occupation can be seen in Prussian and later medieval occupation in the nearby village of Staświny, and reflects a sustained focus on the lakes and surrounding landscapes in this part of the Lake District.

Other sequences emphasize the persistence of woodland throughout much of the medieval period within more peripheral and marginal locations. At Zielone Bagno there are indications of increasing human activity from *c*. 1000, although values for cereals and pasture/meadow plants are consistently low, along with only a small increase in microscopic charcoal, with more significant impacts on woodland not apparent until the seventeenth century (Figs 11.19 and 11.20). The soils around Zielone Bagno are predominantly nutrient-poor sandy podzols (Wacnik et al. 2012, Fig. 2) unsuited for intensive cereal cultivation, with pine-dominated woodland well-established on the sandy soils surrounding wetlands dominated by birch and alder. The predominance of birch pollen in the Wojnowo and Miłkowski sequence over a protracted period of time has been argued to reflect repeated clearings of woodland around the lakes, preventing succession to more closed canopy woodland (Wacnik et al. 2012), a pattern not apparent at Zielone Bagno, where the pollen shows a significant degree of stability in vegetation composition.

The settlement evidence from the landscape surrounding Zielone Bagno supports the palynological picture of a rather sparsely inhabited, still heavily wooded, and marginal landscape. There are traces of Roman Iron Age period settlements within the vicinity of Zielone Bagno, with two scatters of pottery identified 1.5 km east and a distinct concentration of activity, including a Roman cemetery 3–4 km east. Traces of Migration Period settlement are likewise suggested by finds of pottery sherds, though the quantities are small and the chronological precision of the artefacts low. Settlements on the shores of Nietlice swamp, *c*. 5 km south-west, disappeared in the sixth/seventh centuries at the end of the Migration Period and further support the picture of low settlement densities in the surrounding landscape. Written evidence suggests that the existing settlements near Zielone Bagno at Jagodne Małe (1.5 km north-west) and Górkło (2.8 km south-west) are most likely of Prussian origin with the pollen evidence presumably reflecting land-use associated with these settlements. Values for cultivated (up to 1.4%), weed (<1%), and ruderal (0.4–1.6%) taxa during the late medieval period are within the range recorded for the preceding Prussian phase. Although there is a documented increase in settlements in the wider landscape during the mid-sixteenth century, including the nearby settlement at Borki (1.2 km south), most — such as Jagodne Wielkie (3.3 km north-north-east), Paprotki (4.7 km east-north-east), and Marcinowa Wola (7.5 km east) — are located further from Zielone Bagno and are unlikely to register in the pollen profile.

Pollen from Wojsak (Fig. 11.18), 1.8 km north of the castle at Giżycko, similarly suggests the continuance of a heavily wooded landscape during the early medieval period, with an increase in ruderal and cereal pollen not occurring until the fourteenth/fifteenth centuries. Maps from the seventeenth century show a largely cleared landscape around Giżycko, but with evidence for the maintenance of woodland, typically on marginal land unsuited for agriculture, such as at Wojsak. At Lake Łazduny, similar heavy clearance of woodland did not occur much before *c*. 1450 (Wacnik et al. 2012). In the latter two cases, the increase in pollen of cereal, ruderal, and open-ground taxa reflect a renewed phase of colonization well-documented in the Great Masurian Lake District from the mid-fifteenth century.

The clearance of woodland around Lakes Wojnowo and Miłkowski was rapid, with the almost complete removal of woodland by *c*. 1100 and the development of fields and pastures, replacing the previous pattern of shifting slash and burn cultivation (Wacnik et al. 2012). The impact of the Teutonic Order on these patterns of land-use is difficult to detect in those pollen sequences where woodland was previously removed during the early medieval period, and it serves to emphasize a pattern of continuity in the Staświny micro-region. However, the palaeoenvironmental evidence for continuity in land-use within parts of the Masurian Lake District sits in stark contrast with documentary sources, particularly Peter von Dusburg's *Chronicle*. Dusburg describes the defeat of the Galindians by a neighbouring tribe who subsequently devastated their territory, leaving it a depopulated wasteland that was eventually colonized by a vast and dense forest that remained untouched until the arrival of the Teutonic Order (*CTP* III, 4). These events probably relate to a war that ended *c.* 1047 with Poland regaining control over neighbouring Masovia (allied with the Galindians), retained in oral tradition and referred to by Dusburg almost three centuries later.

However, palaeoenvironmental studies from adjoining lake districts support the more complex picture of land-use emerging from the Masurian Lake District (Fig. 11.19). Palynological analysis from Lake Salęt, located within the Mrągrowo Lake District, demonstrates that there is already significant clearance of woodland and intensive agricultural activity from *c.* 1150 to 1350, contrary to archaeological and documentary evidence suggesting abandonment of the settlements around the lake. There is, however, evidence for some regeneration of forests following the crusades between 1350 and 1450, in turn followed by a gradual intensification from the mid-fifteenth century involving intensive exploitation of woodland resources (Szal et al. 2013, 2014). Similar contradictions between the pollen and settlement evidence is apparent from Lake Wigry, located within the Suwałki-Augustów Lake District. Here pollen reveals regeneration of woodland within the first half of the first millennium AD, with decreasing levels of human indicators, a picture at odds with the archaeological evidence for intensive colonization at this time. The subsequent phase (cal. AD 1274–1601) is characterized by decreasing woodland and an increase in cereal cultivation, likewise at odds with the archaeological evidence that the area was uninhabited between the thirteenth and fifteenth centuries (Kupryjanowicz 2007). However, recent palynological analysis from Lake Linówek, located 5 km north of Lake Wigry, shows a measure of consistency between the pollen and cultural evidence: the pollen shows a decrease in human activity from the mid-thirteenth century, with regeneration of woodland, corresponding to documented depopulation of the areas as a result of the crusades. However, this was an unstable frontier environment prior to the crusades with frequent attacks by Slavic groups. The nearby stronghold at Jegliniec was abandoned sometime in the twelfth century as a result of attacks by Ruthenians. Further east, within the Knyszyńska Forest close to the border between Belarus and Lithuania, pollen analysis from Kładkowe Bagno shows that clearance of woodland for arable and pastoral activities did not occur until the fifteenth century; intensive colonization and cereal cultivation did not occur until the sixteenth century (Kupryjanowicz 2004), following successive invasions from the thirteenth to fifteenth centuries that had led to significant disruption of the open settlement pattern.

Conclusion

Important conclusions can be made on the basis of the palynological data from Prussia. The first of these is that significant impacts on woodland occurred in all areas during the medieval period prior to the arrival of the Teutonic Order (Fig. 11.19). Within the Kulmerland, clearance of oak/hornbeam-dominated woodland from the seventh/eighth centuries can be associated with initial Slavic colonization, with increasing intensity in agrarian land-use occurring with Polish control from the tenth century. The pattern is broadly similar within Pomesania to the north; increasing landscape openness is apparent from the eighth/ninth centuries with the appearance of the first Pomeranian strongholds but with evidence for significant continuity of woodland in more sparsely populated areas with limited agricultural potential. Significant impacts on woodland also occurred in the Great Masurian and adjoining Lake Districts resulting from the activities of indigenous tribal communities; this occurred more than a century prior to the arrival of the Teutonic Order and three centuries or more before sustained colonization in the mid-fifteenth century.

Consequently, the arrival of the Teutonic Order in Prussia does not correspond with a clearly identifiable horizon of intensifying land-use in the pollen record; indeed, the picture in many sequences appears to be one of broad continuity from early to late medieval. Within the Kulmerland, increasing arable and pastoral land-use occurs over the twelfth and thirteenth centuries. The initial decades of the crusades when there is likely to have been some political instability are characterized by a measure of continuity in land-use. Many of the larger pitched battles would have taken place along the edges of the Kulmerland as the crusading armies moved north, whilst the impact of periodic Prussian raids and incursions are likely too small to show up in the pollen record. Sufficient stability clearly existed, therefore, to provision the castles and new colonies that were founded and prospered during the initial decades of the crusades. More puzzling is the decline in arable land-use from the late thirteenth century, occurring at a time of relative stability when there is significant archaeological and documentary evidence for increasing German colonization across the Kulmerland. One suggestion is that this reflects a greater focus on stock-rearing, and the statistical modelling of the pollen data emphasizes the significant contribution of pastures and meadows within the landscape. However, agricultural produce, particularly rye, became an increasingly important export to Western Europe and Scandinavia, with shipments of grain up the Vistula recorded from the early fourteenth century (Hybel 2002). The construction of mills and granaries to process and store grain emphasize the organized system of provisioning, and inventory records from castles list large quantities of stored grain. Conversely, pollen sequences and modelling, particularly from the Kulmerland, do not show a noticeable increase in the quantities of either arable or pastoral land-use during the late medieval period when compared to the century preceding the crusades. However, the Kulmerland appears to have been heavily settled by the early thirteenth century, and it is possible that subsequent German colonization following the crusades was not accompanied by any increase in the amount of land under pastoral or arable use, and perhaps involved the more efficient exploitation of this land.

The twelfth and thirteenth centuries saw significant clearance of deciduous woodland across the lower Vistula, most clearly visible in pollen sequences from the Kulmerland. Timber, particularly oak, was a key resource and was the primary material used in the construction of castles and towns from the thirteenth century before being augmented or largely replaced with brick buildings during the fourteenth century. The extensive use of timber in early towns is emphasiszed at Elbląg, where excavations revealed the remains of timber-paved streets and piled oak logs used in foundations to support timber framed buildings (Nawrolska 2004). This is likely to have placed significant pressure on timber resources, particularly mature hardwoods, already denuded in those areas that witnessed increasing Polish colonization from the eleventh century. Even with the increasing use of brick construction from the fourteenth century, wood will have remained an important resource in the production of charcoal for the firing of bricks. Moreover, timber became an important trading commodity, with increasing quantities of timber being exported from the Baltic to north-west Europe (Haneca et al. 2005; Ważny 2002, 2005). Baltic timber was particularly prized for ship construction but was also used as a constructional timber to support rising populations which the local woodlands of north-west Europe, by then much denuded, could not supply.

Documentary records detailing the timber trade are rare before the sixteenth century, but records show the Teutonic Order bought almost one and a half million pieces of timber between *c.* 1389 and 1415, mostly from Masovia in the form of wainscots (high-quality timber boards). Dendro-provenancing of oak-timbers used in ships, buildings, and art-historical objects suggest a distinct correlation with the regional chronologies of northern Poland around Gdańsk and to the south within the Masovian lowlands (Haneca et al. 2005). Clearly, sufficient timber resources were available in the lower Vistula to support both domestic needs and foreign export markets, despite the majority of pollen studies suggesting significant reduction in woodland. However, it is clear from other pollen sequences, particularly those around the Forest of Sztum, that areas of woodland persisted during the early medieval and were actively maintained and managed during the late medieval period. Written sources clearly indicate the Order recognized the importance of preserving and managing woodland as a valuable resource. Within the Forest of Sztum, hardwood species such as oak and hornbeam remained significant components of the canopy throughout the medieval period, probably reflecting the careful management of the woodland to maintain a supply of mature hardwoods. It is probable that every commandery had areas of woodland maintained in a similar fashion, taking advantage of land already of limited agricultural potential. The lack of microscopic charcoal from sediment sequences suggests that there was very low fire activity in the studied landscapes, suggesting little use of fire in management of either agricultural land or woodland (Fig. 11.20).

Trees felled from the mid-fifteenth century show a stronger correlation with the regional chronologies of the Podlasie lowlands of eastern Poland (Haneca et al. 2005). It is no coincidence that this change in supply occurs after the Thirteen Years' War (1454–66) and the Second Peace of Thorn (1466), when large tracts of woodland along the Lithuanian border would have been safe to exploit. It is also at this time that sustained colonization occurs within the Great Masurian Lakelands, and many pollen sequences show evidence for renewed woodland clearance with some intensification of agrarian land-use. The Great Masurian and adjoining lake districts along the Lithuanian border remained unstable frontier zones well into the fifteenth century, and despite sustained colonization they have always remained somewhat more sparsely populated, retaining much of that feeling of a frontier environment. This contrasts with the Kulmerland, Marienburg, Christburg, and Elbing commanderies that rapidly developed into core heartlands of the *Ordensland*, particularly during the fourteenth century, and which saw intensive land-use, rural colonization, and urbanization.

Works Cited

Primary Source

CTP Scholz, K. and Wojtecki, D., trans. 1984. *Chronik des Preussenlandes/Peter von Dusburg*, Darmstadt: Wissenschaftliche Buchgesellschaft

Secondary Works

Badura, M. 2011. *Rośliny Użytkowe w Dawnym Gdańsku*, Studium Archeobotaniczne, Gdańsk: Wydawnictwo Uniwersytetu Gdańskiego

Barford, P. M. 2001. *The Early Slavs: Culture and Society in Early Medieval Eastern Europe*, New York: Cornell University Pres

Bennett, K. D. 1984. 'The Post-Glacial History of *Pinus sylvestris* in the British Isles', *Quaternary Science Reviews* 3, 133–55

Bernadzki, E., Bolibok, L., Brzeziecki, B., Zajączkowski, J., and Żybura, H. 1998. 'Composition Dynamics of Natural Forests in Białowieża National Park, North-Eastern Poland', *Journal of Vegetation Science* 9, 229–38

Blaauw, M., and Christen, A. 2011. 'Flexible Palaeoclimate Age-Depth Models Using an Auto-Regressive Gamma Process', *Bayesian Analysis* 6, 457–74

Bogucki, M., and Jagodziński, M. F., eds. 2012. *Studia nad Truso: Janów pomorski stan. 1. Wyniki ratowniczych badań archeologicznych w latach 2007–2008* (vol. I), Elbląg: Muzeum Archeologiczno-Historyczne

Brown, A. D., and Pluskowski, A. G., 2011. 'Detecting the Environmental Impact of the Baltic Crusades on a Late-Medieval (13th–15th Century) Frontier Landscape: Palynological Analysis from Malbork Castle and Hinterland, Northern Poland', *Journal of Archaeological Science* 38, 1957–66

Buko, A. 2005. *Archeologia Polski wczesnośredniowiecznej*, Warsaw: Trio (repr. in 2008)

Chęć, A., and Gancewski, J., 2009. 'Urząd leśniczego w państwie zakonu krzyżackiego –funkcjonowanie, znaczenie dla gospodarki', in *Las w kulturze polskiej, t. VI*, ed. W. Łysiak, Poznań: Eco, 47–56

Chudziak, W. 1994. 'Podstawy klasyfikacji chronologicznej grodzisk', in *Wczesnośredniowieczne grodziska Ziemi Chełmińskiej*, ed. J. Chudziakowa, Toruń: Universytetu Mikołaja Kopernika, 24–280

——, 1996. *Zasiedlenie strefy chełmińsko-dobrzyńskiej we wczesnym średniowieczu (VII–XI wiek)*, Toruń: Wydawnictwo Naukowe Uniwersytetu Mikołaja Kopernika

Długokęcki, W., and Haftka, M. 2000. 'Między Santyrem a Malborkiem. Z dziejó osadnictwa wzdłuż północno-zachodniej krawędzi Pojezierza Iławskiego w XIII w', in *Mazowsze-Pomorze-Prusy*, ed. D. Śliwiński, Gdańsk: Officina Ferberiana, 75–88

European Environment Agency, 2007. *European Forest Types: Categories and Types for Sustainable Forest Management Reporting and Policy*, Copenhagen: EAA

Falinski, J. B. 1986. *Vegetation Dynamics in Temperate Lowland Primeval Forests: Ecological Studies in Białowieża Forest* (= *Geobotany* 8)

Filbrandt–Czaja, A., and Noryśkiewicz, B. 2003. 'Osadnictwo na pograniczu słowiańsko–pruskim we wczesnym średniowieczu w świetle analizy pyłkowej', in *Pogranicze polsko–pruskie i kryżackie*, ed. K. Grążawski, Brodnica: Włocławek, 57–65

Forest Europe, UNECE, and FAO, 2011. *State of Europe's Forests 2011: Status and Trends in Sustainable Forest Management in Europe*, Oslo: Forest Europe

Gałka, M., Tobolski, K., Zawisza, E., and Goslar, T. 2014. 'Postglacial History of Vegetation, Human Activity and Lake-Level Changes at Jezioro Linówek in Northeast Poland, Based on Multi-Proxy Data', *Vegetation History and Archaeobotany* 23, 123–52

Haftka, M. 2010. *Zamki krzyżackie: Dzierzgoń-Przezmark-Sztum*, Gdańsk: Wydawnictwo 'Marpress'

Haneca, K., Ważny, T., van Acker, J., and Beeckman, H. 2005. 'Provenancing Baltic Timber from Art Historical Object: Success and Limitations', *Journal of Archaeological Science* 32, 261–71

Hellman, S., Gaillard, M.-J., Bunting, J., and Mazier, F. 2009. 'Estimating the Relevant Source Area of Pollen in the Past Cultural Landscape of Southern Sweden — A Forward Modelling Approach', *Review of Palaeobotany and Palynology* 151, 259–71

Hjelmroos, M., 1981. 'The Post-Glacial Development of Lake Wielkie Gacno, NW Poland: The Human Impact on the Natural Vegetation Recorded by Means of Pollen Analysis and 14C Dating', *Acta Palaeobotanica* 21, 129–44

Hybel, N. 2002. 'The Grain Trade in Northern Europe before 1350', *Economic History Review* 45, 219–47

Jagodziński, M. F. 2004. 'Czy Elbląg był kontynuacją Truso? Jeszcze jeden głos w dyskusji', in *Archaeologia et historia urbana*, eds R. Czaja, G. Nawrolska, M. Rębkowski, and J. Tandecki, Elbląg: Muzeum w Elblągu, 95–105

Karczewski, M. 2008, 'Zmiany poziomu lustra wody w jeziorach mazurskich w ciągu ostatnich dwóch tysięcy lat w świetle źródeł archeologicznych i historycznych', in *Botanical Guidebooks, No. 30: Polska północno–wschodnia w holocenie. Człowiek i jego środowisko. Holocene of North — East Poland. Man and his Environment*, eds A. Wacnik and E. Madeyska, Kraków: Instytut Botaniki im. W. Szafera PAN, 47–75

Kowalik, P. 2008. 'Flood Control in the Vistula River Delta (Poland)', *Environmental Biotechnology* 4/1, 1–6

Kupryjanowicz, M. 2004. 'The Vegetation Changes Recorded in Sediments of Kładkowe Bagno Bog in Puszcza Knyszyńska Forest, North-Eastern Poland', *Acta Palaeobotanica* 44, 175–93

——, 2007. 'Postglacial Development of Vegetation in the Vicinity of the Wigry lake', *Geochronometria* 27, 53–66

Lamentowicz, M., Cedro, A., Gałka, M., Goslar, T., Miotk-Szpiganowicz, G., Mitchell, E. A. D., and Pawlyta, J. 2008. 'Last Millennium Palaeoenvironmental Changes from a Baltic Bog (Poland) Inferred from Stable Isotopes, Pollen, Plant Macrofossils and Testate Amoeba', *Palaeogeography, Palaeoclimatology, Palaeoecology* 265, 93–106

Latałowa, M., Święta-Musznicka, J., and Pędziszewska, A. 2009. 'Źródła paleobotaniczne do rekonstrukcji wczesnych etapów rozwoju Gdańska', in *Środowisko–Człowiek–Cywilizacja* (vol. II), eds L. Domańska, P. Kittel, and J. Forysik, Poznań: Bogucki Wydawnictwo Naukowe, 175–185

Makohonienko, M. 2004. 'Late Holocene Period of Increasing Human Impact', in *Late Glacial and Holocene History of Vegetation in Poland Based on Isopollen Maps*, eds M, Ralska-Jasiewiczowa,M. Latałowa, K. Wasylikowa, K. Tobolski, E. Madeyska, H. E. Wright, and C. Turner, Kraków: Polish Academy of Sciences, 411–16

Makowiecki, D. 2006. 'Wybrane zagadnienia ze studiów nad gospodarką zwierzętami we wczesnośredniowiecznych kompleksach grodowych Pomorza, Wielkopolski i Dolnego Śląska', in *Stan i potrzeby badań nad wczesnym średniowieczem w Polsce — 15 lat później*, eds W. Chudziak and S. Moździoch, Toruń: Wydawn. Uniwersytetu Mikołaja Kopernika w Toruniu, 123–50

——, 2007. 'Gospodarka zwierzętami w strefie chełmińskiego pogranicza słowiańsko pruskiego i krzyżackiego w średniowieczu', in *Pogranicze polsko-pruskie i krzyżackie*, ed. K. Grążawski, Włocławek: Lega, 23–37

Nawrolska, G. 2004. 'Remarks on Infrastructure of the Old Town of Elbląg', in *Lübecker Kolloquium zur Stadtarchäologie im Hanseraum IV: die Infrastruktur*, eds R. Dunckel, M. Gläser, U. Oltmanns, and J. Scheschkewitz, Lübeck: Schmidt-Römhild, 303–22

Nitychoruk, J., Bińka, K., and Szymanek, M. 2012. *Sytuacja geologiczna i interpretacja paleośrodowiskowa obszaru dziedzińca Muzeum Arhceologiczno — Historycznego w Elblągu, Bulwar Zygmunta Augusta 11, 82–300 Elbląg*, Warszawa (unpublished report)

Nitychoruk, J., Szymanek, M., Bińka, K., Kasprzycka, M., and Zbucki, Ł. 2016. 'Changes of Sedimentation in the Drużno Lake Based on Geoarchaeological Data from the Teutonic Fortress in Elbląg, North Poland', *Acta Geologica Polonica* 66/1, 85 –98.

Noryśkiewicz, A. M. 2004a. 'Analiza pyłkowa osadów biogenicznych terasy zalewowej Wisły w profile Chełmno/Rybaki', in *Wczesnośredniowieczny zespół osadniczy Kałdusie: Studia przyrodniczo–archeologiczne*, ed. W. Chudziak, Toruń: UMK, 143–50

——, 2004b. 'Przemiany w szacie roślinnej okolic Uścia w okresie ostatnich dwóch tysięcy lat', in *Wczesnośredniowieczny zespół osadniczy Kałdusie. Studia przyrodniczo–archeologiczne*, ed. W. Chudziak, Toruń: UMK, 151–63

——, 2005. 'Preliminary Results of Study on Vegetation History in the Linje Mire Region Using Pollen Analysis', *Monographiae Botanicae* 94, 118–33

——, 2013. *Historia roślinności i osadnictwa ziemi Chełminskiej w późnym Holocenie: Studium palinologiczne*, Toruń: Wydawnictwo Naukowe Uniwersytetu Mikolaja Kopernika

Peterken, G. 1981. *Woodland Conservation and Management*, London: Chapman and Hall

Pidek, I. A., Filbrandt–Czaja, A., Noryśkiewicz, A., Noryśkiewicz, B., and Räsänen, S. 2009. 'New Methods Applied to Interpretations of Pollen Data in the Holocene — Selected Examples from the Last Decade', *Ecological Questions* 11, 49–64

Pluskowski, A. G. 2013. The *Archaeology of the Prussian Crusade: Holy War and Colonisation*, London: Routledge

Poliński, D. 2003. *Późnośredniowieczne osadnictwo wiejskie w ziemi chełmińskiej*, Toruń: UMK

Prygała, J. 1975. 'The Reconstruction of Agriculture and Breeding Economy in Plock Mazovia at the Decline of Antiquity', *Archaeologia Polona* 16, 71–106

Rackham, O. 2003. *Ancient Woodland: Its History, Vegetation and Uses in England*, Colvend: Castlepoint Press

Ralska-Jasiewiczowa, M., Miotk-Szpiganowicz, G., Zachowicz, J., Latałowa, M., and Nalepka, D. 2004. '*Carpinus betulus L.*– Hornbeam', in *Late Glacial and Holocene History of Vegetation in Poland Based on Isopollen Maps*, eds M. Ralska–Jasiewiczowa, M. Latałowa, K. Wasylikowa, K. Tobolski, E. Madeyska, H. E. Wright, and C. Turner, Kraków: Polish Academy of Sciences, 69–78

Scott, T. 2002. *Society and Economy in Germany, 1300–1600*, Basingstoke: Palgrave

Święta-Musznicka, J., Latałowa, M., Badura, M., and Gołembnik, A. 2013. 'Combined Pollen and Macrofossil Data as a Source of Reconstructing Mosaic Patterns of the Early Medieval Urban Habitats — A Case Study from Gdańsk, N. Poland', *Journal of Archaeological Science* 40, 637–38

Szal, M., Kupryjanowicz, M., Wyczółkowski, M. 2013. 'Puszcza Galindzka — kto i kiedy przyczynił się do jej wylesienia', in *Różnorodność biologiczna — od komórki do ekosystemu*, eds I. Ciereszko and A. Bajguza, Białystok: Polskie Towarzystwo Botaniczne, 227–43

Szal, M., Kupryjanowicz, M., Wyczółkowski M., and Tylmann, W. 2014. 'The Iron Age in the Mrągowo Lake District, Masuria, NE Poland: The Salęt Settlement Microregion as an Example of a Long-Lasting Human Impact on Vegetation', *Vegetation History and Archaeobotany* 23, 419–37

Vera, F. W. M. 2000. *Grazing Ecology and Forest History*, Wallingford: CABI

Vuorela, I. 1986. 'Palynological and Historical Evidence of Slash–and-Burn Cultivation in South Finland', in *Anthropogenic Indicators in Pollen Diagrams*, ed. K.-E. Behre, Rotterdam: Balkema, 53–64

Wacnik, A., Goslar T., and Czernik, J. 2012. 'Vegetation Changes Caused by Agricultural Societies in the Great Masurian Lake District', *Acta Palaeobotanica* 52, 59–104

Wacnik, A., Kupryjanowicz, M., Mueller-Bieniek, A., Karczewski, M., and Cywa, K. 2014. 'The Environmental and Cultural Contexts of the Late Iron Age and Medieval Settlement in the Mazurian Lake District, NE Poland: Combined Palaeoebotanical and Archaeological Data', *Vegetation History and Archaeobotany* 23, 419–37

Ważny, T. 2002. 'Baltic Timber in Western Europe: An Exciting Dendrochronological Question', *Dendrochronologia* 20, 313–20

——, 2005. 'The Origin, Assortments and Transport of Baltic Timber', in *Constructing Wooden Images*, eds C. Van de Velde, H. Beeckman, J. Van Acker, and F. Verhaeghe, Brussels: Brussels University Press, 115–26

Zachowicz, J., and Kępińska, U. 1982. 'The Late-Glacial and Holocene Vegetational History of the Żuławy Region, N. Poland: A Biostratigraphic Study of Lake Drużno Sediments', *Acta Palaeobtanica* 22, 141–161

——, 1987. 'Palaeoecological Development of Lake Drużno (Vistula Deltoic Area)', *Acta Palaeobotanica* 27, 227–49

12. Farming, Hunting, and Fishing in Medieval Prussia: The Zooarchaeological Data

Daniel Makowiecki, Mirosława Zabilska-Kunek, Krish Seetah, Marc Jarzebowski, and Aleksander Pluskowski

Introduction

The region which became redefined as the Prussian *Ordensland* after the crusades, encompassing the territories of the Prussia tribes, Pomerelia (eastern Pomerania), and the Kulmerland, was geographically varied. In the west, its ecosystems were defined by the Lower Vistula; in the north, the Vistula Delta had created an estuarine landscape between the Gulf of Gdańsk and the Vistula Lagoon, with expansive wetlands stretching across the floodplains to present-day Lake Drużno. The landscape of Prussian territories located further east of the Vistula was dominated by the presence of a dense network of smaller and larger lakes, formed in the late Pleistocene and early Holocene (see also Chapter 10).

Research excavations led by successive generations of archaeologists on both sides of the Vistula have resulted in the recovery of numerous assemblages of faunal remains (Map 12.1, Table 12.1). These became the basis for studies on the importance of animals in the cultural and economic developments of Kałdus (e.g., Makowiecki and Makowiecka 1999; Makowiecki 2010a) and Pomerelia (e.g., Kubasiewicz 1977; Sobociński 1979). Most recently studies were undertaken at Malbork (Ger. Marienburg) (Maltby et al. 2009; Pluskowski et al. 2010), Grudziądz (Ger. Graudenz) (Makowiecka et al. 2012) and Wąbrzeźno (Ger. Breisen) (Makowiecka and Makowiecki 2014). As a result of excavations carried out within the

Map 12.1: The distribution of sites with faunal assemblages included within this study in the Lower Vistula, Kulmerland, and Prussia. Map by D. Makowiecki and A. Pluskowski.

Table 12.1: Sites with zooarchaeological assemblages included within this study.

Key: M — Modern Time; EM — Early Medieval; LM — Late Medieval; ME — Medieval; K — Kulmerland; VE — Vistula East; VW — Vistula West; PW — Prussia West; PE — Prussia East; GB — Gdańsk Bay, Slavic; S/P — Slavic/Prussian; BiC — Bishop's castle (episcopal); C — Castle; Ch — Church; Mon — Monastery; H — Harbour; PR — Teutonic Order procurator's residence; SE — S — Stronghold, settlement; T — Town.

Miejscowość	Chronology	Function	Region	NISP	Reference
Bachotek, site 2	13th–half 14th	S	S/P	395	Makowiecki and Makowiecka 1999
Baldram, site 1	10th–11th	S	VE	1783	Sobociński (1965)
Barczewo, site 1	14th–M	T	PE	637	Makowiecki (2011)
Biała Góra	2nd half 12th–beg. 15th	SE	VE	3823	Pluskowski et al. 2014
Bobrowo, site 1	9th–1st half 11th	S	K	545	Makowiecki and Makowiecka 1999
Brodnica–Michałowo, site 1	2nd half 13th	S	K	1882	Makowiecki and Makowiecka 1999
Chmielno	10th–13th	S	VW	1996	Sobociński 1979
Ciepłe, site I	14th	S	VW	177	Sobociński 1979
Ciepłe, site IV	11th	S	VW	1120	Sobociński 1979
Elbląg — Old Town	2nd half 13th	T	PW	1565	Makowiecki (2003a)
Elbląg — Castle	end 13th–19th	C	PW	1731	Makowiecki (2011)
Ełk — Castle, site 8	15th–16th	C	PE	955	Makowiecki (2012a)
Gdańsk (Lastadia)	ME–M	T	GB	20,235	Makowiecki (2014a)
Gdańsk, site 1	11th–13th	S	GB	43,017	Krysiak 1955, 1956, 1967; Kubasiewicz 1977; Nogalski 1984; Makowiecki and Gotfredsen 2002; Makowiecki 2003b
Gdańsk, site 10	13th	T	GB	863	Kubasiewicz 1977
Gdańsk, site 10	10th–13th	SE	GB	76	Kubasiewicz 1977
Gdańsk, site 2	1275–1350	T	GB	12,115	Kubasiewicz 1977
Gdańsk, site 4	1255–1390	T	GB	5574	Kubasiewicz 1977
Gdańsk, Sadowa St.	14th–18th	T	GB	66	Makowiecki and Makowiecka 2014a
Gdańsk, Długie Ogrody	15th–M	T	GB	11,358	Makowiecki and Makowiecka 2014a
Gdańsk, Church of St John	1382–18th	T	GB	599	Makowiecki and Makowiecka 2014a
Gdańsk, Hevelius Square	EM–19th	T	GB	19,683	Makowiecki and Makowiecka 2014a
Gdańsk, Św. Ducha 43	1335–18th	T	GB	22,040	Makowiecki 2015
Gdańsk, Kładki 24 St.	15th–17th/18th?	T	GB	18,675	Makowiecki and Makowiecka 2014a
Gdańsk, Młyńska 21 St.	12th–M	T	GB	964	Makowiecki and Makowiecka 2014a
Gdańsk, Mniszki St.	2nd half 14th–17th	T	GB	163	Makowiecki and Makowiecka 2014a
Gdańsk, Na Piaskach St.	2nd half 14th–17th	T	GB	1444	Makowiecki and Makowiecka 2014a
Gdańsk, Olejarna St.	1220–1550	T	GB	12,453	Makowiecki and Makowiecka 2014a
Gdańsk, Piwna St.	2nd half 13th–13th/14th	T	GB	104	Makowiecki and Makowiecka 2014a
Gdańsk, Rajska St.	14th–16th/17th	T	GB	4829	Makowiecki and Makowiecka 2014a
Gdańsk, Seredyńskiego St.	17th	T	GB	195	Makowiecki and Makowiecka 2014a

Miejscowość	Chronology	Function	Region	NISP	Reference
Gdańsk, Wałowa St – 'Bursa'	Late ME–M	T	GB	2023	Makowiecki and Makowiecka 2014a
Gdańsk, Great Armoury	15th/16th	T	GB	160	Makowiecki and Makowiecka 2014a
Gdańsk, Great Mill	11th–14th	T	GB	59	Makowiecki and Makowiecka 2014a
Gdańsk, Granary Island	14th–18th	T	GB	50,651	Makowiecki and Makowiecka 2014a
Gdańsk, Green Gate	14th–18th	T	GB	1508	Makowiecki and Makowiecka 2014a
Giżycko — Castle	mid-15th–1st half 17th	C	PE	101	Makowiecki and Makowiecka 2014a
Gniew — Old Town	2nd half 11th–13th	SE	VW	140	Sobociński and Makowiecki 1992
Gniew — Old Town	13th–M	T	VW	15,614	Sobociński and Makowiecki 1992
Gniew, site 3	2nd half 13th–M	H	VW	3269	Sobociński 1992a
Gniew, site 3	2nd half 13th–M	SE-T	VW	53	Sobociński 1992a
Gniew, site 5	end 9th–10th/11th	S	VW	1573	Sobociński 1992b
Gniew, site 5	11th	SE	VW	1123	Sobociński 1992b
Gorzędziej	10th–13th	S	VW	799	Łuka 1985
Grążawy, site 1	11th/12th–half 13th	S	S/P	1002	Makowiecki 2017
Grążawy, site 2	11th/12th–half 13th	SE	S/P	515	Makowiecki 2017
Grudziądz	14th–1st half 16th	T	K	399	Makowiecki 2010a
Grudziądz, site 1	LM–M	C	K	3342	Makowiecka et al. 2012
Grudziądz, site 1	EM	SE	K	1577	Makowiecka et al. 2012
Gutowo, site 1	12th/13th	S	S/P	564	Makowiecki 2007a; Grążawski 2009
Janów Pomorski, site 1	end 8th–1st half 10th	SE	PW	16,032	Makowiecki 2012b; Makowiecki and Makowiecka 2012
Jedwabno, site 1	1st half 11th–LM	S	K	1435	Makowiecki 1994
Jedwabno, site 2	7th/8th–11th/12th	SE	K	1429	Makowiecki 1994
Jeziorko	10th–13th	SE	PE	155	Krysiak 1950
Kałdus, site 1	2nd half 10th–1st half 13th	SE	K	388	Makowiecki 2010b
Kałdus, site 2	mid 7th–1st half 13th	SE	K	26,352	Makowiecki 2010b
Kałdus, site 3	half 7th–LM	S	K	50,010	Makowiecki 2010b
Lembarg, site 6	12th	S	K	1612	Makowiecki and Makowiecka 1999
Lidzbark Warmiński, site XIX	2nd half 14th–M	BiC	PE	391	Makowiecki 2012c
Lubiszewo	9th–10th	S	VW	393	Sobociński 1979
Malbork	LM–M	C	VE	4967	Maltby et al. 2009
Malbork	EM	SE	VE	158	Maltby et al. 2009
Mała Nieszawka — Castle	14th–1st half 15th	C	VW	9207	Sobociński 1991; Nogalski 1991, Nogalski et al. 1992
Mikołajki — Marketplace	15th	SE	PE	748	Makowiecki and Zabilska-Kunek 2012
Napole, site 1	2nd half 10th–beg. 13th	S	K	4139	Makowiecki 2001
Napole, site 1B	2nd half 7th–beg. 13th	SE	K	7061	Makowiecki 2001
Napole, site 6	2nd half 7th–beg. 13th	S	K	4676	Makowiecki 2001a

Miejscowość	Chronology	Function	Region	NISP	Reference
Olsztyn Municipal Forest	8th–9th	SE	PE	788	Makowiecki 2013
Olsztynek — Castle	half 14th–18th	C	PE	434	Makowiecki 2007b
Osiek Rypiński, site 1	12th	S	S	1363	Makowiecki and Makowiecka 1999
Otomin	9th–1st half 10th	S	VW	309	Kubasziewicz 1965
Owidz, site 1	9th–12th	S	VW	8281	Sobociński 1989
Papowo Biskupie	13th/14th–15th/16th	C	K	603	Makowiecki 2007c
Pelplin — Maciejewo	8th–9th	S	VW	131	Sobociński 1979
Pień, site 1	1st half 15th	PR	K	2038	Makowiecka and Makowiecki 2008
Pień, site 2	2nd half 10th–beg. 11th	SE	K	129	Makowiecki 2003c
Płochocin, site 6	2nd half 11th–2nd half 12th	SE	VW	178	Bienias et al. 2008
Płochocinek, site 14	2nd half 11th–2nd half 12th	SE	VW	1302	Bienias et al. 2008
Podzamcze, site 1	9th–13th	S	VE	1079	Kruszona 1990
Poganowo, site 4	11th–12th	SacP	PE	5096	Wyczółkowski and Makowiecki 2009
Pruszcz Gdański	2nd half 10th–11th	S	VW	88	Sobociński 1979
Skarszewy	9th–10th	S	VW	698	Sobociński 1979
Słoszewy, site 1	2nd half 14th–1st half 15th	C	K	163	Sobociński 1975
Sopot, site 1	8th–10th	S	GB	1584	Sobociński and Makowiecki 1996
Staświny, site 1	EM (10th–13th)	S	PE	25,442	Makowiecki, Makowiecka, and Zabilska-Kunek 2014
Staświny, site 73/163	15th	SE	PE	586	Makowiecki, Makowiecka, and Zabilska-Kunek 2014
Szczuka, site 1	2nd half 11th/1st half 12th	S	S	678	Makowiecki and Makowiecka 1999
Szurpiły	9th–13th	S	PE	1851	Piątkowska-Małecka 2009
Świecie	14th/15th	SE	VW	422	Sobociński 1968
Tolkmicko, site 1	14th–M	T	PW	265	Makowiecka and Makowiecki 2013
Topolno	10th–13th	S	VW	2129	Kranz 1974
Toruń — Castle	EM–M	C	K	4117	Makowiecki and Makowiecka 2013
Toruń — Castle	EM	SE	K	1451	Makowiecki and Makowiecka 2013
Toruń, site 15	mid-13th–M	T	K	901	Makowiecki and Makowiecka 2013
Toruń, site 132	end 13th–M	Mon	K	1535	Makowiecki and Makowiecka 2013
Toruń, site 360 (Church of St Jacob)	14th–M	Ch	K	1661	Makowiecki and Makowiecka 2013
Toruń, site 537	half 14th–half 15th	T	K	9	Makowiecki and Makowiecka 2013
Toruń, site 578	half 14th–half 15th	T	K	456	Makowiecki and Makowiecka 2013
Wąbrzeźno, site 1	14th–M	BiC	K	4875	Makowiecka and Makowiecki 2014
Węgry, site 1	2nd half 11th–mid-12th	S	PW	6429	Waluszewska–Bubień 1976; Sobociński 1978; Makowiecki 2015
Zwiniarz, site 1	12th–12th/13th	S	S/P	1693	Makowiecki 2007d
Żmijewo 3, 4	7th–8th	S/SE	K	301	Makowiecki and Makowiecka 1999

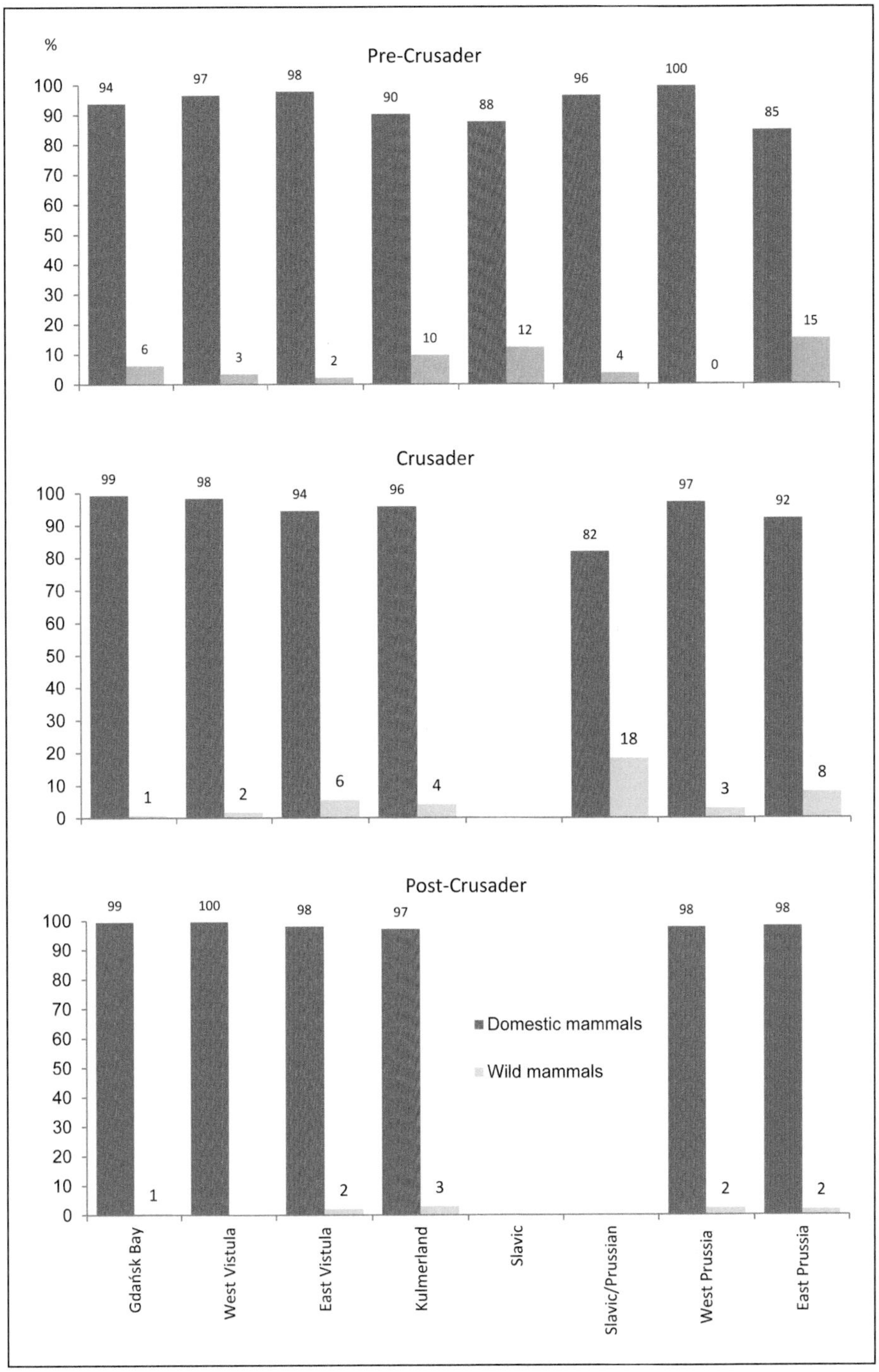

Figure 12.1: Percentage distribution of domestic and wild mammals by region and historical phase. Figure by D. Makowiecki.

information presented here derives from unpublished studies, the result of smaller, independent projects conducted by various Polish scientific institutions and museums (Table 12.1).

The zooarchaeological data presented in this chapter derives from a range of different categories of archaeological sites. The majority of assemblages derive from settlements developing before the arrival of the Teutonic Knights and the Prussian Crusade. These consisted of early medieval strongholds (*grody*) — the regional political-cultural centres — and their adjacent rural settlements. Amongst them, the most dynamic in terms of development and the best-known include Gdańsk (Ger. Danzig) and Culm (Kałdus). From the area of Prussia, such centres were undoubtedly Truso, *lauks* Staświny, or Olsztyn (Ger. Allenstein). Data from the time corresponding to the beginnings and the full development of the *Ordensland* derives from towns (Gdańsk, Toruń (Ger. Thorn), Gniew (Ger. Mewe), and Elbląg), castles (Elbląg, Malbork, Grudziądz, Toruń, Mała Nieszawka (Ger. Nessau), Giżycko, Ełk), the residences of bishops (Wąbrzeźno, Lidzbark Warmiński (Ger. Heilsberg)), and settlements founded or developed further under the auspices of the Order (Biała Góra). A unique case was provided by Poganowo (Site 4) (Ger. Groß Bürgersdorf), where the faunal assemblage consisted almost entirely of the remains of horses, interpreted as the depositional result of the ritual practices of the local Prussian community (Wyczółkowski and Makowiecki 2009). According to radiocarbon dates obtained from five samples of bone from this assemblage, the site functioned from *c.* AD 1003 to 1176.

In terms of their spatial and chronological contexts, the zooarchaeological data are quite heterogeneous.

framework of this present study, data was also collected from Biała Góra (Ger. Weißenberg) (Pluskowski et al. 2014; Sawicki et al. 2015) and Elbląg (Elbing) (Table 12.1). Far less data was available from sites in (tribal) Prussia. Much of this was obtained as a result of this study, which included faunal assemblages from Święta Góra (Ger. Heiliger Berg), Staświny (Ger. Stasswinen), Giżycko (Ger. Lötzen), and Ełk (Ger. Lyck). Much of the

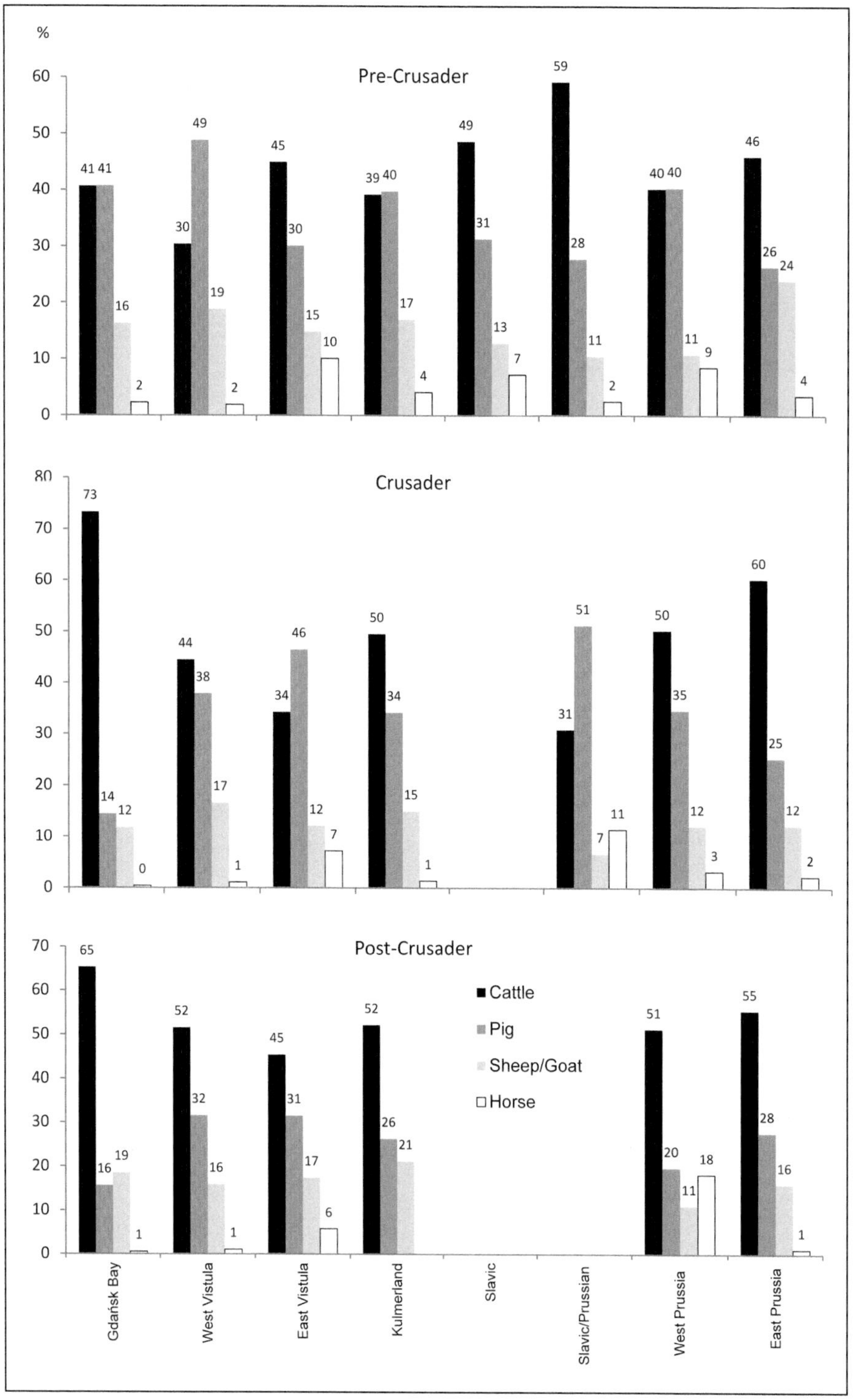

Figure 12.2: Percentage distribution of domestic mammals by region and historical phase. Figure by D. Makowiecki.

Given the quantity of data, it was impossible to present each assemblage individually within the confines of this chapter. Therefore, only the most abundant datasets that enabled the construction of diachronic and spatial trends based on historical and social contexts have been included. Due to the larger volume of data, this chapter also differs in structure from the presentation of comparable trends in animal use in Livonia (Chapter 6). Here, the data has been synthesized for the three main regions represented by Pomerelia, the Kulmerland, and Prussia, and subdivided further by phases of 'social time'. A new synthesis of pre-crusade animal exploitation in Prussia is also presented, the first study confronting zooarchaeological data with the more commonly used historical and ethnographic accounts. Both spatial and temporal categories are of course simplified approximations, but they provide models which can be tested with the analysis of new material. The most complex assemblages were derived from castles, where in most cases the archaeological stratigraphy had been affected by complex taphonomic processes linked to architectural and functional changes, reflecting their protracted occupational history. However, despite these difficulties, the general trends were relatively clear and consistent. The chronologies for all the primary sites included within this study were supported by multi-proxy dating (see Chapter 3).

Animal Use before the Crusades

Trends in animal exploitation by the Slavic and Prussian populations before the arrival of the Order are numerous assemblages derived from strongholds and settlements, developing from around the mid-seventh through to the early thirteenth centuries (Figs 12.1–12.2). The most complete zooarchaeological datasets derive primarily from Gdańsk, Kałdus, Truso, and Święta Góra, whilst the most important assemblages relating to strongholds were obtained from centres located in the Kulmerland. Together, these provide a solid basis for characterizing the region before it was conquered by the Order. In addition,

Table 12.2: NISP of species by region in the Pre-Teutonic Order phase

Animals / Regions	**Gdańsk Bay**	**Vistula West**	**Vistula East**	**Kulmerland**	**Slavic**	**Slavic/Prussia**	**Prussia West**	**Prussia East**	**Total**
Cattle — *Bos primigenius* f. *taurus*	20,889	4025	881	22,286	782	1377	2532	1555	54,327
Pig — *Sus scrofa* f. *domestica*	20,933	6697	591	22,621	504	647	2544	906	55,443
Sheep/Goat — *Ovis ammon* f. *aries*/*Capra aegagrus* f. *hircus*	8336	2555	291	8600	206	208	676	762	21,634
Sheep — *Ovis ammon* f. *aries*	41			792		10	6	25	874
Goat — *Capra aegagrus* f. *hircus*	18			265	0	27	3	16	329
Horse — *Equus ferus* f. *caballus*	1174	179	198	2336	116	58	538	100	4699
Dog — *Canis lupus* f. *familiaris*	351	137	2	332	10	3	56	13	904
Cat — *Felis silvestris* f. *catus*	181	1		153	1		9		345
Hare — *Lepus europaeus*	95	72		242	2	7	8	17	443
European rabbit — *Oryctolagus cuniculus*				5				1	6
Red squirrel — *Sciurus vulgaris*	3	1		18				6	28
European water vole — *Arvicola terrestris*	1	5						14	20
Black rat — *Rattus rattus*	3								3
Rat — *Rattus* sp.								1	1
Beaver — *Castor fiber*	236	17		96	2	2		36	389
Grey wolf — *Canis lupus*	16			3					19
Grey wolf/Dog — *Canis lupus* / *Canis lupus* f. *familiaris*				1					1
Red fox — *Vulpes vulpes*	31	5		36		1		6	79
Brown bear — *Ursus arctos*	22			111				6	139
European polecat — *Mustela putorius*	5			1				1	7
Marten — *Martes* sp.				8			2	3	13
Badger — *Meles meles*	12			10		1		3	26
Mustelids — Mustelidae				1			2	7	10
Least weasel — *Mustela nivalis*	1			1					2
European otter — *Lutra lutra*	47			6		1			54
Wild cat — *Felis silvestris*	15			4					19
Wild/domestic cat — *F. silvestris* / *F. silvestris* f. *catus*				1			1		2
Eurasian lynx — *Felis lynx*	3	2		3				5	13
Wild boar — *Sus scrofa*	685	145		2228	64	25	4	91	3242
Wild boar/Pig — *Sus scrofa*/*Sus scrofa* f. *domestica*				5			2	4	11
Red deer — *Cervus elaphus*	449	113	39	2082	116	28	10	188	3025
Elk — *Alces alces*	1033	13		160	19	3	2	32	1262
Roe deer — *Capreolus capreolus*	374	47	6	1024	15	12	1	179	1658
Aurochs — *Bos primigenius*	222	55		93	8	8		7	393
Aurochs/Cattle — *Bos primigenius*/*Bos primigenius* f. *taurus*				3			1	6	10
Aurochs/European bison — *Bos primigenius*/*Bison bonasus*				4			1	4	9
European bison — *Bison bonasus*	50			3				2	55
Western European hedgehog — *Erinaceus europaeus*						1			1
Seals — Phocidae	28	8							36
Grey seal — *Halichoerus gryphus*	94								94
Common porpoise — *Phocoena phocoena*	1								1
Birds — Aves	1423	316	17	2869	18	16	106	277	5042
European pond terrapin — *Emys orbicularis*				6					6
Fish — Pisces	1316	113	1	2897	2	31	5689	12,248	22,297
Bivalvia	3			30				1	34
European crayfish — *Astacus* sp.				2					2
Identified	58,091	14,506	2026	69,338	1865	2466	12,193	16,522	17,707
Unidentified	1543	4327	940	22,563	176	1273	9722	10,616	51,160
Total	**59,634**	**18,833**	**2966**	**91,901**	**2041**	**3739**	**21,915**	**27,138**	**228,167**

Table 12.3: NISP of bird species in the Pre-Teutonic Order phase

Bird taxa / Site	**Gdańsk**	**Gniew, site 3 & 5**	**Kaldus site 1& 2**	**Janów Pomorski, site 1**	**Malbork**	**Węgry, site 1**	**Olsztyn Las Miejski**	**Staświny, site 1**	**Grążawy**	**Gutowo, site 1**	**Zwiniarz, site 1**
Birds — *Aves* unidentifiable	20			58	1		1	197	1	1	5
Black-throated diver — *Gavia arctica*	1										
Diwer — *Gavia* sp.	1										
Cormorant — *Phalacrocorax carbo*	3		1								
Bittern — *Botarus stellaris*			1								
Grey heron — *Ardea cinerea*	3		4								
Stork — *Ciconia* sp.	2		1								
Mute swan — *Cygnus olor*	13		1								
Greylag goose — *Anser anser*									1		
Greylag goose/Domestic goose — *Anser anser/ A. anser* f. *domestica*	30	18				8					
Goose — *Anser* sp.	66		127		1			1			
Teal — *Anas crecca*			2								
Mallard — *Anas platyrhynchos*			1						1		1
Mallard/Domestic duck — *Anas platyrhynchos/ A. platyrhynchos* f. *domestica*	8	2				5					
Anas sp.	87		31					1			
Garganey — *Anas querquedula*						1					
Long-tailed duck — *Clangula hyemalis*		1									
Common goldeneye — *Bucephala clangula*			1								
White-tailed eagle — *Haliaeetus albicilla*	9										
Sparrowhawk — *Accipiter nisus*			2								
Goshawk — *Accipiter gentilis*	7		15								
Lesser-spotted eagle — *Aqulia pomarina/ Pandion haliaetus*			1								
Golden eagle — *Aquila chrysaetos*						2					
Kestrel — *Falco tinnunculus*			1								
Black grouse — *Tetrao tetrix*	9		21					1			
Capercaillie — *Tetrao urogallus*	10		38					4			
Black grouse/capercaillie — *Tetrao* sp.			1								
Partridge — *Perdix perdix*	4		2								
Quile — *Coturnix coturnix*		2									
Domestic chicken — *Gallus gallus* f. *domestica*	992	21	734	7		25	11	58	1	3	2
Crane — *Grus* sp.			27					1			
Moorhen — *Gallinula chloropus*			1								
Coot — *Fulica atra*	3		1								
Woodcock — *Scolopax rusticola*		2	1								
Stock dove — *Columba oenas*			5								
Dove — *Columba livia* f. *domestica*			1								
Dove — *Columba* sp.	21										
Barn owl — *Tyto alba*	1										
Short-eared owl/long-eared owl — *Asio flammeus/Asio otus*			1								
Jay — *Garrulus glandarius*			1								
Raven — *Corvus corax*	1		2								
Carrion crow — *Corvus corone*			6								
Jackdaw — *Corvus monedula*			2								
Corvus sp.			8								
Total	**1291**	**45**	**1041**	**65**	**2**	**41**	**12**	**263**	**4**	**4**	**8**

Table 12.4: NISP of fish species in the Pre-Teutonic Order phase (data from Makowiecki 2003b, 2010, and unpublished).

Fish taxa / Region	**Gdańsk Bay**	**Vistula West**	**Kulmerland**	**Prussia West**	**Prussia East**
Pike — *Esox lucius*	340	3	283	151	1022
Carp — *Cyprinidae*	1625		1122	1526	955
Roach — *Rutilus rutilus*	174	2	7	8	812
Roach/Rudd — *Rutilus rutilus/Scardinius erythrophthalmus*					4
Eurasian daces — *Leuciscus* sp.	83				
Chub — *Leuciscus cephalus*	5				1
Ide, orfe — *Leuciscus idus*	16		6		2
Rudd — *Scardinius erythrophthalmus*	14		1		26
Asp — *Aspius aspius*	147		28	1	
Tench — *Tinca tinca*	303		51		88
Barbel — *Barbus barbus*	1				
Bleak — *Alburnus alburnus*				2	23
White bream — *Blicca björkna*	42				7
Bream — *Abramis brama*	1470	2	62	90	495
Zarthe — *Vimba vimba*	92				
Sichel — *Pelecus cultratus*	12	1		2	
Crucian carp — *Carassius carassius*	13		8		16
Carp — *Cyprinus carpio* f. *domestica*	1				
Catfish — *Silurus glanis*	121	1	114		29
Perches — Percidae			6	32	15
Perch — *Perca fluviatilis*	237	2	46	102	535
Pikeperch — *Lucioperca lucioperca*	3005	4	39	137	16
Burbot — *Lota lota*	3	1	1		
Salmo sp.	3		5		
Pollan — *Coregonus lavaretus*	1				
Sturgeon — *Acipenser* sp.	5853	9	355	824	
Eel — *Anguilla anguilla*	5		1	34	
Herrings — Clupeidae	361				
Herring — *Clupea harengus*	1		24	420	
Allis (allice) — *Alosa alosa*	1				
Cod — *Gadus morhua*	15				
Flatfishes — Pleuronectidae	42			1	
Identified	13,986	25	2159	3330	4046
Unidentified	5670	24	745	1399	1767
Total	**33,642**	**74**	**5063**	**8059**	**9859**

the assemblage from Poganowo (site 4) provides important data on the pre-crusade period horse population in Prussia.

The Kulmerland, the Lower Vistula, and Gdańsk Bay

The assemblages from these regions inhabited by Slavs in the pre-crusade period indicate a diverse range of fauna were exploited (Tables 12.2–4). These peoples were very familiar with livestock and poultry, with hunting wild mammals and birds. They also fished in nearby lakes, rivers, and the bays of the Baltic Sea. In the majority of settlements and regions, a regular set of species was exploited. Amongst the domestic species were cattle (*Bos taurus*), sheep (*Ovis aries*), goats (*Capra hircus*), pigs (*Sus scrofa*), horses (*Equus caballus*), dogs (*Canis l. familiaries*), cats (*Felis catus*), and chickens (*Gallus gallus*). The first four, along with chicken, formed the basis of subsistence. The other species had different roles: horses were mainly bred for riding, and their meat was also consumed to a small extent. The dog and the cat which appear in smaller quantities in these assemblages (Table 12.2), as in other regions of Europe at that time, were kept by the Slavs as companions (Benecke 1994a).

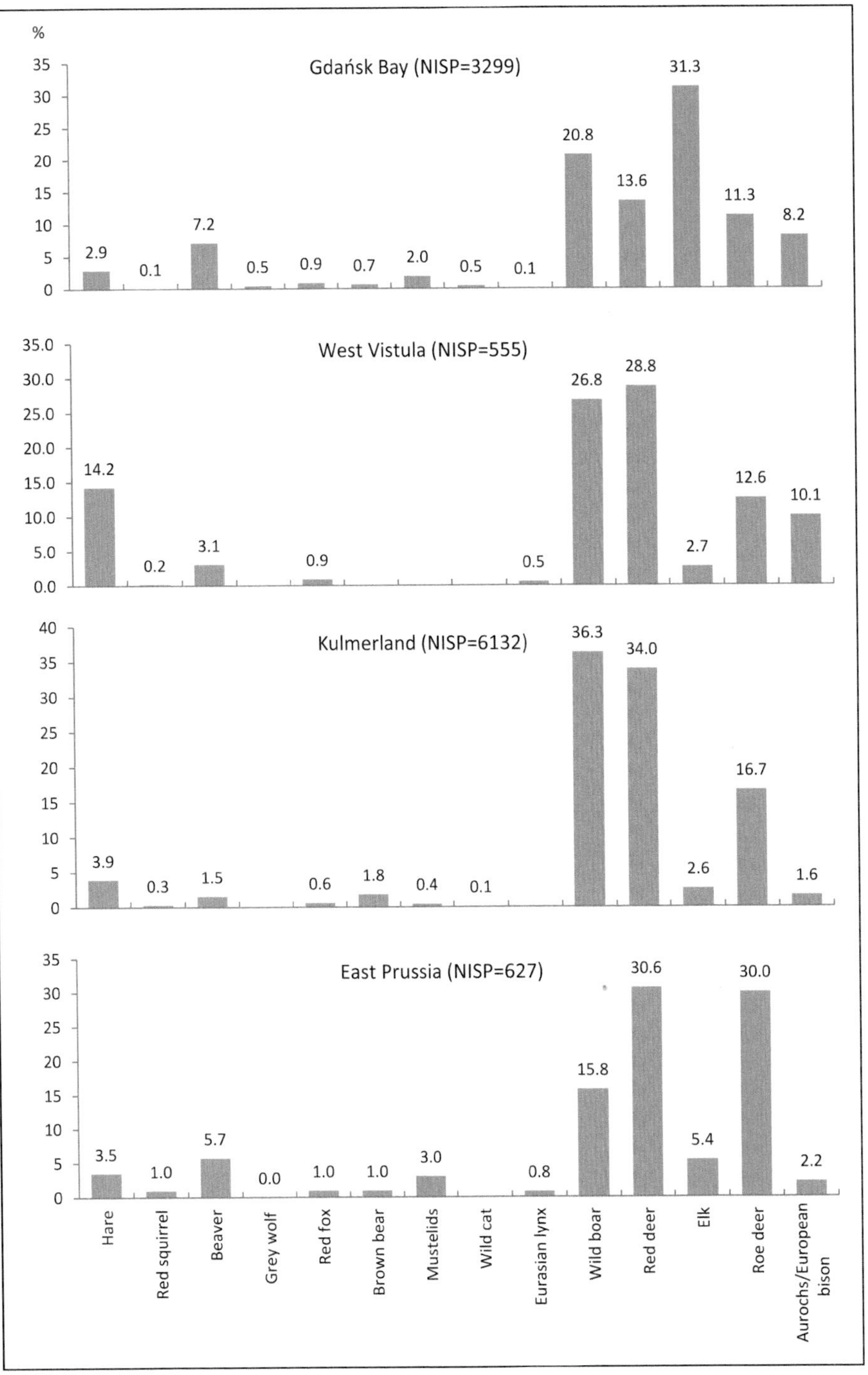

Figure 12.3: Percentage distribution of wild mammals by region in the Pre-Teutonic Order phase. Figure by D. Makowiecki.

Several categories of wild mammals were hunted and trapped. Amongst them were species which were categorized in Polish medieval sources as *Animalia minute* and *Animalia superior* (Samsonowicz 1991) and in German sources as Niederwild and Hochwild respectively (see Chapter 15). The value of the former was linked to their meat and fur (Makowiecki 2010b), and they included the hare (*Lepus europaeus*), red squirrel (*Sciurus vulgaris*), beaver (*Castor fiber*), red fox (*Vulpes vulpes*), European polecat (*Mustela putorius*), pine marten (*Martes martes*), badger (*Meles meles*), least weasel (*Mustela nivalis*), European otter (*Lutra lutra*), wild cat (*Felis silvestris*), grey wolf (*Canis l. lupus*), and Eurasian lynx (*Lynx lynx*). The remains of rabbits (*Oryctolagus cuniculus*) and European water voles (*Arvicola amphibious*) in pre-crusade contexts must be regarded as intrusive, and potentially the remains of black rat (*Rattus rattus*) and other *Rattus* species. The second group included species defined by their large size, referred to today as game. These were wild boar (*Sus scrofa*), red deer (*Cervus elaphus*), roe deer (*Capreolus capreolus*), elk (*Alces alces*), aurochs (*Bos primigenius*), and bison (*Bison bonasus*). The Slavs living along the Baltic zone also exploited marine mammals, specifically the grey seal (*Halichoerus grypus*) and harbour porpoise (*Phocoena phocoena*), whose remains have been found in Gdańsk (Table 12.2).

The quantified data of wild mammals indicates that the relative proportions between species of *Animalia superiora* and *Animalia minuta* were similar in all the regions, where the remains of the former constitute a significantly higher proportion than the latter (Fig. 12.3). However, the individual regions differ in terms of species compositions. The most distinctive is the region around the Gulf of Gdańsk, where assemblages are dominated by the remains of elk and wild boar, followed by red deer and roe deer. The remains of beaver dominate wild mammal assemblages within the other regions, whilst the representation of hare are relatively limited. However, different proportions are found in the Lower Vistula: here, the dominant species are red deer and wild boar, with quantities of hare higher than any other region, surpassing even roe deer, whilst the presence of elk is relatively limited. Contrasts between spe-

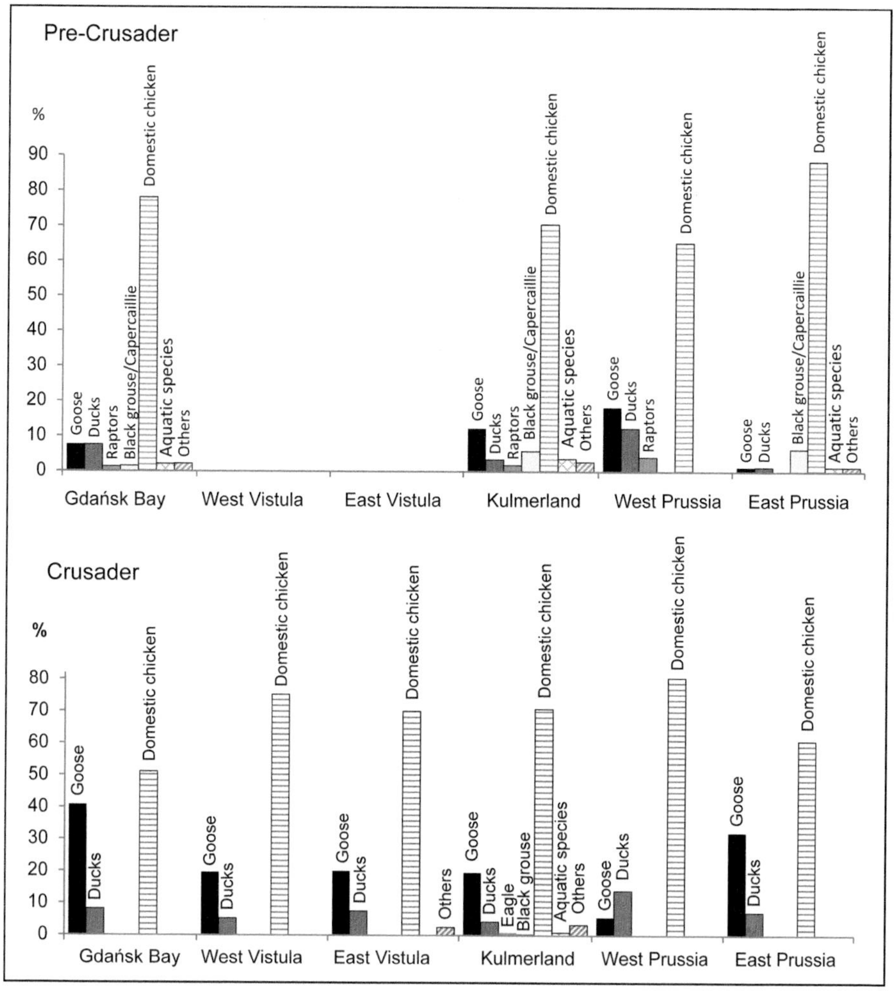

Figure 12.4: Percentage distribution of birds by region and historical phase.
Figure by D. Makowiecki.

cies within the Kulmerland are not so pronounced. The dominant species are wild boar and to a lesser extent red deer. In comparison, numbers of roe deer are significantly smaller, although representing a slightly higher proportion than in the previous region. The importance of elk is almost identical to that on sites west of the Vistula, while the proportion of hare is around 10% less. Around half of the wild mammal assemblage consists of beaver.

The number of exploited bird species was equally extensive and mainly based on the assemblages from Gdańsk (Nogalski 1984) and Kałdus (Makowiecki et al. 2007; Makowiecki 2010b) (Table 12.3). Domestic fowl were also important for the provision of meat and eggs (Fig. 12.4). This species was bred for these purposes throughout the entire western Slavic range, as in Western Europe (Benecke 1994; Serjeantson 2009), and it is possible that geese were also reared, as in the case of the Polabian Slavs (Benecke 1988, 34; Prummel 1993, 82). Amongst the domestic bird species pigeons were also bred, and whilst the positive identification of their domestic variant is difficult, *Columba* has been identified in both Gdańsk and Kałdus. The breeding of domestic pigeons is attested in written sources; according to Arabic accounts, the Slavs caught live wild pigeons and then tamed them. The 'Drewlanie' (a group of eastern Slavs) kept large numbers of pigeons in dovecotes (Lewicki 1954, 468), and the remains of *Columba* have been identified at the majority of centres of the early Piast state (Bocheński et al. 2012). In Kałdus one bone was even positively identified as a domestic pigeon (Table 12.3).

The number of bird species that were hunted was quite extensive (Nogalski 1984; Makowiecki and Gotfredsen 2002; Makowiecki 2010b). Undoubtedly, the most valuable were capercaillie (*Tetrao urogallus*) and black grouse (*Tetrao tetrix*), which inhabited large and old woodlands. In Kałdus their meat was more often consumed by the elite community inhabiting the stronghold than by the population of the adjacent settlement (Makowiecki 2010b, 202). In the case of waterfowl the most important species were wild geese, in particular greylag goose and duck, especially mallard (*Anas platyrhynchos*). These species represent several types of habitat, with the majority deriving from aquatic ecosystems, followed by woodlands, and significantly far less from open landscapes and least exploited were synanthropic species (*Corvidae*).

Ichthyological data derived again primarily from Gdańsk (Makowiecki 2003b) and Kałdus (Makowiecki 2010b), indicate that many different species of fish were exploited in the pre-crusade period (Table 12.4). Within the freshwater range this included cyprinids, as well as pike (*Esox lucius*), perch (*Perca fluviatilis*), and pike-perch or zander (*Sander lucioperca*) (Fig. 12.5). Migratory species included sturgeon (Acipenseridae), salmon/sea trout (*Salmo salar / Salmo trutta*), and eel (*Anguilla anguilla*), while marine species mostly consisted of herring (*Clupea harengus*) (Table 12.4). With the rise of the Polish state, herring became the

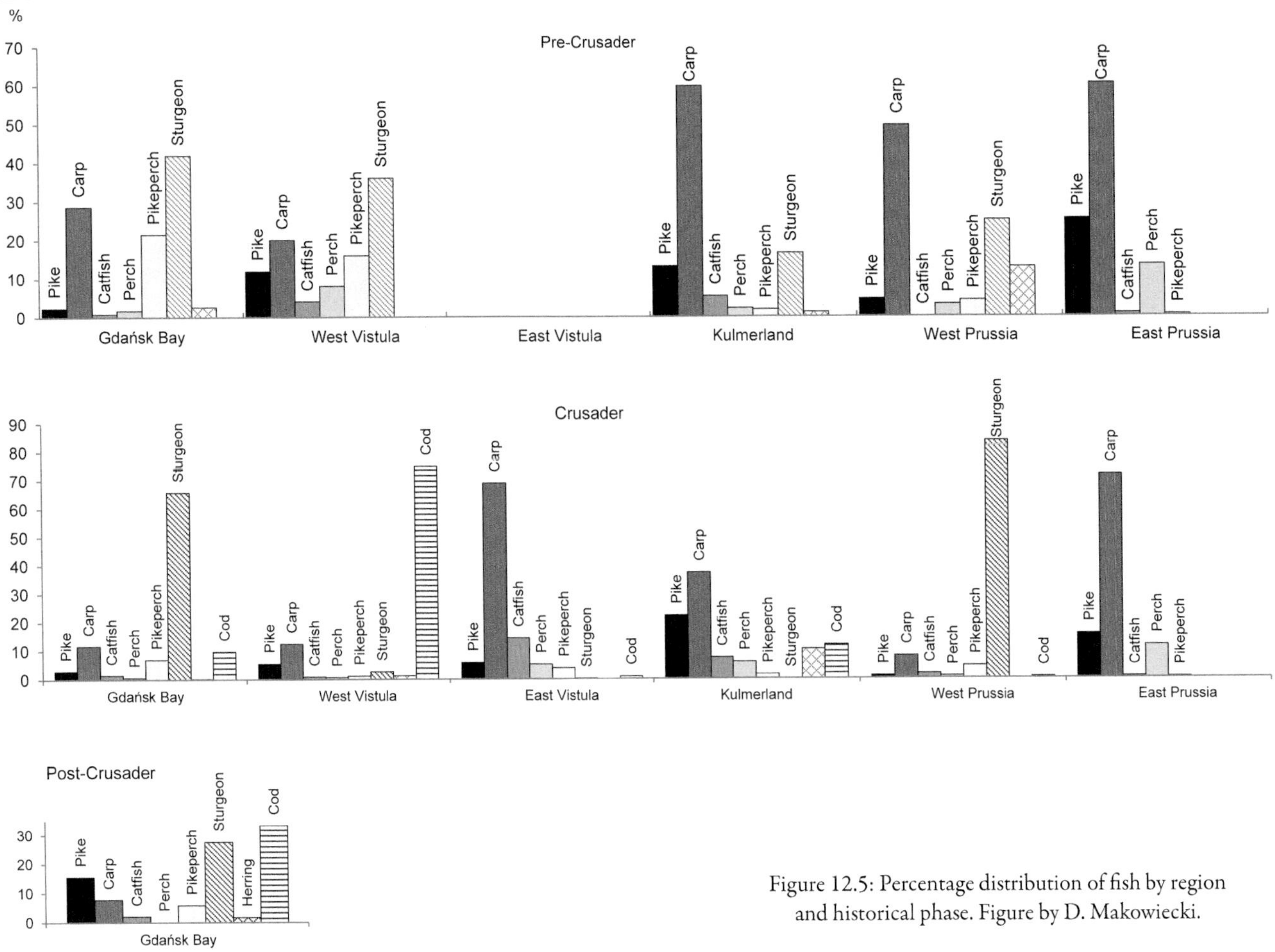

Figure 12.5: Percentage distribution of fish by region and historical phase. Figure by D. Makowiecki.

only marine species of any commercial significance (Makowiecki 2003b, 2008a), and it is not surprising to find its remains were transported to Culm, the principal centre of the Kulmerland (Makowiecki 2010b). In addition to these vertebrates, European pond turtle (*Emys orbicularis*) were also caught, as well as invertebrates represented by crabs (*Astacus* sp.) and mussels (*Unio* sp./ *Anodonta* sp.). Their remains have been found only on some sites, turtles and crabs at Kałdus, and additionally mussels at Gdańsk. On the basis of ethnographic analogues, the latter may have served as feed for pigs and chickens (Kawecki 1988, 205).

Livestock

Livestock husbandry clearly underpinned the economy of every type of settlement. Cattle were the most important in terms of meat provision, followed by pigs (reared for meat and fat) and then small ruminants, sheep, and goats (Fig. 12.2). In the period preceding the emergence of the Polish state, the consumption of pork was significantly lower in settlements within the area corresponding to Polish territory (Makowiecki 2006a). In this respect, the high proportion of pig remains on sites in the Kulmerland is the direct result of adopting breeding strategies from Greater Poland (Pol. Wielkopolska), where this type of husbandry formed the basis of meat consumption, much more so than cattle (Makowiecki 2007e). This region was the core of the emerging Piast state, which from the late tenth century also included the Kulmerland (Chudziak 2003).

On the basis of the age of slaughter profiles for cattle (Fig. 12.6), pig (Fig. 12.7), sheep/goat (Fig. 12.8), and horse (Fig. 12.9), it is clear the regular husbandry practices were defined for each species. There were, however, differences within each stock. Whilst meat was obtained from animals of different ages, in the case of some this played an insignificant role. The mortality profile for horses indicates a completely different role: the majority of individuals were killed when *c.* six to eight years old, but a large proportion of the equine population lived

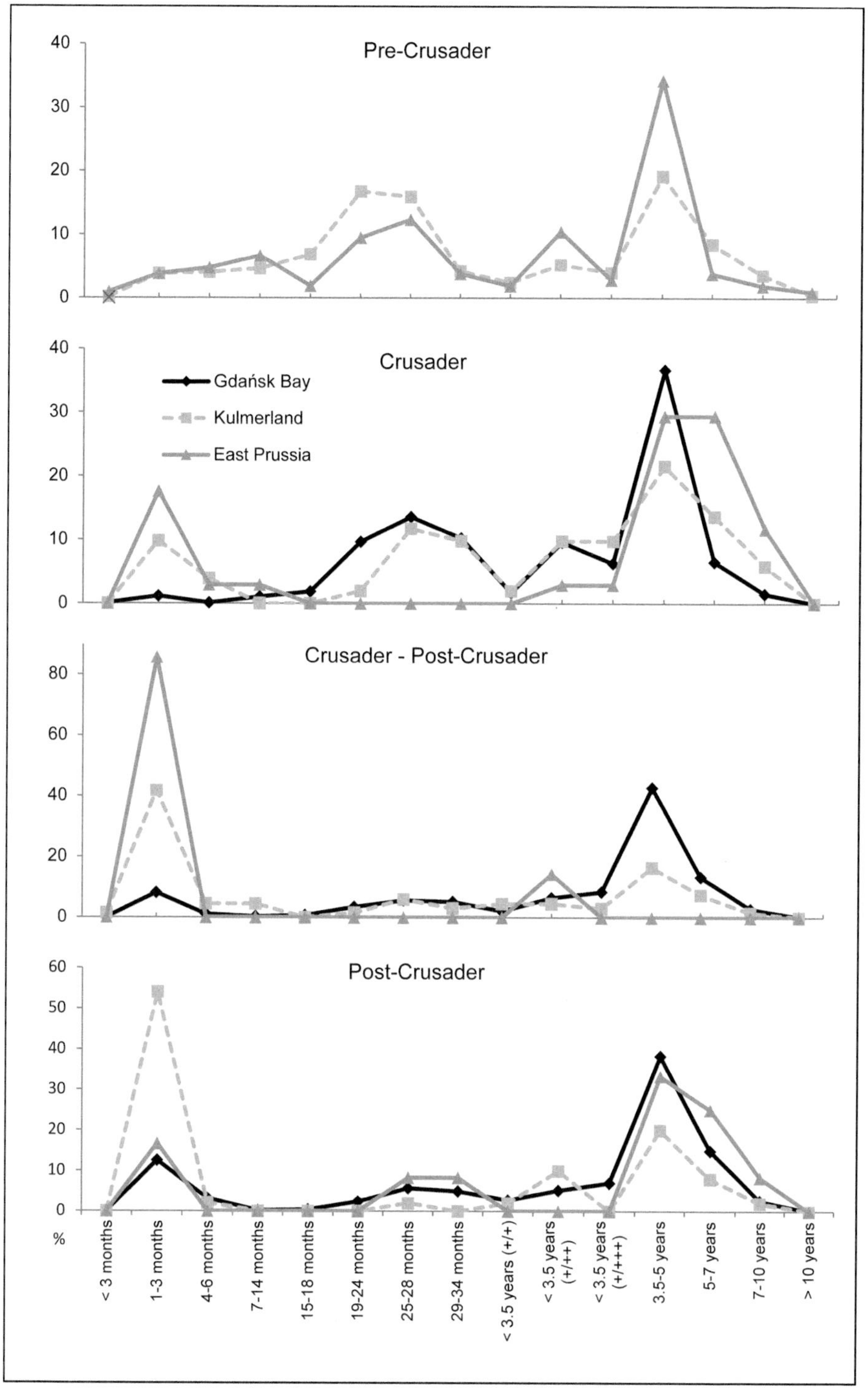

Figure 12.6: Age distribution of cattle by region and historical phase. Figure by D. Makowiecki.

until twenty, indicating these animals were used for as long as possible. This clearly required a level of care and investment which was only profitable when horses were used for more than simply meat exploitation.

Cattle were principally bred for meat and to a lesser extent for dairy produce. The zooarchaeological data contrasts with references in contemporary chronicles which praise the 'milk yields' of Slavic cows and refer to 'butter from cows' (Labuda 1999, 128n). Of course, in the early Middle Ages this product would have been available after calving; however, its significance should not be overstated. First, the relative dominant proportion of juvenile and immature/sub-adult animals to adults is so significant that it can be assumed that only a small portion of the adult herd could have been potentially exploited for dairy produce. Second, the gender composition of the adult herd at Kałdus indicates that there were almost as many males (bulls and castrates) as cows (Makowiecki 2010b, 79). Moreover, the average shoulder height of these animals was 103–08 cm, and only limited quantities of milk could be extracted from such small animals. Presently, the dairying cycle occupies 305 days. One estimate assumes that small medieval cows during a lactation period of 90 to 120 days provided around 400 to 600 kg of milk (Benecke 1994b, 133). From this a calf required around 250–350 kg, with a surplus of only 150–250 kg. As a result, obtaining milk from a cow was heavily dependent on what proportion of calves were designated for further breeding. The mortality profile for cattle shows that calves slaughtered up to three months of age accounted for a small proportion of the herd; they relied on milk, as their stomachs were not developed enough to take roughage. The use of cattle for meat is also reflected in the fact that the total percentage of juveniles and sub-adults is almost the same as adult and mature animals (Fig. 12.6).

Cattle are a versatile species with multiple uses aside from meat and dairy products. They were used in traction, as evident by characteristic pathological changes in their skeletons, including the degeneration of the articular joints of both extremities (Makowiecki 2010). Of all

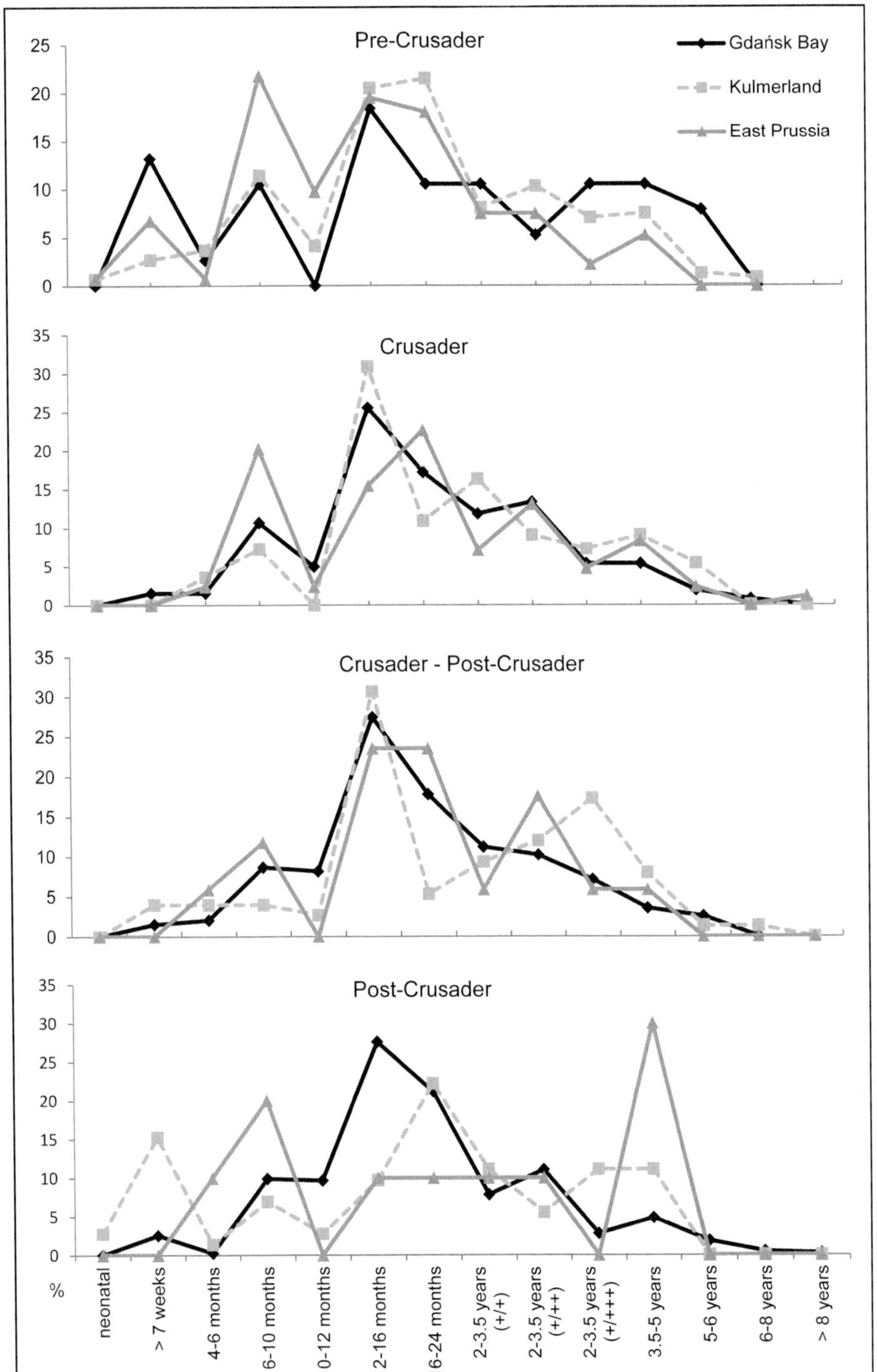

Figure 12.7: Age distribution of pig by region and historical phase. Figure by D. Makowiecki.

Wild Mammals, Birds, and Fish

The small percentage of wild mammal remains indicates that this group played a minor role in daily nutrition (Fig. 12.1). However, there would also have been a season in which the exploitation of wild species, particularly through hunting, played a significant role in provisioning. This crucial period would have been the entirety of winter and early spring, when deer, wild boar, and elk would be hunted (Makowiecki 2010b). It can also be assumed that the hunting of species designated as *Animalia superiora* was more an expression of elite privilege and prestige than economic necessity, as in other regions of Europe where hunting became closely aligned with aristocratic identity (Samsonowicz 1991). This group of mammals also played an important symbolic role within Slavic culture. The accumulation of skulls and mandibles of deer, elk, wild boars, and bears (*Ursus arctos*) is well attested in the peripheral areas of the stronghold at Kałdus, situated on the escarpment overlooking the Vistula (Makowiecki 2010b, 74). It is possible that the paws of bears, in particular, reflected a special type of culinaria, and they were found in one of the designated ritual sites within the settlement complex (Makowiecki 2010b, 159). The elites controlling the strongholds also appear to have hunted with hawks. The remains of females of this species have been found in the assemblages of both the stronghold and the settlement at Kałdus (Makowiecki et al. 2007), as well as at other early medieval *sedes regni principalis* associated with the early Polish state, particularly within Greater Poland (Waluszewska-Bubień 1980; Makowiecki et al. 2007, 2014). Of the various quarry species hunted with hawks, the most commonly represented in these assemblages is hare.

the species, cattle had the highest prevalence of degenerative pathologies. On the basis of ethnographic analogues, it is also possible to conclude that their manure was regularly used to fertilize cultivated fields (Kowalska-Lewicka 1980). Their bones also represented an important raw material for the production of tools and other artefacts, more so than any other domestic mammal.

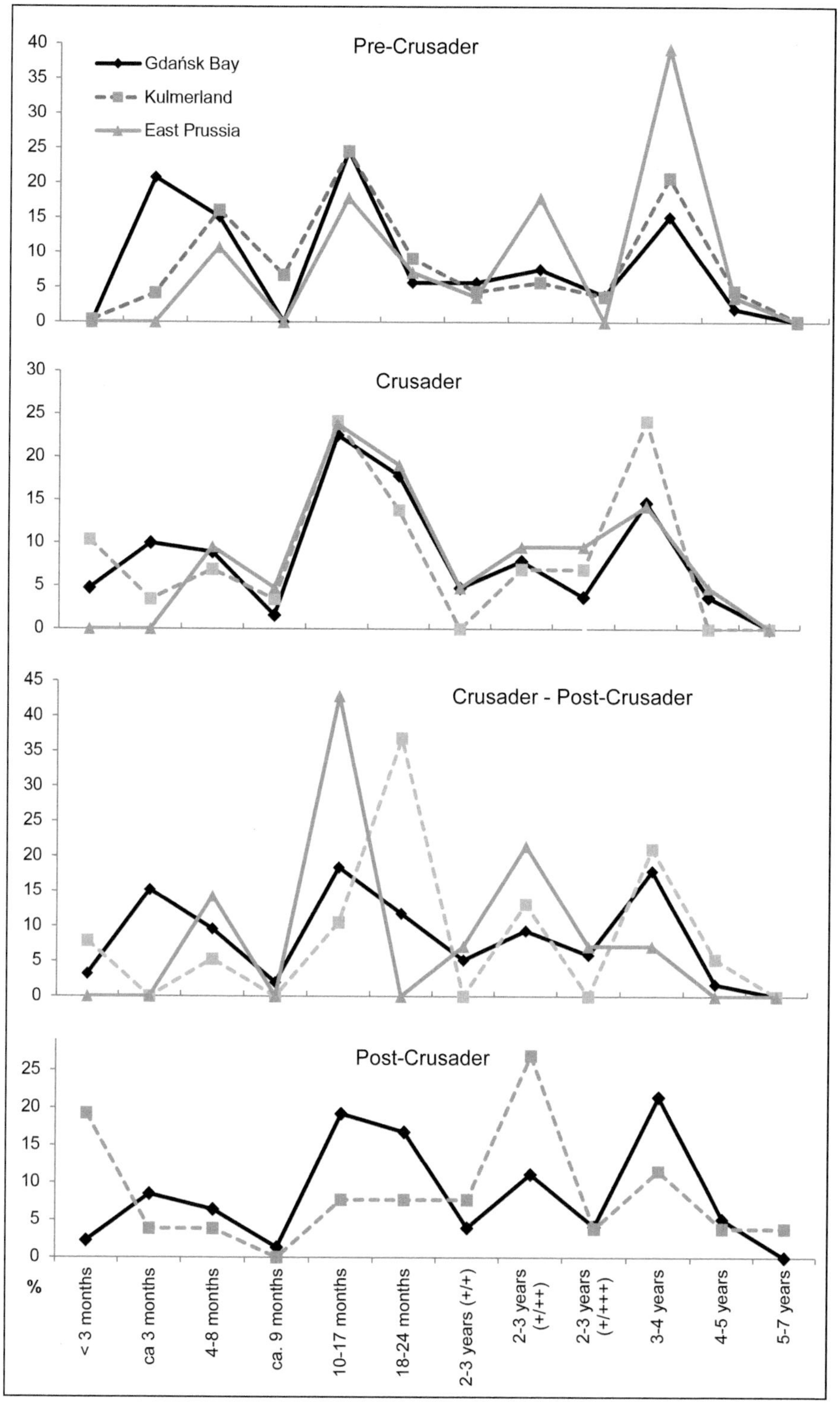

Figure 12.8: Age distribution of sheep/goat by region and historical phase. Figure by D. Makowiecki.

rearing was widespread and important: chickens appear frequently and abundantly in assemblages, and their meat appears to have been available to all social groups. The presence of medullary bone, with completely filled medullary canals, indicates a prolonged cycle of egg production in female birds. As a result, hens were maintained not only for breeding but also for egg production which were used as food. Wild birds were exploited where locally available, particularly geese (e.g., Greylag goose (*Anser anser*)), ducks, capercaillie, black grouse, and partridges.

Similarly, the low representation of fish remains does not reflect their limited economic and alimentary role. Clearly in winter they were of relatively minor importance; however, from early spring until early summer, fishing, in the absence of suitable quantities of meat from livestock, provided important dietary supplements. This was also a time when fish gathered in shoals and would have been easier to catch (Makowiecki 2003b). The dietary importance of fish is reflected by the presence of herring at Kałdus. This species, salted and smoked, had been traded from at least the eleventh century. It was exported from a few settlements on the Baltic coast to the main centres of the Piast state (Makowiecki 2003b). From this time there is also evidence for intensive sturgeon fishing at coastal centres such as Gdańsk. The scale of this exploitation is clearly marked by the fact that over 70% of the fish assemblages from the stronghold and its adjacent settlement consist of sturgeon bones. There were also important centres of sturgeon fishing on the Vistula, exemplified again by Kałdus, where sturgeon represented nearly 30% of the total fish assemblage (Makowiecki 2010b, 197). Within the complex, larger quantities of sturgeon meat were being con-

Whilst the remains of birds only make up a small percentage of the materials included within this study, some species clearly played a significant role in the economic activities and daily alimentation of settlements, especially for those of lower social status. Chicken-

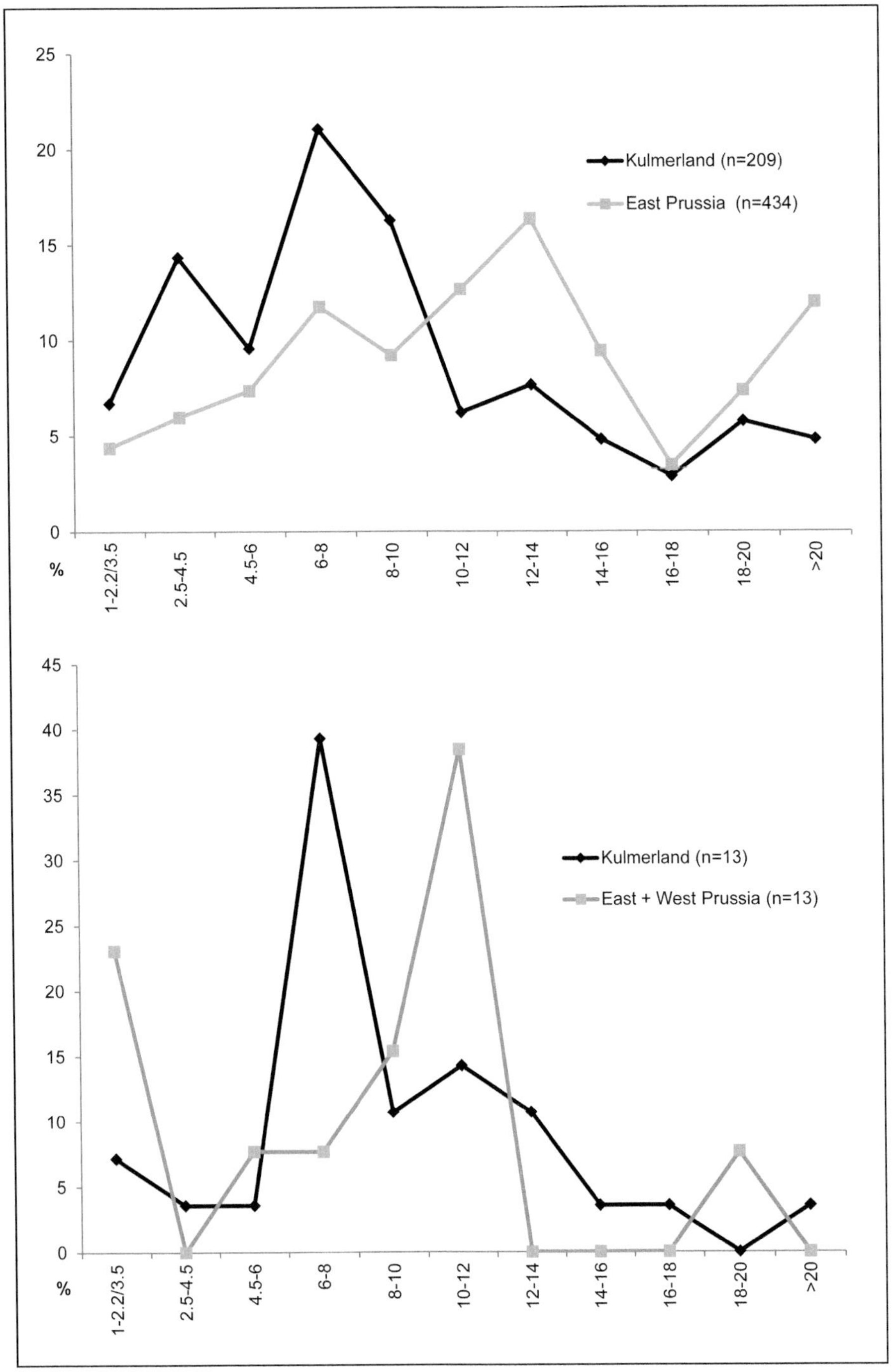

Figure 12.9: Age distribution of horse by region and historical phase. Figure by D. Makowiecki.

sumed inside the stronghold, especially in the fortified centre, which was probably the seat of the Polish castellan. Here, the remains of sturgeon accounted for *c.* 80% of the fish bone assemblage (Makowiecki 2010b, 168).

In the same way as there was geographic variation across the regions considered within this study, so in each region there is evidence for different strategies to obtain food. The most significant differences can be seen between centres situated on either side of the Vistula. These, in turn, were different from the two main regions: the Kulmerland with its political centre at In Culmine, and the Gulf of Gdańsk with the proto-urban centre at Gdańsk dominating the region. These differences are based on varying percentages of cattle, pig, sheep/goat, and horse remains (Fig. 12.2). In both the Gulf of Gdańsk and the Kulmerland the importance of cattle and pigs was very similar, and can be connected with the animal husbandry regimes being promoted in the *sedes regni principales* of the early Piast state in Greater Poland, particularly Poznań, Ostrów Lednicki, Gniezno, and Giecz. These centres, alongside the presence of princely and military classes, were major concentrations of population, with adjacent settlements or *suburbia* (*podgrodzie*) populated by artisans, farmers, and ancillary groups. In this respect, the centres of In Culmine and Gdańsk had very similar socio-economic structures, directly shaped by or strongly influenced by the Piasts. Wild mammals were also comparably exploited in both regions, with variations in species representation reflecting local availability. Due to its coastal location, the population of Gdańsk was able to hunt marine mammals, seals and porpoises, the remains of which are not reported in Kałdus or any other strongholds in the Kulmerland. Variations in the representation of elk between the two centres can also be linked to local environmental contexts (Makowiecki and Satch 2007). In Gdańsk, the species represents almost 30% of all wild fauna, while in Kałdus its remains constituted only 3%. The exploitation of the other major wetland species — beaver — also varies between the two centres, with a significantly higher percentage in Gdańsk than in Kałdus. As in the case of elk, this can be linked to the extensive network of larger and smaller watercourses and wetlands in the Vistula Fens, and the absence of such environments within the vicinity of Kałdus. Here, in the absence of elk, red deer, roe deer, and wild boar were hunted more extensively. Centres in Prussia had different hunting practices, dominated by the exploitation of red deer and roe deer, and to a much lesser extent wild boar. Within the category of *Animalia minute*, beaver and hare dominate, whilst the limited representation of carnivores was comparable at all the sites.

Lower-ranking centres located on either side of the Vistula also had different patterns of livestock husbandry. On the eastern side, cattle rearing was more important; and on the western side, husbandry focused on pigs. This variability can be explained by the greater cultural and ethnic connectivity between the Slavs on the western side of the Vistula with the Slavs within adjacent Greater Poland. In the centres on the eastern side of the river, animal husbandry regimes may have been influenced by the proximity of the Prussian population. This can be inferred on the basis of data from East Prussia, and largely from Święta Góra, where cattle were the dominant species and pigs played a decidedly smaller role. It is also worth noting that at the stronghold of Węgry, which in the eleventh century was a Slavic (Pomerelian) centre and most likely became the residence of a castellan, situated in the borderland of Pomerelia/Pomesania, contained a high proportion of pig remains comparable to other centres of the Piast state located on the eastern side of the river, including Kałdus.

Prussia

In contrast to the abundance of data from the Vistula region, faunal data for Prussia is derived from only a few sites (Map 12.1). These are also situated in different regions of Prussia, and at a considerable distance from each other. In the north is Janów Pomorski (Site 1), located on Lake Drużno, which from the end of the eighth to tenth centuries was a flourishing trading centre described by Wulfstan as Truso, inhabited by Prussians and Scandinavians (Jagodziński 2010). The area was subsequently described as the edge of the Prussian territory of Pogesania. In central Prussia, corresponding to the tribal territory of Warmia, a settlement within the area of Olsztyn Las Miejski was occupied from the eighth to ninth centuries, where excavations recovered a faunal assemblage representing food waste. To the north-east, in the Great Masurian Lake District and corresponding to the tribal territory of Galindia, is *lauks* Staświny, including the site of Święta Góra (see Chapter 10). Here, the assemblage considered within this study derived from the late ninth through to the thirteenth/fourteenth centuries. Nearby, to the west, is the stronghold of Jeziorko, with a settlement inhabited from the tenth to the mid-thirteenth century (Krysiak 1950). Finally, further to the east within the territory of Sudovia is the stronghold of Szurpiły, dating from the ninth to thirteenth centuries (Gręzak and Piątkowska-Małecka 2009).

As in the region of the Lower Vistula, the territories inhabited by the Prussian tribes had a high degree of biodiversity (Tables 12.2–4). Alongside stock keeping, wild mammals, birds, and fish were exploited. Meat and fat was primarily derived from cattle, pigs, and to a much lesser extent from small ruminants, with more emphasis on sheep than goats. In addition, horses, dogs, and cats also feature in the faunal assemblages. Horses were utilized for meat, but they were almost certainly bred primarily for riding. However, at this time there were also populations of wild horses in parts of Prussia, whose meat according to the traveller Erasmus Stella, writing in the later fifteenth century, was 'sometimes willingly consumed by the population' (Pruski 1959, 8). It is therefore possible that some horse remains in these assemblages may have derived from wild individuals. According to Fritz Rünger (1925; Pruski 1959, 9), wild horses lived in the vicinity of Lyck, Angerburg (Pol. Węgorzewo), Ragnit (Rus. Neman), and Taplacken (Rus. Talpaki), and were kept in Königsberg (Rus. Kaliningrad), Marienburg (Pol. Malbork), and Stuhm (Pol. Sztum). The dog, as in the case of the Slavs, was most likely kept as a companion, whilst the small number of cat remains may be associated with the initial phase of dissemination of the species in this part of Europe, which took place only during the Period of Roman Influence (Krajcarz et al. 2015).

Species corresponding to both *Animalia superior* and *Animalia minute* were hunted, although it is not known how the Prussians categorized wild fauna. Based on the number of skeletal remains, the first group was the most important in economic terms and consisted of red deer, roe deer, and wild boar. Elk, aurochs, bison, and brown bear were hunted less frequently. Of the other group, the most important were beaver, hare, and squirrel, whilst the most represented carnivores were the fox, marten, polecat, badger, and lynx. The remains of European rabbit and the European water vole within these assemblages should be regarded as intrusive, potentially along with the remains of black rat and *Rattus* in general (Table 12.2). These species were prized for their furs used for lining luxury garments (Duda 1965; Samsonowicz 1991). Given that 10% of the assemblage from Święta

Góra derived from wild mammals, it seems likely that hunting was relatively important to the tribes of eastern Prussia, whilst in western Prussia it may not have been so significant.

Of the birds, the most important was domestic fowl, whilst capercaillie, grouse, and waterfowl, that is, wild geese and ducks, were hunted (Table 12.3). Judging by the vast quantity of fish remains, nearly 18,000 fragments (a detailed analysis was carried out for only a sample of this assemblage) recovered from Święta Góra and 2300 fragments from Pasym (Dąbrowska 1967; Dąbrowski 1976; Makowiecki 2003b, 131), fishing was extremely important at Prussian lakeside settlements. The majority of fish that were caught were cyprinids, and of these the most frequently occurring were bream, roach, and tench (Fig. 12.5). Of the other groups of freshwater fish, the most important was pike, followed by perch. Prussian fish assemblages indicate that in the early Middle Ages, fish consumption was based on local species, not on imported marine fish. Only in centres located closer to coastal bays, such as Truso, is there evidence for imported herring from at least the first half of the ninth century (Makowiecki 2012b), a species also found at the stronghold of Węgry located on the eastern bank of the Nogat (Makowiecki 2015). From the 1060s/70s until the 1160s, this centre secured the Vistula borderland of the Piast state against Prussian incursions, and as a likely castellany organized colonization within its territory (Haftka and Wadyl 2015). At that time, the trade in herring was well established throughout the territory of the Polish state, constituting an important component of Christian fasting culture. Therefore, its extensive consumption at Węgry suggests both the presence of a Christian culture and the functionality of the trade route from Gdańsk, along the river to the Kulmerland. Conversely, the absence of herring at inland Prussian sites indicates that such connective routes, and most likely a market, were absent. Given the size of sturgeon and the large proportion of its remains within the fish assemblage from Truso, it was clearly one of the most important dietary components for the inhabitants of the trading emporium. Sturgeon were probably caught in the Vistula Lagoon, where their exploitation is already known from the Early Iron Age (Makowiecki 2003b) and would continue until modern times. In this respect, Truso was similar to Gdańsk and Kałdus, exploiting the naturally occurring abundance of this species in the adjacent waters of the Baltic Sea and Vistula Delta.

Mortality Profiles

Based on the age of the slaughter of cattle, it is clear that beef was obtained from adult animals, represented by two main age categories: around 2–2.5 years and 3.5–5 years (Fig. 12.6). The relative proportion of these two age groups was similar, suggesting that cattle were raised primarily for meat. As in the case of the Slavs, dairy products played a minor role; and in the case of Prussian cattle, most animals were small in size, which can be identified with the 'low, short-horned' type (Żuk-Piechota and Towarnicka 1978). The same type of cattle was bred by the neighbouring Slavs (Makowiecki 2001b, 2010).

In the case of pig husbandry, the mortality profile is similar to that of the Gdańsk Bay and Kulmerland, indicative of the procurement of meat from animals at different stages of development, which can be subdivided into three major groups: the youngest individuals slaughtered at *c.* ten months; animals aged *c.* twelve to twenty-four months; and individuals aged over two years. In the case of pigs, increases in body fat take place after four months combined with a decrease in muscle growth (Prawocheński 1958, 170). It can therefore be concluded that individuals from the oldest groups were exploited primarily for fat, younger animals were valued for their succulent meat, and a proportion of sows were kept for breeding purposes. Ethnographic data indicates that in some regions of Poland, even in the second half of the seventeenth and eighteenth centuries, pigs weighing around 60–70 kg were killed for their meat, whilst heavier animals, weighing up to 100 kg, were specifically exploited for their fat (Makowiecki 2001b, 94). It is of course possible that the slaughter regime used by the Prussian tribes also arose from specific culinary needs.

In comparison to pig and cattle, the mortality profiles of small ruminants are even more diverse, with four definable stages of slaughter. The first three age groups were characterized by a similar percentage of slaughtered animals. The fourth group, which contained the largest number of individuals, was derived from the least useful part of the herd. So the dynamics of Prussian alimentary culture stemmed not only from the exploitation of livestock for meat but for meat with variable flavours which was most likely cooked differently. Maintaining individuals also resulted in better quality fleeces (wool attached to skin used as a garment) as well as a greater weight of meat, whilst older animals could provide a regular annual supply of wool which was processed into yarn.

The mortality profile for horses was completely different to those of the other domesticates. Their life expectancy was longer and they fulfilled different roles. A large proportion of individuals lived for more than twenty years, and the majority of the equine population lived for over ten years. Conversely, individuals killed between 2.5–3.5 years accounted for the lowest percentage of the population. In the subsequent age groups, age at death was raised to *c.* twelve to fourteen years.

The data presented above indicates that animal husbandry regimes were not consistent amongst the Prussian tribes. In eastern Prussia the hunting of wild mammals was relatively important, as reflected by the high proportion of the remains of those species (Fig. 12.3). In western Prussia, the ratio of wild to domestic does not exceed 1%, although it is important to note this is based on two sites, one of which was the commercial emporium of Truso. Here, the limited representation of wild mammals almost certainly reflects the trading context of Truso, and by the settlement's own internal social organization. In all likelihood, similar factors determined variations in alimentary cultures linked to differential husbandry regimes. In this respect, it is difficult to explain the low quantities of wild mammals within the assemblage from Węgry, the borderland centre of the Piast state located *c.* 35 km south-west of Truso. However, the proportions of domestic mammals at both centres were the same, with comparable proportions of cattle and pigs. It is also different in relation to the centres of eastern Prussia, where numbers of cattle significantly outnumber those of pig. So it is possible to suggest there were different alimentary cultures operating in both regions. Some aspects, however, were similar: for example, the dominant role of chickens and fish is evident in both, with differences in the latter reflecting local availability.

Animals in Ritual Contexts

External Christian sources, such as Caesarius of Heisterbach and the chronicler of the Teutonic Order Peter von Dusburg, noted that animals played a role in the religious practices of the pre-Christian Prussians. Pope Innocent III had imagined that the Prussians greeted their gods in the form of wild animals, the earliest narratives of the Teutonic Order refer to horse sacrifice, and later accounts suggest the continuation of animal sacrifice within domestic contexts in eastern Prussia (Okulicz-Kozaryn 1983, 231–32; see also concluding discussion). The only archaeologically visible aspect of the religious role of animals is represented by special or structured deposits of whole or partial skeletons, with some animals deliberately killed within a religious context. The main 'cult animal' which characterized the ritual practices of the Prussian tribes was the horse (for a recent summary and references, see Bliujienė 2009; Pluskowski 2013, 77–81). The ritualistic killing and deposition of horses in cemeteries is evident from the first century AD in Masuria and from the second century AD in Sambia (Zinoviev 2009; Karczewski, 2011, 161–91), and in the Migration Period is adopted more or less simultaneously in Lithuania and regions corresponding to southern Latvia (Jaskanis 1966), as far west as the cemetery of the Elbląg Group at Nowinka, dating from the fifth/sixth to eighth centuries (Kontny et al. 2011). By the eleventh century these cemetery deposits appear to be confined to Sambia and Natangia (Shiroukhov 2012). The regional differences could reflect the varying importance of equestrian culture and social structure (Bliujienė and Butkus 2009), and the burial of riding equipment such as spurs, bits, and harness fittings in Prussian graves may also reference horse burial by proxy. The most consistent ritual appears to have been living burial, which is represented by similarly positioned inhumed horses and is also described in later written sources. Horse meat was also consumed in settlements (as at Święta Góra), although the only Prussian site to date where horse butchery and consumption has a convincing ritual context is Poganowo (Wyczółkowski and Makowiecki 2009; see below).

Confronting Zooarchaeological and Historical Sources for a New Thesis of Animal Exploitation in Pre-Crusade Prussia

The role of animals in early medieval (pre-crusade) Prussia has only been comprehensively discussed in two monographs by Okulicz-Kozaryn (1983, 1997). The first was written as a popular book for a general audience, whilst the second monograph, also addressed to a wider audience, slightly broadened the scope of knowledge on the subject. In both cases the author based her thesis largely on the writings of chroniclers such as Wulfstan (ninth century) (Okulicz-Kozaryn 1983, 32; 1997, 159) Helmold (twelfth century) (Popłoński 1862), Peter von Dusburg (early fourteenth century), Jan Długosz (fifteenth century), Simon Grunau (sixteenth century), and the ethnographer and historian Matthaeus Praetorius (seventeenth century). Okulicz-Kozaryn also drew on early modern ethnographies of the 'Old Prussians' (Fischer 1937).

In these various sources, the Prussians are described as maintaining herds of cattle, sheep, and goats in val-

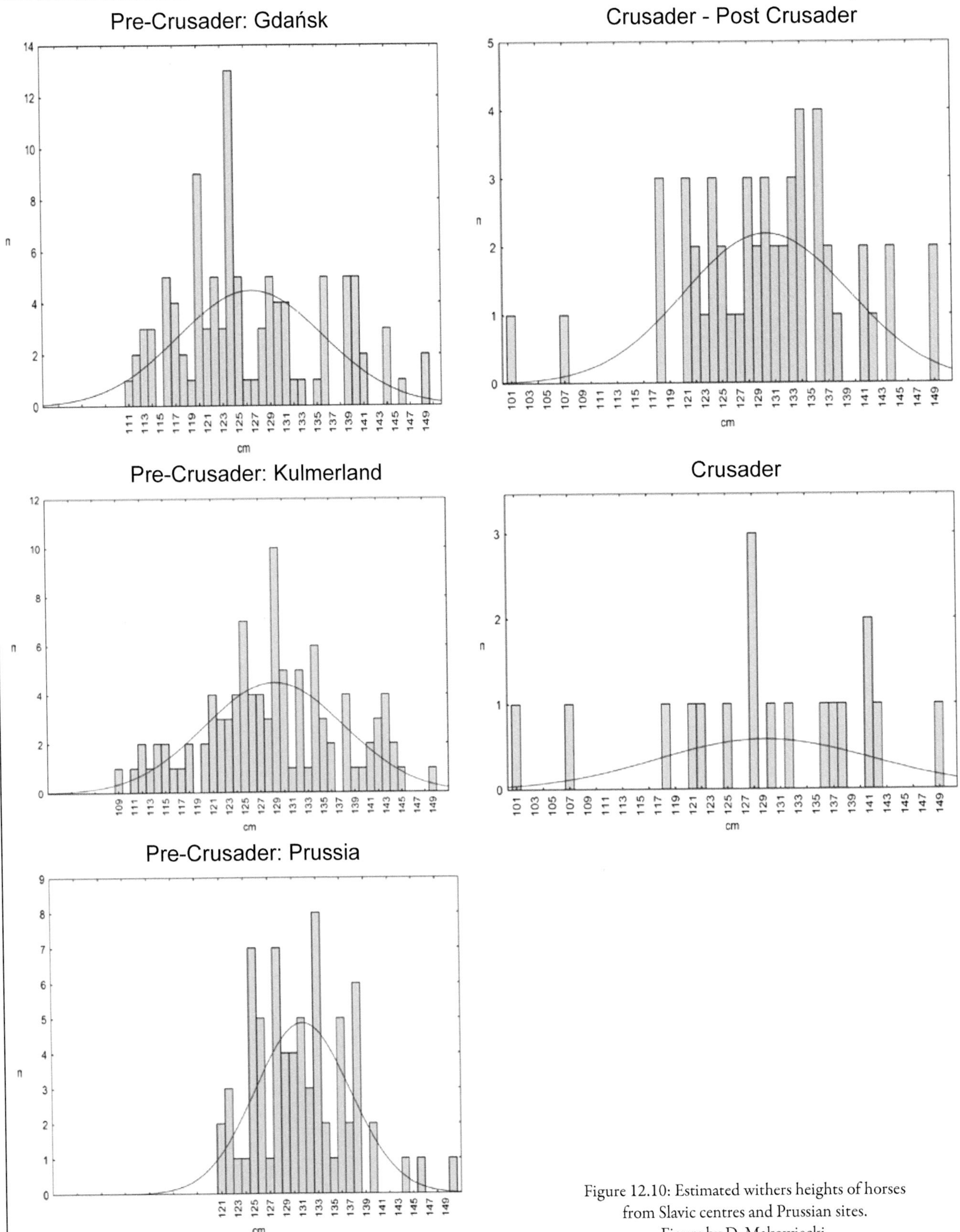

Figure 12.10: Estimated withers heights of horses from Slavic centres and Prussian sites. Figure by D. Makowiecki.

ley meadows and woodland clearings. The importance of beef consumption is noted by Helmold, who wrote that the Prussians 'subsist on cattle meat', whilst Peter von Dusburg referred to the seizure of cattle and horses by the Sudovians from the Galindians, and later Długosz refers to numerous cattle herds (Okulicz-Kozaryn 1983, 80, 134, 165). In the fifteenth and sixteenth centuries, Prussian men would purchase cattle for their wives from their future fathers-in-law (Fischer 1937, 22). The zooarchaeological data presented above clearly validates these accounts; however, the uses of livestock in pre-crusade Prussian society must be qualified further. First, it is clear that the importance of these species was not uniform across the entire region. There were settlements in which pigs were as important as cattle, as in western Prussia, and it is clear that pig rearing was also extremely important in eastern Prussia. Differences could also result from variations in social and economic organization in centres such as Truso, which are absent in the other study areas, although a comparable example is provided by the Sambian emporium of Wiskiauten (Rus. Mokhovoye). Sheep, in turn, were far more important than goats in all the assemblages included within this study, also a characteristic of neighbouring Slavic husbandry regimes (Makowiecki 2001b, 2010).

Livestock were exploited not only for meat but also for a range of secondary products, including the use of oxen within plough teams (Fischer 1937, 8); cattle, sheep, and goat for dairy products; and, in the case of sheep, wool. Okulicz-Kozaryn (1983, 166) highlighted the importance of milk in Prussian society, which would have been drunk fresh on a daily basis. Most milk was sour and provided the basis for cheese, cream, and butter. The zooarchaeological data, in contrast, does not provide clear evidence for the prevalent cultural role of dairy products. The mortality profiles for cattle and small ruminants indicate they were primarily reared for meat and fat. This is also supported by the ratio of males to females in adult cattle and goats (Makowiecki 2010b). Therefore, dairy products were more likely to be a by-product of rearing herds. Milk would have been only available for short periods from lactating females, and after raising juveniles, the surplus would have been comparatively small; it is also possible that mare's milk was set aside for human consumption. In general, it can be assumed that milk was a valued commodity, but it was not as common as it would become in the post-medieval period. In the case of sheep's wool, there is solid archaeological evidence that it was used in the production of clothing, exemplified by the collection of textiles from Równina Dolna (Antoniewicz 1958, 150). However, based on the size of individuals reared by the Prussians and neighbouring Slavs, it can be assumed that these were primitive forms, similar in size and wool coverage to the modern *wrzosówka* breed of sheep (Nowicki et al. 1995). Individuals of this breed provide 2 kg of lanolin wool each year. Only during the period of the Order's rule, when there is a noted increase in the size of sheep, is it possible to speculate that improvements were connected with the aim of producing better-quality wool (Fig. 12.15).

The historical sources all emphasize the importance of horses for riding, although like oxen they could also be used for ploughing. Consumption of horse meat is clearly evident, but the majority of chroniclers noted the consumption of mare's milk (Popłoński 1862, 7). This custom became more widespread between the ninth and thirteenth centuries, but was no longer noted in the sixteenth century, when it appears to have been superseded by a drink made from mixing the blood of cattle and pigs with milk, a beverage also mentioned by Adam of Bremen in the late eleventh century (Fischer 1937, 12). The zooarchaeological data indicates that horses were commonly found in settlements, and the presence of butchered horse bones amongst food waste suggests their meat was occasionally consumed, just as it was amongst the neighbouring Slavs (Makowiecki 2001b, 2010). Given the presence of wild (or feral) horses in the woods of eastern Prussia, it is also possible their meat was consumed. The special status of horses is reflected by the assemblage from Poganowo. The site itself is difficult to characterize at present, but given the regular deposition of disarticulated horse remains, it is plausible they fulfilled a ritual function. The fact that the discrete equine population consisted of larger animals than found in Slavic assemblages may suggest a more intensive breeding culture. It is also possible these horses were bred for longer, since Slavic herds were mostly culled before individuals reached ten years of age, while in Prussia the majority died aged twelve to fourteen years, with a clearly higher proportion of horses ages over twenty years (Fig. 12.10). The special status of the horse was also underlined by its role within Prussian elite culture (Okulicz-Kozaryn 1983, 82, 151), exemplified by the funerary practices most prevalent in the Sambian Peninsula between the eleventh and thirteenth centuries (Shiroukhov 2012).

Alongside cattle and horses, there are recurring references to chickens and ducks in the written sources pertaining to the Prussians. Certain types of poultry appear to have had complex social roles beyond the production of meat and eggs. In seventeenth-century ethnographies, cockerels were used in magical contexts, and particularly during thanksgiving ceremonies held after the threshing of grain. At this point the men of the household killed a black or white cockerel that had been bred that year, then cooked and ate it (Okulicz-Kozaryn 1983, 113). Newly married couples were given cockerel meat to eat on their wedding night in order to induce fertility, and the following day they had to finish the rest of the meat before eating another meal. It was believed that eating the roasted testicles of a goat, bull, or bear would have the same effect (Fischer 1937, 23). The zooarchaeological data certainly confirms the fundamental importance of domestic fowl in Prussia, although it is difficult to gauge the role of ducks before the arrival of the Order. Whilst it is possible that newly hatched eggs of wild ducks were procured, a practice attested in Poland into the seventeenth century, domesticated ducks are only clearly evident from the fifteenth century (Makowiecki and Gotfredsen 2002). The zooarchaeological data indicates that after chickens, geese were more important than ducks during the period of the Order's rule, probably as a result of breeding geese within domestic contexts. The most commonly hunted wild birds were capercaillie and black grouse, clearly abundant in the woodlands of Prussia and neighbouring Slavic regions (Waluszewska-Bubień 1984; Nogalski 1984; Makowiecki 2010b; Bocheński et al. 2012).

In terms of herd management strategies, it is clear that animals were slaughtered at different stages of their life. Previously it was thought that mostly young animals were killed aged around a year or a little older. This was based on nineteenth-century ethnographic descriptions of husbandry practices amongst the Poleszucy and Belarusians, when annual culling removed all animals that were not required for reproduction along with those that had remained unsold. As a result, dried meat was a daily product rather than a rarity, a regular component of soups and sausages (Okulicz-Kozaryn 1983, 165–66). The zooarchaeological data, in particular the mortality profiles of the main domesticates, verifies the use of this analogue for Prussian animal husbandry to a certain extent. The herd management strategies differed for each species, reflecting their particular value. Animals were killed from different age groups, but some were more prevalent than others, in turn linked to specific choices. One reflected the partial culling of a herd after the grazing season, when the animals were in their best condition in terms of meat yield and calorific value. This system of slaughter was used by the Western Slavs in the centres of Greater Poland, evident from later analogues relating to seasonality, as well as mortality profiles (Makowiecki 2001b, 2006b). Cattle and swine were slaughtered in the autumn, whilst small ruminants were killed throughout the year, even during the grazing season. In this way, animal protein was available at all times. Ethnographic data suggests that pork and fat, containing less water, provided the basis for longer-lasting supplies, whilst meat from younger pigs would have been consumed fresh. Given the fact that the mortality profiles of cattle, sheep/goats, and pigs for Greater Poland, the Kulmerland, and Prussia are similar, it is likely that the breeding strategies used by the Slavs and Prussians were very similar.

The woods and lakes of Prussia provided ideal habitats for deer, wild boar, aurochs, bison, mustelids, and squirrels. Wetlands were also favoured by beavers and otters, whilst the numerous lakes, rivers, and estuarine bays of the south-eastern Baltic region supported abundant populations of fish. It is therefore not surprising to see references to hunting and fishing in a number of written sources (Fisher 1937, 6, 7; Okulicz-Kozaryn 1983, 32, 50). According to Peter von Dusburg, the Prussians even knew how to farm fish: the inhabitants of the stronghold of Ragnet supported themselves from a fish pond (Fischer 1937, 7). The Prussians living in coastal areas caught fish, mostly herring (Okulicz-Kozaryn, 1983, 137). Okulicz-Kozaryn (165) wrote that 'large quantities' of fish bones and scales have been found at Prussian sites, but does not provide any further details.

Hunting was thought to be of limited economic significance, which is confirmed by wild mammals representing only 4–6% of assemblages from Prussia. Furbearing mammals were more important, particularly beavers, squirrels, and martens, which the Sudovians exchanged with the Rus' for grain during times of famine. At one point they were paying a tribute to the Danish crown with the pelts of black martens and squirrels, and *Knýtlinga Saga* states that in *c.* 1120, a Sambian merchant offered the Danish king eight thousand white furs. Adam of Bremen and Helmold expressed surprise that the Prussians did not particularly value leather and fur, exchanging them for commercially less valuable linens and wool (Okulicz-Kozaryn 1983, 149). Elk were lured onto the ice with aspen bark and then killed with sticks, and bison were pursued (Fischer 1937, 6; Okulicz-Kozaryn 1983, 150). It is difficult to verify the notion that the Prussian *nobiles* used falcons and specialized dogs for hunting (Okulicz-Kozaryn 1983, 82, 151).

Table 12.5: NISP of mammal species recorded in castles.

Mammal taxa / Sites	Elbląg	Malbork	Grudziądz	Wąbrzeźno (Bishop's)	Papowo Biskupie	Toruń	Mała Nieszawka	Olsztynek	Słoszewy site 1	Lidzbark Warmiński (Bishop's)	Ełk	Giżycko
Cattle — *Bos primigenius* f. *taurus*	692	1077	1350	1191	155	776	3599	158	9	142	646	37
Pig — *Sus scrofa* f. *domestica*	512	705	645	688	189	1265	3207	82	8	102	82	6
Sheep/goat — *Ovis ammon* f. *aries* / *Capra aegagrus* f. *hircus*	127	469	384	478	57	353	901	49	4	14	52	21
Sheep — *Ovis ammon* f. *aries*	28		331	43	29	51		10		4	25	6
Goat — *Capra aegagrus* f. *hircus*	25		12	7		6		4		1	6	2
Horse — *Equus ferus* f. *caballus*	66	170	5	26	6	8	85	13	59	5	5	
Dog — *Canis lupus* f. *familiaris*	2	31	9	144	3	51	23		69		6	
Cat — *Felis silvestris* f. *catus*	9	15	18	5	3		28	1				
European mole — *Talpa europaea*					1							
Hare — *Lepus europaeus*	1	16	15	82	1	18	27	5			2	
European rabbit/Domestic rabbit — *Oryctolagus cuniculus*/*O. cuniculus* f. *domestica*				1	1							
Red squirrel — *Sciurus vulgaris*						3	1					
Rat — *Rattus* sp.				2		4						
Beaver — *Castor fiber*	1		1			1						
Grey wolf — *Canis lupus*									4		2	
Red fox — *Vulpes vulpes*			1	4		2	6				2	2
Brown bear — *Ursus arctos*	2				1						2	
European polecat — *Mustela putorius*		3										
Mustelids — *Mustelidae*					1							
European otter — *Lutra lutra*		1	1									
Wild boar — *Sus scrofa*	9		9	10		10		3		3	2	1
Wild boar/Pig — *Sus scrofa*/*S. scrofa* f. *domestica*	2					2						
Red deer — *Cervus elaphus*	24	7	8	4	1	13	15	13		62	17	
Elk — *Alces alces*	3		1					2		5	5	
Roe deer — *Capreolus capreolus*	14	15	19	21	1	38	167	17		2	15	
Aurochs — *Bos primigenius*	1									1	8	
Aurochs/European bison — *Bos primigenius*/*Bison bonasus*	2										3	
Bos sp.	3			1							4	
European bison — *Bison bonasus*										1	5	
Total	**1523**	**2509**	**2809**	**2707**	**449**	**2601**	**8059**	**357**	**153**	**342**	**889**	**75**

Table 12.6: NISP of bird species in castles.

Bird taxa / Sites	**Elbląg**	**Malbork**	**Grudziądz**	**Wąbrzeźno**	**Papowo Biskupie**	**Toruń**	**Mała Nieszawka**	**Olsztynek**	**Lidzbark Warmiński**	**Ełk**	**Giżycko**	**Total**
Birds — *Aves* unidentifiable	10	105	72	389	6	77	65	3	1	6	3	**737**
Grey heron — *Ardea cinerea*							1					**1**
Heron — *Ardea* sp.						1						**1**
White stork — *Ciconia ciconia*							1					**1**
Whooper swan — *Cygnus cygnus*							1					**1**
Anatidae				5								**5**
Domestic goose — *Anser anser* f. *domestica*						1		1				**2**
Greylag goose/gomestic goose — *Anser anser/ A. anser* f. *domestica*	1		58	58	3	15		2			2	**139**
Goose — *Anser* sp.	1	105	8	255	1	23	60			5	7	**465**
Teal — *Anas crecca*				1			3					**4**
Mallard — *Anas platyrhynchos*		20		23			1					**44**
Mallard/domestic duck — *Anas platyrhynchos/ A. platyrhynchos* f. *domestica*	1		2			3						**6**
Domestic duck — *Anas platyrhynchos* f. *domestica*			1									**1**
Anas sp.	1	4		20	2	2	16					**45**
Garganey/teal — *Anas querquedula/Anas crecca*				2								**2**
Tufted duck — *Aythya fuligula*				1								**1**
Aythya sp.				1								**1**
Long-tailed duck — *Clangula hyemalis*				2								**2**
White-tailed eagle — *Haliaeetus albicilla*						2	4					**6**
Marsh harrier — *Circus aeruginosus*						1						**1**
Eagle — *Aqulia* sp.		1										**1**
Turkey — *Meleagris gallopavo*		1				1					1	**3**
Black grouse — *Tetrao tetrix*				1		1						**2**
Capercaillie — *Tetrao urogallus*		9					6					**15**
Partridge — *Perdix perdix*		1				3	4					**8**
Domestic chicken — *Gallus gallus* f. *domestica*	21	92	21	307	13	154	230	5	7	7	4	**861**
Peacock — *Pavo cristatus* f. *domestica*							2					**2**
Galliformes				13								**13**
Crane — *Grus grus*				4		1	1					**6**
Stock dove — *Columba oenas*							3					**3**
Woodpigeon — *Columba palumbus*							1					**1**
Magpie — *Pica pica*							1					**1**
Carrion crow — *Corvus corone*							1					**1**
Carrion crow/rook — *Corvus corone / C. frugilegus*				1								**1**
Jackdaw — *Corvus monedula*				1								**1**
Corvus sp.				1								**1**
Crow — Corvidae		10		1								**11**
Total	**35**	**348**	**162**	**1086**	**25**	**285**	**401**	**11**	**8**	**18**	**17**	**2396**

The evidence from Greater Poland and the Kulmerland suggests that goshawks rather than falcons were more suited to hunting in densely wooded landscapes, and the remains of the former have been found on numerous sites (Makowiecki 2006a, 2010; Makowiecki et al. 2014; Bochenski et al. 2015). Falcons were better suited for hunting in more open terrain.

The limited representation of *Animalia superior* or game points to their limited role in the alimentary culture of the Prussians. It has been supposed this reflected the fact that only the skin and most valuable joints of meat were brought back to settlements after a hunt (Okulicz-Kozaryn 1983, 151). But the representation of anatomical elements from red deer, wild boar, elk, and roe deer indicates the presence of whole carcasses at sites: both legs, head, and even elements of the spine are present. This is particularly evident for a large number of species in the assemblages from Święta Góra and Olsztyn, and in the Kulmerland at Kałdus (Makowiecki 2010b). There is, however, consistency between the range of species listed in historical sources and identified in faunal assemblages, with deer, aurochs, bison, bear, wild boar, elk, wolf, beaver, and fox (Fischer 1937; Okulicz-Kozaryn 1983, 1997). The proportion of fur-bearing species, particularly beaver, squirrel, and marten, recorded in eastern Prussian sites confirms the commercial value of pelts referred to in medieval sources. Higher percentages of mustelids, squirrels, and beavers have only been found at sites in the Gulf of Gdańsk, where the local woods and wetlands were also clearly exploited for fur.

Animals in the Castles and Towns of the Medieval Ordensland

Reconstructing trends in animal exploitation within the Teutonic Order's Prussian territories is based on faunal assemblages derived from both castles and towns (Table 12.1, Map 12.1). This enables a comparative analysis, with a working assumption that if the brothers of the Order can be considered as the social elite, their food should differ markedly from that of other social groups (see Ervynck 2004). The brethren of the Order maintained a distinctive dietary regime specified in their Rule (see Chapter 17), although castles were occupied by a diverse range of social groups who were not required to follow these proscriptions. At the same time, the economic systems of post-crusade Prussia were determined by the Order, with the result that the husbandry and dietary regimes of castles and towns should, at least in part, be similar. This study included data from assemblages of different sizes recovered from eleven castles, including those belonging to the Prussian episcopate (Table 12.5). These were located in different parts of the former *Ordensland*: in the north-west from Elbląg, in the west from Malbork, on the Vistula from Grudziądz, Toruń, and Mała Nieszawka; in central Prussia from Wąbrzeźno (the episcopal castle), Papowo Biskupie (Ger. Bischöflich Papau), Olsztynek (Ger. Hohenstein), and Lidzbark Warmiński (the episcopal castle); and in eastern Prussia from Giżycko and Ełk. In a few cases material also derived from adjacent towns, namely Elbląg, Grudziądz, and Toruń (Map 12.1, Table 12.6). By far the most numerous assemblages derived from Gdańsk and Gniew, with smaller collections from Tolkmicko (Ger. Tolkemit) and Barczewo (Ger. Wartenburg).

One of the major obstacles to effective comparative analysis is the complex (and at times ambiguous) stratigraphy of castle and urban archaeological sites. In the case of the former, only select portions of faunal assemblages could be assigned to the period of rule under the Order, and some post-medieval assemblages have no exact definition of the upper part of their chronological range. The exception is the castle at Mała Nieszawka which functioned until its dismantlement in 1424, and also in Toruń where the Order's castle was destroyed in 1454. The assemblage from the excavations of the castle in Elbląg can also be assigned primarily to the Order's period of occupation, and it was possible to distinguish the Order's phase during excavations at the castle in Ełk. In contrast, synchronizing the better-dated urban assemblages with the period of the Order's rule is less contentious.

Zooarchaeological Data

Considering the zooarchaeological data for the entire period of the Teutonic Order's rule in Prussia (1230/83–1466/1525) across all the regions, it is possible to identify some general trends. As in the pre-crusade period, proportions of livestock consistently far outnumber those of wild species (Fig. 12.1); however, in the Kulmerland and on the eastern side of the Vistula, the percentage of wild mammals is much higher than in the Gulf of Gdańsk and the western side of the Vistula. Similar differences can be observed between western and eastern Prussia, where the former has a much higher percentage of domestic mammals, whilst the latter has a higher percentage of wild mammals reaching as much as 8% of the overall material.

Similar patterns are evident in the representation of livestock (Fig. 12.2). Here, numbers of cattle dominate

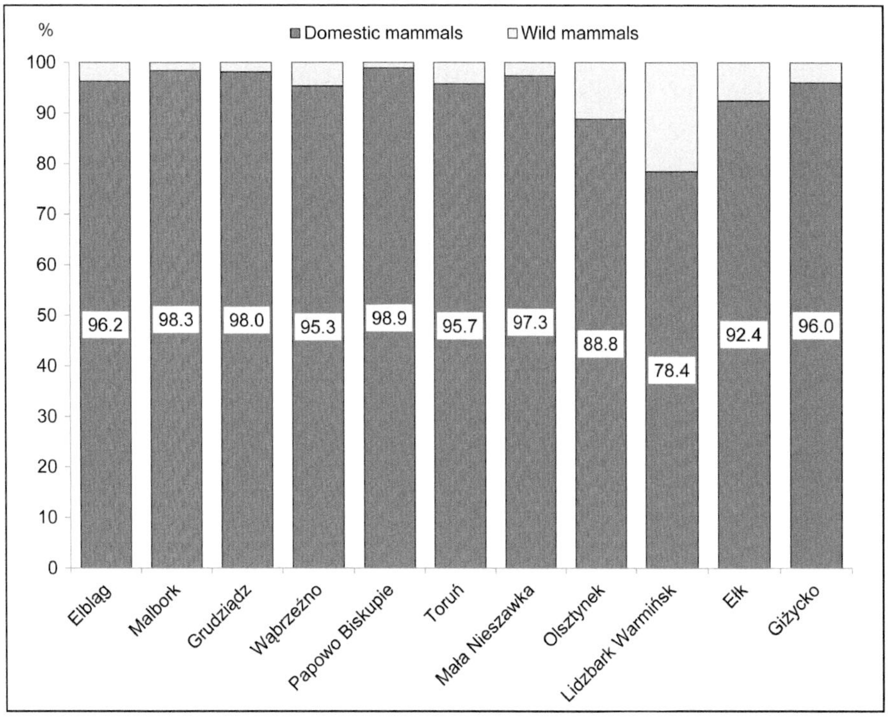

Figure 12.11: Percentage distribution of domestic and wild mammals in castle assemblages. Figure by D. Makowiecki.

Lower Vistula, where pigs outnumber cattle, followed by sheep/goat and horse. The number of bird species is considerably smaller than in the pre-crusade period, dominated again by chicken, but in almost all regions also characterized by substantial numbers of geese and ducks, with smaller numbers of the woodland-dwelling capercaillie and black grouse (Fig. 12.3). The assemblages also contain no traces of hawks and only occasional examples of white-tailed eagles (*Haliaeetus albicilla*).

In the case of fish, each of the regions is different in terms of the relative representation of individual taxa (Fig. 12.4). In the Gulf of Gdańsk, the most abundant species is sturgeon, with marine species dominated by cod (*Gadus morhua*) and freshwater by cyprinids. On the western side of the Vistula, the dominant species is cod, while on the eastern side cyprinids and catfish are prevalent. In the Kulmerland the majority of remains derive from cyprinids and pike, with substantial quantities of cod, sturgeon, and catfish (*Silurus glanis*). In western Prussia sturgeon dominates, whilst eastern Prussia, where cyprinids dominate, is the only region where cod remains are absent.

over pigs and small ruminants, with the lowest percentages represented by horses. The Gulf of Gdańsk stands out as a particularly distinctive region, where the remains of cattle reach their highest value, exceeding 70% of the total assemblage and clearly dominating over pigs. Contrasting trends are visible in the eastern side of the

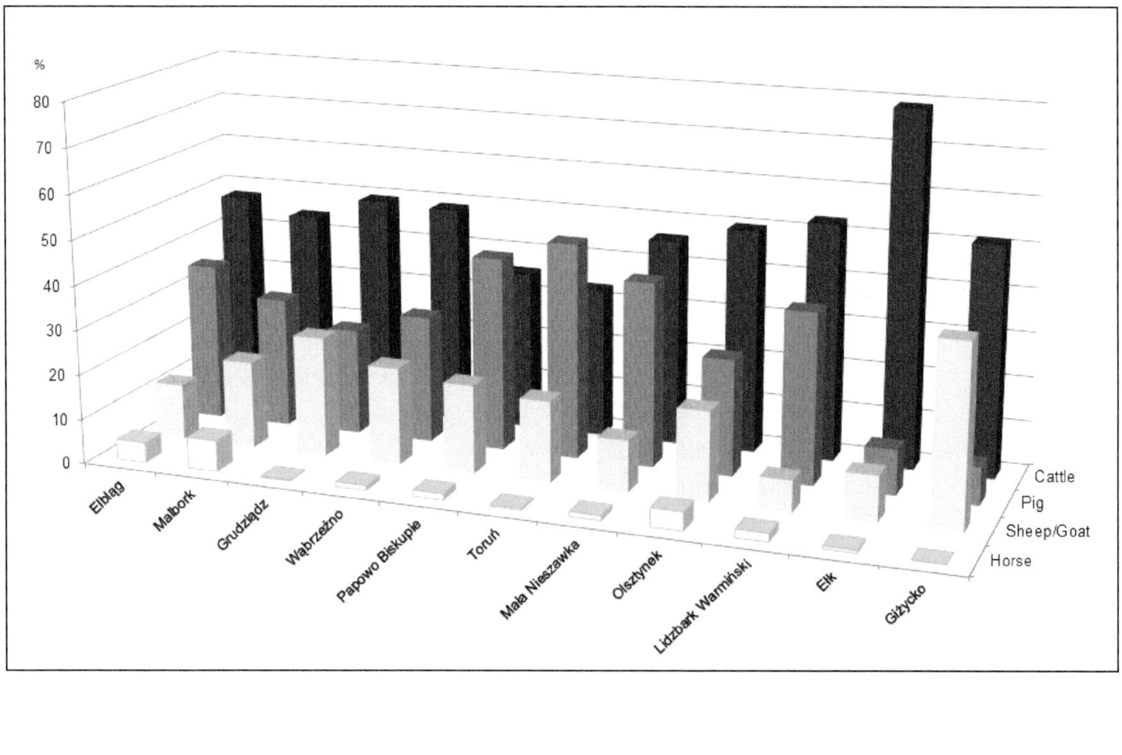

Figure 12.12: Percentage distribution of domestic mammals in castle assemblages. Figure by D. Makowiecki.

Castles

The meat supplies of castles derived from domestic mammals, wild mammals, birds, and fish, and it is clear that food at these sites was characterized by extensive variety. Livestock were far more important than wild mammals (Figs 12.11–12), although the proportion of wild mammal remains is significantly higher in central and eastern Prussia, than in the castles of the Vistula region and western Prussia. This can be plausibly connected with the relative biodiversity of these different regions (see Chapter 11) and exemplified by assemblages from two castles situated in the Great Wilderness: Ełk and Klaipėda (Ger. Memel). At Klaipėda, the earliest zooarchaeological material dates to the last quarter of the fourteenth century, and wild species represent just over 10% of the assemblage, decreasing in importance until after 1521. The range of species included aurochs/bison, elk, red deer, bear, and possibly wild boar. The assemblage from the latter half of the fifteenth century includes aurochs/bison, elk, red deer, wild boar beaver, and fox (Žulkus and Daugnora 2010). There is limited evidence for fur processing at castles, although written sources indicate that alongside the highly prized mustelids and squirrels, hare pelts were also handled by the Order, mostly derived from subservient settlements (Delort 1978, I, 52; see also Chapter 15). Castle communities can also be differentiated by their variable dietary profiles; however, in almost all the staple meat was beef, with higher proportions consumed in the eastern Prussian castles than those located in the west. Only in Papowo Biskupie and Toruń was pork more important than beef, as in the preceding Slavic period. It is possible this also reflects the continuation or legacy of the regional pre-crusade economic system, where pigs were exploited alongside cattle for food. Their popularity could also reflect the fact that pigs were readily housed in smaller enclosed spaces, within castle precincts and towns (see also Scull et al. 2019).

The meat of birds was consumed at all castles, primarily derived from chickens, geese (wild/domestic), and wild ducks. Both groups of *Anseriformes* must have thrived in the favourable habitats provided by lakes, alongside which many of the Order's castles were built. Species such as the greylag goose and mallard, most frequently identified in these assemblages, would have been available from late spring until the late autumn. In the woods maintained by the Order (see Chapter 11), capercaillie and black grouse were hunted. The presence of chickens, as the most abundant domesticated bird, can be linked to both imports and to rearing flocks within castle precincts. Exotic birds were also consumed and included peafowl (*Pavo cristatus*) and, much later, turkey (*Meleagris gallopavo*). The former was identified in the late medieval contexts at the castle of Mała Nieszawka, and the species had already been imported from the eleventh century to centres of the Piast state (Makowiecka et al. 2005; Makowiecki et al. 2014). Turkey remains were found in the outer bailey of the castle at Malbork within the modern assemblage dating from the nineteenth onwards (Maltby et al. 2009) as well as the castles in Giżyko and Toruń. In the case of the latter site, the bones were found in late medieval contexts indicating they had been mixed with early modern material. In Toruń, turkey remains have been found in the vicinity of the church of St James and dated to the last quarter of the seventeenth century and first half of the eighteenth century. Additional undated remains were found within the vicinity of the Dominican friary. The species was disseminated across Europe from the sixteenth century (Benecke 1994b, 392); and from at least the early seventeenth century, and perhaps even earlier, it was consumed by the residents of the Pomeranian castles at Stare Drawsko and Człuchów, where the bones derive from contexts broadly dated to the fourteenth–seventeenth/eighteenth centuries. At the same time, turkey was served on the tables of the Górka family in Szamotuły in Greater Poland, the Branicki's residence in Tykocin, and consumed by the townspeople of Gdańsk (Makowiecki and Gotfredsen 2002).

Given the prevalence of lakes in the region, it is not surprising that fish were an important part of the Order's diet, which also emphasized fasting foods. Local cyprinids dominate castle assemblages, followed by pike, perch, and zander. In castles located on the Vistula, such as at Mała Nieszawka and Toruń, sturgeon contributed significantly to nutrition, as these were fish of considerable size, usually with a length of 100–200 cm (Makowiecki 2003, 2008b). Sturgeon are regularly noted in the Order's castle inventories: there are even records of live individuals at the court of the Pomesanian bishop in Riesenberg (Pol. Prabuty) and at the Pomesanian Chapter's castle in Marienwerder (Pol. Kwidzyn). The second most often consumed species at these castles was European catfish, whilst more luxurious species were represented by salmon/trout, as evident at Toruń. Marine species have also been identified in castle assemblages, consisting of herring and cod. Isotopic studies have demonstrated the latter were caught in the fisheries of the Baltic Sea, indicating the Order promoted the develop-

Table 12.7: NISP of fish in castles.

Fish taxa / Sites	**Elbląg**	**Malbork**	**Grudziądz**	**Wąbrzeźno**	**Papowo Biskupie**	**Toruń**	**Mała Nieszawka**	**Lidzbark Warmiński**	**Ełk**	**Giżycko**	**Total**
Fish – Pisces		1	4	6	1	8		1			**21**
Pike – *Esox lucius*	1	2		33		19	111			1	**167**
Carp – *Cyprinidae*		3		18		22			1		**44**
Roach – *Rutilus rutilus*							11				**11**
Asp – *Aspius aspius*						1	1				**2**
Tench – *Tinca tinca*				11							**11**
Bream – *Abramis brama*	2			2		110	32				**146**
Blue bream – *Abramis ballerus*						1					**1**
Crucian carp – *Carassius carassius*							3				**3**
Catfish – *Silurus glanis*	4					7	5				**16**
Percidae	1								1		**2**
Perch – *Perca fluviatilis*				17		4	9				**30**
Pikeperch – *Lucioperca lucioperca*	2					4	13				**19**
Salmo sp.						1					**1**
Atlantic sturgeon – *Acipenser* sp.	2	2		1		16	31				**52**
Eel – *Anguilla anguilla*						1					**1**
Herring – *Clupea harengus*						7					**7**
Cod – *Gadus morhua*	2	1		1		10	561				**575**
Total	**14**	**9**	**4**	**89**	**1**	**211**	**777**	**1**	**2**	**1**	**1109**

ment and regulation of a local fishing industry (Barrett et al. 2008; Orton et al. 2019). On the basis of the faunal data, castles in eastern Prussia were exclusively reliant on freshwater fish, primarily pike, perch, and cyprinids (Table 12.7).

Towns

The majority of material from urban sites derives from Gdańsk (31,000 fragments) and Gniew (7700 fragments), with the size of other town assemblages ranging from 40 to 1500 fragments. To date, and as a result of its size, the most representative data has been obtained from Gdańsk. All assemblages from the town are dominated by domestic mammals, reaching almost 99% of the total (Fig. 12.13). Cattle dominate overall, followed by pigs, small ruminants, and horses (Table 12.8). In all the towns the presence of dogs was documented, whilst cats were only present in Gdańsk, Gniew, and Elbląg. The most diverse range of wild animals is also evident in Gdańsk, where red deer and roe deer dominate the list of species. However, in the case of red deer more than half of the remains consist of antler fragments, whilst roe deer are represented by more equal proportions of antler and bone. The latter stands out as the most important of the wild species in Gdańsk, and the second most important in Gniew. Wild boar and hare were also hunted. These assemblages have smaller quantities of beaver, elk, and bear, with individual fragments of squirrel and aurochs. Again, these reflect opportunistic exploitation. The Prussian towns of Thorn, Königsberg, and particularly Danzig, were the most important outlets for the fur trade in the south-eastern Baltic, although it is difficult to ascertain how proportions of pelts derived from their patriomonies, from the Order's network, or beyond; most likely they represented a combination of all of these sources areas, with the majority deriving from major

Table 12.8: NISP of mammals in towns of the Teutonic Order phase.

Mammal / Site	Gdańsk	Elbląg – Old Town	Gniew – Old Town	Grudziądz	Toruń	Tolkmicko	Barczewo	Total
Cattle – *Bos primigenius* f. *taurus*	21099	359	3194	233	850	21	32	**25788**
Pig – *Sus scrofa* f. *domestica*	4167	208	2573	23	29	7	21	**7028**
Sheep/Goat – *Ovis ammon* f. *aries*/*Capra aegagrus* f. *hircus*	2527	58	1587	6	9	2	5	**4194**
Sheep – *Ovis ammon* f. *aries*	693	9			12	1		**715**
Goat – *Capra aegagrus* f. *hircus*	186	6		2	11			**205**
Horse – *Equus ferus* f. *caballus*	111	2	77	2	7	3	2	**204**
Dog – *Canis lupus* f. *familiaris*	125		59			1		**185**
Cat – *Felis silvestris* f. *catus*	44	1	68					**113**
Hare – *Lepus europaeus*	11		14					**25**
Red squirrel – *Sciurus vulgaris*	1		1					**2**
Beaver – *Castor fiber*	7		1					**8**
Grey wolf/dog – *Canis lupus*/*Canis lupus* f. *familiaris*	1							**1**
Red fox – *Vulpes vulpes*			1					**1**
Brown bear – *Ursus arctos*	2							**2**
Wild boar – *Sus scrofa*	28		1					**29**
Red deer – *Cervus elaphus*	89		6		3	2		**100**
Elk – *Alces alces*	4							**4**
Roe deer – *Capreolus capreolus*	81	1	29	3				**114**
Aurochs – *Bos primigenius*	1							**1**
Seals – Phocidae	7							**7**
Common porpoise – *Phocaena phocaena*	7							**7**
Total	**30556**	**1500**	**7670**	**269**	**921**	**37**	**66**	**41019**

fur-producing regions. In 1400–02, 60,000 furs were shipped from Novgorod to Königsberg, and from there onto Bruges (Delort 1978). Marine mammals are represented by the remains of seals and harbour porpoises, in both cases consisting of only seven bones. In the abundant assemblage of bird bone, domestic fowl is the most frequently occurring species, with smaller quantities of geese, including greylag and domesticated forms. These are followed by wild ducks, represented by mallard and velvet scoter (*Melanitta fusca*), with birds of prey limited to white-tailed eagles (Table 12.9).

The range of fish species is more extensive, and, as in the case of the castle assemblages, these represent varying aquatic ecosystems (Table 12.10). The most numerous are freshwater species, including cyprinids of which bream (*Abramis brama*) is the most frequently occurring in faunal assemblages. Carp (*Cyprinus carpio*) is also found in these assemblages, a foreign species which only appears after the crusades and whose dissemination in Prussia can be attributed to the Order, as well as to the relatively limited number of monastic institutions. Predatory freshwater species consist of perch, pike, and zander; the latter is by far the most numerous in the collections of Gdańsk and Elbląg, with few remains found in Mikołajki, in the eastern part of Prussia. Another characteristic feature of the assemblages from Gdańsk and Elbląg is the dominating presence of sturgeon remains. In the latter town, on a site at Bednarska 23, a building dating to *c.* 1260–80 included the remains of sturgeon fins and crania, alongside fragments of wine pitchers from Saintonge suggesting the property belonged to a higher status of fisherman (Marcinkowski 2006). Saltwater fish are represented by cod, herring, and flatfish; however, they only form a small component of these assemblages. Freshwater fish clearly played the most important role in the provisioning of towns and would have been the most abundant and accessible within the

Table 12.9: NISP of birds in towns in Teutonic Order phase.

Bird Taxa / Site	**Gdańsk – Św. Ducha St. 43**	**Gdańsk – Młyńska St. 21**	**Gdańsk – Olejarna St.**	**Gdańsk – Piwna St.**	**Gdańsk – Grenary Island**	**Gdańsk, Green Gate**	**Elbląg – Old Town**	**Gniew – Old Town**	**Barczewo, site 1**	**Total**
White stork – *Ciconia ciconia*								1		**1**
Greylag goose – *Anser anser*			1		1	2				**4**
Domestic goose – *Anser anser* f. *domestica*						1				**1**
Greylag goose/Domestic goose – *Anser anser*/*A. anser* f. *domestica*	22		5		39			17	2	**95**
Goose – *Anser* sp.			16		13				1	**30**
Mallard – *Anas platyrhynchos*	1		6		9	1		1		**18**
Mallard/Domestic duck – *Anas platyrhynchos*/ *Anas platyrhynchos* f. *domestica*									1	**1**
Anas sp. – *Anas* sp.	1		3		4		3		1	**12**
Velvet scoter – *Melanita fusca*			1							**1**
White-tailed eagle – *Haliaeetus albicilla*			2							**2**
Capercaillie – *Tetrao urogallus*			1					1		**2**
Domestic chicken – *Gallus gallus* f. *domestica*	24		151		70	9	8	32		**294**
Carrion crow – *Corvus corone*								1		
Carrion crow/rook – *Corvus corone*/*Corvus frugilegus*			1							**1**
Unidentified birds	3	2	28	7	11	3	14	13	1	**82**
Total	**51**	**2**	**215**	**7**	**147**	**16**	**25**	**66**	**6**	**535**

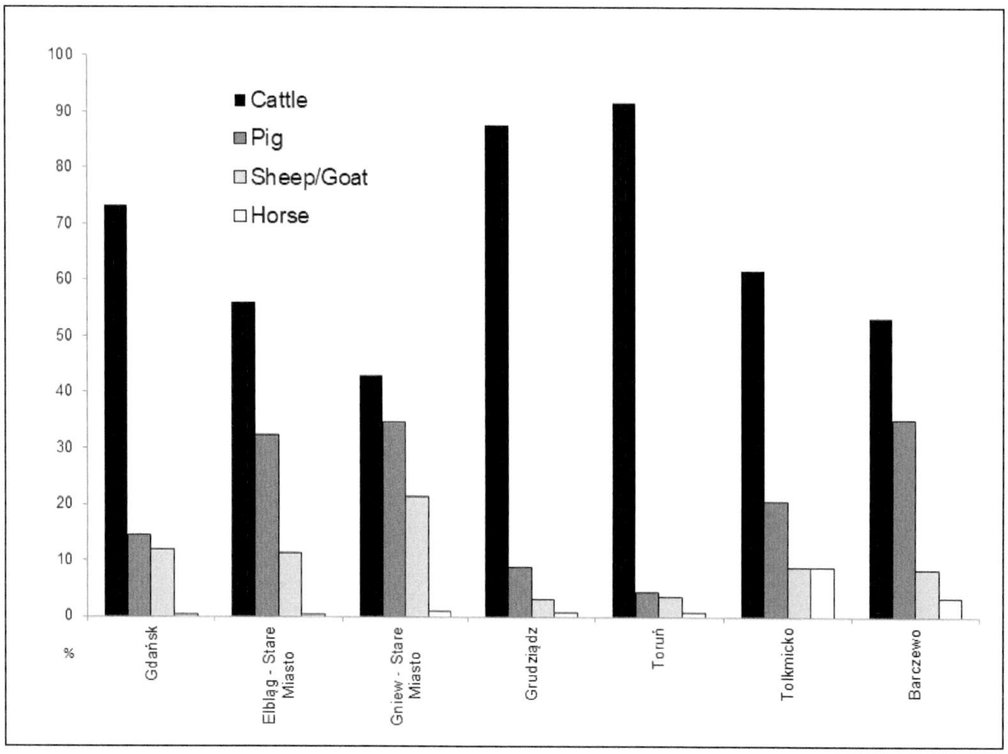

Figure 12.13: Percentage distribution of domestic mammals in town assemblages. Figure by D. Makowiecki.

Table 12.10: NISP of fish in towns in Teutonic Order phase.

Fish Taxa / Site	**Elbląg, Bednarska St.**	**Gdańsk – Green Gate**	**Gdańsk, Olejarna St.**	**Biała Góra (colony)**	**Mikołajki, 14–15th cent.**	**Total**
Pike – *Esox Lucius*	9	1	13	32	283	**352**
Carp – Cyprinidae	20	5	31	269	1538	**1899**
Roach – *Rutilus rutilus*		1	1	10	99	**113**
Ide – *Leuciscus idus*			1	5		**7**
Roach/Rudd – *Rutilus rutilus/Scardinius erythrophthalmus*					28	**28**
Rudd – *Scardinius erythrophthalmus*				3	6	**9**
Asp – *Aspius aspius*	2	1	1	1	1	**8**
Tench – *Tinca tinca*	14		5	22	7	**53**
Bream – *Abramis brama*	25		9	51	24	**118**
Blue bream – *Abramis ballerus*	2					**2**
Sichel – *Pelecus cultratus*				1		**1**
Crucian carp – *Carassius carassius*	3			4	1	**8**
Carp – *Cyprinus carpio* f. *domestica*				1		**1**
Catfish – *Silurus glanis*	13	1	6	78	2	**107**
Perch – *Percidae*	1	2	1	18	323	**348**
Perch – *Perca fluviatilis*	6	2	1	28	387	**427**
Pikeperch – *Lucioperca lucioperca*	37	3	30	21	10	**134**
White fish – *Coregonus albula*					9	**9**
Salmo sp.	1					**1**
Sturgeon – *Acipenser* sp.	694	1	310	1		**1317**
Eel – *Anguilla anguilla*				3		**3**
Herring – *Clupea harengus*		1				**2**
Cod – *Gadus morhua*	4	13	33	5		**101**
Flatfish – *Pleuronectidae*		1	2	1		**7**
Identified	831	32	444	554	2718	**5055**
Unidentified fish	137	29	13	430	4818	**5248**
Total	**968**	**61**	**457**	**984**	**7536**	**10524**

estuarine waters of the Vistula Lagoon, Lake Drużno, the Vistula, and the Masurian lakes. Whilst marine fish were often consumed by the townspeople of Gdańsk, they rarely reached smaller settlements within the broader region, such as Biała Góra. It is also possible that settlements bordering the Masurian lakes, such as Mikołajki, supplied the Order's houses with fish. The assemblage from this site consisted not only of very small cyprinids but also naturally larger species such as pike. Here, individuals of this species measured *c.* 40–50 cm, while data from other medieval sites indicates caught fish were usually 10 cm longer (Makowiecki 2003b). Catfish also played an important role in these assemblages. In Gdańsk, although no faunal assemblages are available from the Order's castle, documents indicate that after the annexation of the town, the community of fisherman living in the district of the ducal stronghold was relocated to Osiek (Hackelwerk). They provided the Order's castle with certain species of fish including sturgeon, trout, salmon, and herring, continuing earlier ducal traditions (Paner 1998–99, 110).

Table 12.11: Selected metrical data of cattle, pig, wild boar, sheep, and horse.

Stage — Region	N	Mean	Minimum	Maximum	Range	sd	V%
CATTLE — Talus: height at the withers (coefficient 1.83 by Calkin 1970)							
Pre-crusader	150	103.6427	92.80000	118.6000	25.80000	5.268819	5.083639
Crusader	56	102.3893	90.00000	117.3000	27.30000	5.239038	5.116783
Post-crusader	147	107.7354	91.70000	139.4000	47.70000	7.961550	7.389912
CATTLE — Metacarpal: height at the withers (coefficient 6.12 by Calkin 1960)							
Pre-crusader	91	104.5780	92.80000	118.3000	25.50000	5.854662	5.598368
Crusader	132	103.1553	90.60000	119.0000	28.40000	4.790613	4.644078
Post-crusader	318	108.1416	91.70000	136.3000	44.60000	7.261200	6.714531
PIG and WILD BOAR — height at the withers, calculated by different element (coefficients by Teichert 1969)							
Pre-crusaders	272	81.34468	60.50200	119.7424	59.24040	14.19329	17.44833
Crusaders	112	78.49740	59.96500	108.9978	49.03280	8.43856	10.75011
Post-crusaders	87	78.74625	67.62600	105.0560	37.43000	5.82740	7.40022
PIG — height at the withers, calculated by different elements (coefficients by Teichert 1969)							
Pre-crusaders	185	72.49226	60.50200	88.87320	28.37120	4.394650	6.062233
Crusaders	103	76.72146	59.96500	91.08450	31.11950	5.904998	7.696671
Post-crusaders	86	78.44033	67.62600	91.22880	23.60280	5.110757	6.515472
SHEEP — Metacarpal: GL							
Early medieval Poland (Schramm 1967; Makowiecki 2001b, 2010a, 2014, 2016)	83	125.4819	114.0000	139.3000	25.30000	5.745309	4.578595
Pre-crusader	35	125.4860	108.8000	146.0000	37.20000	8.411905	6.703461
Crusader	72	129.2681	114.7000	146.7000	32.00000	7.678905	5.940296
Post-crusader	135	124.9956	110.4000	144.9000	34.50000	6.482973	5.186563
SHEEP — Metacarpal: height at the withers, calculated by different elements (coefficients by Teichert 1975)							
Pre-crusader	63	61.08095	53.20000	71.40000	18.20000	3.447386	5.643962
Crusader	157	62.92229	55.10000	71.70000	16.60000	3.650423	5.801478
Post-crusader	267	61.35169	51.80000	75.30000	23.50000	3.665879	5.975189
HORSE — height at the withers, calculated different element (coefficient by Kissewalter 1888 (in Driesch and Boessneck 1974) and by table in Vitt 1952)							
Pre-crusader	255	129.1088	109.6000	150.4000	40.80000	8.31939	6.443700
Crusader	18	129.5556	101.6000	149.6000	48.00000	12.27902	9.477804
Crusader — Post-crusader	51	130.3745	101.6000	149.6000	48.00000	9.29240	7.127469
Pre-crusader: Gdańsk Bay	103	127.0155	111.3000	149.3000	38.00000	9.19299	7.237692
Pre-crusader: Kulmerland	99	129.1798	109.6000	149.8000	40.20000	8.80542	6.816410
Pre-crusader: Prussia	72	131.5243	121.2000	150.4000	29.20000	5.90903	4.492727
Early medieval: Pomerania W (Figurska 2011)	54	134.2907	119.0000	147.6000	28.60000	6.98541	5.201706
Early medieval: Great Poland (Figurska 2011)	142	132.8542	107.9000	146.4000	38.50000	8.19937	6.171701
Early medieval: Lower Silesia (Figurska 2011)	104	134.9529	112.0000	148.5000	36.50000	7.11631	5.273184

Changes in the Size of Animals

Statistical analysis of osteometric data indicates that in the West Slavic regions, sizes of cattle, pigs, sheep, and goats were characterized by minor differentiations (Schramm 1967a, 1967b, 1976; Lasota-Moskalewska et al. 1987; Makowiecki 2006b). Population diversity is only evident in the case of the horse using estimated withers (i.e., shoulder) heights, with some notable regional differences: individuals from Gdańsk and the Kulmerland were on average lower than horses in Greater Poland and Lower Silesia (Makowiecki and Makowiecka 1999; Makowiecki 2006b).

Cattle could be categorized as a small form of the short-horned type (Schramm 1976; Makowiecki 2006), and a similar type was bred in the Gulf of Gdańsk (Kubasiewicz 1977) and the Kulmerland (Makowiecki and Makowiecka 1999; Makowiecki 2010b). Osteometric data for cattle from Prussia also points to the breeding of the same morphotype (Żuk-Piechota and Towarnicki 1978; Piątkowska-Małecka 2009). The range of these animals' withers heights was 90–115 cm, with an average height of *c.* 104 cm. Heights were calculated from measurements of the GL (greatest length) metacarpal and GLl (greatest lateral length) talus, using coefficients of 6.12 and 1.83 estimated by Calkin (1960, 1970), whilst the widths of metacarpals were also synchronized independently with temporal changes. As a result, it is clear that the size of cattle changed in subsequent periods (Table 12.11, Fig. 12.14). In the early Middle Ages, the bovine population was close to a normal distribution, but the histogram illustrating variations in metatarsal shaft width points towards a certain heterogeneity which can plausibly be related to sexual dimorphism. This indicates an equal share of females to males, along with the presence of castrates.

Following the crusades, these characteristics did not change (Table 12.11). However, the majority of individuals stood at 102–03 cm, smaller than the pre-crusade average of 104–05 cm. In terms of the index of metacarpal widths, the average rose to 16–17 from the earlier value of 14–15. This may reflect some change in the size of cattle, and it also seems that the proportion of cows to bulls, with clear examples of oxen, was larger. However, it is only after the period of the Order's rule that the average height of cattle increases by *c.* 5 cm. The range between the smallest and largest individuals also increases, as indicated by the standard deviation and coefficient of variation (Fig.12.14), and the index of diaphyseal (i.e., bone shaft) thickness also increased. The graphic representation of withers heights indicates a multi-modal distribution of characteristics (Fig. 12.14), with two clear sub-populations. The first has an average value of *c.* 16–17 and the second *c.* 19–20. It is therefore clear that in the post-Order period, there were new forms of cattle. One of them was the local population, derived from early types of cattle, whilst the other consisted of significantly larger individuals. Their appearance may be the result of introducing new stock, a process initiated under the Order's rule. However, local cattle may have also been 'improved' as a result of changes in grazing and fodder. It is important to bear in mind that data from the post-Order phase for cattle is largely derived from Gdańsk, and the evident improvements could reflect the grazing opportunities provided by the Vistula Fens. At this time, there was also incremental deforestation within the region, evident in palynological profiles (see Chapter 11), but also in the increasing appearance of hare in faunal assemblages. The conversion of woodland into pasture would have contributed to the development of local livestock husbandry, including that of small ruminants. A particularly high proportion of the latter (almost 30%) is evident in the faunal assemblage from the Order's castle at Grudziądz (Makowiecka et al. 2012). The persistence of earlier livestock types was not confined to the former *Ordensland*. In the villages of Greater Poland, in the sixteenth and seventeenth century, it was still possible to encounter cattle that were the same size as those reared in the earlier Piast state (Makowiecki and Makowiecka 2007).

There is comparatively little osteometric data for pigs from Prussia. More detailed information is only available from the Kulmerland (Makowiecki 2006b, 2010). However, the Slavs living in the Vistula and Odra regions bred small-sized animals, which were noticeably smaller than wild boar (Lasota-Moskalewska et al. 1987; Makowiecki, 2006b). Their heights ranged from 60 to 82 cm, with an average of *c.* 70 cm, and statistical analysis indicates a clear bimodal distribution of this value (Makowiecki 2006b), at least for the study area during the pre-crusade period (Fig. 12.15). The histogram shows a clearly bimodal distribution, in contrast to a natural population, with an extensive range of heights from *c.* 60 to 119 cm, and an average of 81.3 cm. The population is characterized by a very high standard deviation and coefficient of variation. After the crusades, the average height becomes smaller, with a decreased range and other measures of variation. In the post-Order period, the average height did not change; however, the range, standard deviation, and coefficient of variation were reduced even further. In both the medieval and post-medieval phases

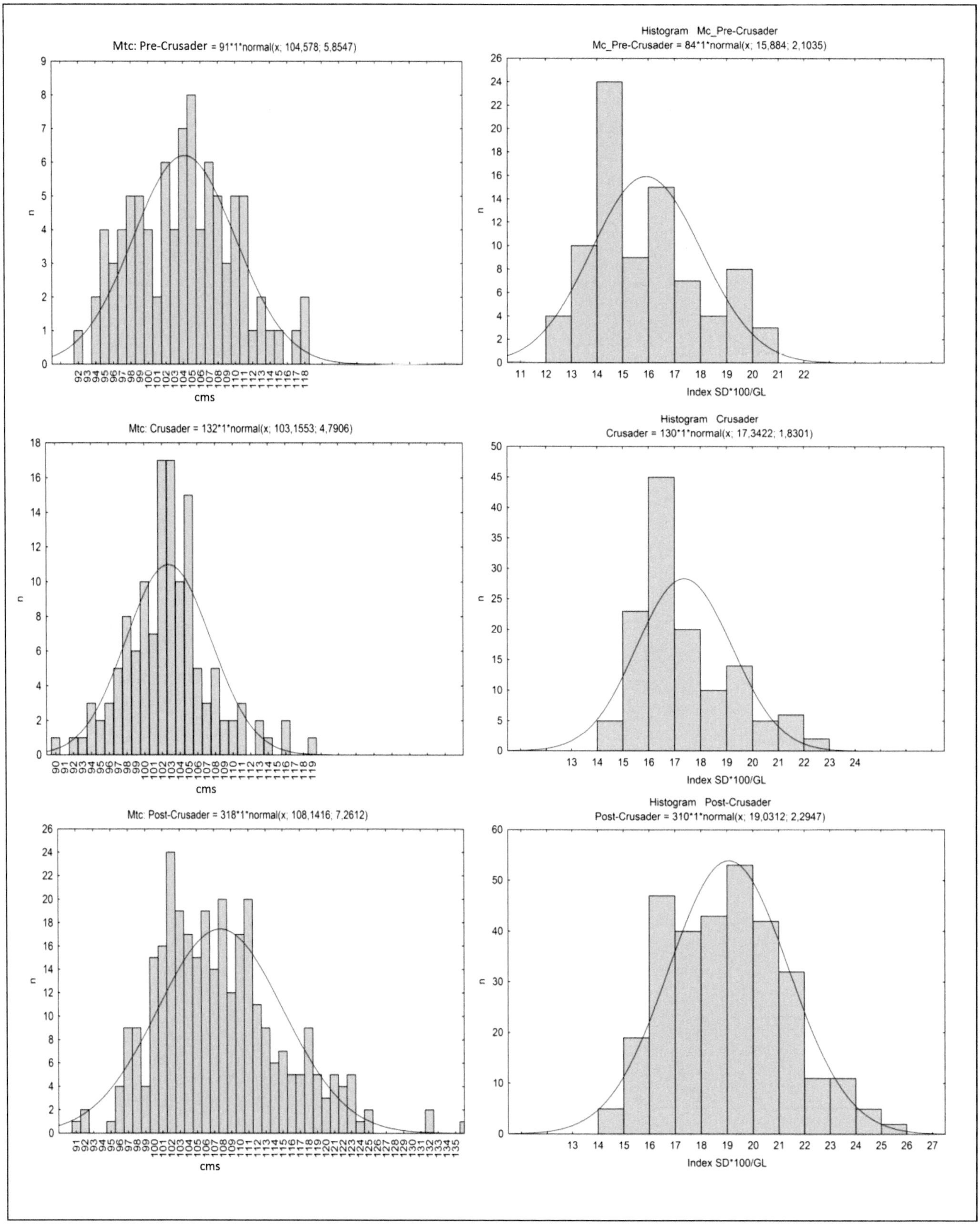

Figure 12.14: Changes in cattle size as indicated by metacarpal measurements.
Figure by D. Makowiecki.

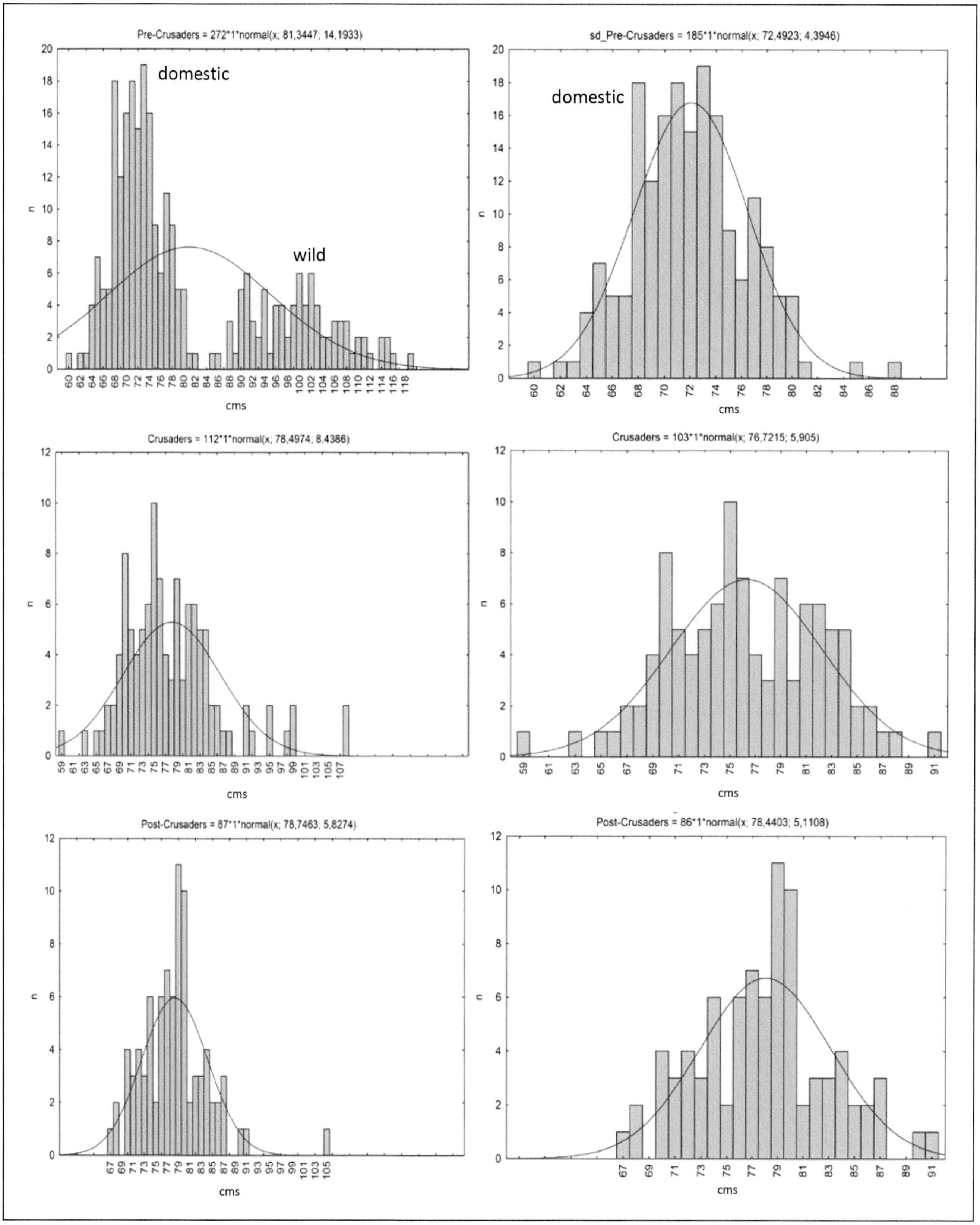

Figure 12.15: Changes in pig size. Withers height calculated according to Teichert 1969 on the basis of different skeletal elements: left - domestic and wild forms; right - domestic form. Figure by D. Makowiecki.

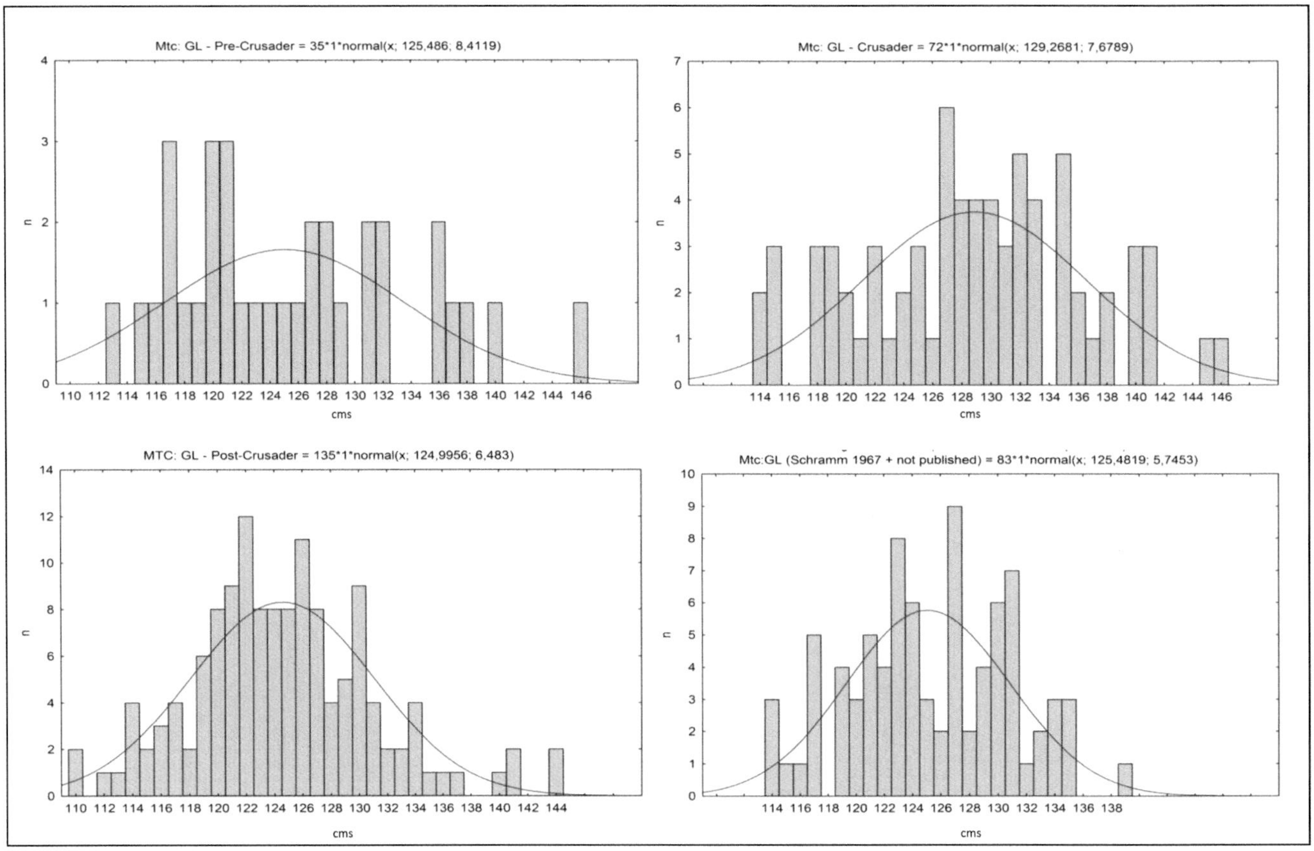

Figure 12.16: Changes in sheep size as indicated by the greatest length (GL) of the metacarpal. Figure by D. Makowiecki.

the histograms indicate close to normal distributions, with an almost complete absence of large individuals corresponding to early medieval wild boars. The boundary between the two populations, domestic and wild, represented by the value of 83 cm, is also absent. As a result, a diachronic model can be suggested where individuals from the pre-crusade period reached heights of 60–82 cm (Fig. 12.15), with an average of *c.* 72 cm (Table 12.11). Subsequently this value increased, with an average of *c.* 76.7 cm during the period of the Order's rule, and 78.5 cm in the following period.

Schramm's (1967b) analysis of sheep from early medieval Polish sites demonstrated variations in withers heights between each centre. Her histogram collating the greatest lengths (GL) of metacarpals, supplemented by more recent data, mainly from Greater Poland (Makowiecki 2001b and unpublished) reinforces these observations (Fig. 12.16). Against this background, individuals from the study area in the pre-crusade period are broadly similar: this is clear from the average value, standard deviation, and coefficient of variation (Table 12.11). The histogram representing the population from the period of the Order's rule shows an increase in size, with a higher average greatest length (Table 12.11). In the post-Order period this average is reduced to a value comparable to the pre-crusade period. On this basis, it is possible to suggest that during the later medieval period larger individuals were bred than in the pre-crusade period.

Osteometric data for horses is, to date, limited from the Order's period, when there is extensive horse breeding documented in the Prussian *Ordensland* (Table 12.11). Larger datasets are available for the pre-crusade period from Gdańsk, the Kulmerland, and Prussia. Based on the length of long bones, shoulder heights were calculated according to Vitt (1952) and Kissewalter (1888, in Driesch and Boessneck 1974), with the result that on average, horses from Gdańsk were the smallest (Fig. 12.10), medium-sized individuals characterize the population of the Kulmerland, whilst the largest came from Prussia, and predominantly from Poganowo. The largest horses within the total range only appear during the period of the Order's rule, whilst the average value is comparable to the pre-crusade horse population from the Kulmerland. These estimates can also be compared with early medieval horses from Greater Poland, western

Pomerania, and Lower Silesia (Figurska 2011), with the result that the average height is higher. The difference between the smallest individuals from medieval Gdańsk and the largest from early medieval Lower Silesia was as much as 8 cm. Comparative analysis of histograms for the regional populations indicates that the group with the most morphological integrity was represented by the horses from Poganowo. In most of the regional populations, it was possible to identify a number of size categories; however, in the case of Poganowo there were no animals smaller than 120 cm.

To summarize, it is clear that during the time of the Order's rule provisioning was still based on stocks derived from the preceding period. There were, however, examples of larger individuals, which may be connected with the influx of German settlers and the potential introduction of new stock. At the same time, changes in fodder may also have contributed to these 'improvements' (Scull et al. 2019). There was no clear increase in the size of horses during the medieval period, which may reflect the limited number of faunal assemblages recovered from castles. A large proportion of the bones used for calculating withers heights derived from towns, and therefore most probably came from working horses rather than warhorses. At the same time, written sources indicate that the proportion of the horse population bred for warfare was comparatively small (see also the discussion on horses in the Order's inventories in Chapter 15).

Conclusion

When considering the overall trends in animal exploitation in Prussia before and after the crusades, several distinct features can be highlighted. First, the Prussian tribes were not solely reliant on livestock but supplemented the animal-based part of their diet with wild mammals, birds, and fish. At the same time, the scale of animal exploitation did not differ markedly from neighbouring Slavic regions. Both societies hunted the same species and fished largely the same fish, facilitated by relatively high degrees of local biodiversity. Provisioning systems were effective enough to ensure sufficient demographic stability, enabling the Prussians to conduct military operations against their Christian neighbours and, indeed, to oppose the incursions of Polish magnates in the twelfth century.

The arrival of the Teutonic Order prompted new developments in animal husbandry and related alimentary culture. The zooarchaeological data confirms the increases in livestock husbandry and pasture highlighted in earlier scholarship (Okulicz-Kozaryn 1983, 34). But it is important to emphasize the persistence of pre-crusade husbandry regimes, which included the maintenance of indigenous livestock herds and the continuation of hunting and fishing. Differences are evident between western and eastern Prussia, both in the exploitation of domestic and wild species. The most significant changes occurred in livestock husbandry; whilst the older system was dominated by pigs and cattle, and the latter were solely important under the new regime. There were also changes in herd management strategies, particularly in the case of cattle with an increase in the numbers of calves being slaughtered. New tools were also introduced for processing meat (Seetah et al. 2013).

During the post-Order period, calves and adults were the two main age groups that were sent to butchers, and these changes may have been linked to changing demands for deadstock, including leather. In the case of cattle, this trend can also be linked to the emergence of different varieties from which larger quantities of milk could be obtained. These animals could have potentially been brought by settlers from the eastern provinces of the Holy Roman Empire, within the context of the waves of colonization encouraged by the Order and Prussian episcopate during the crusading period, but particularly from the late thirteenth through to the mid-fourteenth centuries. This may also account for the appearance of larger cattle, pigs, and sheep during the later medieval period. At the same time, the cultural role of livestock was retained amongst surviving indigenous Prussian communities. In Sambia, calf sacrifices are documented in 1520 and 1531, a goat sacrifice is attested in 1570, and there are ethnographic references to the sacrifice of white pigs, alongside the black and white cockerels mentioned above (Okulicz-Kozaryn 1983, 113). It is of course difficult to project these practices directly back into the pre-crusade period, but given the Christian rejection of animal sacrifice (Ullucci 2012), these were most likely continuations or revivals of earlier customs.

Finally, an intensification in the fur trade is evident after the crusades. Whilst Prussians were clearly handling furs in the pre-crusade period, the Order's network facilitated the expansive trade, with its major ports linking neighbouring fur-producing regions, especially Poland in the case of Prussia, with Western European markets. From the mid-fourteenth century, the trade in furs down the Lower Vistula from Thorn to Danzig shadowed the shipment of grain. Some of these furs may have derived from the Order's own lands, as indicated by the occasional presence of fur-bearing mammal remains in castle and town assemblages, although, as in the case of Livonia, it is highly likely the majority derived from rural settlements, particularly in eastern Prussia, which remain largely unexcavated (see also Chapter 6). The importance of beaver fur, in particular, handled through Prussian ports, is emphasized in French and English documents from the end of the fourteenth to the early fifteenth century. The quantities, usually derived from only a handful of species, were vast — in 1396, 160,000 pieces of fur passed through Königsberg onto Bruges. By the fifteenth century, the Order procured pelts for its own officials; in 1417, a thousand furs are documented at Marienburg for the Grand Master's use, where the presence of imported leopard pelts is also noted (Delort 1978, I, 112–13, 595). This contrasts with the austere image of the Order's brethren which traditionally favoured monochrome clothing. In fact, this did not apply to the highest offices of the Order, as silk was regularly purchased for the Grand Master's garments and for his chapel, and was even used in their funerary robes as evident from the burials within the crypt of the cathedral at Kwidzyn (Grupa 2009).

In the case of the horse, which played such an iconic role during the crusading period, it is harder to identify any improvements due to the limitations of the existing dataset; however, the inventories of the Order's castles refer to many different types of horses with specifically assigned roles, alongside the two broad divisions between the larger imported warhorses and the smaller, indigenous *schweik* (Ekdahl 1991, 1998; see also Chapter 15). This variety appears to be represented by the range of differently sized horses in later medieval Prussian faunal assemblages. The limited numbers of large horses correspond to the evidence for small numbers of warhorses kept at the Order's castles, although the quantities in the written and zooarchaeological evidence are not readily comparable. In the pre-crusade period there were also different sizes of horses, rather than just the one indigenous type, which can plausibly be linked to different functions. As in Great Poland, the smallest were more suited as pack animals than for riding (Makowiecki 2001b, 99), and greater numbers of larger horses amongst the Prussians points to a more pronounced equestrian culture than practised by the Slavs. The smaller horses in Gdańsk may reflect not only the less intensive breeding of the species but also the need for a larger number of animals to transport goods within the trading hub. These reflections must be regarded as a hypothesis which can tested with future studies, particularly concerning the identification of wild horses that appear to have survived in these regions until the mid-seventeenth century (Fischer 1937, 6). The ritual killing of horses which defined Prussian culture in the pre-crusade period appears to end in the thirteenth century, with only deposits of horse equipment attested in the fourteenth and fifteenth centuries (Shiroukhov 2012, 250).

Whilst chickens remained the most important poultry exploited during the period of the Order's rule, there is also a clear increase in the consumption of geese, as well as ducks. Woodland galliforms continued to be hunted, but exotics such as the peafowl were procured for consumption by certain castle communities. The presence of partridge in castle assemblages may be connected with increasing deforestation within their associated territories, which would have resulted in more suitable habitats for this species. Trends in fish consumption also changed with the introduction of cod, increasingly derived from a local fishing industry alongside stockfish imported from greater distances. Sturgeon continued to feature as an important species, but it was not as widely consumed as in the pre-crusade period and was mainly at sites located in the Gulf of Gdańsk and the Vistula Lagoon. Its importance was particularly pronounced in Sambia: the convent at Königsberg had an officer dedicated to their procurement (Długokęcki 2009, 486). This diachronic shift can be readily situated within the well-documented decrease in sturgeon fishing (Makowiecki 2003b) (Fig. 12.5). There were also pronounced differences between the types of fish caught in eastern and western Prussia, with species from local lakes and rivers defining the fishing culture of the former region.

In summary, this chapter has provided a model for changes in animal exploitation as a result of the creation of the Christian theocratic polity in Prussia. This can be tested and modified with future analyses of faunal assemblages from well-dated contexts, as well as more expansive analyses of existing material, in particular DNA and stable isotopic studies. Whilst it has not been possible to include all the regions of Prussia, largely as a result of the availability of material, it would be essential in the future to consider how the regions of Sambia and Sudovia, both extremely important in the early medieval geopolitical context of the south-eastern Baltic, and with contrasting social outcomes as a result of the crusades, fit into the general model presented above.

Works Cited

Primary Sources

Popłoński, J. 1862. *Helmolda kronika sławiańska XII wieku, przełożona z języka łacińskiego na polski przez Jana Popłońskiego*, Warsaw: W drukarni K. Kowalewskiego przy ulicy Królewskiej 1065

Secondary Works

Antoniewicz, J. 1958. 'Prusowie we wczesnym średniowieczu i zarys ich kultury materialnej', in *Szkice z dziejów Pomorza. Pomorze średniowieczne*, ed. G. Labuda, Poznań: Wydawnictwo Poznańskie, 121–59

Barrett J., Johnstone, C., Harland, J., Van Neer, W., Ervynck, A., Makowiecki, D., Heinrich, D., Hufthammer, A. K., Bødker Enghoff, I., Amundsen, C., Christiansen, J. S., Jones, A. K. G., Locker, A., Hamilton-Dyer, S., Jonsson, L., Lõugas, L., Roberts, C., and Richards, M. 2008. 'Detecting the Medieval Code Trade: A New Method and First Results', *Journal of Archaeological Sciences* 35, 850–61

Benecke, N. 1988. *Archäozoologische Untersuchungen an Tierknochen aus der Frühmittelalterlichen Siedlung von Menzlin*, Schwerin: Museum für Ur- und Frühgeschichte

——, 1994a. *Archäozoologische Studien zur Entwicklung der Haustierhaltung in Mitteleuropa und Südskandinavien von den Anfängen bis zum ausgehenden Mittelalter*, Berlin: Akademie

——, 1994b. *Der Mensch und seine Haustiere: Die Geschichte einer jahrtausendealten Beziehung*, Stuttgart: Theiss

Bienias, D., Bojarski, B., Chudziak, W., Janowski, A., Makowiecki, D., Poliński, D., and Weinkauf, M. 2008. 'Katalog', in *Pozostałości wczesnośredniowiecznego osadnictwa w strefie przebiegu autostrady A-1 w województwie kujawsko-pomorskim (byłe województwo bydgoskie). Katalog źródeł archeologicznych*, eds J. Olczak and W. Chudziak, Toruń: Wydawnictwo Naukowe Uniwersytetu Mikołaja Kopernika, 55–543

Bliujienė, A., ed. 2009. *The Horse and Man in European Antiquity* (= *Archaeologia Baltica* 11)

Bliujienė, A., and Butkus, D. 2009. 'Burials with Horses and Equestrian Equipment on the Lithuanian and Latvian Littorals and Hinterlands (from the fifth to the Eighth Centuries)', *Archaeologia Baltica* 11, 149–63

Bocheński, Z., Bocheński, Z. M., and Tomek, T. 2012. *A History of Polish Birds*, Kraków: Institute of Systematics and Evolution of Animals, Polish Academy of Sciences

Bocheński, Z., Tomek, T., Wertz, K., and Wojenka, M. 2015. 'Indirect Evidence of Falconry in Medieval Poland as Inferred from Published Zooarchaeological Studies', *International Journal of Osteoarchaeology*, doi: 10.1002/oa.2457

Calkin, V. I. 1960 'Izmencivost metapodij i jejo znacenje dlja izucenja krupnogo rogotogo skota drevnosti', *Bjulleten Moskovskogo Obšcestva Ispytatelej Prirody, Otdel Biologii* 65/1, 109–26

——, 1970. *Drevnejše domašnje životnye vostocnoj Evropy*, Moskva: Izdatelstvo Nauka

Chudziak, W. 2003. *Wczesnośredniowieczna przestrzeń sakralna in Culmine na Pomorzu Nadwiślańskim*, Toruń: Wydawnictwo Naukowe Uniwersytetu Mikołaja Kopernika

Dąbrowska, T. 1967. 'Analiza morfologiczna i systematyczna resztek kostnych ryb z wykopalisk wczesnośredniowiecznego grodziska „Okrągła Góra' — Ostrowie koło Pasymia', *Zeszyty Naukowe Wyższej Szkoły Rolniczej w Olsztynie* 23/557, 69–88

Dąbrowski, J. 1976. 'Dalsza morfologiczna i systematyczna ocena resztek kostnych ryb z grodziska "Okrągła Góra"', *Zeszyty Naukowe Akademii Rolniczo-Technicznej w Olsztynie* 6, 85–102

Delort, R. 1978. *Le commerce des fourrures en occident á la fin du Moyen âge: (vers 1300 – vers 1450)*, 2 vols, Rome: École Française de Rome

Długokęcki, W. 2009. 'Społeczeństwo wiejskie', in *Państwo zakonu krzyżackiego w Prusach: Władza i społeczeństwo*, ed. M. Biskup et al., Warsaw: Państwowe Wydawnictwo Naukowe, 460–94

Driesch, A. von den and Boessneck, J. 1974. 'Kritische Ammerkungen zur Widerristhöhenberechnung aus Längenmassen vor- und frühgeschichtlicher Tierknochen', *Säugetierkundliche Mitteilungen* 22, 325–48

Duda, I. 1965. *Towaroznawstwo surowych skór futerkowych*, Łódź: Przegląd Skórzany

Ekdahl, S. 1991. 'Das Pferd und seine Rolle im Kriegswesen des deutschen ordens', in *Das Kriegswesen der Ritterorden im Mittelalter*, ed. Z. H. Nowak, Toruń: Wydawnictwo Uniwersytetu Mikołaja Kopernika, 25–47

——, 1998. 'Horses and Crossbows: Two Important Warfare Advantages of the Teutonic Order in Prussia', in *The Military Orders*, II: *Welfare and Warfare*, ed. H. Nicholson, Farnham: Ashgate, 119–51

Ervynck, A. 2004. '*Orant, pugnant, laborant*: The Diet of the Three Orders in the Feudal Society of Medieval North-Western Europe', in *Behaviour behind Bones: The Zooarchaeology of Ritual, Religion, Status and Identity*, eds S. J. O' Day, W. Van Neer, and A. Ervynck, Oxford: Oxbow, 215–23

Figurska, J. 2011. 'Koń w średniowiecznej kulturze mieszkańców dorzecza Wisły i Odry' (unpublished master's thesis, Institute of Archaeology, Nicolaus Copernicus University, Toruń)

Fischer, A. 1937. *Etnografia dawnych Prusów: Rozprawa z pracy zbiorowej Dzieje Prus Wschodnich*, Gdynia: Kasa im. Mianowskiego - Instytut Popierania Nauki

Grążawski, K. 2009. *Ziemia lubawska na pograniczu słowiańsko-pruskim w VIII–XIII w., Studium nad rozwojem osadnictwa*, Olsztyn: Wydawnnictwo Uniwersytetu Warmińsko-Mazurskiego w Olsztynie

Gręzak, A., and Piątkowska-Małecka, J. 1999. 'Stan badań nad zwierzęcymi szczatkami kostnymi Ze stanowisk badanych przez Kompleksową Ekspedycję Jaćwieską', in *Bałtowie i ich sąsiedzi*, eds A. Bitner-Wróblewska and G. Iwanowska, Warsaw: Państwowe Muzeum Archeologiczne, 155–72

Grupa, M. 2009. 'Tkaniny z krypty północnej w Kwidzynie', in *Katedra w Kwidzynie — tajemnica krypt*, eds M. Grupa and T. Kozłowski, Kwidzyn: Kwidzyńskie Centrum Kultury, 149–75

Haftka, M., and Wadyl, S. 2015. *Węgry. Zespół osadniczy na pograniczu Pomorsko-Pruskim w XI–XII w.*, Malbork: Muzeum Zamkowe w Malborku

Jagodziński, M. F. 2010. *Truso: Między Weonodlandem a Witlandem*, Elbląg: Muzeum Archeologiczno-Historyczne w Elblągu

Jaskanis, J. 1966. 'Human Burials with Horses in Prussia and Sudovia in the First Millennium of our Era', *Acta Baltico-Slavica* 4, 29–65

Karczewski, M. 2011. *Archeologia środowiska zachodniobałtyjskiego kręgu kulturowego na pojezierzach*, Poznań-Białystok: Bogucki Wydawnictwo Naukowe

Kawecki, Z. 1988. *Zoologia stosowana*, 3rd edn, Warsaw: PWN

Kontny, B., Okulicz-Kozaryn, J., and Pietrzak, M. 2011. *Nowinka, Site 1: The Cemetery from the Late Migration Period in the Northern Poland*, Gdańsk: Archaeological Museum in Gdańsk & Warszawa University

Kowalska-Lewicka, A. 1980. *Hodowla i pasterstwo w Beskidzie Sądeckim*, Wrocław: Zakład Narodowy im. Ossolińskich

Krajcarz, M., Makowiecki, D., Krajcarz, M., Masłowska, A., Baca, M., Panagiotopoulou, H., Romańska, A., Bednarczyk, J., Gręzak, A., and Sudoł, M. 2015. 'On the Trail of the Oldest Domestic Cat in Poland: An Insight from Morphometry, Ancient DNA and Radiocarbon Dating', *International Journal of Osteoarchaeology*, doi: 10.1002/oa.2471

Kranz, I. 1974. 'Ogólna charakterystyka materiału kostnego zwierzęcego z wykopalisk wczesnośredniowiecznego Topolna pow. Świecie', *Pomorania Antiqua* 5, 385–96

Kruszona, W. 1990. 'Zwierzęcy materiał kostny z wykopalisk w miejscowości Podzamcze, woj. elbląskie', *Roczniki Akademii Rolniczej w Poznaniu* 221, *Archeozoologia* 16, 9–20

Krysiak, K. 1950. 'Charakterystyka materiału zwierzęcego ze stanowiska wczesnośredniowiecznego w Jeziorku, pow. Giżycko', *Materiały Wczesnośredniowieczne* 2, 227–33

——, 1955. 'Szczątki zwierzęce z wykopalisk w Gdańsku w latach 1950-1951', *Studia Wczesnośredniowieczne* 3, 238–51

——, 1956. *Wyniki badań nad materiałem z wykopalisk w Gdańsku*, Prace Wrocławskiego Towarzystwa Naukowego, series B, 78, Wrocław: Państwowe Wydawnictwo Naukowe

——, 1967. 'Szczątki zwierzęce z wykopalisk w Gdańsku', *Gdańsk wczesnośredniowieczny* 6, 7–47

Kubasiewicz, M. 1965. 'Szczątki zwierzęce z wczesnośredniowiecznego grodziska w Otominie', *Pomorania Antiqua* 1, 309–315

——, 1977. *Badania archeozoologiczne na terenie Gdańska IX–XIV wieku*, Gdańsk wczesnośredniowieczny, 9, Gdańsk: Zakład Narodowy im. Ossolińskich

Labuda, G. 1999. *Słowiańszczyzna starożytna i wczesnośredniowieczna*, Poznań: Wydawnictwo *Poznańskiego* Towarzystwa Przyjaciół Nauk

Lasota-Moskalewska, A., Kobryń, H., and Świeżyński, K. 1987. 'Changes in the Size of the Domestic and Wild Pig in the Territory of Poland from the Neolithic to the Middle Ages', *Acta Theriologica* 32/5, 51–81

Lewicki, T. 1954. 'Średniowieczne źródła arabskie i perskie o hodowli zwierząt domowych u Słowian', *Kwartalnik Historii Kultury Materialnej* 2/3, 444–69

Łuka, L. J. 1985. 'Ziemia gdańska w okresie wczesnośredniowiecznym (od VII w. do połowy X w.)', in *Historia Gdańska. Tom 1 do roku 1454*, ed. E. Cieślak, Gdańsk: Wydawnictow Morskie, 27–68

Makowiecka, M. 2011. 'Zwierzęce szczątki kostne z Barczewa, stan. 1', unpublished report, Barkowo: Pracownia Archeologiczno-Konserwatorska M.K. Kocińskiej

Makowiecka, M., and Makowiecki, D. 2008. 'Wyniki badań archeozoologicznych z miejscowości Pień, gm. Dąbrowa Chełmińska, stan. 1 (sezony 2004–2006)', unpublished report, Toruń: Instytut Archeologii, Uniwersytet Mikołaja Kopernika w Toruniu

Makowiecka, M., and Makowiecki, D. 2013. 'Analiza archeozoologiczna szczątków z Placu Wolności (stan. 1) w Tolkmicku', unpublished report, Toruń: Instytut Archeologii, Uniwersytet Mikołaja Kopernika w Toruniu

Makowiecka, M., and Makowiecki, D. 2014. 'Wyniki badań zwierzęcych szczątków kostnych', in *Zamek biskupów chełmińskich w Wąbrzeźnie w świetle badań archeologiczno-architektonicznych. Studia i materiały*, ed. M. Wiewióra, Toruń: Wydawnictwo Uniwersytetu Mikołaja Kopernika, 177–201

Makowiecka, M., Makowiecki, D., Müller, R., and Wussow, J. 2005. 'Wstępne wyniki badań archeozoologicznych z Ostrowa Tumskiego', in *Poznań we wczesnym średniowieczu* (vol. v), ed H. Kóčka-Krenz, Poznań: Wydawnictwo Poznańskie, 179–90

Makowiecka, M., Makowiecki, D., and Zabilska, M. 2012. 'Badania archeozoologiczne', in *Zamek w Grudziądzu w świetle badań archeologiczno-architektonicznych. Studia i materiały*, ed. M. Wiewióra, Toruń: Wydawnictwo Instytut Archeologii UMK, 302–51

Makowiecki, D. 1994. 'Wyniki badań archeozoologicznych szczątków kostnych z wczesnośredniowiecznego grodziska (stanowisko 1) i osady (stanowisko 2) w Jedwabnie, województwo toruńskie', in *Studia nad osadnictwem średniowiecznym ziemi chełmińskiej. Wybrane zagadnienia i materiały*, ed. J. Olczak, Toruń: Uniwersytet Mikołaja Kopernika, 187–217

——, 1996. 'Ekspertyza archeozoologiczna zwierzęcych szczątków kostnych z grodziska wczesnośredniowiecznego w Sopocie, stan. 1 (sezon 1995)',), unpublished report, Toruń: Instytut Archeologii, Uniwersytet Mikołaja Kopernika w Toruniu

——, 2001a. 'Zestawienie inwentaryzacyjne szczątków kostnych z Napola 6 (PL0571), Napola 1 (PL0571/2), Napola 1B (PL0571/1), unpublished report, Toruń: Instytut Archeologii, Uniwersytet Mikołaja Kopernika w Toruniu

——, 2001b. *Hodowla oraz użytkowanie zwierząt na Ostrowie Lednickim w średniowieczu: Studium archeozoologiczne*, Poznań: Biblioteka Studiów Lednickich VI

——, 2003a. 'Zwierzęce szczątki kostne z badań na Starym Mieście w Elblągu — działka przy ul. Bednarskiej 2', unpublished report, Elbląg: Muzeum Archeologiczno-Historyczne w Elblągu

——, 2003b. *Historia ryb i rybołówstwa w holocenie na Niżu Polskim w świetle badań archeoichtiologicznych*, Poznań: IAE PAN

——, 2003c. 'Analiza archeozoologiczna szczątków kostnych z miejscowości Pień, gm. Dąbrowa Chełmińska, stanowisko 2 (sezon 2003)', unpublished report, Toruń: Instytut Archeologii, Uniwersytet Mikołaja Kopernika w Toruniu

——, 2006a. 'Archaeozoology's Contribution to the Improvement of Historians' Conceptions of Subsistence Economy and Environment in Early Medieval Poland — Selected Problems', *Beiträge zur Archäozoologie und Prähistorischen Anthropologie* 5, 77–82

——, 2006b. 'Wybrane zagadnienia ze studiów nad gospodarką zwierzętami we wczesnośredniowiecznych kompleksach grodowych Pomorza, Wielkopolski i Dolnego Śląska', in *Stan i potrzeby badań nad wczesnym średniowieczem w Polsce — 15 lat później*, eds W. Chudziak and S. Moździoch, Toruń: Wydawn. Uniwersytetu Mikołaja Kopernika w Toruniu, 123–50

——, 2007a. 'Zwierzęce szczątki kostne z miejscowości Gutowo, stan. 1, gm. Lubawa', unpublished report, Toruń: Instytut Archeologii, Uniwersytet Mikołaja Kopernika w Toruniu

——, 2007b. 'Zwierzęce szczątki kostne z zamku w Olsztynku, stan. 2', unpublished report, Toruń: Instytut Archeologii, Uniwersytet Mikołaja Kopernika w Toruniu

——, 2007c. 'Zestawienie inwentaryzacyjne szczątków zwierzęcych z zamku w Papowie Biskupim', unpublished report, Toruń: Instytut Archeologii, Uniwersytet Mikołaja Kopernika w Toruniu

——, 2007d. 'Zwierzęce szczątki kostne z grodziska - stan. 1 w miejscowości Zwiniarz (sezon 2005) Zestawienia tabelaryczne', unpublished report, Toruń: Instytut Archeologii, Uniwersytet Mikołaja Kopernika w Toruniu

——, 2007e. 'Animal Subsistence Economy in the Early Medieval Stronghold Complexes of Western Slavs — Comparative Studies of Pomerania, Great Poland and Lower Silesia', in *Eurasian Perspectives on Environmental Archaeology: Annual Conference of the Association for Environmental Archaeology (AEA) September 12–15, 2007, Poznań, Poland*, eds M. Makohonienko, D. Makowiecki, and J. Czerniawska, Poznań: Bogucki Wydawnictwo Naukowe, 100–06

——, 2008a. 'Exploitation of Early Medieval Aquatic Environments in Poland and Other Baltic Sea Countries: An Archaeozoological Consideration', in *Atti Delle Settimane LV, 2: L'acqua nei secoli altomedievali, Spoleto, 12–17 aprile 2007, Settimane di Studio Della Fondazione Centro Italiano di Studi Sull'Alto Medioevo*, Spoleto: Fondazione Centro italiano di studi sull'alto medioevo, 753–75

——, 2008b. 'Sturgeon Fishing on Polish Lowland during Holocene', in *Archéologie du poisson: 30 ans d'archéo-ichtyologie au CNRS. Hommage aux travaux de Jean Desse et Nathalie Desse-Berset*, eds P. Béarez, S. Grouard, and B. Clavel, Antibes: Éditions APDCA, 327–39

——, 2010a. 'Zwierzęce szczątki kostne z badań na Rynku w Grudziądzu', unpublished report, Toruń: Instytut Archeologii, Uniwersytet Mikołaja Kopernika w Toruniu

——, 2010b. *Wczesnośredniowieczna gospodarka zwierzętami i socjotopografia in Culmine na Pomorzu Nadwiślańskim: Studium archeozoologiczne*, Toruń: Wydawnictwo Naukowe Uniwersytetu Mikołaja Kopernika

——, 2011. 'Elbląg - Zamek. Zestawienie inwentaryzacyjne szczątków kostnych', unpublished report, Toruń: Instytut Archeologii, Uniwersytet Mikołaja Kopernika w Toruniu

——, 2012a. 'Ełk – Zamek. Badania archeozoologiczne 2012', unpublished report, Toruń: Instytut Archeologii, Uniwersytet Mikołaja Kopernika w Toruniu

——, 2012b. 'Badania archeoichtiologiczne szczątków ze stanowiska Janów Pomorski 1', in *Janów Pomorski, stan. 1. Wyniki ratowniczych badań archeologicznych w latach 2007–2008, Elbląg, I/3*, eds M. Bogucki and B. Jurkiewicz, Elbląg: Muzeum Archeologiczno-Historyczne w Elblągu, 302–23

——, 2012c. 'Analiza zwierzęcych szczątków kostnych z badań wykopaliskowych w miejscowości Lidzbark Warmiński, stan. XIX, woj. warmińsko-mazurskie', unpublished report, Toruń: Instytut Archeologii, Uniwersytet Mikołaja Kopernika w Toruniu

——, 2013. 'Zwierzęce szczątki kostne z miejscowości Olsztyn – Las Miejski, stan. CIV (AZP 23-61/46 (sezon 2006, 2013)', unpublished report, Toruń: Instytut Archeologii, Uniwersytet Mikołaja Kopernika w Toruniu

——, 2014. 'Studia archeozoologiczne nad znaczeniem wczesnośredniowiecznej i średniowiecznej fauny łęczyckiego grodu', in Początki Łęczycy. Tom I, Archeologia środowiskowa średniowiecznej Łęczycy. Przyroda-Gospodarka-Społeczeństwo, eds R. Grygiel and T. Jurek, Łódź: Muzeum Archeologiczne i Etnograficzne w Łodzi, 261–437

——, 2015. 'Fauna wczesnośredniowieczna i jej znaczenie u ludności zespołu osadniczego w Węgrach', in *Węgry. Zespół osadniczy na pograniczu Pomorsko-Pruskim w XI–XII w.*, eds M. Haftka and S. Wadyl, Malbork: Muzeum Zamkowe w Malborku, 236–55

——, 2016. Zwierzęta średniowiecznego i nowożytnego Poznania oraz okolic: Podstawy archeozoologiczne. Ekologia Historyczna Poznania, Tom 3. Poznań: Wydawnictwo Bogucki

——, 2017. 'Wyniki analizy archeozoologicznej zwierzęcych szczątków kostnych z badań wykopaliskowych (gród i osada)', in Castrum Zvitinense: wczesnośredniowieczny zespół osadniczy w Grążawach nad Drwęcą, ed. K. Grążawski, Brodnica-Olsztyn: Multi B. Grzybowska - T. Siekierski, 185–221

Makowiecki, D., and Gotfredsen, A. B. 2002. 'Bird Remains of Medieval and Post-Medieval Coastal Sites at the Southern Baltic Sea, Poland', *Acta Zoologica Cracoviensia* 45, 65–84

Makowiecki, D., and Makowiecka, M. 1999. 'Gospodarka hodowlano-łowiecka w dorzeczu środkowej Drwęcy we wczesnym średniowieczu: Studium archeozoologiczne', *Studia nad osadnictwem średniowiecznym ziemi chełmińskiej* 3, 27–60

——, 2007. 'Ocena archeozoologiczna kości z osady późnośredniowiecznej i nowożytnej, in Żerniki, gm. Kórnik, stan. 25. Osadnictwo pradziejowe, wieś średniowieczna i folwark nowożytny', in *Archeostrada: Studia i Materiały z badań Wykopaliskowych na autostradzie A2 — Odcinek Wielkopolski* (vol. I), eds E. Pawlak and P. Pawlak, Poznań: Wydawnictwo Poznańskie, 197–221

——, 2012. 'Wyniki badań archeozoologicznych ze stanowiska Janów Pomorski 1', in *Janów Pomorski, Stan 1. Wyniki ratowniczych badan archeologicznych w latach 2007–2008* (vol. I), eds M. Bogucki and B. Jurkiewicz, Elbląg: Muzeum Archeologiczno-Historyczne w Elblągu, 275–312

——, 2013. 'Archeozoologiczna baza danych Torunia', unpublished database, Toruń: Instytut Archeologii, Uniwersytet Mikołaja Kopernika w Toruniu

——, 2014a. 'Archeozoologiczna baza danych Gdańska', unpublished database, Toruń: Instytut Archeologii, Uniwersytet Mikołaja Kopernika w Toruniu

——, 2014b. 'Wyniki badań zwierzęcych szczątków kostnych', in *Zamek biskupów chełmińskich w Wąbrzeźnie w świetle badań archeologiczno-architektonicznych: Studia i materiały*, ed. M. Wiewióra, Toruń: Wydawnictwo Naukowe Uniwersytetu Mikołaja Kopernika, 177–202

Makowiecki, D. Makowiecka, M., and Zabilska-Kunek, M. 2014. 'Archeozoologiczna baza danych dla miejscowości Satświny, stan. 1 i 73/163', unpublished database, Toruń: Toruń: Instytut Archeologii, Uniwersytet Mikołaja Kopernika w Toruniu

Makowiecki, D., and Stach, A. 2007. 'Ssaki wolno żyjące w holocenie Polski — aspect środowiskowy i gospodarczy', *Studia interdyscyplinarne nad środowiskiem i kulturą w Polsce*, eds M. Makohonienko, D. Makowiecki, and Z. Kurnatowska, Poznań: Bogucki Wydawnictwo Naukowe, 155–70

Makowiecki, D., Tomek, T., and Bocheński, Z. M. 2014. 'Birds in Early Medieval Greater Poland: Consumption and Hawking', *International Journal of Osteoarchaeology*, 358–64, doi: 10.1002/oa.2366

Makowiecki, D., Wussow, J., and Müller, R. 2007. 'Knochenfunde von Vögeln aus der Siedlung Kałdus im Weichseltal', *Beiträge zur Archäozoologie und Prähistorischen Anthropologie* 6, 111–20

Makowiecki, D., and Zabilska-Kunek, M. 2012. 'Wyniki badań zwierzęcych szczątków kostnych z miejscowości Mikołajki', Toruń: Instytut Archeologii, Uniwersytet Mikołaja Kopernika w Toruniu

Maltby, M., Pluskowski, A. G., and Seetah, K. 2009. 'Animal Bones from an Industrial Quarter at Malbork, Poland: Towards an Ecology of a Castle Built in Prussia by the Teutonic Order', *Crusades*, 8, 191–212

Marcinkowski, M. 2006. 'Dom poławiacza jesiotrów z połowy XIII wieku na Starym Mieście w Elblągu', in *Nie tylko archeologia: Księga poświęcona pamięci Eugeniusza 'Gwizolona' Wilgockiego*, eds E. Cnotliwy, A. Janowski, K. Kowalski, and S. Słowiński, Szczecin: Muzeum Narodowe w Szczecinie, 235–49

Müldner, G., Britton, K., and Ervynck, A. 2014. 'Inferring Animal Husbandry Strategies in Coastal Zones through Stable Isotope Analysis: New Evidence from the Flemish Coastal Plain (Belgium, 1st–15th Century AD)', *Journal of Archaeological Science* 41, 322–32

Nogalski S. 1984. *Szczątki ptaków wczesnośredniowiecznych Pomorza*, Szczecin: Wydaw Nauk. WSP

——, 1991. 'Szczątki kostne ptaków domowych z wykopalisk w Małej Nieszawce (woj. toruńskie)', *Roczniki Nauk Rolniczych w Poznaniu* 227; *Archeozoologia* 16, 17–24

Nogalski, S., Sałaciak, A., and Kosińska, B. 1992. 'Szczątki kostne ptaków dziko żyjących z wykopalisk w Małej Nieszawce (woj. toruńskie)', *Roczniki Nauk Rolniczych w Poznaniu* 237; *Archeozoologia* 17, 79–87

Nowicki, B., Jasek, S., Maciejowski, J., Nowakowski, P., and Pawlina, E. 1995. *Atlas ras zwierząt gospodarskich*, Warsaw: Wydawnictwo Naukowe PWN.

Okulicz-Kozaryn, Ł. 1983. *Życie codzienne Prusów i Jaćwięgów w wiekach średnich (IX–XIII w.)*, Warsaw: Państwowy Instytut Wydawniczy

——, 1997. *Dzieje Prusów*, Wrocław: Fundacja na Rzecz Nauki Polskiej

Orton, D., Rannamäe, E., Lõugas, L., Makowiecki, D., Hamilton-Dyer, S., Pluskowski, A. G., O'Connell, T., and Barrett, J. 2019. 'The Teutonic Order's Role in the Development of a Medieval Eastern Baltic Cod Fishery', in *Ecologies of Crusading, Colonization, and Religious Conversion in the Medieval Baltic: Terra Sacra II*, ed. A. Pluskowski, Environmental History of the North Atlantic World, 3, Turnhout, Brepols (forthcoming)

Paner, H. 1998–99. 'Z badań archeologicznych Gdańska', in *Z otchłani wieków Pomorza Gdańskiego: zeszyt okazjonalny dla upamiętnienia Tysiąclecia Gdańska: 997 urbs Gyddanyzc — 1997 Gdańsk*, ed. E. Choińska-Bochdan, Gdańsk: Stowarzyszenie Naukowe Archeologów Polskich, 95–119

Piątkowska-Małecka, J. 2009. 'Stan badań nad zwierzęcymi szczątkami kostnymi ze stanowisk badanych przez Kompleksową Ekspedycję Jaćwieską', in *Bałtowie i ich sąsiedzi, Marian Kaczyński in memoriam*, eds A. Bitner-Wróblewska and G. Iwanowska, Warsaw: Państwowe Muzeum Archeologiczne w Warszawie, 155–72

Pluskowski, A. G. 2013. The *Archaeology of the Prussian Crusade: Holy War and Colonisation*, London: Routledge

Pluskowski, A. G., Maltby, M., and Seetah, K. 2010. 'Potential Osteoarchaeological Evidence for Riding and the Military Use of Horses at the Castle in Malbork, Poland', *International Journal of Osteoarchaeology* 20/3, 335–43

Pluskowski, A. G., Sawicki, Z., Shillito, L.-M., Badura, M., Makowiecki, D., Zabilska-Kunek, M., Seetah, K., and Brown, A. 2014. 'Biała Góra: The Forgotten Colony in the Medieval Pomeranian-Prussian Borderlands', *Antiquity* 88/341, 863–82

Prawocheński, R. 1958. *Hodowla świń*, Warsaw: Państwowe Wydawnictwo Rolnicze i Leśne

Prummel, W. 1993. *Die Tierknochenfunde unter besonderer Berücksichtigung der Beizjagd, Starigard/Oldenburg*, Hauptburg der Slawen in Wagrien 4, Offa-Bücher 74, Neumünster: Wachholtz

Pruski, W. 1959. *Dzikie konie wschodniej Europy*, Warsaw: PWRiL

Rünger, F. 1925. 'Herkunft, Rasenzugehürigkeit, Züchtung und Haltung der Ritterpferde des Deutschen Ordens: Ein Beitrag zur Geschichte der ostpreussischen Pferdezucht und der deutschen Pferdezucht im Mittelalter', *Zeitschrift für Tierzüchtung und Züchtungsbiologie* 2/3, 211–308

Samsonowicz, A. 1991. *Łowiectwo w Polsce Piastów i Jagiellonów*, Wrocław: Zakład Narodowy im. Ossolińskich

Sawicki, Z., Pluskowski, A. G., Brown, A., Badura, M., Makowiecki, D., Shillito, L.-M., Zabilska-Kunek, M., and Seetah, K. 2015. 'Survival at the Frontier of Holy War: Political Expansion, Crusading, Environmental Exploitation and the Medieval Colonising Settlement at Biała Góra, North Poland', *European Journal of Archaeology*, 18/2, 282–311

Schramm, Z. 1967a. 'Kości długie a wysokość w kłębie u kozy', *Roczniki Wyzszej Szkoly Rolniczej w Poznaniu* 36, 89–105

——, 1967b. 'Szczątki kostne wczesnośredniowiecznej owcy i kozy z wykopalisk północno-zachodniej Polski', *Roczniki Wyzszej Szkoly Rolniczej w Poznaniu* 36, 135–74

——, 1976. *Bydło wczesnośredniowieczne na ziemiach Wielkopolski i Kujaw: Studium morfologiczne na podstawie wykopaliskowego materiału kostnego z Kruszwicy, Międzyrzecza i Santoka*, Poznań: Akademia Rolnicza w Poznaniu

Scull, C., Müldner, G., and Makowiecki, D. 2018. 'Carbon and Nitrogen Stable Isotope Evidence for Animal Husbandry and Environmental Change in the Medieval Kulmerland', in *Ecologies of Crusading, Colonization, and Religious Conversion in the Medieval Baltic: Terra Sacra II*, ed. A. Pluskowski, Environmental History of the North Atlantic World, 3, Turnhout: Brepols (forthcoming)

Seetah, K., Pluskowski, A. G., Makowiecki, D., and Daugnora, L. 2013. 'New Technology or Adaptation at the Frontier? Butchery as a Signifier of Cultural Transitions in the Medieval Eastern Baltic', *Archaeologia Baltica* 20, 84–101

Serjeantson, D. 2009. *Birds*, Cambridge: Cambridge University Press

Shiroukhov, R. 2012. 'Prussian Graves in the Sambian Peninsula with Imports, Weapons and Horse Harnesses, from the Tenth to the Thirteenth Century: The Question of the Warrior Elite', *Archaeologia Baltica* 18, 224–55

Sobociński, M. 1965. 'Szczątki kostne z grodziska w miejscowości Baldram, stan. 1', unpublished report, Toruń: Instytut Archeologii, Uniwersytet Mikołaja Kopernika w Toruniu

——, 1968. 'Materiał kostny zwierzęcy z wykopalisk w Świeciu nad Wisłą', *Pomorania Antiqua* 2, 321–34

——, 1975. 'Szczątki kostne zwierząt z grodziska średniowiecznego w miejscowości Słoszewy, pow. Brodnica', *Acta Uniwersitatis Nicolai Copernici, Archeologia* 5, 137–53

——, 1978. 'Szczątki zwierzęce z osady i grodziska wczesnośredniowiecznego z miejscowości Węgry', *Roczniki Nauk Rolniczych w Poznaniu* 103, *Archeozoologia* 4, 61–72

——, 1979. 'Zwierzęce szczątki kostne z osady z przełomu epoki brązu i żelaza oraz z osady otwartej i grodzisk wczesnośredniowiecznych Pomorza Wschodniego', *Pomorania Antiqua* 9, 251–93

——, 1989 'Zwierzęcy materiał kostny z grodziska wczesnośredniowiecznego w miejscowosci Owidz, woj. gdańskie: Roczniki Akademii Rolniczej w Poznaniu 198', *Archeozoologia* 13, 127–53

——, 1990. 'Zwierzęce szczątki kostne z grodziska wczesnośredniowiecznego w Sopocie, woj. Gdańskie', Roczniki Akademii Rolniczej w Poznaniu 221, Archeozoologia 15, 55–65

——, 1991. 'Zwierzęcy materiał kostny z wykopalisk w Małej Nieszawce (woj. toruńskie)', *Roczniki Akademii Rolniczej w Poznaniu, 227. Archeozoologia* 16, 57–69

——, 1992a. 'Zwierzęce szczątki kostne z wykopalisk w Gniewie nad Wisłą. Stanowisko 3, podzmacze-port', *Roczniki Nauk Rolniczych w Poznaniu 237, Archeozoologia* 17, 123–47

——, 1992b. 'Zwierzęce szczątki kostne z wykopalisk w Gniewie nad Wisłą. Stanowisko 5, gród i osada wczesnośredniowieczna', *Roczniki Nauk Rolniczych w Poznaniu* 237; *Archeozoologia* 17, 149–60

Sobociński, M., and Makowiecki, D. 1992. 'Zwierzęce szczątki kostne z wykopalisk w Gniewie nad Wisłą. Stanowisko 2 - Stare miasto', *Roczniki Nauk Rolniczych w Poznaniu* 237; *Archeozoologia* 17, 160–95

Teichert, M. 1969. 'Osteometrische Untersuchungen zur Errechnung der Widerristhöhe bei vor- und frühgeschichtlichen Schweinen', *Kühn – Archiv* 38/3, 237–92

——, 1975. 'Osteometrische Untersuchungen zur Berechnung der Widerristhöhe bei Schafen', in *Zooarchaeological Studies*, ed. A. Clason, Amsterdam: New Holland, 51–69

Ullucci, D. C. 2012. *The Christian Rejection of Animal Sacrifice*, Oxford: Oxford University Press

Vitt, V. O. 1952. 'Horses of the Pazyryk Kurgans [Lošadi Pazyrykskikh Kurganov]', *Sovetskaja Archeologija* 16, 163–205

Waluszewska-Bubień, A. 1976. 'Szczątki kostne ptaków z wykopalisk w miejscowości Węgry', *Roczniki Akademii Rolniczej w Poznaniu* 84; *Archeozoologia* 2, 159–64

——, 1980. 'Szczątki kostne ptaków z wykopalisk w Kałdusie (województwo toruńskie)', *Roczniki Akademii Rolniczej w Poznaniu* 121; *Archeozoologia* 6, 159–64

——, 1984. 'The Avifauna of the Early Middle Ages against a Background of Archaeozoological Materials from a Number of Polish Settlement Sites', in *Archaeozoology*, I: *Proceedings of the Third International Archaeozoological Conference Held 23–26th April 1978 at the Agricultural Academy, Szczecin-Poland*, ed. M. Kubasiewicz, Szczecin: Agricultural Academy, 243–55

Wyczółkowski M., and Makowiecki, D. 2009. 'Horse Sacrifices in Prussia in the Early Middle Ages: Ritual Area in Poganowo Site IV, Olsztyn Province (Poland)', *Archaeologia Baltica* 11, 295–303

Zinoviev, A. V. 2009. 'Horses from Two Burials in Samland and Natangen (Second Century, Kaliningradskaia Province, Russia)', *Archaeologia Baltica* 11, 50–55

Żuk-Piechota, E., and Towarnicki, R. 1978. 'Wykopaliskowe resztki kostne wczesnośredniowiecznego bydła kręgu kultury mazurskiej z grodziska Okragła Góra', *Zeszyty Naukowe Akademii Rolniczo Technicznej w Olsztynie, Zootechnika* 14, 3–25

Žulkus, V., and Daugnora, L. 2010. 'What did the Order's Brothers Eat in the Klaipėda Castle? (The Historical and Zooarchaeological Data)', *Archaeologica Baltica* 9, 74–87

13. Exploiting Plants: Macrobotanical Remains from Prussia

Aleksander Pluskowski, Monika Badura, and Marc Jarzebowski

Introduction

This chapter presents the most important results of the plant macrofossils analysis from primary and secondary archaeological sites linked to the project, and consideres this data with the broader archaeological, cultural, and historical context of plant exploitation within medieval Prussia, building on data presented in earlier chapters. An integrated study of select Teutonic Order convents and towns in Prussia can be found in volume II, which considers the culinary role of plants in more detail (Badura and Możejko 2019), with material from Livonian sites published in interim form (Valk et al. 2013) and in preparation for future publication. The analysis of plant macrofossils directly complements the palynological studies of vegetation change in the surrounding landscape (see Chapter 11), by providing direct, synchronous evidence of plant use within related occupation contexts. Plants defined as 'useful' appear in archaeological contexts as the by-products of storage (linked to taxation and trade), food preparation, consumption, and medicinal use, with both cultivated and wild taxa represented (Prance and Nesbit 2005; Dincauze 2006). They are valuable indicators of the economic status — the purchasing power — of individual households. Evidence for their cultivation, gathering, and trade may be apparent in the physical, charred, waterlogged, or mineralized macro-remains of plants in occupation deposits. The interpretation of these remains is often problematic: the integrity of samples and ecological precision are key concerns. In castle and urban archaeological contexts, macrobotanical findings may originate from a variety of ecological zones surrounding a town, resulting from a range of human activities, forming over several years or decades and frequently disturbed and mixed as a result of later taphonomic processes. Also, the use and value of plants could also change over time, with diachronic perspectives best represented in towns such as Gdańsk, where plant macrofossils have been recovered from multiple urban contexts and have been the focus of archaeobotanical studies since the 1950s, and systematically from 1998 (Badura 2011).

The greater part of this chapter focuses on those 'useful plants' derived from the case study sites included within the project, situated within a broader context. The presentation of the results is subdivided into territories — the Prussian tribal territory of eastern Galindia and later the central part Lötzen (Pol. Giżycko) procuratorship, followed by sites in the Lower Vistula Valley and delta region: the commanderies of Graudenz (Pol. Grudziądz), Marienburg (Pol. Malbork), Danzig (Pol. Gdańsk), and Elbing (Pol. Elbląg). The purpose of analysing the plant macro-fossil assemblages was twofold: first, to determine whether there were any differences between sites reflecting variable lifestyles; and second, to contribute to an understanding of the grain trade that historians have characterized as the most important aspect of the Prussian economy in the second half of the fourteenth century (Sarnowsky 1993, 435–36). The main indicators of variable social access to plants are the presence or the lack of exotic species, as well as the relative abundance of edible plant remains. The working hypothesis was that members of the Order had access to expensive plant products, such as overseas spices and fruit, while communities from subordinate settlements, including towns, ate mainly flour-based products, vegetables, as well as fruit and berries that were easily accessible within the countryside. The presence of grain was also expected at the Order's sites, both for the needs of the community and as traces of stored produce intended for export. In 1307, grain measures for Prussia were codified at a meeting of the general chapter in Engelsburg, and subsequently these became used in all inventories and

Figure 13.1: Profiles of features (a) 98 and (b) 99 at Biała Góra (site 1) showing sampling locations. Photos by M. Badura.

accounts. The largest quantity was the *last*; the amount of grain carried in a vessel which consisted of 60 *scheffel* (bushel). A *scheffel* consisted of four *viertel* (quarter) and a *viertel* of four *metzen*. The latter was described as the amount which could be held by a man with both hands for the purposes of grinding. The size of a *scheffel* as the basic grain measure varied and was subject to continuous attempts at standardization (Witthöft 1981, 37). Although useful for understanding the relative measures recorded in fiscal documents, this also highlights the importance placed on the regulation of grain by the Order. At the same time, grain played a fundamental role in the diet of the population of medieval Prussia. The most commonly consumed products were cereals and other farinaceous plants ground into flour, which provided the basis for worst and better quality bread. The latter was largely associated with the wealthier social strata. Given the commercial value of flour, it is not surprising to see references to the construction and control of mills littering the fiscal documents of the Teutonic Order (for Livonian examples of the Order's regulation of mills see Chapter 8; see also below) (Kubicki 2012). Less affluent communities relied on groats and other dishes based on less processed grains (Rutkowska-Płachcińska 1978). The brewing of beer also required regular crops of barley for malt as well as hop (Behre 1999). Finally, fodder for animals, which included various types of crops (Scull et al. 2019), was equally if not more important from the perspective of the Order's provisioning requirements; an essential requisite for the success of crusading in Prussia and the maintenance of the vast herds of cattle, sheep, goats, pigs, and horses within the Order's commanderies (see Chapter 12).

Galindia and the Procuratorship of Lötzen

Święta Góra (Heilsberg)

There is extensive evidence for the expansion of agriculture from the eleventh century in parts of western Prussia and its borderlands (Kulmerland, Pomesania, and Pogesania) but very little evidence in the form of plant macro-fossil assemblages. The few studies that have been conducted, including within the framework of this project, have derived from the tribal territory of Galindia, where palynological studies have demonstrated the expansion of land under cultivation from the eleventh century (see Chapter 11). Buildings at Prussian settlements have been tentatively identified as granaries, and certain pits as silos for storing grain and other food stuffs (Okulicz-Kozaryn 1983, 101, 106–08). In a few instances plant macrofossils have been found in these pits. Plant materials recovered during early excavations at the stronghold of Jeziorko included a small amount of charred remains of cereals, including einkorn (*Triticum monococcum*), spelt (*Triticum spelta*), club white (*Triticum compactum*) (at present considered a subspecies of *T. aestivum*), and barley (*Hordeum* sp.) (Zabłocki 1950). Charcoal fragments were also recovered, which included common hornbeam (*Carpinus betulus*), black alder (*Alnus glutinosa*), and *Populus* sp.. Other tree species — oak (*Quercus robur*, *Q. petrea*), European ash (*Fraxinus excelsior*), pine (*Pinus silvestris*) — were represented by single charcoal pieces. The final phase of the site has been dated from the tenth to twelfth centuries, and potentially into the mid-thirteenth century (Antoniewicz 1950; Jaremek 2003). A slightly more substantial assemblage was recovered from a series of pits at the settlement of Święta Góra, 15 km east of Jeziorko. Sixty-one samples were taken from twenty-five features covering various cultural phases, including the Migration Period, the Viking Age, and the final phase of Prussian occupation into the early thirteenth century within the area of the settlement.

Samples were also taken from a trench in the nearby village of Staświny (Ger. Stasswinnen), with a continuous occupation sequence from the early medieval period through to the fifteenth and sixteenth centuries. Unfortunately, the samples collected from the latter site did not contain any archaeobotanical materials. Samples from the Święta Góra settlement contained badly preserved, charred diasporas of cereals and weeds (both segetal and ruderal) (Table 13.1). The range of species indicates access to permanent rotational field systems, corresponding to the evidence for intensive cultivation apparent from palynological data from nearby lakes, including Lake Wojnowo located immediately adjacent to the site (see Chapter 11; see also Wacnik et al. 2012). Remains of barley (*Hordeum vulgare*) were present mostly in features dated to the Migration Period. In refuse pits dated to the early medieval period, single remains of barley as well as rye (*Secale cereale*), wheat (*Triticum aestivum*), and millet (*Panicum miliaceum*) were noted. Postholes discovered at the centre of the site contained only few plant remains. A charred seed of pea (*Pisum sativum*) suggests that this vegetable was known in the vicinity of Święta Góra from at least the Migration Period. Although in the Middle Ages collected species, alongside cultivated plants, were very important sources of food and medicine, in the samples studied from Święta Góra only one charred diaspora of raspberry (*Rubus idaeus*) were found.

With the conquest of the region by the Teutonic Order, this part of Galindia came under the control of a procurator based at the castle of Lötzen. Although no plant materials have been recovered from the castle, the earliest inventory dating to 1420 makes no mention of grain, with the only stored food consisting of mead, beer, salt, cheese, and flour (*GAD*, 197). Whilst the list may be incomplete, the same inventory lists rye, wheat, oat, barley, and malt in the granaries, mills, and bakeries at other castles within the vicinity, at Barten (Pol. Barciany), Rastenburg (Pol. Kętrzyn), Seehesten (Pol. Szestno), Rhein (Pol. Ryn), and Lyck (Pol. Ełk). It is therefore likely that the castle at Lötzen did not contain a granary at this time, with stores located in nearby settlements within the procuratorship. While there is no grain in the 1420 list, oat dominates the 1507 and 1508 inventories, as well as bushels of unspecified grain, with rye only featuring in the second 1507 inventory. The 1513 and 1514 inventories have much smaller amounts and focus on grain for sowing. Almost all references to seed grain in the *Großes Ämterbuch* are from the sixteenth century, with the first mention appearing in the inventory of the castle at Rhein in 1490 (*GAD*, 198), and this may reflect a relatively later date for the agrarian functions of these eastern Prussian castles. It therefore seems likely that in the fourteenth and early fifteenth centuries the Order used these sites for organizing colonization and focusing on the exploitation of woodlands, and the presence of grain can then be linked to a stabilization in the settlement pattern. This is also suggested by references to flour being processed within castles such as Lötzen throughout the fifteenth century, reflecting local consumption with a reliance on mills within the procuratorship, in turn supplied by settlements such as Stasswinnen. Within the procuratorship of Lötzen, in the absence of any archaeological evidence to date, historical references to mills are relatively late. The first reference to a mill within the vicinity of the procurator's castle is from 1475, followed by another in 1507. Other mills were located at Groß Konopken, Wobelick, Jucha, Stradauen, Klein Skomentnen, and *zcum Eisenberg*, the latter perhaps referring to an iron mill (Białuński 2005, 49–51, 205). In the sixteenth century, as the process of colonization became more stable in the Prussian-Lithuanian frontier, agricultural production intensified and appears to have become more centrally organized.

The Commandery of Graudenz

Grudziądz Castle (Ger. Graudenz)

Excavations at the castle in Grudziądz provided the opportunity to sample contexts with organic preservation for palaeoenviromental data. Four samples were taken from three layers in the north and south-facing profiles in the eastern part of the castle (Ar 0650/6660). One sample was also taken from the deepest layer of the canal excavated in the northern part of the castle (Carson

Table 13.1: Plant remains from Święta Góra (settlement); table includes features dated to the early Middle Ages (the Prussian culture) from which archaeobotanical data were obtained: RP - refuse pits; PH - postholes; n - number of samples; * - collected plant, abundance (determined separately for each of the site); + single; ++ numerous

TAXA	RP.14, n-6	RP.15, n-3	RP.16, n-5	RP.18, n-3	RP.21, n-3	RP.26, n-3	RP.28, n-7	RP.40, n-5	PH.32, n-1	PH.44, n-1	PH.45, n-1	PH.48, n-1	PH.49, n-1
Cultivated Plants													
Cerealia indet.; cereals	+	+	+	+	+	+	++	++	+	-	+	-	-
Avena sp.; oat	-	-	-	-	-	-	+	-	-	-	-	-	-
Hordeum vulgare L.; barley	-	+	+	+	+	-	+	+	+	-	-	-	+
Panicum miliaceum L.; millet	+	+	-	-	+	+	+	+	-	-	+	-	-
cf. *Pisum sativum* L.; pea	+	-	-	+	-	+	-	-	-	-	-	-	-
Secale cereale L.; rye	+	+	+	-	+	-	+	+	+	-	-	-	-
Triticum aestivum L.; wheat	-	+	-	-	++	-	+	+	+	-	-	-	-
Triticum sp.; wheat	+	-	-	-	-	-	-	-	-	-	-	-	-
Triticum sp./*Secale cereale;* wheat/rye	-	-	+	-	-	-	-	-	-	-	-	-	-
Segetal and Ruderal Weeds													
Agrostemma githago L.; corn-cockle	+	-	-	-	-	-	+	-	-	-	-	-	-
Chenopodium album L.; white goosefoot	++	++	++	+	++	++	++	-	-	+	-	-	+
Chenopodium cf. *album* L.; white goosefoot	-	-	-	-	-	-	-	+	-	-	-	-	-
Fallopia convolvulus (L.) Á. Löve; black-bindweed	+	+	+	-	+	+	-	+	-	+	-	-	-
Galium aparine L.; cleavers	-	-	-	-	-	-	-	-	-	-	+	-	-
Galium spurium L.; false cleavers	+	+	+	-	+	+	+	+	-	-	-	-	-
Polygonum lapathifolium L.; pale persicaria	+	+	+	+	-	-	-	+	+	-	+	-	-
Polygonum minus Huds./*mite* Schrank; pygmy smartweed/tasteless water-pepper	-	-	+	-	-	-	-	-	-	-	-	-	-
Polygonum persicaria L.; lady's-thumb	-	-	-	-	+	-	-	-	-	-	-	-	-
cf. *Rhinanthus serotinus* (Schönh.) Oborný var. apterus (Fr.) Hyl.; greater yellow-rattle	-	-	+	-	-	-	+	+	-	-	-	-	-
Rumex acetosella L.; sheep's sorrel	-	-	-	-	+	-	-	-	-	-	-	-	-
Rumex crispus L.; curled dock	+	-	-	-	-	-	-	-	-	-	-	-	-
Setaria pumila (Poir.) Roem. & Schult.; yellow bristle-grass	-	-	-	-	-	-	-	-	+	-	-	-	-
Setaria cf. *pumila* (Poir.) Roem. & Schult.; yellow bristle-grass	-	+	-	-	-	-	-	-	-	-	-	-	-
Spergula arvensis L. s.l.; corn spurey	+	-	-	-	-	-	-	-	-	-	-	-	-
Forest Plants													
Carpinus betulus L.; hornbeam	-	-	-	+	-	-	+	-	-	-	-	-	-
Rubus idaeus L. *; raspberry	-	-	+	-	-	-	-	-	-	-	-	-	-
Other Plants													
Alchemilla sp.; lady's mantle	-	-	+	-	-	-	-	-	-	-	-	-	-
Atriplex sp.; orache	+	+	+	-	-	-	-	-	-	-	-	-	-
Boraginaceae indet.; borawe family	-	-	-	-	-	-	-	-	+	-	-	-	-
Brassicacea indet.; cabbage family	+	-	-	-	-	-	-	-	-	-	-	-	-

TAXA	RP.14, n-6	RP.15, n-3	RP.16, n-5	RP.18, n-3	RP.21, n-3	RP.26, n-3	RP.28, n-7	RP.40, n-5	PH.32, n-1	PH.44, n-1	PH.45, n-1	PH.48, n-1	PH.49, n-1
Carex sp.; sedge	-	+	+	-	-	-	+	-	-	-	-	-	-
Caryophyllaceae indet.; pink family	-	-	+	-	-	-	+	+	-	-	-	-	-
Chenopodium sp.; goosefoot	++	++	++	++	++	++	++	++	+	-	+	+	+
Fabaceae indet.; legume family	-	-	+	-	-	-	+	-	-	-	-	-	-
Galium cf. *palustre* L.; common marsh-bedstraw	-	-	-	-	-	+	-	-	-	-	-	-	-
Galium sp.; bedstraw	+	-	-	-	+	++	+	-	-	-	+	-	-
Luzula sp.; wood-rush	-	-	+	-	-	-	-	-	-	-	-	-	-
Poaceae indet.; grass family	+	-	-	-	-	+	+	-	-	-	+	-	-
Polygonaceae indet.; knotweed family	-	-	-	-	-	-	+	-	-	-	-	-	-
Polygonum sp.; knotweed	-	-	-	-	+	+	+	+	-	-	-	-	-
Ranunculus cf. *acris* L.; meadow buttercup	-	-	-	-	+	-	-	-	-	-	-	-	-
Rosaceae indet.; rose family	-	+	+	-	-	-	-	-	-	-	-	-	-
Setaria sp.; bristle-grass	+	-	-	+	-	+	-	+	-	-	-	-	-
Stellaria sp.; starwort	-	+	-	-	-	-	-	-	-	-	-	-	-
VARIA undetermined	+	-	+	+	-	-	-	+	+	+	-	-	-

et al. 2012). Unfortunately, very few plant remains were preserved and the assemblage sheds little light on activities within the castle, or the preceding Slavic settlement. The first two samples contained only one whole and several fragmentary black elder seeds (*Sambucus nigra*), a species that grows in many variable habitats but that prefers fertile soils. Sample 4 contained two Fabaceae seeds (legume family) and a charred seed that could possibly be a cereal grain. There were also a small number of molluscs present within the assemblage, and in particular ribbed grass snail (*Vallonia costata*) widely distributed on open grasslands, particularly on calcareous soils but also preferring rocks areas and often associated with human occupation sites. The sample from the canal included a charred plant bud, a charred elder seed, and one seed of *Asteraceae* (daisy family).

In summary, the macrobotanical data indicated the presence of nearby open grassland and marshy environments, corresponding with the location of the early medieval settlement in close proximity to the Vistula, and engaged in various forms of fishing, animal husbandry, and (most likely) cereal cultivation. There were also significant quantities of fish scales in the samples that were indicative of intensive processing. The analysis of phytoliths from the same samples corroborated the macrobotanical data, consistently suggesting the presence of grasses in the immediate surroundings. Documentary sources indicate that the convent and town was an important centre for the collection and redistribution of grain, and provide an important counterbalance to the extremely limited palaeoenvironmental data recovered from the site. The faunal assemblage was significantly more abundant (see Chapter 12).

The Commandery of Marienburg

Malbork Castle (Ger. Marienburg)

The commandery of Marienburg saw the expansion of agricultural land following the crusades (see Chapter 11), and the convent was also situated on the Nogat which ultimately linked the southern Prussian commanderies, as well as the Kingdom of Poland, with Elbing and the Baltic coast. In this respect, plant fragments recovered from archaeological contexts at the castle may have derived from both local and distant sources. There is no doubt that the convent housed significant granaries at the end of the fourteenth century, and the archaeological deposits date from this later period. They are derived from two moats associated with the *Vorburg* or outer bailey and so cannot provide information on the socio-topographic complexity of the site (see Chapter 10). The castle complex would have had multiple areas of food processing, waste disposal, and a socially strati-

Table 13.2: Plant remains from Malbork (outer moat); abundance: + single, ++ numerous

TAXA	MAL-100 82–88 cm	MAL-101 88–94 cm	MAL-102 94–100 cm	MAL-103 100–106 cm	MAL-104 106–112 cm	MAL-105 112–118 cm
USEFUL PLANTS						
Cereals						
Panicum miliaceum L.; millet	-	+	-	-	-	-
Vegetables and Spices						
Anethum graveolens L.; dill	+	-	-	-	-	-
Brassica cf. *rapa* L.; turnip	-	-	-	+	-	-
Brassica nigra L.; black mustard	+	-	+	-	-	-
Oil and Fibre Plants						
Cannabis sativa L.; hemp	-	-	+	-	-	-
Papaver somniferum L.; opium poppy	-	-	-	+	-	-
Fruits and Berries						
Ficus carica L.; fig	+	+	++	+	+	-
Fragaria vesca L.; wild strawberry	++	++	++	++	++	+
Prunus sp.; sloe/cherry	-	+	-	-	-	-
Pyrus sp.; Pear	-	+	+	+	+	-
Rubus caesius L.; European dewberry	-	+	-	+	-	-
Vaccinium myrtillus L.; bilberry	+	+	+	++	-	+
NON-USEFUL PLANTS						
Segetal and Ruderal Weeds						
Agrostemma githago L.; corn-cockle	+	+	+	+	+	+
Anthemis arvensis L.; corn chamomile	-	-	+	-	-	-
Anthriscus sylvestris L.; cow parsley	+	+	+	+	+	+
Arctium lappa L.; greater burdock	-	-	+	+	-	+
Carduus crispus L.; welted thistle	+	+	+	+	+	-
Centaurea cyanus L.; cornflower	+	+	++	++	+	-
Conium maculatum L.; hemlock	+	+	+	+	+	+
Chenopodium album L.; white goosefoot	+	+	+	+	+	-
Chenopodium hybridum L.; maple-leaved goosefoot	-	-	-	+	-	-
Echinochloa crus-galli L.; cockspur grass	-	+	+	-	-	-
Fallopia convolvulus (L.) Á. Löve; black-bindweed	+	+	+	+	-	-
Galium aparine L.; cleavers	-	-	-	-	+	-
Hyoscyamus niger L.; henbane	-	-	-	-	-	+
Lapsana communis L.; nipplewort	-	+	+	-	-	-
Leonurus cardiaca L.; motherwort	+	-	+	-	-	-
Neslia paniculata (L.) Desv.; ball-mustard	-	-	+	-	+	-
Plantago major L.; greater plantain	-	-	-	+	-	+
Polygonum aviculare L.; knotgrass	-	+	-	-	-	+
Polygonum hydropiper L.; water pepper	+	-	+	+	-	-
Polygonum lapathifolium L.; pale persicaria	++	+	+	+	+	+
Raphanus raphanistrum L.; wild radish	-	+	+	-	-	-

TAXA	MAL-100 82–88 cm	MAL-101 88–94 cm	MAL-102 94–100 cm	MAL-103 100–106 cm	MAL-104 106–112 cm	MAL-105 112–118 cm
Rumex acetosella L.; sheep's sorrel	+	+	+	+	+	+
Rumex crispus L.; curled dock	+	+	+	+	-	-
Scleranthus annuus L.; annual knawel	-	-	+	-	-	-
Setaria pumila (Poir.) Roem. & Schult.; yellow bristle-grass	+	-	-	-	-	-
Solanum nigrum L. emend. Mill.; black nightshade	+	+	+	+	-	-
Sonchus arvensis L.; perennial sow-thistle	-	-	+	-	-	-
Sonchus asper (L.) Hill; prickly sow-thistle	-	-	-	+	-	-
Stellaria media L.; common chickweed	-	+	+	+	+	-
Taraxacum officinale (L.) H. Wigg; dandelion	-	-	-	+	-	-
Thlaspi arvense L.; field penny-cress	+	+	+	-	-	-
Urtica dioica L.; common nettle	+	+	+	+	+	+
Urtica urens L.; small nettle	+	-	-	+	-	+
Viola arvensis Murray/*tricolor* L.; field pansy/wild pansy	-	-	-	-	-	+
Plants from Riversides, Marshes and Rushes						
Alisma plantago-aquatica L.; common water-plantain	+	-	-	-	-	-
Bidens tripartita L.; trifid bur-marigold	-	-	+	-	-	-
Eleocharis palustris (L.) Roem.&Schult./ *uniglumis* (Link) Schult.; common spike-rush/slender spike-rush	-	-	+	-	-	-
Lemna cf. *trisulca* L.; ivy-leaved duckweed	+	-	-	+	-	-
Oenanthe aquatica L.; fine-leaved water dropwort	+	-	-	-	-	+
Potamogeton sp.; pondweed	+	-	-	+	-	-
Ranunculus flammula L.; lesser spearwort	-	-	+	-	-	-
Ranunculus sceleratus L.; celery-leaved buttercup	+	+	+	+	+	+
Rumex maritimus L.; golden dock	+	+	++	++	+	+
Plants from Fresh Meadows and Pastures						
Lychnis flos-cuculi L.; ragged robin	-	-	-	+	-	-
Prunella vulgaris L.; self-heal	+	-	-	-	-	-
Ranunculus acris L.; ragged-robin	+	+	-	-	-	-
Ranunculus repens L.; creeping buttercup	+	+	+	-	+	+
Scirpus sylvaticus L.; wood club-rush	+	-	-	-	-	-
Other Plants						
Atriplex sp.; orache	+	+	-	+	-	-
Boraginaceae indet.; borawe family	-	-	-	-	+	-
Brassicacea indet.; cabbage family	+	+	+	-	-	+
Carex sp.; sedge	+	+	+	+	+	-
Chenopodium sp.; goosefoot	+	+	-	++	+	+
Galeopsis sp.; hemp-nettle	-	-	+	-	-	-
Poaceae indet.; grass family	+	-	+	++	-	-
Potentilla sp.; cinquefil	+	-	-	-	-	-
Ranunculus sp.; buttercup	-	-	-	-	-	+

continued overleaf

Table 13.2: Plant remains from Malbork (outer moat) *cont.*

TAXA	MAL-100 82–88 cm	MAL-101 88–94 cm	MAL-102 94–100 cm	MAL-103 100–106 cm	MAL-104 106–112 cm	MAL-105 112–118 cm
Rosaceae indet.; rose family	-	-	-	+	-	-
Rumex sp.; dock (sorel)	-	-	-	-	-	+
Sonchus sp.; sow-thistle	-	-	-	+	-	-
Stellaria graminea L./*palustris* Retz.; lesser stitchwort/marsh stitchwort	-	-	+	-	-	-
Stellaria sp.; starwort	-	-	+	-	-	-
VARIA undetermined	-	-	-	+	-	-
ANIMALS						
Insects	-	+	-	+		-
Daphnia sp.	+	-	+	+	-	+
Bryozoa	+	-	-	-	-	-

fied household with differential access to foodstuffs. This is clearly indicated in the written sources (Jóźwiak and Trupinda 2007; Badura and Możejko 2019); however, taken together they represent a diverse assemblage which includes exotic taxa. The plant macrofossils from the headquarters of the Teutonic Order are therefore most useful at the level of the 'site'.

The first discrete assemblage of plant material was obtained from the north-western corner of the waterlogged moat of the middle castle during rescue excavations in 2005, below the northern end of the range housing the Great Refectory (Simmons 2006). Three samples were taken from the lowest waterlogged layers of the moat, in close proximity to a timber pile. The preservation of the plant material in these samples was generally very good, with the presence of small and fragile seed types. The sample from the deepest stratigraphic unit, consisting of a black, artefact-rich humic soil, contained the greatest abundance and diversity of material, a reflection of the effects of waterlogged preservation. A pollen sample taken from the same context also contained abundant, well-preserved pollen grains. The overlying stratigraphic units which were sampled were both humic layers but contained relatively less material. In general, the macrofossil remains represented plants favouring disturbed habitats, such as white goosefoot (*Chenopodium album*), common chickweed (*Stellaria media*), or henbane (*Hyoscyamus niger*). Common nettle (*Urtica dioica*) was recorded in the uppermost stratigraphic level, again indicative of disturbance but also suggesting nitrogen enrichment of the soil perhaps deriving from manure. As in the case of the site in Grudziądz, the remains of elder represent scrub plants growing on waste and moist ground, in this case most likely growing on moist soils along the banks of the Nogat or nearby scrubland. Fragments of plants which prefer marshy habitats, such as rushes (*Juncus* sp.), sedges (*Carex* sp.), and marsh stitchwort (*Stellaria palustris*), could have derived from pastures, meadows, or other places with standing water, especially in winter. The single seeds of both radish (*Raphanus sativus*) and cabbage (*Brassica oleraceae*) were present in the sample recovered from the deepest part of the moat, and it is possible that these seeds represent cultivated foods. Cabbage is regularly noted in the documents of the Order and was a staple food in the medieval southern Baltic (Badura and Możejko 2019).

The second assemblage of plant material was obtained from a trench across the moat outside the curtain wall in the north-east corner of the outer bailey. AMS dating of material from the lowest layers of the moat indicated it was deposited in the fifteenth century (1410–1520, 88.2% probability) and could have derived from a number of different sources, including various latrines in the castle and the residues of mills situated along the connecting canal, the Mühl Graben (Pol. Malborska Młynówka). Six samples were taken from the moat, and the majority of the plant remains recovered from these materials had been extremely well preserved in the waterlogged conditions (Table 13.2). The assemblage that could be linked to human consumption, and therefore reached the moat via the castle's latrines, contained a large number of diaspors of edible plants, both cultivated and collected. Species of edible fleshy fruits and berries included bilberry (*Vaccinium myrtillus*) and wild strawberry (*Fragaria vesca*). Cereals were represented by millet (*Panicum miliaceum*), while vegetables

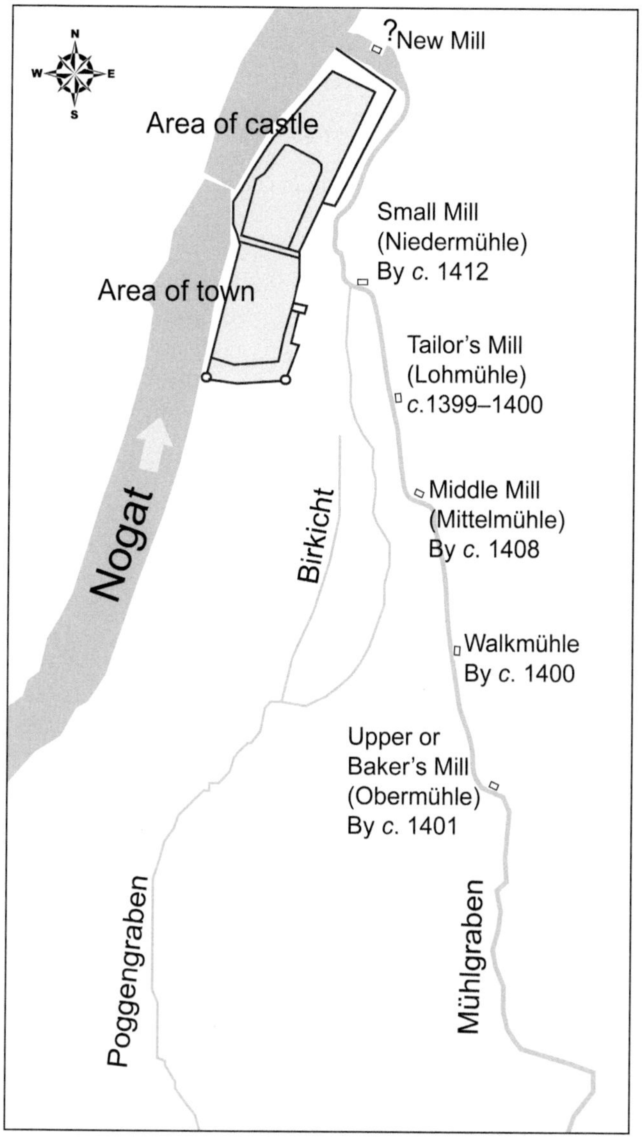

Map 13.1: Documented mills lining the mühlgraben in relation to the town and convent (after Długokęcki 2004, 28).

and spices consisted of dill (*Anethum graveolens*), turnip (*Brassica* cf. *rapa*), and black mustard (*Brassica nigra*). Intact seeds of opium poppy (*Papaver somniferum*) indicated that they were used as flavouring and not for extracting oil. Fruits of hemp (*Cannabis sativa*) provide evidence for the presence of fibrous plants, but this was also an important source of oil. Most of the species listed above are mentioned in the documents of the Order as products that were stored and used within the castle. In samples from the moat the remains of figs (*Ficus carica*) were found. This exotic plant in the form of dried fruits were imported to the towns in Prussia also thanks to the Teutonic Order. As the import of figs required a lot of effort, only the most affluent and influential groups of people had access to them. The convent's registers indicate that significant quantities of figs were used by the community (Długokęcka and Długokęcki 2010). Evidence for grain processing within the vicinity of the castle was represented by fragmented diaspores of segetal weeds, for example: cornflower (*Centaurea cyanus*), corn-cockle (*Agrostemma githago*), black bindweed (*Fallopia convolvulus*), and annual knawel (*Scleranthus annuus*), ultimately derived from the mill-lined canal which connected with the moat. These plants are species typical of winter-sown crops and could be associated with rye or wheat.

The earliest evidence for mills within the commandery of Marienburg dates to 1347 and is associated with the advocate's residence at Herrengrebin (Grabiny-Zameczek); however, the construction of mills is most closely associated with the process of colonization, especially in the Vistula fenlands (see Kubicki 2012 for the latest overview of milling in the Prussian *Ordensland*). In the privileges established for the settlements of Woltersdorf (1318), Marienau (1321), and Selchow (1328), the Order maintained complete control over the mills, their locations, and the roads leading up to them. In the colonizing settlements of Alt Münsterberg (1323), Liebschau (1341), and Tiege (1345), the Order owned the land where the mills were located. When the town adjacent to the convent of Marienburg was granted an island in the Nogat, the Order reserved the right to build a mill there. The organization of mills, which included regulating payments to millers, was standardized across the *Ordensland* by Grand Master Dietrich von Altenburg (1335–41). The mills within any given commandery were overseen by a *Mühlmeister* ('mill master', and at least two such officials are mentioned in the Marienburg convent in 1404), but their construction and maintenance was largely devolved to advocates and procurators, with the cluster nearest to the convent lining the Mühl Graben, all of which are first mentioned in the early fifteenth century (Map 13.1). The canal stretched for 12 km from Lake Damerau (Pol. Dąbrówka) to the castle, linking up a hydrological network that connected a number of the lakes within the southern half of the commandery (see Chapter 15). At least fourteen water-powered mills (including one only accessible by boat), twenty-seven windmills, six mills powered by animals, and a few others, up to an overall total of as many as fifty, were built within the commandery in the fourteenth and fifteenth centuries (Długokęcki 2004). These installations provide the broader context for the intensification of grain production within the commandery that accompanied the process of colonization.

Table 13.3: Plant remains from Biała Góra (settlement); TR - trench, RP - refuse pits, n - number samples; *- non-charred, **- mineralized, abundance: + single, ++ numerous

TAXA	TR.8 n-9	RP.94 n-6	RP.97 n-3	RP.98 n-3	RP.99 n-3	RP.102 n-2	RP.103 n-2	RP.105 n-1	RP.105a n-3	RP.116 n-2
Cultivated Plants										
Avena sp.; oat	-	+	-	-	-	-	-	-	+	-
Cerealia indet.; cereals	++	++	-	+	+	-	+	-	+	-
Hordeum vulgare L.; barley	+	+	-	-	+	-	-	-	-	+
cf. *Hordeum vulgare* L.; barley	+	-	-	-	-	-	-	-	-	-
Panicum miliaceum L.; millet	-	+	-	+	-	-	-	-	+	+
Pisum sativum; pea	-	+	-	-	-	-	-	-	-	-
cf. *Pisum sativum*; pea	+	-	-	-	-	-	-	-	-	-
Secale cereale L.; rye	+	+	-	-	+	+	-	-	+	-
cf. *Secale cereale* L.; rye	+	+	-	-	+	-	-	-	+	+
Triticum aestivum L.; wheat	+	+	-	-	+	-	-	-	-	-
Triticum sp./*Secale cereale* L.; wheat/rye	-	-	-	-	-	+	-	-	-	-
Segetal and Ruderal Weeds										
Agrostemma githago L.; corn-cockle	+	-	-	-	+	-	-	-	-	-
Atriplex nitens Schkuhr; garden orache	+*	-	-	-	-	-	-	-	-	-
Chenopodium album L.; white goosefoot	++	+	-	-	-	+	+	+	+	+
cf. *Echium vulgare* L.; viper's bugloss	-	+*	-	-	-	-	-	-	-	-
Fallopia convolvulus (L.) Á. Löve; black-bindweed	+	-	-	+	-	+	-	-	-	-
Galeopsis tetrahit L.; hemp-nettle	-	+*	-	-	-	-	-	-	-	-
Galium aparine L.; cleavers	+	+	-	-	-	+	-	+*	-	-
Galium aparine L./*tricornutum* Dandy; cleavers/corn clavers	+	-	-	-	-	-	-	-	-	-
Galium spurium L.; false cleavers	+	-	-	+	-	+	-	-	-	+
Lithospermum arvense L.; field gromwell	-	-	-	-	-	-	-	-	+**	-
Polygonum aviculare L.; knotgrass	+*	-	-	+*	-	+	-	-	-	+
Polygonum hydropiper L.; water pepper	-	-	-	-	-	-	-	-	+	-
Polygonum lapathifolium L.; pale persicaria	+	+	-	+	+	+	-	-	-	-
cf. *Rhinanthus serotinus* (Schönh.) Oborný var. *apterus* (Fr.) Hyl.; greater yellow-rattle	+	-	-	-	-	-	-	-	-	-
Rumex cf. *crispus* L.; curled dock	-	-	-	+	-	-	-	-	-	+
Setaria pumila (Poir.) Roem. & Schult.; yellow bristle-grass	-	+	-	-	-	-	-	-	-	-
Setaria viridis (L.) P. Beauv. /*verticillata* (L.) P. Beauv.; green bristle-grass/rough bristle-grass	+	-	-	-	-	-	-	-	-	-
Stellaria media L.; common chickweed	-	+*	-	-	-	-	-	-	-	-
Vicia hirsuta (L.) Grey; hairy tare	+	-	-	-	+*	+*	-	-	-	-

TAXA	TR.8 n-9	RP.94 n-6	RP.97 n-3	RP.98 n-3	RP.99 n-3	RP.102 n-2	RP.103 n-2	RP.105 n-1	RP.105a n-3	RP.116 n-2
Forest Plants										
Corylus avellana L.; hazel	-	+	-	-	-	-	-	-	-	-
Rubus idaeus L.; raspberry	+**	+**	-	-	-	-	-	-	-	-
Sambucus nigra L.; elder	-	-	-	-	-	-	-	+*	-	-
Other Plants										
Boraginaceae indet.; borawe family	-	-	-	-	-	-	+*	-	-	-
Carex sp.; sedge	-	-	-	-	-	-	-	-	-	-
Chenopodium sp.; goosefoot	++	+	+	-	-	+	-	-	+	+
Fabaceae indet.; legume family	+	-	-	-	-	-	+	-	-	-
Galium palustre L.; common marsh-bedstraw	-	+*	-	-	-	-	-	-	-	-
Poaceae indet.; grass family	+	-	-	-	-	+	-	-	-	-
Vicia sp.; vetch	-	-	-	-	-	-	-	-	-	-
VARIA undetermined	+	+	-	+	+	-	-	-	-	-

Biała Góra (Ger. Weißenberg)

The site of Biała Góra, located *c.* 15 km south of Malbork, has been interpreted as having the character of a frontier colony in the twelfth and thirteenth centuries, and then a less intensive settlement that endured into the fifteenth century. Archaeobotanical investigations at the site focused on the assessment of plant materials recovered from excavated features, and a preliminary report has already been published (Pluskowski et al. 2014; Sawicki et al. 2015, 299). A total of forty-four bulk samples were taken from twelve features (pits) within three trenches, predominantly from Trench 8 (see Chapter 10). All samples contained mineral materials preserved above ground water level, and most of the preserved materials were charred.

Samples from eight features contained substantial quantities of plant diaspores, the most abundant of which were traces of cereals, including grains of rye and barley (Table 13.3). Both species were present in larger quantities within the feature identified as having a domestic/economic function within Trench 8 and feature 94, a pit. Single grains of wheat were found in features 94, 99, and Trench 8, and millet (*Panicum miliaceum*) was found in features 94, 98, 105a, and 116 (Fig. 13.1). It seems that these latter two species did not play a significant role at the site, and there were no residues of rachis or glumes. Apart from these crops, single charred seeds of pea (*Pisum sativum*) were observed in feature 94. Wild useful plants were represented by only two charred shell fragments of hazelnut (*Corylus avellana*) and one mineralized seed of raspberry (*Rubus idaeus*). As with many other assemblages containing cereals, there was a relatively high proportion of weeds. The majority of these represented segetal flora which is found growing alongside crops, and included false cleavers (*Galium spurium*), and cleavers/corn cleavers (*G. aparine/tricornutum*). Plants that could belong to both segetal and ruderal flora were represented by single diaspores of curlytop knotweed (*Polygonum lapathifolium*), common chickweed (*Stellaria media)*, and a significant number of fruits from the goosefoots (*Chenopodium*) genus.

The range of cereals and weeds present at the site indicates access to permanent rotational field systems or perhaps to imported grain, although the intensity of crop production is difficult to gauge. Finds of sickles and quern stones certainly provide evidence for harvesting and cereal processing, and palynological profiles from the vicinity suggest continuity in cereal cultivation on either side of the crusading period (see Chapter 11). The fruits and nuts present at the site could also have been sourced locally, most likely from the fringes of the nearby woodland. Charcoal remains found in the features at Biała Góra indicate that in the twelfth and thirteenth centuries pine (*Pinus*), hornbeam (*Carpinus*), and oak (*Quercus*) were also exploited.

Tczew (Ger. Dirschau)

At the turn of the fourteenth century, Tczew was the site of a Pomeranian settlement and stronghold: the residence of a castellan. With the annexation of Pomeralia in 1308, the town came under the control of the Teutonic Order and was managed by an advocate. Although the

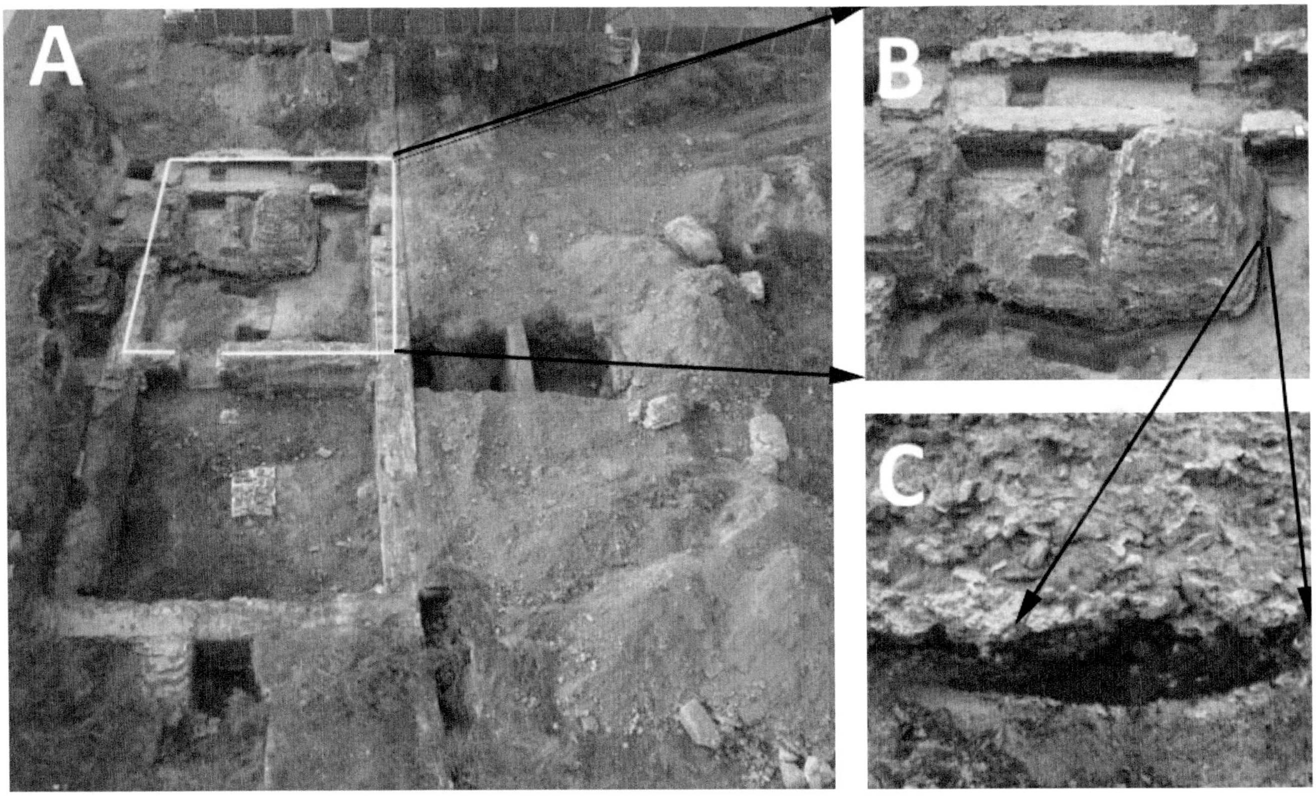

Figure 13.2: Tczew, area 418/14 showing the location (c) of the charred layer. Original photo by M. Jonakowski.

town lies on the western bank of the Vistula and therefore on the edge of the commandery of Marienburg, the advocate of Dirschau became subservient to the Grand Master's convent from 1323 (Długokęcki 1998). The Vistula was also the principal communication route for the movement of commodities from the southern commanderies of the *Ordensland* up to the Baltic coast, particularly from the Kulmerland. This is reflected in one of the most important assemblages of plant macrofossils recovered from the Teutonic Order's territories, albeit not directly associated with a convent or fortified residence. In 2010, excavations near the Vistula foreshore at Tczew uncovered the foundations of substantial medieval brick buildings which appear to have functioned as warehouses for storing produce before shipment (Jonakowski 2009/10). A thick horizon of charred seeds was identified throughout the uncovered complex, which appears to have been abandoned after being damaged by fire (Fig. 13.2). Seeds dated by AMS indicated the structure must have been destroyed between *c.* 1395 and 1475 (93.6% probability). Six samples were taken from across the burnt horizon in the northern and eastern parts of the structure Fig. 13.3).

The assemblage was represented largely by cereals and weeds, as would be expected from a granary or an area where grains could be stored (Table 13.4). Cereals were represented by five species: rye (*Secale cereal*), wheat (*Triticum aestivum*), barley (*Hordeum vulgare*), millet (*Panicum milaceum*), and oat (*Avena* cf. *sativa*) (Fig. 13.3). Rye dominated the samples taken from the northern part of the room; wheat occurred mainly in samples from the eastern part; grains of millet, barley, and oats were present as a small admixture within the sample taken from the western side of the room. A small percentage of the assemblage consisted of the remains of weeds. Most of these represented segetal and ruderal flora can be found in the vicinity of developing settlements. Sample 3 included a considerable number of contemporary diasporas of ruderal plants, which was the result of a very thin layer, partially uncovered. These included lady's-thumb (*Polygonum persicaria*), and white goosefoot (Chenopodium album). The presence of burnt fragments of corn cockle (*Agrostemma githago*), annual knawel (*Scleranthus annuus*) and sheep's sorrel (*Rumex acetosella*) may indicate that the charred cereals were harvested in the winter, as these weeds are present today amongst winter cereals.

Amongst those remains which could not be clearly identified to species or genus, there was a significant quantity of charred seeds of the *Fabaceae* family, most

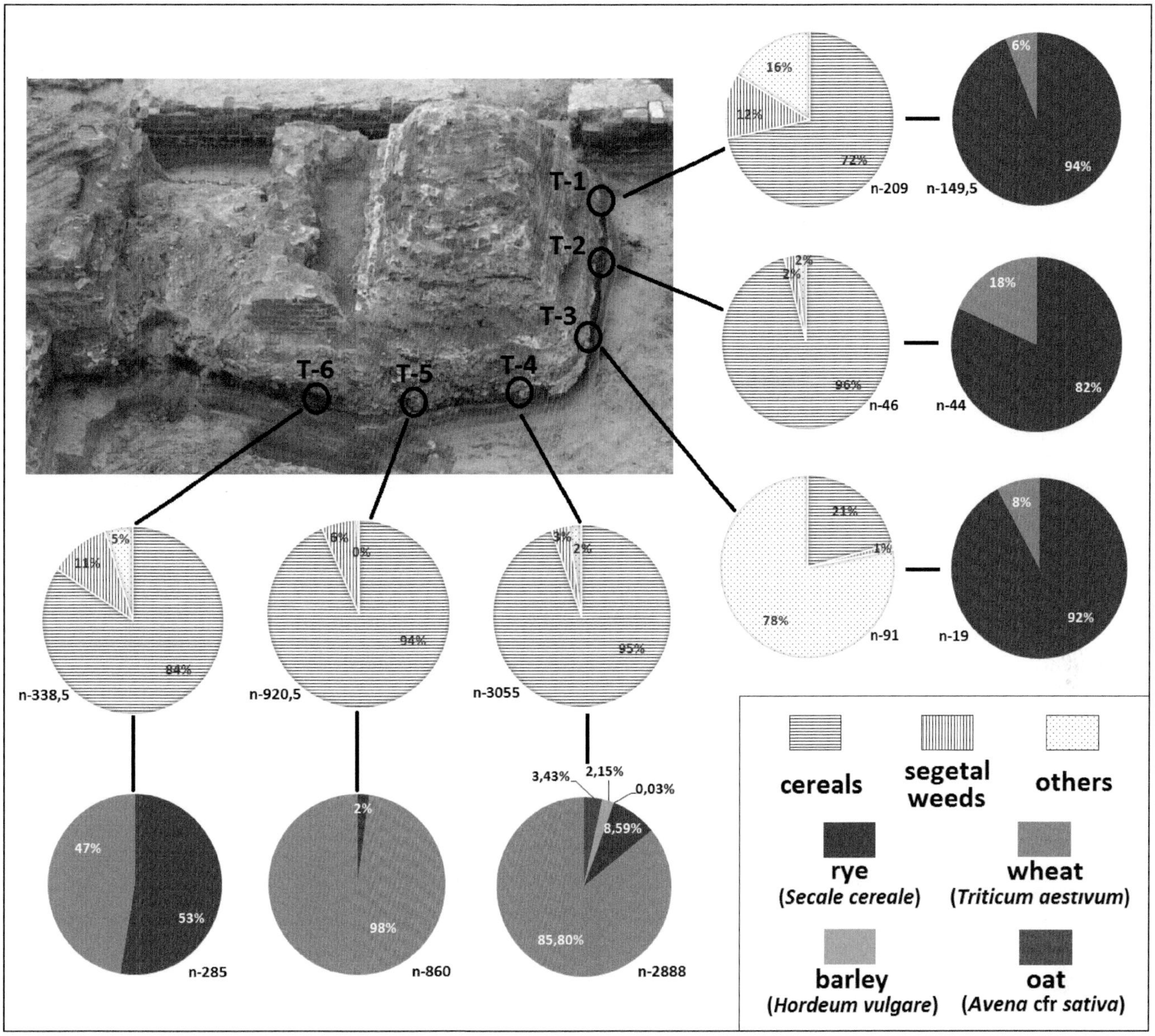

Figure 13.3: The spatial distribution of cereals and other plants at the destroyed granary in Tczew; n = number of remains. Figure by M. Badura.

likely representing the genus *Vicia*. Several species of *Vicia* (e.g., *V. tetrasperma*, *V. hirsuta*, *V. cracca*, *V. angustifolia*) are weeds found amongst cereals (Tymrakiewicz 1976; Lityńska-Zając 2005), and the represented weeds were mainly plants whose fruit and seeds had comparable dimensions or weight to that of cereal grain. Only a few samples contained segetal remains from plants producing lightweight seeds and fruit, such as sheep's sorrel. This indicates the grain had been probably cleaned and prepared for milling or sowing (Engelmark 1989; Henriksen and Robinson 1996). The varied distribution of percentages of different species within the charred horizon suggests the segregated storage of grain within the room: rye was stored against the northern wall, wheat against the eastern, with the mixture of oats, barley, and millet likely to have been accidental.

The fruits of sedges (*Carex elata*, *C. hirta*, *C. pallescens*, *C. panicea*, and *C. spicata*) and marsh-bedstraw (*Galium palustre*) represent a range of bog plants and rushes. They almost certainly reflect the presence of wetlands within the vicinity, habitats that can be connected with the Vistula and its floodplain, whilst plant communities reflecting meadows and pastures are represented by self-heal (*Prunella vulgaris*). Since the fruits of sedges and marsh-bedstraw had been charred, it can be assumed that the communities which included these plants were located in very close proximity to the burning building. Yet despite the proximity of these wetlands, none of

Table 13.4: Plant remains from Tczew (destroyed building); n - number of samples;
*- non-charred; abundance: + single, ++ numerous, +++ abundant

TAXA	N part			E part		
	T-1 n-1	T-2 n-1	T-3 n-1	T-4 n-1	T-5 n-1	T-6 n-1
Cereals						
Avena cf. *sativa* L.; oat	-	-	-	++	-	-
Cerealia indet.; cereals	+	+	+	+++	++	++
Hordeum vulgare L.; barley	-	-	-	++	-	-
Panicum miliaceum L.; millet	-	-	-	+	-	-
Secale cereale L.; rye	++	++	+	++	+	+
Triticum aestivum L.; wheat	+	+	I	+++	+++	+
Segetal Weeds						
Agrostemma githago L.; corn-cockle	+	-	-	++	-	-
Camelina sativa L.; gold-of-pleasure	-	-	-	-	-	+
Centaurea cyanus L.; cornflower	-	-	-	+	+	+
Galium aparine L./*tricornutum* dandy; cleavers/rough corn	-	-	-	+	-	-
Galium spurium L.; false cleavers	+	-	-	+	-	+
Lithospermum arvense L.; field gromwell	+	-	-	+	++	+
Lolium temulentum L.; darnel	-	-	+	-	-	-
Papaver argemone L.; prickly poppy	-	-	-	-	-	+
Rumex acetosella L.; sheep`s sorrel	+	+	-	+	-	+
Rumex crispus L.; curled dock	-	-	-	-	-	+
Scleranthus annuus L.; annul knawel	-	-	-	+	-	-
Vicia tetrasperma (L.) Schreb.; smooth tare	-	-	-	-	+	-
Ruderal Weeds						
Chenopodium album L.; white goosefoot	-	-	+	-	-	-
Erysimum cheiranthoides L.; treacle-mustard	-	-	++*	-	-	-
Lactuca serrolia L.; prickly lettuce	-	-	+*	-	-	-
Polygonum aviculare L.; knotgrass	-	-	+*	-	-	-
Polygonum lapathifolium L.; pale persicaria	+	-	-	-	-	+
Polygonum persicaria L.; lady's thumb	-	-	-	+	-	-
Senecio vulgaris L.; groundsel	-	-	+*	-	-	-
Setaria pumila (Poir.) Roem. & Schult.; yellow bristle-grass	-	+*	-	-	-	-
Taraxacum officinale F. H. Wigg.; dandelion	-	-	-	+*	-	+
Plants from Fresh Meadows and Pastures						
Prunella vulgaris L.; self-heal	-	-	-	-	-	+
Plants from Riversides, Marshes and Rushes						
Carex elata All.; tufted sedge	+	-	-	+	-	+
Carex hirta L.; hairy sedge	-	-	-	+	-	-
Carex pallescens L.; pale sedge	+	-	-	-	-	-
Carex panicea L.; carnation sedge	-	-	-	+	-	-

TAXA	N part			E part		
	T-1 n-1	T-2 n-1	T-3 n-1	T-4 n-1	T-5 n-1	T-6 n-1
Carex spicata Huds.; spiked sedge	+	-	-	-	-	-
Carex sp.; sedge	+	-	+*	+/+*	-	+
Galium palustre L.; common marsh-bedstraw	-	-	-	-	-	+
Other Plants						
Asteraceae indet.; aster family	-	-	+*	-	+	-
Betula sect. *albae*; white birches	-	-	+*	-	-	-
Boraginaceae indet.; borawe family	-	-	-	+*	-	-
Brassica sp.; cabbage	-	-	-	+	-	-
Bromus sp.; brome	+	-	-	-	+	+
Chenopodium sp.; goosefoot	+	-	-	+	-	+
Fabaceae indet.; legume family	-	-	-	++	+	+
Galium sp.; bedstraw	+	-	-	-	-	-
Lolium sp.; ryegrass	-	-	-	-	-	+
Poaceae indet.; grass family	+	-	-	-	-	+
Polygonum sp.; knotweed	-	-	-	+	-	-
Sonchus sp.; sow-thistle	-	-	+*	-	-	-
Rumex sp.; dock (sorrel)	+	-	-	-	-	-
VARIA undetermined	+	-	+*	+*	-	+

these seeds displayed signs of germination (swelling or the presence of an embryo), indicating the room where the grain was stored was well insulated.

The Commandery of Danzig

Gdańsk (Ger. Danzig)

Following the siege and annexation of the town in 1308, the Teutonic Order reorganized the urban districts, built a convent, and promoted trade. It was at this time that the port suburb situated on an island in the canalized Motława, the so-called 'Granary Island', was developed. Gdańsk was already an important centre for the export of grain in the twelfth century, and there is tentative evidence for activity on the 'Granary Island' in the thirteenth century: but following the imposition of the Order's rule and the increase in trade with the Hanse, a series of granaries, warehouses, shipyards, and, from the mid-fourteenth century, a bathhouse were constructed. In the fourteenth century, 130 granaries are documented alongside a civic court, ash yard, pitch yard, and stores for timber and other goods. Despite a devastating fire in 1424, the district was quickly rebuilt and by the early sixteenth century there were 340 buildings on the island (Paner 1999, 388–89).

Over the last few years, numerous assemblages of plant macrofossils have been recovered from across the various historical districts of Gdańsk, enabling a series of detailed diachronic botanical study of the environmental conditions and economy within the town (Latałowa et al. 2007; Badura 2011; Możejko and Badura 2013; Badura et al. 2015; Święta-Musznicka et al. 2013). A significant proportion of the botanical remains from Gdańsk are derived from 'useful' plants, recovered from latrines, granaries, domestic buildings, and related occupation deposits. Amongst them, findings on the Granary Island confirm the importance of the town as a major transit port for the export of grain throughout the Baltic and North Sea (Badura 2011). Rye and wheat were the most important types of grain passing through the town for both local consumption and export, and historical sources indicate that in the thirteenth century most of the grain exported from Gdańsk came from Kuyavia, an area of northern-central Poland around the middle Vistula River basin and the upper River Noteć, but, by the fifteenth century, the source of grain had shifted to Masovia in the north-eastern part of the Kingdom of Poland, bordering with the *Ordensland* (Latałowa et al. 2007). The town represented the end point of the Vistula trade route, and was one of the most important hubs for exporting grain produced within the commanderies of the *Ordensland*, as well as from neighbouring regions, in particular the Kingdom of Poland. This is also confirmed by the presence of weed species in fourteenth-century deposits on the Granary Island typical of the limestone soils of southern Poland (e.g. *Galium tricornutum*) (Badura 2011, 95). The low number of weeds present in the town's grain assemblages is indicative of highly efficient grain preparation and cleaning, as in the case of Tczew, whilst the weed species that were present suggest the prevalence of winter crops in the grain trade. The impact of the Teutonic Order's economic policy is therefore represented by an intensification in the grain trade, with Gdańsk fully connected within the Hanseatic networks that emerged during the crusading period.

In addition to exported cereals, the plant macro-fossil assemblages also provide evidence for local consumption and the purchasing power of urban households. Many are widely found staples, such as cabbage, turnips, and various fruits, but the presence of buckwheat (*Fagopyrum esculentum*) in fourteenth- and fifteenth-century deposits ranks amongst the earliest in Northern Europe, and would have been used in making groats and beer. The widespread presence of hop fruits from the fifteenth century can also be linked to the town's brewing industry. The presence of some fruits is only attested from this period, such as red currant (*Ribes rubrum*) from the fourteenth century and black currant (*Ribes nigrum*) and gooseberry (*Ribes uva-crispa*) from the fifteenth century. Wild strawberries and bilberries are also evident from the fourteenth century. In contrast to the recovery of material from the Old Town of Elbląg, the remains of exotics and expensive commodities (rice *Oryza sativa*, black pepper *Piper nigrum*, cardamom *Elettaria cardamomum*) in Gdańsk are largely found from the end of the fifteenth century. These are slightly later than in German Hanseatic centres, although some such as rice are documented within the Order's documentation already in the fourteenth century, whilst one of the most prevalent exotic species imported into the Order's territories — fig — is present in fourteenth-century latrine deposits (Badura 2011, 190–91). This suggests there may have been subtle differences in access to exotic commodities between different towns within the *Ordensland*, reflecting variable purchasing power amongst communities of burghers, as well as the Order's convents. It is also possible that the chronological and historical differences between Pomeranian towns which came under the Order's rule in the fourteenth century, and those colonies founded on former Prussian lands in the thirteenth century (especially those promoted by mercantile interests) resulted in divergent economic trends. Within the various urban districts of Gdańsk, the main town stands out as the wealthiest on the basis of access to luxury plant goods. This district saw the most investment in infrastructure from the Order and its burghers, whilst the Old Town remained comparatively poorer with the earlier Slavic settlement persisting into the fifteenth century (Paner 1998–99, 2001).

Direct evidence of what plants were consumed on contemporary ships comes from the unique fifteenth-century 'Copper Wreck' (Pol. *Miedziowiec*), sunk off the coast of Gdańsk, from which an exceptionally well-preserved plant remains was recovered during excavations in 1971–75 led by the Polish Maritime Museum (Badura 2014; Badura et al. 2015). The find included amongst others bulbs of onion (*Alium cepa*) and garlic (*Allium sativum*), seed of a field bean (*Vicia faba* var *minor*), as well as the ear of rye fragment.

The Commandery of Elbing

Elbląg Town and Castle (Elbing)

The convent and town of Elbing was situated at the north-eastern end of the Vistula trade route, accessible via the Nogat which ran through the centre of the commandery of Marienburg. The colony played a pivotal role during the crusading period, but afterwards declined as access to the Vistula Lagoon became increasingly difficult. Excavations in the waterlogged soils of the Old Town have produced a range of plant macrofossil assemblages, which have been the focus of investigations for over a decade (Badura et al. 2004; Latałowa et al. 1998, 2003, 2007, 2009). Compared to Gdańsk, however, cereal remains are comparatively rare, despite historical sources indicating the importance of the town in the grain trade (Czaja 2009). Exotic plant remains, in particular nutmeg (*Myristica fragrans*) from Indonesia and figs from the Mediterranean, point towards longer-distance trade during the thirteenth to fifteenth centuries (Latałowa et al. 2007). A few fruits of carrot (*Daucus carota*) and a larger assemblage of onion seeds were found in fourteenth-century contexts within the town. Both species are rarely reported in medieval European sites, with the exception of carrots in Kraków and onions (both seeds and bulb) in Gdańsk (Latałowa et al. 2007;

Figure 13.4: Charred cereals from excavations in the courtyard of the Archaeology and History Museum of Elbląg (Ger. Elbing). Photo by M. Badura.

Badura et al. 2013). As with Gdańsk, there is evidence for the widespread use of dill (*Anethum graveolens*), black mustard (*Brassica nigra*), caraway (*Carum carvi*), wild marjoram (*Origanum vulgare*), celery (*Apium graveolens*), parsley (*Petroselinum crispum*), and black pepper (*Piper nigrum*). All of this evidence points to the significant purchasing power of the town's patricians during the fourteenth century.

Excavations in the courtyard of the castle in 2012 also produced an abundant range of plant material, dating to the second half of the thirteenth and the first half of the fourteenth centuries. Samples were collected from both trenches, in which the consecutive contexts represented the early occupation and destruction phases of the castle and the associated moat (Fig. 13.4). In both trenches there were archaeological layers with charred plant material, dating to the second quarter of the fourteenth century (Trench 1/context 57, Trench 2/context 17) (Table 13.5). Although the trenches were some distance from each other, comparative analysis of their archaeology indicated the charred layer was the result of the same fire. This was confirmed by archaeobotanical analysis from both trenches: contexts 57 and 17 were characterised by identical botanical compositions. Samples included a significant quantity of charred oats (*Avena sativa*), alongside the burnt remains of segetal weeds, as well as ruderal species. This particular assemblage can be interpreted as the remnants of fodder, most likely for horses (see Chapters 12 and 14), and the oats appear to have been stored in a container outside the destroyed building. The remainder of the samples were waterlogged. In Trench 1, a mixture of plant elements from different vegetation types was preserved, providing evidence for the range of habitats within the immediate vicinity of the castle. Some samples (dated to the fourteenth century) contained species representing nearby wetlands, particularly bulrush (*Typha* sp.), rushes (Juncus spp.) and celery-leaved buttercup (*Ranunculus sceleratus*), with some fragments derived from ruderal plants, such as common nettle (*Urtica dioica*), as well as segetal weeds. Cereals were represented by single charred kernels of wheat (*Triticum aestivum*) and barley (*Hordeum vulgare*). Other materials were dominated by mostly ruderal weed species, as well as remains of fresh/wet meadows plants such as meadow buttercup (*Ranunculus acris)*, creeping buttercup (*R. Repens)*, self-heal (*Prunella vulgaris*), ragged robin (*Lychnis flos-cuculi)*, and common valerian (*Valeriana officinalis*). Single seeds of *Hypericum maculatum* indicated the presence of dry meadows and pastures. Apart from cereals, cultivated species were represented by single seeds of cabbage (*Brassica oleraceae*), turnip (*Brassica rapa*), and fruits of caraway (*Carum carvi*). In addition, the remains of wild useful plants (hazel, raspberry, and bilberry) were only present in small quantities. Hop (*Humulus lupulus*), one of the most important flavouring and preservative agents used in brewing, was noted in material dating to the fourteenth century. At other medieval sites located within the area of the Baltic coastline, the fruits of hop are present in varying abundance. Not all finds of single remains of this plant reflect its use in brewing. As hop occurs naturally in more or less disturbed flood-plain forests, it is possible that such communities developed naturally within the vicinity of the castle and were the sources of the remains identified in the samples. With the exception of context 17, other materials from Trench 2 were connected with the edge of the moat (most probably a regulated, natural watercourse) and included a substantial amount of wood. Bark and small branches (fascines) could perhaps be linked to the construction and maintenance of the earthwork (Latałowa et al. 2007; Badura et al. 2013). Plant remains in these samples were scarce and represented mostly those species which tolerated wet areas and nitrates.

Table 13.5: Plant remains from Elbląg: C - context; n - number of samples; *- charred remains, abundance: + - single, ++ - numerous, +++ - abundant

	Trench 1					Trench 2							
	C.57		C.57/1	C.57/2	C.18	below C.26	C.26	C.25	C.23	C.18b	C.11	above C.11	C.17
TAXA	El.1/12	El.5/12	El.2/12	El.3/12	El.4/12	El.8/13	El.9/13	El.10/13	El.11/13	El.12/13	El.13/13	El.15/13	El.16/13
USEFUL PLANTS													
Cereals													
Avena sativa L.; oat	-	+++*	-	+*	-	-	-	-	-	-	-	-	+++*
Avena sp.; oat	-	-	-	-	-	-	-	-	-	-	-	-	
Cerealia indet.; cereals	-	++	-	-	-	-	-	-	-	-	+	-	
Hordeum vulgare L.; barley	-	-	+	-	-	-	-	-	-	-	+	-	++
Panicum miliaceum L.; millet	-	-	-	-	+	-	-	-	-	-	-	-	
Secale cereale L.; rye	-	-	-	-	-	-	-	-	-	-	-	-	+
Triticum aestivum L.; wheat	-	-	+*	-	-	-	-	-	-	-	-	-	+*
Vegetables and Spices													
Brassica oleraceae L.; cabbage	-	-	-	-	-	-	+	-	-	-	-	-	-
Brassica rapa L.; turnip	-	-	-	-	++	-	-	-	-	+	-	-	+*
Brassica cf. *rapa* L.; turnip	-	-	-	-	-	-	-	-	-	-	-	-	+*
Carum carvi L.; caraway	-	-	+	-	-	-	-	-	-	-	-	-	-
Fruits and Nuts													
Corylus avellana L.; hazel	-	-	+	-	+	-	-	-	-	-	-	+	-
Fragaria vesca L.; wild strawberry	+	-	-	-	+	-	-	-	-	+	-	-	-
Rubus idaeus L.; raspberry	-	-	+	-	+	-	-	-	-	-	-	-	-
Vaccinium myrtillus L.; bilberry	-	-	-	-	-	-	-	-	-	+	-	-	-
NON-USEFUL PLANTS													
Segetal and Ruderal Weeds													
Agrostemma githago L.; corn-cockle	-	++*	+	+*	+	-	-	-	-	+	+*	+	+++*
Anthemis cotula L.; stinking chamomile	-	-	-	+	-	-	-	-	-	-	-	-	-
Bromus secalinus L.; rye brome	-	-	-	-	-	-	-	-	-	-	-	-	+*
Centaurea cyanus L.; cornflower	-	-	-	+	+	-	-	-	-	-	-	-	+*
Chenopodium album L.; white goosefoot	-	-	+	-	++	-	+	+	-	+	+*	+	+*
Cirsium oleraceum (L.) Scor.; cabbage thistle	-	-	-	-	+	-	-	-	-	-	-	-	-
Fallopia convolvulus (L.) Á. Löve; black bindweed	-	+*	-	+*	+	-	-	+	-	+	-	-	++*
Galeopsis tetrahit L.; hemp-nettle	-	-	-	-	+	-	+	+	-	-	-	-	+
Galium aparine L.; cleavers	-	+*	-	+*	-	-	-	-	-	-	-	-	+
Galium aparine L./*tricornutum* Dandy; cleavers/corn clavers	-	-	-	-	-	-	-	-	-	-	-	-	+*
Galium cf. *aparine* L.; cleavers	-	-	-	-	+	-	-	-	-	-	-	-	
Galium spurium L.; false cleavers	-	++*	-	-	+	-	-	-	-	-	-	-	++*

	Trench 1					Trench 2							
	C.57		C.57/1	C.57/2	C.18	below C.26	C.26	C.25	C.23	C.18b	C.11	above C.11	C.17
TAXA	**El.1/12**	**El.5/12**	**El.2/12**	**El.3/12**	**El.4/12**	**El.8/13**	**El.9/13**	**El.10/13**	**El.11/13**	**El.12/13**	**El.13/13**	**El.15/13**	**El.16/13**
Galium tricornutum Dandy; corn clavers	-	-	+	-	-	-	-	-	-	-	-	-	+*
Melandrium album (Mill.) Garcke; white campion	-	-	-	-	+	-	-	-	-	-	-	-	-
Myosotis arvensis (L.) Hill; field forget-me-not	-	-	-	-	+	-	-	-	-	-	-	+	-
Neslia paniculata L.; ball-mustard	-	-	-	-	+	-	-	-	-	-	-	-	-
Plantago lanceolata L.; ribwort plantain	-	-	-	-	-	-	-	-	-	-	-	-	+*
Plantago major L.; greater plantain	-	-	-	-	+	-	-	-	-	-	-	-	-
Polygonum aviculare L.; knotgrass	-	-	+	+	+	-	-	-	-	+	-	+	+*
Polygonum persicaria L.; lady's thumb	-	+*	-	-	+	-	+	-	-	+	-	+	-
Polygonum cf. *persicaria* L.; lady's thumb	-	-	-	-	-	-	-	-	-	-	-	-	+*
Polygonum hydropiper L.; water pepper	-	+*	+	+*	+	-	-	-	-	-	-	-	+*
Polygonum lapathifolium L.; pale persicaria	-	*	+	+*	+++	+	+	+	+	-	-	-	+*
Polygonum minus Huds.; pygmy smartweed	-	-	-	-	-	-	-	-	-	-	-	-	+*
Raphanus raphanistrum L.; wild radish	-	-	-	-	-	-	-	-	-	+	-	-	-
Rhinanthus serotinus (Schönh.) Oborný var. *apterus* (Fr.) Hyl.; greater yellow-rattle	-	-	-	+*	-	-	-	-	-	-	-	-	-
Rumex acetosella L.; sheep`s sorrel	-	++*	+	++*	+*	-	+*	-	-	+	-	+	++*
Rumex crispus L.; curled dock	-	-	-	-	+	-	+	-	-	-	-	-	+*
Setaria pumila (Poir.) Roem. & Schult.; yellow bristle-grass	-	+*	-	-	+	+	-	-	-	-	-	-	+*
Setaria viridis (L.) P. Beauv. /*verticillata* (L.) P. Beauv.; green bristle-grass/rough bristle-grass	-	+*	-	-	-	-	-	-	-	-	-	-	-
Silene gallica L.; small-flowered catchfly	-	-	-	-	-	-	-	-	-	-	-	-	+*
Silene vulgaris (Moench) Garcke; bladder campion	-	-	-	-	+	-	-	-	-	-	-	-	-
Spergula arvensis L. s.l.; corn spurrey	-	+	-	-	-	-	-	-	-	-	-	-	-
Spergula arvensis L. subsp. *vulgaris;* corn spurrey	-	-	-	-	+	-	-	-	-	-	-	+	+++*
Spergula arvensis L. subsp. *maxima* (Weihe) O. Schwarz; corn spurrey	-	+	-	-	-	-	-	-	-	-	-	-	+*
Stellaria media L.; common chickweed	+	-	+	-	++	-	+	+	-	+	-	-	-
Thlaspi arvense L.; field penny-cress	-	-	-	-	+	-	-	-	-	-	-	-	-
Urtica dioica L.; common nettle	-	-	+	+	+	+	-	-	-	-	-	-	-
Vicia sativa L.; common vetch	-	-	-	-	-	-	-	-	-	-	-	-	++*
Viola arvensis Murray/*tricolor* L.; field/wild pansy	-	-	-	-	+	-	-	-	-	-	-	-	-

	Trench 1					Trench 2							
	C.57		C.57/1	C.57/2	C.18	below C.26	C.26	C.25	C.23	C.18b	C.11	above C.11	C.17
TAXA	El.1/12	El.5/12	El.2/12	El.3/12	El.4/12	El.8/13	El.9/13	El.10/13	El.11/13	El.12/13	El.13/13	El.15/13	El.16/13
Plants from Fresh Meadows and Pastures													
Filipendula ulmaria (L.) Maxim.; meadowsweet	-	-	-	-	+	-	-	-	-	-	-	-	-
Lychnis flos-cuculi L.; ragged robin	-	-	-	+	+	-	-	-	-	-	-	-	-
Potentilla anserina L.; silverweed	-	-	-	-	-	-	+	-	-	-	-	-	-
Prunella vulgaris L.; self-heal	-	+*	+	+	+	-	-	-	-	-	-	-	-
Ranunculus acris L. s. str.; meadow buttercup	-	-	-	+*	++	+	+	-	-	+	-	1.0	+*
Ranunculus repens L.; creeping buttercup	+	-	-	**	++	+	+	-	+	++	+	2.0	+*
Scirpus sylvaticus L.; wood club-rush	+	-	+	+*	++	-	-	-	-	-	-	-	+*
Stellaria graminea L.; lesser stitchwort	-	-	-	-	+	-	-	-	-	-	-	-	-
Valeriana officinalis L.; common valerian	-	-	-	+	-	-	-	-	-	-	-	-	-
Plants from Riversides, Marshes, and Rushes													
Carex elata All.; tufted sedge	-	-	-	-	+	+	+	-	-	+	-	-	-
Carex elata All./*nigra* Reichard; tufted sedge/common sedge	-	-	-	-	+	+	-	-	-	-	-	-	-
Carex ovalis Gooden.; oval sedge	-	-	-	-	+	-	+	-	-	-	-	-	-
Carex pallescens L.; pale sedge	-	-	-	-	+	-	+	+	-	-	-	+	-
Galium palustre L.; common marsh-bedstraw	-	-	-	-	+	-	-	-	-	+	-	-	-
Galium cf. *palustre* L.; common marsh-bedstraw	-	-	-	-	-	-	-	-	-	-	-	+	-
Humulus lupulus L.; hop	-	-	+	-	+	-	-	-	-	-	-	-	-
Juncus articulatus L. emend. K. Richt./*bulbosus* L.; jointed rush/bulbous rush	-	-	-	-	+	-	-	-	-	-	-	-	-
Juncus cf. *bufonius* L.; toad rush	-	-	+	-	+	-	-	-	-	-	-	-	-
Juncus conglomeratus L. emend. Leers/effusus L.; compact rush/soft rush	-	-	+	-	++	-	-	-	-	-	-	-	-
Juncus cf. *filiformis* L.; thread rush	-	-	-	-	+	-	-	-	-	-	-	-	-
Juncus sp.; rush	-	-	++	-	++	-	-	-	-	-	-	-	-
Lycopus europaeus L.; gypsywort	-	-	-	-	+	-	-	-	-	-	-	-	-
Mentha cf. *aquatica* L.; water mint	-	-	-	-	+	-	-	-	-	-	-	-	-
Ranunculus flammula L.; lesser spearwort	-	-	-	+	+	-	-	-	-	+	-	-	+*
Ranunculus scelerathus L.; celery-leaved buttercup	-	-	-	-	+	-	-	-	-	-	-	-	-
Stellaria palustris Retz.; marsh stitchwort	-	-	+	-	-	-	-	-	-	-	-	-	-
Typha sp.; bulrush	-		++	-	-	-	-	-	-	-	-	-	-

TAXA	Trench 1					Trench 2							
	C.57		C.57/1	C.57/2	C.18	below C.26	C.26	C.25	C.23	C.18b	C.11	above C.11	C.17
	El.1/12	El.5/12	El.2/12	El.3/12	El.4/12	El.8/13	El.9/13	El.10/13	El.11/13	El.12/13	El.13/13	El.15/13	El.16/13
Forest Plants													
Betula sect. *albae*; white birches	-	-	-	-	+	-	-	-	-	-	-	-	-
Carex spicata Huds.; spiked sedge	-	-	-	+	+	+	-	-	-	-	-	-	-
Ranunculus auricomus L. s.l.; goldilocks buttercup	-	-	-	-	-	+	-	-	-	-	-	-	-
Other Plants													
Brassica sp.; cabbage	-	-	-	-	-	-	-	+	-	-	-	-	-
Bromus sp.; brome	-	-	-	-	-	-	-	-	-	-	-	-	+*
Carex sp.; sedge	-	-	+	+*	+	+	+	-	-	-	-	+	+*
Caryophyllaceae indet.; pink family	-	-	-	-	+	-	-	-	-	-	-	-	
Centaurea sp.; cornflower	-	-	-	-	-	-	-	-	-	-	-	-	+*
Chenopodium sp.; goosefoot	+	-	-	+*	++	+	+	+	-	+	-	+	+*
Cirsium sp.; thistle	-	-	-	-	+	-	-	-	-	-	-	-	-
Cyperaceae indet.; sedge family	-	-	-	-	+	-	-	-	-	-	-	-	-
Fabaceae indet.; legume family	-	++*	-	+*	-	-	-	-	-	-	-	-	-
Galeopsis sp.; hemp nettle	-	+*	-	-	+	-	-	-	-	-	-	-	+*
Galium sp.; bedstraw	-	+*	-	-	-	-	-	-	-	-	-	-	+
Galium sp.; bedstraw	-	-	-	-	+	-	-	-	-	-	-	-	-
Glyceria sp./*Melica* sp.; sweet-grass/melic	-	-	-	-	-	-	-	-	-	-	-	+	-
Hypericum maculatum L.; imperforate St John's wort	-	-	+	-	-	-	-	-	-	-	-	-	-
Lamiaceae indet.; mint family	-	-	-	-	+	-	-	-	-	-	-	-	-
Poaceae indet.; grass family	-	+*	++	+*	+	-	-	-	-	-	-	-	+*
Polygonum sp.; knotweed	-	-	-	-	-	-	-	-	-	-	-	-	+*
Ranunculus sp.; buttercup	-	-	-	-	++	-	+	-	-	+	-	-	+*
Setaria sp.; bristle grass	-	-	-	-	-	-	-	-	-	-	-	-	+*
Stellaria graminea L./*palustris* Retz.; lesser stitchwort/marsh stitchwort	-	-	-	-	-	-	-	-	-	-	-	-	+*
Vicia sp.; vetch	-	-	-	-	-	-	-	-	-	-	-	-	+*
VARIA undetermined	-	-	-	+*	-	-	-	-	-	-	+*	-	+*

Conclusion

The plant macro-fossil assemblages from medieval sites in northern Poland, within the former territory of the Teutonic Order, provide a social and diachronic snapshot of plant exploitation on either side of the crusading period (Tables 13.1–13.5; Fig. 13.5). The range of contexts from which plant material has been recovered is very diverse and includes refuse pits in settlements, castle moats, and granaries, as well as more ambiguous occupation deposits. No silos were identified at any of the sites considered in this study, although discrete concentrations of plant material could be interpreted as being stored in containers or one type or another. Considered together, these assemblages provide supporting evidence for the vegetation changes that are

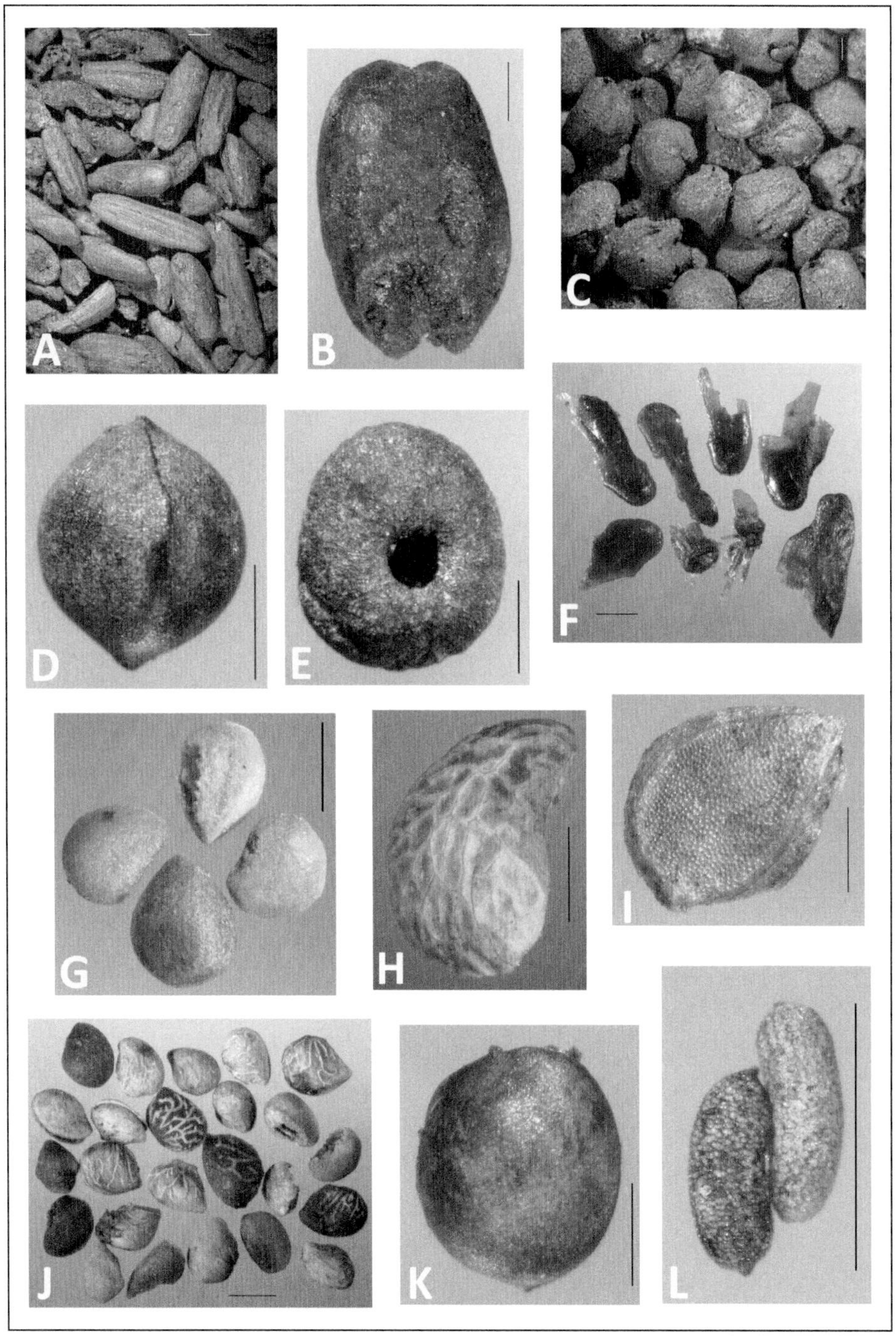

Figure 13.5: Plant remains from sites mentioned in the text.
A: *Avena sativa* (Elbląg),
B: *Triticum aestivum* (Tczew),
C: *Agrostemma githago* (Tczew),
D: *Fallopia convolvulus* (Biała Góra),
E: *Galium aparine/tricornutum* (Biała Góra),
F: *Centaurea cyanus* (Malbork),
G: *Ficus carica* (Malbork),
H: *Rubus caesius* (Malbork),
I: *Ranunculus repens* (Elblag),
J: *Fragaria vesca* (Malbork),
K: *Humulus lupulus* (Elblag),
L: *Hypericum maculatum* (Elblag).
Photos and figure by Monika Badura.

visible in the landscape and, in particular, the agrarian surplus generated within the Teutonic Order's territories in Prussia. There are also clear differences between settlement types which provide a model of socially variable access to plant resources that can be tested further with future analyses. The most abundant range of species are encountered in towns and castles which include several exotic species, whilst assemblages from smaller settlements are less diverse, represented by cereals and legumes many of which were probably locally sourced. For example, both convents and towns were markets for dried or candied fruit, particularly figs (*Ficus carica*) and grapes (raisins) (*Vitis vinifera*), species which are not found on rural settlements or before the crusading period in Prussia. The evident parity between the Order's convents and the more affluent urban households indirectly reflects the growth in the political and commercial power of towns by the fifteenth century, which ultimately enabled the Prussian Union to challenge the rule of the Order in the 1440s and finance military opposition during the Thirteen Years' War (1454–66).

The Order's internal hierarchy was mirrored in differential access to expensive fruits and vegetables: the most diverse and abundant quantities were documented in Marienburg, with increasingly diminutive amounts recorded in other convents. However, even the smallest residences of the Order's officials had access to a diverse range of expensive foodstuffs. The accounts of the castle of Lötzen indicate that peas, mustard, poppy seeds, hemp, millet, pearl barley, buckwheat, sauerkraut, and onions were used in the kitchen. Imported products, categorized as *crewed/krude* or 'confectionery', also appear to have been bought in for special occasions, such as Lenten feasts. Expenses for small quantities of sugar, raisins, figs, almonds, rice, caraway, pepper, and cloves are listed in the castle's accounts in 1507 and 1508 (*OF*, fols 11^r, 96^r). The convent of Brandenburg, to which the procurator of Lötzen was usually subservient for much of the existence of the *Ämter*, had more abundant supplies within its stores (*GAD*, 210 ff.), whilst the convent of Memel (Lith. Klaipėda), despite its precarious location on the frontier, had access to a diverse range of imported commodities due to its connection with maritime trade routes. Its inventories from 1404 to 1447 record the presence of rye, barley, oats, malt, hops, flour, peas, and salt, as well as the more exotic onions, figs, raisins, almonds, mustard, and vinegar (Willoweit 1969, 79–80).

It is also possible to contribute towards a diachronic understanding of the Order's impact on vegetation and plant exploitation. This can be linked to the cultural phases outlined in Chapter 2. The evidence from the recent excavations in Elbląg suggests that the Order's *castra* built during the crusading period, and the more durable castles constructed in the first half of the fourteenth century, included facilities for storing grain. The first references to such stores only appear in documents from the later fourteenth century (Jóźwiak and Trupinda 2012). Agrarian production for local consumption by people and livestock, and to generate a taxable surplus, was a fundamental component of the process of militarized colonization in the thirteenth century. The extent to which this was implemented varied between commanderies, although it is clear that frontier colonies such as Biała Góra were involved in sustained agricultural activities already before the crusading period, as evident across the frontier castellanies of the Polish state. Although the resolution of the archaeological record at the site is coarse, the intensity of activity in the thirteenth century suggests this process was barely interrupted by the crusades. Palynological data indicates agricultural land-use became increasingly intensive from the thirteenth century in the Lower Vistula Valley (see Chapter 11), continuing a pattern begun in the eighth and ninth centuries and reflecting the progressive colonization of the landscape by successive Pomeranian and German groups. The intensification in agricultural production can be linked to requirements to provision both rural and urban settlements, such as Biała Góra, and the Order's fortified *castra* and later *Konventsburgen*.

The second chronological phase can be associated with surplus grain production supplying the well-documented export trade in grain to Western Europe and Scandinavia (Hybel 2002). This is best attested along the valley of the Lower Vistula. The movement of grain (amongst other commodities, such as timber and fur) down the Vistula is attested at various points along

Figure 13.6: Fifteenth-century granary buildings in Toruń (Piekary 2) (Ger. Thorn). Photo by A. Pluskowski.

the river. The route was punctuated by four important commanderies — Thorn, Graudenz, Marienburg, and Danzig. Granary buildings have survived in each of these towns. In modern Toruń they are best represented by the granary buildings in the southern part of the Old Town, close to the river, with the best-preserved example situated on Piekary 4 and dating to the second half of the fifteenth century (Fig. 13.6). During the period of the Order's rule, there were thirty such buildings documented here (Czacharowski 1983, 82). The Order became a major supplier grain to Western Europe as regional economies diversified, refocused on pastoral farming or suffered from famines (see Chapter 11). In the fourteenth century, Flemish towns — and in the fifteenth century, Dutch towns — were the main consumers of Prussian and Polish grain, representing more than 50% of all the products exported by the Order. The trading connections between Prussia and Poland saw the movement of grain from Kuyavia up the Vistula, purchased by Hanseatic merchants in Thorn, a town which itself had a diverse population of Poles and Germans. The short-lived commandery of Nessau (Pol. Nieszawa) also provided direct access to the farmlands of Kuyavia in the fourteenth century. After Bruges, the primary market for export was England: during the Great Famine under the reign of Henry VI, the Order provided large quantities of grain; conversely in 1389 grain was shipped from British ports to Danzig (Samsonowicz 1994, 142). This period is also associated with the noticeable appearance of exotics. This may partly reflect the fragmentary nature of the evidence: archaeological and written data significantly increases from the latter half of the fourteenth century, but when contextualized within a broader timeframe and combined with the evidence of plant macrofossils, there is a noticeable rise in both abundance and distribution of these plants, corresponding to the rising affluence of the Order's convents and urban populations.

The final phase, associated with the fiscal decline of the Teutonic Order in the fifteenth century, is represented by three sites within the Lower Vistula region: the granary at Tczew, the moats at Malbork Castle, and the urban contexts of Gdańsk. However, in eastern Prussia, which remained in the hands of the Order after 1466, there is documentary evidence suggesting the intensification of grain production and its storage at castles. The grain trade also continued after the disintegration of the *Ordensland*, and there is certainly no interruption in the palynological indicators of cultivation in the post-medieval period (see Chapter 11). From 1562, *The Book of Sound Dues* recorded the tolls on cargo vessels passing through the Sound of Denmark, demonstrating that significant quantities of grain (as well as timber) were being exported through Danzig to Western Europe (Haneca et al. 2005). The archaeobotanical record from Gdańsk for the post-medieval period corresponds to a widely documented flourishing urban centre with an elevated economic status, represented by greater access to luxury goods (Badura 2011, 192–96). Within this longue durée perspective the period of the Order's rule can be characterized as one of growth, following the brief period of destruction during the siege of the city. This corresponds to the significant programme of urban development promoted by the Order (Pluskowski 2013, 221–28).

There are of course some important caveats to the social and diachronic model presented above. First, the taphonomic complexity of castle and urban sites has resulted in highly fragmented and unevenly represented assemblages of plant materials. In the case of Malbork Castle, material has only been recovered from very small, keyhole excavations and only in suitably preserved contexts. The written record for the convent is much richer than any archaeological dataset and indicates, for example, the consumption of imported rice and other exotics which have yet to be confirmed archaeologically (Długokęcka and Długokęcki 2010). The absence of any systematic excavations, let alone environmental sampling, in the Old Town of Malbork has prevented any meaningful comparisons with the convent. In contrast, the Old Town of Gdańsk has been one of the most intensively studied in north Poland from the perspective of environmental archaeology, but relatively little data has been obtained from the Order's castle. When it comes to the presence of exotics, there is a significant disparity between the plant macro-fossil record and the written sources. Almonds (*Amygdalus communis*), for example, are first attested archaeologically in north Poland in the seventeenth century but feature in the documents of the Teutonic Order from the fourteenth century (Badura and Możejko 2019). The documented intensification of viticulture under the Order's rule in Prussia is also largely absent from the archaeological record to date (Tandecki 1991).

Whilst the medieval grain trade itself is archaeologically visible, alongside the purchasing power of the convents and towns of the Prussian *Ordensland*, the agricultural activities of rural settlements — and indeed, markets — remain poorly known. This is particularly true of settlements governed under the segregating Prussian Law in the eastern Prussian commanderies. These communities were an integral part of the agrarian machine constructed by the Order within the conquered territories, but at present the indigenous Prussian population which survived the onslaught of the crusades is only visible through observing changes in vegetation and micro-charcoal through the lens of palynology, the dietary regimes of livestock as revealed by trends in stable isotopes (Scull et al. 2019), and excavated tools associated with harvesting, threshing, and milling. What is clear is that the changes visible in the landscape are staggered chronologically, and that agricultural production on the frontier remained small-scale and limited until the sixteenth century. In this respect, the experience and impact of the Teutonic Order's rule in eastern Prussia is more comparable to parts of Livonia than to the productive and densely settled heartlands of the *Ordensland*. Future archaeological investigations in medieval sites within this region would benefit from a systematic sampling strategy with the aim of recovering more plant macrofossils, particularly from rural settlements. This is essentially for clarifying and testing the social and diachronic models presented above.

Finally, it is important not to forget the religious dimensions of the agrarian world, whether Christian, indigenous, or some creolized hybrid. Matthaeus Praetorius, writing on the customs of the Prussians of Nadruvia in the seventeenth century, described the ritual killing and consumption of black and white cockerels amongst Prussian families, a ceremony conducted only by men after the drying and threshing of grain (Okulicz-Kozaryn 1983, 113). It is easy to dismiss such accounts as either exaggeration or fabrication (see Bojtár 1999), but the ritual spheres of agrarian society in medieval Prussia would have contributed to a sense of segregation between the colonists and colonized, and to the persistence of indigenous customs and identity (see Szczepański 2019). This certainly appears to have been the case in Livonia and would equally apply to parts of central and particularly eastern Prussia. Despite the effective integration of the Teutonic Order's provisioning networks that connected the smallest farmstead with the most powerful convents, in a society that was predominantly rural, this may be part of the multifaceted reason for why the crusades failed to eliminate pre-Christian praxis in the eastern Baltic.

Works Cited

Manuscripts

Berlin-Dahlem, Geheimes Staatsarchiv Preußischer Kulturbesitz, *Ordensfolianten*

Primary Sources

GAD Ziesemer, W., ed. 1968. *Das große Ämterbuch des Deutschen Ordens: Mit Unterstützung des Vereins für die Herstellung und Ausschmückung der Marienburg*, Wiesbaden: M. Sändig (repr. 2009)

Secondary Works

Antoniewicz, J. 1950. 'Wyniki prac wykopaliskowych na grodzisku i podgrodziu w Jeziorku, pow. Giżycko w 1950 r.', *Materiały Wczesnośredniowieczne* 2, 235–56

Badura, M. 2011. *Rośliny użytkowe w dawnym Gdańsku*, Studium Archeobotaniczne, Gdańsk: Wydawnictwo Uniwersytetu Gdańskiego

——, 2014. 'The Plant Remains', in *The Copper Ship: A Medieval Shipwreck and its Cargo*, ed. W. Ossowski, Gdańsk: National Maritime Museum in Gdańsk, 419–28

Badura, M., Latałowa, M., Jarosińska, J., and Święta, J. 2004. 'Rośliny użytkowe w średniowiecznych i nowożytnych materiałach archeobotanicznych z miast północnej Polski (Kołobrzeg, Gdańsk, Elbląg)', in *Archeologia et Historia Urbana*, eds R. Czaja, G. Nawrolska, M. Rębkowski, and J. Tandecki, Elbląg: Muzeum w Elblągu, 277–86

Badura, M., and Możejko, B. 2019. 'The Plant Element in the Diet of the Inhabitants of Gdańsk, Elbląg and Malbork during the Rule of the Teutonic Order: The Historical and Archaeobotanical Perspective', in *Ecologies of Crusading, Colonization, and Religious Conversion in the Medieval Baltic: Terra Sacra II*, ed. A. Pluskowski, Environmental History of the North Atlantic World, 3, Turnhout: Brepols (forthcoming)

Badura, M., Możejko, B., and Ossowski, W. 2013. 'Bulbs of Onion (*Allium cepa L.*) and Garlic (*Allium sativum L.*) from the 15th-Century Copper Wreck in Gdańsk (Baltic Sea): A Part of Victualling?', *Journal of Archaeological Science* 40/11, 4066–72

Badura, M., Możejko, B., Święta-Musznicka, J., and Latałowa, M. 2015. 'The Comparison of Archaeobotanical Data and the Oldest Documentary Records (14th–15th Century) of Useful Plants in Medieval Gdańsk, Northern Poland', *Vegetation History and Archaeobotany* 24/3, 441–54

Behre, K.-E. 1999. 'The History of Beer Additives in Europe: A Review', *Vegetation History and Archaeobotany* 8, 35–48

Białuński, G. 2005. *Siedlungswesen im Bereich der Großen Masurischen Seen vom 14. bis zum 18. Jahrhundert — Ämter Lötzen und Rhein*, Hamburg: Selbstverlag des Vereins für Familienforschung

Bojtár, E. 1999. *Foreword to the Past: A Cultural History of the Baltic People*, Budapest: Central European University Press

Carson, S., Shillito, L.-M., Brown, A., and Pluskowski, A. G. 2012. 'Environmental Assessment of Samples from the Castle Site at Grudziądz, Poland', in *Zamek w Grudziądzu w świetle badań archeologiczno-architektonicznych: Materiały i studia*, ed. M. Wiewióra, Toruń: Wydawnictwo Naukowe Uniwersytetu Mikołaja Kopernika, 351–57

Czacharowski, A. 1983. 'Toruń średniowieczny (do roku 1454)', in *Toruń dawny i dzisiejszy: zarys dziejów*, ed. M. Biskup, Warsaw: Państwowe Wydawnictwo Naukowe, 31–131

Czaja, R. 2009. 'Urbanizacja kraju', in *Państwo zakonu krzyżackiego w Prusach: Władza i społeczeństwo*, ed. M. Biskup et al., Warsaw: Państwowe Wydawnictwo Naukowe, 177–230

Dincauze, D. F. 2006. *Environmental Archaeology: Principles and Practice*, Cambridge: Cambridge University Press

Długokęcka, O., and Długokęcki, W. 2010. 'The Sumptuous Use of Food at Castle Marienburg (Malbork) at the Start of the Fifteenth Century', *Acta Poloniae Historica* 102, 101–26

Długokęcki, W. 1998. 'Zakon krzyżacki wobec Tczewa w początkach XIV wieku', in *Historia Tczewa*, ed. W. Długokęcki, Tczew: Kociewski Kantor Edytorski, 43–47

——, 2004. 'Młynarstwo w komturstwie malborskim w XIII–XV w.', in *Młyny w Malborku i okolicy od XIII do XIX wieku*, eds W. Długokęcki, J. Kuczyński, and B. Pospieszna, Malbork: Muzeum Zamkowe w Malborku, 7–36

Engelmark, R. 1989. 'Weed-Seeds in Archaeological Deposits', in *Models, Experiments and Interpretations: Approaches to Swedish Prehistory*, eds T. B. Larsson and H. Lundmark, British Archaeological Reports, 500, Oxford: Archaeopress, 179–87

Haneca, K., Ważny, T., van Acker, J., and Beeckman, H. 2005. 'Provenancing Baltic Timber from Art Historical Object: Success and Limitations', *Journal of Archaeological Science* 32, 261–71

Henriksen, P. S., and Robinson, D. 1996. 'Archaeobotany as a Source of Information about Past Agrarian Practices', in *Proceedings of the Sixth Nordic Conference on the Application of Scientific Methods in Archaeology: Esbjerg 1993*, eds V. Mejdahl and P. Siemen, Esbjerg: Esbjerg Museum, 159–67

Hybel, N. 2002. 'The Grain Trade in Northern Europe before 1350', *Economic History Review* 45, 219–47

Jaremek, A. 2003. 'Badania w osadzie przygrodowej w Jeziorku stan. I.', *Swiatowit* 5/46 (B), 251–54

Jonakowski, M. 2009/10. *Tczew, ul. Zamkowa, działka nr 747/2: Sprawozdanie archeologiczne z przeprowadzonych prac* (unpublished report)

Jóźwiak, S., and Trupinda, J. 2007. *Organizacja życia na zamku krzyżackim w Malborku w czasach wielkich mistrzów (1309–1457)*, Malbork: Muzeum Zamkowe w Malborku

Kubicki, R. 2012. *Młynarstwo w państwie Zakonu krzyżackiego w Prusach w XIII–XV wieku (do 1454 r.)*, Gdańsk: Wydawnictwo Uniwersytetu Gdańskiego

Latałowa, M., Badura, M., and Jarosińska, J. 2003. 'Archaeobotanical Samples from Non-Specific Urban Contexts as a Tool for Reconstructing Environmental Conditions (Examples from Elbląg and Kołobrzeg, Northern Poland', *Vegetation History and Archaeobotany* 12, 93–104

Latałowa, M., Badura, M., Jarosińska J., and Święta-Musznicka, J. 2007. 'Useful Plants in Medieval and Post-Medieval Archaeobotanical Material from the Hanseatic Towns of Northern Poland (Kołobrzeg, Gdańsk and Elbląg)', *Studies in Archaeology and History (National Museum (PNM), Copenhagen)* 12, 39–72

Latałowa, M., Badura, M., Jarosińska, J., Święta-Musznicka, J., and Pińska, K. 2009. 'Archeobotaniczne przesłanki dotyczące pokarmów roślinnych w Gdańsku, Elblągu i Kołobrzegu na tle danych z innych ośrodków hanzeatyckich północnej Europy (XIII–XVIII w.)', in *Stan badań archeologicznych miast w Polsce*, eds H. Paner, M. Fudziński, and Z. Borcowski, Gdańsk: Muzeum Archeologiczne w Gdańsku, 167–82

Latałowa, M., Jarosińska, J., and Badura, M. 1998. 'Elbląg średniowieczny w świetle dotychczasowych materiałów archeobotanicznych', *Archeologia Polski*, 43, 147–66

Lityńska-Zając, M. 2005. *Chwasty w uprawach roślinnych w pradziejach i wczesnym średniowieczu*, Kraków: Instytut Archeologii i Etnologii Polska Akademia Nauk

Możejko, B., and Badura, M. 2013. 'Między źródłem pisanym a archeobotaniką. Rośliny użytkowe w późnośredniowiecznym Gdańsku', *XVII Sesja Pomorzoznawcza (2) Od wczesnego średniowiecza do czasów nowożytnych*, eds H. Paner and M. Fudziński, Gdańsk: Muzeum Archeologiczne w Gdańsku, 581–89

Okulicz-Kozaryn, Ł. 1983. *Życie codzienne Prusów i Jaćwięgów w wiekach średnich (IX–XIII w.)*, Warsaw: Państwowy Instytut Wydawniczy

Paner, H. 1998–99. 'Z badań archeologicznych Gdańska', *Z otchłani wieków Pomorza Gdańskiego: zeszyt okazjonalny dla upamiętnienia Tysiąclecia Gdańska: 997 urbs Gyddanyzc — 1997 Gdańsk*, ed. E. Choińska-Bochdan, Gdańsk: Stowarzyszenie Naukowe Archeologów Polskich, 95–119

——, 1999. 'Archaeological Evidence for Trade in Gdańsk from the 12th to 17th Century', in *Lübecker Kolloquium zur Stadtarchäologie im Hanseraum*, II: *Der Handel*, eds R. Duncke, M. Gläser, and U. Oltmanns, Lübeck: Schmidt-Römhild, 387–402

——, 2001. '10th- to 17th-Century Domestic Architecture in Gdańsk', in *Lübecker Kolloquium zur Stadtarchäologie im Hanseraum*, III: *Der Hausbau*, eds B. Dahmen, M. Gläser, U. Oltmanns, and S. Schindel, Lübeck: Schmidt-Römhild, 491–509

Pluskowski, A. G. 2013. *The Archaeology of the Prussian Crusade: Holy War and Colonisation*, London: Routledge

Pluskowski, A. G., Sawicki, Z., Shillito, L.-M., Badura, M., Makowiecki, D., Zabilska-Kunek, M., Seetah, K., and Brown, A. 2014. 'Biała Góra: The Forgotten Colony in the Medieval Pomeranian-Prussian Borderlands', *Antiquity* 88/341, 863–82

Prance, G., and Nesbitt, M. 2005. *The Cultural History of Plants*, London: Routledge

Rutkowska-Płachcińska, A. 1978. 'Pożywienie, napoje', in *Historia kultury materialnej Polski*, II: *Od XIII do XV wieku*, ed. A. Rutkowska-Płachcińska, Wrocław: Zakład Narodowy im. Ossolińskich, Wydawnictwo Polskiej Akademii Nauk, 247–79

Samsonowicz, H. 1994. 'Europejski handel Krzyżaków w XIV–XV wieku', *Komunikaty Mazursko-Warmińskie* 2–3, 137–45

Sarnowsky, J. 1993. *Die Wirtschaftsführung des Deutschen Ordens in Preußen (1382–1454)*, Cologne: Böhlau

Sawicki, Z., Pluskowski, A. G., Brown, A., Badura, M., Makowiecki, D., Shillito, L.-M., Zabilska-Kunek, M., and Seetah, K. 2015. 'Survival at the Frontier of Holy War: Political Expansion, Crusading, Environmental Exploitation and the Medieval Colonising Settlement at Biała Góra, North Poland', *European Journal of Archaeology* 18/2, 282–311

Scull, C., Müldner, G., and Makowiecki, D. 2019. 'Carbon and Nitrogen Stable Isotope Evidence for Animal Husbandry and Environmental Change in the Medieval Kulmerland', in *Ecologies of Crusading, Colonization, and Religious Conversion in the Medieval Baltic: Terra Sacra II*, ed. A. Pluskowski, Environmental History of the North Atlantic World, 3, Turnhout: Brepols (forthcoming)

Simmons, E. 2006. *Preliminary Assessment of Waterlogged Plant Remains from MZM-Malbork Castle, Poland* (unpublished report)

Święta-Musznicka, J., Latałowa, M., Badura, M., and Gołembnik, A., 2013. 'Combined Pollen and Macrofossil Data as a Source of Reconstructing Mosaic Patterns of the early medieval Urban Habitats — A Case Study from Gdańsk, N. Poland', *Journal of Archaeological Science* 40, 637–38

Szczepański, S. 2019. 'Surkapurn – Kreken; The Phenomenon of a Name, the Phenomenon of a Place: The Organization of Sacred Space in Prussia using Archaeology and Other Sources', in *Ecologies of Crusading, Colonization, and Religious Conversion in the Medieval Baltic: Terra Sacra II*, ed. A. Pluskowski, Environmental History of the North Atlantic World, 3, Turnhout: Brepols (forthcoming)

Tandecki, J. 1991. 'Weinbau im mittelalterlichen Preußen', *Beiträge zur Geschichte Westpreußens* 12, 83–99

Tymrakiewicz, W. 1976. *Atlas chwastów*, Warsaw: Państwowe Wydawnictwo Rolnicze i Leśne

Valk, H., Rannamäe, E., Brown, A., Pluskowski, A., Badura, M., and Lõugas, L. 2013. 'Thirteenth-Century Cultural Deposits at the Castle of the Teutonic Order in Karksi', *Arheoloogilised välitööd Eestis/Archaeological Field Work in Estonia* 2012, 73–92

Wacnik, A., Goslar, T., and Czernik, J. 2012. 'Vegetation Changes Caused by Agricultural Societies in the Great Masurian Lake District', *Acta Palaeobotanica* 52, 59–104

Willoweit, G. 1969. *Die Wirtschaftsgeschichte des Memellandes*, I: *Wissenschaftliche Beiträge zur Geschichte und Landeskunde Ost-Mitteleuropas; Auftrage des Johann Gotfried Herder-Instituts, 85/1*, Marburg: Lahn

Witthöft, H. 1981. 'Scheffel und Last in Preußen', *Blätter für Deutsche Landesgeschichte* 117, 335–72

Zabłocki, J. 1950. 'Szczątki roślinne ze stanowiska wczesnośredniowiecznego w Jeziorku, powiat Giżycko, wydobyte w r. 1950', *Materiały Wczesnośredniowieczne* 2, 211–25

14. Settlement Life in Prussia at the Microscopic Scale and the Impact on their Territories

Rowena Banerjea

The application of a geoarchaeological approach has enabled the extent of the impact of the process of colonization associated with the crusades to be examined at the microscopic scale, with specific reference to the development of occupation and use of settlements, moats and the flow of effluent, and the geochemical traces of sites in their immediate landscapes. The integration of thin-section micromorphology, phytolith analysis, and inorganic geochemical analysis of sediments from settlements and castles provides high-resolution insights into the lifestyles of inhabitants, and the spatial and diachronic use of settlement space (e.g., Banerjea et al. 2017; Matthews et al. 1997; Shillito et al. 2014) with specific reference to key cultural transitions associated with the crusading period. The site at Święta Góra (Ger. Heiliger Berg) provided the opportunity to examine the longue durée nature of activities and settlement use, and in particular the important transitional period within the tribal landscape from the Migration Period through to the later Middle Ages. As a result of excavations at Biała Góra, this multi-proxy geoarchaeological approach was applied to examine the nature of occupation at this rural colony in relation to the transition between early and later medieval phases (Pluskowski et al. 2014). The excavations within the courtyard of the castle in Elbląg (Ger. Elbing) presented a unique opportunity to study the early colonizing activities driven by the Order and its allies, in what became the theocracy's principal administrative centre in the latter half of the thirteenth century until its role was replaced by Marienburg, and the development and changing use of the castle's earliest *Vorburg* over time, linking in with the palynological (see Chapter 11), zooarchaeological (see Chapter 12), and archaeobotanical evidence from this site (see Chapter 13).

The Development of Occupation and Use of Settlements

Święta Góra

Święta Góra (site 1, 'Sacred Hill') is a settlement on a hill adjacent to Lake Staświny, Masuria, located in the tribal territory of Galindia. There are three or possibly four phases of occupation of the site (Table 14.1): the earliest dates from the Early Iron Age (fifth to first century BC), the second started in the developed part of Migration Period (*c.* second part of the fifth century AD) and lasted until the end of seventh/early eighth century, phase 2a; the third phase is dated to the developed part of the early Middle Ages, the ninth to tenth centuries, phase 2b; and the last phase of occupation is connected with the settlement from the early Middle Ages until the time of the Order's rule, the tenth to the thirteenth/fourteenth centuries. The excavations and the application of a geoarchaeological approach are particularly important for understanding the nature of the occupation at the site as, in comparison to earlier periods, the early medieval archaeology of Masuria remains poorly understood (Nowakiewicz and Rzeszotarska-Nowakiewicz 2013, 9). The spatial and diachronic geoarchaeological analysis of several pits from each of the phases contributes to the understanding of this site and challenges the accuracy of historical sources and the interpretations of the archaeological evidence from Masuria at this time. Peter von Dusburg (*CTP* III, 4) refers to a story of a time of major upheaval in this part of Prussia before the crusades and describes the territory of Galindia as desolate and depopulated into the early fourteenth century. However, written sources including papal bulls from the second half of the thirteenth century refer to the rivalry between the Teutonic Order and the Polish Kuyavian prince

Table 14.1 The chronological framework for the pit features at Święta Góra

Phase	Date	Feature (s)
1	Early Iron Age (fifth to first century BC). Western Balt Barrows Culture	23
2	a) Late Migration Period (second half of fifth century until the end of seventh–beginning of eighth century AD)	13, 19, 24, 25, 30, 31, 39, 41, 43, 47
	b) Early Middle Ages (ninth century until the tenth century AD) 'Prussian Culture'	14, 15, 16, 26, 28, 40, 44, 45, 49
3	Early Middle Ages, last phase of the 'Prussian Culture' (late tenth/early eleventh century until the thirteenth/fourteenth century AD).	17, 18, 21, 32

Casimir regarding control of Galindian (e.g., Powierski 2001, 9–88), whilst the archaeological and palaeoenvironmental evidence suggests this region was not completely abandoned (Nowakiewcz and Wróblewski 2003; Nowakiewicz and Rzeszotarska-Nowakiewicz 2013, 10).

Samples for both micromorphological and geochemical analysis were collected from vertical profiles through pit feature 40 to examine similarities and/or changes in the nature of the occupation of the site associated with the Early Iron Age (or Western Balt Barrows Culture) and the early Middle Ages (or Prussian Culture). Geochemical samples from several other pit features were collected from profiles, which represent occupation deposits from the Late Iron Age to the first half of the thirteenth century (Table 14.1). Geochemical results for P (phosophorus), Ca (calcium), Cu (copper), and Zn (zinc) are the most interesting of all the XRF data from the Święta Góra pits (Appendix C), and their enrichment factors have been classified as very high (> the mean elemental enrichment factor at Święta Góra), high (>2.95), medium (2.50–2.94), and low (1.5–2.49). The mean enrichment factor at Święta Góra for P is 7.30, 4.95 for Ca, and 3.67 for Zn. The key geochemical results are summarized in relation to each phase (Figs 14.1–4). The results from the pit features from each phase are discussed in relation to their spatial and diachronic use, and the range of activities that are represented by the refuse in the pits and particularly their intensity of use.

Spatial and Diachronic Use of Pits

Feature 23 is the only pit with that is associated with Phase 1 (Fig. 14.1) and shows high to very high enrichments of P, Ca, and Zn between the depths of 28–67 cm, and a general decrease in element enrichments above 28 cm (Fig. 14.1), which could indicate that the intensity of use of the pit started to decline towards the end of the life of this feature.

The features associated with the Migration Period, Phase 2a (Fig. 14.2), are generally shallow features with single fills. The geochemical analysis highlights nuances in the intensity of use of specific pits and/or groups of pits. Features 19, 30, and 43 are have the highest geochemical enrichment factors of P, Ca, Cu, and Zn of the Migration Period group, and notably Feature 19 has enrichments of P that are nineteen times the amount of the background control samples (Fig. 14.2). The later material within Feature 30 has greater elemental enrichments, particularly in P, Ca, and Zn (Fig. 14.2), than those lower down the profile towards the base, which may indicate the deposition of coprolite or bone material towards the end of the use of this pit. Features 13, 31, and 39 also have high enrichment factors of P

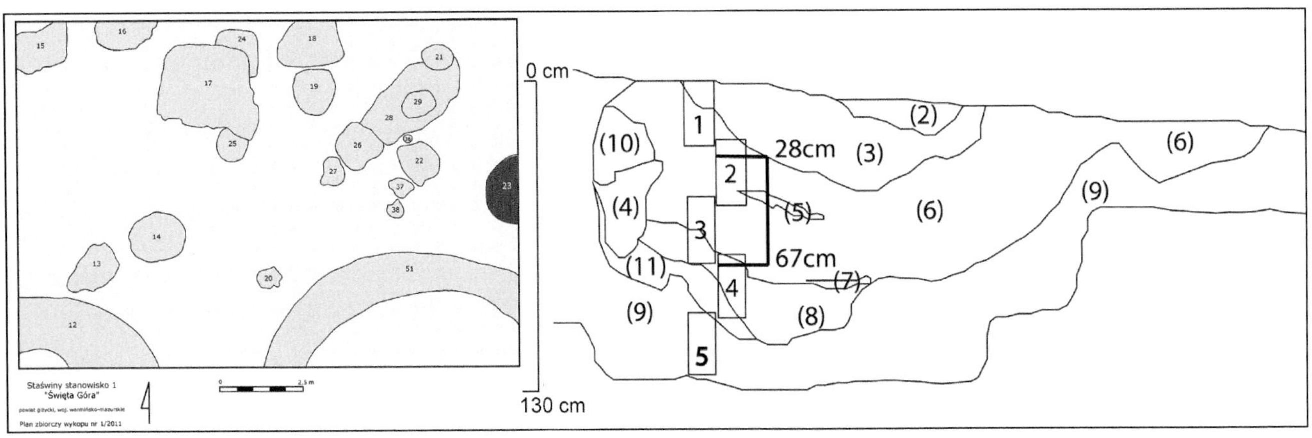

Figure 14.1: Left: Location of Feature 23 (dark grey) on the site plan of Święta Góra of all pit features (site plan by M. Karczewski and M. Karczewska). Right: Profile drawing of Feature 23 showing the location of monolith samples collected and sub-sampled for geochemical analysis. Figure by author.

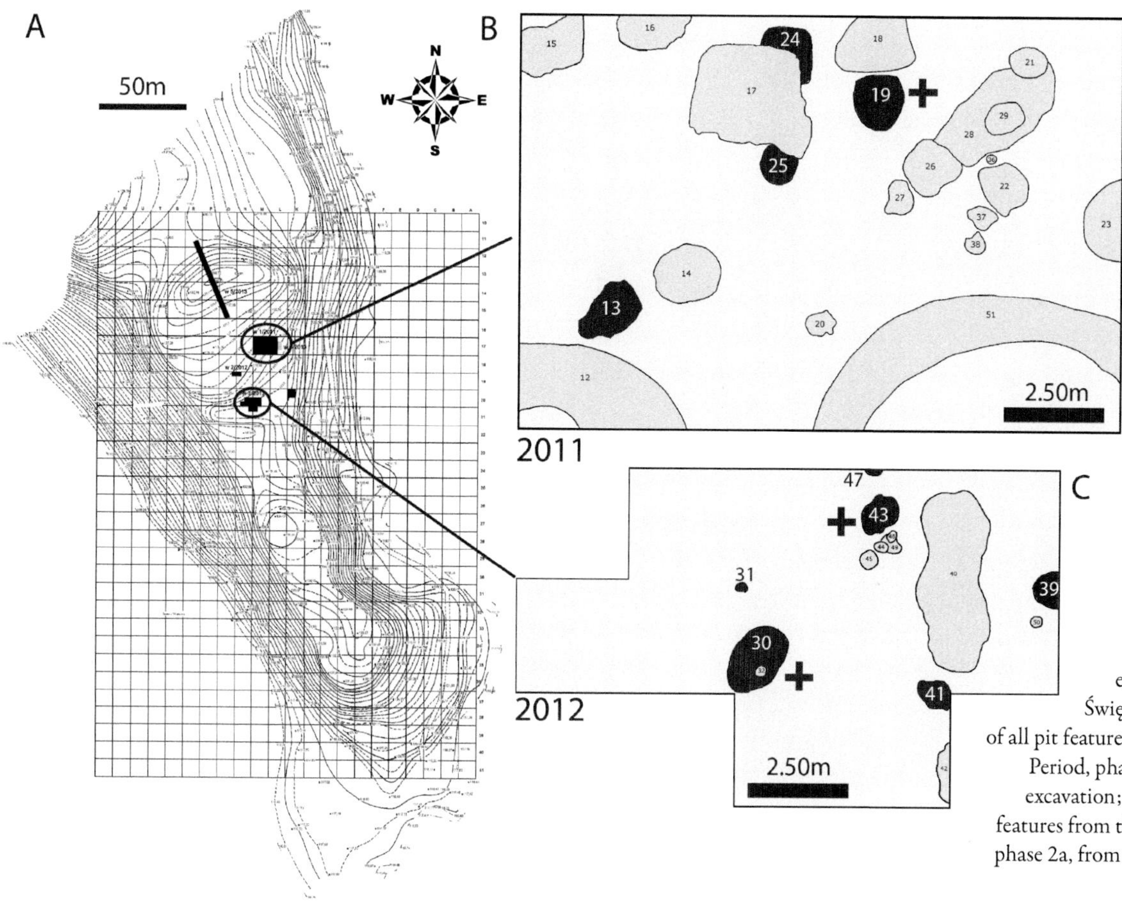

Figure 14.2: (a) The location of the 2011 and 2012 excavation trenches at Święta Góra; (b) location of all pit features from the Migration Period, phase 2a, from the 2011 excavation; (c) location of all pit features from the Migration Period, phase 2a, from the 2012 excavation. Figures by author.

and Ca, but not to the same extent as Features 19, 30, and 43, which suggests that they had a different use to these features and/or that they were not as intensively used. Feature 47 is the least intensively used with particularly low geochemical enrichments (Fig. 14.2).

As with Phase 2a, the geochemical analysis of features associated with Phase 2b highlights nuances in the intensity of use of specific pits and/or groups of pits from the early Middle Ages. Features 28 and 40, the largest pits associated with this phase (Fig. 14.3), have the greatest element enrichment factors, specifically P and Ca enrichments are up to fifteen times the amount of the background control samples (Fig. 14.3). The other pits from Phase 2b, Features 14, 15, 16, 26, 44, 45, 48, and 49 (Fig. 14.3), have high enrichment factors of P but unlike Features 28 and 40 did not have high enrichment factors of other elements such as Ca, Cu, and Zn, which again suggests that they had a different use to these Features 28 and 40, and/or that they were not as intensively used.

Feature 40 had an unusual shape and dimensions for a refuse pit; it is a wide (3.70 m), narrow (1.50 m), and shallow (0.60 m) pit, and its purpose was unclear from excavation alone. During excavation, four units were identified within Feature 40, and these were further separated into microstratigraphic units as result of micromorphological analysis (Banerjea et al. 2015b, Fig. 2; Fig. 14.5), of which microstratigraphic units 2a, 2b, and 1 all indicate that massive dumping events took place within a very short time period. If these fills had micro-laminated bedding structures, then it could be assumed that periodic accumulation of refuse took place (Banerjea et al. 2015a; Shillito and Matthews 2013). Massive dumping events may have taken place at the end of a prolonged period of occupation at the site, perhaps when the site was cleared. Micromorphological analysis has shown that Field Context 3 comprises laminations of mineral-rich sediment; microstratigraphic unit 3 [2], which are interspersed with thin discard deposits; microstratigraphic unit 3 [1], which are only 1–2 cm thick (Fig. 14.5). This sequence of events suggests that the sterile, mineral-rich sediment may have washed into the pit during periods when it was not in use, perhaps indicating a period in towards the tenth century when the site was temporarily unoccupied. The thin discard events, microstratigraphic unit 3 [1], may indicate very short periods of sporadic occupation.

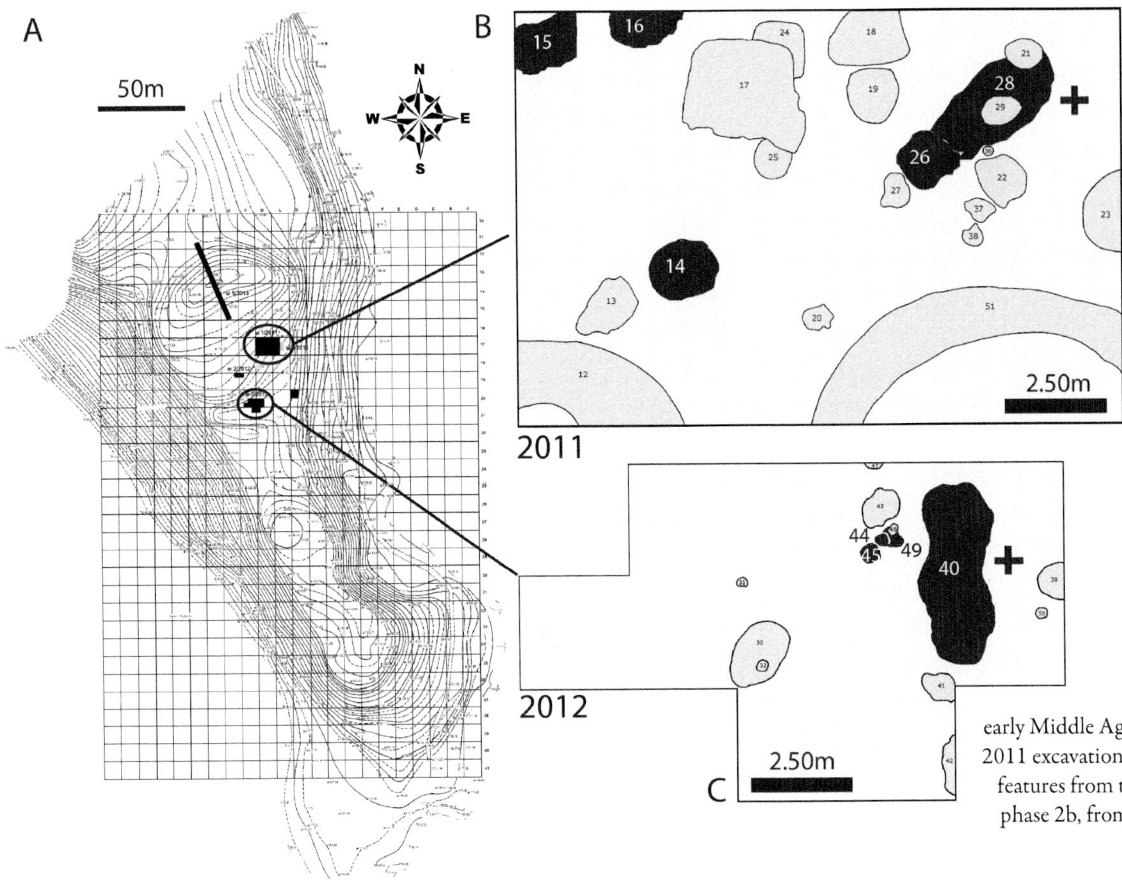

Figure 14.3:
(a) The location of the 2011 and 2012 excavation trenches at Święta Góra;
(b) location of all pit features from the early Middle Ages, phase 2b, from the 2011 excavation; (c) location of all pit features from the early Middle Ages, phase 2b, from the 2012 excavation.
Figures by author.

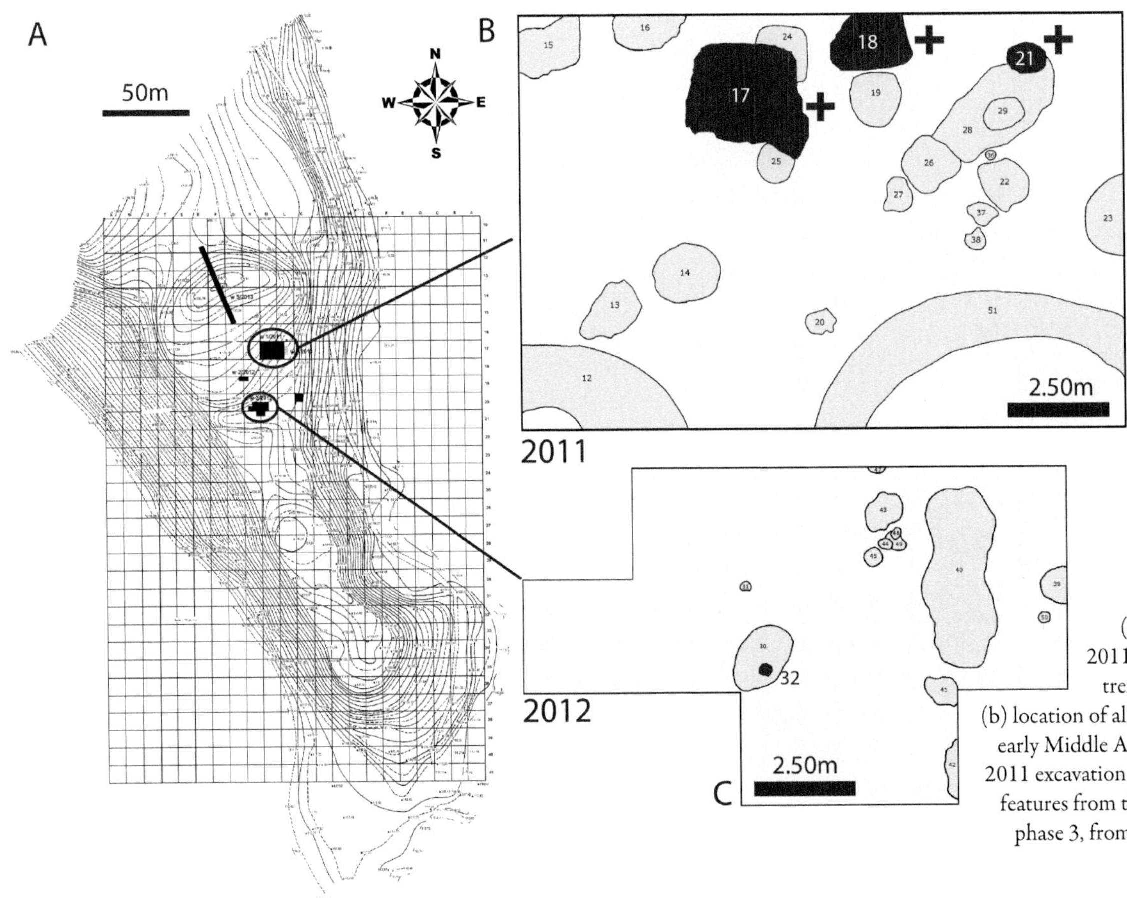

Figure 14.4:
(a) The location of the 2011 and 2012 excavation trenches at Święta Góra;
(b) location of all pit features from the early Middle Ages, phase 3, from the 2011 excavation; (c) location of all pit features from the early Middle Ages, phase 3, from the 2012 excavation.
Figures by author.

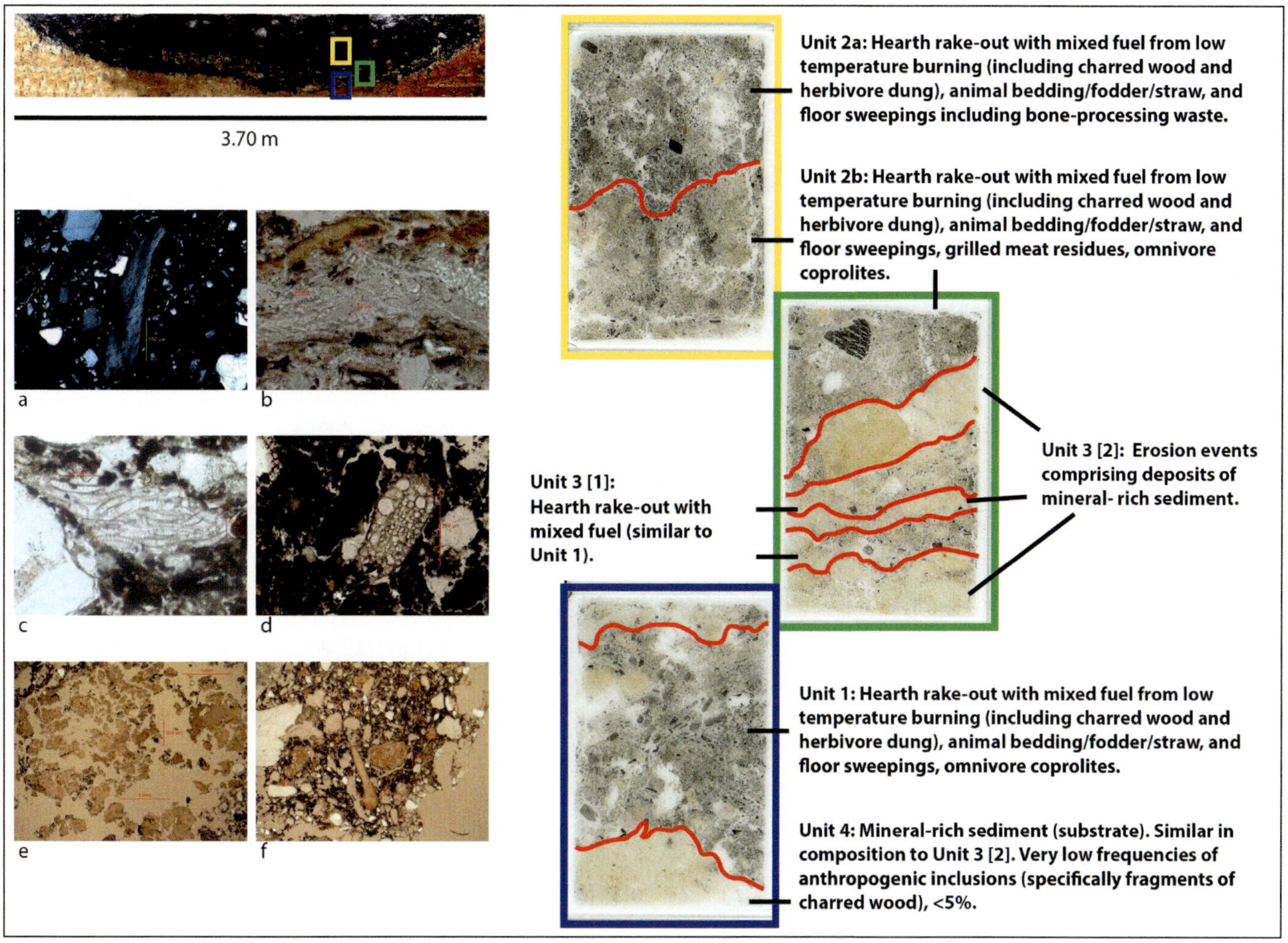

Figure 14.5: Top left: the location of micromorphology samples 295 (yellow), 296 (green), and 297 (blue) on the profile of Feature 40, Święta Góra. Left: (a) Photomicrographs of non-burnt bone, discard deposit 2a; (b) phytoliths, discard deposit 2a; (c) diatoms, discard deposit 1; (d) melted silica, discard deposit 3[1]; (e) omnivore coprolite fragments in discard deposit 2a; (f) discard deposit 1. Right: Interpretations of microstratigraphic units within samples 295 (yellow), 296 (green), and 297 (blue). Figures by author.

The inclusions within the discard deposits (microstratigraphic units 2a, 2b, 3 [1], and 1) most probably represent activities in spaces/buildings surrounding Feature 40 and have been deposited as a result of their maintenance. The lower discard deposits (microstratigraphic units 3 [1] and 2b) represent nearby cooking activities that involved the grilling of meat (see Chapter 12 for evidence of processing carcasses at the site). The deposition of unburnt bone fragments in discard deposit (microstratigraphic unit 2a), the final fill of Feature 40, suggests that the local activities around the pit may have changed, perhaps to butchery or bone processing (Banerjea et al. 2015b).

Much of the refuse has been burnt, which serves both as a method of disposal and fuel. Herbivore dung is amongst the burnt material. The use of dung as fuel is not uncommon, and its good burning properties are well established (Sillar 2000). The occurrence of herbivore dung indicates that livestock may have been penned nearby throughout the time-period that Feature 40 was in use. The origin (species) of the omnivore coprolites is unclear, although they bear a strong micromorphological likeness to those from humans (Macphail and Goldberg 2010, 607; Shillito et al. 2011, 1873).

In Feature 40, geochemical samples 305 A and B, and 299 A and B, which approximately relate to microstratigraphic units 2b and 1 respectively (Fig. 14.5), also have high enrichment indexes of Cu. These samples also showed the highest (microstratigraphic unit 2b) and second highest (microstratigraphic unit 1) enrichment indexes of Ca (9.05 highest, 6.60 second highest) and P (14.50 highest, 10.64 second highest), and samples 305 A and B also showed the highest Zn enrichments, 8.19 (Fig. 14.5). These high geochemical enrichments most probably reflect the inclusion of coprolite/latrine waste,

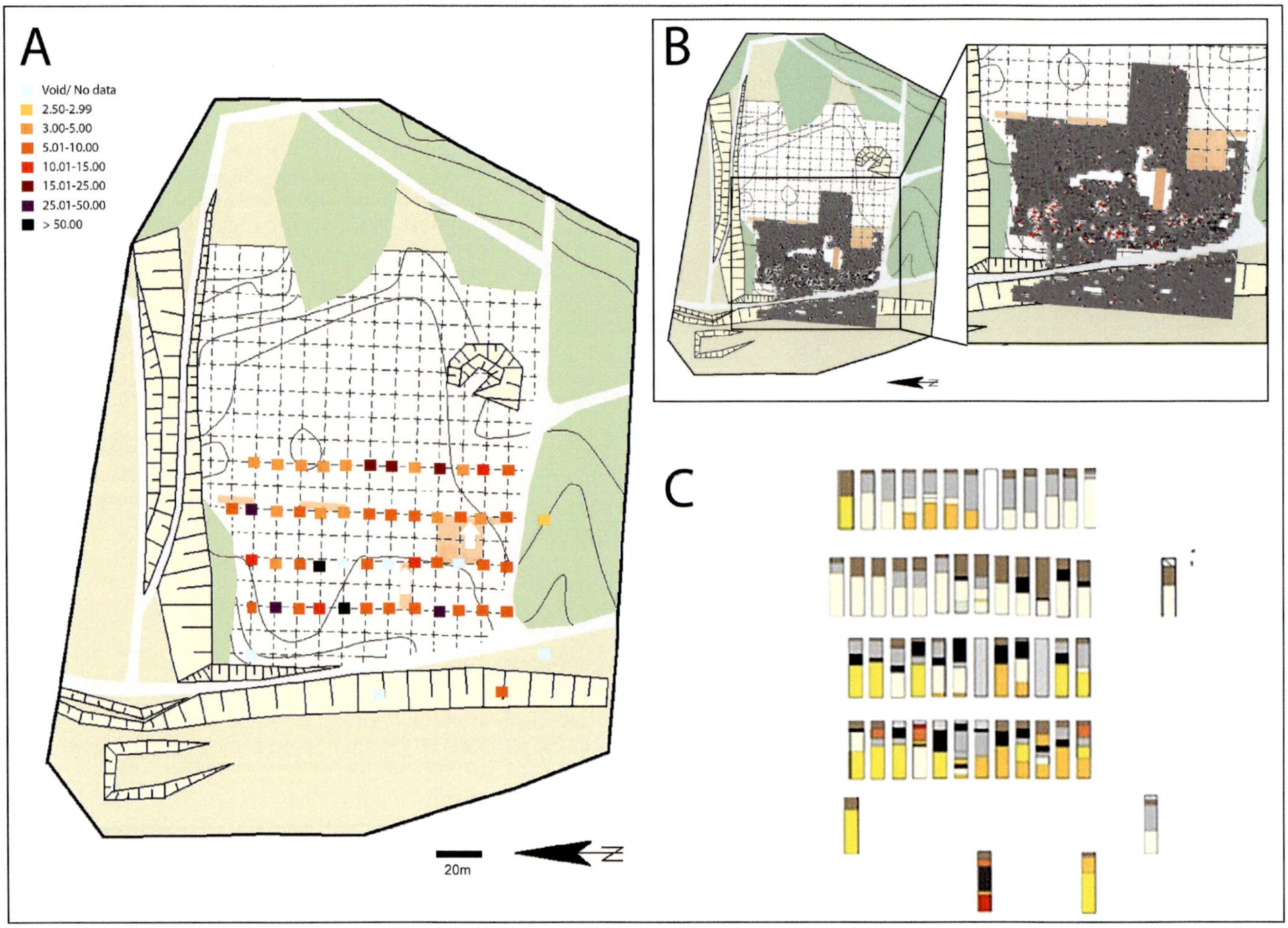

Figure 14.6: (a) Site plan of Biała Góra 3 showing the location of cores collected for geochemical analysis and their corresponding phosphate enrichment factors from the 'cultural layer', and excavation trenches (light orange); (b) geophysical anomalies within the 'cultural layer' highlighted by gradiometer survey; (c) Coring transect (after Pluskowski et al. 2014) showing the 'cultural layer' (black and grey), which varied in depth between 200–500 mm. Figures by author.

bone, hearth ash, dung, and plant materials (Fig. 14.5) in these units (Banerjea 2008; Dockrill et al. 1994; Hutson and Terry 2006; Middleton and Price 1996; Pierce et al. 1998; Robinson 2006, 212; Robinson 2011, 288; Terry et al. 2004; Wilson et al. 2005). The other pit features from the early Middle Ages, Phase 2 (Fig. 14.3), have high enrichment factors of P, but no particularly important enrichments of Ca, Cu, and Zn (Fig. 14.3), which may again suggest a different use or a less intensive use of these features.

In Phase 3, during the second half of the early Middle Ages, Features 17, 18, and 21 have the highest enrichment factors of P, Ca, Cu, and Zn (Fig. 14.4), which most probably can be attributed to excrement in these features, as coprolites were recovered from Features 17 and 18. Posthole Feature 32 has high levels of P, Ca, and Zn, which may represent ashes, perhaps from a burnt post.

Summary of Activities and Use of Pits

With the exception of Feature 23, which is a deeper pit and much earlier in date than the other features, and Feature 40, which contained several microstratigraphic units that indicated that massive dumping events took place within a very short time period, all other pits are fairly shallow with single fills. The geoarchaeological results suggest an interpretation of sporadic, short-term use of the site throughout the Migration Period and the early Middle Ages, Phases 2a, 2b, and 3. The microstratigraphic analysis of Feature 40 shows periods of intense use, followed by periods of disuse, which is consistent with this interpretation.

Phases 2a, 2b, and 3 all have groups of pits that either differed from the others in terms of the type of refuse that was deposited, or in their intensity of use. Those with particularly enriched levels of P, Ca, and Zn may contain coprolite material, as Features 17, 18, and 40 have indicated. As the micromorphological analysis of Feature 40 has shown, geochemical enrichments of pit fills could be attributed to the deposition of refuse from a range of activities such as the cooking and preparation of meat, the deposition of hearth ashes, which can include burnt herbivore dung, and the inclusion of coprolites.

Biała Góra

Biała Góra (Ger. Weißenberg) represents a unique example of an excavated medieval rural colony in the Vistula borderlands that sustained itself into the fourteenth century, which was occupied during the transitional period between Pomeranian and the Order's rule (Pluskowski et al. 2014; Sawicki et al. 2015). Biała Góra 3 was first discovered in 1972 during a field survey and was subject to excavations in 2007 and 2008 by the Department of Archaeology at the Castle Museum, Malbork, and in 2011 within the research framework of the Ecology of Crusading project (Pluskowski et al. 2014). Biała Góra 3 was founded in the late twelfth or early thirteenth century and falls in between the two expansionist phases. The excavations produced a range of material culture that demonstrate links with both Pomeranian and German colonists, including building materials of the kind associated with the monastic and military orders (Pluskowski et al. 2014). The diverse range of material culture at Biała Góra is unusual for a rural site as the composition of assemblage is otherwise associated with urban communities such as Gdańsk (Ger. Danzig) and Kołobrzeg (Ger. Kolberg) (Pluskowski et al. 2014).

Samples for micromorphology and geochemical analysis (Fig. 14.6a) were collected from the 'cultural layer' at Biała Góra 3. The 'cultural layer' was an organic-rich, dark grey sand deposit immediately below the current topsoil, and varied in depth between 200–500 mm (Fig. 14.6c). It contained significant quantities of artefacts and merged into various pit features. Micromorphological analysis has identified that the 'cultural layer' formed by similar processes to a rural 'dark-earth' under short grassland (Pluskowski et al. 2014). The study and definition of 'dark earth' has been a much debated issue during the last decades, and has seen its definition explored beyond characterizing thick, dark, homogenous units within urban stratigraphy that mark the transition between the late Roman and early medieval periods in the United Kingdom (Macphail et al. 2003; Yule 1990) to understand its origin and formation (Borderie et al. 2015; Nicosia and Devos 2014), and to understand the differences and similarities between European and Amazonian 'dark earth' (Arroyo-Kalin et al. 2003; Nicosia and Devos 2014). 'Dark earth' is not a scientific soil classification term, but as 'archaeological jargon' it is useful for identifying these diverse deposits on account of their common features (Nicosia and Devos 2014). Micromorphology has played an integral role in numerous interdisciplinary scientific studies that have examined the genesis and formation of 'dark earth' (Borderie et al. 2015; Devos et al. 2013). 'Dark earths' develop as a result of pedogenesis, which reworks the underlying parent material of occupation materials and archaeological strata, and provide a lens through which the former activities and land-use within an urban area can be ascertained (Borderie et al. 2015; Nicosia and Devos 2014); however, 'dark earths' need to be studied on a case-by-case basis as there is not one single model that explains them all, and, importantly, interpretations must be made on the basis of context (Goldberg and Macphail 2006, 271–73; Nicosia and Devos 2014).

It is possible that the 'cultural layer' represents a short period of intensive occupation that comprises a shallow spread of middening dating from, on the basis of the artefacts that were recovered, the mid-thirteenth to mid-fourteenth centuries. Microscopic occupation residues within the 'cultural layer' are altered by weathering and comprise highly degraded bone fragments, ceramic material (probably brick fragments), degraded mortar, and degraded charcoal within black amorphous fine material. Experimental research has demonstrated that occupation deposits in a temperate climate, which have been exposed to wind and rain as a result of failing roofs, can turn to soil within two and a half years (Banerjea et al. 2015a); therefore, it is plausible that pedogenesis within external occupation deposits in a temperate climate will be more rapid than those occupation deposits that have been more sheltered by the remains of a building.

The spatial distribution of phosphate enrichment (Fig. 14.6a) indicates areas of the 'cultural layer' (or 'dark earth' formation) that coincide with anomalies highlighted by gradiometer survey (Fig. 14.6b). Geochemical results for P, Ca, and Zn are the most interesting of all the pXRF data from Biała Góra 3 (Appendix D), and their enrichment factors have been classified as very high (> 5.00), high (>2.95), medium (2.50–2.94), and low (1.5–2.49). Enrichment factors have been calculated using surface geology control samples from Pleistocene sands, alluvial sands, silts, and clays, and Aeolian sands from around Biała Góra 3. The enrichment factors of P are predominantly very high (>5.00), with enrichments

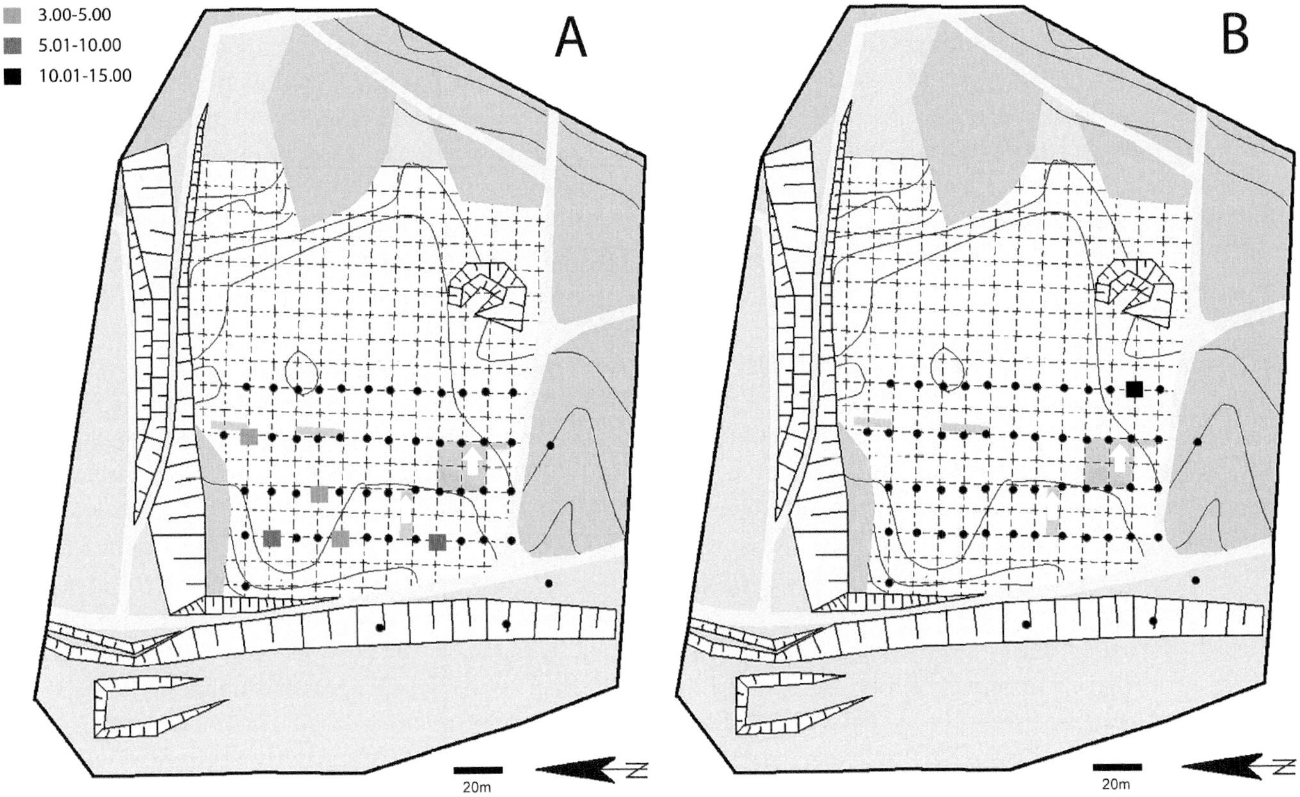

Figure 14.7: (a) Site plan of Biała Góra 3 showing the location of cores collected for geochemical analysis and their corresponding calcium enrichment factors from the 'cultural layer'; (b) site plan of Biała Góra 3 showing the location of cores collected for geochemical analysis and their corresponding zinc enrichment factors from the 'cultural layer'. Figures by author.

of up to 83.5 times the background geology in places. All enrichment factors, with the exception of three low or medium factors, phosphate enrichments, have been plotted incrementally using colour coding (Fig. 14.6a). Those enrichments of P that are greater than twenty-five times the background geology occur within samples from cores in an area of the 'cultural layer' that produced geophysical anomalies (Fig. 14.6b) that result from clusters or individual metal objects, and/or caches of brick and burnt stone. The metal objects included iron knives, iron and bronze rivets, iron boat sintels, military and equestrian equipment, buckles, belt fittings, and pieces of lead (Pluskowski et al. 2014). The enrichments of Ca that are high or very high correspond with some of those cores that had enrichments of P that are greater than twenty-five times the background geology (Fig. 14.7a), and the very high enrichment of Zn corresponds with a very high, but less than fifteen times the background, enrichment factor of P in the south-east corner of the site (Fig. 14.7b). It is probable that the geochemistry has identified depositional hot spots where, in addition to metal objects, other refuse that has since decayed was also dumped. The geochemistry probably represents microscopic occupation residues within the 'cultural layer' that have been altered by weathering and comprise highly degraded bone fragments, ceramic material (probably brick fragments), degraded mortar, and degraded charcoal within black amorphous fine material (Pluskowski et al. 2014). As at Święta Góra, high geochemical enrichments of P, Ca, and Zn most probably reflect the inclusion of coprolite/latrine waste, bone, hearth ash, dung, and plant materials in the 'cultural layer', and perhaps also Ca indicating the presence of decayed mortar (Banerjea 2008; Dockrill et al. 1994; Hutson and Terry 2006; Middleton and Price 1996; Pierce et al. 1998; Robinson 2006, 212; Robinson 2011, 288; Terry et al. 2004; Wilson et al. 2005).

Elbląg

In the absence of comprehensive excavation and preservation of early occupation phases, relatively little is known about the internal layout of the Order's thirteenth-century *castra* in the eastern Baltic (see Chapter 1). However, distinct spaces are specified in the Order's Rule, and written sources provide information on the variety of spaces within some of the larger convents from the fourteenth century (Jóźwiak and Trupinda 2012). The excavations that were conducted in 2013 ahead of development at Elbląg (Ger. Elbing) have produced the only example where the timbers have survived in situ from multiple phases of rebuilding. An exceptionally well-preserved and well-stratified sequence of organic-rich deposits that were interspersed with depositions of alluvial sediment was revealed. Timbers at the base of this sequence produced a dendrochronology date of *c.* AD 1245, which means that the lowest deposits within this sequence represent the very earliest occupation and use of the *Vorburg*, or outer bailey, of the early *castrum*. The organic-rich layers were attributed to external occupation that had regularly become inundated with flood waters; however, samples were collected for micromorphological, phytolith and geochemical analysis as their formation and composition was not fully understood through excavation alone. Phytolith samples and geochemical samples for ICP-OES analysis were collected from each stratigraphic unit within the profile (Table 14.2). Micromorphology samples were collected from the main stratigraphic units, which included the organic-rich deposits (Table 14.2). Samples 1–3 were collected with an overlap to prevent a break in the stratigraphic sequence (Fig. 14.8). Sample 4 was collected from lower alluvium and the organic-rich layer (Fig. 14.8). In addition to bulk elemental analysis, the resin-impregnated, 1 cm-thick micromorphological slabs (off-cuts) were analysed using an ITRAX XRF Core-scanner (Banerjea et al. 2017). This has enabled the in situ analysis of fluctuations in elemental concentrations to be examined at a higher resolution, 200 µm intervals, within this profile.

Sediment Formation Processes within the Earliest Outer Bailey

Micromorphological analysis has further identified additional microstratigraphic units within the sequence of contexts that were identified during the excavation of Trench 2 in 2013 (Table 14.2; Fig. 14.8). The analysis shows that context 26 at the base of the sequence is three microstratigraphic units comprising mixed lenses of trampled herbivore dung and alluvium. These mixed lenses (26i–26iii) most probably formed as a result of animals trampling in the earliest outer *bailey* while low level flooding took place, which deposited alluvial sediment that was trampled by animals. These lenses are then overlain by a more substantial deposit of alluvium, context/microstratigraphic unit 25, which formed as a result of more substantial flooding of this area of the earliest *Vorburg*. On top of this alluvial deposit, an earth-based occupation surface formed, microstratigraphic unit 23b, which has been reworked, most probably due to churning under damp conditions. The occupation surface contains occupation detritus such as fragments of charred wood and phytoliths, and it shows substantial effects of weathering, decay, and bioturbation processes (Table 14.2), including the occurrence of iron sulphide framboids and perhaps indicates a dry period in the area. Sedimentary framboidal pyrite commonly occur in marine and perimarine environments and form at the oxic-anoxic interface. The formation of these sulphides requires a source of ferrous iron (Fe(II)), a source of sulphur and a reducing environment (Kattenberg and Aalbersberg 2004, 234). The occupation surface is overlain by more alluvium, unit 23a, although this is less substantial in thickness than unit 25. These flooding episodes may relate to inter-tidal estuarine regimes linked to a previous stage when the castle was located on the riverbank (Nitychoruk et al. 2012).

Units 20b and 20c represent two possible floor-raising attempts. Unit 20b seems to be a 'brush wood' levelling platform that was laid when the lower constructional deposit, unit 20c, began to decay. The 'brush wood' (branches from small trees and shrubs) material may have been laid so that herbivores could be kept in the area. The 'brush wood' platform is overlaid by herbivore dung that is in situ, unit 18b, and the top of the dung layer has begun to degrade under wet conditions to form a dopplerite stabling crust, unit 16a4 (Stolt and Lindbo 2010, 385–86). The area is then inundated again with alluvial sediment, unit 16a3, and when this has stabilised leaf litter, unit 16a2, accumulated. The microstratigraphic sequence ends with a deposit of discarded material, unit 16a1, that represents the abandonment of the area. This discarded material comprises brick, daub, and burnt bone inclusions, as well as organic components such as leaves, charred cereals, wood, bark, and charred wood (Table 14.2).

The Characterization and Development of Activities within the Earliest Outer Bailey

Integrated micromorphological, phytolith, and geochemical analyses have identified traces of dung from livestock within the early occupation phases of the *Vorburg* (Table 14.2). The micromorphological evidence comprises fragments of herbivore dung pellets in units 26i–iii that contain calcareous faecal spherulites indicative of the presence of caprids (Canti 1999), and in situ

Table 14.2: Summary of integrated bulk geochemistry, phytolith, and microstratigraphic results (surface to base), Elbląg. Geochemical enrichment factors >2.95 are considered high.

Field Context Number	Key bulk geochemistry enrichment factors	Summary of phytoliths	Monolith Sample Number	Microstrat classification	Microstrat unit	Particle size
16a	Ca: 5.00 P: 6.49 Sr: 3.27	Single cells: Monocots 97% Dicots 3% Multi-cells: Monocots 95% Dicots 5% Multi/single cell ratio 1:16.2 Mainly leaf/stem of tall grass species with husks well represented	1	Discard/ abandonment debris	16a1	Loamy sand
				Decayed leaf litter	16a2	Sand/ loamy sand
				Alluvium	16a3	Sandy clay loam
				In situ stabling crust	16a4	Peat
18b	P: 5.41	Single cells: Monocots 99% Dicots 1% Multi-cells: Monocots 95% Dicots 5% Multi/Single cell ratio 1:7.4 Mainly leaf/ stem of tall grass species.	1/2	*In situ* oxidised dung	18b	Loamy sand/ fine sand
20b	Ca: 4.64 K: 4.76 P: 7.15 Sr: 3.56	Single cells: Monocots 88% Dicots 12% Multi-cells: none Mainly leaf/stem of tall and short grass species, also tree/shrub leaves.	2	Constructiona/ levelling wood	20b	Very coarse gravel-sized wood fragments (horizontal extent beyond slide).
20c	K: 5.08 P: 5.15	Single cells: Monocots 69% Dicots 31% Multi-cells: Monocots 100% Multi/ Single cell ratio 1: 179 Lowest overall count of phytoliths. Mainly leaf/stem of tall grass species, and tree/shrub leaves.		Decayed constructional/ levelling wood	20c	Loamy sand
23	K: 3.39, Ti: 3.34	Single cells: Monocots 85% Dicots 15% Multi-cells: Monocots 100% Multi/ Single cell ratio 1: 94.5 Mainly leaf/stem of tall grass species.	2/3	Alluvium	23a	Sandy clay/ Sandy clay loam
				Reworked occupation surface	23b	Sandy clay loam
25	P: 4.15	Single cells: Monocots 89% Dicots 11% Multi-cells: Monocots 75% Dicots 25% Multi/Single cell ratio 1: 36.4 Mainly leaf/stem of tall grass species.	3/4	Alluvium	25	Sandy clay loam/loamy sand
26	Ca: 5.68, Cu: 3.89, P: 18.6, Sr: 5.94, Zn: 4.84	Single cells: Monocots 64% Dicots 36% Multi-cells: Monocots 100% Multi/ Single cell ratio 1: 5.5 Mainly leaf/ stem of tall grass species with husks represented, also dicots from twigs/ branches/foliage.	4	Mixed: trampled dung and alluvium	26i	Loamy sand
				Mixed: alluvium and trampled dung	26ii	Sandy clay loam
				Mixed: trampled dung and alluvium	26iii	Loamy sand

nclusion categories	Key inclusion types	Key post-depositional alterations
ological 38–39%, Micro-artefacts/agg 17%, arch <2%, Organic/plant 43–44%, crofossil 0%	Brick, Daub, Charred cereal grains, Wood, Charred wood, bark, Leaves (shrubs)	Silty clay coatings, unlaminated 5–10%, Organic staining 10–20%, Iron sulphides 2–5%
ological 52–59%, Micro-artefacts/agg -17%, Bioarch 0%, Organic/plant 31%, crofossil 0%	Phytoliths in dung, Wood, Leaves, Grass seeds, Amorphous organics	Silty clay coatings, unlaminated 5–10%, Vivianite neomineral formation 5–10%, Organic staining 5–10%.
ological 58–63%, Micro-artefacts/agg 7%, Bioarch <5%, Organic/plant 28–32%, crofossil <2%	Mainly quartz and glauconite, Burnt bone, Ferruginous and amorphous organics	Silty clay coatings, unlaminated 2–5%, Vivianite neomineral formation 5–10%, Organic staining 5–10%, Iron sulphides 2–5%
ological 28–29%, Micro-artefacts/agg 5–6%, arch <5%, Organic/plant 62–63%, crofossil <2%	Leaves, Ferruginous and amorphous organics, Phytoliths, including dendritics, Grass seeds, *Ascaris* spp, parasite eggs.	Organic staining>20%, Iron sulphides 2–5%, Dopplerite >20%
ological 58–60%, Micro-artefacts/agg , Bioarch <2%, Organic/plant 36–38%, crofossil <5%	Leaves, Bark, Ferruginous and amorphous organics, Phytoliths, including dendritics, Grass seeds, Charred seeds — probably blackberry, Ascaris spp, parasite eggs.	Silty clay coatings, unlaminated 5–10%, Organic staining 5–10%.
ological 16%, Micro-artefacts/agg 0%, arch 0%, Organic/plant 84%, Microfossil 0%	Mainly deciduous wood, Bark, Amorphous organics.	Organic staining 5–10%. Iron sulphides 2–5%
ological 26–30%, Micro-artefacts/agg 6, Bioarch <2%, Organic/plant 60–62%, rofossil 0%	Mainly amorphous organics, Wood, Bark	Organic staining 10–20%, Iron sulphides, <2%
logical 68–88%, Micro-artefacts/agg Bioarch 0%, Organic/plant 12–30%, rofossil <2%	Mainly quartz, Coprolite (small fragment), Organics	Silty clay coatings, microlaminated 5–10%, unlaminated 2–5%, Vivianite neomineral formation, 10–20%, Organic staining 5–10%
logical 66–76%, Micro-artefacts/agg 6–20%, rch 0%, Organic/plant 12–18%, rofossil <2%	Mainly quartz, Daub, Charred wood, Ferruginous and amorphous organics	Silty clay coatings, microlaminated 5–10%, unlaminated 5–10%, Vivianite neomineral formation, 5–10%, Organic staining 10–20%, Iron sulphides 2–5%
logical 70–74%, Micro-artefacts/agg 15%, rch 0%, Organic/plant 11–15%, rofossil 0%	Mainly quartz and glauconite	Microlaminated silty clay coatings, 5–10%, Vivianite neomineral formation, 10–20%
logical 25%, Micro-artefacts/agg 2%, rch 6%, Organic/plant 63%, Microfossil 4%	Amorphous organics, Non-charred seeds, Leaves, Bark, Charred wood, Fish gills	Organic staining, >20%, Iron sulphides, 5–10%, Dopplerite, 2–5%.
logical 76%, Micro-artefacts/agg 3%, rch 2%, Organic/plant 15%, Microfossil 4%	Mainly quartz and glauconite	Organic staining, 5–10%.
logical 28%, Micro-artefacts/agg 15%, rch 0%, Organic/plant 46%, Microfossil 11%	Herbivore dung, Faecal spherulites, Ascaris spp, parasite eggs, Phytoliths including dendritics	Organic staining, 5–10%, Sodium silicate neoformation, 5–10%

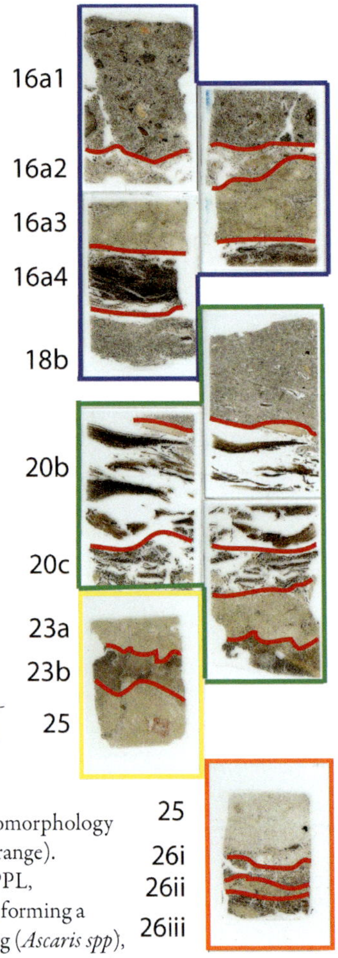

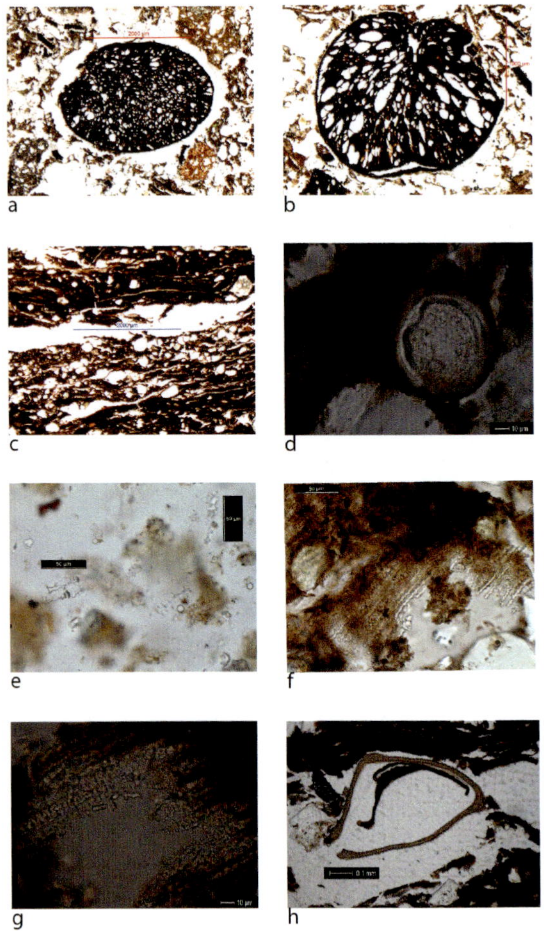

Figure 14.8: Left: The location of micromorphology samples 1 (blue), 2 (green), 3 (yellow), and 4 (orange) on the profile through the earliest *Vorburg*, Elbląg. Centre: slide scans and their associated microstratigraphic units within micromorphology samples 1 (blue), 2 (green), 3 (yellow), and 4 (orange). Right: (a) photomicrographs of charred seeds, PPL, including: (b) cereal grains; (c) unit 16a1, dung forming a dopplerite crust, PPL; (d) unit 16a4, parasite egg (*Ascaris spp*), PPL; (e) unit 16a4, single dendritic cells, grass husk phytoliths, and spherical (or globular psilate) phytoliths from broad-leafed foliage and the bark of twigs; (f) and (g) unit 26 phytoliths within herbivore dung, PPL (f and g), unit 18b. Figures by author.

cattle or horse dung (units 16a4 and 18b), quite possibly horse given the absence of calcareous faecal spherulites (Canti 1999), in the form of elongated organic fragments, which are oriented with a parallel alignment to the basal boundary, and contain roundworm (*Ascaris* spp) parasite eggs (Fig. 14.8; Pichler et al. 2014). As observed in both modern and archaeological organic-rich dung deposits, where livestock feeds primarily on grasses, their dung is enriched in fibrous organic and inorganic particles that upon trampling rearrange perpendicular to the direction of the force resulting in the formation of sub-parallel microlaminations, which preserve as linear arrangements of opal phytoliths that remain articulated after the ingested organic grass fibers have degraded (Shahack-Gross 2011, 210). A stabling 'crust' comprising dopplerite (unit 16a4) has formed at the surface of the in situ cattle/horse dung layer, and shows comparable microstratigraphic (Fig. 14.8) and geochemical (Table 14.2) attributes with those that were identified within experimental stabling deposits, which was an organic-rich horizon with high levels of P and a strong association between Ca, K, and P consistent with the identification of hydroxyapatite by XRD (Macphail et al. 2004, 179–80).

The dung traces in microstratigraphic units 26i–iii suggest that the earliest animals to be kept in this area were caprids; whereas later, in microstratigraphic units 18b and 16a4, the dung traces either belong to cattle or horse. Cattle, and sheep/goat are present in the zooarchaeological assemblage from the castle (Chapter 12). The ITRAX XRF data shows peaks of P and Zn within the in situ stabling crust and dung (Banerjea et al. 2017). The geochemical spot samples (1–7) show element enrichments throughout the profile, presented here in the form of an enrichment factor (Table 14.2), a ratio calculated against the average for the local geology (see Chapter 2). Sample 7, which corresponds to the stabling crust layer (microstratigraphic unit 16a4), shows enriched levels of P, Ca, and K, as also observed

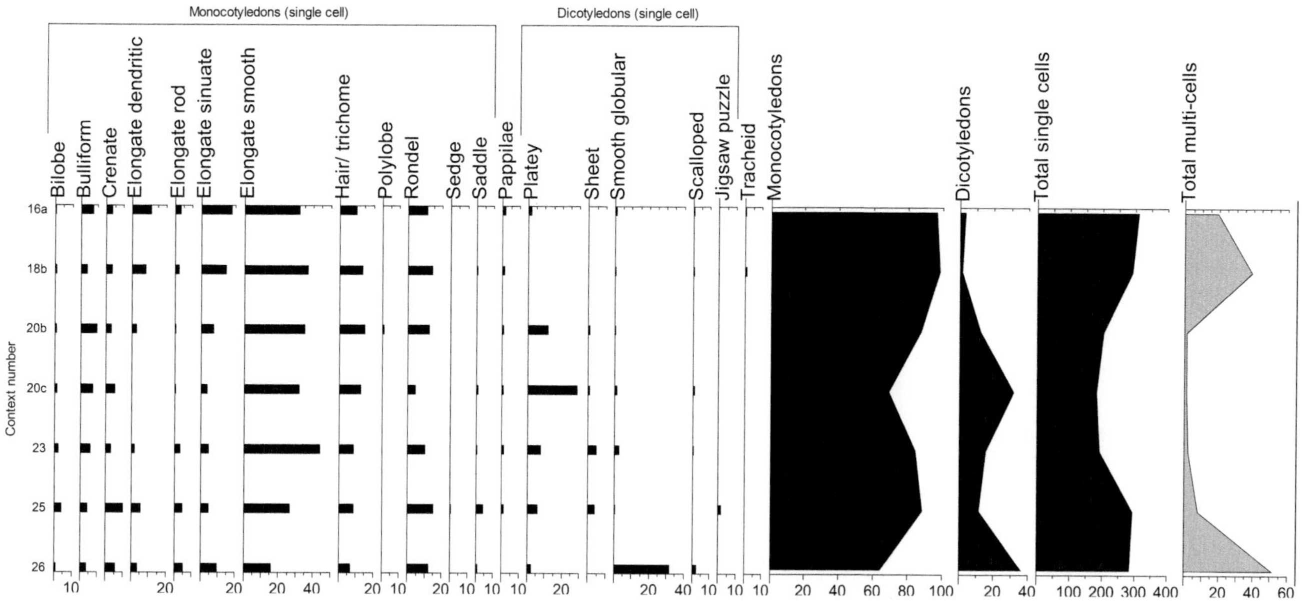

Figure 14.9: Percentages of phytolith types within samples collected from contexts within the profile through the earliest *Vorburg*, Elbląg. Figure by author.

experimentally (Macphail et al. 2004, 179–80), which are, respectively, 6.49, 5.0, and 2.6 times the background geology, as well as trace element enrichments (Cu, Ni (nickel), Pb (lead), and Zn), which can occur in cess layers (Wilson et al. 2005).

Changes and differences in livestock alimentation have been identified in the phytolith assemblages (Fig. 14.8) from the in situ dung (16a and 18b) and the trampled lenses of herbivore, caprid pellets (26). The phytolith assemblage from the trampled lenses of herbivore, caprid pellets at the base of the sequence comprises arboreal (dicotyledon) and grass (monocotyledon) phytoliths (Fig. 14.9; Table 14.2), particularly spherical (or globular psilate) forms that occur with broad-leafed foliage and the bark of twigs (Delhon et al. 2008; Piperno 2006). Dicotyledon phytoliths are under represented in the phytolith record as monocotyledons can produce up to twenty times more phytoliths than dicotyledons (shrubs/trees), and so a direct comparison between monocots and dicots does not represent the true ratio of grasses to trees/shrubs (Tsartsidou et al. 2007). Therefore, woody species are very likely to be much more prolific than indicated in the phytolith record, and could indicate the use of leaf fodder in addition to grasses (Delhon et al. 2008) during the earlier occupation at the castle, represented in unit 26 (Fig. 14.9).

Later on, higher up the profile in units 16a and 18b (Fig. 14.8), the morphology of the dung changes, resembling that of cattle or horse, and the phytolith assemblage predominantly comprises monocotyledons from grasses, particularly elongate smooth cells from the leaf/stem (Fig. 14.9; Table 14.2). The occurrence of dendritic phytoliths from wild and/or domesticated cereal husks, which include multi-cell formations in thin-section that bear a strong morphological likeness to oat, barley, and rye (Fig. 14.9), may represent the use of grain as fodder. The short cell assemblage is dominated by rondels, which are representative of C3 grasses, which are those species that prefer moist and wet conditions in well-watered habitats (Vincentini et al. 2008). Bilobes and saddle phytoliths also occur, but in very low frequencies (Fig. 14.9). Bilobes are produced by the C4 panicoideae grasses (Twiss 1992; Barboni et al. 1999) and represent tall grasses such as reeds, and saddles are produced by C4 chloridoideae grasses and represent either short grasses or common field weeds. Papillae are also represented in units 16a and 18b (Fig. 14.9). Papillae occur in the husk of grasses (wild and cereal). The range of phytolith morphologies within dung units 16a and 18b from wetland species, field grasses and weeds, and cereal grasses suggest that horses (or possibly cattle) were both put out to pasture in an area close to water and also fed on grain, as indicated by the presence of multi-cell dendritic forms from cereal husks.

Multi-proxy analysis of a thirteenth century horse stable within a medieval bailey in Veselí nad Moravou, Czech Republic (Dejmal et al. 2014), provides an interesting comparative case study for the potential horse stabling evidence, and livestock alimentation at Elbląg. There are similarities between the Elbląg and Veselí sequences in the formation of the stabling layers, and the location of the stable close to water as evident by flooding depositing alluvium. At Veselí, horses were fed meadow grasses as well as woody vegetation, millet, oat, and less commonly hemp, wheat, and rye (Dejmal et al. 2014), which is similar to the mixed fodder evidence from units 16a and 18b at Elbląg. A 'collapsed hurdle fence' was identified at Veselí (Dejmal et al. 2014), and it is plausible that the 'brush wood' platform at Elbląg could also be collapsed fencing.

The Geochemistry of Moats: Effluent and Pollution at Malbork

Artificial or natural moats protecting the Order's fortified houses in Prussia and Livonia had functions beyond those of security, the most important of which was to flush effluent from castle latrines. The date of construction of the outer moat at Malbork (Ger. Marienburg) is well established historically to *c.* 1320–40; a sample from 71.5–72 cm was dated to 430±30 BP (*c.* 1410–1520, 88.2%), consistent with the historically attested use of the moat in the final phase of the Order's rule. However, an earlier date obtained from seeds from 94 to 100 cm of 899±29 BP (*c.* 1040–1212) predates the construction of the castle, suggesting the likely erosion and inclusion of earlier material into the moat (see Chapter 11). Twelve spot samples were collected from a core taken through the moat outside the north-eastern walls of the castle's outer bailey. The results of the XRF geochemical analysis (Appendix E) show no enrichments trace elements (Ni, Cu, Pb, Zn, Sr (strontium)) that are associated with human activities such as the deposition of cess, industrial, domestic waste; however, there is an isolated peak in P enrichment (3.65) at a depth of 0.8–0.84 m, which may indicate an accumulation of effluent or organic material towards the surface of a series of organic rich silt layers (Cook et al. 2014; Wilson et al. 2005). Pollen from the outer moat adjacent to the inner curtain wall focused on a series of organic rich silt layers between 72 and 117 cm. These are considered to reflect the ponding of fine sediment within a low energy environment, separated by sand bands deposited under higher energy environments, perhaps related to periodic flushing of the moat, which may be reflected by fluctuations in Zirconium (Zr) enrichments during these depths, with higher enrichments reflecting coarser, sandier horizons (Appendix E). Pollen within the moat (see Chapter 11) is characterized by high non-arboreal pollen values, particularly grasses but also a range of cultivated plants and members of the mustard and carrot family perhaps reflecting herbs and vegetables. Areas of pasture and disturbed ground are suggested by pollen of grasses, ribwort plantain, buttercups, and the daisy and lettuce family. Arboreal pollen is represented mainly by pine growing within a predominantly open landscape, with stands of alder, willow, and birch more likely to be growing in wetter areas. The lack of significant geochemical enrichments within the moat sediments could suggest that there were no significant inputs of effluent into the moat at this time from nearby large scale agricultural, industrial, or domestic activities.

Geochemical Traces of Sites in their Territories

Radzyń Chełmiński and the Kulmerland

The close association between the lakeside and later medieval settlements at Radzyń Chełmiński (Ger. Rehden) presents the ideal opportunity to reconstruct past vegetation and land-use dynamics within a well-documented archaeological, historical, and cultural context. Previous studies of lake sediments have emphasized how the geochemical record can be influenced by a variety of within-lake and outside-lake processes, requiring careful consideration when interpreting the geochemical data (Brown et al. 2015). Identifying the traces of specific activities from the geochemical record in lake sediments can be problematic due to factors such as: element loading of lake sediments as a result of atmospheric and erosion inputs: lateral fluxes of P; and within-lake processes such as the recycling of trace elements, redox cycling, and the quantities of biogenic silica (Brown et al. 2015). However, interpretation of the geochemical data are strengthened in this case study by both the close proximity of the lake to the castle and the use of Ti (titanium), which is both chemically stable and unaffected by biological transformations (Lomas-Clarke and Barber 2007; De Vleeschouwer et al. 2009), as a Ti ratio as a soil mineral tracer, whereby the anthropogenic fraction of trace element enrichment can be calculated by subtracting the background fraction from the total enrichment (Boyle 2001); the remaining fraction represents the natural trace element supply. Three broad phases of increasing landscape impact are visible in the pollen data, augmented with multi-element geochemistry, dating from the eighth/ninth, tenth/eleventh, and thirteenth centuries, reflecting successive phases of Slavic and German colonization within the Kulmerland. These integrated data display a strong correlation in evidence for the effects of increasing colonization and agricultural activity in the surrounding landscape over the course of the medieval period. Although the late medieval castles, towns, and rural settlements of the Kulmerland are the most visible traces of medieval colonization and economic intensification, the most significant phase of ecological activity in the pollen and geochemical record occurs during the early medieval period associated with Polish control of the Kulmerland from the latter half of the tenth century. This is surprising given the close proximity of the pollen sequence to the castle and town established shortly after the start of the crusades. Instead, the pollen and geochemical records suggest that the crusaders and German colonists entered an already significantly altered landscape, with the late medieval period characterized instead by broad stability in the intensity of land-use (Brown et al. 2015).

A Summary of the Impact of the Crusades in Prussia at the Microscale

The Initial Arrival of the Teutonic Order

The geoarchaeological analyses presented above have illustrated that the early occupation associated with the arrival of the Teutonic Order varied in its intensity depending on the geographical region. The Kulmerland, considered to be the first frontier and an unstable region for many decades into the thirteenth century, remained under the Order's control after the initial phase of German colonization (1230–*c.* 1343) until 1466 (Pluskowski 2013, 112–14). The pollen and geochemical records from Castle Lake at Radzyń Chełmiński suggest that the crusaders and German colonists entered an already significantly altered landscape, with the late medieval period characterized instead by broad stability in the intensity of land-use (Brown et al. 2015). These integrated datasets suggest that the most significant phase of ecological activity in the pollen and geochemical record occurred during the early medieval period, associated with Polish control of the Kulmerland from the latter half of the tenth century. At Radzyń, political authority was centred on the stronghold overlooking the south-western shore of the lake.

In Masuria, although the archaeological and palaeoenvironmental evidence suggests that parts of Galindia were not completely abandoned in the pre-crusade period (Pluskowski 2013, 58), the geochemical and micromorphological evidence from pits suggests that the intensity of the occupation and range of activities at Święta Góra does not change with the arrival of the Teutonic Order in the vicinity, but continues to be fairly low key, sporadic, perhaps seasonal, occupation. This geoarchaeological evidence along with the zooarchaeological evidence, which shows that that animals were slaughtered, butchered, cooked, and consumed up on the hill of Święta Góra, including large quantities of fish (Chapter 12), and the absence of archaeological evidence for substantial structures may suggest an intermittent use of the site for feasting or hunting purposes, with the main settlement focus located in nearby Staświny. The nature of this type of occupation is comparable with the use of specific sites such as Neolithic causewayed enclosures, some of which may have served as places of aggregation for a dispersed population, with deposits of human remains, finds of non-local artefacts, whole or partial animals, and the residues of feasts. These were also frequently located close to water; for example, in Denmark such monuments were located in prominent locations overlooking river valleys or fjords (Bradley et al. 2015, 99–100). Here, it is not argued that the site at Święta Góra is a causewayed enclosure, but rather that the evidence suggests that the site may have been used in a similar way, and represents a continuity in specific places in the landscape holding significance in the way in which some causewayed enclosures did. It is arguable that the site at Święta Góra was one location, a focal point, within a wider pre-Christian sacred landscape comparable with the use of natural places during the early medieval period in Britain and Scandinavia (Semple 2010). Natural places have an archaeology because they acquired significance in the minds of people and retained this significance through possessing special meaning and becoming a focal point, as well as attracting the construction of monumental architecture and occupation around them (Bradley 2000). A sacred place can achieve this significance through its natural distinctive environs, accessibility, visibility, surrounding archaeology, and also through its contrast with the wider natural setting (Semple 2010, 25). The hill-top location of Święta Góra, its proximity to Lake Staświny, as well as the absence of identifiable buildings (to date) support the interpretation that this site served as an outdoor meeting place away from bounded or roofed spaces (Semple 2010, 42). Hill-tops are widely accepted as an indisputable aspect of pre-Christian sacred locales (Semple 2010, 27), and the nature of the occupation, particularly the absence of evidence for permanent or prolonged inhabitation, at Birutė Hill in Palanga, western Lithuania, is an example of a comparable focal point that was visited, most probably by people living in surrounding settlements (Žulkus 2007, 58–59). Watery places represent special loci in the landscape for pre-Christian populations, from prehistoric periods through to the early Middle Ages (Vaitkevičius 2004; Lund 2010; Semple 2010; Bradley et al. 2015).

The dates and timescales of post-depositional soil forming processes, based on experimental archaeology observations (Banerjea et al. 2015a), indicate that the cultural layer (the 'dark earth' type layer) at Biała Góra represents a short period of intensive occupation (Pluskowski et al. 2014). Soil forming processes could have begun within two and a half years between the mid-thirteenth to mid-fourteenth centuries, suggesting a point at which this area of the site was no longer in use, and may be consistent with a gradual shrinkage of this rural colony which was finally abandoned by the late fourteenth or early fifteenth century (Pluskowski et al. 2014). Nearby at Elbląg, the microstratigraphic sequence from the earliest *Vorburg* shows persistent occupation relating to animal husbandry in this area, despite repeated flooding episodes (as evident by deposits of alluvium) from AD 1245 until the destruction layers at the top of the sequence, unit 16a1 (Fig. 14.7), which may be related to an intensive fire destroying the initial timber structures at the beginning of the fourteenth century, after the earlier obliteration of the town by fire in *c*. 1288 (Fonferek forthcoming). The micromorphological evidence shows the presence of caprids in this location early on, which are subsequently replaced by the stabling of larger herbivores, probably horses.

The Use of Settlements and Castles

Multi-proxy geoarchaeological analyses have identified animal husbandry, on-site subsistence, and refuse disposal practices that relate to pre-crusade Pomeranian colonization and the arrival of the Order at Biała Góra (Sawicki et al. 2014), the early colonizing activities of the Order at Elbląg, and the presence of the indigenous population at Święta Góra until at least the early phase of contact with the Order, or incoming colonists. The evidence for animal husbandry and alimentation from the earliest *Vorburg* at Elbląg provides revealing insights into the types of livestock that arrived with the Order, how they were foddered, and about the use of castle wards during the period of active crusading. The in situ evidence for livestock within this area lends support to descriptions of pastoral farming regimes in later written sources, that record animals being taken from rural sites into towns and castle baileys 'on the hoof' to be slaughtered by butchers on site (Pluskowski 2013, 321). In the earliest trampled lenses at the base of the profile, woody species are very likely to be much more prolific and could indicate the use of leaf foddering of caprids in addition to grasses (Delhon et al. 2008) during the earliest colonizing occupation at the castle. The presence of caprid dung provides complimentary evidence for the importance of sheep and goats, which feature in the Order's written sources but are otherwise relatively poorly represented in faunal assemblages in the Vistula Delta region, or the zoning of space for livestock stabling given the prevalence of cattle and pigs in these assemblages (see Chapter 12).

In the later phase, further towards the top of the profile within the earliest *Vorburg* at Elbląg there is a change in the morphology and composition of the dung deposits, which is probably from horses. Documentary sources indicate that stud horses were more often stabled in the outer baileys of castles to ensure their safety (Ekdahl 1998), and horseshoes were recovered from mid-thirteenth-century contexts during excavation (Fonferek forthcoming). In addition to the documentary and material culture, comparison with a thirteenth-century horse stable within a Moravian medieval bailey in Veselí nad Moravou, Czech Republic (Dejmal et al. 2014), provides an interesting comparative case study for the potential horse stabling evidence in terms of the similar formation of the stabling layers, a comparable location of the stable close to water as evident by flooding depositing alluvium, and mixed grain fodder and pasture diet that is also evident from units 16a and 18b at Elbląg; charred oats in the macrobotanical evidence have been interpreted to be fodder (see Chapter 13). At Elbląg, the earliest evidence for caprids and the later evidence for horses both show the importance of stabling animals in close proximity to an existing water source, which is also observed at Karksi Castle (Ger. Karkus) in Livonia (see Chapter 7).

There is evidence at Święta Góra, specifically from the early medieval phase of occupation (Phase 2b, late ninth to tenth century) that herbivores were kept on site during the sporadic episodes of occupation and, as the zooarchaeological evidence shows, were also slaughtered and butchered there (Chapter 12). Diatoms within the herbivore dung indicate that nearby Lake Staświny most probably served as their water source. These sporadic episodes of low-key, domestic occupation occurred throughout the Migration Period and the early Middle Ages, Phases 2a, 2b, and 3 (Table 14.1). Phases 2a, 2b, and 3 all have groups of pits that either differed from the others in terms of the type of refuse that was deposited, or in their intensity of use. Those with particularly enriched levels of P, Ca, and Zn, may contain coprolite material, as Features 17, 18, and 40 have shown. As the micromorphological analysis of Feature 40 has shown, geochemical enrichments of pit fills could be attributed to the deposition of refuse from a range of activities such as the cooking and preparation of meat, the deposition of hearth ashes, which can include burnt herbivore dung, and the inclusion of coprolites. Most of the pits are shallow features with single fills, which could suggest that individual pits were dug and filled with domestic refuse, as well as coprolites, for each episode of short-term occupation.

At Biała Góra, the geoarchaeological occupation evidence relates to the abandonment of part of the site. It is possible that the 'cultural layer' (or 'dark earth') represents a short period of intensive occupation that comprises a shallow spread of middening dating from, on the basis of the artefacts that were recovered, mid-thirteenth to mid-fourteenth centuries, which was continuously reworked into the Crusading period; prior to this spread of midden material, refuse appears to have been deposited in pits, representing a change in waste management strategies from a more to less organized system (Schiffer 1987; Simpson and Barrett 1996; Needham and Spence 1997; Simpson 1998). Geochemical results for P, Ca, and Zn are the most interesting of all the pXRF data from Biała Góra 3. The spatial distribution of phosphate enrichment (Fig. 14.6a) (greater than twenty-five times the background geology) indicates areas of the 'cultural layer' (or 'dark earth' formation) that coincide with anomalies highlighted by gradiometer survey (Fig. 14.6b The geophysical anomalies (Fig. 14.6b) that result from clusters or individual metal objects, and/or caches of brick and burnt stone. The enrichments of Ca that are high or very high correspond with some of those cores that had enrichments of P that are greater than twenty-five times the background geology (Fig. 14.7a), and the very high enrichment of Zn corresponds with a very high, but less than fifteen times the background, enrichment factor of P in the south-east corner of the site (Fig. 14.7b). It is probable that the geochemistry has identified depositional hot spots where, in addition to metal objects, other refuse that has since decayed was also dumped. The geochemistry probably represents microscopic occupation residues within the 'cultural layer' are altered by weathering and comprise highly degraded bone fragments, ceramic material (probably brick fragments), degraded mortar, and degraded charcoal within black amorphous fine material (Pluskowski et al. 2014).

Works Cited

Primary Sources

CTP Scholz, K. and Wojtecki, D., trans. 1984. *Chronik des Preussenlandes/Peter von Dusburg*, Darmstadt: Wissenschaftliche Buchgesellschaft

Secondary Works

Arroyo-Kalin, M., Neves, E. G., Petersen, J., and Bartone, R., 2003. 'Geoarchaeological Approaches to the Study of Amazonian Anthropic Dark Earths', in *Second International Conference on Soils and Archaeology Pisa, 12th–15th May, 2003, Extended Abstracts*, ed. G. Boschian, Pisa: Felici editore, 7–9

Banerjea, R. Y. 2008. 'Experimental Geochemistry: A Multi-Elemental Characterisation of Known Activity Areas', *Antiquity* 82, 318 (Project Gallery)

Banerjea, R. Y., Badura, M., Kalējs, U., Cerina, A., Gos, K., Hamilton-Dyer, S., Maltby, M., Seetah, K., and Pluskowski, A. 2017. 'A Multi-Proxy, Diachronic and Spatial Perspective on the Urban Activities within an Indigenous Community in Medieval Riga, Latvia', *Quaternary International* 460, 3–21

Banerjea, R. Y., Bell, M., Matthews, W., and Brown, A. 2015a. 'Applications of Micromorphology to Understanding Activity Areas and Site Formation Processes in Experimental Hut Floors', *Journal of Archaeological and Anthropological Sciences* 7, 89–112

Banerjea, R., Karczewski, M., Karczewska, M., and Pluskowski, A. 2015b. 'The Role of Geoarchaeology in Understanding the Formation and Function of Święta Góra (Staświny), Masuria, Poland: Interim Results from Feature 40', in *Materiały do Archeologii Warmii i Mazur*, eds. S. Wadyl, K. Karczewski, and M. Hoffmann, Warszwa, Białystok: Instytut Archaeologii Uniwersytetu Warsawskiego/Instytut Historii i Nauk Politycznych Uniwersytetu w Białymstoku, 221–31

Barboni, D., Bonnefille, R. A. A., and Meunier, J. D. 1999. 'Phytoliths as Paleoenvironmental Indicators, West Side Awash Valley, Ethiopia', *Palaeogeography, Palaeoclimatology, Palaeoecology* 152, 87–100

Borderie, Q., Devos, Y., Nicosia, C., Cammas, C., and Macphail, R. I. 2015. 'Les terres noires dans l'approche géoarchéologique des contextes urbains', in *La géoarchéologie française au XXIe siècle*, eds. N. Carcaud and G. Arnaud-Fassetta, Paris: CNRS, 217–26

Boyle, J. F. 2001. 'Inorganic Geochemical Methods in Palaeolimnology', in *Tracking Environmental Change Using Lake Sediments*, II: *Physical and Geochemical Methods*, eds W. M. Last and J. P. Smol, Dordrecht: Kluwer Academic, 83–141

Bradley, R. J. 2000. *The Archaeology of Natural Places*, London: Routledge

Bradley, R., Haselgrove, C., Vander Linden, M., and Webley, L. 2015, *The Later Prehistory of North-West Europe: The Evidence of Development-Led Fieldwork*, Oxford: Oxford University Press

Brown, A., Banerjea, R., Wynne, A. D., Stivriņš, N., Jarzebowski, M., Shillito, L.-M., and Pluskowski, A. 2015. 'The Ecological Impact of Conquest and Colonisation on a Medieval Frontier Landscape: Combined Palynological and Geochemical Analysis of Lake Sediments from Radzyń Chełmiński, Northern Poland', *Geoarchaeology* 30/6, 511–27

Canti, M. G. 1999. 'The Production and Preservation of Faecal Spherulites: Animals, Environment and Taphonomy', *Journal of Archaeological Science* 26, 251–58

Cook, S. R., Clarke, A., Fulford, M., and Voss, J. 2014. 'Characterising the Use of Urban Space: A Geochemical Case Study from Calleva Atrebatum (Silchester, Hampshire, UK) Insula IX during the Late First/Early Second Century AD', *Journal of Archaeological Science* 50, 108–16

Dejmal, M., Lisà, L., Nývltová, M. F., Bajer, A., Petr, L., Kočàr, P., Kočàrova, R., Nejman, L., Rybníček, M., Sůvova, Z., Culp, R., and Vavrčík, H. 2014. 'Medieval Horse Stable: The Results of Multi Proxy Inter Disciplinary Research', *PLoS ONE* 9 (3) e89273

Delhon, C., Martin, L., Argant, J., and Thiébault, S. 2008. 'Shepherds and Plants in the Alps: Multi-proxy Archaeobotanical Analysis of Neolithic Dung from "La Grande Rivoire" (Isére, France)', *Journal of Archaeological Science* 35, 2937–52.

De Vleeschouwer, F., Piotrowska, N., Sikorski, J., Pawlyta, J., Cheburkin, A., Le Roux, G., Lamentowicz, M., Fagel, N., and Maquoy, D. 2009. 'Multiproxy Evidence of "Little Ice Age" Palaeoenvironmental Change in a Peat Bog from Northern Poland', *The Holocene* 19, 625–37

Devos, Y., Nicosia, C., Vrydaghs, L., and Modrie, S. 2013. 'Studying Urban Stratigraphy: Dark Earth and a Microstratfied Sequence on the Site of the Court of Hoogstraeten (Brussels, Belgium): Integrating Archaepedology and Phytolith Analysis', *Quaternary International* 315, 147–66

Dockrill, S. J., Bond, J. M., Milles, A., Simpson, I. A., and Ambers, J. 1994. 'Tofts Ness, Sanday, Orkney: An Integrated Study of a Buried Orcadian landscape', in *Whither Environmental Archaeology?*, eds R. Luff and P. Rowley-Conwy, Oxford: Oxbow, 115–32

Ekdahl, S. 1998. 'Horses and Crossbows: Two Important Warfare Advantages of the Teutonic Order in Prussia', in *The Military Orders*, II: *Welfare and Warfare*, ed. H. Nicholson, Farnham: Ashgate, 119–51.

Fonferek, J. 2019. 'The Environmental Setting of the Earliest Teutonic Order Stronghold in Elbląg', in *Ecologies of Crusading, Colonization, and Religious Conversion in the Medieval Baltic: Terra Sacra II*, ed. A. Pluskowski, Environmental History of the North Atlantic World, 3, Turnhout: Brepols (forthcoming)

Goldberg, P., and Macphail, R. I. 2006. *Practical and Theoretical Geoarchaeology*, Oxford: Blackwell

Hutson, S. R., and Terry, R. E. 2006. 'Recovering Social and Cultural Dynamics from Plaster Floors: Chemical Analyses at Ancient Chunchucmil, Yucatan, Mexico', *Journal of Archaeological Science* 33, 391–404

Jóźwiak, S., and Trupinda, J. 2012. *Krzyżackie zamki komturskie w Prusach*, Toruń: Wydawnictwo Naukowe Uniwersytetu Mikołaja Kopernika

Kattenberg, A. E., and Aalbersberg, G. 2004. 'Archaeological Prospection of the Dutch Perimarine Landscape by means of Magnetic Methods', *Archaeol. Prospect* 11, 227–35

Lomas-Clarke, A. H., and Barber, K. E. 2007. 'Human Impact Signals from Peat Bogs — A Combined Palynological and Geochemical Approach', *Vegetation History and Archaeobotany* 16, 419–29

Lund, J. 2010. 'At the Water's Edge', in *Signals of Belief in Early England: Anglo-Saxon Paganism Revisited*, eds M. Carver, A. Sanmark, and S. Semple, Oxford: Oxbow, 49–66

Macphail, R. I., Cruise, G., Allen, M. J., Linderholm J., and Reynolds, P. 2004. 'Archaeological Soil and Pollen Analysis of Experimental Floor Deposits; With Special Reference to Butser Ancient Farm, Hants, UK', *Journal of Archaeological Science* 31, 175–91

Macphail, R. I., Galinié, H., and Verhaeghe, F. 2003. 'A Future for Dark Earth?', *Antiquity*, 77, 349–58

Macphail, R. I., and Goldberg, P. 2010. 'Archaeological Materials', in *Interpretation of Micromorphological Features of Soils and Regoliths*, eds G. Stoops, V. Marcelino, and F. Mees, Amsterdam: Elsevier, 589–622

Matthews, W., French, C. A. I., Lawrence, T., Cutler, D. F., and Jones, M. K. 1997. 'Microstratigraphic Traces of Site Formation Processes and Human Activities', *World Archaeology* 29/2, 281–308

Middleton, W. D., and Price, T. D. 1996. 'Identification of Activity Areas by Multi-Element Characterization of Sediments from Modern and Archaeological House Floors Using Inductively Coupled Plasma-Atomic Emission Spectroscopy', *Journal of Archaeological Science* 23, 673–87

Needham, S., and Spence, T. 1997. 'Refuse and Formation of Middens', *Antiquity* 71, 77–90

Nicosia, C., and Devos, Y. 2014. 'Urban Dark Earth', in *Encyclopedia of Global Archaeology* (vol. XI), ed. C. Smith, New York: Springer, 7532–40

Nitychoruk, J., Bińka, K., and Szymanek, M. 2012. *Sytuacja geologiczna i interpretacja paleośrodowiskowa obszaru dziedzińca Muzeum Archeologiczno — Historycznego w Elblągu, Bulwar Zygmunta Augusta 11, 82–300 Elbląg*, Warszawa (unpublished report)

Nowakiewicz, T., and Rzeszotarka-Nowakiewicz, A. 2013. *Antiquities of Mrągowo Lake District: An Archaeological Guide*, Mrągowo: Mrągowo Municipality

Nowakiewcz, T., and Wróblewski, W. 2003. ''Legenda galindzka' Piotra z Dusburga w świetle badań archeologicznych', in *Antyk i barbarzyńcy*, eds A. Bursche and R. Ciołek, Warsaw: IA UW, 313–22

Pichler, S. L., Pümpin, C., Brönnimann, D., and Rentzel, P. 2014. 'Life in the Proto-Urban Style: The Identification of Parasite Eggs in Micromorphological Thin Sections from the Basel-Gasfabrik Late Iron Age Settlement, Switzerland', *Journal of Archaeological Science* 43, 55–65

Pierce, C., Adams, K. R., and Stewart, J. D. 1998. 'Determining the Fuel Constituents of Ancient Hearth Ash Via ICP-AES Analysis', *Journal of Archaeological Science* 25/6, 493–503

Piperno, D. 2006. *Phytoliths: A Comprehensive Guide for Archaeologists and Palaeoecologists*. Lanham, MD: Alta Mira Press

Pluskowski, A. G. 2013. *The Archaeology of the Prussian Crusade: Holy War and Colonisation*, London: Routledge

Pluskowski, A. G., Sawicki, Z., Shillito, L.-M., Badura, M., Makowiecki, D., Zabilska-Kunek, M., Seetah, K., and Brown, A. 2014. 'Biała Góra: The Forgotten Colony in the Medieval Pomeranian-Prussian Borderlands', *Antiquity* 88/341, 863–82

Powierski, J. 2001. *Prusowie, Mazowsze i sprowadzenie krzyżaków do Polski* (vol. II.1), Malbork: Muzeum Zamkowe w Malborku

Robinson, M. 2006. 'The Macroscopic Plant Remains', in *Life and Labour in Late Roman Silchester: Excavations In Insula IX Since 1997*, ed. M. G. Fulford, A. Clarke, and H. Eckardt, Britannia Monograph Series 22, London: Society for the Promotion of Roman Studies

——, 2011. 'The Macroscopic Plant and Invertebrate Remains', in *Silchester, City in Transition: The Mid-Roman Occupation of Insula IX, c AD 125–250/300*, eds M. G. Fulford and A. S. Clarke, Britannia Monograph, 25, London: Promotion of Roman Studies, 281–93

Sawicki, Z., Pluskowski, A. G., Brown, A., Badura, M., Makowiecki, D., Shillito, L.-M., Zabilska-Kunek, M., and Seetah, K. 2015. 'Survival at the Frontier of Holy War: Political Expansion, Crusading, Environmental Exploitation and the Medieval Colonising Settlement at Biała Góra, north Poland', *European Journal of Archaeology*, 18/2, 282–311

Schiffer, M. B. 1987. *Formation Processes of the Archaeological Record*, Albuquerque: University of New Mexico Press

Semple, S. 2010. 'In the Open Air', in *Signals of Belief in Early England: Anglo-Saxon Paganism Revisited*, eds M. Carver, A. Sanmark, and S. Semple, Oxford: Oxbow, 21–48

Shahack-Gross, R. 2011. 'Herbivorous Livestock Dung: Formation, Taphonomy, Methods for Identification, and Archaeological Significance', *Journal of Archaeological Science* 38, 205–18

Shillito, L.-M., Balazs Major, B., Almond, M., Anderson, E., and Pluskowski, A. 2014. 'Micromorphological and Geochemical Investigation of Formation Processes in the Refectory at the Castle of Margat (Qal'at al-Marqab), Syria', *Journal of Archaeological Science* 50, 451–59

Shillito, L.-M., Bull, I. D., Matthews, W., Almond, M. J., Williams, J. M., and Evershed, R. P. 2011. 'Biomolecular and Micromorphological Analysis of Suspected Faecal Deposits at Neolithic Çatalhöyük, Turkey', *Journal of Archaeological Science* 38/8, 1869–77

Shillito, L.-M., and Matthews, W. 2013. 'Geoarchaeological Investigations of Midden Formation Processes in the Early to Late Ceramic Neolithic Levels at Çatalhöyük, Turkey ca. 8550–8370 cal BP', *Geoarchaeology* 28, 25–49

Sillar, B. 2000. 'Dung by Preference: The Choice of Fuel as an Example of How Andean Pottery Production is Embedded within Wider Technical, Social and Economic Practices', *Archaeometry* 42, 43–60

Simpson, I. 1998. 'Early Land Management at Tofts Ness, Sanday, Orkney: The Evidence of Thin Section Micromorphology', in *Life on The Edge: Human Settlement and Marginality*, eds C. M. Mills and G. Coles, Oxford: Oxbow, 91–98

Simpson, I., and Barrett, J. 1996. 'Interpretation of Midden Formation Processes at Robert's Haven, Caithness, Scotland Using Thin Section Micromorphology', *Journal of Archaeological Science* 23, 543–56

Stolt, M. H., and Lindbo, D. L. 2010. 'Soil Organic Matter', in *Interpretation of Micromorphological Features of Soils and Regoliths*, eds G. Stoops, V. Marcelino, and F. Mees, Amsterdam: Elsevier, 369–90

Terry, R. E., Fernández, F. G., Parnell, J. J., and Inomata, T. 2004. 'The Story in the Floors: Chemical Signatures of Ancient and Modern Maya Activities at Aguateca, Guatemala', *Journal of Archaeological Science* 31, 1237–50

Tsartsidou, G., Lev-Yadun, S., Albert, R.-M., Miller-Rosen, A., Efstratiou, N., and Weiner, S. 2007. 'The Phytolith Archaeological Record: Strengths and Weaknesses Evaluated on a Quantitative Modern Reference Collection from Greece', *Journal of Archaeological Science* 34, 1262–75

Twiss, P. C. 1992. 'World Distribution of C3 and C4 Grass Phytoliths', in *Phytolith Systematics: Emerging Issues*, eds G. Rapp Jr and S. C. Mulholland, New York: Plenum, 113–28

Vaitkevičius, V. 2004. *Studies into the Balts' Sacred Places*, BAR international Series, 1228, Oxford: Archaeopress

Vincentini, A., Baber, J. C., Aliscioni, S. S., Giussani, L. M., and Kellogg, E. A. 2008. 'The Age of the Grasses and Clusters of Origins of C4 Photosynthesis', *Global Change Biology* 14/12, 2963–77

Wilson, C. A., Davidson, D. A., and Cresser, M. S. 2005. 'An Evaluation of Multi-Element Analysis of Historic Soil Contamination to Differentiate Space Use and Former Function in and around Abandoned Farms', *Holocene* 15, 1094–1099

Yule, B. 1990. The 'Dark Earth' and Late Roman London', *Antiquity* 64, 620–28

Žulkus, V. 2007. *Palanga in the Middle Ages: Ancient Settlements*, Vilnius: Versus Aureus

15. Reorganizing the Prussian Landscape

Marc Jarzebowski, Aleksander Pluskowski, Tomasz Nowakiewicz, and Maciej Karczewski

The process of reorganizing conquered tribal territories began almost immediately with the Teutonic Order's entry into the Kulmerland in the 1230s, and continued through to the conquest of Sudovia in 1283. The latter region was depopulated and would become the frontier with Lithuania, known as *Die Große Wildnis* or Great Wilderness (Mager 1960, 145ff.; Białuński 2009). The annexation of Pomerelia and the relocation of the Order's headquarters to Marienburg in 1309 marked the beginning of a process of administrative centralization. Territorial reorganization then continued throughout the fourteenth century, as new districts were created run primarily by advocates (sing. *Vogt*) and procurators (sing. *Pfleger*). The specific details of the primary case study sites are outlined in Chapter 10, and full coverage of the chronological development of the various categories of *Ämter* within the Prussian *Ordensland* can be found in Biskup et al. 2009.

This chapter compliments Chapter 8 ('Reorganizing the Livonian Landscape') with how the organization of the landscape in Prussia changed over the course of the thirteenth century. Where the two regions differ is in the relative representation of inventories from the Order's castles, which are far more abundant for Prussia and enable select, systematic comparisons of individual convents and the residences of lesser officials. Territorial organization before the crusades is considered first, followed by how the Order reorganized and assigned value to the landscape, and then a comparison of the role of different categories of castles in contrasting heartland and frontier landscapes as expressed in their surviving inventories: Rehden (Pol. Radzyń Chełmiński) and Stuhm (Pol. Sztum), followed by Nessau (Pol. Mała Nieszawka) and Lötzen (Pol. Giżycko). The full range of data relating to the ecological changes initiated by the crusades in Prussia, defined as the cultural landscape, are considered together in Chapter 16 and compared with Livonia in Chapter 17.

Territories before the Crusades

The term 'Prussia' is used to refer to the territories occupied by the westernmost Baltic language-speaking group, but would become expanded to refer to the entirety of the Order's territories in this region, including the Kulmerland and Pomerelia, which in the eighteenth century would be defined as 'West Prussia' (*Westpreußen*). The antiquity of the individual Prussian tribal territories outlined by Peter von Dusburg remains unclear (see Chapter 10; for a full discussion, see Bojtár 1999). Claudius Ptolemy had referred to Galindians and Sudines in the second century AD, whilst by the later eleventh century when Adam of Bremen was using the term *Prussi* he specified Semland. Sambia, Samland, and its derivatives were sometimes used to denote a different group to the Prussians and at other times to include Prussians in the twelfth and thirteenth centuries (Okulicz-Kozaryn 1997, 8). The Kievan chronicler Nestor mentioned the Jatviagi in the first quarter of the twelfth century, a name which would feature more frequently in Rus' and Polish sources in the thirteenth century. These were referred to as Sudovians by Peter von Dusburg, although opinion remains divided on whether this easternmost group spoke a Prussian dialect or had its own language (Bojtár 1999, 158; see also Engel et al. 2006). Indeed, both Galindians and Sudovnians may have been separate groups included within the Order's geographic understanding of Prussia, due to similarities in customs and language (Powierski 2003, 123). Although strongholds had been a distinct form of settlement in Prussia (including Sudovia) since the Bronze Age and Early Iron Age, in the last quarter of the first millennium AD they became more visibly linked to a broader settlement cluster. The character of territories is best understood from the settlement pattern, rather than differences in material culture between the documented tribal groups (see Pluskowski 2013 for a regional summary).

Settlement was particularly clustered on the Sambian Peninsula, on either side of the Masurian Lake District and around the Vistula Delta and Lagoon (Długokęcki 2009a, 27–32), although this distribution may also reflect the regional preferences of Prussian settlement studies in the twentieth century (Nowakiewicz 2010, 487n1). The archaeologically documented strongholds of Prussia were almost certainly organised in relation to the basic territorial unit referred to as the *pulko* (Lat. *territorium*; Rus. *vlost*; Pol. *włość*), which varied in size and had an estimated population of around a thousand inhabitants (Łowmiański 1985, 63). Various interpretations of *pulko* and other subdivisions, particularly the *lauks*, defined by Gerard Labuda and Marian Biskup (1986, 71–72) as the economic hinterland of the *pulko*, have been suggested on the basis of later sources (reviewed in Nowakiewicz, 2010). Later social stratification separated the Prussian aristocracy or *nobiles* from the peasantry, but indigenous political structures had been more complex and included leaders, slaves, merchants (most visible in Sambia: Antoniewicz 1958, 141–49), and assemblies (*wayde*), which in the thirteenth century took the form of war councils (*kariowaite*; Długokęcki 2006, 35). The principal 'core' Prussian territory included within this study has been Galindia, which has also been one of the most intensively studied Prussian regions in recent years.

Territorial units within Galindia varied spatially and temporally. In its western part, early medieval settlement clustered around Lake Sałęt. The centre of this cluster was the ring fort and associated open settlement at Szestno-Czarny Las (site 3). The stronghold was probably occupied from the tenth century into the eleventh century and supported by a network of smaller settlements, which communicated with each other through waterways, particularly the lake. Its territory, as can be reconstructed archaeologically, was defined by a series of smaller strongholds with little evidence for occupation and no identified dwelling structures, suggesting they functioned as 'sentinel' fortifications. The northernmost chain of these structures corresponds to the documented border between Galindia and Barta, perhaps representing a state of hostility between these groups during the time of their construction. The presence of linear earthworks extending east–west within wooded and marshy terrain in close proximity to these mounds may have been intended to protect settlements or define territory, although their chronology remains ambiguous. Social and military organization beyond the central stronghold was focused on mound sites, referred to as 'high houses', interpreted as residences of the Prussian nobility (Nowakiewicz 2010). In some cases, as at Boże (site 1), these were attached to an open settlement; elsewhere, as at Wyszembork, they were defined by a more commercial character. The stronghold at Szestno-Czarny Las lay at the centre of an international trade network, as suggested by the presence of artefacts from Scandinavia, Ruthenia, Slavic regions, and the Near East. Extensive numbers of iron Thor's hammers also point to the presence of Scandinavians, perhaps functioning as mercenaries for the Prussian aristocracy (Wróblewski et al. 2003). The settlement complex has been interpreted as a *pulko* (Nowakiewicz 2010, 497).

This *pulko* appears to have declined in the mid-eleventh century, with no functioning strongholds in this region in the twelfth century (Nowakiewicz 2010, 492). This collapse coincides with the documented devastation of the war ending in 1047, which saw the reincorporation of Masovia (allied with the Galindians) into the Piast state. Its impact was so traumatic that it appears to have been preserved in the oral culture of Galindia and was subsequently noted down by Peter von Dusburg, who stated the region had been left completely desolate (Nowakiewcz and Wróblewski 2003). However, other written sources (such as papal bulls) indicate this region continued to be of political interest and populated, and the Kuyavian prince Casimir in the second half of the thirteenth century sought to Christianize the Galindians, despite opposition from the Teutonic Order (Powierski 2001, 9–88). The settlement at Ruska Wieś (site 13) on the eastern banks of Lake Sałęt was potentially occupied into the thirteenth and fourteenth centuries (Auch and Nowakiewicz, 2009). Further east the stronghold and associated settlement of Jeziorko may have been occupied into the crusading period (Antoniewicz 1950, 253), whilst the Staświny settlement complex occupied into the thirteenth or fourteenth century has been interpreted as a *lauks* (see Chapter 10). In summary, the settlements of Galindia, and more broadly in Prussia, are understood as discrete clusters represented by a variety of sites associated with discrete topographic units — in the case of Szestno-Czarny Las and Staświny these were focused on lakes. In contrast to the overlordships of Estonia, there is little evidence for comparable regional power centres in Galindia in the early thirteenth-century. The connection, if any, between

Staświny and the stronghold of Jeziorko, *c.* 15 km apart, remains unknown, and they may have been associated with completely independent territorial units.

Settlement dynamics in the Slavic-Prussian borderlands, of which two are included within this study, has been traced through contrasting forms of material culture. The presence of Slavic groups colonizing the southern fringes of Galindia coincides with the appearance of new strongholds, burial rites and distinctive 'Slavic' pottery (Nowakiewicz 2006, 169). Slavic influence continued to be felt into the eleventh century, as suggested by a limestone casting mould from Barczewko used for making crescent-shaped pendants. This was a period of flourishing metalwork in Galindia, as indicated by locally made hoards of silver objects (Antoniewicz 1955). Following the incorporation of Masovia into the Polish state, the political authority of the new Christian regime was articulated by a new phase of stronghold construction, alongside impressive linear earthwork defences that redefined the borderland. In the ninth century, Masovian borderland settlements had reached Grzebsk, Działdowo, and Orzyc; from the ninth to thirteenth centuries, the extent of this colonization stretched beyond Goniądz. The nearest Galindian settlements were located around Pasym and Szczytno some 40 km away and separated by an expansive belt of woodland, parts of which is still visible today (Białuński 1999, 8). Neighbouring Lubavia and Sasna also saw Masovian colonization from the early eleventh century, whilst the Kulmerland had been colonized by Slavic groups from the eighth century, and came under Polish dominion in the second half of the tenth century. In the borderland between Pomerelia and Pogesania, Slavic settlement spread across the Vistula Delta and Iława Lake District from the eighth to thirteenth centuries, whilst the main Prussian settlement clusters were centred on Dzierzgoń and the Elbląg uplands east of Lake Drużno.

The main tool of colonization was the stronghold, which initially functioned as a border fortification and subsequently organized clusters of settlements and taxation, and provided a measure of security. The largest strongholds which dominated regional groups of smaller fortifications were the residences of castellans, who represented the authority of the dukes of Masovia (e.g., Kałdus) and the dukes of Gdańsk (e.g., Węgry; Haftka and Wadyl 2015) (Buko 2005, 283–84). As in the case of the central Prussian strongholds, the seats of castellans are defined by high-status material culture, a military presence, specialized manufacturing, trade, and consumption patterns. The frontier between the Kulmerland and Pomesania was equally permeable, with the strongholds at Grudziądz (Ger. Graudenz) and Radzyń Chełmiński, around 17 km apart, organizing the northern cluster of Slavic settlement in the twelfth century. By the thirteenth century there were Masovian knights and peasants present in the Kulmerland (Biskup 2002, 132), and the network of strongholds which developed has been interpreted as a response, in particular, to the threat from neighbouring Prussians (Poliński 2003, 130).

Polish investment in borderland colonization was twinned with missionary activity, aimed at Christianizing Pomeranian, Masovian, and ultimately neighbouring Prussian areas. Kałdus and Biała Góra (if identified with Zantyr) may have been bases for missions into Prussia, and there is documentary evidence for the establishment of parishes in the borderlands before the arrival of the Order. At the same time, Polish dukes and magnates were quick to adopt the Christian rhetoric of holy war established by the First Crusade (1095–99). Bolesław III's campaigns against the Pomeranians in 1109, Bolesław IV's in Prussia in 1147 and 1166, and Casimir I's wars against the Sudovians in 1191–92 were all portrayed as crusading engagements in early Polish narratives sources such as the *Gesta Principum Polonorum* and the *Chronicon Polonorum* (von Güttner Sporzyński 2008, 14), even if they were not officially sanctioned as such by the papacy (Śliwiński 2004, 53). This was followed by another crusade in 1222–23 and, from 1216 to 1230, Prussian reprisals followed these incursions, including an assault on Pomerelia that destroyed the Cistercian monastery at Oliwa and the Norbertine convent in Żuków. There is even evidence for the temporary occupation of the borderland by Prussian groups, including the Kulmerland. This provides a context for the abandonment of strongholds and the collapse of the settlement pattern in the borderlands with Prussian territories. In the Masovian/Galindian border, the stronghold at Stare Łączyno was occupied into the mid-thirteenth century by which time the majority of archaeologically identified settlements around the River Wkra (a tributary of the Narew) had also been abandoned (Powierski 2001, 171–72, 192). In the Kulmerland and Lubavia, many settlements were abandoned or destroyed by the early thirteenth century (Poliński 2003, 133). In the Vistula Delta region, the centre at Węgry was abandoned but the settlement pattern did not completely collapse, and the site of Biała Góra endured (or was reoccupied) into the crusading period. Whilst Polish and Pomerelian magnates encouraged the Hospitallers, Templars, and

Knights of Calatrava to operate in the borderlands with grants of lands (Starnawska 2001, 420–22), along with the foundation of a local military order by Bishop Christian with estates provided by Konrad of Masovia, the *Fratres Milites Christi* (later the Order of Dobrin; Białunski 1999, 87), this proved ineffective and set the scene for Duke Konrad's enlistment of the Teutonic Order to secure the Kulmerland (Śliwiński, 2004, 62).

Redefining the Conquered Landscape and the First Phase of Colonization

By 1235, the Teutonic Order, with papal support, was directing the crusades against the Prussians and had also established themselves in the former Masovian castellany of the Kulmerland, with additional territory across the Vistula in Kuyavia. The (re)conquest of the territory was secured with *castra*, the most strategic of which was at Rehden on the border with Prussian territories. By this time the Order had founded three towns: Thorn (Pol. Toruń), Kulm (Pol. Chełmno), and perhaps already Rehden (between 1234 and 1239), whilst the towns of Marienwerder (1234) (Pol. Kwidzyn) and later Elbing (1237) (Pol. Elbląg) and Königsberg (Rus. Kaliningrad) were founded by various leaders of crusading hosts. In the 1240s, Lübeckian merchants planned to establish a town in Sambia in collaboration with the Order, where in exchange for fiefs the burghers would provide military service (Dygo 2009b). Elbing had been founded with the Lübeck Law, but the Order introduced its own town charter — the Kulm Law (Ger. *Kulmer Handfeste*) — which became the most widely applied in Prussia. This charter, first issued in Latin on 28 December 1233 in Kulm, emulated the Magdeburg Law, but specified the allocation of land and resources in a slightly different way. The sole copy of the original charter was destroyed during a fire in Kulm and was reissued on 1 October 1251. Both forms of the charter had been granted to only two towns at this time: Thorn and Kulm, but from the last decades of the thirteenth century it became widely used for both towns and smaller settlements. The basic unit of property defined in the Kulm Law was the *mansus* (consisting of around 16.8 hectares), adopted from Flemish land divisions used in Silesia, which in turn had derived from earlier Carolingian designations of farmland (Długokęcki 2009b, 462). In addition to dealing with monetary regulations, military obligations, and hereditary rights, the Kulm Law designated access to resources, specifying the Order's ownership of lakes and rivers, whilst granting fishing and milling rights to the burghers, and the obligation of providing the right shoulder of any game animal to the Order, with the exception of beavers which belonged to the Knights (Zdrójkowski 1983, 19). Beavers were not specified in the Magdeburg Law, and the multiple references to the species in the Kulm Law indicates their locally recognized importance (Militzer 1997, 336). Stipulated taxes consisted of a tithe of a measure of wheat and rye, and an annual tax of one-half mark per *Hufe* (interchangeable with Lat. *mansus* and Ger. *Hube*). The towns founded during the Prussian Crusade were therefore important components in the developing administration of the conquered territories, and some acquired substantial patrimonies. Their military resources bolstered the security of the Order's territories, whilst their taxes ensured the maintenance of the Order's garrisons and from the 1250s until 1291 contributed to the Order's defence of the Holy Land.

The Order's organization of conquered territory became more complex as further *castra* were established with associated territories, provisioning networks and subjects to manage. In the first few decades of the Prussian Crusade, the territories were governed by a *preceptor* and his deputy, the equivalent of a provincial master, with a named marshal and provisors, comparable to castellans (Dygo 2009b, 73). Although the first commandery is mentioned at Elbing in 1246, the introduction of a standardized system of organization has been attributed to the activities of the Master of the Order's German province, Eberhard von Sayn. During his visit to Prussia in 1251–52, shortly after the suppression of the First Prussian Uprising, he renewed the charters of Kulm and Thorn, and facilitated the designation of the convent at Elbing as the provincial headquarters of the Order in Prussia, alongside the foundation of the commanderies of Balga and Christburg. This has been regarded as the start of a more clearly defined hierarchical system of territorial organization in Prussia (Jóźwiak 2001; Jóźwiak and Trupinda 2012, 33). By this time the Kulmerland was clearly defined as a singular territory, bounded by the Vistula, Osa (Ger. Ossa), and Drwęca (Ger. Drewenz) rivers (Dygo 2009b, 59). Polish knights were granted territories here as vassals of the Order, with newly defined privileges and obligations defined under the Polish Law (Ger. *Polnisches Recht*, Lat. *Ius Polonicum*), a customary law but one that was codified in law books alongside the Magdeburg, Lübeck, Kulm, and Prussian laws (Päsler 2003, 235). Alongside the subdivision of the Kulmerland into commanderies, a provincial master was also appointed to govern the region, a post that is first named in 1248 and only abolished in the third decade of the fourteenth century.

Advocates were initially introduced to manage judicial and economic matters in Sambia and Natangia, enabling commanders to focus on organizing military resources. Galindia was one of the last regions to come under the Order's control, following a brief period of territorial contestation with the Duke of Kuyavia, and whilst outposts were constructed here, further subdivisions of territory, largely delegated to procurators, would accompany colonization from the mid-fourteenth century. The surviving Prussian nobility, once officially converted to Christianity, were redefined as neophytes and vassals with the obligation to provide military service and pay tithes in relation to the size of their fiefs. This changed after the Great Prussian Uprising (1260–74), when grants of fiefs to loyal converts increased, and included the reassignment of Prussian families and villages. In this respect, pre-crusade social structures were systematically reshaped by the Order (Dygo 2009a, 103–04). These territorial changes over the course of the thirteenth century were accompanied by the imposition of a new sacred topography, not only with a diocesan network of cathedrals, churches, and monasteries, but also with the establishment of the Order's convents, which were centres of religious life and loci of relics.

Organizing and Regulating Territories in Prussia from c. 1309–1525: The Written Evidence

The fourteenth century is defined by two major trends in the organization of territory; the restructuring of the Order's administration and a protracted and regionally variable process of colonization and settlement reorganization. The establishment of the Order's headquarters in Marienburg in 1309 resulted in increasing administrative centralization, alongside the importance of the elected Grand Master as the political figurehead of the Order, which was otherwise run as a corporation. From the 1320s, semi-autonomous advocateships and procuratorships were created, directly subservient to the grand masters and designed to bolster Marienburg's income. The centralized bureaucracy included regular visitations of castles and stock-checks of their arms, armour, and supplies. From the mid-fourteenth century, commanders in eastern Prussia and Pomerelia delegated the management of discrete districts within their territories to procurators, centred on their smaller castles. The third office responsible for territorial management was the equivalent of a bailiff (*Kammer*, sometimes translated as 'chamberlain'), who governed *Kammerämter* (Pol. sing. *komornictwo*) which included between fifteen and forty-five villages (Białuński 1996, 14). The office of forest master, the *Waldmeister* (who governed *Waldämter*), was also created to manage densely wooded territories. All four offices were deployed on an ad hoc basis by grand masters and commanders, depending on the diachronic vicissitudes of managing specific regions. In the densely settled Kulmerland there were additional irregular districts; individual villages, farms, or other small territories which had separate administrative status (Jóźwiak 2001, 2009b, 137). All of this indicates the Order adopted a reflexive response to managing the landscapes it annexed.

In contrast to Livonia, the Order's (and indeed, episcopal) lands in Prussia were occupied by three main groups: migrant Germans; migrant Poles, as well as those who had settled and survived in the borderlands in the twelfth century (particularly Masovians and Pomerelians); and indigenous Prussians. Some of the colonists were knights, and the surviving Prussian *nobiles* redefined as neophytic vassals held a comparable status, but the majority of settlers and settled were peasants. The Order remained the most important landowner on its Prussian territories, with only small fiefs handed to knights and their families who were obliged to provide military service. The majority of fiefs consisted of 10–20 *Hufen*; however, between the 1230s and 1330s, there were examples of particularly generous grants of land to secular knights. The most widely cited in the scholarly literature was the case of Dietrich von Tiefenau, who in 1236 had been granted a stronghold along with three hundred *mansi* in conquered Pomesania, and his estate expanded in 1239 and again in 1242. By the later 1280s, the Stange family had acquired the Tiefenau lands which amounted to *c.* 1400 *Hufen*, split between the territory of the Pomesanian episcopate and the Order. Such grants had clearly been effective in consolidating the security of the conquered territories during the period of active crusading. However, once the Prussians had been subjugated, the Stange lordship was clearly perceived as threatening, and the Order retrieved those parts of the estate situated within its own lands. The remainder gradually shrank and was further subdivided amongst descendants. Another estate of 1400 *Hufen* situated in the former territory of Sasna, within the commandery of Christburg (and from 1341, Osterode, Pol. Ostróda) was short-lived. Subsequent examples of large fiefs consisted of a few hundred *Hufen* with one or more villages (Jóźwiak 2009a, 222–26).

The fiscal and political decline of the Order was paralleled by the increasing prosperity of the knightly class. By the mid-fifteenth century, knightly families in the southern and western parts of the *Ordensland* became more actively involved in producing a surplus from their estates for sale, and particularly grain from the 1440s (Jóźwiak 2009c, 455).

On the majority of the Order's territory, the only intermediate authorities were the 'locators' of settlements (Ger. *Scholtis* or variants), and in Pomerelian and Prussian villages 'elders'. These community leaders, along with a 'commune' (Ger. *Gemeinde*, Pol. *gmina*, which developed late in Prussian villages) administered settlements on behalf of the Order, or their episcopal or knightly lords (Długokęcki 2009b, 465, 471). The peasants who populated these villages were defined according to their ethnic background, as well as the status of the settlement. Settlements founded with the Kulm charter were entrusted with an area ranging from 20 to over 100 *mansi*, with an average of 50–60 (roughly 840–1000 hectares), with units of 2 *mansi* best suited to the use of the heavier 'German plough' pulled by two horses or four oxen; a standard of ploughland is first evident in the Lower Vistula in the 1340s and by the end of the fourteenth century had been introduced as far east as Sambia (Długokęcki 2009b, 460–63, 472). Earlier, indigenous subdivisions of land appear to have been retained in some cases, as evident from the appropriation of Prussian *lauks*, and settlements with indigenous Prussians could retain ownership of ancestral lands, albeit redefined under a new system of taxation (Okulicz-Kozaryn 1983, 102). However, in many cases the settlement charters created new territorial units. Some villages were designated as specializing in particular economic activities, primarily fishing and bee-keeping, of which the largest number were found in the commandery of Marienburg (Długokęcki 2009b, 465). Town patrimonies contributed to the management of the landscape within the *Ordensland* and varied dramatically in size. Amongst the largest was held by Elbing (the Old Town) consisting of 1147 *Hufen*, whilst Thorn only had around 335; smaller towns had on average 65. Patrimonies included a variety of land-use owned and leased by burghers, whilst settlements within the territories of the larger towns paid dues to their councils. Here, rural communities were legally defined as urban and economically as peasantry (Czaja 2009, 195–96).

In this respect the environmental changes evident in the territory of any given castle largely reflect, as in Livonia, the agency of its varied rural population and adjacent town. However, the castle represented the organizing force behind these changes, which were fundamental to the maintenance of its community and lordship. Colonization was organized and directed from the top, and in this respect the process of settlement reflects the broader agenda of the theocracy. As in Livonia, a sense of intensive regulation is evident from the Order's rule in Prussia. Dues from subjects in the form of taxes, rents, and sales (including the enforcement of selling to the Order at fixed prices), as well as those procured by the Order's own households, were subdivided between the maintenance of the district (i.e. the castle and its household) and the presiding official. The latter often had separate or additional quarters, and even their own stores which are specified in the inventories. Part of the produce collected in this way was then designated for sale at local and international markets. The system overall was extremely successful at producing a financial surplus, and in *c.* 1400 the Order's annual expenditure in Prussia of over 36,000 marks was sustained by an income of over 40,000 marks (a good warhorse was worth 12–18 marks *c.* 1400; Rünger 1925, 235). Throughout much of the fourteenth century, the convent at Marienburg maintained itself through its own commandery and the lands of its subservient advocates and procurators. By the end of the fourteenth century, it also began to draw on the reserves of other commanderies, partly through irregular payments of the so-called *Geschoßem* (Jóźwiak 2009b, 132–37). From 1457, the Order's centralized administration was relocated to Königsberg and a decade later governed a significantly reduced territory.

This pivotal function of the various castles constructed by the Order in Prussia, directly connected with the administration of a defined territory and its resources, can be compared at a high resolution using sequences of inventory records and related documents. Variations between castles can be linked to their size, associated territory, settlements, and particularly the types of local resources that were available. This variability will be demonstrated by a consideration of the supply of different types of castles in four contrasting territories: the commanderies of Rehden (a heartland) and Nessau (a frontier), the advocateship of Stuhm (a heartland), and the procuratorship of Lötzen (a frontier). It is important to note that only a few categories of ecofacts recovered from castle excavations map onto these data, alongside palynological profiles from their associated with territories, all of which are compared in Chapter 16.

Organizing a Heartland: The Commandery of Rehden

The first mention of the castle in Rehden is in 1234, the adjacent town was founded by 1239, and the first commander is mentioned in 1251 (Torbus 1998, 595). Clearly the convent played a frontline and then an organizational role during the Prussian Crusade, but no further information is available from the written record about how this was articulated. The town was destroyed by the Prussians in 1277, and its reconstruction is marked by Grand Master Konrad von Thierberg's renewal of the town's *Handfeste* in March 1285. The revised charter refers to the needs of provisioning the convent of Rehden in several ways. The burghers had to deliver grain (see below), and they were permitted to catch fish in three lakes: Bliese (Pol. Jezioro Blizno), Okrengel (unknown), and Boberau (Pol. Jezioro Bobrowo). They had to share this privilege with the brethren, they were not permitted to use drag nets, and any caught fish could only be sold within the town of Rehden. The lake to the east of the castle (Schloss See) was explicitly excluded from the list. Concerning the more generic references to the exploitation of gold, silver, and other mines, as well as salt and beavers, the privilege replicated the generic terms established at Kulm (*PrUB* I–II, 457; Froelich 1868, 245 f.). The decades of 1310–40, when the conventual complex was rebuilt in brick, were crucial for the development of the commandery when the intensification of resource exploitation can be linked to the growth of settlement. Resource allocation was carefully controlled by the convent. For example, the charter for the village of Rehwalde (Rywałd) founded in 1312 by the commander of Rehden, granted permission to its inhabitants to catch fish in Lake Bliese (*PrUB* II, 51; Frölich 1868, 267). In *c.* 1365, commander Petzold von Korwis granted the entire lake to the town of Rehden, but in 1438 the commander, Johann von Erlebach, took it back for the exclusive use of the convent (*OBA* 7518, 1438 Oct 19). These modifications of accessibility to the lake illustrate its economic importance within a commandery in which freshwater resources were present, but less abundant than the eastern Prussian districts such as Lötzen's *Ämter*.

The twenty-two Rehden inventories dating from 1377 until 1449 are recorded in the *Großes Ämterbuch* (*GAD*). They cover a relatively short time span of the Order's rule in Prussia, but in quite a dense sequence which varies between a few months and twelve years. This is roughly the same period as covered by the smaller Stuhm inventories (see below), which total only fourteen in number. The first thirteen Radyzn inventories from 1377 until 1422 are taken from *Ordensfoliant* (*OF*) 130, the *Großes Ämterbuch* manuscript. The inventory from February 1434 has a secondary tradition in *OBA* 13: the 1435, 1446, and 1449 inventories are not in *OF* 130, but only in the *OBA*, and the 1437 inventory is from the *Großes Zinsbuch* (*OF* 131). Most of the inventories were handover certificates produced when a new commander came into office. They were sent in letter form to Marienburg and then copied into the *Ämterbuch* in the Grand Master's registry. In this respect both traditions (*OF* 130 and *OBA*) served the same purpose, and all of Rehden's inventories are relatively homogenous. Only the 1446 inventory was produced during a visitation of the convent, and this explains its different composition. The visitations at this time were more concerned with mapping the remaining manpower and military resources of the Order (see also Chapter 9). For example, the Rehden inventory names the convent brethren with their horses, but does not list livestock. The inventories therefore provide two levels of diachronic information: the details of resource exploitation and provisioning within the commandery, and the organizational paradigm of the Order. The diachronic status of the convent is also evident with the Battle of Grunwald (Ger. Tannenberg) as a defining moment in the long-term fiscal decline of the Order. The castle was briefly occupied by a Polish garrison in 1410 and was taken back by the Order after a siege. However, its economic status never recovered as the inventories from 1410 to 1449 show, a situation that would only begin to change when the castle and the entire region came under Polish sovereignty as a result of the Second Peace of Thorn in 1466. Indeed, it appears that after 1410 the castle's trading network, which had been growing in the previous two decades, virtually collapsed (Gancewski 2001, 31, tab. 31). For comparative purposes, the inventories are subdivided into grain, beverages, and livestock.

Grain (Fig. 15.1)

The grain measures in the Order's inventories referred to a system defined in 1307 at the meeting of the general chapter in Engelsburg (Pol. Pokrzywno). It consisted of *last*, *scheffel* (bushel), *viertel* (quarter), and *metze*. A *last*, which originally referred to freight, being the amount of grain carried in a vessel, consisted of 60 *scheffel*, a *scheffel* of 4 *viertel*, and a *viertel* of 4 *metzen*. A *metze* was described as the amount which could be grabbed by a man with his two hands for grinding. The size of a *schef-*

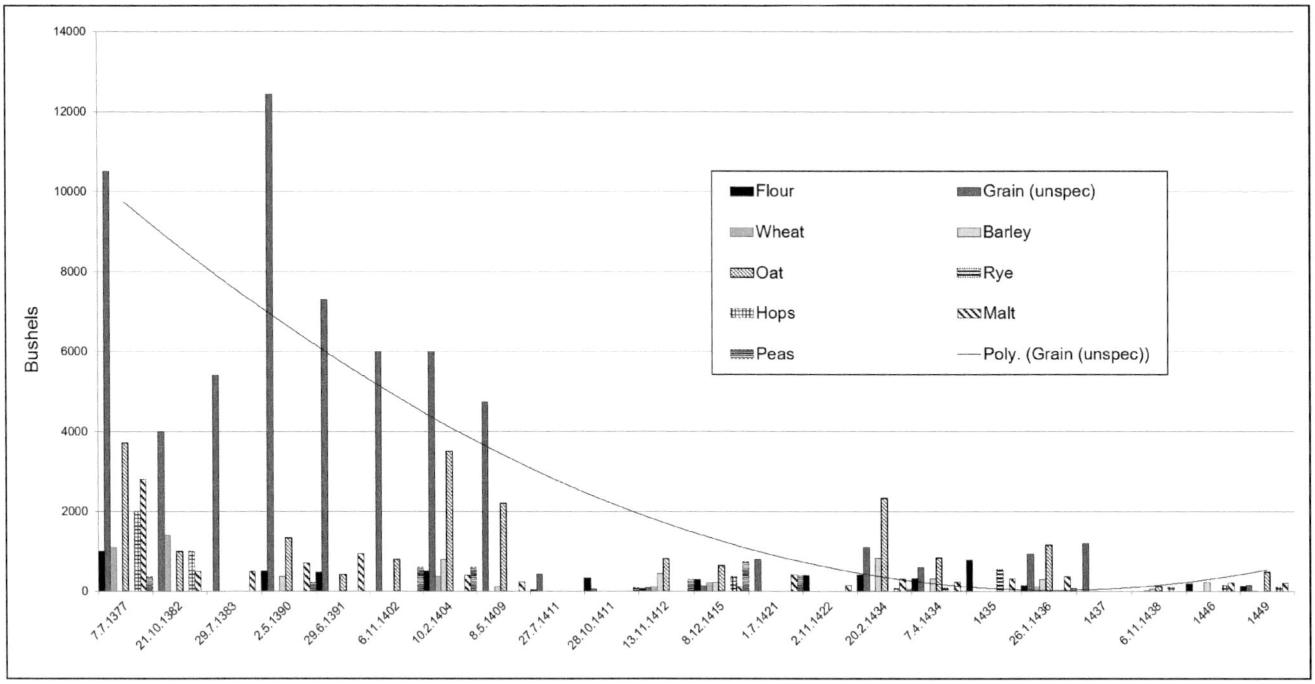

Figure 15.1: Grain measures at the convent of Rehden (Pol. Radzyń Chełmiński), with the polynomial trend for grain showing an overall decrease from the late fourteenth to the mid-fifteenth century. Figure by M. Jarzebowski.

fel as the basic grain measure varied and was subject to continuous attempts at standardization. It was a measure of capacity rather than of weight and equalled *c.* 54 litres. Taking the old 5 to 6 weight/volume ratio for grain, this would be *c.* 45 kg. As the density of flour is different (one litre has a weight of *c.* 600 g), a *scheffel* of flour would have a weight of about 32 kg (Witthöft 1981).

The 1285 town privilege for Rehden, written in Latin, states that the burghers had to deliver to the convent each year on St Martin's day (November 11) a specific amount of four varieties of grain, namely two kinds of wheat (*triticum* and *siligo*), barley (*hordeum*), and oat (*avena*) (Frölich 1868, 245). Rye (*secale*) is not mentioned here, which corresponds to its absence in the convent's inventories, where it is only mentioned once in 1435, although it is important to note the inventories only begin around nine decades after the 1285 charter, and the huge quantities of unspecified grain that are listed may have also included rye; moreover, another charter for Rehden dating to 1352 and written in German lists all four species ,including rye (*OBA*, 62). The mills in the area, which had to pay contributions to the commander, were primarily windmills. In 1449, the Grand Master granted the miller Niclas two windmills, free timber and rods (for the vanes), as well as financial support in case of their destruction by storm or warfare. One of the mills was situated in front of the convent, the other in front of the town. Niclas, in exchange, had to contribute 52.5 bushels of flour and 1 *last* of malt each year to the convent and to grind grain and malt, as and when it was required, assisted by a servant of the commander (*OBA*, 10079, 1449 Oct 19; included in *OBA*, 63). Mills were also situated in Bogusch (Pol. Boguszewo), Schwetz (Pol. Swiecie), Neumühl (Pol. Nowy Młyn — 'New Mill'), and Klein Sagen (Pol. Gołębiewko) (Frölich 1868, 47 ff., 80, 227, 311).

The diachronic chart (Fig. 15.1) illustrates the massive difference in the total amounts of grain which were stored at Rehden before and after 1410, with the quantity of unspecified grain peaking at more than 670 modern tonnes in 1390. Named categories of grain appear consistently throughout but in much smaller quantities. Oat appears most regularly, while wheat is only mentioned in eight cases and rye in only two inventories. Malt and peas appear often, but taken together there are no significant regularities or developments in the use of these plants. Comparing the diachronic trends in named and unspecified grain, it is possible to suggest the former were for the consumption of the convent (and its livestock), and the latter primarily intended for export. Documents from 1450 to 1454 refer to the export of rye and in one case wheat (Gancewski 2001, 95, Table 27). The vast amounts of unspecified grain in the last decades of the fourteenth century correspond to the period when this commodity represents one of the most significant exports from the Prussian *Ordensland* (Sarnowsky 1993). At the same time the subdivision of space within the conventual complex was overwhelmingly dedicated to production. Of the sixteen service areas referred to in the inventories, storage spaces, including a granary and an attic, accounted for 31.3% of space, whilst production spaces, which included two mills (the commandery had a further five) and a bakery, accounted for 68.7% (Gancewski 2001, 46).

Beverages

Beverages and other fluids, along with semi-dry products (butter, honey, and cheese) and dry food, which was stored and transported in similar containers, were measured in barrels and tonnes. A barrel (*Fass*) was the larger entity and in northern Germany could contain between two and four tonnes (*Tonne*; referring to the recorded medieval weight which was inconsistent, rather than modern equivalents) (Ziegler 1977, 304–09). In the Hanseatic area and in Prussia a barrel/tonne ratio of 1/3.5 was predominant (Sattler 1887, 272); in litres this would have been *c.* 421.8/120.8 or 451.2/128.9 (Ziegler 1977, 306, 308). In order to facilitate comparisons between inventories, a barrel has been converted to 422 litres and a tonne to 120.5 litres. Although there is a clear distinction between barrels and tonnes in the inventories, it is important to note that the containers, which were counted by officials in the cellars, may not have corresponded to any standard norms.

The quantities of beverages listed in the Rehden inventories include larger and smaller barrels, barrels of two tonnes, tonnes, and *standen*, which makes conversions into litres and a graphic representation of the amounts in a chart impossible. The lists include mead, ale, old ale, March ale, ship ale, *collatio* ale (consumed during a light meal in the afternoon), and convent ale, the latter reflecting the castle's role as a religious house. Ale was also imported from Danzig (Pol. Gdańsk), Elbing, Bromberg (Pol. Bydgoszcz), and, significantly further away, from Wismar, a major centre of beer production in the Baltic (Blanckenburg 2001, 90–108), alongside wine. The 1434 inventory also mentions mugwort ale, which was frequently stored in other castles during the fifteenth and sixteenth centuries. Thus the Rehden con-

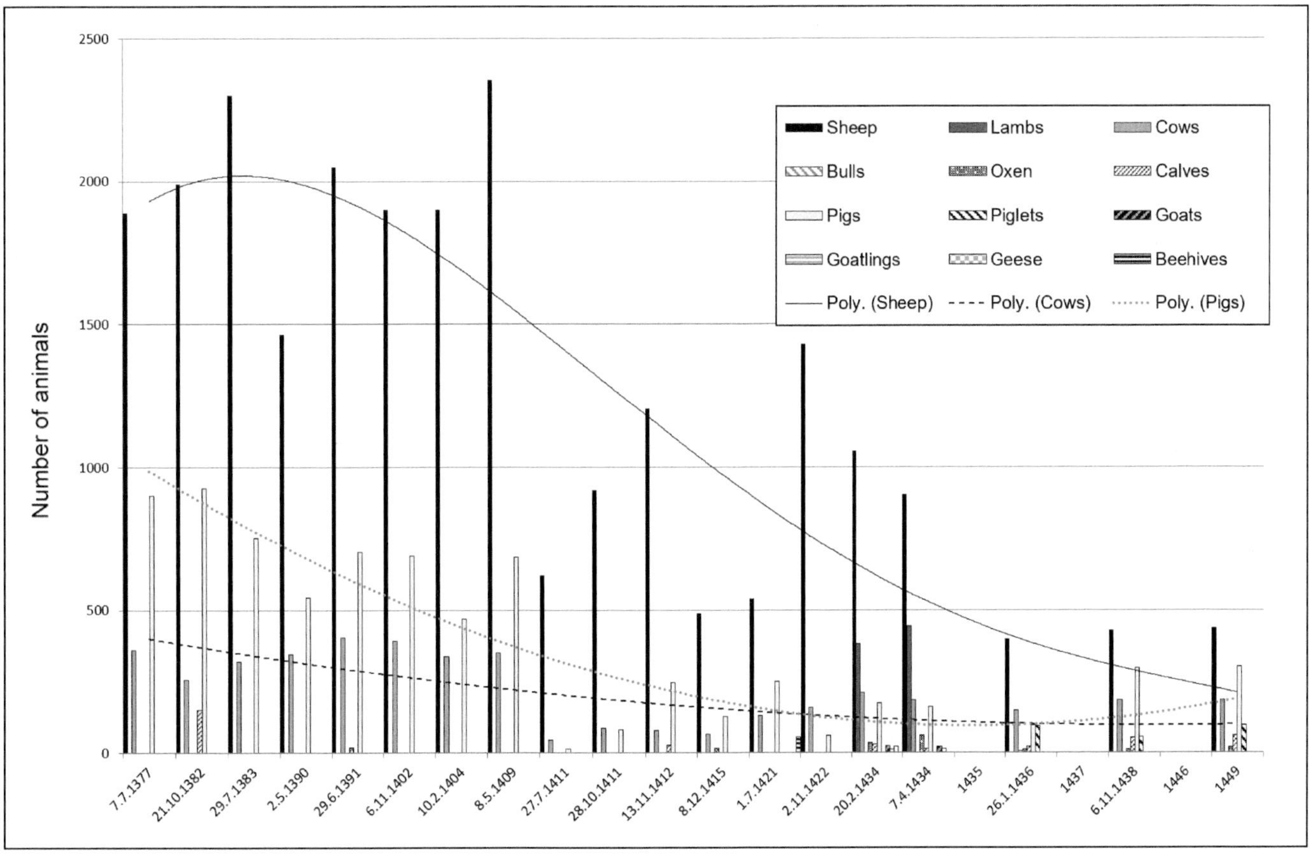

Figure 15.2: Numbers of livestock owned by the convent of Rehden (Pol. Radzyń Chełmiński), with the polynomial trends for sheep, cattle, and pigs showing an overall decrease from the late fourteenth to the mid-fifteenth century. Figure by M. Jarzebowski.

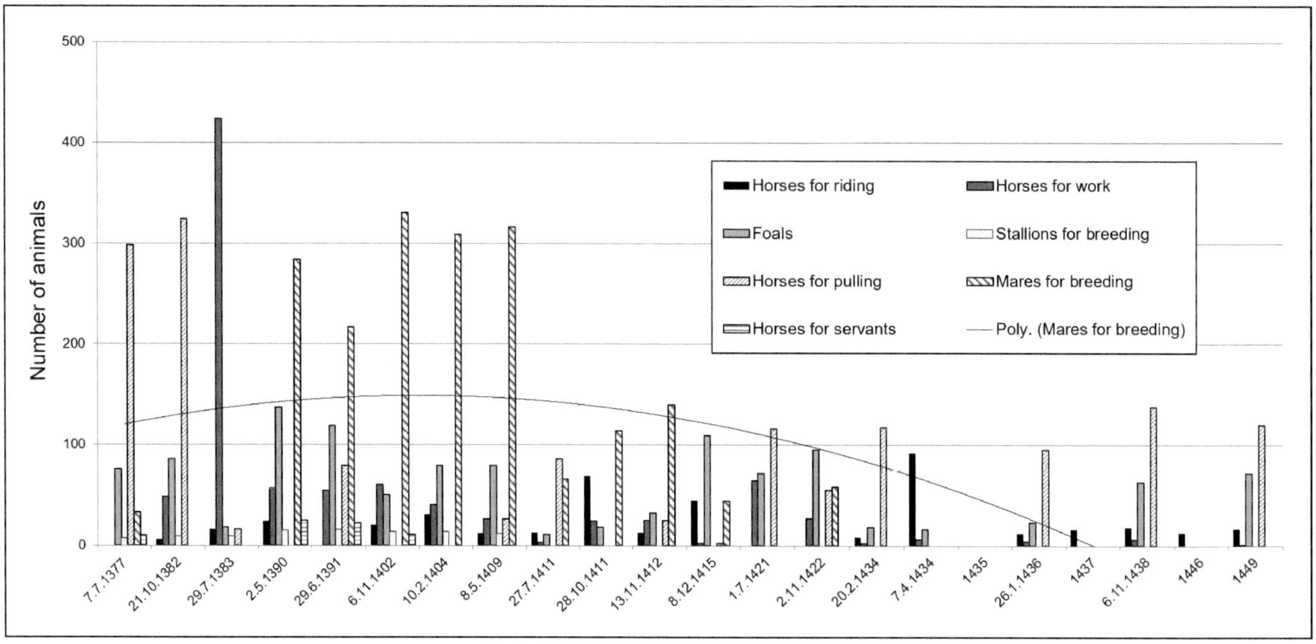

Figure 15.3: Numbers and types of horses owned by the convent of Rehden (Pol. Radzyń Chełmiński), with the polynomial trend for breeding mares showing an overall decrease from the late fourteenth to the mid-fifteenth century. Figure by M. Jarzebowski.

vent relied on a broad variety of beverages with and without their place of origin specified, both locally produced (from grain and honey) and imported. This former is supported by references to a brewery, a malt house, a cellar for the storage of honey, and a possible apiary within the castle's grounds (Blanckenburg 2001, 46).

Livestock (Fig. 15.2)

Livestock were described according to their age (young/old, of the same year, two years, three years, etc.) sex (especially cattle), and purpose (for breeding or fattening, oxen for pulling). The broadest diversity of categories in the inventories related to horses, as the most important animals for agriculture, administration, transport and military purposes (see below).

The earliest inventory for Rehden from 1377 dates from a time when the convent's commanders increased their engagement with agriculture and stockbreeding. The castle itself had stables and pigsties within its precinct (Gancewski 2001, 46), but the majority of herds were kept on farms within the commandery. In 1365, Commander Petzold von Korwis acquired the village of Bliesen (Pol. Blizno), and the place is listed, alongside Bogusch (Pol. Boguszewo) and Schwetz (Pol. Świecie), as a cattle farm in the Rehden inventories from 1391 (Frölich 1868, 42, 315). When Commander Engelhard Rabe von Wildenstein sold the farm at Klein Sagen (Pol. Gołębiewko) in 1386, he obliged its inhabitants to deliver timber from the Order's beech forest to the convent, as well as to cut grass and to bring it to the nearest farm to be used as fodder (Frölich 1868, 81). Whilst the first four (and some of the later) inventories summarize numbers of livestock, including horses, with formulas such as 'on all farms' or *summa summarum*, the 1391 inventory and some of the later texts refer to them by name. Not all of them have been clearly identified, but they included the cattle farm of Rehden, Bogusch, Schwetz, Semyn (Pol. Sumin?), Konyad (Pol. Konojady), Raczlow, Blise (Pol. Blizno), and Kobilbude. From 1411, the commanders of Rehden also used the farm at Aldenhof, which was part of the village of Melno (Pol. Mełno) (Frölich 1868, 213), and in 1434 a farm is referred to at Paulsdorf (Pol. Przesławice?) and in 1449 at a place called Lynkorn (Pol. Łąkorek?). As in the case of Stuhm, sheep dominated the herds present on the Order's farms, with about two thousand animals listed between 1377 and 1409, followed by pigs, which make up around half or a third of the numbers compared to sheep. Of the 722 pigs which are mentioned in the 1391 inventory, 433 are noted as being grazed in the 'beech forest' (Mager 1960, 359). Cattle is in third place, with 250–400 animals listed until 1410. After this period of rapid decline, sheep numbers recovered and reached between 500 and 1500 again in the 1420s and early 1430s. The most detailed inventories are the two from

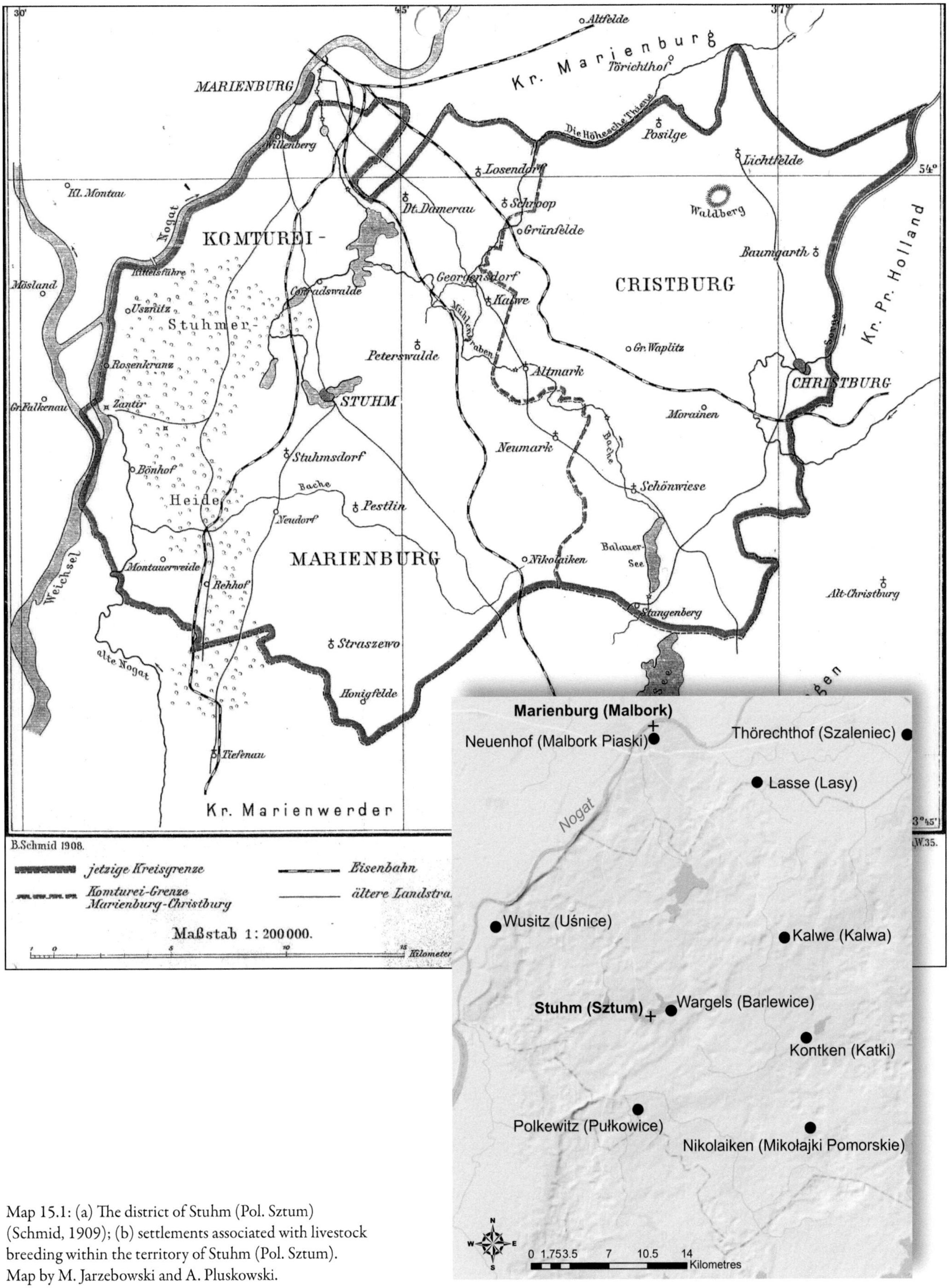

Map 15.1: (a) The district of Stuhm (Pol. Sztum) (Schmid, 1909); (b) settlements associated with livestock breeding within the territory of Stuhm (Pol. Sztum). Map by M. Jarzebowski and A. Pluskowski.

1434, which mention — alongside sheep, pigs, and cattle — lambs, calves, piglets, goats, geese, and beehives. At Paulsdorf in 1436 there was a *Mietkuh*, which is only noted again on the farm at Roschau (Pol. Rościszewo) owned by the castle at Dirschau, in 1415 and 1418 (*GAD*, 722, 724). The charter for Miller Niclas from 1449 states that he was granted pasture land for twelve cows, two horses, and twenty pigs, which were all to be kept together with the commander's livestock (*OBA* 10079, 1449 Oct 19; included in *OBA* 63). The sources differentiate livestock belonging to the Order from those which were the property of local farmers, although as subjects they were required to pay for the privilege (Boockmann 1983, 570). Numbers of horses belonging to Rehden follow a similar trend, with the substantial breeding population decreasing significantly after 1410 (Fig. 15.3) (see also Chapter 12).

In summary, the inventories of Rehden's convent, within one of the more densely settled commanderies of the Prussian *Ordensland*, point to a sustained focus on the production of grain in the decades before 1410. This would have derived from the convent's subjects, which included the adjacent town. The breeding of sheep, presumably for wool, was also important within the context of the commandery and appears to have been maintained after 1410.

Managing the Centre of Prussia: The Advocateship of Stuhm

The commandery of Marienburg has attracted extensive attention from scholars, and its relatively abundant documentary sources have been synthesized most recently by Jóźwiak and Trupinda (2007). The convent itself was headed by a commander (and also by his deputy *hauskomtur*), but the Grand Master clearly made regular decisions concerning the administration of the commandery. This was subdivided into smaller administrative districts, the most important of which relating to this study was the advocateship governed from the castle at Stuhm, which encompassed a large portion of the southern part of the commandery (Map 15.1a), whilst the adjacent Forest of Stuhm was managed by at least one *Waldmeister* who was usually subservient to Marienburg, but at times under the administration of the advocate of Stuhm.

The southern part of the Marienburg's commandery can be subdivided into two areas. The first is associated with the convent and adjacent town which was granted its charter in 1286, not long after the construction of the castle. The expansion of the castle complex in the first half of the fourteenth century was accompanied by major hydrological works in the landscape to the north and north-east of the Forest of Stuhm (Ger. Stuhmer Heide), alongside an intensive process of colonization, especially in the floodplain of the Nogat and Vistula. In contrast, the early settlement of the Stuhm district between the late thirteenth and mid-fourteenth century was limited; only a few foundation charters have survived, specifically for Conradswalde (1284) (Pol. Koniecwałd; *PrUB* I–II, 446), Pestlin (1295) (Pol. Postolin; *PrUB*, I–II, 633), Kalwe (1297) (Pol. Kalwa; *PrUB* I–II, 672), Neumark (1336) (Pol. Nowy Targ; *PrUB* III, 82), and Willenberg (1374) (Pol. Wielbark; Semrau 1928, 202). Stuhm itself would remain a village until it was granted a town charter by the Grand Master in 1416, a means of bolstering the defence of Marienburg's territory (Biskup 1956). Aside from the usual regulations concerning tithes and work duties, there are very few special obligations relating to the castle. The villagers of Conradswalde were allowed to catch fish in the lakes and brooks within their allotted territory, and — for their own use only — to lead cattle onto the heath for pasture, as well as to cut wood for building material and fuel, except for large trees, which were used for beehives (Biskup 1956, 80). The name of the settlement linked the woodland with the locator Conrad, who had been granted 60 *Hufen* in 1284 (Semrau 1928, 79). This can be related to the role of the castle, which never gained major economic or political importance. Its principal role appears to have been strategic; the castle was situated on a strip of land flanked by two lakes, controlling the road from Marienburg to Marienwerder and ultimately through to the Kulmerland. Its secondary role as a tool of the Grand Master was the management of its defined *Ämter*, which included directly contributing to the maintenance of the Marienburg convent.

Whilst its topographic location was comparable to the similarly-sized castle of Lötzen (see below), the advocateship of Stuhm had significantly fewer lakes. This is also reflected in the castle's inventories, which hardly mention any stocks of freshwater fish. A description of the Stuhm eldership from 1565 mentions only five lakes, the White Lake (Jezioro Białe, today called Sztum Lake), the Periotha (Parlety), the Goreje (near Braunswalde), the Virgin Lake (Pol. Jezioro Przezwane Pannie), the Slugza (formerly Ger. Neuer See), and the Damerau Lake (Pol. Jezioro Dąbrówka) (Semrau 1928, 5–7). Correspondingly, there were also few smaller streams, which limited the construction of water mills. Canalization in the landscape to the south of Marienburg enabled a series of mills to be constructed leading up to the town, whilst the wetlands of the Vistula floodplain contained many more (see Chapter 13). During the conflict in 1410, Stuhm Castle was captured by Polish troops but regained by the Order after a few weeks. This was followed by a period of repair and extension work, which was unusual during this phase of stagnation, and almost certainly reflects the continued use of Marienburg (which resisted the Polish siege) as the headquarters of the Order (Torbus 1997, 180ff.). As with many other castles, this break is clearly recognizable in the sequence of inventories. During 1454, the first year of the war against Poland and the Prussian Confederation, Stuhm Castle was seized again, but handed back to the Order after its victory at the battle of Konitz (Pol. Chojnice) and remained in its possession until 1466 (Torbus 1997, 173).

Managing Woodland

The southern-central part of the Marienburg commandery was dominated by the Forest of Stuhm and the Stuhm Heath, which reached along the eastern banks of the rivers Weichsel and Nogat from the area between Tiefenau (Pol. Tychnowy) and Rehhof (Pol. Ryjewo) to Willenberg near Marienburg, with its eastern borders at Pestlin (Pol. Postolin) and Stuhm (Schmid 1909, 344). The *Waldamt* of Bönhof (Pol. Benowo), situated within the heath, was probably established during the construction of Stuhm Castle in *c*V1331. The Bönhof and Stuhm inventories in the *Marienburger Ämterbuch* mention horses for work in the forest (*Waldschweiken*) and refer to the production of timber (*AMH*, 212, 215, 305), charcoal production on the heath and at Braunswalde (Pol. Gościszewo), as well as tar and honey (*PrUB* I–II, 446; Semrau 1928, 80). According to the inventories, honey was kept at Bönhof, and the product is also mentioned in charters such as the one for Conradswalde from 1284. The place-names Bönhof (*Bön* derived from *Bienen*, 'bees') and Honigfelde (Pol. Trzciano, 'honey field') also have clear connections with the production of honey and wax. At the same time, the *Waldmeister* at Bönhof managed pastures which were probably situated within the heath and along the borders of the forest, and Bönhof was one of several cattle farms which served the castle at Stuhm (see below). Despite the presence of the extensive corridor of woodland which had remained more or less intact from the pre-crusade period, consisting mainly of oak, lime, beech, alder and — especially between Tiefenau and Rehhof — of spruce trees (see Chapter 11), levels of biodiversity had dropped. The absence of game in the kitchen inventories of Stuhm Castle, except for flitches of wild boar in 1437 (though the unspecified meat may have also partly been game), suggests that the population of wild ungulates within the *Ämter* may have been limited, although an absence of local game is only noted in the eighteenth century (Semrau 1928, 16–20).

The Park of Stuhm

The castle at Stuhm is best known as the so-called 'summer residence' of the Grand Master from at least the last decades of the fourteenth century, which may be connected with the construction of a park mentioned in the *Tresslerbuch*. The park would have been a recognizable status symbol, created at a time when the Grand Master's court was adopting increasingly secular aristocratic traits. But the park's function may have been the provision of game, rather than providing a theatrical setting for hunting, especially if wild ungulates were already being depleted within the commandery. The hunting activities of the Order's higher officials are rarely mentioned in the extant written sources. In October 1403, the Grand Master crossed the Nogat on the so-called *Kittelsferry* near Wernersdorf (Pol. Pogorzała Wieś) to go hunting on the 'geese meadow' (Ger. *Gänsewiese*), situated between the Nogat and the Vistula (*MTB*, 270). The limited role of game in the diet of the Marienburg convent is also evident in the faunal assemblages recovered from the castle complex (Maltby et al. 2009; see Chapter 12). A wall painting of a hunting scene in the convent's high castle, dated to *c.* 1380–90, may have reflected this secular tendency or was also plausibly part of a larger and more explicitly religious iconographic programme (Jóźwiak and Trupinda 2007, 133). Game was more likely to have been procured by the Order's subjects, which included specialists. The *Tresslerbuch* lists expenses for servants transporting deer to the park at Stuhm, for the erection of fences and for the keeper of the *Meerkuh* and *Meerochse*. Treichel (1899, 31) interprets *Meerochse* as *Moorochse* (moss oxen) and therefore as elk, which would be possible, as two *Meerkühe* were sent to the Grand Master by the commander of Balga from the wooded wetlands in eastern Prussia. By the end of the fourteenth century, while elk was regularly encountered in the east, they were comparatively rarer in Marienburg's commandery and required specialist handlers to maintain, as the *Tresslerbuch* differentiates

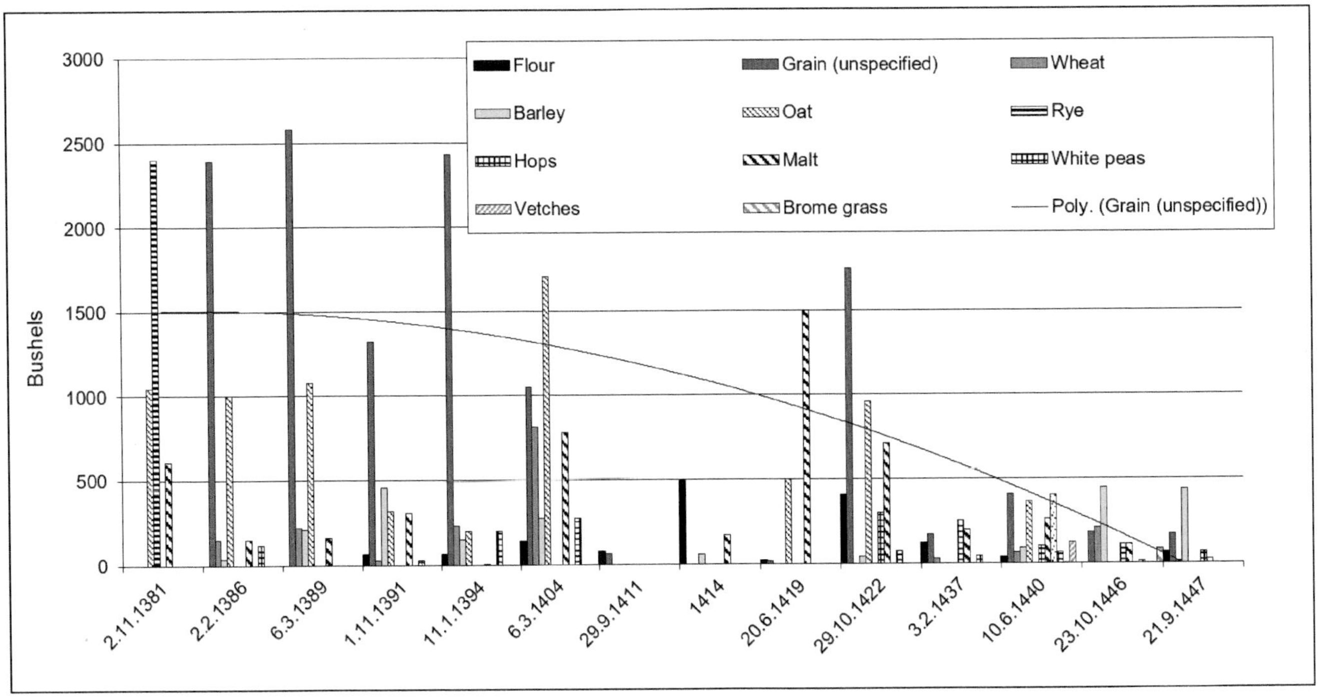

Figure 15.4: Grain measures at the castle of Stuhm (Pol. Sztum), with the polynomial trend for grain showing an overall decrease from the late fourteenth to the mid-fifteenth century. Figure by M. Jarzebowski.

the keepers of the animals in general (i.e., deer) from those responsible for the *Meerkühe*. The park at Stuhm can therefore be situated within the context of the provisioning networks established to sustain the Order's headquarters, with game playing a small but conspicuously high-status component of the diet of the convent's highest officials and their more important guests.

The Inventories of Stuhm

The inventories for Stuhm are included in the *Marienburger Ämterbuch* rather than the *Großes Ämterbuch*, as the district fell under the general administration of the Grand Master, delegated in turn to the advocate. The *Marienburger Ämterbuch* (*OF* 129) is a paper manuscript of 308 pages, covering the period from 1375 to 1442, and it has been the main source for Walter Ziesemer's edition from 1916, supplemented by texts from *OF* 132 and other material. The first ten inventories from Stuhm are from *OF* 129, the 1422 and the 1437 texts too, but with secondary traditions in the *Ordensbriefarchiv*, of which the most detailed is the 1440 inventory. The 1446 and 1447 lists are from *OF* 132. Together the inventories span a period of almost seventy years, including the most prosperous period for the Order in Prussia during the late fourteenth and early fifteenth centuries. However, the break caused by Grunwald and the first Peace of Thorn in 1411 is clearly visible in the diachronic trends of the castle's stores. No inventories have survived from the period of the conflict between the Order and the Prussian Confederation, the Thirteen Years' War from 1453, or from the early sixteenth century. The inventories summarize the mobile property, food provisions, and livestock of the castle, alongside its economic facilities consisting of farms, mills, stores, and workshops, along with the farms and mills in district governed by the castle.

In addition to the farm located in front of the castle, the others were at Wargels (Pol. Barlewice), Wusitz (Pol. Uśnice) (both mentioned in 1391–1447), Laase (Pol. Lasy) (1394–1447), and only until the beginning of the fifteenth century at Neuenhof (Pol. Malbork Piaski) (1394–1404) and from *c.* 1440 at Nikolaiken (or Niclasdorf; Pol. Mikołajki Pomorskie) (1440–47) (Semrau 1928, 32, 37, 86, 107ff., 156, 182ff.). Other places came under the supervision of the advocate of Stuhm only temporarily, such as Kontken (Pol. Kątki) (1404), Thörechthof (Pol. Szaleniec) (1419), Polkewitz (Pol. Pułkowice) (1440), and Kalwe (Pol. Kalwa) (Kalesim, 1440) (Semrau 1928, 82, 123ff., 156) (Map 15.1b). The fact that the inventories refer to various places rather than solely the castle, especially in the case of livestock including horses, may lead to some distortion when this data is combined. While the 1389 and 1391 inventories list the livestock from 'all of the farms and in (or in front of) the house (i.e., Stuhm Castle) itself', the later inventories specify livestock for each farm.

Grain (Fig. 15.4)

The overview of the provisioning of Stuhm Castle with grain represented in Fig. 15.4a–c (for 1381–1411 and 1414–47) clearly shows the break resulting from the conflict with Poland and the consequences of 1410, which included laying waste to Marienburg's commandery during the siege of the convent. The frequent references to large amounts of unspecified grain (and flour) makes it difficult to draw precise conclusions about what varieties were stored at the castle, but it seems that rye lost its importance during the late fourteenth century, whereas all other types of grain were stored throughout the entire period in irregular amounts, such as wheat, oat, hop, and (white) peas, which clearly had the status of grain; it also seems that barley was of limited importance. Seasonal differences are also difficult to detect. It is possible the list from October 1422 represents the grain stocks immediately after harvesting, while the much smaller quantities in the February 1437 inventory reflects the consumption of stored grain during the winter. However, the amounts in the first three inventories from 1381, 1386, and 1389 are quite similar, although they reflect the status of November, February, and March respectively, so it is not possible to identify any diachronic pattern. With its close proximity to Marienburg, the agricultural importance of the Vistula and Nogat region, as well as the highly developed infrastructure of the Order's heartland linked to the fundamental role of trade, there was no need to store consistent amounts of grain at Stuhm. The requirements of the castle clearly varied according to the situation, but it did not house a convent of brethren who would have depended on a regular supply. The provisioning needs of the advocate's household would have been comparatively small.

The largest quantity of grain features in the 1404 inventory, which may be connected to the construction of a new granary between 1400 and 1403 documented in the *Tresslerbuch* (*MTB*, 51, 94, 137, 210.). The inventory from 1440 also lists grain which was stored for sowing, though this information is clearly incomplete (Semrau 1928, 156). It seems that the first few decades of the fifteenth century was a period of agricultural development within the advocateship. The 1440 inventory is the only one which mentions vetches (i.e., peas) amongst the grain. This is a rare instance, as the *Großes Ämterbuch* refers to vetches only once, also in 1440, in Schwetz (Pol. Świecie) (*GAD* 630, line 24). In the *Marienburger Ämterbuch* vetches are mentioned also in 1394 at Lesewicz (Pol. Lasowice), a few kilometres north of Marienburg, and in 1419 at Pomyn, a farm belonging to Mösland (Pol. Międzyłęż), west of Stuhm

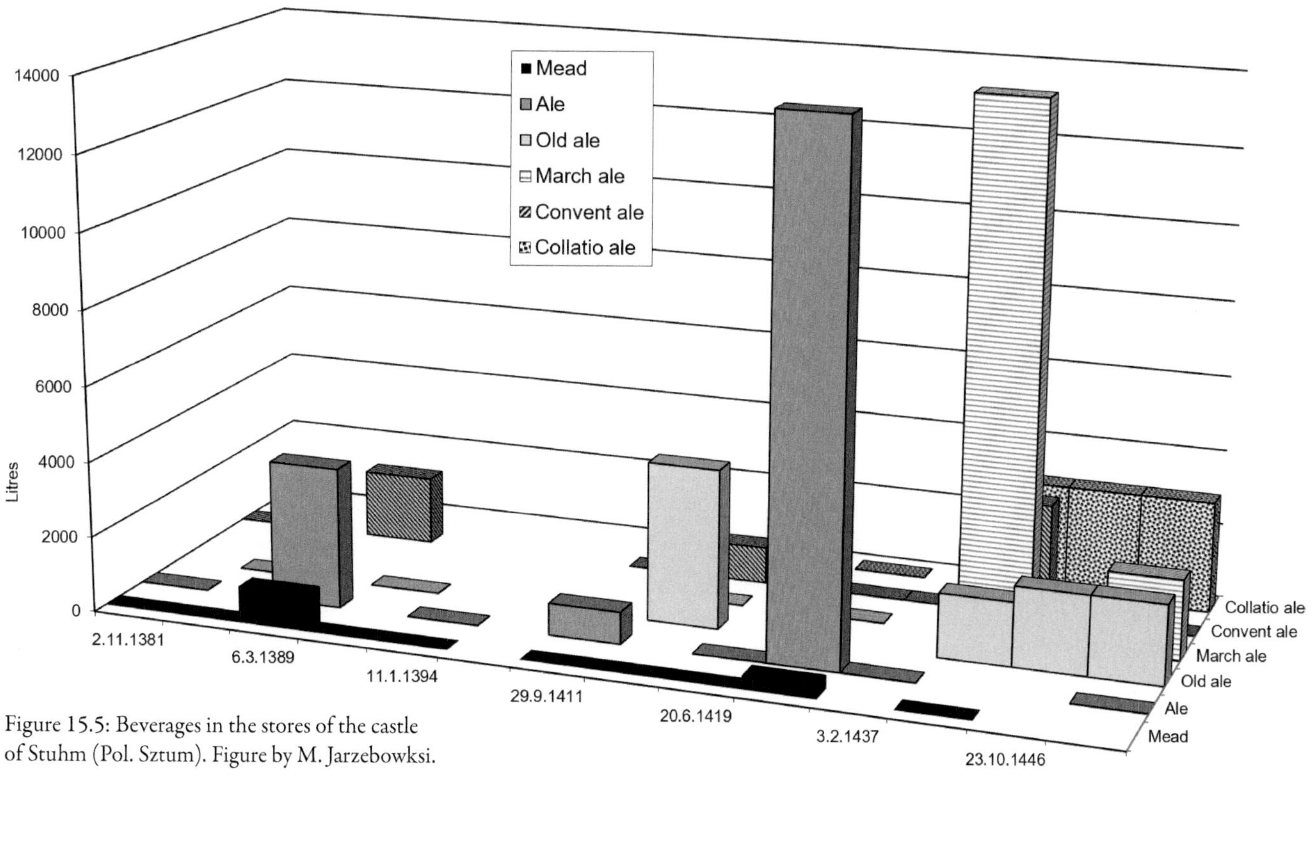

Figure 15.5: Beverages in the stores of the castle of Stuhm (Pol. Sztum). Figure by M. Jarzebowksi.

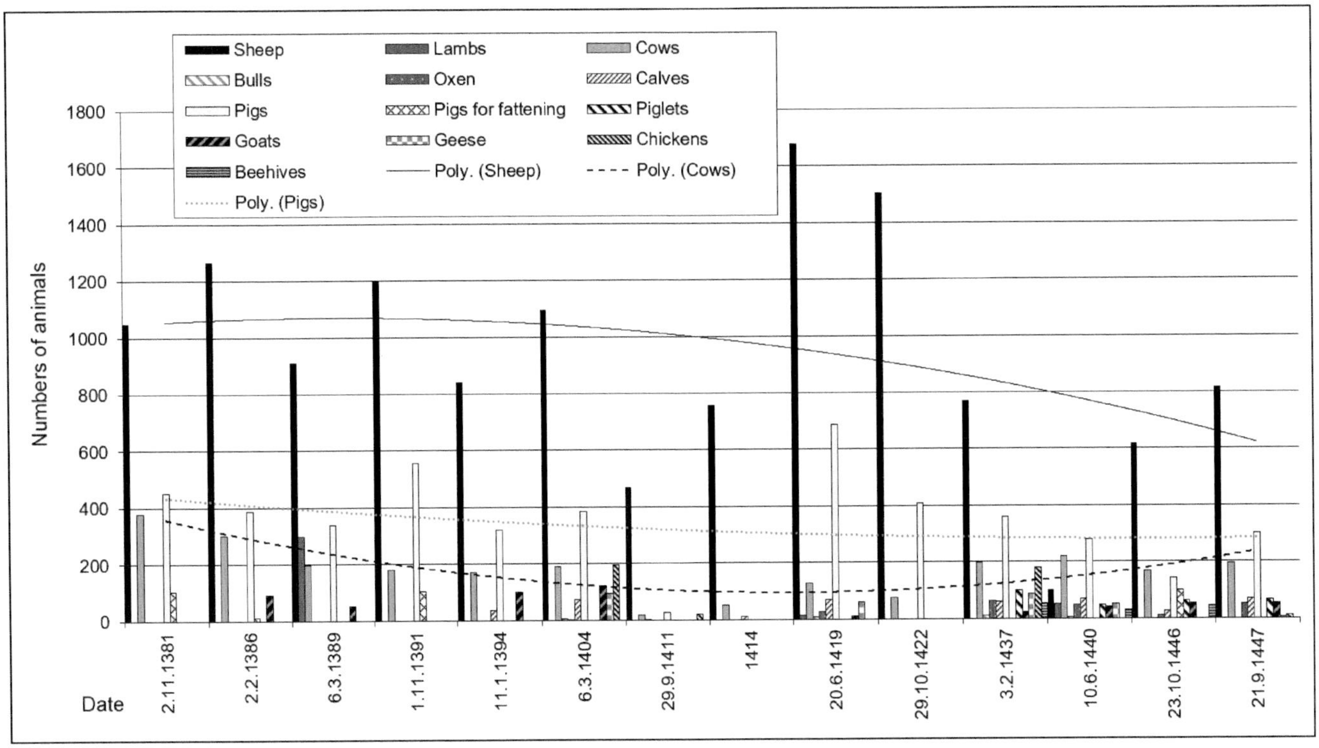

Figure 15.6: Numbers of livestock owned by the castle of Stuhm (Pol. Sztum), with the polynomial trends for sheep, cattle, and pigs showing an overall decrease in sheep from the late fourteenth to the mid-fifteenth century and a renewal in cattle stocks after 1410. Figure by M. Jarzebowski.

(*MA*, 67, 85), which may indicate that vetches were only grown along the Vistula. The 1446 Stuhm inventory registers two more species which appear exclusively here: brome grass (*bromus*), used for feeding livestock, and spelt (*Triticum spelta*). A number of mills were situated along the canal between Stuhm and Marienburg, which was probably built during the first half of the fourteenth century. These mills were governed by Marienburg's mill master, while the advocate of Stuhm supervised the windmills at Stuhm, Pirswitten (Pol. Pierzchowice, documented in 1411–46) and Tillendorf (Pol. Tulice) (documented in 1437 / 1446), as well as the watermill at Schroop (Pol. Szropy, documented in 1446) (Semrau 1928, 121, 145, 180).

Beverages (Fig. 15.5)

Half of the inventories make no mention of beverages at all. The 1422 list has an extraordinary amount of ale (more than 13,000 litres), the 1440 inventory registers the same amount of March ale, and the other reserves of old ale and *collatio* ale only feature in regular amounts in 1440, 1446, and 1447. It is remarkable that no imported beer from nearby cities such as Danzig and Elbing is mentioned. Wine was produced locally, as a wine garden is mentioned in the early fifteenth century (*MTB*, 94; Treichel 1900, 26ff.), and it may be that the advocate's household could not afford to import wine after its other expenditures. In April 1457, the captain of Stuhm Castle (during the events of the Thirteen Years' War), wrote to the Grand Master that he would willingly host him, but that he had to bring his own beer and any other food he required (*OBA* 14856).

Livestock (Fig. 15.6)

As in the case of Nessau, sheep dominate the inventories of Stuhm, but in contrast the second most abundant group of livestock was pigs rather than cattle. Numbers of pigs are around one-third to a half of those of sheep, and this proportion as well as the total numbers seem to have been barely affected by times of crisis, except during the years 1411 and 1414, which followed the Order's defeat and the attack on the commandery in 1410. Just a few years later, in 1419, the inventory lists the largest number of sheep on the farms belonging to Stuhm Castle, with more than 1600 animals and numbers of cattle occupying third place amongst livestock. The four inventories from 1437 to 1447 are the most specific in terms of details. They are not (exclusively) from the *Marienburger*

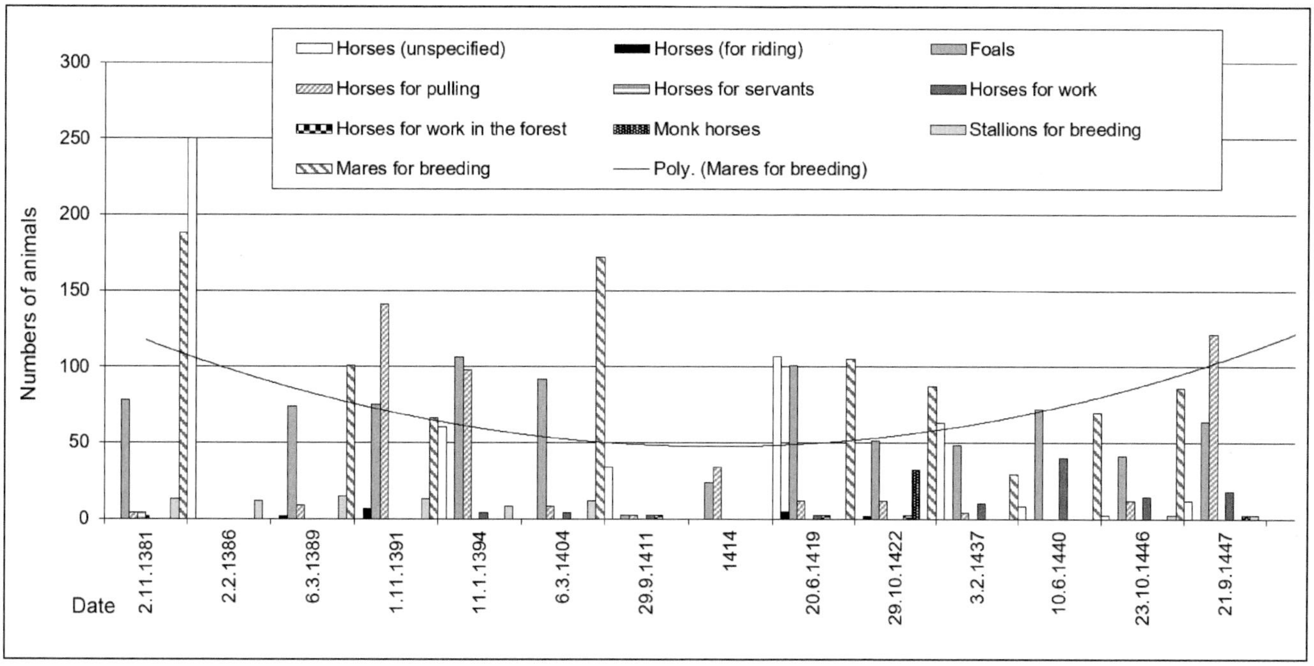

Figure 15.7: Numbers and types of horses owned by the castle of Stuhm (Pol. Sztum), with the polynomial trend for breeding mares showing a renewal after 1410. Figure by M. Jarzebowski.

Ämterbuch (*OF* 129), but the 1437 list in Ziesemer's edition is supplemented from the *Ordensbriefarchiv*, which is also the source for the 1440 inventory, while the 1446 and 1447 inventories are from *OF* 132. They also mention geese, chickens, and ducks. They list pigs, pigs for fattening, and piglets separately, they also have goats (such as the small number in the 1419 list), beehives and, in 1440 at the farm at Niclasdorf, two peafowl. Peafowl are also listed in the inventories of the *Großes Ämterbuch*, but only for 1488 in Osterode (Ostróda) and 1513/1514 for Angerburg (Pol. Węgorzewo). Peacock feathers — as décor in churches — appear in 1416/1422 at Mewe (Pol. Gniew), between 1416 and 1442 in Schönsee (Pol. Kowalewo Pomorskie), in 1428 at Strasburg (Pol. Brodnica), in 1437 at Brandenburg (Rus. Uschakovo) and Schlochau (Pol. Człuchów), between 1437 and 1447 in Rastenburg (Pol. Kętrzyn), and in 1508/1510 again in Angerburg. The peacock was an established Christian symbol of eternity and was popularly used in some regions of Europe in the fifteenth century to refer to dynastic longevity. It is entirely possible this meaning was promoted by the Order during a time when their rule in Prussia was being contested (see Chapters 12 and 16).

To summarize, the purpose of the advocateship of Stuhm was to support Marienburg in the management of a substantial portion of its commandery, which included a mixture of land use. Emphasis was placed on grain production, forestry, which was also delegated to Bönhof, and herds of sheep and pigs. The castle's own needs and finances were also much smaller compared to those of a convent. After 1410, there was renewed investment in the castle reflecting the continued importance of the advocateship in the delegated management of the heartland of the Order's Prussian territories. This is also reflected in the short-lived renewal of horse breeding (Fig. 15.7).

Organizing a Frontier: The Commandery of Nessau

For much of its existence, the commandery of Nessau was on the frontier between the *Ordensland* and Polish lands, and along with Thorn facilitated access to Kuyavia (Torbus 1998, 555). There is some discrepancy between the first mention of a commander of Nessau in 1255 and the earliest reference to a castle in 1348, whilst excavations at Mała Nieszawka indicated a structure established in the fourteenth century (see Chapter 10). The inventories cover the period from 1376 to 1435, and the commandery appears to have been highly productive before 1410; the charters for its subservient villages indicate that standard tax contributions were paid to the convent in the form of grain, hens, eggs, wax, geese, and hemp (Märcker 1899, 634–39, 642). At the same time, more specific obligations were included in some

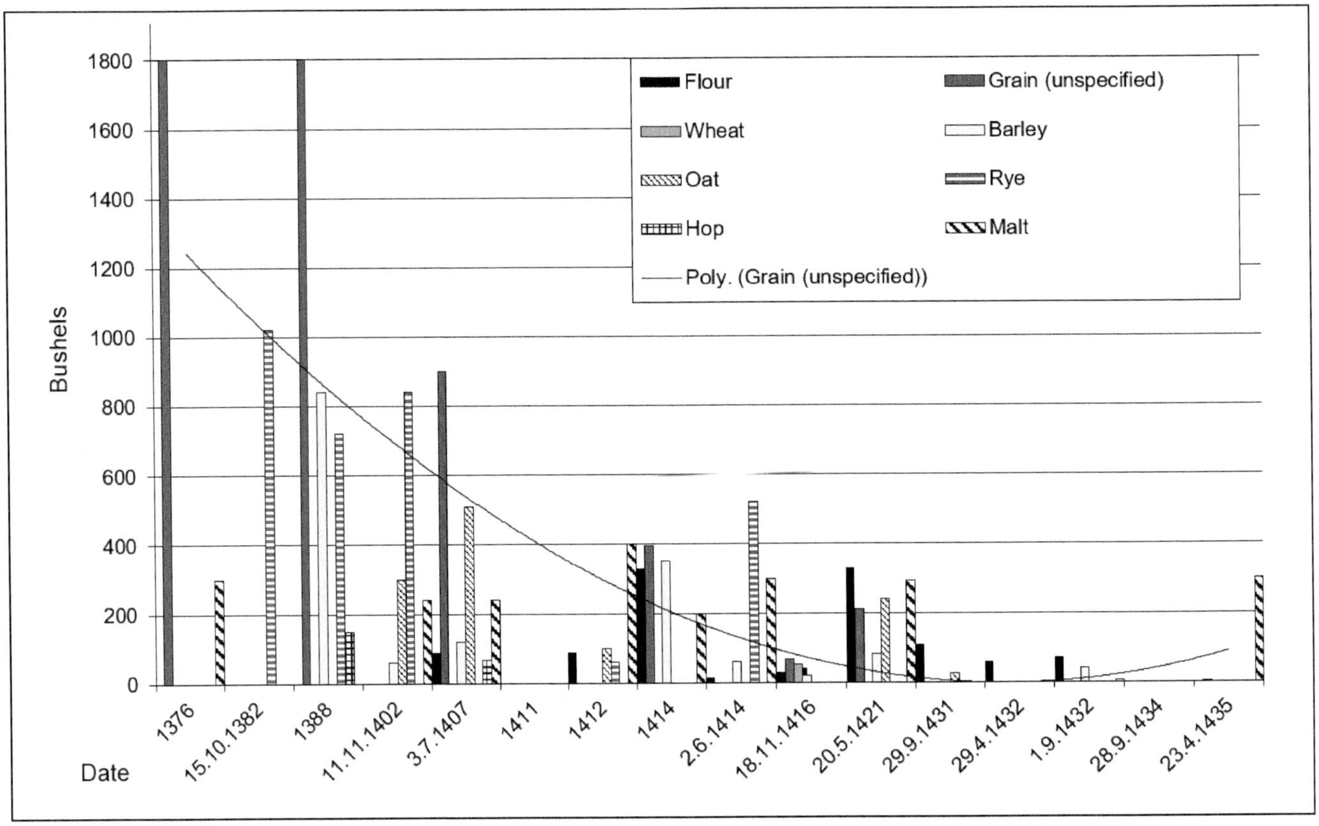

Figure 15.8: Grain measures at the convent of Nessau (Pol. Mała Nieszawka), with the polynomial trend for grain showing an overall decrease from the late fourteenth to the early fifteenth century. Figure by M. Jarzebowski.

of the charters, alongside privileges that indicate a level of dependency between the convent and its subservient settlements. The inhabitants of Kostbar (Pol. Kozibór) received permission in 1364 to lead their cattle into the *Heide und Felde* (heath and field) of the castle, as long as this did not result in any damage (Märcker 1899,322, 620). According to a charter from 1390, each farmer from the village of Kostrin was allowed to lead two pigs onto the Order's fields, to fish in the Vistula with a rod, to collect green wood on the Order's heath for building houses and dried wood for fuel, and to sell wood from the heath on the market, of which a portion of the proceeds had to be paid to the Order (Märcker 1899, 323, 634ff.). The villagers were also obliged to assist the Order in hunting roe deer for eight days each year. In 1409 the villagers of Nieder-Nessau received their charter from the commander which included the obligation to deliver hay to the castle (with the help of all the gardeners), to transport grain and, as in the case of Kostrin, to assist with hunting for eight days each year. All the inhabitants were expected to bring firewood to the brick barn and each had to deliver a day's portion of firewood to the castle once a year. The charters of Gross and Klein Stab contain similar regulations, with some additions. The inhabitants had to deliver grain and building timber, cut grass and provide hay for the castle. The women were also requested to wash and shear the sheep (Märcker 1899, 540, 640). This is only a fragmentary glimpse of how the commandery functioned in its last decades, but it suggests that any instability in the settlement pattern or access to the Vistula that the convent could not mitigate, would compromise its sustainability. As a result the prosperity of the convent was inextricably to the volatility of the frontier, and particularly the conflict with Poland-Lithuania in the early decades of the fifteenth century.

Grain (Fig. 15.8)

The diachronic evidence for grain stored in the castle, measured in *last* and *Scheffel* (bushel), reflects the three phases of the convent of Nessau, the first from the late fourteenth century until 1410, associated with large quantities of grain (the 85 *last* or 5100 bushels in the 1376 list cannot be represented in scale), a short recovery with lower quantities after 1411 until the destruc-

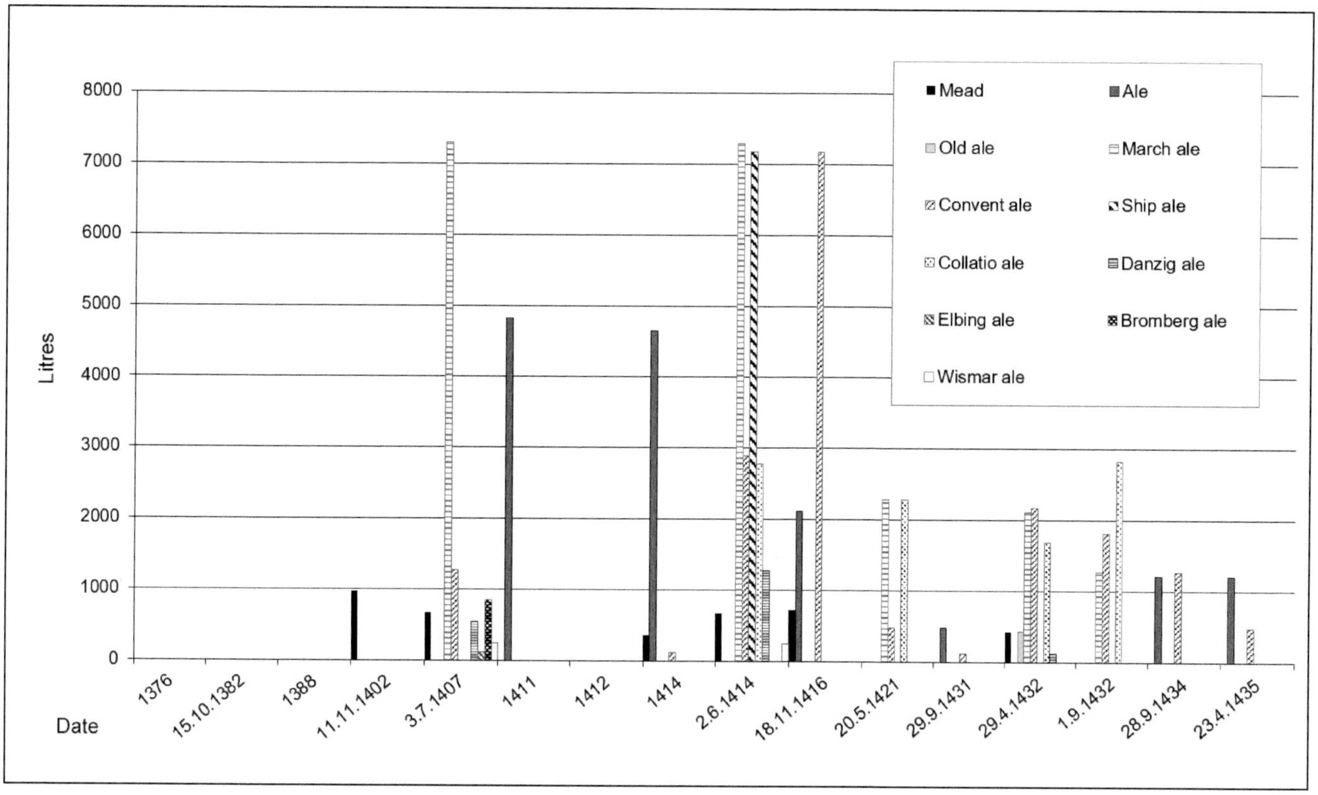

Figure 15.9: Beverages in the stores of the convent of Nessau (Pol. Mała Nieszawka). Figure by M. Jarzebowski.

tion of the castle, and the brief aftermath at Neu Nessau (Pol. Dybów). Apart from this phased trend, there are no regular patterns in grain storage at the convent, perhaps only when comparing the 1402 and 1407 inventories which show similar amounts of grain, although they were recorded in different seasons (November and July respectively). The listings of unspecified grain makes it additionally difficult to compare the different species. The irregularities may reflect the fact that grain was easily available throughout the year thanks to the Order's infrastructure of storage, transport, and trade with neighbouring Kuyavia and Masovia. However, it is clear that grain was also produced within the commandery.

The 1382 inventory mentions 'the mill', and other mills in the commandery are referred to in charters, namely the *Sandmühle* (1380) (Märcker 1899, 488, 632ff.), the *Herrenmühle* (1405) (Märcker 1899, 293, 641ff.), the *Eynnokenmühle* (impossible to localize today, but part of Ober-Nessau) (Märcker 1899, 234), and the *Haselmühle* (Märcker 1899, 287). In 1405, the commander of Nessau sold the *Herrenmühle* near Kostbar to Hans Wopetz and his heirs. They had to deliver 13½ *last* of rye each year and the convent could also use the mill for free. It must have been a substantial mill, as the miller received permission to add a sixth wheel. He was also granted a meadow near the Order's 'horse garden', use the Order's meadow for pasturing livestock, free fishing with small nets for personal use in the mill pond, and wood for fuel and building material for the mill and its wheels, with the exception of alder. The *Großes Zinsbuch* (Thielen 1958, 24, no. 87) mentions several mills with their respective income for the Order: the *Große Wirdelaw* (with a tax contribution of 10 *last* of grain), the *Kleine Wirdelaw* (8 marks), the *Herrenmühle* (12 *last* of rye and 12 fattened pigs), and the *Haselmühle* (30 marks and 4 bushels of wheat), all of which were deserted. The status of milling at this time — in *c.* 1414/1422 — can be clearly linked with ongoing warfare with Poland. The Nessau inventories also list additional amounts of grain, which were given to the convent's officials as separate tax contributions. (These amounts are not included in Fig. 15.8.) The 1414 inventory mentions 5 *last* of wood-oats (*Waldhafer*), which had to be delivered to feed the dogs of the Order's *Waldmeister* (Grimm and Grimm 1984, XXVII, col. 1142ff.). *Waldhafer* is also recorded in the *Großes Ämterbuch* for Osterrode (1446) alongside *Hühnerhafer* (chicken oats) and *Wiesenhafer* (meadow oats), and at Strassburg (1396), Tuchel (1396/1397/1404), and Schlochau (1382–1411) next to *Zinshafer* and *Frauenhafer*.

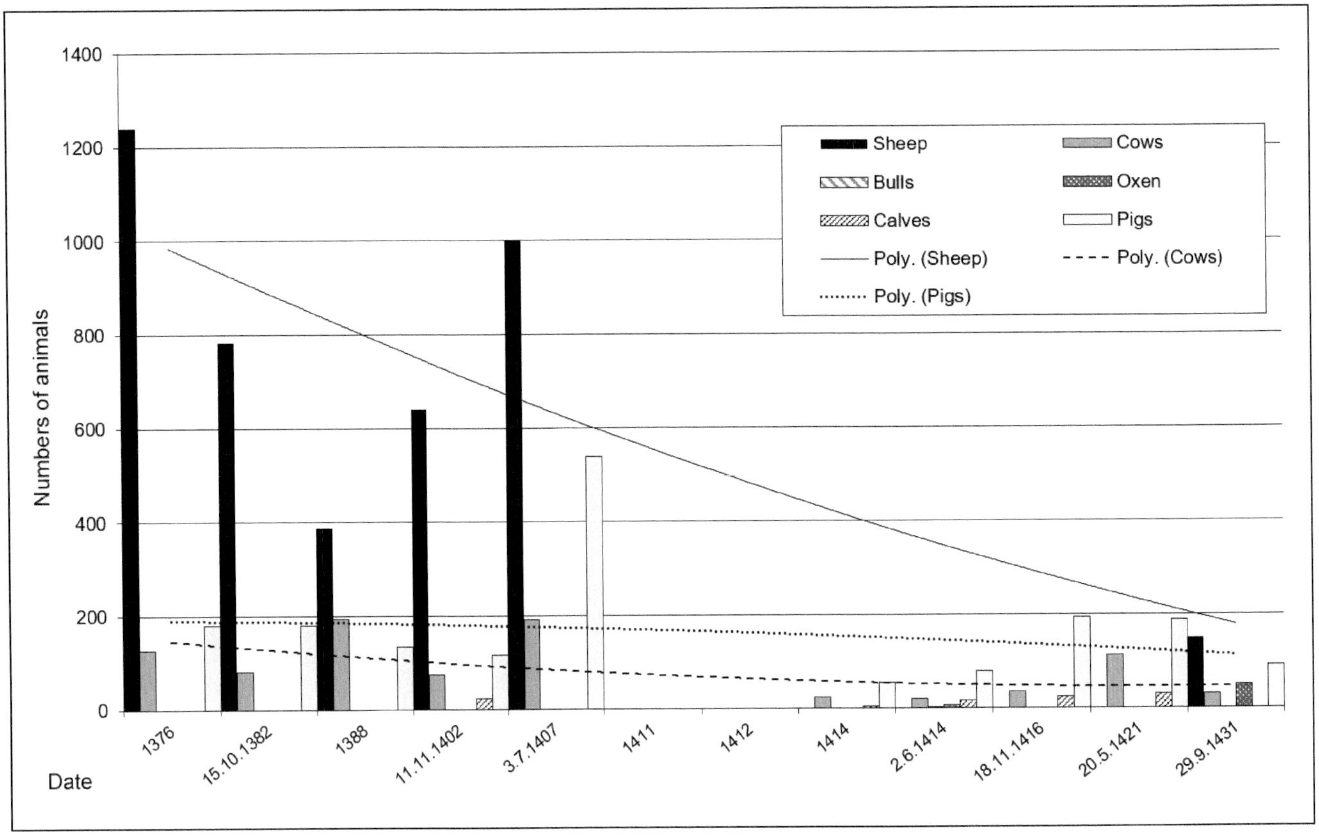

Figure 15.10: Numbers of livestock owned by the convent of Nessau (Pol. Mała Nieszawka), with the polynomial trends for sheep, cattle and pigs showing an overall decrease from the late fourteenth to the early fifteenth century. Figure by M. Jarzebowski.

Beverages (Fig. 15.9)

Beverages and other fluids were measured in barrels and tonnes. For the purposes of comparative measurements, a barrel has been converted into 422 litres and a tonne into 120.5 litres (39). Somewhat surprisingly, the diachronic trend in the storage of beverages at Nessau is antithetical to those for grain and livestock (including horses). There are hardly any beverages mentioned for the first phase (1376–1410), with the exception of 1407, while there were considerable quantities during the second phase between the treaties of Thorn and Melno (1411–22). Also the cellars of Neu Nessau in the third phase were far from empty. It is difficult to find probable explanations for this pattern, but some other interesting observations can be made in comparison to Lötzen (see below). While at Lötzen different kinds of herbal beers were consumed alongside ale made from grain, there are more varieties of imported beer at Nessau, not only from Danzig and Elbing (as in Lötzen in 1507) but also from nearby Bromberg (Pol. Bydgoszcz) and Wismar, with the broadest range of these four varieties in 1407. It is clear that Nessau, like Rehen, which had a similar level of imports, benefitted from its connection with the Vistula. It suggests that convents, irrespective of their location, had the purchasing power to bring in imported commodities whilst smaller castles, even if situated in the very heart of the *Ordensland*, as in the case of Stuhm, typically found this beyond their means. The Nessau inventories also include the types of beverage reflecting the castle's status, specifically the so-called convent and *collatio* beers. The inventories from 1416, 1432, and 1434 also list the beverages in the cellars of the commander and the convent separately, reflecting the social stratification of alimentation.

Livestock (Fig. 15.10)

The inventories of Nessau's cattle farm are dominated by sheep and cattle. The banks of the Vistula and the surrounding eyots, which appear several times in the charters of the Nessau commandery, provided valuable pasture. The break in 1410 is clearly visible in Fig. 15.10, when the castle was taken by the Poles who also seized the livestock within its vicinity. Sheep, which were predominant from the last quarter of the fourteenth century, do

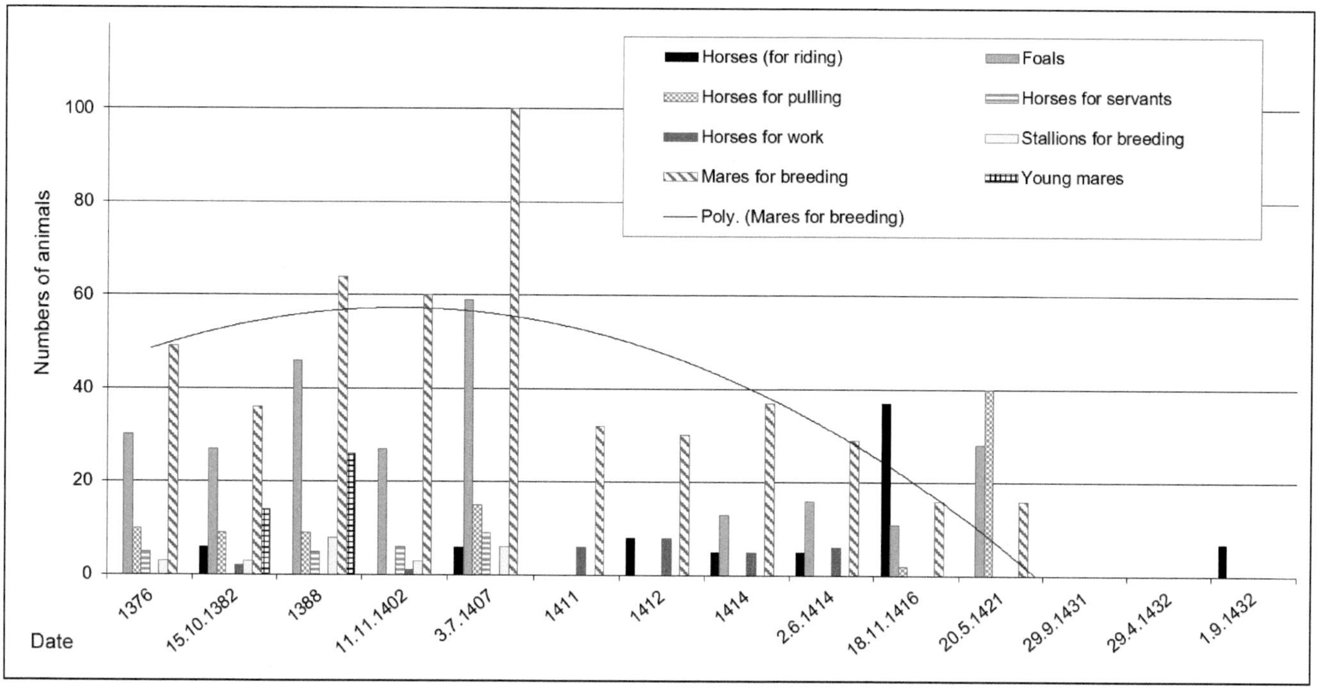

Figure 15.11: Numbers of horses at the convent of Nessau (Pol. Mała Nieszawka), with the polynomial trend for breeding mares showing an overall decrease from the late fourteenth to the early fifteenth century. Figure by M. Jarzebowski.

not appear again until 1431. Cattle and pigs, which are grouped only in the 1382 list (132 pigs under the care of the swineherd, forty pigs for fattening, eight pigs in the pigsty) were also important, while other animals such as goats are not listed in the inventories. The 1407 inventory has the largest number of livestock, which corresponds with the information from other sources that the years before 1410 were the most prosperous for the commandery (Märcker 1899, 9–11). In the period after the first Peace of Thorn, the number of cattle is much smaller than before 1410, while the number of pigs increases slowly to its former level. From the five inventories from 1431 to 1435, which refer to Neu Nessau, only one mentions livestock at all. Livestock were primarily stabled at Nessau's cattle farm (*Karwan*), but the 1416 inventory lists the *Karwan* and Kufros (Pol. Kowróz), whilst the 1421 inventory refers to the *Karwan*, the cattle farm at Birgelau (Pol. Bierzgłowo) and another *Karwan*. This data is summarised together in Fig. 15.10.

Horses (Fig. 15.11)

The detailed classification of horses according to sex, age, and function in the inventories reflects their importance for the Order (see Chapter 12) and the fact that Nessau was a major site for horse breeding. The lands along the Vistula provided fertile pasture ideally suited to rearing horses (Rünger 1925, 255, 303). The nomenclature used in the inventories is complex; the term *Pferd* (horse) appears frequently, but rather than a collective term it refers to animals used for riding and travelling, ranking in their quality from cavalry horses to work horses (Rünger 1925, 213). *Kobel* is an archaic word of unknown origin for 'mare', which, depending on the context, can refer to a mare for breeding or for other purposes. The 1382 inventory mentions 'Kobeln, die zu Rosse gehen' (i.e., mares brought together with studs), young *Kobeln* (probably mares which were not old enough for breeding), and *Kobelchen* (mare foals). The male equivalent of *Kobel* is *Ross*. This is also an archaic general term for 'horse' and especially for a military horse, but in the Prussian sources a *Ross* usually designated a male horse for breeding (Rünger 1925, 214). More difficult is the term *Hengst* (stallion). As Rünger shows, this is — in the Prussian sources — not a term specifying the sex, but it refers to the animal's function as a cavalry horse. Nevertheless, they were typically males, both stallions and geldings. Foals are usually listed according to their age: that is, if they were born that year, the year before, or two, three, or four years ago. Usually their sex is not specified, although sometimes their intended purpose is listed. A *Hengstfohlen*, for example, is a male animal envisaged for military service, a *Kobelchen* (i.e., little mare) a female foal envisaged for breeding. For the sake of clarity, all the foals from different age groups have been placed into one category in Fig. 15.11. Work horses are specified according to their tasks such as pulling carts or ploughs (*Wagenpferde, Pflugpferde*) or being used by servants (*Knechtpferde*). Sometimes they are referred to as *Schweiken*, which is an old Prussian term for plough-horses, probably derived from *sweikys* (healthy, strong) (Rünger 1925, 225). The Order sometimes used these for riding but primarily for work purposes, such as pack animals. *Briefschweiken*, for example, were courier horses. But there was also a connotation of breed, as *Schweiken* were the smaller native horses which had ultimately been bred from wild horses and clearly distinct from the breeds introduced by the Teutonic Order (Rünger 1925, 219). Fig. 15.11 shows continuous horse breeding at Nessau's cattle farm between 1376 and 1410 with the regular presence of studs. The inventory from 1407 which has the largest number of animals, parallels the abundance of other livestock and the prosperity of the district before 1410. Between the first Peace of Thorn and the destruction of the castle in 1423, horse breeding continued but at a lower intensity, with no stallions being continuously registered on the farm. Moreover, during the short occupation of Neu Nessau by the Order until 1435, horses played no role at all.

Food

It is important to understand that from the medieval and early modern perspective, volume was more important than weight, as it was more relevant for transport and storage. Conversion to modern weights, therefore, requires calculating density. Most kinds of food, whether semi-dry, such as butter, or dry, such as peas, were measured as fluids in barrels and tonnes, and also frequently in bushels. Other items such as pieces of meat (flitches, slabs and skewers; today, a flitch or slab would be *c.* 4–6 kg), fish, and cheese are recorded in the inventories in absolute numbers, usually as a *schock* referring to sixty pieces, and it would be difficult to determine average weights for each units. However, it is possible to compare proportions of these different foods.

Meat was regularly kept in Nessau's kitchen, except immediately after the Order's defeat in 1410. Despite the presence of a *Waldmeister* and stipulations within charters issued to subservient villages regarding participation in hunts, game is never explicitly mentioned in the inventories. The majority of meat is unspecified, with only species-specific references to beef (including from oxen), pork, and mutton, which were also documented as pieces, skewers, barrels, or in total numbers of complete animals. The inventories from 1407 and from April 1432 (the latter one referring to Neu Nessau) have the broadest range of meat, but this is not indicative of any development, all food stocks in these two lists are abundant and diverse. Bacon is mentioned in five of the sixteen inventories, and sausages only twice. Concerning fish, there was no regularity in the supply of local freshwater fish such as pike, carp, bream, and zander, whilst imported marine species — herring, cod, and flatfish — were readily available via the Vistula trading network. Salt, milk, and fat products, alongside various seed and grain products such as poppy, hemp, millet, turnip, and different porridges were regularly present. Spices such as pepper and saffron were exceptional and are only listed in 1407, which, according to the inventory, was the most prosperous year for the convent of Nessau.

In summary, the convent of Nessau focused on agriculture and cattle rearing, which defined the economy of the Lower Vistula region in the fourteenth and fifteenth centuries. The only wild animal products mentioned in the Nessau inventories are fish and roe deer. This corresponds with the relatively low levels of woodland within the commandery; pasture land, fields, and heath are regularly referred to, while the term *wald* is hardly ever mentioned, as it is in the 1414 inventory, when the *Hegewald* which belonged to the convent and was worth 2000 marks was destroyed by the Poles (Märcker 1899, 11). However, indirect references to woodland are evident from the frequent mentions of firewood and timber for building, as well as references to the *Waldzins*. The charters defined this heath as a place where trees grow, rather than as a dense woodland. It was also a place where beehives were maintained, although they are mentioned only once in the charter for Kostbar in 1364, situated in the 'heath and field' (Märcker 1899, 630). The importance of the heath is reflected in an undated letter written by the commander of Thorn to the Grand Master, probably Michael Küchmeister (1414–22). The commander, along with his retinue, had started cutting trees in the heath, as he had been permitted to by the Grand Master, but in the second year they were prevented from doing so by the commander of Nessau. The commander of Thorn complained that unless he could have the trees cut during the current period of good weather period, his castle would suffer from a shortage of timber (*OBA* 28529). This indicates that Nessau's heath was also important as a source of wood for the convent and town on the other side of the Vistula, and that such valuable resources could even be contested between commanderies. Some trees were specifically reserved as boundary markers. In the 1404 charter for the village of Gross Stab, the associated estate is described with eight spruces as boundary markers (Märcker 1899, 638), whilst the 1405 charter for the Herrenmühle excluded free access to alder for fuel and building material for the miller, as alder and oak trees had been designated as boundary markers.

Organizing Territory in the Wilderness: The Procuratorship of Lötzen (Pol. Giżycko)

The fact that the Order would come to fortify the area between Lakes Niegocin (Ger. Löwentin) and Kisajno (Ger. Kissain) indicates a break with the earlier network of autonomous (or semi-autonomous) power centres, and a new emphasis on securing the main passes between the largest lakes in the region. This resulted in a cohesive defence line, marked by a series of castles which were nonetheless associated with different commanderies. The creation of a procuratorship centred on Lake Niegocin resulted in the definition of a completely new territorial unit, which related to the boundaries of adjacent commanderies, and perhaps the amount of land required to support the Order's centre at Lötzen. The territory included German colonies situated to the west, and indigenous settlements elsewhere, such as the one at Staświny. After 1466, more colonies were established in the south and east of the procuratorship, and by 1540 most of the settlers were Polish (Długokęcki 2009c, 378; see also Białuński 1996). The earliest inventory for the castle is quite late, dating to 16 October 1420, when the *Ämter* of Barten (Pol. Barciany), Leunenburg (Pol. Sątoczno), Rastenburg (Pol. Kętrzyn), Rhein (Pol. Ryn), Lötzen, Seehesten (Pol. Szestno), and Lyck (Pol. Ełk) were inventoried upon the resignation of the commander of Rhein (a convent at the time), Luppolt Reitenbach. The text is as follows (*GAD*, 197):

> Leczen: First, on the cattle farm: twelve cows and one bull, also three calves, ½ schock less two old goats (i.e., twenty eight), thirteen younger goats, eighteen pigs for breeding, young and old (*Faselschwein*, i.e. pigs not used for fattening), six pigs for fattening, twenty one piglets, two oxen for pulling; kitchen: two large pots (*gropen*), five cauldrons, small and large; brewery: three troughs, one cauldron for brewing; parapet walk: five stone guns, two plumb guns, one leather sack for gun powder, eight crossbows, two bolt crossbows; harness: seven mail shirts, four bonnets, one helmet, two pairs of iron arm guards, two pairs of iron front arm guards, half a tonne of gun powder, forty schock [i.e. 2400] arrows, six schock [i.e. 360] arrows with sockets (*getollete*); cellar: five tonnes of mead, three tonnes of beer, one barrel of convent beer, three steel pitchers, one tablecloth, one towel, two new scythes, ten bushels of salt, one and a half *last* of flour, two house cheeses, three(?) cheeses (*iseksse*), one grid, one meat axe. The Schoneberg cully: three horses; Jocub von Nicclusdorff, three horses.

The provision of livestock and food is very basic, and smaller than the resources documented at neighbouring castles: the cattle farm of Seehesten housed sixty horses for ploughing, forty-five cattle, two hundred sheep, and ninety pigs; the one in Rastenburg sixty horses for ploughing, eighteen horses for pulling, hunting, and for couriers, sixty cattle, five calves, ninety sheep, forty goats, 106 pigs, and fifty-three piglets. The farm in Barten housed 357 sheep, fifty-seven horses for pulling, four foals, eighty-nine cattle, eighteen calves, and 120 pigs. Farms and stables belonging to convents were much larger. An inventory from Balga in 1431 lists almost 1300 horses and foals of different ages and for different purposes, as well as 428 cattle, 150 calves, 1263 pigs (both young and old), and 1704 sheep, both young and old (*GAD*, 163). The only stored food mentioned in the Lötzen inventory from 1420 is mead, beer, salt, cheese, and flour. However, it is important to note that this list may be incomplete, and the absence of grain is striking. The same inventory lists rye, wheat, oat, barley, and malt in the granaries, mills, and bakeries at Barten, Rastenburg, Seehesten, Rhein, and Lyck, and it may well be that in the case of Lötzen the grain was stored elsewhere and for whatever reason was not included in the inventory.

The provisioning of Lötzen castle was largely based on products from its immediate territory. This is illustrated by the confirmation of the founding charter (*Handfeste*) for the adjacent village of Neuendorf (named Lötzen from the sixteenth century). The charter of the original foundation, drawn up some time after 1437, was probably lost during the period of warfare between 1453 and 1466, which had a significant impact on the castle its district (Białuński 2005, 19, 205). However, Lötzen remained within the truncated *Ordensland* after the Second Peace of Thorn and most of the related historical data derives from this final phase of the Order's rule. At the time when the parish of Lötzen was formed in 1475 (Karczewska et al. 2005, 108), the substitute charter from highlights the economic interrelationship between the village and the castle, referring to fishing, honey production, and hunting. Neuendorf, in particular, was designated as a 'bee master' settlement (Meye 1976, 4ff., 20, 25), along with Ortelsburg (Pol. Szczytno) (1360), Johannisburg (Pol. Pisz) (1376), and Lyck (1425). The

Neuendorf charter granted the inhabitants the right to exploit beehives in the woods, but they had to deliver the due part of the honey, which has been interpreted as half of the total amount, to the Order and to sell the rest of it to them (minus what they needed for their own use) for 2½ marks per tonne. The procedure for cultivating honey was also specified in a clause; if a beehive in the woods fell from its tree, the one who had put it up was allowed to place it in their garden. Fishing was allowed for the villagers' own use, but explicitly not for sale and only within a strictly defined geographic area, consisting of the banks of Lake Löwentin and within the fields and meadows around Neuendorf. The use of rods and small nets is not explicitly mentioned, but features in other charters and was confirmed in the city charter for Lötzen from 1612, defined as the 'old custom'. In 1475 it was strictly prohibited for the villagers to block streams and rivers within their borders. Charters for other villages did not allow fishing at all, but the inhabitants of Neuendorf had to deliver fish to the procurators at both the castles of Barten and Lötzen (*c.* 35 km apart as the crow flies), as often as requested. However, the castle households were not solely dependent on such deliveries, as they also undertook fishing themselves and employed two fishermen in *c.* 1513. The sixteenth-century Lötzen inventories mention different kinds of nets, fabrics, ropes, axes (for breaking ice), and boats, and the 1513 text starts with the *Inventory of the Fishery at Letzenn* (*GAD*, 252). This is perhaps why no stocks of cured fish are mentioned in the 1420 inventory, and given the local availability of freshwater fish provided by the Order's subjects and castle households, it is unsurprising to find no mentions of imported saltwater fish until the sixteenth century (see also Chapter 12).

Hunting was usually only granted to freemen and rarely to settlement leaders or locators (Ger. *Schulze*). Exceptions were the 'bee master' villages which were more dependent on exploiting the resources of woods and wetlands. Hunted animals was subdivided in high (*Hochwild*; aurochs, stag, elk, bear, and boar) and low species (*Niederwild*; wild horse, lynx, wolf, roe deer, hare, and wild birds), comparable to neighbouring Poland (see Chapter 12). The inhabitants of Lötzen were permitted to keep two large dogs, which were used for hunting *Hochwild*. According to Meye's interpretation of the regulations, the villagers were permitted to hunt only by assignment and predominantly to protect the fields against damage caused by game. They were allowed to keep the meat except for the front portion (*der Vorderbug*), which they had to deliver to the Order's procurator, to whom they also had to sell skins from both *Hochwild* and *Niederwild* for fixed prices. The price list from 1475 provides some indication of the biodiversity (or perceived biodiversity) of the procuratorship, which is also reflected in the earlier faunal assemblage from Święta Góra:

- Bear (large, i.e., male) skin: 5 good *skot*
- Aurochs (bull) skin: 1 good *skot* (per foot in length)
- Elk skin: 5 good *skot*
- Stag skin: 1 good *firdung*
- Wild horse skin: 1 good *firdung*
- Aurochs (cow) skin: 4 *skot*
- Marten skin: 4 good *skot*
- Beaver skin: 9 minor *skot* (if found at own expense) or 4 good *skot* (if caught while being on duty for the lord)
- Lynx skin: 5 good *skot*
- Wolf skin: 5 good *skot*
- Polecat skin: 2 good *schillinge*
- Otter skin: 4 good *skot*
- Otter (female) skin: 2 good *skot*
- Fox skin: 1 *firdung*
- Bear (common, i.e., female) skin: 4 *skot*

Specific uses for some of the skins are provided; aurochs skins were to be used for making bellows, red deer skins for delivery to Königsberg, and cattle skins were to be handed to the shoemaker (*OF* 175, fols 9[r], 53[r]). Hazel grouse and black grouse also feature in Lötzen's account books (*OF* 175, fols 11[v], 96[v]). Game products such as salted meat (*Wildpret*) continuously feature in the Lötzen inventories from the beginning of the sixteenth century, and there are also references to hunting gear (as with fishing equipment), to nets for catching black grouse, roe deer (also stated as suitable for stags), and hares, some of them in a good, others in a bad condition (*GAD*, 35 lines 37ff.; 252 lines 26ff.; 253, lines 26ff.; 254, lines 23–25, Appendix 11, lines 25ff.). A number of written sources relating to Lötzen have survived from the early sixteenth century, perhaps reflecting the effective and strictly supervised administration of Dietrich von Bobenhausen, who was appointed governor of the castle in 1507. His contract with the Grand Master states that he should take care of and register (and not for his own advantage) the income from tax collection, judicial trials, honey, fishing, martens, beavers, and 'all other fur animals whatever names they have'. In the same year the procurator received and bought fifty-five martens (*OF*

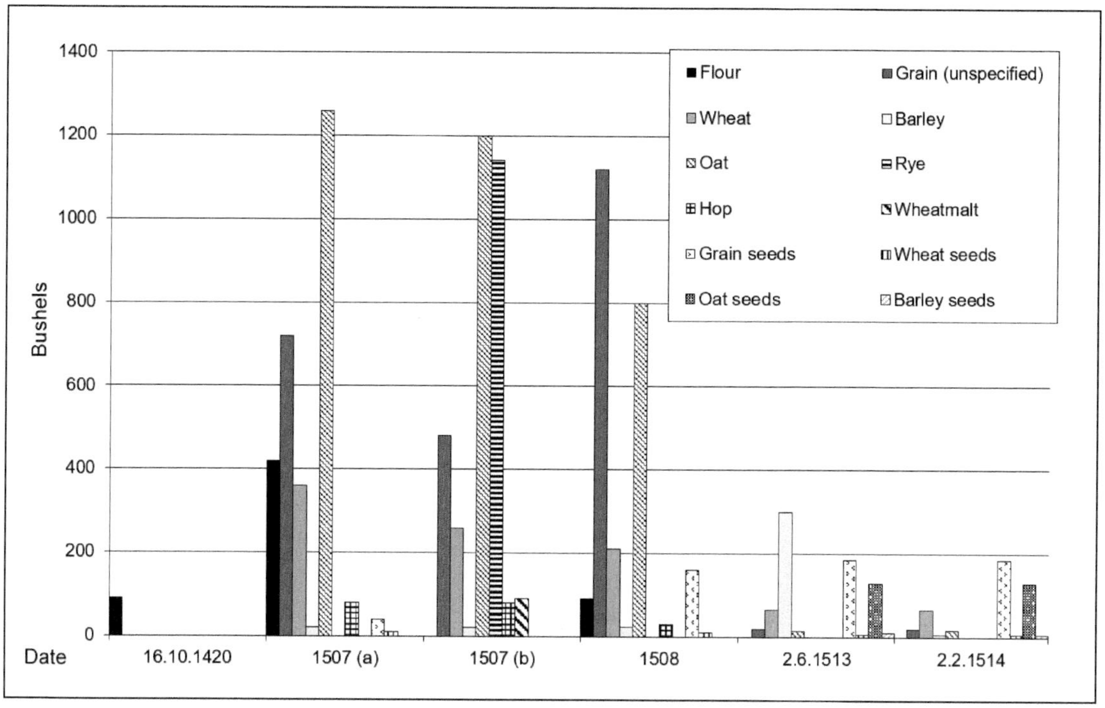

Figure 15.12: Grain measures at the castle of Lötzen (Pol. Giżycko). Figure by M. Jarzebowski.

175, fol. 21r), which had to be delivered several times as fines for threatening another person, for giving false evidence or for appearing at the castle armed with a gun (*OF* 175, fol. 7v, fol. 52r). His duties also included preventing anyone from illegally exploiting the woodlands and to provide accounts of his sales of pike and honey to merchants. The requirements of his office, especially money, salt, two horses for riding, and one servant, would be sent to him by the Grand Master from Königsberg, whilst roe deer were regularly delivered from Lötzen at this time (*OF* 175, fol. 23v; *OBA* 19183).

Most of Dietrich's accounts have survived, including those covering the years from 1507 to 1513 (*OBA* 19631; *OF* 175). They contain valuable information supplementing the evidence from the inventories and enable a more systematic comparison of the diachronic use of resources in the final phase of the Order's rule before secularization in 1525. The governor's duties were more elaborately specified in *c.* 1513 in a document which lists the castle's staff; namely, a writer, a cook, a procurator (*Schäffer*) ('who should also function as cellar servant'), a baker, a hunter and a hunter boy, a brewer, a carriage servant, a servant and a maidservant on the farm, a shepherd, a servant for the governor, another servant boy, and two fishermen. At this time the castle's household consisted of fifteen people, and four were resident on the farm, as also mentioned in the 1507 and 1508 inventories (*GAD*, 251). These two early inventories were compiled during the Order's official visitations. In Ziesemer's edition of *GAD* they are followed by the inventory from 2 June 1513, when the governor's office was handed over from Dietrich von Bobenhausen to Hans Breitenbach. Added to this is another inventory, which remains undated, but documents the transfer of office from Jacob Reiff, called Walter, to Dietrich von Bobenhausen. Reiff was in office from about 1489 to 1507, thus this inventory must also date from 1507. As a result, the order of all inventories for the comparative studies below has to be changed to 1420, 1507 (a), 1507 (b), 1508, 1513, followed by another inventory from 2 February 1514, which was not included by Ziesemer and which is published here for the first time.

The inventories are complex documents, with inconsistent specifications of weights, measures, and numbers. This limits the possibilities for quantitative comparisons, except for grain, beverages in the cellar, and livestock on the farm. The other categories of food stored in the pantry and kitchen cannot be easily translated into comparable quantities, and are tabulated with the original measures as noted in the inventories.

Grain (Fig. 15.12)

Flour was recorded in 1420, 1507 (a), and 1508. The castle had its own baker, and flour would be ground in the nearby mill, first recorded in the 1475 charter (Meye 1976, 28; Białuński 2005, 205). A 1507 account states that 'the miller in front of the castle supplies his

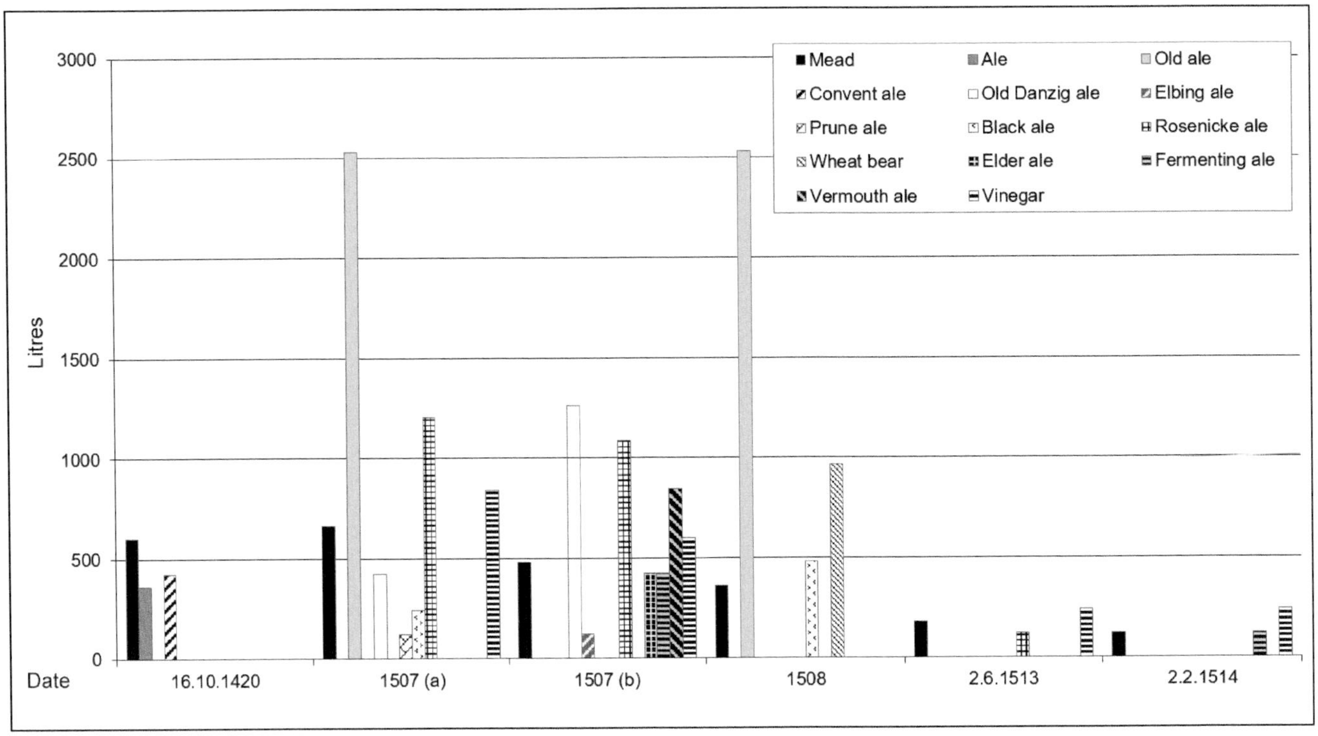

Figure 15.13: Beverages in the stores of the castle of Lötzen (Pol. Giżycko). Figure by M. Jarzebowski.

mill with all things necessary'. Other mills were located *zcum Eisenberg*, at Groß Konopken (Pol. Konopki Wielkie), Wobelick, Jucha (Pol. Stare Juchy), Stradauen (Pol. Straduny), and at Klein Skomentnen (Pol. Skomętno Małe) (*GAD*, 251ff.). While there is no grain in the 1420 list, oat dominates the 1507 and 1508 inventories, as well as bushels of unspecified grain, whilst rye is only present in the second 1507 inventory. The 1513 and 1514 inventories have much smaller amounts and focus on grain for sowing. Almost all references to seed grain in the *Großes Ämterbuch* are from the sixteenth century. Its first mention is at Rhein in 1490 (*GAD* 198, line 13), and this corresponds to other evidence for later agricultural intensification in the region.

Beverages (Fig. 15.13)

Beer was the most common beverage in the castle cellars and consisted of convent ale; old ale; ale imported from Elbing, Danzig, and Rastenburg (*OF* 175, fols 12ᵛ, 97ʳ.); dark beer; and raisin beer. In 1507 the governor bought a tonne, when he hosted the grand commander (*Großkomptur*), one of the leading officials of the Order in Prussia (*OF* 175, fol. 12ᵛ). The inventories also register the presence of ale, which was still in the process of fermenting, as well as prune ale, elder ale, and vermouth ale. The last three do not appear in the *Großes Ämterbuch* before the sixteenth century, and other inventories from this later period contain an even broader variety of types: ales made from centaury, cherries, raspberries, blackthorn (sloe), lavender, rue, salvia, mugwort, and laurel (e.g., Ortelsburg (Pol. Szczytno) in 1513; *GAD*, 120). The inventories indicate that wine was not regularly kept in the cellars of Lötzen castle, but only acquired for special occasions, such as 23 *stof* (around 27 litres) in 1508 during the visit of the commander of Königsberg (*OF* 175, fol. 97ʳ.).

Livestock (Fig. 15.14)

The Lötzen inventories mention horses for pulling, hunting, and fishing, whilst the horses attributed to higher ranking staff were probably intended for riding. Other inventories also list horses for delivering mail (*Briefsweiken*). Cattle and pigs clearly dominated the castle's farm in the sixteenth century in relatively consistent numbers. The 1508 inventory includes 32 *stuck kukelber*, that is, cows and calves, while there were only half as many cows in 1420. In the early fifteenth century, the extent of fields and meadows around Lötzen would have been limited, and it was easier to feed pigs with pannage from the more abundant woods. Another characteristic of the 1420 inventory is the large number of goats, which is absent in any of the subsequent registers. When

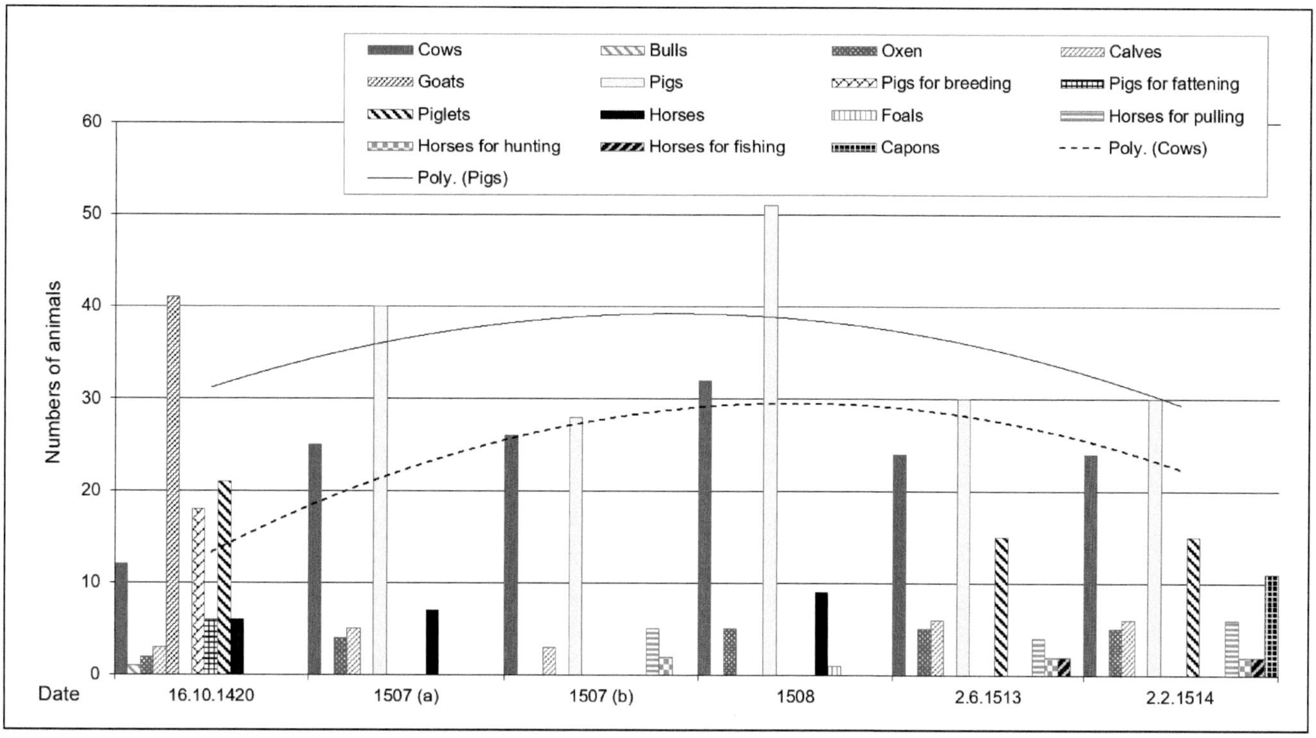

Figure 15.14: Numbers of livestock owned by the castle of Lötzen (Pol. Giżycko), with the polynomial trend for pigs and cattle showing slight fluctuations on either side of *c.* 1507. Figure by M. Jarzebowski.

comparing numbers of goats at other castle farms, this appears to be a regional trend. According to the *Großes Ämterbuch*, goats were still kept in the sixteenth century at some places (although not at every farm), but their numbers were almost always significantly smaller than those of horses, cattle, sheep, and pigs. For example, to the south-west at Ortelsburg, between 1507 and 1520, there were 15–36 goats, 21–28 horses, 70–83 cows, 51–70 pigs and 85–150 sheep. Chickens, geese, and ducks are also documented on the Order's farms, but the only fowl in the Lötzen inventories are 11 capons — cockerels fattened for consumption, in 1514. An inn-keeper at Lötzen also paid an interest of 6 marks, 4 capons, and 4 bushels of oat in 1507 (*OF* 175, fol. 5^{r}). The *Großes Ämterbuch* also only lists capons at Schaken (Pol. Szaki) (1514), Angerburg (1513, 1514), and Nessau (1407). The 1507 and 1508 accounts list expenses for chickens, geese, and hazel hens (*OF* 175, fols 11^{v}, 96^{v}).

Food

At Lötzen, game clearly dominates the castle's stores, and in this way differs from the Order's castles in more cultivated and less densely wooded regions. Bacon was also regularly available. Fish, readily available from the vast lake within the procuratorship, only appear in limited quanities in the inventories. Pike and bream represent locally caught fish, while flatfish was imported from coastal areas. It is remarkable that herring, which was a widely available commercial product, does not appear in the lists at all. This suggests that the supply of fish in the Lötzen procuratorship was primarily based on the exploitation of local lakes and rivers, and that imported marine species were a delicacy for the procurator's table. Alongside flatfish, the Lötzen account books specify that in 1507 and 1508 small amounts of dried herring (1 bushel each) were received from Königsberg for Lent. Other species mentioned in the account books are catfish and crabs.

Certain commodities recur in all the inventories. Cheese, which appears frequently and in various forms in the inventories (such as house cheese, small cheese) of the *Großes Ämterbuch*, seems to be underrepresented at Lötzen. This may be connected with the small size of the adjacent farm, with its limited number of cows where cheese is recorded in 1420. Plant products are recorded more regularly, namely peas, mustard, poppy seeds, hemp, millet, pearl barley, buckwheat, various different types of porridge, onions and sauerkraut. There is also a category of imported goods called *crewde/krude*, that is, confectionary. In 1507 and 1508, expenses for sugar, raisins, figs, almonds, rice, caraway, pepper, and cloves are listed, only in small amounts. The raisins were bought for Lent (*OF* 175, fols 11^{r}, 96^{r}.), and as no such luxury goods are mentioned in any of the Lötzen inventories, it is probable that they were only acquired for special occasions.

The overall impression provided by the data from Lötzen is of a small castle that was self-sufficient, relying almost entirely on resources from its territory to sustain its household. Occasional imports provided by the Order's network were intended for special occasions, most likely for the table of the castle's procurators and their guests. By the time of the earliest surviving inventory, the castle was still within a relatively volatile frontier region. Ahead of 1410, it had served as a mustering point and depot for arms and armour, and previously it had been used as a base for some of the Order's *Reisen* into Lithuania. This volatility is also represented by the relatively late foundations of the first churches within the procuratorship. By the early sixteenth century, several decades after the truncation of the Prussian *Ordensland*, the castle's most important role alongside the organization of settlement and taxation was the procurement of fur from its subjects. The provisioning of the castle also depended on wild natural products, even as late as the early sixteenth century. This, along with the emphasis placed on honey and wax production, reflects a highly specialized economy tailored to the local ecological context, which was very different from the other castles considered within this chapter. Allowances were made for the settlements within the procuratorship, enabling them to exploit the resources of landscapes dominated by woods and wetlands. The most important relationship was articulated with the settlement adjacent to the castle, but it is clear that the landscape transformations evident within the vicinity were the direct results of the activities of German and Masovian colonists, alongside the remaining Galindian communities who were gradually losing their distinctive identity.

Conclusion: Landscapes of Regulation in Prussia

This chapter has considered the reorganization of an important component of the Prussian cultural landscape; the management of resource exploitation. All existing centres of tribal authority were destroyed, abandoned or appropriated by the Order during the crusading period. This also included borderland regions which had been previously under the control of Polish castellans, as well as subsequently acquired Christian territories governed by convents in Pomerelia and Kuyavia. The crusades did not result in the complete destruction of the indigenous Prussian aristocracy, although in some regions large numbers had been killed, but pre-crusade power structures were replaced with a new, hierarchical, interregional superstructure. This consisted of the incremental imposition of the commandery system, which defined the reorganization of territories in the latter half of the thirteenth century. These districts largely ignored the bounds of tribal districts as outlined in the earliest written sources of the Teutonic Order. Instead, the location of convents was determined by the events of the Prussian Crusade, which included early grants within the territory of the Duke of Masovia, and the extension of territorial bounds by subsequent historical processes. At the level of individual settlements and their associated territories, it appears that where the indigenous settlement pattern remained intact, it was appropriated and accentuated with colonization. From the very onset of the Prussian Crusade, colonies were founded with clearly defined privileges that guaranteed the effective functioning of provisioning systems and the maintenance of the Order's sovereignty. Colonists were encouraged to settle the land in order to ensure the convents could sustain themselves and provide effective security against pagan Lithuania, and the generous grants of land and composite town foundations that defined the crusading period were replaced by tighter control over territory in the fourteenth century.

The inventories presented above provide a highly detailed, albeit chronologically fragmented, insight into the functioning of fully developed territories and how these changed in 1410 in the Lower Vistula region, and after 1466 in eastern Prussia. A comparison of two convents and two smaller castles in both frontiers and heartlands indicates that status, as well as the specific ecological context, played important roles in determining how these communities were provisioned and how resource exploitation was managed through subservient settlements. There is little difference in how effectively castles were sustained in the heartlands and frontiers of the Ordensland, with the conflicts in the early fifteenth century affecting the frontier commandery of Nessau as much as the heartland commandery of Rehden and the advocateship of Stuhm. The importance of grain production, in particular, has long-been recognized in the historiography of the Prussian *Ordensland* as the principal cause of the Order's booming economy in the latter decades of the fourteenth century. It is clear that the castles of the Lower Vistula played a pivotal role in the procurement of this grain, whether from their own subjects or from long-distance trade. The grain trade along the Vistula intensified in the early fourteenth century, punctuated by Thorn and Danzig as the major centres of redistribution. In eastern Prussia, relatively more abundant game was valued and sent on via the Order's networks to other convents, especially Königsberg, whilst fur was procured for the Hanseatic market. In contrast, amber played a fundamental role in the economies of the coastal commanderies, especially in the south-eastern part of the Sambian Peninsula. From a fiscal perspective, this was the most important luxury product obtained from the Order's remaining Prussian territories at the start of the sixteenth century, accounting for 57.6% of the Grand Master's total budget in 1501/02 (Jahnke 2015, 211).

Works Cited

Primary Sources

AMH Ziesemer, W., ed. 1911. *Das Ausgabebuch des Marienburger Hauskomturs für die Jahre 1410–1420*, Königsberg: Thomas & Oppermann

Froelich, X. 1868. *Geschichte des Graudenzer Kreises*, Graudenz: self-published

GAD Ziesemer, W., ed. 1968. *Das große Ämterbuch des Deutschen Ordens: Mit Unterstützung des Vereins für die Herstellung und Ausschmückung der Marienburg*, Wiesbaden: M. Sändig (repr. 2009)

Grimm, J. and Grimm, W. 1984. *Deutsches Wörterbuch*, 33 vols, München: Deutscher Taschenbuch (orig. pub. 1854–1971)

MA Ziesemer, W., ed. 1916. *Das Marienburger Ämterbuch, mit Unterstützung des Vereins für die Herstellung und Ausschmückung der Marienburg*, Danzig: Kafemann

MTB Joachim, E., ed. 1896. *Das Marienburger Tresslerbuch der Jahre 1399–1409*, Königsberg: Thomas and Oppermann

PrUB *Preußisches Urkundenbuch* (*Prussian Records Book*), available online at <http://www.spaetmittelalter.uni-hamburg.de/Urkundenbuch/ [accessed December 2017]

Sattler, C., ed. 1887. *Handelsrechnungen des Deutschen Ordens*, Leipzig: Duncker & Humblot

Thielen, P. G., ed. 1958. *Das große Zinsbuch des Deutschen Ritterordens (1414–1438)*, Marburg: N. G. Elwert

Secondary Works

Antoniewicz, J. 1950. 'Wyniki prac wykopaliskowych na grodzisku i podgrodziu w Jeziorku, pow. Giżycko w 1950 r.', *Materiały Wczesnośredniowieczne* 2, 235–56

Antoniewicz, J. 1955. 'Niektóre dowody kontaktów słowiańsko-pruskich w okresie wczesnośredniowiecznym w świetle źródeł archeologicznych', *Wiadomości Archeologicznych* 23, 233–77

——, 1958. 'Prusowie we wczesnym średniowieczu i zarys ich kultury materialnej', in *Szkice z dziejów Pomorza. Pomorze średniowieczne*, ed. G. Labuda, Poznań: Wydawnictwo Poznańskie, 121–59

Auch, M., and Nowakiewicz, T. 2009. 'Ruska ceramika z Ruskiej Wsi. Przyczynek do studiów nad interregionalnymi kontaktami ziem pruskich u schyłku wczesnego średniowiecza', in *Bałtowie i ich sąsiedzi. Marian Kaczyński in Memoriam*, eds. A. Bitner-Wróblewska and G. Iwanowska, Warsaw: Państwowe Muzeum Archeologiczne w Warszawie, 481–93

Białuński, G. 1996. *Osadnictwo regionu Wielkich Jezior Mazurskich od XIV do początku XVIII wieku — starostwo leckie (giżyckie) i ryńskie*, Olsztyn: Ośrodek Badań Nauk. im. Wojciecha Kętrzyńskiego

——, 1999. *Studia z dziejów plemion pruskich i jaćwieskich*, Olsztyn: Ośrodek Badań Nauk. im. Wojciecha Kętrzyńskiego

——, 2005. *Siedlungswesen im Bereich der Großen Masurischen Seen vom 14. bis zum 18. Jahrhundert — Ämter Lötzen und Rhein*, Hamburg: Selbstverlag des Vereins für Familienforschung

——, 2009. *Bevölkerung und Siedlung im ordensstaatlichen und herzoglichen Preussen im Gebiet der 'Großen Wildnis' bis 1568*, Hamburg: Selbstverlag des Vereins für Familienforschung

Biskup, M. 1956. 'Sztum', *Studia z historii budowy miast polskich* 6/2, 237–52

Biskup, M. 2002. 'Etniczno-demograficzne przemiany Prus Krzyżackich w rozwoju osadnictwa w średniowieczu', in *Opera Minora: Studia z dziejów zakonu Krzyżackiego*, Toruń: Towarzystwo Naukowe w Toruniu, 129–50

Biskup, M., and others, eds. 2009. *Państwo zakonu krzyżackiego w Prusach: Władza i społeczeństwo*, Warsaw: Państwowe Wydawnictwo Naukowe

Blanckenburg, C. von. 2001. *Die Hanse und ihr Bier. Brauwesen und Bierhandel im hansischen Verkehrsgebiet*, Köln: Böhlau

Bojtár, E. 1999. *Foreword to the Past: A Cultural History of the Baltic People*, Budapest: Central European University Press

Boockmann, H. 1983. 'Die Vorwerke des Deutschen Ordens in Preußen', in *Die Grundherrschaft im späten Mittelalter* (vol. I), ed. H. Patze, Sigmaringen: Jan Thorbeke, 555–76

Buko, A. 2005. *Archeologia Polski wczesnośredniowiecznej*, Warsaw: Trio (repr. in 2008)

Czaja, R. 2009. 'Urbanizacja kraju', in *Państwo zakonu krzyżackiego w Prusach: Władza i społeczeństwo*, ed. M. Biskup et al., Warsaw: Państwowe Wydawnictwo Naukowe, 177–230

Długokęcki, W. 2006. 'Uwagi o genezie i rozwoju wczesnośredniowiecznych Prus do początków XIII wieku', *Pruthenia* 2, 9–54

——, 2009a. 'Prusy w starożytności i we wczesnym średniowieczu', in *Państwo zakonu krzyżackiego w Prusach: Władza i społeczeństwo*, ed. M. Biskup et al., Warsaw: Państwowe Wydawnictwo Naukowe, 25–50

——, 2009b. 'Społeczeństwo wiejskie', in *Państwo zakonu krzyżackiego w Prusach: Władza i społeczeństwo*, ed. M. Biskup et al., Warsaw: Państwowe Wydawnictwo Naukowe, 460–94

Dygo, M. 2009a. 'Ekspansja terytorialna i władztwo terytorialne (1249–1309)', in *Państwo zakonu krzyżackiego w Prusach: Władza i społeczeństwo*, ed. M. Biskup et al., Warsaw: Państwowe Wydawnictwo Naukowe, 79–104

——, 2009b. 'Początki i budowa władztwa zakonu krzyżackiego (1226–1309)', in *Państwo zakonu krzyżackiego w Prusach: Władza i społeczeństwo*, ed. M. Biskup et al., Warsaw: Państwowe Wydawnictwo Naukowe, 53–78

Engel, M., Iwanicki, P., and Rzeszotarska-Nowakiewicz, A. 2006. '"Sudovia in qua Sudovitae": The New Hypothesis about the Origin of Sudovian Culture', *Archeologia Lituana* 7, 184–211

Gancewski, J. 2001. *Rola zamków krzyzackich w ziemi chelminskiej od polowy XIV wieku do 1454 roku*, Olsztyn: Ośrodek Badań Naukowych im. Wojciecha Kętrzyńskiego

Güttner Sporzyński, D. von. 2008. 'Recent Issues in Polish Historiography of the Crusades', in *The Military Orders*, IV: *On Land and by Sea*, ed. J. Upton-Ward, Farnham: Ashgate, 13–23

Haftka, M., and Wadyl, S. 2015. *Węgry. Zespół osadniczy na pograniczu Pomorsko-Pruskim w XI–XII w.*, Malbork: Muzeum Zamkowe w Malborku

Jahnke, C. 2015. 'The Baltic Trade', in *A Companion to the Hanseatic League*, ed. D. J. Harreld, Leiden: Brill, 194–240

Jóźwiak, S. 2001. *Centralne i terytorialne organy władzy zakonu krzyżackiego w Prusach w latach 1228–1410: rozwój, przekształcenia, kompetencje*, Toruń: Wydawnictwo Uniwersytetu Mikołaja Kopernika

——, 2009a. 'Dobra rycerskie', in *Państwo zakonu krzyżackiego w Prusach: Władza i społeczeństwo*, ed. M. Biskup et al., Warsaw: Państwowe Wydawnictwo Naukowe, 217–30

——, 2009b. 'Podziały administracyjne', in *Państwo zakonu krzyżackiego w Prusach: Władza i społeczeństwo*, ed. M. Biskup et al., Warsaw: Państwowe Wydawnictwo Naukowe, 132–37

——, 2009c. 'Rycerstwo-szlachta', in *Państwo zakonu krzyżackiego w Prusach: Władza i społeczeństwo*, ed. M. Biskup et al., Warsaw: Państwowe Wydawnictwo Naukowe, 450–59

Jóźwiak, S., and Trupinda, J. 2007. *Organizacja życia na zamku krzyżackim w Malborku w czasach wielkich mistrzów (1309–1457)*, Malbork: Muzeum Zamkowe w Malborku

——, 2012. *Krzyżackie zamki komturskie w Prusach*, Toruń: Wydawnictwo Naukowe Uniwersytetu Mikołaja Kopernika

Karczewska, M. et al.. 2005. *Miłki: Monografia krajoznawcza gmniy mazurskiej*, Białystok/Miłki: Wydawnictwo Kwadrat

Labuda, G., and Biskup, M. 1986. *Dzieje Zakonu Krzyzackiego w Prusach*, Gdańsk: Wydawnictwo Morskie

Łowmiański, H. 1985. *Geografia polityczna Bałtów w dobie plemiennej*, Poznań: Lituano-Slavica Posnaniensia.

Mager, F. 1960. *Der Wald in Altpreussen als Wirtschaftsraum*, Köln: Böhlau

Maltby, M., Pluskowski, A. G., and Seetah, K. 2009. 'Animal Bones from an Industrial Quarter at Malbork, Poland: Towards an Ecology of a Castle Built in Prussia by the Teutonic Order', *Crusades*, 8, 191–212

Märcker, H. 1899. *Geschichte der ländlichen Ortschaften und der drei kleineren Städte des Kreises Thorn*, Quellen und Darstellungen zur Geschichte Westpreußens, 2, Danzig: Th. Bartling

Meye, H. 1976. 'Die erneuerte Handfeste von Neuendorf vor Lötzen vom Jahre 1475', in *Lötzener Fragmente: Nachrichten aus verlorenen Quellen 1700–1900*, ed. R. Heling, Hamburg: Selbstverlag Verein für Familienforschung in Ost- u. Westpreussen e., 15–28 (orig. pub. 'Lötzener Zeitung', *Aus der Heimat* 1936 (supplement), 4ff)

Militzer, K. 1997. 'Jagd und Deutscher Orden', in W. Rösener, *Jagd und höfische Kultur in Mittelalter*, Göttingen: Vandenhoeck & Ruprecht, 325–63

Nowakiewicz, T. 2006. 'Galindia in the Viking Age – New Shape of the Culture', in *Transformatio Mundi: The Transition from the Late Migration Period to the Early Viking Age in the East Baltic*, ed. M. Bertašius, Kaunas: Kaunas University of Technology, Department of Philosophy and Cultural Science, 161–72

Nowakiewicz, T. 2010. 'Some Remarks on Settlement Systems of Early Medieval Prussians: The Case of Northern Galindia', in *Worlds Apart? Contacts across the Baltic Sea in the Iron Age*, eds U. L. Hansen and A. Bitner-Wróblewska, Copenhagen: Det kgl. Nordiske Oldskriftselskab, 487–504

Nowakiewcz, T., and Wróblewski, W. 2003. '"Legenda galindzka" Piotra z Dusburga w świetle badań archeologicznych', in *Antyk i barbarzyncy*, eds A. Bursche and R. Ciołek, Warsaw: IA UW, 313–22

Okulicz-Kozaryn, Ł. 1983. *Życie codzienne Prusów i Jaćwięgów w wiekach średnich (IX–XIII w.)*, Warsaw: Państwowy Instytut Wydawniczy

——, 1997. *Dzieje Prusów*, Wrocław: Fundacja na Rzecz Nauki Polskiej

Päsler, R. G. 2003. *Deutschsprachige Sachliteratur im Preußenland bis 1500: Untersuchungen zu ihrer Überlieferung*, Köln: Böhlau

Pluskowski, A. G. 2013. *The Archaeology of the Prussian Crusade: Holy War and Colonisation*, London: Routledge

Poliński, D. 2003. *Późnośredniowieczne osadnictwo wiejskie w ziemi chełmińskiej*, Toruń: UMK

Powierski, J. 2001. *Prusowie, Mazowsze i sprowadzenie krzyżaków do Polski* (vol. II.1), Malbork: Muzeum Zamkowe w Malborku

——, 2003. *Prussica 1*, Malbork: Muzeum Zamkowe w Malborku

Rünger, F. 1925. 'Herkunft, Rassenzugehörigkeit, Züchtung und Haltung der Ritterpferde des Deutschen Ordens: Ein Beitrag zur Geschichte der ostpreussichen Pferdzucht und der deutchen Pferdzucht im Mitelalter', *Zeitschrift für Tüthrzüchtung und Züchtungsbiologie* 2/3, 211–308

Sarnowsky, J. 1993. *Die Wirtschaftsführung des Deutschen Ordens in Preußen (1382–1454)*, Cologne: Böhlau

Semrau, A. 1928. 'Die Orte und Fluren im ehemaligen Gebiet Stuhm und Waldamt Böhnhof (Komturei Marienburg)', *Mitteilungen des Coppernicus-Vereins für Wissenschaft und Kunst zu Thorn* 36, 5–7

Śliwiński, B. 2004. 'The Christianisation of Prussia: The Polish Contribution until the Introduction of the Teutonic Order', in *Christianization of the Baltic Region*, ed. J. Gąssowski, Pułtusk: Baltic Research Center in Frombork, 39–63

Starnawska, M. 2001. 'Military Orders and the Beginning of Crusades in Prussia', in *The Crusades and the Military Orders: Expanding the Frontiers of Medieval Latin Christianity*, eds Z. Hunyadi and J. Laszlovszky, Budapest: Department of Medieval Studies, Central European University, 417–28

Torbus, T. 1997. 'Die Ordensburg von Stuhm (Sztum) im ehemaligen Ordensland Preußen', in *Der frühe Schloßbau und seine mittelalterlichen Vorstufen*, Forschungen zu Burgen und Schlössern, 3 München: Deutscher Kunstverlag, 171–84

——, 1998. *Die Konventsburgen im Deutschordensland Preussen*, Munich: Oldenbourg

Treichel, A. 1899. 'Der Thiergarten von Stuhm', *Zeitschrift des Historischen Vereins fuer den Regierungsbezirk Marienwerder* 37, 27–35

Witthöft, H. 1981. 'Scheffel und Last in Preußen', *Blätter für Deutsche Landesgeschichte* 117, 335–72

Wróblewski, W., Nowakiewicz, T., and Bogucki, M. 2003. '*Terra desolata*: Wczesnośredniowieczna Galindia w świetle badań mikroregionu Jeziora Salęt', in *Studia Galindzike* (vol. I), ed. W. Wróblewski, Warsaw: Instytut Archeologii Uniwersytetu Warszawskiego, 157–80

Zdrójkowski, Z. 1983. *Zarys dziejów prawa chełmińskiego 1233–1862*, Toruń: Wydawnictwo Uniwersytetu Mikołaja Kopernika

Ziegler, H. 1977. 'Flüssigkeitsmaße, Fässer und Tonnen in Norddeutschland', *Blätter für Deutsche Landesgeschichte* 113, 276–337

16. The Environmental Impact of the Conquest of Prussia

Aleksander Pluskowski

Without a doubt, what the Lord spoke through Isaiah would be fulfilled through them (the brothers of the Teutonic Order): 'The Lord will have mercy on Zion and will take pity on its ruin. He will turn its deserts into a paradise and its wildernesses into the Lord's garden.' (*CTP* II, 10)

The narratives of the Prussian Crusade are saturated with biblical imagery. They paint a picture of a land devastated by warfare, with regions reduced to a desolate wilderness, comparable to the biblical *vastum*. This also served to remind its audiences — in Peter von Dusburg's case the papal curia, in Nicolaus von Jeroschin's the brothers of the Order — that the Prussian Crusade was above all a spiritual war. In the early fourteenth century, when Dusburg was writing, the ongoing battle against paganism was also conceptualized as an apocalyptic struggle, and by invoking Isaiah (51. 3), the German priest linked the activities of the Order with the divine restoration of harmony. Seen in this light, the Order's rule, on behalf of the Virgin Mary as Queen of Heaven, was part of the divine plan (Fig. 16.1). Nicolaus, translating Peter's *Chronicon*, ensured that future recruits would understand the Order's role in the conquest of Prussia, and its ongoing war with Lithuania. By the time his narrative was being compiled, the process of colonization and the reorganization of the landscape was defining theocratic rule in the conquered tribal territories. With subsequent regulated settlements populated largely by internal migrants, colonization had become a way of life in Prussia. In effect, the Order's prophetic mission was being fulfilled with strikingly visible precision. But if much of the thirteenth century was conceptualized as a time of destruction and desolation across Prussia and its borderlands, to be followed by an era of regeneration, to what extent is the impact of the Order's rule reflected in the palaeoenvironmental record? This chapter is a synthesis of all the data for the Prussian case studies presented in the previous chapters; pollen (Chapter 11), animal bones (Chapter 12), plant macrofossils (Chapter 13), sediments (Chapter 14), and written sources (Chapter 15). Additional data has been included from a range of secondary sites. As in the case of the synthetic chapter on Livonia, these are integrated here to focus on two key questions: how did patterns of exploitation change and why?

Figure 16.1: A limestone roof boss with the Virgin enthroned from the church of the high castle at Malbork (Ger. Marienburg), *c.* 1340 (Pospieszna 2010, I, 1.2.13).

As in the case of Livonia, environmental exploitation in Prussia has traditionally been understood in economic terms. The broadly accepted economic history of medieval Prussia can be quickly summarized as follows (see Arnold 1989; Sarnowsky 1993). The crusades led by the Teutonic Order (1230–83) were accompanied by protracted phases of colonization that intensified from the last decades of the thirteenth century until the mid-fourteenth century, and continued in eastern Prussia into the second half of the fifteenth century. The intro-

Figure 16.2: The 'Golden Gate' of the Church of the Blessed Virgin Mary in the high castle of Malbork (Ger. Marienburg). Photo by M. Elander.

duction of a heavier plough and rotational field systems facilitated the intensification of grain production, which provided the basis for imposed taxation; the Order's well-documented rural rent economy. The annexation of Pomerelia in 1309 gave the Order complete control over the Lower Vistula and the trading city of Danzig (Pol. Gdańsk). The export of surplus grain to Western Europe defined the Order's financial gains from the latter decades of the fourteenth century until 1410, at which point there was a gradual and irreversible fiscal decline. This was exacerbated by ongoing conflicts with Poland-Lithuania, including the loss of control over key towns which rebelled against the Order, culminating in the truncation of the Prussian *Ordensland* in 1466. The final phase of the Order's rule before secularization in 1525 was no longer dominated by the export of grain, but by timber, amber, and fur.

As in the case of Livonia, it is important to emphasize how the Order's Prussian branch expressed its role and how it was perceived by others. In contrast to Livonia, this was much more explicitly linked to the defence of *ecclesia* and the ongoing holy war with Lithuania. There is also more surviving evidence for the promotion of the Marian cult in Prussia, which underlined the Order's sacred mandate (Dygo 1989, 2009). Convents themselves were doorways to paradise — literally, if we accept the most recent interpretations of the unusually well-preserved iconographic programme of the 'Golden Gate' of Marienburg (Błażejewska 2010; Kutzner 2010) (Fig. 16.2). But the activities of the Church Militant in Prussia were tempered by necessary engagements with neighbouring secular powers. A survey of the foreign policy of the Order's Prussian branch indicates widespread involvement with the Hanse, the Danish crown, and the Holy Roman Empire, and points to an expansionist agenda with the purchase of the Neumark and Gotland (Tandecki 2009, 122). After the annexation of Pomerelia, the limited periods of conflict with Poland in the fourteenth century focused on the borderland, particularly Kuyavia and the Dobrin Land.

However, the most investment by far in terms of time, financial expenditure, and ideological emphasis was on the ongoing war with Lithuania, which Peter von Dusburg (*CTP* III, 221) stated began immediately after the subjugation of Prussia in 1283. Königsberg became the principal mustering centre and point of departure for the *Reisen*, with campaigns organized from 1304 and intensifying from 1360 through to 1410. The *Reisen* were expensive for both the Order and its guests (in some cases ruinous), and a percentage of the Order's annual income was spent on maintaining the castles, garrisons, stocks of arms and armour, as well as related provisioning requirements on the frontier of eastern Prussia for this purpose. At the same time, participants had to cover their own costs, which included the provision of services and loans from the Order, and the *Reisen* represented an important economic component of the *Ordensland* (Paravicini 1995, 163–316). Whilst the Order was not actively engaged in missionary activity, its foreign policy regarding Lithuania — along with neighbouring Catholic polities — was framed in the rhetoric of Christianization. In the mid-fourteenth century, the Polish king Casimir the Great initiated a failed crusade against Lithuania and an Imperial legation was briefly established with the aim of Christianizing the Lithuanian nobility in return for a cessation of further attacks (Tandecki 2009, 116). The Order's polemic at the time may not have emphasized miracles, relics, and religiosity in the same way as Dusburg did for the benefit of the papal curia, but its treaties with the Lithuanian Grand Dukes, which included territorial demands in Samogitia, emphasized the acceptance of Christianity.

Historians recognise an internal crisis within the Order following the establishment of the Bishopric of Vilnius, accompanying the political union with Poland. The Order responded by attempting to discredit the effectiveness of this conversion and focused on the occupation of Samogitia, again emphasizing their intent to Christianize the region. The terms of the treaty of Salynas (Ger. Sallinwerder) in 1398 with Grand Duke Vytautas also twinned territorial demands (which included part of Sudovia under Lithuanian control, as well as a claim to Pskov) with the promotion of Christianity in Lithuania. With Vytautas' assistance, the Order was able to subjugate Samogitia in 1400, the management of which was delegated to an advocate (*Vogt*) directly under the Grand Master's authority. The Order's rule in Samogitia was quickly challenged, and following Vytautas' lobby of the papal curia, a bull was issued in 1403 forbidding the Knights from attacking and annexing Lithuanian territories. Nonetheless, the Order retained its hold on Samogitia, confirmed by the Bohemian king Wenceslaus IV on 15 February 1410 with the specification that Poland could not aid Samogitia, Lithuania, or Rus', whose were declared 'unbelievers'. The Order's ensuing war with Poland-Lithuania was justified by its polemicists that Lithuanians (and in particular Samogitians) were still pagans and Poles, by allying with them, should be considered culpable. The issues were fiercely debated at the Council of Constance (1414–18), where Paulus Vladimiri, leading the Polish delegation, challenged the Order's ideology of forced conversion. This has been regarded as a seminal event in the development of a legal doctrine promoting co-existence between Christians and non-Christians (Alexandrowicz 1963), although the argument was rejected by the Order's supporters, who argued that pagans were justifiable targets.

The Order had also been enlisted by the Holy Roman Emperor to fight the Hussites (the Polish delegation in Constance had unsuccessfully defended Jan Hus, who was burnt at the stake in 1415), and whilst the conflict was also framed in crusading rhetoric, it blurred with the ongoing war with Poland. Between 1466 and 1525, the Order was subservient to the Polish crown, although in 1493, resisting incorporation into the Kingdom of Poland, the Order emphasized the spiritual character of its organization and territories. In 1495, the Bishop of Ermland even suggested the Order should be relocated from Prussia to Podolia (under the control of the Polish crown at the time) to fight the Turks, and the Order briefly became involved in this new crusading frontier two years later. In 1512, within the matrix of continuing political intrigue, an agreement (ultimately rejected by the Grand Master) referred to the Order's role of fighting heathens (Biskup 2009, 278, 282). In 1523, when Luther urged the Order to renounce vows of celibacy, he described the Order as unique, differing from all other orders in that it was founded for the purpose of waging war against infidels (*LW* 45, 133). It therefore seems that until the moment of secularization the Order retained a distinctive external identity, irrespective of the decline in the spiritual life of its membership, which was linked to the ideology of Christian holy war. This identity was increasingly challenged, in relation to conflicting political agendas in Samogitia, but perceived as viable for confronting the threat of Turkish expansion.

In summary, it is impossible to untangle the Order's actions from Imperial, Polish, and Lithuanian politics even during the crusading period, let alone in the fourteenth and fifteenth centuries, and there is evidence for an agenda of territorial expansion within the Prussian branch of the Order. At the same time, the consistent justification of the war against Samogitia and Lithuania as a means of imposing Christianity was not supported by the papacy in the late fourteenth and fifteenth centuries. Therefore, it is possible to suggest that the Order's investment in military resources, alongside the maintenance of convents and other houses, was ultimately promoted in relation to an internal ideology of a righteous holy war against pagans and their allies, tempered (and from the Order's perspective 'frustrated') by regional *Realpolitik* and securing the territorial integrity of the *Ordensland*. As a result, this blurred the Order's foreign policy with wars against Catholic Poland-Lithuania, alongside other military actions such as the assault on the Vitellian Brotherhood on Gotland.

With this adjusted framework in mind, the results of each of the primary case studies are now summarized in turn, followed by a discussion characterizing the Order's ecological impact on Prussia subdivided according to periods of 'social time' (see Chapter 3). Detailed references can be found in earlier chapters. Both Livonia and Prussia are compared in Chapter 17, where the study's initial objectives outlined in Chapter 1 are addressed. The various caveats and critiques of the individual datasets have been presented in the previous thematic chapters and are not repeated here. What these summaries present are necessarily abbreviations, but, as in the case of Livonia (Chapter 9), the aim is to present a general model which can be challenged and developed with future research.

A Long-Term Focus for Colonization: Radzyń Chełmiński-Rehden

The stronghold at Radzyń Chełmiński (Ger. Rehden) was established to secure the northern frontier of the Kulmerland and, as in the case of other *grody*, to manage the process of colonization and its related infrastructure. Not as important as the centre at Kałdus, the stronghold at Radzyń and its projected territory in the twelfth/early thirteenth century is comparable to the other smaller strongholds forming a defensive ring around the Kulmerland (Poliński 2003, 130). The pollen data indicates clearance of the immediate landscape began in the seventh century, but intensified from the mid-eleventh century with a marked increase in values for cultivated, weed, ruderal, and grassland taxa. The establishment of the stronghold was accompanied by the development of an intensive system of permanent rotational field systems involving the cultivation of both winter and summer crops, primarily rye, oat-wheat, and hemp, as well as the clearance of woodland for pasture. Although there is an absence of corresponding faunal and plant macro-botanical data from this site, the changes in the surrounding landscape match those of Węgry (Ger. Wengern), and represent a typical ecological profile for a frontier stronghold of the early medieval Piast state. Although there are evident changes in land-use already from the Roman Period in the vicinity of Radzyń, the establishment of the stronghold corresponded to a marked intensification of cultivation and deforestation as reflected in both the pollen and geochemical record of the adjacent lake. This is also evident within the surrounding territory associated with the settlement cluster on the frontier; at Rywałd (Ger. Königlich Rehwalde) there is also a distinct increase in pollen of cereals, particularly rye and wheat/oat from the mid-eleventh century, alongside increasing quantities of ruderals, weed, and grassland taxa. This extent of arable land and pasture noticeably increases during the twelfth century, followed by a clear increase in cereal pollen during the early thirteenth century.

The border between the Kulmerland and Pomesania was retrospectively described by Peter von Dusburg (*CTP* III, 12) as a 'wilderness' when the Order build their fortress at Rehden in 1234 , which had served as a 'gateway' (Lat. *introitus*) to Prussia. The transition from the early medieval stronghold to the convent is not entirely clear, but it appears that a *castrum* was established within the immediate vicinity during the thirteenth century and then replaced with the brick convent after the cessation of hostilities in 1283. This period saw the replacement of a number of timber and earth *castra* in the Kulmerland with brick castles (Poliński 2007), resulting in the densest concentration of fortified structures within the Order's territories by the end of the thirteenth century (Mroczko 1974, 299). The establishment of a commandery at Rehden saw the convent develop into one of the most powerful and important in medieval Prussia. The pre-crusade settlement pattern appears to have collapsed before the Order's arrival, but colonists were encouraged to settle in the commandery and contribute to the maintenance of the convent. A few secular knightly families were also permitted to construct or reuse strongholds at the fringes of the commandery. The impact of this second phase of colonization is evident within the vicinity of the convent and town of Rehden, and around small settlements such as Königlich Rehwalde. However, it differs from the earlier Slavic phase and points to the reorganization of the immediate landscape.

The commanders of Rehden tailored their regulations to the topographic situation. Despite the presence of multiple wetlands within the commandery, the number of lakes was limited and access to these was carefully regulated. The shortage of suitable rivers also meant that more mills had to be wind-powered. Above all, the regional importance of grain was recognized during the refoundation of the town, with abundant quantities stored in the convent in the latter decades of the fourteenth century. Despite this emphasis on grain production and processing, there is a decline in cultivated taxa from the mid-/late thirteenth century within the vicinity of the convent, with ruderal species decreasing marginally during the fourteenth century, implying a reduction in the intensity of agriculture. Instead, a significant part of the landscape around the convent and town appears to have been reassigned as pasture. Rehden's inventories indicate the presence of substantial herds by 1377, some of which would have been located in other parts of the commandery but clearly large numbers of animals were also housed in at least one of castle's two outer baileys. Elsewhere in the convent's territory, there is further evidence for the conversion of land to pasture to meet the demands of intensified livestock husbandry. Within the vicinity of Rywałd, there is a reduction in cereal cultivation by the late thirteenth century, with meadows and pastures dominating the landscape into the post medieval period. Increasing quantities of pine again point to the opening up of the landscape and the reduction of commercially valuable species in surviving woodlands. Does this diachronic trend reflect the hiatus between the early and later medieval settlement patterns here, followed by a deliberate reassignment of landuse? The principal areas of grain production may have been situated in the northern parts of the commandery, which saw the highest density of settlement in the late medieval period. The remaining woodlands within the commandery were further depleted of commercially valuable broad-leaf trees and recolonized by pine, a trend found across the Lower Vistula region (see below).

After 1410, numbers of livestock, including horses, as well as cereals and other supplies, dramatically decreased. The notable exception was sheep, with relatively substantial herds maintained in the 1420s, although these only reached 60% of earlier numbers of animals in one year. From the 1430s, the decline in activity at the castle is much more pronounced. At the same time, the devastation that is usually attributed to the incessant conflicts between 1410 and 1466 is not clearly visible in the landscape. The town of Rehden endured and there is evidence for sustained cultivation throughout the fifteenth century which contrasts with the dramatically reduced status of the convent. At this time, there is an increase in oat-wheat pollen within the vicinity of the castle, suggesting a renewed emphasis on localized cultivation. It is possible that the fragile authority of Rehden's convent at this time and the dramatic reduction in local herds, as reflected in the inventories, may have prompted the conversion of local pastures.

It is also important to situate the convent of Rehden within the broader regional context of the Kulmerland. The Vistula continued to link the Kulmerland with Danzig, as it had done during the earlier centuries of Polish rule in the borderland. But the geopolitical situation of the site changed dramatically between the twelfth and fourteenth centuries: it transformed from a frontier outpost to a centre managing a productive heartland. The north-eastern border with the territory of the Pomesanian episcopate was of little political consequence as the chapter had been incorporated into the Order. The quickest means of exporting grain from Rehden would have been via neighbouring Graudenz (Pol. Grudziądz), where the town had invested in impressive granaries fronting the river. The Kulmerland remained a volatile region throughout the crusading period (Pluskowski 2013, 118), the target of a number of Prussian and Lithuanian attacks, but ultimately the transition from the period of Polish rule to that of the Order can be characterized as relatively smooth. The range of data indicates that the colonists settling the region after the cessation of hostilities appropriated the existing agrarian and pastoral infrastructure. This most likely involved the movement of livestock and farming technology from neighbouring regions such as Kuyavia, as most of the pre-crusade settlement pattern with the exception of Kałdus appears to have collapsed as a result of Prussian incursions.

This is evident in the continued breeding of small, short-horned cattle, which defined the region's stocks into the post medieval period. Beef became the staple meat for the castles and towns of the Kulmerland, although in at least two cases — Bischöflich Papau (Pol. Papowo Biskupie) and Thorn (Pol. Toruń) — the consumption of pork was more important, perhaps reflecting the adoption this distinctive element of the Piast state's economic system, alongside access to suitable woodland pasture as specified at Rehden. The presence of sturgeon at the convents of Thorn and Nessau (Pol. Mała Nieszawka), where at the latter peafowl was also identified, is reminiscent of the high-status role of these species in earlier Polish centres. However, if there was continuity in the types of animals bred in the Kumerland, there were changes in their quantities, how herds were fed and their meat processed. Isotopic studies from sites within the region indicate a broad shift

Figure 16.3: (a) Standardized cattle rib cuts and cleaver marks on limb bones recovered from excavations of the Order's castle at Mała Nieszawka (Ger. Nessau); (b) cleaver and saw marks on bone from excavations at the Order's castle in Toruń (Ger. Thorn). Photos by K. Seetah.

in livestock diets on either side of the crusading period (Scull et al. 2019), indicative of changes in cereal production linked to the diversification of crop yields, a form of intensification clearly promoted by the Order. This is also confirmed by the range of cereal species documented in the convent's inventories. At the same time, the incremental replacement of woodland with pasture is indirectly reflected in a higher proportion of sheep and goat in assemblages, as evident in the faunal assemblage of Graudenz and listed in the inventories of Nessau in the last quarter of the fourteenth century. These animals were also typically larger in size and would have supplied local wool industries. Within the Kulmerland, the 'new town' of Thorn became known as a major centre of (average quality) cloth production, exporting its products to other regions of the *Ordensland* (Czacharowski 1983, 43), and cut marks on bone from the town and castle indicate that heavy cleavers had been introduced for processing carcasses, alongside evidence for specialization and standardization. Cleavers and standardized butchery also typify how meat was prepared within the precinct of the nearby convent of Nessau (Fig. 16.3). Sturgeon remained a popular food for the regional elites in the fourteenth and fifteenth centuries, but the Vistula now also facilitated the import of new species into the region, in particular cod which consistently features in the stores and waste pits of the Order's convents.

From a Frontier to the Heartland of Prussia: Malbork–Marienburg

In the early medieval period the landscape within the vicinity of Malbork (Ger. Marienburg) was dominated by a series of strongholds at Węgry, Baldram, and Kalwa. The immediate landscape of Węgry, the northernmost stretch of the forest of Stuhm, was progressively cleared and cultivated from the eleventh century, intensifying a trend already evident from the eighth century. The palynological evidence from Gościszewo (Ger. Braunswalde) located *c.* 3 km north of Węgry shows significant clearance of woodland from the eleventh century (Brown and Pluskowski 2011), accompanied by the development of cultivated fields and pasture. In all the settled areas, there is evidence for the cultivation of both winter and summer crops alongside hemp, with a distinct peak in rye pollen within the vicinity of Lake Rozlewisko before the onset of the Prussian Crusade. This can be linked to the presence of nearby strongholds at Stary Targ (Ger. Altmark) and Kalwa. Further south, in the vicinity of Benowo (Ger. Bönhof), *c.* 14 km north-west of Baldram, there is evidence for the impact of early Slavic colonization with the gradual decline of hornbeam from the eighth century, and particularly from the eleventh and twelfth centuries. Whilst there is evidence for an increase in pasture during the eleventh century, arable and pastoral activity appears to have been limited, represented by rye, wheat/oat, and barley along with intermittent weed and ruderal pollen taxa. Here, the relatively limited evidence for late medieval cultivation probably also reflects the location of the site deeper within the forest. The settlement at Biała Góra (Ger. Weißenberg), which straddles the chronology of the early and later medieval period, indicates that the full range of habitats within the immediate vicinity of the site was exploited, with the maintenance of permanent rotational field systems and the importance of the Vistula is emphasized by the presence of imported cod in the post-crusade period. Whilst cereal production at the site was almost certainly intended for local consumption, pelts of fur-bearing mammals were most likely prepared for export via the Order's trading network. One of the earliest examples of brick production in the region has been identified here (Fig. 16.4), alongside new kiln firing technology which saw the production of large quantities of 'greyware'. The continued occupation of the settlement, or at least its reoccupation if it can be identified as the historically documented Zantyr (see Pluskowski et al. 2014; Sawicki et al. 2015), belies the fundamental territorial changes which transformed this early medieval frontier into the heartland of the Teutonic Order's polity. From an ecological perspective, the frontier can be characterized as progressively settled and exploited before it becomes a militarized zone during the crusading period.

Figure 16.4: One of the brick and stone 'caches' excavated at Biała Góra (site 3). Photo by Z. Sawicki.

The early medieval power centres were replaced by the convent of Marienburg, later supported by the residences of advocates and procurators (sing. *Pfleger*) established within its commandery. The associated changes to the physical landscape were far more extensive here than virtually anywhere else in the Prussia and Livonia, perhaps with the exception of the convent and urban cluster of Königsberg (Rus. Kaliningrad). The dramatic expansion of the Marienburg complex over the course of the fourteenth century saw the levelling of the plateau and the establishment of brickyards processing vast quantities of clay. Although this may have derived from more distant sites, a proportion appears to have derived from local sources, perhaps obtained during the excavation of the canal and moat systems linking the castle with the lakes to the south (Hayward 2019). At this time the town was also fortified with a set of brick walls constructed on stone foundations (in its southern stretch consisting of four parallel walls), replacing the late thirteenth-century timber and earth defences (Duda and Kryzia 2011, 191–92). Substantial quantities of pine were used in the western side of the middle castle (particularly under the Great Refectory) in the form of piles driven into the waterlogged soil to provide stable foundations (as, for example, in Elbing, Danzig, and Königsberg) (Mierzwiński 2010, 169, 171).

The heavily fortified convent and town were also associated with significantly higher provisioning require-

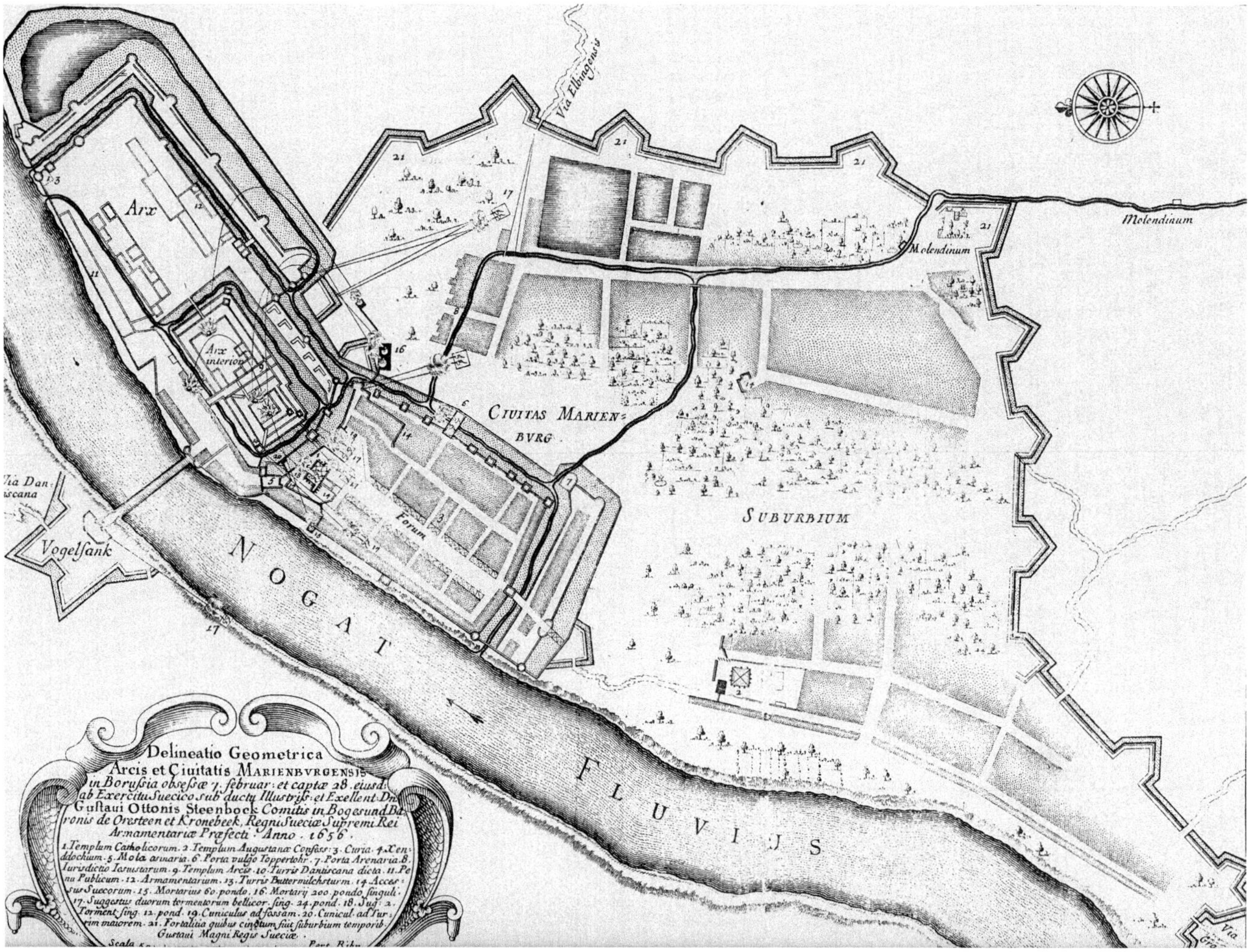

Map 16.1: Plan of the town and castle of Marienburg in 1656 (Pol. Malbork) (after Czaja and Czacharowski 2003, Map 8).

ments than the earlier castellany. Cultivation is evident from the establishment of the convent: the pollen profile from the high castle courtyard suggests this space functioned as a garden for growing herbs and vegetables in the first occupation phase (dated by AMS (68.8%) to *c.* 1267–1311). Along with remains of cultivated fruits and vegetables recovered from moats and documented in the convent's stores, gardens, and orchards located appear to have been located around and within the castle precinct at least by the end of the fourteenth century (Jóźwiak and Trupinda 2007). The immediate vicinity of the castle and town was a mosaic of landuse. Pollen from the outer moat indicates an open agricultural landscape in the fifteenth century, reiterated in the earliest maps of the castle and town (Map 16.1) — today a built-up landscape resulting from urban expansion in the second half of the twentieth century. The castle housed its own herds of cattle, sheep, goats, pigs, and particularly horses. Written sources from the later fourteenth century point to the prevalence of sheep at the convent (Jóźwiak and Trupinda 2007, 426), all of which would have required substantial quantities of pasture and fodder. The limited faunal assemblage from the Order's period indicates that livestock were slaughtered and butchered within the outer bailey, where deadstock was also processed for manufacturing a range of artefacts, including crossbows (Maltby et al. 2009); a crossbow-maker's workshop was also identified during excavations in the outer bailey (Dąbrowska 2007). The rare inventory from the convent's crossbow workshop indicates that a large number of varied animal-based ingredients were required for manufacturing this weapon, and given the stockpiles of crossbows documented in the Order's castles from the mid-fourteenth to fifteenth centuries, particularly in the Kulmerland and Marienburg (Gancewski 2001, 77), substantial quantities of deadstock would have been required. This was an iconic industry associated with the period of the Order's rule; by the end of the fifteenth century, crossbows,

Figure 16.5: Views around the modern village of Parpary (Ger. Parpharen) showing the maintenance of cultivated and cleared land, which rapidly come to typify the landscape following the establishment of the colony. Photos by M. Elander.

although still in use, were becoming replaced by hand-held firearms (Nowakowski 1994, 102–03).

The convent was not solely reliant on animals and plants from within its own fortified precincts. Much of the grain, meat, livestock, and fish stored and consumed at Marienburg derived from its commandery, and it is not surprising, given its status and purchasing power, that the convent was the most important consumer of exotic products in the *Ordensland* (Długokęcka and Długokęcki 2010). The town would have provided the obvious source of artisanal products, and it is likely that the relatively small amounts of good quality skin required for Marienburg's chancellery were sourced from beyond the convent's walls on an ad hoc basis (see Pluskowski et al. 2019). The remains of weed species, associated with winter-sown crops in the outer bailey moat, most probably derived from the mills lining the connecting Mühl Graben. Indeed, the lack of suitable water channels in the southern part of the commandery almost certainly prompted the dramatic hydraulic works which connected the lakes between Marienburg and the Stuhm. This resulted in a more industrialized, rather than aesthetic landscape, a process mirrored in the north-western part of the commandery — the colonization of the Vistula fenlands in the first half of the fourteenth century was associated with the construction of mills. Indeed, the important changes in the territory of Marienburg can be subdivided geographically into two regions; the south-west dominated by the plateau and Forest of Stuhm (Pol. Sztum), and the Vistula fenlands. The process of clearing the edges of the wooded expanse continued from the early fourteenth century, with a significant decrease in hornbeam and notable increases in cereals, weeds, ruderal plants, and grasslands around Parpary (Fig. 16.5). The notable increase in rye pollen in the Gościszewo sequence from the fourteenth century (Brown and Pluskowski 2011), is also evident within the vicinity of Lake Rozlewisko, where hornbeam and oak are almost completely cleared during the fourteenth and fifteenth centuries giving way to a predominantly agricultural landscape. Permanent field-systems of winter and summer crops were established along with meadows and pastures. The forest itself was carefully managed; commercially valuable hardwoods such as oak were maintained into the sixteenth century, when there is a reduction in pollen of all arboreal taxa and a significant increase in pollen of rye and higher values for grassland and ruderal communities in the north-western side of the forest. In the south-western side, the process of colonization corresponds to a slight expansion in arable and pasture in the surrounding landscape by the mid-fourteenth century. However, the general integrity of the forest has been retained into the present day.

In contrast to Rehden, the exploitation of local resources within Marienburg's commandery was delegated to a series of procurators, advocates, and lesser officials. The advocates of Stuhm provided the Grand Master's table with fresh game, timber was supplied from the substantial corridor of woodland managed by the *Waldmeister* of Bönhof and the procurators governing stretches of the Vistula's floodplain ensured the Order maintained control over milling. All of these officials were involved in the maintenance of what amounted to substantial herds of livestock until decline set in after 1410. At the intra-regional level, the largest meetings of the general chapter at Marienburg were — uniquely — provisioned with fish from virtually *all* the other Prussian convents, but the networks which supplied the Grand Master's court were also provisioning the Order's most important houses in other parts of

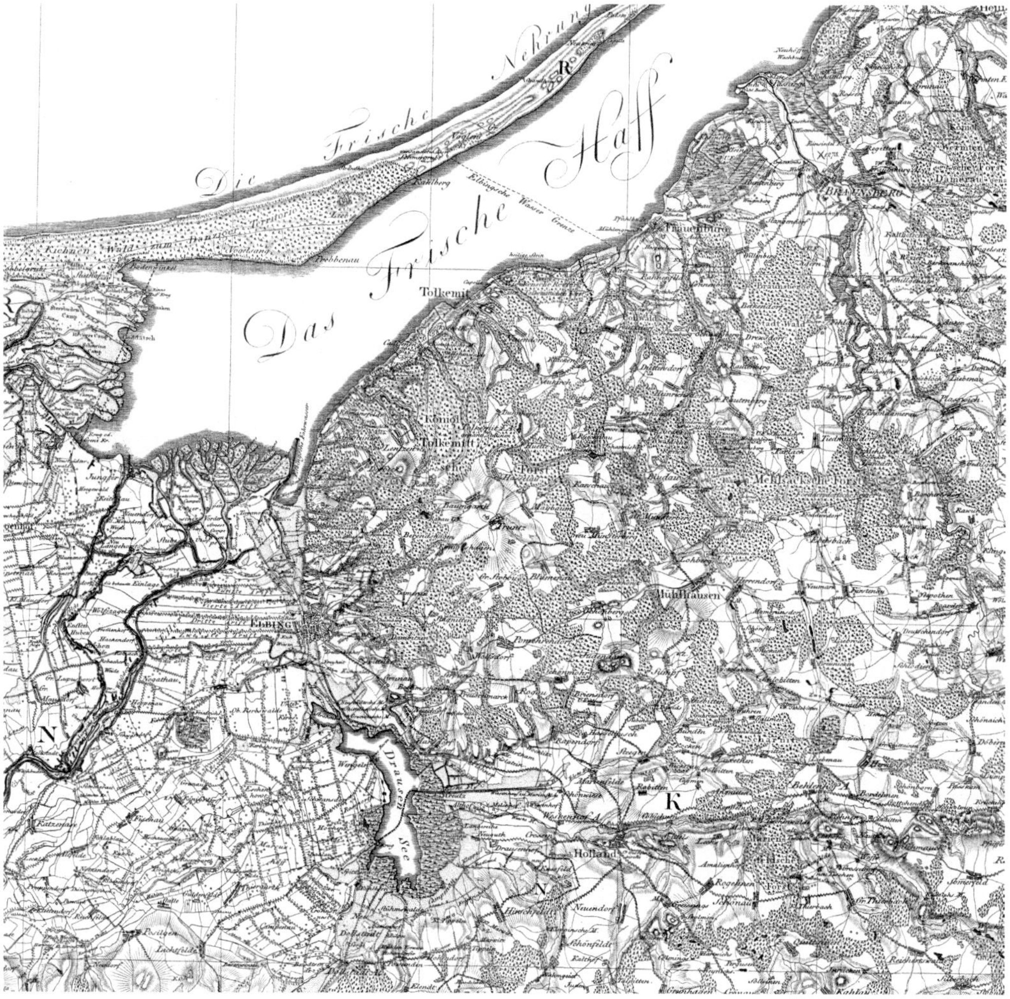

Map 16.2:
The Elbląg uplands (a) on Schroetter's map from 1810 (Staatsbibliothek zu Berlin – Preußischer Kulturbesitz, Kart. N 1020; image reproduced with permission); (b) today (Google Earth, 2016 DigitaalGlobe, MGGP Aero and CNES/Astrium); (c) a high contrast digital elevation model showing the difference between the floodplain and uplands (ESRI).

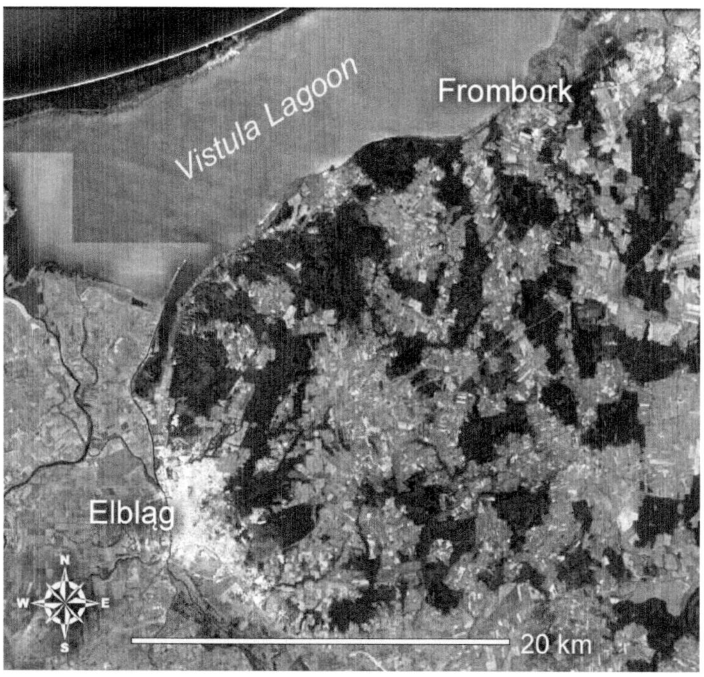

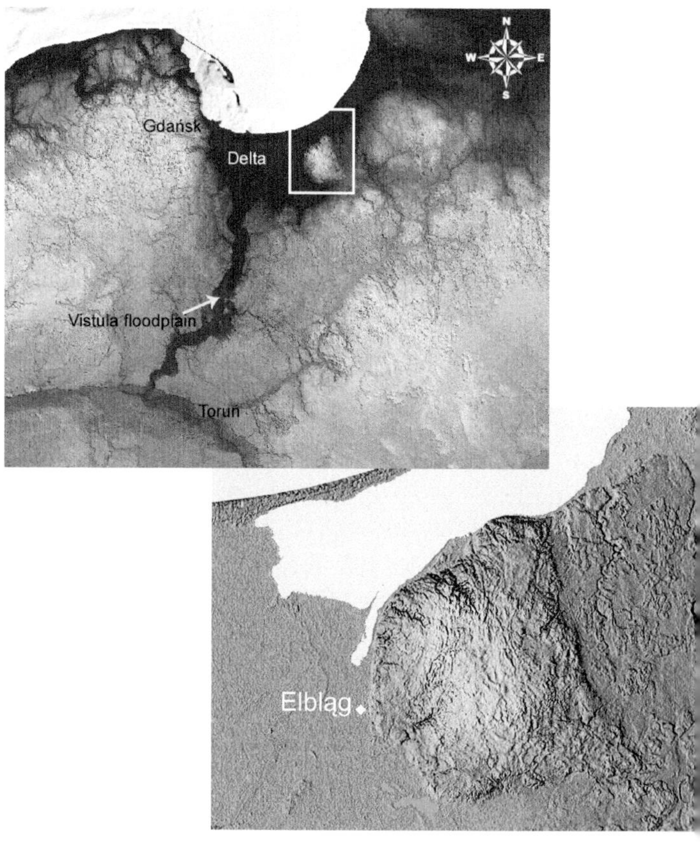

the eastern Baltic. Taking the commandery as a whole, with its fifty mills, twenty-five specialized farms, riverine embankments, canalized lakes, settled by a multi-ethnic (but largely German and Pomerelian) population, with a visible Christian infrastructure already by the mid-thirteenth century, the transformation from a frontier to a heartland could not have been more dramatic.

The Early Crusading Frontier: Elbląg-Elbing

Palaeoenvironmental studies of the vicinity of Elbląg (Ger. Elbing), particularly those carried out within the framework of the 'Adalbertus' programme, have been largely concerned with tracking the impact of the settlement at Janów Pomorski (Truso), which was occupied into the early eleventh century. Subsequently the region is dominated by strongholds in the Elbląg uplands and at Dzierzgoń (Ger. Christburg) (Długokęcki 1992; Szczepański 2016). The sequence from Lake Drużno included within this study indicates the presence of a substantial mixed-deciduous woodland in the vicinity during the Migration Period, with clearance commencing several centuries before the occupation of Truso, alongside increasing values for grasses, ruderal, and cereals. Unfortunately, due to the absence of a secure chronology it is difficult to link the abandonment of Truso with the notable decline in rye pollen in this sequence, although woodland decreases significantly during the upper portion corresponding to the medieval period, with pollen from thirteenth- and fourteenth-century deposits within the courtyard of the outer bailey of the Order's castle suggesting a predominantly open environment in the immediate vicinity. Truso was certainly sustained by grain, substantial herds of livestock, as well as locally available marine and wild mammal resources. Aurochs (or perhaps bison) may have been procured locally or obtained from denser corridors of woodland further east, but hunting played a minimal role at the settlement (Makowiecki and Makowiecka 2012, 297). At this time, a broad river connected Lake Drużno with the Vistula Lagoon which, in turn, was accessible via the Baltic Sea through a break in the spit (Kasprzycka 1999, 172). When the Order arrived and established their *castrum* on an island in the River 'Elbing', the waterways were still navigable, and crusading armies were following well-established routes. However, Lake Drużno had begun to shrink along with the waterways connecting it with the Vistula Lagoon, and notable changes in the local environment are evident following the construction of the Order's castle and adjacent town.

During the crusading period, the castle's *Vorburg* clearly functioned as a secure area for agricultural storage and for housing, slaughtering and butchering livestock. Some of the structures within the early castle's outer bailey were constructed from old and substantial oaks, much larger than any used in the town. It is entirely possible that, given the size of these trees, they may have been considered sacred by the local Pogesanians and their incorporation into the castle's superstructure during the crusading period was a deliberate religious act. Whether these derived from one of the 'sacred woods' documented in nearby toponyms, such as Heiligenwalde (Pol. Święty Gaj), south of Lake Drużno, a village which at the end of the eighteenth century was close to two areas of woodland, or whether they were simply cut down for their size, will invariably remain speculation. In the earliest phase of the castle, greater quantities of arboreal pollen indicate a more significant presence of pine, birch, alder, oak, and hornbeam in the nearby vicinity, with very small quantities of cereals, than evident in the upper part of the Lake Drużno sequence. Of course, the castle and lake represent two very different pollen catchment zones, but the construction of the convent and the two towns of Elbing would have necessitated the removal of much of the surrounding woodland. This is particularly attested in the earliest phases of the Old Town where the majority of buildings were constructed from timber (Nawrolska 2004). The rebuilding of the town and convent in brick marks a significant break from earlier architectural traditions, and a brickyard is documented early on in association with the Dominican monastery, later given to the town (Nawrolska 2003).

The provisioning systems of the town in the thirteenth century have yet to be investigated in more detail, with most of the data deriving from the fourteenth and fifteenth centuries. However, the organization of territory and access to resources was clearly fundamental to maintaining the colony in the decades following its establishment. The town was granted the Lübeck Law in *c.* 1240 (confirmed by the Order in 1246), with its rights listed in eighty-two articles, but this number had doubled by 1275 and increased to 256 by 1295 (Domagała 2004, 92). Alongside a regular food supply, access to timber for housing was as important as clay for bricks and ceramic vessels. At least four residential tower houses dating to the 1260s and 1280s had stone foundations, and along with the presence of imported luxuries such as jugs from Saintogne and nutmeg, they indicate the rapid development of social stratification within a generation of establishing the colony (Nawrolska 2008, 512). There is more evidence of imported foodstuffs in the fourteenth and fifteenth centuries along with other luxury commodities such as glassware, reflecting the significant purchasing power of the town's patricians. However, much of the town's provisioning almost certainly derived from its own extensive patrimony. This territory, eventually covering an area of *c*. 200 km^2, was the largest urban patrimony in late medieval Prussia, mostly located west of the town within the Vistula Fens (Nawrolska 2008, 518). The town's patrician class also supported the Order's ongoing wars against Lithuania, at least until 1410. The discovery of seventeen spurs, dating from the latter half of the thirteenth century through to the 1390s, reflects the provision of military support for the Order's *Reisen* as documented in the *Elbinger Kriegsbuch* in 1321, 1360, and 1377. The loss of the town's military elite in 1410 correlates with the absence of fifteenth-century spurs in the Old Town (Marcinkowski 2003, 186).

This study has largely focused on the transition from the pre-crusade period to the early phase of the Elbing convent at a relatively coarse resolution. However, the historically attested process of colonization within the commandery over the course of the fourteenth century can be expected to have been accompanied by similar processes of locally variable landscape transformations, as elsewhere in the Lower Vistula and western Prussian regions. One important area for future research would be to investigate the impact of the crusades in the Elbląg uplands, which appear substantially wooded in the earliest maps of the region and into the present day (Map 16.2). The uplands were controlled by a series of strongholds which were systematically destroyed by crusading armies. Once the region was pacified, colonists were encouraged to settle, but to what extent were these extensive woods maintained as a source of timber for the Elbing convent, were they exploited for export and was this distinctive landscape used to define the border with the neighbouring Ermland episcopate?

From a Tribal Core to a Managed Wilderness: Giżycko-Lötzen

In the pre-crusade period, the focus of settlement within the vicinity of Giżycko is most evident to the south, associated with the Staświny-Święta Góra settlement complex, interpreted as a *lauks*. Four stages of anthropogenic impact on the local environment have been identified by Wacnik et al. (2012) from sequences obtained from Lakes Wojnowo and Miłkowski. The fourth stage is characterized by extensive clearance of woodland around the lakes and the appearance of intensive agricultural activity from *c*. 1100. This was also confirmed by a pollen sequence from Lake Staświny which had a much smaller catchment, showing intensifying cultivation and deforestation from *c*. 1000. Woodland was almost completely removed from the vicinity by the early twelfth century, accompanied by the development of permanent rotational field systems and pastures. At Święta Góra the remains of rye, wheat, and millet were found in early medieval contexts. Nonetheless substantial woodland survived beyond the immediate vicinity of the settlement, corresponding to a high degree of biodiversity in the faunal assemblage from Święta Góra. It is unclear how this *lauks* related to the territory of the nearby stronghold at Jeziorko (Ger. Jeziorken), but it is possible the two may have been connected politically. Future excavations and palynological studies ought to focus on identifying how settlement clusters related to each other around these series of lakes. Along with evidence for significant clearance of woodland and intensive agricultural activity from *c*. 1150 to 1350 around Lake Salęt, this indicates the Galindian settlement pattern was not completely obliterated in the eleventh and twelfth centuries. Peter von Dusburg's description of Galindia as desolate and depopulated in the mid-thirteenth century is clearly a comparison to the more densely settled parts of Prussia. It is also non-specific and settlement from the mid-thirteenth to fifteenth century remained unevenly distributed across Galindia. The region remained part of the broad frontier zone with Lithuania and the process of colonization did not begin in earnest here until the later fifteenth century.

There is significant evidence for continuity in this landscape following the crusades, and the region is not reorganized until the procuratorship is established in *c*. 1340. This takes place during a new phase of colonization in the Masurian Lake District driven by Grand Master Dietrich von Altenburg; however, it is clear that much of the indigenous population remained intact, and in the case of Staświny there is direct archaeological evidence for settlement continuity. The intensity of agricultural land-use is sustained in the Lake Wojnowo, Miłkowski, and Staświny sequences into the late medieval period, although this still appears to have been relatively localized. In contrast, the landscape to the north of the castle at Lötzen remained heavily forested until the process of colonization began in earnest, resulting in gradual clearance with woods eventually confined to marginal lands. The vicinity of Lake Łazduny was only significantly cleared from the mid-fifteenth century, again corresponding to trends in colonization (Wacnik et al. 2012). What this suggests is that the surviving indigenous settlement pattern maintained a relatively localized impact on the landscape, and more significant change would take place with the wave of colonization following the establishment of the procuratorship, and particularly once the border with Lithuania was stabilized. The period between 1420, when Lötzen castle had no granary, and the early sixteenth century, when grain is documented in its stores, corresponds to the general increase in cultivated land associated with intensified colonization. In this respect, the infrastructure of exploitation in this macro-region can be synchronized with the development of the post-crusade settlement pattern, where tenuerial relationships between the procurator and Masovian and German colonists may have been more clearly defined than with the indigenous population in the fourteenth century. At this time levels of biodiversity within the procuratorship were still high — the list of fur-bearing species from Lötzen castle is more diverse than the earlier wild mammal assemblage from Święta Góra — and the final phase of the Order's rule in this region (1466–1525) is characterized by an emphasis on fur, wax, and honey. The castle's stores of game were also higher than in the Order's houses in *Ämter* generating large quantities of grain. The faunal assemblage from the village of Staświny has comparatively limited biodiversity, but given its size it cannot be taken as representative of the full range of species exploited by such communities on an ad hoc basis. Fish continued to be obtained from local lakes and rivers until the sixteenth century, when the first saltwater species are noted in the castle inventories. It is therefore possible to characterize this macro-region between the crusades and secularization as an incrementally managed wilderness.

Further east, the settlement pattern in the territory of Sudovia may well have collapsed following Rus' attacks and the final campaign of the Teutonic Order in 1283, although it is poorly known archaeologically. Jegliniec, the best known Sudovian stronghold, was certainly abandoned before the crusades, but the pollen evidence suggests some communities endured in the midst of the 'Great Wilderness'. In the Suwałki-Augustów Lake District, the late thirteenth to seventeenth centuries are defined by decreasing woodland and an increase in cereal cultivation, whilst in Knyszyńska Forest an expansion in arable and pasture accompanying the clearance of woodland is evident from the fifteenth century, intensifying a century later (Kupryjanowicz 2004). The Order's fortified outpost at Lyck, established at the start of the fifteenth century, was sustained by its own livestock herds alongside the relatively high biodiversity of its associated territory. In summary, it is possible to correlate the shifting of the eastern Prussian frontier with the stability of the settlement pattern and trends in environmental exploitation. It was difficult to impose an effective system of territorial exploitation here before the fifteenth century. Up to this point the process of colonization can be defined as 'militaristic', aimed at sustaining the Order's outposts and *Reisen* into Lithuania. At the same time, the topography of the Lake District limited the possibilities of agrarian expansion. Although the indigenous population survived in settlements like Staświny and maintained earlier patterns of exploitation, they were not mobilized within the procuratorship until the new phase of colonization. Whilst contact with new technologies (e.g., the presence of greyware vessels) is evident, the recording of sacred site toponyms suggests their function was recognized by the theocratic administration. This is a region where future excavations of indigenous settlements and, in particular, cemeteries, will shed important new light on the nature of cultural transformation following the crusades. In the absence of a Christian infrastructure before the fifteenth century, the working hypothesis must be that indigenous lifestyle and religious beliefs were retained, perhaps as long as the Prussian language survived in this region. Environmental exploitation intensified only with the influx of substantial numbers of migrants.

Discussion: Contextualizing and Defining the Impact of the Teutonic Order in Prussia

The Early Medieval Period

The final phase of indigenous Prussian culture following the collapse of the south-eastern Baltic emporia, from the eleventh to early thirteenth centuries, is defined by two major trends in settlement in Prussia and its borderlands: first, general expansion focused around strongholds with localized episodes of contraction; and second, Slavic colonization in the southern and western frontiers, accentuated by the development of the Polish state. If Peter von Dusburg's retrospective account of the collapse of the Galindian political system finds its archaeological parallel in the abandonment of settlements in north-western part of the region in the twelfth century (Nowakiewicz and Wróblewski 2003), then enough of the population survived to maintain continuous pressure on woodland and sustain permanent rotational field systems into the thirteenth century. This is particularly evident along the belt of lakes stretching south of Lake Niegocin. The territorial role of strongholds, such as Jeziorko, in this final pre-crusade phase of Galindian society, remains obscure, but their political reach would have been limited by the dispersed nature of settlement and the topography of the Masurian Lakelands. The stronghold, located on a morainic outcrop, was within a few kilometres of a major route that would eventually link the Order's castles at Rhein (Pol. Ryn) and Lötzen, but in contrast to other strongholds which dominated their surroundings, this settlement complex was relatively secluded (Jaremek 2003).

In western Pomesania, the construction of Pomerelian strongholds is accompanied by the first significant expansion in cultivation, alongside evidence for significant continuity of woodland in more sparsely populated areas where agricultural potential was limited. On the fringes of Pomesania-Pogesania, the ecological impact of Truso is multi-scalar, with evidence for local landscape transformation and the exploitation of distant hinterlands. Truso's ecological profile is comparable to Gdańsk albeit the latter has a far larger dataset from both the town and its territory. The presence of, for example, herring at these emporia, as well as at frontier strongholds such as Węgry (Makowiecki 2012, 2015), is indicative of similar procurement networks. However, the intensity of fish consumption in Pomerelian centres would have met the ideological needs of a recently established Christian fasting culture, and the more pronounced emphasis on cattle rearing in Prussia and pig-rearing in the Polish-Prussian frontiers may also have reflected contrasting culinaria, given that suitable grazing was available across the region. The abandonment of Truso in the early eleventh century did not result in a break in cultivation and deforestation within the vicinity of Lake Drużno. This can be explained by the continuation of settlement around Dzierzgoń and the Elbląg uplands, where strongholds replaced the role of the former emporium (Szczepański 2016). Over the course of the twelfth century, the earlier commercial relationships between Pomerelians and Prussians were overshadowed by ideological confrontation, represented in the militarization of the Christian frontier, the contraction of Pomerelian settlement, and evidence for Prussian westward expansion, with raids reaching as far as Gdańsk (Sawicki et al. 2015). The Pomerelian settlements within the Pomesanian frontier were largely abandoned at this time, with exceptions such as Biała Góra, which may have endured due to its sheltered location and connection with the Nogat.

Further south, the Kulmerland, consolidated as a Polish castellany in the twelfth century, began to be cleared of oak-hornbeam dominated woodland from the seventh and eighth centuries, with increasing intensity in cultivation and livestock husbandry from the tenth century, associated with the expansion of the Piast state. By the early thirteenth century this was a highly developed and well-connected region, centred on the castellan's stronghold at Kałdus. To summarise, important levels of environmental exploitation are evident across Prussia, and its borderlands are linked to the expansion of settlement and focused on strongholds. Variability reflected accessibility to resources and the regional organization of environmental exploitation, which in the case of western Pomesania and the Kulmerland were linked to the devolved superstructure of the Polish state. In the parts of Prussia surveyed in this study, there is no evidence for such massive regional organizational structures after Truso. The other important difference was, of course, linked to religion.

The relationship between cult sites and the treatment of the physical environment in pre-Christian Prussia briefly requires clarification. All descriptions of Prussian religious beliefs were written by external, Christian observers and as such contain bias, misunderstanding, and simplification. But there is a consistent emphasis on attaching sacred meaning to very specific elements of the landscape: individual trees, groves, fields, lakes, rivers, and hills (Szczepański 2019). This is supported by studies of toponyms recorded in the documents of the Teutonic Order and Prussian episcopates. Archaeologically, the most visible sites of cult activity are cemeteries, although some glacial erratics are also associated with special deposits (Pluskowski 2013, 66–81). Cult sites were an important territorial component in pre-crusade Prussia, typically located on borders (Szczepański 2011, 135), although they have not been easy to identify in all the case study regions. In the Christian borderlands of Prussia, moreover, the diachronic use of these sites is difficult to reconstruct at a high resolution. Toponyms offer a snapshot of use as the landscape was coming under the control of the Order, and the evidence of cemeteries indicates a variable chronology, itself linked to fluctuations in settlement. The fact that earlier cemeteries continued to be remembered and used by the indigenous population is clear (Szczepański 2019), although in the case study regions included here they are poorly known archaeologically. Indeed, with the exception of cemeteries in the Sambian Peninsula, the best example of a post-crusade Prussian cemetery is, to date, solely represented by the site near Równina Dolna (Ger. Unter Plehnen) within the former tribal territory of Barta, incorporated into the commandery of Balga (Odoj 1958).

Animals were also attributed supernatural agency by external observers, with a particular emphasis on the afterlife. The Treaty of Christburg refers to a role for sacrificed horses in the afterlife, whilst Vincentius, the Bishop of Kraków, writing in the late twelfth century in his *Chronica Polonorum* dismissed beliefs in the transmigration of human souls into the bodies of animals (Plezia 1994, 167). Sacrifices of animals to deities are occasionally mentioned, whilst Pope Innocent III imagined the Prussians greeted their gods in the form of wild animals (Okulicz-Kozaryn 1983, 231–32). The variable treatment of horses evident in the archaeological record suggests practices varied across Prussia. Not all were deposited in cemeteries: some were slaughtered and butchered within settlements (as at Staświny-Święta Góra), and not all were consumed. Moreover, by the 1230s, the deposition of horses appears to have been confined to cemeteries in Natangia and Sambia (Shiroukhov 2012). Later ethnographies also record animal sacrifice (of chickens, pigs, goats, and calves) in eastern Prussian regions. It is entirely possible that pre-crusade missionary activity, although generally considered unsuccessful, had contributed to the abandonment of public animal sacrifice in western Prussia. This is not the place for an extended discussion of pre-Christian Prussian religion (for the most recent summary, see Kawiński and Szczepański 2016), but the impact of the process of Christianization on the landscape, the treatment of animals and plants, is invariably linked to responses to the physical environment. The physical transformation of the landscape can therefore be envisaged as part of the missionary process (Szczepański 2019). It is clear that Prussian communities did not flinch from expanding their ecological niches and exploiting all available habitats within their territories, and so sacred meaning was attached to very discrete elements of the natural world.

The Crusading Period

The alleged devastation caused by Prussian attacks in the late twelfth/early thirteenth century corresponds to the abandonment of Polish strongholds in the borderlands. Peter von Dusburg describes how the Kulmerland was a wasteland, and alongside the human cost of the subsequent crusading campaigns, many other regions were laid to waste, along with the regular seizure of horses, livestock, and supplies. The environmental record suggests that, if this was indeed the case in the early decades of the thirteenth century, it was comparatively short-lived and did not interrupt the ecological impact of the previous phase of settlement. The resolution of the palaeoenvironmental data is relatively coarse, but it is clear that some settlements were abandoned during this time for at least several decades. In the Kulmerland, key strongholds such as Rehden were reused by the Teutonic Order, and the final phase of the Kałdus complex has been identified with the Order's short-lived *castrum* of Potterberg (Chudziak 2003, 179). The Order then established Kulm to the north, which would be granted town rights in 1233. With it came a new paradigm for organizing the governance of conquered territory as a means of sustaining the Order's houses: taxation was imposed from the very onset and extracted from incoming migrants and conquered indigenous communities alike. A castle garrisoned by the Order was attached to the colony by the 1240s (Mroczko 1974). With the possible exception of Thorn, the use of brick as a constructional material only began in earnest after the cessation of hostilities in the 1280s. However, the crusading period in the Kulmerland is associated with an intensive phase of urbanization and the development of associated provisioning networks. These either drew on the remnants of the Polish castellany or were resupplied from neighbouring Kuyavia. Given the chronological resolution of the Order's castles in the Kulmerland, it is difficult to separate the middle decades of the thirteenth century from fourteenth- and even fifteenth-century occupation phases. In any event, the palynological data indicates broad continuity in land-use.

Further north, in the former territory of the stronghold of Węgry, there is a dramatic break between the early and later medieval settlement pattern, with the exception of Biała Góra. The site's chronology could easily mask the short interludes in occupation which are documented for Zantyr. In any event, in the latter half of the thirteenth century and into the early fourteenth century, the settlement is fully connected with its surrounding landscape and local networks, although nothing compared to Elbing's access to imported luxury commodities. The site was certainly stable enough to enable brick production to begin, although there was only tentative evidence for the use of brick in structures. The Order established a bridgehead at Elbing, only to find itself fighting on two fronts across the Vistula throughout the 1240s. Nonetheless, between the two Prussian uprisings, the military campaigns organized by the Order had secured much of western and central

Prussia with strategically located *castra* and accompanying settlements. Archaeological data from the Old Town of Elbing indicates the early presence of a wealthy mercantile class. The convent had been designated as the headquarters of the Order in Prussia, although there is virtually no information on how it differed from other convents at the time. Its fortified outer bailey, laid out on an island with substantial timber structures, was used for housing livestock, processing meat and deadstock, and for storing grain. The extensive use of timber in the urban colonies established by crusading hosts is exemplified here, where excavations revealed the remains of timber paved streets and piled oak logs used in foundations to support timber-framed buildings (Nawrolska 2004). This renewed pressure on timber resources within the vicinity of the town and castle. The destruction of Elbing by fire in 1288 indicates its vulnerability: the central part of Prussia (broadly represented by the tribal territories of Warmia and Barta) remained a relatively volatile frontier, exposed to Lithuanian incursions, until the mid-fourteenth century. Nonetheless, the rebuilding of Elbing indicates a remarkable level of resilience and investment in what was the most important centre in the eastern Baltic during the second half of the thirteenth century after Riga.

In eastern Prussia, the community using Święta Góra endured during the crusading period, and the limited data from Staświny indicates no break in occupation between the early and later medieval periods. This was almost certainly the case in many settlements across eastern and parts of central Prussia where the population survived or was not relocated. All indigenous political centres, however, appear to have been abandoned. Although the Order's *castra* were not always built on earlier sites, it appears that earlier systems of provisioning — linking strongholds with their network of subservient settlements — were appropriated by the colonizing authorities. This is particularly visible in the continuation of local stocks of cattle, pigs, sheep, goats, and horses. In the case of all the thirteenth-century colonies, and despite the Order's evident concern with controlling hunting, there is no evidence that game played a significantly elevated role in the 'war economy' of the crusading period; the meat and skin supply of the colonies was almost entirely derived from livestock. In this respect, the economic systems of indigenous Prussia did not collapse but were realigned to serve convents, outposts and towns. Major changes would come with rural colonization from the end of the thirteenth century.

Peter von Dusburg's account of the Prussian Crusade paints a picture of half a century of Christian radicalism, punctuated by martyrdom on the battlefield and the power of relics. The destruction of sacred sites is referred to only once in his narrative (the village of Romene in Aukštaitija, within Lithuanian territory), although it is clear that forced conversions and obligations to construct churches were imposed (Szczepański 2013). Crusading armies targeted centres of tribal authority — strongholds — rather than sacred sites located at the fringes of territories. Cemeteries, with their focus on public cult linked to the identity of the militarized elite, were more obvious targets. This is perhaps the subtext behind the provisions of the Treaty of Christburg, which targeted ritual specialists alongside a string of mortuary practices. In the case of Pomesania, the fragmentary evidence for cult sites corresponds with the distribution of Prussian settlement in the eastern parts of the region, and the fact that these were named and recognized by the theocracy provides a window onto the Order's *Realpolitik* (Szczepański 2011). The survival of some substantial oaks within the Prussian *Ordensland* raises questions as to why such commercially valuable trees were not exploited. One explanation may be they were left to function as boundary markers; alternatively, if they had a sacred function, they were tolerated for the benefit of the Order's indigenous vassals. The recognition of sacred woods in Curonian charters already in the 1230s provides an early precedent for this (see Chapter 8). The narratives of the Prussian Crusade highlight the biographies of individual leaders and castle garrisons and it is clear that responses to the conquered 'other' varied on the ground, despite the official agenda of conversion promoted by the papacy at this time.

The Middle Ages

The imposition of the commandery system during the thirteenth and early fourteenth centuries saw the reorganization of the political configuration of Prussia and neighbouring Pomerelia. The Order's castles became the lynchpins of the rural economy of conquered Prussia, serving as centres of agrarian administration, storage, and redistribution (Arszyński 2013, 73). These castles were clearly self-sufficient. Their faunal assemblages indicate that livestock were slaughtered, butchered, and processed within their fortified precincts, rather than simply brought in as joints of meat. They had access to mills either within their precincts or close by, and all were stocked with the aim of sustaining their garrisons for

Table 16.1: Anatomical distribution of wild mammal bones from the Order's phase in Ełk (Ger. Lyck) castle; *Bos* spec. could be domestic or wild.

	Hare (*Lepus europaeus*)	**Brown bear (*Ursus arctos*)**	**Fox (*Vulpes vulpes*)**	**Wolf (*Canis lupus*)**	**Wild boar (*Sus scrofa*)**	**Red deer (*Cervus elpahus*)**	**Roe deer (*Capreolus capreolus*)**	**Elk (*Alces alces*)**	**Aurochs (*Bos primigenius*)**	**Aurochs/Bison**	**Bison (*Bison bonasus*)**	**Bos spec.**
Skull and antler							1					
Skull						1	1	1				
Mandible			1	1			1					
Thoracic vertebra					1				1			1
Rib					1		1		1			1
Scapula	1			1		1	2				1	1
Humerus	1						1			1		
Radius						1	1		2			
Ulna								1				
Radius and ulna								1	1			
Metacarpal		1				2						
Pelvis						2						
Femur						2			1			
Tibia			1			4	2	1	1	2	3	1
Calcaneus		1					1		1			
Astragalus						1					1	
Metatarsal						3	4					
Phalanx								1				
Total	**2**	**2**	**2**	**2**	**2**	**17**	**15**	**5**	**8**	**3**	**5**	**4**

periods of up to a year. Virtually all of the Order's houses had access to luxury foods and condiments: convents regularly stored saltwater fish, expensive fruits, and vegetables, which begin to be purchased by the urban patrician class in the largest Prussian towns from the fifteenth century. This underlies how the character of provisioning at the Order's houses was not simply about meeting nutritional requirements. In the easternmost outpost of Lyck at the start of the fifteenth century, whole carcasses of red deer, elk, roe deer, and even aurochs were processed within the castle precinct (Table 16.1). It is likely, given the evidence for the internal mobility of commodities, that choice cuts were cured and sent onto the convent of Balga, or elsewhere within the Order's network. However, even though biodiversity was relatively high in the surrounding wilderness, the garrison clearly depended largely on cattle for their meat and deadstock. At Lötzen, which governed another territory with high biodiversity, game was regularly stored but its quantities were small in comparison to other staples. Hunting may have been practised as a form of military training; the presence of skinned wolf bones from a mature individual would have been of limited commercial value.

Further west, with little evidence of game species in the faunal assemblages recovered from the Order's castles, game was more likely to have been supplied through the tenuerial obligation specified in the Kulm Law or via the Order's internal network. Here, the consumption of game can be situated within the contemporary culture of both secular and ecclesiastical lordship which was uniquely blended in the military orders. The park at Stuhm was a short-lived expression of the aristocratic character of the Grand Master's court at the end of the fourteenth century. The fact that in 1440 two peacocks are documented at the farm at Niclasdorf, subservient to the advocate of Stuhm, suggests these were not simply occasional imports but were bred, selectively and erratically, on the Order's lands, at least in the Lower Vistula. The growing popularity of displaying peacocks at royal and princely banquets in the later fourteenth and fifteenth centuries was matched by their use as a symbol of immortality, particularly dynastic longevity. However, the use of peacock feathers in Prussian churches provides an additional context for the Order's own use of the plumage, which combined contemporary aristocratic feasting culture with a well-established religious symbol. When coupled with the impressive display of the Order's substantial cavalry in the first decade of the fifteenth century, the use of peacock feathers in this public setting could readily emphasize the longevity of the Order's rule, particularly at a time when the Knights' authority in Prussia was being contested.

If the Order's castles were self-sufficient communities which conveyed a clear sense of lordship through their purchasing power, dietary habitats, and bureaucratic regulations, not to mention their use of a militarized form of monastic architecture, the sustainability was inextricably linked to the stability of their subservient settlements, in both frontier and heartland *Ämter*. The entire rationale behind promoting colonization, which in turn complicated the management and bureaucratic structures of the Order's Prussian branch, was to develop local resource exploitation (alongside military capability) in order to sustain the Order's houses. This is also expressed in the regular and precisely defined grants of multiple privileges to tenants, such as access to lakes, rivers, meadows, woodland, the ownership of mills, and access to their ponds. The environmental transformations in Prussia which characterize the fourteenth and fifteenth centuries can therefore be directly connected with how the process of colonization was articulated by the Order. Across the whole of the *Ordensland*, the landscape became a shifting mosaic of varying ratios of arable, pasture, and woodland. Grain production, the principal pillar of the late medieval Prussian economy, is best attested along the Vistula between Nessau/Thorn and Danzig. Not all grain derived from commanderies flanking the river: significant quantities of rye and wheat passing through Danzig were sourced from Masovia.

But for any given commandery, grain production was only one component of its economic system, and the importance of livestock husbandry is clearly evident at convents such as Rehden and Nessau. Of course, both were closely linked as substantial quantities of fodder were required for sustaining these herds, with horses alone amounting to an estimated 16,000 animals on the Order's Prussian lands by the end of the fourteenth century (Ekdahl 1998). With the exception of officials designated with the management of specific resources — such as fish or mills — the 'improvements' that are visible in plant and animal husbandry between the early and later medieval periods reflect the agency of the colonists and existing settlers. The persistence of archaic forms in the Kulmerland reflects the adoption of surviving or neighbouring herds, whilst gradual changes in size can be linked to changes in fodder and variable breeding strategies. This could result in smaller animals, as in the case of pigs, as well as larger ones, as in the case of cattle. There is an interesting disparity between the regular dominance of cattle in zooarchaeological assemblages from the Order's sites, and the documented numerical superiority of sheep herds associated with convents such as Rehden and Marienburg. Clearly wool production for local consumption was important whilst beef was a more important component in the conventual and urban diet, but animal resources were also distributed across the commandery. Assemblages recovered from castles therefore only represent one portion of their resource base, but provide valuable insight into the diet of the household and their use of deadstock.

Clearance of woodland for agriculture provided a supply of timber. Commercially valuable species, particularly oak, were used in the construction of castles and towns from the thirteenth century, after which timber was augmented or largely replaced with brick. Nonetheless, timber continued to be used for a range of structures, even for fortified constructions in the early fifteenth century such at Lyck and Pien (Pol. Pień). But even with the increasing use of brick in construction from the fourteenth century, wood remained an important resource in the production of charcoal for the firing of bricks, and this is reflected in the Order's regulations over access to woodland. Timber was also increasingly exported from the Baltic to north-west Europe (Haneca et al. 2005; Ważny 2002, 2005). Baltic hardwood timber was particularly prized for ship construction, as it met demands for cheap building which became increasingly difficult from the denuded woodlands of north-west Europe. Dendro-provenancing of oak-timbers used in ships, buildings, and art-historical objects has demonstrated a clear correlation with the regional chronologies of supply. By the fourteenth century, the commanderies located to the east of the Lower Vistula were the main source of exported hardwoods, a shift in supply from the vicinity of Gdańsk (Haneca et al. 2005). As in the case of grain, timber was also sourced by the Order from neighbouring regions. Documentary records relating to the timber trade are rare before the sixteenth century, but records indicate the Order bought almost 1.5 million pieces of timber between *c.* 1389 and 1415, mostly sourced from Masovia in the form of wainscots (high-quality timber boards). Clearly, sufficient timber resources were available in these commanderies to support both domestic needs and those of foreign markets, despite the palynological evidence pointing to significant reductions in woodland. However, it is clear from other pollen sequences that substantial corridors of woodland persisted into the late medieval period and were actively maintained and managed by the Order. The best example of this is the Forest of Stuhm within Marienburg's commandery, where hardwood species such as oak and hornbeam remained significant components of the canopy throughout the period of the Order's rule, reflecting the careful management of the woodland to maintain a supply of mature hardwoods. The wood was also important for the production of honey and wax, which represented extremely lucrative exports alongside grain, but were also required in the production of beverages and candles. It is likely that every commandery had defined areas of woodland which were maintained in a similar fashion, taking advantage of land with recognizably limited agricultural potential. In the absence of palynological data it is possible to see this on the earliest detailed maps of Prussia, where in some cases discrete areas of woodland are named after the convent — perhaps indicative of earlier ownership.

The persistence of large expanses of woodland also has implications for the endurance of pre-Christian cult sites in the form of sacred groves and trees. These are exclusively known from surviving toponyms recorded in contemporary sources and suggest that a number of 'holy woods' were tolerated by the Order, particularly in eastern Prussia. Given the limited parochial infrastructure before the late fifteenth century in this region, it is not surprising that elements of pre-Christian practises survived as an important component of indigenous identity. The evidence for this is very fragmentary; in the case of the procuratorship of Lötzen, the best example of this is represented by Stasswin (Pol. Staświny) and its associated Prussian toponyms, as well as nearby Heiliger Berg (Pol. Święta Góra). Future excavations of fourteenth-century cemeteries within the region will provide an invaluable insight into the persistence of indigenous identity and the impact of religious conversion.

From an economic perspective, the system of territorial management appears to have functioned effectively into the fifteenth century, at which point the impact on the population of the wars with Poland-Lithuania saw a fall in productivity which contributed to the Order's fiscal decline. For example, the farmers of Nikolaiken (or Niclasdorf, Pol. Mikołajki Pomorskie), south-east of Stuhm, could not produce enough to pay their tithes regularly, which is perhaps why the Order established a farm there in *c.* 1440 in an attempt to shore up the district's resource base (Semrau 1928, 103–07). Within a number of commanderies in what was left of the Order's territories in Pomesania and Upper Prussia after 1466, more than 50% of estates were still abandoned by the start of the sixteenth century. Mills were left unattended, and flood defences were not maintained as effectively as in the past; in 1505, the waters of the Vistula burst through the embankment at Marienwerder (Długokęcki 2009, 377). This historically documented decline over the course of the fifteenth century does not interrupt the continuation of cultivation in the study areas as evident in the palynological data, which in turn reflects localized variability. Settlements such as Parpahren (Pol. Parpary) continued to function, and with no visible break in land-use the most extensive period of woodland clearance in the vicinity began in the sixteenth century. In the case of Rehden, the drop in grain and animals at the convent after 1410 is more likely to reflect a reduction in territorial control and purchasing power, rather than the collapse of the associated settlement pattern. Further east, the truncation of the Prussian *Ordensland* in 1466 resulted in the focus of promoting colonization shifting to the south-east of the remaining territory. Here, particularly in the procuratorships of Lötzen, Rhein, and Lyck, this corresponded to the expansion of cultivated land, the intensification of grain production, and an increase in agrarian storage at the Order's castles; by the sixteenth century, the procurator's castle at Lötzen was storing grain alongside game and fur. There is also a shift in the regional timber supply to eastern Prussia and Lithuania, underpinning the economy of the final phase of the Order's rule before secularization, alongside fur and amber.

The Post Medieval Period

The long-term impact of the Order's rule can be considered with two case studies where there is sufficient multi-proxy data from the later fifteenth century through to the partitions of the Polish-Lithuanian Commonwealth in 1772–95, which saw the historical territories of the *Ordensland* incorporated into the Kingdom of Prussia. This forms a convenient period of 'social time' which

Map 16.3: Henry Zella's Map of Prussia printed in 1573 (after Jäger 2008).

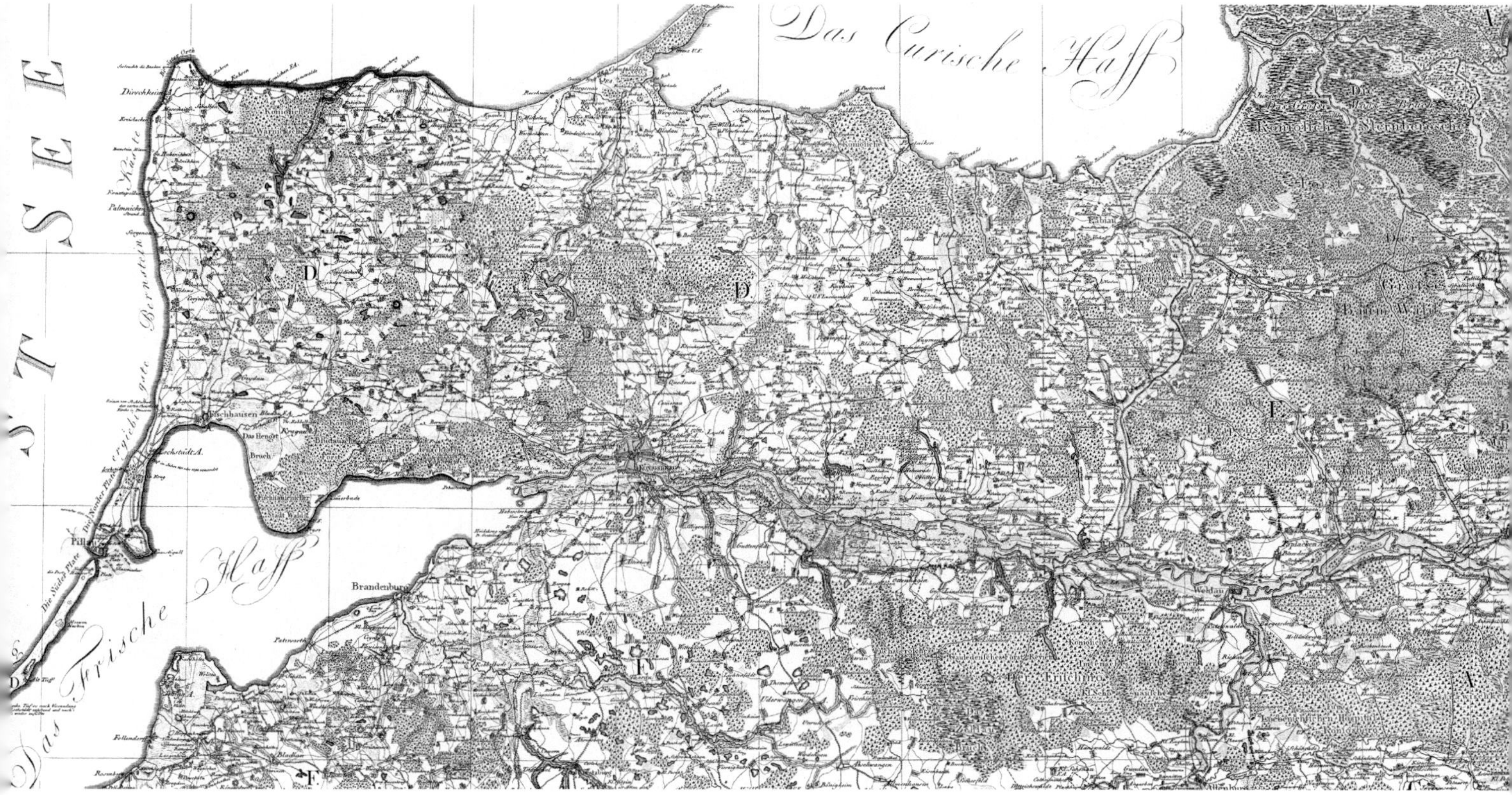

Map 16.4: Schrötter's map (published 1810) of the Sambian Peninsula, showing the extensive belts of woodland around and especially to the east of Königsberg (Rus. Kaliningrad), corresponding to the formerly designated 'Great Wilderness'. Staatsbibliothek zu Berlin – Preußischer Kulturbesitz, Kart. N 1020; image reproduced with permission.

can be compared with the previous chronological divisions. The two treaties of Thorn in 1411 and 1466 saw the Order cede its western Prussian territories to the Kingdom of Poland. The conventual castles obtained by the Polish crown were redefined as the centres of elderships (*starostwa*). The settlements in their associated territories had been variously affected by the ongoing conflicts between the Order and Poland-Lithuania. The castle and town at Rehden survived more or less intact. The castle at Marienburg was also intact and had already been in Polish possession from 1457, but the town had been largely destroyed in the process. The former headquarters of the Order were then used as one of several royal residences in the expanded kingdom. The household of the castle was dramatically reduced, and Polish kings used the residence infrequently. The election of the voivode and prince of Transylvania, Stefan Batory, to the Polish throne by the landed nobility in 1575–1576 provoked a rebellion from Gdańsk which had supported the candidature of Emperor Maximilian II. Batory established a military camp at Marienburg and from there successfully crushed the rebellion by the end of 1577. These were unusual episodes which clearly would have increased the burden on local provisioning networks, and the faunal assemblage corresponding to the early post-medieval period indicates the continued use of the outer bailey for slaughtering and processing livestock, with cattle as the dominant species. In contrast to the medieval period, there is relatively more evidence for hunting: roe deer, red deer, hare, polecat, and capercaillie point to the continued exploitation of the nearby forest, whilst the presence of cod and sturgeon signifies the enduring role of the Vistula. The comparable diversity index for both medieval and post-medieval assemblages reinforces this continuity (Maltby et al. 2009). Subsequent residents of the castle would inherit the problem of a falling water table, already affected by the use of four wells within the complex. With the subsequent drop in the water table by almost two metres in the mid-nineteenth and early-twentieth centuries following the regulation of the River Nogat, the castle's pine foundations began to rapidly weaken, resulting in the subsidence of the western side of the middle castle by the 1980s, which necessitated rapid conservation measures (Mierzwiński 2010).

Six of the modelled pollen sequences from Prussia extend into the post-medieval period. At this time the picture of land-use is variable, in part reflecting the different landscape contexts and histories of the sites concerned. There is general evidence for continuity in agricultural land-use, particularly with the Kulmerland and Marienburg sequences. The landscape within the vicinity of Rehden remains relatively stable in terms of agricultural land-use during the late fifteenth to seventeenth centuries. Around Rywałd, there is evidence for the renewed clearance of broad-leaved woodland in the seventeenth and eighteenth centuries, accompanied by a substantial increase in pine, but with no substantial reduction in woodland coverage. The Parpary sequence, in particular, shows evidence for increasing intensity in arable land-use and a sharp decline in broad-leaved woodland from the sixteenth century, suggesting a significant denudation of the forest during the period of Polish rule. The sequence from Benowo, although located further within the forest, also shows increasing arable landuse, and a change in the structure of the woodland with a greater proportion of coniferous species. Within the Masurian Lakelands there are phases of intensive cultivation within the Zielone Bagno sequence during the late seventeenth to early nineteenth centuries, but the general picture from nearby pollen sequences suggests a high degree of continuity in land-use over the course of the post-medieval period (Wacnik et al. 2012). Henrico Zellio's map of Prussia from 1573 is of little use as a document of Cartesian space (Map 16.3), but it highlights the persistence of the eastern wilderness, with areas of woodland around Lötzen and Lyck, and south-east of Johannisburg (Pol. Pisz). Extensive belts of woodland are documented in this region in the nineteenth and early twentieth centuries, and have persisted into the present day. However, in the eastern part of the Sambia, the cartographic data indicates woodland was extensively cleared between the early nineteenth and late twentieth century (Map 16.4).

Conclusion

The Prussian Crusade both interrupted and accentuated existing trends in environmental exploitation in the borderlands and heartlands of Prussia. The Order's military campaigns against the Prussians lasted half a century, but at the regional level their impact was relatively short-lived. The Prussian population remained partially intact, especially in central and eastern regions, although the archaeological record points to the widespread abandonment of strongholds by or during the thirteenth century. As in the case of Livonia, some Prussian nobles retained their status as the Order's vassals. In contrast to Livonia, however, the Order was able to manage the conquered territories more effectively through its policy of encouraging colonists. Phases of colonization can be linked to trends in environmental exploitation at both local and regional levels, in turn linked to the reconfiguration of heartlands and frontiers. The earliest impact can be associated with Slavic/Polish colonization of the western and southern borderlands and the development of Prussian settlements in the interior. Following the crusades, despite the Order's ideology of renewal and the imposition of a commandery system, the process of environmental exploitation had to be synchronized with a chronologically protracted and regionally uneven phase of colonization. This process was ultimately driven by grand masters, commanders, and lesser officials, and therefore settlement patterns in conquered Prussia can be regarded as directly reflecting the Order's agenda.

This may go some way to explain why in some regions of Prussia indigenous communities were resettled or reorganized. This action provided the Order with an opportunity to impose clear conditions of tenancy, reaffirm their authority and promoted the intensive exploitation of local resources. The relatively late intensification of cultivation and deforestation in eastern Prussia, despite the widespread survival of the indigenous population, can therefore be attributed to the comparatively later mobilization of incoming settlers. Differences in language, religion, and lifestyle — reinforcing the projection of the pagan 'other' in the ideology of the theocracy — would have presented the officials of the Order's eastern Prussian castles with some daunting barriers when exercising lordship or enforcing church attendance. It is almost certainly the reason why native locators were used for reorganizing Prussian villages. This would also explain the time lag in the introduction of a Christian infrastructure, which accompanied the establishment of colonizing settlements, sometimes many years after their foundation. In this region of Prussia, the environment has served as a valuable proxy of human activity, and it can now be consolidated with targeted excavations with the express aim of investigating how discrete clusters of settlements — indigenous and colonizing — and their cultural landscapes were affected by the Order's territorial policies.

Works Cited

Primary Sources

CTP Scholz, K. and Wojtecki, D., trans. 1984. *Chronik des Preussenlandes/Peter von Dusburg*, Darmstadt: Wissenschaftliche Buchgesellschaft

LW Luther, M. 1883–2009. *D. Martin Luthers Werke: kritische Gesamtausgabe*, Weimar: H. Böhlau

Plezia, M., trans., and ed. 1994. *Magistri Vincentii dicti Kadłubek Chronica Polonorum*, Kraków: Secesja

Secondary Works

Alexandrowicz, C. H. 1963. 'Paulus Vladimiri and the Development of the Doctrine of Coexistence of Christian and Non-Christian Countries', *British Yearbook of International Law* 39, 441–48

Arnold, U. (ed.) 1989. *Zur Wirtschaftsentwicklung des Deutschen Ordens im Mittelalter*, Marburg: Elwert

Arszyński, M. 2013. 'Zamki i umocnienia zakonu krzyżackiego i hierarchii kościelnej w Prusach', in *Zakon krzyżacki w Prusach i Inflantach: podziały administracyjne i kościelne w XIII–XVI wieku*, eds R. Czaja and A. Radzimiński, Toruń: Wydawnictwo Naukowe Uniwersytetu Mikołaja Kopernika, 55–80.

Biskup, M. 2009. 'Polityka zewnętrzna zakonu krzyżackiego', in *Państwo zakonu krzyżackiego w Prusach: Władza i społeczeństwo*, ed. M. Biskup et al., Warsaw: Państwowe Wydawnictwo Naukowe, 233–331

Błażejewska, A. 2010. 'Złota brama', k. XIII. in *Fundacje artystyczne na terenie Państwa Krzyżackiego w Prusach*, ed. B. Pospieszna, Pelplin: Wydawnictwo Bernardinum, 19

Brown, A. D., and Pluskowski, A. G. 2011. 'Detecting the Environmental Impact of the Baltic Crusades on a Late-Medieval (13th–15th Century) Frontier Landscape: Palynological Analysis from Malbork Castle and Hinterland, Northern Poland', *Journal of Archaeological Science* 38, 1957–66

Chudziak, W. 2003. *Wczesnośredniowieczna przestrzeń sakralna in Culmine na Pomorzu Nadwiślańskim*, Toruń: Wydawnictwo Naukowe Uniwersytetu Mikołaja Kopernika

Czacharowski, A. 1983. 'Toruń średniowieczny (do roku 1454)', in *Toruń dawny i dzisiejszy: zarys dziejów*, ed. M. Biskup, Warsaw: Państwowe Wydawnictwo Naukowe. 31–131

Czaja, R., and Czacharowski, A. 2003. *Malbork: Atlas historyczny miast polskich*, Toruń: Wydawnictwo Uniwersytetu Mikołaja Kopernika

Dąbrowska, M. 2007. 'Badania archeologiczno-architektoniczne na terenie zamku niskiego w Malborku w latach 1998–2004', in *XV Sesja Pomorzoznawcza*, ed. G. Nawrolska, Elbląg: Muzeum Archeologizno-Historyczne w Elblągu, 303–16

Długokęcka, O., and Długokęcki, W. 2010. 'The Sumptuous Use of Food at Castle Marienburg (Malbork) at the Start of the Fifteenth Century', *Acta Poloniae Historica* 102, 101–26

Długokęcki, W. 1992. *Osadnictwo na Żuławach w XIII i na początku XIV wieku*, Malbork: Muzeum Zamkowe w Malborku

——, 2009. 'Odbudowa osadnictwa, kolonizacja i przemiany wsi w państwie krzyżackim (1411–1525)', in *Państwo zakonu krzyżackiego w Prusach: Władza i społeczeństwo*, ed. M. Biskup et al., Warsaw: Państwowe Wydawnictwo Naukowe, 375–83

Domagała, T. 2004. 'Zagubiony Elbląski kodeks prawa lubeckiego z 1240 roku: Próba identyfikacji i rekonstrukcji tekstu', in *Archaeologia et historia urbana*, eds R. Czaja, G. Nawrolska, M. Rębkowski, and J. Tandecki, Elbląg: Muzeum w Elblągu, 89–93

Duda, Z., and Kryzia, K. 2011. 'Mury obronne miasta Malbork', *Górnictwo i Geoinżynieria*, 35/2, 191–98

Dygo, M. 1989. 'The Political Role of the Cult of the Virgin Mary in Teutonic Prussia in the Fourteenth and Fifteenth Centuries', *Journal of Medieval History* 15, 63–80

——, 2009. 'Ideologia panowania zakonu niemieckiego w Prusach', in *Państwo zakonu krzyżackiego w Prusach: Władza i społeczeństwo*, ed. M. Biskup et al., Warsaw: Państwowe Wydawnictwo Naukowe, 357–69

Ekdahl, S. 1998. 'Horses and Crossbows: Two Important Warfare Advantages of the Teutonic Order in Prussia', in *The Military Orders*, II: *Welfare and Warfare*, ed. H. Nicholson, Farnham: Ashgate, 119–51

Gancewski, J. 2001. *Rola zamków krzyzackich w ziemi chelminskiej od polowy XIV wieku do 1454 roku*, Olsztyn: Ośrodek Badań Naukowych im. Wojciecha Kętrzyńskiego

Haneca, K., Ważny, T., van Acker, J., and Beeckman, H. 2005. 'Provenancing Baltic Timber from Art Historical Object: Success and Limitations', *Journal of Archaeological Science* 32, 261–71

Hayward, K. M. J. 2019. 'Resources for Castle Building in Prussia and Livonia', in *Ecologies of Crusading, Colonization, and Religious Conversion in the Medieval Baltic: Terra Sacra II*, ed. by A. Pluskowski, Environmental History of the North Atlantic World, 3, Turnhout: Brepols (forthcoming)

Jäger, E. 2008. *Entlang der Weichsel und der Memel: historische Landkarten und Stadtansichten von Ost- und Westpreuben, Polen und dem Baltikum. Eine Ausstellung des Kulturzentrums Ostpreuben in Ellingen/Bay = Z biegiem Wisły i Niemna: historyczne mapy iwidoki miast Prus Wschodnich i Zachodnich, Polski oraz krajów bałtyckich: wystawa Centrum Kultury Prus Wschodnich w Ellingen/Bawaria*, Ellingen/Bay: Kulturzentrum Ostpreuben

Jaremek, A. 2003. 'Badania w osadzie przygrodowej w Jeziorku stan. I.', *Swiatowit* 5/46 (B), 251–54

Jóźwiak, S., and Trupinda, J. 2007. *Organizacja życia na zamku krzyżackim w Malborku w czasach wielkich mistrzów (1309–1457)*, Malbork: Muzeum Zamkowe w Malborku

Kasprzycka, M. 1999. 'Tło paleogeograficzne osadnictwa Żuław Elbląskich w pierwszym tysiącleciu naszej ery', in *Adalbertus* (vol. v), ed. P. Urbańczyk, Warsaw: Instytut Archeologii i Etnologii Polskiej Akademii Nauk, 84–93

Kawiński, P., and Szczepański, S. 2016. *Szkice o religii Prusów*, Olsztyn: Towarzystwo Naukowe Pruthenia

Kupryjanowicz, M. 2004. 'The Vegetation Changes Recorded in Sediments of Kładkowe Bagno Bog in Puszcza Knyszyńska Forest, North-Eastern Poland', *Acta Palaeobotanica* 44, 175–93

Kutzner, M. 2010. 'Propaganda władzy Zakonu Niemieckiego w sztuce na terenie Państwa Zakonnego w Prusach', in *Fundacje artystyczne na terenie Państwa Krzyżackiego w Prusach*, ed. B. Pospieszna, Pelplin: Wydawnictwo Bernardinum, 238–59

Makowiecki, D. 2012. 'Badania archeoichtiologiczne szczątków ze stanowiska Janów Pomorski 1', in *Janów Pomorski, stan. 1. Wyniki ratowniczych badań archeologicznych w latach 2007–2008, Elbląg, I/3*, eds M. Bogucki and B. Jurkiewicz, Elbląg: Muzeum Archeologiczno-Historyczne w Elblągu, 302–23

——, 2015. 'Fauna wczesnośredniowieczna i jej znaczenie u ludności zespołu osadniczego w Węgrach', in *Węgry: Zespół osadniczy na pograniczu Pomorsko-Pruskim w xi–xii w.*, eds M. Haftka and S. Wadyl, Malbork: Muzeum Zamkowe w Malborku, 236–55

Makowiecki, D., and Makowiecka, M. 2012. 'Wyniki badań archeozoologicznych ze stanowiska Janów Pomorski 1', in *Janów Pomorski, Stan 1. Wyniki ratowniczych badan archeologicznych w latach 2007–2008* (vol. i), eds M. Bogucki and B. Jurkiewicz, Elbląg: Muzeum Archeologiczno-Historyczne w Elblągu, 275–312

Maltby, M., Pluskowski, A. G., and Seetah, K. 2009. 'Animal Bones from an Industrial Quarter at Malbork, Poland: Towards an Ecology of a Castle Built in Prussia by the Teutonic Order', *Crusades*, 8, 191–212

Marcinkowski, M. 2003. 'Ostrogi ze Starego Miasta w Elblągu', *Pomorania Antiqua* 19, 177–91

Mierzwiński, M. 2010. 'Prace ratownicze i konserwatorskie w zachodnim skrzydle zamku średniego w Malborku w latach 1990–2010', in *Wielki refektarz na zamku średnim w Malborku: dzieje-wstrój-konserwacja*, ed. J. Trupinda, Malbork: Muzeum Zamkowe w Malborku, 169–214

Mroczko, T. 1974. 'Ruch budowlany na ziemi chełmińskiej w xiii wieku', in *Sztuka i ideologia xiii wieku*, ed. P. Skubiszewski, Wrocław: Zakład Narodowy im. Ossolińskich, 281–332

Nawrolska, G. 2003. 'Dominikanie w średniowiecznym Elblągu', in *Dominikanie: Gdańsk–Polska–Europa*, eds D. A. Dekański, A. Gołembnik, and M. Grubki, Gdańsk: Bernardinum, 321–33

——, 2004. 'Remarks on Infrastructure of the Old Town of Elbląg', in *Lübecker Kolloquium zur Stadtarchä ologie im Hanseraum*, iv: *Die Infrastruktur*, eds R. Dunckel, M. Gläser, U. Oltmanns and J. Scheschkewitz, Lübeck: Schmidt-Römhild, 303–22

——, 2008. 'A Way of Life. Luxury in a Medieval Town', in *Lübecker Kolloquium zur Stadtarchäologie im Hanseraum*, vi: *Luxus und Lifestyle*, eds C. Kimminus-Schneider, D. Mührenberg, M. Schneider, and C. Sheehan, Lübeck: Schmidt-Römhild, 509–27

Nowakiewcz, T., and Wróblewski, W. 2003. '"Legenda galindzka" Piotra z Dusburga w świetle badań archeologicznych', in *Antyk i barbarzyncy*, eds A. Bursche and R. Ciołek, Warsaw: IA UW, 313–22

Nowakowski, A. 1994. *Arms and Armour in the Medieval Teutonic Order's State in Prussia*, Łódź: Oficyna Naukowa MS

Odoj, R. 1958. 'Sprawozdanie z prac wykopaliskowych przeprowadzonych w Równinie Dolnej, pow. Kętrzyn, w 1956 i 1957 roku', *Rocznik Olsztyński* 1/1, 117–56

Okulicz-Kozaryn, Ł. 1983. *Życie codzienne Prusów i Jaćwięgów w wiekach średnich (ix–xiii w.)*, Warsaw: Państwowy Instytut Wydawniczy

Paravicini, W. 1995. *Die Preussenreisen des europäischen Adels* (vol. ii), Sigmaringen: Thorbecke

Pluskowski, A. G. 2013. *The Archaeology of the Prussian Crusade: Holy War and Colonisation*, London: Routledge

Pluskowski, A. G., Brown, A. D., Banerjea, R., Makowiecki, D., Seetah, K., Rannamäe, E., Badura, M., Jarzebowski, M., Kreem, J., and Kļaviņš, K., 2019. 'From the Convent to the Commandery: The Pivotal Role of the Environment in Defining the medieval Baltic *Ordensland*', *Quellen und Studien zur Geschichte des Deutschen Ordens* (in press)

Pluskowski, A. G., Sawicki, Z., Shillito, L–M., Badura, M., Makowiecki, D., Zabilska-Kunek, M., Seetah, K., and Brown, A. 2014. 'Biała Góra: The Forgotten Colony in the Medieval Pomeranian-Prussian Borderlands', *Antiquity* 88/341, 863–82

Poliński, D. 2007. 'Wczesne warownie krzyżackie w Prusach w kontekście miejscowych obiektów obronnych w umocnieniach drewniano-ziemnych', *Archaeologia Historica Polona* 16, 41–61

Pospieszna, B. ed. 2010. *Fundacje artystyczne na terenie Państwa Krzyżackiego w Prusach*, 2 vols, Malbork: Muzeum Zamkowe w Malborku

Sarnowsky, J. 1993. *Die Wirtschaftsführung des Deutschen Ordens in Preußen (1382–1454)*, Cologne: Böhlau

Sawicki, Z., Pluskowski, A. G., Brown, A., Badura, M., Makowiecki, D., Shillito, L.-M., Zabilska-Kunek, M., and Seetah, K. 2015. 'Survival at the Frontier of Holy War: Political Expansion, Crusading, Environmental Exploitation and the Medieval Colonising Settlement at Biała Góra, North Poland', *European Journal of Archaeology*, 18/2, 282–311

Scull, C., Müldner, G., and Makowiecki, D. 2019. 'Carbon and Nitrogen Stable Isotope Evidence for Animal Husbandry and Environmental Change in the Medieval Kulmerland', in *Ecologies of Crusading, Colonization, and Religious Conversion in the Medieval Baltic: Terra Sacra ii*, ed. by A. Pluskowski, Environmental History of the North Atlantic World, 3, Turnhout: Brepols (forthcoming)

Semrau, A. 1928. 'Die Orte und Fluren im ehemaligen Gebiet Stuhm und Waldamt Böhnhof (Komturei Marienburg)', *Mitteilungen des Coppernicus-Vereins für Wissenschaft und Kunst zu Thorn* 36, 5–7

Shiroukhov, R. 2012. 'Prussian Graves in the Sambian Peninsula with Imports, Weapons and Horse Harnesses, from the Tenth to the Thirteenth Century: The Question of the Warrior Elite', *Archaeologia Baltica* 18, 224–55

Szczepański, S. 2011. 'Sakralizacja obszaru pogranicza na przykładzie Pomezanii pruskiej', *Pruthenia* 6, 129–56

——, 2013. 'Chomor Sancti Adalberti (1249): a możliwości lokalizacji terenowej wybranych kościołów pomezanii', *Komunikaty Mazursko-Warmińskie* 1/279, 19–45

——, 2014. *Pomezania pruska: Dzieje osadnictwa w XIII–XV wieku*, Olsztyn: Rozprawy i Materiały Ośrodka Badań Naukowych im. Wojciecha Kętrzyńskiego w Olsztynie

——, 2019. 'Surkapurn – Kreken: The Phenomenon of a Name – The Phenomenon of a Place. The Organisation of Sacred Space in Prussia Using Archaeology and Other Sources', in *Ecologies of Crusading, Colonization, and Religious Conversion in the Medieval Baltic: Terra Sacra II*, ed. by A. Pluskowski, Environmental History of the North Atlantic World, 3, Turnhout: Brepols (forthcoming)

Tandecki, J. 2009. 'Polityka zewnętrzna', in *Państwo zakonu krzyżackiego w Prusach: Władza i społeczeństwo*, ed. M. Biskup et al., Warsaw: Państwowe Wydawnictwo Naukowe, 110–23

Wacnik, A., Goslar, T., and Czernik, J. 2012. 'Vegetation Changes Caused by Agricultural Societies in the Great Masurian Lake District', *Acta Palaeobotanica* 52, 59–104

Ważny, T. 2004. 'Zastosowanie dendrochronologii w badaniach starego miasta w Elblągu', in *Archaeologia et Historia Urbana*, eds R. Czaja, G. Nawrolska, M. Rębkowski, and J. Tandecki, Elbląg: Muzeum w Elblągu, 377–82

——, 2005. 'The Origin, Assortments and Transport of Baltic Timber', in *Constructing Wooden Images*, eds C. Van de Velde, H. Beeckman, J. Van Acker, and F. Verhaeghe, Brussels: Brussels University Press, 115–26

LIVONIA and PRUSSIA

A Comparison

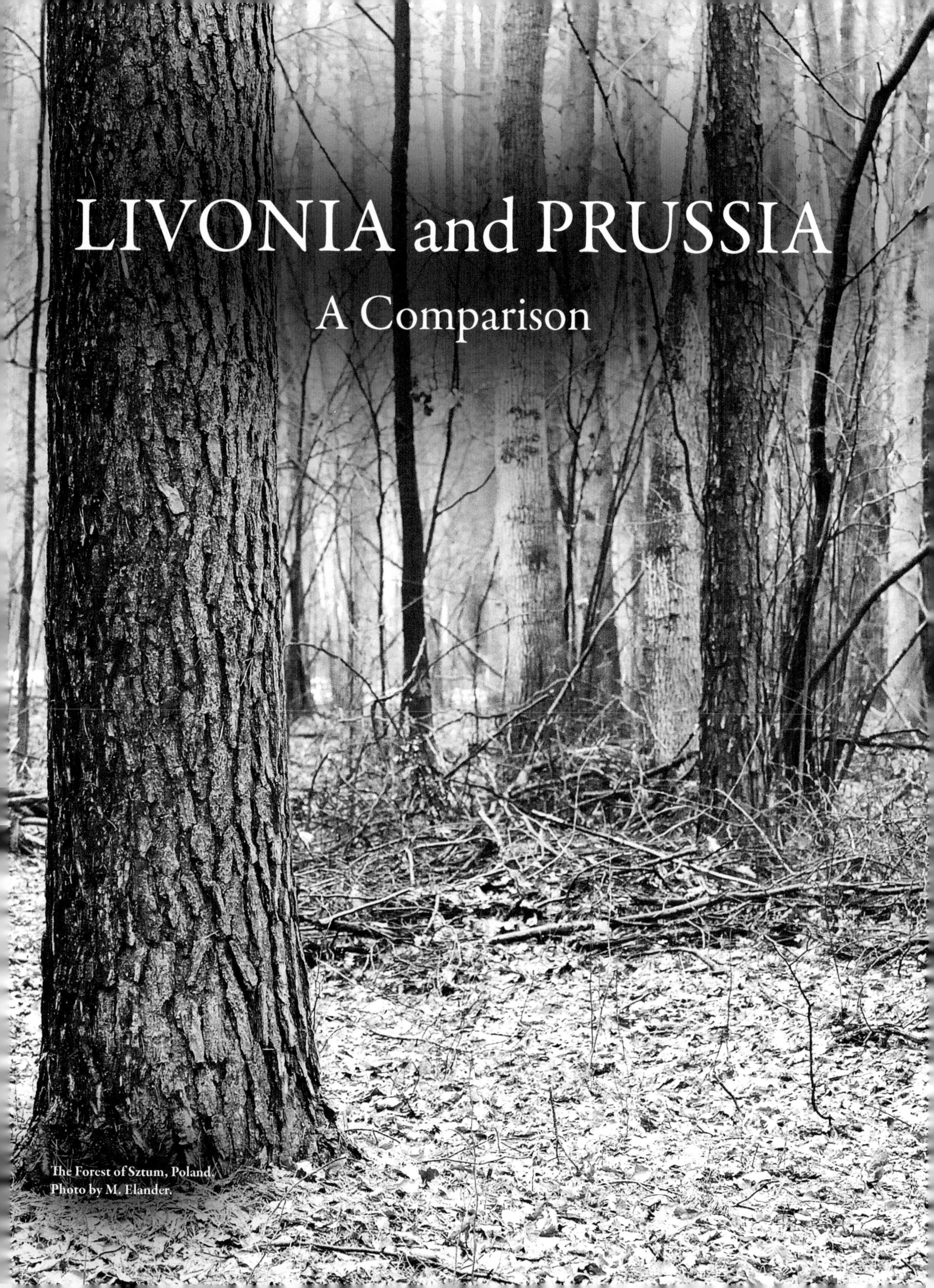

The Forest of Sztum, Poland.
Photo by M. Elander.

17. The Environmental Impact of the Crusades in the Eastern Baltic

Aleksander Pluskowski

Introduction: 'Clash of Cultures'

The immediate and long-term impact of the crusades on the indigenous cultures of the eastern Baltic has been mapped by over a century of scholarship. Recent historiography has drawn attention to the disparity between how the crusades — and the conquered territories — were envisaged compared with their historical realities, the difference between 'the doing of crusading and the writing of crusades' (Bull 2003, 15; Tamm 2009, 35; 2011; 2013). Medieval Christians viewed the Baltic Crusades as a polarized struggle with the pagan 'other', a concept promoted by the Teutonic Order which redefined itself as the protector of *eccelsia* in the north-eastern frontier of Latin Christendom (Kutzner 1995; Pospieszny 2010; Kwiatkowski 2013). This laid the foundations for the anachronisms of ethnic conflict promoted in nationalist historiography that have long been abandoned by historians (Eihmane 2009, 45), although which continued to dominate the cultural-historical archaeological paradigm of the eastern Baltic (e.g., Mugurēvičs 1990). The essentialist image of the pagan frontier advocated by the conquering theocracy can be contrasted with the realities of diverse encounters between the colonizing and indigenous populations.

The conquest of Livonia was facilitated by inter-tribal feuding, and a complex regional struggle for influence between Rus', Lithuanian, and incoming European powers (Kala 2001; Tamm 2009; Selart 2015). The conquest of Prussia was similarly facilitated by fragmented tribal polities and complicated by the expansionist agendas of Polish, Pomerelian, Danish, and Lithuanian magnates. However, the Teutonic Order's involvement in both regions changed the geopolitical dynamic and resulted in the creation of virtually unique theocratic dominions. Only the Duchy of Estonia did not follow this trajectory until its acquisition by the Order in 1346 (Kala 2007). Given the consistent use of crusading to define the Order's role in the Baltic, and the mobilization of its subjects and resources towards sustaining its houses and lands (Kļaviņš 2009), these territories, otherwise referred to in the scholarly literature as the *Ordensland*, can usefully be called 'crusader states', a term that has been more commonly used in relation to the Latin East (e.g., Barber 2012). Both colonists and indigenous communities contributed to the military and financial resources that enabled the Order to maintain its ongoing war with Lithuania throughout the fourteenth century and the various conflicts with neighbouring polities into the sixteenth century. The Order's crusading role was diluted by its entanglement in foreign politics, even during the thirteenth century, but nonetheless this raison d'être was publicly promoted into the early sixteenth century.

Previous scholarship has drawn a distinction between Prussia, where a larger proportion of the indigenous population was annihilated during the crusades, and Livonia, where casualties were comparatively smaller. Where local power structures survived the onslaught of the crusades they appear to have been maintained, clearly as a means of facilitating the Order's rule. This was also linked to divergent patterns of colonization in both regions, which accompanied and developed in the wake of the crusades. Where the local population survived intact, there was limited colonization by incoming migrants or very protracted migration. In Livonia, this was restricted to towns and castles until German vassals began to establish themselves in rural manors, leading historians and archaeologists to characterise medieval society as culturally polarized between these Christian enclaves (which included indigenous 'reagents') and the 'pagan' countryside. The conceptualization of rurality as 'other' by urban populations (Cloke and Little 1997), would have been accentuated by differences in language, religion, technology and customs. Indeed, in both Livonia and Prussia, the clearest material examples of enduring indigenous identities, connected in part to native expressions of spirituality, can be found in rural settlements and cemeteries. This ensured that Livonia remained 'a profoundly dualist society' (Šnē 2008, 99), as did parts of central and eastern Prussia (Kawiński and Szczepański 2016). But dualism was diluted with forms of hybridity that remain to be fully understood, and the incoming culture was not absent from the countryside. It was represented by the detailed regulation of land; by the presence of loyal vassals and village leaders who had converted to Christianity; by the enactment of obligations to provide taxes, tithes, and military services; by the construction of chapels; and by provisioning networks extending beyond urban patrimonies. Investment in rural parochial infrastructure, as well as tax collection, provisioning, feudal obligations, and judicial functions, depended on a level of coexistence. This suggests that while the urban-castle/rural divide is a legitimate reflection of the type of legal apartheid visible from the mid-fourteenth century and linked to the development of local German elite identity (Kala 2006; 2009), the relationship between castles, towns, and rural settlements was more of a variable continuum. The Baltic *Ordensland* was a compromise between crusading ideology and pragmatic *Realpolitik* (exemplified by trade; Mažeika 1994), the variability of which can only be understood through a comparative approach of its individual territorial units.

The focus on the environment as a lens for scrutinizing the cultural changes associated with the Baltic Crusades has provided a new contribution to our understanding of this variability. It has demonstrated the value of highly localized and multi-scalar studies for contextualizing the varied impact of a supranational political and military force. The detailed management of the landscape became *the* principal means of sustaining the conquering regime, just as it had underpinned pre-crusade polities. There are some hints that colonization was tied to an ideology of legitimized exploitation, and whilst the motivations of individual colonists were not recorded, it is evident that migrants in the thirteenth century understood the dangers (Piskorski 2004, 337). But this was not about the occupation of empty space as imagined by the early German historians of *Ostsiedlung*: rather, it concerned the justification of Christian rule framed in the essentialist rhetoric of the struggle with paganism and the defence of Christendom. This had already been implied by the Archbishop of Magdeburg's call for holy war against the Polebian Slavs in 1108, which emphasized the value of natural resources, linking crusading with the seizure of territory (Trupinda 1999, 127; Constable 2008, 197–214). In the early thirteenth century, this connection was applied to the lands of excommunicated Occitan magnates. In Iberia, calls for crusades were inextricably linked to the occupation of territory. From a contemporary, external Christian perspective, the peoples of the eastern Baltic lived in a land full of abundant resources, but their erroneous worship of animals, trees, lakes, rivers, plants, as well as their pagan ferocity, had hindered the exploitation of this 'paradise' (Tamm 2009). In Prussia, this became more explicitly tied to the functionality of the Teutonic Order. Here, in the early decades of the fourteenth century, as narratives justifying and celebrating the Order's actions were being compiled, and as colonization was well underway, the Order's defence of Latin Christianity in the north-eastern Europe became twinned with its role as a restorer of the war-ravaged land.

This chapter summarizes the data and conclusions presented for Livonia and Prussia in the rest of this volume, returning to the key questions set out in the first chapter. As with any ambitious research programme, these have raised further questions: but having presented the full suite of data in the previous sixteen chapters, it is possible to characterize the ecological impact of the crusades on the societies of the eastern Baltic, to consider whether this impact can be understood as the result of the implementation of a 'European' system, separating it from pre-crusade indigenous systems (see Chapter 2), and whether it constitutes a defining moment in the ecological as well as social history of north-eastern Europe. These questions are addressed in turn, incorporating the suite of data from the Livonian and Prussian macroregions considered within this study.

1. How Were Local Environments in the Eastern Baltic Region Transformed from the Thirteenth to Fifteenth Centuries with the Establishment of Castles and their Territories?

New Territorial Paradigms

The transformation of local environments in the eastern Baltic was directly linked to the successful imposition of new paradigms of territorial organization. Although the largest military orders were supranational institutions, the management of their estates in crusading frontiers and European heartlands was not standardized. The Teutonic Order's development was ultimately connected with Imperial patronage and Frederick II's involvement in the Latin East. However, the Order's experiences in the Transylvanian Burzenland (1211–25), Prussia (from 1230), and Livonia (from 1237), as well as obtaining estates within Imperial territories and beyond, provided unique logistical challenges within a few decades of militarization. The Order became highly adaptive in all the frontier regions it was actively involved in and adopted local administrative systems to govern its properties. In the Holy Land, there was a tendency to adopt existing institutions for tax collection, using intermediaries between lords and vassals, and the Order employed scribes here (Morton 2009, 175).

In the eastern Baltic, as with all 'first crusades', the Order's transformation was arguably more profound. Here, the organization of conquered territory which had initially begun with the formation of Livonia saw the adoption of the fief as a basic unit linked to vassalage. Indeed, the Sword Brothers were themselves effectively episcopal vassals, a status inherited by the Livonian branch of the Order in 1237 and a subsequent cause of political tension. The crusading period was characterized by the parcelling out of land to both indigenous and migrant vassals; however, fiefs granted to indigenous vassals were often existing parcels of land and so perpetuated pre-crusade landscape divisions linked to ownership by specific lineages. The theocracy also took over many existing centres of power, all of which were physically remodelled. What changed was how ownership was defined — the introduction of written documents as forms of territorial validation came to replace oral confirmations — along with the reconfiguration of regional power, a network which became focused on Riga, and a reflexive response to governing conquered territories where local political structures had been replaced by the Order's administration.

The best examples of this, considered within this study, were the commanderies of Wenden (Latv. Cēsis) and Fellin (Est. Viljandi), where the native aristocracies were either subsumed as allies or destroyed. This necessitated the creation of new power centres in areas that were previously peripheral, as with the construction of the castles of Arrasch (Latv. Āraiši) and Karkus (Est. Karksi). Both were strategically located within the communication network that linked the Order's territories. Both governed districts where the indigenous population increased between the thirteenth and mid-sixteenth centuries, and where colonists remained confined to the Order's convents and their associated towns. As a result of the absence of local power structures, there were no indigenous vassals in these territories in comparison to the Lower Daugava or parts of Curonia, where control of the landscape was delegated to successive generations of loyal indigenous families. In the final phase of the Order's rule in Livonia, the new emphasis on Wenden and the establishment of the Master's district reflected the increasing political autonomy of the Livonian Master, as well as continuing political tensions in Riga. It also coincided with a shift in the ownership of land from indigenous to German vassals, although the Order maintained the same level of control over its territories until secularization.

In Prussia, the Order's early involvement and direct leadership in the crusading enterprise saw the establishment of a different territorial blueprint. Following the acquisition of the Kulmerland and territories along the Lower Vistula, the Polish castellany system was replaced with convents governing commanderies, again located on or close to existing centres of power, as at Rehden (Pol. Radzyń Chełmiński) and Graudenz (Pol. Grudziądz). Further north, the stronghold at Węgry was not reoccupied, and attention focused on the Pomerelian outpost of Zantyr, Bishop Christian's base, plausibly linked to the site at Biała Góra. From the very onset, these convents had to be sustained. The population of the borderlands appears to have been heavily depleted by the early thirteenth century, and so colonists were required to support the Order's garrisons. At the same time, the collaborative foundations of urban settlements reflected a host of geopolitical and commercial agendas, whose legacy was the development of urban patriciates in major towns, as in Thorn (Pol. Toruń) and Elbing (Pol. Elbląg) (Jensen 2001). Additional security was provided by indigenous and migrant vassals in exchange for territory. This form of organization was far from novel by the 1230s, as already evident in Livonia. The brethren of the Order had experience of colonists in the Levant, in Transylvania, as well as familiarity with the well-established culture of colonization in the eastern provinces of the Holy Roman Empire. The latter provided the model for the Kulm Law which enabled settlement and resource exploitation — the two were inextricably linked — to be rapidly organized.

As the crusades progressed through Prussia, some tribal centres were occupied and remodelled, as in the case of Balga, but many were destroyed or abandoned and not reoccupied. Elbing was constructed on a geographical and cultural frontier. By that time, the regional centre of power had shifted from former Truso to the forested uplands, as well as to the territory to the south that would be reorganised under the convent of Christburg (Pol. Dzierzgoń). Marienburg's (Pol. Malbork) location in the 1280s was on the frontier with Pomerelia, one that would change dramatically in 1309 and result in a shift in central authority from Elbing. Both convents had been constructed in a frontier landscape which would become the heartland of the Order's Prussian territories. This reconfiguration was accompanied by a sustained wave of colonization from the 1280s through to the mid-fourteenth century. In eastern Galindia, indigenous power centres were not reoccupied: instead, when the Order came to establish its administrative centre at Lötzen (Pol. Giżycko), it was linked to the strategic defence of passes in the Masurian Lake District. This reflected the management of a frontier that was precarious not just in terms of military threats from Lithuania but also in the organization of the conquered population. These factors arguably prevented the effective management of the landscape by the theocracy until the increase in the number of colonists within the procuratorship in the fifteenth century, whose communities were easier to incentivise and regulate. In some cases, as at Staswinnen (Pol. Staświny), the gradual introduction of 'Masovian' ceramics at this time suggests a substantial level of interaction between the indigenous population and incoming colonists.

In Prussia, the Order could not translate its experience of building stone castles, which it had used in Transylvania much to the alarm of the Hungarian nobility (Laszlovszky and Soós 2001), due to the lack of suitable building material. Instead, timber and earth *castra* or strongholds, used by neighbouring Slavs, Prussians, and Germans alike, were eventually augmented by brick and smaller quantities of field stones; as an exception, some castles such as Bischöflich Papau (Pol. Papowo Biskupie) were built largely from field stones. The logistical requirements of construction were therefore vastly different from the Holy Land. When the Order acquired the Sword Brothers' properties in Livonia in 1237, a comparable system of land organization was already in place, stone was in plentiful supply, and masons were available in the fortified colonies. Brick also became popular here, albeit on a smaller scale than in Prussia. The absence of the Kulm Law underlines the relatively autonomous nature of the Order's Livonian branch at this time, within a supra-regional corporate structure that was distinct from the Templars and Hospitallers. The Order's territories in Prussia and Livonia were effectively 'quasi-independent dominions', which were also recipients of supplies from the West. The Order's power was theoretically centred on the Levant, but in fact shared with the provincial masters of Prussia, Livonia, and Germany, who acted with a great degree of autonomy (Morton 2009, 166–67). This was reflected in a variable policy of support for the Holy Land, and sustained investment in the Baltic. It was also reflected in the more autonomous behaviour of Livonian officials in the fourteenth century, compared to the centralised hierarchy in Prussia dominated by the Grand Master, a tendency which facilitated the privatization of some convents in the final decades of the Order's rule in Livonia. But even in Prussia where grand masters played an important role in directing the process of colonization, the agency of commanders (and convent chapters) is visible in how the landscapes under their direct control were managed.

The reorganization of territory in the eastern Baltic was directly connected with the introduction of literacy and record-keeping. This was revolutionary, although it remained confined to a small section of the population. It brought with it new ways of thinking, a new precision concerning territorial organization and the ownership of resources, and it required new resources: parchment, paper, and wax tablets. The use of writing in urban administration increased in the second half of the thirteenth century but proliferated during the second half of the fourteenth century when paper increasingly superseded parchment, coinciding with the gradual replacement of Latin by Middle Low German as the main language of recording keeping in the eastern Baltic by the end of the fourteenth century, itself linked to the growth of local German identity. Alongside the increasing complexity of urban bureaucracy, this can be situated within the context of the Hanse as a Baltic phenomenon (if not a cohesive confederation), which saw a dramatic increase in the organization and volume of long-distance trade (Salminen 2004, 146–47; Jahnke 2015).

Intensifying Cultivation

A recurring theme in medieval environmental historiography, most recently highlighted by Hoffmann (2014, 140) in relation to the Order's promotion of intensified agriculture in Prussia, has been the concept of farming as part of the ideological package of Christian civilization, replacing the (perceived) disorder of the pre-Christian world. On the one hand, this was partly how the Order's role in the conquered territories was

Figure 17.1: The Teutonic Order's canal at Jurkowice. Photo by M. Elander.

validated: on the other, it is clear that existing systems of cultivation, woodland management (particularly the use of 'bee-trees') and animal husbandry were adopted by the conquerors. Rotational farming was already present in the Polish-Prussian borderlands and north-west Estonia before the crusades, and in Livonia and parts of eastern Prussia existing agricultural practices took several centuries to change.

The main areas of grain production in the fourteenth century were in the two regions which had functioned as pre-crusade borderlands and became heartlands of the Prussian *Ordensland*; the Kulmerland and the Lower Vistula. These regions had already seen intensive cultivation before the crusades, linked to the expansion of the Polish state. Following the crusades, intensive waves of colonization from the last decades of the thirteenth century through to the mid-fourteenth century saw the introduction of new ploughs, corresponding measurements of farmland, and diversification in the number of cultivated crop species. But this was focused in very specific areas of commanderies, and grain production had to meet local needs before it could be exported. In the case of Rehden, for example, the refounding of the town included an obligation to supply a variety of grain to the convent. The landscape within the vicinity of the convent was dominated by pasture, and the other principal grain-producing settlements were most probably located in the northern part of the territory. The convent became an important storehouse for grain, which would have been shipped down the Vistula via Graudenz through to Danzig. Merchants may have picked up further stocks of grain en route from adjacent districts, such as those that had been prepared in the waterfront warehouse at Dirschau before it was destroyed by fire. The Vistula route, also used for transporting other bulk commodities such as timber, extended to the heart of the Polish realm in Kraków, with the main interfaces between the *Ordensland* and Polish territories managed by the frontier convents of Nessau (Pol. Mała Nieszawka) and Thorn. In Marienburg, the northern edges of the great corridor of woodland — the Forest of Stuhm (Pol. Sztum) — were slowly cleared by colonies such as Parpharen (Pol. Parpary), as they had been in the pre-crusade period by Pomerelian colonists organized from the stronghold of Węgry. The colonization of the floodplains of the Nogat and Vistula accentuated the earlier settlement pattern and was consolidated by substantial embankments and hydraulic modifications. Here in the absence of suitable environmental data, grain production is represented in the landscape by the widespread construction of mills. In eastern Prussia, despite the survival of indigenous communities, intensification is only evident following the influx of colonists and the establishment of new, regulated settlements.

Grain production required granaries and mills, with specialists to manage them. Processing is evident on pre-crusade sites in all the case study regions, represented by quernstones and storage pits, but from the crusading period there is evidence for substantial installations and for the refinement of cereal processing. Indeed, mills formed an integral component of every major castle, and whilst archaeological traces are limited to traces of processed grain, this is clearly stated in the written sources. In 1406, when the commander of Ragnit (Rus. Neman) informed the Grand Master of his prospection of sites for a new stronghold on the Samogitian frontier, he suggested connecting two rivers to ensure there was enough water to power a mill, essential for the needs of the garrison (Arszyński 2013, 563–66). The Order also oversaw the construction of substantial urban mills, such as in Riga and Danzig (Pol. Gdańsk), as well as canals functioning as mill runs within landscapes with limited waterways, most extensively at Marienburg (Fig. 17.1; see Chapter 13 and Map 13.1). In both Livonia and Prussia, the ownership of mills was carefully regulated. Investment in this infrastructure maps onto the regional chronology of cultivation. In eastern Prussia, in the procuratorship of Lötzen, intensified cultivation is only evident from the fifteenth century, corresponding to sustained colonization and the proliferation of mills and grain storage. In this respect, there is a disparity between the management of colonizing settlements and indigenous communities. The latter, although conquered and taxed, were more difficult to govern and perhaps even resisted top-down attempts at maximizing their productivity. In Riga, the virtual absence of cleaver technology in the Liv quarter suggests that if technological transference within multicultural towns was obstructed by economic and/or cultural differences, then the dissemination of other cultural elements to rural communities would have been even more challenging. This may also have ensured the persistence of indigenous rituals linked to harvesting, embedded in a non-Christian mythological framework (Pompeo 2000).

Patterns of Deforestation

Peter von Dusburg's narrative of the Prussian Crusade is punctuated with descriptions of regions laid to waste and reduced to 'deserts'. This is how the Kulmerland following Prussian raids in the early thirteenth century is described, and later the border between the Kulmerland and Pomesania. Subsequent episodes of regional violence, such as the Order's annexation of Pomerelia in 1308–09, as well as short-lived outbreaks of plague from 1373 to 1410 also took their toll (Jäger 1981, 230). In the fifteenth century, significant expanses of ploughed fields within the advocateship of Karkus lay fallow, perhaps reflecting a localized population collapse. Woodland could regrow quite quickly during periods of depopulation; even as late as 1582, the colonization of formerly cultivated fields by oak, beech, birch, and spruce was noted in eastern Prussia (Mager 1960). In Livonia, the impact of the war with Russia (1558–83) devastated the population; the description of Wenden's estates in the late sixteenth century is a catalogue of abandoned farmsteads and fields. Palynological evidence for the regrowth of woodland associated with historically documented depopulation is uneven, and may reflect a perceived rather than an actual collapse of the settlement pattern. This is the case for those parts of the 'Great Wilderness' of the east Prussian frontier which have been studied, although it remains to be seen whether territories in Sudovia and perhaps even the Elbąg uplands — the frontier between the territory of the Order and the episcopate of Ermland — were partially reforested following the destruction of indigenous power centres in the region. The earliest detailed maps of these regions suggest that extensive woodland may have survived, if it had not regrown. The persistence of cultivation and the lack of regrowth of woodland between the twelfth and fifteenth centuries within the central parts of the procuratorship of Lötzen can be linked to enduring indigenous communities, before the onset of intensive colonization from the mid-fifteenth century. However, the impact of the Livonian War is evident in the landscapes around Arrasch, Wenden, and Fellin.

Where the Order promoted cultivation, this was often twinned with deforestation, with woodland deliberately cleared for arable. This resulted in a corresponding drop in biodiversity, although in a number of regions this decline had already begun before the crusades. This relationship in shifting landuse can be seen in all the commanderies surveyed in this study. But the rate of deforestation was highly variable, and there is extensive evidence for the persistence of large corridors of woodland, with some located in recognizably marginal areas with poor agricultural potential. The first major phases of deforestation had already taken place before the crusades in the landscapes around Rehden, Marienburg, Lötzen, Riga, and Arrasch. Within the territories of Fellin, Wenden, and Karkus the crusading authorities managed a mosaic landscape with substantial tracts of woodland which have occupied the same marginal locations into the present day. Timber was recognized as a valuable commodity by the theocracy, perhaps more so than preserving discrete woodlands as habitats for bees producing invaluable wax, since individual trees or lines of trees were all that was required to sustain hives, and there is evidence for management in all the commanderies. In the same respect, woodlands were not preserved as habitats for game or for fur-bearing mammals, but those districts with a high degree of biodiversity were exploited accordingly. As a result, between 1466 and 1525 the economy of the shrunken Prussian *Ordensland* was dominated by woodland products and luxuries such as amber, reflecting the loss of the western grain-producing territories and a reliance on landscapes dominated by wetlands, coastlines, and woods.

The exploitation of timber in both regions was connected with rivers, which enabled large quantities of lumber to be shipped to down the Daugava to Riga and the Vistula to Danzig. This affected the regionally uneven impact on woodlands within both Livonia and Prussia, which also reflected the emphasis on exploiting commercially valuable hardwoods, rather than conifers. This explains why substantial corridors of woodland survived in landscapes that were intensively settled, such as in the Lower Vistula (e.g., in the southern part of Marienburg's commandery) and Daugava (e.g., in the patchwork of territories on both sides of the river around Riga). Woodland in eastern Prussia and Livonia was more difficult to access and would only be exploited in the post-medieval period; in the case of the former region this was only possible after the frontier with Lithuania had stabilized. But whilst the hardwood component of many of the case study regions was depleted, the continuation of a long-term pattern of exploitation, it was not completed extirpated, and oaks and hornbeams continued to grow, in some cases deliberately preserved as isolated boundary markers, in others as elements of managed woodlands. There is evidence for the deliberate preservation of discrete, sacred woodlands in Curonia, and the tolerance of sacred groves and trees in other parts of the eastern Baltic represents an unusual and ad hoc policy to territorial management by individual convents.

Organizing the Coastal Zone

The Order established a number of convents close to the coast of the Vistula Lagoon (Elbing, Balga, Brandenburg (Pol. Pokarmin), Königsberg (Rus. Kaliningrad)), on the Baltic coast itself (Memel, Lith. Klaipėda; after 1346 Reval (Est. Tallinn)), whilst Riga and Danzig were directly connected to the sea through their waterways. The commandery of Marienburg also encompassed a significant stretch of coastline where the Vistula Delta drained into the Baltic Sea, whilst a large portion of the Curonian Lagoon was controlled by the convent at Ragnit. The episcopates of Ermland and Samland also controlled stretches of the coastline. These provided the Baltic theocracies with opportunities to reorganise two crucial coastal resources: fish and amber.

The impact on fishing is clearly evident following the crusades. Convents provided new markets for imported marine species, which are not found in the Order's smaller houses in the Prussian interior until the fifteenth or sixteenth centuries. In Livonia, the presence of marine species in smaller castles such as Arrasch and Karkus reflected their connectivity with coastal centres, as well as the elevated status of the latter, whilst their absence from Lötzen reflects poorer connectivity and perhaps even the comparatively lesser status of the Order's procurator before the later fifteenth century. Fishing in the sea, rivers, and lakes was carefully regulated in both regions, and much of the disputation in Riga between the various political groups included fishing rights. Marine fishing was particularly restricted. Isotopic studies indicate that the Order fostered the development of a local fishing industry, sourcing fresh cod from Baltic waters. At the same time, access to marine species by other groups within Prussia and Livonia remained limited until the fifteenth century (Orton et al. 2019). However, even in the fifteenth and early sixteenth centuries, access to littoral waters was regulated. In 1445, the commander of Windau wrote to the Bishop of Courland to inform him that his subjects may not harvest hops in the Order's territory, since the old custom gave the land lord the first right to the harvest. The bishop's subjects were also forbidden from fishing on the Order's stretch of the seashore (*LUB* x, 164). The regulation of fishing was mirrored by the control of amber.

The coastal zones of the southern Baltic contained abundant quantities of tertiary amber, which had been widely exported in the Roman Period and the Viking Age. Amber artefacts regularly feature as high-status grave goods in the south-eastern Baltic at this time, and the material may have been attributed with magical or symbolic qualities (e.g., Bliujienė 2003, 63; Žulkus 2002). Amber was collected and worked under the auspices of the dukes of Gdańsk, and amber artefacts were manufactured throughout the early medieval period. After the Prussian Crusade, the Order sought to regulate the collection of amber, of which the richest concentrations were located on the coastline of the Sambian Peninsula, where its collection was subcontracted to the Sambian Bishop in 1264. The international market for amber was already vibrant by the first decade of the fourteenth century, and paternoster guilds such as the one in Lübeck documented from 1302 were its main consumers; amber had also reached Italy by the end of the thirteenth century (King 2013, 760). In the fourteenth and fifteenth centuries, the collection of amber appears to have been the responsibility of commanders, advocates, and perhaps even 'fish masters'. The Sambian coastline remained the main source of amber for the Order, and an official documented as the *Borstein Herr* in 1327 at Lochstädt (Rus. Pawlowo) and later as the 'amber master' (*Bernsteinmeister*) reflected the Order's typical administrative response to the localized abundance of the fossilized resin in the south-western corner of the Peninsula. A similar function was fulfilled by episcopal officials at the castle at Fischhausen (Rus. Primorsk), located a few kilometres away (Hermanowski 1996, 98, 185). Regional control of amber collection and mining became centred on Königsberg, where fines were introduced for the possession of unworked amber in 1394 (Hagen 1824). Substantial quantities must have also been available in the Vistula Delta region, for the Order subcontracted amber collection rights to the fishermen of Danzig in 1312 and the abbey of Oliwa in 1342. Amber collection noticeably intensified following the secularization of the Order's Prussian branch (Willoweit 1969a; 1969b, 89–109), and whilst this period is also associated with an increase in the numbers of luxury amber products, the export of amber by the latter decades of the fourteenth century was well organized and substantial enough to generate a significant income for the Order (Jankhe 2015).

In the fourteenth century, it appears that amber was largely used for the manufacture of rosaries in some of the Order's centres and also in Western European workshops, although by the 1370s and 80s, more sophisticated artefacts were being manufactured. From the turn of the fifteenth century, there are several extant amber carvings which have been attributed to workshops in either Marienburg (where a goldsmith and amber carver is mentioned in the *Tresslerbuch*) or Königsberg. These are all religious in character and consist of a sculpture of the Virgin, a figure of St Catherine attached to a walrus skull chandelier, and a nativity scene, as well as three mounted *Vera* icons (popularized by the crusading movement) which may have been produced in Bruges (King 2013). Large numbers of amber beads and partially processed fragments of amber were found in excavations at the castle in Klaipėda, within occupation layers dating to the end of the fifteenth/start of the sixteenth centuries. The evidence for the storage of amber in the Order's Livonian houses is focused on Curonia, again reflecting the presence of local deposits. The inventory for the commandery of Goldingen in 1341 referred to one barrel of amber in the bailiff's room (*in camera camerarii*) worth 13 Curonian marks, while the advocate of Kandau was responsible for a quantity of amber worth 10 Curonian marks (Bauer 1924, 182–83). In 1538, when plots of land were exchanged between the commander of Windau and the vassal Gerdt Doenhow, this included part of the coastline with all the rights except for the mining of amber, which remained the sole privilege of the commander (Bauer 2003, 1929).

The Order's monopoly over amber has been typically interpreted in terms of export; however, it remained available to the indigenous population, whether through collection or purchase, and was included within jewellery (beads, cross pendants) that was ultimately deposited in graves. It is possible that amber cross pendants and rosaries were manufactured in towns, but other forms of jewellery may have been produced in rural settlements. This suggests that raw amber was also traded or collected by indigenous communities, and not solely intended for export. It is noticeable that relatively small amber objects are found in fourteenth- and fifteenth-century graves, whereas the largest pieces were used for devotional sculptures, of which little has survived from the period of the Order's rule.

2. *How Were Wild and Domestic Plants and Animals Used by Crusading Institutions in the Eastern Baltic, and to What Extent Did This Reflect Ideology and Adaptation to Local Ecology?*

The use of plants and animals by the Order in the eastern Baltic reflected a combination of subsistence reflecting adaptation to local ecology and the implementation of religious and political ideology. The crusading period was defined by the strategic mobilization of resources; Riga was a regular assembly point for crusading armies because of its concentrations of provisions, which could, if necessary, be moved to military frontiers. Quantities of meal, malt, meat, and other food had been shipped to the frontier castle of Mitau from Riga in advance of the winter crusade in the final years of the conquest of Semigallia (*LRC*, 11032). Thorn, Elbing, and Königsberg would function in similar ways. The narratives of the crusades regularly refer to the seizure of livestock and grain in both Livonia and Prussia, and agrarian production for provisioning people and livestock, and to generate a taxable surplus, was a fundamental component of the process of militarised colonization in the thirteenth century. The ownership of territory was renegotiated several times in Livonia, whilst the application of the Kulm Law in Prussia provided the Order with relatively more control over resources. The appropriation of existing provisioning systems is therefore an important aspect of the crusading period, although these had to be realigned to serve the needs of the Order's houses and towns, as well as existing and new rural settlements.

Plants

The staple food in the Order's houses, as well as the towns and rural settlements of Livonia and Prussia, was derived from cereals. Grain, particularly barley and hops, was also essential for brewing beer, and various crops were required to feed livestock. Given the growth in population in both regions after the crusades, as well as the exponential increase in livestock — particularly horses — the Order's grain requirements were vast. The diversification of crop species following the crusades is reflected in some of the isotopic dietary profiles of livestock in the Kulmerland (Scull et al. 2019). Whilst grain is poorly represented in the archaeological record of the Order's sites, inventories indicate substantial stores in both Livonian and Prussian castles, intended to sustain households for several seasons, as well as contain surpluses for export. Much of this would have derived from the produce of commander-

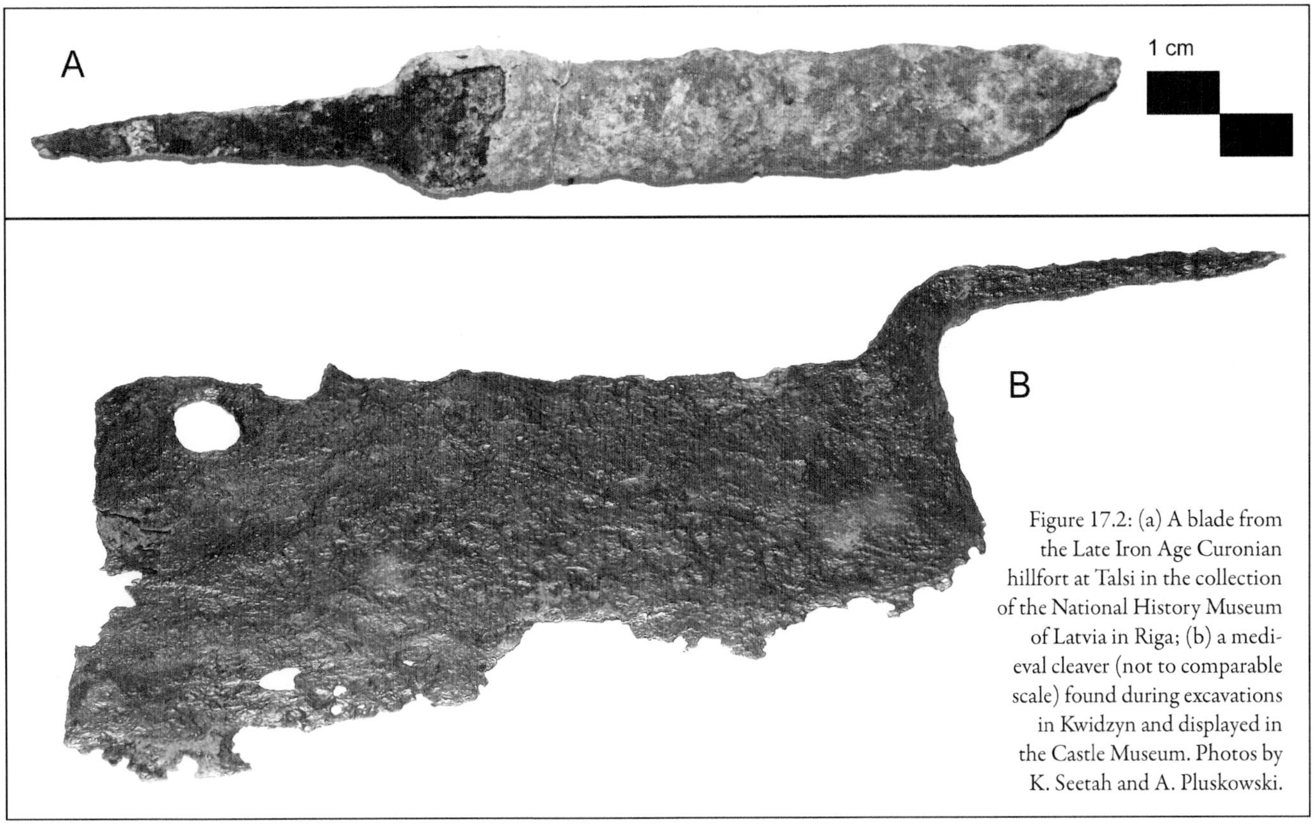

Figure 17.2: (a) A blade from the Late Iron Age Curonian hillfort at Talsi in the collection of the National History Museum of Latvia in Riga; (b) a medieval cleaver (not to comparable scale) found during excavations in Kwidzyn and displayed in the Castle Museum. Photos by K. Seetah and A. Pluskowski.

ies, although in Livonia, in particular, there is evidence for the concentration of grain production in the central parts of the Order's territories. But the Order's houses were also consumers of expensive plant products, particularly imported spices and fruit, while communities in towns and villages were reliant on flour-based products, vegetables, and locally available fruit and berries. The rich diets of the theocratic elites have been confirmed by isotopic studies of the grand masters buried in Kwidzyn (Ger. Marienwerder) (Szostek et al. 2009) and high-status individuals from the cathedral in Tartu (Ger. Dorpat) (Malve pers. comm). There is also evidence for the use of plants for medicinal uses, and herbaria are documented in some of the Order's houses. The palynological profile from the courtyard of the high castle in Malbork suggests that a garden was most likely cultivated in this space during the first phase of the convent. It is important to note that an important, and often understated, function of the Order was the provision of welfare. Convents included infirmary complexes, and hospitals were established in a number of towns within the *Ordensland*, although the Order was not the sole sponsor of these institutions (Pospieszny 1998; Pluskowski 2013, 236). The provision of medicinal plants would have therefore been extremely important.

Livestock

The Order oversaw the maintenance of composite husbandry regimes, inheriting existing Polish traditions in the Kulmerland and Lower Vistula, and Prussian, Balt, and Finno-Ugric practices in the conquered territories. In the Late Iron Age/early Middle Ages, strongholds may have operated specific types of husbandry regimes, or where faunal assemblages contained low species diversity and younger animals of prime stock, operated command economies involving the redistribution of livestock acquired in rent and tribute (O'Connor 2014). The faunal assemblages from the Order's houses in both Livonia and Prussia consistently indicate access to good quality meat, processed with heavy cleavers (Fig. 17.2), although these were socially stratified communities with the brethren only forming one component.

The Order's conventual diet combined access to expensive commodities with the alimentary ideology embedded within the Rule, which was modelled on that of the Templars and Dominicans (Trupinda 2002, 11). In 1244, the Order petitioned Pope Innocent IV to disregard certain aspects of their Rule which were perceived as redundant, including the provision of vegetarian meals on Wednesdays following a vigil, and replacing meals consisting of vegetables and pulses ordained on two to three days a week with eggs and cheese. During this time, the Templars and Hospitallers were grappling for control of the Teutonic Order, and this shift in dietary regulations was one means of preserving a visible degree of independence. By 1264, the new dietary regime had been clarified: the brethren were to eat meat three days each week, and Friday and Wednesday were treated as fasting days, alongside Lent. The Rule also stated that every tenth loaf (and all broken loaves) was given as alms to the poor, prayers were to be said before and after meals, and silence was observed whilst eating, except for the reading of passages from the Bible or the Order's own prescribed texts. Communal meals in the Order's convents took place in refectories where the ideology of holy war and the defence of Christendom would be reinforced through decorative schemes, the best examples of which are known from Marienburg. By the mid-fourteenth century, there is evidence for decreasing interest in, and diversions from, the Order's earlier regulations. However, given the relative dietary freedom of the Order's regulations this would not be evident in any faunal assemblages.

What is striking is that in both Prussia and Livonia, there is limited evidence of improvements in the indigenous livestock, including the iconic horse (Chapters 6 and 12). Sheep in Prussia appear to have become slightly larger between the pre-crusade and late medieval period, and the identification of two sub-populations of cattle can be interpreted as continuity in native stock alongside the introduction of new animals or perhaps the improvement of existing stock (see Chapter 12). This means that whilst some animals may have been imported from other parts of Europe, particularly the larger warhorses during the crusading period, the Order simply appropriated existing provisioning systems, which included herds maintained by its conquered subjects in both regions. The use of horses was directly linked to adaptations to the topography, as well as extreme winters and summers in the eastern Baltic. Winter became a preferred time for moving armies across the frozen rivers and lakes when 'the ice could bear a hundred armies' (*LRC*, 1644–45). Horses, rather than carts, became the preferred means of transporting provisions across the marshy and forested borderlands into Lithuania, and given the documented numbers involved in some of the *Reisen*, convoys would have stretched for miles. By the end of the fourteenth century, there were an estimated *c.* 16,000 horses on the Order's lands in Prussia (Ekdahl 1998), the end result of a definitive breeding programme which drew upon local stocks whilst maintaining a relatively small number of the largest stallions for warfare. This composition of this equine population can be glimpsed qualitatively in the range of estimated horse sizes from the scattered faunal assemblages in both Livonia and Prussia, where large individuals represent a small proportion of the measurable bones. This can be confirmed with future studies, particularly based on assemblages from castles where there is documentary evidence for keeping warhorses. The written sources, however, suggest that horse mobility within and between the two regions was much more significant than in the preceding period. There may have been discrete populations, but the Order's (and indeed, urban and episcopal) herds were regularly supplemented by raids into Lithuania, and by trade with neighbouring regions.

What changed, as evident from the isotopic studies of livestock diet in the Kulmerland, was the knock-on effect of new agrarian regimes which emphasized rye (Scull et al. 2019). This in turn, affected fodder, and there is also some evidence that urbanization saw more pigs confined in castle wards and towns, a trend that was not visible in the dietary signatures of this species in early medieval centres. Significant changes in the size of livestock would only appear in Prussia in the post-medieval period, coinciding with the influx of Dutch migrants. In Livonia, indigenous forms were retained until the nineteenth century, and therefore the changes can be disconnected from any medieval breeding regimes. However, whilst the sizes of animals did not change substantially, the appropriated provisioning systems were reconfigured to serve castles, towns, as well as the growing rural populations of Livonia and Prussia. This resulted in a new emphasis on rearing cattle for an expanding meat market, and in regions where pig husbandry had been prevalent, as in the early medieval Kulmerland, a retention of existing husbandry practices. But the animal economy had been broadened: castles were important consumers of meat, dairy produce, wool, and to a small extent parchment. Moreover, the quantity of animals must have increased, although it is difficult to provide statistically meaningful comparisons with the early medieval period. There is also a disparity between the zooarchaeological data and documented numbers of livestock in some of the Order's castles, which emphasize underrepresented animals such as sheep, and also indicate diachronic variations in herd numbers, particularly in the fifteenth century. Sheep would have been important in both Livonia and Prussia for the production of low-grade wool for local markets, including the Order's houses.

Map 17.1: The relative representation of wild species (%, averaged across relevant phases) in faunal assemblages at castle sites associated with the Baltic theocracy included within this study. Game documented in inventories has not been included as the quantities are not comparable. Map by author.

Wild Mammals and Birds

Hunting large carnivores — specifically wolf, bear, lynx, and lion — for the common good was permitted, although brethren were forbidden to use dogs and had to employ hunters to pursue game and trap for furs. Monastic orders were forbidden to hunt, but for the military orders it also served as valuable training for warfare and acquiring knowledge of the local landscape. The crossbow was a decisive weapon against the Prussian tribes, and shooting wild birds was recognized within the Rule as a useful means of developing skill with projectiles. The limited exploitation of wild fauna is comparable to medieval secular high-status sites from other parts of Europe, including the castles and provincial headquarters of the military orders (Ervynck 1992, 2004; Pluskowski 2010). A survey of the Order's sites in Livonia and Prussia indicates that game played a small but consistent role in both convents and the residences of lesser officials (Mugurēvičs, 2002) (Map 17.1). The evidence from the castles at Lötzen, Memel, and Lyck suggests that a greater diversity of game was available in frontier territories with higher levels of biodiversity, although even here the majority of meat consumed by these households derived from cattle. It is also entirely possible that those few remains that have been identified were procured by members of castle households and the Order's subjects. This is clearly expressed in the obligations of the Kulm Law, and most vividly in the fifteenth-century regulations concerning the sale of wild mammal pelts to the procurator of Lötzen.

Although hunting should be understood as most useful for training retinues for warfare, it was clearly a conspicuous expression of high status consumption. The Order's network facilitated the movement of wild meat from districts where it could be readily acquired to the headquarters — Marienburg and later Königsberg. The aristocratic character of the Grand Master's court in the late fourteenth century was further emphasized by the creation of the park at Stuhm, facilitating the provision of fresh game that complemented stocks of fresh fish bred in the ponds located north of the outer bailey. The presence of wild birds is even more limited at the Order's sites, and it seems that falconry was not widely practised; in the fifteenth century, there is some limited evidence for the keeping of hawks and specialized hunting hounds amongst the higher officials of the Order (Militzer 1997). Whilst game was used to express lordship, the acquisition of fur was largely driven by commercial interests. The late medieval international fur market was one of the defining features of the Hanse, and there is fragmentary evidence that substantial quantities of fur could be sourced from Livonia and Prussia. However, these were not the principal fur-producing regions in eastern Europe, and most of the furs passing through Riga, Reval, and Danzig came from neighbouring regions. The wearing of fur-lined garments within the Order became a feature of the highest officials at the end of the fourteenth century, and included leopard pelts which can be regarded as one of several exotic commodities imported into Marienburg in the late fourteenth and early fifteenth centuries.

Fish

The crusades introduced a fasting culture into eastern Baltic societies, but it is difficult to determine how effectively this was implemented amongst local populations. The theocracy led by example, with fish consistently documented at both regional branches of the Order's houses. Fish were recognized as a valuable resource from the very onset of the crusading period, as indicated in regulations established in Riga and set out in the Kulm Law, and fishing represented a fundamental component of territorial management for every *Ämter*. As with livestock, the Order appropriated existing fishing practices — and indeed, fishing communities — and freshwater species remained the staple for houses governing territories with large numbers of lakes and rivers. Coastal houses of course exploited littoral waters, but the presence of imported marine fish in a number of the Order's convents is consistent with the purchasing power of these high-status communities. Cod, in particular, appears to have been favoured by German colonists as evident in the faunal assemblages from Gdańsk and Kołobrzeg (Ger. Kolberg) before the crusades (Makowiecki 2003), whilst the documentation of at least two types of carp within the Order's inventories, alongside the presence of the species in faunal assemblages, represents the first evidence for the widespread consumption of this species in the south-eastern Baltic. These imports, along with ponds supplying freshwater fish to some of the Order's convents and occasional evidence for beaver consumption, indicates that fasting in the Order's houses could be experienced as a high-status feasting activity (Mänd 2004). In this respect, the Order was continuing existing trends, at least in the early medieval Vistula castellanies, where the increase in sturgeon fishing coincides with the formation of the Polish state (Cios 2009, 76), exemplified by Kałdus where sturgeon remains represent 12% of the entire fish assemblage between the tenth and twelfth centuries, at a time when stocks were already decreasing as a result of overfishing (Makowiecki 2008). The continued exploitation of the species put further pressure on the populations migrating up the Vistula, although Baltic sturgeon continued to be caught in the Curonian Lagoon and nearby waters into the seventeenth century (Žulkus and Daugnora 2010, 79). Fishing was therefore more intensive where it was associated with clusters of major urban centres and convents, unevenly distributed across the Order's territories. As in the case of the exploitation of livestock and wild mammals, the exploitation of fish combined ideological requirements with local adaptations.

Raw Materials

The castles and towns in the *Ordensland* developed industries utilizing a range of raw materials derived from plants and animals. Inventories of the Order's houses in the fifteenth century were predominantly concerned with documenting stocks of food and arms. Both required natural products, and with the exception of good quality iron which had to be imported, the principal components for crossbows (antler, bone, horn — notably from goats —, linen (derived from flax) hemp, sinews, and wood) were sourced locally. Armouries are documented in both convents (Jóźwiak and Trupinda 2012) and the residences of lesser officials, whilst archaeological evidence for crossbow production is evident in Malbork, Viljandi (Luik 2015), and Cēsis, and bone and antler waste from Karksi may have been linked to this. Evidence for the processing of goat horns is also evident in the Liv quarter in Riga and both the castle and town of Cēsis. Other essentials, such as leather, parchment, and tallow for candles, would have also been sourced from deadstock, whilst the procurement of wax from hives was regulated in both Livonia and Prussia. Brewing played an important role in all of the Order's convents, although certain beverages were also imported. Overall this indicates the presence of specialists and interconnected industries situated within the Order's precincts and adjacent towns, with mercantile guilds first documented in Livonia from the early thirteenth century and subsequently in Prussia. In the former region, these guilds admitted artisans from the mid-thirteenth century, which then evolved into corporations of craftsmen in the mid-fourteenth century (Czaja 2005). In Prussia, the documentation for specialist guilds can be quite late. For example, the butchers of Memel only founded a guild in 1627 (Sembritzki 1926, 166).

3. *To What Extent Did the Crusading Movement in the Eastern Baltic Frontier Represent a Cohesive Force of Ecological Transformation? Is There Evidence for Standardization?*

The replacement of pre-crusade power centres with fortified convents and smaller castles, connected by an interregional network, represented a clear shift from independent territories to a set of interdependent territories, governed by a religious corporation with an administration to rival any royal court in Christendom. The regularity of its conventual castles, the most impressive symbols of authority in the medieval eastern Baltic, reinforced the Order's image of unity, hierarchy, religiosity, and military superiority (Jensen 2009, 150). But the Order was not a static organization, with evident tensions, changes, declines, and regular attempts at reform and standardization. The Grand Master was elected as the leading representative of the Order, but de facto power rested in the hands of ambitious commanders and convent chapters. As a result, the economy of the *Ordensland* functioned on two tiers: the autonomous economies of individual convents and the integrated economy of the Order's territories as controlled by Marienburg (Czaja and Nowak 2013, 22). This was particularly true in Livonia, physically separated from the Order's territories in Prussia: here, despite the presence of a provincial master, commanders retained a high degree of independence and political agency, although it is important to bear in mind the collegiate context of a convent — decisions were also mediated by chapters and the commander's deputies, the *hauskomturs*.

The key differences between the two branches were determined by the early history of the Order's involvement in both regions. In Prussia, the Order quickly incorporated all but the Ermland Cathedral Chapter into its ranks and exercised complete sovereignty. In Livonia, the Order only had comparable authority in Curonia, inheriting the subservient status of the Sword Brothers. In 1254, the representative of the Grand Master gave the oath of obedience to the respective Livonian bishops, swearing that the Order would respect all the liabilities it had in Livonia with respect to the episcopates (*LUB* I, 277; VII, 3024b). This was most probably reiterated in 1264 (*LUB* I, 429, 2745); however, with the consolidation of territories in both Livonia and Prussia by the 1290s coinciding with the loss of Acre, and particularly following the annexation of Pomerelia and the relocation of the Order's headquarters to Marienburg in 1309, the subjugation of the Livonian branch to episcopal authority became disputed. The Livonian brethren became associated with periodic and excessive demonstrations of force, and the most significant change came in 1393 when the Cathedral Chapter of Riga was incorporated into the Order, although conflict between the Livonian theocracies would continue into the fifteenth century (Selart 2011). Nonetheless, the effects of centralized organization are more visible in fourteenth-century Prussia than in Livonia, exemplified by the relative control of trade. In the latter region, commanders and advocates dealt with merchants directly, whilst in Prussia regional trade was overseen by the Quartermaster (Ger. *Trappier*) based in the convent of Christburg (adjacent to Marienburg), with representatives in the Hanseatic *Kontores* in Bruges and London.

Over the course of the fifteenth century, the separation between Prussia and Livonia became increasingly marked until Plettenberg effectively ruled as a sovereign in his own land. It is only at this point that the Livonian masters came into their own as the equivalent of princely rulers (Kreem 2002), coinciding with the reorganization of territories in central Livonia which saw the realignment of power to the convent at Wenden. This level of heterogeneity across the *Ordensland* is not surprising,

Figure 17.3: Building materials used in the Order's houses: (a) field stones and brick at Papowo Biskupie (Ger. Bischöflich Papau); (b) brick at Radzyń Chełmiński (Ger. Rehden); (c) field stones, dolomite, and brick at Cēsis (Ger. Wenden). Photos 17.3a and b by M. Elander; photo 17.3c by A. Pluskowski.

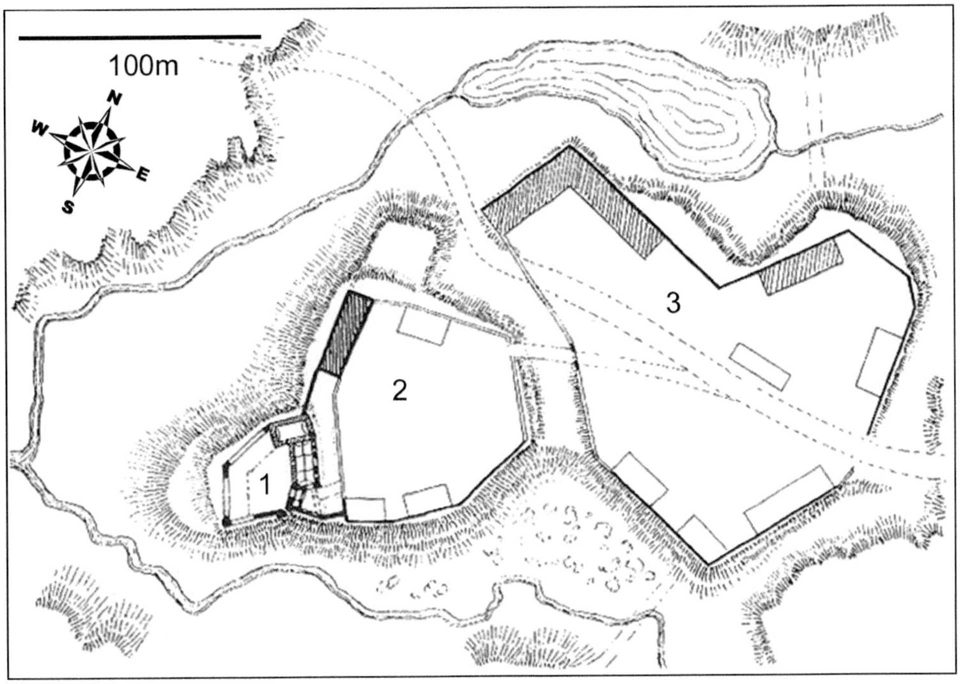

Figure 17.4:
Plan of the convent at Engelsburg (Pol. Pokrzywno), with the (a) *Oberstes Haus* (high house or castle); (b) *Mittlere Haus* (middle house or castle); (c) *Vorburg* (outer bailey) marked (after Steinbrecht in Haftka 1999, 238).

and all supranational religious organizations in medieval Europe struggled with regulating the behaviour of their provincial communities. Outside of Prussia and Livonia, the Order functioned differently. Its houses varied in their design, often reflecting local architectural styles (e.g., Vairo 2014), and from the mid-fourteenth century the Order's German Master became increasingly treated as a member of the Imperial court (Biskup 2009, 291). These organizational differences provide the backdrop for contextualizing the impact of the Order's individual houses.

Standardization within the Order's houses

The Order's houses adopted a broadly consistent approach to the management of agrarian products and animals within their precincts. Whilst architectural forms appear to have been largely standardized (particularly in the Order's Prussian convents constructed from the end of the thirteenth century), variations in the layout and spatial organization of the Order's castles can be closely related to their specific function, with changes reflecting new financial and strategic investment in expanded roles for these structures. Building materials also varied regionally (Fig. 17.3). The constructional history of each castle, in turn linked to territorial management, therefore reflects the gradual and regionally varied administrative structure of the *Ordensland*. Moreover, the decision-making process in castle construction and expansion was negotiated between the Order's officials, architects, and masons (Arszyński 2010, 42). However, there were recurring features which were essential for the sustainability of both large and smaller houses. Convents had important storage functions, spatially subdivided between the 'high castle' (*Haupt Schloss*) and 'outer bailey' (*Vorburg*) (Jóźwiak and Trupinda 2012; Pluskowski et al. 2019). Convent stores, typically located in cellars and attics, were for regular, short-term use, whilst the main storage areas were located in outer baileys (Gancewski 2001, 51). The use of these outer wards for storage and animal penning is evident from the earliest phases of these sites, as in the case of Elbing and Karkus. However, recent scholarship has highlighted that outer baileys were more than simply areas for service buildings; in the late fourteenth century these also contained residences for officials, refectories, and chapels (Jóźwiak and Trupinda 2012). Marienburg, although seemingly atypical by virtue of its massive size and role as the Order's headquarters, can in fact be envisaged as a convent attached to two substantial outer baileys. The so-called 'middle castle' initially functioned as the outer bailey for the castle in the last two decades of the thirteenth century, and its later spatial character, which included refectories and the 'Grand Master's palace', is mirrored in a number of other major convents such as Elbing (Mierzwiński 2010; Jóźwiak and Trupinda 2012). The full extent of the conventual complex at Engelsburg (Pol. Pokrzywno) can also be divided into three principal spaces — the *Oberstes Haus* (high house or castle), *Mittlere Haus* (middle house or castle), and *Vorburg* (outer bailey) (Fig. 17.5).

The spatial configurations of smaller castles varied across Livonia and Prussia and can be related to the status of the residing officials, the functions of the houses and levels of investment. The outer bailey of the castle of Arrasch was most likely devoted to livestock and patches of cultivation, and in the period of Polish rule the site would become the focus of a *folwark* or specialized farm. In the absence of any archaeological data, written sources indicate that the procurator's castle in Lötzen was attached to a farm. The advocate's castle at Stuhm (Pol. Sztum) was served by an outer bailey which was later converted into a town, and also managed — uniquely — a park in the late fourteenth century, reflecting its proximity to the Grand Master's court. Despite its size, the written sources for Stuhm indicate relatively low productivity, with a focus on maintaining the garrison and serving the requirements of nearby Marienburg. All three castles had relatively specialized administrative roles and were subservient to their respective convents. In contrast, the advocates of Karkus were virtually on par with Livonian commanders, and the substantial outer bailey of their castle had, in all probability, multiple functions similar in character to those of a convent. Analyses of faunal assemblages have confirmed that livestock were regularly slaughtered, butchered, and processed in the outer baileys of all the Order's houses, both at convents and lesser residences. There is evidence for stabling within the Order's *Vorburgen*, documented in Riga, Fellin, Marienburg, Nessau, and Rehden, and identified archaeologically in Elbląg and Cēsis. It is also likely the outer baileys of Karkus and Arrasch contained penning and stabling areas. In the case of castles situated next to towns, this suggests they relied on their own supplies of meat and deadstock. In a number of instances, castles were reliant on services from adjacent settlements, such as milling in fifteenth-century Lötzen.

Standardization in Environmental Exploitation

The Order's houses were sustained by the settlements in their territories, a level of integration between sites and their territories that mirrored the relationship between towns and their patrimonies, one of Blomkvist's defining features of Europeanization (2005, 701). When this was threatened or collapsed, these houses could still sustain themselves on a limited scale or by relying on imports from the Order's network, which mapped onto the broader Hanseatic network. This can be seen in the successful maintenance of frontier castles such as Lyck, Ragnit, Memel, and Nessau. It can also be seen in the internal mobility of commodities across the Order's two Baltic branches. In 1417, the commander of Windau (Latv. Ventspils) sent the Grand Master hunting hawks (for which he asked payment), along with a request for a keg of wine from Thorn as the grape harvest in Curonia had been poor that year (*LUB* v, 2168). The surviving records of decisions taken by the Order's officials indicate a shared culture of managing territory in both Livonia and Prussia. This is most clearly expressed in the use of settlement laws, alongside the detailed regulation of mills, fishing rights, and access to woodland and pasture. When grand masters sought to standardize the management of territory across both regions, as they did in the fifteenth century, this focused on ensuring the sustainability of the Order's houses. But there were factors which diluted attempts at corporate homogeneity. The inventories of Prussian and Livonian castles, which provide snapshots of the distribution of resources from the late fourteenth through to sixteenth centuries, indicate strategic uses of the hierarchies of houses in both regions. They also underline the variable uses of the landscapes governed by these houses which, in turn, reflected their capabilities at mobilizing their subjects.

Where new regional centres were created, as at Riga and Marienburg, their associated territories saw significant transformation. In Riga this took the form of localized deforestation and the intensive regulation of the surrounding landscape where indigenous settlement had remained intact, whilst in Marienburg this was reflected by the establishment of multiple farms and mills linked to extensive colonization, with the governance of the commandery supported by a host of lesser officials. In the case of Fellin and Wenden, which replaced existing power centres, there is little evidence for dramatic change, and the indigenous settlement pattern does not appear to have been disrupted. In fact, the populations of these regions increased during the medieval period, and the most significant disruption came with the devastation of the Livonian War in the later sixteenth century. The convent of Rehden replaced an earlier power centre at the frontier of the Kulmerland, but with no evidence for continuity. The smaller castles of Lötzen, Arrasch, and Karkus were all established at new sites linked to the reorganization of communication routes following the crusades. In the case of Arrasch, there was no continuity from the nearby Viking Age fortified lake settlement which had been abandoned several centuries before the castle was built. In all of these cases, the territories associated with the Order's houses were new creations which developed incrementally and changed in the later period of the Order's rule. The evidence for pre-crusade territories is tentative at best, but whilst the commandery of Fellin may have mapped onto a substantial part of the preceding indigenous overlordship, the commandery of Wenden significantly expanded beyond the earlier Vendish core, although the Gauja Valley remained an important frontier. The commandery of Marienburg incorporated a frontier region straddled by the Vistula and Nogat, and its north-eastern boundaries were shaped according to the neighbouring commanderies of Elbing and Christburg.

In the general absence of excavations of the domestic and industrial spaces of medieval rural settlements within the Order's territories, with important exceptions such as Mārtiņsala and Biała Góra, the agency of indigenous communities and colonists is largely visible through their documented obligations, and from palynological studies of their immediate landscapes. The latter enables the impact of these settlements to be contextualised. The deforestation evident between Węgry and the modern northern border of the Forest of Sztum can be readily connected with the activities of the early medieval settlement cluster within the vicinity of the stronghold, and therefore reflects the impact of establishing the castellany as a definitive cultural expression of early medieval Pomeranian colonization in the Vistula Delta region. However, the later deforestation of the north-eastern edge of the forest can be specifically linked to the activities of the colonists in Parpharen, in much the same way as the changing vegetation within the vicinity of Rywałd (Ger. Königlich Rehwalde) reflects the highly localized strategies of landuse adopted by the colonists in the early decades of the fourteenth century. Small increases in livestock size in western Prussia coincide with improvements in fodder. In eastern Prussia, the indigenous population is more visible, and there is evidence for continuity in the exploitation of local environments, in terms of cultivation, livestock husbandry, hunting, and fishing. The focused palynological work conducted within the macro-region of Staświny reflects the activities of a string of early medieval settlements identified by archaeological traces, with clear evidence for continuity in land-use into the post-crusade period. Even with the establishment of the procuratorship of Lötzen, there is little evidence of intensification until the second half of the fifteenth century — a century after its foundation — coinciding with the installation of a mill at the castle, and following the relocation of the Order's headquarters from distant Marienburg to the relatively closer centre at Königsberg. In Livonia, the environmental changes associated with the territories of Fellin, Karkus, Wenden, and Arrasch reflect the agency of indigenous communities much more clearly.

In summary, the replacement of pre-crusade power centres and the restructuring of territories was enabled by the use of a consistent administrative unit run by commanders and advocates in the thirteenth century, but exactly how the new territories were run and the impact of the Order's rule on subjugated landscapes can only be understood on a case-by-case basis. In this respect is it also heterogeneity, rather than homogeneity, that characterizes the ecological impact of the Christian theocracy in the eastern Baltic.

4. How Did the Process of Christianization, Facilitated by the Baltic Crusades, Change the Way Indigenous Groups Conceptualized the Natural World?

The pre-Christian societies of the eastern Baltic invested multiple aspects of the natural world with spiritual agency (Vaitkevičius 2004). These sites can be linked to the distribution of settlement, and they clearly formed a component of territorial units; in the case of Prussia at least, they appear to have been typically located on boundaries (Szczepański 2011). The crusading period was also a time of conflict over the spiritual meaning of the landscape. To Christian commentators, the topography was as polarized as the struggle against paganism, with Christian colonies represented by town and convents, protected by relics and divine dedications, juxtaposed against the spiritually dangerous forests and marshes where the indigenous population hid and ambushed. Paganism is best understood as an expression of 'otherness' in the medieval eastern Baltic, a category of difference in relation to the type of Christian identity introduced during the crusading period (cf. Palmer 2007). The crusades facilitated more than just the establishment of a Latin diocesenal system, with parishes, churches, pilgrimage, priests and tithes. They created polities dominated by a religious elite (with the exception of the Danish Duchy of Estonia until the mid-fourteenth century), which ruled, from the perspective of the conquered population, on behalf of a new deity. Designating the conquered territories as the 'land of Mary' may have been intended to raise support for the early crusades in Livonia, but the promotion of the Virgin as a figure of authority by the Teutonic Order reinforced the conquest as a spiritual battle (Dygo 1989). At the same time, the proliferation of the symbol of the new religion — par-

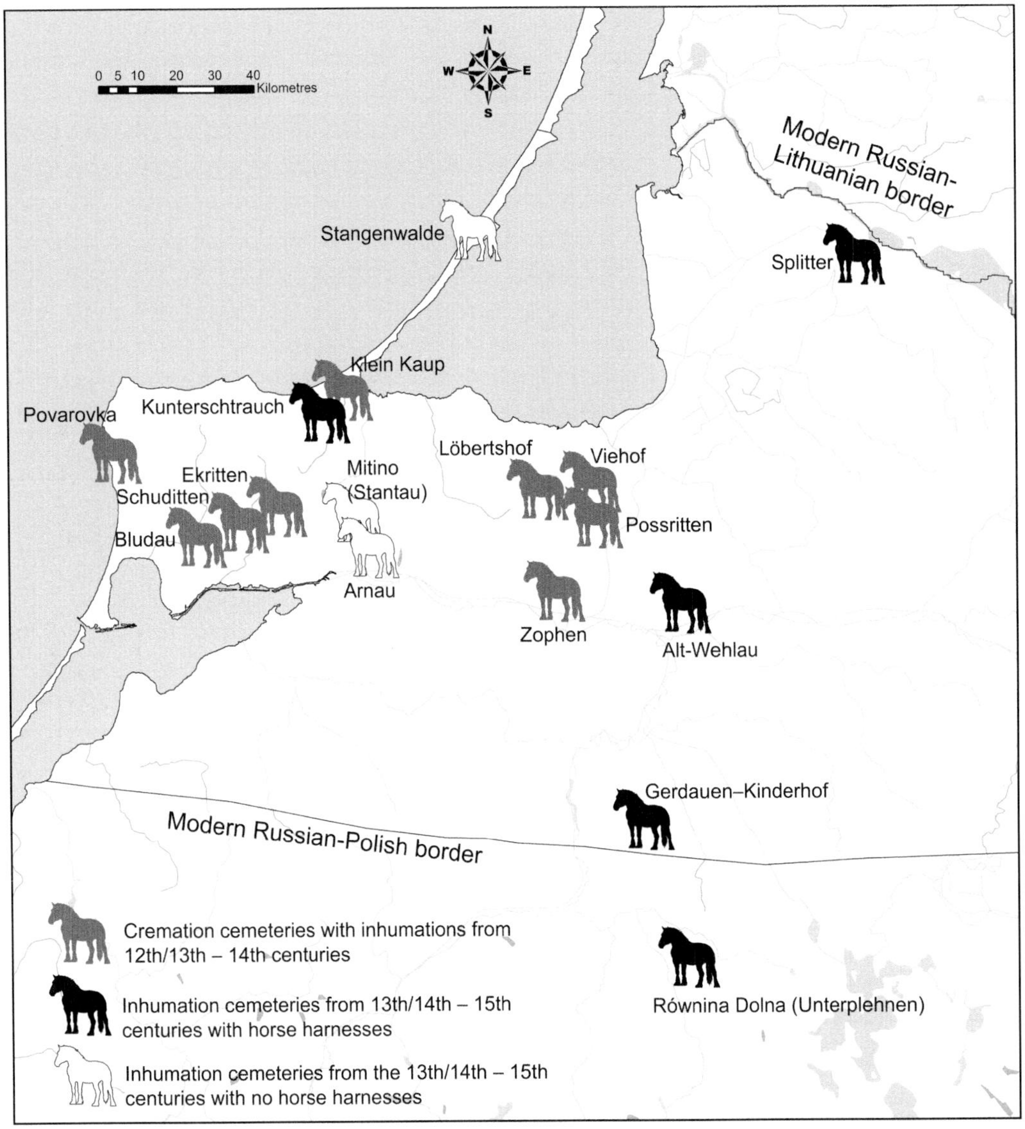

Map 17.2: Archaeologically known cemeteries with inhumations, weapons, and horse harnesses from the twelfth/thirteenth to fifteenth centuries in Prussia. Horse deposits are found only in pre-crusade cemeteries dating to the twelfth and up to the first half of the thirteenth century (after Shiroukhov 2012, Figure 15).

ticularly as a boundary marker — would have had a double meaning in the conquered territories; the emblem of the theocracy and the most important symbol of the new religion. For the Teutonic Order the symbolism was more nuanced, and the cult of the cross, intimately connected with crusading, provided a direct reference to the Holy Land. The use of crosses cannot be untangled from new definitions of property introduced after the crusades. The same conceptual transformation is evident in Masovia in the fourteenth century, where the explicit marking of boundaries is clearly linked with the proliferation of German settlement laws and a more precise understanding of spatial organization, ultimately resulting in an unprecedented culture of litigation over land ownership (Myśliwski 2002). In Livonia, the proliferation of crosses in the landscape — as boundary markers, cemetery markers, and features of local shrines — was inextricably linked to the process of Christianization and the gradual visual assertion of a Christian pastoral infrastructure (Valk 2019). Yet despite the reorganization of all the conquered territories into dioceses, and the eventual establishment of bishoprics in Lithuania and Samogitia as recognized (if disputed) markers of Christianization, there is variable evidence for the persistence of pre-Christian sacred sites and values attached to pre-crusade worldviews. Indeed, there is relatively little evidence for the systematic destruction of sacred sites in Livonia or Prussia during the crusading period, and where it does occur it almost certainly did so within a political context.

The sites most closely connected with the memory of pre-crusade political structures were cemeteries, and in both Livonia and Prussia rural cemeteries continued to be used by indigenous communities with gradual changes in burial rites (Szczepański 2019; Valk 2019). There is some evidence for the Christian theocracy attempting to deter local communities from using earlier cemeteries which were not associated with churches nor consecrated, but their main impact appears to have been on the abandonment of public animal sacrifice and deposition at cemeteries. The most important of these was the horse, whose spiritual role may have been connected with the political status of the equestrian elites. The distribution of horse deposits in the eastern Baltic varied temporally and spatially, and in the twelfth and thirteenth centuries they appear to have been restricted to the Sambian Peninsula and its fringes in eastern Prussia (Jaskanis 1966; Shiroukhov 2012, 2015), and amongst certain Latgalian and southern Liv communities in the Daugava region (Brūzis and Spirģis 2011). Lithuanian horse deposits are also unevenly distributed and appear to have been connected to the centralized authority of the Grand Dukes, reinforcing the political symbolism of public ritual animal killing (Bertašius and Daugnora 2001; Bertašius 2006). The public sacrifice of horses (and people) at cemeteries was singled out in the Treaty of Christburg in 1249 and represented the limits of tolerance by the conquering Christian authorities. The abandonment of horse deposits in cemeteries and their replacement by proxies suggests a more concerted effort to ban animal sacrifice in line with Christian moral codes, both in Livonia and Prussia (Map 17.2) (Shiroukhov 2012). Whilst indigenous power structures were accommodated by the new theocracy, it is clear that public animal sacrifice was a step too far. The practice continued in Lithuania, reinforcing its pagan otherness in the eyes of Christian observers, and in the conquered territories it was reduced to domestic ceremony which endured beyond the Reformation. The political symbolism of the horse continued under the Order's rule, and the connection between warhorses and convents reinforced the military authority of the Christian regime. However, the earlier religious connection was broken, and may have only persisted with the continued consumption of horse meat by indigenous communities, as in the case of the Liv quarter in Riga. This lingering Christian taboo characterized pagan otherness more broadly in Northern Europe, although it was clearly ignored in many regions (Poole 2013). Until more faunal assemblages from medieval indigenous rural settlements in Livonia and Prussia are examined, it will be difficult to determine how widespread this practice was after the crusades, or how long it persisted. Whilst domestic animal sacrifice is documented in both Livonia and Prussia in the post-medieval period, there are some examples of animal deposits within hybrid cultural contexts in the Lower Daugava in the thirteenth century. The calf deposit in the doorway of Vecdole Castle and the placement of horse and goat skulls under the rebuilt north wall of Riga has been typically linked to the actions of indigenous craftsmen, but could reflect — or at the very least resonated with — the foundation deposits or *Bauopfern* that were part of the colonizing culture. It is unclear if animals were deliberately killed for this purpose, or if these deposits were drawn from deadstock.

Evidence for the persistence of pre-crusade religious praxis in the regions considered within this study is variable. The Kulmerland and Lower Vistula Valley had been colonized by Christians following the expansion of the Polish state in the tenth century; the placement of the altar of the church at Kałdus on top of an alleged offering site represents this earlier phase of religious conversion (Chudziak 2003). The Order effectively encountered an already Christianized landscape in these borderlands. In eastern Galindia, there is only indirect evidence of the persistence of earlier meanings in the form of toponyms, although if the interpretation of Święta Góra as a cult site is accepted, the presence of greyware ceramics in its final phase documented to date, suggests an enduring role into the crusading or post-crusading period. Here, the absence of a Christian infrastructure until the fifteenth century would have enabled indigenous spirituality to endure with little incentive for interference from the procurator at Lötzen. In Livonia, the situation is also related to the relative survival and distribution of the indigenous population. In Riga, where two pre-crusade Liv settlements have been identified, there is evidence for the continuation of indigenous burial rites in the thirteenth and into the fourteenth century (Caune 2007). However, as in the case of Estonian towns (Valk 2004), these were eventually abandoned in favour of Christian eschatological uniformity. The cathedral of Riga appears to have also been constructed on the location of a Liv cult site (Caune 1992). Given its role in the crusading period and subsequently as the principal centre of Christian Livonia, this may also explain the general absence of sacred sites within the town's highly regulated patrimony.

However, the absence of evidence for non-Christian sites in the commandery of Wenden may reflect the devastation of the indigenous population during the Livonian War, which would have resulted in a hiatus in local traditions. But at the end of the twelfth century this was also a borderland between different tribal territories

— the Gauja Livs, the Vends, and the Latgalians to the south — and sacred sites appear to have been located in relation to central places, which, within the vicinity of Wenden are solely represented by Riekstu Hill. The Vendish community here was quickly Christianized and allied with the Sword Brothers in the early years of crusading. Further north, the establishment of new power centres at Fellin and Karkus did not result in the obliteration of sacred sites within their broader territories; in contrast the number appears to have proliferated with the growth of settlement (Valk 2019). The majority of toponyms related to indigenous sacred sites have been identified in Curonia, where there is clear evidence for their toleration as the earliest territorial boundaries were being established by the conquering theocracies. Pre-Christian practices, focused on sacred natural sites as well as cemeteries, appear to have persisted amongst indigenous rural populations which were poorly provided with parochial support during the period of the Order's rule (Kala 2009, 190), but the simultaneous existence of chapels and non-Christian sacred sites within the same landscapes indicates the presence of parallel religious systems (Valk 2019). The persistence of pre-crusade meanings may also be evident in the continued use of carnivore claw and tooth pendants, which are usually interpreted as having an amuletic function in the Late Iron Age. They have even been found at the Order's sites, such as at Āraiši, and linked to the presence of an indigenous community (Apals 1995). No systematic survey of species-specific artefacts has been conducted for the later medieval period except in western Lithuania, where the deposition of bear claw pendants in late fourteenth- and fifteenth-century graves has been interpreted as a form of religious hybridity, rather than a straightforward continuation of pre-Christian symbolism (Svetikas 2008).

Conclusion: Ecological 'Event Horizons' and Ending Medieval Baltic Singularities

The crusades in the eastern Baltic were defined by escalating violence and Christian radicalism (Provost-Bien 2013), but also by variable encounters between Latin Europeans and Balt, Finno-Ugric, and Rus' communities. They resulted in the creation of distinctive, composite polities largely governed by theocracies, of which the most powerful was the Teutonic Order. The simultaneous rise of late medieval German and Scandinavian mercantile groups as a commercial force in the Baltic Sea region can, in part, be connected with the war economy of the crusades (Wienberg 2004). This alliance between *ecclesia* and *mercatura* is integral to the model of Europeanization (Blomkvist 2004, 2005), where trends in environmental exploitation are understood as the result of the implementation of the Catholic world system, with the export of raw materials rather than commodities from the eastern Baltic defining the entire region as economically and culturally peripheral (Mortensen and Bisgaard 2013, 16). However, just as the supra-regional integrity of the Hanse has been challenged in favour of blended international, supra-regional, and regional trading spheres (Jahnke 2015, 240), so the implementation of the new world system — in Livonia and Prussia represented by the Teutonic Order — consisted of adaptations and innovations which integrated various pre-crusade systems, from the remnants of Polish castellanies and the hereditary territories of Liv vassals to the conservative lifestyles of Saccalian peasants, who continued to recognize a landscape imbued with supernatural agency. This resulted in a cultural hybrid, a composite society under a monolithic political aegis with extensive regional variation.

One of the defining features of this hybridity was the uneven impact of the Order's houses across Livonia and Prussia on the landscape, despite the imposition of a standardized and hierarchical administrative structure. The incremental process of military conquest in both Livonia and Prussia resulted in variable transitions from indigenous heartlands to frontiers, the creation of new cores and the persistence of those where the Order had directly replaced the pre-crusade elites and reused their monumental centres. In some cases, this accentuated earlier trends in environmental exploitation, in other cases it did not result in intensification until a century or more after their establishment. Only in pre-crusade frontier regions was the Order able to implement a significantly new design on the landscape, exemplified by the commandery of Marienburg. The earlier crusading centre in the Baltic, Riga, accentuated existing trends in environmental exploitation, whilst the impact of the much later (and relatively short-lived) central role of Wenden reflected the mobilization of resources from its satellite districts. Blomkvist's (2005) revised model of Europeanization and Gaimster's (2005, 2007) Hanseatic archaeology emphasized the commercial and cultural dependence of the eastern Baltic periphery on the Western European core, but the *Ordensland* was neither a passive periphery, nor a replica of an Imperial province. Indeed, Piskorski (2004) has argued that processes of colonization in Central and Eastern Europe should not be regarded as much a part of 'German history', but rather considered on their own terms. The theocratic authorities of the eastern Baltic answered directly to the papacy and maintained lobbies in the papal court for this purpose; however, in reality they were de facto rulers within their own territories, conducting their own domestic and foreign policies (Eihmane 2012). The function of the Order's houses, and the theocracy as a whole, was not to support an imperial core but to sustain its own communities with the aim of effectively maintaining its own sovereignty. This, in turn, was validated by the Order's redefined role as the defender of Latin Christendom against pagan Lithuania and schismatic Rus' principalities. The economic strength of the Templars and Hospitallers — intended to sustain their houses in the Levant, which in turn supported the security of the crusader states, as well as infrastructure for pilgrims and local communities — was founded upon their western estates. The Teutonic Order also needs to be understood in this context.

William Urban (2001, 67) characterized the military orders during the crusading period in the eastern Baltic as highly adaptable to the challenges of local climate and geography. The Order's system of environmental exploitation which developed after the cessation of hostilities can also be characterized as adaptive. Its houses became largely standardized in terms of their management of agrarian resources and animals, with little difference between heartland and frontier sites. This reflected the paradigm of self-sufficiency consistently promoted within the Order's hierarchy. The principal differences were between convents and lesser residences; a question of relative status and purchasing power, with the former having greater access to luxury commodities, although even the Order's smallest communities were able to import products such as marine fish and ale. There was far more variability in how individual territories were governed. This was related to the nature of settlement — the interplay between existing and incoming communities — the possibilities of cultivation provided by local soils, as well as the constraints of topography. The result was that many convents ruled over mosaic landscapes, with access to variable combinations of cultivated land, woodland, and wetland. As a general tendency, intensification in cultivation was accompanied by deforestation which targeted commercially valuable hardwoods, although substantial expanses of woodland remained in all the commanderies surveyed in this study. Smaller districts, run by advocates, procurators, and forest masters, tended to be more specialized, particularly in the eastern Prussian frontier.

The term 'event horizon' has been convincingly used by James Barrett (et al. 2004) to describe the figurative moment, in time, when the beginnings of commercial marine fishing in north-western Europe became visible in the archaeological and historical records, dated to *c.* AD 1000. When contextualized within the four main periods of 'social time' (see Chapter 1), the crusades in the eastern Baltic were clearly a force of major cultural change, but from an ecological perspective their impact was temporally and spatially variable. This does not amount to the thirteenth century as a clear and unilineal 'ecological event horizon' in the history of north-eastern Europe. The variable levels of impact across the Order's territories in Livonia and Prussia clearly reflect a host of interrelated factors, but the relationship with subjects working in the fields, pastures and woodlands would have been pivotal. This exemplifies the adaptation of the conquering theocracy to the eastern Baltic, a complex and at times contradictory interaction between ideology and pragmatism. What remains less recognized, perhaps because of its enduring perception as a periphery, is how the eastern Baltic is situated within the general trends of the 'Great Transition' of late medieval Europe (Campbell 2016). This must await future pan-European/Eurasian syntheses, but it is hope this study has drawn attention to the socio-ecological resilience of societies in the eastern Baltic.

Finally, this study is situated within the recent milieu of international scholarship that has stressed the cultural importance of crusading in the Baltic region, challenging economically deterministic models of cultural transformation in the context of Europeanization. The most recent synthesis on the Hanse has argued for the need to reconsider traditional German-centred historiographic perspectives, which had neatly subdivided the historical trajectories of lands to the east and west of the Elbe. This study has contributed to breaking up the historical singularity of Prussia and Livonia, countering traditional scholarly regionalism in the Baltic (Jahnke 2015, 206). In the process of comparative analysis, it has contributed to a critique of these polities as political, social, and cultural singularities, and highlighted the benefits of situating local case studies within an interregional framework. This critique has been most pronounced in studies of the political history of the medieval eastern Baltic, which have demonstrated spatial and temporal variability driven by a host of local, regional, and international factors (e.g., Kala 2007). Historians of the military orders have also been concerned with the international organization of the Templars and Hospitallers, and how this varied between provinces and at different times, in contrast to archaeologists who have consistently focused on individual sites or discrete regions (Pluskowski 2018). The rejection of historical singularities also applies to the pervasive concept of the eastern Baltic as a frontier. Both postcolonial theory and the revised world-systems model highlight how frontiers created new categories of people: archetypal frontier authorities, merchants and colonists (Naum 2010, 107, 125), and this is a useful way of understanding the crusading period in the eastern Baltic. However, the locations of frontiers shifted during and after the crusading period and their ecological profiles were not consistent; compare for example, the productivity of the commandery of Nessau with the procuratorship of Lötzen or the eastern segments of the commanderies of Balga and Brandenburg by the early fifteenth century. Considering the entirety of the eastern Baltic as a frontier is useful for understanding its externally perceived situation within the context of Latin Christendom, but not when attempting to explain cultural and ecological heterogeneity across the region. The most important contribution of this study is therefore to promote the importance of multi-scalar research in furthering our understanding of supra-regional phenomena, using the environment as a multidisciplinary lens on cultural transformation.

Works Cited

Primary Sources

Bauer, A. 1924. *Die Wartgutsteuresliste der Komturei Goldingen: Mitteilungen aus der livländischen Geschichte*, Rīga: Gesellschaft für Geschichte und Altertumskunde zu Riga

Bauer, A. 2003. Transcripts from the Sources of the History of Manors in Courland collected by Albert Bauedr, digitized by the Herder Institute in Marburg (the date refers to the completion of the digitization project)

LRC Smith, J. C. and Urban, W., trans. 2006. *The Livonian Rhymed Chronicle*, London: Routledge Curzon

LUB Bunge, F. G. von, ed. 1853–57. *Liv-, Est- und Kurländisches Urkundenbuch nebst regesten* (vols i–iii), 13 vols to date, Reval: Kluge and Ströhm

Trupinda, J., ed. 2002. *Reguła Zakonu Szpitala Najświętszej Marii Panny Domu Niemieckiego w Jerozolimie*, Malbork: Muzeum Zamkowe w Malborku

Secondary Works

Apals, J. 1995. 'Āraišu mūra pils arheoloģiskā izpēte', *Cēsu novada vēsture* 1, 59–76

Arszyński, M. 2010. 'Architektura warowna Zakonu Krzyżackiego w Prusach', in *Fundacje artystyczne na terenie państwa krzyżackiego w prusach*, ed. B. Pospieszna, Pelplin: Wydawnictwo Bernardinum, 7–45

——, 2013. 'Budownictwo', in *Zakon krzyżacki w Prusach i Inflantach: podziały administracyjne i kościelne w xiii–xvi wieku*, eds R. Czaja and A. Radzimiński, Toruń: Wydawnictwo Naukowe Uniwersytetu Mikołaja Kopernika, 545–73

Barber, M. 2012. *The Crusader States*, New Haven: Yale University Press

Barrett, J. H., Locker, A. M., and Roberts, C. M. 2004. 'The Origin of Intensive Marine Fishing in Medieval Europe: The English Evidence', *Proceedings of the Royal Society* B 271, 2417–21

Bertašius, M. 2006. 'From Social Expression to Institutionalized Religion. Making an Attempt of central Lithuania', in *Transformatio Mundi: The Transition from the Late Migration Period to the Early Viking Age in the East Baltic*, ed. M. Bertašius, Kaunas: Kaunas University of Technology, 65–71

Bertašius, M., and Daugnora, L. 2001. 'Viking Age Horse Graves from Kaunas Region (Middle Lithuania)', *International Journal of Osteoarchaeology* 11, 387–99

Biskup, M. 2009. 'Polityka zewnętrzna zakonu krzyżackiego', in *Państwo zakonu krzyżackiego w Prusach: Władza i społeczeństwo*, ed. M. Biskup et al., Warsaw: Państwowe Wydawnictwo Naukowe, 233–331

Bliujienė, A. 2003. 'Lithuanian Amber Artifacts from the Roman Iron Age to Early Medieval times', in *Amber in Archaeology*, eds C. W. Beck, I. B. Loze, and J. M. Todd, Rīga: Institute of the History of Latvia Publishers, 47–71

Blomkvist, N. 2004. 'The Medieval Catholic World-System and the Making of Europe', in *The European Frontier: Clashes and Compromises in the Middle Ages*, ed. J. Staecker, Lund: Almqvist and Wiksell, 15–34

——, 2005. *The Discovery of the Baltic: The Reception of a Catholic World-System in the European North (ad 1075–1225)*, Leiden: Brill

Brūzis, R., and Spirģis, R. 2009. 'The Ogresgala Čabas Cemetery and Horse Sacrifice', *Archaeologia Baltica* 11, 283–94

Bull, M. 2003. 'Views of Muslims and of Jerusalem in Miracle Stories, *c.* 1000–*c.* 1200: Reflections on the Study of First Crusaders' Motivations', in *The Experience of Crusading* (vol. I), eds M. Bull and N. Housley, Cambridge: Camridge University Press, 13–38

Campbell, B. M. S. 2016. *The Great Transition: Climate, Disease and Society in the Late-Medieval World*, Cambridge: Cambridge University Press

Caune, A. 1992. 'Arheoloģiskās liecības par senāko apdzīvotību Rīgas Doma baznīcas apkārtnē', *Latvijas Vēstures Institūta Žurnāls* 3, 21–38

——, 2007. *Pētījumi Rīgas archeoloģijā*, Rīga: LVIA

Chudziak, W. 2003. *Wczesnośredniowieczna przestrzeń sakralna in Culmine na Pomorzu Nadwiślańskim*, Toruń: Wydawnictwo Naukowe Uniwersytetu Mikołaja Kopernika

Cios, S. 2009. 'What Do We Know about the Extinction of Sturgeon in Poland?' in *Fishes–Culture–Environment through Archaeoichthyology, Ethnography and History*, eds D. Makowiecki, S. Hamilton Dyer, I. Riddler, and N. Trzaska, Poznań: BWN, 72–80

Cloke, P., and Little, J. 1997. 'Introduction: Other Countrysides?' in *Contested Countryside Cultures: Otherness, Marginalisation and Rurality*, eds P. Cloke and J. Little, London: Routledge, 1–18

Constable, G. 2008. *Crusaders and Crusades in the Twelfth Century*, Farnham: Ashgate

Czaja, R. 2005. 'Patrician Guilds in Medieval Towns on the Baltic Coast', *Acta Poloniae Historica* 92, 31–51

Czaja, R., and Nowak, Z. H. 2013. 'Państwo zakonu krzyżackiego w Prusach — próba charakterystyki', in *Zakon krzyżacki w Prusach i Inflantach: podziały administracyjne i kościelne w XIII–XVI wieku*, eds R. Czaja and A. Radzimiński, Toruń: Wydawnictwo Naukowe Uniwersytetu Mikołaja Kopernika, 1–27

Dygo, M. 1989. 'The Political Role of the Cult of the Virgin Mary in Teutonic Prussia in the Fourteenth and Fifteenth Centuries', *Journal of Medieval History* 15, 63–80

Eihmane, E. 2009. 'The Baltic Crusades: A Clash of Two Identities', in *The Clash of Cultures on the Medieval Baltic Frontier*, ed. A. V. Murray, Farnham: Ashgate, 37–51

——, 2012. 'Livonia and the Holy See in 13th–Early 15th Centuries: Centre and Periphery of Christendom', in *The Image of the Baltic: A Festschrift for Nils Blomkvist*, eds M. F. Scholz, R. Bohn, and C. Johansson, Visby: Gotland University Press, 23–40

Ekdahl, S. 1998. 'Horses and Crossbows: Two Important Warfare Advantages of the Teutonic Order in Prussia', in *The Military Orders*, II: *Welfare and Warfare*, ed. H. Nicholson, Farnham: Ashgate, 119–51.

Ervynck, A. 1992. 'Medieval Castles as Top-Predators of the Feudal System: An Archaeozoological Approach', *Château Gaillard* 15, 152–59

——, 2004 '*Orant, pugnant, laborant*: The Diet of the Three Orders in the Feudal Society of Medieval North-Western Europe', in *Behaviour behind Bones: The Zooarchaeology of Ritual, Religion, Status and Identity*, eds S. J. O' Day, W. Van Neer, and A. Ervynck, Oxford: Oxbow, 215–23

Gaimster, D. 2005. 'A Parallel History: The Archaeology of Hanseatic Urban Culture in the Baltic *c.* 1200–1600', *World Archaeology* 37/3, 408–23

——, 2007. 'The Baltic Ceramic Market 1200–1600: Measuring Hanseatic Cultural Transfer and Resistance', in *Cultural Exchange in Early Modern Europe*, IV: *Forging European Identities, 1400–1700*, ed. H. Roodenburg, Cambridge: Cambridge University Press, 30–58

Gancewski, J. 2001. *Rola zamków krzyżackich w ziemi chełmińskiej od połowy XIV wieku do 1454 roku*, Olsztyn: Ośrodek Badań Naukowych im. Wojciecha Kętrzyńskiego

Haftka, M. 1999. *Zamki krzyżackie w Polsce: Szkice z dziejów*, Malbork: Wydawnictwo CONSORT i Muzeum Zamkowe w Malborku

Hagen, K. G. 1824. 'Geschichte der Verwaltung des Börnsteins in Preussen: Von dem Herrn Medizinal- und Regierungsrath Hagen. Erster Abschnitt. Von der Zeit des Ordens bis zur Regierung König Friedrich I', *Beiträge zur Kunde Preussens* 6/1, 1–41

Hermanowski, G. 1996. *Ostpreussen Wegweiser durch ein Unvergessenes Land*, Augsburg: Flechsig

Hoffman, R. 2014. *An Environmental History of Medieval Europe*, Cambridge: Cambridge University Press

Jäger, D. 1981. 'The International Background – Late Medieval Agrarian Crisis and Deserted Settlements in Central Europe', in *Danish Medieval History: New Currents*, eds N. Skyum-Nielsen and N. Lund, Copenhagen: Museum Tuscunalum Press, 223–37

Jäger, E. 2008. *Entlang der Weichsel und der Memel: historische Landkarten und Stadtansichten von Ost- und Westpreuben, Polen und dem Baltikum. Eine Ausstellung des Kulturzentrums Ostpreuben in Ellingen/Bay = Z biegiem Wisły i Niemna: historyczne mapy i widoki miast Prus Wschodnich i Zachodnich, Polski oraz krajów bałtyckich: wystawa Centrum Kultury Prus Wschodnich w Ellingen/Bawaria*, Ellingen/Bay: Kulturzentrum Ostpreuben.

Jahnke, C. 2015. 'The Baltic Trade', in *A Companion to the Hanseatic League*, ed. D. J. Harreld, Leiden: Brill, 194–240

Jaskanis, J. 1966. 'Human Burials with Horses in Prussia and Sudovia in the First Millennium of our Era', *Acta Baltico-Slavica* 4, 29–65

Jensen, K. V. 2001. 'Denmark and the Second Crusade: The Formation of a Crusader State?', in *The Second Crusade: Scope and Consequences*, eds J. Phillips and M. Hoch, Manchester: Manchester University Press, 164–79

——, 2009. 'Sacrilization of the Landscape: Converting Tress and Measuring Land in the Danish Crusades against the Wends', in *The Clash of Cultures on the Medieval Baltic Frontier*, ed. A. V. Murray, Farnham: Ashgate, 141–50

Jóźwiak, S., and Trupinda, J. 2012. *Krzyżackie zamki komturskie w Prusach*, Toruń: Wydawnictwo Naukowe Uniwersytetu Mikołaja Kopernika

Kala, T. 2001. 'The Incorporation of the Northern Baltic Lands into the Western Christian World', in *Crusade and Conversion on the Baltic Frontier, 1150–1500*, ed. A. V. Murray, Aldershot: Ashgate, 3–20

——, 2006. 'Languages in a Medieval North European City: An Example from Medieval Tallinn', in *Frontiers in the Middle Ages*, eds O. Merisalo and P. Pahta, Louvain-la-Neuve: Fédération Internationale des Institutes d'Études Médiévales, 585–603

——, 2007. 'Northern Estonian Church Life in the Thirteenth and Fourteenth Century: The Religious Environment in which the Pirita Monastery Emerged', *Studies on Art and Architecture (Kunstiteaduslikke Uurimusi)* 4, 57–59

——, 2009. 'Rural Society and Religious Innovation: Acceptance and Rejection of Catholicism among the Native Inhabitants of medieval Livonia', in *The Clash of Cultures on the Medieval Baltic Frontier*, ed. A. V. Murray, Farnham: Ashgate, 168–90

Kawiński, P., and Szczepański, S. 2016. *Szkice o religii Prusów*, Olsztyn: Towarzystwo Naukowe Pruthenia

King, R. 2013. 'Rethinking the Oldest Surviving Amber in the West', *The Burlington Magazine* (Sculpture) 1328 /155, 756–62

Kļaviņš, K. 2009. 'The Significance of the Local Baltic People in the Defence of Livonia (Late Thirteenth–Sixteenth Century)', in *The Clash of Cultures on the Medieval Baltic Frontier*, ed. A. V. Murray, Farnham: Ashgate, 321–40

Kreem, J. 2002. *The Town and its Lord: Reval and the Teutonic Order (in the Fifteenth Century)*, Tallinn: Tallinna Linnaarhiiv

Kutzner, M. 1995. 'Propaganda władzy w sztuce Zakonu Niemieckiego w Prusach', in *Sztuka w kręgu zakonu krzyżackiego w Prusach i Inflantach*, ed. M. Woźniak, Toruń: Uniwersytet Mikołaja Kopernika, 32–85

Kwiatkowski, K. 2013. '*Christ ist erstanden*... and Christians win! Liturgy and the Sacralization of Armed Fight against Pagans as Determinants of the Identity of the Members of the Teutonic Order in Prussia', in *Sacred Space in the State of the Teutonic Order in Prussia*, ed. J. Wenta, Toruń: Wydawnictwo Naukowe Uniwersytetu Mikołaja Kopernika, 101–30

Laszlovszky, J., and Soós, Z. 2001. 'Historical Monuments of the Teutonic Order in Transylvania', in *The Crusades and the Military Orders: Expanding the Frontiers of Medieval Latin Christianity*, eds Z. Hunyadi and J. Laszlovszky, Budapest: Central European University, Department of Medieval Studies, 319–36

Luik, H. 2015. 'Bone-Workers in Medieval Viljandi, Estonia: Comparison of Finds from Downtown and the Order's Castle', in *Everyday Products in the Middle Ages: Crafts, Consumption and the Individual in Northern Europe, c. AD 800–1600*, eds G. Hansen, S. P. Ashby, and I. Baug, Oxford: Oxbow, 91–109

Mager, F. 1960. *Der Wald in Altpreussen als Wirtschaftsraum*, Köln: Böhlau

Makowiecki, D. 2003. *Historia ryb i rybołówstwa w holocenie na Niżu Polskim w świetle badań archeoichtiologicznych*, Poznań: IAE PAN.

——, 2008. 'Sturgeon Fishing on Polish Lowland during Holocene', in *Archéologie du poisson: 30 ans d'archéo-ichtyologie au CNRS. Hommage aux travaux de Jean Desse et Nathalie Desse-Berset*, eds P. Béarez, S. Grouard, and B. Clavel, Antibes: Éditions APDCA, 327–39

Mänd, A. 2004. 'Beaver Tails and Roasted Herring Heads: Fast as Feast in Late-Medieval Livonia', *Medium Aevum Quotidianum* 50, 5–12

Mažeika, R. 1994. 'Of Cabbages and Knights: Trade and Trade Treaties with the Infidel on the Northern Frontier, 1200–1390', *Journal of Medieval History* 20/1, 63–76

Mierzwiński, M. 2010. 'Prace ratownicze i konserwatorskie w zachodnim skrzydle zamku średniego w Malborku w latach 1990–2010', in *Wielki refektarz na zamku średnim w Malborku: dzieje-wystrój-konserwacja*, ed. J. Trupinda, Malbork: Muzeum Zamkowe w Malborku, 169–214

Militzer, K. 1997. 'Jagd und Deutscher Orden', in W. Rösener, *Jagd und höfische Kultur in Mittelalter*, Göttingen: Vandenhoeck & Ruprecht, 325–63

Mortensen, L. B., and Bisgaard, L. 2013. 'Introduction: Late Medieval Urban Civilization and its North European Variant', in *Guilds, Towns and Cultural Transmission in the North, 1300–1500*, eds L. Bisgaard, L. B. Mortensen and T. Pettitt, Odense: University Press of Southern Denmark, 7–30

Morton, N. E. 2009. *The Teutonic Knights in the Holy Land: 1190–1291*, Woodbridge: Boydell

Mugurēvičs, Ē. 1990. 'Interactions between Indigenous and Western Culture in Livonia in the 13th to 16th Centuries', in *From the Baltic to the Black Sea*, eds D. Austin and L. Alcock, London: Routledge, 168–78

——, 2002. 'Forest Animals and Hunting in Medieval Livonia', in *Core-Centre-Periphery: Medieval Europe 2*, eds G. Helmig, B. Scholkmann, and M. Untermann, Basel: Folio Verlag Dr G. Wesselkamp, Archäologische Bodenforschung Basel-Stadt, 177–81

Myśliwski, G. 2002. 'Boundaries and Men in Poland from the Twelfth to the Sixteenth Century: The Case of Masovia', in *Medieval Frontiers: Concepts and Practices*, eds D. Abulafia and N. Berend, Cambridge: University of Cambridge Press, 217–37

Naum, M. 2010. 'Re-emerging Frontiers: Postcolonial Theory and Historical Archaeology of the Borderlands', *Journal of Archaeological Method and Theory* 17, 101–31

O'Connor, T. 2014. 'Livestock and Animal Husbandry in Early Medieval England', *Quaternary International* 346/30, 109–18

Palmer, J. 2007. 'Defining Paganism in the Carolingian World', *Early Medieval Europe* 15/4, 402–25

Piskorski, J. M. 2004. 'The Medieval Colonization of Central Europe as a Problem of World History and Historiography', *German History* 22/3, 323–43

Pluskowski, A. G. 2010. 'The Zooarchaeology of Medieval "Christendom": Ideology, the Treatment of Animals and the Making of Medieval Europe', *World Archaeology* 42/2, 201–14

——, 2012. *The Archaeology of the Prussian Crusade: Holy War and Colonisation*, London: Routledge

——, 2018. 'The Archaeology of the Military Orders: The Material Culture of Holy War', *Medieval Archaeology* 62.1, 105–34

Pluskowski, A. G., Brown, A. D., Banerjea, R., Makowiecki, D., Seetah, K., Rannamäe, E., Badura, M., Jarzebowski, M., Kreem, J., and Kļaviņš, K., 2019. 'From the Convent to the Commandery: The Pivotal Role of the Environment in Defining the Medieval Baltic *Ordensland*', *Quellen und Studien zur Geschichte des Deutschen Ordens* (in press)

Pompeo, L. 2000. 'Etnografia umanistica ne L'epistola sulla religione ed i sacrifici degli antichi prussiani di Jan Sandecki Malecki (Meletius)', *Studia Mythologica Slavica* 3, 63–74

Poole, K. 2013. 'Horses for Courses? Religious Change and Dietary Shifts in Anglo-Saxon England', *Oxford Journal of Archaeology* 32/3, 319–33

Pospieszny, K. 1998. 'Program i forma architektoniczna krzyżackich infirmerii zamkowych w Prusach', in *Szpitalnictwo w dawnej Polsce*, eds M. Dąbrowska and J. Kruppé, Warsaw: Wydawnictwo PAN, 109–33

——, 2010. 'I.2.13. Główny zwornik z "madonną tronującą" z kościoła zamkowego w Malborku, ok. 1340', in *Fundacje artystyczne na terenie Państwa Krzyżackiego w Prusach*, ed. B. Pospieszna, Pelplin: Wydawnictwo Bernardinum, 35–36

Provost-Bien, L. 2013. 'L'envers de la conversion et de la croisade: la réponse païenne au phénomène de conversion religieuse comme justification de la radicalisation', *History in the Making Review* 2/1, 26–40

Salminen, T. 2004. 'Rescripta gratis? Professional Writing in the Administration of Late Medieval Merchant Cities in the Baltic Sea Area', in *Der Ostseeraum und Kontinentaleuropa 1100–1600: Einflussnahme, Rezeption, Wandel*, eds D. Kattinger, J. E. Olesen, and H. Wernicke, Schwerin: Helms, 145–150

Scull, C., Müldner, G., and Makowiecki, D. 2019. 'Carbon and Nitrogen Stable Isotope Evidence for Animal Husbandry and Environmental Change in the Medieval Kulmerland', in *Ecologies of Crusading, Colonization, and Religious Conversion in the Medieval Baltic: Terra Sacra II*, ed. A. Pluskowski, Environmental History of the North Atlantic World, 3, Turnhout: Brepols (forthcoming)

Selart, A. 2011. 'The Use and Uselessness of the Chronicle of Henry of Livonia in the Middle Ages', in *Crusading and Chronicle Writing on the Medieval Baltic Frontier*, eds M. Tamm, L. Kaljundi, and C. S. Jensen, Farnham: Ashgate, 345–63

——, 2015. *Livonia, Rus' and the Baltic Crusades in the Thirteenth Century*, Leiden: Brill

Sembritzki, J, 1926. *Geschichte der Königlich Preussischen See- und Handelsstadt Memel*, Memel: Siebert

Shiroukhov, R. 2012. 'Prussian Graves in the Sambian Peninsula with Imports, Weapons and Horse Harnesses, from the Tenth to the Thirteenth Century: The Question of the Warrior Elite', *Archaeologia Baltica* 18, 224–55

——, 2015. 'Contacts between Prussians and Curonians in the 11th–Early 13th Centuries, According to the Archaeological Data', in *Materiały do Archeologii Warmii i Mazur*, 1, eds M. Hoffmann, M. Karczewski, and S. Wadyl, Warszawa-Białystok: Instytut Archeologii Uniwersytetu Warszawskiego i Instytut Historii i Nauk Politycznych Uniwersytetu w Białymstoku, 255–73

Svetikas, E. 2008. 'XIV a. Pabaigos-XV a. amuletai iš apkaustyto okio nago lietuvos didžiojoje kunigaikštystėje ir kaimyniniuose kraštuose', *Lietuvos Archeologija* 34, 171–210

Szczepański, S. 2011. 'Sakralizacja obszaru pogranicza na przykładzie Pomezanii pruskiej', *Pruthenia* 6, 129–56

——, 2019. 'Surkapurn - Kreken: The Phenomenon of a Name - the phenomenon of a Place. The Organisation of Sacred Space in Prussia using Archaeology and Other Sources', in *Ecologies of Crusading, Colonization, and Religious Conversion in the Medieval Baltic: Terra Sacra II*, ed. A. Pluskowski, Environmental History of the North Atlantic World, 3, Turnhout: Brepols (forthcoming)

Szostek, K., Kozłowska, A., and Kralisz, A. 2009. 'Pochówki osobników z krypty północnej w prezbiterium katedry w Kwidzynie w świetle paleoserologicznych i fizykochemicznych analiz materiału kostnego', in *Katedra w Kwidzynie — tajemnica krypt*, eds M. Grupa and T. Kozłowski, Kwidzyn: Kwidzyńskie Centrum Kultury, 87–114

Tamm, M. 2009. 'A New World into Old Words: The Eastern Baltic Region and the Cultural Geography of Medieval Europe', in *The Clash of Cultures on the Medieval Baltic Frontier*, ed. A. V. Murray, Farnham: Ashgate, 11–36.

——, 2011. 'Inventing Livonia: The Name and Fame of a New Christian Colony on the Medieval Baltic Frontier', *Zeitschrift für Ostforschung* 60, 186–209

——, 2013. 'How to Justify a Crusade? The Conquest of Livonia and New Crusade Rhetoric in the Early Thirteenth Century', *Journal of Medieval History* 39/4, 431–55

Trupinda, J. 1999. *Ideologia krucjatowa w kronice Piotra z Dusburga*, Gdańsk: Officina Ferberiana

Urban, W. 2001. 'The Frontier Thesis and the Baltic Crusade', in *Crusade and Conversion on the Baltic Frontier 1150–1500*, ed. A. V. Murray, Aldershot: Ashgate, 45–71

Vairo, G. R. 2014. 'Originality and Adaptation: The Architecture of the Teutonic Order in Italy', in *Archaeology and Architecture of the Military Orders*, eds M. Piana and C. Carlsson, Aldershot: Ashgate, 193–218

Vaitkevičius, V. 2004. *Studies into the Balts' Sacred Places*, BAR international Series, 1228, Oxford: Archaeopress

Valk, H. 2004. 'Artefacts in Estonian Urban Churchyards: Reflections on Different Traditions and Ideologies', in *Der Ostseeraum und Kontinentaleuropa 1100–1600*, eds D. Kattinger, J. E. Olesen, and H. Wernicke, Schwerin: Helms, 103–12

——, 2019. 'Sacred Landscapes of Medieval Livonia: Lands of the Teutonic Order in Southern Estonia', in *Ecologies of Crusading, Colonization, and Religious Conversion in the Medieval Baltic: Terra Sacra II*, ed. by A. Pluskowski, Environmental History of the North Atlantic World, 3, Turnhout: Brepols (forthcoming)

Wienberg, J. 2004. 'Medieval Gotland: Churches, Chronologies and Crusades', in *The European Frontier: Clashes and Compromises in the Middle Ages*, ed. J. Staecker, Lund: Almqvist and Wiksell, 285–98

Willoweit, G. 1969a. 'Der Bernstein im Memeland', *Memelar Dampfboot* 20, 275–76

——, 1969b. *Die Wirtschaftsgeschichte des Memellandes*, I: *Wissenschaftliche Beiträge zur Geschichte und Landeskunde Ost-Mitteleuropas; Auftrage des Johann Gotfried Herder-Instituts, 85/1*. Marburg: Lahn

Žulkus, V. 2002. 'The Balts: Economy And Society', in *The Neighbours of Poland in the 11th Century*, ed. P. Urbańczyk, Warsaw: Institute of Archaeology and Ethnology, Polish Academy of Sciences, 169–210

Žulkus, V., and Daugnora, L. 2010. 'What did the Order's Brothers eat in the Klaipėda Castle? (The Historical and Zooarchaeological Data)', *Archaeologica Baltica* 9, 74–87

Appendices

A. Accelerated Mass Spectrometry (AMS) Radiocarbon Dates from Poland, Latvia, and Estonia

Alexander Brown

Table A.1: AMS radiocarbon dates from Poland. **A** — Lake*, Peat Bog**, Occupation Deposits***.
B — Lab numbers are GU except Wk +. **C** — All calibrated date ranges expressed at 2σ 95.4% probability.

Site A	Lab No B	Material Dated	Depth (cm)	Age (BP)	δ13C (‰)	Age Range (cal. AD / BC) C
Rehden Commandery and Kulmerland Sites						
Radzyń[1] *	27673	Gyttja bulk	34–35	595±35	-27.3	AD 1296–1412
Radzyń[1]	26408	Gyttja bulk	37–38	625±30	-26.7	AD 1290–1399
Radzyń[1]	24506	Gyttja bulk	41–42	690±30	-26.9	AD 1260–1320 (67.6%) AD 1350–90 (27.8%)
Radzyń[1]	27674	Gyttja bulk	44–45	765±35	-27.1	AD 1212–88
Radzyń[1]	26407	Gyttja bulk	47–48	805±30	-26.2	AD 1178–1274
Radzyń[1]	26406	Gyttja bulk	51–52	890±30	-27.6	AD 1041–1109 (35.2%) AD 1116–17 (60.2%)
Radzyń[1]	26405	Gyttja bulk	55–56	925±30	-27.6	AD 1026–1180
Radzyń[1]	26404	Gyttja bulk	61–62	1000±30	-28.2	AD 982–1052 (70.1%) AD 1081–1129 (19.2%) AD 1134–53 (6.1%)
Radzyń[1]	24507	Gyttja bulk	70–71	1035±30	-26.8	AD 890–920 (5.6%) AD 940–1040 (89.8%)
Radzyń[1]	26409	Gyttja bulk	79–80	1060±30	-26.5	AD 896–924 (16.2%) AD 939–1024 (79.2%)
Radzyń[1]	24508	Gyttja bulk	95–96	1505±30	-27.6	AD 430–90 (11%) AD 530–640 (84.8%)
Radzyń[2] *	31523	Gyttja bulk	29–30	485±34	-27.8	AD 1398–1459
Radzyń[2]	31524	Gyttja bulk	35–36	946±34	-27	AD 1020–1164
Radzyń[2]	32578	Gyttja bulk	35–36	906±29	-27.1	AD 1036–1206
Radzyń[2]	31525	Gyttja bulk	42–43	966±34	-28.8	AD 1015–1160
Radzyń[2]	32579	Gyttja bulk	42–43	1007±26	-27.1	AD 983–1046 (86.1%) AD 1093–1121 (7.9%) AD 1141–48 (1.4%)
Radzyń[2]	31526	Gyttja bulk	52–53	867±34	-27.6	AD 1045–95 (17.9%) AD 1120–42 (6.1%) AD 1147–1256 (71.4%)
Radzyń[2]	31527	Gyttja bulk	60–61	1105±34	-27.9	AD 875–1019
Radzyń[2]	31528	Gyttja bulk	70–71	1587±34	-28.4	AD 406–551
Radzyń[2]	31529	Gyttja bulk	88–89	1850±34	-27.4	AD 80–239
Radzyń[3]	32609	Gyttja bulk	20–21	97±29	-27.4	AD 1682–1735 (27.1%) AD 1806–1930 (68.3%)
Radzyń[3]	32610	Gyttja bulk	30–31	499±29	-26.8	AD 1399–1449
Radzyń[3]	32611	Gyttja bulk	40–41	583±29	-27.1	AD 1300–69 (65.3%) AD 1381–1416 (30.1%)

Table A.1: AMS radiocarbon dates from Poland. **A** — Lake*, Peat Bog**, Occupation Deposits***. **B** — Lab numbers are GU except Wk +. **C** — All calibrated date ranges expressed at 2σ 95.4% probability. *(cont.)*

Site A	Lab No B	Material Dated	Depth (cm)	Age (BP)	δ13C (‰)	Age Range (cal. AD / BC) C
Radzyń[3]	32612	Gyttja bulk	50–51	609±29	-28.1	AD 1296–1404
Radzyń[3]	32613	Gyttja bulk	60–61	674±29	-27.3	AD 1272–1318 (56.1%) AD 1353–90 (39.3%)
Radzyń[3]	32614	Gyttja bulk	70–71	941±29	-27.2	AD 1026–1158
Radzyń[3]	32615	Gyttja bulk	75–76	1026±29	-29	AD 904–18 (1.6%) AD 966–1043 (92.6%) AD 1106–18 (1.2%)
Radzyń[3]	32616	Gyttja bulk	80–81	1038±29	-27.4	AD 901–21 (4.2%) AD 954–1034 (91.2%)
Radzyń[3]	32617	Gyttja bulk	90–91	1034±29	-26.1	AD 901–21 (3.3%) AD 962–1037 (92.1)
Radzyń[3]	32618	Gyttja bulk	100–01	1245±29	-25.7	AD 680–780 (69.2%) AD 788–874 (26.2%)
Radzyń[3]	32619	Gyttja bulk	110–11	1302±27	-26	AD 660–727 (65%) AD 737–69 (30.4%)
Radzyń[3]	32620	Gyttja bulk	120–21	1482±29	-28.5	AD 538–644
Rywałd *	32596	Gyttja bulk	34–35	147±29	-28.2	AD 1667–1710 (16.3%) AD 1717–83 (30.4%) AD 1796–1890 (31.9%) AD 1910–49 (16.7%)
Rywałd	32597	Gyttja bulk	46–47	498±29	-26.8	AD 1400–49
Rywałd	32598	Gyttja bulk	58–59	467±29	-26.9	AD 1410–60
Rywałd	32599	Gyttja bulk	68–69	571±29	26.2	AD 1305–65 (58.5%) AD 1384–1421 (36.9%)
Rywałd	32600	Gyttja bulk	75–76	798±27	-23.3	AD 1190–1275
Rywałd	32601	Gyttja bulk	80–81	785±29	-23.6	AD 1206–80
Rywałd	32602	Gyttja bulk	88–89	1005±29	-22.6	AD 980–1050 (78.7%) AD 1085–1125 (13.3%) AD 1137–50 (3.3%)
Rywałd	32603	Gyttja bulk	100–01	1200±29	-20	AD 716–44 (5.7%) AD 766–895 (88.6%) AD 929–40 (1.2%)
Rywałd	32604	Gyttja bulk	115–16	1372±**29**	-21	AD 610–85
Rywałd	32605	Gyttja bulk	145–46	1830±29	-28	AD 87–108 (3.1%) AD 120–252 (92.3%)
Grudziądz ***	25760	Animal bone	G1–73/09	235±30	-20.7	AD 1528–44 (2.3%) AD 1633–83 (48.3%) AD 1737–56 (2.6%) AD 1761–1804 (32.6%) AD 1936–55 (8.6%)
Grudziądz	25761	Animal bone	G2–74/09	270±25	-21.3	AD 1521–77 (35.3%) AD 1582–91 (1.7%) AD 1622–68 (53.7%) AD 1782–97 (4.7%)
Grudziądz	25762	Animal bone	G3–71/09	225±30	-21.6	AD 1640–84 (40.4%) AD 1735–1806 (42.1%) AD 1931–55 (13.0%)

Table A.1: AMS radiocarbon dates from Poland. **A** — Lake*, Peat Bog**, Occupation Deposits***.
B — Lab numbers are GU except Wk +. **C** — All calibrated date ranges expressed at 2σ 95.4% probability. *(cont.)*

Site A	Lab No B	Material Dated	Depth (cm)	Age (BP)	δ13C (‰)	Age Range (cal. AD / BC) C
Marienburg Commandery						
Malbork Castle Moat ***	23112	Bark	71.5–72	430±35	-25.4	AD 1410–1520 (88.2%) AD 1590–1620 (7.2%)
Malbork high castle ***	30189	Mammal bone fragment	73	690±29	-22.5	AD 1267–1311 (68.8%) AD 1359–88 (26.6%)
Parpary **	29810	Peat bulk	4–4.5	Modern fraction	-29.4	Modern fraction
Parpary	29802	Peat bulk	14–14.5	33±33	-29.6	AD 1694–1728 (21.2%) AD 1812–1919 (74.2%)
Parpary	31206	Peat bulk	17–17.5	407±29	-29.6	AD 1433–1522 (82.5%) AD 1578–81 (0.3%) AD 1591–1620 (12.5%)
Parpary	32584	Peat bulk	18–18.5	162±26	-28.9	AD 1664–99 (16.5%) AD 1722–1818 (50.5%) AD 1833–79 (9.1%) AD 1916– (19.3%)
Parpary	31207	Peat bulk	20–20.5	160±27	-28.9	AD 1665–1706 (16.7%) AD 1720–87 (38.2%) AD 1792–1819 (10.8%) AD 1832–81 (11.1%) AD 1915–54 (18.5%)
Parpary	29803	Peat bulk	22–22.5	454±33	-28.7	AD 1410–86
Parpary	31208	Peat bulk	24–24.5	561±27	-28.7	AD 1310–60 (49.7%) AD 1386–1426 (45.7%)
Parpary	31209	Peat bulk	26–26.5	546±26	-28.5	AD 1317–53 (32.6%) AD 1389–1433 (62.8%)
Parpary	31210	Peat bulk	28–28.5	690±29	-28.3	AD 1267–1311 (68.8%) AD 1359–88 (26.6%)
Parpary	29804	Peat bulk	30–30.5	658±33	-27.9	AD 1276–1326 (46.7%) AD 1343–95 (48.7)
Parpary	31211	Peat bulk	32–32.5	657±30	-28	AD 1278–1324 (46.2%) AD 1346–94 (49.2%)
Parpary	31212	Peat bulk	34–34.5	929±29	-28.2	AD 1025–1170
Parpary	32586	Peat bulk	42–43	1221±27	-27.2	AD 695–702 (1.6%) AD 708–46 (17.7%) AD 764–885 (76.1)
Parpary	32587	Peat bulk	50–51	2059±27	-27	166 BC–AD 2
Benowo **	29823	Peat bulk	20–20.5	28±33	-28.1	AD 1694–1728 (20.7%) AD 1812–1919 (74.7%)
Benowo	32581	Peat bulk	25–26	255±29	-26.7	AD 1521–92 (20.2%) AD 1620–75 (56.6%) AD 1777–1800 (16.2%) AD 1941– (2.4%)
Benowo	31199	Peat bulk	28–29	418±29	-28.2	AD 1428–1515 (88.2%) AD 1600–18 (7.2%)
Benowo	31200	Peat bulk	33–34	375±29	-28.2	AD 1446–1526 (58.9%) AD 1556–1633 (36.5%)

Table A.1: AMS radiocarbon dates from Poland. **A** — Lake*, Peat Bog**, Occupation Deposits***.
B — Lab numbers are GU except Wk +. **C** — All calibrated date ranges expressed at 2σ 95.4% probability. *(cont.)*

Site A	Lab No B	Material Dated	Depth (cm)	Age (BP)	δ13C (‰)	Age Range (cal. AD / BC) C
Benowo	31201	Peat bulk	36–37	432±29	-28.1	AD 1420–95 (92.3%) AD 1601–15 (3.1%)
Benowo	312.2	Peat bulk	40–41	487±29	-28.5	AD 1405–50
Benowo	27664	Peat bulk	44–45	840±35	-28.4	AD 1052–80 (4.9%) AD 1128–33 (0.5%) AD 1152–1268 (90%)
Benowo	32582	Peat bulk	45–46	788±26	-27.2	AD 1212–77
Benowo	31203	Peat bulk	48–49	922±29	-28.9	AD 1027–1180
Benowo	31204	Peat bulk	53–54	981±32	-28.2	AD 992–1155
Benowo	29824	Peat bulk	58–59	955±33	-26.9	AD 1010–1159
Benowo	31205	Peat bulk	65–66	1128±29	-28.1	AD 783–88 (0.6%) AD 816–43 (3.7%) AD 859–990 (91.1%)
Benowo	27665	Peat bulk	72–73	1100±35	-28	AD 881–1019
Benowo	29825	Peat bulk	99–100	1438±33	-26.9	AD 565–656
Lake Rozlewisko *	32580	Peat bulk	52–53	458±29	-27.8	AD 1414–67
Lake Rozlewisko	31223	Peat bulk	63–63.5	568±29	-28.5	AD 1305–64 (56%) AD 1384–1424 (39.4%)
Lake Rozlewisko	31224	Peat bulk	69.5–70	646±29	-28.7	AD 1281–1328 (42.5%) AD 1342–95 (52.9%)
Lake Rozlewisko	29842	Peat bulk	72–72.5	754±26	-29.2	AD 1223–84
Lake Rozlewisko	29843	Peat bulk	80–80.5	1269±25	-29.9	AD 668–780 (94%) AD 793–803 (1.4%)
Lake Rozlewisko	29844	Peat bulk	90–90.5	1448±29	-29.5	AD 564–651
Biała Góra ***	29832	Charred *Secale cereale* seeds	F94, 120a	649±28	-22.7	AD 1281–1325 (43.1%) AD 1344–94 (52.3%)
Biała Góra	29833	Charred *Cerealia* seeds	F94, 120b	713±28	-26.2	AD 1255–1305 (87.5%) AD 1364–85 (7.9%)
Biała Góra	29834	Charred *Cerealia* seeds	F94, 116	649±29	-21.2	AD 1281–1326 (43.2%) AD 1343–95 (52.2%)
Biała Góra	29835	Charred Secale cereal, Triticum aestivum, Agrostemma githago seeds	F99, 187	816±28	-25.1	AD 1171–1268
Biała Góra	29836	Charred *Galium spurium, Triticum/ Secale* cereale, *Fallopia convolvulus* seeds	F102, 189	570±26	-23	AD 1307–63 (57.3%) AD 1385–1421 (38.1%)
Tczew ***	29830	Charred *Secale cereale* seeds	Tczew T1	460±33	-21.5	AD 1409–76
Tczew	30872	Charred *Secale cereale* seeds	Tczew 5E	478±39	-22.6	AD 1328–41 (1.8%) AD 1395–1475 (93.6%)

Table A.1: AMS radiocarbon dates from Poland. **A** — Lake*, Peat Bog**, Occupation Deposits***.
B — Lab numbers are GU except Wk +. **C** — All calibrated date ranges expressed at 2σ 95.4% probability. *(cont.)*

Site A	Lab No B	Material Dated	Depth (cm)	Age (BP)	δ13C (‰)	Age Range (cal. AD / BC) C
Brandenburg Commandery/Lötzen Procturatorship						
Lake Staświny*	30863	Pollen concentrate	220–22	1715±34	-29.1	AD 246–405
Lake Staświny	34965	Pollen concentrate	250–52	2566±29	-30.5	807–748 BC (75.4%) 685–67 BC (5.8%) 614–587 BC (11.4%) 581–58 BC (2.9%)
Lake Staświny	34966	Pollen concentrate	300–02	3172±27	-29.9	1501–1407 BC
Staświny Village ***	29859	Wood		461 ± 26	-23	AD 1414–58
Swięta Góra ***	29860	Elk bone	SG12/1	1121±29	-22.3	AD 827–41 (1.4%) AD 864–993 (94%)
Swięta Góra	29861	Cattle bone	SG12/4	1126±29	-21.2	AD 783–87 (0.5%) AD 818–43 (3%) AD 860–990 (91.9%)
Swięta Góra	29862	Pig bone	SG12/5	1191±28	-21.1	AD 723–40 (2.2%) AD 770–897 (90.7%) AD 923–40 (2.5%)
Swięta Góra	29863	Cattle bone	SG12/8	1119±29	-21.3	AD 827–40 (1.1%) AD 865–994 (94.3%)
Swięta Góra	34729	Charred cereal grain	253, Ft. 31	1087±36	-26.3	AD 888–1019
Swięta Góra	34730	Charred cereal grain	255, Ft. 32	1221±30	-24.7	AD 692–748 (21.9%) AD 762–887 (73.5%)
Zielone Bagno **	32591	Peat bulk	44–45	205±26	-28.7	AD 1649–84 (27.1%) AD 1736–1806 (50.1%) AD 1933. (18.3%)
Zielone Bagno	32592	Peat bulk	48–49	265±26	-28.5	AD 1521–76 (29.3%) AD 1584–91 (0.9%) AD 1625–69 (57.2%) AD 1781–99 (8.0%)
Zielone Bagno	32593	Peat bulk	51–52	201±29	-28.1	AD 1647–86 (24.9%) AD 1731–1809 (51.2%) AD 1927– (19.3%)
Zielone Bagno	32594	Peat bulk	60–61	772±26	-28	AD 1219–79
Zielone Bagno	31438	Peat bulk	70–71	2027±30	-28.3	153–40 BC (1.7%) AD 111 BC–54 BC (93.7%)
Zielone Bagno	32595	Peat bulk	74–75	1854±25	-28	AD 86–231
Wojsak **	33983	Peat bulk	30–31	791±29	-28.7	AD 1191–1279
Wojsak	33984	Peat bulk	63–64	2183±29	-28.8	361–171 BC
Wojsak	31439	Gyttja bulk	72–73	2480±30	-29.1	769–502 BC (90.2%) 494–87 BC (0.7%) 463–49 BC (1.5%) 442–17 BC (3.0%)
Giżycko Castle ***	34885	Goat tooth	Context?	329±37	-21.2	AD 1468–1645
Giżycko Castle	34886	Cow longbone	Context?	379±37	-21.4	AD 1442–1529 (56.3%) AD 1544–1635 (39.1%)

Table A.2: AMS radiocarbon dates from Latvia, Cēsis commandery. **A** — Lake*, Peat Bog**, Occupation Deposits***. **B** — Lab numbers are GU except Poz +. **C** — All calibrated date ranges expressed at 2σ 95.4% probability.

Site A	Lab No B	Material Dated	Depth (cm)	Age (BP)	δ13C (‰)	Age Range (cal. AD / BC) C
Trikāta *	29840	*Ranunculus* sp, 1 × *Carex*, leaf fragments	481–82	158±26	-27.6	
Trikāta	61639 +	Wood	483–84	175±30	?	AD 1656–97 (18%) AD 1725–1815 (51.2%) AD 1835–78 (6%) AD 1916– (20.1%)
Trikāta	61640 +	*Picea abies* needle and seed wing fragment	545–46	915±30	?	AD 1030–1189
Trikāta	61641 +	*Picea abies* needle fragment	577–78	1345±30	?	AD 641–715 (86.4%) AD 744–65 (9%)
Trikāta	61642 +	Wood	689–90	2485±30	?	776–488 BC
Trikāta	61643 +	Wood	723–24	2990±30	?	1374–56 BC (2.6%) 1301–1118 BC (92.8%)
Blanku Mire **	29808	Peat bulk	53–54	171±33	-28.1	AD 1655–1706 (18.2%) AD 1720–1819 (49.3%) AD 1832–81 (9.4%) AD 1915–54 (18.5%)
Blanku Mire	31214	Peat bulk	65–66	379±29	-28.1	AD 1445–1525 (61.7%) AD 1557–1632 (33.7%)
Blanku Mire	27666	Peat bulk	80–81	515±35	-28.2	AD 1320–50 (31.1%) AD 1391–1448 (82.3%)
Blanku Mire	31216	Peat bulk	83–84	578±29	-28.7	AD 1302–67 (62.5%) AD 1382–1419 (32.9%)
Blanku Mire	31217	Peat bulk	86–87	847±29	-28.7	AD 1056–76 (3.2%) AD 1154–1261 (92.2%)
Blanku Mire	29809	Peat bulk	90–91	932±33	-28.7	AD 1023–1174
Blanku Mire	31218	Peat bulk	98–99	1339±29	-28.4	AD 645–715 (83.8%) AD 744–68 (12.1%)
Blanku Mire	29810	Peat bulk	129–30	2031±33	-28.5	161–34 BC (4.9%) 116 BC–AD 53 (90.5%)
Blanku Mire	29811	Peat bulk	165–66	2573±29	-27.3	810–749 BC (75.6%) 688–66 BC (11.4%) 641–592 BC (7.6%) 577–69 BC (0.7%)
Blusu Mire **	29812	Peat bulk	31–32	143±33	-27.8	AD 1668–1782 (44.3%) AD 1979–1891 (35.1%) AD 1908–47 (15.9%)
Blusu Mire	32583	Peat bulk	40–41	236±25	-27	AD 1533–36 (0.3%) AD 1636–80 (55.6%) AD 1763–1801 (33.3%) AD 1938– (6.3%)
Blusu Mire	31220	Peat bulk	49–50	480±29	-28.9	AD 1409–50
Blusu Mire	27667	Peat bulk	52–53	545±35	-28.4	AD 1310–60 (38.2%) AD 1386–1438 (57.2%)

Table A.2: AMS radiocarbon dates from Latvia, Cēsis commandery. **A** — Lake*, Peat Bog**, Occupation Deposits***.
B — Lab numbers are GU except Poz +. **C** — All calibrated date ranges expressed at 2σ 95.4% probability. *(cont.)*

Site A	Lab No B	Material Dated	Depth (cm)	Age (BP)	δ13C (‰)	Age Range (cal. AD / BC) C
Blusu Mire	29815	Peat bulk	55–56	581±33	-27.8	AD 1298–1370 (63.6%) AD 1380–1419 (31.8%)
Blusu Mire	31221	Peat bulk	58–59	745±29	-28.6	AD 1223–88
Blusu Mire	31222	Peat bulk	61–62	833±29	-28.5	AD 1159–1264
Blusu Mire	29813	Peat bulk	64–65	1092±33	-27.6	AD 891–1016
Blusu Mire	29814	Peat bulk	75–76	2038±33	-27.6	164–129 BC (8.2%) 120 BC–AD 29 (85.1%) 39–50 BC (2.1%)
Nineris Mire **	29817	Peat bulk	25–26	98±33	-28.4	AD 1681–1738 (27.1%) AD 1757–62 (0.7%) AD 1803–1937 (67.7%)
Nineris Mire	23349	Peat bulk	55–56	205±30	-25.9	AD 1640–90 (27.4%) AD 1730–1810 (50.4%) AD 1920–60 (17.6%)
Nineris Mire	29818	Peat bulk	75–76	244±33	-27.7	AD 1522–74 (11.5%) AD 1627–83 (48.6%) AD 1737–1804 (28%) AD 1936–55 (7.3%)
Nineris Mire	31227	Peat bulk	110–11	883±29	-27.8	AD 1042–1107 (28.8%) AD 1117–1220 (66.6%)
Nineris Mire	29820	Peat bulk	125–26	1010±33	-26.9	AD 904–13 (1%) AD 971–1052 (77.2%) AD 1082–1127 (13.1%) AD 1135–53 (4.2%)
Nineris Mire	31228	Peat bulk	145–46	1248±29	-28.2	AD 679–870
Nineris Mire	29821	Peat bulk	175–76	1350±33	-26.6	AD 621–718 (85.7%) AD 743–69 (9.7%)
Nineris Mire	29822	Peat bulk	199–200	1426±33	-27.6	AD 570–660
Rāmuli **	33985	Peat bulk	20–21	112±27	-28.5	AD 1681–1739 (27.6%) AD 1755–63 (1.3%) AD 1803–1938 (66.6%)
Rāmuli	29806	Peat bulk	40–41	614±33	-27.9	AD 1293–1404
Rāmuli	33986	Peat bulk	48–49	1148±29	-27.6	AD 776–94 (7.7%) AD 800–973 (87.7%)
Rāmuli	33987	Peat bulk	80–81	4377±27	-27.4	3090–53 BC (11.3%) 3032–2912 BC (84.1%)
Riekstu Hillfort ***	25391	Wood charcoal	Trench B	1375±30	-26.5	AD 607–85
Cēsis Castle, outer bailey ***	30869	Horse bone	Trench M <63> (325)	407±39	-22.5	AD 1429–1524 (71.4%) AD 1558–1631 (24%)
Āraiši Castle, western range ***	30865	Pig MC4	Trench 2, Layer 4	582±39	-22.1	AD 1297–1421
Āraiši Castle, western range	30866	Mammal bone fragment	Trench 2, Layer 5	356±39	-21.5	AD 1452–1636
Āraiši Castle, western range	30867	Pig MC4	Trench 2/3, Layer 6	686±39	-22.4	AD 1260–1325 (60.1%) AD 1345–94 (35.3%)

Table A.2: AMS radiocarbon dates from Latvia, Cēsis commandery. **A** — Lake*, Peat Bog**, Occupation Deposits***. **B** — Lab numbers are GU except Poz +. **C** — All calibrated date ranges expressed at 2σ 95.4% probability. *(cont.)*

Site A	Lab No B	Material Dated	Depth (cm)	Age (BP)	δ13C (‰)	Age Range (cal. AD / BC) C
Āraiši Castle, western range	30868	Pig radius	Trench 3/4, Layer 3	244±39		AD 1520–93 (17.2%) AD 1619–84 (42.0%) AD 1735–1806 (28.1%) AD 1931–55 (8.2%)
Āraiši Castle, outer bailey**	32622	Peat bulk	40–41	62±29	-29.3	AD 1693–1728 (23.1%) AD 1812–1920 (72.3%)
Āraiši Castle, outer bailey	32623	Peat bulk	60–61	608±27	-29.1	AD 1296–1404
Āraiši Castle, outer bailey	32624	Peat bulk	80–81	864±29	-28.9	AD 1048–85 (11.7%) AD 1124–1137 (2.6%) AD 1150–1252 (81.1%)

Table A.3: AMS radiocarbon dates from Estonia, Karksi district. **A** — Lake*, Peat Bog**, Occupation Deposits***. **B** — Lab numbers are GU except Wk +, Tin ++. **C** — All calibrated date ranges expressed at 2σ 95.4% probability.

Site A	Lab No B	Material Dated	Depth (cm)	Age (BP)	δ13C (‰)	Age Range (cal. AD / BC) C
Univere **	29807	Peat bulk	56–57	197±26	-29.1	AD 1651–85 (23.4%) AD 1733–1808 (52.8%) AD 1928–54 (19.1%)
Univere	31226	Peat bulk	90–91	1583±30	-29.5	AD 414–545
Äriküla **	23341	Peat bulk	21–22	50±30	-27.9	AD 1690–1730 (21.6%) AD 1810–1920 (70.4%) AD 1950–60 (3.5%)
Äriküla	23342	Peat bulk	41–42	395±30	-29	AD 1430–1530 (73%) AD 1570–1630 (22.4%)
Äriküla	23343	Peat bulk	51–52	540±30	-28.6	AD 1310–60 (29.4%) AD 1380–1440 (66%)
Äriküla	23344	Peat bulk	61–62	840±30	-27.9	AD 1050–80 (1.6%) AD 1150–1270 (93.8%)
Äriküla	24855 +	Peat bulk	102–04	1039±30	-26.9	AD 890–920 (6.8%) AD 940–1040 (88.6%)
Karksi high castle ***	28488	Hazelnut	Tr. 1 204	724±26	-27.8	AD 1239–99 (93.9%) AD 1371–79 (1.5%)
Karksi high castle	28487	Hazelnut	Tr. 1 207	697±26	-27.5	AD 1265–1306 (78.1%) AD 1363–85 (17.3%)
Karksi high castle	28486	Leather	Tr. 1 209	771±26	-28	AD 1219–79
Karksi high castle	28485	Peat bulk	Tr. 1 215	694±29	-29.2	AD 1265–1310 (72.2%) AD 1360–88 (23.2%)
Karksi high castle	28484	Peat bulk	Tr. 1 229	767±28	-28.7	AD 1220–80
Karksi high castle	28483	Birch bark	Tr. 1 <14>	754±28	-29.3	AD 1221–85
Karksi high castle	30638	Birch bark	Tr. 1 <15>	677±27	-29.3	AD 1272–1315 (58.5%) AD 1355–89 (36.9%)
Karksi high castle	3486 ++	Secondary used timber	Tr. 1. <18>	386±55	?	AD 1436–1638

Table A.4: Dates either excluded or not presented as accompanying pollen sequences not subsequently analysed. All lab codes GU except Wk + R after la code = rerun sample.

Site A	Lab No B	Material Dated	Depth (cm)	Age (BP)	δ13C (‰)	Age Range (cal. AD / BC) C
Węgry *	31229	Peat bulk	95–96	8699±28	-28.6	7784–7601 BC
Tczew	29831	Charred Triticum aestivum seeds	Tzcew T4	1664±33	-23.4	AD 257–300 (9.4%) AD 317–436 (82.9%) AD 490–509 (2.1%) AD 518–29 (1.1%)
Koniecwałd **	35066	Peat bulk	52–53	1070±29	-28.6	AD 896–928 (20.3%) AD 941–1021 (75.1%)
Koniecwałd	34732R	Peat bulk	70–71	2793±37	-28.3	1028–839 BC
Koniecwałd	35067R	Gyttja bulk	73–74	3222±36	-28.1	1609–1581 BC (7.5%) 1532–1424 BC (87.9%)
Lake Drużno *	30864	Pollen concentrate	200–02	2995±34	-25	1377–37 BC (8%) 1321–1124 BC (87.4%)
Malbork Castle moat	31437	Various seeds	94–100	899±289	-27.3	AD 1040–1110 (42.2%) AD 1116–1212 (53.2%)
Trikāta LV *	27675	Gyttja bulk	450–51	1115±35	-32.1	784–87 (0.3%) 825–42 (2.0%) 862–1016 (93.1%)
Trikāta LV	27676	Gyttja bulk	475–76	1200±35	-32.2	693–748 (10.3%) 765–897 (82.6%) 923–41 (2.6%)
Trikāta LV	27677	Gyttja bulk	500–01	1535±35	-30.1	430–599
Trikāta LV	27678	Gyttja bulk	525–26	1735±35	-31.8	230–402
Trikāta LV	27679	Gyttja bulk	550–51	2800±35	-33.5	1045–890 BC (88.9%) 881–45 BC (6.5%)
Trikāta LV	27680	Gyttja bulk	575–76	2840±35	-33.5	1116–913 BC
Trikāta LV	29839	*Spergula arvensis*	477–78	1918±26	-25	22–133
Lake Nineris *	29826	Gyttja	988–89	972±33	-28.2	1014–1157
Lake Nineris	27668	Gyttja	1000–01	905±35	-27.7	1036–1210
Lake Nineris	29827	Gyttja	1045–46	805±33	-26.6	1172–1275
Lake Nineris	27669	Gyttja	1086–87	745±35	-26.8	1217–94
Lake Nineris	29828	Gyttja	1106–1107	1546±33	-29.7	427–591
Lake Nineris	29829	Gyttja	1150–51	3762±33	-29.4	2289–2123 BC (82.6%) 2092–43 BC (12.8%)
Bille **	34731	Peat bulk	36–37	4167±36	-28.8	2883–2631 BC
Blusu Mire **	29816	Peat bulk	139–40	4878±33	-27.8	3712–3633 BC 3552–3541 BC
Lake Kosilase *	27670	FRE/ HARD WATER ERROR	154–55	805±35	-31.2	AD 1170–1275
Lake Kosilase	27671	FRE/ HARD WATER ERROR	200–01	1475±35	-31.8	AD 535–652
Lake Viljandi *	27672	FRE / HARD WATER ERROR	900–01	1145±35	-31.7	AD 780–93 (4.1%) AD 804–980 (91.3%)
Karksi 7 km south **	G31602	Peat bulk	28–29	1084±34	-28.7	AD 893–1017
Karksi 7 km south	34733	Peat bulk	36–37	1571±36	-28.5	AD 407–565

Table A.4: Dates either excluded or not presented as accompanying pollen sequences not subsequently analysed. All lab codes GU except Wk + R after la code = rerun sample. *(cont.)*

Site A	Lab No B	Material Dated	Depth (cm)	Age (BP)	δ13C (‰)	Age Range (cal. AD /BC) C
Karksi 7 km south	31603	Peat bulk	100–01	3842±34	-28.2	2459–2202 BC
Viljandi Lake Edge **	Wk-24857	Peat bulk	90–92	848±30	-27.8	AD 1050–80 (4.4%) AD 1150–1270 (91%)
Viljandi Lake Edge	Wk-24858	Peat bulk	204–06	2481±30	-27.3	973–1049 BC (82.4%) 1085–1125 BC (10.4%) 1137–50 BC (2.6%)
Pärsti **	31597	Peat bulk	68–69	2126±34	-28.1	351–03 BC (10.8%) 210–48 BC (84.6%)
Vöistre **	31600	Peat bulk	55–56	8329±34	-28.4	7511–7315 BC
Vöistre	31601	Peat bulk	104–05	8019±34	-28.3	7064–6822 BC
Heimtali **	31598	Peat bulk	40–41	2657±34	-27.6	896–792 BC
Heimtali	31599	Peat bulk	80–81	4069±34	-27.7	2855–2489 BC

B. URANIUM SERIES DATING

Stuart Black

Background

Reviews of U-Th disequilibria methods on bone can be found elsewhere (e.g., Schwarcz et al. 1998; Pike et al. 2005; Ivanovich and Harmon 1992), but briefly: U is taken up by animals during life from groundwater and ingested via food, both of which have a relatively low content of dissolved U (up to 150 ± 45 µg/l; 55 ± 45 µg/kg; Ivanovich and Harmon 1992; FSA 2004). Dissolved thorium levels and food concentrations, however, are very low (<8 × 10^{-5} µg/kg; Ivanovich and Harmon 1992; FSA 2004), and uptake from solution and food are negligible. The dating method of bone is based on the gradual in growth of ^{230}Th ($t_{1/2}$ = 75.2 ka) from the decay of its parents ^{238}U ($t_{1/2}$ = 4.468 × 10^9 years). Since bone when newly deposited contains only small amounts of ^{238}U (and daughter products; Hamilton 1971), virtually all the U found in archaeological bones are as a result of deposition post-death and subsequent in growth of the daughter radionuclides (Harley and Fisenne 1990; Hedges and van Iclinken 1994; Rae and Ivanovich 1986). Thus, in order to determine the age of the bone sample,the following expressions are solved for *t* (age):

$$\left[\frac{^{230}Th}{^{238}U}\right] = 1 - e^{-\lambda_{231}t} \qquad (1)$$

Where the isotopes indicated are the activities (Bq kg) of the isotope and is the decay constant (Ln $2/t_{1/2}$) of the isotope in question (e.g. λ_{230} = decay constant of ^{230}Th; Table B.1). The above equation can be used if U is taken up quickly, shortly after deposition and is referred to as the Early Uptake (EU) model. An alternative uptake model is the Linear Uptake model (LU), which is assumes continuous U uptake and is defined as:

$$\left(\frac{^{230}Th}{^{238}U}\right) = 1 - \frac{1}{\lambda_{230}t}\left(1 - e^{-\lambda_{230}t}\right) \qquad (2)$$

However, in the case of the ^{238}U decay chain, ^{234}U which is present in excess (i.e., $^{234}U/^{238}U>1$) in groundwater and, hence, in deposited bone, over longer periods of time will attribute more decays for the ingrowth of the daughter isotope and the above equation is not appropriate. This excess fraction decays away with time, but, below 30 ka, this effect can be disregarded (and equation 2 can be used). Above 30 ka, the age can be derived from the following equation (Ivanovich and Harmon, 1992).

$$\left[\frac{^{230}Th}{^{238}U}\right] = \left(1 - e^{-\lambda_{230}t}\right) + \frac{\lambda_{230}}{\lambda_{230} - \lambda_{234}}\left[\left(\frac{^{234}U}{^{238}U}\right) - 1\right] x\left(1 - e^{-(\lambda_{230}-\lambda_{234})t}\right) \qquad (3)$$

Where, ^{230}Th and λ_{230} are the activity concentration and decay constant of ^{230}Th, respectively. The remaining fraction is supported by ^{238}U ($t_{1/2}$ = 4.468 Ga; Ivanovich and Harmon 1992). The practical age limit for this method is 350 ka for $^{230}Th/^{238}U$. The method requires that the system has been closed since the deposition, that is, no digenetic contribution or loss of U isotopes, and that there was no significant incorporation of ^{230}Th into the bone. There are various sources of error related to departures from these conditions, which are discussed in Ivanovich and Harmon (1992).

Equation 3 can be solved for t if the ($^{234}U/^{238}U$) and ($^{230}Th/^{238}U$) ratios are determined. Pike et al. (2005) also developed a diffusion–adsorption model for U-series in bones. This U-D/A model predicts the behaviour of the U-series when compared to analyses across a single bone, and can then be used to calculate accurate ages for bones and suggests that unless cross-sectional analyses of the bones are conducted certain inaccuracies accrue as a result of misinterpretation of the U uptake.

Table B.1: Gamma ray intensities and efficiencies for detectors used in this study

Radionuclide	Energy (keV)	γ Intensity (%)	Efficiency* (%)	Interfering γ	Interfering γ Energy (keV)	Interfering γ Intensity (%)
^{238}U-Series						
^{234}Th	63.3	3.81	45	-		-
^{234m}Pa	1001.0	0.82	9	-		-
^{234}U	53.2	0.119	38	^{214}Pb	53.2	1.10
^{234}U	120.9	0.041	22	^{223}Ra	122.2	0.054
^{230}Th	67.7	0.376	55	-		-
^{226}Ra	186.1	3.28	19	^{235}U	185.7	2.4
^{214}Pb	53.2	1.10	38	^{234}U	53.2	0.119
^{214}Pb	241.9	7.46	17	^{234}Ra	240.8	3.9
^{214}Pb	295.1	19.2	16	-		-
^{214}Pb	351.9	37.1	15	-		-
^{214}Bi	609.3	46.1	13	-		-
^{214}Bi	1120.2	15.0	7	-		-
^{214}Bi	1764.5	15.9	5	-		-
^{210}Pb	46.5	4.05	31	-		-
^{235}U-Series						
^{235}U	163.3	0.21	20	-		-
^{235}U	185.5	2.40	19	^{226}Ra	186.1	3.28
^{235}U	205.3	0.21	18	-		-
^{232}Th-Series						
^{228}Ac	911.1	28.0	10	-		-
^{224}Ra	240.8	3.9	17	^{214}Pb	241.9	7.46
^{212}Pb	238.6	43.6	17	-		-
^{212}Pb	727.3	6.65	11			
^{208}Tl	583	86.0	12	-		

* Based on a combination of NIST-SRM 4356, 3159, 3164, and 1646 in a hydroxyapatite sample (Fisher ultra-pure $Ca_{10}(PO_4)_6OH_2$ made into in a cylinder shape to match the dimensions of the analysed bone fragments). Decay constants used during this study are: $^{238}U = 1.55125 \times 10^{-10}\ yr^{-1}$; $^{235}U = 9.8485 \times 10^{-10}\ yr^{-1}$; $^{230}Th = 9.1952 \times 10^{-6}\ yr^{-1}$; $^{226}Ra = 4.332 \times 10^{-4}\ yr^{-1}$; $^{210}Pb = 0.0311387\ yr^{-1}$.

Methodological Approach

Gamma Spectrometry

U-series dating by gamma spectrometry has been reported previously by Yokoyama and Nguyen (1980), Barton and Stringer (1997), Berzero et al. (1997), Simpson and Grun (1998) and Schwarcz et al. (1998). This study was carried out at the University of Reading for ^{230}Th, ^{238}U, ^{234}U, ^{235}U, ^{226}Ra, ^{210}Pb and, ^{228}Ra measured directly by γ-spectrometry using the peaks identified in Table B.1 on the assumption that the short-lived daughters will be in equilibrium with their parent isotopes. However, diffusion loss of the intermediate daughter ^{222}Rn (between ^{226}Ra and ^{214}Pb) from fine-grained material can affect ^{214}Pb activities; to overcome this, all samples were sealed in airtight plastic bags. Samples were counted on a Harwell Instruments, Broad Energy, BE5030 high purity germanium coaxial photon detector. This detector has an ultra-low background set up (detector and cryostat)

Table B.2: U-series results for the NIST-SRM bone ash (4356). Parentheses indicate uncertified values.

	^{210}Pb (mBq g)	^{226}Ra (mBq/g)	^{230}Th (mBq/g)	^{232}Th (mBq/g)	^{238}U (mBq/g)	^{234}U (mBq/g)	U (µg/kg)	Th (µg/kg)
Reported Value	(20)	14.5 ± 1.1	0.52 ± 0.03	0.98 ± 0.03	0.63 ± 0.02	0.64 ± 0.02	50.6 ± 1.6	242.6 ± 7.4
This Study (n = 5)	20.5 ± 0.8	14.4 ± 1.5	0.55 ± 0.05	1.00 ± 0.06	0.66 ± 0.04	0.67 ± 0.09	51.0 ± 1.9	244.1 ± 5.8

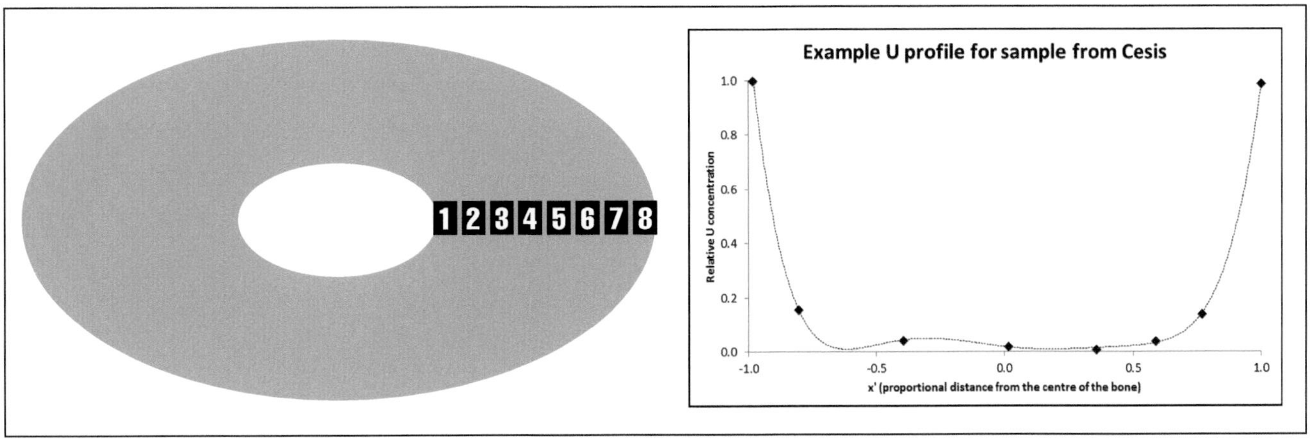

Figure B.1: A) Plot showing a diagrammatic cross section of a bone sample with an example of sub-samples taken; B) Relative U concentration versus proportional distance from the centre of the bone plot to show relative uptake timescale of the uranium.

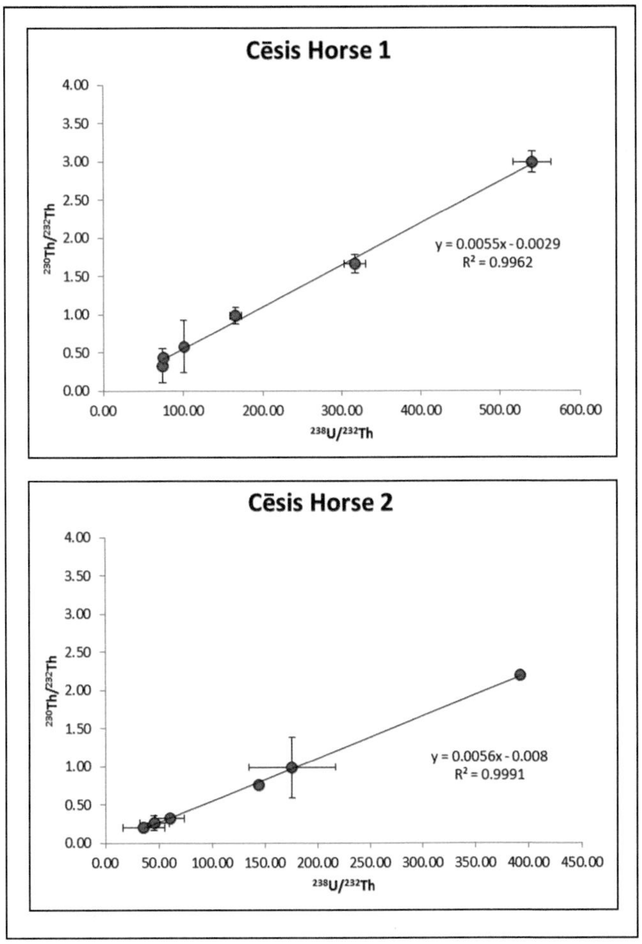

Figure B.2: ^{238}U/^{232}Th versus 230Th/^{232}Th isochron plot for Cēsis horse samples 1, (A), and 2, (B).

with a 0.5 mm thick carbon-epoxy window and remote detector chamber. Detector specifications were FWHM @ 5.9 keV = 0.45 keV, FWHM @ 1.3 MeV = <1.2 keV. To keep self-absorption differences negligible, standard samples were used to calibrate the detectors using material of similar density and dimensions (10.5 × 4.5 cm; volume = 167 cm^3; 0.29 g cm^3) of the bone materials analysed. The standard was made using phosphate rock standard (NBS 964 (now NIST SRM 964) spiked with NIST SRMs 3164 and 3159), which was configured into the correct shape via use of a low density epoxy resin. A secondary standard was also made in the form of a disc (80 mm diameter) from the same material to which the detector had been calibrated previously.

The (^{230}Th/^{238}U) activity ratio was determined from the activities at the 67.7 keV and 63.3 keV γ-ray peaks. In addition, the activity of the (^{226}Ra(^{214}Pb)/^{230}Th), using the 295, 352, and 67.7 keV γ-rays, and the (^{226}Ra (^{214}Bi)/^{238}U (^{234m}Pa)) ratios using the 609 and 1764 keV γ-rays for ^{214}Bi and 1001 keV γ-ray for ^{234m}Pa were also determined.

Whole samples were counted for approximately ten to sixty days each in order to reduce the uncertainties by accumulating a large number of counts in each analyte peak. Most analyte peaks were> 10,000 net counts (i.e., <1% uncertainty). External reproducibility was checked using international standards (NIST-SRM 4356), the results of which can be seen in Table B.2.

Table B.3: U-series dates from Cēsis Castle and Hillfort samples

Sample Location	Sub-sample Number	Reference Number	U-series Date (Years BP)	Uncertainty (+/- years)
Cēsis Castle	1	Ces 1	446	37
	2	Ces 1	491	39
	3	Ces 1	479	35
	4	Ces 1	363	31
	5	Ces 1	467	40
	6	Ces 1	414	39
Cēsis Castle	1	Ces 2–4	444	38
	2	Ces 2–4	471	36
	3	Ces 2–4	458	40
	4	Ces 2–4	460	41
	5	Ces 2–4	420	39
	6	Ces 2–4	456	36
Cēsis Castle	1	Trench M	292	30
Cēsis Hillfort	1	Trench 12	743	49
	2	Trench 12	734	55
	3	Trench 12	705	58
	4	Trench 12	800	59
	5	Trench 12	798	53
	6	Trench 12	777	56
	7	Trench 12	736	55
	8	Trench 12	717	50
	9	Trench 12	815	57
	10	Trench 12	744	52
Cēsis Hillfort	1	Trench 12a	747	51
	2	Trench 12a	739	50
	3	Trench 12a	719	56

Mass Spectrometry

Small sub-samples (100–500 mg) were also taken from the bones for destructive analysis for determination of the $^{234}U/^{238}U$, $^{235}U/^{238}U$, and $^{230}Th/^{232}Th$ ratios. These were undertaken using a Thermo-fisher iCAPQ Inductively Coupled Plasma Mass Spectrometer. The mass ratio of the 234/38 is low (<1%) and 230/32 very low (<0.1%), but the counts were increased by running the mass spectrometer in isotope ratio mode using ten replicate analyses, an increased dwell time (100 ms), together with an average of forty-five passes per replicate sample for 234/38 and increased replicates for 230/32. This brought the uncertainty of the ratios to within a tolerable level (<1.5% for 234/38 and <2% for 230/32). External reproducibility was checked using international standards (NIST SRM 3164) and by monitoring the (235/38) ratios in the samples to be within the naturally abundant ratio (137.5). Uranium, thorium, barium, and a range of trace elements were also determined via mass spectrometry using the same instrument.

Quality Assurance

Accuracy of the gamma spectrometry data was assessed in several ways: 1) by running several bone samples that were known to be older than 752, 000 years (Pleistocene mastodon teeth from the Kennet Valley in the United Kingdom, which showed $^{230}Th = ^{226}Ra = ^{210}Pb$ within uncertainty (± 0.98%); (2) by running several NIST (SRM) international reference materials. NIST SRM 4356, 3159, 3164, which were within 0.64–0.98% specific activities for all nuclides peaks.

Table B.4: U-series dates from Poganowo

Sub-sample Number	Reference Number	U-series Date (Years BP)	Uncertainty (+/- years)
1	AP-POG-173	845	45
2	AP-POG-173	857	49
3	AP-POG-173	850	55
4	AP-POG-173	841	48
5	AP-POG-173	829	47
6	AP-POG-173	816	43
1	AP-POG-148	855	50
2	AP-POG-148	866	54
3	AP-POG-148	851	51
4	AP-POG-148	831	49
5	AP-POG-148	838	48
6	AP-POG-148	825	45
1	AP-POG-133	913	56
2	AP-POG-133	902	57
3	AP-POG-133	910	48
4	AP-POG-133	923	51
5	AP-POG-133	905	53
6	AP-POG-133	921	50
1	AP-POG-741	955	49
2	AP-POG-741	971	47
3	AP-POG-741	950	52
4	AP-POG-741	960	53
5	AP-POG-741	931	56
6	AP-POG-741	950	55
1	AP-POG-1438	1016	60
2	AP-POG-1438	1005	67
3	AP-POG-1438	1020	63
4	AP-POG-1438	1012	65
5	AP-POG-1438	1005	68
6	AP-POG-1438	990	54

Determination of ^{232}Th by mass spectrometry is very accurate (<0.1% uncertainty); however, determination of the 230/32 mass ratio using a single collector instrument poses problems of detecting enough of the low mass abundance 230 and long count times can lead to instrumental drift. Whole samples were analysed on the mass spectrometer and on the gamma detector such that a comparison of the calculated ^{230}Th concentrations could be compared. There was a clear linear correlation between the two independent sets of data indicating that the mass spectrometry data was indeed accurate and that little mass drift was occurring during analysis.

Table B.5: U-series dates from Tartu

Sub-sample Number	Reference Number	U-series Date (Years BP)	Uncertainty (+/- Years)
1	LOS-M-169–1	768	38
2	LOS-M-169–2	760	42
3	LOS-M-169–3	782	47
4	LOS-M-169–4	765	33
1	LOS-M-171–1	726	40
2	LOS-M-171–2	730	37
3	LOS-M-171–3	721	35
4	LOS-M-171–4	741	37
1	LOS-M-175–1	749	34
2	LOS-M-175–2	751	40
3	LOS-M-175–3	730	38
4	LOS-M-175–4	734	34
1	LOS-M-181–1	391	58

Table B.6: U-series dates from shells from Lake Drużno

Depths (cm)	Reference Number	U-series date (years BP)	Uncertainty (+/- years)
129	Sample 129 1	283	80
129	Sample 129 2	305	51
129	Sample 129 3	239	81
129	Sample 129 4	264	83
130.5	Sample 130.5 1	253	75
130.5	Sample 130.5 2	283	41
130.5	Sample 130.5 3	259	72
130.5	Sample 130.5 4	289	87
145.5	Sample 145.5 1	298	34
145.5	Sample 145.5 2	314	40
145.5	Sample 145.5 3	310	34
145.5	Sample 145.5 4	307	64
153	Sample 153 1	363	98
153	Sample 153 2	369	33
153	Sample 153 3	382	35
180	Sample 180 1	389	108
180	Sample 180 2	393	96
180	Sample 180 3	412	54
180	Sample 180 4	391	73

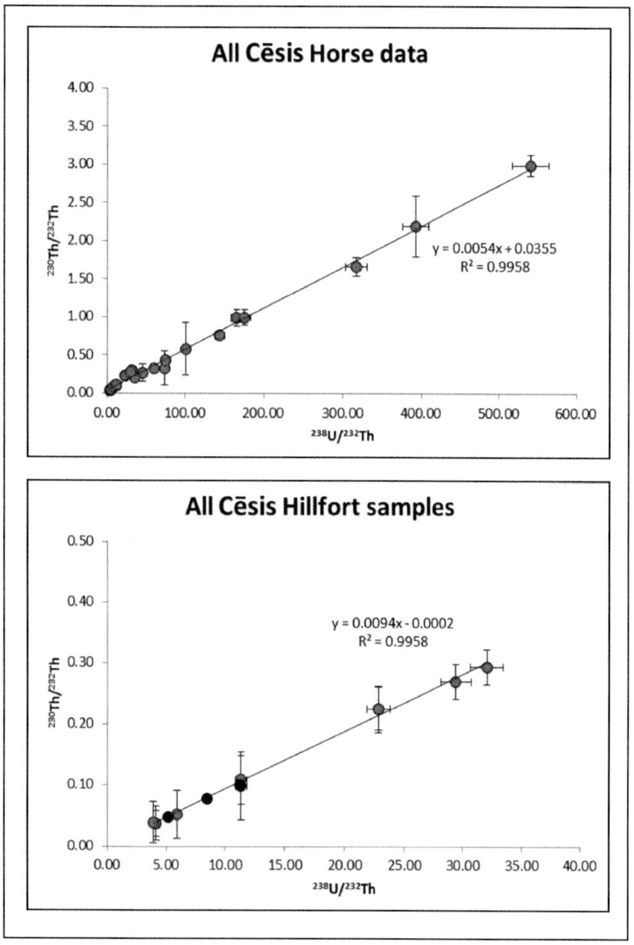

Figure B.3: $^{238}U/^{232}Th$ versus $230Th/^{232}Th$ isochron plot for all Cēsis horse samples combined, (A) and Cēsis and all Riekstu Hillfort samples, (B).

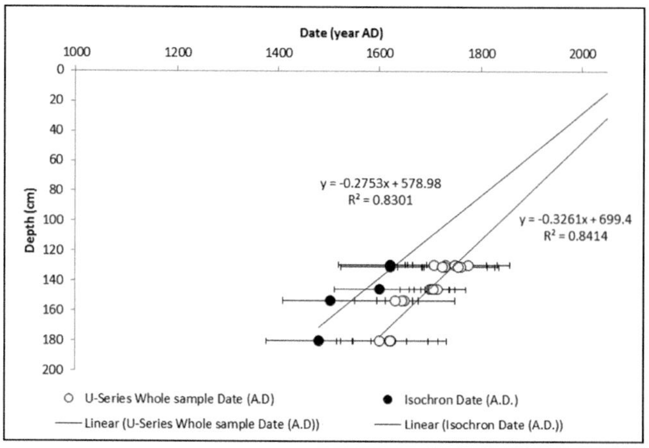

Figure B.4: U-series ages with depth for Cēsis horse samples from Trench M. Both isochron ages (corrected) and whole samples (uncorrected) are displayed for reference. A best-fit line for both distributions is also presented.

Table B.7: U-series isochron ages from shells from Lake Drużno

Depths (cm)	Reference Number	U-series Date (Years BP)	Uncertainty (+/- Years)
129	Isochron 129	391	102
130.5	Isochron 130.5	391	98
145.5	Isochron 145.5	413	90
153	Isochron 153	511	94
180	Isochron 180	533	104

Age Determination — U-Th

The U-Th ages determined using the early (EU) models modified in Pike et al. (Pike and Hedges 2001; Pike et al. 2002, 2005) are presented in Tables B.3–B.7. The data show the samples to be all young (less than a thousand years since death) and so a simple early uptake model was used in all cases. The samples were drilled out across the bone and analyzed using the methods above. EU profiles were used to determine a simple U-Th age using equation 3 which was solved for t using the measured ratios of ^{238}U, ^{234}U and, ^{230}Th concentrations.

Isochrons were also constructed for some samples to check the integrity of the EU model ages. Sub-samples of the same age from the same sample will show variations in $^{238}U/^{232}Th$ or $^{234}U/^{232}Th$, but the $^{230}Th/^{232}Th$ will only vary as a function of time and therefore plots of $^{238}U/^{232}Th$ versus $^{230}Th/^{232}Th$ will produce linear correlations which can be used to determine the age. Correlated errors were reduced by calculating isochron ages in ISOPLOT v4.15 (Ludwig, 2008).

Works Cited

Barton, J. C., and Stringer, C. B. 1997. 'An Attempt at Dating the Swanscombe Skull Bones Using Non-Destructive Gamma-Ray Counting', *Archaeometry* 39, 205–16

Berzero, A., Carmella-Crespi, V., and Cavagna, P. 1997. 'Direct Gamma-Ray Spectrometric Dating of Fossil Bones: Preliminary Results', *Archaeometry* 39, 189–203

FSA 2004. Food Standards Agency – Scottish Environment Protection Agency Report RIFE-9, Radioactivity in Food and the Environment, 2003

Harley, N. H., and Fisenne, I.M. 1990. 'Distribution and Alpha Radiation Dose from Naturally Occurring U, Th and Ra in the Human Skeleton', *Health Physics* 58, 515–18

Hedges, R., and Klinken, G. J. van, eds 1994. Special issue on bone diagenesis, *Journal of Archaeological Science* 22/2, 147–340

Ivanovich, M., and Harmon, R. S., eds 1992. *Uranium Series Disequilibrium Applications to Environmental Problems*, Oxford: Oxford University Press

Ludwig, K. R. 2008. *Isoplot User's Manual*, Berkley: Berkeley Geochronology Center

Pike, A. W. G., Eggins, S., Grun, R., Hedges, R. E. M., and Jacobi, R. M. 2005. 'U-series Dating of the Late Pleistocene Fauna from Wood Quarry (Steetley) Nottinghamshire, UK', *Journal of Quaternary Science* 20 /1, 59–65

Pike, A. W. G., and Hedges, R. E. M. 2001. 'Sample Geometry and Uranium Uptake in Archaeological Teeth: Implications for U-Series and ESR Dating', *Quaternary Science Reviews* 20, 1021–25

Pike, A. W. G., Hedges, R. E. M., and Calsteren P. van. 2002. 'U_series Dating of Bone Using the Diffusion-Adsorption Model', *Geochimica et Cosmochimica Acta* 66, 4273–86

Rae, A. M., and Ivanovich, M. 1986. 'Successful Application of Uranium Series Dating of Fossil Bone', *Applied Geochemistry* 1, 419–26

Schwarcz H. P., Simpson, J. J., and Stringer, C. B. 1998. 'Neanderthal Skeleton from Tabun: U-series Dating by Gamma Spectrometry', *Journal of Human Evolution* 35, 635–45

Simpson, J. J., and Grun, R. 1998. 'Non-destructive Gamma Spectrometric U-series Dating', *Quaternary Geochronology* 17, 1009–22

Yokoyama, Y., and Nguyen, H.-V. 1980. 'Direct and non-Destructive Dating of Marine Sediments, Manganese Nodules, and Corals by High Resolution Gamma-Ray Spectrometry', in *Isotope Marine Chemistry*, ed. K. Saruhashi, Tokyo: Uchida-Rokaku, 235–65

C. Enrichment Factors for Major and Trace Elements within Pit Deposits, Święta Góra, Poland

Rowena Banerjea

		Majors										Trace										
Feature	Sample name	Na EF	Mg EF	Al EF	Si EF	P EF	K EF	Ca EF	Ti EF	Mn EF	Fe EF	V EF	Cr EF	Co EF	Ni EF	Cu EF	Zn EF	Rb EF	Sr EF	Y EF	Zr EF	Pb EF
Phase 1																						
23 (Tr1)	SG12 273A	0.80	0.83	0.80	0.93	10.07	0.85	5.52	0.86	3.50	0.83	0.29	0.80	1.00	1.33	3.22	4.40	0.99	1.71	1.00	0.95	0.79
23 (Tr1)	SG12 273B	0.78	0.80	0.78	0.93	10.00	0.84	5.49	0.86	3.50	0.83	1.29	0.67	1.00	1.22	3.22	4.38	0.99	1.71	1.00	0.91	0.71
23 (Tr1)	SG12 275A	0.72	0.90	0.80	0.87	11.52	0.84	6.30	0.89	3.75	0.86	0.82	0.77	1.00	1.44	4.78	5.83	1.00	1.76	1.00	0.83	0.79
23 (Tr1)	SG12 275B	0.72	0.88	0.79	0.87	11.41	0.84	6.26	0.91	3.75	0.87	2.29	0.77	1.00	1.44	4.78	5.81	1.01	1.81	1.00	0.93	0.71
23 (Tr1)	SG12 277A	0.65	0.98	0.70	0.73	21.11	0.84	12.56	0.91	4.75	0.88	1.41	0.77	1.17	1.67	6.00	7.10	1.09	2.81	1.13	0.89	0.64
23 (Tr1)	SG12 277B	0.63	0.98	0.70	0.73	21.30	0.84	12.67	0.89	4.75	0.88	1.35	0.77	1.33	1.67	6.00	7.17	1.11	2.83	1.07	0.90	0.71
23	STAS 231 0–2	1.02	1.00	0.93	1.00	3.44	0.92	2.39	0.97	2.00	0.98	1.76	1.07	0.83	1.11	1.56	2.24	1.08	1.36	1.27	1.00	1.00
23	STAS 231 4–6	0.95	0.95	0.89	0.98	4.33	0.90	2.75	0.94	2.25	0.92	1.71	0.87	0.83	0.89	1.78	2.52	1.03	1.39	1.20	0.98	1.14
23	STAS 231 8–10	1.03	0.95	0.91	1.01	2.96	0.91	2.28	0.94	1.75	0.95	1.65	0.90	0.83	0.78	1.22	1.98	0.97	1.20	1.13	0.88	1.14
23	STAS 231 12–14	0.98	1.00	0.95	1.00	3.26	0.93	1.97	1.00	1.75	0.98	1.88	1.23	0.83	1.00	1.33	2.00	1.07	1.27	1.33	1.00	1.43
23	STAS 231 18–18	0.97	0.95	0.91	1.00	3.78	0.90	2.33	0.91	1.75	0.93	1.71	1.37	1.17	1.11	1.56	2.36	1.07	1.36	1.33	0.98	1.07
23	STAS 231 20–22	1.02	0.88	0.88	1.02	3.33	0.90	2.25	0.94	1.75	0.97	1.82	0.93	0.83	0.89	1.44	2.38	1.08	1.33	1.33	1.04	1.07
23	STAS 231 24–26	0.97	0.91	0.89	1.00	3.81	0.89	2.43	0.94	1.75	0.92	1.76	0.83	0.83	0.78	1.44	2.38	1.05	1.34	1.27	1.03	1.07
23	STAS 231 28–30	0.85	0.85	0.78	0.93	7.00	0.82	4.77	0.89	3.00	0.86	1.65	0.77	0.83	1.00	2.78	4.26	0.99	1.71	1.13	0.94	1.21
23	STAS 231 32–34	0.83	0.72	0.71	0.94	7.37	0.77	4.91	0.80	3.25	0.82	1.59	0.63	0.83	0.89	3.22	4.86	0.93	1.73	1.07	0.93	1.07
23	STAS 231 36–38	0.88	0.76	0.75	0.97	6.22	0.80	3.85	0.83	3.00	0.83	1.47	0.70	0.83	0.78	2.56	4.33	0.97	1.59	1.20	0.95	1.00
23	STAS 231 40–42	0.87	0.83	0.80	0.96	6.30	0.82	3.75	0.89	2.75	0.87	1.71	0.83	0.00	0.78	2.33	4.36	1.00	1.57	1.13	0.92	1.00
23	STAS 231 42–44	0.82	0.84	0.77	0.95	6.56	0.80	4.29	0.86	2.75	0.86	1.59	0.93	1.00	0.67	2.67	4.36	1.01	1.60	1.13	0.94	1.00
23	STAS 231 44–46	0.90	0.85	0.81	0.97	5.93	0.84	3.53	0.89	2.50	0.89	1.65	0.77	0.83	0.67	2.11	4.12	1.05	1.53	1.33	0.96	1.07
23	STAS 231 46–48	0.85	0.80	0.75	0.96	6.33	0.80	3.94	0.83	3.00	0.84	1.41	0.73	0.83	0.56	2.33	4.24	0.95	1.51	1.00	0.85	1.00
23	STAS 231 50–52	0.83	0.79	0.77	0.97	6.15	0.82	3.60	0.89	3.00	0.86	1.59	0.67	0.83	0.78	2.33	4.31	1.01	1.57	1.13	0.96	1.00
23	STAS 231 54–56	0.83	0.83	0.77	0.96	6.59	0.81	3.95	0.86	3.00	0.87	1.65	0.80	0.83	0.89	2.67	4.55	1.00	1.63	1.20	0.96	1.07
23	STAS 231 58–60	0.82	0.85	0.78	0.93	7.52	0.81	4.79	0.89	3.00	0.88	1.71	0.80	0.00	1.00	2.78	5.05	1.03	1.67	1.13	0.92	0.86

		Majors										Trace										
Feature	**Sample name**	**Na EF**	**Mg EF**	**Al EF**	**Si EF**	**P EF**	**K EF**	**Ca EF**	**Ti EF**	**Mn EF**	**Fe EF**	**V EF**	**Cr EF**	**Co EF**	**Ni EF**	**Cu EF**	**Zn EF**	**Rb EF**	**Sr EF**	**Y EF**	**Zr EF**	**Pb EF**
23	STAS 231 61–63	0.68	0.67	0.65	0.91	8.04	0.70	5.36	0.77	4.75	0.90	1.82	0.77	1.00	1.11	5.11	7.57	0.87	1.79	1.07	0.82	1.07
23	STAS 231 62–64	0.80	0.77	0.73	0.94	7.48	0.78	4.59	0.86	3.25	0.89	1.71	0.87	0.83	0.89	3.33	5.60	0.92	1.51	0.87	0.75	1.07
23	STAS 231 65–67	0.63	0.73	0.65	0.94	8.19	0.68	4.73	0.77	5.50	0.96	1.94	0.80	1.00	1.33	5.67	7.79	0.74	1.41	0.80	0.63	1.00
23	STAS 231 69–71	0.92	1.35	0.94	0.96	2.67	0.93	3.91	1.06	1.75	1.06	2.24	1.17	1.00	1.00	1.44	2.79	1.04	1.40	1.47	1.00	1.14
23	STAS 231 73–75	0.93	1.91	0.95	0.88	1.07	0.96	7.55	1.06	1.25	1.00	2.06	1.07	0.83	0.67	0.00	1.55	1.03	1.61	1.33	0.92	0.93
23	STAS 231 75–77	0.55	0.71	0.60	0.95	7.44	0.64	4.40	0.69	5.50	0.93	2.06	0.70	1.17	1.22	5.11	6.90	0.61	1.09	0.53	0.50	1.00
23	STAS 231 77–79	0.92	2.26	0.88	0.89	0.93	0.92	7.76	0.97	1.25	0.86	1.94	1.07	0.83	0.67	0.56	1.57	0.97	1.51	1.13	0.85	1.00
23	STAS 231 79–81	0.55	0.74	0.61	0.95	7.70	0.66	4.46	0.74	5.25	1.00	2.18	0.67	1.17	1.11	4.56	6.69	0.82	1.51	0.93	0.72	1.14
23	STAS 231 81–83	0.97	1.63	0.91	0.92	1.22	0.94	6.45	1.00	1.25	0.91	1.82	1.10	0.00	0.78	0.67	1.76	1.01	1.57	1.27	0.95	0.93
23	STAS 231 83–85	0.75	0.97	0.82	0.97	5.22	0.81	3.32	0.91	3.25	1.13	2.29	0.93	1.17	1.00	2.67	4.48	0.91	1.36	1.00	0.77	1.14
23	STAS 231 87–88	0.98	1.38	1.10	0.99	1.67	1.02	1.75	1.17	1.00	1.23	2.41	1.27	0.83	0.89	0.00	1.83	1.13	1.19	1.47	1.06	1.21
23	STAS 231 91–93	0.97	2.16	0.98	0.85	1.22	1.01	8.68	1.26	1.25	1.19	2.53	1.30	1.00	0.89	0.67	1.83	1.13	1.81	1.67	1.06	1.00
23	STAS 231 95–97	0.88	2.07	0.80	0.87	1.41	0.88	8.13	0.97	1.25	0.88	1.71	0.87	0.00	0.00	0.00	1.36	0.93	1.63	1.20	0.95	1.07
Phase 2																						
13 (Tr 1)	SG12 258A	0.98	0.66	0.83	1.00	6.11	0.92	2.35	0.97	2.50	0.68	0.00	0.57	1.00	0.67	1.67	2.48	0.99	1.33	0.93	1.04	0.57
13 (Tr 1)	SG12 258B	0.97	0.66	0.82	1.00	6.11	0.90	2.33	0.94	2.50	0.68	1.12	0.73	1.00	0.67	1.56	2.48	0.97	1.33	0.93	1.01	0.64
19 (Tr1)	SG12 270A	0.65	0.78	0.71	0.77	19.04	0.76	10.67	0.83	5.00	0.86	1.76	0.73	1.17	1.44	4.78	6.71	0.93	2.59	1.00	0.83	0.79
19 (Tr1)	SG12 270B	0.65	0.78	0.71	0.77	19.04	0.76	10.71	0.86	5.00	0.86	2.00	0.80	1.00	1.44	4.89	6.74	0.93	2.64	1.07	0.81	0.71
25 (N17A2)	SG11 131	0.93	0.90	0.91	0.96	6.70	0.94	2.85	1.00	2.00	0.88	1.18	0.80	1.00	0.89	1.67	2.52	1.07	1.36	1.07	1.02	0.79
30 (Tr 3)	SG12 264A	0.82	0.99	0.87	0.87	8.67	0.90	6.04	0.97	3.00	0.95	0.65	0.80	1.17	1.11	3.11	4.40	1.04	1.99	1.13	0.95	0.86
30 (Tr 3)	SG12 264B	0.82	0.98	0.86	0.87	8.52	0.90	5.95	0.97	3.00	0.94	1.24	1.07	1.17	1.22	3.00	4.33	1.03	1.97	1.13	0.92	0.79
30 (Tr 3)	SG12 266A	0.90	1.05	1.00	0.95	5.96	1.01	3.01	1.11	2.50	0.94	0.00	0.97	1.17	1.22	2.00	2.69	1.12	1.44	1.00	1.00	0.79
30 (Tr 3)	SG12 266B	0.90	1.06	1.00	0.95	6.04	1.02	3.03	1.09	2.50	0.94	0.71	1.10	1.33	1.11	2.00	2.71	1.12	1.43	0.93	1.05	0.86
31 (Tr 3)	SG12 254A	0.87	1.03	0.93	0.94	6.30	0.96	4.00	1.03	2.50	0.97	1.47	0.90	1.17	1.11	2.22	3.26	1.08	1.64	1.07	1.01	0.79
31 (Tr 3)	SG12 254B	0.87	1.05	0.94	0.94	6.37	0.97	4.03	1.03	2.50	0.97	0.76	0.83	1.17	1.22	2.22	3.29	1.08	1.64	1.07	0.96	0.79
39 (Tr3)	SG12 278A	0.95	1.20	1.07	0.92	6.48	1.06	3.08	1.14	2.00	1.07	0.65	1.17	1.17	1.11	1.89	2.90	1.09	1.50	1.27	1.00	0.79
39 (Tr3)	SG12 278B	0.95	1.20	1.08	0.93	6.37	1.06	3.06	1.17	1.75	1.08	0.41	1.43	1.33	1.22	1.89	2.93	1.08	1.50	1.33	1.00	0.93
41 (Tr3)	SG12 286A	1.00	1.09	1.01	0.93	5.00	1.05	3.26	1.11	1.75	0.91	0.65	1.03	1.17	1.00	1.44	2.02	1.09	1.41	1.00	1.04	0.79
41 (Tr3)	SG12 286B	1.00	1.12	1.02	0.93	5.07	1.06	3.29	1.14	2.00	0.92	0.00	0.87	1.33	1.11	1.33	2.00	1.08	1.41	1.00	1.01	0.93
43 (Tr3)	SG12 284A	0.98	1.06	0.99	0.82	8.85	1.05	4.99	1.14	3.00	1.18	1.06	1.00	1.33	1.44	3.44	3.79	1.20	2.09	1.33	0.92	0.86
43 (Tr3)	SG12 284B	0.98	1.02	0.97	0.83	8.63	1.04	4.91	1.14	3.00	1.17	1.76	1.03	1.33	1.44	3.44	3.76	1.18	2.01	1.33	0.89	0.93
47 (Tr3)	SG12 294A	1.15	0.79	0.95	1.02	1.96	1.03	0.74	1.09	1.25	0.75	-0.29	0.93	1.00	0.67	0.00	0.86	1.00	1.11	0.80	1.03	0.79
47 (Tr3)	SG12 294B	1.17	0.78	0.95	1.03	1.89	1.02	0.73	1.03	1.25	0.74	1.18	0.90	0.83	0.67	0.44	0.86	1.00	1.11	0.80	1.07	0.64

		Majors										Trace										
Feature	Sample name	Na EF	Mg EF	Al EF	Si EF	P EF	K EF	Ca EF	Ti EF	Mn EF	Fe EF	V EF	Cr EF	Co EF	Ni EF	Cu EF	Zn EF	Rb EF	Sr EF	Y EF	Zr EF	Pb EF
Phase 2a																						
14 (M17D3)	SG11 148	1.03	0.70	0.82	0.93	4.85	0.90	2.86	0.91	2.00	0.78	0.76	0.63	1.00	0.67	1.78	2.33	1.00	1.43	1.00	0.86	0.79
14 (M17D3)	SG11 149	1.08	0.93	0.88	0.86	6.48	0.90	4.68	1.23	3.00	1.20	1.94	0.97	1.33	1.11	2.67	3.05	1.03	2.07	1.07	0.94	0.71
14 (M17D3)	SG11 150	1.00	0.95	0.97	0.94	4.81	1.02	2.75	1.06	2.00	0.87	0.59	1.00	1.17	0.89	1.78	2.36	1.04	1.34	1.00	0.94	0.71
14 (M17D3)	SG11 151	0.92	1.14	1.11	1.02	1.93	1.07	0.66	1.11	0.75	1.17	1.00	1.13	1.17	1.11	0.78	0.83	1.16	1.03	0.87	0.99	1.00
14 (Tr 1)	SG12 260A	1.05	0.90	0.90	0.88	6.81	0.94	3.92	1.09	2.75	0.96	0.71	0.87	1.17	0.89	2.22	2.81	1.03	1.89	1.13	0.98	0.71
14 (Tr 1)	SG12 260B	1.07	0.90	0.91	0.88	6.93	0.94	3.95	1.06	2.75	0.95	1.06	0.83	1.17	0.89	2.11	2.79	1.04	1.90	1.07	1.02	0.71
14 (Tr 1)	SG12 262A	1.02	0.77	0.90	0.95	5.07	0.98	2.43	1.11	2.00	0.76	0.00	0.93	1.00	0.78	1.33	2.14	1.04	1.33	1.00	1.00	0.79
14 (Tr 1)	SG12 262B	1.03	0.78	0.91	0.96	5.15	0.99	2.44	1.09	2.00	0.76	0.00	0.87	1.17	0.78	1.44	2.12	1.04	1.34	1.00	1.02	0.79
15 (M17C1)	SG11 115	1.00	0.81	0.87	0.93	5.78	0.94	3.45	1.00	2.75	0.84	0.76	0.83	1.00	1.11	2.11	2.86	1.00	1.41	0.87	0.88	0.79
15 (M17C1)	SG11 116	1.02	0.87	0.93	0.97	3.93	0.98	2.03	1.09	2.00	0.81	0.65	1.00	1.17	0.89	1.33	1.88	0.93	1.09	0.87	0.89	0.71
15 (M17C1)	SG11 117	1.07	0.95	1.01	0.99	2.41	1.03	1.13	1.09	1.25	0.88	0.94	1.03	1.17	0.89	0.78	1.10	1.03	1.09	0.87	0.96	0.79
16 (M17D1)	SG11 121	1.15	0.51	0.82	1.04	2.11	0.96	0.70	0.94	0.75	0.54	0.06	0.70	0.83	0.33	0.33	0.86	0.93	1.07	0.80	1.06	0.57
16 (M17D1)	SG11 122	1.00	0.77	0.86	0.95	5.59	0.95	2.54	0.94	2.50	0.82	0.35	0.83	1.17	0.78	1.89	2.38	1.00	1.31	0.80	0.93	0.79
16 (M17D1)	SG11 123	0.98	0.88	0.89	0.89	7.63	0.96	4.25	1.00	2.50	0.90	1.18	0.97	1.00	1.00	2.44	3.12	0.99	1.61	0.93	0.82	0.79
16 (M17D1)	SG11 124	1.02	1.00	0.98	0.97	4.52	1.00	2.00	1.06	2.25	0.99	0.71	1.07	1.17	0.89	1.56	1.88	1.11	1.33	1.13	0.99	0.86
16 (M17D1)	SG11 125	1.03	0.95	0.96	0.99	3.74	1.00	1.68	1.06	2.00	0.97	1.18	0.87	1.17	1.00	1.33	1.43	1.09	1.27	1.00	0.99	0.79
26 (N17C2)	SG11 133	0.78	0.83	0.75	0.83	6.22	0.76	4.05	0.74	2.00	0.68	0.94	0.63	1.00	0.78	2.00	2.17	0.54	0.84	0.60	0.46	0.43
26 (N17C2)	SG11 134	1.07	1.00	0.76	1.02	3.11	0.88	4.74	0.83	1.75	0.88	2.06	0.80	1.17	1.00	1.67	1.90	1.00	1.56	1.00	0.95	0.79
26 (N17C2)	SG11 135	0.82	1.14	0.83	0.83	10.41	0.86	8.95	1.00	3.50	1.00	1.59	0.97	1.50	1.22	3.78	5.19	1.01	2.37	1.20	0.92	0.79
28a (N17C1/C2)	SG11 162	0.88	0.65	0.68	0.86	12.15	0.76	8.43	0.71	4.25	0.77	2.06	0.60	1.17	1.11	3.67	5.52	0.91	2.36	0.93	0.73	0.64
28a (N17C1/C2)	SG11 163	0.65	0.81	0.69	0.72	15.89	0.77	12.49	0.86	5.25	0.90	2.18	1.33	1.33	1.44	5.22	6.81	0.97	2.94	1.07	0.79	0.79
28a (N17C1/C2)	SG11 164	0.77	0.80	0.64	0.69	13.96	0.76	15.40	0.74	4.75	0.82	2.24	0.63	1.33	1.33	5.56	6.83	1.03	3.50	1.07	0.74	0.57
28b (N17C1/C2)	SG11 168	0.82	0.64	0.68	0.91	9.93	0.77	6.34	0.77	4.50	0.79	2.24	0.67	1.17	1.00	3.78	5.57	0.89	2.04	1.00	0.89	0.57
28b (N17C1/C2)	SG11 169	0.65	0.84	0.69	0.79	12.44	0.75	10.86	0.80	5.00	0.87	2.53	0.93	1.00	1.33	4.67	6.31	0.91	2.73	1.07	0.81	0.79
28b (N17C1/C2)	SG11 170	0.78	1.16	0.72	0.70	10.30	0.82	14.95	0.83	4.25	0.91	1.65	0.90	1.00	1.22	4.56	5.60	1.01	3.20	1.13	0.78	0.71
44 (Tr3)	SG12 288 A	1.03	0.84	0.95	1.00	3.48	1.03	1.45	1.09	1.75	0.78	-0.35	1.03	1.33	0.78	1.11	1.55	1.04	1.13	0.87	0.91	0.71
44 (Tr3)	SG12 288B	1.03	0.76	0.91	1.00	4.48	0.98	1.86	1.00	1.75	0.72	0.18	0.90	0.83	0.78	1.44	1.88	1.01	1.30	1.00	1.08	0.71
45 (Tr3)	SG12 281B	0.95	0.98	0.97	0.94	6.67	1.00	3.10	1.09	2.25	0.92	0.47	0.97	1.00	1.22	1.89	2.79	1.07	1.44	1.00	1.00	0.93

		Majors										Trace										
Feature	**Sample name**	**Na EF**	**Mg EF**	**Al EF**	**Si EF**	**P EF**	**K EF**	**Ca EF**	**Ti EF**	**Mn EF**	**Fe EF**	**V EF**	**Cr EF**	**Co EF**	**Ni EF**	**Cu EF**	**Zn EF**	**Rb EF**	**Sr EF**	**Y EF**	**Zr EF**	**Pb EF**
45 (Tr3)	SG12 282A	0.95	1.01	0.99	0.94	6.85	1.01	3.17	1.11	2.25	0.94	1.59	0.93	1.33	1.11	2.00	2.88	1.07	1.46	1.07	1.04	0.86
49 (Tr3)	SG12 291A	1.15	0.51	0.82	1.03	2.11	0.96	0.70	0.91	0.75	0.53	-0.06	0.57	1.00	0.33	0.33	0.86	0.95	1.09	0.80	1.01	0.57
49 (Tr3)	SG12 291B	1.07	0.77	0.92	1.00	4.56	1.00	1.89	1.06	1.75	0.74	0.18	0.87	1.00	0.78	1.22	1.93	1.03	1.30	0.93	1.12	0.64
Phase 3																						
17 (M17D1, N17A1)	SG11 214	1.00	0.55	0.75	1.06	4.74	0.86	2.27	0.86	2.25	0.81	1.76	0.63	1.00	0.78	1.89	2.71	1.00	1.43	1.00	0.96	0.64
17 (M17D1, N17A1)	SG11 215	0.90	0.73	0.83	1.00	6.04	0.90	2.53	0.94	2.75	0.84	2.00	0.83	1.17	1.00	2.22	2.79	1.04	1.44	1.07	0.99	0.64
17 (M17D1, N17A1)	SG11 216	0.90	0.65	0.79	1.01	6.22	0.87	2.66	0.91	2.75	0.82	0.82	0.87	1.00	0.78	2.56	3.07	1.01	1.29	0.93	0.94	0.71
17 (M17D1, N17A1)	SG11 217	0.87	0.72	0.78	0.93	11.56	0.85	4.92	0.91	3.25	0.82	1.71	0.77	1.17	0.89	2.44	3.60	1.01	1.60	1.00	0.98	0.86
17 (M17D1, N17A1)	SG11 218	0.85	0.74	0.76	0.90	10.59	0.85	6.01	0.83	3.25	0.84	1.35	0.87	1.00	1.00	3.44	4.57	1.04	1.83	1.07	0.90	0.64
17 (M17D1, N17A1)	SG11 219	0.80	0.74	0.75	0.89	11.63	0.84	6.65	0.89	3.50	0.83	1.41	0.83	1.17	1.11	3.56	4.43	0.97	1.83	1.00	0.88	0.79
17 (M17D1, N17A1)	SG11 220	0.85	0.65	0.72	0.93	10.26	0.81	6.25	0.83	3.25	0.76	1.00	0.80	0.83	0.89	3.22	4.10	0.97	1.79	1.00	0.92	0.64
17 (M17D1, N17A1)	SG11 221	0.87	0.94	0.79	0.91	7.63	0.88	6.52	0.91	3.50	0.83	1.00	0.73	1.17	0.89	3.00	4.14	1.01	1.73	1.07	0.90	0.79
17 (M17D1, N17A1)	SG11 222	0.88	1.21	0.82	0.89	6.44	0.89	6.82	1.09	3.75	0.96	1.76	0.90	1.17	1.22	3.11	4.12	1.09	1.73	1.13	0.84	0.71
18 (Tr 1)	SG12 268A	0.88	1.01	0.94	0.95	7.52	0.95	3.41	1.03	2.75	0.90	0.00	0.93	1.00	1.00	2.11	3.12	1.05	1.46	1.20	1.01	0.71
18 (Tr 1)	SG12 268B	0.88	1.01	0.94	0.95	7.48	0.95	3.38	1.00	2.75	0.90	0.00	0.93	1.17	0.89	2.11	3.10	1.05	1.44	1.13	0.99	0.71
18 (Tr 1)	SG12 272A	0.93	0.99	0.91	0.90	8.22	0.97	5.41	1.00	3.00	0.93	1.18	0.97	1.00	1.22	2.67	3.36	1.09	1.77	1.13	0.92	0.71
18 (Tr 1)	SG12 272B	0.92	0.99	0.90	0.90	8.26	0.97	5.44	1.03	3.00	0.93	0.65	0.97	1.50	1.11	2.78	3.33	1.08	1.81	1.13	0.93	1.00
21 (N17D1)	SG11 139	0.67	0.74	0.69	0.88	12.67	0.74	6.24	0.80	5.00	0.82	1.53	0.80	1.33	1.22	4.22	6.21	0.83	1.83	0.87	0.76	0.57
21 (N17D1)	SG11 140	0.80	1.15	0.72	0.88	9.00	0.80	8.56	0.86	3.25	0.83	1.76	0.63	1.00	1.00	2.78	4.26	0.95	2.04	1.07	0.90	0.64
21 (N17D1)	SG11 141	0.90	1.87	0.78	0.84	1.11	0.88	11.54	0.91	1.25	0.76	0.41	0.97	0.83	0.78	0.78	0.86	0.92	1.74	1.13	0.95	0.50
32 (Tr3)	SG12 256A	0.90	1.00	0.93	0.95	6.33	0.96	4.03	1.00	2.50	0.97	1.76	1.00	1.00	1.22	2.33	3.36	1.08	1.64	1.07	1.02	0.64
32 (Tr3)	SG12 256B	0.90	1.02	0.94	0.95	6.37	0.96	4.06	1.03	2.50	0.97	1.06	0.97	1.17	1.22	2.33	3.36	1.08	1.63	1.07	1.03	0.86

D. PPM and Enrichment Factors for P, Ca, and Zn, 'Cultural Layer', Biała Góra

Rowena Banerjea

Sample	P	P EF	Ca	Ca EF	Zn	Zn EF
Core1 S122 A	2966.39	5.41	3210.50	0.40	34.54	1.07
Core1 S122 B	3063.58	5.58	3087.53	0.39	33.22	1.03
Core1 S122 C	2979.68	5.43	3208.31	0.40	27.42	0.85
Core2 S123 A	2713.27	4.95	3072.32	0.39	23.62	0.73
Core2 S123 B	2605.69	4.75	3081.39	0.39	23.20	0.72
Core2 S123 C	2834.03	5.17	2985.90	0.38	32.54	1.01
Core3 S124 A	3000.48	5.47	4931.55	0.62	37.33	1.16
Core3 S124 B	3294.58	6.01	5014.85	0.63	39.98	1.24
Core3 S124 C	3123.35	5.69	4795.19	0.60	33.68	1.04
Core4 S125 A	2244.52	4.09	2786.52	0.35	45.41	1.41
Core4 S125 B	2005.32	3.66	2651.55	0.33	31.02	0.96
Core4 S125 C	2139.07	3.90	2747.02	0.35	40.74	1.26
Core5 S126 A	2968.21	5.41	3299.26	0.42	38.37	1.19
Core5 S126 B	3014.16	5.49	3353.10	0.42	37.55	1.16
Core5 S126 C	2874.71	5.24	3221.66	0.41	34.69	1.07
Core6 S127 A	2777.33	5.06	3171.22	0.40	34.43	1.07
Core6 S127 B	2691.40	4.91	3130.94	0.39	39.99	1.24
Core6 S127 C	2789.38	5.09	3248.03	0.41	26.31	0.81
Core7 S128A A	4197.83	7.65	5911.07	0.74	31.48	0.97
Core7 S128A B	4166.52	7.60	5906.97	0.74	38.04	1.18
Core7 S128A C	4495.93	8.20	5799.04	0.73	31.11	0.96
Core7 S128b Control A	2701.53	4.92	2832.05	0.36	40.20	1.24
Core7 S128b Control B	2650.06	4.83	2883.03	0.36	45.48	1.41
Core7 S128b Control C	2879.85	5.25	2840.66	0.36	30.78	0.95
Core8 S129 A	2488.54	4.54	2181.81	0.27	21.08	0.65
Core8 S129 B	2724.39	4.97	2243.28	0.28	21.90	0.68
Core8 S129 C	2596.78	4.73	2139.08	0.27	0.00	0.00
Core9 S130 A	2507.02	4.57	2243.83	0.28	51.07	1.58
Core9 S130 B	2597.01	4.73	2261.39	0.28	46.11	1.43
Core9 S130 C	2591.52	4.72	2407.69	0.30	48.50	1.50
Core10 S131 A	2992.35	5.46	2023.29	0.25	0.00	0.00
Core10 S131 B	3018.38	5.50	2037.90	0.26	17.10	0.53
Core10 S131 C	3045.80	5.55	2049.25	0.26	20.53	0.64

Sample	P	P EF	Ca	Ca EF	Zn	Zn EF
Core11 S132 A	2145.20	3.91	2156.44	0.27	22.62	0.70
Core11 S132 B	2255.02	4.11	2087.85	0.26	18.67	0.58
Core11 S132C	2238.89	4.08	2079.43	0.26	22.83	0.71
Core12 S133 A	23,494.72	42.83	28,978.53	3.65	23.60	0.73
Core12 S133 B	23,108.28	42.13	28,938.04	3.64	23.44	0.73
Core12 S133 C	3885.12	7.08	4091.68	0.51	40.54	1.25
Core13 S134 A	3854.08	7.03	4441.33	0.56	24.17	0.75
Core13 S134 B	4068.31	7.42	4355.65	0.55	29.79	0.92
Core13 S134 C	4036.03	7.36	4499.55	0.57	29.17	0.90
Core14 S135 A	4009.86	7.31	6324.45	0.80	21.23	0.66
Core14 S135 B	4096.18	7.47	6416.29	0.81	32.65	1.01
Core14 S135 C	3820.19	6.96	6297.93	0.79	0.00	0.00
Core15 S136 A	6249.42	11.39	17,613.22	2.22	347.33	10.75
Core15 S136 B	6223.00	11.34	17,358.71	2.18	337.98	10.46
Core15 S136 C	6557.61	11.95	17,574.37	2.21	334.74	10.36
Core16 S137 A	2386.54	4.35	2284.24	0.29	0.00	0.00
Core16 S137 B	2393.42	4.36	2170.48	0.27	15.86	0.49
Core16 S137 C	2225.55	4.06	2200.03	0.28	0.00	0.00
Core17 S138 A	13,668.05	24.92	15,420.50	1.94	24.55	0.76
Core17 S138 B	12,712.96	23.18	15,903.62	2.00	23.70	0.73
Core17 S138 C	12,325.01	22.47	16,580.93	2.09	25.26	0.78
Core18 S139 A	2027.74	3.70	1692.03	0.21	0.00	0.00
Core18 S139 B	2176.81	3.97	1682.86	0.21	20.75	0.64
Core18 S139 C	2081.54	3.79	1581.24	0.20	24.47	0.76
Core19 S140 A	12,146.15	22.14	10,613.38	1.34	23.80	0.74
Core19 S140 B	10,737.08	19.57	9918.95	1.25	23.55	0.73
Core19 S140 C	1591.48	2.90	1409.31	0.18	21.99	0.68
Core20 S141 A	9539.65	17.39	10,285.81	1.29	22.49	0.70
Core20 S141 B	10,407.37	18.97	10,250.88	1.29	22.77	0.70
Core20 S141 C	10,370.89	18.91	9920.19	1.25	22.85	0.71
Core21 S142 A	1682.55	3.07	1461.91	0.18	22.91	0.71
Core21 S142 B	1724.27	3.14	1530.80	0.19	16.84	0.52
Core21 S142 C	1698.55	3.10	1510.01	0.19	20.64	0.64
Core22 S143 A	2279.31	4.16	2258.73	0.28	19.72	0.61
Core22 S143 B	2118.26	3.86	2151.01	0.27	15.83	0.49
Core22 S143 C	13,169.13	24.01	14,541.52	1.83	23.34	0.72
Core23 S144 A	2133.09	3.89	1947.36	0.25	0.00	0.00
Core23 S144 B	2034.67	3.71	1924.23	0.24	0.00	0.00
Core23 S144 C	12,555.23	22.89	13,871.71	1.75	23.18	0.72
Core24 S145 A	2304.06	4.20	2432.48	0.31	17.48	0.54
Core 24 S 145 B	2283.04	4.16	2473.02	0.31	18.89	0.58
Core24 S145 C	2245.80	4.09	2440.66	0.31	16.34	0.51
Core25 S146 A	2520.30	4.59	2594.18	0.33	17.34	0.54
Core25 S146 B	2584.43	4.71	2523.73	0.32	25.11	0.78
Core25 S146 C	2328.09	4.24	2474.47	0.31	26.24	0.81

Sample	P	P EF	Ca	Ca EF	Zn	Zn EF
Core26 S147 A	3224.38	5.88	5899.31	0.74	51.76	1.60
Core26 S147 B	3108.94	5.67	5899.20	0.74	44.16	1.37
Core26 S147 C	3187.51	5.81	5722.96	0.72	45.62	1.41
Core27 S148 A	4590.46	8.37	10,371.03	1.31	88.05	2.72
Core27 S148 B	4492.93	8.19	10,186.36	1.28	82.88	2.56
Core27 S148 C	4575.06	8.34	10,363.72	1.30	90.40	2.80
Core29 S149 A	2992.26	5.45	4590.82	0.58	43.91	1.36
Core29 S149 B	3107.18	5.66	4502.93	0.57	36.90	1.14
Core29 S149 C	3120.69	5.69	4577.79	0.58	39.11	1.21
Core30 S150a A	7675.13	13.99	13,494.81	1.70	74.76	2.31
Core30 S150a B	7535.58	13.74	13,205.80	1.66	68.78	2.13
Core30 S150a C	7605.81	13.87	13,424.01	1.69	73.21	2.27
Core30 S150b Control A	5572.41	10.16	6946.11	0.87	73.23	2.27
Core30 S150b Control B	5676.21	10.35	6874.94	0.87	63.53	1.97
Core30 S150b Control C	5779.86	10.54	6910.71	0.87	60.40	1.87
Core30 S150c Control A	11,971.20	21.82	7894.28	0.99	0.00	0.00
Core30 S150c Control B	12,106.74	22.07	7952.08	1.00	0.00	0.00
Core30 S150c Control C	12,210.91	22.26	7913.03	1.00	0.00	0.00
Core32 S151 A	3457.45	6.30	3856.28	0.49	60.00	1.86
Core32 S151 B	3487.52	6.36	3709.01	0.47	57.93	1.79
Core32 S151 C	3550.04	6.47	3721.73	0.47	57.11	1.77
Core34 S153a A	4421.74	8.06	5118.66	0.64	52.35	1.62
Core 34 S153a B	4542.13	8.28	5011.65	0.63	48.15	1.49
Core 34 S153a C	4682.79	8.54	5064.63	0.64	56.90	1.76
Core34 S153B A	5698.84	10.39	5429.48	0.68	25.88	0.80
Core34 S153B B	5634.86	10.27	5562.83	0.70	24.49	0.76
Core34 S153B C	5590.78	10.19	5523.46	0.70	29.48	0.91
Core34 S153C A	7817.29	14.25	1892.06	0.24	0.00	0.00
Core34 S153C B	7923.51	14.44	1887.34	0.24	0.00	0.00
Core34 S153C C	45,721.74	83.35	12,168.20	1.53	20.78	0.64
Core34 S153d A	28,709.68	52.34	38,614.11	4.86	23.41	0.72
Core 34 S153d B	27,674.38	50.45	38,646.82	4.86	22.99	0.71
Core 34 S153d C	28,389.97	51.75	39,129.88	4.92	23.21	0.72
Core35 S154 A	2711.34	4.94	3519.30	0.44	42.94	1.33
Core35 S154 B	2615.20	4.77	3485.03	0.44	30.13	0.93
Core35 S154 C	2750.48	5.01	3441.96	0.43	32.39	1.00
Core35 S152 A	4348.98	7.93	4186.44	0.53	41.48	1.28
Core35 S152 B	4245.09	7.74	4209.54	0.53	47.39	1.47
Core35 S152 C	4442.27	8.10	4252.78	0.54	46.55	1.44
Core36 S155 A	2570.42	4.69	3928.19	0.49	41.16	1.27
Core36 S155 B	2564.60	4.68	3881.30	0.49	29.50	0.91
Core36 S155 C	2540.46	4.63	3793.13	0.48	36.98	1.14
Core37 S156 A	6132.59	11.18	9742.35	1.23	53.32	1.65
Core37 S156 B	6086.56	11.10	9774.75	1.23	59.95	1.85
Core37 S156 C	5975.28	10.89	9706.10	1.22	62.47	1.93

Sample	P	P EF	Ca	Ca EF	Zn	Zn EF
Core38 S157 A	2821.40	5.14	4090.94	0.51	23.43	0.72
Core38 S157 B	2899.49	5.29	4269.77	0.54	29.08	0.90
Core38 S157 C	2904.85	5.30	4191.80	0.53	29.88	0.92
Core39 S158 A	4504.85	8.21	7538.38	0.95	48.38	1.50
Core39 S158 B	4508.73	8.22	7494.34	0.94	38.54	1.19
Core39 S158 C	4517.98	8.24	7590.80	0.96	36.24	1.12
Core40 S159 A	2885.90	5.26	16,559.64	2.08	51.68	1.60
Core40 S159 B	2922.22	5.33	16,420.81	2.07	43.87	1.36
Core40 S159 C	2870.08	5.23	16,321.66	2.05	42.71	1.32
Core41 S160 A	26,713.74	48.70	62,571.95	7.87	25.34	0.78
Core41 S160 B	26,262.25	47.88	63,048.11	7.93	24.93	0.77
Core41 S160 C	26,586.51	48.47	64,079.91	8.06	24.82	0.77
Core42 S161 A	3504.66	6.39	6485.41	0.82	78.57	2.43
Core42 S161 B	3560.39	6.49	6336.60	0.80	76.57	2.37
Core42 S161 C	3727.74	6.80	6466.78	0.81	71.49	2.21
Core43 S162 A	4937.07	9.00	6932.23	0.87	69.80	2.16
Core43 S162 B	4924.33	8.98	6711.28	0.84	71.36	2.21
Core43 S162 C	5021.16	9.15	6747.70	0.85	74.53	2.31
Core44 S163 A	4779.52	8.71	6411.78	0.81	58.82	1.82
Core44 S163 B	4590.69	8.37	6358.17	0.80	57.20	1.77
Core44 S163 C	4672.65	8.52	6411.82	0.81	61.21	1.89
Core45 S164 A	27,903.85	50.87	31,831.55	4.01	22.87	0.71
Core45 S164 B	27,298.82	49.77	32,531.84	4.09	23.58	0.73
Core45 S164 A	28,250.13	51.50	33,066.15	4.16	22.96	0.71
Core46 S165 A	5902.97	10.76	8613.98	1.08	86.73	2.68
Core46 S165 B	6085.35	11.09	8490.35	1.07	74.34	2.30
Core46 S165 C	6199.00	11.30	8313.21	1.05	81.29	2.52
Core47 S166 A	3385.75	6.17	4638.16	0.58	50.36	1.56
Core47 S166 B	3340.85	6.09	4557.38	0.57	47.00	1.45
Core47 S166 C	3543.80	6.46	4633.22	0.58	47.89	1.48
Core48 S167 A	14,835.45	27.04	41,144.35	5.18	24.07	0.74
Core48 S167 B	15,398.45	28.07	40,767.68	5.13	24.87	0.77
Core48 S167 C	14,835.31	27.04	41,408.32	5.21	24.01	0.74
Core49 S168 A	18,503.16	33.73	48,048.91	6.05	19.76	0.61
Core49 S168 B	3490.91	6.36	7761.41	0.98	53.08	1.64
Core49 S168 C	3526.72	6.43	7628.21	0.96	49.72	1.54
Core50 S164 A	1358.58	2.48	2353.82	0.30	23.21	0.72
Core50 S164 B	1554.42	2.83	2325.15	0.29	17.60	0.54
Core50 S164 C	1379.92	2.52	2316.09	0.29	21.27	0.66
Core52 S170 A	5144.02	9.38	5406.95	0.68	0.00	0.00
Core52 S170 B	5127.32	9.35	5554.79	0.70	0.00	0.00
Core52 S170 C	5172.64	9.43	5517.76	0.69	0.00	0.00
F94 A	4743.91	8.65	6690.15	0.84	29.03	0.90
F94 B	4805.78	8.76	6640.92	0.84	18.45	0.57
F94 C	4671.19	8.52	6773.80	0.85	30.38	0.94

E. Enrichment Factors for Major and Trace Elements within the Moat Sediment at Malbork

Rowena Banerjea

	Majors										Trace										
Sample name	**Na**	**Mg**	**Al**	**Si**	**P**	**K**	**Ca**	**Ti**	**Mn**	**Fe**	**V**	**Cr**	**Co**	**Ni**	**Cu**	**Zn**	**Rb**	**Sr**	**Y**	**Zr**	**Pb**
MAL 11 0.25–0.26	0.39	0.00	0.45	0.89	1.50	0.52	0.33	0.23	0.00	0.30	0.24	0.32	0.00	0.00	0.61	0.36	0.45	0.70	0.22	1.93	1.22
MAL 11 0.34–0.38	0.54	0.00	0.81	1.03	1.80	0.83	0.93	0.63	0.00	0.74	0.69	0.71	0.10	0.55	1.30	0.80	0.90	1.38	0.54	3.94	1.48
MAL 11 0.48–0.52	0.48	0.00	0.98	0.96	1.90	0.99	1.28	0.80	0.00	1.02	0.95	0.84	0.10	0.61	1.07	0.91	1.03	1.29	0.76	4.38	1.65
MAL 11 0.59–0.61	0.61	0.00	1.14	0.90	1.50	1.18	5.08	1.23	0.00	1.50	1.45	0.99	0.16	1.09	1.15	1.40	1.42	1.75	0.99	6.17	2.35
MAL 11 0.64–0.7	0.00	0.00	0.06	0.13	0.90	0.11	0.99	0.03	0.00	0.34	0.00	0.00	0.00	0.00	2.98	0.72	0.38	1.11	0.22	1.76	0.78
MAL 11 0.8–0.84	0.39	0.00	0.38	0.87	3.65	0.48	2.23	0.23	0.00	0.23	0.17	0.00	0.00	0.00	0.69	0.66	0.46	0.87	0.22	1.85	0.70
MAL 11 1.04–1.06	0.43	0.00	0.57	1.04	1.80	0.63	2.08	0.43	0.00	0.62	0.55	0.35	0.00	0.00	1.98	0.94	0.74	1.09	0.45	3.39	1.30
MAL 11 1.16–1.18	0.00	0.00	0.01	0.11	0.05	0.03	0.00	0.00	0.00	0.05	0.00	0.00	0.00	0.00	0.00	0.00	0.28	0.30	0.00	1.07	0.00
MAL 11 1.22–1.24	0.01	0.00	0.11	0.17	0.35	0.21	1.27	0.15	0.00	0.66	0.24	0.09	0.00	0.00	0.00	0.36	0.52	1.21	0.49	3.53	0.61
MAL 11 1.34–1.36	0.00	0.00	0.04	0.14	0.15	0.07	0.00	0.00	0.00	0.09	0.00	0.00	0.00	0.00	0.00	0.00	0.23	0.21	0.00	1.82	0.43
MAL 11 1.58–1.6	0.00	0.00	0.04	0.16	0.75	0.08	0.48	0.00	0.00	0.15	0.00	0.00	0.00	0.00	0.00	0.83	0.29	0.97	0.00	1.82	0.00
MAL 11 1.8–1.82	0.00	0.00	0.05	0.15	0.20	0.07	0.00	0.00	0.00	0.13	0.00	0.00	0.00	0.00	0.00	0.22	0.28	0.25	0.00	1.76	0.00

Index

Environmental Histories of the North Atlantic World

All volumes in this series are evaluated by an Editorial Board, strictly on academic grounds, based on reports prepared by referees who have been commissioned by virtue of their specialism in the appropriate field. The Board ensures that the screening is done independently and without conflicts of interest. The definitive texts supplied by authors are also subject to review by the Board before being approved for publication. Further, the volumes are copyedited to conform to the publisher's stylebook and to the best international academic standards in the field.

Title in Series

Thomas Finan, *Landscape and History on the Medieval Irish Frontier: The King's Cantreds in the Thirteenth Century* (2016)

In Preparation

Ecologies of Crusading, Colonization, and Religious Conversion in the Medieval Baltic: Terra Sacra II, ed. by Aleksander Pluskowski

Philip Slavin, *Experiencing Famine in Fourteenth-Century Britain*